THEORETICAL NUCLEAR PHYSICS VOLUME I: NUCLEAR STRUCTURE

THEORETICAL NUCLEAR PHYSICS VOLUME I: NUCLEAR STRUCTURE

Amos deShalit
Weizmann Institute

Herman Feshbach
Massachusetts Institute of Technology

JOHN WILEY & SONS, INC.
New York • Chichester • Brisbane • Toronto

Library of Congress Cataloging in Publication Data:

Shalit, Amos de, 1926-1969.
Theoretical nuclear physics.

Bibliography: v. 1, p.
CONTENTS: v. 1. Nuclear structure.
1. Nuclear physics—Collected works.
2. Nuclear reactions—Collected works.
I. Feshbach, Herman, joint author. II. Title.

QC771.S48 539.7 73-17165
ISBN O-471-20385-8

Printed in the United States of America

10 9 8 7 6 5

To
Aihud, Avner
and
Sylvia

FOREWORD

The frontiers of science are broad and varied. As the methods of observation are improved and expanded, an immense variety of facts and occurrences are faced and new phenomena are continually discovered. Science tries to understand the phenomena as the consequences of a few basic processes and laws of nature. In this way, insights are gained into the fundamental ordering principles that govern the great variety of observed events.

One can distinguish two different aspects in the research at the frontiers of science: in some research the fundamental laws and forces that govern the relevant phenomena are believed to be known and beyond doubt; the problems in understanding come from the infinite variety of ways in which nature realizes the potentialities of these laws and forces, a variety that by far exceeds what the human mind could have expected. At these frontiers, ingenuity and insight are applied to create new ideas and concepts in order to recognize and formulate unexpected consequences of known fundamental principles. The frontiers of chemistry, solid state physics, low temperature physics, statistical mechanics, quantum optics, and plasma physics belong into this category.

In other frontier research the situation is different. One is dealing with phenomena where it seems that the known laws of nature are no longer applicable, where one faces conditions that probably are beyond the domain of validity of these principles. One therefore searches for new principles or for a generalization of known laws. Examples of these frontiers are high-energy physics and some fields of astronomy.

Nuclear physics—the science of the structure and the properties of atomic nuclei—has a special position in the edifice of physical science. Its frontiers occupy an intermediate position, partaking of both aspects of frontier research, a fact that imparts a special character and interest to this field. The central concept is the nuclear force that keeps the nucleus together. This force is not part of a well-understood system of natural laws, in contrast to the electromagnetic forces that are responsible for the atomic and molecular structure. The nature and origin of the nuclear force is unknown; it is a realization of the so-called "strong interactions" between elementary particles, interactions that still defy any attempt to obtain a systematic understanding. Its existence is known only for about 40 years. Nuclear force fields are too short ranged to be realized on a macroscopic scale in contrast to electric or gravitational fields.

In the study of these forces, nuclear physics aims at the exploration of new natural laws. Another evidence of this aspect is found in the study of radioactive phenomena, which play such an important role in nuclear phenomena.

Also here processes are faced that are not understood within the framework of known natural laws. They seem to be related to the so-called "weak interactions," between elementary particles. The violation of left-right symmetry was first discovered in a nuclear radioactive decay.

On the other hand, a large part of nuclear physics is devoted to the study of nuclear properties on the basis of a system of laws and forces that is assumed to be valid and known. True enough, our knowledge of the nuclear force and of the weak interactions is derived only from empirical evidence and is in some respects incomplete. Still, many of the most important creative efforts and insights of nuclear physics are directed toward the interpretation of phenomena and processes as consequences of these assumed fundamental laws.

As in so many fields of physics, such efforts and insights were made possible by the invention of new physical concepts and new formulations. Some of them have found applications in different fields of physics too, some were borrowed from other fields. Here are a few examples of such intellectual creations. One is the concept of isotopic spin that was introduced by Heisenberg for the purpose of a more concise description of the similarities between the proton and the neutron in regard to nuclear forces; it led to the important concept of "analogue states" and became a fundamental concept of particle physics. Another example is Bohr's concept of compound nucleus, which is essential for the description of certain nuclear reactions. This concept led, among other things, to the application of thermodynamic concepts to nuclear processes, such as "nuclear temperature" and "evaporation"; today it is used successfully also for the understanding of atomic collisions and elementary particle processes.

A concept that was borrowed from other fields and further developed in nuclear physics is the single-particle model, in which the dynamics of a nucleus is described in terms of the motion of each constituent in the average field of all others. The concept of superconductivity also found its application in nuclear physics. Most of these ideas are ways to cope with the problems of the dynamics of a large number of particles bound together by a strong force. The nuclear physicist faces the same problems in the nucleus as the solid state physicist faces in a metal, a crystal, or a liquid. Both study the dynamics of many particles and their collective motions. The nuclear physicist, however, has to worry about the effects of the overall shape of the object, which play a much more important role in the nucleus because of its small dimensions. These problems gave rise to fruitful concepts such as deformed nuclei, rotating nuclei, and deformation vibrations.

The development in recent years of our understanding of many-body systems is a good example of successful cross-fertilization between fields of physics that study objects as disparate as a nucleus and a chunk of metal. The same intellectual tools are used for the understanding of widely different phenomena. Such concepts as effective interaction, the particle-hole description of excited quantum states, plasma frequencies, spin waves, and so on, find their applications equally well in nuclear and in ordinary matter. It is a proof of the unity and power of modern quantum physics.

A cross-fertilization of a different kind has taken place between nuclear physics and astronomy. Here it is not the conceptual methods but the subject matter that brought these two sciences together. The discovery that stellar energy is supplied by nuclear processes is one of the greatest achievements of modern science. Nuclear processes occur rarely on the earth's surface. Most of the phenomena of nuclear physics studied in the laboratories are "man-made." We need powerful accelerators to induce nuclear excitations. The only exceptions are the decays of natural radioactive substances, which are the last embers still remaining of the great cosmic fire in which terrestrial matter was formed. The natural habitat of nuclear physics is the interior of stars. In the nuclear laboratories, man has created here on earth a cosmic environment in which he studies the processes that are of importance for the universe at large. What we see in the sky, the stars, the novae and supernovae, the galaxies and quasars, and the neutron stars are all manifestations of nuclear effects induced by gravitational compression.

There is a fascination in dealing with nuclear processes, with nuclear matter with its tremendous density; a matter, however, that is inert on earth but is not inert at all in most other large accumulations of matter in the universe. The dynamics of nuclear matter are probably much more essential to the life of the universe than our terrestrial atomic and molecular physics. After all, what is that physics? It deals with the electron shells around nuclei that are only formed at very low temperatures on a few outlying planets where the conditions are just right—where the temperature is not too high, low enough to form those electron shells but high enough to have them react with each other. These conditions are possible only because of the nearness of a nuclear fire. Under the influence of that nuclear fire, self-reproducing units were formed here on earth. And after billions of years of benign radiation from the solar furnace, thinking beings evolved who investigate the processes that may be nearer to the heart of the universe than the daily world in which we live.

The present state of insight into nuclear structure is presented to us in this textbook by two of the most dedicated contributors to this knowledge. Tragically one of them—Amos deShalit—is no longer among us. He was able to finish most of his contributions to this book, which Herman Feshbach welded together into an impressive work. The achievements of nuclear research of half a century are described here in a concise, systematic, and lucid form. May this book be a testimony to the ability of the human mind to clarify the ways in which nature works.

Victor F. Weisskopf

PREFACE

It is our intention in these two volumes to describe the fundamental principles underlying the present understanding of nuclear structure and interactions. It is not our intention to be complete, giving the most recent views on each and every nuclear problem. Instead, we hope that, after studying these volumes, the reader will have gained sufficient insight to enable him or her to profitably turn to the original literature and review articles. Even more, we hope we have been able to transmit the quality of the challenge presented by nuclear phenomena and the nature of the intellectual rewards that attend their study.

Nuclear phenomena are enormously varied. To understand them it has been necessary to call on the entire armamentarium of modern theoretical physics. Concepts and methods from every subdiscipline such as statistical mechanics, thermodynamics, atomic and molecular physics, chemical reactions, the physics of solids and liquids, optics and sound-wave propagation and elementary-particle physics have been borrowed and transformed. The consequent ability to appreciate advances in these fields and even to contribute to them on occasion is one of the dividends of the conscientious study of nuclear theory. However, this wide-ranging approach and the great variety of phenomena to be discussed present a formidable pedagogic problem.

For this reason we have emphasized principles instead of attempting to cover all aspects and to be up to date. The reader is assumed to be moderately sophisticated. The audience we have in mind is typified by the graduate student who might make a career in nuclear physics or more generally in nuclear science. We assume a working knowledge of quantum mechanics and, as far as nuclear physics itself is concerned, a general knowledge of the phenomena as well as the vocabulary and symbols in terms of which these are commonly described, as might be acquired from an introductory course.

In addition we have provided an introductory chapter that discusses in a brief and qualitative fashion many of the important phenomena, facts, and concepts, furnishing thereby a quick first look at the nucleus. Both structure and reactions are reviewed. Of course, these are given a fuller discussion later on in this and the next volume. However, it is often true that these qualitative descriptions are sufficient input for many developments, and it is convenient to be able to refer to them rather than to the fuller but necessarily less transparent descriptions.

The structure of nuclei and the principal features of nuclear reactions are a consequence of the strong nuclear forces. Strong interaction physics here and elsewhere involves two characteristic and interactive features. On the one hand, a systematic study of nuclei is needed to establish regularities in their properties. These regularities can suggest or are suggested by models of nuclear structure that mimic some of the attributes of the real nucleus. Initially the models are rather crude and have a narrow range of applicability. As the field develops, the experiments become more subtle, and the models become more sophisticated and encompassing. One such model, the generalized nuclear shell model, has been able to provide a framework for understanding many of the outstanding features of nuclei, particularly the properties of the ground and low-lying states. This is a remarkable achievement, since the model has been established in spite of the absence of a complete description of the underlying force. For the most part, only a qualitative understanding of that force is required.

In many ways the generalized shell model can be considered as the central subject of this first volume. Why and in what ways is it successful? What are the ad-hoc elements? How is it related to the simpler models, which preceded it historically and which remain very useful in their domains of validity? How can it be justified and, more than that, quantitatively related to the nuclear forces? Upon what features of the latter does it rely?

The first step is taken in Chapter II in which it is shown that the features of the semiempirical law giving the binding energy of nuclei as a function of the number of protons and the number of neutrons follow if the nucleus is thought to consist of noninteracting nucleons constrained to have the observed density. The fact that the nucleons obey the Pauli exclusion principle is of essential importance for this development. The agreement with experiment of this very primitive model is not quantitative, but order-of-magnitude results, correct to within a factor of two, are obtained.

Of course, one is surprised that it is possible to replace (even though roughly) the complex motion of the strongly interacting nucleons by independent motion in a constant potential. In addition, the question remains of how the strong nuclear forces can lead to the observed nuclear densities. There is some discussion of these problems in Chapter II. But it is taken up in earnest in Chapter III in which the properties of an idealized system, infinite nuclear matter, is discussed. The theory of this system is presented in Chapter III in a relatively elementary way. It is quickly realized that the potential in which the particle moves represents the average effect on one of the nucleons by all of the others. The quantitative development of this notion is carried out with the aid of the independent-pair approximation. With this development comes an understanding of the independent-particle nature of nuclear dynamics as well as the role played by various components of the nuclear force and the Pauli principle. A more modern and sophisticated discussion is presented in Chapter VII.

The independent-particle concept is applied to finite nuclei in Chapter IV. Each particle is now assumed to move independently in a central potential. The phenomenon of shells, the necessity for a spin-orbit component in that potential, the values of the spins and magnetic moments of the ground and low-lying states are discussed. Finally the Hartree-Fock method for deriving the "best" central potential is described. The limitations of the model and the significance of the agreement with experiment obtained in the course of the discussion become quite apparent. A possible cure, the residual interactions between the particles, is introduced in Chapter V, and their effect is evaluated with the aid of perturbation theory. The nature of the configurations that can be involved and the way in which their effects can be evaluated in first order are discussed in some detail. It becomes apparent that first-order theory is adequate for nuclei near closed shells but that a more elaborate perturbation theory becomes essential for nuclei at some distance from the closed shells.

However, it was not by simply extending perturbation theory that progress was made. Instead, it was made through a brilliant hypothesis that provided a new insight into nuclear dynamics. This is described in Chapter VI on collective motion. The success of the hypothesis, that some nuclei such as the rare earths are rotators, suggests that there are characteristic modes of motion of nuclei that involve substantial numbers of nucleons moving together. Rotational motion is just one example. Chapter VI discusses others such as vibrations of which the giant dipole is one realization. But now the problem of connecting these phenomenological descriptions with the shell model arises. Its resolution, also discussed in Chapter VI, lies in the behavior of the shell-model potential. We learn that that potential can be deformed, that it can vibrate, and that the corresponding modification in the shell-model orbits can be connected with the observed rotational and vibrational motions.

Up to this point, no attempt has been made in the text to connect the above descriptions of collective and shell motion with the underlying nuclear forces. This is the ultimate goal of the investigations reported in Chapter VII. Vibrations are treated by considering the time-dependent Hartree-Fock method, and also by searching for the appropriate linear combination of particle-hole excitations that can describe a vibration, the so-called RPA method. These methods fail if the nuclear interaction is singular or very strong as is certainly true for small nucleon-nucleon separations. The Brueckner method, which is designed to deal with this problem, is described and applied to the nuclear-matter problem. Finite nuclei are considered next. Recent work reviewed includes (1) the Hartree-Fock method for soft nucleon-nucleon potentials, (2) the Brueckner-Hartree-Fock method that generalizes the method of self-consistent fields to include the case of singular potentials, and (3) the local density approximation in which the infinite nuclear matter results for a given density are assumed to apply in finite nuclei in regions with the same density. Various attempts to relate these results to derive an effective interaction that can be used in shell-model calculations are described, and the present status of these theories is discussed. Much of the discussion of this chapter uses the methods of "second quantization." One model that is most easily described in this language leads to a "superconducting" solution from which the existence of a gap between the ground state of an even-even nucleus and its first excited state can be inferred.

The various models and theories of nuclei described in Chapters II to VI predict not only the properties of the nuclear states but also their radiative and β-decay transitions. These are discussed in Chapters VIII and IX. Chapter IX also contains a review of the nuclear tests of the theory of weak interactions including the "fall of parity," double β-decay, the nuclear tests of CVC and PCAC, and the effect of weak interactions on nuclear forces.

It is obvious that many important subjects have been omitted; the choice of topics that are thought to constitute the basis of nuclear theory is necessarily idiosyncratic. For this reason, references to review articles and books that can supplement the material presented here have been listed at the end of this volume.

The second volume of this book is not yet completed. It will be concerned with two subjects. The first will focus on nuclear forces, their origin in the exchange of bosons and their manifestations in nucleon-nucleon scattering and binding, as well as in the three- and four-body nuclear systems.

The second will be concerned with nuclear reactions and will include reactions induced by high-energy particles, direct reactions, transfer reactions, compound nuclear resonances, doorway state resonances, and the statistical theory of nuclear reactions. Reactions involving electromagnetic probes such as electrons and photons, weak interaction probes, the various neutrinos, as well as those induced by strongly interacting projectiles such as pions, kaons, nucleons and antinucleons, deuterons, alpha particles and finally heavy ions will be discussed.

I am very grateful to a number of my colleagues who took time from their busy lives to carefully read and criticize the original manuscript. Approximately the whole manuscript was read by N. Austern, J. Devaney, R. L. Feinstein, A. Gal, J. Hufner, and I. Talmi. Chapter IX was carefully scrutinized by C. W. Kim and J. Weneser. Discussions with J. Weneser and K. Gottfried resulted in the rewriting of the Appendix on time reversal. T. Lauritsen read the first few chapters while D. A. Bromley contributed suggestions in connection with Chapter VI. R. Jackiw was helpful in developing the description of the Adler-Weisberger relation given in Chapter IX. Not all of their advice was taken, nor did they see the final version. Any errors that inescapably are present in a work of this magnitude are of course my own responsibility. But I hope that with their help, major errors and unclear discussions have been eliminated.

I thank the members of the M.I.T. Center for Theoretical Physics, particularly, M. Baranger, A. K. Kerman, J. Negele, F. Villars, and V. F. Weisskopf. Their attitudes toward physics in general and toward nuclear physics in particular is reflected in the spirit with which this book is written.

We are grateful to Miss Ilana Eisen and Mrs. Lillian Horton for their dedicated assistance in the preparation of the manuscript.

The references listed at the end of this volume are not in any way complete, nor has any attempt been made to ascertain the origin of many of the arguments presented. The preparation of a complete list and the assignment of credit for original discoveries is an enormous task that we did not attempt.

It was in 1960 that Amos deShalit and I decided to write this book. But we were unable to start until 1965 and even then it proceeded slowly because of our many commitments to other projects. Amos came often to M.I.T. to write and to work intensively with me on the manuscript. Collaborating with Amos was an exhilarating experience. My own understanding of nuclear physics grew significantly with each visit as we discussed the plan of each chapter, sharpened up an argument, found a more incisive presentation, all so that the underlying physics would be more strikingly revealed. Amos died in August 1969. I have written elsewhere of the loss to the world, to his country, to science and education, and to his many friends. I want to record here my own profound sense of loss. I hope that the completed volume is faithful to his vision of a book that would not only instruct but also inspire.

Herman Feshbach

CONTENTS

LIST OF SYMBOLS

Symbols		See Chapter
A	Nuclear mass number	I
$\mathbf{A}$	Vector potential	IV, VIII
$\mathcal{A}$	Antisymmetrization operator	III
B	Nuclear binding energy	I
	Collective mass	VI
	Proportional to the square of the matrix element for an electromagnetic transition	VI, VIII
C	Collective force constant	VI
$D_{mm'}{}^{(j)}$	Wigner D matrix	Appendix A
$\mathbf{D}$	Electric dipole operator	VI, VIII
E	Energy	I
$\mathbf{E}$	Electric field	VIII
$F^{(V)}$	Electromagnetic form factor	IX
G	G-matrix	III, VII
	Nucleon-form factor	VIII
H	Hamiltonian	
$\mathbf{H}$	Magnetic field	IV, VIII
I	Total angular momentum	VI
$\mathcal{J}$	Moment of inertia	I, VI
J^{π}	Total angular momentum, in units of $\hbar$, parity π	I
$J^{(A)}$ $J^{(V)}$	Axial and vector current density	IX
K	Kaon, K-meson	I
	Component of the angular momentum along a body-fixed axis	VI
L	Orbital angular momentum in units of $\hbar$	I
L_{λ}	Lepton current	IX
M	Mass, subscript denotes particle	I
	z-component of angular momentum	I
M_T	"3" component of isospin T	I
M_F	Fermi matrix element	IX
M_{GT}	Gamow–Teller matrix element	IX
N	Neutron number	I
P	Polarization	I
P_l	Legendre polynomial of order l	
$\mathbf{P}$	Momentum of the center of mass	VI

Symbols		See Chapter
Q	Quadrupole moment	I
$Q_{lm}^{(E,M)}$	Electric (E) or magnetic (M) multipole moments of order 2^l; dipole ($l = 1$), quadrupole ($l = 2$), octupole ($l = 3$), hexadecupole ($l = 4$)	
Q	Projection operator	III, VII
Q_F	Fermi-projection operator	III, VII
Q_0	Intrinsic quadrupole moment	VI
$\hat{Q}$	Charge operator	IX
R	Nuclear radius	I
R_{nlj}	Radial single-particle wave functions	IV
$\mathbf{R}$	Center of mass	IV, VIII
S	Surface area	II
S_0	s-wave strength function	I
S_n, S_p	Separation energy for neutron and proton	I
S_{12}	Tensor-force operator	I
T	Kinetic energy operator	II
	Temperature	III
	Total isospin	I
$\mathbf{T}$	Isospin operator	I
$T_\kappa^{(k)}$	Spherical tensor	Appendix A
U_k	Occupation amplitude	VII
V	Potential	I
V_k	Occupation amplitude	VII
Y_{lm}	Spherical harmonic	Appendix A, I
$\mathcal{Y}_{jlm}$	Total angular momentum-eigenfunction	IV
Z	Atomic number	I
a	Decoupling parameter	VI
a_μ	Boson operator	VI
	Fermion operator	VII
b	Boundary condition parameter	VII
c	Velocity of light	
c.c.	Complex conjugate	
d	Deuteron	I
	$l = 2$ orbital	IV
e	Charge on proton	
	Electron	I
f	Scattering amplitude	I

Symbols		See Chapter
	Boundary-condition parameter	VII
	$l = 3$ orbital	IV
g	Nucleon–pion coupling constant	I
	Magnetic moment/angular momentum	I
	$l = 5$ orbital	
g_V, g_W, g_S	Vector-form factors	IX
g_A, g_P, g_T	Axial-form factors	IX
h	One-body Hamiltonian	VII
$\hbar$	Planck's constant/2π	
h_l	Spherical Hankel function	
$h_{\alpha\beta}$	Healing distance	III
j_l	Spherical Bessel function	II
j_μ	Four-vector current density $(j, i\rho)$	VIII
$\mathbf{j}$	Current density	VIII
k	Wave number	I
k_F	Fermi momentum	II
$\mathbf{k}$	Momentum/$\hbar$	II
l	Orbital angular momentum	
$\mathbf{l}$	Orbital angular momentum vector	I
m	Mass, subscript denotes particle	
n	neutron	I
	Index of refraction	I
	Number of solutions	II
	Number of nodes in radial wave functions	IV
p	Proton	I
	$l = 1$ orbital	IV
$\mathbf{p}$	Momentum	I
$\mathbf{q}$	Momentum transfer	I
r_0	Scale for nuclear radius, $R = r_0A^{1/3}$	I
r_c	Radius of the core	II
$\mathbf{s}$	Spin	I
	Displacement	II, VI
t	Time	I
	Nuclear surface thickness	I, II
	Square of four momentum transfer	VIII
$\mathbf{v}$	Velocity	I, VI
v	Potential	
w	Transition probability/time	VIII
α	Alpha particle	I
	Additional quantum numbers	IV
	Collective variable	VI

Symbols		See Chapter
	Internal conversion coefficient, electric	VIII
α_k	Quasi-particle operator	VII
$\boldsymbol{\alpha}$	Dirac matrix	Appendix of IX
β	Quadrupole deformation parameter	I, VI
	Internal conversion coefficient, magnetic	VII
	Dirac matrix	Appendix of IX
γ	Gamma ray	
	Shape parameter	VI
γ_μ	Dirac matrix	Appendix of IX
Γ	Full width of a resonance at half maximum	
Δ	Gap energy	VII
δ	Dirac delta function	
	Pairing energy	I
	Phase shift	I
	Variation	IV
	Deformation parameter	VI
δ_{ab}	Kronecker delta, $=0$ if $a \neq b$, $=1$ if $a = b$	
ϵ	E/mc^2 where m is the electron mass	IX
ϵ_F	Fermi energy	II
ϵ	Nuclear excitation energy	I
ϵ_{ABC}	Antisymmetric tensor	VI
$\boldsymbol{\varepsilon}$	Polarization vector	VIII
η	Coulomb scattering parameter	I
	Hartree–Fock single-particle energy	VII
θ	Spherical coordinate	
Θ	Unit function	VII
λ	Wavelength/2π	I

Symbols		See Chapter
Λ	Lambda baryon	I
	z-component of orbital angular momentum in deformed potential	VI
μ	Mu meson	I
	1/(Pion compton wavelength)	I
	Nuclear magnetic moment	I
μ_0	Nuclear magneton	VIII
$\boldsymbol{\mu}$	Nuclear magnetic moment operator	I
ν	Neutrino	I
	Neutron	IV
ν_e, ν_μ	Electron neutrino, muon neutrino	IX
$\bar{\nu}$	Antineutrino	I
ξ	Intrinsic coordinate	IV
Ξ	Xi baryon	I
π	Pion $\equiv$ pi meson	I
	Proton	IV
	Parity	VI
ρ	Density	I
	Rho meson	I
ρ_i	Dirac matrix	Appendix of IX
$\rho_{if}^{(1)}$	One-particle density matrix	II
$\rho_{if}^{(2)}$	Two-particle density matrix	II
σ	Cross-section	I
	Signature for electric or magnetic multipole	VIII
$\boldsymbol{\sigma}$	Pauli spin operator	I
Σ	Sigma baryon	I
	Spin projection	VI
$\boldsymbol{\tau}$	Isospin vector operator with components τ_1, τ_2, τ_3	I
ϕ	Spherical coordinate	
$\chi(m)$	Spin state vector with z-component of spin, m	II

Symbols		See Chapter
$\chi(m_s, m_t)$	Spin and Isospin state vector	III
ω	Harmonic-oscillator frequency	IV, VII
Ω	Solid angle Volume Omega baryon z-component of total angular momentum	 II I VI
$\boldsymbol{\Omega}$	Unit vector in radial direction	
*	Superscript complex conjugation	
∇^2	"Del squared" $= \partial^2/\partial x^2 + \partial^2/\partial y^2 + \partial^2/\partial z^2$	
$[A, B]$	$AB - BA$	
$\{A, B\}$	$AB + BA$	
^	When used above an italic = operator When used above a bold face = unit vector	
tr	trace	
		IX
$\langle 0 \rangle$	Average of 0	I
ft		I, IX
CVC	Conserved vector current	IX
OPEP	One-pion exchange potential	IX
PCAC	Partially conserved axial current	IX
V–A	Vector–Axial weak interaction	IX
${}^{A}_{Z}(\ )_N$	Symbol for a nuclear species	I
~	Tilde above a letter refers to time reverse of that quantity	Appendix A
$(-\hat{1})^J$		VI
$\lvert [\alpha\beta] \rangle$	$= \phi_\alpha(1)\phi_\beta(2) - \phi_\alpha(2)\phi_\beta(1)$	III

Symbols		See Chapter
$>_\alpha$	Quasi-particle state	VII
$np\ mh$	n-particle m-hole state	
$\langle lm_l sm_s \| jm \rangle$	Clebsch–Gordan coefficient	Appendix A
$\begin{pmatrix} j_1 & j_2 & j_3 \\ m_1 & m_2 & m_3 \end{pmatrix}$	Wigner 3j symbol	Appendix A
$\begin{Bmatrix} j_1 & j_2 & j_3 \\ j_4 & j_5 & j_6 \end{Bmatrix}$	Wigner 6j symbol	Appendix A
$\begin{Bmatrix} j_1 & j_2 & j_3 \\ j_4 & j_5 & j_6 \\ j_7 & j_8 & j_9 \end{Bmatrix}$	Wigner 9j symbol	Appendix A
$(\alpha jm \| \| T^{(k)} \| \| \alpha' j' m')$	Reduced matrix element	Appendix A
$(\ldots \{ \| \ldots)$	Coefficient of fractional parentage	V
$(\ldots \| \} \ldots)$	Coefficient of fractional parentage	
$s, p, d, f, g, h, i, \ldots$	$l = 0, 1, 2, 3, 4, 5, 6, \ldots$ orbitals	V
${}^{2S+1}L_J$	S = spin, L = orbital angular momentum, J = total angular momentum	
l_j	Single-particle orbital with orbital angular momentum, l, and total angular momentum j	
$(l_j)^n$	n-particle configuration, each particle in a l_j orbital	
$(l_j)^{-1}$	Hole in an otherwise filled level composed of l_j orbitals	V

CHAPTER I

NUCLEAR PHYSICS—INTRODUCTORY REVIEW

These two volumes deal with the theoretical concepts that underlie our understanding of the structure of nuclei and nuclear reactions. One can distinguish two main complementary approaches. In one, attention is focused upon nuclear forces. The ultimate aim is to quantitatively relate these to the properties of nuclei. From this point of view, each nucleus, its excitations, and the way in which these properties shift as the nucleus changes are all manifestations of the nuclear force. The traditional strategy has assumed that nuclear forces would be determined principally from the properties of the two-, three-, and four-nucleon systems together with the fundamental theory of the nuclear force that asserts that this interaction is the consequence of the interchange of the mesons (π, ρ, ω, etc.) among the nucleons. However the possibility that the comparison between nuclear properties and their predicted values may help in choosing among rival theories of nuclear forces should not be forgotten.

The second approach looks toward the discovery and elucidation of the nuclear modes of motion and their associated degrees of freedom. Examples include the nuclear shell model with its single particle states. Another is the rotational model with such dynamical parameters as the moment of inertia. Vibrations are indicated by such phenomena as the giant dipole resonance. The optical model, the compound nuclear resonance, the doorway state resonance are also special modes of nuclear motion. The discovery of these relies upon the observation of regularities in the properties of nuclei from which the nature of a nuclear degree of freedom may be deduced or verified as the case may be. Significantly involved here is the formulation of a model, with a kinetic and interaction energy depending directly upon the degrees of freedom under study. Such a model depends upon a number of parameters whose value is determined from comparison with experiment. If these parameters vary smoothly or understandably over a number of nuclei, the model is said to be a good one. Of course, models may be deduced from the data or they may be postulated and their consequences suggest experimental tests. In any event experiments and calculations involving a number of nuclei and a number of processes are required before the model is established. Originally the models were rather simplistic but by this time rather elaborate and sophisticated schemes have been developed that unify and combine several earlier models.

These two broad areas of effort, (1) the study of nuclear forces and their relation to the properties of nuclei, and (2) the development of widely applicable models meet in the effort to "derive" the models, calculating the model parameters, and determining their range of validity from nuclear forces. The sorts of question that need to be answered, for example, in the case of the shell model include: (1) when is the description of the nucleus in terms of each nucleon moving independently in a potential correct and (2) how is this shell model potential quantitatively determined?

These paragraphs describe very briefly the areas to be discussed in these volumes. The first volume will be devoted mainly to nuclear structure, that is, to the properties of the ground state, the low-lying levels, and the electromagnetic and β-decay transitions that can occur between them. Models including their relation to nuclear forces are discussed. The second volume concerns itself with the origin of nuclear forces, the semiempirical determination of nuclear forces from the nuclear two, three-, and four-body problems, and finally nuclear reactions and the associated reaction models. An important part of the discussion of nuclear reactions is the description of how they can be employed to determine properties of the stationary states of nuclei.

It has been stated that the physics of the twentieth century has been primarily concerned with the quantal structure of matter [Weisskopf (71)]. Each system, an atom or a nucleus or a hadron studied by particle physicists, has a ground state and a spectrum of excited states that are specified by a set of internal quantum numbers, such as spin, in addition to their energy. Transitions can take place between these states. These transitions may involve the emission or absorption of photons, electron-neutrino pairs, mesons of various kinds, and for the relatively highly excited states of atoms, electrons, and of nuclei, neutrons, alpha particles, etc. These and other parallels have been emphasized by Weisskopf.

But of course there are substantial qualitative differences. We can distinguish between the three great classes of matter (1) atoms, molecules, solids, and plasmas, (2) nuclei, and (3) hadrons. Nuclei are those systems that can be formed from A-nucleons, where A is the mass number, of which Z, the atomic number, are protons. No other particles need be involved. The hadrons are the strongly interacting particles including the baryons, that is, nucleons and the "strange"baryons, the Λ, Σ, Ξ, and Ω as well as the mesons such as the π, ρ, K, etc. The concept of nuclei can be generalized by including in its domain all systems with two or more baryons. When these baryons are nucleons, the systems are the ordinary nuclei. When some are the strange baryons as well as nucleons, the systems are called hypernuclei. We now emphasize that these three types of matter, atoms, etc., nuclei, and hadrons differ dynamically. That is, the major forces acting in each type are dissimilar. This is indicated in the table below.

Characteristic Forces for Different Classes of Matter

	Strength of Interaction	Range	Importance of Relativity	Characteristic Excitation Energies	Number of Particles
(1) Atoms	Weak	Long	Little	1–10^5 eV	1 to many
Molecules	Weak	Long	Little	10^{-3}–10^{-1} eV	2 to many
Solids	Weak	Long	Little	10^{-4}–1 eV	very many
(2) Nuclei	Moderate	Short	Some	10^5–10^7 eV	2 to many
(3) Hadrons	Strong	Short	Great	10^7–10^9 eV	?

As the table shows, nuclei are the only systems that consist of a finite number of particles, (the most massive known nucleus consists of 259 particles, the least the deuteron with two) with moderately strong forces acting between the particles. They are a unique form of matter.

Because of the strength of the forces it is not possible to use simple perturbation theory, which is so useful in the theory of atoms. Because there are relatively few particles, the many-body theory developed for solids and quantum fields does not immediately apply. New methods have had to be devised. This process is not complete but even at this point, application of these methods has already been made to the theory of atoms and molecules and their interactions, as well as to solids and plasmas.

There is remarkable commonality in the qualitative nature of the phenomena exhibited by the various forms of matter. But there is an equally remarkable diversity in their underlying dynamics and thus in quantitative aspects, as, for example, is apparent from the column, "Characteristic Excitation Energies." Each field has its own great problems but their solution in one area may still prove of use and significance for the others.

1. NUCLEAR SIZES

Evidence for the existence of nuclei comes now from so many different sources that it is hardly necessary to describe any of them. The picture of the atom as consisting of a small, positively charged, massive nucleus, surrounded by a "cloud" of electrons, is now established beyond any doubt. Together with the powerful tools of quantum mechanics, it forms the backbone of numerous quantitative calculations of the properties of atoms and molecules, and these all agree very well with experiment.

The detailed study of atomic spectra shows that the atom possesses degrees of freedom in addition to those of the electrons that manifest themselves, for instance, through the "splitting" of spectral lines. These so-called hyperfine effects, which modify the behavior of electrons in atoms, [see Herzberg (44)], are most naturally ascribed to detailed properties of the nucleus of the atom. The nucleus emerges from these studies of atomic spectra, as an object of finite dimensions, with a finite distribution of charge and magnetization. Furthermore, it is quite evident from the studies of atomic spectra that a finite angular momentum should be generally associated with the nucleus, and that, because of the interaction between the electrons and the nucleus, it is only the *total* angular momentum—the angular momentum of the electrons plus that of their nucleus—that is generally conserved.

The dimensions of a typical nucleus are about 10^{-12} cm. The density of electrons in the atom changes very little over these dimensions. Yet, in many cases the atomic-spectroscopy data are accurate enough to trace even the effects of the shape of the nuclear charge distribution on the dynamics of its surrounding electrons. It is thus possible to determine the electric quadrupole moment of the nucleus, which measure the extent to which the charge distribution in the nucleus deviates from spherical symmetry and acquires an ellipsoidal shape. Magnetic dipole moments, which reflect the current and spin distributions in nuclei, have generally a more dramatic effect on atomic spectra, and further refined measurements yielded information on nuclear magnetic octupole moments as well.

An important step forward in the elucidation of nuclear charge and current distribution has been made through the study of high energy electron scattering from nuclei [T. deForest and J. D. Walecka (66)]. Because the forces on electrons penetrating the nucleus depend on the details of the charge distribution in the nucleus, and because the electromagnetic interaction is, on the whole, very well understood, it is possible to extract fairly accurate information about nuclear electromagnetic properties from such electron scattering data. Figure 1.1 shows typical results of high energy elastic electron scattering experiments and the fit to the data obtained by using different forms of the charge density. A typical charge density employed is shown in Fig. 1.2. Figure 1.1 gives us an idea of the sensitivity with which such experiments can determine nuclear charge distributions.

The information obtained from high-energy electron scattering complements that which is obtained from the bound electrons. In fact, in the Born

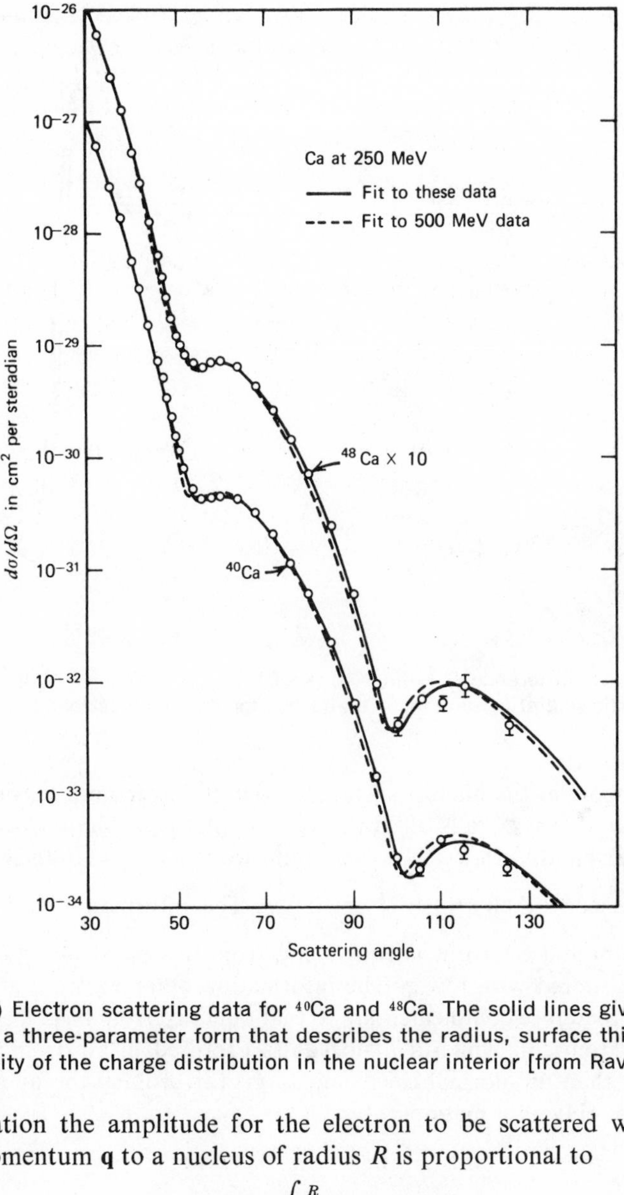

FIG. 1.1. (*a*) Electron scattering data for ^{40}Ca and ^{48}Ca. The solid lines give the fit obtained with a three-parameter form that describes the radius, surface thickness, and nonuniformity of the charge distribution in the nuclear interior [from Ravenhall (67)].

approximation the amplitude for the electron to be scattered while transferring momentum $\mathbf{q}$ to a nucleus of radius R is proportional to

$$f(q) \simeq (1/q)\int_0^R \rho(r)\, r \sin qr\, dr \tag{1.1}$$

Note that $\hbar\mathbf{q} = \mathbf{p}_i - \mathbf{p}_f$ where $\mathbf{p}_i$ is the initial momentum of the electron, $\mathbf{p}_f$ the final momentum. If $qR \gg 1$, that is, for large-momentum transfer, the main contribution to (1.1) comes from regions in which $\rho(r)$ varies rapidly. This

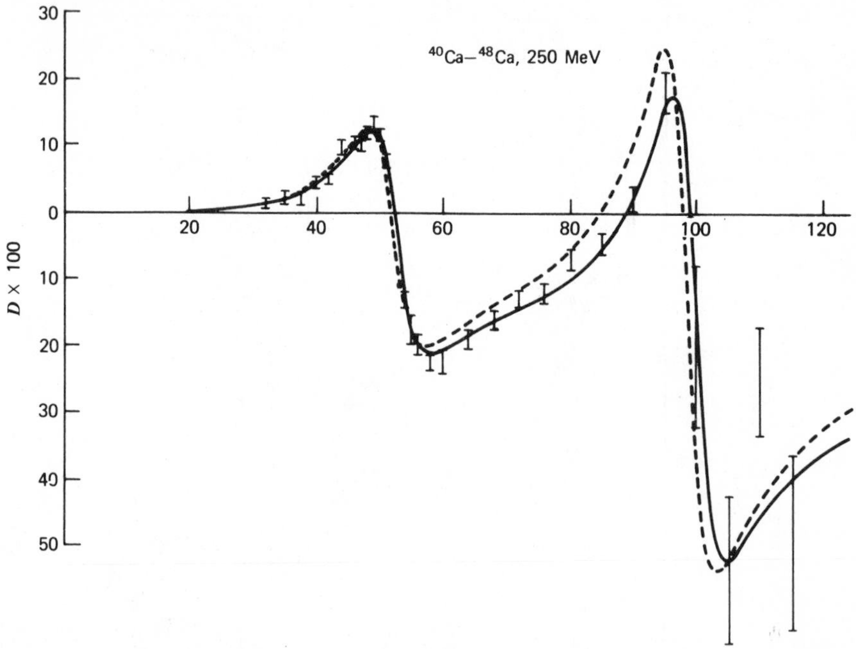

FIG. 1.1 (*b*) The difference $D \equiv [\sigma(^{40}\text{Ca}) - \sigma(^{48}\text{Ca})]/[\sigma(^{40}\text{Ca}) + \sigma(^{48}\text{Ca})]$ between the electron elastic scattering data for ^{40}Ca and ^{48}Ca for 250 MeV electrons [from Ravenhall (67)].

usually happens at the nuclear surface, so that (1.1) becomes sensitive to how fast the charge density falls off at the surface, that is, to the *surface thickness.* On the other hand, for $qR \ll 1$ we can put $\sin qr \approx qr - (1/6)(qr)^3$ and we see that $f(q)$ measures then $\langle r^2 \rangle = \int_0^R r^4\rho(r)\,dr$. The information obtained from energies of bound electrons is equivalent, from this point of view to that of electrons scattered with low q. The information obtained from atomic spectroscopy is therefore complementary to that obtained from large q e-scattering.

Muonic atoms, in which a μ^--meson is captured in an atomic orbit are even better than the normal electronic atoms for probing the nuclear charge distribution. Having a mass of

$$m_\mu = 105.659 \pm 0.002 \text{ MeV} \tag{1.2}$$

μ-mesons are about 200 times heavier than the electron and, hence, have a Bohr radius that is 200 times smaller. A μ-meson therefore probes the nuclear charge distribution from much closer distances and mu-mesic x-rays have greatly helped clarify the charge and current distribution in nuclei [see Sens (67); Devons and Duerdoth (69); and Wu and Wilets (69)].

All atomic spectroscopy data, electron scattering data, mu-mesic x-rays and

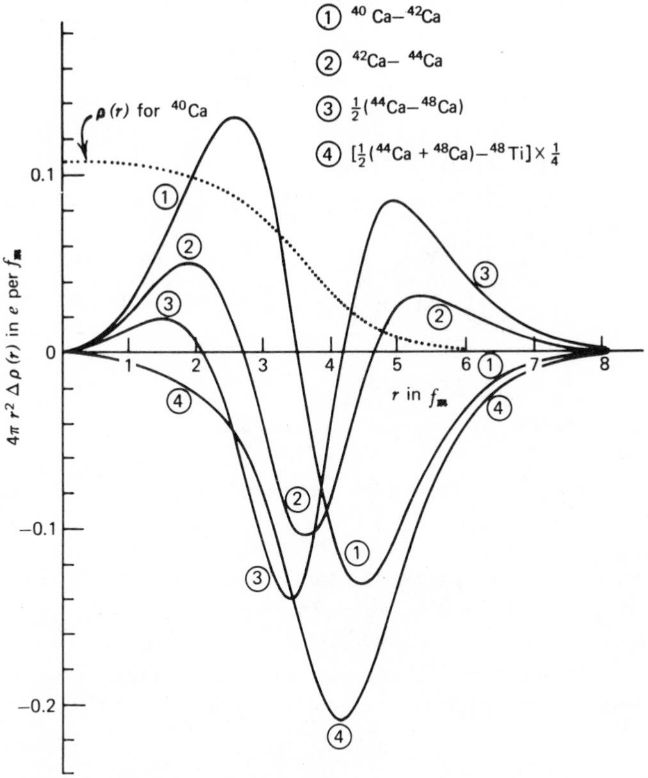

FIG. 1.1 (c) Differences between the charge distributions of ^{40}Ca and other nuclei. The assumed charge distribution for ^{40}Ca is shown by the dotted line [from Ravenhall (67)].

many other experiments are consistent with an A independent nuclear density. Unlike atoms, whose size on the average varies very little with increasing Z, nuclei keep swelling as we increase their mass. Since the average nuclear density is found to be independent of A we can write for the nuclear radius

$$R = r_0 A^{1/3} \tag{1.3}$$

where r_0 was found, empirically, to have the value of

$$r_0 = 1.12 \times 10^{-13} \text{ cm} = 1.12 \text{ fm} \tag{1.4}$$

corresponding to an average nuclear density of

$$\rho = 1.72 \times 10^{38} \text{ nucleons/cm}^3 \tag{1.5}$$

Actually the nuclear density does not change abruptly from its value (1.5) to

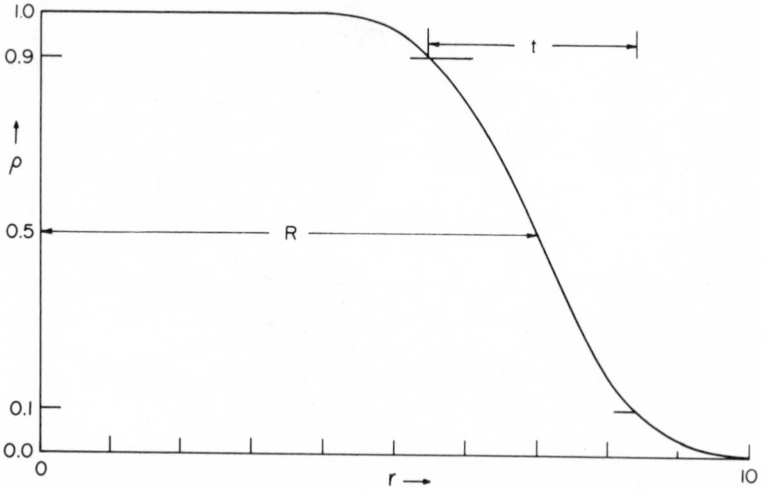

FIG. 1.2. Typical charge density. *R* is the nuclear radius and *t* is the surface thickness.

$\rho = 0$ outside the nucleus; there is a finite region over which ρ goes down to zero. This region is often called the *nuclear surface*; see Chapter II. The width of that region labeled t in Fig. 1.2 is defined to be the distance over which the density drops from 0.9 of its value at $r = 0$ to 0.1 of that value. Empirically t is a constant

$$t \simeq 2.4 \text{ fm} \tag{1.6}$$

2. NUCLEAR MASSES

Nuclear masses are measured, both directly and indirectly, through the energetics of specific reactions leading to the desired nucleus. What is usually tabulated is the mass of the *neutral atom* built around the nucleus in question. It is given in such units, the "unified scale," that make the atom of ^{12}C have exactly the atomic mass 12. In these units, 1 unit $= (1.66043 \pm 0.00002) \times 10^{-24}$ g $= 931.478 \pm 0.005$ MeV/c^2. The mass of the hydrogen atom is

$$M_H = 1.00782522 \text{ units} \tag{2.1}$$

The mass of the ^{16}O atom is

$$M_{^{16}O} = 15.99491494 \pm 0.00000028 \text{ units} \tag{2.2}$$

[In some older tabulations a "physical scale" or a "chemical scale" for atomic masses is used; the former assigns the exact mass 16 to ^{16}O, while the latter assigns it to the natural isotopic mixture of oxygen isotopes; in making precise calculations using tabulated nuclear masses it is thus advisable to make sure which mass unit is used. For complete discussions and tabulations, see E. R. Cohen and Jesse W. M. DuMond (65)].

The bare nuclear mass can be obtained from the atomic mass using the relation

$$M_{\text{nuc}} = M_{\text{ato}} - [Zm_e - B_e(Z)] \tag{2.3}$$

Here m_e is the mass of the electron

$$\begin{aligned} m_e &= (5.48597 \pm 0.00003) \times 10^{-4} \text{ units} \\ &= 511006 \pm 2 \text{ eV} \end{aligned} \tag{2.4}$$

and $B_e(Z)$ is the binding energy of the Z electrons in the neutral atom. A good estimate for $B_e(Z)$ is given by the expression derived from the Fermi–Thomas model of the atom [L. L. Foldy (51)]:

$$B_e(Z) = 15.73Z^{7/3} \text{ eV} \tag{2.5}$$

For the mass of the bare proton one obtains, using (2.3) and the measured binding energy of the electron:

$$\begin{aligned} M_p &= (1.00727663 \pm 0.00000008) \text{ units} \\ &= (938.256 \pm 0.005) \text{ MeV} \end{aligned} \tag{2.6}$$

while the measured mass of the neutron is

$$\begin{aligned} M_n &= (1.0086654 \pm 0.0000004) \text{ units} \\ &= (939.550 \pm 0.005) \text{ MeV} \end{aligned} \tag{2.7}$$

A detailed table of atomic masses, cross-checked by using various nuclear reactions, can be found in a paper by J. H. E. Mattauch, W. Thiele, and A. H. Wapstra, (65) [see also A. H. Wapstra and N. B. Gove (71)]. The standard deviations of the masses given there are generally less than a few parts per million.

Many derived quantities can be obtained from the data on masses. The nuclear binding energy $B(Z,N)$, for instance, is defined through the identity (neglecting small corrections due to the electronic binding energy)

$$M(Z,N) = ZM_H + NM_n - B(Z,N) \tag{2.8}$$

where $M(Z,N)$ is the atomic mass of a nucleus with Z protons and $N = A - Z$ neutrons and M_H is the hydrogen mass and M_n the neutron mass. $B(Z,N)$ is thus the energy that is required to completely break up the nucleus (Z,N) into its A-nucleons; it is given as a *positive* number, as implied by the minus sign in (2.8).

The nuclear binding energy $B(Z,N)$ is, generally speaking, an increasing function of Z and N; a derived quantity—the binding energy per nucleon: $B(Z,N)/(Z + N)$—remains rather constant from $A = 12$ and up. To a good approximation one finds empirically that

$$\frac{B(Z,N)}{A} \approx 8.5 \text{ MeV/nucleon} \tag{2.9}$$

Figure 2.1 shows the more detailed trend of the binding energy per nucleon as a function of A. It can be contrasted with a corresponding plot for the electronic binding energy per electron that, from (2.5), is seen to increase as $Z^{4/3}$.

If the binding of a system of n objects arises from the interaction between every pair of such objects, then, crudely speaking, the *total* binding energy of the system should increase with n in direct proportion to the number of pairs $\frac{1}{2}n(n-1)$; *the binding energy per object* should then increase linearly with n, which agrees, roughly, with the observation in atomic electron binding energies. The totally different behavior of nuclear binding energies, as indicated by (2.9), shows that nuclear interactions possess the *saturation* property; a nucleon in a nucleus interacts at most only with a small, fixed, number of nucleons, which is independent of the size of the nucleus. As the nucleus increases in size its *total* binding energy should then, crudely speaking, increase only linearly with A, leading to an A-independent binding energy per nucleon, as observed.

Result (2.9) is of course quite crude. There have been many attempts to provide a more accurate formula for $B(Z,N)$ or M in terms of a universal function of Z and N. These attempts were motivated on the one hand by the apparent regularities observed in tables of nuclear masses and, on the other, by some general theoretical considerations. They have therefore come to be

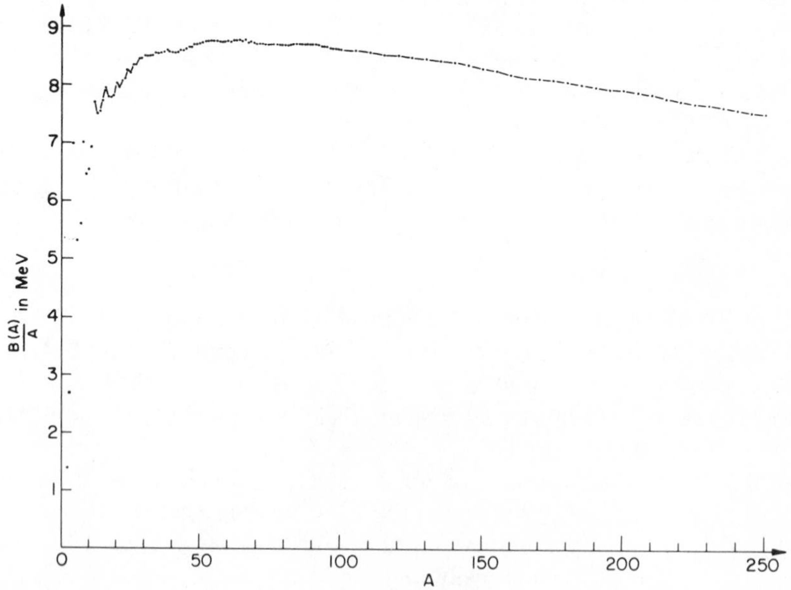

FIG. 2.1. Binding energy per nucleon as a function of A.

known as *semiempirical mass formulae*, the most commonly used being Weizsäcker's formula (see also Section II.3):

$$M = ZM_p + NM_n - a_1A + a_2A^{2/3} + a_3 \frac{Z^2}{A^{1/3}} + a_4 \frac{(Z - N)^2}{A} + \delta(A) \quad (2.10)$$

The numerical value for the coefficients a_i in this formula are given in Eq. II.3.2. The physical picture behind this formula is the following:

The nuclear mass consists mainly of the masses of the Z-protons and N-neutrons that constitute the given nucleus. The next term a_1A is in the form given by (2.9).* Since nuclear densities are to a first approximation independent of A, this term is proportional to the nuclear volume and is therefore called the *volume energy*. Since the nucleons at the nuclear surface contribute less to the total binding, the volume energy a_1A has to be reduced by a term proportional to the nuclear surface area: $+a_2A^{2/3}$. Then comes the Coulomb forces that, being repulsive, again increase the nuclear mass, this time in proportion to the square of the charge Z^2, and in inverse proportion to the nuclear radius, that is, to $A^{1/3}$. Of the last two terms—the a_4 term is the *symmetry energy* and $\delta(A)$ is the *pairing energy*. They have to do with the empirical observation that stable nuclei prefer, in the absence of Coulomb forces, to have as far as possible, equal numbers of protons and neutrons, and that an even number of protons or neutrons seems to be more strongly bound than an odd number of these nucleons.

It should be stressed that semiempirical mass formulae give only an average behavior of nuclear masses, and important systematic deviations of actual masses from the "predictions" of such formulae are observed locally in specific parts of the periodic table. These will be discussed later in Chapters II and III.

3. NUCLEAR FORCES

Having established the fact that nucleons bind themselves to each other in the nucleus, we naturally ascribe this binding to a force that must exist among nucleons. It is definitely a different force from the two forces already known to us—gravitation and electromagnetic—because both are too weak to explain the observed binding. The electromagnetic force is of course repulsive [in Weizsäcker's mass formula (2.10) it contributes to a *reduction* in the total binding energy].

---

*The constant a_i is, however, *not* 8.5 MeV but rather 15.68 MeV. The value 8.5 MeV is a consequence of the effect of all the terms in (2.10) over a limited range in A.

Several things can be said about this new force:

(a) Inside the nucleus the nuclear force is substantially stronger than the electromagnetic interactions, or otherwise most stable nuclei would not exist.

(b) The nuclear force is attractive, or at least has dominant attractive components in it. This, again follows from the fact the nuclei are bound.

(c) The nuclear force has a short range. That its range is shorter than interatomic distances we can conclude from the fact that at the molecular level there seems to be no necessity for forces other than electromagnetic to explain known phenomena. However, we can also put a much lower limit on its range. As we see from Fig. 2.1, the binding energy per nucleon seems to reach its saturation level already around $A = 10$ and, in fact, even for ^{4}He the binding energy per particle is nearly 90% of its average value all through the periodic table. It is thus evident that the nuclear interaction is "exhausted" already at distances of the order of magnitude of the size of the ^{4}He nucleus. We conclude, therefore, that the range of nuclear forces is of the order of magnitude of the radius of that nucleus, that equals approximately the average distance between nucleons in the nucleus, that is, between 1 and 2 fermis.

(d) Nuclear forces have the saturation property; in other words their general character is such that each one of them cannot interact with more than the few nucleons within its range of influence leading, as we mentioned before, to a total binding energy that increases linearly with A.

These qualitative conclusions about the nuclear force can be made more quantitative by a closer study of the two nucleon system. This shall be done in detail in Volume II, but we would like to summarize here some of the more detailed results:

(e) The strength of a force cannot be measured just by the binding energy that it produces. There is another energy to which it should be compared, as we shall now show. Consider the two nucleon system. If we force the nucleons to be within the range of their mutual (nuclear) interaction, we increase their relative momentum because of the uncertainty relation $\Delta p \Delta x \approx \hbar$. That leads to an increase in their kinetic energy in the center-of-mass system

$$E_{\text{kin}} = \frac{(\Delta p)^2}{2(M/2)} \approx \frac{\hbar^2}{M(\Delta x)^2} \tag{3.1}$$

[the reduced mass $M/2$ was used in (3.1) to emphasize that we are dealing with the relative kinetic energy]. The nuclear attraction has to overcome this kinetic energy in order to produce a bound state, so it is therefore natural to ask whether its binding energy is large or small compared to E_{kin} in (3.1.).

From the fact that there is only one bound state known for the two-nucleon system—the deuteron—we conclude that the nuclear interaction is barely strong enough to overcome the kinetic energy that results when the two nucleons come within each other's range. Thus the nuclear interaction is basically a *weak* interaction—that is, weak in comparison with its task of overcoming E_{kin} in (3.1). This situation can be contrasted with the Coulomb interaction in atoms: because of the infinite range of e^2/r it is always possible to go to large enough distances so that the kinetic energy (3.1), which decreases like $1/r^2$, will be small compared to the interaction energy. And, indeed, the hydrogen atom has many bound states!

(f) The nuclear force depends not only on the relative separation of the two nucleons, but also on their intrinsic degrees of freedom—the spin and the charge. Its dependence on the spins of the interacting nucleons can be inferred from the fact that the only bound *n-p* system—the deuteron—has the proton and the neutron spins parallel to each other giving rise to a total angular momentum $J = 1$. There is no known bound state for that *n-p* system where the two spins are antiparallel to each other leading to $J = 0$.

The spin dependence* of the nuclear interaction probably comes in three different ways: a direct *spin-spin interaction* of the form $\boldsymbol{\sigma}_1 \cdot \boldsymbol{\sigma}_2\, V(|\mathbf{r}_1 - \mathbf{r}_2|)$; a *spin-orbit interaction* of the type $[(\boldsymbol{\sigma}_1 + \boldsymbol{\sigma}_2)\cdot(\mathbf{r}_1 - \mathbf{r}_2) \times (\mathbf{p}_1 - \mathbf{p}_2)]\, V(|\mathbf{r}_1 - \mathbf{r}_2|)$ [note that $(\mathbf{r}_1 - \mathbf{r}_2) \times (\mathbf{p}_1 - \mathbf{p}_2)$ is proportional to the relative orbital angular momentum, hence, the name *spin-orbit* interaction for this part of the nucleon nucleon force]; and a *tensor interaction* of the type

$$\{3[\boldsymbol{\sigma}_1\cdot(\mathbf{r}_1 - \mathbf{r}_2)]\,[\boldsymbol{\sigma}_2\cdot(\mathbf{r}_1 - \mathbf{r}_2)] - (\mathbf{r}_1 - \mathbf{r}_2)^2\, \boldsymbol{\sigma}_1\cdot\boldsymbol{\sigma}_2\}\, V(|\mathbf{r}_1 - \mathbf{r}_2|).$$

The evidence for these various components in the nucleon-nucleon interaction comes from many sources, the most direct one being nucleon-nucleon scattering experiments. For the tensor force we have an additional very important evidence from the quadrupole moment of the deuteron, which is found to be positive and nonnegligible [$Q(H^2) = 2.78 \times 10^{-27}\ \text{cm}^2$]. We conclude that the deuteron has a preference for a shape similar to the one shown in Fig. 3.1*a*, rather than oscillating equally between the two shapes (Figs. 3.1*a* and 3.1*b*), which would have led to a zero quadrupole moment. A preference for one shape over the other can come only through an interaction that couples the spins to the vector $\mathbf{r} = \mathbf{r}_1 - \mathbf{r}_2$ and, in the case of the deuteron, this can be achieved only via a tensor force. We may recall that in its spin dependence this is a force similar to the one between two magnetic dipoles, but its strength in the deuteron is too large to be accounted for by the interaction between the magnetic moments of the proton and the neutron.

---

*The vector $\boldsymbol{\sigma}$ is the Pauli spin operator whose components satisfy

$$\sigma_a \sigma_b + \sigma_b \sigma_a = 2\delta_{ab}$$
$$\sigma_a \sigma_b = i\, \sigma_c \qquad a, b, c, \qquad \text{cyclical}$$

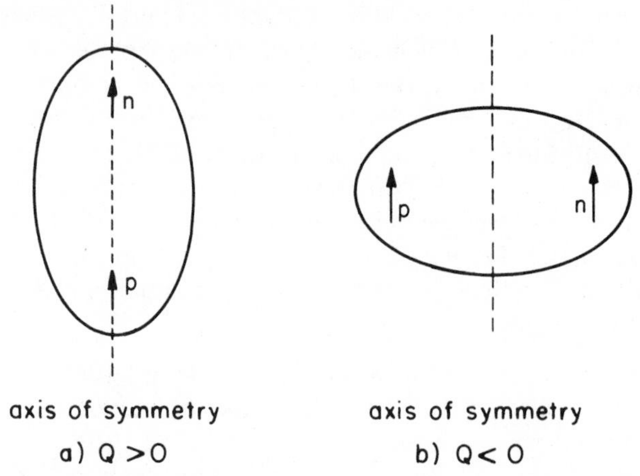

FIG. 3.1. Schematic mass distributions for the deuteron that lead to (a) positive and (b) negative quadrupole moments. The axis of symmetry is in the direction of the total angular momentum.

The dependence of the nuclear interaction on the charge variables is a very interesting one. As will be discussed in greater detail below, it was found that the nuclear *p-p* interaction (i.e., that part of their interaction that cannot be accounted for by the electromagnetic interaction between them) is very nearly the same as that of the *n-n* interaction. Furthermore, if a *p-n* pair is put in a state that is similar to that of a *p-p* or *n-n* pair, its interaction will again be the same as that of the *p-p* or the *n-n* pair. Because the Pauli principle operates for protons or neutrons but does not exclude states for the *p-n* system that are *symmetric* under the exchange of the spin and space variables, the *p-n* pair can be found also in states that have no counterpart in the *p-p* or *n-n* system. In such states its interaction is *different*, and as a matter of fact *stronger*, than that of the *p-p* or *n-n* pairs.

The equality between *p-p* and *n-n* nuclear interactions is known as *charge symmetry*, whereas the fact that in similar space-spin states the *p-n* interaction also turns out to be the same, is known as *charge independence*. Both charge symmetry and charge independence have important consequences with regard to nuclear structure. This will be discussed in detail in Chapter V.

Charge symmetry and charge independence are not satisfied exactly. Deviations from charge independence that ultimately have their origin in electromagnetic effects and the failure of charge symmetry are of the order of a few percent.

(g) An interaction between two nucleons, like every other interaction, involves the exchange of momentum between them. It has been found, however, that the two nucleons can also exchange their charges at the same time.

This is manifested most dramatically in *p-n* scattering experiments that show a peaking of the differential cross section at backward angles (in the center-of-mass system) as drawn in Fig. 3.2. Such peaking at back angles is most easily understood in terms of charge exchange, as indicated in Fig. 3.3; it is difficult to get a backward scattering as large as the forward scattering at such relatively high energies by any other mechanism. The component of the potential responsible for this effect is called the *exchange potential.*

(h) Nucleon-nucleon scattering at higher energies—200 MeV laboratory energy and more—shows that at very small separations the nucleon-nucleon interaction becomes very strongly repulsive. This is often referred to as a *hard core* in the nucleon-nucleon potential, whose radius, it turns out, has to

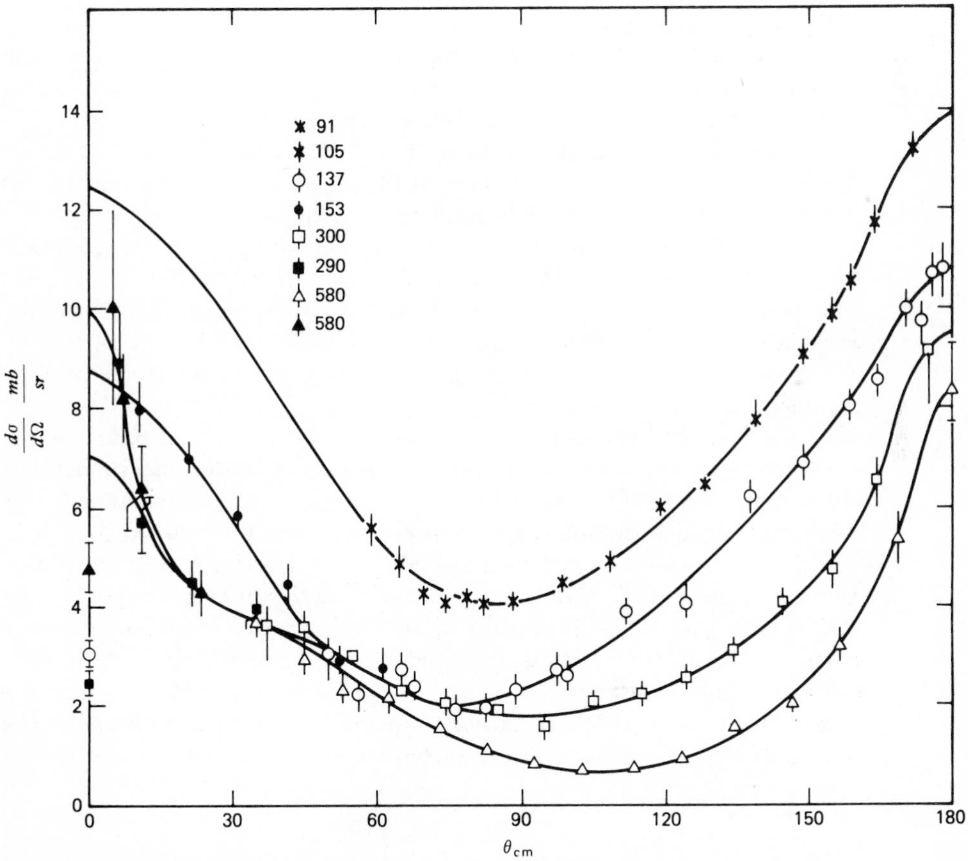

FIG. 3.2. *np*-angular distribution 90 to 580 MeV in the laboratory system. The lines are merely to guide the eye. [from Wilson (63)].

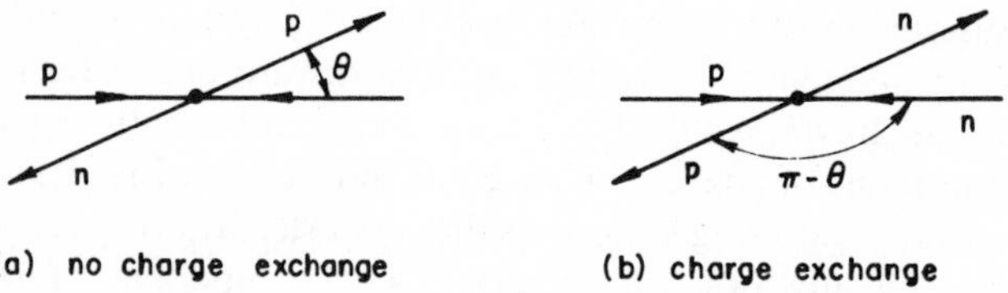

FIG. 3.3. Schematic description of charge exchange scattering; scattering of the original proton through the angle θ looks like a backward scattering by an angle $\pi-\theta$ if charge is exchanged.

be taken as $r_c \sim 0.5$ fm to fit the scattering data. It should be emphasized, however, that all we really know is that the interaction becomes repulsive at such short distances, but whether it really takes the form of an infinite repulsive hard core, or whether it takes other possible forms we do not really know at this stage. It is this repulsive part in the nucleon-nucleon interaction that is in part responsible for the saturation of nuclear forces, although it has been shown [Bethe (71)] that the tensor forces and the exchange potential also play important roles in determining the actual saturation density of nuclei.

Our understanding of nuclear forces is still rather limited. Most probably they arise out of the strong coupling of the nucleons to the various mesons that have been discovered. This coupling is manifested through the prolific production of mesons whenever a nucleon of high energy is decelerated. Like the electromagnetic bremstrahlung emitted by stopped electrons, nucleons emit their characteristic radiation when they are stopped. It is therefore reasonable to assume that nucleons can also emit virtual mesons, and by exchanging them exchange also momentum that gives rise to a force between the nucleons.

In electromagnetism it is easy to derive the important features of the Coulomb force from the analogous exchange of virtual photons. In fact, if a charged particle 1 emits a virtual photon of energy E and momentum E/c, it itself must recoil with momentum $p_1 = -E/c$; when the charged particle 2 absorbs the virtual photon it must also absorb its momentum and recoil with momentum $p_2 = +E/c$; thus an exchange of momentum $\Delta p = p_2 - p_1 = 2(E/c)$ took place. This whole process, since it violates conservation of energy by an amount E, can last only for a time $\Delta t = \hbar/E$, during which the photon can travel only a distance $r = c\,\Delta t = c\hbar/E$. $c\hbar/E$ must therefore be the separation between the two particles when they exchange momentum Δp during a time t. The force between these two particles at this distance is then

$$F = \frac{\Delta p}{\Delta t} = \frac{2(E/c)}{\hbar/E} = \frac{2}{\hbar c}\,E^2 = 2\hbar c\left(\frac{E}{\hbar c}\right)^2 = \frac{2\hbar c}{e^2}\,\frac{e^2}{r^2} \tag{3.2}$$

This force must be multiplied by the probability of absorbing and emitting the photon. From (3.2) we see that this probability must be of the order of the fine structure constant $(e^2/\hbar c)$ in order to obtain the Coulomb law of force.

It is obvious from (3.2) that the force at large distances is communicated by the low energy photons which, because they violate conservation of energy only by very little, can afford to stay outside the sources for a longer time and transfer their momentum over greater distances. However, because their energy is low they also carry small momentum and, hence, the decrease of the force with distance.

When we apply a similar argument to the nuclear force we find that because here the virtually exchanged particle has a mass, there is a *minimum* amount by which energy conservation must be violated: $\Delta E > m_\pi c^2$ where m_π is the mass of the lightest known meson that is strongly coupled to the nucleon:

$$\begin{aligned} m_\pi c^2 &= 139.576 \pm 0.011 \text{ MeV} \qquad \text{(charged pions)} \\ &= 134.972 \pm 0.012 \text{ MeV} \qquad \text{(neutral pion)} \end{aligned} \tag{3.3}$$

That means that there is an upper limit on the time a virtual meson can stay away from its source:

$$\Delta t < \hbar / m_\pi c^2$$

and there is consequently a maximum distance over which it can carry momentum:

$$\Delta r < c\,\Delta t = (\hbar / m_\pi c) \approx 1.4 \times 10^{-13} \text{ cm} \tag{3.4}$$

The range of nuclear forces is thus naturally related to the mass of the lightest observed meson [Wick (37)].

We see from these arguments that the heavier mesons, or virtual transitions that involve the emission of more than one meson, will be effective only at shorter distances, and the complexity of possible mesons that can be exchanged at such shorter distances has thus far prevented the derivation of any reliable force from such fundamental processes. Actually, the situation is even more complex since, although the *range* at which heavy mesons contribute to nuclear forces is smaller, the *strength* of their contribution may be much greater. The rho meson contribution still accounts for nearly 10% of the nuclear forces at 1.4 fm, although its mass, $m_\rho = 765.0$ MeV, is much larger than that of the pion.

The empirical data on the nuclear force at distances of 0.7 to 1.4 fm is also not complete. In fact, nuclear forces that differ considerably from each other in this region fit the data equally well [see Lomon and Feshbach (68)]. Still, the tail of the nucleon–nucleon potential, from distances $\mu^{-1} = \hbar/m_\pi c = 1.4$ fm and up, can be calculated on the basis that one pion exchange is responsible for that part of the interaction. One obtains then the OPEP (One Pion Exchange Potential) interaction that is given by

$$V_{12} = \frac{1}{3}\frac{g^2}{\hbar c}\, m_\pi c^2 (\boldsymbol{\tau}_1 \cdot \boldsymbol{\tau}_2)\left[(\boldsymbol{\sigma}_1 \cdot \boldsymbol{\sigma}_2) + \left(1 + \frac{3}{\mu r} + \frac{3}{(\mu r)^2}\right) S_{12}\right] \frac{e^{-\mu r}}{\mu r} \tag{3.5}$$

where $\boldsymbol{\tau}$ is the isospin operator (see below, Section I.6), and S_{12} is the tensor force operator,

$$S_{12} = \frac{1}{r^2} [3(\boldsymbol{\sigma}_1 \cdot \mathbf{r}) (\boldsymbol{\sigma}_2 \cdot \mathbf{r}) - (\boldsymbol{\sigma}_1 \cdot \boldsymbol{\sigma}_2) r^2] \tag{3.6}$$

and $r = |\mathbf{r}_1 - \mathbf{r}_2|$ is the mutual separation of the two nucleons. The coupling constant $g^2/\hbar c$ is determined by the π-nucleon coupling constant and is found to be

$$g^2/\hbar c = 0.081 \tag{3.7}$$

4. NUCLEAR SEPARATION ENERGIES

Nuclear masses can be used to obtain other important derived quantities besides the binding energies of Section 2. One such quantity is the separation energy S for a nucleon, or a group of nucleons, from the original nucleus. More precisely, the total mass of the nucleus (Z,N) is generally smaller than that of the nucleus $(Z,N-1)$ plus a free neutron; it is necessary to provide the nucleus (Z,N) with a certain amount of energy in order to "ionize" it and make it emit a neutron. We thus define

$$M(Z,N) = M(Z,N-1) + M_n - S_n \tag{4.1}$$

and call S_n the *neutron separation energy* in the nucleus (Z,N). It is seen from (2.8) and (4.1) that

$$S_n = B(Z,N) - B(Z,N-1) \tag{4.2}$$

Figures 4.1 to 4.3 show three sections from the chart of neutron and proton separation energies [N. B. Gove and M. Yamada (68)]. It exhibits a striking regularity: for any fixed number of neutrons (protons) the neutron (proton) separation energy $S_n(S_p)$ increases as the number of protons (neutrons) increases. This regularity obviously represents a refinement over the crude constant-binding-energy-per-nucleon rule (2.9). Indeed, if (2.9) were precisely valid we would have gotten for the neutron separation energy (4.2) the constant value $S_n = 8.5$ MeV. The fact that S_n is not a constant demonstrates the approximate nature of (2.9). The regular behavior of S_n as a function of both N and Z points at some regularity in the deviations of the actual binding energies from their average value (2.9).

Figures 4.1 to 4.3 also show that for a fixed proton (neutron) number, the neutron (proton) separation energy $S_n(S_p)$ *decreases* as the number of neutron (protons) increases. However, this decrease has sharp discontinuities in some regions. Thus while S_n decreases by about 300 keV in going from ^{137}Ce to ^{139}Ce from ^{141}Ce to ^{143}Ce, or from ^{143}Ce to ^{145}Ce, it decreases by nearly seven times that much—about 2 MeV—when we go from ^{139}Ce to ^{141}Ce. Similar dis-

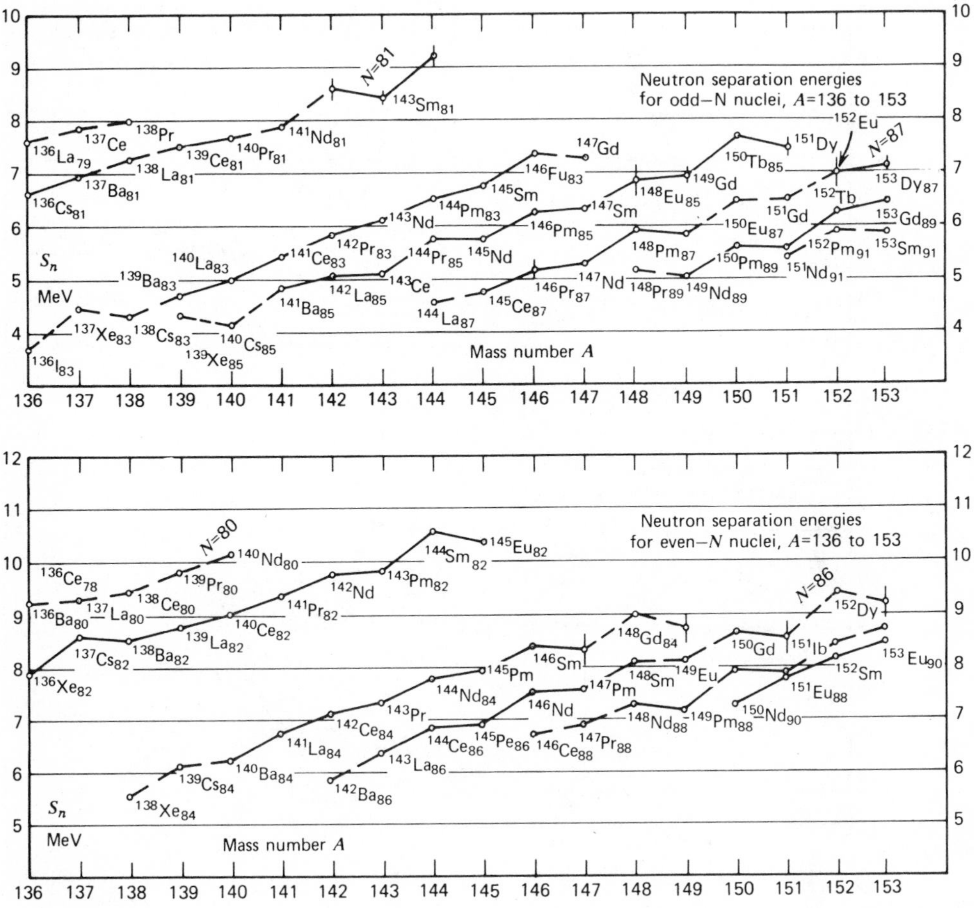

FIG. 4.1. Reproduced from N. B. Gove and M. Yamada, (68).

continuities exist in the chart of the proton separation energies. They can be summarized in saying that the 83rd or 84th nucleon (proton or neutron) is considerably less strongly bound than the 81st or 82nd proton or neutron. Drawing from the analogy with electronic binding energies, one then says that at neutron or proton number 82 a corresponding "shell" is being closed, so that the next nucleon seems to go to the next, less strongly bound, shell.

Similar effects are seen at both neutron and proton numbers

$$2, 8, 20, 28, 50, 82, \text{ and } 126 \tag{4.3}$$

These numbers thus seem to represent especially stable nuclear configurations, and they became known as the nuclear *magic numbers*. As we shall see later,

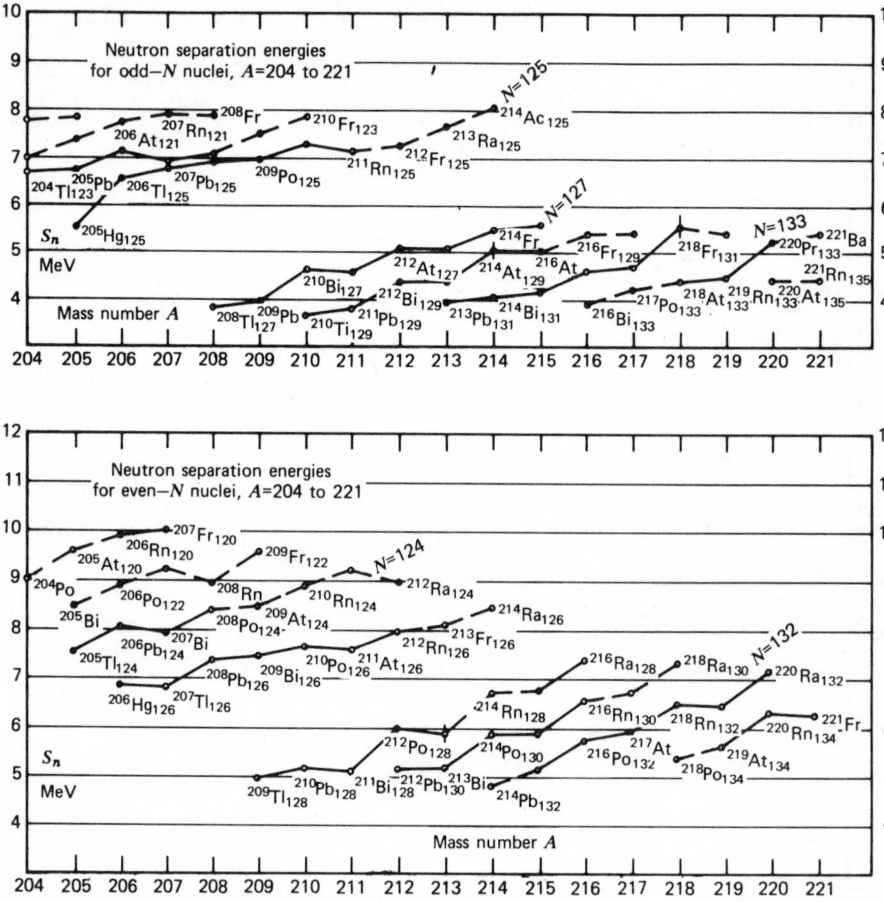

FIG. 4.2. Reproduced from N. B. Gove and M. Yamada, (68).

there are many other nuclear properties that exhibit a characteristic "discontinuity" at these numbers, strongly suggesting their interpretation in terms of closed nuclear shells in the nucleus. The nuclear shell model, which was developed to provide a simple, first order, description of nuclei, is largely based on these findings. It asserts that nuclei can be described as a collection of nucleons moving independently in well-defined orbits. These orbits, or *single-particle levels* as they are often called, are determined by an average smooth potential that takes into account in an average way the mutual interaction among the nucleons.

The neutron and proton separation energies exhibit also a characteristic odd-even structure. In fact, we obtain more regular patterns if we group separately S_n for odd-N (or odd-Z) and for even-N (or even-Z) nuclei. Further-

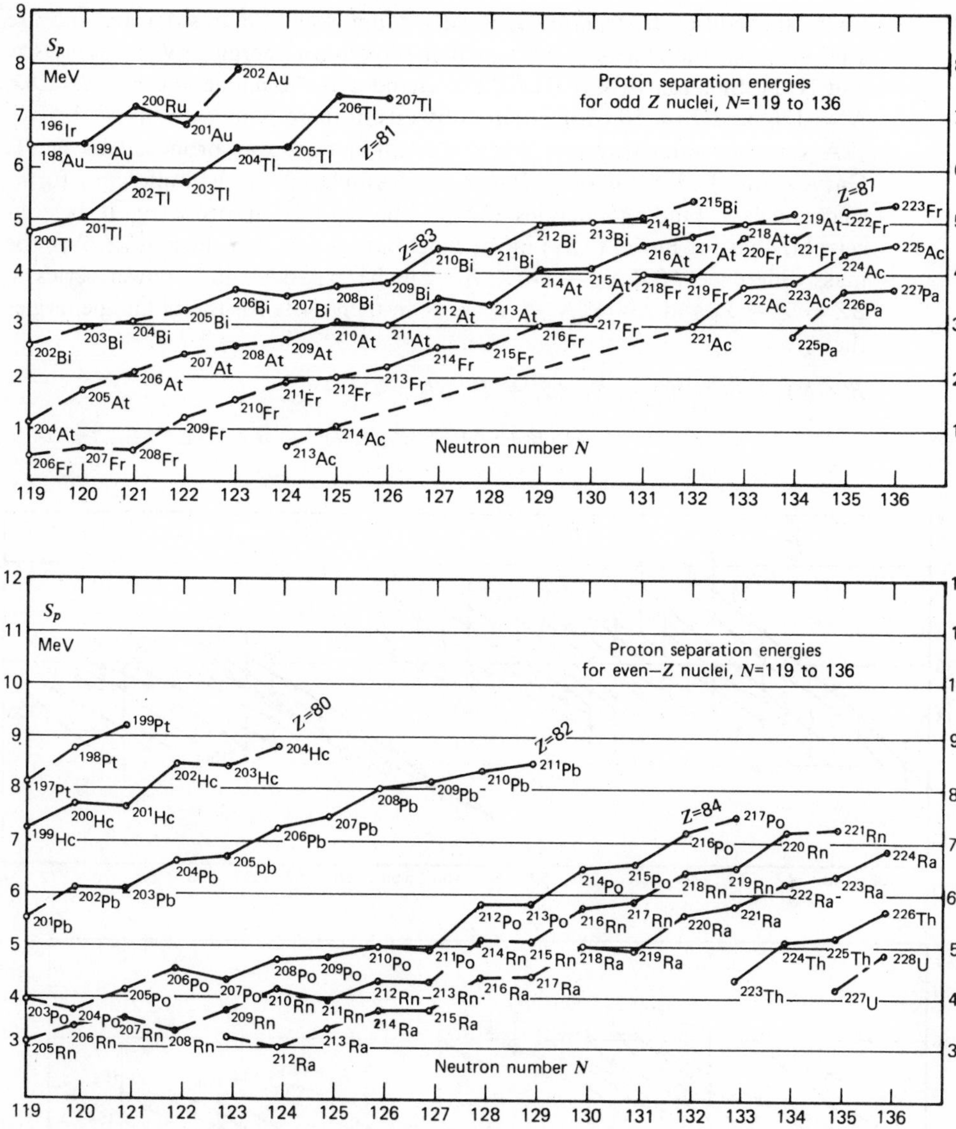

FIG. 4.3. Reproduced from N. B. Gove and M. Yamada, (68).

more, within each group of values of S_n, we note that S_n for a series of isotopes very often changes less in going from an odd-N nucleus to an even one, than in going from an even-N nucleus to the next odd-N nucleus. [The pairing term $\delta(A)$ in the Weizsäcker formula (2.10) is based on this observation.] It is therefore interesting to look also at the *neutron-pair separation* energies defined by

$$S_{2n} = B(Z,N) - B(Z,N-2) \tag{4.4}$$

where the odd-even structure is greatly suppressed. Figures 4.4 and 4.5 are sections from the charts of nuclear pair separation energies [V. A. Kravtsov and N. N. Skachkov (66)]. The sharp changes at the magic number $N = 82$, $N = 126$, and $Z = 50$ (Sn) is quite obvious in these plots as well.

A very revealing way to look at nuclear masses was proposed by G. T. Garvey and I. Kelson (66). It can be formulated in the following form: since nuclear binding energies exhibit the saturation property, it is conceivable that for nuclei (Z,N) in the neighborhood of any nucleus (Z_0,N_0), the mass difference $M(Z,N) - M(Z_0,N_0)$ can be expanded in a power series in $\Delta Z = Z - Z_0$ and $\Delta N = N - N_0$. We can then write for the binding energies that:

$$B(Z,N) = B(Z_0,N_0) + B_{10}\,\Delta Z + B_{01}\,\Delta N + B_{20}(\Delta Z)^2 + B_{02}(\Delta N)^2 + B_{11}\Delta Z\cdot\Delta N + \ldots \quad (4.5)$$

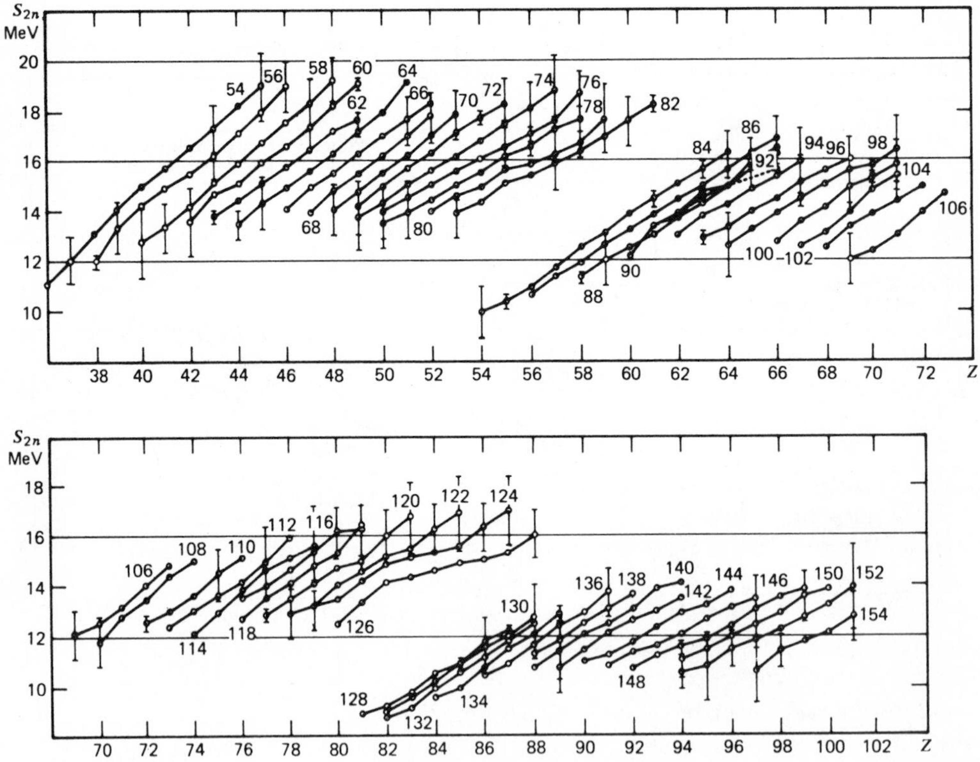

FIG. 4.4. The neutron separation energy $S_{2n}(Z)$ for $N = 54$ to 154. Points for nuclei with the same N-values are connected by line segments; N is indicated at the line. Errors are shown by error bars when greater than 100 keV. Errors of 1 MeV were assigned to binding energies obtained by interpolation. Reproduced from V. A. Kravtsov and N. N. Skachkov, (66).

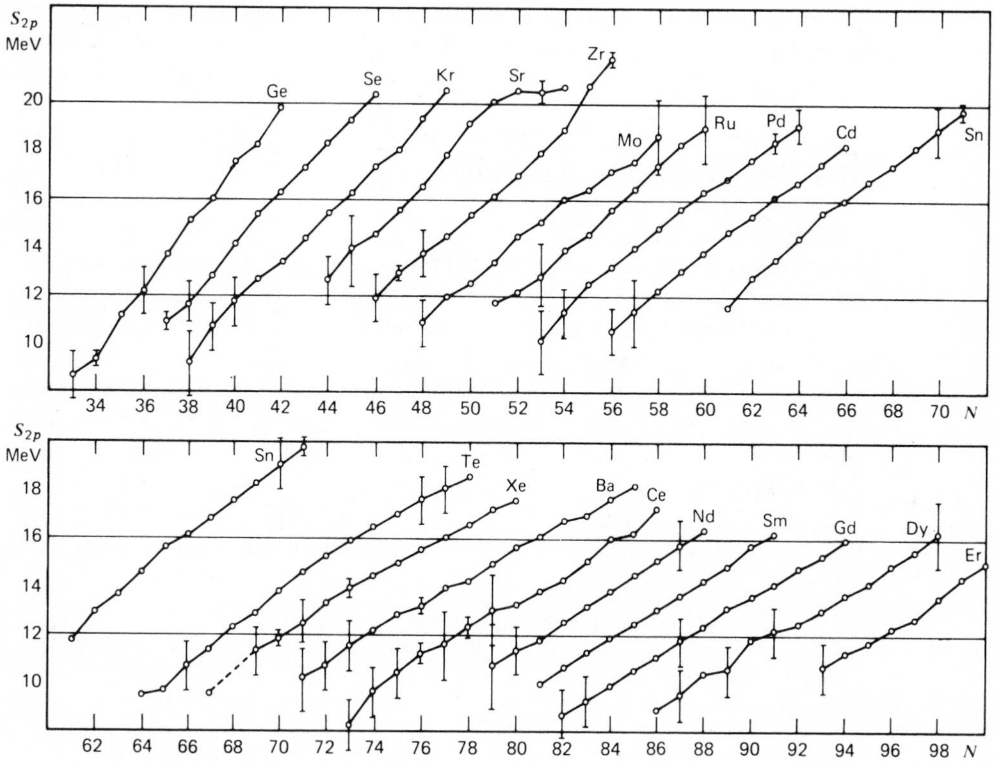

FIG. 4.5. The proton pair separation energy $S_{2p}(N)$ for $Z = 32$ to 68, Ge to Er. Points for nuclei with the same Z-values are connected by line segments. Error are shown by error bars when greater than 100 keV. Errors of 1 MeV were assigned to binding energies obtained by interpolation. Reproduced from V. A. Kravtsov and N. N. Skachkov, (66).

The coefficients B_{10} etc. are "partial derivatives" of $B(Z,N)$ taken at $(Z,N) = (Z_0,N_0)$. Assuming that it is a good approximation to terminate the series (4.5) after the second derivatives, we see that $B(Z,N)$ in the neighborhood of (Z_0,N_0), is given by the six constants: $B(Z_0,N_0)$, B_{10}, B_{01}, B_{20}, B_{02}, and B_{11}. It is therefore obvious that some linear relations can be established between the binding energies $B(Z,N)$ of at least six different nuclei around the nucleus (Z_0,N_0). It follows, for instance, from the expansion (4.5), provided it is terminated after the second derivatives, that

$$B(Z-2, N+2) - B(Z, N) + B(Z-1, N) - B(Z-2, N+1) + B(Z, N+1) - B(Z-1, N+2) = 0 \quad (4.6)$$

and also [see G. T. Garvey et al. (68), (69)]

$$B(Z, N+2) - B(Z-2, N) + B(Z-2, N+1) - B(Z-1, N+2) + B(Z-1, N) - B(Z, N+1) = 0 \quad (4.7)$$

To the extent that such relations actually hold around every point (Z_0, N_0) in the periodic table, one can draw further conclusions about the Z and N dependence of $B(Z, N)$. Indeed, (4.6) is satisfied for every Z and N only if

$$B(Z, N) = g_1(Z) + g_2(N) + g_3(Z + N) \tag{4.8}$$

while (4.7) is satisfied for every Z and N only if

$$B(Z, N) = f_1(Z) + f_2(N) + f_3(N - Z) \tag{4.9}$$

Figure 4.6 represents an attempt by Garvey et al. (68) to fit the available data on binding energies of nuclei to (4.6). $\Delta(Z, N)$ in Fig. 4.6 is the empirical value for the left-hand side of (4.6), and different points in the figure for the same value of $Z + N$ represent different initial pairs (Z, N). Note that the scale of $\Delta(Z, N)$ is given in *keV*; characteristic binding energies that enter into the left-hand side of (4.6) to produce these Δ's are in the neighborhood of 1000 *MeV*. The relative smallness of $\Delta(Z, N)$ thus indicates that (4.6) is fairly well satisfied by actual binding energies of nuclei.

Equation 4.7 is also rather well satisfied by the empirical data. Garvey et al. (68), (69) performed a least-square fit of the empirical binding energies to an expression of the type (4.9):

$$B(Z, N) = 114.0130 + f_1(Z) + f_2(N) + f_3(N - Z) \tag{4.10}$$

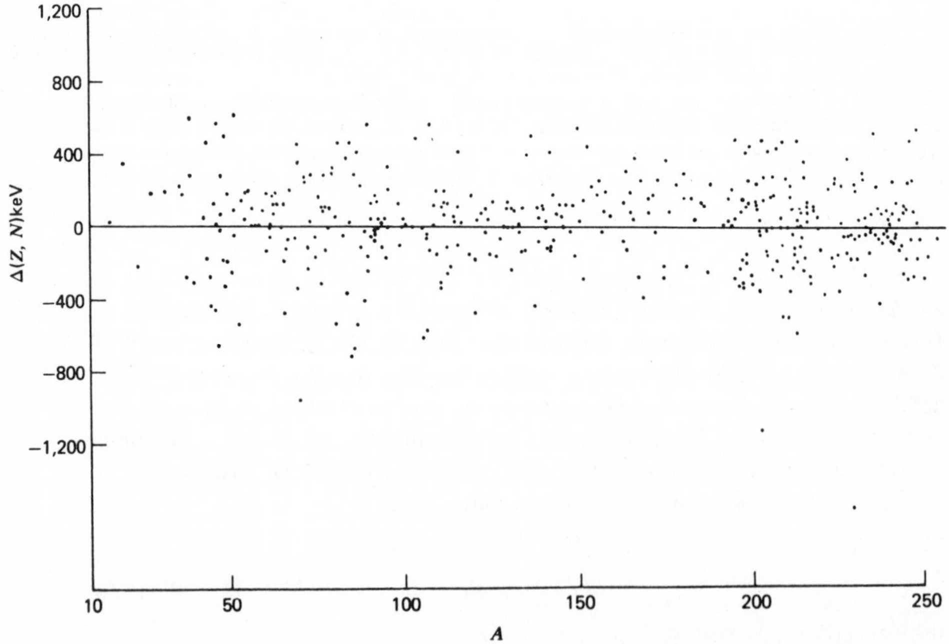

FIG. 4.6. $\Delta(Z, N)$ as a function of A (see text). Reproduced from G. T. Garvey et al., (68).

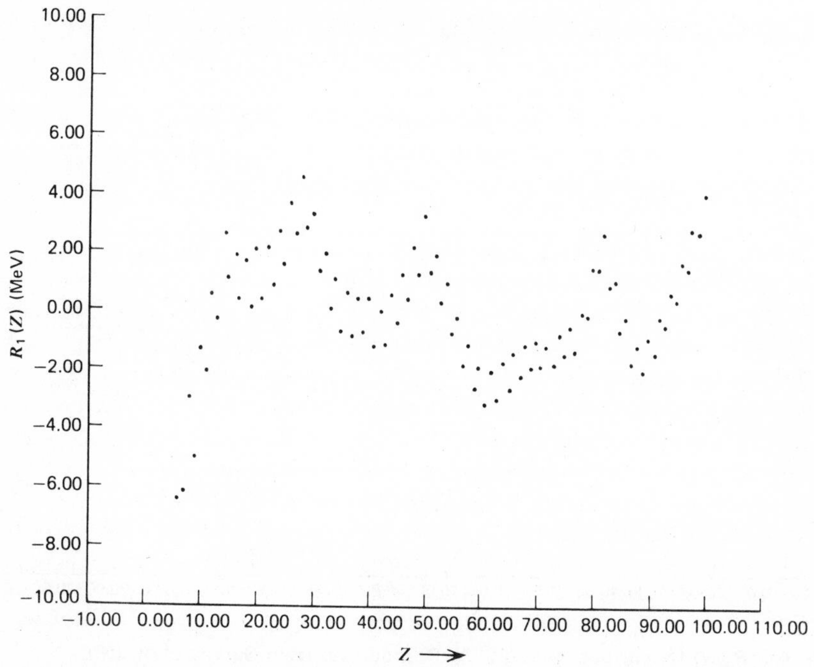

FIG. 4.7. $R_1(Z)$ for various values of Z. Reproduced from Garvey et al. (69).

The numerical results for the f's were found to be

$$f_2(Z) = [-65.4326 + 12.8234Z - 0.14203Z^2 + R_1(Z)]\ \text{MeV} \qquad Z > 6$$

$$f_1(N) = [-113.976 + 9.07899N + 0.551232N^2 + R_2(N)]\ \text{MeV} \quad N > 10$$

$$f_3(N - Z) = [11.2943 - 2.76076(N - Z) - 0.160758(N - Z)^2 + R_3(N - Z)]\ \text{MeV} \qquad N > Z \qquad (4.11)$$

The "residues" $R_1(Z)$, $R_2(N)$, and $R_3(N - Z)$ are relatively small corrections and their empirical values are given in Figs. 4.7 to 4.9. It is interesting to observe the "peaks" in $R_1(Z)$ and $R_2(N)$ at the magic numbers 28, 50, 82, and 126. They indicate again that around these numbers there are significant deviations from the "smooth" behavior of nuclear binding energies found elsewhere in the periodic table. The pairing effect is also clearly indicated by the grouping of the points into parallel curves for the heavier nuclei.

A comparison of (2.10) and (4.11) indicates that although the total nuclear binding energy $B(Z, N)$ turns out to be approximated fairly well by a linear function of $Z + N$, this linearity is less well satisfied by the functions $f_1(Z)$ and $f_2(N)$ *separately*. In particular, it is apparent that $f_1(Z)$, for larger values of Z, even changes its sign, making a negative contribution to the total binding energy of heavier nuclei.

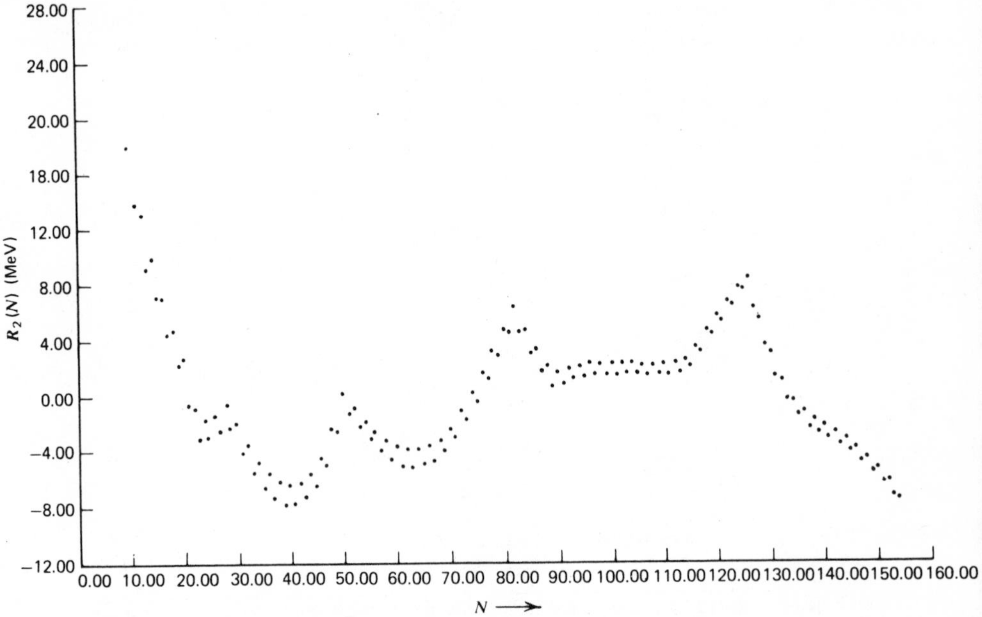

FIG. 4.8. $R_2(N)$ for various values of N. Reproduced from Garvey et al. (69).

This last result is not really surprising, since the repulsive Coulomb interaction between the protons in a nucleus is expected to make heavier nuclei relatively less tightly bound. The repulsive Coulomb energy is, of course, quadratic in Z since it does not possess the saturation property and all $\frac{1}{2}Z(Z - 1)$ pairs of protons are expected to make their contribution to this energy (see Eq. 2.8). The negative coefficient of Z^2 in (4.11) supports this interpretation, and its order of magnitude is consistent with the Coulomb energy of a uniformly charged sphere of radius R, $(3/5)\ (Z^2e^2/R)$, with a typical nuclear radius of $R \sim 5$ fermi.

5. COULOMB FORCES AND MIRROR NUCLEI

The study of the effects of Coulomb forces on nuclear masses has led tô the observation of a number of additional important regularities. One of these is exemplified in Fig. 5.1. Of the three fairly well-studied nuclei of mass $A = 27$: ^{27}Mg, ^{27}Al, and ^{27}Si, ^{27}Al turns out to be the most strongly bound. Yet, when one compares the excited states of ^{27}Al and ^{27}Si one finds great similarity between these two nuclei. In both nuclei excited states are found at about the same excitation energies and the angular momenta associated with these states

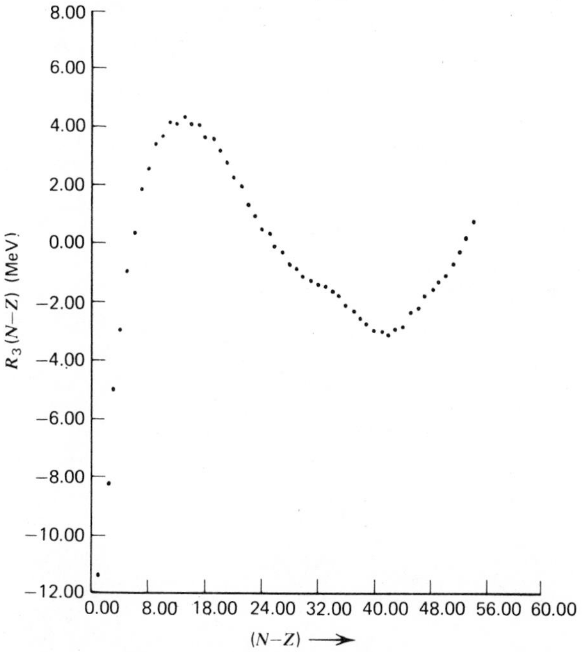

FIG. 4.9. $R_3(N\text{–}Z)$ for various values of $(N\text{–}Z)$. Reproduced from Garvey et al. (69).

are identical. The similarity between these two nuclei is even more striking when we compare them with the third $A = 27$ nucleus—^{27}Mg; the latter exhibits an entirely different sequence of angular momenta and altogether different excitation energies.

^{27}Al and ^{27}Si are often referred to as *mirror nuclei* because the number of neutrons (protons) in one is equal to the number of protons (neutrons) in the other. Many other mirror nuclei are known and a systematic study of these pairs of mirror nuclei reveals that the excitation spectrum of one member of the pair is always very similar to that of the other (see Fig. 5.2). On general grounds we are inclined to expect that an excitation spectrum of a system is intimately connected with its detailed internal structure and, hence, our general conclusion that *mirror nuclei have similar intrinsic structure.*

There is, however, one important difference between the two members of a pair of mirror nuclei: the one with the smaller Z is always found to be more strongly bound, the difference in binding energies between the two members of the pair being well approximated by

$$\Delta E_c \approx \frac{Ze^2}{R} \tag{5.1}$$

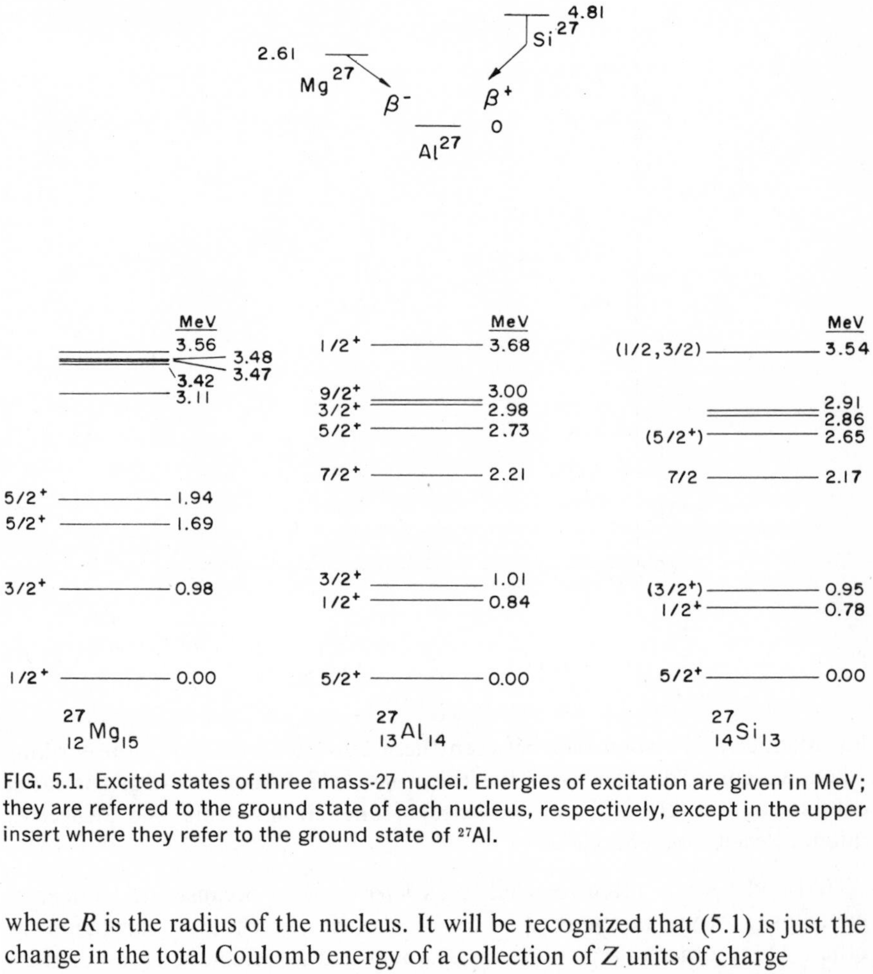

FIG. 5.1. Excited states of three mass-27 nuclei. Energies of excitation are given in MeV; they are referred to the ground state of each nucleus, respectively, except in the upper insert where they refer to the ground state of ^{27}Al.

where R is the radius of the nucleus. It will be recognized that (5.1) is just the change in the total Coulomb energy of a collection of Z units of charge

$$E_c \propto \frac{Z^2e^2}{R}$$

when we change the total charge by one unit.* The following picture therefore emerges:

Mirror nuclei have a similar intrinsic structure determined by the dynamical effects of the nuclear interactions. However the difference in the total charge between two mirror nuclei makes all the states of the member with the smaller charge more tightly bound by nearly the same amount ΔE_c.

*For a uniformly charged sphere of radius R, the constant of proportionality is 3/5. This is an overestimate since it neglects the exchange contribution. [(Cooper and E. Henley (53)].

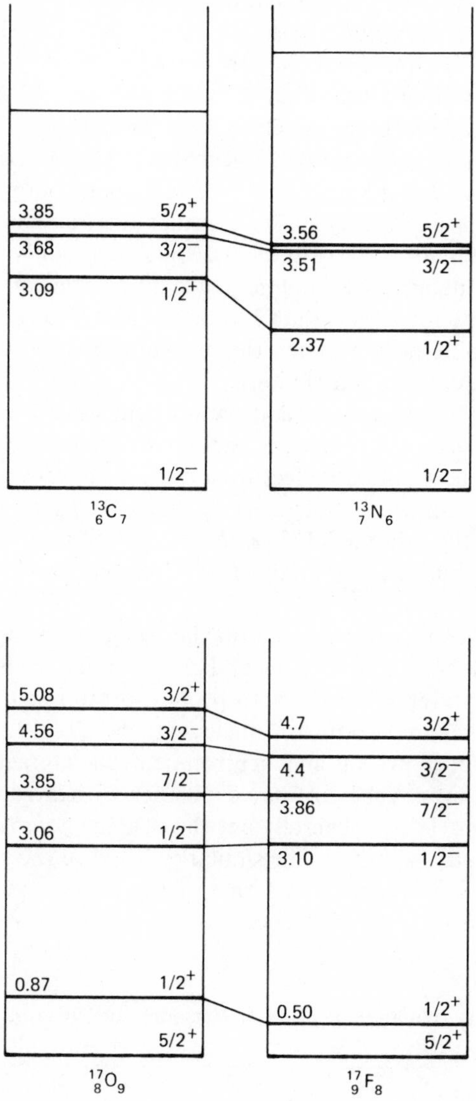

FIG. 5.2. Single-particle levels for mirror nuclei (in MeV). Talmi and Unna (60).

The fact that ΔE_c, which is about 4.8 MeV for the pair ^{27}Al-^{27}Si, remains the same for all excited states of both nuclei is reflected through the near equality of the excitation energies of corresponding states from the ground state (Figs. 5.1 and 5.2). It indicates that the nuclear size hardly changes for excited states of these nucleus.

In the example discussed above and shown in Fig. 5.1, ^{27}Mg has a smaller charge than ^{27}Al; since its size is known to be similar to that of ^{27}Al, its repulsive Coulomb energy must be smaller than that of ^{27}Al. Superficially we may expect then ^{27}Mg to be more tightly bound than ^{27}Al. Actually its ground state is *less* tightly bound than that of ^{27}Al by about 2.61 MeV (see insert in Fig. 5.1). We must conclude, therefore, that the nature of the nuclear forces makes a system of 12 protons and 15 neutrons (^{27}Mg) *less* tightly bound than that of 13 protons and 14 neutrons (^{27}Al) although both contain the same number of nucleons. Actually, if we consider the fact that Coulomb forces alone would have made ^{27}Al by about 4.5 MeV less tightly bound than ^{27}Mg, we conclude that nuclear forces alone make 12 protons and 15 neutrons less tightly bound by about 7 MeV than 13 protons and 14 neutrons (the observed 2.61 MeV mass difference between the ground states of ^{27}Mg and ^{27}Al plus the effect of the Coulomb interaction).

We are caught here in a somewhat paradoxical situation. On the one hand the similarity between the spectra of the mirror nuclei ^{27}Al and ^{27}Si suggests that, apart from the Coulomb effects, the dynamical structure of nuclei is the same whether the A-nucleon system is composed of n_1 protons and n_2 neutrons or vice versa. On the other hand if $(Z, N) \neq (Z', N')$ we see a marked difference between the nuclei (Z, N) and (Z', N') despite the fact that $Z + N = Z' + N'$.

The resolution of this paradox lies in the recognition that the existence of two distinct nucleons—the neutron and the proton—leads to three distinct classes of nuclear interactions: proton-proton, proton-neutron, and neutron-neutron. If we compare two mirror nuclei with each other, we are comparing essentially the proton-proton and neutron-neutron interactions; the proton-neutron interactions involve the same number of neutron-proton pairs for both mirror nuclei. This is generally not the case for nonmirror nuclei, as can be clearly seen from Table 5.1. The similarity between the structure of mirror nuclei can then be interpreted in terms of the charge symmetry of nuclear forces (see Section I.3); the difference between the structure of ^{27}Mg and ^{27}Al

TABLE 5.1 Number of Pairs of Nucleons for Different Mass-27 Nuclei.

Nucleus→ / Number of	^{27}Mg	^{27}Al	^{27}Si
Protons	12	13	14
Neutrons	15	14	13
p-p pairs	66	78	91
n-n pairs	105	91	78
p-n pairs	180	182	182

arises from the fact that a *p-n* pair can exist in states that are not allowed to a *p-p* or *n-n* pair and have a characteristically stronger interaction in these states (see Section I.3).

6. CHARGE SYMMETRY AND CHARGE INDEPENDENCE OF NUCLEAR FORCES

The two known nucleons—the neutron and the proton—differ from each other very significantly in that one is electrically charged and the other is electrically neutral. Yet in many other respects they markedly resemble each other. First, their masses are nearly equal:

$$\frac{M_n - M_p}{1/2(M_n + M_p)} \approx 1.4 \times 10^{-3} \tag{6.1}$$

Second, both are fermions having an internal spin $s = 1/2$. Their intrinsic magnetic moments are indeed different:

$$\begin{aligned} \mu_p &= 2.792782 \pm 0.000017 \text{ nuclear magnetons} \\ \mu_n &= -1.913148 \pm 0.000066 \text{ nuclear magnetons} \end{aligned} \tag{6.2}$$

But if we take into account the fact that as elementary fermions they are supposed to have the intrinsic magnetic moments

$$\mu_p^{(0)} = 1 \text{ nm} \qquad \mu_n^{(0)} = 0 \text{ nm} \tag{6.3}$$

we see that the deviations of the actual moments from those expected for elementary structureless fermions are again nearly equal in magnitude:

$$\mu_p - \mu_p^{(0)} \approx 1.79 \text{ nm} \qquad \mu_n - \mu_n^{(0)} \approx -1.91 \text{ nm} \tag{6.4}$$

The similar structure of mirror nuclei discussed in Section 5 brings out another similarity between the two nucleons: the charge symmetry of the *n-n* interaction and the nuclear part of the *p-p* interaction (we stress the nuclear part in the *p-p* interaction since the similarity between the structure of mirror nuclei comes out only after correcting for the Coulomb interaction as explained previously).

It is convenient for these reasons, as well as for other reasons that will become clear later, to consider the neutron and the proton as two states of the same entity—the nucleon. This is to be understood in the same sense that positively polarized electrons and negatively polarized electrons are two states of the same entity—the electron. We should then write a nucleon's wave function as a function that has two components in the "nucleon's space," much the same as the electron's wave function has two components in the spin space. Thus we write

$$\psi(\mathbf{r}) = \begin{pmatrix} \psi_+(\mathbf{r}) \\ \psi_-(\mathbf{r}) \end{pmatrix} \tag{6.5}$$

and adopt the following convention. If the nucleon wave function $\psi(\mathbf{r})$ is that of a proton its structure (6.5) will be

$$\psi_p(\mathbf{r}) = \begin{pmatrix} \psi_+(\mathbf{r}) \\ 0 \end{pmatrix} \tag{6.6}$$

whereas if it is a wave function of a neutron its structure will be

$$\psi_n(\mathbf{r}) = \begin{pmatrix} 0 \\ \psi_-(\mathbf{r}) \end{pmatrix} \tag{6.7}$$

Thus the nucleon wave function (6.5) describes a nucleon with the probability amplitude ψ_+ of being a proton and ψ_- of being a neutron. It should be remembered that since the nucleon is a fermion with $s = \frac{1}{2}$, each of the functions $\psi_+(\mathbf{r})$ and $\psi_-(\mathbf{r})$ in (6.5) has two components in spin space, which we may label with arrows as $\psi_{+\uparrow}(\mathbf{r})$ and $\psi_{+\downarrow}(\mathbf{r})$. It is too cumbersome to denote explicitly all the components of a wave function, and we shall usually drop the indices, or the symbols, which are not relevant to the specific point we wish to make. The reader is advised, however, to develop the habit of keeping in mind the full structure of these wave functions.

Because of the analogy of (6.5) with the spin-space structure of wave functions, it is common to call $\psi_+(\mathbf{r})$ and $\psi_-(\mathbf{r})$ the two components of the nucleon wave function $\psi(\mathbf{r})$ in the charge space, or in isospace. For now this is a pure formality, and we can conceive only of states of the form (6.6) or (6.7) in which one of the components vanishes. However, as we shall soon see, the actual similarity between the proton and the neturon makes it sensible to talk of a nucleon state that has mixed properties of a proton and a neutron.

To be able to make full use of the isospace formalism we should introduce also some operators operating in this space.

There are only four independent operators in the space of dichotomic variables, that is, in a space of two-component wave functions like (6.5). These four independent operators can be chosen in different ways, but it is most convenient to use the three hermitian Pauli matrices and the identity operator. We thus introduce the operators

$$1 = \begin{pmatrix} 1 & 0 \\ 0 & 1 \end{pmatrix} \qquad \tau_1 = \begin{pmatrix} 0 & 1 \\ 1 & 0 \end{pmatrix} \qquad \tau_2 = \begin{pmatrix} 0 & -i \\ i & 0 \end{pmatrix} \qquad \tau_3 = \begin{pmatrix} 1 & 0 \\ 0 & -1 \end{pmatrix} \tag{6.8}$$

and proceed to formulate our findings and conjectures in terms of these operators.

We notice first that the operators τ_i satisfy the commutation relations of the Pauli spin matrices:

$$\tau_1\tau_2 = i\tau_3 \qquad \tau_2\tau_3 = i\tau_1 \qquad \text{and} \qquad \tau_3\tau_1 = i\tau_2 \tag{6.9}$$

Second, it is obvious that the operator τ_3 tells us whether a given wave function describes a proton or a neutron. When operating on the wave functons (6.6) and (6.7), τ_3 yields:

$$\begin{aligned} \tau_3\psi_p(\mathbf{r}) &= +\psi_p(\mathbf{r}) \\ \tau_3\psi_n(\mathbf{r}) &= -\psi_n(\mathbf{r}) \end{aligned} \tag{6.10}$$

We can interpret (6.10) by saying that the proton belongs to the eigenvalue of $+1$ of τ_3, whereas the neutron belongs to the eigenvalue -1.

If τ_{i3} is the τ_3 operator that operates on the ith nucleon, then it follows from (6.8) that the charge of the ith nucleon is given by

$$e_i\psi(i) = e\,\frac{1 + \tau_{i3}}{2}\,\psi(i) \tag{6.11}$$

where e is the unit of charge, that is, of a positron. The charge of a nucleon in units of e is thus the eigenvalue of the operator

$$q_i = \tfrac{1}{2}(1 + \tau_{i3})$$

and the total charge of a system of A-nucleons, in units of e, is given by the operator

$$Q = \frac{1}{2}\sum_{i=1}^{A}(1 + \tau_{i3}) \tag{6.12}$$

The operator representing the Coulomb interaction of a system of A-nucleons can now be written as:

$$H_c = \frac{e^2}{4}\sum_{i<j}^{A}\frac{(1 + \tau_{i3})(1 + \tau_{j3})}{|\mathbf{r}_i - \mathbf{r}_j|} \tag{6.13}$$

When written in the form (6.13) the Coulomb interaction takes place, formally, between *all* the nucleons including the neutrons. The operators τ_{i3} (6.10) see to it that actual contributions come from the protons only.

The total charge of a system of nucleons is known to be conserved. Thus if ψ is an A-nucleon wave function belonging to an eigenvalue E of the Hamiltonian H, then

$$Q\psi = \frac{1}{2}\left[\sum_{i=1}^{A}(1 + \tau_{i3})\right]\psi = Z\psi$$

is also an eigenfunction of H belonging to the same eigenvalue. In other words every nuclear wave function must satisfy

$$H(Q\psi) = Q(H\psi) \tag{6.14}$$

where Q is the operator (6.12). It follows that every nuclear Hamiltonian must satisfy

$$HQ - QH = [H, Q] = 0$$

or, since $Q = (1/2)A + T_3$ where $T_3 \equiv \frac{1}{2}\sum_{i=1}^{A} \tau_{i3}$,

$$[H, T_3] = 0 \tag{6.15}$$

It is seen immediately that the Coulomb interaction H_c, given by (6.13) satisfies (6.15). It is less trivial to see that the OPEP potential (3.5) commutes with Q, since the operator $\boldsymbol{\tau}_i \cdot \boldsymbol{\tau}_k \equiv \tau_{i1}\tau_{k1} + \tau_{i2}\tau_{k2} + \tau_{i3}\tau_{k3}$ does not commute separately with either τ_{i3} or τ_{k3}:

$$\begin{aligned} [\boldsymbol{\tau}_i \cdot \boldsymbol{\tau}_k, \tau_{i3}] &= -i\tau_{i2}\tau_{k1} + i\tau_{i1}\tau_{k2} \\ [\boldsymbol{\tau}_i \cdot \boldsymbol{\tau}_k, \tau_{k3}] &= -i\tau_{i1}\tau_{k2} + i\tau_{i2}\tau_{k1} \end{aligned} \tag{6.16}$$

However, it is still true that

$$[\boldsymbol{\tau}_i \cdot \boldsymbol{\tau}_k, \tau_{i3} + \tau_{k3}] = 0 \tag{6.17}$$

so that Q does commute with OPEP.

We notice in passing that (6.16) actually says that the *individual* charge of each nucleon separately is not conserved by OPEP; only the *total* charge is conserved. OPEP may therefore lead to charge exchange between nucleons and, in fact, it does whenever it exchanges a charged pion between a proton and a neutron as in charge exchange scattering (Section I.3).

We also note that the isospin operators $\boldsymbol{\tau}_i$ and $\boldsymbol{\tau}_k$, which behave like vectors in isospace, appear in (3.5) in the *scalar* product $\boldsymbol{\tau}_i \cdot \boldsymbol{\tau}_k$; it thus commutes with the operator $T_3 = \sum \tau_{i3}$ that represents rotations around the 3-axis in isospace.

The statement that nuclear forces are charge symmetric can also be formulated in terms of the isospin operators. In fact we notice that the operator τ_1 has the property of converting a neutron state into a proton state and vice versa:

$$\begin{aligned} \tau_1\psi_p(\mathbf{r}) &= \psi_n(\mathbf{r}) \\ \tau_1\psi_n(\mathbf{r}) &= \psi_p(\mathbf{r}) \end{aligned} \tag{6.18}$$

As a consequence the operator

$$\Pi_\tau(1, \ldots, A) \equiv \tau_1(1)\tau_1(2)\tau_1(3) \ldots \tau_1(A) \tag{6.19}$$

operating on $\psi(1, \ldots, A)$ will yield a wave function in which every neutron is converted into a proton and vice versa. If nuclear forces are charge symmetric, the energy of the system is unchanged by this operation:

$$H(\Pi_\tau\psi) = E(\Pi_\tau\psi) = \Pi_\tau(H\psi) \tag{6.20}$$

If this result is to be valid for any eigenfunction of H, it follows that

$$[H, \Pi_\tau] = 0 \tag{6.21}$$

It is easy to verify that $\boldsymbol{\tau}(1)\cdot\boldsymbol{\tau}(2)$ does commute with Π_τ and, hence, we can conclude that OPEP, (3.5), is charge symmetric. The Coulomb interaction, (6.13) does not commute with Π_τ;

$$[H_c, \Pi_\tau] \neq 0 \tag{6.22}$$

This is not surprising since the Coulomb energy will obviously change (except for $N = Z$) if all protons are changed into neutrons and vice versa.

We have seen that charge conservation amounts to the statement (6.15) that H commutes with T_3; charge symmetry requires that it commutes with Π_τ (Eq. 6.21). We shall now show that charge independence requires that H commutes with T_1, T_2, and T_3, that is, with the whole isovector $\mathbf{T}$:

$$[H, \mathbf{T}] = 0 \tag{6.23}$$

To see, however, how this comes about we have to be more specific on what we mean by saying that "charge independence implies equal *n-p* and *n-n* interactions when both pairs are in the same space-spin states." The *p-p* or *n-n* system is restricted by the Pauli principle to antisymmetric states only

$$\psi_{nn}(\mathbf{r}_1 s_1, \mathbf{r}_2 s_2) = -\psi_{nn}(\mathbf{r}_2 s_2, \mathbf{r}_1 s_1) \tag{6.24}$$

where s_i stands for the spin coordinate of the ith particle. No such restriction exists for the *p-n* system, but it is possible to break up any *p-n* wave function into a part that behaves like (6.24) and a part that is symmetric with respect to the exchange of particles 1 and 2.

Suppose the neutron-proton wave function is $\psi(\mathbf{r}_p s_p, \mathbf{r}_n s_n)$. We may remove the explicit charge labels from $\mathbf{r}$ and s on the charge of the nucleon by introducing the basis vectors of a spin space to be referred to as *isospin space*

$$\chi_p = \begin{pmatrix} 1 \\ 0 \end{pmatrix} \qquad \chi_n = \begin{pmatrix} 0 \\ 1 \end{pmatrix} \tag{6.25}$$

Then

$$\psi(\mathbf{r}_p s_p, \mathbf{r}_n s_n) \equiv [\psi(\mathbf{r}_1 s_1, \mathbf{r}_2 s_2]\, \chi_p(1)\chi_n(2) \tag{6.26}$$

The isospin wave function $\chi_p(1)\chi_n(2)$ tell us that particle 1 is a proton, particle 2 is a neutron. We may now break $\psi(\mathbf{r}_1 s_1, \mathbf{r}_2 s_2)$ up into two parts that are symmetric or antisymmetric against space and spin exchange, respectively, so that

$$\psi(\mathbf{r}_1 s_1, \mathbf{r}_2 s_2) \equiv \frac{1}{\sqrt{2}} [\psi^{(0)}(\mathbf{r}_1 s_1, \mathbf{r}_2 s_2) + \psi^{(1)}(\mathbf{r}_1 s_1, \mathbf{r}_2 s_2)] \tag{6.27}$$

where

$$\psi^{(0)}(\mathbf{r}_1 s_1, \mathbf{r}_2 s_2) \equiv \frac{1}{\sqrt{2}} [\psi(\mathbf{r}_1 s_1, \mathbf{r}_2 s_2) + \psi(\mathbf{r}_2 s_2, \mathbf{r}_1 s_1)]$$

$$\psi^{(1)}(\mathbf{r}_1 s_1, \mathbf{r}_2 s_2) \equiv \frac{1}{\sqrt{2}} [\psi(\mathbf{r}_1 s_1, \mathbf{r}_2 s_2) - \psi(\mathbf{r}_2 s_2, \mathbf{r}_1 s_1)] \tag{6.28}$$

If the *p-n* wave function in (6.26) depends spatially only on the relative separation of *p* and *n* (and not, say, on their separate coordinates, which would have been the case for a *p-n* pair in an external Coulomb field), then it really makes no difference whether particle 1 is the proton and particle 2 the neutron or vice versa. We could therefore use linear combinations of both situations. Nature chose to have the antisymmetric combination of the isospin wave functions go with the symmetric space-spin wave function $\psi^{(0)}$, while the symmetric combination will be associated with $\psi^{(1)}$: (we could have just as well chosen, at this stage of our discussion the opposite combination, but we shall see later that charge exchange forces introduce a *physical* difference between the two choices. Both combinations are eigenvectors of $\Pi_\tau(1, 2)$ but with eigenvalues of opposite sign).

$$\psi_{pn}^{(0)}(\mathbf{r}_1 s_1, \mathbf{r}_2 s_2) = \tfrac{1}{2}[\psi_{pn}(1,2) + \psi_{pn}(2,1)]\,[\chi_p(1)\,\chi_n(2) - \chi_n(1)\,\chi_p(2)] \tag{6.29}$$

$$\psi_{pn}^{(1)}(\mathbf{r}_1 s_1, \mathbf{r}_2 s_2) = \tfrac{1}{2}[\psi_{pn}(1,2) - \psi_{pn}(2,1)]\,[\chi_p(1)\,\chi_n(2) + \chi_n(1)\,\chi_p(2)] \tag{6.30}$$

Equation 6.30 can be supplemented now with the *p-p* and *n-n* wave functions written in the same formalism

$$\begin{aligned}\psi_{pp}^{(1)}(\mathbf{r}_1 s_1, \mathbf{r}_2 s_2) &= \psi_{pp}(1,2)\,\chi_p(1)\,\chi_p(2) \\ \psi_{nn}^{(1)}(\mathbf{r}_1 s_1, \mathbf{r}_2 s_2) &= \psi_{nn}(1,2)\,\chi_n(1)\,\chi_n(2)\end{aligned} \tag{6.31}$$

In (6.29) to (6.31) 1 and 2 stand for the space and spin coordinates of particles 1 and 2, ψ_{pp} and ψ_{nn} being antisymmetric [see (6.24)].

By inspecting (6.29) to (6.31) we can now formulate a *Generalized Pauli Principle*: two nucleon wave functions have to be antisymmetric with respect to the simultaneous exchange of their space, spin, and isospin coordinates. We further notice that the isospin wave functions that go with the three wave functions $\psi_{pn}^{(1)}$, $\psi_{nn}^{(1)}$, and $\psi_{pp}^{(1)}$ are the analogs of the ordinary spin combinations for the $S = 1$ wave functions of two spin-$\frac{1}{2}$ particles. We therefore introduce the *total isospin* operator

$$\mathbf{T} = \mathbf{t}_1 + \mathbf{t}_2 \tag{6.32}$$

where $\mathbf{t} = (1/2)\boldsymbol{\tau}$ and rewrite (6.29) to (6.31) in the form

$$\begin{aligned}\psi_{pn}^{(0)}(\mathbf{r}_1 s_1, \mathbf{r}_2 s_2) &= \frac{1}{\sqrt{2}}\,[\psi_{pn}(1,2) + \psi_{pn}(2,1)]\,\chi(12, T = 0, M_T = 0) \\ \psi_{pn}^{(1)}(\mathbf{r}_1 s_1, \mathbf{r}_2 s_2) &= \frac{1}{\sqrt{2}}\,[\psi_{pn}(1,2) - \psi_{pn}(2,1)]\,\chi(12, T = 1, M_T = 0) \\ \psi_{pp}^{(1)}(\mathbf{r}_1 s_1, \mathbf{r}_2 s_2) &= \psi_{pp}(12)\,\chi(12, T = 1, M_T = +1) \\ \psi_{nn}^{(1)}(\mathbf{r}_1 s_1, \mathbf{r}_2 s_2) &= \psi_{nn}(12)\,\chi(12, T = 1, M_T = -1)\end{aligned} \tag{6.33}$$

where $\chi(12,TM_T)$ is a two-particle isospin wave function of the two nucleons corresponding to total isospin T and a total three-component $T_3 = M_T$.

Charge independence now says that the nucleon-nucleon interaction is the same for all two particle $T = 1$ systems. The *p-n* system can be either in the state $\psi_{pn}^{(0)}$ or $\psi_{pn}^{(1)}$. In the state $\psi_{pn}^{(1)}$, its energies are the same as in the corresponding states $\psi_{pp}^{(1)}$ and $\psi_{nn}^{(1)}$ of the *p-p* and *n-n* system. It is implied that as a function of the two space and spin coordinates 1 and 2, the three functions $\psi_{pn}^{(1)}$, $\psi_{pp}^{(1)}$, and $\psi_{nn}^{(1)}$, determined by the nuclear Hamiltonian, are identical with each other. This is of course true only in the absence of charge independence breaking interactions such as the Coulomb interaction (6.13).

We see from (6.33) that with our choice of isospin wave functions charge independence says two things:

(a) The two nucleon states are split according to their total isospin.

(b) For a given total isospin, the energies are independent of M_T.

These two conclusions are similar to the well-known conclusions about the space-rotational invariance of the Hamiltonian of closed systems which make J a good quantum number and the energies degenerate with respect to $J_z = M$. Thus, in a similar way, charge independence amounts to the invariance of the Hamiltonian with respect to any rotation in isospin, and we can formulate it in the form of (6.23).

It is worthwhile to note that charge independence *does not* require that $\psi^{(0)}$ and $\psi^{(1)}$ have the same energies. In fact, if we consider OPEP and note that

$$\begin{aligned}\langle 12, TM_T|(\boldsymbol{\tau}_1\cdot\boldsymbol{\tau}_2)|12, TM_T\rangle &= 4\langle 12, TM_T|(\mathbf{t}_1\cdot\mathbf{t}_2)|12, TM_T\rangle \\ &= 2\langle 12, TM_T|T^2 - \mathbf{t}_1{}^2 - \mathbf{t}_2{}^2|12, TM_T\rangle \\ &= 2[T(T+1) - \tfrac{3}{4} - \tfrac{3}{4}] = \begin{cases} -3 & T = 0 \\ +1 & T = 1 \end{cases}\end{aligned} \tag{6.34}$$

we see that it leads to contributions of an opposite sign for $T = 0$ and $T = 1$. OPEP is attractive for $T = 0$ and $S = 1$ (like the ground state of the deuteron) and repulsive for $T = 1$ and $S = 1$. We note that the levels in the mirror nuclei ^{27}Al and ^{27}Si that match (see Section 5) have the same total isospin, $T = \frac{1}{2}$ their values of M_T differing while the low lying levels of ^{27}Mg belong to $T = 3/2$.

To see further the operational significance of this state of affairs let us consider, for instance, the nucleus ${}^{21}_{9}\mathrm{F}_{12}$ with 9 protons and 12 neutrons. Because of the charge symmetry of nuclear forces its structure should be similar to that of ${}^{21}_{12}\mathrm{Mg}_9$ with 12 protons and 9 neutrons. This indeed is found to be true when one compares the excitation energies of the first few states in these two nuclei. However, consider now the nucleus ${}^{21}_{10}\mathrm{Ne}_{11}$ with 10 protons and 11 neutrons. Formally it can be obtained from ^{21}F by changing one of the neutrons into a proton. Because of the equality of the *n-n* and *n-p* forces when the spin and spatial part of the wave functions are the same, there ought to be states in ^{21}Ne that mimic the states in ^{21}F; in other words we expect to find in ^{21}Ne states whose spins and energy spacings are identical with those of ^{21}F. The *p-n* system can also be found in states, like $\psi_{pn}^{(0)}$ in (6.29), which have no counterpart in the *p-p* and *n-n* system. We should therefore expect ^{21}Ne to have also states that have no counterpart in ^{21}F. But the important conclusion we have reached is that every state in ^{21}F should have its counterpart in ^{21}Ne.

Figure 6.1 [reproduced from Butler et al. (68)] shows the actual experimental situation for the mass 21 nuclei. $^{21}_{9}F_{12}$ and $^{21}_{12}Mg_{9}$, being a pair of mirror nuclei, show a similarity in their structure; so do $^{21}_{10}Ne_{11}$ and $^{21}_{11}Na_{10}$, being another pair of mirror nuclei. The spectrum of ^{21}Ne is, as expected, very different from that of either ^{21}F or ^{21}Mg. However the lowest three states in ^{21}F have counterparts in ^{21}Ne and ^{21}Na that have the same spins and parities assigned to them, and whose energy spacing is very similar to that of the corresponding states in ^{21}F and ^{21}Mg. Thus the states in ^{21}Ne and ^{21}Na are composed of states that are identical in their spatial and spin structure to the states of ^{21}F and ^{21}Mg, plus other states that result from the *p-n* combinations that have no counterpart in the *p-p* or *n-n* system.

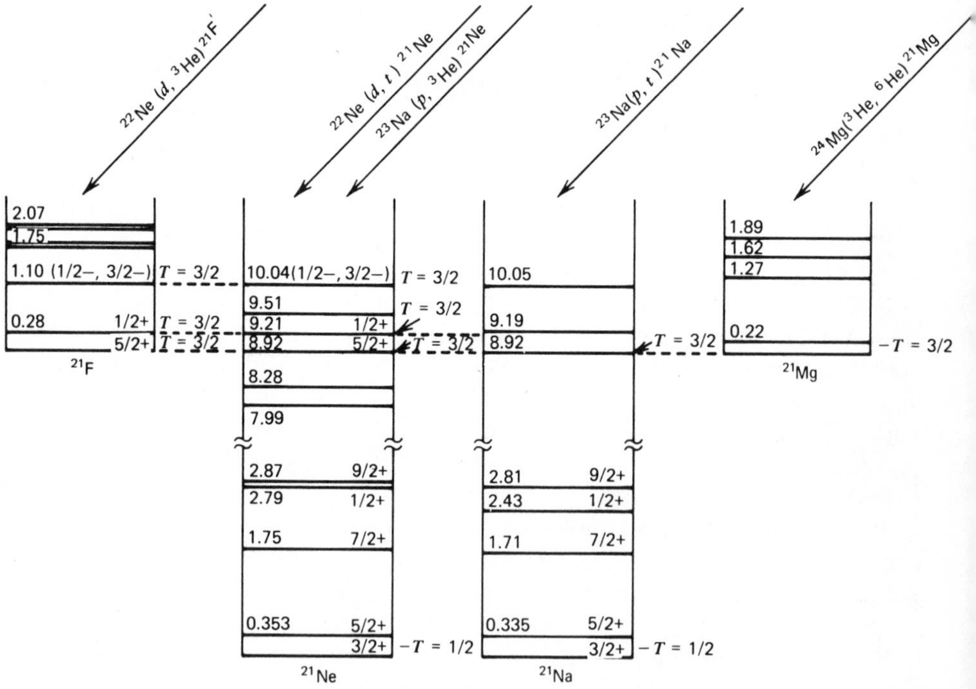

FIG. 6.1. Energy level diagrams for the members of the mass-21 isospin quartet showing the positions of the $T = 3/2$ levels in each nucleus. For clarity, the ground state energies of mirror-nuclei have been equated, and many of the excited levels of Ne and Na below 10-MeV excitation have been deleted [Butler, Cerny, and McCarthy, (68)].

The fact that nuclear forces are independent of the nature of the nucleons in two nucleon states that are antisymmetric with respect to the exchange of the spin and space coordinates, that is, the phenomenon of *charge independence*, thus manifests itself in many parts of the periodic table. In the last few years there have been also extensive studies of the consequence of charge independence for heavy nuclei, and it was possible to find many cases in which a sequence of excited states in a nucleus (Z,N) correspond to the sequence of the lowest states of a nucleus $(Z-1, N+1)$. Such *analog states* in the neighboring nuclei offer important tools for the detailed study of their structure [see, for instance, Fox and Robson (66) and Anderson, Wong, and McClure (62)].

Charge independence implies charge symmetry, but not vice versa. It thus puts a stronger limitation of the nature of nuclear forces than that implied by charge symmetry alone, and enhances the similarity between the two nucleons. Actually with the discovery of other elementary particles it was found that the proton and the neutron are two members of a bigger family of particles, with which they share many common properties. The other members of the *p-n* family are the three Σ-particles (Σ^+, Σ^0, and Σ^-), the Λ^0 and the two Ξ-particles (Ξ^- and Ξ^0). We shall not discuss here the nature of the similarities between all eight members of this family, but refer the reader to appropriate studies in elementary particle physics [see, for instance, Gasiorowicz (67)].

One application of the generalized Pauli principle that will be important for later chapters is to the two-body nucleon system. According to that principle the space-spin wave function for $T = 1$ two-particle system must be antisymmetric, for the $T = 0$ system, symmetric. Consider the $T = 1$ system first (e.g., a two-proton system). Suppose the spin wave function is a singlet and therefore antisymmetric. Then the space dependence must be even. More explicitly in the expression

$$\psi(T = 1, S = 0) = \phi(\mathbf{r}_{12})\, \chi_\tau(1, M_T)\, \chi_\sigma(0,0)$$

$$\phi(\mathbf{r}_{12}) = \phi(\mathbf{r}_{21}) \tag{6.35}$$

If ϕ is expanded in a series of Legendre functions

$$\phi(\mathbf{r}_{12}) = \sum \frac{u_l(r)}{r} P_l(\cos\theta)$$

where $|\mathbf{r}_{12}| \equiv r$

$$\phi(\mathbf{r}_{21}) = \sum \frac{u_l(r)}{r} (-)^l P_l(\cos\theta)$$

It follows from (6.35) that only the even l's survive (i.e., u_l for l odd are zero). Thus only even orbital angular momenta are involved. In other words, if $T = 1$, $S = 0$, then l is even. The states allowed are thus $^1S_0$, $^1D_2$, $^1G_4$, etc. In the same fashion it is possible to determine the states for which $T = 1$, $S = 1$. In this case l must be odd.

Similar discussion may be made in the $T = 0$ two-body case. The results are summarized in Table 6.1.

TABLE 6.1 Two-Body States Allowed by the Generalized Pauli Principle

$T = 1$ $S = 0$	$T = 1$ $S = 1$	$T = 0$ $S = 1$	$T = 0$ $S = 0$
${}^1S_0$	${}^3P_{0,1,2}$	${}^3S_1$	${}^1P_1$
${}^1D_2$	${}^3F_{2,3,4}$	${}^3D_{1,2,3}$	${}^1F_3$
${}^1G_4$	${}^3H_{4,5,6}$	${}^3G_{3,4,5}$	${}^1H_5$
etc.	etc.	etc.	etc.

7. OTHER REGULARITIES OF NUCLEAR SPECTRA

The accumulation of data on nuclei and their excited states resulted in the discovery of several outstanding regularities that greatly helped to clarify what otherwise would have been a hopelessly complicated collection of data. Most of the regularities concern the spin sequences of various nuclear levels, their parities and their excitation energies, and we shall discuss these first.

The spin of a nuclear level, that is, its total angular momentum, can be determined directly from the multiplicity of hyperfine structures in atomic spectra or from the splitting of atomic beams in inhomogenous magnetic fields. It can also be inferred indirectly from the study of radiations and nuclear reactions involving the particular level in question. It was thus found that without any exception the ground states of even-even nuclei all have $J = 0$. Furthermore in many cases it is possible to determine also the parity of the nuclear state studied, that is, whether its wave function is odd or even under space reflection. All ground states of even-even nuclei turn out to have positive parity, without a single known exception.

It is surprising that these universal rules have not yet found a simple, model independent, interpretation. To be sure each of the models proposed for the elucidation of nuclear structure reproduces these empirical rules in a straightforward manner (or else it would have not been proposed to begin with), but what is still lacking is a *direct* theoretical proof that the ground state of a nuclear Hamiltonian for an even-even system must have angular momentum $J = 0^+$.

The derivation of the rule $J = 0^+$ in the framework of the various models will be discussed in the following chapters. Here we shall mention only briefly the common underlying idea: all the Hamiltonians that we use for the description of nuclei have the property that if ψ is an eigenstate belonging to an eigenvalue E, then its time-reversed conjugate state $\tilde{\psi}$ is also an eigenstate belonging to the same eigenvalue E. $\tilde{\psi}$, loosely speaking, is obtained from ψ by reversing the velocities and spins of all the particles (for more exact definition, see Appendix A.3). Hence, if in ψ one of the nucleons moves with momentum **k**, it will move with momentum $-\mathbf{k}$ in $\tilde{\psi}$. In forming the ground state of a nucleus, the nuclear wave function should allow each nucleon to find another nucleon with which it can interact most strongly, since the overall interaction between nucleons is attractive. For two particles to interact strongly their wave functions (or densities) should overlap as much as possible. The maximum overlap would occur if both of them were in the same state. But just this ideal situation is forbidden by the Pauli exclusion principle. It turns out that the maximum overlap consistent with the exclusion principle is obtained when a particle pairs off with another particle in a time-reversed state to its own (say, a particle rotating clockwise with one rotating counterclockwise). The angular momentum of a pair of such particles vanishes, and the parity of the pair is always positive, since any state and its time-reversed state have the same parity. In an even-even nucleus the opportunity exists for all particles to pair off in such $J = 0^+$ pairs, and the only state one can get from adding together any number of pairs with 0^+ angular momenta, is again a state with $J = 0^+$. The α-particle is a good example where the two protons and the two neutrons each couple their spins to $S = 0$, all of them moving in the same spatial state that is its own time conjugate. This is then the best overlap one can hope for and, hence, the large binding energy of He^4. The remarkable regularity found for the ground states of all even-even nuclei, at least in this way of looking at it, depends therefore strongly on the validity of the Pauli principle for the nucleons and on the fundamentally attractive nature of their mutual interaction.

It should be stressed, however, that the arguments described above may be misleading in their simplicity. In fact, for an antisymmetric wave function it is not possible to ascribe well-defined properties to one pair of nucleons. The intricacies of dealing with antisymmetric wave functions are greatly responsible for the more complicated arguments that the different nuclear models have to present for the explanation of the remarkable regular behavior of ground states of even-even nuclei. This will become clearer in the subsequent chapters.

Another important regularity found in even-even nuclei, which is nearly equally universal, is concerned with their first excited state: its spin and parity is in nearly all cases $J = 2^+$. There are few exceptions to this rule, such as ${}^4_2He_2$ whose first excited state at 20 MeV has $J = 0^+$, ${}^{16}_8O_8$ with a $J = 0^+$ first excited state at 6.05 MeV, ${}^{40}_{20}Ca_{20}$ with a $J = 0^+$ state at 3.35 MeV, ${}^{90}_{40}Zr_{50}$ with a $J = 0^+$ state at 1.752 MeV, and ${}^{208}_{82}Pb_{126}$ with its first excited state at 2.615 MeV having $J = 3^-$. It is noteworthy that all these exceptional nuclei are doubly magic nuclei, that is, nuclei for which both the proton number and the neutron number belong to the series of magic numbers (Eq. 4.3). (The 40 protons in ${}^{90}_{40}Zr_{50}$ have other features of a magic number as well; see discussion following Fig. 8.1.) If the magic numbers stand for closed shells, as suggested earlier, then the regularity referred to can be formulated in the following way: the lowest possible excitation of nonclosed shell configurations of even-even nuclei has $J = 2^+$.

Again, the explanation of this regularity is far from trivial in the various theories that describe nuclear excitations. What is crucial to all these explanations is the validity of the Pauli principle for the nucleons in the nucleus and the short range of the interparticle interaction. No model independent explanation of this regularity is available yet, nor is there a simple explanation for the very regular behavior of the excitation energy of the first 2^+ state in even-even nuclei as a function of Z and N. Figure 7.1, reproduced from O. Nathan and S. G. Nilsson (65), shows this regularity and indicates its relation to the magic numbers. It is seen that $E(2^+)$, the excitation energy of the first excited state in even-even nuclei, increases sharply as either Z or N approaches a magic number, and no 2^+ state is observed as a *first* excited state when both Z and N are magic numbers.

Whereas the regular behavior of the ground states and first excited states of even-even nuclei is uniform throughout the periodic table, there are other features that show up as a regular behavior in preferred regions of the periodic table. One of the first such regularities to be observed is the existence of "*islands of isomerism*" for odd-even nuclei when the odd nucleon number is just below the magic numbers 50, 82, or 126.

Isomers are long-lived excited states, and it is now established that their long lifetime is connected with the fact that their angular momentum is sub-

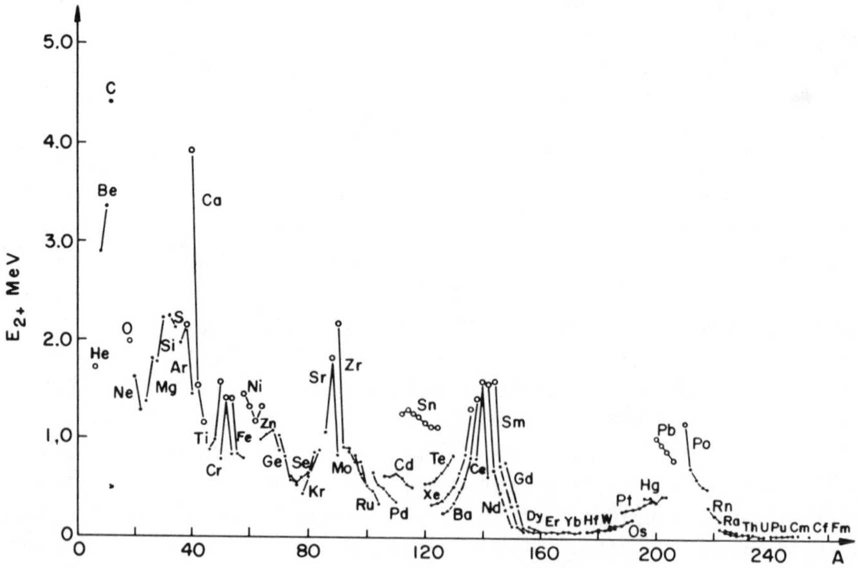

FIG. 7.1. The energy E_{2+} of the first excited 2^+ state in even–even nuclei [taken from O. Nathan and S. G. Nilsson (65)].

stantially different from that of any of the states they can energetically decay into. Angular momentum, like energy and linear momentum, must be conserved in any decay process of one nuclear level into another. The energy available to the decay product is determined by the energy difference between initial and final states, and this energy is shared between the decay product and the recoiling nucleus in order to conserve total momentum. When the decay product is a photon and neglecting the nuclear recoil in this case, its momentum is

$$P_\gamma = E/c$$

If this photon is emitted from a point at a distance R from the nuclear center, it can carry away at most L units of angular momentum where

$$L \simeq ER/\hbar c$$

If we recall that $\hbar c = 197.32$ MeV fm we see that for photons of energy $E \sim 1$ MeV to carry away a few units of angular momentum with ease, they will have to be emitted at about 100 fm from the nuclear center, which is way outside the nucleus! Thus the whole radiation process goes through the quantum mechanical "tails" of the photon and nuclear wave functions, and it is no surprise that if the photon has to carry away four units of angular momentum, the probability of it doing so is extremely small, leading to very long lifetimes—even as long as a few days or weeks—for the decay of the isomeric states.

The regular appearance of islands of isomerism near magic numbers finds its natural explanation within the framework of the nuclear shell model. In this model, which we shall discuss in greater detail in Chapters IV and V, as we build up heavier nuclei, the nucleons fill in one orbit after the other. It is a particular feature of the ordering of the orbits in the shell model that sets of orbits are grouped together leading to the phenomenon of a shell structure and the corresponding magic numbers. It, furthermore, turns out that near the top of such shells there are always orbits with relatively high values of the angular momentum lying close to orbits with low angular momentum. This results from the relatively strong *spin-orbit interaction* in the nucleon-nucleon force. The realization of the importance of this spin-orbit interaction led Mayer (49) and independently Haxel, Jensen, and Suess (49), to the formulation of the nuclear shell model, and opened the way to a great progress in our understanding of nuclear structure.

The average nuclear potential, which determines the orbits of the individual nucleons, has a shape that follows roughly the mass distribution of the nucleons. It is often approximated by a parabolic potential, and the levels and shells in the harmonic oscillator potential thus form a natural starting point for the classification of nuclear levels. Figure 7.2 shows the sequence of proton orbits in the nuclear shell model, and one sees from it that with 81 protons, for

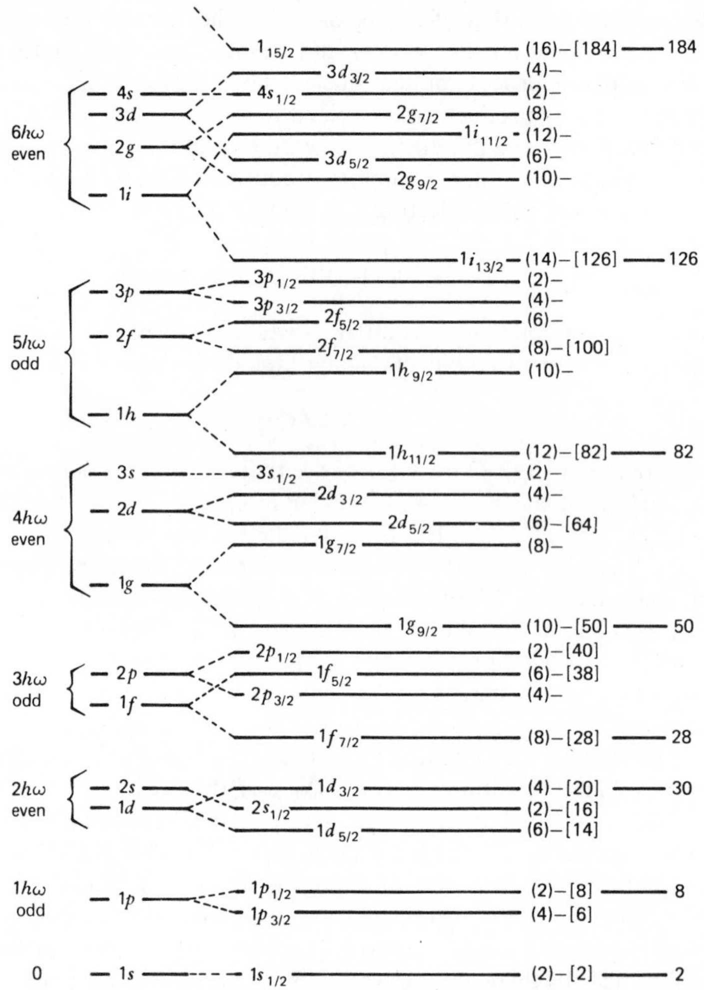

FIG. 7.2. Approximate level pattern for protons. The spin-orbit splitting is adjusted in such a way that the empirical level sequence is represented. For convenience the oscillator-level grouping and the parities of these groups are indicated at the left side of Fig. 7.2. Round brackets (2), (4), etc. and square brackets [2], [6], etc. indicate the level degeneracies and the total occupation numbers. In the $6\hbar\omega$ oscillator group the splittings are not drawn in a proper scale, the $3d$ splitting is too large. A more accurate drawing would have confused the picture too much. [Taken from Mayer and Jensen (55)].

instance, one can have all levels filled up to the end of the fourth shell, except that the $3s_{1/2}$ level will have just one proton in it. (Although the $1h_{11/2}$ orbit lies higher than the $3s_{1/2}$, it is a general feature of the shell model that for *pairs* of nucleons it is actually lower even than the $2d_{3/2}$ level. See Chapter V on the shell model for further clarification of this point). This will then be the ground state configuration of 81 protons and its spin will be $J = 1/2^+$. To obtain an excited state we can "lift" a proton from the $2d_{3/2}$ orbit and make it fill the $3s_{1/2}$ orbit; there will result an excited state with $J = 3/2^+$. The next excited state will be gotten by lifting a proton from the filled $1h_{11/2}$ orbit and make it fill the "hole" in the $2d_{3/2}$ orbit. The resulting excited state will have $J = 11/2^-$. We now see that if a nucleus with 81 protons is to be deexcited from the $J = 11/2^-$ state by the emission of a photon, this photon has to carry away at least 4 units of angular momentum if the nucleus settles in the excited state $J = 3/2^+$, or even 5 units of angular momentum if it goes straight to the ground state. The $J = 11/2^-$ state will consequently live for a relatively long time and will show up as an isomer. Such isomers, involving a transition from a $11/2^-$ level to a $3/2^+$ level, are indeed very common for odd Z or odd N before the magic number 82.

Figure 7.3, reproduced from M. Goldhaber and A. W. Sunyar (65), shows the distribution of observed isomers over the periodic table. The absence of any isomers right after magic numbers is very striking. It is equally important to note that isomerism shows up at the same nucleon number for both odd-Z and odd-N nuclei. It appears that the rules governing nuclear structure apply separately for protons and for neutrons, as if the nucleus were composed of two separate liquids—a proton liquid and a neutron one—that are held together by forces that do not upset the intrinsic structure of each of them separately.

In the framework of the various models that we shall discuss, this special feature of nuclear structure finds a very natural interpretation. There is a common starting point to all these models: the effect of the mutual interaction of the nucleons is approximated by an overall potential that keeps all the nucleons together, something like the central Coulomb field that keeps all the electrons together in the atom. This overall potential is assumed to take care, in an average way, of the bulk of the nucleon-nucleon interaction, and the remaining unaccounted for residues of the interaction are then invoked to explain the details of nuclear structure. As in the case of atoms, the Pauli principle is very instrumental in determining the "packing" of the nucleons inside this overall potential. Since, the Pauli principle applies separately for protons and for neutrons one sees that the structure of nuclei will be dominated by the numbers Z and N separately, rather than by $A = Z + N$.

Odd-even nuclei show other remarkable regularities. If one looks at a whole series of odd-Z–even-N isotopes: (Z, N), $(Z, N + 2)$, $(Z, N + 4)$, etc. or a whole series of odd-N even-Z isotopes: (Z, N), $(Z + 2, N)$, $(Z + 4, N) \ldots$, one often finds that the excitation spectra of the nuclei in such series resemble each other very much. An example is shown in Fig. 7.4 where the excited states of odd-A isotopes of Tl are shown together with their spin assignments. It seems that the addition of pairs of neutrons to an odd-A isotope of Tl does very little to the general nature of its low-energy excitations. This result, which holds for many other sequences of isotopes and isotones is again consistent with the observation made with regard to the $J = 0^+$ rule for the spins of the

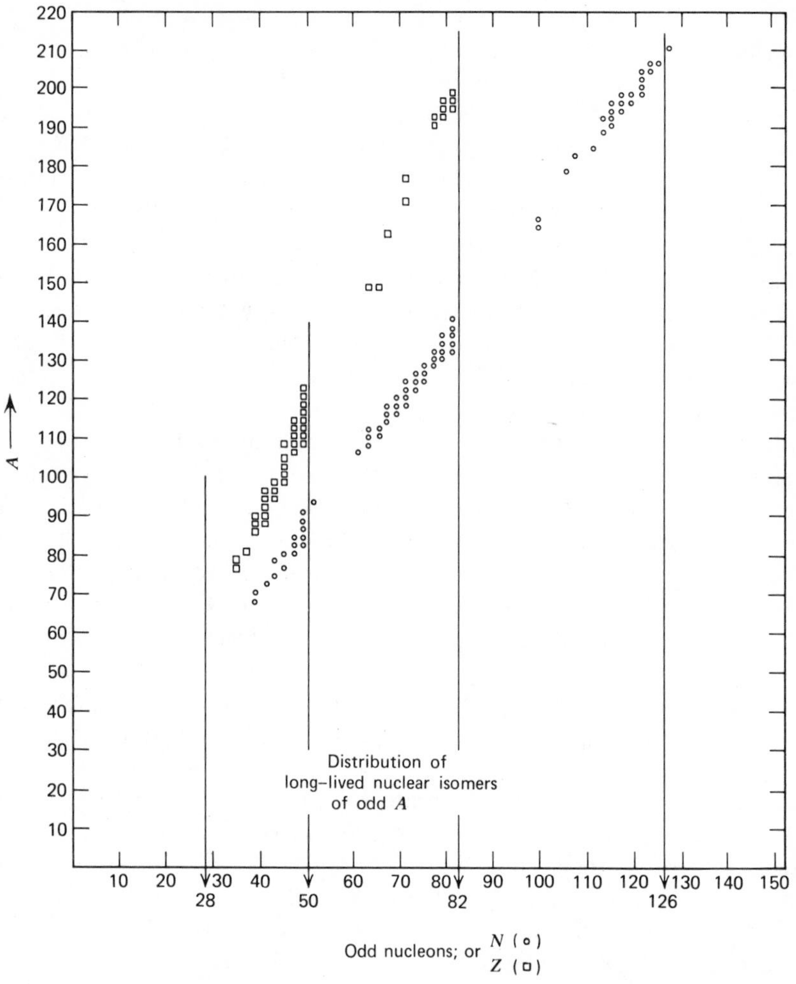

FIG. 7.3. The distribution of long-lived isomers of odd mass number A, plotted against the number of odd nucleons (N or Z) [taken from Goldhaber and Sunyar (65)].

ground states of even-even nuclei. In fact, if pairs of identical nucleons do tend to pair off in states that are time-reversed conjugates of each other, then they add no angular momentum or parity to a "host" nucleus, and the spin sequence of its low-lying excited states will remain unchanged after the addition of the paired-off nucleons.

Another very remarkable regularity that led to extensive studies of various collective features of nuclear motion is connected with the spectra of nuclei in the so-called *regions of large deformations*, ($150 \lesssim A \lesssim 180$; $220 \lesssim A \lesssim$

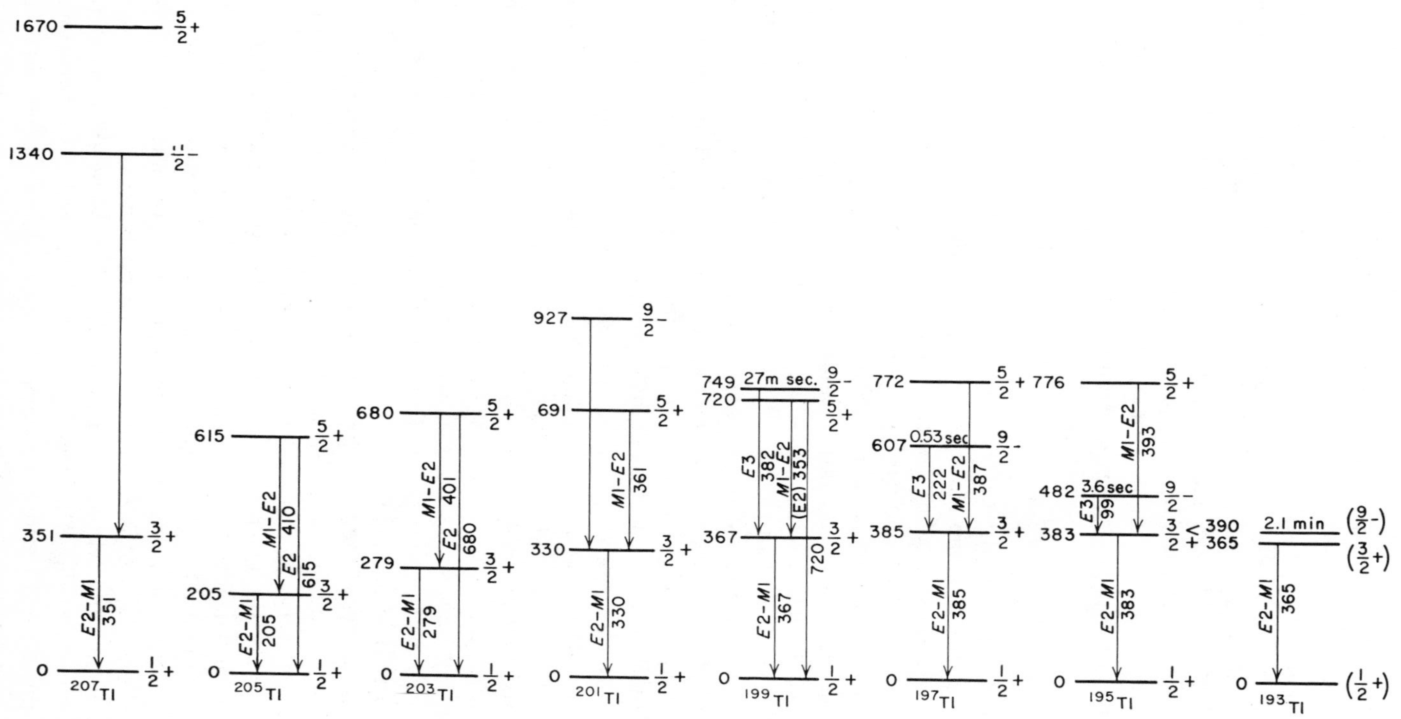

FIG. 7.4. Level schemes for the odd-*A* isotopes of Tl [taken from Diamond and Stephens (63); the 9/2 level in ^{201}Tl is based on a measurement of Conlon (67)].

250). In between magic numbers nuclei tend to acquire an elongated shape with a large positive quadrupole moment. The excitation spectra of nuclei in this region then shows the characteristic features of rotational motion. Such rotational motion are encountered in molecular physics where nonspherical molecules with axial symmetry can absorb energy by increasing their rate of rotation around an axis perpendicular to their symmetry axis. This shows up as a band of excited states with angular momenta J increasing with excitation, the excitation energy itself being approximately proportional to $J(J + 1)$. Figure 7.5 shows several examples of such *nuclear* spectra observed in various regions of the periodic table where in first approximation:

$$E_J \simeq \frac{1}{2\mathcal{J}} J(J + 1)$$

The picture that emerges from attempts to put together the obvious evidence for shell structure and the equally impressive evidence for collective motions of the nucleus as a whole can be stated as follows.

The effects of the mutual interaction that keep the nucleons together in a nucleus can be simulated by an overall average potential in which these nucleons move with very little residual interaction among them. This average potential in itself, being produced by the nucleons, has some dynamical features of its own. In particular, when this average potential turns out to have a nonspherical shape, its axes rotate in space, slowly compared to the motion of the nucleons inside it; the nucleons in this deformed potential are then dragged along and are thus performing a collective rotational motion on top of their nearly independent faster motion in the average potential well.

The average overall potential can undergo other slow changes, such as shape oscillations or volume oscillations, each time giving rise to another type of collective motion of the nucleons moving in it. Thus excited states have been observed that probably correspond to vibrations of the nucleus manifesting themselves through surface waves at the surface of the nucleus. It is possible to identify many such states that arise out of one such phonon excitation and, in several cases, states arising from the excitations of two or more phonons have been also observed.

These modes of excitation like the rotational modes seem to be common to many nuclei and to exhibit a marked regularity in the dependence of their excitation energy on mass and atomic numbers. Perhaps the best documented of these are the *giant dipole resonance*, observed either by the absorption process in which the incident projectile is a photon and the major reaction involves the production of a neutron or sometimes two nucleons, or by the (p, γ) process in which the photon is produced. In either case a broad resonance (see Fig. 7.6) is observed. The general behavior of these resonances, their energy and width, is shown in Fig. 7.7. These resonances were first understood by Goldhaber and Teller (48) as collective vibrations of the neutrons against the protons, giving rise to a large electric dipole moment. The quantum numbers of the resonance in this case can be calculated by

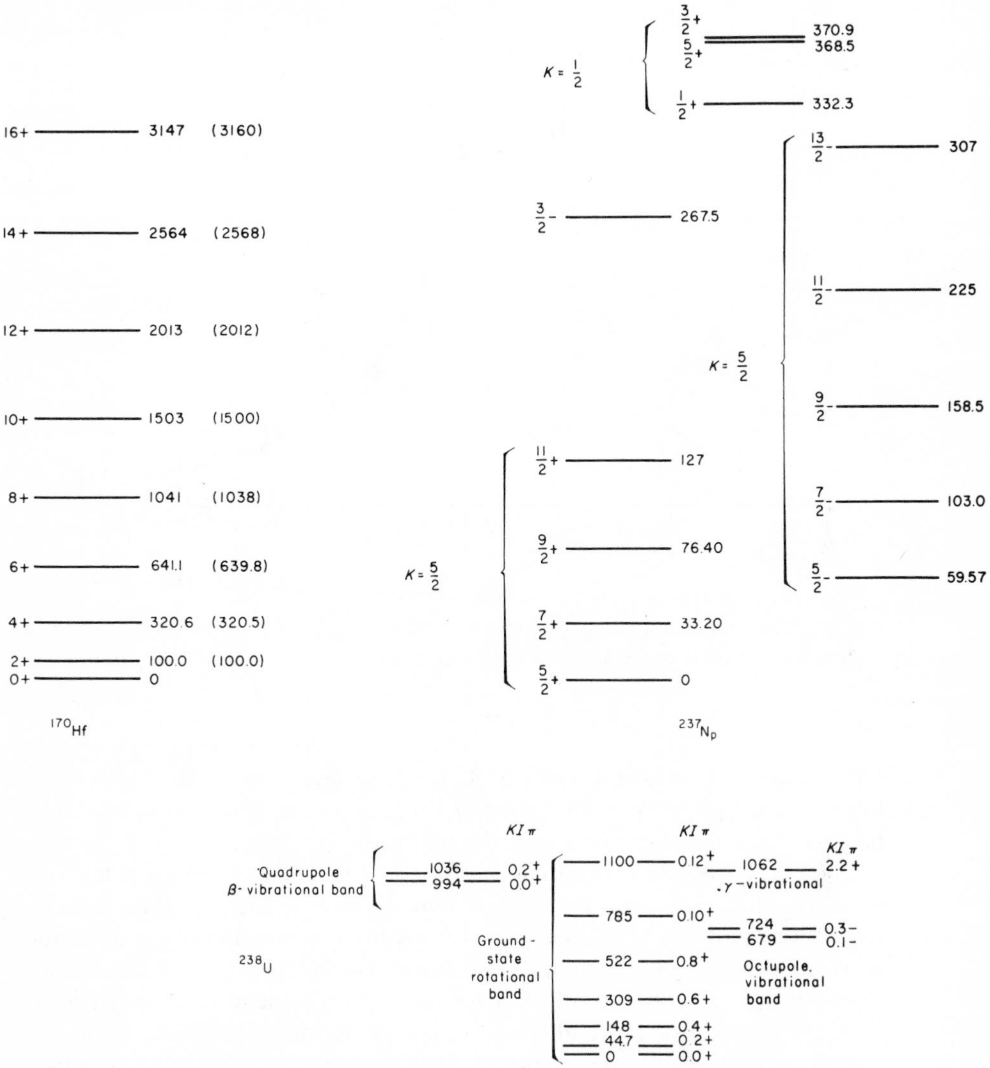

FIG. 7.5. Numbers in brackets are calculated energies based on a collective model of Davidov and Chaban for the rotational excitation of the levels in ^{170}Hf. The excitation energies actually deviate slightly from the $J(J + 1)$ rule expected for a rotational band [Data taken from F. S. Stephen et al. (65)].

assigning to the photon a unit isospin, spin, and odd parity. We shall postpone the discussion of these excitations to a later chapter. For the present it should be pointed out that their occurrence in a wide variety of nuclei suggests that collective motion of a vibratory type is a universal form of nuclear motion.

Other "vibrations" have been discovered. Giant resonances built upon other radiative multipoles (e.g., the magnetic dipole) have been reported.

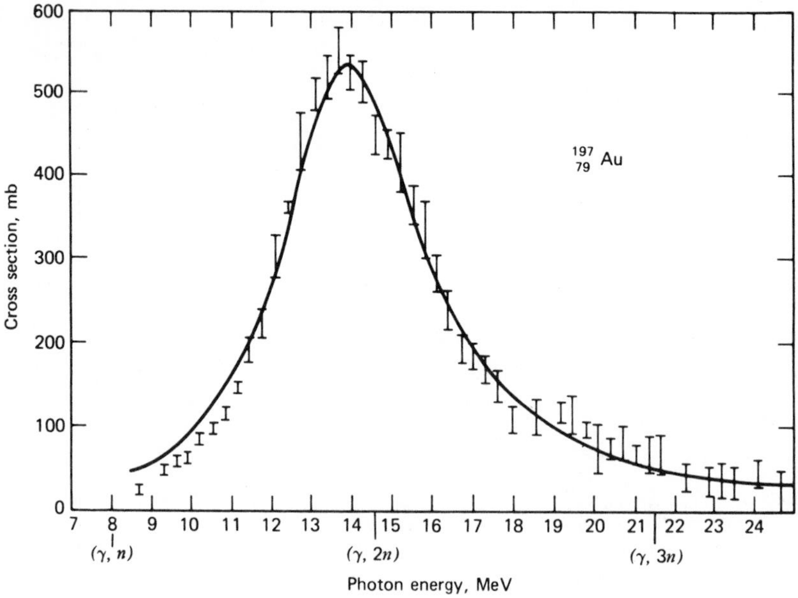

FIG. 7.6. The photonuclear cross section for ^{197}Au $\sigma = \sigma(\gamma, n) + \sigma(\gamma, 2n) + \sigma(\gamma, np)$. The resonance energy is about 13.90 MeV, the width of the resonance 4.2 MeV [from Fultz et al. (62)].

Other examples include a series of 3^- levels in heavy nuclei (the 3^- level in ^{208}Pb at nearly 3 MeV is an example). In light nuclei the 3^- level in ^{16}O at 6.14 MeV is an example of a level thought to be vibratory.

In fact it is also clear now that one and the same nucleus can show features in its excitation spectrum that may require different models for their description. If one plots the total energy of A-nucleons as a function of the deviation of their overall shape from spherical symmetry, a curve similar to the one shown in Fig. 7.8*a* may result. For low energies, E_0, the nucleus may prefer, then, the spherical shape, and characteristic excitations will involve vibrations and single particle excitations. For higher excitations, say E_1, the nucleus may prefer a shape that deviates appreciably from spherical symmetry giving rise to characteristic rotational spectra. A possible example of such a situation is found in ^{16}O.

For other values of A the situation may be just reversed, as shown in Fig. 7.8*b*: the lowest energies may prefer a nonspherical nucleus with its characteristic rotational spectra, and the higher excitations may have spherical symmetry.

The two groups of nuclear states, all in the same nucleus, one centered around spherically symmetric shapes and the other around deformed states, may sometimes be so different from each other that a direct transition from

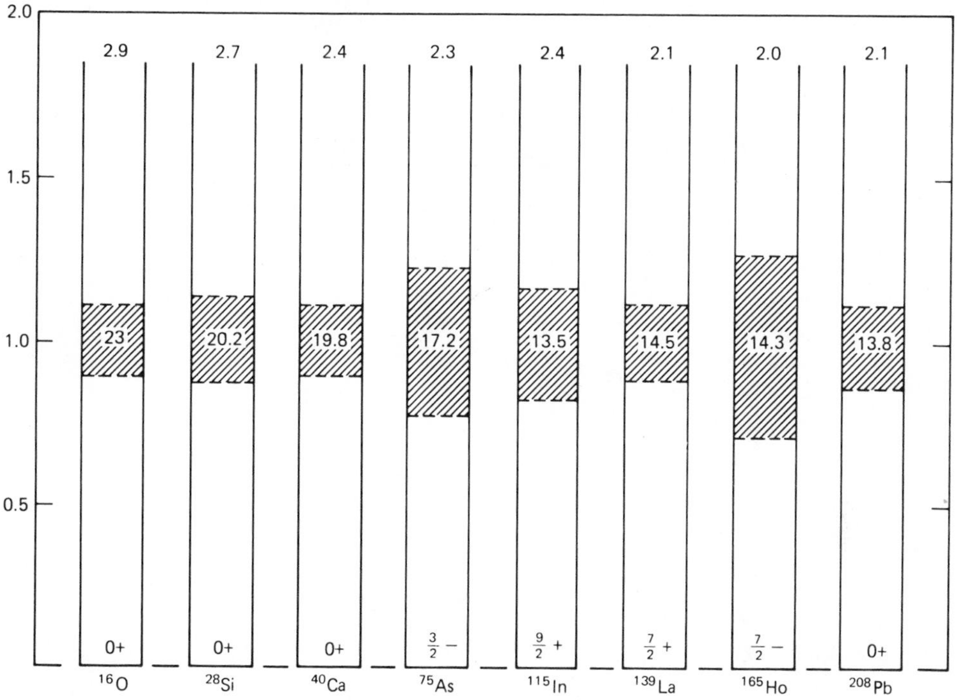

FIG. 7.7 The location and width of giant dipole resonance for a number of nuclei. The shaded area gives the width, the enclosed number the resonance energy (MeV) [from Fuller (66)].

one to the other becomes greatly hindered. One talks then about *shape isomers* [Hill and Wheeler (53) and V. M. Strutinsky (67)], that is, excited nuclear states that owe their long lifetime to the fact that they correspond to a nuclear shape that is very different from that of any of the states they can possibly decay into. The decay of these states therefore involves a simultaneous transition by several nucleons at a time, which drastically cuts its probability. In fact, it is possible that some of these shape isomers are responsible for the fast spontaneous fission observed in some of the Am isotopes. Such nuclei in their ground states are energetically unstable against a breakup into two smaller fragments and can undergo spontaneous fission from their ground state with a characteristic lifetime of 10^5 to 10^7 years. However, sometimes excited states are observed at an excitation of 2 to 3 MeV that undergo spontaneous fission with a lifetime of 1 msec or less and, furthermore, these excited states do not seem to decay via a γ-emission back to the ground state. It is possible that we see here extreme shape isomers that are much closer in their shape to the fissioning configuration of the nucleus than to its ground state, giving rise to this tremendous change of 14 orders of magnitude in the

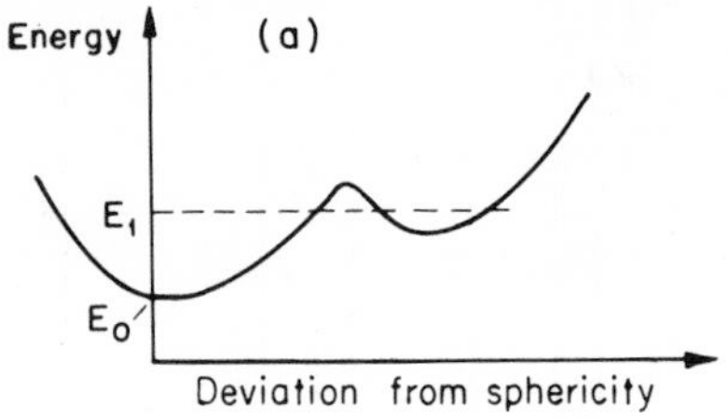

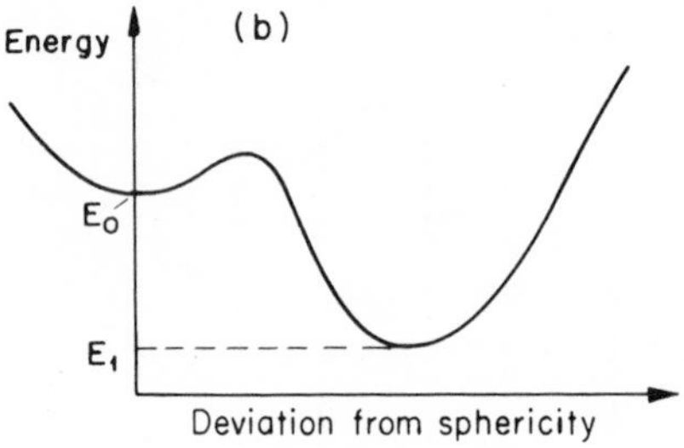

FIG. 7.8. Two possible situations for the dependence of the energy of *A*-particles on the overall shape of their mass distribution.

spontaneous fission lifetime. Some further evidence in favor of this interpretation was given by the work of Migneco and Theobold (68), who studied subthreshold fission in Pu (see below).

We therefore see that the collective properties of nuclei reflect the special features of the single particle levels available for the specific number of nucleons one is considering. Near magic numbers there are few possibilities for the particles to distribute themselves among various levels if they want to stay at the low energies and a spherical shape is preferred. Between magic numbers or at higher excitations the availability of several close-lying levels makes it possible for the particles to accommodate themselves better in non-spherical-shaped nuclei, and some collective features show up. This way one can satisfy the need for both independent particle motion as required by shell structure and other phenomena, as well as the accumulating evidence for the existence of several modes of collective motion in nuclei.

The fact that the average potential is, in the last analysis, just a convenient way of representing the bulk of the mutual interactions of the nucleons reflects itself in a coupling between the collective motions of the average potential and the motion of the nucleons inside it. Very interesting features of nuclear structure find their natural explanation through this coupling, as we shall see in later chapters. Among other things the inertial parameters that determine the collective motion (such as the moment of inertia for collective rotations) are sensitive to the details of the motion of the nucleons in the potential, including the effects of the small residual interaction on this motion. The elucidation of such details is a good measure of the success of the models, and we defer their study to the chapter on collective motion.

8. ELECTROMAGNETIC PROPERTIES OF NUCLEI

A very useful way of studying nuclear properties involves their interaction with outside electromagnetic fields. Two features of the electromagnetic interaction contribute to making it so productive in the study of nuclei: first, it is a relatively well understood interaction and therefore allows rather clear conclusions to be drawn. Second, because the electromagnetic interaction is weak compared to nuclear interactions, an electromagnetic "probe" that is used to study nuclear properties disturbs the nucleus very little. One is therefore measuring properties of the free, undisturbed nucleus. These features are to be contrasted with the study of nuclei using nuclear reactions, where the uncertainties concerning the nuclear interaction and the strong perturbation of the target nucleus by the projectile complicate the analysis markedly.

The simplest way of studying nuclei by means of electromagnetic interactions is to measure their interaction with a known static field. If we consider an electric field given by the electric potential $V(\mathbf{r})$, then its interaction energy with a nucleus is given by

$$E = e\langle\psi^*_{JM}(1, \ldots, A)|\sum \frac{1+\tau_{i3}}{2} V(\mathbf{r}_i)|\psi_{JM}(1, \ldots, A)\rangle \tag{8.1}$$

where $\psi_{JM}(1, \ldots, A)$ is the nuclear wave function characterized by the total angular momentum J and its projection M along an arbitrary z-axis. It is convenient to expand $V(\mathbf{r}_i)$ around the center of mass of the nucleus since the wave functions ψ_{JM} are usually referred to this point. Furthermore, for reasons that will become clear soon, it is convenient to expand $V(\mathbf{r}_i)$ in terms of spherical harmonics $Y_{lm}(\theta,\phi)$ (θ and ϕ are the spherical angles taken with respect to the z-axis mentioned above) rather than in terms of the conventional Taylor expansion. Taking the center of mass of the nucleus as the point $\mathbf{r} = 0$, and choosing an arbitrary direction for the z-axis, we then obtain

$$V(\mathbf{r}_i) = \sum_{l=0}^{\infty} \sum_{m=-l}^{l} V_{lm}(r_i)\, Y_{lm}(\theta_i\phi_i) \tag{8.2}$$

where θ_i and ϕ_i are the polar coordinates of the point $\mathbf{r}_i$ and $V_{lm}(r_i)$ is given by

$$V_{lm}(r) = \int Y^*_{lm}(\theta\phi)\, V(\mathbf{r})\, d(\cos\theta)\, d\phi \tag{8.3}$$

The functions $V_{lm}(r)$ play the role of the l-th term in the Taylor expansion of $V(\mathbf{r})$, and it can be shown that to a good approximation they are given by

$$V_{lm}(r) \simeq r^l V_{lm} \tag{8.4}$$

where V_{lm} depends, of course, on the detailed funtional form of $V(\mathbf{r})$, but is independent of r. V_{lm} is simply related to the derivatives of $V(\mathbf{r})$ at the origin.

Introducing (8.4) into (8.1) we obtain

$$E = \sum_{l=0} \sum_{m=-l}^{l} V_{lm} \cdot Q_{lm} \tag{8.5}$$

where

$$Q_{lm} = e\langle\psi_{JM}(1, \ldots, A)| \sum_i \frac{1 + \tau_{i3}}{2} r_i^l Y_{lm}(\theta_i \phi_i) |\psi_{JM}(1, \ldots, A)\rangle \tag{8.6}$$

The quantities Q_{lm} in (8.6) are often referred to as the *static electric multipole moments* of the nucleus in the state ψ_{JM}. As we see from (8.5) their knowledge determines uniquely the interaction of the nucleus with any outside static electric field.

Using some well-known properties of the spherical harmonics (see Appendix A,, A.2.44,) we can draw some further conclusions about Q_{lm}. First, the matrix element $\langle JM| Y_{lm} |J'M'\rangle$ vanishes unless $M = m + M'$ and $|J - J'| < l < J + J'$. Applying these results to (8.6) we conclude that

$$Q_{lm} = 0 \qquad \text{for} \qquad m \neq 0$$

$$\text{and for} \qquad l > 2J \tag{8.7}$$

Equation 8.7 actually says that for a nucleus with angular momentum J the highest nonvanishing multipole can be of the order $l = 2J$. Thus nuclei with $J = 0$ have no dipole ($l = 1$) or quadrupole ($l = 2$) moments, and nuclei with $J = 1/2$ have no quadrupole moments. Second, we know that under the transformation $\mathbf{r}_i \rightarrow -\mathbf{r}_i$ (space reflection), the spherical harmonics undergo the transformation $Y_{lm} \rightarrow (-1)^l Y_{lm}$ (see Appendix A, A.2.32). If the eigenfunction $\psi_{JM}(1, \ldots, A)$ have a definite parity, it then follows that

$$Q_{lm} = 0 \qquad \text{for odd } l \tag{8.8}$$

Combining (8.7) and (8.8) we see that the electric, static interaction of a nucleus in a state of angular momentum J is determined completely by the $[J + 1]$ numbers

$$Q_{00}, Q_{20}, Q_{40}, \ldots, Q_{2[J],0}$$

where $[X]$ is the largest integer that is smaller than or equal to X. These remarks assume parity conservation of nuclear forces. Weak nonparity conserving forces may exist (see Chapter IX).

Q_{00} has a very simple meaning; since $Y_{00} = 1/\sqrt{4\pi}$ we see from (8.6) that Q_{00} is just proportional to the total charge, that is,

$$Q_{00} = \frac{Ze}{\sqrt{4\pi}} \tag{8.9}$$

Q_{20} defines the *quadrupole moment* of the nucleus. For historical reasons a somewhat different normalization has been adopted for the quadrupole moment; it is proportional to the expectation value of Q_{20} for the special magnetic substate $M = J$ or, more precisely,

$$q = \sqrt{\frac{16\pi}{5}}\,\frac{1}{e}\,Q_{20}|_{M=J} = \langle JM = J|\sqrt{\frac{16\pi}{5}}\sum_{i=1}^{A}\frac{1+\tau_{i3}}{2}\,r_i^2\,Y_{20}(\theta_i\phi_i)|JM = J\rangle \tag{8.10}$$

(Note the omission of the charge e; quadrupole moments are measured in cm^2 or in barns where 1 barn $= 10^{-24}\ \text{cm}^2$). Q_{40} measures the *hexadecupole moment* of the nucleus, etc.

The various electric multipole moments provide us with a measure of the complexity of the charge density of the nucleus. Thus, if the density has complete spherical symmetry all multipoles, except the monopole, vanish identically. On the other hand, a strongly deformed nucleus may have large values of $|q|$, as can be more clearly seen if we introduce into (8.10) the identity

$$\sqrt{\frac{16\pi}{5}}\,r^2\,Y_{20}(\theta,\phi) = r^2(3\cos^2\theta - 1) = 3z^2 - r^2 \tag{8.11}$$

The quadrupole moment then takes on the form

$$q = Z[\langle 3z^2\rangle - \langle r^2\rangle] = Z[2\langle z^2\rangle - \langle x^2 + y^2\rangle] \tag{8.12}$$

where $\langle A\rangle$ stands for the average value of the operator A_i taken over all *protons* of the system.

Taking a nucleus with axial symmetry, with uniform charge density and choosing the z-direction along the axis of symmetry, we see that $\langle x^2\rangle = \langle y^2\rangle$ and therefore, from (8.12)

$$q = 2Z[\langle z^2\rangle - \langle x^2\rangle]$$

If ΔR is the difference between the extension of the nucleus along the z-direction and along the x-direction, and if $\Delta R \ll R$, then q reduces to

$$q = \frac{4}{5}ZR\Delta R = \frac{4}{5}ZR^2\cdot\frac{\Delta R}{R} \tag{8.13}$$

The parameter

$$\beta = \frac{4}{3}\sqrt{\frac{\pi}{5}}\,\frac{\Delta R}{R} \sim 1.06\,\frac{\Delta R}{R}$$

is known as the quadrupole deformation parameter; in terms of β

$$q \approx \frac{3}{\sqrt{5\pi}} ZR^2\beta \tag{8.14}$$

For a prolate charge distribution $\langle z^2 \rangle > \langle x^2 \rangle$ and q can be large and positive; for an oblate charge distribution $\langle z^2 \rangle < \langle x^2 \rangle$ and q may be large (in magnitude) and negative. For a spherical charge distribution $\langle z^2 \rangle = \langle x^2 \rangle$ and q vanishes.

In actual cases one finds, for instance, for ^{17}O a quadrupole moment of $q(^{17}\text{O}) = -0.0205$ barns leading to $\Delta R/R = -0.03$—indeed a very small deviation from sphericity. The largest measured quadrupole moment is that of ^{175}Lu: $q(^{175}\text{Lu}) = 5.68$ barn, which leads to $\Delta R/R \approx 0.23$.

The transition electric charge density shows up in various features of the electromagnetic radiation emitted by nuclei. Classically a system whose charge density varies with time will emit electromagnetic radiation. If, in particular, the charge density changes only its quadrupole moment, without at the same time changing the other moments, the system will be emitting a pure quadrupole radiation that has a characteristic pattern for its intensity distribution in space. We shall derive the quantum mechanical laws for the emission of electromagnetic radiation in Chapter VIII, but it is obvious that nuclei with large quadrupole moments should emit quadrupole radiation more readily than those with small quadrupole moments. There are other factors that affect the rate of radiation of any given multipole, such as the energy of the emitted radiation and the size of the system. If one corrects for both these factors one obtains a measure of the intrinsic ability of the particular system one is studying to emit the particular radiation. Such studies for electric quadrupole radiations are shown in Fig. 8.1, reproduced from P. H. Stelson and L. Grodzins (65). They show very striking regularity that is again connected with the magic numbers: nuclei close to magic numbers emit electric quadrupole radiation less readily than nuclei between them. The static quadrupole moments show a similar behavior—they are large between magic numbers and small around them. These two similar patterns are quantitatively related to each other, as we shall show in Chapter VI on the collective model of deformed nuclei.

Note that in the data described in Fig. 8.1, proton number $Z = 40$ behaves as if it were a magic number. Nuclei with Z close to 40 emit quadrupole radiation less readily than those further removed from this region. We noted already before that $^{90}_{40}\text{Zr}_{50}$ is one of the very few even-even nuclei whose first excited state is not a $J = 2^+$ state, all other being doubly magic nuclei. The data on the emission of quadrupole radiation is consistent therefore with the pattern observed in the first excited states of even-even nuclei.

Electric multipole moments provide us with information on the charge density in nuclei. Similar information on the current and magnetization densities is provided by the magnetic multipole moments. These moments measure the interaction of a nucleus with an external magnetic field in much

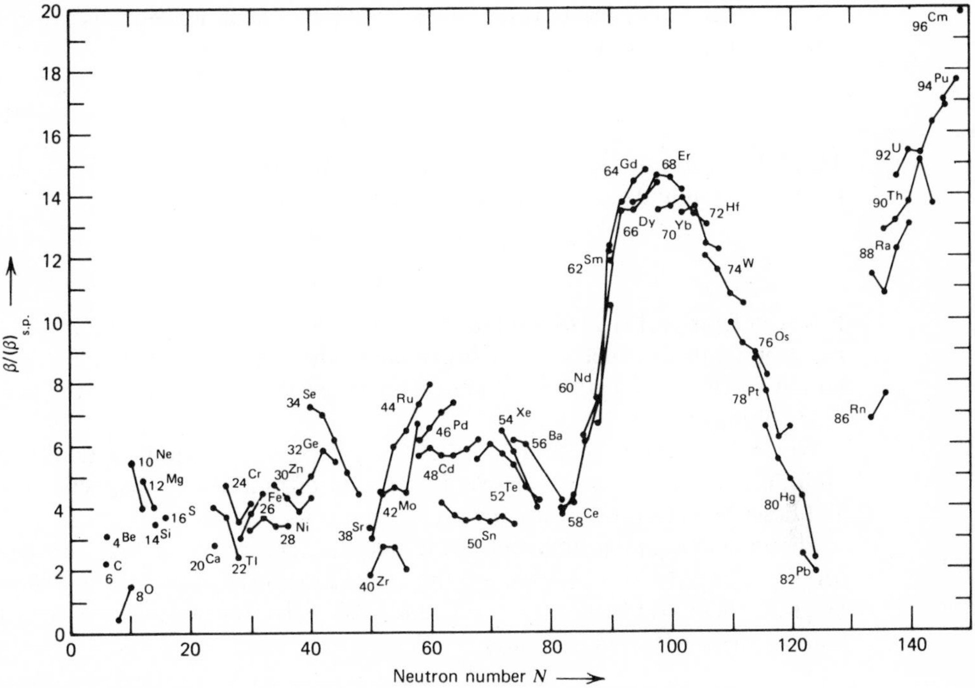

FIG. 8.1. The ratio as a function of neutron number N, of the adopted value of β, the quadrupole deformation parameter as obtained from quadrupole transitions, to $(\beta)_{s.p.}$ the value expected from the single-particle model [reproduced from Stelson and Grodzins 65)].

the same way that the static electric moments measure its interaction with an external electric field.

Two factors contribute to the magnetic multipole moments of nuclei: one comes from the currents of the protons and the other from the intrinsic magnetic moments of the protons and the neutrons. Both are of the same order of magnitude. In fact, it is well known [see, for instance, K. Gottfried, *Quantum Mechanics* (66), p. 313], that if $\mathbf{l}$ is the operator for the orbital angular momentum in units of $\hbar$ of a charged particle of mass M, its current will produce an effective magnetic dipole whose operator is

$$\boldsymbol{\mu}_l = \frac{e\hbar}{2Mc}\mathbf{l} \tag{8.15}$$

where $e\hbar/2M_pc$—the nuclear magneton, has the numerical value

$$\frac{e\hbar}{2M_pc} = 3.1525 \times 10^{-18}\ \text{MeV/gauss} \tag{8.16}$$

On the other hand, the intrinsic magnetic dipole of the nucleons is given in terms of their spins by

$$\boldsymbol{\mu}_s = g^{(s)} \frac{e\hbar}{2M_p c} \mathbf{s} \tag{8.17}$$

M_p is the proton mass and $g^{(s)}$ are the *spin g-factors* whose numerical values for the proton and the neutron are

$$\begin{aligned} g_p^{(s)} &= 5.585564 \pm 0.000034 \\ g_n^{(s)} &= -3.82630 \pm 0.00013 \end{aligned} \tag{8.18}$$

It is seen from (8.15) to (8.18) that as long as the orbital angular momenta of the protons in the nucleus is of order unity, the orbital and the spin contribution to the total magnetic dipole moment of the nucleus are of the same order of magnitude.

We shall postpone to Chapter VIII the complete derivation of the higher multipole operators describing the interaction of nuclei with magnetic fields and describe here briefly only the dipole moment.

The magnetic dipole moment is the only nucleon property that is needed for the complete specification of the interaction of nuclei with a *uniform* magnetic field. According to (8.15) and (8.17) the appropriate operator in units of the nuclear magneton is:

$$\boldsymbol{\mu} = \sum_{i=1} (g_i^{(l)}\, \mathbf{l}_i + g_i^{(s)}\, \mathbf{s}_i) \tag{8.19}$$

where $g_i^{(s)}$ are given by (8.18), and the *orbital g-factor* is given by

$$g^{(l)} = \begin{matrix} 1 & \text{for protons} \\ 0 & \text{for neutrons} \end{matrix} \tag{8.20}$$

Note that each term in (8.19) depends upon the coordinate of a single nucleon. Because of pion exchange, currents flow between the nucleons inside the nucleus. These exchange currents discussed in Chapter VIII will contribute terms in $\boldsymbol{\mu}$ depending on at least the coordinates of two nucleons. Their effect appears to be small.

Using the isospin formalism, (8.19) is sometimes written also in the following form:

$$\begin{aligned} \boldsymbol{\mu} &= \sum \frac{1+\tau_{i3}}{2}\mathbf{l}_i + \left(\frac{1+\tau_{i3}}{2} g_p^{(s)} + \frac{1-\tau_{i3}}{2} g_n^{(s)}\right)\mathbf{s}_i \\ &= \sum \frac{1}{2}\mathbf{l}_i + \frac{1}{2}(g_p^{(s)} + g_n^{(s)})\,\mathbf{s}_i + \frac{1}{2}\tau_{i3}[\mathbf{l}_i + (g_p^{(s)} - g_n^{(s)})\,\mathbf{s}_i] \\ &= \frac{1}{2}\sum (\mathbf{l}_i + \mathbf{s}_i) + \frac{1}{2}(g_p^{(s)} + g_n^{(s)} - 1)\sum \mathbf{s}_i \\ &\quad + \frac{1}{2}\sum \tau_{i3}[\mathbf{l}_i + (g_p^{(s)} - g_n^{(s)})\,\mathbf{s}_i] \end{aligned}$$

Noting that $\sum (\mathbf{l}_i + \mathbf{s}_i) = \mathbf{J}$ is the total angular momentum of the nucleus and putting in numerical values from (8.18) we obtain

$$\boldsymbol{\mu} = \left[\frac{1}{2}\mathbf{J} + 0.38 \sum_{i=1}^{A} \mathbf{s}_i\right] + \left[\frac{1}{2}\sum \tau_{i3}(\mathbf{l}_i + 9.41\mathbf{s}_i)\right] \tag{8.21}$$

The first term is often referred to as the isoscalar part of $\boldsymbol{\mu}$, since it involves no dependence on the isospin operators, whereas the second term is referred to as the isovector part. We notice the considerably bigger contribution of the nucleon spins to the isovector part of the magnetic moment, as compared to their contribution to the isoscalar part. This will turn out to be important in the study of some magnetic dipole radiations (see Chapter VIII).

What the physicist refers to as the *magnetic moment* of a system in a state $\psi(JM)$, as distinct from its magnetic moment operator, is the expectation value of μ_z in the state $\psi(J, M = J)$:

$$\mu = \langle J, M = J|\mu_z|J, M = J\rangle \tag{8.22}$$

For a single nucleon moving in a central field—the so-called Schmidt case—μ can be very easily evaluated. Assuming the nucleon moves in an orbit with orbital angular momentum l, and that its spin $\mathbf{s}$ couples with $\mathbf{l}$ to form a total angular momentum $\mathbf{j}$, we obtain

$$\mu_{\text{sch}} = \langle lj, m = j|g^{(l)}l_z + g^{(s)}s_z|lj, m = j\rangle \tag{8.23}$$

We can evaluate these matrix elements by noting that

$$|ljm\rangle = \sum_{mlm_s} (lm_l, \tfrac{1}{2}m_s|jm)|lm_l\rangle|\tfrac{1}{2}m_s\rangle \tag{8.24}$$

where $(lm_l, (1/2)m_s|jm)$ is the Clebsch-Gordan coefficient for the addition of $\mathbf{l}$ and $\mathbf{s}$ to form $\mathbf{j}$ (see Appendix A, *A*.2.57) while $|lm_l\rangle$ and $|(1/2)m_s\rangle$ are the eigenfunctions of the orbital and spin motions separately. Since

$$l_z|lm_l\rangle = m_l|lm_l\rangle \qquad s_z|\tfrac{1}{2}m_s\rangle = m_s|\tfrac{1}{2}m_s\rangle$$

we obtain, from (8.23) and (8.24), using the orthogonality of the wave functions $|lm_l\rangle$ and $|(1/2)m_s\rangle$,

$$\mu_{\text{sch}} = \sum_{m_l m_s} |(lm_l, \tfrac{1}{2}m_s|j, m = j)|^2\,(m_l g^{(l)} + m_s g^{(s)}) \tag{8.25}$$

The relevant Clebsch-Gordan coefficients are:

$$\text{for} \quad j = l + \tfrac{1}{2} \qquad (lm_l, \tfrac{1}{2}m_s|j, m = j) = \begin{cases} 1 & \text{for} \quad m_l = l \quad m_s = +\tfrac{1}{2} \\ 0 & \text{for all other cases} \end{cases} \tag{8.26}$$

$$\text{for} \quad j = l - \tfrac{1}{2} \qquad (lm_l, \tfrac{1}{2}m_s|j, m = j) = \begin{cases} -\sqrt{\dfrac{1}{2j+2}} & \text{for} \quad m_l = l - 1, \; m_s = \tfrac{1}{2} \\ +\sqrt{\dfrac{2j+1}{2j+2}} & \text{for} \quad m_l = l, \; m_s = -\tfrac{1}{2} \\ 0 & \text{for all other cases} \end{cases}$$

Substituting from (8.26) into (8.25) we obtain:

$$\mu_{\text{sch}} = \begin{cases} lg^{(l)} + \tfrac{1}{2}g^{(s)} & \text{if} \quad j = l + \tfrac{1}{2} \\ j/(j+1)[(l+1)\,g^{(l)} - \tfrac{1}{2}g^{(s)}] & \text{if} \quad j = l - \tfrac{1}{2} \end{cases} \tag{8.27}$$

Equation 8.27, which is often referred to as the *Schmidt value* or *the single particle value* for the magnetic moment, plays an important role in the study of nuclear structure. The reason for this is the following: if, as we were led to conclude, pairs of identical nucleons are paired off in states that are time-reversed conjugates of each other, then such pairs make no contribution to the total magnetic moment of the nucleus; both the orbital and the spin contributions of one member of the pair are cancelled by those of the other member. In an odd-even nucleus one is thus left with the sole contribution of the odd, unpaired nucleon. To the extent that it can have well defined values of l and j, its magnetic moment will be given by the Schmidt relations (8.27).

Experimentally one finds that the situation is very close to what we have just described. Since the g-factors for protons and neutrons are so different from each other, it is advisable to study separately the odd-Z–even-N and odd N–even Z nuclei. Plotting then the magnetic moments as a function of the total angular momentum—the so-called Schmidt diagram—one obtains Fig. 8.2. The full lines in these diagrams represent the Schmidt values for the magnetic moments as derived from (8.27) by inserting the appropriate values for the g-factors. It is clearly seen there that magnetic moments of odd-A nuclei tend to fall into two groups, those that could be identified with the Schmidt line for $j = l + \tfrac{1}{2}$, and those that belong to $j = l - \tfrac{1}{2}$. Furthermore, whereas the moments of odd-Z nuclei tend, on the whole, to increase with j, those of odd-N nuclei do not show this tendency. This feature of the Schmidt formula is related, as can be easily seen, to the different values of g_l for protons and neutrons: the increase of μ with j for odd-Z nuclei reflects the increasing contribution to μ of the orbital current of the proton; for odd neutron nuclei this contribution vanishes on account of the zero charge of the neutron.

There are other interesting patterns observed in the behavior of nuclear magnetic moments; we shall mention here just one of them that follows from (8.21) if one assumes that nuclear states have well defined values for the isospin.

Taking, for instance, nuclear states with $T = 0$, and recalling that the

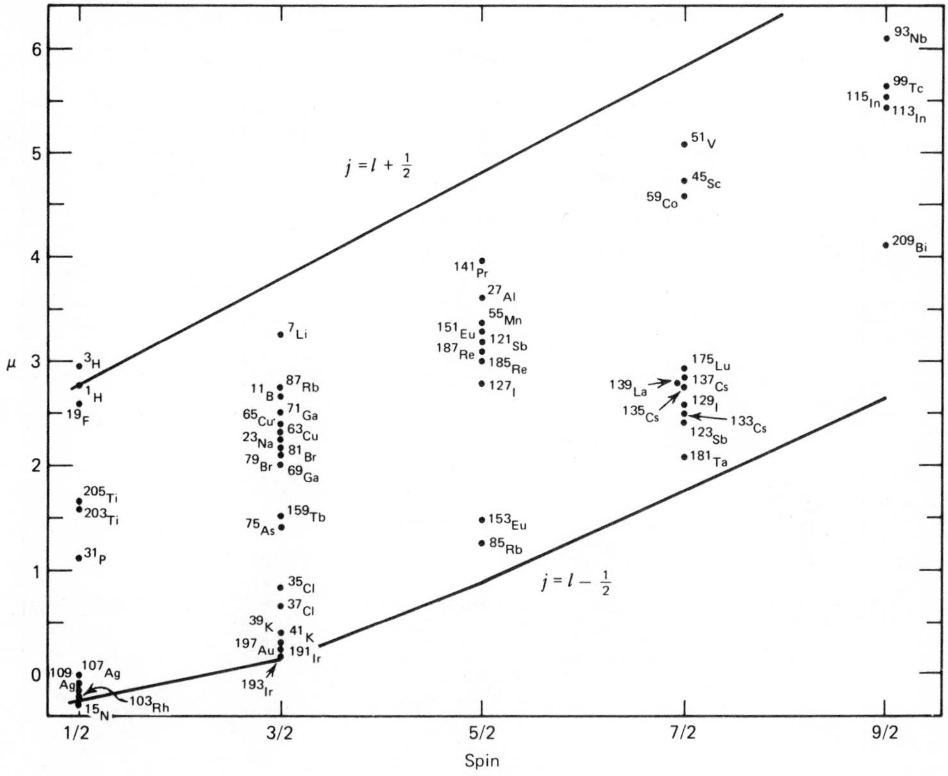

(a)

FIG. 8.2. (a) Schmidt diagram for odd-proton nuclei [from Blin-Stoyle (56)].

expectation value of any component of an isovector vanishes in states with $T = 0$, we obtain from (8.21) and (8.22) that

$$\mu(T = 0) = \langle J, M = J | \tfrac{1}{2}J_z + 0.38 \sum_{i=1} s_{iz} | J, M = J \rangle$$

$$= \tfrac{1}{2}J + 0.38 \langle J, M = J | \sum s_{iz} | J, M = J \rangle \qquad (8.28)$$

or

$$g(T = 0) = \tfrac{1}{2} + 0.38/J \langle J, M = J | \sum s_{iz} | J, M = J \rangle$$

In the deuteron ground state, which is the simplest $T = 0$ nuclear state, we know that most of the angular momentum $J = 1$ is provided by the two spins. In this case we can put $\langle | \Sigma s_{iz} | \rangle = 1$ and predict a magnetic moment $\mu =$ 0.88 nm; the actual value is $\mu = 0.85742$ nm. The difference between the two

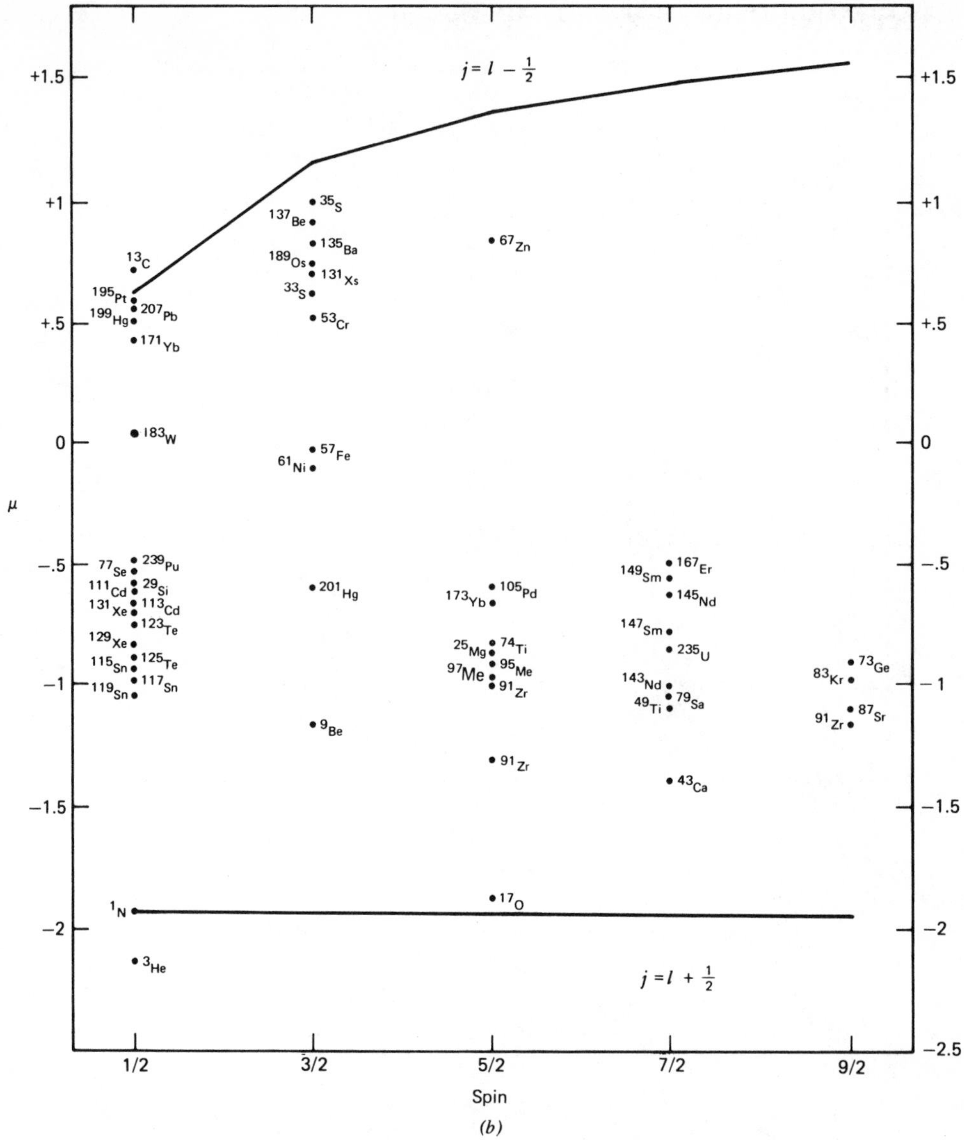

FIG. 8.2—Continued (*b*) Schmidt diagram for odd-neutron nuclei [from Blin–Stoyle (56)].

values may very well be due to the fact that the expectation value of the total spin is somewhat less than 1, there being a slight contribution to J from orbital angular momentum as well (Volume II).

In other heavier nuclei, the evaluation of $\langle|\Sigma s_{iz}|\rangle$ may require more detailed knowledge of the nuclear wave function, but our picture of pairs of identical nucleons pairing off in states that are time-reversed conjugates of each other (see Section 1.7) leads us to expect that at most two nucleons—an odd proton and an odd neutron—will be contributing to $\langle|\Sigma s_{iz}|\rangle$ in $T = 0$ nuclei. We thus expect to have $-1 < \langle|\Sigma s_{iz}|\rangle < +1$ leading to the limits ($J \neq 0$)

$$0.22 \text{ nm} < g(T = 0) < 0.88 \text{ nm} \tag{8.29}$$

This limit is indeed found to be obeyed by all $T = 0$ nuclear states whose magnetic moments were measured, as can be seen in Table 8.1 [reproduced from J. Bleck et al. (69)] that gives the experimental g-factors of these nuclei (^{2}H, ^{6}Li, ^{10}B, ^{22}Na, ^{22}Na*, ^{14}N, and ^{38}K).

One can go even one step further and consider pairs of mirror nuclei. For corresponding states in such pairs the isoscalar part of the magnetic moment

TABLE 8.1 Compilation of the Experimental Values for Isoscalar g-Factors [from Bleck et. al. (69)]

Mirror Nuclei or Doubly-Odd Nucleus		g_j (experimental)
^{2}H	(1^+)	0.857
^{3}He, ^{3}H	$(\frac{1}{2}^+)$	0.851
^{19}Ne, ^{19}F	$(\frac{1}{2}^+)$	0.742
^{6}Li	(1^+)	0.822
^{10}B	(3^+)	0.600
^{11}C, ^{11}B	$(\frac{3}{2}^-)$	0.552(3)
^{17}O, ^{17}F	$(\frac{5}{2}^+)$	0.566
^{18}F	(5^+)	0.568(3)
^{19}Ne, ^{19}F	$(\frac{5}{2}^+)$	0.573(3)
^{21}Ne, ^{21}Na	$(\frac{3}{2}^+)$	0.575
^{22}Na	(3^+)	0.582
^{22}Na	(1^+)	0.536(6)
^{13}C, ^{13}N	$(\frac{1}{2}^-)$	0.380
^{14}N	(1^+)	0.404
^{15}O, ^{15}N	$(\frac{1}{2}^-)$	0.436
^{35}Ar, ^{35}Cl	$(\frac{3}{2}^+)$	0.485
^{37}Ar, ^{37}K	$(\frac{3}{2}^+)$	0.38 (7)
^{38}K	(3^+)	0.458

should be the same, while the isovector part should be equal in magnitude and opposite in sign. This is clearly seen if one recalls that the protons and neutrons exchange their roles when one goes from one member of a mirror pair to the other, so that all we have to do in (8.21) is change the sign of τ_{i3}. It follows, therefore, that the average value for the magnetic moments of corresponding states in mirror nuclei with $T = \frac{1}{2}$ is again given by

$$\tfrac{1}{2}[\mu(T = \tfrac{1}{2} \quad T_3 = \tfrac{1}{2}) + \mu(T = \tfrac{1}{2} \quad T_3 = -\tfrac{1}{2})] = \tfrac{1}{2}J + 0.38 \langle |\textstyle\sum s_{iz}| \rangle$$

so that the same considerations apply to this average as to the moments of $T = 0$ nuclear states. Comparison with the shell model yields agreement within the limits given by (8.29).

The regular behavior of nuclear magnetic moments thus seems to corroborate our previous findings and add some new insight into the nature of nuclear structure. The pairing off of pairs of identical nucleons shows up most dramatically in the Schmidt diagrams, as does the independent behavior of neutrons and protons. On the other hand the fact that the empirical moments do not lie on the Schmidt lines probably indicates the approximate nature of the picture we have been developing thus far. The understanding of the deviations of magnetic moments from the Schmidt lines in terms of refinements to the picture of paired-off nucleons has thus been the subject matter of many interesting studies in nuclear structure, some of which will be discussed in later chapters.

We notice an interesting fundamental difference between the general behavior of electric quadrupole moments and magnetic dipole moments. Whereas the single-particle values for the magnetic moments of nuclei, (8.27), do turn out to describe actual moments fairly well, the experimental values of the electric quadrupole moment, (8.12), often exceed the contribution from an average single particle, indicating a constructively coherent addition of the contributions from all particles. This difference stems from a fundamental difference in the nature of the two relevant operators. The magnetic dipole operator leads to opposite contributions from a state ψ and its time-reversed state $\tilde{\psi}$ since both the currents and the spins change sign when we go from ψ to $\tilde{\psi}$. It is thus an odd operator with respect to the transformation $\psi \to \tilde{\psi}$, and one member of paired-off nucleons cancels the contribution of the other. The quadrupole operator's contribution on the other hand does not change when we switch from ψ to $\tilde{\psi}$; the two *add up* their contributions. If now the nuclear deformation also assures some correlations between the motion of the various pairs, a coherent buildup of the quadrupole moment can, and actually does, take place.

We see, therefore, that there are odd operators, like the magnetic dipole moment, which naturally respond primarily to the few unpaired nucleons, whereas there are even operators, like the electric quadrupole moment, which are more sensitive to collective correlations between the various pairs. By cleverly using such operators one can separate the study of the independent particle features of nuclear structure from those of the collective degrees of freedom.

9. SYSTEMATICS OF OTHER DECAY MODES OF NUCLEI

The richness of nuclear phenomena makes it possible to inspect the structure of the same nucleus from many angles. We mentioned nuclear masses and their systematics, and have briefly glanced at some of the electromagnetic properties of nuclei and their relevance to nuclear structure. Another fruitful way to investigate nuclear structure uses the weak interaction of nucleons with the leptonic field. The fundamental reactions that underlie this whole field are the beta decay of the neutron and the μ-capture by protons

$$n \to p + e^- + \bar{\nu}_e \tag{9.1}$$

$$\mu^- + p \to n + \nu_\mu \tag{9.2}$$

where ν_e is the neutrino that goes with β-decay ($\bar{\nu}_e$ is the antineutrino; see Chapter IX) and ν_μ is the neutrino that goes with μ-decay. Both are spin-$\frac{1}{2}$ particles and are believed to have zero mass [actual present limits are $m(\nu_e) <$ 60 eV, $m(\nu_\mu) <$ 1.6 MeV]. They are believed to be different particles, in that the neutrino that leads to (9.6) (see below) does not lead at the same time to a positron production. Both above reactions are exothermic and have been observed. Under suitable energetic conditions other reactions, which follow from (9.1) and (9.2), can take place and they have also been observed. Thus from (9.1) we can derive a reaction that will involve the capture of electrons on protons

$$p + e^- \to n + \nu_e \qquad \text{in complex nuclei} \tag{9.3}$$

This can be seen in complex nuclei whenever the conversion of a proton into a neutron is energetically favorable. In fact, if the balance of energy is favorable enough, one finds even spontaneous positron emission

$$p \to n + e^+ + \nu_e \qquad \text{in complex nuclei} \tag{9.4}$$

Inverse reactions involving neutrino absorption have been observed for both the electron-neutrino ν_e and the μ-neutrino ν_μ

$$\nu_e + n \to p + e^- \tag{9.5}$$

$$\bar{\nu}_\mu + p \to \mu^+ + n \tag{9.6}$$

Although the fundamental reactions in all these weak interactions take place with individual nucleons, when these nucleons are bound in the nucleus the rate at which these reactions proceed is sensitive to nuclear structure. Since in these reactions a nucleon of one kind always winds up as a nucleon of the other kind, that is, the nucleon isospin changes, the rate depends on the ease with which the final nucleus can accommodate the second kind of nucleon.

In the study of beta decays it has been customary, for historical reasons, to measure the rate of decay by a quantity called log *ft*. Here *t* stands for the half-life (in seconds) of the transition, and *f* is a universal function that corrects for the effects of the available energy, Coulomb repulsion, etc. on the transition rate. We shall deal with all these questions in detail in the Chapter IX; here we want just to point out that a *large* log *ft* value stands for a *low* transition probability.

As in the case of γ-emission, β-emission proceeds less readily if the electron and the neutrino have to carry away angular momentum with them. Beta decays have therefore been classified according to the angular momentum (and parity) that they carry away into transitions that are "allowed," "1st forbidden," "2nd forbidden," etc. In Fig. 9.1 we show just the distribution of observed log *ft*-values of the various types of transitions, as compiled by

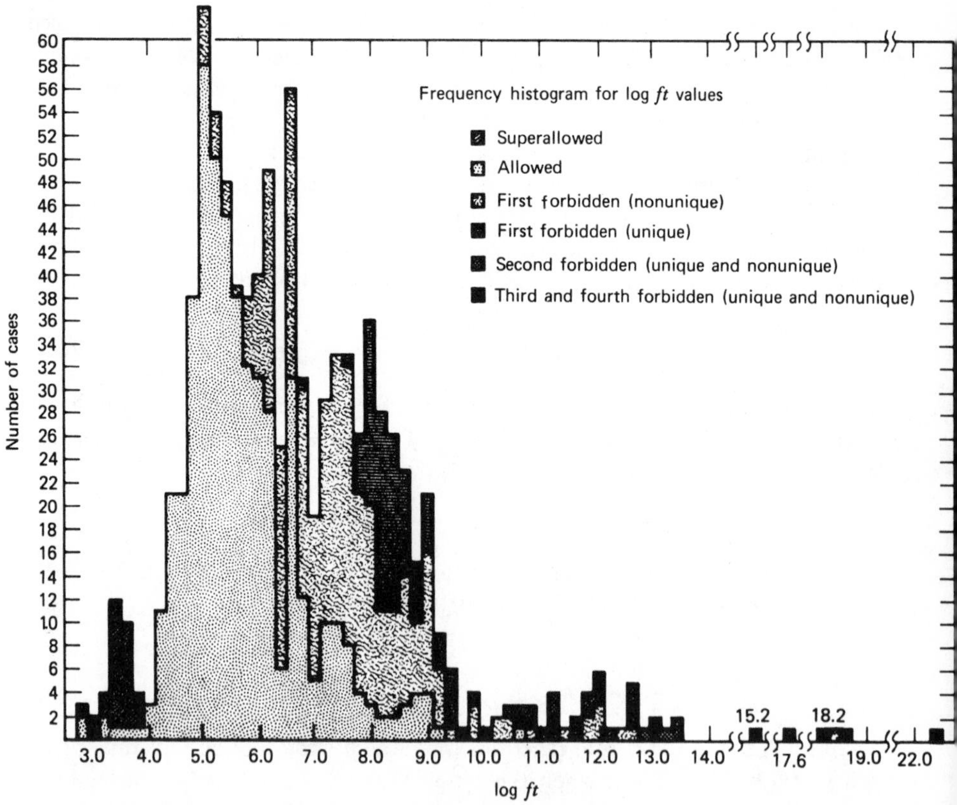

FIG. 9.1. Frequency histogram for log *ft*-Values. Reproduced from Nuclear Data Sheets 5-5-109 (1963).

Gleit, Tang, and Coryell (63). One notices a group of very fast (low log *ft*) transitions; these are the so-called "superallowed" transitions that take place between two members of an isomultiplet. The members of an isomultiplet, it will be recalled, have a very similar structure as evidenced by the similarity in their spectra as described in Section 6. Had it not been for the Coulomb interaction they would have all been degenerate in their energies. The Coulomb repulsion is enough to make the high Z members of the isomultiplet less stable than the low Z members, and positron decay can take place. The fact that the superallowed beta decays are considerably faster than the regular allowed transitions is a very strong additional proof for the similarity between the wave functions of the various members of an isomultiplet. It fits in nicely with the observed similarity of the spectra of mirror nuclei.

Negative muon capture by nuclei is more than just another example of electron capture. Because of the much larger mass of the muon, nuclear matrix elements for correspondingly larger values of neutrino momenta are involved. Thus it should be possible to extract additional information on the nuclear wave functions from these experiments.

Nuclei are known also to decay spontaneously via the emission of other nuclei. The best known cases are those of α-decay and fission. All these decays owe their presence to the Coulomb repulsion that makes it energetically advantageous for a nucleus to break up into two smaller parts. The total energy of such nuclei as a function of the separation between the two fragments may look something like Fig. 9.2. The energy does have a minimum at zero separation, that is, when the two fragments are together and form the initial nucleus; a slight increase in the separation increases the energy of the system because one has to overcome the attractive nuclear interactions without gaining much from the slower decrease in the repulsive Coulomb interaction. As the separation increases further, however, the main work in overcoming the short range nuclear forces has been done, and if the eventual gain in the Coulomb interaction is big enough, one may wind up at a lower total energy. The lifetime for such decays is then determined by the available gain in energy and by the width and structure of the potential barrier. Since the probability for barrier penetration decreases exponentially with its increasing

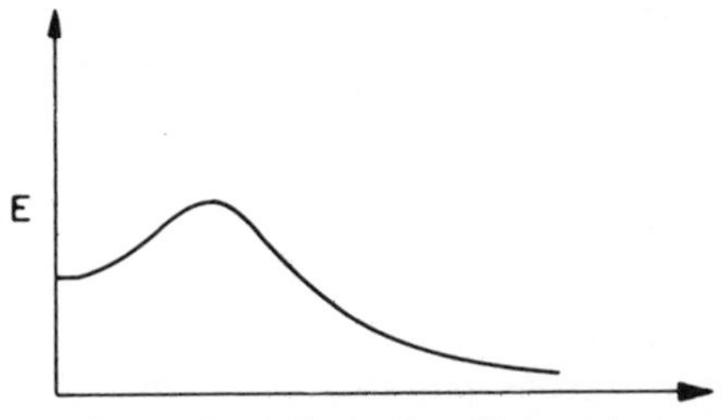

FIG. 9.2. Total energy of fissioning nuclei as a function of the separation between the two fragments.

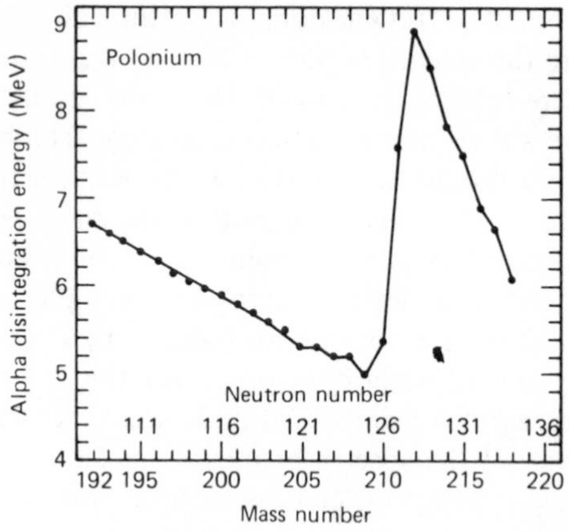

FIG. 9.3. Alpha disintegration energies of polonium isotopes (From Hyde, Perlman, and Seaborg (64)].

height [see Messiah (61)], we expect big variations in the lifetime for α-emission or spontaneous fission, as indeed is found to be true.*

Both α-emission and fission again reflect the magic numbers very vividly. Figure 9.3, taken from E. K. Hyde, I. Perlman, and G. T. Seaborg (64), shows the huge jump in the disintegration energies of the polonium isotopes as one passes the neutron number 126. This, of course, is nothing but another way of looking at the masses of the specific nuclei involved in these processes, but it does dramatize the effect of the magic number 126.

10. THE STUDY OF NUCLEI BY MEANS OF NUCLEAR REACTIONS

Although the study of unstable nuclei has been and continues to be an important source, by far the greatest amount of detailed information on nuclear structure and nuclear dynamics is obtained from nuclear-reaction studies. The methods employed may involve the application of general conservation principles such as the conservation of momentum and energy. It may require, in the case of certain classes of experiments, the use of an understanding and associated semiempirical analysis based upon systematic study of many individual cases. Nuclear information that is obtained in these several ways will include properties of nuclear states such as their energy, angular momentum, parity, and isospin. It may also yield detailed information on the nuclear wave function. Indeed, it has been one of the accomplishments of nuclear physics that specific reactions that are sensitive to particular properties of the wave function have been discovered and developed. Clearly the nature of the

*Recent experiments indicate that there is a second valley and peak in the curve of Fig. 9.2.

nuclear dynamics plays a central role. In this and later chapters we shall briefly review by example the characteristic features of nuclear reactions that have not only been useful for the determination of important nuclear data but also for the elucidation of the interaction dynamics between the incident and emergent projectiles with the target or residual nucleus.

In a typical two-body reaction $A(a,b)B$

$$A + a \rightarrow b + B$$

a is the incident projectile, A the target, b the emergent projectile, and B the residual nucleus. Experimentally the incident projectile is one of an incident beam of particles that often have acquired their energy in an accelerator or are a secondary or tertiary particle produced when a primary accelerated particle strikes a target. An exception is of course neutron beams extracted from reactors.

The kinematic variables usually observed include the energy and momentum of the incident and emergent projectiles. From these measurements and application of the conservation of energy and momentum, the excitation energy of residual nucleus can be determined. In this way the energy of a state in that nucleus can be obtained. The accuracy depends naturally upon the accuracy with which the incident and emergent particle energies and the momenta have been measured. Figure 10.1 shows a characteristic spectrum of protons scattered from ^{90}Zr. The peak at zero excitation energy gives the elastically scattered protons in which the target nucleus remains in its ground state while the numbered groups are located at the energies of the excited states of the target.

For another example of the use of a conservation principle, consider a nucleus excited by inelastic α-scattering, and assume that this is done at high enough energies to assure that isospin-conserving nuclear forces, rather than electromagnetic, dominate in the process. The α-particle has isospin $T = 0$ (see Section I.6). Since isospin is conserved by nuclear interactions, an inelastic α-scattering can lead only to such states in the nucleus that have the same isospin as that of the ground state. Protons, on the other hand, have $T = 1/2$, and their inelastic scattering can therefore lead to states with $T = T_0 \pm 1$, and T_0 if T_0 is the isospin of the target's ground state. By thus comparing levels that can be reached by α and p inelastic scattering, it is possible to determine the isospins of many levels.

The parity of the excited state can very often be determined by nuclear reactions. Thus, if an α-particle is inelastically scattered giving rise to an excited state, in transferring to the nucleus the necessary angular momentum it should also transfer parity to make up for the possible parity differences between the ground and the excited state. The α-particle can transfer angular momentum; orbital angular momentum wave functions are the spherical harmonics $Y_{lm}(\theta,\phi)$; these have a parity that is determined by l: $(-1)^l$. It therefore follows that when an α-particle transfers l units of angular momentum it also transfers a parity of $(-1)^l$. If we consider an inelastic α-scattering

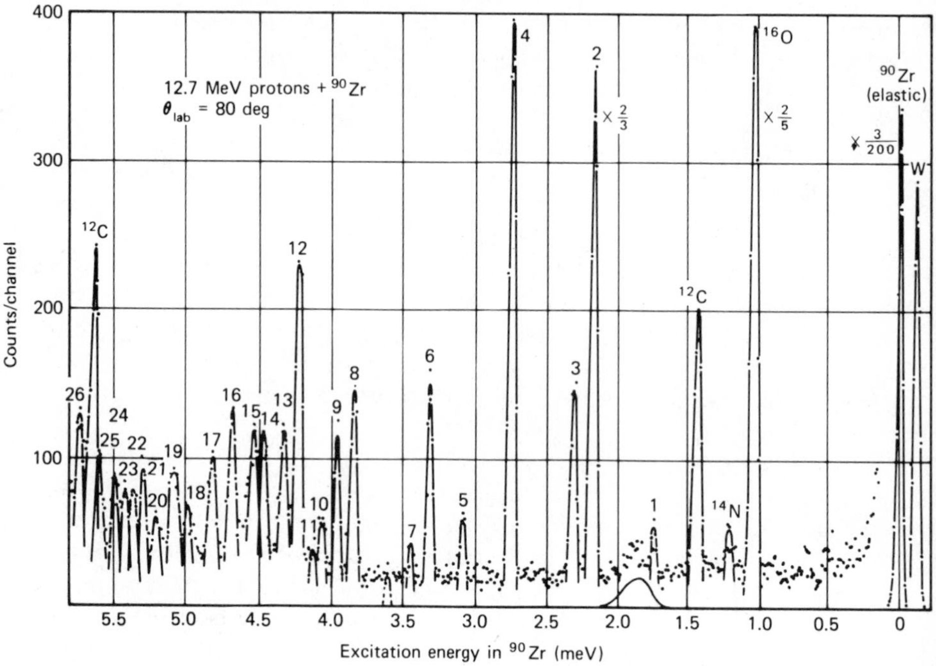

10.1. Pulse height spectrum of protons scattered from ^{90}Zr. Proton groups associated with scattering from contamination nuclei are labeled with the symbol for that nucleus. The numbered groups correspond to proton excitation of levels in ^{90}Zr [taken from Dickens, Eichler and Satchler (67)].

from even-even nuclei, whose ground state has $J = 0^+$ we find then that an α-particle can excite a level with spin J only if it has the so-called *natural parity* $(-1)^J$. States with $J = 2^+$ or $J = 3^-$ can thus be excited by inelastic α-scattering; those with $J = 2^-$ or 3^+ cannot. Figure 10.2, taken from Segel et al., shows the absence of an α-decay from the 2^- state in ^{16}O to the ground state of ^{12}C thus confirming the parity assignment to the level at 8.88 MeV in ^{16}O.

Not only are we interested in the presence or absence of a level but also in the probability for the process leading to a particular final state that is usually expressed in terms of a cross section for the process. This describes what would have been the yield in number of particles per second of that particular reaction if a single nucleus were bombarded with a flux of one projectile per cm^2 per sec. A cross section for a reaction is therefore defined by yield/flux. If we imagine an area equal to the cross section placed across the incident beam, the yield is given by the number of incident particles striking that area. Since the dimensions of nuclei are of the order of 10^{-12} cm, we expect nuclear

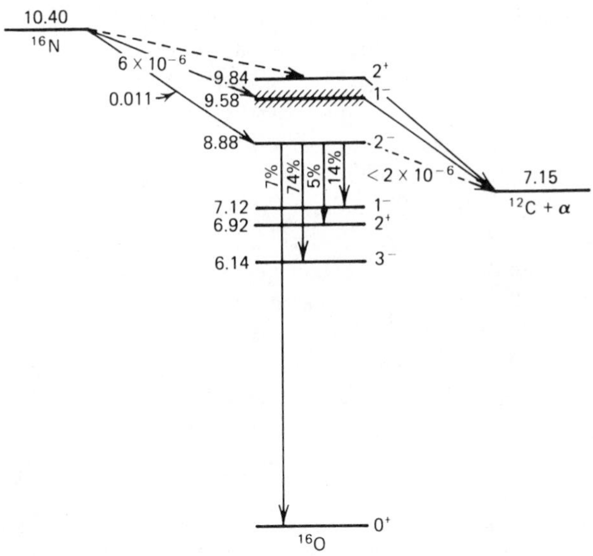

FIG. 10.2. Energy-level diagram showing the relevant states in the ^{16}N (β^-) $^{16*}O$ (α) ^{16}C decay chain [taken from Segel, Olness and Sprenkel (61)].

cross sections to be of the order of 10^{-24} cm^2 ($=1$ barn). Although this is a good first orientation in nuclear cross sections, we encounter of course cross sections that are much smaller, either because the specific reaction we have in mind is a rare event among other, more common, reactions, or because the specific projectile interacts only weakly with the nucleus. We also encounter considerably larger cross sections, even as high as 10^3 barns, when the projectile "resonates" with the nucleus in one sense or another.

It is common to talk about *differential cross sections* when some kinematic restrictions are put on the reaction products, such as specifying their direction of emission, or their energy (when this is not uniquely determined by the momentum of the projectile, such as in reactions leading to three or more reaction products).

We also encounter polarization measurements; these are usually given by the ratio

$$P = \frac{N_\uparrow - N_\downarrow}{N_\downarrow + N_\uparrow}$$

where $N_\uparrow$ can be the number of nuclei or other emergent particles observed at specified conditions (energy, direction, etc.) with spin up, and similarly for $N_\downarrow$. We note that polarization as defined above varies between plus and minus one, so that it is always given as a percentage $-100\% < P < +100\%$.

These cross sections are sensitive to the nuclear dynamics. This is demonstrated by the very different probability with which different processes excite a nuclear level in the residual nucleus. An example is shown in Fig. 10.3. The height of each line indicate the differential cross section for the excitation of each level. We see that most levels are excited by inelastic proton scattering, but that the interaction dynamics is much more selective in inelastic deuteron and the stripping (*d,p*) reaction. The "specificity" observed in this case is thought to be a consequence of the small probability for composite particles to penetrate into the interior of the target nucleus. As a consequence their interaction is localized at the surface. Hence a transfer of given amount ΔJ of angular momentum must be accompanied by linear momentum transfer q so that $qR \sim \hbar\Delta J$. This requirement, together with the conservation of energy, then fixes, classically, the direction into which the projectile is inelastically scattered. Quantum mechanically these directions are not sharply determined, but it is obvious that for a given inelasticity, the higher ΔJ is, the greater will be the average scattering angle. We may thus expect that the complete (angular) differential cross section for such a reaction is a very good tool in the determination of the spin of the excited state. The angular distribution will reveal, if we are dealing with a process in which a specific angular momentum is transferred, what that value is. This is illustrated in Fig. 10.4. The value of

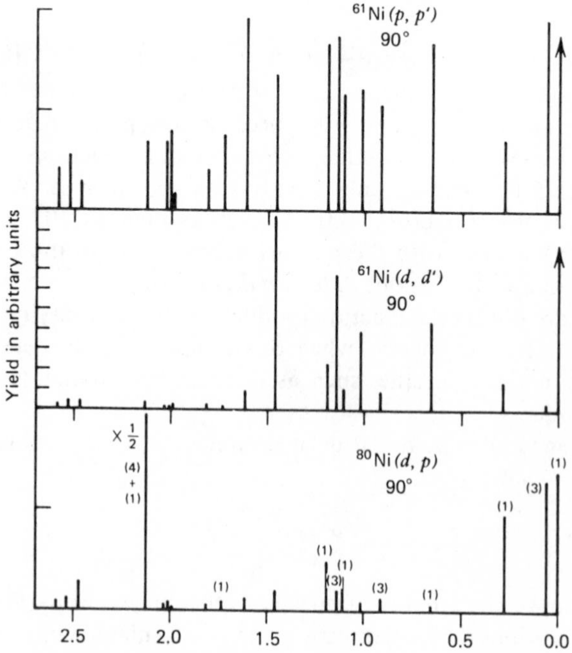

FIG. 10.3. The excitation of levels in ^{61}Ni (MeV) using different reactions [from Cosman, Schramm, Enge, Sperduto and Paris (67)].

the transferred angular momentum is denoted by l. We see that each value of l is characterized by a characteristic angular distribution. Note also the first maximum occurs at angle that is greater for the larger l.

Knowing the transferred l gives us information on the nature of the excited state of the residual nucleus involved. It must be a "surface" state, that is, its amplitude must peak at the nuclear surface. Combining this with some

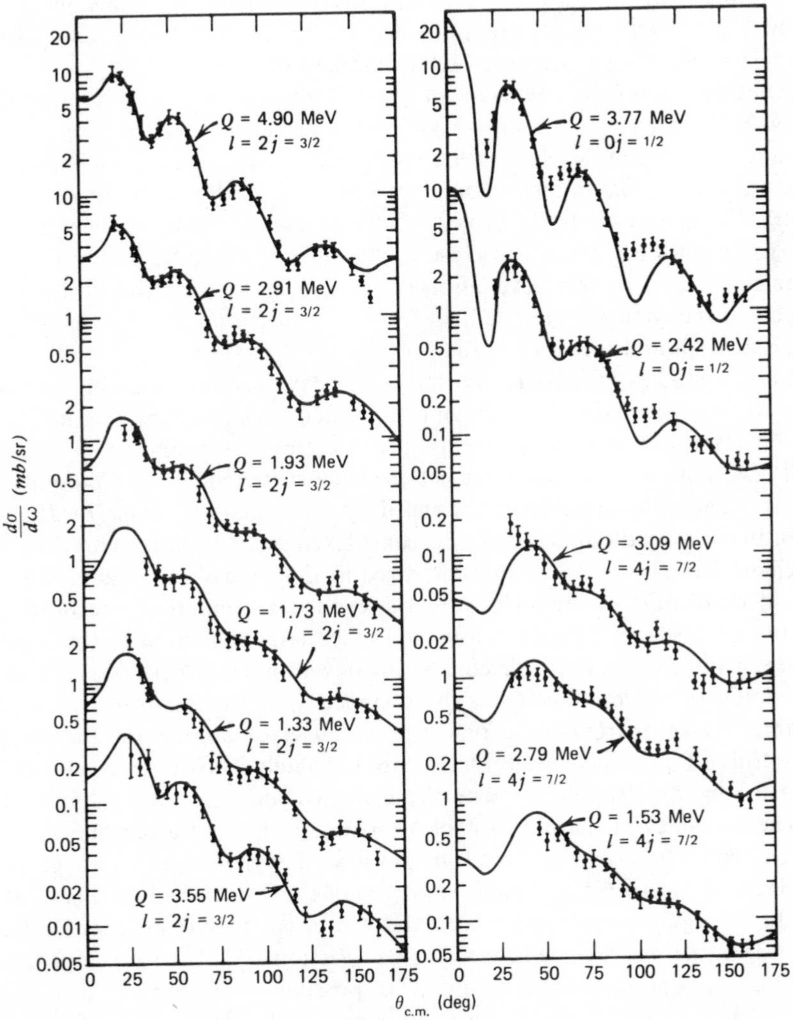

FIG. 10.4. The angular distribution for the $^{90}Zr(d, p)$ reaction with 12 MeV deuterons in which the neutron transferred to ^{90}Zr carries an orbital angular momentum of $l = 0, 2, 4$ [from G. R. Satchler (66)].

knowledge of the nucleus can take us much farther. For example, if the nucleus is a closed shell similar to ^{90}Zr the stripping (d,p) reaction will most likely deposit the transferred neutron in a single-particle state in the open shells. In this way one can then determine the energy of the levels that are principally single-particle levels.

This method can be generalized to examine two-particle states in, for example, a (^{3}H, p) reaction or a four-particle state in, say, a (^{16}O, ^{12}C) reaction, and so on.

Another sort of surface states are the collective vibrations mentioned earlier. These are also preferentially excited by composite projectiles like the deuteron or alpha particle or a heavy ion like ^{16}O.

Another example of the use of a probe with a given specificity are excitations induced by a low energy high charge projectile. Such a projectile will preferentially excite levels that are easily excited by an electromagnetic field (these are levels that in their decay have a relatively high probability for γ-emission). The proper choice of projectile and its energy can be made so that the projectile does not come too close to the nucleus (because of the Coulomb repulsion), and whatever excitation is observed must therefore be due to the effects of the changing electric field felt by the target nucleus as the projectile approaches it and recedes from it.

Careful studies of inelastic scattering of heavy projectiles make it possible also to measure quadrupole moments of excited nuclear states. Two projectiles of different mass and charge are used with their energies chosen so that both will follow the same classical trajectory when scattered by the Coulomb field of the nucleus (assumed to be produced by a point charge *Rutherford scattering*). Both projectiles may excite the same level, say a 2^+ state in an even-even nucleus. Due, however, to the difference in their mass and charge, one will spend more time in the vicinity of the excited nucleus than the other (the lifetime of the excited state is assumed to be long compared to the collision time); it will thus be more affected by the deviations of the electric field, in the proximity of the nucleus, from the assumed spherical symmetry of a point charge. By comparing the actual differential cross section for the inelastic scattering of two such projectiles at appropriately chosen energies, we can measure the quadrupole moment of the excited state.

The examples of the use of nuclear structure that are given above are all taken from "low energy" nuclear physics, the projectiles being generally nucleons or nuclei whose kinetic energy is of the order of a few tens of MeV. Higher energy projectiles have been and will be extensively used for studies of nuclear structure. We mentioned earlier (Section I.1) that from high energy electron-nuclear elastic scattering it is possible to determine the nuclear charge density. But there are many more examples. These will be discussed in Section I.15. And more fully in Volume II.

One should also mention that the availability of intense beams of various elementary particles provide new types of reactions for the study of nuclear structure. Thus pions, for instance, interact strongly with nuclei; they cannot be absorbed by single nucleons (except if another particle, say a photon, is emitted) because energy and momentum cannot be conserved, but they can be absorbed on an interacting nucleon pair $\pi + N + N \rightarrow N + N$. Pion absorption in nuclei, which is not accompanied by high energy γ-rays (nonradiative absorption) is therefore a good tool for the study of pair interactions in nuclei.

Since pions have a unit isospin the excitation of new isospin states not accessible to nucleon, α-particle, or deuteron excitation becomes possible. For example, in the charge exchange reaction

$$A + \pi^+ \rightarrow \pi^- + B$$

it is possible to explore states that differ in isospin from that of the ground state by as much as two units.

A few other elementary particles that may soon become available in the future as nuclear probes are given in the table below with some of their relevant properties. Of particular interest is the production or absorption of kaons that will yield nuclei one of whose particles will be a strange baryon or "hyperon" like the $\Lambda°$. These nuclei are referred to as hypernuclei. Since the $\Lambda°$ differs from the nucleon it is not inhibited by the Pauli exclusion principle and

TABLE 10.1 Possible Elementary Particle Probes[a]

Particle	T	J	π (parity)	Mass (MeV)	Mean Life in Seconds or Width (Γ)
$\pi^\pm$	1	0	−	139.576 ± .011	2.6×10^{-8}
π^0	1	0	−	134.972 ± .012	0.84×10^{-16}
$K^\pm$	$\frac{1}{2}$	0	−	493.84 ± .011	1.24×10^{-8}
K^0	$\frac{1}{2}$	0	−	497.79 ± .015	$K_{short} = .86 \times 10^{-10}$
					$K_{long} = 5.17 \times 10^{-8}$
η	0	0	−	548.8 ± 0.6	$\Gamma = 2.63 \pm 0.59$ keV
ρ	1	1	−	765 ± 10	$\Gamma = 125 \pm 20$ MeV
ω	0	1	−	783.9 ± 0.3	$\Gamma = 11.4 \pm 0.9$ MeV
Λ	0	$\frac{1}{2}$	+	1115.59 ± 0.06	2.52×10^{-10}
Σ^+	1	$\frac{1}{2}$	+	1189.42 ± .42	$.800 \times 10^{-10}$
Σ^0	1	$\frac{1}{2}$	+	1192.51 ± 0.10	$<1.0 \times 10^{-14}$
Σ^-	1	$\frac{1}{2}$	+	1197.37 ± 0.07	1.49×10^{-10}
Ξ^0	$\frac{1}{2}$	$\frac{1}{2}$	+	1314.7 ± 0.7	3.03×10^{-10}
Ξ^-	$\frac{1}{2}$	$\frac{1}{2}$	+	1321.31 ± 0.17	1.66×10^{-10}
Ω^-	0	$\frac{3}{2}$	+	1672.5 ± .5	$1.3^{+0.4}_{-0.3} \times 10^{-10}$

[a] These data taken from Particle Data Group (71) where a more complete listing can be found.

thus will be able to occupy particle states not available to a neutron or proton. It can thus serve as a unique probe of nuclear structure.

11. THE DETERMINATION OF NUCLEAR SIZES BY NUCLEAR REACTIONS

We mentioned in Section I.1 that x-ray data, mu-mesic x-rays, and *e*-scattering experiments can give us fairly good information on nuclear sizes and shapes. Strictly speaking these experiments are indicative only of the nuclear charge's size and shape parameters, but not of the nuclear-mass parameters. There is no a priori reason why the nuclear charge distribution should follow its mass distribution, especially in heavy nuclei, where the number of protons is significantly different from that of neutrons. In fact, the separate determination of the size and shape parameters for the charge and the mass distributions in nuclei can be a good test of our detailed understanding of nuclear structure.

Size reflects itself most dramatically in nuclear reactions. It was through its effect on α-scattering that Rutherford was first able to formulate his famous model for the atom and give the correct order of magnitude for the size of the atom's nucleus.

The differential cross section for the scattering of one point charge Z_1e by another one with charge Z_2e, is given by the well-known Rutherford formula

$$\frac{d\sigma}{d\Omega} = \left[\frac{Z_1Z_2e^2}{2mv^2}\frac{1}{\sin^2\theta/2}\right]^2 \tag{11.1}$$

where m is the reduced mass of the target-projectile system and v is their relative velocity. As far as the angular dependence is concerned, (11.1) is a monotonically decreasing function of $\theta(0 < \theta < \pi)$, and low energy scattering of α-particles from heavy nuclei follows (11.1) very closely. However, at higher energies significant deviations from (11.1) show up.

As is well known, there is a connection in classical physics between the scattering angle and the distance of closest approach of the projectile's trajectory, given by

$$D = \frac{\eta}{k}\left[1 + \frac{1}{\sin\theta/2}\right] \tag{11.2}$$

where k is the wave number and where η, the Coulomb scattering parameter, is given by

$$\eta = \frac{mZ_1Z_2e^2}{\hbar^2k} = \frac{Z_1Z_2e^2}{\hbar v} \tag{11.3}$$

As one sees from (11.2) for a given energy the large angle scattering probes small distances, and for a given angle it is the higher energy that probes smaller distances. Thus the angle at which significant deviations from (11.1) start to show up corresponds to values of D at which the Coulomb interaction fails because of the additional nuclear interactions. D is then of the order of nuclear dimensions and, hence, the usefulness of nuclear scattering and reactions as a tool in determining the size of the nucleus. The energies one has to use for that purpose must be larger than the Coulomb energy between the target and the projectile when they "touch" each other; in fact, one sees from (11.2), by putting $\sin \theta/2 \sim 1$ that the projectile's energy is given by

$$\frac{\hbar^2 k^2}{2m} \approx \frac{Z_1 Z_2 e^2}{D} \tag{11.4}$$

The right-hand side of (11.4) is the Coulomb energy of the projectile when it is a distance D from the target, and in order to be sensitive to the target's size we must have

$$D < r_0(A_1^{1/3} + A_2^{1/3}) \tag{11.5}$$

and consequently

$$\frac{\hbar^2 k^2}{2m} > E_{\text{Coul}} = \frac{Z_1 Z_2 e^2}{r_0(A_1^{1/3} + A_2^{1/3})} \tag{11.6}$$

Here A_1 and A_2 are the masses of the target and the projectile.

A beautiful manifestation of the nuclear size as reflected in the Coulomb scattering is shown in Fig. 11.1 [taken from McIntyre et al. (60)] that shows the angular distribution (in center of mass system) of ^{13}N in the reaction $^{197}Au + {}^{14}N \rightarrow {}^{198}Au + {}^{13}N$. Since the absorption of ^{14}N by Au is very strong, the reaction can take place only when ^{14}N is outside the ^{197}Au nucleus. On the other hand, since a transfer of a neutron has to take place from ^{14}N to ^{197}Au, ^{14}N cannot be too far from ^{197}Au for the reaction to occur. Thus this particular reaction is fairly well localized at mutual distances of slightly more than

$$D = r_0(A_1^{1/3} + A_2^{1/3}) \approx 10 \text{ fm} \tag{11.7}$$

The reaction product should therefore come off at angle θ so that (11.2) will be satisfied with a value of D slightly bigger than (11.7). An inspection of Fig. 11.1 does show a characteristic peaking of the differential cross section at a specific angle, which becomes smaller as the energy increases. Introducing the angle θ_{max} at which the differential cross section attains its maximum value into (11.2) with the appropriate energy gives consistently a value of $D = 12$ to 13 fm. It is thus in excellent agreement with the expectation based on (11.7). [See also A. Dar (65) for a more detailed study of this particular experiment.]

Our description of the ^{197}Au (^{14}N, ^{13}N) ^{198}Au reaction was made essentially in terms of classical concepts. This was justified because of the very short wave length associated with nitrogen nuclei at these energies (around 0.1 fm). In many cases of interest, such as lighter projectiles at lower energies, the

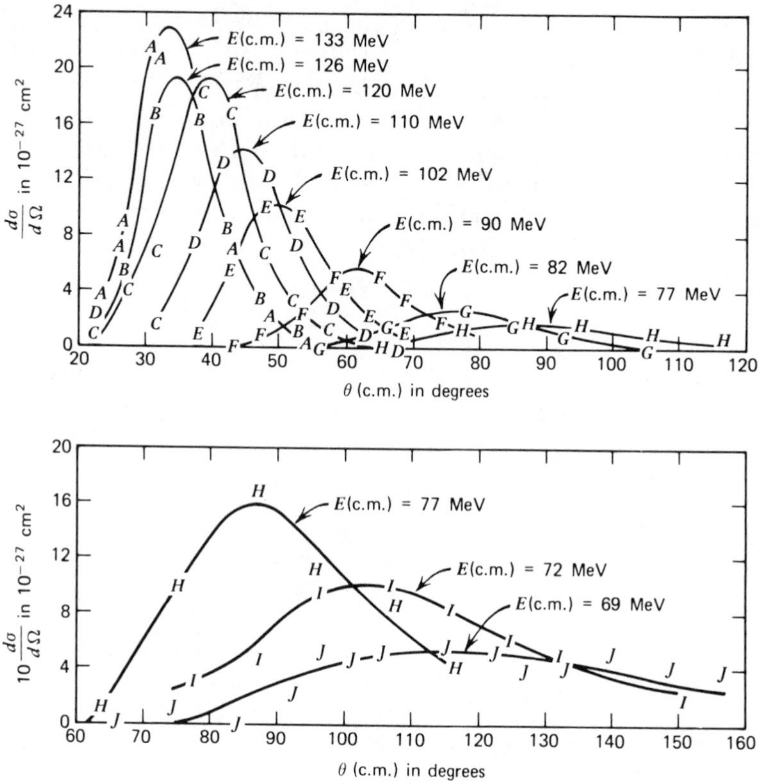

FIG. 11.1. Summary of angular distribution measurements as a function of θ in the center-of-mass system for the reaction for $^{197}Au + ^{14}N \rightarrow ^{198}Au + ^{13}N$. The letters are plotted to indicate measured differential cross-section values. A different letter is assigned for each bombarding energy. The curves are smooth lines drawn through the experimental data. The lower portion of the figure shows the cross sections plotted with a magnification of 10 to reveal more details [taken from McIntyre, Watts and Jobes (60)].

wave length of the projectile becomes of the order of magnitude of the dimensions of the system and more caution has to be exercised in the use of classical concepts.

Figure 11.2 [taken from F. G. Perey (66)] shows the differential cross section from the scattering of α-particles of 43 MeV on ^{58}Ni and ^{24}Mg. What one observes now, both in the elastic scattering and the inelastic scattering, is a characteristic oscillation of the differential cross section with angle, which is nothing but a manifestation of diffraction of the α-wave around these nuclei. In fact, since α-particles are fairly strongly absorbed by nuclei (in the sense that their coming out as α-particles after passing through a nucleus is highly improbable), we can use the familiar concepts of optics in order to describe

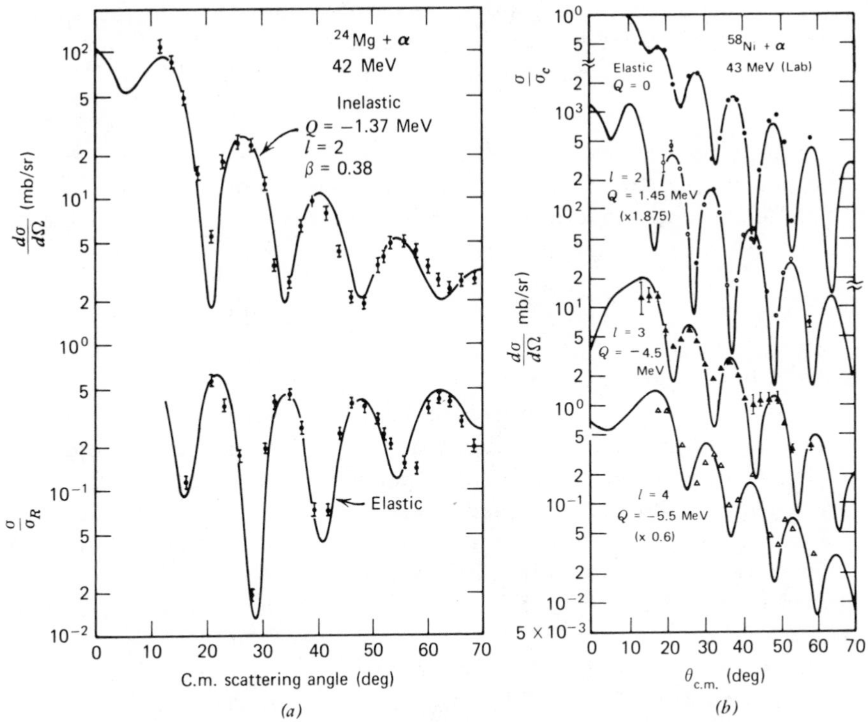

FIG. 11.2. Elastic and inelastic α-particle scattering [taken from F. G. Perey (66)].

these reactions. The nucleus is considered to be a black sphere. Such an object will in the short wave length limit diffract an incident plane wave, the resulting diffraction pattern showing minima of intensity at angles θ satisfying

$$2kR \sin \theta/2 \approx n\pi \qquad n \text{ is an integer} \tag{11.8}$$

Although one needs a more refined theory of the elastic and inelastic processes to pin down the exact positions of the maxima and minima in the differential cross section, two things emerge from (11.8):

(a) The spacing in angle $\Delta\theta$ between successive maxima at small angles is given by

$$\Delta\theta \approx \frac{\pi}{kR} \tag{11.9}$$

and can thus serve to determine the nuclear radius (or to be more specific but still quite approximate—the sum of the radii of the target nucleus and the α-particle).

(b) The consistency of the diffractive description of these processes can be tested by measuring the energy dependence of the angles of minima and maxima in the cross sections; this should be such that $k\sin\theta/2$ remains constant.

An inspection of Fig. 11.2 shows that for 43 MeV α-scattering of ^{58}Ni the spacing in angle between two successive maxima is about 11°, whereas in ^{24}Mg for α's of about the same energy it is about 13°. According to (11.9) these two numbers should be in the ratio of

$$\frac{R(\alpha\text{-}^{24}\mathrm{Mg})}{R(\alpha\text{-}^{58}\mathrm{Ni})} = \frac{(24)^{1/3} + (4)^{1/3}}{(58)^{1/3} + (4)^{1/3}} = 0.83$$

which is in very good agreement with observation.

We thus see that some rather detailed features of the differential cross section can be understood in terms of the general principles of quantum mechanics and, in this case, a single parameter describing the size of the target. Only the much more detailed study of the cross sections is going to give us more specific information about the structure of the target nucleus.

There are, however, cases in which even an inspection of the cross sections at the level we have employed thus far can give us interesting information on the reaction mechanism.

An interesting example in this connection is shown in Fig. 11.3, taken from W. von Oertzen et al., (68). Shown here is the angular distribution for the reaction

$$^{12}\mathrm{C} + {}^{16}\mathrm{O} \rightarrow {}^{12}\mathrm{C} + {}^{16}\mathrm{O}$$

over a wide range of angles at two different energies. In the forward direction one sees the characteristic Rutherford scattering of two charges off each other. Scattering at small angles results from collisions with a large impact parameter,* and we expect to see the effects of nuclear forces and the structure of the nucleus only as we proceed to larger angles. Indeed, as one sees from Fig. 11.3, the differential cross section starts to deviate from the smooth exponentially decreasing Rutherford scattering at angles of about 40°, in the center-of-mass system and a characteristic diffraction pattern shows up. However, instead of the expected continuously decreasing cross section as we go to larger angles, we find an increase in the cross section at backward angles (in the center-of-mass system) with a marked diffraction structure in it too. This suggests that we may have here two processes that are schematically shown in Fig. 11.4. In process (a) ^{12}C and ^{16}O collide, exchange momentum via the nuclear force and scatter at an angle θ; in process (b), during their collision

---

*The impact parameter is defined to be the perpendicular distance between a ray in the incident direction and a parallel ray passing through the center of the nucleus.

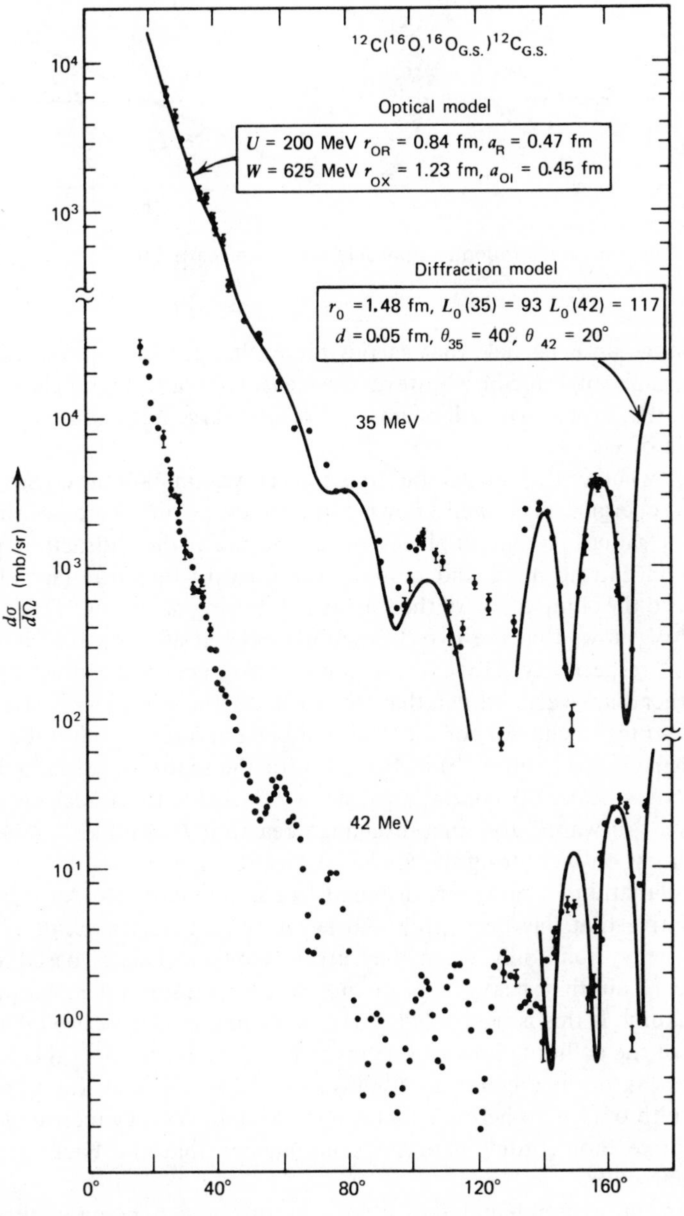

FIG. 11.3. The elastic scattering of ^{16}O on ^{12}C at 35 and 42 MeV [taken from von Oertzen et al. (68)].

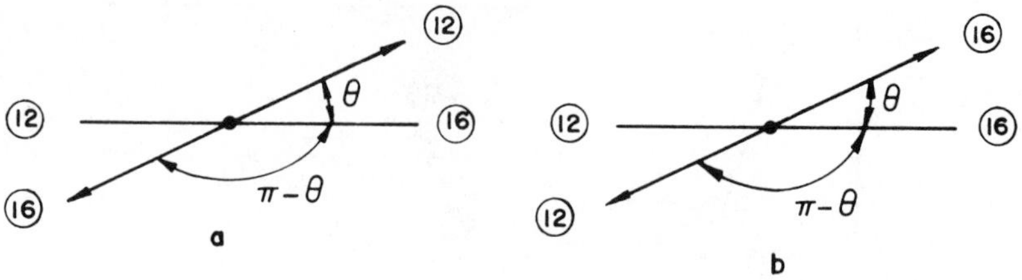

FIG. 11.4. Schematic description of α-particle exchange scattering.

they exchange an α-particle that carries the exchanged momentum and proceed at the same angle θ. Since however the measured scattering angle is always referred to ^{12}C, process (b) will be interpreted as a large angle scattering of ^{12}C at an angle of π-θ.

To check whether this interpretation is correct we compare the cross sections at different energies. As is well known from optics, as one decreases the wave length of a radiation that impinges on an obstacle, the diffraction pattern tends to concentrate more and more in the forward direction (see also Eq. 11.9). Indeed the comparison of the forward scattering of ^{12}C on ^{16}O at 35 MeV and 42 MeV shows that the two diffraction peaks at 80° and 100° moved to 60° and 80°, respectively. However, we see that the backward diffraction peaks move farther backward, rather than forward, as the energy increases. This seemingly strange behavior for a diffraction pattern finds its natural explanation in terms of mechanism (b) in Fig. 11.4 for the scattering at large angles: the diffraction peaks do continue to move *forward* with increasing energy, except that "forward" for an α-exchange reaction looks like "backward" when we insist on measuring the angles of the ^{12}C nucleus.

This explanation is, however, disputed [for a summary see Austern (71)]. An alternative that has been proposed suggests that nuclear scattering will occur for heavy ions only when they are in contact. This is based on the qualitative result that heavy ions do not interpenetrate and maintain their identity readily. If this is true only a small range in the relative orbital angular momenta of the colliding ions will be involved in the reaction. If this is true a backward peak in the angular distribution should appear and should have an angular width of $(1/L)$ where L equals $pR/\hbar$. As E the energy increases so will p, the relative momentum, with the consequence that the backward peak becomes narrower.

Thus far the experimental data does not distinguish between these two explanations. But in their resolution we shall learn much about the reaction mechanism. We shall learn with what probability an α-particle can be transferred from ^{16}O to ^{12}C, what states in the residual nucleus are most readily excited, etc.

12. ENERGY DEPENDENCE AND TIME DELAY IN NUCLEAR CROSS SECTIONS; RESONANCES

The cross sections for the processes discussed in the preceding section for the most part vary smoothly with energy. An example is given in Fig. 12.1 that shows that the proton angular distribution for the $^{40}Ca(d,p)$ reaction to a particular excited state of ^{41}Ca both in magnitude and shape remain remarkably stable over a bombarding energy range from 7 MeV to 12 MeV. On the other hand some cross-sections are not at all smooth and exhibit a very complex structure. Figure 12.2, reproduced from the Brookhaven neutron cross sections table (65), shows the energy dependence of the neutron cross section for ^{235}U at very low energies. The resonances that one observes in Fig. 12.2 are characteristic of the general features of low energy neturon scattering for all nuclei. We see that some of the resonances are sharper whereas others are broader; some are "intense" whereas others are "weak"; the density of resonances may be higher in one region and lower in another, etc. Figure 12.3 shows other interesting features of some resonances. We see that between the resonances the cross section is smooth and almost constant. This is called the background scattering. On approaching some of the resonances from their low energy side, one observes a destructive interference with the background scattering. This interference indicates that the two processes proceed coherently; the fact that it is seen with some resonances and not with others suggests that there are some quantum numbers characterizing the resonance that may not be the same as the "smooth" background scattering that becomes visible between the resonances in that region. Only resonances that have the same quantum numbers as the background scattering can interfere with it.

In the vicinity of an isolated resonance the cross section $\sigma(E)$ takes on the Breit-Wigner form

$$\sigma(E) = \frac{4\pi}{k^2}\sin^2\delta = \frac{4\pi}{k^2}\left|\sin\sigma - e^{i\sigma}\frac{\Gamma/2}{E - E_R + i\Gamma/2}\right|^2 \qquad (12.1)$$

where δ is the phase shift and $\hbar^2k^2/2m = E$. The "background" scattering cross section is given by $4\pi/k^2 \sin^2\sigma$. The second term in (12.1) gives the resonant scattering amplitude, the resonance occurring at $E = E_R$, assuming that the only physical process possible is elastic scattering.* The full width of the resonance at half maximum is Γ. The presence of interference between the background amplitude and the resonance amplitude is clear.

---

*If there are other processes the resonance amplitude should be multiplied by the relative probability for elastic scattering. Equation 12.1 holds when only one orbital angular momentum is resonating.

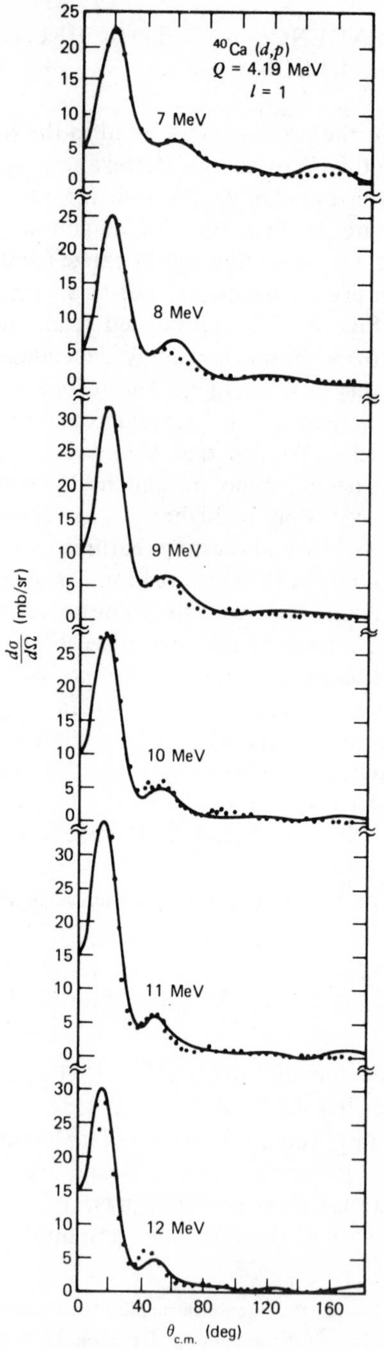

FIG. 12.1. The ^{40}Ca (d, p) reaction for bombarding energies in the range 4.19 MeV to 12 MeV exciting a level in ^{41}Ca, the solid lines are theoretical [from L. L. Lee et al. (64)].

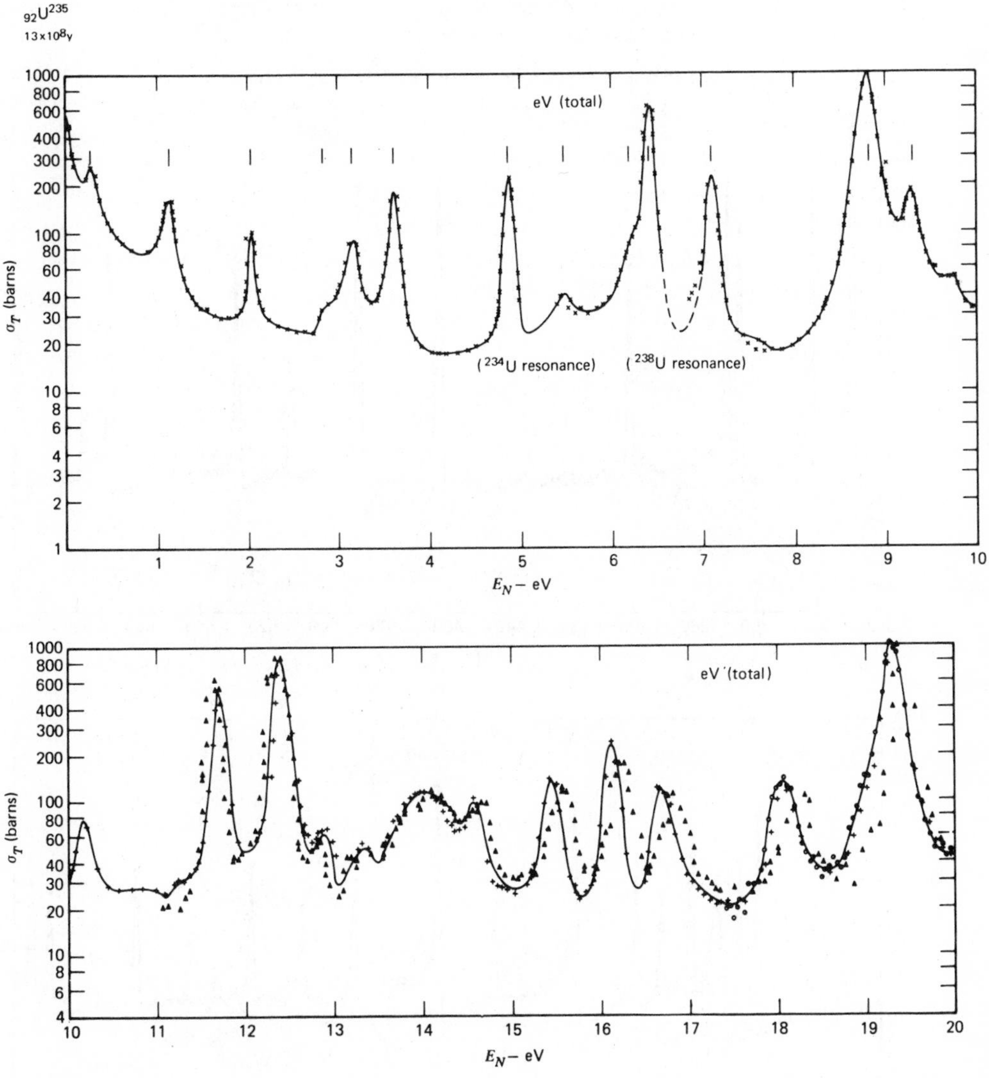

FIG. 12.2. Neutron total cross section for ^{235}U. Taken from BNL-325 2nd edition, Supplement No. 2(65).

Resonances characterize the low energy behavior of almost all processes, both elastic and inelastic, and much has been learned about them. To be sure they characterize the behavior of complex systems in fields other than nuclear physics. They are known in classical physics, atomic physics, and high-energy physics. In many cases they represent special dynamical conditions around the energy E_R for the specific process one is studying. In other cases they may

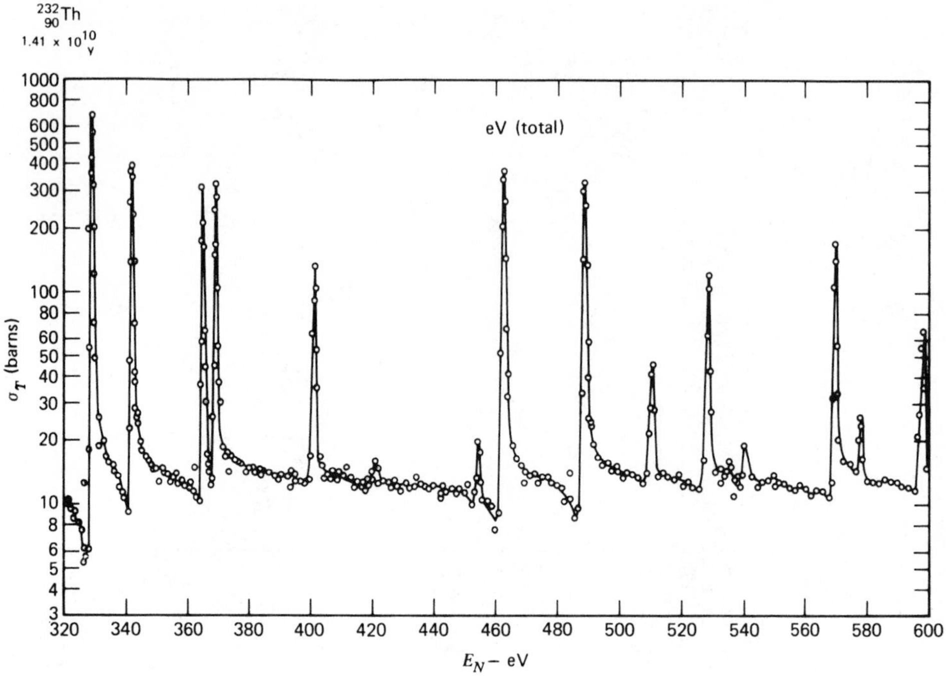

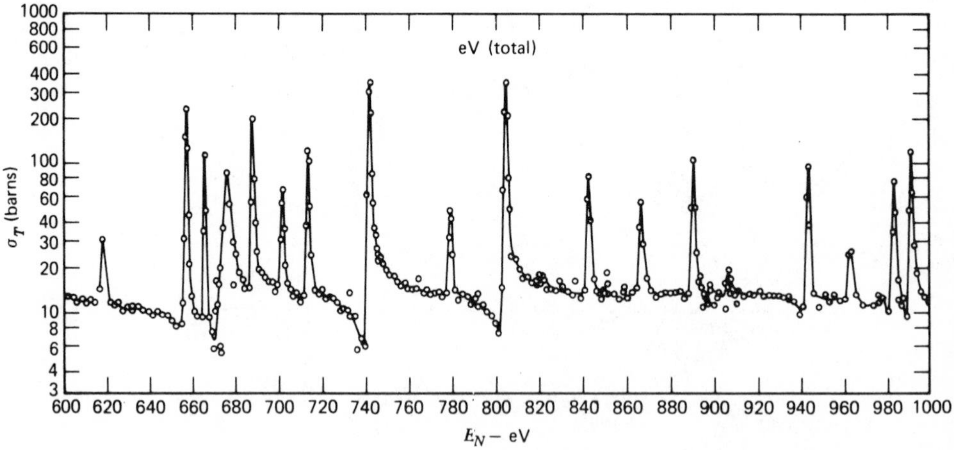

FIG. 12.3. Neutron total cross section for ^{232}Th. Taken from BNL-325 2nd edition, Supplement No. 2(65).

reflect the size or the shape of the system. Thus it is well known [see Messiah (61)] that the scattering from a potential shows characteristic peaks, as a function of energy, for such values of E that make an integral number of wave lengths sit within the potential. The resulting *shape resonances* are rather broad, their width being of the order of a fraction of $(1/m)\,(\hbar/R)^2$, that is, about several MeV for nuclei. It is clear that the resonances of the sort shown in Figs. 12.2 and 12.3, whose widths are several orders of magnitude less than 1 MeV, cannot be described by simple potential scattering.

At higher bombarding energy fluctuations in the cross section are seen if sufficient energy resolution in the incident beam and detectors is employed. This is illustrated in Fig. 12.4. The fluctuations are not nearly as rapid as those that appear at low energy and cannot be described in terms of resonances. They are referred to as Ericson fluctuations [T. Ericson (60, 63)].

If we return to resonances described by (12.1), because of their very narrow widths, these resonances are thought to reflect the existence of almost stationary states of the compound system C;

$$A + a \rightarrow C \rightarrow a + A$$

A stationary state has an infinite lifetime while these states have a lifetime equal to $\hbar/\Gamma$, which although finite is large compared to the time it would take particle a to make one traversal of the target nucleus. According to this

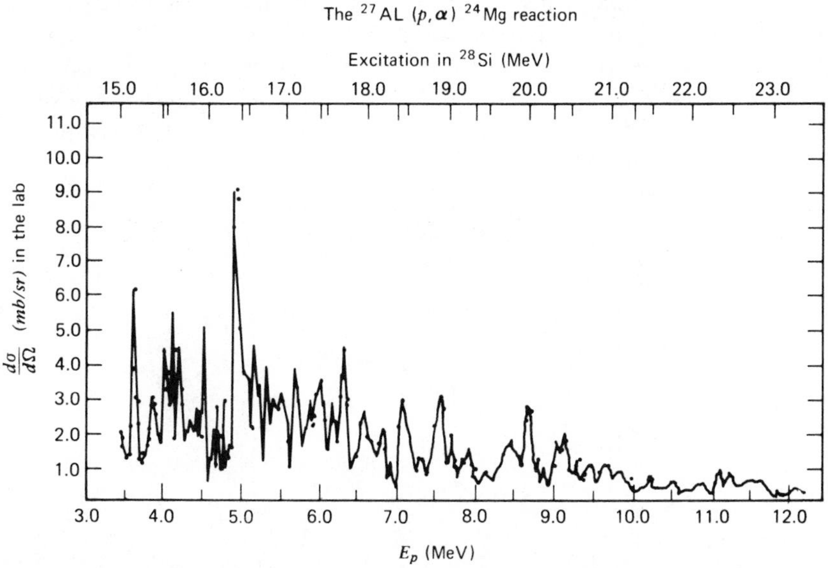

FIG. 12.4. Excitation curve for the reaction ^{27}Al (p, α) Mg to the ground state of Mg observed at 90° to the beam [from G. Temmer (64) quoted by H. Feshbach (64)].

picture it makes many traversals remaining inside the nucleus for a time of the order of $\hbar/\Gamma$ before being reemitted. The resonance is thus referred to as a *compound nuclear resonance*, the state as a *compound nuclear state*.

As Friedman and Weisskopf (55) have emphasized a considerable insight can be gained by studying the time dependence of nuclear reactions. Toward this end let us study the time delay that occurs during a reaction. [A more complete treatment will be given in Volume II, Chapter XI. See also N. Austern (70)]. According to the above discussion a resonance reaction should be characterized by a relatively long time delay.

For simplicity consider a spherical wave packet expanding from the origin ($r = 0$). In the absence of any interaction, the form of a component of such a wave at a distance r is

$$\frac{1}{r}\, e^{i[(kr-(E/\hbar)\,t)]} \tag{12.2}$$

where $k = p/\hbar$, $E = \hbar^2k^2/2m$, and $t = 0$ corresponds to the wave packet localized at $r = 0$. The net effect of the interaction is to introduce a change of phase δ at r equal to the radius R of the interaction so that (12.2) is replaced by

$$\frac{1}{r}\, e^{i[kr+\delta(E)-(E/\hbar)\,t]} \qquad r \geq R \tag{12.3}$$

The phase shift δ can be identified with the δ in (12.1). The time it takes for the *wave packet* to make the journey from $r = 0$ to $r = R$ is taken to be the time for the maximum of the wave packet amplitude to arrive at R. The maximum of the wave packet occurs when there is constructive interference. Two components of the wave packet with wave numbers k and k' interfere constructively at $r = R$ at a time t when their phase difference is zero. The time t at which this occurs at $r = R$ is:

$$(k - k')\,R + \delta(E) - \delta(E') - \frac{1}{\hbar}\,(E - E')\,t = 0$$

Solving for t and taking the limit of $E \to E'$ yields:

$$t = \hbar R\,\frac{dk}{dE} + \hbar\,\frac{d\delta}{dE}$$

or

$$t = \frac{R}{v} + \hbar\,\frac{d\delta}{dE} \tag{12.4}$$

The first term gives the time for the wave packet to go from $r = 0$ to $r = R$ without the presence of an interaction so that $\hbar\; d\delta/dE$ measures the time delay because of the interaction. The total time delay τ for an incident spherical wave is twice that given by (12.4) since the wave must propagate from $r = R$ to the origin and then back again. Hence

$$\tau = 2\hbar \frac{d\delta}{dE} \tag{12.5}$$

Expression 12.1 is obtained if

$$\tau = \frac{\hbar\Gamma}{(E - E_R)^2 + \Gamma^2/4} \tag{12.6}$$

Integrating (12.5) with this value of τ gives

$$\delta = \sigma - \tan^{-1} \frac{\Gamma}{2(E - E_R)}$$

where σ is a constant of integration. The final expression for $\sigma(E)$ in (12.1) can be obtained by substituting this value for δ in the first of the equations in (12.1). From (12.6) we see that at $E = E_R$, $\tau = 4\hbar/\Gamma$ as expected (except for the qualitatively unimportant numerical factor of 4).

But there is much more that one can learn from (12.5). We note that time delay is relatively larger whenever δ varies rapidly with energy. Or from the dependence of the cross section on δ, (see Eq. 12.1) the time delay is large whenever the cross section varies rapidly with energy. This condition is of course satisfied at the resonances in Figs. 12.2 and 12.3. However, for the (d,p) reaction of Fig. 12.1 or the background scattering between resonances the cross section varies slowly with energy so that in these cases the time delay is short. Such a short time delay characterizes the collision between systems with at most few internal degrees of freedom, collisions that can be described in terms of a potential. For background scattering the potential is, say, $U(p)$ where p gives the coordinate of the incident particle relative to the center of mass of the target. The potential description is more complicated for the (d,p) reaction since there is more than one channel involved and would require too great a digression at this point to describe it.*

The description of the background scattering in terms of a potential is reminiscent of the independent particle description of the nuclear bound states. As in that case the exact interaction between the incident particle, p, and the nucleons $(1, \ldots, A)$ of the target nucleus is replaced by a smooth interaction

---

*However "for the record" in this case the effective Schrödinger equation has the following coupled equation form:

$$[E - T_D - U(p,n)]\, \psi_D(p,n) = \int U(p,n;p')\, \psi_p(p')\, dp'$$

$$[E - T_p - U(p)]\, \psi_p(p) = \iint U(p;p',n')\, \psi_D(p',n')\, dp'\, dn'$$

It is left as a problem for the reader to ascertain the physical significance of the quantities involved. Or he can refer to Volume II [See also Austern (70) p. 84f.]

$U(p)$. The complex many-body interaction is thus approximated by a simple two-body potential. Of course this does not apply to an energy region in which the resonance amplitude dominates.

In introducing such a potential it is, of course, conjectured that a more precise and detailed treatment of nuclear reactions, including scattering, will take into account that part of the exact interaction that is not accounted for by $U(p)$, and that this residual interaction can be treated as a perturbation. This conjecture is based primarily on the success of the shell model and the collective model in describing properties of bound states.

There is, however, an important difference between the situation with low-lying bound states and scattering states. As we shall see in Chapter III, the success of the various independent particle models is due primarily to the fact that nuclear interactions are basically weak (in the sense described in Section I.3), and the Pauli principle blocks many of the states into which bound nucleons could scatter under the influence of this weak interaction. Thus pairs of nucleons in bound states, when they collide, have hardly any other choice than to scatter in the forward direction only, which is tantamount to saying that they move independently of each other.

When a proton with a positive energy hits a nucleus, it is no longer true that its interaction with the individual nucleons can have only little effect on them. If the proton has enough energy it can kick a nucleon from the nucleus into unoccupied states in the continuum; it can excite the target nucleus into one of its low-lying states; it can be captured into the target nucleus and emit a γ-ray, etc. Thus the arguments that could justify the use of an independent particle approximation for bound states are less valid when it comes to scattering states.

Here, however, comes an important generalization of the bound-state average potential in the assumption that all these processes that do lead to meaningful interactions of the projectile with the target nucleons can also be described by means of an average potential $U(p)$ acting on the projectile coordinates only. In fact, we notice that all the processes we described above, such as knocking out a nucleon, etc. necessarily involve a loss of energy for the projectile; therefore, if we consider only projectiles of the initial energy—that is, if we stay in the elastic channel—these processes are equivalent to an attenuation of the intensity of the elastic beam. The ordinary Schrödinger equation with its Hermitian operators does not allow for such loss of particles from a wave function. If, however, we relax the condition of hermiticity of the operators, we can describe the processes that lead to the disappearance (as well as appearance) of particles. The generalization of the shell model to scattering and reactions thus takes the form of allowing the average potential $U(p)$ to be complex.

$$U(p) = V(p) + iW(p) \tag{12.7}$$

Indeed, if we write down the time-dependent Schrödinger equation with $H = T_p + U(p)$

$$i\hbar\frac{\partial\psi}{\partial t} = H\psi, \quad -i\hbar\frac{\partial\psi^*}{\partial t} = \psi^* H^\dagger$$

we find that

$$\frac{\partial}{\partial t}[\psi^*\psi] = \frac{1}{i\hbar}[\psi^* H\psi - \psi^* H^\dagger \psi]$$
$$= \frac{2}{\hbar}[\psi^* W\psi] \tag{12.8}$$

($[\psi^* W\psi]$ is just the product of ψ^* and $W\psi$, *not* integrated over space coordinates). Therefore a choice of $W(p)$ that makes the right-hand side of (12.8) negative will lead to a depletion of the beam intensity with time, that is, to absorption.

13. ENERGY AVERAGES, OPTICAL MODEL, AND INTERMEDIATE STRUCTURE

However the description of the nucleon-nucleus interaction in terms of an absorptive potential is still incomplete since it omits any description of the compound nuclear resonances. To obtain a detailed theoretical description of each of many thousands of resonances is of course impractical and probably not worthwhile. One resorts therefore to statistical methods. Of these the simplest is smoothing the fluctuations by taking an energy average. This may be accomplished experimentally by using sufficiently poor resolution or by numerically averaging good resolution data. In either event the energy region over which the average is made ΔE must be much larger than the width of the resonances, Γ, and also much larger than the energy spacing, D, between them. By making ΔE sufficiently large one can obtain cross sections that vary slowly with the energy. An example of the result of such a smoothing procedure is shown in Fig. 13.1.

In principle it then becomes possible to describe each cross section in terms of a potential model. However, from the point of view of theory this is awkward since the cross section depends quadratically upon the wave function. It is therefore customary following Feshbach, Porter, and Weisskopf (54) to energy average the wave function ψ and thus the scattering amplitude rather than the cross section; that is, to define

$$\langle\psi(E)\rangle \equiv \int dE_0\, \rho(E,E_0)\, \psi(E_0) \tag{13.1}$$

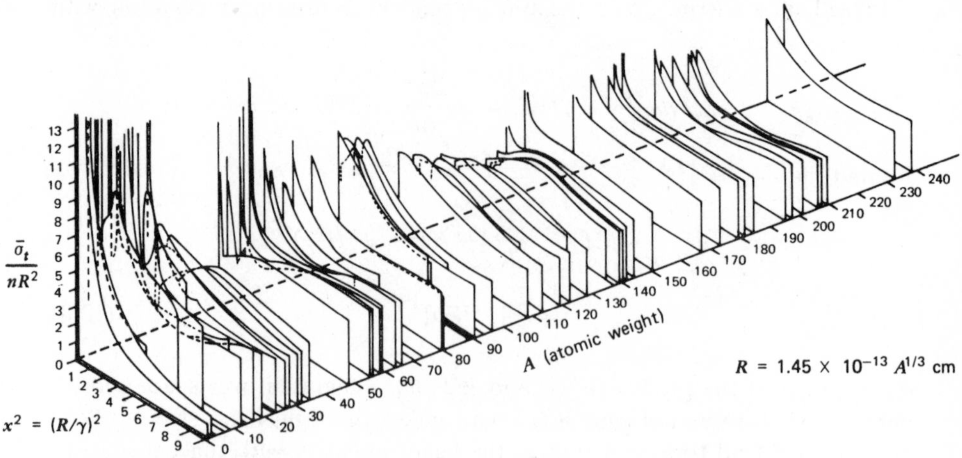

FIG. 13.1. Observed neutron total cross sections as a function of energy and mass number compiled by Feshbach et al. (54) from experiments by Barschall (52), Miller et al. (52), Walt et al. (53), Okazaki et al. (54), Nerenson and Darden (53, 54), Cook and Bonner (54), and Neutron Cross-Sections, *U. S. AEC Report*, AECU-2040.

where ρ is a weighting factor normalized to unity. Such a superposition forms a wave packet and permits thus a sequential description of the scattering process [Friedman and Weisskopf (55)]. The various components of the wave packet will be delayed inside the nucleus differently. Those with energies at which the background scattering dominates will be promptly emitted while those near resonance energies will be much delayed. (The lifetime of a resonant state whose width is 1 eV is 6.6×10^{-16} sec. The time it takes a particle moving in a potential of depth 50 MeV to travel a distance of 5 fm, the radius of a medium mass nucleus, is about 5.5×10^{-23} sec). Thus the magnitude of the prompt amplitude will be less than that of the incident amplitude. As far as the prompt amplitude is concerned an absorption appears to have taken place. It is not a true absorption in the sense that the missing portion will eventually be emitted after a meantime of $(\hbar/\Gamma)$. It is important to remember that the potential can relate only to the prompt amplitude. It therefore must be absorptive, that is, have a negative imaginary part as is true for the true absorption (see Eq. 12.8 and below).

We are thus led to the following result. The energy averaged wave function in the elastic channel is the solution of a two-body Schrödinger equation with a complex potential whose imaginary part is negative definite. Part of the resultant absorption is a consequence of real processes that deplete the amplitude in elastic channel as described in the discussion leading to (12.7). The remainder is a false absorption originating in the energy average and describes that part of the amplitude that is not prompt because of compound nuclear resonances. Thus the cross section calculated from the energy averaged wave function does not yield the energy averaged cross section. This is obtained only if one adds on the effect of the delayed amplitude. We shall postpone the discussion of how that is done to Volume II. The reader should also consult Friedman and Weisskopf (55).

The Schrödinger equation with the complex potential described above is referred to as the optical model Schrödinger equation, although some authors prefer the phrase "complex potential" to "optical." The equation has the following form

$$\left\{\nabla^2 + \frac{2m}{\hbar^2}\,[E - (V + iW)]\right\}\langle\psi\rangle = 0 \tag{13.2}$$

If $\langle\psi\rangle$ were a classical wave it would correspond to the propagation through a medium with an index of refraction n given by

$$n^2 = 1 - \frac{1}{E}\,(V + iW) \tag{13.3}$$

the index of refraction for an absorptive medium. The term "optical" applies most accurately at high energies where the real inelastic processes dominate the absorption, and there is no need to make an energy average. It is of course always possible to describe exactly the passage of a nucleon through a nuclear medium in terms of an index of refraction. But that index of refraction is generally a complicated many-body energy-dependent operator. [See Feshbach (58) where the generalized optical potential is derived or Feshbach (62) where it is related to the effective Hamiltonian described there.] Only at high energies or upon energy averaging does it reduce to the simple form (13.3).

In the absence of a delayed amplitude and inelastic processes, the potential in (13.2) is real. At low nucleon energies it is this real potential that can be regarded as the extrapolation of the potential of the independent particle model to positive nucleon energies. Such a real potential will have "shape" elastic resonances* described earlier. Within this shape resonance the amplitude of the nuclear wave function inside the nucleus is enhanced with a corresponding enhancement of the formation of a complex many-body compound nuclear resonance at the resonance energy.† This means that the widths of compound nuclear resonances occurring within the shape resonance are also enhanced. This is indicated schematically in Fig. 13.2. In this figure the size of the widths is indicated by the height of the lines placed at the resonance

---

*The shape resonance is closely related to the giant resonance of Lane, Thomas, and Wigner (55).

†For a description of how the compound nuclear state develops from the simple single nucleon wave function see Section 14.

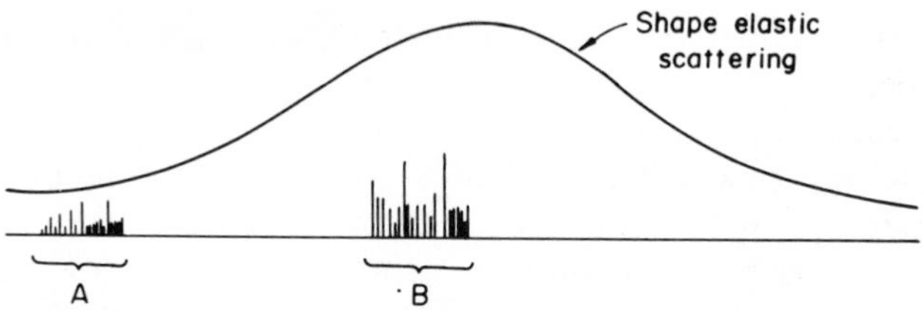

FIG. 13.2. Schematic picture of the cross section for shape elastic scattering in the neighborhood of a shape resonance. The heights of the vertical lines indicate the width of compound nuclear resonances at that value of the energy. Two regions one outside the shape resonance (A) the other inside (B) are shown.

energy. The resonances in region B are enhanced because of the shape resonance with respect to resonances in energy region A.

Upon averaging the real potential will be modified but, more important, quantitatively W in (13.2) is no longer zero. The consequent absorption calculated from (13.2) is proportional to the average probability of forming a compound nucleus. This is proportional to the fraction of the energy domain occupied by the resonances, that is, to the strength function S_o where

$$S_o \equiv \frac{\langle \Gamma \rangle}{D} \tag{13.4}$$

where $\langle \Gamma \rangle$ is the average width while D the average energy between resonances. One can expect that S_o is large within a shape resonance and correspondingly smaller outside. Observationally this phenomena is most easily seen, by comparing different nuclei that have differing radii. The radii of some of these nuclei will be just appropriate at the energy in question for a shape resonance in the wave function. Figure 13.3 taken from Seth (66) shows the experimental variation of the neutron strength function $\langle \Gamma \rangle / D$ with A, together with the calculated curves using appropriate optical models. In this calculation s-resonances ($l = 0$) of the neutron optical potential were considered, so that only corresponding resonances had to be included. The peaks in the dependence of $\langle \Gamma \rangle / D$ on A are due to the fact that at these values of A the size of the nucleus just happens to be able to accommodate an integral number of neutron wave lengths within it.

The addition of W has a second effect: the elastic cross section calculated from (13.2) no longer has an observable shape resonance. This is because W is of the order of several MeV, which must be added to the width, Γ_{SP}, of the shape resonance, and the height of the resonance must be lowered by roughly $\Gamma_{SP}/(\Gamma_{SP} + 2W)$.

In actual applications one assumes a certain functional form for the potential $U(p)$, with a few parameters built into it, and then tries to get the values of the parameters that will best reproduce the experimental data. For α-scatter-

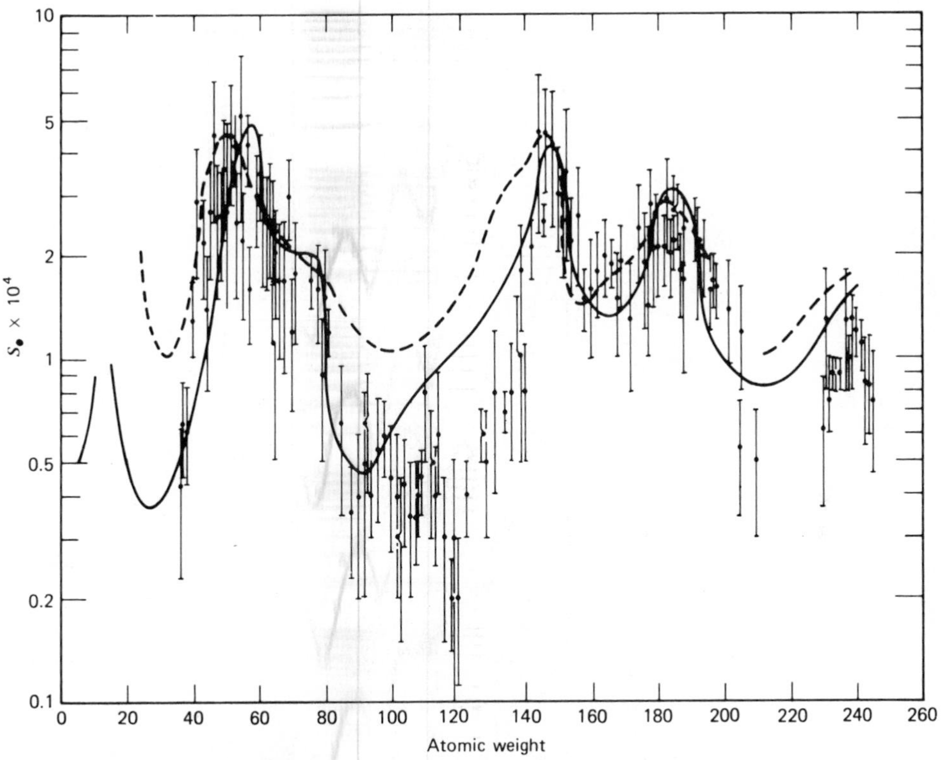

FIG. 13.3. *S*-wave neutron strength function plotted against atomic weight, *A*. ——————Optical model prediction of Chase et al. (58); - - - - - - - - - - - - - - - - - - Optical model prediction of Buck and Perey (62), [taken from Seth (66)].

ing, for instance, McFadden and Satchler (66) used for $U(\alpha)$ the following expression:

$$U(r_\alpha) = -V(1 + e^x)^{-1} - iW(1 + e^{x'})^{-1} + V_c(r) \qquad (13.5)$$

where

$$x = \frac{(r_\alpha - r_0 A^{1/3})}{a} \qquad x' = \frac{(r_\alpha - r'_0 A^{1/3})}{a'} \qquad (13.6)$$

and $V_c(r)$ is the Coulomb field of a uniformly charged sphere with a radius $r_c A^{1/3}$, $r_c = 1.3$ fm. The parameters that were used to fit the data were taken to be V, W, $r_0 = r_0'$, and $a = a'$ (the data did not justify the use of the freedom to make $r_0 \neq r_0'$, $a \neq a'$). The data included the measured elastic scattering differential cross sections for 24.7 MeV α's on various nuclei from O to U. Figure 13.4 shows a characteristic fit of the calculated differential cross section to the measured one.

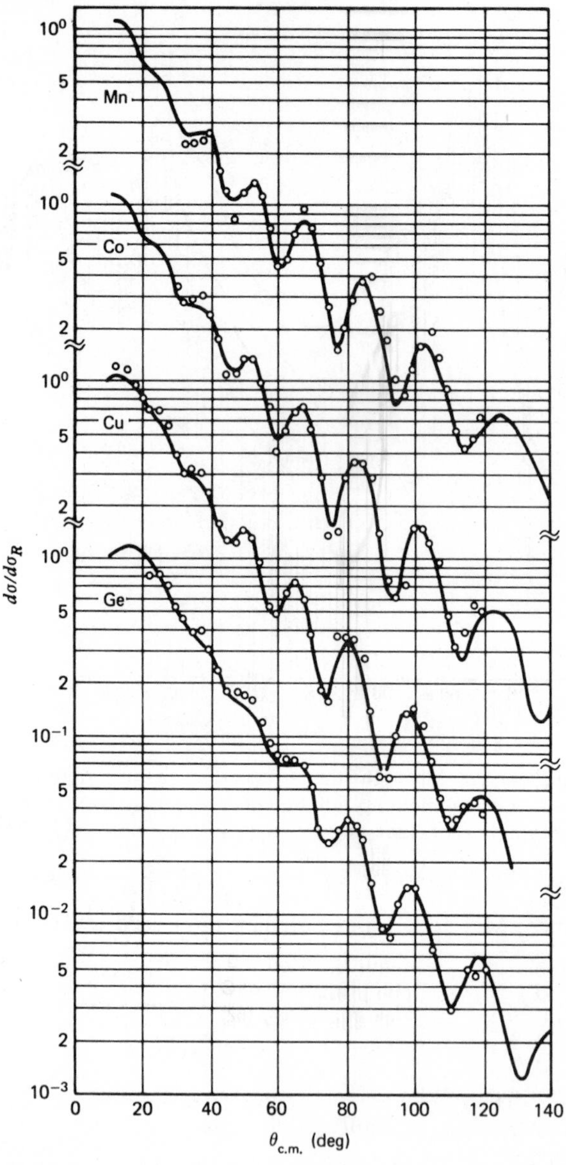

FIG. 13.4. Scattering of 24.7 MeV α-particles. Comparison of experimental data with the prediction of the optical model [from McFadden and Satchler (66)].

For protons and neutrons the optical potential is usually more complicated since additional terms should be added to account for the strong spin orbit interaction. These may again be present in both the real and the imaginary part of the potential. Although this means that there are more parameters that can be used to fit the data, one should remember that there is also more data to be fit, since the polarization of the elastically scattered nucleons can also be measured. Figure 13.5 [taken from Rosen et al. (65)] shows a fit to the polarization data on 14.5 MeV protons scattered off various nuclei from ^{12}C to ^{120}Sn. The optical potential was made to fit the differential cross sections for proton scattering off these various nuclei and average empirical parameters

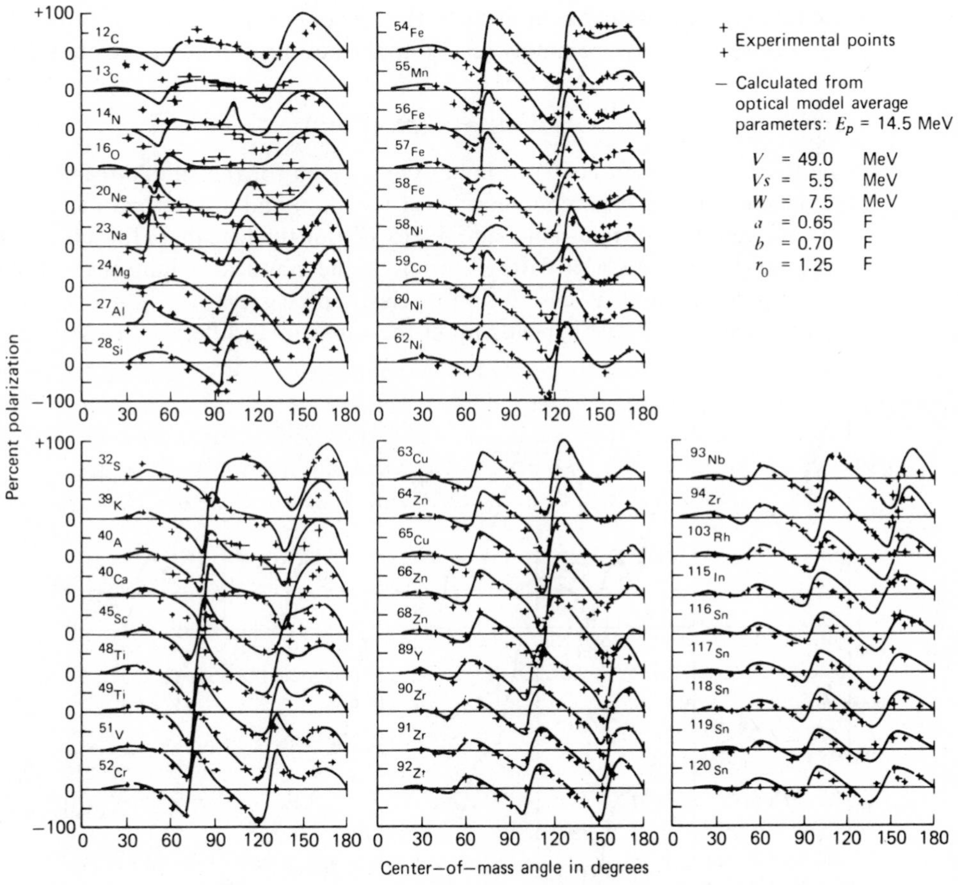

FIG. 13.5. Comparison of predictions of optical model potential with 14.5 MeV polarization data, using average parameters. Taken from Rosen, Beery, Goldhaber, and Auerbach (65).

thus derived were then used to calculate the polarization with the results shown in Fig. 13.5. The fit for the proton scattering is shown in Fig. 13.6.

The optical potential in its present form was originally introduced by Feshbach, Porter, and Weisskopf (54) to explain the regularities observed by Barschall (52) in the elastic scattering of neutron by a wide variety of nuclei. Its great usefulness stems from the fact that its parameters are found to vary slowly and smoothly with the atomic weight A and the projectile's energy E. Its success in reproducing differential cross sections and polarizations must therefore be tied to a more profound aspect of nuclear structure that makes such a description possible. This question will be dealt with at greater length in Volume II; here we would like just to mention that the physical meaning of the optical potentials goes even beyond the scope of their first presentation: extensive studies by various authors on the best optical potentials for deuterons have shown that they are simply related to the best proton and neutron optical potentials:

$$U(d) \approx U(p) + U(n) \tag{13.7}$$

One would not expect (13.7) to hold exactly, even for the simple reason that the deuteron is a special combination of a proton and a neutron, and this is not reflected on the right-hand side of (13.7). But the fact that (13.7) holds in some approximation is a further proof of the physical significance of the optical potential and its description of elastic processes.

The discussion above depends upon the substantial difference between the

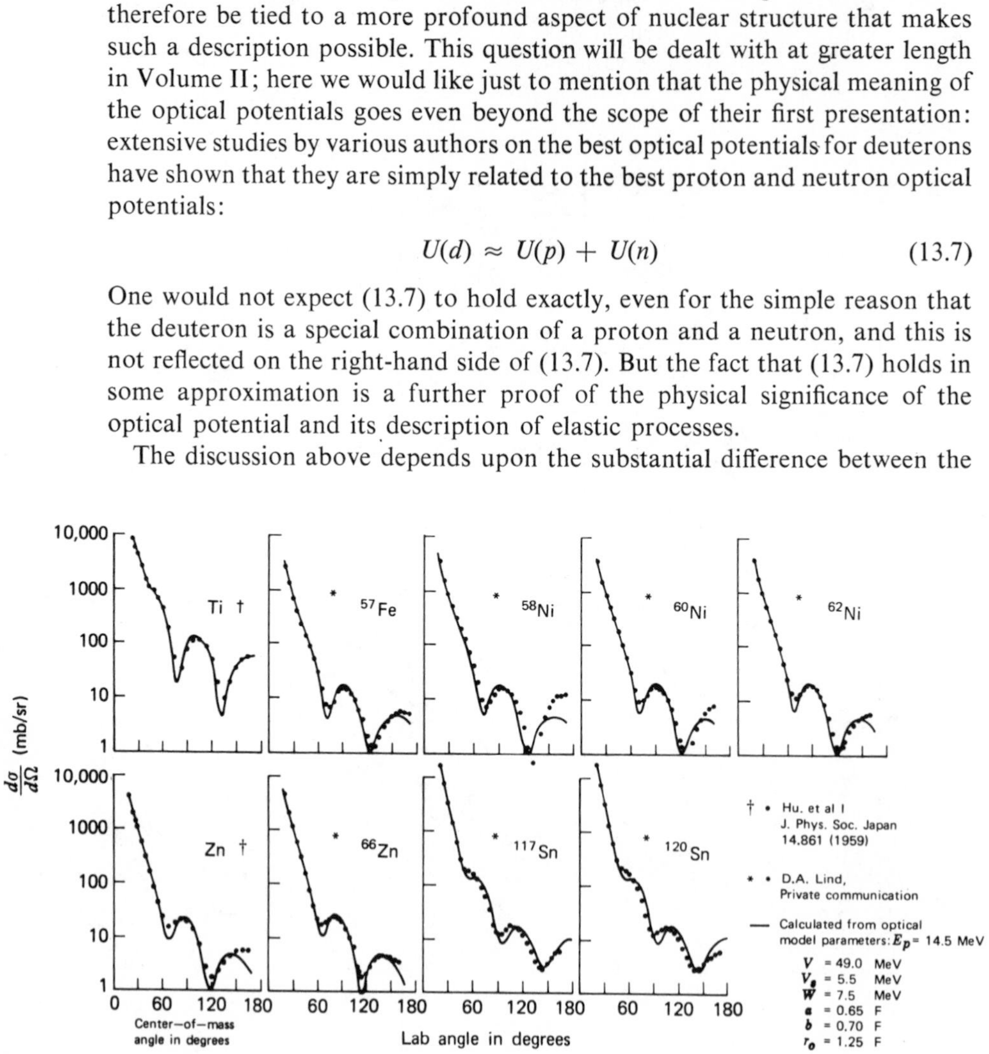

FIG. 13.6. Comparison of optical model predictions with differential elastic scattering cross sections at 14.5 MeV. Taken from Rosen, Beery, Goldhaber, and Auerbach (65).

time needed for a particle to traverse a nucleus and the delay time for particles at or near a resonance energy. In the example quoted above, these times and the corresponding widths were 6.6×10^{-16} sec, $\Gamma_{CN} = 1$ eV and 5.5×10^{-23} sec, $\Gamma_{SP} = 12$ MeV. The question naturally arises as to whether there are intermediate situations with characteristic times lying between these two extreme values. Such an intermediate situation would be indicated by the presence of structures in the cross sections whose width Γ_d is such that

$$\Gamma_{SP} \gg \Gamma_d \gg \Gamma_{CN} \tag{13.8}$$

It would be observed as the energy interval ΔE over which the cross section is averaged increases from the very small value required to make the compound nuclear resonances or Ericson fluctuations visible. If over some range in ΔE satisfying

$$\Gamma_{CN} < \Delta E \ll \Gamma_{SP}$$

structure (that is, energy dependence in the average cross section) is observed, the structure is referred to as *intermediate structure*; the widths of the observed peaks are then Γ_d, which must be larger than ΔE and of course Γ_{CN}. An example of this phenomenon is shown in Figs. 13.7*a* and Fig. 13.7*b*. Figure 13.7*a* is the partially resolved cross section, the individual peaks representing single or groups of compound nuclear resonances. Upon averaging the cross section Fig. 13.7*b* results. This average cross section also resonates, the resonance in this case is an *isobar analog resonance* [see Auerbach, Hufner, Kerman, and Shakin (72)]. This is an example of a general class of resonances, the *doorway state resonance* [Block and Feshbach (63), Feshbach, Kerman, and Lemmer (67)] that are responsible for intermediate structure. Characteristically its width is much larger than the compound nucleus width but much smaller than Γ_{SP}. Another example is the giant dipole resonance described earlier that exhibits a similar behavior with change in resolution [see P. P. Singh et al. (65)]. In Fig. 13.8 one sees the successive emergence of intermediate structure as the energy resolution ΔE is increased. Still another occurs in fission (Fig. 13.9) where we observe selective enhancements of the fission cross section. In these regions the cross section breaks up into narrow fluctuations.

In each of these cases the relation between the good resolution results and the intermediate structure is similar to that between the good resolution data and the shape resonance. In the absence of coupling to the compound nuclear resonances or rapid fluctuations in the cross section, the cross section would exhibit a resonance. Within that region the widths of the compound nuclear resonances, because they do in fact couple with the doorway state, are enhanced. This phenomenon is illustrated in Fig. 13.9, the fission cross section being greatly enhanced in the region of the doorway state resonances but being practically zero elsewhere. Upon averaging over the compound nuclear resonances an effective absorption is introduced that increases the width of the doorway state resonance just as the width of the shape resonance was increased

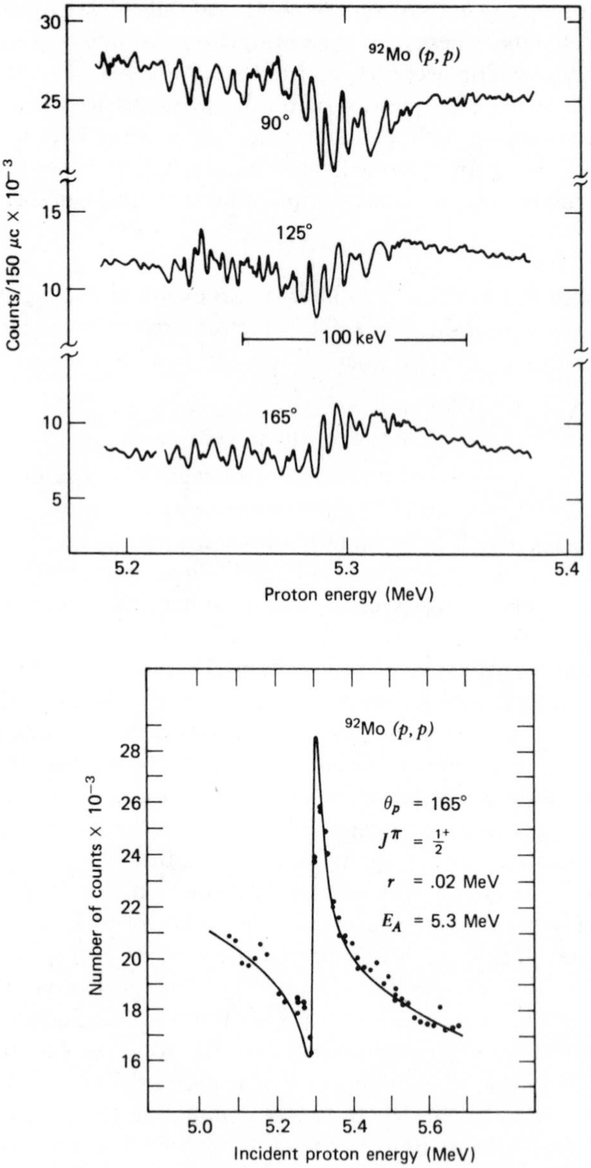

FIG. 13.7. (*a*) Elastic scattering of protons by ^{92}Mo in the neighborhood of the *s*-wave isobar analog resonance at proton energy of 5.3 MeV. (*b*) Curve obtained by averaging the data of (*a*) [from Richard, Moore, Robson, and Fox (64)].

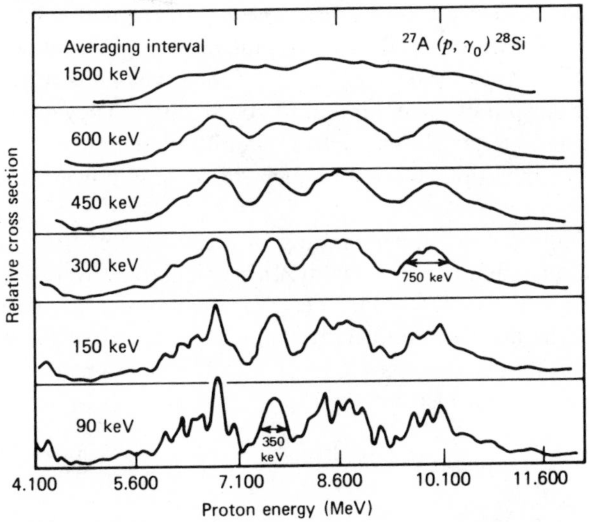

FIG. 13.8. Photocapture of protons by ^{27}Al to the ground state of ^{28}Si. The data is presented for various energy resolutions [taken from P. P. Singh et al. (65)].

by energy averaging; the additional width is referred to as the *spreading width*, Γ_W. If Γ_W is too large the resonance may spread so far and its magnitude reduced so greatly as not to be visible. Thus the relative value of Γ_W and Γ_d are critical. Γ_W depends directly upon the coupling between the doorway state and the compound nuclear resonances. If this coupling is weak, Γ_W will be relatively small and the doorway resonance will be narrow even after the energy

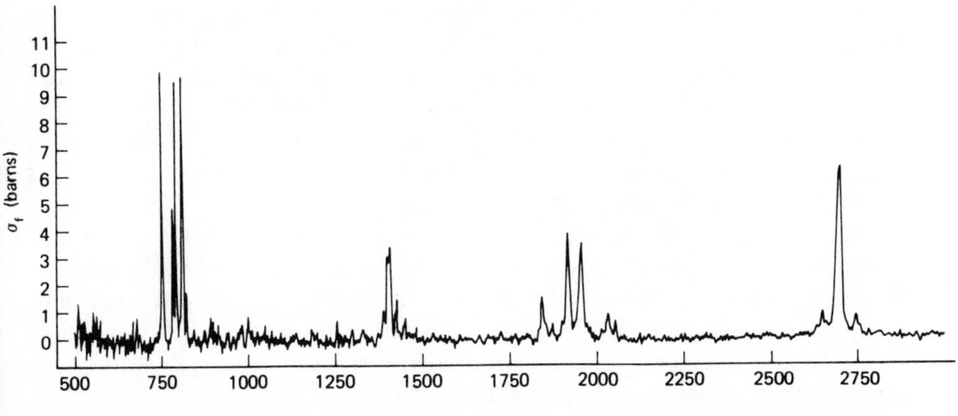

FIG. 13.9. The sub threshold fission cross section ^{240}Pu (n,f) [taken from Migneco and Theobold (68)].

average is made. As we shall see in the case of isobar analog resonances, the coupling is weak because of the approximate validity of isospin conservation. Here the isospin of the analog state, as the doorway state is referred to, T_A is larger than the isospin of the compound nuclear states. They are coupled only in virtue of those parts of the nuclear Hamiltonian that are not invariant against rotations in isospin space, of which the most important example is the Coulomb interaction.

It is worthwhile for the understanding of intermediate structure to discuss the isobar analog case more thoroughly. In particular, let us consider the case of elastic proton-nucleus scattering. Let Z be the atomic number of the target nucleus and N the neutron number. It is a general rule that the isospin of the ground state of this nucleus is $(N - Z)/2$. The combined system, proton plus target, can be in two isospin states,

$$T_> = \frac{N - Z + 1}{2} \qquad T_< = \frac{N - Z - 1}{2}$$

Consider the $T_>$ component. The essential point is that the compound system with $Z + 1$ protons and N neutrons does have a lowest $T_>$ state that for heavier nuclei is in the continuum. This, through the action of nuclear forces, can couple to the incident channel, proton plus nucleus, with a consequent resonance. Most important for the width is the fact that these $T_>$ states are isolated. There is, for example, a lowest $T_>$ state, the other states nearby in energy are all $T_<$. These can couple only because of the isospin symmetry breaking interactions such as the Coulomb interaction. Since the nondiagonal components of these forces are relatively weak, the spreading width will be correspondingly small and the resonance narrow and observable.

The reason that these $T_>$ states in the $(Z + 1, N)$ nucleus are relatively isolated is because they are the "isobar analogs" of the low-lying $T_>$ levels in the $(N + 1, Z)$ nucleus. This nucleus is called the *parent* nucleus. For example the ground state of this nucleus is a $T_>$ state and there is generally a healthy separation in energy of this state from the next highest $T_>$ state. The T_3 value is $-(N + 1 - Z)/2$. The analog of this ground state in the $(Z + 1, N)$ nucleus is the member of the $T_>$ isospin multiplet with $T_3 = -(N - Z - 1)/2$. If isospin symmetry were exact these two states in the $(N + 1, Z)$ and $(N, Z + 1)$ nuclei would have the same energy. But because of the Coulomb force the level in the $(N, Z + 1)$ nucleus lies higher. We have described such a case earlier (Fig. 6.1). In heavy nuclei this increase in energy is enough to make the analog level unbound and therefore accessible to excitation by proton scattering from a (Z, N) nucleus. The $T_<$ levels can have a considerably smaller energy because in addition to the repulsive Coulomb energy, there is an effectively attractive compensating nuclear term. Since a $T_<$ state involves a wave function that is more symmetric in space and spin than a $T_>$ state, it can readily take greater advantage of the short range nuclear force and will be more deeply bound.

Thus the first $T_>$ state is at an energy very much* above the ground state for the $T_<$ levels. The advantages of having a relatively simple state with positive energy that emits particles are many; we shall return to this point briefly.

For the moment we wish to emphasize the more general picture for which this example of the isobar analog resonance can serve as a prototype. The wave function for the system can be expanded in terms of wave functions of increasing complexity. The simplest component is that given by the initial state ψ_0; in the above example, this is the $T_>$ combination of the proton plus the target nucleus in its ground state, a "single-particle" state. The next most complex component is the doorway state ψ_d, in this case the analog state. The relative motion of the proton and the target that is the single-particle state ψ_0 is described, for example, by some single-particle potential analogous to the independent-particle model.† The doorway state is coupled to ψ_0 by the residual interaction; that is what remains of the nucleon-nuclear Hamiltonian after the single-particle potential is subtracted. ψ_d is a more complex state than ψ_0 because it involves some excitations of the target nucleus. But these are chosen to be the simplest variety possible, such as particle-hole states in which a particle in the target is excited leaving a "hole" behind; or it might be a vibrational or rotational excitation. Which it is will depend upon the target nucleus and the nature of the incident projectile. In the case just discussed the analog was such a simple state. This two-channel system, ψ_0 and ψ_d can under the appropriate conditions have resonances of both the broad shape resonance variety and of the much narrower doorway kind.

Of course, ψ_0 and ψ_d do not in general give a complete description of the exact wave function. There generally needs to be many components of greater and greater complexity before a compound nuclear state can be adequately described. Both the shape and doorway state resonance are meaningful only if ψ_0 and ψ_d couple not too strongly to the remainder of the wave function. The coupling causes the "strength" of the doorway state to be shared among a number of compound nuclear states over an energy range of the order of Γ_W, the spreading width. If the coupling is too large Γ_W will be too large and the doorway state resonance will not be observable.

Let us return briefly to the isobar analog state. There is a considerable advantage to having such a simple state at high energy where it can be excited by an incident proton. Because of this high energy processes in which the analog resonance decays to several final states are possible. The branching ratio to a final state would indicate whether the analog state was similar to that final state. For example, consider the inelastic scattering of a proton by ^{208}Pb.

---

*The symmetry energy involved here is about 90 $(N - Z)/(N + Z)$ MeV with the result that there are many $T_<$ states at the first $T_>$ energy level.

†There are many ways to choose ψ_0 and the corresponding potential. This is one example. For others see Feshbach, Kerman, and Lemmer (67) or Volume II of this work.

The isobar-analog states are in ^{209}B, the parent nucleus is ^{209}Pb. Several analog states are found [Zaidi et al. (68), Bromley and Weneser (68)]. All these decay by proton emission into the 3^- first excited state of ^{208}Pb, while decay to the second and third levels, 5^- and 4^- respectively, occurs from only one of the analog resonances, the one corresponding to a $g_{9/2}$ neutron level in the parent nucleus. It is thus possible to conclude that in ^{208}Pb the 3^- is a superposition of many neutron particle-neutron hole configurations while the 5^- and 4^- are principally a neutron hole plus a neutron in a $g_{9/2}$ orbit.

14. THE MICROSCOPIC STRUCTURE OF THE NUCLEUS AND ITS RELATION TO NUCLEAR REACTIONS

We picture the nucleus as consisting of nucleons moving more or less independently in a common potential. A particle p of energy E approaches the nucleus (Fig. 14.1*a*) interacting with each one of the nucleus via the interaction $v(pi)$. As a result of this interaction the projectile p may fall down into one of the allowed states lifting up a particle i_1 of the target nucleus, thereby conserving total energy (Fig. 14.1*b*). A further interaction with particle i_2 may bring p farther down in the nuclear well, lifting i_2 to an excited state. The process continues until one of the particles is lifted to a positive energy and comes out of the nucleus as a reaction product, leaving the rest of the nucleus at its ground state or at one of its excited states.

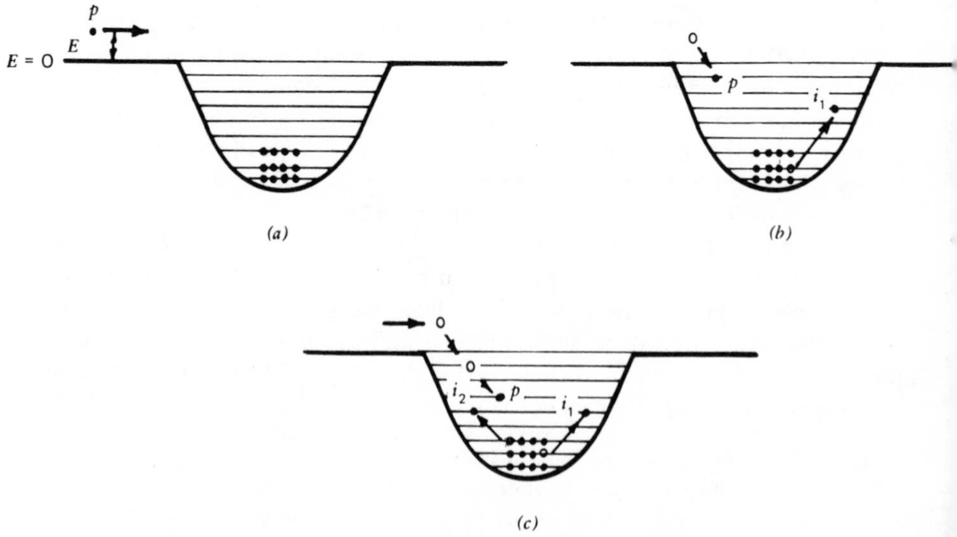

FIG. 14.1. The reaction process: a schematic microscopic picture.

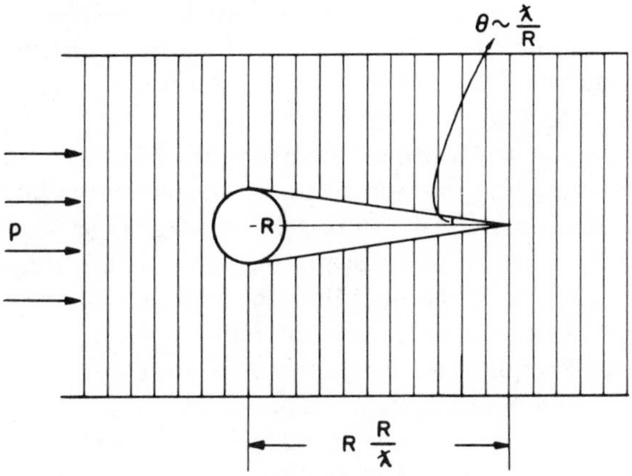

FIG. 14.2. Diffraction scattering by an absorbing sphere.

The process we have just described is actually that of the formation of the compound nucleus. It can readily proceed if the initial energy of the projectile coincides with that of the level in the compound nucleus, and the more complex the excitation involved, the sharper is the energy dependence of the process. Sharp resonances are compound nucleus states that, in the picture we have just described, have many components representing excitations in the compound nucleus of various nucleons to various states, so that the total excitation energy (including corrections due to the residual interaction) lies around E_R. When the projectile hits the target at the appropriate energy it is "trapped" in one of the components (usually the simpler ones) of this compound state and is carried over through all other components until a reaction product emerges through another component of the resonance wave function.

The kinematics of the system and the strong conservation laws (conservation of charge, baryonic number, total angular momentum, etc.) generally make many otherwise reasonable end products strictly forbidden. One says then that these particular *channels* are closed. There is, however, always at least one channel that is open; the *entrance channel* to the reaction, and elastic scattering is therefore a process that *always* takes place.

Actually, because of the wave character of the projectile, there is another contribution to the elastic scattering that shows up whenever there is a reaction. To see how this comes about let us consider a nucleus that *absorbs completely* the projectile p and an energy E. In the language of the description given above it means that at this energy there is an overwhelming probability that the reaction products will differ from p. Looking now at the *elastic* channel the picture would be something like that shown in Fig. 14.2: a wave of the projectile with wave length $\lambdabar$ impinges on a nucleus with radius R; that part which hits the nucleus is totally absorbed, casting a shadow on the other side of the nucleus (there may be reaction products in this shadow; we are considering only the elastic channel). Because of the wave nature of the projectile, however, this shadow will extend only over distances of the order of $R/(\lambdabar/R)=R^2/\lambdabar$. If we translate it to the particle language, it means that because of the absorption by the nucleus a projectile passing outside of the nucleus will be deflected by an angle $\lambdabar/R$ in order to form the shadow by interference with the incident beam. Such a deflection is interpreted as an elastic scattering, thus showing why there is elastic scattering, even if the compound nucleus prefers to decay into channels other than the entrance channel.

Coming back to our microscopic description of nuclear reactions, Fig. 14.1, we see that if the energy of the projectile does not coincide with one of the resonances of the compound system, many things can still happen. The most trivial process is that in which the projectile p interacts with one or more of the target nucleons and is scattered elastically without exciting these nucleons. The recoil momentum is then absorbed by the nucleus as a whole. This process is probably described to a large extent by the optical potential, although this potential includes also a description of processes in which the target nucleus is first excited and then deexcited while the projectile goes through it. Furthermore, through the imaginary part of the optical potential some account is taken of the effects of reactions (i.e., absorption) on the elastic channel.

Another simple process involves a projectile with energy E, exciting one of the nucleons by an energy $\epsilon < E$, and emerging from the nucleus as an inelastically scattered projectile with its residual energy $E - \epsilon$. This process, as is quite obvious from its description, is not very sensitive to the energy of the incoming projectile, and we do not expect any resonance structure in it.

A third simple process is that of the *knock-out reactions* or *pick-up reactions*. Here a projectile, say a proton comes in and either knocks out a nucleon or a cluster of nucleons with itself being captured in one of the nuclear states [such as in (p, n) or (p, α) reactions], or it picks up a neutron, or a pair of neutrons, or a deuteron, and comes out as a more complex reaction product [such as in (p, d), (p, t), $(p, {}^3\text{He})$ reactions, etc.]. In the same class we have also the stripping reactions, where a complex projectile comes in, leaves part of itself in the nucleus, and emerges as a lighter projectile [reactions like (d, p), (t, p), etc.]. All of these reactions have simple components that proceed very rapidly and involve just one or two collisions. Thus, since Δt is small, they cannot, be associated at the same time according to the uncertainty principle with a small energy spread.

Reactions that proceed in between resonances, that is, that do not show a structure in the energy dependence of their cross sections, are called *direct reactions*. Since they are also associated with fast processes, they do not involve multiple scatterings, and are thus invariably connected with the excitation of simple degrees of freedom in the nucleus. These may be the excitation of a single nucleon from one orbit to another, or the excitation of a collective oscillation of the nucleus, or the picking of one nucleon from a definite level to the continuum, etc.

In some cases, especially if the projectile is energetic enough, the nucleus may be left at quite a high excitation after a direct interaction. It may then continue to emit reaction products also a relatively long time after the direct reaction is over. One talks then about the *evaporation* of such reaction products, and they can indeed be identified by their characteristic spectrum.

Since direct reactions involve only simple degrees of freedom, they offer an excellent tool for the isolated study of these degrees of freedom. In pickup reactions, for instance, it is possible to study the quantum numbers of the orbit from which the nucleon was picked up. If this nucleon was, say, in an s-orbit, with orbital angular momentum $l = 0$, then its pickup does not require the transfer of any angular momentum to the residual nucleus, and the product can come out in the forward direction. If, on the other hand, the picked-up nucleon had a finite orbital angular momentum in its orbit, this has to be provided for by the reaction, and the reaction products will be peaked at an angle that will assure the required momentum transfer q to satisfy $qR \approx l\hbar$. Figure 14.3 shows typical stripping reactions that are peaked in the forward direction when the orbit involved is in $l = 0$, and are peaked at other angles for orbits with a finite angular momentum. The same effect is seen in elastic and inelastic proton scattering as shown in Fig. 14.4.

Direct reactions make it possible to study also other properties of the nuclear wave function. We can always decompose an A-particle wave function in the form:

$$\psi_\alpha(1, \ldots, A) = \sum C^\alpha_{\beta n} \psi_\beta(1, \ldots, A - 1)\, \phi_n(A)$$

Here ψ_β is a complete set of the $(A - 1)$ particle system and $\phi_n(A)$ is a complete set for the one-nucleon wave functions in the original nucleus. By picking out a

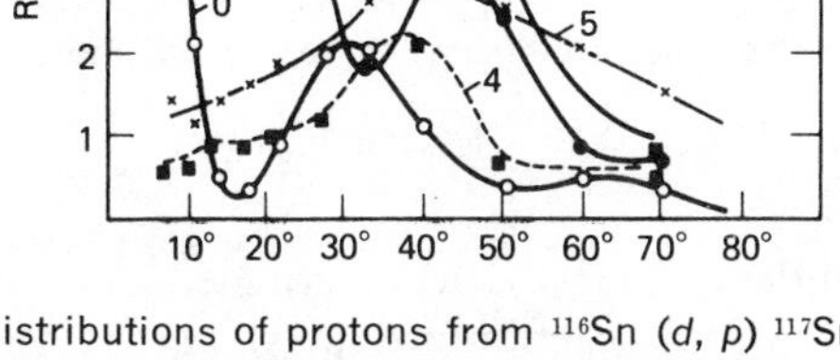

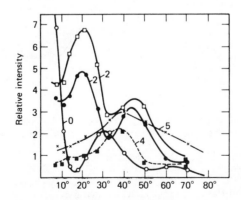

FIG. 14.3. Angular distributions of protons from ^{116}Sn (d, p) ^{117}Sn leading to states of ^{117}Sn with known spins and parities. Figures attached to the curves are l_n, the orbital angular momentum of the stripped neutron. Of the two curves with $l_n = 2$, the upper leads to the $d_{3/2}$ states and the lower to the $d_{5/2}$ state [taken from Cohen and Price (61)].

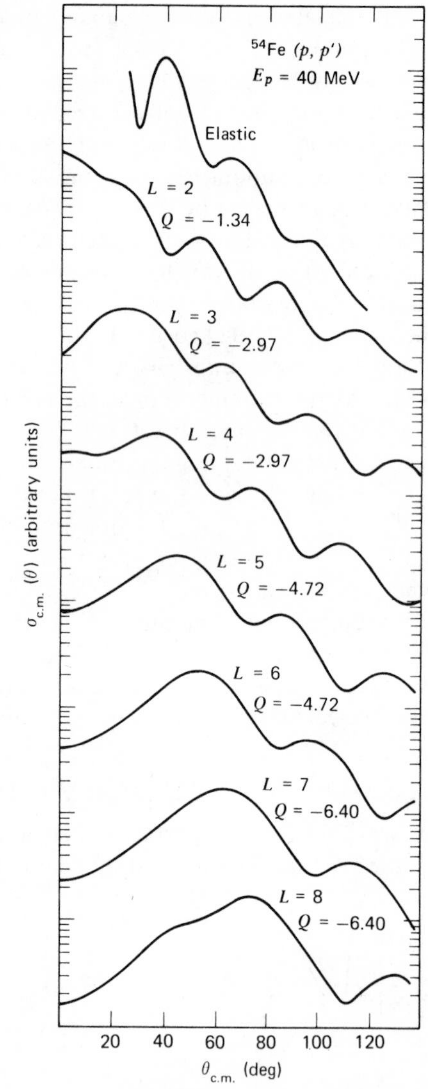

FIG. 14.4. Typical predictions of the optical model for elastic and inelastic angular distributions of 40 MeV protons scattered from ^{54}Fe for various multipole excitations [taken from Blair (66)].

nucleon from a well-defined orbit n (as determined from the angular distribution and polarization of the reaction products), and by observing the β-states in which the final nucleus is left, one can determine the coefficients $|C^{\alpha}_{\beta n}|^2$. These coefficients, generally called *spectroscopic factors*, are easily determined in the various models that serve to describe nuclear structure; their experimental determination is thus of great importance in the study of the range of validity of these models.

The transition from direct reactions, with their characteristic smooth energy dependence, to the compound nucleus reactions, with their characteristic narrow resonances, is not an abrupt one. There is a whole array of intermediate situations where the compound nucleus is starting to be formed, so that the process is still fast enough not to show narrow resonances, but already complex enough to show some preference for one range of energies over another. These intermediate structures show themselves most strikingly when one averages compound nucleus resonances over an appropriate energy range [see Section 13, Feshbach (64) and Griffin (66)].

15. HIGH-ENERGY SCATTERING AND REACTIONS

We know from optics that the limit on the resolving power of any instrument is set by the wave length of the radiation it employs. The same holds true for the study of nuclei by means of various projectiles: if we want to perform experiments that are sensitive to the detailed microscopic structure of the nucleus, we should be using projectiles of short wave lengths, shorter, in fact, than the average separation between nucleons.

The wave length of a particle is given nonrelativistically by

$$\lambda\!\!\!^{-} = \frac{\hbar}{p} = \frac{\hbar}{\sqrt{2ME}}$$

so that $\lambda\!\!\!^{-}$ can be made small either by increasing M or by increasing E. The use of heavy projectiles, such as heavy nuclei, is, however, not practical for the purpose of studying the internal properties of nuclei because of their own structure and because of their strong absorption by the target nucleus. We are therefore left with the possibility of using projectiles—electrons, protons, pions, and other mesons—of high energy and, indeed, with protons of about 1 GeV (=1000 MeV) we can get down to wave lengths of about 0.1 fermi.

Figure 15.1 gives the angular distribution obtained when 1 GeV protons are elastically scattered by ^{4}He. The typical diffraction pattern can be easily explained on the basis of multiple scattering theory. Note the following features. At small angles or small momentum transfers the cross section drops precipitously. Then there is an interference region including a minimum and a maximum after which the cross section continues to fall off with increasing angle but now at a considerably slower rate, roughly one-half of the original. At larger angles there is an indication of another intereference region.

In terms of multiple scattering analysis this can be interpreted as follows: it is known empirically [Bugg et al. (66) Dutton (67)] that the proton-nucleon elastic scattering is given by

$$\frac{d\sigma}{d\Omega} = |f_p(q)|^2 \tag{15.1}$$

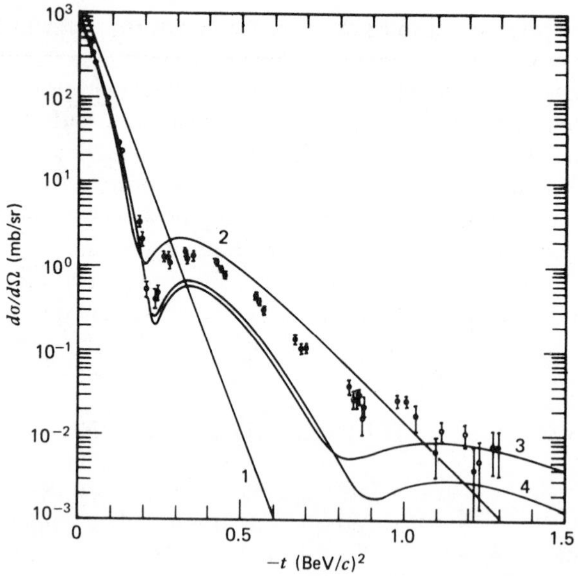

FIG. 15.1. 1 GeV elastic scattering of protons by ^{4}He. The straight line curve includes single scattering. The upper curve with one minimum includes single and double scattering. The remaining curve with two minima includes triple scattering as well [taken from Bassel and Wilkin (68)]. The data is that of Palevsky et al. (67).

where $f_p(q)$ the scattering amplitude for a momentum transfer $\mathbf{q}$ is given by

$$f_p(q) = \frac{(i + \rho)\, p}{4\pi\hbar}\, \sigma e^{-(q/q_0)^2} \tag{15.2}$$

where p is the proton momentum and

$$\rho = -0.3 \qquad q_0 = 0.57\, \frac{\text{GeV}}{\text{c}}$$

and σ has the following values for p-p and p-n scattering:

$$\sigma_{pp} = 47.53 \text{ mb} \qquad \sigma_{pn} = 40.4 \text{ mb}$$

Note the very rough nature of (15.2). It does not include spin dependence and only approximately agrees with not very complete or accurate data.

When a high energy proton impinges on a ^{4}He nucleus and is observed to have transferred momentum $\mathbf{q}$ to ^{4}He, the momentum transfer could have occurred in several ways:

(i) It could have transferred momentum $\mathbf{q}$ in a *single* collision with a ^{4}He nucleon with probability amplitude given by $f_p(q)$ multiplied by the probability $\rho(q)$ that ^{4}He can recoil with that momentum transfer without excitation.

The quantity $\rho(q)$ turns out to be the Fourier transform of the nucleon density $\rho(\mathbf{r})$. The charge density and thereby the nucleon density (certainly for ^{4}He) has been measured in electron-^{4}He elastic scattering and is given by

$$\rho(q) = \left(1 - a^2 \frac{q^2}{\hbar^2}\right)^6 e^{-(q/Q)^2}$$

where

$$a = .316 \text{ fm} \qquad \text{and} \qquad Q = .282 \frac{\text{GeV}}{c}$$

Hence the probability amplitude for single scattering is proportional to

$$\frac{(i + \rho)\, p\sigma}{4\pi\hbar}\left(1 - \frac{a^2q^2}{\hbar^2}\right)^6 e^{-q^2(1/q_0^2 + 1/Q^2)} \tag{15.3}$$

Since $Q \ll q_0$ it is clear that the q dependence and therefore the angular distribution is dominated by the q dependence of the nucleon density.

(ii) It could have transferred momentum (1/2) $\mathbf{q}$ in one collision and (1/2) $\mathbf{q}$ in a second collision. The probability amplitude for this double scattering process is roughly proportional to

$$\frac{[\rho(q/2)\, f_p(q/2)]^2}{R} = \frac{1}{R}\left[\frac{(i + \rho)\, p\sigma}{4\pi\hbar}\left(1 - \frac{a^2q^2}{\hbar^2}\right)^6\right]^2 e^{-(q^2/2)(1/q_0^2 + (1/Q^2)} \tag{15.4}$$

where R is the dimension of the target nucleus. This amplitude consists of an amplitude for the first scattering, a probability amplitude proportional to $(1/R)$ to reach a second nucleon after scattering by the first one and the probability amplitude for the second scattering.

(iii) It could transfer a momentum $\mathbf{q}$ in multiple collisions, each time transferring only a small fraction of $\mathbf{q}$.

All these scattering amplitudes add up, of course, coherently. Although it is true that the probability for multiple scattering decreases with their multiplicity since $\sigma/2\pi R^2 \ll 1$, it is equally true that the multiple scattering decreases less rapidly with momentum transfer q than the single scattering. This can be easily seen by comparing the exponential in (15.4) with that in (15.3). There is therefore a value of q for which double scattering becomes comparable to single scattering. At that point as one can see by comparing (15.3) and (15.4), one expects a destructive interference between the two to take place. At higher values of q the double scattering will predominate with its characteristic slower decrease with q, until it reaches a point where triple scattering becomes comparable to double scattering. Interference between the two amplitudes will take place again, after which the triple scattering amplitude will dominate with its even slower decrease with q, and so on. The data shown in Fig. 15.1 has been analyzed using a model of multiple scattering by Glauber (59, 70), and the full lines represent the theoretical predictions using single, single + double, and single + double + triple scattering amplitudes. No free parameters were used to fit the data, and the p-He^4 cross section (at least at angles that are not too large) is accounted to within the accuracy shown in Fig. 15.1 by the nucleon-nucleon cross section and a ^4He nucleon density that is taken from electron scattering. To the extent that these results are dominated by that density, the agreement is not suprising. But we learn that it is possible from measurements of this type to determine properties of the nuclear-nucleon density. This is more significant for the heavier nuclei where these components differ from those obtained from electron scattering. It is important to learn that the forward angle region is dominated by the density, and that the interference region involves interference between single and double scattering. This suggests that the angular distribution in this latter region through its dependence upon the double scattering will be sensitive to the pair correlations in the target nucleus.

We shall discuss pair correlations at greater length and with greater precision later on and in Volume II. For the present it will suffice to give the rough definition that they measure the probability of finding a nucleon at a displacement **r** away from a given nucleon.

The effect of correlations upon the angular distribution is given in Fig. 15.2. The various curves correspond to differing estimates of the pair-correlation terms. An accurate experiment could decide among these possibilities but only if the elementary nucleon-nucleon amplitude is known with sufficient accuracy.

The second interference region still imperfectly observed will presumably offer opportunities for the measurement of triple correlations.

Inelastic scattering will similarly measure transition densities and correlations. Omitting any dependence on spin and isospin, we find that the transition density $\rho_{fi}(1)$ is

$$\rho_{fi}(1) = \int d(2)\, d(3) \ldots d(A)\, \psi^*_f\,(1, 2, \ldots, A)\, \psi_i\,(1, 2, \ldots, A)\, \delta(\mathbf{R}) \tag{15.5}$$

where all the coordinates refer to the center of mass through the delta function by placing

$$\mathbf{R} = \frac{1}{A} \sum \mathbf{r}_i$$

The pair transition correlations are obtained simply by integrating the integrand in (15.5) over one less coordinate, that from (3) to (A), and clearly triple correlations are obtained by dropping still one more coordinate.

Looking at these various expressions it should be clear that one way of

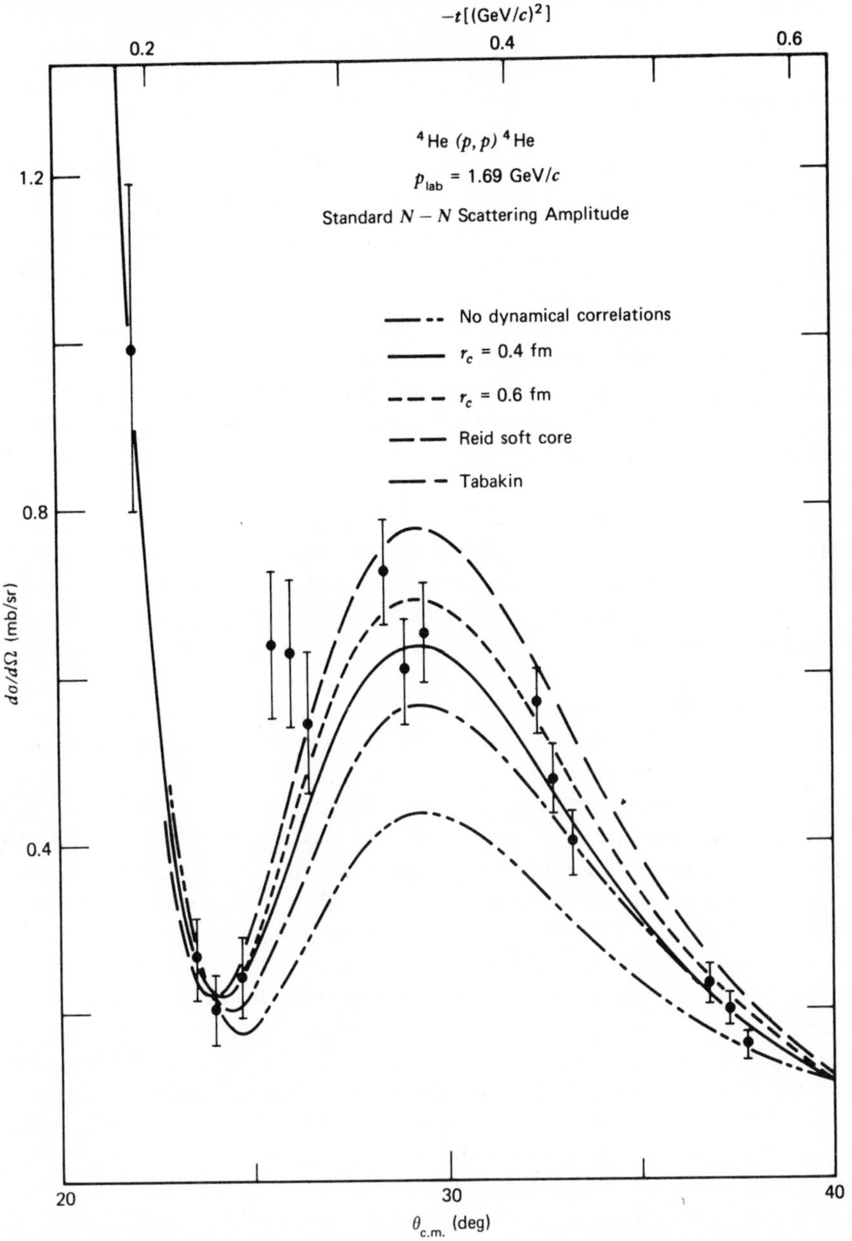

FIG. 15.2. The cross section for the scattering of 1 GeV protons in the interference region. The various curves are for differing forms of the pair correlation function. Two, those labeled by r_c are postulated forms in which roughly speaking the correlation is significant for particle separations of 0.4 fm and 0.6 fm. The other two, the pair correlation is computed from nucleon–nucleon interactions proposed by Reid and Tabakin [from Lambert and Feshbach (72)]. These results were obtained using a multiple scattering series for the optical model potential rather than for the direct calculation of the scattering amplitude.

describing the goal of this research is that it aims to measure the nuclear density matrix:

$$\rho_{fi} \sim \psi^*_f(1, \ldots, A)\, \psi_i(1, \ldots, A)$$

where f can also equal i as is appropriate for elastic scattering. Of course one cannot expect to measure this quantity for all f's or even for one f for all A-coordinates. But even partial measurements such as those of the density, second and perhaps triple correlations will, when combined with other nuclear structure information, be sufficient to choose among proposed nuclear models and nucleon-nucleon interactions as is indicated in Fig. 15.2. Of course one must not forget that information from experiments with other high energy projectiles, such as the photon and the electron, will play as important a role as the nucleon projectile since all are needed in order to disentangle the spin and isospin dependence of ρ_{fi}.

High energy projectiles are not only useful because of their short wave length but also because of their high momentum that allows them by direct collision to "knock out" nuclear nucleons and clusters of particles such as deuterons, ^3H, ^3He, as well as alpha particles. The presence of such clusters is an expression of the existence of correlations in the initial and final A-particle nuclear state.

Let us briefly discuss only one such experiment, the $(p, 2p)$ or (e, ep) in which the incident proton or electron "knocks out" a proton. The energy and momentum of one or of both final particles may be measured as well as the cross section for these final states. If one adopts the simple model for the nucleus in which each particle is moving in a single-particle potential, such experiments would yield the momentum distribution of the nuclear nucleons. It would also yield the energy required to knock the particle out of the nucleus.* Of course these considerations are too simplistic. Some of the energy goes into excitation of the residual nucleus. And of course the independent-particle description provides only a first approximation to the nuclear wave function. Nevertheless, as often happens in nuclear physics, the rough results obtained in this way are very suggestive. Table 15.1 gives the results obtained using the results of an (e, ep) type of experiment, the electrons having an energy of 500 MeV [Moniz et. al. (71)]. The quantity k_F is the Fermi momentum (see Chapter II) that measures the maximum momentum of a nucleon in the nucleus. In Fig. 15.3 the energy required to eject a nucleon in the indicated particle orbit out of the nucleus has been determined in a $(p, 2p)$ experiment [James (70)], the protons having an energy of 385 MeV. This should be compared with the column $\bar{\epsilon}$ in Table 15.1, which represents some rough average of the energies in the illustration. They are comparable. For the light nuclei it

---

*These experiments will be described in greater detail in Volume II, including the special kinematic conditions that are generally employed.

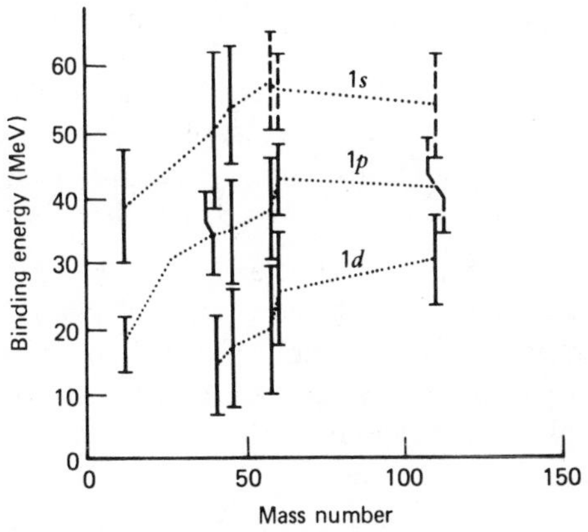

FIG. 15.3. The "binding energy" of single-particle levels as a function of nuclear mass obtained from (*p*, 2*p*) experiments. The incident proton energy is 385 MeV, the target nuclei are ^{12}C, ^{40}Ca, ^{45}Sc, ^{59}Co, ^{58}Ni, ^{120}Sn, ^{208}Pb, and ^{209}Bi [from James (69)].

would appear as if the nucleon ejected in the electron experiment is from the *lp* orbit. In both experiments note the approach of these energies to independence of A for heavy nuclei.

Further details on these and other high-energy reactions and on scattering together, with the possible uses of such experiments will be discussed in Volume II.

TABLE 15.1 Nuclear Fermi Momentum k_F and Average Binding Energy ϵ Determined in (e,ep) Experiment. The Electron Energy is 500 MeV [from Moniz et al. (71)]

Nucleus	k_F (MeV/c)[a]	$\bar{\epsilon}$ (MeV)[b]
$^{6}_{3}Li$	169	17
$^{12}_{6}C$	221	25
$^{24}_{12}Mg$	235	32
$^{40}_{20}Ca$	251	28
$^{58.7}_{28}Ni$	260	36
$^{89}_{39}Y$	254	39
$^{118.7}_{50}Sn$	260	42
$^{181}_{73}Ta$	265	42
$^{208}_{82}Pb$	265	44

[a] The fitting uncertainty in these numbers is approximately ± 5 MeV/c.
[b] The fitting uncertainty in these numbers is approximately ± 3 MeV.

16. SUMMARY

Our understanding of nuclear structure and nuclear reactions is not complete, yet a picture of the nucleus has emerged that combines phenomenological and qualitative considerations with rigorous applications of the laws of quantum mechanics. The principle ingredients that go to make this picture are the following:

(1) It is enough to consider just the degrees of freedom of the nucleons in describing nuclear data. The degrees of freedom of the pions and other mesons, which clearly exist in nuclei since they are known to exist in the nucleon, do not play an important role in the range of energies in which nuclei are studied.

(2) The theoretical description of nuclear phenomena is most probably given a good approximation via the solution of the nonrelativistic Schrödinger equation

$$\left[\sum_i \frac{p_i^2}{2M_i} + \sum_{i<j} v(ij) \right]\psi = E\psi \tag{16.1}$$

with the boundary condition appropriate to the special problem one is dealing with. In (16.1), M_i are the free-nucleon masses, and $v(ij)$ is most probably the free nucleon-nucleon interaction. The latter point is not completely clear at this stage, but there is evidence to show that at least the long range part of $v(ij)$—from 1 fm and up—is the same as that of the free nucleon-nucleon interaction. Much of the nuclear data is less sensitive to the short range part of $v(ij)$.

It is also to be noticed that in (16.1) no three-body force is present. As of now there is no evidence of its importance.

(3) It seems to be a good zeroth-order approximation to replace in (16.1) $\sum_{i<j} v(ij)$ by $\sum_i U(i)$, where the one body potential $U(i)$ is energy dependent and may look rather different for bound and unbound nucleons. The remaining part of the interaction, that is, the residual interaction, can then be treated as a perturbation.

(4) The one body potential $U(i)$, which converts (16.1) from an A-body problem to a one-body problem, may have dynamical collective coordinates in it, representing slow time variations in $U(i)$. The nucleus can then be treated best by dealing first with the collective and individual nucleon coordinates independently and then introducing their coupling as a perturbation.

(5) As far as we know the nucleus does not violate any of the fundamental principles of quantum mechanics, even though most of them were derived from the study of atoms.

The following chapters will be devoted to a more detailed examination of these various points and to an attempt to justify them at least to some extent. In so doing we shall have to develop the appropriate machinery and formalism, some of which, it turns out, is also useful for other fields of physics. Our main emphasis, however, will be on the particular approximations that render such formulations most useful in the realm of nuclear physics.

CHAPTER II
THE NUCLEUS AS A FERMI GAS

Nuclei, like atoms, can be completely broken apart. They can be ionized by tearing off one or more nucleons and they can also be taken apart completely. Unlike the atom, where after complete ionization the atomic nucleus remains, the complete "ionization" of the nucleus of all its constituent nucleons leaves nothing behind. That is, it is possible, by using the minimum required energy, to completely disintegrate a nucleus into a certain number, say Z-protons and N-neutrons. It is in this sense that the nucleus is said to be composed of Z-protons and N-neutrons. All measurements of the minimum energy, $B.E.$, required to break up such a nucleus are consistent with the expression

$$M(Z,N) = ZM_p + NM_n - \frac{B.E.}{c^2} \tag{0.1}$$

$M(Z,N)$ is the mass of the nucleus. M_p and M_n are the proton and neutron mass respectively, and c is the velocity of light. Relation (0.1) has not been checked directly except for a few light nuclei. Complete ionization of most nuclei is very difficult experimentally. However, it is not necessary, because through mass spectroscopy and by means of nuclear reactions it is possible to compare the masses and binding energies of nuclei that are not far apart and, thereby, (0.1) is verified indirectly.

1. THE NUCLEAR WAVE FUNCTION AND THE ROLE OF MODELS

The conclusion, drawn from (0.1) that the nucleus is composed of Z-protons and N-neutrons, finds additional support in the fact that the total charge of the nucleus is found to be Ze with a very high accuracy (e is the positive charge of the proton); furthermore it is found from molecular spectra that nuclei with Z-protons and N-neutrons satisfy a Bose-Einstein or a Fermi-Dirac statistics according to whether $A = Z + N$ is even or odd. Since both the proton and the neutron separately satisfy the Fermi-Dirac statistics, this result is exactly what one expects of a system of Z-protons and N-neutrons.

On the other hand the neutrons and protons in a nucleus are rather closely packed: the average distance between nucleons in the nucleus is comparable to the size of the nucleon. One may wonder if such close packing of nucleons can still be considered simply as an assembly of Z-protons and N-neutrons. More rigorously we can formulate this question as follows.

We propose to describe a nucleus by a wave function ψ; to the extent that a nucleon is described by five coordinates (three for its location in space $\mathbf{r}$, one for its spin orientation σ_z ,and one for its charge τ_3), is it sufficient to consider ψ as a function of the five A-coordinates

$$\mathbf{r}_i, \sigma_{iz}, \tau_{i3} \qquad i = 1, \ldots, A$$

or must we consider explicitly additional degrees of freedom as well? It is in this sense that we have to understand the question of whether the nucleus is composed just of Z-protons and N-neutrons, or whether there is "something else" to it. It extends the meaning of the phrase "composed of Z-protons and N-neutrons" far beyond the rigorous implications of (0.1).

The answer to this question is not unique. If we are concerned with processes involving energies of a few hundred MeV or more, then in the description of the single free nucleon we must include the field of π-mesons, and possibly that of other mesons as well. Only as long as we are considering energies below the threshold for π-meson production can we regard the coordinates $\mathbf{r}$, σ_z, τ_3 as sufficient to describe the nucleon degrees of freedom. Similarly it is evident that complex nuclei may also require an explicit description of their mesonic field once we are considering processes involving energies beyond roughly one hundred MeV. However, at lower energies it is still an open question whether the close packing of the nucleons in the nucleus requires the explicit consideration of the meson field. Most treatments of nuclear structure assume that mesonic effects at most change some of the intrinsic properties of the nucleon (its magnetic moment and more generally possibly also the distribution of its charge and magnetic density). Sometimes more complicated corrections are also considered, such as the current due to the virtual exchange of mesons among the nucleons—the so-called *exchange current* (see Chapter VIII). But for the most part, the theories of nuclear structure that we shall discuss in this volume, for energies below meson production threshold, describe the nucleus in terms of a wave function of the type:

$$\psi(\mathbf{r}_1, \sigma_{1z}, \tau_{13}, \mathbf{r}_2\sigma_{2z}\tau_{23}, \ldots, \mathbf{r}_A\sigma_{Az}, \tau_{A3})$$

The central problem in nuclear physics is that of characterizing the motion of these A-nucleons under the influence of their mutual forces. We need hardly mention that the general A-body problem for an arbitrary mutual interaction has not been solved exactly for $A \geq 3$ and that we must satisfy ourselves with approximate solutions only. However the nuclear A-body problem becomes even more complicated on account of our rather limited knowledge of the nuclear force and its apparent great complexity. Had we had a fundamental field theory describing the strong nuclear interactions, we could have perhaps extracted from it also the exact nucleon-nucleon interaction to be used in the nuclear A-body problem. In the absence of such a theoretically deduced interaction we must satisfy ourselves with a more-or-less phenomenological nucleon-nucleon force derived from nucleon-nucleon scattering experiments. The limitations on our knowledge of such a phenomenological force were described in Chapter I.

There is a further uncertainty that concerns the possible existence of many-body forces, especially among the strongly interacting particles. Nucleon-nucleon scattering experiments cannot teach us anything about three-body forces among nucleons. Such potentials describe a situation in which the interaction between two particles 1 and 2 depends not only on their mutual position, but also on the presence of a third nucleon in position $\mathbf{r}_3$. A three-body potential is a function of $\mathbf{r}_{12}$ and $\mathbf{r}_{13}$ that cannot be broken up into a sum of two-body potentials. It might have a form such as $V(r_{12} + r_{23} + r_{31})$ or $V(r_{12} \cdot r_{23} \cdot r_{31})$. They are thus completely absent in two-nucleon systems. In an $A > 2$ nucleon system, however, their contribution may become significant.

In view of these fundamental difficulties special methods had to be developed to minimize the sensitivity of the calculation of the nuclear property under study to uncertainties in the basic input. Generally these methods consist in the formulation of a "model," a simple solvable physical system that mimics some of the attributes of the real nucleus. The hope is that there exists a range of phenomena for which the omitted features are not of critical importance. Since no attempt, at least initially, is made to connect these models with nuclear forces, they will involve a number of parameters that will characterize the simplifications made regarding nuclear interactions, nuclear sizes and shapes, etc. Models thus involve built-in parameters whose values are determined from experiment. A model is valid and thus also very useful when these parameters vary slowly over the appropriate range of experimental conditions. In some cases such models can be justified more or less rigorously; in many cases they are justified only intuitively to begin with and are later checked through their agreement with experiment. In this way a number of models were proposed for the study of nuclear structure and nuclear reactions, some of them claiming a relatively broad range of validity, others limiting themselves to a selected set of phenomena within a selected set of nuclei.

A well-known nuclear model of this type, which we shall discuss in detail later, is the "collective model." In its simplest form it asserts that certain nuclei are cigar-shaped rather than spherical, and, as a result they exhibit rotational bands in their spectra (see Chapter I, page 46). The energy of a level in a band, relative to the lowest level of the band, can be described by a simple relation:

$$E(J) = \frac{1}{2\mathcal{J}} J(J + 1) \tag{1.1}$$

Here J is the angular momentum of the state considered and $\mathcal{J}$ is a parameter—referred to as the "nuclear moment of inertia"—whose value is determined by the experimental values of $E(J)$. Although the model in its simplest version does not give an absolute value for $E(J)$, it predicts for any given nucleus the *ratios* of various excitation energies. Thus, if a set of excited states of the nucleus is found to agree with (1.1), they can all be characterized by a single number, the moment of inertia $\mathcal{J}$. A more detailed theory, when developed, will have then only to relate the moment of inertia $\mathcal{J}$ to the nuclear forces, thereby automatically explaining the energies of a whole set of excited states.

As we see from this example, a model can be considered as a way of summarizing, through a few parameters, a large amount of experimental information. Evidently the model is convenient only if its predictions are relatively easy to obtain. A model should always reduce to soluble problems, and very often even to a relatively simple soluble problem. It is therefore bound to have a limited applicability and fail at least in some aspects. We should be prepared to find different, and even opposing, models being applicable in different regions of the periodic table, or even to encounter two different models being used for the elucidation of different properties of the same nucleus. It is comforting to know, however, that the better models do have a surprising broad range of validity.

The next step—that of a more complete theory—has then two tasks. First, it has to explain the success of the particular model in its range of validity. Second, it has to relate the experimentally observed values of the parameters of the model to the forces, or other basic features, of the more complete theory.

It is also obvious that in working out models we shall often try to reduce as much as possible the number of ad-hoc assumptions, sometimes even at the cost of being able to describe only gross features of some nuclei rather than their detailed properties. We shall now investigate one such model, the so-called Fermi gas model, which considers the nucleus as a collection of non-interacting fermions whose density is fixed somehow from the outside. The purpose of this model is to clarify which of the nuclear properties is due just to the Pauli exclusion principle at the given density, and how much of these properties remains, so to speak, to be explained by the detailed dynamics of the interacting nucleons. This model is *not* a theory of nuclear structure; for one thing it does not even pretend to explain why nuclei have exactly their observed density; but it does tell us that if we understand the factors that determine the density—some of the other properties of nuclei follow simply from the Pauli principle.

2. THE FERMI GAS MODEL

In our attempt to account for some nuclear properties with this simple and crude model we shall concentrate initially on the parameters empirically observed in the Weizsäcker formula (see Section I.2). This mass formula, it will be recalled, gives the masses of nuclei as a function of their charge and mass number, and is composed of different terms whose plausibility is argued for on some general principles. It is for this reason that we expect to be able to derive the parameters of the formula from a very simple model and obtain a qualitative order of magnitude agreement with experiment.

Consider a system of a number of neutrons and protons put in a cubic box of linear dimension a. The Schrödinger equation for a single particle in this box reduces to

$$-\frac{\hbar^2}{2M}\nabla^2\psi = E\psi \tag{2.1}$$

We impose the boundary conditions corresponding to an infinite potential barrier at the surface of the box:

$\psi(x,y,z) = 0$ on the faces of the box, that is, whenever at least *one* of the following conditions holds:

$$x = 0, \quad y = 0, \quad z = 0, \quad x = a, \quad y = a, \quad z = a \tag{2.2}$$

The action of the walls of the box replaces to some extent the average interaction between the nucleons. By adjusting the size of the box we can obtain the observed nucleon density in nuclei. The solution of (2.1) with the boundary conditions (2.2) is given by

$$\psi(x,y,z) = A\cdot\sin k_x x\cdot\sin k_y y\cdot\sin k_z z \tag{2.3}$$

provided

$$k_x a = n_x\pi \qquad k_y a = n_y\pi \qquad \text{and} \qquad k_z a = n_z\pi \tag{2.4}$$

where n_x, n_y, and n_z are all positive integers, and A is a normalization factor. [Negative integers lead to solutions equivalent to (2.3).] Each set of positive integers (n_x, n_y, n_z) defines a different solution corresponding to an energy

$$E(n_x,n_y,n_z) = \frac{\hbar^2}{2M}(k_x^2 + k_y^2 + k_z^2) = \frac{\hbar^2}{2M}\mathbf{k}^2 \tag{2.5}$$

Equation 2.5 together with the restriction (2.4) on the values of k_x, k_y, and k_z represent the quantization of a single particle in a box. It is obvious from (2.5), as well as from (2.3), that

$$\mathbf{k} \equiv (k_x, k_y, k_z)$$

is the momentum (divided by $\hbar$) of the particle in the box.

Remark. Actually the eigenstates in a cubical box are characterized by the line along which the particle moves and the magnitude of its momentum, that is, by the ratio $k_x : k_y : k_z$ and by $\mathbf{k}^2$. We shall refer to $\mathbf{k} = (k_x,k_y,k_z)$ loosely as the momentum of the particle in the box, understanding that actually we are dealing with a standing wave composed of **k** and −**k**.

Because of the Pauli principle a given momentum state can be occupied at most by four nucleons, two neutrons with opposing spins, and two protons with opposing spins. The lowest energy of a nucleus will then be obtained by filling the lowest possible energy states in the box. The values of this lowest energy will be dependent upon the number of available states, and it is the determination of this quantity that we shall now consider.

Consider then the space of the vectors **k**; because of (2.4), for every cube of sides π/a [and volume $(\pi/a)^3$] there is, in this space, just one point that represents a permissible solution of the type (2.3). (Fig. 2.1) The number of permissible solutions $n(k)$ with the magnitude of **k** between k and $k + dk$ is then given by

$$dn(k) = \frac{1}{8}\cdot 4\pi k^2 dk \cdot \frac{1}{(\pi/a)^3} \tag{2.6}$$

In this expression $4\pi k^2 dk$ is the volume of the spherical shell in k-space with radii k and $k + dk$; only one-eighth of this shell is considered since only positive values of k_x, k_y, and k_z are required to span all different solutions, and $(\pi/a)^3$ is the volume, in k-space per permissible solution.

The total number of permissible states up to energy $\epsilon = (\hbar k)^2/2M$ is then given by

$$n(k) = \frac{4\pi}{3}\frac{k^3}{8(\pi/a)^3} \tag{2.7}$$

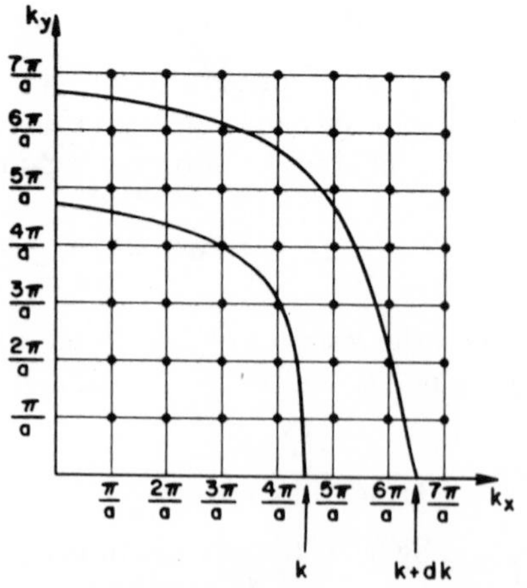

FIG. 2.1. Each point at the intersection of the thin lines represents an allowed pair of values (k_x, k_y). To find the number of pairs allowed between k and $k + dk$, find the number of points in the region enclosed by the heavy lines; this number is its area divided by the area per point. Equation 2.6 is obtained from a similar three-dimensional diagram.

Since each momentum state **k** can accommodate, according to the Pauli principle, two neutrons and two protons, we shall obtain the lowest energy state of A noninteracting nucleons put in a box of linear dimension a if we take half of them to be neutrons and half to be protons. The highest occupied momentum state k_F will then be given according to (2.7) by:

$$\frac{A}{4} = \frac{4\pi}{3}\frac{k_F^3}{8(\pi/a)^3} \qquad \text{or} \qquad A = \frac{2\Omega}{3\pi^2}k_F^3 \tag{2.8}$$

where $\Omega = a^3$ is the volume of the box. It follows that under these conditions the momentum of the highest occupied state depends only on the *density* $\rho = A/\Omega$ of nucleons in the box, and is given by:

$$\rho = \frac{2}{3\pi^2}k_F^3 \tag{2.9}$$

We have disregarded thus far the interaction among the nucleons (except when it determines their density). Within this limit the momentum distribution per unit volume in momentum space is a step function with a constant value for $k < k_F$ and zero for $k > k_F$ (Fig. 2.2). This momentum distribution is referred to as the *Fermi distribution*. In high energy scattering experiments it is possible to measure the momentum distribution of nucleons in a nucleus (see Section I.15) and although one does not observe a sharp cutoff at $k = k_F$, the probability of finding in the nucleus nucleons with momenta $k > k_F$ decreases very rapidly with k.

From the observed density of nuclei, $\rho = 1.72 \times 10^{38}$ particles/cm^3, which, as is known, is practically the same for all nuclei with $A \gtrsim 12$, we obtain

$$k_F = 1.36 \text{ fm}^{-1} \tag{2.10}$$

with a corresponding energy of

$$\epsilon_F = 38 \text{ MeV} \tag{2.11}$$

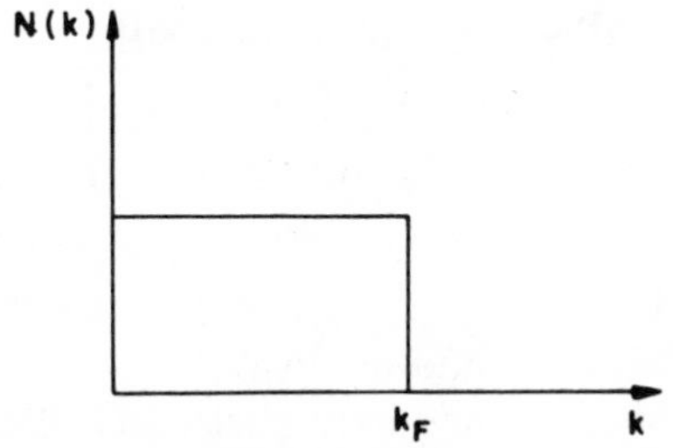

FIG. 2.2. The momentum distribution of nucleons in the ground state of the Fermi gas model. k_F is the Fermi momentum. $N(k)\,d\mathbf{k}$ equals number of states with momenta between **k** and **k** + d**k**.

k_F and ϵ_F are called, respectively, the *Fermi momentum* and *Fermi energy* of the degenerate gas model. The *average* kinetic energy of the fermions in the box is less than ϵ_F, and is given by:

$$\langle T \rangle = \frac{1}{A/4}\int_0^{\epsilon_F} \epsilon \frac{dn}{d\epsilon}\, d\epsilon = \frac{3}{5}\,\epsilon_F \tag{2.12}$$

$$= 23 \text{ MeV for nuclei of the observed density}$$

It is also convenient to introduce a *nuclear radius* R through $(4/3)\,\pi R^3\,\rho = A$. Since ρ is experimentally found to be independent of A it follows that

$$R = r_0 A^{1/3} \tag{2.13}$$

With the value of ρ given above r_0 becomes

$$r_0 = 1.12 \text{ fm} \tag{2.14}$$
$$k_F r_0 = 1.52$$

It should be stressed that there are different ways to define a nuclear radius, since the nucleus does not have sharp boundaries. Equation 2.13 corresponds to the radius of the sphere that would have included all the mass of a nucleus if its density were uniform. In comparing radii it is very important to make clear which radius is being considered.

3. THE WEIZSÄCKER MASS FORMULA

Let us now use the Fermi gas model to try and understand the order of magnitude of some of the empirical nuclear parameters. To this end we recall the Weizsäcker semiempirical mass formula:

$$M(A) = ZM_p + NM_n - a_1 A + a_2 A^{2/3} + a_3 \frac{Z^2}{A^{1/3}} + a_4 \frac{(Z-N)^2}{A} + \delta(A) \tag{3.1}$$

This empirical formula gives the mass of a nucleus with Z-protons and $N = A - Z$ neutrons in terms of some empirically determined parameters a_i. Their best values are [Myers and Swiatecki (66)].

$$\begin{aligned}
a_1 &= 15.68 \text{ MeV (coefficient of volume energy)}\\
a_2 &= 18.56 \text{ MeV (coefficient of surface energy)}\\
a_3 &= 0.717 \text{ MeV (coefficient of coulomb energy)}\\
a_4 &= 28.1 \text{ MeV (coefficient of the symmetry energy)}
\end{aligned} \tag{3.2}$$

$$\delta(A) = \left\{\begin{array}{ll} 34\,A^{-3/4} & \text{MeV for odd-odd nuclei}\\ 0 & \text{MeV for odd-even nuclei}\\ -34\,A^{-3/4} & \text{MeV for even-even nuclei}\end{array}\right\}\text{(coefficient of pairing energy)}$$

Since we have found that $\langle T \rangle = 23$ MeV at the observed nuclear density, the value of $a_1 = 15.68$ MeV for the average (volume) binding energy per particle indicates that the "nuclear box" in the Fermi gas model represents an average potential energy of $\langle U \rangle = -15.68 - \langle T \rangle \approx -39$ MeV. Scattering of neutrons on complex nuclei provides another estimate of this quantity. If the interaction energy is U, then we conclude from (2.12) that the average (volume) binding energy per particle should be 23 MeV + U. Empirical values of U are scattered around $U \approx -40$ MeV leading to a value of the (volume) binding energy per particle of -17 MeV that is not too far from the value quoted above.

Next we shall try to calculate the symmetry energy term within the framework of the Fermi gas model. As we can see from (3.1), this term measures the increase in the mass M that occurs when the number of neutrons no longer equals the number of protons. We therefore consider the energy required to take protons from the highest filled proton levels and transform them into neutrons in the lowest unfilled neutron level. Consider a nucleus with A-nucleons, half neutrons, and half protons. Let λ represent the fraction of protons changed into neutrons, so that after the change is made we have

$$Z' = \frac{A}{2}(1-\lambda) \qquad N' = \frac{A}{2}(1+\lambda) \qquad \text{or}$$

$$\Delta N = -\Delta Z = \frac{\lambda A}{2} \tag{3.3}$$

The energy consumed in this change is

$$\Delta \text{E} = 2\int_{(A/4)}^{A/4(1+\lambda)} \epsilon\, dn - 2\int_{A/4(1-\lambda)}^{A/4} \epsilon\, dn \tag{3.4}$$

Introducing

$$F(n) \equiv \int_0^n \epsilon\, dn$$

$$\begin{aligned} \Delta E &= 2\left\{F\left[\frac{A}{4}(1+\lambda)\right] - 2F\left(\frac{A}{4}\right) + F\left[\frac{A}{4}(1-\lambda)\right]\right\} \\ &= 2\left(\frac{A\lambda}{4}\right)^2\left(\frac{d^2F}{dn^2}\right)_{n=A/4} = \frac{1}{8}(A\lambda)^2\left(\frac{d\epsilon}{dn}\right)_{n=A/4} \end{aligned} \tag{3.5}$$

But

$$\epsilon = \epsilon_0 n^{2/3}$$

where ϵ_0 is a constant. When $n = A/4$, ϵ equals ϵ_F. Hence

$$\frac{1}{\epsilon}\frac{d\epsilon}{dn} = \frac{2}{3n}$$

so that

$$\left(\frac{d\epsilon}{dn}\right)_{n=A/4} = 8\,\epsilon_F/3A \tag{3.6}$$

Hence

$$\Delta E = \frac{1}{3} \frac{\epsilon_F (N' - Z')^2}{A} \simeq 13 \frac{(N' - Z')^2}{A} \text{ MeV} \tag{3.7}$$

Comparison of the empirical value for the coefficient of $(Z - N)^2/A$ in the Weizsäcker formula with (3.7) shows that the Pauli principle accounts for about one-half of the observed symmetry energy.

Problem. Assuming that a nucleus consists of two Fermi gases, one of neutrons and the other of protons, show that the energy of the nucleus is

$$E = \frac{3}{10} \epsilon_F \left[\left(\frac{2N}{A} \right)^{2/3} N + \left(\frac{2Z}{A} \right)^{2/3} Z \right] \tag{3.8}$$

where $\epsilon_F \simeq 38$ MeV is obtained from (2.8). Derive (3.7) starting from (3.8).

Another nuclear property we want to calculate concerns the nuclear surface energy. If we refer to (2.6) and Fig. 2.1. we see that for any given value of k_z, say, we should have not counted those states in which k_x or $k_y = 0$, since in this case $\psi \equiv 0$. The number of such states in a shell dk at k is from (2.6)

$$3 \cdot \frac{1}{4} \frac{(2\pi k)\, dk}{(\pi/a)^2} \tag{3.9}$$

The correct number of permissible states between k and $k + dk$ is then

$$\begin{aligned} dn(k) &= \frac{k^2 a^3}{2\pi^2} \left(1 - \frac{3\pi}{ak} \right) dk \\ &= \Omega \frac{k^2}{2\pi^2} \left(1 - \frac{\pi}{2} \frac{S}{\Omega} \frac{1}{k} \right) dk \end{aligned} \tag{3.10}$$

where we have chosen to introduce the surface $S = 6a^2$ of the cube in order to write the correction to $dn\,(k)$ as a characteristic surface term proportional to S/Ω. It can be shown that $dn\,(k)$ is indeed a function of S/Ω only, so that (3.10) is rigorously valid for any shape of the nucleus. For a recent discussion see Balian and Bloch (70, 71). Using this expression we shall be able to deduce a surface correction to the nuclear energy as well as the surface thickness of large nuclei. In fact, let us expand the binding energy per particle in powers of S/Ω. We can write then:

$$\frac{E}{A} = \frac{4 \int_0^{k_F} \epsilon(k) \dfrac{dn\,(k)}{dk}\, dk}{4 \int_0^{k_F} dn\,(k)} = a_0 + b_0 \frac{S}{\Omega} + \dots \tag{3.11}$$

where $a_0 = (3/5)\, \epsilon_F$ and $b_0 = (\pi/8)\,(1/k_F)\, a_0$. The coefficient b_0 is simply related to the surface energy coefficient a_2 in the Weizsäcker mass formula (3.1) .From (3.11) we get

$$E = a_0 A + b_0 (S/\Omega) A + \dots$$

We have for the nuclear radius

$$R = r_0 A^{1/3} \qquad \text{and} \qquad \rho = \frac{A}{\Omega} = \frac{3}{4\pi r_0^3}$$

Hence

$$\frac{S}{\Omega} A = \frac{3}{r_0} A^{2/3}$$

Comparison with (3.1) then yields immediately

$$a_2 = \frac{3b_0}{r_0} = \frac{9\pi}{40 k_F r_0} \epsilon_F \tag{3.12}$$

Substituting (2.11) into (3.12) and using the observed constant density of nuclei, one obtains a surface energy of 18 $A^{2/3}$ MeV. This time the Fermi gas model gives nearly the correct value for an empirical parameter since a_2 is 18.56 MeV.

In a similar way we can apply the surface correction to ρ:

$$A = 4\int_0^{k_F} dn\,(k) = \Omega\left(a_0' + b_0' \frac{S}{\Omega}\right)$$

leading to an average density of the form

$$\rho = \frac{A}{\Omega} = a_0' + b_0' \frac{S}{\Omega} \tag{3.13}$$

The existence of a term proportional to S/Ω in (3.13) indicates that ρ cannot be uniform throughout the nucleus. A step function for $\rho(r)$ would have led to an average density independent of S/Ω.

To interpret the surface term $b_0'(S/\Omega)$ in (3.13) in terms of more familiar concepts, it is reasonable to assume for $\rho(r)$ a trapezoidal function as shown in Fig. 3.1. Nothing in the derivation of (3.13) really tells us that the term in S/Ω results from nucleons actually located at the surface, but since it is known from electron scattering that $\rho(r)$ has approximately a trapezoidal shape, our assumption is the natural one to make. Using this density distribution, that is,

$$\rho(r) = \begin{cases} \rho_0 & r \leq R \\ \rho_0 \dfrac{R - r + t}{t} & R \leq r \leq R + t \\ 0 & r \geq R + t \end{cases}$$

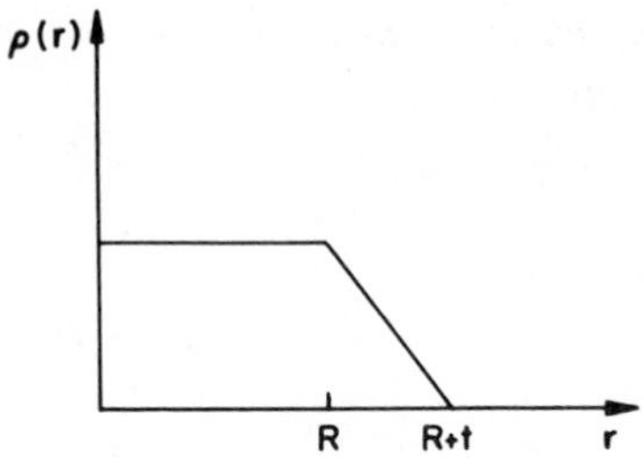

FIG. 3.1. The trapezoidal charge distribution $\rho(r)$ assumed in deriving the surface thickness t.

we have to first order in $t(S/\Omega)$:

$$A = 4\pi \int_0^\infty \rho(r)\, r^2\, dr = \frac{4}{3}\pi R^3 \rho_0 \left[1 + \frac{1}{2}\frac{S}{\Omega}t\right] \tag{3.14}$$

On the other hand the average density ρ_0 is given by [see Eq. (3.9)]

$$\rho_0 = \frac{A}{\Omega} = \frac{4}{\Omega}\int_0^{k_F} dn\,(k) = \frac{4}{3}\frac{k_F{}^3}{2\pi^2}\left(1 - \frac{3\pi}{4k_F}\frac{S}{\Omega}\right)$$

The left-hand side of (3.14) is independent of S/Ω; hence,

$$\rho_0\left(1 + \frac{1}{2}\frac{S}{\Omega}t\right) = \frac{4}{3}\frac{k_F{}^3}{2\pi^2}\left(1 - \frac{3\pi}{4k_F}\frac{S}{\Omega}\right)\left(1 + \frac{1}{2}\frac{S}{\Omega}t\right)$$

should be independent of S/Ω to first order. t is thus fixed at a value of

$$t = \frac{3}{2}\frac{\pi}{k_F} \approx 3.5 \text{ fm} \tag{3.15}$$

The quoted empirical value of the surface thickness is the distance between the radii at which $\rho = 0.9$ and 0.1 of the central value. This thickness is predicted by the Fermi gas model (3.15) to be $0.8 \times 3.5 = 2.8$ fm. Actual measurements of nuclear shapes yield a surface thickness of more like 2.5 so that again the Fermi degenerate gas model gives only an approximation to the effect. It is, however, interesting to note that the density of a finite system must fall off gradually instead of abruptly. In actual nuclei the distance over which the density almost falls to zero is close to the distance corresponding to free, noninteracting fermions with the same average density.

It should be pointed that general principles of quantum mechanics do not permit the density to fall sharply to zero. In the Fermi gas model all the wave functions are zero at R and rise to their first maximum a quarter of a wave length away. If one assumes that the closest maximum is dominated by the high k's, that is, for $k \simeq k_F$, that maximum occurs at approximately $r/2k_F \simeq$ 1.15 fm away from $r = R$.

Our description of nuclei in terms of a degenerate Fermi gas is, of course, very crude. For simplicity we carried out the quantization in a cube of dimensions a and, through the introduction of its volume $\Omega = a^3$ and its surface $S = 6a^2$, switched over to a spherical box with the same volume and surface. This is not always a rigorous procedure particularly when the potential at the surface is finite. Strictly speaking we should have carried out the quantization in a spherical, rather than cubical, geometry. However, for our present purpose of first orientation what we have done thus far is sufficient. Clearly, the nucleus is more complex than just a collection of noninteracting fermions forced to stay within a certain volume. The comparison of our estimate with experiment did show us, nevertheless, that the Pauli principle accounts qualitatively for some of the trends observed in nuclei. This success suggests that we may use, as an initial approximation, this picture of nucleons moving independently, that is, without any mutual interaction, in a certain potential well. Various effects of the interaction among the nucleons will then be included as corrections of various orders. To the extent that this approach turns out to be successful, as seems now to be the case, we would like to understand why a system of nucleons, which are in fact strongly interacting fermions, can be approximated well enough by a system of noninteracting fermions confined to a given volume. We shall see that this seemingly paradoxical situation can be anticipated for a system of nucleons in its lowest energy states, and that both the Pauli principle, which the nucleons have to obey, as well as some specific features of their interaction play an important role.

4. THE NUCLEAR WAVE FUNCTION IN THE FERMI GAS MODEL

Before we proceed with this program let us investigate some further consequences of the close packing of nucleons and their relation to observed nuclear characteristics. Formally, if we had a precise prescription for the derivation of nuclear wave functions, our task of understanding nuclear structure will have been achieved. When we claim that the Fermi degenerate gas model reproduces some empirical data fairly well, we are implying that the wave function appropriate for the Fermi gas can approximately replace the real nuclear wave function. But does that mean also that the two wave functions are very similar to each other? Or can very different wave functions reproduce empirical data equally well?

A wave function $\psi(\mathbf{r}_1, \mathbf{r}_2, \ldots, \mathbf{r}_A)$ of A-particles (for convenience we have omitted spin and isospin coordinates) gives us complete information on the A-nucleon system. For instance, it can tell us what is the probability of finding particle number 5 in a given location, if the locations of particles numbers $1, \ldots, 4$ are fixed and if we average over the positions of particles $6, 7, \ldots, A$. But when $\psi(\mathbf{r}_1, \mathbf{r}_2, \ldots, \mathbf{r}_A)$ is used to calculate nuclear properties we generally use only a very small part of the information contained in it.

For example, a typical calculation may involve the rate at which the nucleus radiates as it goes from an initial state to a final state. The operator, whose matrix elements determine this rate, is symmetric in all the nucleons and involves a sum of operators each concerning one nucleon at a time. Physically, this means that the nucleus acquires the ability to radiate through the ability of each one of its nucleons to do so. We can write then the operator $\hat{\Omega}$ that gives rise to the radiation in the form

$$\hat{\Omega} = \sum_{k=1}^{A} \omega(k) \tag{4.1}$$

and the transition amplitude becomes

$$T_{fi} = \langle f|\hat{\Omega}|i\rangle = \sum_{k=1}^{A} \int \psi_f^*(\mathbf{r}_1, \mathbf{r}_2, \ldots, \mathbf{r}_A)\, \omega(k)\, \psi_i(\mathbf{r}_1, \mathbf{r}_2, \ldots, \mathbf{r}_A)\, \mathbf{dr}_1 \ldots \mathbf{dr}_A \tag{4.2}$$

where ψ_i is the initial state and ψ_f the final state wave function. If we use the isospin formalism and treat all nucleons equally, then ψ is an antisymmetric function. Since $\hat{\Omega} = \sum_{k=1}^{A} \omega(k)$ is a symmetric operator, we can evaluate (4.2) for one of the particles, say particle 1, and multiply the result by A.

Problem. Prove it using the permutation operator on ψ_i and ψ_f.

Thus we obtain

$$T_{fi} = A\int \psi_f^*(\mathbf{r}_1, \mathbf{r}_2, \ldots, \mathbf{r}_A)\, \omega(1)\, \psi_i(\mathbf{r}_1. \mathbf{r}_2, \ldots, \mathbf{r}_A)\, d\mathbf{r}_1 \ldots d\mathbf{r}_A \tag{4.3}$$

In (4.3) the operator acts on the coordinates of particle 1 only; the integration over $d\mathbf{r}_2 \ldots d\mathbf{r}_A$ can therefore be carried out *irrespective* of the form and content of the operator ω. We therefore introduce the *one-particle transition density matrix* defined by:

$$\rho_{fi}^{(1)}(\mathbf{r},\mathbf{r}') = \int \psi_f^*(\mathbf{r},\mathbf{r}_2,\mathbf{r}_3, \ldots, \mathbf{r}_A)\, \psi_i(\mathbf{r}',\mathbf{r}_2,\mathbf{r}_3, \ldots, \mathbf{r}_A)\, d\mathbf{r}_2 \ldots d\mathbf{r}_A \tag{4.4}$$

Equation 4.3 can now be written as:

$$T_{fi} = A\int d\mathbf{r}\, \delta(\mathbf{r} - \mathbf{r}')\, d\mathbf{r}'\, \omega(\mathbf{r}')\, \rho_{fi}(\mathbf{r},\mathbf{r}') = A\ tr(\hat{\omega}\rho_{fi}^{(1)}) \tag{4.5}$$

The tr (trace) operation can be carried out in any representation. The quantity $\omega(\mathbf{r})\, \delta(\mathbf{r} - \mathbf{r})$ can be considered as the matrix element of an operator $\hat{\omega}$. It is diagonal in the $\mathbf{r}$ representation as indicated by $\delta(\mathbf{r} - \mathbf{r}')$. In general it could be a nonlocal operator $\omega(\mathbf{r},\mathbf{r}')$, in which case T_{fi} will be given again by

$$T_{fi} = A\int \mathbf{dr}\, \omega(\mathbf{r}',\mathbf{r})\, \rho_{fi}^{(1)}(\mathbf{r},\mathbf{r}')\, dr' = A \text{ tr } (\hat{\omega}\hat{\rho})$$

Problem. Show that

$$\hat{\rho}^{(1)} = \delta(\mathbf{r} - \mathbf{r}_1)\,\delta(\mathbf{r}' - \mathbf{r}_1')\,\delta(\mathbf{r}_2' - \mathbf{r}_2)\,\delta(\mathbf{r}_3 - \mathbf{r}_3') \ldots \delta(\mathbf{r}_A - \mathbf{r}_A')$$

Hint: Evaluate $\int \psi_f^*(\mathbf{r}_1,\mathbf{r}_2, \ldots, \mathbf{r}_A)\,\hat{\rho}^{(1)}\,\psi_i(\mathbf{r}_1',\mathbf{r}_2', \ldots \mathbf{r}_A')\,d\mathbf{r}_1 d\mathbf{r}_2 \ldots d\mathbf{r}_A \cdot$

$d\mathbf{r}_1' d\mathbf{r}_2' \ldots d\mathbf{r}_A'$ to obtain (4.4).

Problem. Suppose $\phi(\mathbf{r})$ form an orthogonal normalized complete set of functions. Show directly from the integral that

$$T_{fi} = A \sum_{nm} \langle \phi_n(\mathbf{r}) | \omega(\mathbf{r},\mathbf{r}')\,\phi_m(\mathbf{r}') \rangle \langle \phi_m(\mathbf{r}'') | \rho_{fi}(\mathbf{r}'',\mathbf{r}''')\,\phi_n(\mathbf{r}''') \rangle$$

If the particles have spin intrinsic degrees of freedom the *tr* operation is meant to take a sum over the spin variables as well. An operator $\hat{\omega}$ that is independent of the spin variables will then be understood to multiply the unit operator 1 in spin space. The extension to isospin variables proceeds along the same lines.

Remark. The slightly pedantic way of writing T_{fi} in this form has to do with the possibility that $\omega(\mathbf{r}')$ involves a differential operator; in this case one has to make sure that it operates on the ψ_i component of ρ_{fi} only, and not on both the ψ_i and the ψ_f^* components. For operators $\omega(\mathbf{r}')$ that are simply functions of $\mathbf{r}'$, we can use the usual transition density $\bar{\rho}_{fi}^{(1)}(r) = \rho_{fi}^{(1)}(r,r)$ and reduce (4.5) to the more familiar form

$$T_{fi} = A \int d\mathbf{r}\,\bar{\rho}_{fi}^{(1)}(\mathbf{r})\,\omega(\mathbf{r})$$

More generally, if we have any operator $\hat{\Omega}$, its matrix elements in the A-particle r-representation can be written in the form

$$\langle \mathbf{r}_1,\mathbf{r}_2, \ldots, \mathbf{r}_A | \hat{\Omega} | \mathbf{r}_1',\mathbf{r}_2', \ldots, \mathbf{r}_A' \rangle$$

If $\hat{\Omega}$ is a symmetric sum of one body operators, like (4.1), then we have

$$\begin{aligned} &\langle \mathbf{r}_1,\mathbf{r}_2, \ldots, \mathbf{r}_A | \hat{\Omega} | \mathbf{r}_1',\mathbf{r}_2', \ldots, \mathbf{r}_A' \rangle \\ &= A \langle \mathbf{r}_1 | \hat{\omega} | \mathbf{r}_1' \rangle\,\delta(\mathbf{r}_2' - \mathbf{r}_2), \ldots \delta(\mathbf{r}_A - \mathbf{r}_A') \end{aligned} \tag{4.6}$$

If the operator $\hat{\Omega}$ is local, then we have also

$$\langle \mathbf{r}_1 | \omega | \mathbf{r}_1' \rangle = \delta(\mathbf{r}_1 - \mathbf{r}_1')\,\omega(\mathbf{r}_1')$$

and hence the form of (4.5).
The generalization of (4.6) to two- or more- body operators is straightforward.

It is seen therefore that for the evaluation of matrix elements of operators of the type of $\hat{\Omega}$—for a sum of single-particle operators—it is sufficient to utilize only a small part of the information contained in ψ: that included in $\rho_{fi}^{(1)}(\mathbf{r},\mathbf{r}')$, as given by (4.4). Two sets of wave functions that differ from each other in their more detailed structure, but yield similar functions for $\rho_{fi}(\mathbf{r},\mathbf{r}')$ will provide an equally good "explanation" for the matrix element $\langle f|\hat{\Omega}|i\rangle$.

Many of the measurable quantities in nuclear physics are indeed represented by single-particle operators like $\hat{\Omega}$. The operator measuring charge distribution, for instance, is

$$\rho = \sum_k \frac{e}{2}(1 + \tau_{k3})$$

that of the kinetic energy of the nucleus is

$$T = \sum_i -\frac{\hbar^2}{2M}\nabla_i^2$$

etc. There are, however, some operators of importance that cannot be written in this form. The one best known is the two-particle interaction $v(\mathbf{r}_1 - \mathbf{r}_2)$, or more generally v_{12}. In typical situations we may have to evaluate matrix elements of the type

$$G_{fi} = \langle f|\sum_{k<l} g_{kl}|i\rangle = \sum_{k<l}\int \psi_f^*(\mathbf{r}_1, \ldots, \mathbf{r}_A)\, g_{kl}\, \psi_i(\mathbf{r}_1, \ldots, \mathbf{r}_A)\, d\mathbf{r}_1 \ldots d\mathbf{r}_A \tag{4.7}$$

Here g_{kl} is an operator involving the coordinates of both particles k and l in a way that cannot be reduced into a simple sum: $g_{kl} \neq \omega(k) + \omega(l)$. Again using the symmetry of $G = \sum g_{kl}$ with respect to interchange of particles and the antisymmetry of $\psi(\mathbf{r}_1, \ldots, \mathbf{r}_A)$ we can cast (4.7) into the form

$$G_{fi} = \frac{A(A-1)}{2}\int \psi_f^*(\mathbf{r}_1, \ldots, \mathbf{r}_A)\, g_{12}\, \psi_i(\mathbf{r}_1, \ldots, \mathbf{r}_A)\, d\mathbf{r}_1 \ldots d\mathbf{r}_A \tag{4.8}$$

This expression cannot be evaluated if all we know about ψ is the one-particle density matrix $\rho_{fi}(\mathbf{r},\mathbf{r}')$. But we still need to use only a small part of the information contained in ψ to evaluate (4.8). We define now the *two-particle density matrix*

$$\rho_{fi}^{(2)}(\mathbf{r}_1,\mathbf{r}_2;\mathbf{r}_1',\mathbf{r}_2') = \int \psi_f^*(\mathbf{r}_1,\mathbf{r}_2,\mathbf{r}_3, \ldots, \mathbf{r}_A)\, \psi_i(\mathbf{r}_1',\mathbf{r}_2',\mathbf{r}_3, \ldots, \mathbf{r}_A)\, d\mathbf{r}_3 \ldots d\mathbf{r}_A \tag{4.9}$$

in terms of which G_{fi} can be written as

$$G_{fi} = \frac{A(A-1)}{2}\int d\mathbf{r}_1 d\mathbf{r}_2\, \delta(\mathbf{r}_1 - \mathbf{r}_1')\, \delta(\mathbf{r}_2 - \mathbf{r}_2')\int g_{12}(\mathbf{r}_1',\mathbf{r}_2')\, \rho_{fi}^{(2)}(\mathbf{r}_1,\mathbf{r}_2;\mathbf{r}_1',\mathbf{r}_2')\,d\mathbf{r}_1'\, d\mathbf{r}_2' \tag{4.10}$$

Physically it is very easy to understand results (4.5) and (4.10). If we have a single-particle operator, to which each particle contributes independently of the others, all we want to know from the wave function $\psi(\mathbf{r}_1, \ldots, \mathbf{r}_A)$ is the average one-particle behavior that it represents. Similarly, if we have a two-particle operator, to which each *pair* of particles contributes independently of all other particles, all we need to know is the average behavior of a *pair* in $\psi(\mathbf{r}_1,\mathbf{r}_2, \ldots, \mathbf{r}_A)$; correlations among three or more particles are of no significance to the evaluation of the matrix elements of $G = \sum_{k<l} g_{kl}$.

When the potential between the nucleons is of the two-body type

$$V = \sum_{k<l} v(\mathbf{r}_k, \mathbf{r}_l) \equiv \sum_{k<l} v_{kl}$$

it becomes possible to write the expectation value of the nuclear Hamiltonian $\langle H \rangle$ in terms of the one- and two-particle density matrix $\rho^{(1)}$ and $\rho^{(2)}$

$$\langle H \rangle = A \operatorname{tr}\left(-\frac{\hbar^2}{2M} \nabla_1^2 \rho_{00}^{(1)}\right) + \frac{A(A-1)}{2} \operatorname{tr} v_{12}\rho_{00}^{(2)} \tag{4.11}$$

In this form the Hamiltonian is independent of representation so that one can directly consider it in either momentum or coordinate space. Returning to the original point made in opening this section, we see that the nuclear binding energies depend only upon certain weighted averages of one- and two-body densities. It is thus possible for a wave function whose three-body and higher order densities are in considerable error to still yield reasonable values for the expectation value of H, that is, for the energy. It is even possible for the one- and two-body densities to be incorrect if the potential in (4.11) emphasizes those regions where they are correct or if errors in the potential compensate for errors in the densities. The latter is presumably true for the Fermi gas model!

5. ONE- AND TWO-PARTICLE DENSITIES

To obtain a more complete picture of the usefulness of the Fermi gas model we want, therefore, to compute the one- and two-particle density matrices that go with it. For simplicity let us first deal with the special case $i \equiv f$. To avoid dealing with surface effects it is convenient to replace the boundary conditions (2.2) by periodic boundary conditions $\phi(-a) = \phi(a)$, $\phi'(-a) = \phi'(a)$ and write the normalized wave function for a single particle in the form:

$$\phi_\alpha(\mathbf{r}_l) = \frac{1}{\sqrt{\Omega}} e^{i\mathbf{k}_\alpha \cdot \mathbf{r}_\alpha} \chi_l(m_s^{(\alpha)}, m_\tau^{(\alpha)}) \tag{5.1}$$

where $\mathbf{k}_\alpha = (\pi/a)\,(n_{\alpha x},n_{\alpha y},n_{\alpha z})$, the n_α's being both positive and negative integers. $\chi_l(m_s{}^{(\alpha)},m_\tau{}^{(\alpha)})$ is the spin and isospin wave function that goes with the state α, so that $\chi(1/2,1/2)$ stands for a proton with spin up, $\chi(-1/2,1/2)$ a proton with spin down, etc. The χ's, which can be considered as four-rowed columns, satisfy the orthogonality relations

$$\chi^\dagger(m_s,m_\tau)\,\chi(m_s',m_\tau') = \delta(m_s,m_s')\,\delta(m_\tau,m_\tau'). \tag{5.2}$$

The complete ϕ's satisfy the orthogonality relation

$$\int \phi_\alpha^\dagger(\mathbf{r})\, \phi_\beta(\mathbf{r})\, d^3\mathbf{r} = \delta(\mathbf{k}_\alpha,\mathbf{k}_\beta)\, \delta(m_s^\alpha,m_s^\beta)\, \delta(m_\tau^\alpha,m_\tau^\beta) \tag{5.3}$$

when the integration is limited to the cube of volume Ω. The set $\mathbf{k}_\alpha$ is a discrete set under these conditions (of a finite box Ω) and

$$\delta(\mathbf{k}_\alpha,\mathbf{k}_\beta) = \delta(k_{\alpha x},k_{\beta x})\cdot\delta(k_{\alpha y},k_{\beta y})\cdot\delta(k_{\alpha z},k_{\beta z})$$

is then an ordinary Kronecker delta function.

A normalized antisymmetric state of A noninteracting particles in the box is obtained by forming the Slater determinant

$$[\Phi_\alpha] \equiv \Phi_{\alpha_1,\ldots,\alpha_A}(1,2,\ldots,A) = \frac{1}{\sqrt{A!}} \begin{vmatrix} \phi_{\alpha_1}(\mathbf{r}_1)\ \phi_{\alpha_2}(\mathbf{r}_1),\ \ldots,\ \phi_{\alpha_A}(\mathbf{r}_1) \\ \phi_{\alpha_1}(\mathbf{r}_2)\ \phi_{\alpha_2}(\mathbf{r}_2),\ \ldots,\ \phi_{\alpha_A}(\mathbf{r}_2) \\ \cdot \\ \cdot \\ \cdot \\ \phi_{\alpha_1}(\mathbf{r}_A)\ \phi_{\alpha_2}(\mathbf{r}_A),\ \ldots,\ \phi_{\alpha_A}(\mathbf{r}_A) \end{vmatrix} \tag{5.4}$$

If $\alpha_1, \alpha_2, \ldots, \alpha_A$ are the A lowest states in the box, then (5.4) gives the wave function ψ_0 of the ground state of the degenerate Fermi gas.

It is a simple matter to calculate $\rho_{00}^{(1)}$ if $A = 4n$, and each state is fully occupied by two protons and two neutrons. Then following the definition (4.4),

$$\rho_{00}^{(1)} = \langle m_s m_t | \rho_{00} | m_s' m_t' \rangle$$

where

$$\hat{\rho}_{00}^{(1)}(\mathbf{r},\sigma_z\tau_3;\mathbf{r}'\sigma_z\tau_3') = \frac{\mathbb{1}}{A\Omega} \sum_{l=1}^{n} e^{i\mathbf{k}_l\cdot(\mathbf{r}'-\mathbf{r})} \tag{5.5}$$

where $k_l(l = 1, \ldots, n = A/4)$ runs over all occupied states, and the unit 4×4 matrix $\mathbb{1}$ in spin and isospin variables implies that in (4.5) the trace of $\omega(\mathbf{r}'\boldsymbol{\sigma}'\boldsymbol{\tau}')$ has to be taken also with respect to the spin and isospin variables. The diagonal elements of ρ_{00} (that is, $\mathbf{r}' = \mathbf{r}$), after taking the trace with respect to spin and isospin variables (tr $\mathbb{1} = 4$), are given simply by

$$\bar{\rho}_0(\mathbf{r}) = \frac{1}{\Omega} \tag{5.6}$$

Thus the Fermi degenerate gas model represents a constant single-particle density, independent of $\mathbf{r}$.

Remark. In the Fermi gas model as defined by the boundary conditions (2.2), which implied the *vanishing* of ψ on the walls of the box, the density in the box does not *come* out to be constant. Near the walls all the single-particle wave functions vanish, and since the highest wave number associated with a single-particle wave function is k_F, the single-particle density will not rise to its constant average value in the middle of the box until we go a distance $\sim 1/k_F$ away from the walls. Our result (5.6) of a constant density *all through* the nucleus is a direct consequence of the *periodic* boundary conditions. Since these boundary conditions represent a nucleus that can be "repeated" in all directions an infinite number of times, surface effects are washed out. The present treatment of the Fermi gas is meant to be just for first orientation and we shall disregard these details. Later we shall treat the more realistic situation of independent particles in a spherical "box" with greater detail and rigor.

There are many wave functions, even quite complex ones, that lead to an essentially constant nuclear density, expecially if we do not come too close to the nuclear surface. If an $\hat{\Omega}$-type operator derives the important contribution to its matrix elements from the interior of the nucleus, a Fermi gas model and more sophisticated and realistic wave functions will then give an equally good description of $\hat{\Omega}$. It is important to bear this result in mind in order to distinguish later whether an agreement between theory and experiment is, or is not, meaningful.

The two-particle density matrix for the Fermi gas is equally easy to obtain. For simplicity we drop for a moment the spin and isospin dependence and assume that the indices α_i in (5.4) refer just to momentum states. With $\phi_k(\mathbf{r}) = (1/\sqrt{\Omega})\, e^{i\mathbf{k}\cdot\mathbf{r}}$ we then get for the diagonal ($\mathbf{r}_1 = \mathbf{r}_1$, $\mathbf{r}_2 = \mathbf{r}_2'$) part of the two-particle density matrix, by expanding the determinant (5.4),

$$\rho_{00}(\mathbf{r}_1\mathbf{r}_2,\mathbf{r}_1\mathbf{r}_2) \equiv \bar{\rho}(\mathbf{r}_1,\mathbf{r}_2)$$

$$= \frac{1}{A(A-1)}\frac{1}{2\Omega^2}\sum_{k_j,k_i}[2 - e^{i(\mathbf{k}_i\cdot\mathbf{r}_1+\mathbf{k}_j\cdot\mathbf{r}_2-\mathbf{k}_i\cdot\mathbf{r}_2-\mathbf{k}_j\cdot\mathbf{r}_1)} - e^{-i(\mathbf{k}_i\cdot\mathbf{r}_1-\mathbf{k}_j\cdot\mathbf{r}_1-\mathbf{k}_i\cdot\mathbf{r}_2+\mathbf{k}_j\cdot\mathbf{r}_2)}]$$

$$= \frac{1}{A(A-1)}\frac{1}{\Omega^2}\sum_{k_j,k_i}[1 - \cos 2\mathbf{k}_{ij}\cdot\mathbf{r}] \tag{5.7}$$

where

$$\mathbf{k}_{ij} = \tfrac{1}{2}(\mathbf{k}_i - \mathbf{k}_j) \qquad \text{and} \qquad \mathbf{r} = \mathbf{r}_i - \mathbf{r}_j \tag{5.8}$$

and the sums are over occupied states only ($A/4$ of them). Equation 5.7 determines the pair correlation in the Fermi gas. Since both $\mathbf{k}_i$ and $\mathbf{k}_j$ are limited in their magnitude by the Fermi momentum k_F, we see from (6.8) that the relative momentum $\mathbf{k}_{ij}$ satisfies

$$|\mathbf{k}_{ij}| = |\tfrac{1}{2}(\mathbf{k}_i - \mathbf{k}_j)| \leq k_F$$

Thus if we look at separations $\mathbf{r}$ that are small compared with $1/k_F$, $\bar{\rho}(\mathbf{r}_1,\mathbf{r}_2)$ tends to zero.

We can evaluate $\sum_{i,j}$ (1-cos $2\mathbf{k}_{ij}\cdot\mathbf{r}$) in (5.7) directly, but since we are interested in that part of the sum which is independent of the size of the box, we can replace summation by integration over all allowed momenta:

$$\sum(1-\cos 2\mathbf{k}_{ij}\cdot\mathbf{r})\rightarrow\left(\frac{\Omega}{8\pi^3}\right)^2\int \mathbf{dk}_i d\mathbf{k}_j(1-\cos 2\mathbf{k}_{ij}\cdot\mathbf{r})$$

$$=\left(\frac{\Omega}{8\pi^3}\right)^2\int d\mathbf{k}_i\, d\mathbf{k}_j[1-\cos(\mathbf{k}_i\cdot\mathbf{r}-\mathbf{k}_j\cdot\mathbf{r})]$$

$$=\left(\frac{\Omega}{8\pi^3}\right)^2\int d\mathbf{k}_i\, d\mathbf{k}_j[1-\cos(\mathbf{k}_i\cdot\mathbf{r})(\cos\mathbf{k}_j\cdot\mathbf{r})]$$

because sin $\mathbf{k}_i\cdot\mathbf{r}$ leads to a vanishing integral.

Now taking $\mathbf{r}$ as the z-axis in polar coordinates in k-space:

$$\int d\mathbf{k}\cos(\mathbf{k}\cdot\mathbf{r}) = 2\pi\int k^2\, dk\, d(\cos\theta)\cos(kr\cos\theta)$$

$$=4\pi\int_0^{k_F}k^2dk\,\frac{\sin kr}{kr} = -4\pi\,\frac{1}{r}\,\frac{d}{dr}\left[\int_0^{k_F}\cos kr\, dk\right]$$

$$=-4\pi\,\frac{1}{r}\,\frac{d}{dr}\left[\frac{\sin k_Fr}{r}\right]=\frac{4\pi}{r^2}\left[\frac{\sin k_Fr}{r}-k_F\cos k_Fr\right]$$

We therefore obtain

$$\bar{\rho}(\mathbf{r}_1,\mathbf{r}_2) = \frac{[(4/3)\,\pi k_F{}^3]^2}{(2\pi)^6A(A-1)}\left\{1-\left[\frac{3}{(k_Fr)^2}\left(\frac{\sin k_Fr}{k_Fr}-\cos k_Fr\right)\right]^2\right\} \tag{5.9}$$

$$\approx\frac{1}{\Omega^2}g_-(x)\qquad x\equiv k_Fr$$

where

$$g_-(x)\equiv 1-\left[\frac{3}{x^2}\left(\frac{\sin x}{x}-\cos x\right)\right]^2 \tag{5.10}$$

This function vanishes for $x = 0$, and for small values of x acts like $g_-(x)\sim(1/5)\,x^2$; for large values of x, $g_-(x)\rightarrow 1$. The effects of the Pauli principle on the correlations between two fermions in a degenerate Fermi gas are clearly exhibited by (5.9). There is a "repulsion" of two identical particles expressed by the vanishing of the probability of finding them both at the same place, and it gradually disappears as the distance between the particles becomes big compared with $1/k_F$ [note that already for $r = (\pi/2)\,(1/k_F)$, $g_-(x) = 0.77$]. This "Pauli repulsion" has nothing to do with the forces between the nucleons and does not result from any dynamical considerations. It is merely a manifestation of the Pauli principle, and exists in any system of fermions irrespective of whether the real forces between them are attractive or repulsive.

In deriving (5.7) we have disregarded the spin and isospin coordinates of the nucleons, and assumed the nuclear wave function to be antisymmetric with respect to the exchange of spatial coordinates only. According to the Pauli principle this is true only if the two particles we have been considering have a symmetric spin and isospin wave function. If the spin and isospin wave function is antisymmetric, the spatial wave function has to be symmetric with respect to the exchange of the two particles to satisfy the Pauli principle. Equation 5.7 is then replaced by

$$\bar{\rho}_{\text{sym}}(\mathbf{r}_1,\mathbf{r}_2) = \frac{1}{A(A-1)\,\Omega^2} \sum_{i,j} [1 + \cos 2\mathbf{k}_{ij}\cdot\mathbf{r}] \tag{5.11}$$

This, then leads to a symmetric correlation

$$\bar{\rho}_{\text{sym}}(\mathbf{r}_1,\mathbf{r}_2) \approx \frac{1}{\Omega^2}\, g_+(k_F r) \tag{5.12}$$

where

$$g_+(x) = 1 + \left[\frac{3}{x^2}\left(\frac{\sin x}{x} - \cos x\right)\right]^2 \tag{5.13}$$

Unlike $g_-(x)$, $g_+(x)$ remains finite for small values of x and in fact

$$\lim_{x\to 0} g_+(x) = 2$$

The particles are thus "attracted" to each other if they are in a spatially symmetric state.

It is no accident that

$$\tfrac{1}{2}[g_-(x) + g_+(x)] = 1$$

A situation in which two particles have equal chance of being found in a spatially symmetric state and spatially antisymmetric state leads neither to "Pauli repulsion" nor to "Pauli attraction."

The spin wave function $\chi_1(s_z)$ and $\chi_2(s_z)$ of two particles can be combined to form three symmetric combinaitons

$$\chi_1\left(+\frac{1}{2}\right)\chi_2\left(+\frac{1}{2}\right),\ \frac{1}{\sqrt{2}}\left[\chi_1\left(+\frac{1}{2}\right)\chi_2\left(-\frac{1}{2}\right) + \chi_1\left(-\frac{1}{2}\right)\chi_2\left(+\frac{1}{2}\right)\right],$$

$$\chi_1\left(-\frac{1}{2}\right)\chi_2\left(-\frac{1}{2}\right)$$

and one antisymmetric combination

$$\frac{1}{\sqrt{2}}\left[\chi_1\left(+\frac{1}{2}\right)\chi_2\left(-\frac{1}{2}\right) - \chi_1\left(-\frac{1}{2}\right)\chi_2\left(+\frac{1}{2}\right)\right].$$

These will be recognized as the three states of the triplet and the one singlet state. For isospin we have similarly three symmetric states of the isotriplet and one antisymmetric state for the isosinglet. For the combined spin and isospin wave function there are therefore 10 symmetric wave functions (the nine formed from the triplet and isotriplet plus the one formed from the singlet and the isosinglet) and six antisymmetric wave functions. When we average the correlation function between two nucleons over their spins and isospins $g_-(x)$ gets a weight of 10/16 and g_+ a weight of 6/16. We thus obtain for $\bar{\rho}_F(\mathbf{r}_1,\mathbf{r}_2)$ in a degenerate Fermi gas of nucleons, averaged over spin and isospin,

$$\bar{\rho}_F(\mathbf{r}_1,\mathbf{r}_2) = \text{constant}\left\{1 - \frac{1}{4}\left[\frac{1}{(k_F r)^2}\frac{\sin k_F r}{k_F r} - \cos k_F r\right]^2\right\} \tag{5.14}$$

where

$$r = |\mathbf{r}_1 - \mathbf{r}_2|$$

Remark. In the Fermi gas model the diagonal two-particle density matrix is the same for the symmetric isospin-spin state formed from triplet isospin, triplet spin, and from the singlet isospin, singlet spin states. A similar result holds for the singlet isospin, triplet spin and triplet isospin, singlet spin combinations. For more general wave functions (e.g., when $N \neq Z$) than that of the Fermi gas model, four two-particle density matrices are required

The preponderance of spatially antisymmetric pairs over spatially symmetric ones in a system of fermions in its ground state has very important consequences. Physically it tells us that spin-1/2 fermions, even when they possess an additional intrinsic degree of freedom (charge, in our case), have a tendency, on the whole, to stay apart. There is a net repulsion. This tendency is there *before* the consideration of any dynamical effects. Since attractive short range forces should be particularly effective in spatially symmetric states of the pair of interacting nucleons, the preponderance of spatially antisymmetric pairs in an ensemble of fermions can help prevent a collapse of this system. The extent to which it can do so is the subject of Section 7 of this chapter.

6. COULOMB ENERGY (C.E.) IN THE FERMI GAS MODEL

The effect of the "Pauli repulsion" and "Pauli attraction" is illustrated by the simple but important example of the Coulomb energy for the nuclear ground state. According to (4.11) this is given by

$$\text{C.E.} = \frac{A(A-1)}{2}\int \bar{\rho}_{00}(\mathbf{r}_1,\mathbf{r}_2)\, v_{12}(\mathbf{r}_1,\mathbf{r}_2)\, d\mathbf{r}_1\, d\mathbf{r}_2 \tag{6.1}$$

where v_{12} is given by (I.6.13).

$$v_{12} = \frac{e^2}{r_{12}}\left[\frac{1+\tau_3(1)}{2}\right]\left[\frac{1+\tau_3(2)}{2}\right]$$

$$r_{12} \equiv |\mathbf{r}_1 - \mathbf{r}_2|$$

The isospin factors guarantee that the interaction takes place only between protons. Since a pair of protons are in a symmetric isospin state, the spatial wave function is symmetric when they are in a singlet spin state and antisymmetric when the protons are in a triplet spin state. Hence the spatially antisymmetric $\bar{\rho}_{00}$ has a weight of 3/4, the symmetric a weight of 1/4. Thus

$$\text{C.E.} = \frac{Z(Z-1)}{2\Omega^2}\int d\mathbf{r}_1\, d\mathbf{r}_2\, v_{12}(\mathbf{r}_1,\mathbf{r}_2)\left(\frac{3g_- + g_+}{4}\right)$$

or

$$\text{C.E.} = \frac{Z(Z-1)}{2\Omega^2}\int d\mathbf{r}_1\, d\mathbf{r}_2\, v_{12}(\mathbf{r}_1,\mathbf{r}_2)\left\{1 - \frac{1}{2}\left[\frac{3}{2}\left(\frac{\sin x}{x} - \cos x\right)\right]^2\right\} \tag{6.2}$$

where $x = k_F r_{12}$.

The first term is just the classical Coulomb energy of a sphere (we assume a spherical nucleus) and so yields $3/5\ Z^2/R$ when $Z \gg 1$. The second term gives the effects of the Pauli principle. They can be traced back to the exchange Coulomb integrals if the wave function (5.4) is directly used in (6.1) for $\bar{\rho}_{00}$. It (omitting the minus sign) is therefore referred to as the exchange Coulomb energy [$\equiv$ (C.E.)$_{\text{ex}}$]. Changing variables from $\mathbf{r}_1$ and $\mathbf{r}_2$ to $\mathbf{r}_1$ and $\mathbf{r}$ ($\equiv \mathbf{r}_1 - \mathbf{r}_2$) it becomes

$$\begin{aligned}(\text{C.E.})_{\text{ex}} &= \frac{Z(Z-1)}{4\Omega}\, e^2 \int d\mathbf{r}\left[\frac{3}{x^2}\left(\frac{\sin x}{x} - \cos x\right)^2\right]^2 \\ &= \frac{9\pi Z(Z-1)\, e^2}{\Omega}\int_0^R \frac{r\,dr}{x^2}\left(\frac{\sin x}{x} - \cos x\right)^2 \\ &= \frac{9\pi Z(Z-1)\, e^2}{\Omega k_F^2}\int_0^{k_F R} \frac{dx}{x^3}\left(\frac{\sin x}{x} - \cos x\right)^2\end{aligned}$$

The integral can be evaluated exactly yielding

$$(\text{C.E.})_{\text{ex}} = \frac{9\pi}{4}\,\frac{(Z)(Z-1)\, e^2}{\Omega k_F^2}\,[1 - j_0^2(k_F R) - j_1^2(k_F R)] \tag{6.3}$$

where j_0 and j_1 are the spherical Bessel function [Morse and Feshbach (53), Chapter 11]. These spherical Bessel function terms can be dropped as $k_F R$ becomes large. The square bracket is 0.88 for $k_F R$ of 3, that is, for a nucleus with 2 fm radius! Hence to a very good approximation

$$(\text{C.E.})_{\text{ex}} = \frac{9\pi}{4} \frac{Z^2 e^2}{\Omega k_F^2} \tag{6.4}$$

where again $Z(Z - 1)$ has been replaced by Z^2.

The total Coulomb energy is then

$$\text{C.E.} = \frac{3}{5} \frac{Z^2 e^2}{R} - (\text{C.E.})_{\text{ex}}$$

or

$$\text{C.E.} = .77 \frac{Z^2}{A^{1/3}} \left(1 - \frac{1.0}{A^{2/3}}\right) \text{MeV} \tag{6.5}$$

The exchange energy subtracts simply because of the fact that on the average the exclusion principle keeps the protons apart. For protons the net effect of the "Pauli attraction" and "Pauli repulsion" is repulsion. We note that the exchange effect decreases with increasing A. This is because the "range" of the Pauli repulsion is $(1/k_F)$ and this involves a smaller fraction of the nuclear volume as A and therefore the nuclear volume increases.

Formula (6.5) should be compared with the semiempirical result of Myers and Swiatecki (66):

$$\text{C.E.} = \frac{.717\, Z^2}{A^{1/3}} \left(1 - \frac{1.69}{A^{2/3}}\right) \tag{6.6}$$

Again the Fermi gas model is surprisingly close. The coefficient of the $A^{-2/3}$ term is considerably larger than the Fermi gas model value. It has been pointed out that the reason for this discrepancy is in part a consequence of the diffuse nature of the nuclear surface. The calculation leading to (6.5) assumed that the nucleus had a sharp radius whereas a more reasonable assumption, as we know from electron scattering, is that the proton density falls from 0.9 of its value at the center of the nucleus to 0.1 of that value in about 2.4 fm. The charge density has the shape shown in Fig. I.1.2. This effect [Seeger (67)] and others are reviewed by J. Jänecke (69) to whom the reader is referred for further details. It should also be remarked that the calculation also neglects the finite size of the protons and neutrons (see Chapter VIII). This and other electromagnetic effects are discussed by Auerbach, Hufner, Kerman, and Shakin (72). The effect of the finite neutron size is discussed by Bertozzi et al. (72).

7. THE NUCLEON-NUCLEON FORCE AND NUCLEAR STABILITY—QUALITATIVE

It has become evident from our discussions up to this point that although some properties of nuclei do not require a detailed knowledge of the nuclear force for their qualitative elucidation, we have to analyze the effects of the nuclear forces if we want a more thorough understanding of nuclear structure. In the Fermi gas model the equilibrium density of nuclei is taken as a parameter, but if we want to understand why the density is roughly constant and predict its experimental value, we must make use of the properties of the nuclear force. In order to obtain a more quantitative determination of nuclear parameters, like those of the Weizsäcker formula, than that provided by the Fermi gas model, we must take into account in a more detailed way the interactions among the nucleons.

In the introduction to this chapter we have already referred to some of the difficulties connected with the formulation of a dynamical Schrödinger equation for the nucleus. For the sake of our present discussion we shall make the very plausible postulate that the nucleus is governed by the nonrelativistic Schrödinger equation (see Chapter I):

$$\left(\sum_{i=1}^{A} T_i + \sum_{i<j=1}^{A} v_{ij}\right)\psi(1, \ldots, A) = E\psi(1, \ldots, A) \tag{7.1}$$

with boundary conditions on ψ that are appropriate to the problem we are dealing with. Furthermore, we assume that the interaction v_{ij} is *independent* of A; in other words the same basic interaction is to be taken for the analysis of ^{208}Pb, say, as is taken for the analysis of two-nucleon scattering. Of course, inside a large nucleus collisions between nucleons occur over a range of relative momenta, and in the presence of the rest of the nucleus. Therefore, among other things, the calculation of the dynamics of a large nucleus at a given total energy involves a knowledge of two-nucleon system over a wide range of energies and momenta.

Let us first try to estimate the range of relative momenta in collisions between two nucleons in a nucleus. The probability density of finding in a degenerate Fermi gas two particles each with momentum of magnitude k relative to the motion of their center of mass is proportional to the integral

$$\int d\mathbf{k}_1\, d\mathbf{k}_2\, \delta(\tfrac{1}{2}|(\mathbf{k}_1 - \mathbf{k}_2)| - k)$$

since each particle is uniformly spread over the various space elements in momentum space. The evaluation of this integral yields for the desired probability [see Gomes et al. (58)]:

$$P(k)\, d\mathbf{k} = \frac{6 d\mathbf{k}}{\pi k_F{}^3}\left\{1 - \frac{3}{2}\frac{k}{k_F} + \frac{1}{2}\left(\frac{k}{k_F}\right)^3\right\} \qquad k \leq k_F \tag{7.2}$$

The distribution of relative momenta corresponding to (7.2) is shown in Fig. 7.1.

The most probable relative momentum is approximately $k_0 \simeq 0.6\, k_F$. At the observed nuclear density this corresponds to a relative kinetic energy of (note the use of reduced mass $M/2$)

$$\epsilon_0 = \frac{\hbar^2(0.6\, k_F)^2}{2(M/2)} = 27 \text{ MeV}$$

In other words, the most probable collisions between nucleons in the nucleus correspond to nucleon-nucleon scattering at 30 MeV in the center-of-mass system, or 60 MeV in the laboratory system, that is, the system in which one nucleon is initially at rest.

The maximum relative momentum is k_F (that is, when the two nucleons have equal and opposite momenta, both equal to k_F); because of the reduced mass in the relative motion this corresponds to a *relative* kinetic energy that is *twice* the Fermi energy of a nucleon, that is, about 80 MeV in the center-of-mass system. It takes 160 MeV incident protons to reproduce these conditions in nucleon-nucleon scattering when the target nucleon is at rest (laboratory system).

We see therefore that in the description of nuclear dynamics we shall depend primarily on the nucleon-nucleon interaction up to energies of $\sim$200 MeV in the laboratory system. In other words, potentials v_{ij} that can be made to fit the nucleon-nucleon data up to laboratory energies of 200 MeV should be sufficient for the description of the dynamics of complex nuclei as well. This conclusion relies heavily on the picture of independent-particle motion in the nucleus with no correlations beyond those implied by the Pauli principle. To the extent that we shall find later that additional important correlations exist between nucleons in the nucleus, we shall have to review this conclusion. In particular, if, as seems to be the case, there are significant short range correlations, that is, if the nuclear two-body density $\rho(\mathbf{r}_1,\mathbf{r}_2)$ is significantly different from the Fermi gas model density $\bar{\rho}_F(\mathbf{r}_1,\mathbf{r}_2)$ for small values of $r = |\mathbf{r}_1 - \mathbf{r}_2|$, then higher relative momenta will be involved in the collisions between nucleons in the nucleus. In that event the potentials used for the study of the

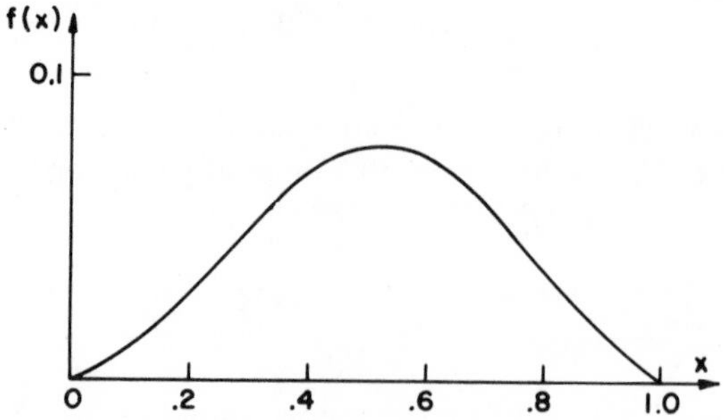

FIG. 7.1. The function $f(x) = x^2(1 - \frac{3}{2}x + \frac{1}{2}x^3]x < 1$ that describes the distribution of relative momenta in the Fermi gas model.

dynamics of nuclei should be made to fit the nucleon-nucleon scattering data at higher energies as well.

In principle, even this may not be sufficient. There is another refinement, whose effects thus far have not been extensively studied, and which should be mentioned at this stage. To solve the Schrödinger equation (7.1) for the bound states we can proceed by choosing an arbitrary orthonormal set of antisymmetric function Φ_α of the A-nucleons satisfying the proper boundary conditions, and then diagonalize the matrix $\langle\Phi_\alpha|\sum T_i + \sum_{i<j} v_{ij}|\Phi_\beta\rangle$. In computing these matrix elements we shall encounter matrix elements of, say v_{12} taken between two states of the particles 1 and 2. The statement that v_{12} is independent of A means that these matrix elements could be taken from the two nucleon scattering data. However, because of conservation of energy, the two-nucleon scattering can supply, *in principle*, only a special subset of the matrix elements $\langle\alpha|v_{12}|\beta\rangle$: those in which the two-particle states $|\alpha\rangle$ and $|\beta\rangle$ correspond to the *same* total energy, that is, these matrix elements are on the energy shell. In the A-particle problem, energy is conserved of course for the complete A-particle system, but in computing

$$\langle\Phi_\alpha|\sum_{i<j=1}^{A} v_{ij}|\Phi_\beta\rangle$$

there is no guarantee that matrix elements of v_{12} will occur only between states having the same total energy for particle 1 and 2 *alone*. Quite the contrary, the Hamiltonian for particles 1 and 2 alone does not commute generally with v_{12}, We are therefore sure to encounter in the A-particle problem matrix elements of v_{12} between states of different energies for these two particles. Thus, whereas the nucleon-nucleon scattering determines only matrix elements on the "energy shell"* of the two particles, the A-nucleon problem also requires the knowledge of the interaction "off the energy shell."

Notice that whenever we write the interaction v_{12} in a certain *functional form* we are actually giving a prescription for the connection of matrix elements *off* the energy shell with those *on* the energy shell. For example, suppose we were to assume that $v_{12} = v(|\mathbf{r}_1 - \mathbf{r}_2|)$. For the sake of illustration we disregard the spin and isospin dependence. What we say is: look for a function $v(|\mathbf{r}_1 - \mathbf{r}_2|)$ that reproduces the data on nucleon–nucleon scattering, and use the *same function* for the evaluation of off-energy shell matrix elements. Implicit in this prescription is the assumption that a potential of the functional form $v(|\mathbf{r}_1 - \mathbf{r}_2|)$, is really adequate to describe the nucleon–nucleon interaction both on and off the energy shell. This need not be the case. The interaction may, for instance, depend on the relative momenta of the two particles; or there may be other complications. A more fundamental theory of the inter-

*"Energy shell" refers to the subspace, in momentum space for which $E_1 + E_2 = \text{constant}$.

actions among particles will eventually give us a unique prescription for all matrix elements of v_{ij} or its equivalent, both on and off the energy shell. Until such a theory becomes available we shall satisfy ourselves with some simple forms of v_{ij}, guided by whatever other information we may have, and keeping in mind the above limitations (see Section VII.17).

Let us now come back to the evaluation of nuclear binding energies and their equilibrium density. We have already seen, (2.12), that the average kinetic energy in a Fermi gas is given by

$$\langle T \rangle = \tfrac{3}{5}\epsilon_f = \frac{3}{5}\cdot\frac{\hbar^2 k_F{}^2}{2M}$$

Using (2.9) to express k_F in terms of the nuclear density ρ, we obtain

$$\langle T \rangle = \frac{3}{10}\left(\frac{3\pi^2}{2}\right)^{2/3}\frac{\hbar^2}{M}\rho^{2/3} \tag{7.3}$$

The increase of the kinetic energy with density is easy to understand if we recall the "Pauli repulsion" among nucleons. Because of this repulsion an increase in density is accompanied by an increase in the density of nodes in the wave function and therefore an increase in its curvature; thus the average curvature is of the order of magnitude of the distance between the nucleons, that is, $\rho^{-1/3}$. The kinetic energy is quadratic in the inverse curvature of the wave function and, hence, the factor $\rho^{2/3}$.

To obtain an estimate of the equilibrium density of nuclei we have to know also the density dependence of the potential energy. We can estimate this in the following way: take the interaction between two nucleons to be represented by an attractive square well of range b and depth $-V_0(V_0 > 0)$. At a given density there is a probability p for a nucleon to be within the range of forces of another prescribed nucleon. The total contribution to the potential energy will therefore be

$$\langle V \rangle = \frac{-A(A-1)}{2}pV_0 \tag{7.4}$$

The probability p can be easily estimated for large nuclei; indeed to the extent that we can neglect surface effects it is just the ratio of the interaction volume to the total volume, that is,

$$p = \frac{(4/3)\pi b^3}{\Omega} = \frac{4}{3}\pi\frac{1}{A}b^3\rho \qquad \text{for} \qquad \Omega \gg b^3 \tag{7.5}$$

For finite nuclei p can be expressed in terms of the θ function defined by:

$$\theta(x) = \begin{cases} 1 & \text{if} \quad x > 0 \\ 0 & \text{if} \quad x < 0 \end{cases} \tag{7.6}$$

Then

$$p(b, R) = \frac{1}{\Omega^2} \int \theta(b - |\mathbf{r}_1 - \mathbf{r}_2|) d\mathbf{r}_1 \, d\mathbf{r}_2 \tag{7.7}$$

Note that $d\theta/dx = \delta(x)$, so that (7.7) can be obtained by integrating (7.2), after proper transformations. Upon integration we obtain

$$p(b, R) = \begin{cases} \left(\frac{b}{R}\right)^3 \left[1 - \frac{9}{16}\frac{b}{R} + \frac{1}{32}\left(\frac{b}{R}\right)^3\right] & \text{if} \quad R > \frac{b}{2} \\ 1 & \text{if} \quad R < \frac{b}{2} \end{cases} \tag{7.8}$$

where R is the nuclear radius—that is, $\Omega = (4/3)\pi R^3$. For $R \to \infty$, (7.8) goes over to (7.5) as expected.

Introducing first the approximate form of p, (7.5), into (7.4) we find that

$$\langle V \rangle = -\tfrac{2}{3}\pi \cdot Ab^3 V_0 \rho \qquad \text{for} \qquad R \gg b \tag{7.9}$$

where we have taken $A \gg 1$. The total energy, in this approximation, takes the form

$$\langle E \rangle = \frac{3A}{5}\left(\frac{3\pi^2}{2}\right)^{2/3} \frac{\hbar^2}{2M} \rho^{2/3} - \tfrac{2}{3}\pi A b^3 V_0 \rho \tag{7.10}$$

where the total average kinetic energy was taken to be A times the average kinetic energy (7.3). We see that $\langle E \rangle$ in (7.10) has no minimum as a function of ρ. Thus, in this approximation the lowest energy is $\langle E \rangle = -\infty$; the nucleus has an infinite density and would collapse to a point. This is inconsistent with the approximation (7.5) that assumed $R \gg b$. We therefore employ the exact expression (7.8) for $p(b, R)$. Notice that it deviates from the approximate form (7.5) as soon as b is smaller than $2R$. Using (7.5) in this region shows $\langle E \rangle$ increasing with large R to a maximum. For $R < b/2$ and using (7.8) to calculate $\langle E \rangle$ we find that $\langle E \rangle$ increases with decreasing R. $\langle E \rangle$ will thus have a minimum at $R \approx b/2$. See Fig. 7.2. As can be seen from the approximate form of E, (7.10), it also has a maximum at some large value of R.

We reach the following conclusion: if the density dependence of $\langle V \rangle$ is given by (7.4), where the density dependence of p is given by (7.8), then nuclei will fall apart if their density is below a certain critical low density, and once they are compressed beyond this critical density they will collapse into a sphere of radius $\sim b/2$, b being the range of the nuclear attractive interaction. The density will then become $\rho \sim \dfrac{A}{(4/3)\pi(b/2)^3}$ and the total energy $\langle E \rangle$, (7.10) will be proportional to A^2.

This conclusion is in violent disagreement with experiment. Nuclei are experimentally known to have more or less a constant *density*, but certainly not

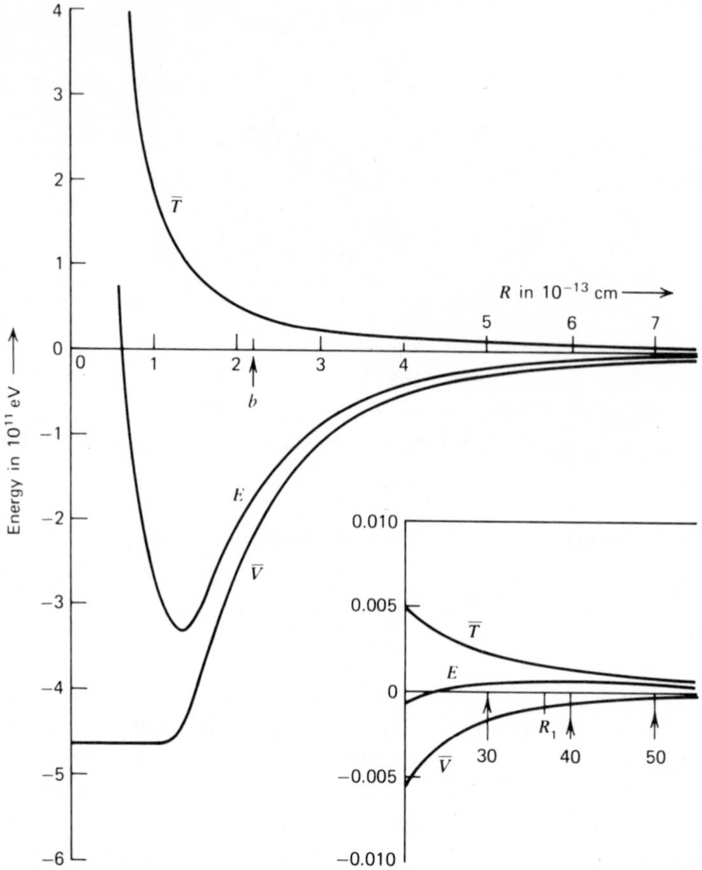

FIG. 7.2. A schematic picture of the average potential energy $\bar{V} = \langle V \rangle$, kinetic energy $\bar{T} = \langle T \rangle$ and total energy $E = \langle E \rangle$ as a function of nuclear radius R. R_1 is the point in which E attains its maximum value [taken from Blatt and Weisskopf (52)].

a constant radius. Also, their binding energy is roughly proportional to A, but certainly not proportional to A^2. Something must be wrong with our arguments, and it is not too difficult to see where it went wrong. Indeed, as long as we assume that the nucleon–nucleon interaction is purely attractive within a range b, we are bound to have a collapsed nucleus of this radius. The Pauli repulsion does tend to keep nucleons apart, but having only a $\rho^{2/3}$ dependence it is not sufficient to overcome the gain in potential energy, which is proportional to ρ, if all the nucleons are within the range of their forces. It is also clear why such nuclei will not collapse to sizes smaller than b; a decrease in size beyond b is accompanied by a further increase in kinetic energy without a compensating gain in potential energy.

It is clear that the attractive square well adopted for the nucleon–nucleon interaction is at fault. We must take into account the repulsive parts of the nuclear interaction. The equilibrium density observed in nuclei must be the combined effect of the Pauli repulsion, the dominant nucleon–nucleon attraction, and the particular features of the nucleon–nucleon repulsion. In fact, the analysis of nucleon–nucleon scattering does show (see Section (*I*.3)) the presence of a strong repulsion when the nucleons are sufficiently close. Its clearest manifestation is the vanishing of the ^1S-phase shift at a laboratory energy of about 200 MeV. As we shall show below, this leads to equilibrium density that is independent of A. Indeed, we saw that as the density increases, k_F increases as well. The distribution of relative momenta in a Fermi gas extends up to k_F, so that an increase in density leads to the introduction of higher components of relative momenta into the wave function. If a part of the nuclear force becomes repulsive at certain high relative momenta, it will not pay energy-wise for the nucleus to increase its density further, and we can therefore expect that the nucleus will stop collapsing at a corresponding density. At any rate, since the two-particle relative momentum distribution in a system of A-fermions depends on the *density*, rather than the *size*, of the nucleus, we see why the onset of nuclear repulsion at a given relative momentum would lead to stable nuclei of a given *density* rather than of a given *size*.

Let us try to estimate this density somewhat more quantitatively. The main effect of a repulsive core on the many-nucleon system is the introduction of a further correlation between the nucleons. The Pauli repulsion gave two identical nucleons a zero probability of being at exactly the same place, but a repulsive core extends this absolutely "forbidden zone" to a finite sphere around the center of each nucleon. The nucleus now looks in a way, like a Swiss cheese with "holes" around each of the nucleons. For a given density, the nuclear wave function has got now less space to spread itself over, and it is therefore forced into a bigger curvature. This increases the kinetic energy counteracting the tendency of the attractive interaction to make the nucleus collapse.

We have already noticed that the kinetic energy is proportional to $\rho^{2/3}$, or inversely proportional to r_0^2, where $2r_0$ is the average spacing between nucleons [$(4\pi/3)r_0^3$ is the volume per particle]. If r_c is the radius of the repulsive core then the kinetic energy should go to infinity as the spacing between the nucleons becomes equal to r_c.

We thus take for the kinetic energy of the A-particles the approximate form that reduces to (7.3) when r_c is zero!

$$T = A\langle \bar{T} \rangle = \frac{\alpha A}{(r_0 - r_c)^2} \qquad \text{where} \qquad \alpha = \frac{3}{5}\frac{\hbar^2}{2M}\left(\frac{9\pi}{8}\right)^{2/3} \approx 30 \text{ MeV (fm)}^2 \tag{7.11}$$

*A more careful analysis shows that $T = \alpha A/(r_0 - 0.8r_c)^2$; we shall, however, disregard it here. [See Huang and Yang (57)].

For the average potential energy we take again (7.4) with p given by (7.8). Noting that $R = r_0 A^{1/3}$ we obtain the expression for the average potential energy (we put $A - 1 \approx A$):

$$\langle V \rangle = -V_0 \frac{A}{2}\left(\frac{b}{r_0}\right)^3 \left[1 - \frac{9}{16}\frac{b}{R} + \frac{1}{32}\left(\frac{b}{R}\right)^3\right] \tag{7.12}$$

To obtain now the equilibrium density, we want to minimize $\langle T \rangle + \langle V \rangle$ with respect to r_0. This leads to the equation:

$$\frac{3V_0 b^2}{4\alpha}\left(\frac{b}{r_0}\right)\left[1 - \frac{3}{4}\frac{b}{r_0 A^{1/3}} + \frac{1}{16}\left(\frac{b}{r_0 A^{1/3}}\right)^3\right]\left[1 - \left(\frac{r_c}{r_0}\right)\right]^3 = 1 \tag{7.13}$$

The product $V_0 b^2$ is known from the low-energy scattering data, and to good approximation

$$V_0 b^2 = 100 \text{ MeV (fm)}^2$$

With the value of α given by (7.11) we can proceed to solve (7.13) numerically for r_0 in terms of r_c, b, and A. With $b = 1.8$ fm and $r_c = 0.4$ fm one then obtains values of r_0 that range between 0.9 fm and 1.5 fm as A changes from 4 to 216. The experimental value of r_0 is around 1.2 fm. We see therefore that even this crude model for the average kinetic-potential energies is sufficient to establish a "scale" for nuclear sizes based on the radius of the hard core, and the interaction radius. The average spacing between nucleons or more precisely the constant r_0 that relates the nuclear radius to its atomic number $R = r_0 A^{1/3}$, turns out to be about three times the hard core radius.

A close inspection of (7.13) allows us also to establish various limits on V_0, b, and r_c that determine whether a bound state exists at all. If there is a (real) solution to (7.13), then there are generally two solutions: the smaller solution $r_{0<}$ corresponds to the equilibrium density and the bigger solution represents that value of $r_{0>}$ beyond which the nucleons will not form a bound state. In other words, if we were to bring nucleons together with an average separation bigger than $r_{0>}$ and leave them there, they will fall apart so as to reduce their kinetic energy; if, however, we were to compress them so that their average spacing became smaller than $r_{0>}$, the attractive forces between them will overcome the repulsive effects of the kinetic energy, and the nucleons will eventually settle at a density determined by $r_{0<}$.

Extensive numerical calculations [see Bethe (71) for a survey of this work] have shown that the effects of the repulsive core and the exclusion principle do not suffice to establish the correct value of equilibrium density. The effect of the tensor force and the exchange potential turns out to be the required additional critical elements. As a function of the density, the contribution of the tensor force to the binding energy decreases with increasing density. This phenomenon is well known from early calculations comparing ^{2}H and ^{3}H binding [Inglis (39), Pease and Feshbach (52)]. In the absence of tensor forces the ground state of ^{2}H is $^3S_1$. The tensor force coupled the $^3S_1$ with a $^3D_1$ state. A very similar situation prevails for ^{3}H. The ground state is principally $^2S_{1/2}$ that by virtue of the tensor force couples with $^4D_{1/2}$ states. This coupling is, however, much less effective in ^{3}H because it is a system much smaller than ^{2}H. As a consequence the unperturbed D state is located at a much greater energy in ^{3}H than in ^{2}H, the angular momentum barrier being much more effective in ^{3}H. Thus the contribution of the tensor force to the binding energy of the denser system ^{3}H is relatively smaller.

Much the same effect occurs in nuclear matter with the Pauli exclusion principle playing a role similar to that of the angular momentum barrier. In this case the tensor forces can couple the ^3S states in nuclear matter, only to unfilled ^3D levels lying above k_F. As the density increases so does k_F and, therefore, the energy denominator of the second-order perturbation theory increases with increasing density. Hence the tensor force becomes less effective as the density of nuclear matter increases.

This effect is illustrated in Fig. 7.3. The ^1S contribution to the binding energy of nuclear matter does not involve the tensor force. As k_F and the density increase, its contribution increases in magnitude. It does not saturate. Contrast this behavior with that of the ^3S whose growth with k_F parallels that of the ^1S for low k_F but as k_F grows the rate of increase decreases and finally the ^3S contribution turns around and begins to decrease. This is the effect of the tensor forces.

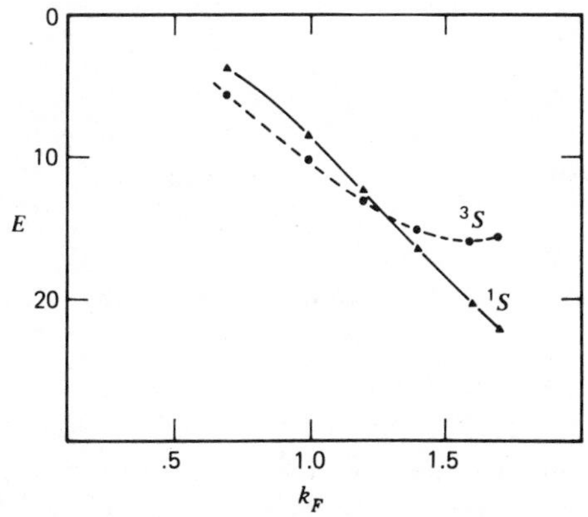

FIG. 7.3. Contribution to the binding energy per particle in nuclear matter of the ^1S and ^3S states as a function of the Fermi momentum k_F according to Sprung (quoted by Bethe (71)].

The effect of the two-body exchange potential can be illustrated by choosing the appropriate terms from OPEP (Eq. I. 3.5):

$$V_{\sigma\tau} = \frac{1}{3}\frac{g^2}{\hbar c} m_\pi c^2 (\boldsymbol{\sigma}_1 \cdot \boldsymbol{\sigma}_2)(\boldsymbol{\tau}_1 \cdot \boldsymbol{\tau}_2) Y(\mu r_{12}) \tag{7.14}$$

where

$$Y(x) \equiv e^{-x}/x$$

For spin singlets $(\boldsymbol{\sigma}_1 \cdot \boldsymbol{\sigma}_2) = -3$, spin triplets $= 1$. Therefore for the two-body state ($T = 1$, $S = 0$) such as the ${}^1S_0$ (Table I.6.1)

$$V_{\sigma\tau} = -3\left(\frac{1}{3}\frac{g^2}{\hbar c} m_\pi c^2\right) Y(\mu r_{12}) \qquad (T = 1,\ S = 0)$$

an attractive force. The other T, S combinations are shown in Table 7.1.

TABLE 7.1 Value of $\boldsymbol{\sigma}_1 \cdot \boldsymbol{\sigma}_2\ \boldsymbol{\tau}_1 \cdot \boldsymbol{\tau}_2$ for Various Values of the Total Spin S and Isospin T

(S, T)	State	Value of $\boldsymbol{\sigma}_1 \cdot \boldsymbol{\sigma}_2\ \boldsymbol{\tau}_1 \cdot \boldsymbol{\tau}_2$	Nature of Potential
$S = 0, T = 1$	${}^1S_0$, ${}^1D_2$, etc.	-3	Attractive
$S = 1, T = 1$	${}^3P_{0,1,2}$, etc.	1	Repulsive
$S = 0, T = 0$	${}^1P_1$, ${}^1F_2$, etc.	9	Repulsive
$S = 1, T = 0$	${}^3S_1$, ${}^3D_{1,2,3}$, etc.	-3	Attractive

From Table 7.1 it is clear that the ${}^1P_1$ state is strongly repulsive. This feature remains when the full nuclear force is considered, and is borne out by the experimental nucleon–nucleon scattering. These terms thus contribute an additional repulsive energy that is required for saturation at the observed value of the density. The contribution of both the S and P states to the potential energy per particle is shown in Fig. 7.4, where the repulsive nature of the latter is clearly seen.

In conclusion, we see that saturation and the particular value of the density realized in nuclei is the consequence of many effects, that of the Pauli exclusion principle, the repulsive core, the behavior of the tensor force contribution as a function of density, and the repulsive nature of the potential acting between particles in relative P states. These compensate for the attractive forces such as those in the S states that by themselves would lead to a collapse of the nucleus to very small radii.

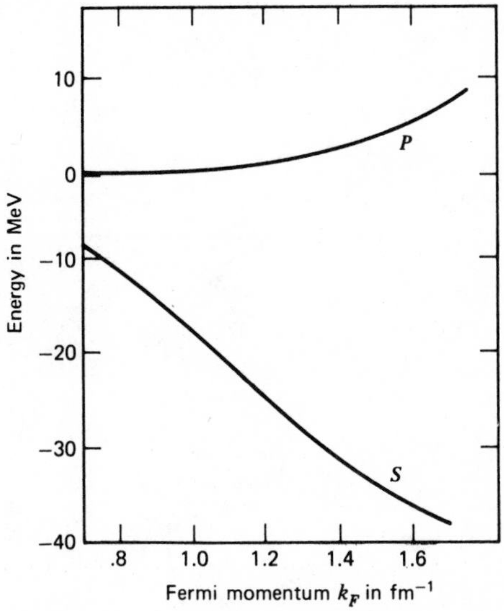

FIG. 7.4. Contribution of the *S* and *P* states to the binding energy per particle in nuclear matter as a function of the Fermi momentum k_F according to Sprung [quoted by Bethe (71)].

CHAPTER III

THE PROPERTIES OF NUCLEAR MATTER

It is most remarkable that, in spite of the relatively strong nucleon-nucleon interaction, a description of the nucleus as consisting of noninteracting nucleons moving in a common potential provides a rough but reasonable approximation to nuclear properties. In the preceding chapter the simplest possible potential, that of a box with infinitely repulsive walls was used. Even in that case we found that, by choosing the radius of the box so as to obtain the observed density, the prediction of various nuclear properties were at most a factor of two away from their observed values. In later chapters a more sophisticated common potential will be used but the surprising conclusion remains: that the major effect of the strong nuclear force is to fix the density. The nuclear forces and the Pauli principle act so as to provide the common potential well that with the action of the Pauli principle determines the nuclear density. The residual interaction, the difference between the nucleon–nucleon interaction and the common potential, may be treated as a perturbation. Although the perturbation is not always "small," the underlying picture of the nucleons moving independently remains. To be sure these "small effects are responsible for many of the striking features of nuclear structure and their analysis forms the subject matter of a greater part of this book. In this chapter our primary concern will be with the question:

Why is it that the nucleons in a nucleus exhibit such a high degree of independence?

The discussion in this chapter will emphasize the physical concepts involved; many of the arguments used will be intuitive. A more precise and more rigorous analysis is presented in Chapter VII.

The independent-particle picture and the saturation exhibited by the Weizsäcker semiempirical formula (II.3.1) are closely connected. If a nucleus consists of A noninteracting nucleons moving in a common potential, its binding energy will in the first approximation be proportional to A. It is the nucleon–nucleon interaction acting between *all* the particles which would give rise to a binding energy proportional to $A(A - 1)$. In Section II.7 we gave an intuitive explanation of this phenomenon. In this chapter it is necessary to deepen our understanding and to establish a connection and thereby the reason for the validity of the independent particle picture for nuclei.

The term in the semiempirical mass formula of interest is the one with the a_1 coefficient. It is the one proportional to A. The other terms, the surface energy (a_2), the Coulomb energy (a_3), the symmetry energy (a_4) have been discussed in Chapter II and the origin of their A and Z dependence is reasonably well understood. It would therefore be useful to discuss a physical situation in which the a_1 term is isolated. If the semiempirical mass formula is valid for all A, this will occur for A infinite. In this limit the binding energy per particle equals a_1.

Of course it is possible to question this extrapolation. The extrapolation does seem reasonable in virtue of the excellent fit to the known binding energies of the whole known periodic table obtained with the semiempirical mass formula.

The reason for the finite limit on the charge of stable nuclei is believed to be known: it concerns the competition between the repulsive Coulomb forces and the attractive nuclear forces; since the former have an infinite range wheres the range of nuclear forces is finite, it is obvious that when the nucleus radius gets to be several times the range of nuclear forces, the Coulomb force will overcome the nuclear attraction and the nucleus will split. Nuclei thus become unstable against fission into smaller components and a natural limit to sizes of stable nuclei is set. It is interesting to estimate roughly when nuclei become unstable against fission. We know from the mass formula (Eq. II.3.1, the a_1 term) that the nuclear binding energy per particle is about 16 MeV. It is obvious that once the repulsive Coulomb energy per particle becomes of the same order of magnitude, nuclei will certainly become unstable. The Coulomb energy of a uniformly charged sphere of radius R is $(3/5)(Z^2e^2/R) = (3/5)\cdot(Z^2e^2/r_0A^{1/3})$. Hence we expect the limit of stability to satisfy

$$\frac{3}{5}\frac{Z^2e^2}{r_0A^{4/3}} < 16 \text{ MeV}$$

where $r_0 = 1.2$ fm, that is,

$$\frac{Z^2}{A^{4/3}} < 25$$

Actually we shall see later that considerably lower limits can be put on the size of the biggest nuclei that are stable against fission; but we note that for the heaviest known stable nucleus—^{238}U—we have $Z^2/A^{4/3} \sim 6$, well within the limit derived above.

From the above considerations it is apparent that if we are willing to disregard Coulomb effects we are probably justified in supposing the possible existence of very large nuclei. One should, however, bear in mind the possibility that a "correct" mass formula may include additional and more complicated functions of A with very small coefficients. Such terms will not affect the masses of known nuclei but may be important in infinite nuclei. The possible existence of neutron stars makes such considerations not too academic, but we shall not enter into them here.

To stress that we are dealing with a somewhat hypothetical system, a special name has been given to it: *nuclear matter* refers to an infinite system of nucleons—neutrons and protons—interacting with the full nuclear interaction, but with their electromagnetic interaction switched off. To be even more specific, the masses of the neutron and the proton are assumed to be equal, so that in nuclear matter there is no way of telling the difference between a neutron and a proton: they are just the two possible states of a nucleon. The particles in nuclear matter are thus nucleons, which have two internal degrees of freedom—spin $\boldsymbol{\sigma}$ and isospin $\boldsymbol{\tau}$; each nucleon can be found in one of four internal states (spin up—isospin up, spin up—isospin down, etc.), and the total wave function is required to be antisymmetric with respect to the exchange of all the nucleons.

1. THE INDEPENDENT-PARTICLE APPROXIMATION

Formally the problem is best defined as that of N-nucleons in a box of volume Ω, satisfying the Schrödinger equation

$$\left(\sum_{i=1}^{N} T_i + \frac{1}{2}\sum_{i\neq j}^{N} v(ij)\right)\Psi_\alpha(1, \ldots, N) = E_\alpha\Psi_\alpha(1, \ldots, N) \quad (1.1)$$

where $\Psi_\alpha(1, \ldots, N)$ is required to satisfy periodic boundary conditions, and is taken to be normalized in the box:

$$\int \Psi_\alpha^*(1, \ldots, N)\Psi_{\alpha'}(1, \ldots, N)d(1)\ldots d(N) = \delta(\alpha, \alpha') \quad (1.2)$$

The integral in (1.2) stands for integration over space coordinates and summation over spin and isospin variables. The results for whatever quantity is calculated are then evaluated in the limit $N \to \infty$ and $\Omega \to \infty$ taken so that the density remains unchanged

$$\lim_{\substack{N\to\infty \\ \Omega\to\infty}} \rho\left(\equiv\frac{N}{\Omega}\right) = \rho_0 \quad (1.3)$$

It is in this limit that we shall discuss the various properties of nuclear matter.

In more refined treatments of infinite quantum mechanical systems in their ground states there is another limit to worry about, that of $T \to 0$ where T the temperature of the system. Our physical systems are all studied at finite temperatures, and are therefore to be found distributed among various states according to the Boltzmann distribution. Due, however, to the smallness of the Boltzmann constant

$$k = 8.6 \times 10^{-5} \text{ eV } (\text{deg K})^{-1}$$

we see that even at room temperature the probability of a state 1 eV above the ground state to be occupied as a result of thermal equilibrium is about e^{-40} and thus negligibly small. Therefore in calculating various properties of real nuclei, it is perfectly legitimate to go to the limit $T \rightarrow 0$ at any stage. For large nuclei, however, the excited states lie at lower energies the larger the system is. For a system of linear dimensions of 1 cm, the characteristic excitations lie at an energy of $\Delta E = (1/M)(\hbar/1 \text{ cm})^2 \sim 10^{-17}$ eV. Since now $kT > \Delta E$ even for temperatures of 10^{-6} °K, the proper procedure really is to evaluate the various properties of the system at *finite* temperature, go first to the limit $\Omega \rightarrow \infty$, and only then to the limit $T \rightarrow 0$. In our discussions in this chapter we shall disregard the temperature and calculate nuclear matter properties in the *ground state* of (1.1) for finite Ω; this actually means that we take the limit $T \rightarrow 0$ first. To the extent that the two limiting processes are interchangeable that really does not matter, but one must watch out for strange effects that may be due to the noninterchangeability of the limits $\Omega \rightarrow \infty$ and $T \rightarrow 0$.

Returning to (1.1) we begin by asking for its solution in the independent-particle approximation. When the interaction $v(ij) = 0$ this solution consists of independent particles; the spatial dependence of each is described by a plane wave. To simplify notation we shall refer to the different quantized states allowed in the box by greek letters $\alpha, \beta, \ldots$ where each letter stands for both the momentum associated with that state as well as the z-component of the spin and third component of the isospin:

$$\alpha \equiv (\mathbf{k}_\alpha, m_{s\alpha}, m_{\tau\alpha}) \tag{1.4}$$

In the Fermi gas approximation a single-particle wave function is then given by

$$\phi_\alpha(k) \equiv \frac{1}{\sqrt{\Omega}} \chi(m_{s\alpha}, m_{\tau\alpha}) \exp(i\mathbf{k}_\alpha \cdot \mathbf{r}_k) \tag{1.5}$$

where $\chi(m_{s\alpha}, m_{\tau\alpha})$ is the spin-isospin wave function. These functions form a complete orthonormal set. The N-particle wave function is given, still in this approximation, by the Slater determinant

$$[\Phi_\alpha(1, \ldots, N)] = \frac{1}{\sqrt{N!}} \det |\phi_{\alpha_1}(1), \phi_{\alpha_2}(2), \ldots . \phi_{\alpha_N}(N)| \tag{1.6}$$

where $\alpha_1, \ldots, \alpha_N$ are the occupied states (the N-lowest states if $[\Phi_\alpha]$ is to be the ground state of the system). The bracket will be used to indicate antisymmetrization.

We consider now the independent-particle approximation in the presence of the interaction. The independent-particle wave function is still of the form (1.6). Because we are considering an "infinite" nucleus, it follows immediately that the single particle wave functions $\phi_\alpha(k)$ are given, even in the presence of an interaction, by (1.5). This is a major simplification, one which is characteristic for nuclear matter. The reason lies in the translational invariance of the nuclear Hamiltonian in this case. This means that the Hamiltonian does not change when the origin of the coordinate system is shifted. It is thus possible to find a solution of (1.1) that upon shifting the origin remains unchanged except for a change in phase that does not depend upon the particle coordinates. This condition for a product wave function can be satisfied only by a product of exponentials. The Pauli principle then immediately leads to the antisymmetrized wave function (1.6). Under the transformation

$$\mathbf{r}'_\alpha = \mathbf{r}_\alpha + \boldsymbol{\delta}_\alpha$$

this wave function is multiplied by the factor

$$\exp\left(i \sum_\alpha \mathbf{k}_\alpha \cdot \boldsymbol{\delta}_\alpha\right)$$

Thus the independent-particle wave function is unchanged even in the presence of an interaction. What does change is the relation between the single-particle energy and the momentum $\mathbf{k}_\alpha$ (see Eq. 1.10 below) and of course the energy of the system as a whole. (See (1.13) below.) In the following discussion we shall describe the nature of these changes.

In the no interaction case, ϕ_α is a solution of

$$T\phi_\alpha = \epsilon'_\alpha \phi_\alpha \tag{1.7}$$

where T is the kinetic energy operator and

$$\epsilon'_\alpha = \frac{\hbar^2}{2M} k_\alpha{}^2 \tag{1.8}$$

In the presence of an interaction, ϕ_α will satisfy a single particle Schrödinger equation with a translationally invariant single particle Hamiltonian. This equation must be of the form

$$T\phi_\alpha + \int_{-\infty}^{\infty} U(\mathbf{r} - \mathbf{r}')\phi_\alpha(\mathbf{r}')d\mathbf{r}' = \epsilon_\alpha \phi_\alpha \tag{1.9}$$

It is easy to verify that the plane wave (1.5) is a solution of this equation and

$$\epsilon_\alpha = \epsilon'_\alpha + U(\alpha) \tag{1.10}$$

where

$$U(\alpha) = \int \exp(-i\mathbf{k}_\alpha \cdot \mathbf{r}) U(\mathbf{r})\, d\mathbf{r}$$

Obviously U depends upon k_α so that the relation between the energy ϵ_α and k_α is no longer necessarily quadratic as in (1.8).

It remains to determine $U(\alpha)$. This is done by inserting the wave function (1.6) into the Schrödinger equation for the system (1.1), multiplying from the left by

$$\Phi^*(2, 3, \ldots, A) \equiv \phi^*_{\alpha_2}(2)\phi^*_{\alpha_3}(3)\ldots\phi^*_{\alpha_A}(A)$$

and integrating over coordinates (2) to (A):

$$\int d(2)d(3)\ldots d(A)\Phi^*(2, 3, \ldots, A)[\Sigma T_i + \tfrac{1}{2}\Sigma v_{ij}]\Psi_\alpha(1, 2, \ldots, A)$$
$$= E_\alpha \int d(2)d(3), \ldots, d(A)\Phi^*(2, 3, \ldots, A)\Psi_\alpha(1, 2, \ldots, A)$$

Note that Φ is *not* antisymmetrized and that the normalization of Ψ_α is chosen to be:

$$\int d(2)d(3)\ldots d(A)\Phi^*\Psi_\alpha = \phi_{\alpha_1}(1)$$

It is now simple to recover (1.10) and determine the potential energy term. It is

$$U(\alpha_1)\phi_{\alpha_1}(1) = \int d(2)d(3)\ldots d(A)\Phi^*\left(\sum_{i=2}^{N} v(i, 1)\right)\Psi_\alpha \quad (1.11)$$

The factor 1/2 no longer appears because both v_{ij} and v_{ji} are included in the original sum.

For local interactions $v(ij)$, $U(\alpha_1)$ cannot depend on the coordinate $\mathbf{r}_1$ because we are dealing with an infinite nucleus, and the potential felt by a particle in the state α_1 is the same irrespective of where it is. But $U(\alpha_1)$ may, and generally does, depend on the momentum $\mathbf{k}_{\alpha_1}$ and the quantum numbers $m_{s\alpha_1}$ and $m_{\tau\alpha_1}$. More precisely, it depends on the remaining quantum numbers $\alpha_2, \ldots, \alpha_N$ of all the other $N - 1$ states, but since we assumed that we were dealing with a given set of occupied states, this is the same as specifying α_1. Returning to (1.10) $\mathbf{k}_\alpha$ is still determined by the periodic boundary conditions of the box, and its allowed values are the same in both (1.8) and (1.10). Thus $U(\alpha)$ is determined once the nucleon density is given. However the energy associated with this state is changed as a result of the interaction. Since the energy ϵ'_α determines the frequency of the time oscillations of the single-particle wave function and $1/k_\alpha$ determines its wavelength, we see that (1.10) is a dispersion formula for a particle moving in a medium in which it interacts with other particles.

The total energy of a Fermi gas is given by

$$E'_\alpha = \sum_{i=1}^{N} \epsilon'_{\alpha i} \quad (1.12)$$

where ϵ'_α is given by (1.8). If we take now the interactions into account in the way outlined in (1.9) we should be more careful. The potential $U(\alpha)$ is that felt by particle in the state α due to all the other particles. If we sum it over all states each particle will appear once as a "source" of the interaction and once as the particle acted upon, and we shall be counting these interactions twice. We therefore find that in the approximation represented by (1.9)

$$E_\alpha = \sum_{i=1}^{N} \epsilon'_{\alpha i} + \frac{1}{2}\sum_{i=1}^{N} U(\alpha_i) \tag{1.13}$$

This energy E_α for the wave function (1.6) is just the expectation value of the Hamiltonian in (1.1) with respect to Ψ_α as can be ascertained by a direct evaluation.

Equation 1.13 is the best we can do with an independent-particle wave function. We have not really proved here that it is the best approximation to the energy (this will be shown in the next chapter when we discuss the Hartree–Fock approximation), but the physical arguments that led us to (1.13) support this conjecture.

2. THE INDEPENDENT PAIR APPROXIMATION; THE BETHE-GOLDSTONE EQUATION

To improve on (1.13) we have to go beyond the independent-particle model. In terms of the formal treatment it means that we cannot any longer define a unique set of states as being fully occupied and the others completely empty. Rather we would like to consider a number of Slater determinants $[\Phi_\alpha]$, $[\Phi_\beta]$, . . . constructed out of different sets of N single-particle states $(\alpha_1, \ldots, \alpha_N)$, $(\beta_1, \ldots, \beta_N)$ etc; we can then produce an improved wave function by taking linear combinations of $[\Phi_\alpha]$, $[\Phi_\beta]$, etc.

The physics behind this step is also quite obvious: the N-particles moving in their orbits collide with each other from time to time [that is the meaning of the existence of an interaction $v(ij)$]. If we confine ourselves from the outset to just one Slater determinant we give the particles no possibility of colliding and being scattered from one state to another, since we put at the disposal of the N-particles exactly N-states. The inclusion of other wave functions Φ_β, . . . in our description of the system is equivalent to opening up the possibility for the particles to collide and scatter into states different from their original ones.

In determining which additional Slater determinants ought to be included, and with what amplitudes, lies the essence of the various approximations to nuclear structure. We shall deal in later chapters with a number of different approximations and discuss their justification. In this chapter we want to concentrate on a specific approximation known as the *Independent Pair Approximation*. This approximation is motivated by two considerations:

(a) The success of the independent-particle model in explaining some of the nuclear data (see Chapter II) suggests that the next important correction has to do with two-particle correlations, ignoring the effects of higher-particle correlations.

(b) The possible existence of a *hard core* in the nucleon-nucleon potential requires a better determination of the two body part of the N-particle wave function. Indeed for this case, $U(\alpha)$ of (1.11) is infinite. In particular we would like the better N-particle wave function $\Phi_\alpha(1, \ldots, N)$ to vanish whenever any two particles come within a distance $r < r_c$ of each other, where r_c is the radius of the hard core. Otherwise, some of the matrix elements involving $v(ij)$ will diverge. We saw in Section II.7 that a wave function that has this restriction built into it may give rise to a nuclear density of a reasonable order of magnitude.

As a bonus, this approximation should also indicate why the independent-particle model turns out to be as successful as it is.

The independent-pair approximation focuses its attention on the motion of a pair. It asserts that except for restrictions originating in the Pauli principle and to be described below, the remainder of the nucleus acts as a "spectator." There are two limiting situations to bear in mind. First, there is the independent-particle wave function for the pair that holds in the absence of the interaction v_{12}. The un-antisymmetrized form is

$$|\alpha\beta\rangle \equiv \phi_\alpha(1)\phi_\beta(2)$$

The antisymmetrized form is

$$\sqrt{2}|[\alpha\beta]\rangle \equiv [\phi_\alpha(1)\phi_\beta(2)] \equiv [\phi_\alpha(1)\phi_\beta(2) - \phi_\beta(1)\phi_\alpha(2)]$$

The interaction will modify $\phi_\alpha(1)\phi_\beta(2)$ by a correlation function, that is a function of (1) and (2) and also of α and β. The determination of this modification is the objective of the discussion that follows. Second, in the absence of the rest of the nucleus, that is, under "free field" conditions, the two-particle wave function satisfies the two-body Schrödinger equation.

Consider the two-particle Schrödinger equation

$$(T_1 + T_2 + v_{12})\psi(12) = E\psi(12) \tag{2.1}$$

We ask how is this equation modified by the fact that the two particles are inside nuclear matter. We want to find a solution of that modified equation*

*In (2.1) $\psi(12)$ is assumed not to be antisymmetrized. If we look for antisymmetrized solutions denoted by $[\psi_{12}]$ then we ask for the modification of $[\phi_\alpha(1)\phi_\beta(2)]$ part of $[\Phi_\alpha]$. For further discussion see problem on p. 165.

that could replace a given two-particle part $\phi_\alpha(1)\phi_\beta(2)$ in the Fermi-gas wave function $[\Phi_\alpha]$ and represent an improvement over $[\Phi_\alpha]$ in the sense referred to at the end of last section. This is essential if there is a hard core in the nucleon-nucleon potential. The Fermi gas wave function $[\Phi_\alpha]$ does not vanish when $|\mathbf{r}_i - \mathbf{r}_j| < \mathbf{r}_c$ and, hence, for interactions with a hard core, $v_{12}\Psi_\alpha$ is infinite whenever $|\mathbf{r}_1 - \mathbf{r}_2| < r_c$. We may think of accounting properly for the two-particle correlations by utilizing somehow the exact scattering solution of (2.1). This solution, being exact, will make sure that $v_{12}\Psi(12)$ does not diverge for any value of $|\mathbf{r}_1 - \mathbf{r}_2|$. However, a two-particle scattering solution is not appropriate: if we expand it in terms of the independent-particle wave functions we find that it generally contains components in all states, including states that are occupied by particles other than the two particles 1 and 2 considered. In other words, in the presence of other nucleons, we cannot consider the interaction between 1 and 2, as just giving rise to a scattering identical with that which would result if the other nucleons were not present. This will violate the Pauli principle and will be of no use for us in our present problem. Actually what we want is to solve (2.1) with the additional restriction that whatever scattering states it leads to will be *outside* of the occupied states. We can even relax this condition somewhat: since we want to replace a two-particle part of Φ_α by a solution of (2.1), we can allow scattering also into the two occupied states in Φ_α that are to be replaced.

Let us concentrate on two occupied states α and β. We define a projection operator $Q_{\alpha\beta}(12)$ so that if $f(12)$ is any two particle function, then

$$Q_{\alpha\beta}(12)f(12) = |\alpha\beta\rangle\langle\alpha\beta|f\rangle + \sum_{\lambda\mu} |\lambda\mu\rangle\cdot\langle\lambda\mu|f\rangle \tag{2.2}$$

where $\lambda\mu$ run over all *unoccupied* states and

$$\langle\rho\kappa|f\rangle = \int \phi_\rho^*(1)\phi_\kappa^*(2)f(12)d(1)d(2)$$

$Q_{\alpha\beta}(12)$ is a projection operator that projects on to the unoccupied states and on to the specific state $\phi_\alpha(1)\phi_\beta(2)$. It is a hermitian operator as can be seen from the equality

$$\langle Qf|g\rangle = \langle f|Qg\rangle \tag{2.3}$$

and

$$Q_{\alpha\beta}^2 = Q_{\alpha\beta}$$

It also commutes with $T_1 + T_2$ and when written in coordinate space it is a nonlocal operator.

It follows from (2.2) that

$$Q_{\alpha\beta} = |\alpha\beta\rangle\langle\alpha\beta| + Q_F$$

$$Q_F \equiv \sum_{\lambda\mu} |\lambda\mu\rangle\langle\lambda\mu| \tag{2.4}$$

We now modify (2.1) so that only the states allowed by the Pauli principle can become involved in the scattering. The resulting equation is known as the *Bethe–Goldstone equation* [Bethe and Goldstone (57)]:

$$[T_1 + T_2 + Q_{\alpha\beta}(12)v(12)]\psi_{\alpha\beta}(12) = E_{\alpha\beta}\psi_{\alpha\beta}(12) \tag{2.5}$$

It is an equation designed for the description of the motion of two interacting fermions, in the presence of other fermions that affect particles 1 and 2 only via the Pauli principle. In the independent pair approximation this is the only way the spectator nucleons affect the two-particle wave function. The insertion of the operator $Q_{\alpha\beta}(12)$ to the left of $v(12)$ is meant to prevent $v(12)\psi_{\alpha\beta}(12)$ from leading to any of the occupied states. Thus if $\phi_{\gamma\delta}(12)$ is a two-particle Slater determinant involving occupied states different from $(\alpha\beta)$ (either both γ and δ are occupied or just one of them is), then it follows from (2.2) that

$$\langle\Phi_{\gamma\delta}(12)|\, Q_{\alpha\beta}v(12)\psi_{\alpha\beta}(12)\rangle = 0 \;\; (\gamma\delta) \text{ occupied and } \neq (\alpha\beta) \tag{2.6}$$

The potential in (2.5) is not hermitian since

$$[Q_{\alpha\beta}v(12)]^{\dagger} = v(12)Q_{\alpha\beta} \neq Q_{\alpha\beta}v(12)$$

This difficulty is only apparent. To see this note first that $\psi_{\alpha\beta}(12)$ may be decomposed into two components

$$\psi_{\alpha\beta} = Q_{\alpha\beta}\psi_{\alpha\beta} + (1 - Q_{\alpha\beta})\psi_{\alpha\beta} \tag{2.7}$$

The equations satisfied by the two components of $\psi_{\alpha\beta}$ in (2.7) can be obtained by inserting (2.7) into (2.5) and operating on both sides of that equation by $Q_{\alpha\beta}$ yielding the equation

$$(T_1 + T_2 + Q_{\alpha\beta}v_{12}Q_{\alpha\beta})Q_{\alpha\beta}\psi_{\alpha\beta} + Q_{\alpha\beta}v_{12}(1 - Q_{\alpha\beta})\psi_{\alpha\beta} = E_{\alpha\beta}Q_{\alpha\beta}\psi_{\alpha\beta} \tag{2.8}$$

Note that the commutativity of T_1 and T_2 with $Q_{\alpha\beta}$ has been used. If we multiply (2.5) from the left by $(1 - Q_{\alpha\beta})$ we obtain [note $(1 - Q_{\alpha\beta})Q_{\alpha\beta} = 0$]

$$(T_1 + T_2)(1 - Q_{\alpha\beta})\psi_{\alpha\beta} = E_{\alpha\beta}(1 - Q_{\alpha\beta})\psi_{\alpha\beta} \tag{2.9}$$

We see that $(1 - Q_{\alpha\beta})\psi_{\alpha\beta}$ satisfies the zero potential equation. Hence its solution is one of the many levels $|\gamma\delta\rangle$ that have the energy $E_{\alpha\beta}$. However the essential point is that the amplitude of the $(1 - Q_{\alpha\beta})\psi_{\alpha\beta}$ component is not determined by (2.9). Choosing that amplitude is thus one of the conditions needed to make the solution of the Bethe–Goldstone equation unique as can be directly seen from (2.9). The choice is made by requiring

$$(1 - Q_{\alpha\beta})\psi_{\alpha\beta} = 0 \tag{2.10a}$$

A consequence of (2.10a) is that in the filled states in $\psi_{\alpha\beta}$ can contain only $|\alpha\beta\rangle$. In that event (2.8) becomes

$$(T_1 + T_2 + Q_{\alpha\beta}v(12)Q_{\alpha\beta})\psi_{\alpha\beta} = E_{\alpha\beta}\psi_{\alpha\beta} \tag{2.10b}$$

and now the potential is hermitian.

The solution of (2.5) is not antisymmetrized since $Q_{\alpha\beta}$ is not symmetric. However it does satisfy the condition

$$\psi_{\alpha\beta}(12) = \psi_{\beta\alpha}(21) \tag{2.10c}$$

Problem. Prove that the antisymmetric combination

$$[\psi_{\alpha\beta}] = \psi_{\alpha\beta}(12) - \psi_{\beta\alpha}(12)$$

satisfies

$$(T_1 + T_2)[\psi_{\alpha\beta}] + [Q_{\alpha\beta}]v_{12}[\psi_{\alpha\beta}] = E_{\alpha\beta}[\psi_{\alpha\beta}] \tag{2.10d}$$

where

$$[Q_{\alpha\beta}] = |[\alpha\beta]\rangle\langle[\alpha\beta] + \Sigma|[\lambda\mu]\rangle\langle[\lambda\mu]|$$

where

$$|[\alpha\beta]\rangle \equiv \frac{1}{\sqrt{2}}[\phi_\alpha(1)\phi_\beta(2) - \phi_\beta(1)\phi_\alpha(2)] \equiv \frac{1}{\sqrt{2}}[\phi_\alpha\phi_\beta]$$

Prove that the condition replacing (2.10a) becomes

$$(1-[Q_{\alpha\beta}])[\psi_{\alpha\beta}] = 0 \tag{2.10e}$$

We have of course not derived (2.5). A derivation will be given in Section 4 and also in Chapter VII.

To summarize, then, we shall be looking for solutions of the Bethe–Goldstone equation (2.5), or its equivalent form (2.10b) that satisfy (2.10a). These solutions will then consist of $\phi_\alpha(1)\phi_\beta(2)$ plus a "correction" that has components only outside of the occupied states. It will thus represent a sort of a scattering state for the pair in the Fermi gas, which takes into account the effects of the Pauli principle by excluding scattering into the occupied states. Note that the antisymmetric solution $[\psi_{\alpha\beta}]$ can be obtained either by solving (2.10d) or by combining two solutions of (2.5), $\psi_{\alpha\beta}(12)$, and $\psi_{\beta\alpha}(12)$.

Since we shall consider (2.5) primarily for potentials $v(12)$ with a hard core, it may be worthwhile to look more closely into the behavior of $\psi_{\alpha\beta}(12)$ for $|\mathbf{r}_1 - \mathbf{r}_2| < r_c$. For such short separations, $v(12)$ on the left-hand side of (2.5) becomes infinite; since the right-hand side is proportional to $\psi_{\alpha\beta}(12)$, which must remain finite, it follows that on the left-hand side of (2.5) $\psi_{\alpha\beta}$ must satisfy

$$\psi_{\alpha\beta}(12) = 0 \qquad \text{for} \qquad |\mathbf{r}_1 - \mathbf{r}_2| < r_c \tag{2.11}$$

If (2.11) is now substituted on the right-hand side of (2.5) it follows also that

$$Q_{\alpha\beta}(12)v(12)\psi_{\alpha\beta}(12) = 0 \qquad \text{for} \qquad |\mathbf{r}_1 - \mathbf{r}_2| < r_c \tag{2.12}$$

The product $Q_{\alpha\beta}(12)v(12)\psi_{\alpha\beta}(12)$ becomes finite for $|\mathbf{r}_1 - \mathbf{r}_2| > r_c$. It is possible to show that at $|\mathbf{r}_1 - \mathbf{r}_2| = r_c$ the product $v(12)\psi_{\alpha\beta}(12)$ has a δ-function discontinuity (see Bethe and Goldstone (57)). Its origin is in the fact that for $|\mathbf{r}_1 - \mathbf{r}_2| < r_c$ both $\psi_{\alpha\beta}(\mathbf{r}_1, \mathbf{r}_2)$ and $\partial\psi_{\alpha\beta}(\mathbf{r}_1, \mathbf{r}_2)/\partial(|\mathbf{r}_1 - \mathbf{r}_2|)$ must vanish. In order to obtain a nontrivial solution to (2.5) it is then necessary for the derivative of $\psi_{\alpha\beta}$ to be discontinuous at $|\mathbf{r}_1 - \mathbf{r}_2| = \mathbf{r}_c$. $(T_1 + T_2)\psi_{12}$ consequently involves a δ-function singularity, leading via (2.5) to the singularity in $v(12)\psi_{\alpha\beta}(12)$.

The reader is referred to the paper of Bethe and Goldstone (57) for the detailed study of the solutions to (2.5). For more recent discussions see Bethe (71), Sprung (72) and Day (67). We shall also return to this question in Section 6. Here we shall satisfy ourselves with just a few comments on these solutions.

The solution $\psi_{\alpha\beta}(12)$ to (2.5) or (2.10b) can be expanded in a series of the functions $\phi_\rho(1)\phi_\sigma(2)$. Because of (2.10a) it will include just one function from the occupied states: $\phi_\alpha(1)\phi_\beta(2)$, so that we obtain

$$\psi_{\alpha\beta}(12) = \phi_\alpha(1)\phi_\beta(2) + \sum_{\lambda\mu} C_{\lambda\mu}^{(\alpha\beta)}\phi_\lambda(1)\phi_\mu(2) \tag{2.13}$$

Looked upon as a two-body Schrödinger equation, (2.10b) gives the scattering of particles 1 and 2 off each other under the influence of the mutilated interaction $Q_{\alpha\beta}v(12)Q_{\alpha\beta}$. In (2.13), $\sum_{\lambda\mu} C_{\lambda\mu}^{(\alpha\beta)}\phi_\lambda(1)\phi_\mu(2)$ then represents the "scattered wave." Since the unoccupied states lie, by assumption, higher in energy than the occupied ones, the "scattered wave" consists of no part that has the same unperturbed energy as $\phi_\alpha(1)\phi_\beta(2)$. Since at large separations of 1 and 2 all components of $\psi(12)$ must have the same energy it follows that the scattered wave can have no amplitude at infinity.

This is a very important result, due entirely to the Pauli principle that blocks, for particles 1 and 2, all states that are energetically degenerate with $\phi_\alpha(1)\cdot\phi_\beta(2)$. As a result of this blocking the interaction between two particles in occupied states can modify their wave function at relatively short mutual distances, but cannot affect it at large relative separations. We shall return to this point later.

Notice that the innocent looking operator $Q_{\alpha\beta}$ causes in reality a very appreciable change in v_{12}. If we neglect for a moment the spin and isospin variables we obtain from (2.4)

$$w(\mathbf{r}_1, \mathbf{r}_2) \equiv Q_{\alpha\beta}v(\mathbf{r}_1, \mathbf{r}_2) = \phi_\alpha(\mathbf{r}_1)\phi_\beta(\mathbf{r}_2)\int \phi_\alpha^*(\mathbf{r}_1')\phi_\beta^*(\mathbf{r}_2')v(|\mathbf{r}_1' - \mathbf{r}_2'|)d\mathbf{r}_1'd\mathbf{r}_2'$$
$$+ \sum_{\mu\lambda} \phi_\lambda(\mathbf{r}_1)\phi_\mu(\mathbf{r}_2)\int \phi_\lambda^*(\mathbf{r}_1')\phi_\mu^*(\mathbf{r}_2')v(|\mathbf{r}_1' - \mathbf{r}_2'|)\,d\mathbf{r}_1'd\mathbf{r}_2' \tag{2.14}$$

Note that $w(\mathbf{r}_1, \mathbf{r}_2)$ is an operator. The integrations on the right-hand side of (2.14) are to be performed only after multiplication by a function of $\mathbf{r}_1'$ and $\mathbf{r}_2'$. Thus even if $v(|\mathbf{r}_1 - \mathbf{r}_2|)$ has a short range and vanishes for $|\mathbf{r}_1 - \mathbf{r}_2| > r_0$, $w(\mathbf{r}_1, \mathbf{r}_2) = Q_{\alpha\beta}v(|\mathbf{r}_1 - \mathbf{r}_2|)$ will generally have a range as large as that of a typical $\phi_\gamma(\mathbf{r}_1)\phi_\delta(\mathbf{r}_2)$ and will not vanish for $|\mathbf{r}_1 - \mathbf{r}_2| > r_0$. The physical reason for this result is obvious: $Q_{\alpha\beta}$ has to see to it that all the components of $v(12)\psi_{\alpha\beta}(12)$ in the occupied states should vanish. It should therefore be effective wherever an occupied state is to be found, and thus leads to a range comparable with that of a typical $\phi_\gamma(\mathbf{r}_1)\phi_\delta(\mathbf{r}_2)$.

Intuitively we would expect $\psi_{\alpha\beta}(12)$ to be a better description of a pair than $\phi_\alpha(1)\phi_\beta(2)$, since, as we have stressed, $\psi_{\alpha\beta}(12)$ includes also corrections resulting from the virtual scattering of the pair into unoccupied states. A formal treatment of this problem, which would also point out which terms in the perturbation series are included by replacing $\phi_\alpha(1)\phi_\beta(2)$ by $\psi_{\alpha\beta}(12)$, will be given in Chapter VII. Here we would just like to make some additional observations on the solutions of the Bethe–Goldstone equation and their relation to the N-fermion wave function.

3. THE TWO-PARTICLE HEALING DISTANCE

Let us consider again an exact solution $\Psi_\alpha(1, \ldots, N)$ of the N-particle problem (1.1), and define $\Phi_\alpha(1, \ldots, N)$ to be a *non*-antisymmetrized independent N-particle wave function of the N-lowest states in the box:

$$\Phi_\alpha = \phi_{\alpha_1}(1)\phi_{\alpha_2}(2)\ldots\phi_{\alpha_N}(N) \qquad \alpha_1 < \alpha_2 \ldots < \alpha_N \tag{3.1}$$

where the notation $\alpha_1 < \alpha_2 \ldots < \alpha_N$ implies that we order the particles according to the order of the states in the box, so that all α_i's are different.

We note that a function $\Phi'_\alpha(1, \ldots, N)$ in which any two states are identical, say $\alpha_1 = \alpha_2$, satisfies:

$$\begin{aligned}\int \Phi'^*_\alpha(1, 2, \ldots, N)\Psi_\alpha(1, 2, \ldots, N)d(1)\ldots d(N) \\ = \int \Phi'^*_\alpha(2, 1, 3, \ldots, N)\Psi_\alpha(1, 2, \ldots, N)d(1)\ldots d(N) \\ = -\int \Phi'^*_\alpha(2, 1, 3, \ldots, N)\Psi_\alpha(2, 1, \ldots, N)d(1)\ldots d(N)\end{aligned} \tag{3.2}$$

where in the last step we used the antisymmetry of $\Psi_\alpha(1, \ldots, N)$ (note that Φ'_α is not antisymmetrized). Since integration variables are dummy variables it follows from (3.2) that

$$\int \Phi'^*_\alpha(1, 2, \ldots, N)\Psi_\alpha(1, 2, \ldots, N)d(1)\ldots d(N) = 0 \qquad \text{for } \alpha_1 = \alpha_2 \tag{3.3}$$

In order to separate out from the exact solution a characteristic one-particle or two-particle behavior, we define the following auxiliary functions

$$\Phi_{\alpha,k} = \frac{\Phi_\alpha}{\phi_{\alpha_k}(k)} \tag{3.4}$$

$$\Phi_{\alpha,kl} = \frac{\Phi_\alpha}{\phi_{\alpha k}(k)\phi_{\alpha l}(l)} \qquad k < l \tag{3.5}$$

Note that in Φ_α particle i occupies the state α_i, and that $\Phi_{\alpha,k}$ is an $(N - 1)$ independent-particle wave function formed from Φ_α by taking away the k-particle; in $\Phi_{\alpha,kl}$ two particles have been taken away, etc. The roman subscripts, $k, l \ldots$, will denote the single particle quantum numbers to be omitted from the set $\{\alpha\}$.

The one-particle structure of $\Psi_\alpha(1, \ldots, N)$ can be given by

$$\psi_{\alpha k}(k) = \int \Phi^*_{\alpha,k}\Psi_\alpha(1, \ldots, N)\, d'\tau \tag{3.6}$$

where $d'\tau$ stands for integration over all the variables except k. $\psi_{\alpha k}$ actually describes the behavior of a single particle in $\Psi_\alpha(1, \ldots, N)$: that particle which, in the Fermi-gas approximation, is assigned to the state α_k. Because of (3.3) we see that $\psi_{\alpha k}(k)$ has no component among the occupied states except $\phi_{\alpha k}(k)$. Indeed, if $\phi_\gamma(k)$ is an occupied state with $\gamma \neq \alpha_k$, then

$$\int \phi^*_\gamma(k)\psi_{\alpha k}(k)d(k) = \int \Phi'^*_\alpha(1, \ldots, N)\Psi_\alpha(1, \ldots, N)d(1)\ldots d(N) = 0$$

since $\Phi'_\alpha(1, \ldots, N)$ is a function of the type (3.1) but with two particles occupying the same state.

We can define similarly the two-particle part of $\Psi_\alpha(1, \ldots, N)$ by

$$[\psi_{\alpha k\alpha l}(kl)] = \int \Phi^*_{\alpha,kl}\Psi_\alpha(1, \ldots, N)d'\tau \tag{3.7}$$

where $d'\tau$ now means integration over all variables but k and l. We use the same notation for the two-particle part of Ψ_α and the Bethe–Goldstone solution; this is done in anticipation of the result that the two are the same within a certain approximation (see Sec. 4). It should be borne in mind that overlap (3.7) can be quite small. It can be seen easily that $[\psi_{\alpha k\alpha l}]$ has no components in the occupied states except for $[\phi_{\alpha k}\phi_{\alpha l}]\cdot [\psi_{\alpha k\alpha l}]$ thus satisfies (2.10e):

$$[Q_{\alpha k\alpha l}][\psi_{\alpha k\alpha l}] = [\psi_{\alpha k\alpha l}] \tag{3.8}$$

Problem. Prove that $[\psi_{\alpha k\alpha l}(kl)] = -[\psi_{\alpha k\alpha l}(lk)]$

We can proceed and define the three-particle part of $\Psi_\alpha(1, \ldots, N)$, etc. It can now be shown [Brenig (57)] that in the approximation

$$v(12)\psi_{\alpha k\alpha l\alpha m}(123) \equiv v(12)[\psi_{\alpha k\alpha l}(12)\phi_{\alpha m}(3) - \psi_{\alpha k\alpha m}(12)\phi_{\alpha l}(3) - \psi_{\alpha m\alpha l}(12)\phi_{\alpha k}(3)] \tag{3.9}$$

$\psi_{\alpha_k \alpha_l}$ satisfies the Bethe–Goldstone equation (2.5) with the projection operator $Q_{\alpha_k \alpha_l}$. The approximation (3.9) implies that the three-particle structure is dominated by the two-particle correlation in the following sense:

The regions of configurations space in which $v(12)$ is appreciable are those for which $|\mathbf{r}_1 - \mathbf{r}_2| < b$ where b is the range of $v(12)$; (3.9) asserts that there is only a small probability also for particle 3 to be then close enough to 1 and 2 to affect their relative wave function. Actually we could have started with (3.9) and derive from it the Bethe–Goldstone equation for $\psi_{\alpha\beta}(12)$ (see Brenig 57). See also discussion in Section 4.

We noticed before, following (2.13), that at large relative separations of the two particles, the Bethe–Goldstone solution satisfies

$$\psi_{\alpha\beta}(12) \to \phi_\alpha(1)\phi_\beta(2) \qquad (3.10)$$
$$|r_{12}| \to \infty$$

Because of the projection operator $Q_{\alpha\beta}(12)$, $\psi_{\alpha\beta}(12)$ can differ from $\phi_\alpha(1) \cdot \phi_\beta(2)$ only at short relative separations. The whole validity of the Bethe–Goldstone approximation can now be examined in terms of the magnitude of r_{12} for which $\psi_{\alpha\beta}(12)$ effectively goes over into $\phi_\alpha(1)\phi_\beta(2)$. This distance has been called the *healing-distance* $h_{\alpha\beta}$, picturing the interaction $v(12)$ as producing a "wound" in $\phi_\alpha(1)\phi_\beta(2)$ by converting it into $\psi_{\alpha\beta}(12)$. [Gomes et al (58)]. For two free particles scattering off each other the healing distance is infinite; the existence of a phase shift in the relative function at $r_{12} \to \infty$ indicates that the wound inflicted on $\phi_\alpha(1)\phi_\beta(2)$ by the interaction does not heal even at infinity. However, when the pair scatters inside a Fermi gas where other states of the same energy are blocked by the Pauli principle, no phase shifts are possible and $\psi_{\alpha\beta}(12)$ heals at a finite relative separation $h_{\alpha\beta}$.

The two-particle part $\psi_{\alpha_k \alpha_l}(kl)$ of the exact solution Ψ_α satisfies the Bethe–Goldstone equation only if the approximation (3.9) is valid. This condition is automatically satisfied if the distances between particles k, l, and m are large enough so that even $\psi_{\alpha_k \alpha_l}(kl) \sim \phi_{\alpha_k}(k)\phi_{\alpha_l}(l)$. In order for (3.9) to be valid also for smaller distances it is required that $\psi_{\alpha_k \alpha_l}(kl)$ differs appreciably from $\phi_{\alpha_k}\phi_{\alpha_l}$ only for such small separations that there is a very small probability of finding particle m within the same region as well. Thus we see that the Bethe–Goldstone solution $\psi_{\alpha\beta}(12)$ can serve as a valid description of the behavior of a pair only if the healing distance turns out to be smaller than the average spacing between particles. Under these conditions the presence of particle 3 will not affect $\psi_{\alpha\beta}(12)$ in those regions in which $\psi_{\alpha\beta}(12)$ reflects the the effects of the interaction $v(12)$. The effects of this interaction can therefore be evaluated ignoring the position of all other particles. This is saying, in other words, that an independent pair approximation is valid.

The magnitude of the healing distance $h_{\alpha\beta}$ depends on the nature of the nucleon-nucleon force, and ultimately so does the average equilibrium interparticle distance r_0. Can a situation be realized in which $h_{\alpha\beta} < r_0$? Does the actual nuclear force lead to a realization of this condition?

To answer these questions we shall need to carry out more detailed calculations, but we may note here some qualitative aspects of this question. We saw in the previous chapter that a purely attractive nuclear force of range b would have lead to a collapse of the N-nucleon system into a sphere of radius $\sim b$. An N-independent density was obtained when we took into account the hard core, the tensor and the exchange potential components of the nucleon-nucleon force. It is the joint effect of all of these elements in the nucleon-nucleon force that determined the equilibrium density. The healing distance depends, of course, on the hard core radius r_c. In fact

$$\psi_{\alpha\beta}(r_{12}) = 0 \qquad \text{for} \qquad r_{12} < r_c \tag{3.11}$$

and since $\phi_\alpha(r_1)\phi_\beta(r_2) \neq 0$ for $r_{12} < r_c$ it is obvious that

$$h_{\alpha\beta} > r_c \tag{3.12}$$

How much $h_{\alpha\beta}$ is bigger than r_c depends on the remaining, that is, the attractive, part of the nuclear interaction. If that interaction is strong, the "wounds" it inflicts on $\phi_{\alpha\beta}$ will be felt over large distances; if, on the other hand, that attractive part is weak we can expect $\psi_{\alpha\beta}$ to heal from the wound inflicted by the hard core over short distances. The nuclear equilibrium density and the healing distance are therefore particularly sensitive to different parts of the nucleon-nucleon force. For nuclear forces the healing distance does seem to be shorter than the average interparticle equilibrium distance. It is due to these circumstances that the independent particle approximation, or its slightly improved version—the independent pair approximation—is valid.

4. ENERGIES AND WAVE FUNCTIONS IN THE INDEPENDENT PAIR APPROXIMATION

Having discussed the possible validity of the independent pair approximation for nuclei, we should proceed to solve the Bethe–Goldstone equation with realistic nucleon-nucleon forces and show that it does lead to a short enough healing distance. Prior to this, however, we want to see what is the relation between the Bethe–Goldstone energies $E_{\alpha\beta}$ (2.5) and the total nuclear energy. This relation will be required for our subsequent discussions.

We start again from the exact N-particle Hamiltonian (1.1) and assume that the set of independent-particle states states ϕ_{α_i}, (1.5), can serve to construct a zeroth-order approximation wave function. Multiplying (1.1) to the left with Φ_α^* defined in (3.1), and integrating we obtain:

$$E_\alpha \int \Phi_\alpha^* \Psi_\alpha d(1) \ldots d(N) = \int \Phi_\alpha^* [\Sigma T_i + \sum_{i<j} v_{ij}] \Psi_\alpha d(1) \ldots d(N)$$
$$= \sum_{\alpha i} \langle \alpha_i | T\psi_{\alpha i} \rangle + \sum_{\alpha_i < \alpha_j} \langle \alpha_i \alpha_j | v [\psi_{\alpha_i \alpha_j}] \rangle \tag{4.1}$$

where $\psi_{\alpha i}$ and $\psi_{\alpha_i\alpha_j}$ are the one-particle, (3.6), and the two-particle, (3.7), parts of the exact wave function Ψ_α and $|\alpha_i\rangle \equiv \phi_{\alpha_i}(i)$, $|\alpha_i\alpha_j\rangle \equiv \phi_{\alpha_i}(i)\phi_{\alpha_j}(j)$. The quantum numbers α_i denote occupied states. The matrix elements in (4.1) are:

$$\langle \alpha_i | T\psi_{\alpha_i} \rangle = \int \phi_{\alpha_i}^*(i) T_i \psi_{\alpha_i}(i) d(i) \tag{4.2}$$

$$\langle \alpha_i \alpha_j | v [\psi_{\alpha_i\alpha_j}] \rangle = \int \phi_{\alpha_i}^*(i) \phi_{\alpha_j}^*(j) v(ij) \psi_{\alpha_i\alpha_j}(ij) d(i) d(j) \tag{4.3}$$

We notice that

$$\int \phi_{\alpha_i}^*(i) \psi_{\alpha_i}(i) d(i) = \int \Phi_\alpha^* \Psi_\alpha d(1) \ldots d(N) \tag{4.4}$$

We shall further assume that Ψ_α is normalized in the following fashion

$$\int \Phi_\alpha^* \Psi_\alpha d(1) \ldots d(N) = 1 \tag{4.5}$$

Observing that

$$T_i \phi_{\alpha_i}(i) = \epsilon_{\alpha_i} \phi_{\alpha_i}(i) \tag{4.6}$$

we obtain then from (4.1) and (4.2)

$$E_\alpha = \Sigma \epsilon_{\alpha_i} + \sum_{\alpha_i < \alpha_j} \langle \alpha_i \alpha_j | v [\psi_{\alpha_i\alpha_j}] \rangle \tag{4.7}$$

Equation (4.7) is still an exact expression for the total energy of the N-nucleon system in terms of its two-particle part $[\psi_{\alpha_i\alpha_j}]$.

If we adopt now for $[\psi_{\alpha_i\alpha_j}]$ the solution of the Bethe–Goldstone equation, then it follows from (2.10d) that

$$\langle \alpha_i \alpha_j | v [\psi_{\alpha_i\alpha_j}] \rangle = \langle \alpha_i \alpha_j | [Q_{\alpha_i\alpha_j}] v [\psi_{\alpha_i\alpha_j}] \rangle$$
$$= \langle \alpha_i \alpha_j | (E_{\alpha_i\alpha_j} - T_1 - T_2) [\psi_{\alpha_i\alpha_j}] \rangle$$
$$= \epsilon_{\alpha_i\alpha_j} \langle \alpha_i \alpha_j | [\psi_{\alpha_i\alpha_j}] \rangle \tag{4.8}$$

Here

$$\epsilon_{\alpha\beta} = E_{\alpha\beta} - (\epsilon_\alpha + \epsilon_\beta) \tag{4.9}$$

is the change in the zeroth-order energy $(\epsilon_\alpha + \epsilon_\beta)$ induced by the Bethe–Goldstone interaction $[Q_{\alpha\beta}]v$. The normalization (4.5) of Ψ_α implies also

$$\langle \alpha_i\alpha_j | [\psi_{\alpha_i\alpha_j}] \rangle = 1 \tag{4.10}$$

It therefore follows from (4.7) that

$$E_\alpha = \Sigma\epsilon_{\alpha_i} + \sum_{\alpha_i<\alpha_j} \epsilon_{\alpha_i\alpha_j} \tag{4.11}$$

where a summation over all occupied states is implied. Equation 4.11 provides the connection between the energy *change* for the two-particle system obeying the Bethe–Goldstone equation, and the corresponding correction to the zeroth-order energy of the N-particle system. Note that in (4.11) it is $\epsilon_{\alpha_i\alpha_j}$ that is summed upon, and not $E_{\alpha_i\alpha_j}$. The independent pair approximation should not be taken to imply the existence of $N(N-1)/2$ independent pairs in the nucleus. It is only in the *correction* to the total energy that each pair makes its contribution independently.

The task of constructing the N-particle wave function once the Bethe–Goldstone wave function is known remains. This has been carried out by deShalit and Weisskopf (58) in the independent pair approximation. Following their procedure define a two-particle correlation factor $f_{\alpha\beta}$ through

$$\psi_{\alpha\beta}(12) \equiv (1 + f_{\alpha\beta}(12))\phi_\alpha(1)\phi_\beta(2) \tag{4.12}$$

where $\psi_{\alpha\beta}(12)$ is the un-antisymmetrized Bethe–Goldstone wave function and $f_{\alpha\beta}(12) = f_{\beta\alpha}(21)$.

We note (2.11) that if the interaction $v(12)$ has a hard core of radius r_c, the correlation factor satisfies

$$f_{\alpha\beta}(12) = -1 \qquad \text{for} \qquad |\mathbf{r}_1 - \mathbf{r}_2| < r_c \tag{4.13}$$

Also, since $\psi_{\alpha\beta}(12) \to \phi_{\alpha\beta}(12)$ for $|\mathbf{r}_1 - \mathbf{r}_2| \to \infty$, we conclude that

$$f_{\alpha\beta}(12) \to 0 \qquad \text{for} \qquad |\mathbf{r}_1 - \mathbf{r}_2| \to \infty \tag{4.14}$$

Comparing (4.12) with the Bethe–Goldstone wave function (2.13) yields

$$f_{\alpha\beta}(12)\phi_\alpha(1)\phi_\beta(2) = Q_F\psi_{\alpha\beta}(12) = \Sigma C_{\lambda\mu}^{(\alpha\beta)}\phi_\lambda(1)\phi_\mu(2)$$

Thus $f_{\alpha\beta}(12)$ operating on the Fermi gas wave function leads to unoccupied states. It follows that

$$\int \phi_\delta^*(1)f_{\alpha\beta}(12)\phi_\alpha(1)d(1) = 0 \tag{4.15}$$

if ϕ_δ is an occupied level. Although the properties (4.13) to (4.15) are common to all factors $f_{\alpha\beta}$, yet for different values of $(\alpha\beta)$ we generally obtain different factors $f_{\alpha\beta}(12)$. In some approximations [Jastrow (55)] the dependence of $f_{\alpha\beta}(12)$ on $(\alpha\beta)$ is neglected.

We can now write down an N-particle wave function that takes account of the independent pair approximation, by using the two-particle correlation factor $f_{\alpha\beta}(12)$. To this end we define

$$\Psi'_\alpha(1, \ldots, N) = \mathcal{A}\{[1 + f_{\alpha_1\alpha_2}(12)][1 + f_{\alpha_1\alpha_3}(13)] \ldots [1 + f_{\alpha_2\alpha_3}(23)] \ldots \cdot [1 + f_{\alpha_{N-1}\alpha_N}(N-1, N)]\, \phi_{\alpha_1}(1)\phi_{\alpha_2}(2) \ldots \phi_{\alpha_N}(N)\} \quad (4.16)$$

where $\mathcal{A}$ is the antisymmetrization operator. The structure of Ψ'_α is quite transparent. It starts with the independent-particle wave function $\phi_{\alpha_1}(1) \ldots . \phi_{\alpha_N}(N)$ and through the use of the $N(N-1)/2$ correlation factors guarantees that each pair is properly correlated.

We shall show that if

$$\psi_{\alpha\beta}(12) = [1 + f_{\alpha_1\alpha_2}(12)]\phi_{\alpha_1}(1)\phi_{\alpha_2}(2)$$

satisfies the Bethe–Goldstone equation (2.5), Ψ'_α is an approximate solution of the many-body Schrödinger equation (1.1). We shall also show that in the same approximation Ψ'_α has a two-particle part equal to the antisymmetrized $[\psi_{\alpha\beta}(12)]$. The essential approximation can be described upon expansion of the product of the correlation factors in (4.16).

The leading term is 1; then comes a term

$$\sum_{\alpha_i < \alpha_j} f_{\alpha_i\alpha_j}(i, j);$$

the next term contains products of two f's, and of these there are two types: "linked" products in which one of the particles appears twice like $f_{\alpha_1\alpha_2}(12) \cdot f_{\alpha_2\alpha_3}(23)$, and "unlinked" products like $f_{\alpha_1\alpha_2}(12)f_{\alpha_3\alpha_4}(34)$, etc. The approximation we shall adopt is that of neglecting all linked products of f's; that is, we assume that

$$f_{\alpha_1\alpha_2}(12)f_{\alpha_2\alpha_3}(23) \ll f_{\alpha_1\alpha_2}(12) \quad (4.17)$$

The physical argument behind this approximation is identical to the one behind (3.9); the product $f_{\alpha_1\alpha_2}(12)f_{\alpha_2\alpha_3}(23)$ can only be significantly different from zero if all three particles are found simultaneously within a volume of order h^3 where h is the healing distance; if the healing distance is short compared with the average internucleon distance the probability for this to happen is small.

To determine the extent to which Ψ'_α satisfies the many-body Schrödinger equation, let us begin by evaluating

$$V\Psi'_\alpha \equiv \sum_{i \neq j} v(ij)\Psi'_\alpha = \mathcal{A} \sum_{i \neq j} v(ij)[1 + f_{\alpha_i\alpha_j}(ij)]\Phi_\alpha + \mathcal{A} \sum_{i \neq j} v(ij)[1 + f_{\alpha_i\alpha_j}(ij)]F^{(1)}_{\alpha,ij} \quad (4.18)$$

where

$$F^{(1)}_{\alpha,ij} = [1 + f_{\alpha_1\alpha_2}(12)][1 + f_{\alpha_1\alpha_3}(13)] \ldots [1 + f_{\alpha_{N\text{-}1}\alpha_N}(N-1, N)] - [1 + f_{\alpha_i\alpha_j}(ij)] \quad (4.19)$$

For example

$$F^{(1)}_{\alpha,12} = \sum_{i<j}{}' f_{\alpha_i\alpha_j} + \tfrac{1}{2} \sum_{i<j,k<l}{}'' f_{\alpha_i\alpha_j} f_{\alpha_k\alpha_l} + \ldots$$

where the prime on the first sum requires the omission of the term $f_{\alpha_1\alpha_2}$. The double prime in the second sum requires the omission not only of the term $f_{\alpha_1\alpha_2}$ but also of terms for which (α_i, α_j) equals (α_k, α_l). The important point is that $F^{(1)}_{\alpha,12}$ depends at least linearly on the correlation amplitude $f_{\alpha_i\alpha_j}$.

We turn to the first term of (4.18) to be designed $(V\Psi_\alpha')_1$. Because of the antisymmetrization operator it follows that in the evaluation of

$$v(ij)[1 + f_{\alpha_i\alpha_j}(ij)]\Phi_\alpha$$

one can drop all terms that lead to filled states except for the state in which particles i, j are in states $\phi_{\alpha_i}\phi_{\alpha_j}$. Therefore

$$(V\Psi_\alpha')_1 = \mathcal{A} \sum_{ij} Q_{\alpha_i\alpha_j} v(ij)[1 + f_{\alpha_i\alpha_j}(ij)]\Phi_\alpha$$

We now replace the potential energy term using the Bethe–Goldstone equation

$$\begin{aligned}(V\Psi)_{\alpha 1}' &= \mathcal{A} \sum_{i<j} (E_{\alpha_i\alpha_j} - T_i - T_j)\Phi_\alpha [1 + f_{\alpha_i\alpha_j}(ij)] \\ &= (E_0 - T)\mathcal{A} \sum_{i<j} (1 + f_{\alpha_i\alpha_j}(ij)\Phi_\alpha \\ &\qquad + \mathcal{A} \sum_{i<j} \epsilon_{\alpha_i\alpha_j}(1 + f_{\alpha_i\alpha_j}(ij)]\Phi_\alpha\end{aligned}$$

where $T = \Sigma T_i$, $E_0 = \Sigma \epsilon_{\alpha_i}$ and we have used (4.9). Inserting this value of $(V\Psi_\alpha')_1$ into (4.18) one obtains

$$\begin{aligned}V\Psi_\alpha' = (E_0 - T)\Psi_\alpha' + \sum_{i<j} \epsilon_{\alpha_i\alpha_j}\Psi_\alpha' - (E_0 - T)\mathcal{A}F_\alpha{}^{(2)}\Phi_\alpha \\ + \mathcal{A} \sum_{i<j} [v(ij) - \epsilon_{\alpha_i\alpha_j}]\Phi_\alpha(1 + f_{\alpha_i\alpha_j})F^{(1)}_{\alpha,ij}\end{aligned}$$

where

$$F^{(2)}_\alpha \equiv [1 + f_{\alpha_1\alpha_2}(12)][1 + f_{\alpha_1\alpha_3}(13)] \ldots [1 + f_{\alpha_{N\text{-}1,N}}(N-1, N)] - 1 - \sum_{i<j} f_{\alpha_i,\alpha_j} \quad (4.20)$$

Finally using (4.11)

$$\begin{aligned}(E_\alpha - T - V)\Psi_\alpha' = (E_0 - T)\mathcal{A}F^{(2)}_\alpha\Phi_\alpha \\ - \mathcal{A} \sum_{ij} [v(ij) - \epsilon_{\alpha_i\alpha_j}]\Phi_\alpha(1 + f_{\alpha_i,\alpha_j})F^{(1)}_{\alpha,ij} \quad (4.21)\end{aligned}$$

Now the remarkable result of de Shalit and Weisskopf (58) obtained is that the error term on the right involves only linked products of $f_{\alpha_i\alpha_j}$ as in (4.17). Let us take an example, a particular unlinked term in $F_\alpha^{(2)}$, say the $f_{\alpha_1\alpha_2}(12)\cdot f_{\alpha_3\alpha_4}(34)$ term (see Eq. 3.5):

$$\begin{aligned}(E_0 - T)f_{\alpha_1\alpha_2}(12)f_{\alpha_3\alpha_4}(34)\Phi_\alpha &= (E_{\alpha_1} + E_{\alpha_2} - T_1 - T_2)\psi_{\alpha_3\alpha_2}f_{\alpha_3\alpha_4}\Phi_{\alpha;12}\\ &\quad + (E_{\alpha_3} + E_{\alpha_4} - T_3 - T_4)\psi_{\alpha_3\alpha_4}f_{\alpha_1\alpha_2}\Phi_{\alpha;34}\\ &= (Q_{\alpha_1\alpha_2}v_{12} - \epsilon_{\alpha_1\alpha_2})\psi_{\alpha_1\alpha_2}f_{\alpha_3\alpha_4}\Phi_{\alpha;12}\\ &\quad + (Q_{\alpha_3\alpha_4}v_{34} - \epsilon_{\alpha_3\alpha_4})\psi_{\alpha_3\alpha_4}f_{\alpha_1\alpha_2}\Phi_{\alpha;34}\end{aligned}$$

or

$$\begin{aligned}(E_0 - T)f_{\alpha_1\alpha_2}(12)f_{\alpha_3\alpha_4}(34)\Phi_\alpha &= (v_{12} - \epsilon_{\alpha_1\alpha_2})(1 + f_{\alpha_1\alpha_2})f_{\alpha_3\alpha_4}\Phi_\alpha\\ &\quad + (v_{34} - \epsilon_{\alpha_3\alpha_4})(1 + f_{\alpha_3\alpha_4})f_{\alpha_1\alpha_2}\Phi_\alpha\end{aligned}$$

It is clear that this term is cancelled by the unlinked components of the second term on the right-hand side of (4.21). If therefore we, in the spirit of the independent pair approximation, neglect all linked $f_{\alpha_i\alpha_j}$, the right-hand side of (4.21) is zero so that $(E_\alpha - T - V)\Psi_\alpha' = 0$ if linked $f_{\alpha_i\alpha_j}$ terms are neglected. Hence in the independent pair approximation Ψ_α' satisfies the Schrödinger equation. The question now arises as to the relation between Ψ_α' and the Bethe–Goldstone solutions. Toward this end we now compute the two-particle part $\psi_{\alpha_k\alpha_l}(kl)$, (3.7), of Ψ_α' term by term. The leading term $\psi^{(0)}$ with no f's at all will give rise to

$$[\psi^{(0)}_{\alpha_k\alpha_l}(kl)] = \int \Phi^*_{\alpha,kl}\mathcal{A}[\phi_{\alpha 1}(1)\phi_{\alpha 2}(2)\ldots\phi_{\alpha_N}(N)]d'\tau = [\phi_{\alpha_k}\phi_{\alpha_l}] \qquad (4.22)$$

The terms linear in $f_{\alpha\beta}$ will give rise to

$$\begin{aligned}[\psi^{(1)}_{\alpha_k\alpha_l}(kl)] &= \int \phi^*_{\alpha,kl}\mathcal{A}\{[\Sigma f_{\alpha_i\alpha_j}(ij)]\phi_{\alpha 1}(1)\phi_{\alpha 2}(2)\ldots\phi_{\alpha_N}(N)\}d'\tau\\ &= f_{\alpha_k\alpha_l}(kl)[\phi_{\alpha_k}\phi_{\alpha_l}]\end{aligned} \qquad (4.23)$$

Use has been made of (4.15) from which one can obtain the result that $f_{\alpha\beta}$, when operating on a Fermi-gas wave function Φ_β, can lead only to nonoccupied states; since in $\Phi_{\alpha,kl}$ only the pair state (kl) is unoccupied, only the corresponding term in $\Sigma f_{\alpha_i\alpha_j}$ gives rise to a nonvanishing contribution.

Higher terms, ignoring linked products of f's, will all give vanishing contributions to $\psi_{\alpha_k\alpha_l}(kl)$. Indeed, due to (4.15)

$$\int \Phi^*_{\alpha,kl}\mathcal{A}[\sum_{(ij)\neq(m,n)} f_{\alpha_i\alpha_j}(ij)f_{\alpha_m\alpha_n}(mn)\phi_{\alpha 1}(1)\ldots\phi_{\alpha_N}(N)]d(1)\ldots d(N) = 0$$

since for any two different pairs $\alpha_i\alpha_j$ and $\alpha_m\alpha_n$, at least one of them is occupied in $\Phi_{\alpha,kl}$.

To summarize in the approximation of neglecting linked products of the f's, if the Bethe–Goldstone wavefunction $\psi_{\alpha\beta}$ is related to $f_{\alpha\beta}$ by

$$\psi_{\alpha\beta} = [1 + f_{12}(\alpha\beta)]\phi_\alpha(1)\phi_\beta(2)$$

then

(1) Ψ_α' given by (4.16) is a solution of the many-body problem and

(2) the antisymmetrized $\psi_{\alpha\beta}$, $[\psi_{\alpha\beta}]$, is the two-particle part of Ψ'_α.

When two particles are relatively close the dependence of the N-body wave function upon their coordinates is given by the Bethe–Goldstone two-particle wave function $\psi_{\alpha\beta}$ multiplied into $\Phi_{\alpha;\alpha\beta}$ and antisymmetrized. In these circumstances the pair can be thought of as interacting in the presence of the other $N-2$ particles, as one might expect in an independent pair approximation.

The validity of the independent pair approximation hinges very critically on the properties of the three-particle correlations. These should show a small probability for three particles to be found simultaneously within a volume of order h^3. If the average volume per particle is r_0^3 (see Section II.2) then, for the independent pair approximation to be valid we must have

$$\left(\frac{h^3}{r_0^3}\right) \ll 1 \tag{4.24}$$

Since r_0 and h are of the same order of magnitude, it is important to estimate the healing distance h fairly well, to ascertain the validity of this approximation. We note that even for $h = (1/2)r_0$, (4.24) still holds. The independent pair approximation is therefore a low-density approximation.

5. INDEPENDENT PAIR AND THE INDEPENDENT-PARTICLE MODEL

In the preceding sections, the independent pair approximation has been described. The relation of the solutions of the Bethe–Goldstone equation to the many-body nuclear wave function was established and the corresponding nuclear energy determined. Solving the nuclear many-body problem is accordingly reduced to solving the Bethe–Goldstone equation. In this section we shall sketch a simple but important improvement on this procedure.

Recall that asymptotically $\psi_{\alpha\beta}$ approaches $\phi_\alpha\phi_\beta$ where

$$\phi_\alpha = \frac{1}{\sqrt{\Omega}} \exp{(i\mathbf{k}_\alpha \cdot \mathbf{r})} \tag{5.1}$$

The energy corresponding to ϕ_α is taken to be

$$\epsilon_\alpha = \frac{\hbar^2}{2m} k_\alpha^2$$

However this expression for the energy does not take into account the effect of the interaction of the particle in level α with all the other particles in the nucleus. It seems reasonable that if this interaction is taken into account in the definition of ϕ_α a better approximation to the nuclear wave function will result. This additional energy can be, as we shall shortly see, related to the Bethe–Goldstone solutions whose asymptotic behavior is expressed in terms of ϕ_α, which as we have just proposed depends upon the Bethe–Goldstone solutions. An iterative procedure suggests itself.

Let ϕ_α be a solution of the equation

$$(T + U_\alpha)\phi_\alpha = \epsilon_\alpha\phi_\alpha$$

where U_α is a single-particle potential that takes into account the interaction with the other particles. [Note that ϕ_α will still be given by (5.1). Why?] We shall shortly show how this potential is related to the Bethe–Goldstone wave function. The Bethe–Goldstone equation is modified as follows:

$$[T_1 + T_2 + U_\alpha(1) + U_\beta(2) + Q_{\alpha\beta}v_{12}][\psi_{\alpha\beta}] = E_{\alpha\beta}[\psi_{\alpha\beta}] \qquad (5.2)$$

As a first step U_α and U_β could be placed equal to zero, $[\psi_{\alpha\beta}]$ determined, and a first approximation to U_α and U_β obtained. These would then be inserted into (5.2), a new $[\psi_{\alpha\beta}]$ obtained, and the second approximation to U_α and U_β obtained.

Let us now obtain the relation between U_α and the Bethe–Goldstone solution. Toward this end we introduce the G operator by the equation

$$\langle\alpha'\beta'|G(12)|[\alpha\beta]\rangle = \langle\alpha'\beta'|v_{12}[\psi_{\alpha\beta}]\rangle \qquad (5.3)$$

where $|\alpha'\beta'\rangle \equiv \phi_{\alpha'}(1)\phi_{\beta'}(2)$, and $|[\alpha\beta]\rangle = \phi_\alpha(1)\phi_\beta(2) - \phi_\beta(1)\phi_\alpha(2)$. In terms of G the energy correction (4.9) $\epsilon_{\alpha\beta}$ is

$$\epsilon_{\alpha\beta} = \langle\alpha\beta|v_{12}[\psi_{\alpha\beta}]\rangle = \langle\alpha\beta|G(12)|[\alpha\beta]\rangle$$

The total energy of the N-particle system is then

$$E = \sum_\alpha \frac{\hbar^2 k_\alpha{}^2}{2m} + \Sigma\epsilon_{\alpha_i\alpha_j} = \langle\Phi_\alpha|[\Sigma T_i + \sum_{i<j} G(ij)]|[\Phi_\alpha]\rangle \qquad (5.4)$$

where $[\Phi_\alpha]$ is the product wave function Φ_α antisymmetrized and satisfying the normalization condition $\langle\Phi_\alpha|[\Phi_\alpha]\rangle = 1$. It is tempting to think of $G(ij)$ as an effective interaction. However it should be remembered that $G(ij)$ gives this energy only when its expectation value with respect to antisymmetrized plane wave state $[\Phi_\alpha]$ is taken.

The one-particle energy is defined as in (1.11)

$$U(\alpha_1)\phi_{\alpha_1}(1) = \langle\alpha_2\alpha_3\ldots\alpha_N|\sum_{j=2} G_{1j}|[\Phi_\alpha]\rangle \qquad (5.5)$$

This is the U_α to be inserted in (5.2). In terms of $U(\alpha_i)$ the energy (5.4) is

$$E = \sum_\alpha \frac{\hbar^2 k_\alpha^2}{2m} + \frac{1}{2}\sum U(\alpha) = \sum \epsilon_\alpha - \frac{1}{2}\sum U(\alpha) \tag{5.6}$$

Equation 5.5 relates the effective interaction energy and, therefore, the single-particle energy ϵ_α, and the Bethe–Goldstone wave function. The latter being a solution of (5.2) depends upon $U(\alpha_i)$.

These relationships are often expressed in terms of an integral equation for G. To obtain this let us first write (5.2) as an integral equation

$$[\psi_{\alpha\beta}] = [\phi_\alpha\phi_\beta] + \frac{1}{E_{\alpha\beta} - H_0} Q_{\alpha\beta} v_{12} [\psi_{\alpha\beta}]$$

where $H_0 = T_1 + T_2 + U(\alpha_1) + U(\alpha_2)$. Multiplying both sides of this equation by $Q_{\alpha\beta} v_{12}$, we then get

$$G = Q_{\alpha\beta} v_{12} \left[1 + \frac{1}{E_{\alpha\beta} - H_0} G \right] \tag{5.7}$$

In terms of this equation, the iterative procedure consists of solving (5.7) for a given choice of $U(\alpha)$ and $U(\beta)$, then determining a new value of $U(\alpha)$ from (5.5), which is inserted into (5.7) and a new iterative cycle begun.

We shall show later (Chapter VII) how this self-consistency procedure is carried out for nuclear matter. The method is greatly simplified by the fact that the single-particle wave function in nuclear matter $(1/\sqrt{\Omega}) \exp(i\mathbf{k}_\alpha \cdot \mathbf{r})$ is independent of the single-particle potential $U(\alpha_i)$. The former is completely determined by the periodic boundary conditions, and the latter affects only the energies. For finite nuclei the analogous single particle potential has both coordinate and momentum dependence, so that *both* the wave functions and the energies change when the single-particle potential is modified. This makes it extremely complicated to carry out in finite nuclei such self-consistent calculations taking into account the independent pair approximation.

The relation between G and v, as given by (5.7) depends on the single-particle energy spectrum, that is, on ϵ_α and ϵ_β. The latter, in addition to being dependent on $U(\alpha)$ as discussed above, depends also on the density of the system, since it is this density that determines the value of the Fermi momentum k_F and therefore also which are occupied states and which are unoccupied. To see this more explicitly we solve (5.7) formally by iterations, and obtain

$$G = (Q_{\alpha\beta} v) + (Q_{\alpha\beta} v) \frac{1}{E_{\alpha\beta} - H_0} (Q_{\alpha\beta} v)$$
$$+ (Q_{\alpha\beta} v) \frac{1}{E_{\alpha\beta} - H_0} (Q_{\alpha\beta} v) \frac{1}{E_{\alpha\beta} - H_0} (Q_{\alpha\beta} v) + \ldots \tag{5.8}$$

The nonlocal character of G is now manifest. For example the second term in (5.8), in the coordinate representation, is:

$$\sum |12\rangle\langle 12| Q_{\alpha\beta} v \frac{1}{E_{\alpha\beta} - H_0} Q_{\alpha\beta} v |1'2'\rangle\langle 1'2'|$$

where the summation extends over all two-particle states. The occurrence of $Q_{\alpha\beta}$ means that in taking matrix elements of G the intermediate states could involve only states above the Fermi momentum in addition to $|\alpha\beta\rangle$. For instance, up to second order in v, $\langle\alpha\beta|G|\alpha\beta\rangle$ is given by

$$\langle\alpha\beta|G|\alpha\beta\rangle = \langle\alpha\beta|v|\alpha\beta\rangle + \langle\alpha\beta|v|\alpha\beta\rangle \frac{1}{E_{\alpha\beta} - (\epsilon_\alpha + \epsilon_\beta)} \langle\alpha\beta|v|\alpha\beta\rangle$$

$$+ \sum_{\lambda\mu > k_F} \langle\alpha\beta|v|\lambda\mu\rangle \frac{1}{E_{\alpha\beta} - (\epsilon_\lambda + \epsilon_\mu)} \langle\lambda\mu|v|\alpha\beta\rangle$$

where the summation over $\lambda\mu > k_F$ indicates that both states should be unoccupied. It is obvious that $\langle\alpha\beta|G|\alpha\beta\rangle$ depends on k_F and therefore on the density ρ of the system. It follows that $U(\alpha_i)$, the single-particle potential, will also depend on the density, and we shall indicate it explicitly by writing $U(\alpha_i) \equiv U(\alpha_i; \rho)$. The dependence of the average kinetic energy per particle on the density was given in (II. 7.3). We obtain therefore for the ρ-dependence of the total energy E:

$$E(\rho) = N \frac{3}{10} \left(\frac{3\pi^2}{2}\right)^{2/3} \frac{\hbar^2}{M} \rho^{2/3} + \tfrac{1}{2} \sum_{\alpha_i < k_F} U(\alpha_i, \rho) \tag{5.9}$$

where the ρ-dependence of the potential energy comes both through the explicit dependence of $U(\alpha_i)$ on ρ, and through limiting the summation to occupied states only $\alpha_i < k_F$.

The total energy $E(\rho)$ as a function of ρ may have a minimum for certain $\rho = \rho_0$. If it does, ρ_0 will correspond to the equilibrium density, and $E(\rho_0)/N$ will be the binding energy per particle. Both will include pair interaction effects in the approximation of the independent pair model.

In Chapter II we obtained the equilibrium density for interactions with a repulsive core through the effects on the kinetic energy. In (5.9) it looks as if the whole effect is in the potential energy. This is really just a matter of convenience since it will be recalled that $U(\alpha_i)$ was obtained from a solution of the Bethe–Goldstone equation. The latter includes the effects of a repulsive core in its eigenvalues $E_{\alpha\beta}$, and it makes no difference whether this is ascribed to a modification in the kinetic energy or to a change in the potential energy.

In the next section we shall discuss the actual solution of the Bethe–Goldstone equation and the nuclear matter parameters derived therefrom.

6. THE SOLUTION OF THE BETHE-GOLDSTONE EQUATION

The solution of the Bethe–Goldstone equation (2.5)

$$[T_1 + T_2 + Q_{\alpha\beta}(12)v(12)]\psi_{\alpha\beta}(12) = E_{\alpha\beta}\psi_{\alpha\beta}(12) \tag{2.5}$$

is made complicated by several factors. First, the operator $Q_{\alpha\beta}(12)$ is a nonlocal operator as seen from (2.4). In fact,

$$Q_{\alpha\beta}(\mathbf{r}_1, \mathbf{r}_2; \mathbf{r}_1'\mathbf{r}_2') = \phi_{\alpha\beta}(\mathbf{r}_1, \mathbf{r}_2)\phi^*_{\alpha\beta}(\mathbf{r}_1', \mathbf{r}_2') + \sum_{\lambda,\mu} \phi_{\lambda\mu}(\mathbf{r}_1, \mathbf{r}_2)\phi_{\lambda\mu}(\mathbf{r}_1', \mathbf{r}_2') \tag{6.1}$$

where the summation (λ, μ) is carried over all unoccupied states $(\lambda, \mu) > k_F$. The solution of differential equations with nonlocal potentials involves dealing with integro-differential equations, and these, as a rule, are more complicated than differential equations.

Comparing the Bethe–Goldstone equation with the ordinary Schrödinger equation for two particle scattering it looks like the Bethe–Goldstone equation has an additional complication, since unlike the free two particle equation, it involves also an eigenvalue problem; a solution of (2.5) satisfying the proper boundary conditions exists only for a specific value of $E_{\alpha\beta}$. Fortunately this particular difficulty is easy to remove in practice. We notice that the order of magnitude of $\langle\alpha\beta|v|\psi_{\alpha\beta}\rangle$ is given by:

$$\begin{aligned}\langle\alpha\beta|v|\psi_{\alpha\beta}\rangle &\approx \frac{1}{\Omega^2}\int \exp\,[-i(\mathbf{k}_\alpha\cdot\mathbf{r}_1 + \mathbf{k}_\beta\cdot\mathbf{r}_2)]v \exp\,[i(\mathbf{k}_\alpha\cdot\mathbf{r}_1 + \mathbf{k}_\beta\cdot\mathbf{r}_2)]\\ &\approx v_0\frac{b^3}{\Omega}\end{aligned} \tag{6.2}$$

where the integration is limited to $|\mathbf{r}_1 - \mathbf{r}_2|$ outside the hard core, b is the range of v, and v_0 its depth. Thus for $\Omega \to \infty$ the matrix element in (6.2) tends to zero. Comparing (6.2) with (4.8) we conclude that

$$\epsilon_{\alpha\beta} = 0\left(\frac{1}{\Omega}\right) \tag{6.3}$$

and thus

$$\epsilon_{\alpha\beta} \to 0 \qquad \text{as} \qquad \Omega \to \infty \tag{6.4}$$

This result need not mean that the corrections to the N-particle energy vanish, since there we have a sum $\Sigma_{\alpha_i<\alpha_j}\,\epsilon_{\alpha_i\alpha_j}$ (see Eq. (4.11)] and the number of terms in this sum being of order N^2, it follows that

$$\sum \epsilon_{\alpha_i\alpha_j} \sim N^2\frac{1}{\Omega} = N\rho \tag{6.5}$$

For $N \to \infty$ and $\Omega \to \infty$ with a constant density N/Ω, (6.5) shows that the correction to the *energy per particle*: $1/N \sum_{\alpha_i<\alpha_j} \epsilon_{\alpha_i\alpha_j}$, remains finite. For the solution of the Bethe–Goldstone equation that deals with an isolated pair at a time, the relation (6.3) can be of great help. We can write (2.5) in the form

$$[T_1 + T_2 - (\epsilon_\alpha + \epsilon_\beta)]\psi_{\alpha\beta} = -Q_F v\psi_{\alpha\beta} + \epsilon_{\alpha\beta}\psi_{\alpha\beta} - \langle\alpha\beta|v\psi_{\alpha\beta}\rangle\phi_\alpha\phi_\beta$$
$$= -Q_F v\psi_{\alpha\beta} + \epsilon_{\alpha\beta}(\psi_{\alpha\beta} - \phi_\alpha\phi_\beta) \quad (6.6)$$

Here the projection operator Q_F includes just the projection on the unoccupied Fermi states, so that

$$Q_{\alpha\beta} = |\alpha\beta\rangle\langle\alpha\beta| + Q_F \quad (6.7)$$

The term proportional to $(\psi_{\alpha\beta} - \phi_\alpha\phi_\beta)$ in (6.6) vanishes for $|\mathbf{r}_1 - \mathbf{r}_2| \gg h$ where h is the healing distance; for smaller values of $|\mathbf{r}_1 - \mathbf{r}_2|$ it is of order $1/\Omega$ compared to the other terms. We can therefore neglect it in solving for $\psi_{\alpha\beta}$, and are thus left with the approximate Bethe–Goldstone equation

$$[T_1 + T_2 - (\epsilon_\alpha + \epsilon_\beta)]\psi'_{\alpha\beta} = -Q_F v\psi'_{\alpha\beta} \quad (6.8)$$

In view of the condition $\psi_{\alpha\beta} \to \phi_\alpha\phi_\beta$ for $|\mathbf{r}_1 - \mathbf{r}_2| \to \infty$ and since $(\epsilon_\alpha + \epsilon_\beta)$ is also the energy in the state ϕ_α (6.8) is equivalent to a scattering problem with a potential $Q_F v$. The eigenvalue problem of the original Bethe–Goldstone equation thus causes no difficulties; we can solve the simple approximate (6.8), and to a very good approximation we obtain then $\epsilon_{\alpha\beta}$ by evaluating

$$\epsilon_{\alpha\beta} \approx \langle[\phi_\alpha\phi_\beta]v|\psi_{\alpha\beta}\rangle$$

There is, however, another feature of (6.8) that causes serious difficulties for its solution. Usually a two-particle equation is solved by transforming into center-of-mass coordinate and relative coordinate; in these coordinates the equation separates because the interaction depends on the relative coordinate only. Furthermore, since the interaction is a scalar and depends only on the magnitude of the relative coordinate, the relative angular momentum is a constant of motion and it is possible to write down separate equations for each partial wave. This is no longer true for the Bethe–Goldstone equation, as can be seen from the following argument:

Let us define for each pair $\alpha\beta$ its relative momentum and center of mass momentum:

$$\mathbf{k}_{\alpha\beta} = \tfrac{1}{2}(\mathbf{k}_\alpha - \mathbf{k}_\beta) \quad (6.9)$$
$$\mathbf{P}_{\alpha\beta} = \mathbf{k}_\alpha + \mathbf{k}_\beta$$

and similarly

$$\mathbf{r} = \mathbf{r}_1 - \mathbf{r}_2 \qquad \mathbf{R} = \tfrac{1}{2}(\mathbf{r}_1 + \mathbf{r}_2) \quad (6.10)$$

We can then write a typical product in $[Q_{\alpha\beta}]$ (6.1), in the following form:

$$\tfrac{1}{2}[\phi_\alpha(\mathbf{r}_1)\phi_\beta(\mathbf{r}_2)][\phi_\alpha^*(\mathbf{r}_1')\phi_\beta^*(\mathbf{r}_2')] = \frac{1}{2\Omega^2}\{\exp[i(\mathbf{k}_\alpha\cdot\mathbf{r}_1 + \mathbf{k}_\beta\cdot\mathbf{r}_2)]$$

$$- \exp[i(\mathbf{k}_\beta\cdot\mathbf{r}_1 + \mathbf{k}_\alpha\cdot\mathbf{r}_2)]\}$$

$$\times \{\exp[-i(\mathbf{k}_\alpha\cdot\mathbf{r}_1' + \mathbf{k}_\beta\cdot\mathbf{r}_2')] - \exp[-i(\mathbf{k}_\beta\cdot\mathbf{r}_1' + \mathbf{k}_\alpha\cdot\mathbf{r}_2')]\}$$

$$= \frac{1}{\Omega^2}\exp[i\mathbf{P}_{\alpha\beta}\cdot(\mathbf{R} - \mathbf{R}')]$$

$$\cdot[\cos \mathbf{k}_{\alpha\beta}\cdot(\mathbf{r} - \mathbf{r}') - \cos \mathbf{k}_{\alpha\beta}\cdot(\mathbf{r} + \mathbf{r}')] \quad (6.11)$$

When we now substitute (6.11) into (6.1), or into the equivalent expression for $[Q_F]$, a complication is involved in specifying the limits of $\sum_{\lambda\mu}$. In the single-particle representation λ and μ are limited by the requirement that they be unoccupied, that is,

$$k_\lambda^2 > k_F^2 \qquad \text{and} \qquad k_\mu^2 > k_F^2$$

but when we go over to relative momenta, the lower limit on the magnitude of $\mathbf{k}_{\lambda\mu}$ depends on its angle with $\mathbf{P}_{\lambda\mu}$.

This can be most easily seen by plotting the allowed regions for $\mathbf{k}_\lambda$ and $\mathbf{k}_\mu$ (see Fig. 6.1). The allowed region for $\mathbf{k}_{\lambda\mu}$ is from 0 to points outside the intersection of the two spheres of radius k_F centered at $0'$ and $0''$. We see that $k_{\lambda\mu}$ attains its lowest magnitude $\sqrt{k_F^2 - [(1/2)P_{\lambda\mu}]^2}$ when $\mathbf{k}_{\lambda\mu}\cdot\mathbf{P}_{\lambda\mu} = 0$. If, however, $\mathbf{k}_{\lambda\mu} \times \mathbf{P}_{\lambda\mu} = 0$ the lowest allowed magnitude of $k_{\lambda\mu}$ equals its maximum value, $k_F + (1/2)P_{\lambda\mu}$. Thus when the sum of $\mathbf{k}_\lambda$ and $\mathbf{k}_\mu$ is replaced by a

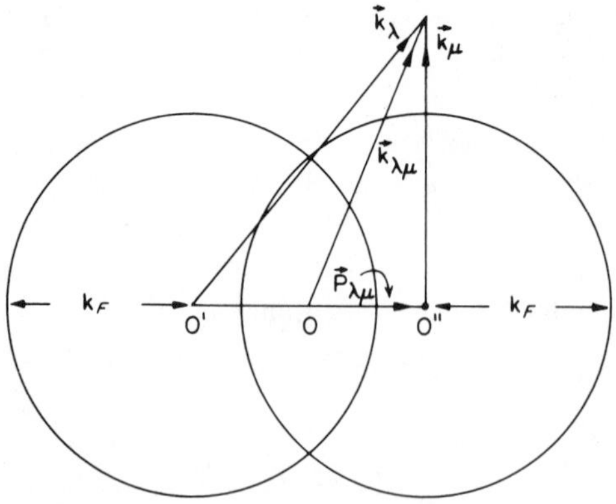

FIG. 6.1. The area outside the two spheres of radius k_F is the allowed region for $\mathbf{k}_{\lambda\mu}$, since both $\mathbf{k}_\lambda$ and $\mathbf{k}_\mu$ have to satisfy $|\mathbf{k}_\lambda| > k_F$, $|\mathbf{k}_\mu| > k_F$. The point 0 bisects the line $0'0'' = P_{\lambda\mu}$.

sum over $\mathbf{k}_{\lambda\mu}$ and $\mathbf{P}_{\lambda\mu}$, the sum over $\mathbf{k}_{\lambda\mu}$ will depend upon $\mathbf{P}_{\lambda\mu}$ in virtue of the lower limit of that sum. It follows therefore that $Q_{\alpha\beta}(\mathbf{r}_1\mathbf{r}_2, \mathbf{r}_1'\mathbf{r}_2')$ is a function not only of the magnitude $|\mathbf{r} - \mathbf{r}'|$ but also of its orientation relative to $\mathbf{P}_{\lambda\mu}$. It is then impossible to separate the Schrödinger equation for the relative coordinate into partial waves of definite angular momenta, and the equation has to be solved for all partial waves simultaneously. One might think that the whole advantage of going over to relative coordinates is thereby lost, and that the problem can be solved just as well in the space of $\mathbf{r}_1$ and $\mathbf{r}_2$. This, however, is not true if $v(12)$ contains a repulsive core, since the need to express this particular $v(|\mathbf{r}_1 - \mathbf{r}_2|)$ in terms of $\mathbf{r}_1$ and $\mathbf{r}_2$ will then make the calculation even more complicated.

In practice the approximation used to avoid this difficulty has been to average Q_F over the angle between $\mathbf{k}_{\lambda\mu}$ and $\mathbf{P}_{\lambda\mu}$. For our purpose here, which is simply to illustrate the nature of the solutions of the Bethe–Goldstone equation, it will suffice to consider the simplest case: $\mathbf{P}_{\alpha\beta} = 0$. Since the Hamiltonian commutes with the total linear momentum, it follows that all the unoccupied two-particle states $\phi_{\lambda\mu}$ into which $v(12)$ can scatter must also have $\mathbf{P}_{\lambda\mu} = 0$; it follows that for such states $\mathbf{k}_\lambda = -\mathbf{k}_\mu$ and hence the states into which scattering can take place are characterized by $|k_\mu| = (1/2)|\mathbf{k}_\lambda - \mathbf{k}_\mu| > k_F$. We can therefore use (6.11) to write the projection operator for $\mathbf{P}_{\alpha\beta} = 0$ in the form of the nonlocal operator

$$[Q_F] = \frac{1}{\Omega^2} \sum_{|k_{\lambda\mu}|>k_F} [\cos \mathbf{k}_{\lambda\mu}\cdot(\mathbf{r} - \mathbf{r}') - \cos \mathbf{k}_{\lambda\mu}\cdot(\mathbf{r} + \mathbf{r}')] \tag{6.12}$$

The approximate Bethe–Goldstone equation (6.8) for the relative coordinate $r = |\mathbf{r}_1 - \mathbf{r}_2|$ then takes the form

$$\left(-\frac{\hbar^2}{M}\nabla_r^2 - \frac{\hbar^2 k_{\alpha\beta}^2}{M}\right)[\psi_{\alpha\beta}(\mathbf{r})] = -\frac{1}{\Omega}\int d\mathbf{r}' \sum_{|k_{\lambda\mu}|>k_F} [\cos \mathbf{k}_{\lambda\mu}\cdot(\mathbf{r} - \mathbf{r}') - \cos \mathbf{k}_{\lambda\mu}\cdot(\mathbf{r} + \mathbf{r}')]v(r')[\psi_{\alpha\beta}(\mathbf{r}')] \tag{6.13}$$

In (6.13) we have made the transformation to the relative coordinate, noting that for $\mathbf{P}_{\alpha\beta} = 0$ only the relative coordinate remains and therefore

$$[\psi_{\alpha\beta}(\mathbf{r}_1, \mathbf{r}_2)] = \frac{1}{\sqrt{\Omega}}\psi_{\alpha\beta}(\mathbf{r})$$

The reduced mass $M/2$ has been introduced, and we have used the fact that $\mathbf{P}_{\alpha\beta} = 0$. We find

$$\epsilon_\alpha + \epsilon_\beta = \frac{\hbar^2}{2M}(k_\alpha^2 + k_\beta^2) = \frac{\hbar^2}{M}k_{\alpha\beta}^2 \qquad \mathbf{k}_{\alpha\beta} = \tfrac{1}{2}(\mathbf{k}_\alpha - \mathbf{k}_\beta) \tag{6.14}$$

Only $1/\Omega$ is left in front of the projection operator in (6.13); when the $d\mathbf{r}'$ integration is carried out it will give just this factor Ω. The summation over $k_{\lambda\mu}$ in (6.13) can be replaced by integration using the relation (II. 2.6)

$$\sum_{k_i} F(\mathbf{k}_i) \to \frac{\Omega}{(2\pi)^3} \int F(\mathbf{k}) d\mathbf{k} \tag{6.15}$$

Equation 6.13 then takes on the form

$$\frac{\hbar^2}{M} [\nabla_r^2 + k_{\alpha\beta}^2][\psi_{\alpha\beta}(\mathbf{r})] = \int_{k_F}^{\infty} \frac{k^2 dk}{(2\pi)^3} \int d\Omega_k \int d\mathbf{r}' [\cos \mathbf{k}\cdot(\mathbf{r} - \mathbf{r}') - \cos \mathbf{k}\cdot(\mathbf{r} + \mathbf{r}')] v(r') [\psi_{\alpha\beta}(\mathbf{r}')] \tag{6.16}$$

with the asymptotic boundary condition

$$[\psi_{\alpha\beta}(\mathbf{r})] \underset{r\to\infty}{\longrightarrow} [\phi_{\alpha\beta}(\mathbf{r})] \equiv \frac{1}{\sqrt{\Omega}} [\exp(i\mathbf{k}_{\alpha\beta}\cdot\mathbf{r}) - \exp(-i\mathbf{k}_{\alpha\beta}\cdot\mathbf{r})]$$

$$= \frac{2i}{\sqrt{\Omega}} \sin k_{\alpha\beta}\cdot\mathbf{r} \tag{6.17}$$

The hard core potential can be introduced into (6.16) rather simply by putting

$$[\psi_{\alpha\beta}(\mathbf{r}')] = 0 \qquad \text{for} \qquad |\mathbf{r}'| < r_c \tag{6.18}$$

and solving (6.16) for $|\mathbf{r}| > r_c$ only.

Having reduced the problem of solving the Bethe–Goldstone equation to that of an explicit integro-differential equation, we shall not go into the details of the various mathematical methods used for its solution, but rather examine some of the results.

Figures 6.2 to 6.5 taken from Gomes, Walecka, and Weisskopf (58) show the radial s-wave solution to the Bethe–Goldstone equation for pairs with total momentum $\mathbf{P}_{\alpha\beta} = 0$ and relative momentum $k_{\alpha\beta}$ as indicated. $u_0(r)$ is related to the wave function $\psi_{\alpha\beta}(r)$ through

$$\psi_{\alpha\beta}(\mathbf{r}) = \sum (2l+1) i^l P_l(\cos\theta) \frac{u_l(r)}{r} \tag{6.19}$$

where θ is the angle between $\mathbf{k}_{\alpha\beta}$ and $\mathbf{r}$. All solutions are for $k_F = 1.48 \times 10^{13}$ cm^{-1} and $r_c = 0.4 \times 10^{-13}$ cm.

Figures 6.2 and 6.3 describe the effect of a hard core all by itself. The broken line (- - -) is the s-wave with no hard core (i.e., no interaction at all)—the unperturbed solution. The line (- · · -) indicates the solution corresponding to the same energy for the free-particle scattering. The full line (———) corresponds to the Bethe–Goldstone solution. We see in these figures very clearly

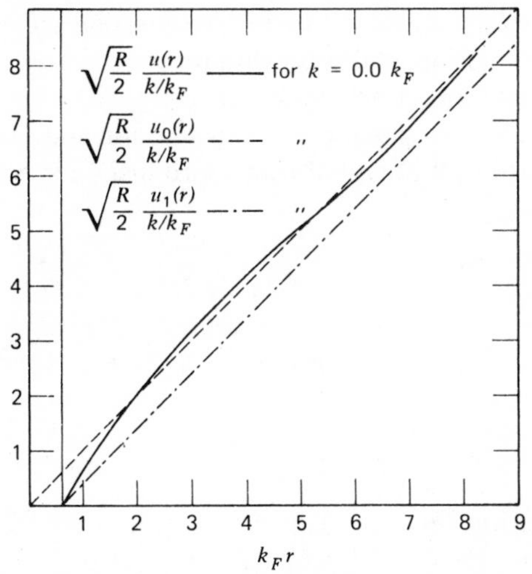

FIG. 6.2. The wave functions of the relative motion in the s-state of two particles for the case of no interaction [$u_0(r)$], for the case of a repulsive core interaction for an isolated pair [$u_1(r)$], and for a pair embedded in a Fermi distribution with $k_F = 1.48\ \mathrm{fm}^{-1}$, [$u(r)$]. The relative momentum $k = 0$ [taken from Gomes, Walecka, and Weisskopf (58)].

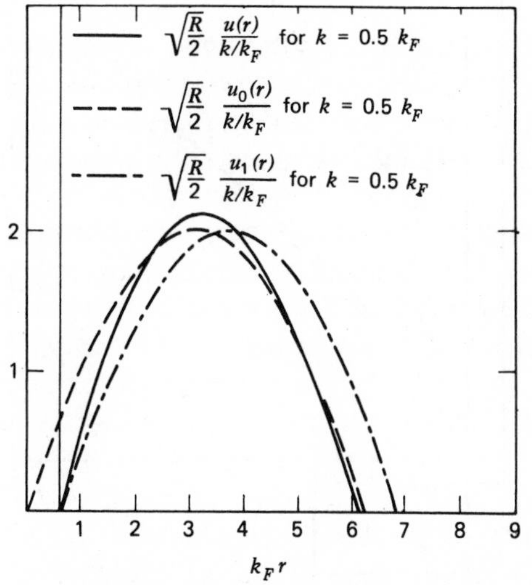

FIG. 6.3. The wave functions of the relative motion in the s-state of two particles for the case of no interaction [$u_0(r)$], for the case of a repulsive core interaction for an isolated pair [$u_1(r)$], and for a pair embedded in a Fermi distribution with $k_F = 1.48\ \mathrm{fm}^{-1}$, [$u(r)$]. The relative momentum $k = 0.5k_F$ [taken from Gomes, Walecka, and Weisskopf (58)].

how this solution starts at $r = r_c$ like the free-particle solution and goes over to the unperturbed solution, and stays with it as r increases. The healing effects of the Pauli principle are clearly recognizable.

Figures 6.4 and 6.5 show the unperturbed solution and the Bethe–Goldstone solution for a square well potential that reproduces correctly the low energy singlet scattering

$$V(r) = \begin{cases} \infty & \text{for} \quad r < r_c = 0.4 \times 10^{-13}\ \text{cm} \\ \dfrac{-\hbar^2\pi^2}{4Mb^2} & \text{for} \quad r_c < r < r_c + b \qquad b = 1.9 \times 10^{-13}\ \text{cm} \\ 0 & \text{for} \quad r < r_c + b \end{cases} \tag{6.20}$$

The average separation between nucleons, d, as indicated in these figures was taken to be 1.66 fm [$d = (\Omega/A)^{1/3}$]. We see that for all practical purposes we can consider the Bethe–Goldstone solution as healed at a distance h such that $h/d \sim 0.75$. It is remarkable that the wave function is nearly healed at a distance considerably smaller than b. Since the corrections to the Bethe–

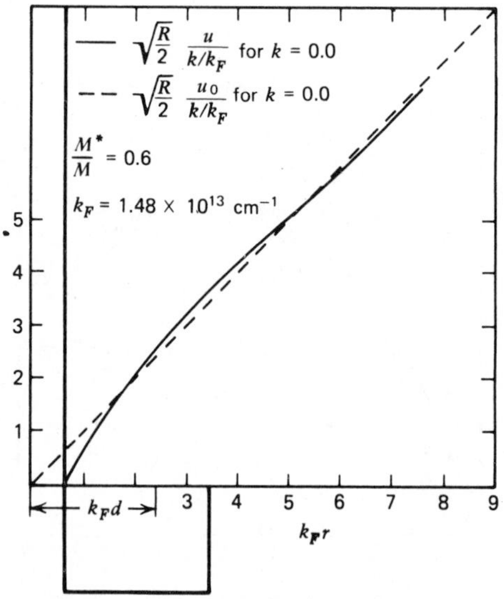

FIG. 6.4. The wave function of the relative motion in the s-state of two particles for the case of no interaction [$u_0(r)$] and for the case of the nuclear interaction [$u(r)$], for a pair embedded in a Fermi distribution $k_F = 1.48\ \text{fm}^{-1}$. The relatlve momentum is $k = 0$. The heavy line indicates the nuclear potential as a function of r. The average distance d of the next neighbor is also indicated [taken from Gomes, Walecka, and Weisskopf (58)].

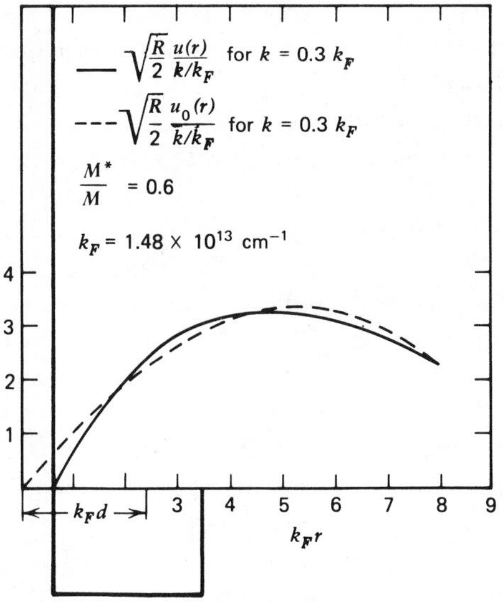

FIG. 6.5. The wave function of the relative motion in the *s*-state of two particles for the case of no interaction [$u_0(r)$] and for the case of the nuclear interaction [$u(r)$] for a pair embedded in a Fermi distribution $k_F = 1.48\ \text{fm}^{-1}$. The relative momentum is $k = 0.3k_F$. The heavy line indicates the nuclear potential as a function of r. The average distance d of the next neighbor is also indicated [taken from Gomes, Walecka, and Weisskopf (58)].

Goldstone energy are proportional to $(h/d)^6$ [see discussion leading to (4.21)], the independent pair approximation can be expected to yield the nuclear potential energy with an accuracy of about 10 to 15%.

The binding energy per particle includes the contributions of both the kinetic and the potential energies. The average kinetic energy per particle in nuclear matter is ~23 MeV (Eq. II.2.12). To obtain the observed average binding energy per particle of ~16 MeV, the average potential energy should be ~39 MeV. We see that a 15% uncertainty in the calculated average potential energy may lead to nearly 40% uncertainty in the binding energy per particle. We should not be surprised, therefore, if calculations based on the Bethe–Goldstone equation miss the binding energy per nucleon by that much.

An uncertainty in the average potential energy per nucleon reflects itself also in an uncertainty in the equilibrium density; we would expect then equilibrium densities to be given by the Bethe–Goldstone equations to an accuracy of about 10 to 15%. Even simplified realistic calculations [Gomes, Walecka, and Weisskopf, (58)] bear this out, giving an equilibrium Fermi momentum that is within 10% or less of the Fermi momentum deduced from the observed density.

Improving the calculations will require going beyond the independent pair approximation, that is, the inclusion of the effects of triple and higher order correlations. It will take us too far afield to discuss these effects here. The reader is referred to the review paper by Bethe (71).

Another effect that should be mentioned here is the dependence of the results on the assumed form of the nucleon–nucleon potential. This at the present writing is a matter of some controversy. Some authors [Lomon (72), Haftel and Tabakin (71), Coester et al. (70)] have found a moderately strong dependence. The point at issue is that there are infinitely many potentials that can fit the nucleon–nucleon data. But this agreement does not mean that the matrix elements of these differing potentials agree, and thus one would expect differing Bethe–Goldstone solutions and energies and thus different values for the binding energy per particle in nuclear matter. The authors mentioned above have found it possible to change the value of the binding energy by several MeV, by varying the nucleon–nucleon potentials. However it is not clear how the triple and higher correlations would be affected. It may be that these would tend to reduce the net dependence on the potential. However some discrimination among potentials should remain and will help eventually to choose between proposed potentials. Additional discussion of this point will be found in Chapter VII.

TABLE 6.1 Contributions to Nuclear-Matter Energy (MeV per Particle [taken from Bethe (71)]

Contribution	$k_F = 1.2$ fm^{-1}	1.36 fm^{-1}	1.6 fm^{-1}
2-body correlations	−9.79	−11.05	−10.20
3-body correlations	(− 1.6)	− 1.76	(− 2.6)
4-body correlations	(− 0.9)	− 1.09	(− 1.5)
3-body forces	(− 0.8)	− 1	(− 1.0)
Minimal relativity	(− 0.35)	− 0.5	(− 0.7)
Total	−13.4	−15.4	−16.0

With these caveats in mind, the "best" results according to Bethe (71) are given in the table 6.1 for the soft core Reid nucleon–nucleon potential. The numbers in parentheses are estimates. The three body forces and "minimal relativity" will be discussed in Volume II. For the present it will suffice to say that current theories of nuclear forces predict three-body potentials, that is, potentials that depend upon the coordinates of three particles, $v(1, 2, 3)$ that cannot be written as a sum of two-body interactions. The "minimal relativity" correction arises when the relativistic description of the nucleon–nucleon interaction is replaced by a nonrelativistic potential [H. Partovi and Lomon

(70)]. Returning to the table note that the corrections to the result cbtained with the independent pair approximation are considerable. The three- and four-body correlation contributions are (1/4) of the two-body value. The total for the experimental $k_F(=1.36 \text{ fm}^{-1})$ is remarkably close to the experimental binding energy per particle of 15.68 MeV.

7. CONCLUSION

The discussion of the two-body interactions in nuclear matter have enabled us to determine the essential factors that determine both the nuclear stability and its independent-particle nature.

The repulsive core, the tensor force, the exchange terms, and the repulsive force in the $^1P_1$ state in the nucleon–nucleon interaction combined with the Pauli principle are the dominant factors preventing the nucleus from collapsing under the action of the attractive part of the interaction. This latter part, although rather weak, would have led to a nuclear collapse because its effect increases with increasing density faster than the opposing effect of the kinetic energy. As it stands the attractive part of the nuclear interaction holds the nucleons together against the repulsive effects of both their kinetic energy and the repulsive components of the nuclear force. But due to the relative weakness of the attractive forces, the equilibrium density established thereby, corresponds to a rather high dilution of the hard cores; the equilibrium relative separation d is four times the radius of the hard core: $d/r_c \sim 4$. The hard cores occupy, therefore, only $\sim 2\%$ of the total nuclear volume.

We have also seen how the Pauli principle induces a certain "rigidity" in the nuclear wave function. Two nucleons colliding with each other inside nuclear matter must have high enough momentum components in their interaction in order for the collision to affect their motion. In fact this interaction should be able to lift the two nucleons virtually above the Fermi energy to unoccupied states, and this requires $v(12)$ to have momentum components of order k_F or bigger. Thus in nuclear matter small momenta are prevented from affecting the two-particle wave function. The attractive part of nucleon–nucleon potential does not have big momentum components in it; in fact it cannot "bend" the free two-nucleon wave function to produce more than one bound state, and even that one is barely bound. It follows that in nuclear matter the attractive part of the nucleon–nucleon interaction is prevented from affecting the two-particle correlations. The repulsive part by itself does create "holes" in the nuclear wave function, but their effect is "propagated" only over a distance $1/k_F \approx 2r_c$ because of the Pauli principle. Since equilibrium density is obtained only at $d \approx 4r_c$, these holes in the nuclear wave function are hardly felt by nucleons other than the pair involved, and we are led to the approximate validity of the independent particle picture.

The independent motion of the nucleons in the nucleus is therefore accidental in a certain sense. It is due to a fortunate combination of circumstances in which the weakness of the nuclear long range attraction, as indicated by the existence of just one bound state in the two-nucleon problem, results in a rather low nuclear density. This relative weakness combines with the Pauli principle to reduce drastically the effect of nuclear forces on two-body correlations in the nucleus. The major correlations in the resulting wave function are those imposed by antisymmetry.

CHAPTER IV

INDEPENDENT-PARTICLE MODELS FOR FINITE NUCLEI

1. SINGLE-PARTICLE POTENTIALS AND CENTER-OF-MASS MOTIONS

Thus far we have considered such nuclear properties that could be understood in terms of very idealized models for the nucleus. We have seen how the particular nature of the nuclear interaction can account for an important part of nuclear characteristics if attention is paid to the Pauli principle. The fact that nuclear forces become repulsive at large relative momenta was shown to be instrumental in determining the equilibrium density of nuclei, and in making this density essentially A-independent. On the other hand the Pauli principle, coupled with the fact that the attractive part of the nuclear interaction barely gives rise to one bound state in the two-nucleon system, was shown to lead to a nearly independent-particle motion of the nucleons in an infinite nucleus (nuclear matter). More precisely, it was shown how and why an independent *pair* approximation can give such a good approximation to the energy of nuclear matter. Moreover, when in the independent pair approximation we replaced the interaction v_{ij} by the corresponding $G(ij)$ matrix we got, formally at least, a very simple structure for nuclear matter: its energy could be calculated by considering a Fermi gas with just "first order corrections" induced by G (Section III.5).

When we consider finite nuclei we naturally try to treat them in a similar way. The most naive approach would tell us to put all the nucleons in a spherical box—that is, in a potential $U(r)$ of some form—and to generate in this way the zeroth-order wave function of a degenerate Fermi gas with a finite number of particles. Then, to include in the nuclear energy the effects of the nuclear interaction, we may attempt to introduce again a G-matrix and consider only its diagonal elements. This should give us the ladder approximation in v_{ij} (see Chapter VII), and thus corresponds to the independent pair approximation.

However here we encounter the first difficulty in carrying out such a program in practice. The equation connecting the *G*-matrix to the interaction v_{ij} depends on the density of the Fermi gas, as was pointed out in Section II.5. For nuclear matter the density is uniform and there is no problem. For a finite nucleus there is a gradient of the density at the nuclear surface, and since for all real nuclei the "surface" includes a good part of the nuclear mass, this gradient cannot be overlooked. It is possible to write down formally an equation for the *G*-matrix also for finite nuclei, but without some assumption of the effects of the density variation on *G* it is technically very hard to solve these equations.

Various approaches have been suggested to overcome this difficulty. For two body forces without infinitely repulsive cores a Hartree–Fock self-consistent field approach promises to be of great value (see Sections IV.19–21). For forces with repulsive cores a *G*-matrix has been used which, for every point $\mathbf{r}$, coincides with the nuclear matter *G*-matter derived for a constant density $\rho = \rho(\mathbf{r})$ (For a more complete discussion see ch. VII.).

The nuclear shell model is a model that escaped the problem of the relation between *G* and v in finite nuclei, by *asserting* that the nucleus can be described to some approximation in the following way:

The nucleons are assumed to move in a single-particle potential $U(i)$, which depends on the nucleon's spatial, spin, and charge coordinates; in addition, one takes into account an *effective residual interaction* v_{ij}, limiting oneself to first, or sometimes second, order perturbation theory only. This very radical assumption was not of course immediately made by nuclear physicists but was proposed by Mayer and Jensen only after a sufficient accumulation of experimental facts [See Mayer (48); Haxel, Jensen and Suess, (49)]. Indeed it is the development of these connections to which we shall devote much of this chapter.

The physical idea behind this model is rather simple: since we are dealing with fermions whose mutual interaction leads to a short healing distance (see Section III.3), it is plausible to assume that in the lowest states collisions between nucleons in the nucleus rarely lead to scattering outside of the forward direction. In a finite nucleus this is equivalent to saying that the probability of an excitation being induced by a nucleon collision is small. The residual effective interaction is thus "weak" and can be treated as a small perturbation. We shall later deal with the question of the plausibility of this argument. In particular we must ask what is the best $U(i)$ and what then is the corresponding effective residual interaction v_{ij}. But before discussing these points we must deal with another connected problem raised by the shell model: an arbitrary single-particle potential, when filled with *A*-fermions leads to a wave function—say a Slater determinant, which *does not* separate into a product of wave functions describing the center-of-mass motion and an internal motion. By taking such a wave function for the description of a given nucleus we mix two things that in nature are clearly separated. Our energies, for instance, instead of reflecting purely the internal forces, have, mixed in a generally complicated manner, both the physical internal energies and the center-of-mass energy. This center-of-mass energy is a reflection of the motion of the center of mass in some external field. The determination of this energy and this field for a given shell-model wave function is a very difficult problem.

This difficulty is connected in an essential way with the fact that we now consider a finite system. In the *infinite* nuclear matter we can use a plane wave for the wave function of a single nucleon, since this is the appropriate wave function for a translationally invariant system. For the same reason momentum is conserved in each collision of the pair, even though we are not dealing with pairs of particles in free space. A finite nucleus is invariant with respect to the translation of its center-of-mass only, and all this teaches us is that a plane wave $\exp\,[i(\mathbf{P}\cdot\mathbf{R}/\hbar)]$ gives the proper dependence of a correct nuclear wave function on the center-of-mass coordinate. (Here $\mathbf{P}$ is the center-of-mass momentum and $\mathbf{R} = (1/A)\sum \mathbf{r}_i$ is the center-of-mass coordinate). This, however, is generally not interesting in the present context since it is the *intrinsic* motion of the nucleons within the nucleus that we are after.

Physically it is desirable to formulate the finite-nucleus problem in terms of the coordinates of the nucleons with respect to their center of mass. $\Psi(\mathbf{r}_1, \mathbf{r}_2, \ldots, \mathbf{r}_A)$ describes then the intrinsic motion of the nucleons. However, in this case the A-vectors $\mathbf{r}_1, \ldots, \mathbf{r}_A$ are not independent, since they satisfy the relation $\sum \mathbf{r}_i = 0$. Unfortunately the handling of the Schrödinger equation with nonindependent variables becomes quite complicated.

Another possible way of dealing with intrinsic motion only is that of introducing $A - 1$ *independent* vectors $\boldsymbol{\xi}_i$, $i = 2, \ldots, A$ to describe the intrinsic coordinates. A possible set of such vectors that is often used in the treatment of three body systems, is defined in the following way:

$$\boldsymbol{\xi}_2 = \mathbf{r}_2 - \mathbf{r}_1,$$

$$\boldsymbol{\xi}_3 = \mathbf{r}_3 - \frac{\mathbf{r}_1 + \mathbf{r}_2}{2},$$

$$\boldsymbol{\xi}_4 = \mathbf{r}_4 - \frac{\mathbf{r}_1 + \mathbf{r}_2 + \mathbf{r}_3}{3}, \ldots,$$

$$\boldsymbol{\xi}_A = \mathbf{r}_A - \frac{\mathbf{r}_1 + \mathbf{r}_2 + \ldots \mathbf{r}_{A-1}}{A - 1}$$

Thus, $\boldsymbol{\xi}_n$ is the vector from the nth particle to the center of mass of the first $n - 1$ particles. This set of variables has one very serious drawback: it is not symmetric in all the nucleons, and the handling of antisymmetrization with these coordinates becomes quite involved.

The nature of the difficulty is quite simple to understand: a free nucleus occupies, strictly speaking, all of space, because its center of gravity is described by a plane wave. There are, of course, important correlations in this nuclear wave function, in that although its center of mass fills space uniformly, the probability of finding *two* nucleons separated by a distance larger than the nuclear diameter approaches zero. But our way of handling systems of many degrees of freedom always calls for their description in terms of independent, or nearly independent, degrees of freedom. The transformation of a system of coupled oscillators into new coordinates that lead to the noninteracting normal mode oscillations is such an example. In addition, to be able to carry out antisymmetrization in a transparent way, it is desirable to visualize the A-particle system as composed of A single-particle systems. The latter have to fill space uniformly in order to match the behavior of the free nucleus, and we are back to the description of nuclei in terms of plane waves. In other words the requirements of antisymmetrization and the available methods for handling many-body problems lead to the obviously inadequate plane-wave function for each nucleon.

2. THE SHELL MODEL WITH HARMONIC-OSCILLATOR POTENTIAL

These remarks hint at a possible way out of this dilemma: In order to be able to approximate the nuclear wave functions by some combination of bound-state, single-nucleon wave functions, we have to bind the whole nucleus to a point in space, but *without affecting its intrinsic structure.* This can be achieved if instead of studying a system obeying the Schrödinger equation

$$H\Psi_\alpha \equiv \left(\sum T_i + \sum_{i<j} v_{ij}\right)\Psi_\alpha(1, \ldots, A) = E_\alpha\Psi_\alpha(1, \ldots, A) \quad (2.1)$$

we choose to study a system of A-particles obeying a modified Schrödinger equation

$$H'\Psi_\alpha \equiv \left[\sum T_i + \sum_{i<j} v_{ij} + U(R)\right]\Psi_\alpha(1, \ldots, A) = E'_\alpha\Psi_\alpha(1, \ldots, A) \quad (2.2)$$

Equation 2.2 differs from (2.1) through the introduction of a potential $U(R)$ acting on the center-of-mass coordinate only. The energy E_α in (2.1) is composed of two parts: one part originates in the kinetic energy of the center-of-mass motion; it is totally independent of the nuclear interactions and is given by $P_\alpha^2/2AM$, where P_α is the center-of-mass momentum. The rest of the energy, $\epsilon_{i\alpha}$, has to do with the effects of the interaction v_{ij} and is called the "internal energy in the state α". $\epsilon_{i\alpha}$ is the part we are generally interested in; unlike the center-of-mass energy, which always assumes a continuous set of values, $\epsilon_{i\alpha}$ has generally a spectrum that in part consists of discrete energy levels. These are the discrete bound states observed in nuclei. The corresponding spectrum of E_α shows a continuous "band" of different center-of-mass energies superimposed on each internal state:

$$E_\alpha = \epsilon_{i\alpha} + \frac{P_\alpha^2}{2AM} \tag{2.3}$$

Turning to (2.2) the center of mass (c.m.) is bound by the potential $U(R)$ (more precisely, if $U(R)$ is chosen to be attractive and strong enough, there are states Ψ_α that correspond to bound states of the center of mass). E_α is therefore given by

$$E_\alpha' = \epsilon_{i\alpha} + E_\alpha^{\text{c.m.}} \tag{2.4}$$

Since $U(R)$ is a known potential, $E_\alpha^{\text{c.m.}}$ is in principle known, like $P_\alpha^2/2AM$ in (2.3), and we can therefore extract the interesting internal energies from (2.4), just as well as from (2.3). The advantage of the modified equation (2.2) is, however, obvious: $\Psi_\alpha(1, \ldots, A)$ is now a *localized* wave function. Its extension depends on the actual size of the nucleus as determined by its internal wave function. To be more precise this situation can be achieved if the potential $U(R)$ ties the center of mass strongly enough. In fact the strength of $U(R)$ can be considered as a free parameter at our disposal, and it can be chosen to fit best the particular approximation that is used to solve (2.2). Such localized wave functions can be approximated by a combination of single-particle wave functions tied to the *same* point in space. We are thus led to the description of the nuclear wave function in terms of independent particles moving in a single-particle potential, that is, in a finite spherical box, or in other words we are led to the shell model.

The single-particle potential in which the nucleons are considered to be bound is often compared to the central potential in the atom that binds the electrons in their orbits. However, in the latter case there is physically a heavy central body—the nucleus—whose separate motion can be handled to a good approximation, and that serves as the source for the binding potential. Such a physical central source does not exist in the nuclear case, and yet we were able to introduce it in a rigorous way. The price to be paid is the necessity of separating from the total energy that part which corresponds to the center-of-mass motion. The introduction of an attractive $U(R)$ makes this task simpler because $E_\alpha^{\text{c.m.}}$ has a discrete spectrum. The separation can easily be done if we can evaluate the exact energy E_α'; but if, due to our approximations, we have only an approximate value for E_α', the separation of the center-of-mass energy from the total energy is not always straightforward.

The choice of $U(R)$ in (2.2) has been left arbitrary thus far, and everything we have said holds true for a wide variety of attractive potentials $U(R)$. There is, however, one choice of U that is particularly simple to handle: that of the harmonic oscillator

$$U(R) = \tfrac{1}{2}AM\omega^2R^2 \qquad \mathbf{R} = \frac{1}{A}\sum_1^A \mathbf{r}_i \tag{2.5}$$

The usefulness of this particular choice for $U(R)$ becomes evident if we note the identity

$$AR^2 = \sum_{i=1}^{A} r_i^2 - \frac{1}{A}\sum_{i<j} (\mathbf{r}_i - \mathbf{r}_j)^2 \tag{2.6}$$

Using (2.6), we can write the modified Schrödinger equation (2.2) for this particular choice of $U(R)$:

$$H'\Psi_\alpha \equiv [\sum_{i=1} (T_i + \tfrac{1}{2}M\omega^2 r_i^2) + \sum_{i<j} v'_{ij}]\Psi_\alpha(1, \ldots, A) = E'_\alpha\Psi_\alpha(1, \ldots, A) \tag{2.7}$$

where the new two body interaction v'_{ij} is given by

$$v'_{ij} = v_{ij} - \frac{M\omega^2}{2A}(\mathbf{r}_i - \mathbf{r}_j)^2 \tag{2.8}$$

Formally (2.7) reads like the Schrödinger equation for A-particles moving in a central harmonic potential with an additional interaction v'_{ij} among them. This already looks like a shell model: a description of the nucleus in terms of particles moving independently in a central potential, except that we do not know whether the residual interaction in (2.8) is small enough to be treated consistently as a perturbation.

Equation 2.7 is still an exact formulation of the nuclear problem and its exact solution for any value of ω will yield the same spectrum of internal energies $\epsilon_{i\alpha}$ independent of ω. Taking different values of ω will lead to different energies E'_α, but after correction for the center-of-mass energy, which in this case is just $(n_\alpha + 3/2)\hbar\omega$, the same internal energies must emerge.

Although the choice of ω in (2.5) or (2.7) is therefore immaterial, yet, when we consider that the exact solution of (2.7) is beyond our means today, and that we have to be satisfied with approximations to (2.7), there is a "best" choice of ω. This can be seen as follows. We want our approximate wave function to be as close as possible to that of an independent-particle model, because the latter can be written down immediately. Suppose then that we disregard $\sum_{i<j} v'_{ij}$ in (2.7). We are left with the problem of A-fermions moving in a central field. To obtain the lowest state, we just calculate the energy levels in the three-dimensional harmonic-oscillator potential and fill the lowest $A/4$ states. A Slater determinant $\Phi(1, \ldots, A)$ of the corresponding wave functions will yield the desired antisymmetric independent-particle wave function. This will not, of course, be the exact eigenfunction of the com-

plete Hamiltonian (2.7), there being three main differences between the approximate Slater determinant Φ and the exact wave function of the ground state Ψ:

(1) Φ contains only the Pauli correlations between the nucleons, that is, those that are produced by antisymmetrization, whereas Ψ has dynamical correlations as well.

(2) The spatial extension of Φ is characterized by ω—the steepness of the harmonic-oscillator potential, whereas that of Ψ is determined by the size of the nucleus A.

(3) The behavior of Φ at large distances from the origin is dictated by the fact that the harmonic oscillator potential keeps increasing to infinity. Thus the nucleon density in Φ falls off as $\exp(-\alpha r^2)$. The exact solution corresponds to a situation where there is no force acting between the nucleons at large relative separations from each other. The nucleon density in Ψ falls off as $\exp(-\beta r)$. Thus there is an important difference in the asymptotic behavior of Φ and Ψ.

As we proceed from the approximate wave function Φ, trying to improve it by taking into account the residual interaction in various approximations, we shall gradually correct for these "defects" in Φ. It will, therefore, be wise to choose ω right from the beginning so that the spatial extension of Φ is similar to that of Ψ. In this case the "burden" put on the successive approximations will be the lightest; all they will have to do is to introduce dynamical correlations into the wave function, and correct the asymptotic behavior without having to change the extension of the wave function. Qualitatively it would seem advisable to choose ω so that the average of some power of r, say the rms radius, calculated with Φ coincides with its measured value. Thus ω is chosen so that

$$\int \Phi_\omega^*(1, 2, \ldots, A)\left[\frac{1}{A}\sum_{i=1}^{A} r_i^2\right]\Phi_\omega(1, 2, \ldots, A)d(1)d(2)\ldots d(A)$$
$$\equiv \langle R^2 \rangle = (r_0 A^{1/3})^2 \quad (2.9)$$

Here Φ_ω is a Slater determinant obtained by filling the lowest $A/4$ states in a harmonic-oscillator potential $(1/2)M\omega^2 r^2$, putting four particles in each state.

For harmonic-oscillator functions the integral on the left of (2.9) is easy to calculate. From the virial theorem one knows that for a single particle in a given harmonic-oscillator orbit nl

$$\tfrac{1}{2}M\omega^2\langle nl|r^2|nl\rangle = \tfrac{1}{2}E(nl) = \tfrac{1}{2}(2n + l - 1/2)\hbar\omega \quad (2.10)$$

where l stands for the orbital angular momentum of the specific orbit and n the number of nodes in the radial wave function (excluding the one at the origin and including the one at infinity). Filling the lowest $A/4$ orbits we find then the following relation between ω and A for large values of A:

$$\hbar\omega = \frac{5}{4}\frac{\hbar^2}{Mr_0^2}\left(\frac{3}{2A}\right)^{1/3} \approx \frac{41}{A^{1/3}} \text{ Mev} \tag{2.11}$$

Remark. To obtain (2.11) we note that the kth shell in the harmonic oscillator can accommodate

$$N(k) = \tfrac{1}{2}(k^2 + k) \text{ particles}$$

Therefore, if we fill all shells up to, and including $k = K$, the total number of particles that can be accommodated is

$$\sum_{k=1}^{K} N(k) = \frac{1}{2}\sum_{1}^{K} (k^2 + k) = \frac{K^3}{6} + \frac{K^2}{2} + \frac{K}{3} = \tfrac{1}{6}K(K+1)(K+2) = \frac{A}{4} \tag{2.12}$$

The energy of the kth shell is

$$E(k) = (k + 1/2)\hbar\omega$$

and therefore each particle there contributes to $< r^2 >$ the amount of

$$\langle r^2 \rangle_k = \frac{E(k)}{M\omega^2} = \frac{\hbar}{M\omega}(k + 1/2) \tag{2.13}$$

For K-filled shells the mean square radius is therefore

$$\begin{aligned}
\overline{\langle r^2 \rangle_k} &= \frac{1}{\sum_{k=1}^{K} N(k)}\frac{\hbar}{M\omega}\sum_k (k + 1/2)N(k) = \frac{2\hbar}{AM\omega}\sum_{k=1}^{K}(k^3 + \tfrac{3}{2}k^2 + \tfrac{1}{2}k) \\
&= \frac{2\hbar}{AM\omega}\left[\frac{K^4}{4} + K^3 + \frac{5K^2}{4} + \tfrac{1}{2}K\right] \\
&= \frac{2\hbar}{AM\omega}\cdot\frac{1}{6}(K+1)\left(\frac{1}{6}K^3 + \frac{K^2}{2} + \frac{K}{3}\right) \\
&= \frac{\hbar}{2AM\omega}K(K+1)^2(K+2) = \frac{3}{4}\frac{\hbar}{M\omega}(K+1)
\end{aligned} \tag{2.14}$$

where we have used (2.12) to make the last step. From (2.12)

$$(K+1)^3 = \tfrac{3}{2}A + K + 1$$

Hence, for large values of A (i.e., $A \gg K$) we can put

$$(K + 1) \approx (\tfrac{3}{2}A)^{1/3} \tag{2.15}$$

[we notice that already for $K = 3A = 40$, so that (2.15) is already good to about 2%]. In terms of the nuclear radius $R = r_o A^{1/3}$ we have

$$\overline{\langle r^2 \rangle_k} = \int r^2 \rho(r) d\tau \bigg/ \int \rho(r) d\tau = \frac{3}{5} R^2 \tag{2.16}$$

where $\rho(r)$ is assumed to be a step function. Hence

$$R^2 = \tfrac{5}{3} \overline{\langle r^2 \rangle_k} = \frac{5}{4} \frac{\hbar}{M\omega} (\tfrac{3}{2}A)^{1/3} = r_0^2 A^{2/3}$$

and therefore

$$\hbar\omega = \frac{5}{4} \frac{\hbar^2}{Mr_0^2} \cdot \left(\frac{3}{2A}\right)^{1/3}$$

3. THE HARMONIC-OSCILLATOR WAVE FUNCTION

The usefulness of the introduction of harmonic-oscillator function for $U(R)$ depends on the extent to which the solutions of the modified Hamiltonian (2.7) can be approximated by solutions of the single-particle harmonic-oscillator Hamiltonian

$$H_0 = \sum_i (T_i + \tfrac{1}{2} M\omega^2 \mathbf{r}_i^2) \tag{3.1}$$

We do not expect the eigenfunctions of H_0 themselves to provide a good description of the nucleus; but if they can constitute a good zeroth-order approximation to the solution of the complete modified Hamiltonian (2.2) then we have made real progress. This depends largely on the particular part of the nuclear wave function we need to know. To demonstrate this let us assume that we require the knowledge of the nuclear wave function at relatively large distances from the nuclear center of mass. We know that the effective nuclear potential is a *finite* one (it requires about 8 MeV to ionize a nucleus) and therefore we expect a quantum mechanical "leakage" of the wave function "outside" the nucleus (i.e., outside the sphere of radius $R = r_0 A^{1/3}$). Particles that interact strongly with the nuclear constituents feel the presence of nucleons already at relatively large distances from the nuclear surface, and the knowledge of the nuclear wave function there may therefore be of real interest. For example, the probability of finding just a *single* nucleon outside the nucleus is of particular importance in the (p, d) pickup reaction. In that case we need to know the behavior of $\Psi_\alpha(\mathbf{r}_1, \mathbf{r}_2, \ldots, \mathbf{r}_A)$ when $|\mathbf{r}_A - \mathbf{r}_i| \gg R$ for $i = 1, 2, \ldots, A - 1$. Let us examine this behavior in more detail.

To obtain this behavior of the exact Ψ we note that it satisfies the Schrödinger equation (2.1), in the absence of $U(R)$:

$$\left(\sum_{i=2}^{A} T_i + \sum_{i\geq 2,\, j>i}^{A} v_{ij} + T_1 + \sum_{j=2}^{A} v_{1j}\right) \Psi_\alpha(1, \ldots, A)$$
$$= E_\alpha{}^{(A)} \Psi_\alpha(1, \ldots, A) \quad (3.2)$$

Recalling the discussion of Section 1, only $3(A - 1)$ spatial variables are independent. We introduce now the exact $(A - 1)$ particle wave function $\Phi_\beta(2, \ldots, A)$ satisfying

$$\left(\sum_{i=2}^{A} T_i + \sum_{i\geq 2,\, j>i}^{A} v_{ij}\right) \Phi_\beta(2, \ldots, A) = E_\beta{}^{(A-1)} \Phi_\beta(2, \ldots, A) \quad (3.3)$$

Multiplying (3.2) by Φ_β^* and integrating over all independent variables except the first we obtain, taking into account (3.3)

$$T_1 \psi_{\beta\alpha}(1) + \int \Phi_\beta^*(2, \ldots, A) \sum_{j=2}^{A} v_{1j} \psi_\alpha(1, 2, \ldots, A) d(2) \ldots$$
$$= [E_\alpha^{(A)} - E_\beta^{(A-1)}] \psi_{\beta\alpha}(1) \quad (3.4)$$

where $\psi_{\beta\alpha}(1)$ describes the motion of a particle in Ψ_α when the other $(A - 1)$ particles are known to be in the state Φ_β, that is,

$$\psi_{\beta\alpha}(1) = \int \Phi_\beta^*(2, \ldots, A) \Psi_\alpha(1, 2, \ldots, A) d(2) \ldots \quad (3.5)$$

In (3.4) we now take Ψ_α to be the ground state of the A-particle system. The bulk of the contribution to the integral in (3.4) comes then from that part of configuration space for which $|\mathbf{r}_i - \mathbf{r}_j| < 2R$; $i, j = 2, \ldots, A$. If we are interested in the large distance behavior of $\psi_{\beta\alpha}(1)$, that is, $\mathbf{r}_1 \to \infty$, then, due to the finite range of v_{1i}, and the finite radii of the $(A - 1)$ and A-particle wave functions the contributions to the integral in (3.4) will be small and we can neglect it in comparison with $(E_\alpha{}^{(A)} - E_\beta{}^{(A-1)})\psi_{\beta\alpha}(1)$. With our choice of the states α and β, $E_\alpha{}^{(A)} - E_\beta{}^{(A-1)}$ is just the ionization energy of the Ath particle. If the ground state of the nucleus A is stable against the emission of the Ath particle then $E_\alpha{}^{(A)} - E_\beta{}^{A-1)} < 0$, and, defining

$$\chi = \sqrt{-\frac{2M}{\hbar^2}(E_\alpha{}^{(A)} - E_\beta{}^{(A-1)})} \quad (3.6)$$

(3.4) leads to the result

$$\psi_{\beta\alpha}(1) \to \text{constant} \exp(-\chi r_1) \quad (3.7)$$

where r_1 is measured from the center of mass of $(A - 1)$.

Remark. Strictly speaking we should be more careful in our definition of Ψ_α and Φ_β; since these are supposed to be exact solutions of their respective Hamiltonians, the center-of-mass motion has to be specified. However it is possible to overcome this difficulty if we define Ψ_α and Φ_β to be just the *intrinsic* part of the nuclear wave function, so that the complete solution of (3.2) is not Ψ_α but exp $[(i/\hbar)\mathbf{P}\cdot\sum \mathbf{r}_i]\Psi_\alpha$, and similarly for Φ_β. Only relative coordinates should then appear in Ψ_α, and that is why we are able to say that in (3.7) r_1 is measured relative to the center of mass $(A-1)$. More precisely, it could be measured relative to any of the $A-1$ particles, but in the asymptotic region in which we are interested this makes no difference.

Equation 3.7 is nothing but the well-known result that quantum mechanical leakage through a potential dies off exponentially, with a characteristic inverse length determined by the height of the potential. In our case this height is replaced, of course, by the binding energy of the particle we are considering.

If we now look back at the wave functions generated by the harmonic oscillator, we see a different asymptotic behavior. In fact the wave function of a particle bound by a harmonic oscillator potential dies off at large distances exp $(-\alpha r^2)$ (α being simply related to ω). This is due to the fact that here we have to do with a potential that grows steadily so that the binding energy of any particle in this potential is infinite. There is nothing strange in this result, since by going over to the modified Hamiltonian and putting the nucleus in a harmonic oscillator potential $U(R)$ we deliberately caused its internal wave function to be multiplied by a harmonic-oscillator eigenfunction that involves a factor exp $(-\alpha R^2)$. We can recover the correct internal wave function from an exact solution of (2.7) by just multiplying the latter by the inverse harmonic oscillator eigenfunction, which would involve a factor exp $(+\alpha R^2)$. Thus if we proceed to solve (2.7) by first putting independent particles in a harmonic oscillator potential and then figuring out *exactly* the effects of v'_{ij}, we shall indeed get a solution in which each particle dies off at large distances as exp $(-\alpha r_i^2)$. But after multiplying this solution by the inverse wave function for the center-of-mass motion, all these exp $(-\alpha r_i^2)$ factors will be cancelled out, and we shall be left with the expected exponential decay exp $(-\chi r_1)$ for the single-particle function $\psi(r_1)$. The trouble, however, is that cancellation to yield the correct exponential behavior depends critically on our having an *exact* solution to (2.7). If, as usually is the case, we have only *approximate* solutions to (2.7) we cannot expect that a cancellation of the exp $(-\alpha r_i^2)$ asymptotic behavior will result from its multiplication by the inverse of the center-of-mass wave function. Thus the method of the harmonic oscillator potential we have outlined will generally lead to unreliable results for such processes in which the *tail* of the nuclear wave function plays the dominant role. For processes in which the *interior* part of the nucleus makes the most significant contributions, the harmonic oscillator wave functions will be more adequate particularly for matrix elements of single-particle operators. The question of the validity of a shell model with a harmonic-oscillator potential depends on the application we have in mind for that model. This situation will repeat itself in the future again and again. The simplifying assumptions that make a model soluble necessarily limit its range of validity. We should be prepared to meet with only a partial validity of our models, and as we shall see one can gain important information about nuclear structure also from the search for phenomena that "contradict" the model.

The harmonic-oscillator center-of-mass potential can really simplify the handling of the nuclear structure problem if it is possible to consider v'_{ij} in (2.7) as a perturbation. Otherwise we are back to the exact problem and we have not gained anything. On the basis of what we have seen thus far we may feel some justification in expecting this to be really the case. First, by choosing ω so as to reproduce in zeroth order the observed size of the nucleus A, we automatically obtain, even before we take v'_{ij} into account, all the Pauli correlations at the pertinent density. This, in itself, as we saw in the case of the Fermi gas, is sufficient to give us a semiquantitative understanding of a number of nuclear properties. Then, from the study of nuclear matter, we expect that by limiting ourselves to the independent pair approximation we shall not be ignoring important factors that determine the details of nuclear structure. Furthermore, as long as we limit ourselves to independent pair descriptions we can make use of the G-matrix formalism and work with the zeroth-order wave functions with an even greater reliability.

4. THE GENERALIZED SHELL MODEL

At this stage it might be proper to go back to our results on the two nucleon system, work out the G-matrix appropriate for the modified two-body interaction v'_{ij}, and carry out with it the computation of nuclear properties.

This, however, turns out not to be the best way for the understanding of nuclear structure. There are some additional important restrictions on the nuclear wave function, which are independent of some, or even all, of the details of the residual interaction v'_{ij}. These restrictions are connected with the conservation laws of angular momentum, of isospin and of parity, and we gain much insight by studying their effects first. Normally there is relatively little information that can be gotten about an A-particle system from the fact that its total angular momentum is conserved. But if the system can be described in zeroeth order as a system of *A-independent* fermions in a spherical box the situation changes drastically. For now the angular momentum of each of the fermions separately is conserved, and this fact together with the conservation of the total angular momentum leaves very little freedom in the A-fermion wave function. Moreover, if v'_{ij} is to be treated as a first-order perturbation only, then, as we shall see, it again is possible to make many general statements about the properties of nuclear levels without invoking the detailed features of v'_{ij}.

Instead of proceeding with the various approximations to (2.7) we therefore propose to study the following model for the nucleus:

The nuclear wave functions $\Psi(1, \ldots, A)$ and the nuclear energies E are to be the eigenfunctions and the eigenvalues of the Hamiltonian

$$H = \sum_{i=1}^{A} [T_i + U(i)] + \sum_{i<j=1}^{A} v(ij) \tag{4.1}$$

where it is understood that:

(a) Ψ is to be antisymmetrized with respect to all its particles (protons and neutrons are considered to be different internal states of the nucleon).

(b) $U(i)$ is an overall attractive potential that is a scalar function of the variables of the ith particle [i.e., it can be only a function of the magnitude of $\mathbf{r}_i$ and of $(\mathbf{l}_i \cdot \boldsymbol{\sigma}_i)$].

(c) $v(ij)$ is a scalar function of the variables of particles i and j, symmetric in i and j [i.e., it could be a function of $|\mathbf{r}_i - \mathbf{r}_j|$, or even a function of $(\boldsymbol{\sigma}_i \cdot \mathbf{r}_i)(\boldsymbol{\sigma}_j \cdot \mathbf{r}_j)$, etc.].

(d) $v(ij)$ is a "weak potential," in the sense that it can be treated as a small perturbation. The unperturbed Hamiltonian is

$$H_0 \equiv \sum_{i=1}^{A} (T_i + U_i) \tag{4.2}$$

The modified Hamiltonian (2.7) is seen to be a special case of (4.1), but thus far we have not shown to what extent v'_{ij} can in fact be treated as a perturbation [see (d)]. For simplicity the prime on v_{ij} has been dropped in (4.1).

The physical picture behind (4.1) is that of nucleons moving in a spherical box nearly independently, and it is motivated by the success of the Fermi-gas model in explaining some general characteristics of nuclei, and by the results we found on the "healing" of the nuclear wave function in nuclear matter. There it was shown that the effects of the interaction of two particles in nuclear matter are felt only when they are close to each other, and that the wave function is "healed" as soon as the separation between the particles increases beyond $\backsim 1/k_F$. A third particle colliding with one of these particles therefore does not feel the effects of their previous interaction.

The scalar property of $U(i)$ and of $v(ij)$ is required in order that the total angular momentum of the system is conserved both for the zeroth-order wave function [when $v(ij)$ is ignored] as well as for the exact wave functions. In fact, if $\mathbf{J} = \sum (\mathbf{l}_i + \mathbf{s}_i)$ is the total angular momentum operator ($\mathbf{l}_i$ and $\mathbf{s}_i$ being the orbital and spin angular momenta of the ith particle), then

$$\left[\sum_{i=1}^{A} [T_i + U(i)], \mathbf{J}\right] = 0$$

implies

$$[[T_i + U(i)], (\mathbf{l}_i + \mathbf{s}_i)] = 0$$

Since the kinetic energy T_i commutes with both $\mathbf{l}_i$ and $\mathbf{s}_i$, it follows that we must have

$$[U(i), (\mathbf{l}_i + \mathbf{s}_i)] = 0 \tag{4.3}$$

Equation 4.3 is equivalent to the statement that $U(i)$ is invariant under rotations, and hence point (b). In a similar way the requirement $[H, \mathbf{J}] = 0$ leads to

$$[v(ij), (\mathbf{l}_i + \mathbf{s}_i + \mathbf{l}_j + \mathbf{s}_j)] = 0 \tag{4.4}$$

which is again equivalent to point (c).

The exact functional form of $U(i)$ and $v(ij)$ and their dependence on $\mathbf{r}_i$, $\mathbf{r}_j$ and the intrinsic variables of the particles i and j will, of course, reflect itself in the eigenfunctions and eigenvalues of (4.1). The interesting thing, however, is that many conclusions can be drawn without having a very detailed knowledge of $U(i)$ and $v(ij)$, and we shall now proceed to derive some of them.

Since $v(ij)$ is to be considered as a perturbation, the zeroth-order wave function is determined by $U(i)$. For any given number A of particles such a wave function can be formed by taking a Slater determinant of appropriate single-particle wave functions. If there is more than one Slater determinant of A-particles that has the same zeroth-order energy, linear combinations of these Slater determinants may be more convenient to work with.

5. THE SINGLE-PARTICLE POTENTIAL

The potential $U(i)$ can depend on $\mathbf{r}_i$, $\mathbf{p}_i$, $\boldsymbol{\sigma}_i$, and τ_{i3}. The only scalars that can be constructed out of these operators and that also conserve parity (i.e., are invariant under reflections) are:

$$\mathbf{r}_i^2,\ p_i^2 \qquad \text{and} \qquad \boldsymbol{\sigma}_i \cdot (\mathbf{r}_i \times \mathbf{p}_i)$$

Therefore $U(i)$ will generally consist of two parts: one that is independent of $\boldsymbol{\sigma}$ (but may depend on p_i^2, and we shall come to that later), and one that is proportional to $\boldsymbol{\sigma}_i \cdot \mathbf{l}_i$ (since $\mathbf{r}_i \times \mathbf{p}_i = \hbar \mathbf{l}$). The former is often referred to as the "central potential." In addition, since we allow a dependence of $U(i)$ also on τ_{i3}, we can have different central potentials for neutrons and for protons. Indeed, the central potential will include a term $(1/2)(1 + \tau_{i3})V_c(\mathbf{r}_i)$, giving rise to a Coulomb potential $V_c(\mathbf{r}_i)$ when the ith particle is a proton and vanishing when it is a neutron.

The radial dependence of the central potential and the spin-orbit potential need not be related to each other, but we can expect some simple general features for both. If $U(i)$ is to give rise to a reasonable zeroth-order wave function, then it should represent the average potential that the nucleon i feels as it traverses the nucleus. Later we shall make this statement more precise, but a few things can be said already at this stage. Since the range b of the nucleon–nucleon interaction is smaller than the nuclear radius R for $A \gtrsim 10$, we can expect that if the nucleon i is inside the nucleus and removed more than a distance b from the nuclear surface there will be equal number of nucleons of all spin orientations on all its sides and it will experience no net force. Thus

$$U(i) = \text{constant} \qquad \text{if} \qquad R - r_i > b \tag{5.1}$$

At the vicinity of the nuclear surface there will be a net force on the nucleon attracting it back into the nucleus. Thus, for the central part

$$\frac{d}{dr} U_{\text{cent}}(i) > 0 \qquad \text{for} \qquad |R - r_i| < b \tag{5.2}$$

Outside the nucleus, a distance b and more from its surface, there is again no nuclear force on the ith particle (there will be a Coulomb force if the particle is charged), so that for that part of $U(i)$ we have

$$U(i) = 0 \qquad \text{for} \qquad r_i - R > b \tag{5.3}$$

[Actually we could have put $U(i) = $ constant in (5.3), and the choice $U(i) = 0$ is just a convenient normalization of the energy.]

Equation 5.1 holds for both the central part and the spin-orbit part; however, in the latter case we can even determine the value of the constant. Indeed, according to (5.1) the spin-orbit potential at the nuclear interior would be

$$U_s(i) = \text{constant}\ \boldsymbol{\sigma}_i \cdot \mathbf{l}_i \tag{5.4}$$

Depending on the sign of this constant, (5.4) would tell us that a situation in which $\boldsymbol{\sigma}_i$ is, say, parallel to $\mathbf{l}_i$ has a lower energy than the one in which it is antiparallel to $\mathbf{l}_i$. But that means that the spin of the ith nucleon experiences a force at the interior of the nucleus tending to make it parallel to the particle's orbital angular momentum. This contradicts our assertion that at the interior of the nucleus the ith nucleon sees equal density of nucleons with spin-up and spin-down on all sides and therefore should experience no net force whatsoever. We conclude therefore that the constant in (5.4) must be zero, and therefore the spin-orbit *potential* must vanish inside the nucleus. Since it obviously vanishes at large distances from the nucleus it follows that the spin-orbit potential is a *nuclear-surface* phenomenon.

We have remarked before that real nuclei consist mostly of a surface. This does not invalidate our conclusions about the central potential $U(i)$ and the spin-orbit potential. There is a region of radius $\backsim(R - b)$ around the center of the nucleus in which $U(i)$ is constant. The *volume* of that region may indeed be a small part of the total nuclear volume, and this is reflected in volume elements in integrations; but in determining a radial wave function the "flat" part of $U(i)$ is very important.

The effectiveness of a potential on a particle is proportional to the time it spends in the potential. The fact that the spin-orbit potential is concentrated at the surface therefore means that its effectiveness relative to the central potential will be roughly proportional to the surface to volume ratio, that is, to $A^{-1/3}$. We shall see later how this effectiveness is determined quantitatively.

Our qualitative considerations therefore call for a central potential of the type shown in Fig. 5.1.

The magnitude of the spin-orbit potential, on the other hand, will have a radial dependence of the type shown in Fig. 5.2.

Due to the fact that $U_s(r)$ has a shape similar to $(d/dr)U_{\text{cent}}(r)$ and since the spin-orbit interaction for an electron in an electric field has the form $U_s(r) = (g/r)(dV_c/dr)\mathbf{l}\cdot\mathbf{s}$, ($V_c \equiv$ Coulomb potential) it is often assumed in nuclear structure calculations that $U_s(r) = \gamma(1/r)(d/dr)U_{\text{cent}}(r)\mathbf{l}\cdot\mathbf{s}$. We shall not make this assumption at this stage.

The central potential shown in Fig. 5.1 does not greatly resemble the harmonic-oscillator potential. Although the harmonic-oscillator potential does have the important feature that the force it exerts on a particle is weak at the center and gets stronger as the particle moves closer to the surface, it becomes much too strong outside the nuclear radius, unlike the potential $U_{\text{cent}}(r)$. In this respect $U_{\text{cent}}(r)$ may be a more realistic potential with which to form zeroth-order wave functions. It does have, however, an important disadvantage over the harmonic-oscillator potential: a Slater determinant of single-particle orbits computed with $U_{\text{cent}}(i)$ cannot be decomposed into a product of a center-of-mass wave function and an internal wave function. It can be shown

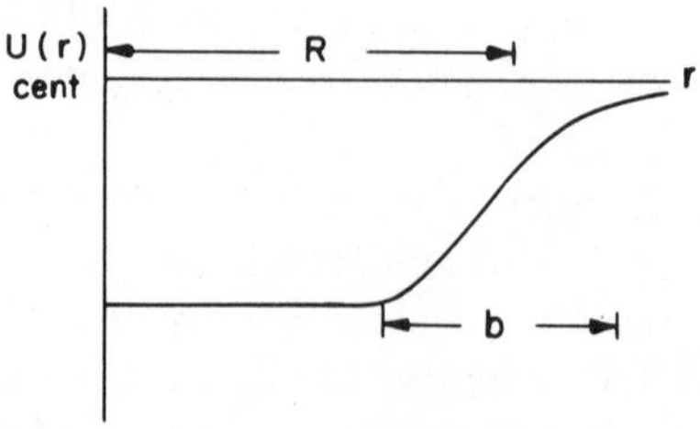

FIG. 5.1. Qualitative features of the central potential.

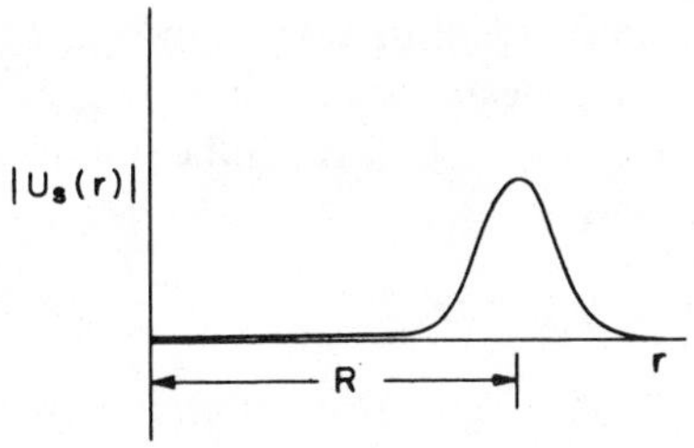

FIG. 5.2. Qualitative features of the radial dependence of the spin-orbit potential.

that this latter property exists only for harmonic-oscillator wave functions. Thus the wave functions generated from $U_{\text{cent}}(i)$ will not have a simple center-of-mass motion and, generally, it will not be possible to isolate from them, cleanly various center-of-mass effects. The final decision as to which potential to prefer must depend on the nature of the problem one is dealing with. This is again a reflection of the fact that we are dealing with a model and not with an exact theory.

6. WAVE FUNCTIONS IN THE SINGLE-PARTICLE POTENTIAL

The bound states of a single particle moving in a potential $U(i)$ are, of course, quantized. They are characterized by four quantum numbers: n—the number of nodes in the radial wave function, l—the orbital angular momentum of the particle, $j(=l \pm 1/2)$—the total angular momentum of the particle, and m—the z-projection of j where the z-axis can be chosen arbitrarily. These results follow immediately if we notice that the equation

$$[T + U_{\text{cent}}(r) + U_s(r)\boldsymbol{\sigma}\cdot\mathbf{l}]\psi(r, \theta, \phi) = E\psi(r, \theta, \phi) \tag{6.1}$$

reduces to

$$\frac{\hbar^2}{2M}\frac{d^2R_{nlj}(r)}{dr^2} + \left[E_{nlj} - U_{\text{cent}}(r) - \lambda_{lj}U_s(r) - \frac{\hbar^2}{2M}\frac{l(l+1)}{r^2}\right]R_{nlj}(r) = 0 \tag{6.2}$$

if we put

$$\psi_{nljm}(r, \theta, \phi) = \frac{1}{r}R_{nlj}(r)\mathcal{Y}_{ljm}(\theta, \phi) \tag{6.3}$$

Here $\mathcal{Y}_{ljm}(\theta, \phi)$ includes also the spin coordinates and is constructed by vector coupling the spin $\mathbf{s}$ to the orbital angular momentum $\mathbf{l}$ to form the total single-particle angular momentum $\mathbf{j}$:

$$\mathcal{Y}_{ljm}(\theta, \phi) = \sum_{m_s, m_l} (1/2 m_s l m_l | jm)\chi_{1/2}(m_s)Y_{lm_l}(\theta, \phi) \tag{6.4}$$

Here $\chi_{1/2}(m_s)$ are the spin eigenfunctions and $Y_{lm_l}(\theta, \phi)$ are spherical harmonics. The coefficient λ_{lj} reflects the fact that the spin-orbit potential is different for $j = l + 1/2$ and $j = l - 1/2$. It can be obtained by operating directly on $\mathcal{Y}_{ljm}$ with $\boldsymbol{\sigma}\cdot\mathbf{l}$. Noting that

$$\boldsymbol{\sigma}\cdot\mathbf{l} = 2\mathbf{s}\cdot\mathbf{l} = \mathbf{j}^2 - \mathbf{l}^2 - \mathbf{s}^2 \tag{6.5}$$

and that $\mathcal{Y}_{ljm}$, by its construction, is a simultaneous eigenfunction of $\mathbf{l}^2$ and $\mathbf{j}^2$, we obtain:

$$(\boldsymbol{\sigma}\cdot\mathbf{l})\mathcal{Y}_{ljm}(\theta, \phi) = [j(j+1) - l\,(l+1) - 3/4]\mathcal{Y}_{ljm}(\theta, \phi) \tag{6.6}$$

Hence

$$\lambda_{lj} = \begin{cases} l & \text{if} \quad j = l + 1/2 \\ -(l+1) & \text{if} \quad j = l - 1/2 \end{cases} \tag{6.7}$$

The radial function R_{nlj} has to satisfy the boundary conditions

$$R_{nlj}(r) \to 0 \qquad \text{for} \qquad r \to \infty \tag{6.8}$$

and $R_{nlj}(0) = 0$ (or more rigorously $R_{nlj}/r \underset{r\to 0}{\longrightarrow} M < \infty$). R_{nlj} may, or may not, have additional zeros between $r = 0$ and $r = \infty$. The quantum number n is related to the number of these zeros (nodes), and in nuclear physics literature it is customary to put it *equal* to their number (excluding the one at zero and including the one at infinity). R_{nlj} is furthermore normalized so that

$$\int_0^\infty |R_{nlj}(r)|^2\, dr = 1 \tag{6.9}$$

Two radial functions with the *same* lj-quantum numbers but different radial quantum numbers are orthogonal as can be easily verified from (6.2). Together with (6.9) we therefore obtain

$$\int R^*_{nlj}(r) R_{n'lj}(r)\, dr = \delta(n, n') \tag{6.10}$$

The orthogonality relation (6.10) *does not* necessarily hold between radial functions with different values of l and/or j. The total wave function ψ_{nljm} does satisfy such orthogonality relation that, however, results from the orthogonality of the angular spin functions $\mathcal{Y}_{ljm}$:

$$\int [\mathcal{Y}^*_{ljm}(\theta, \phi), \mathcal{Y}_{l'j'm'}(\theta\phi)]\, d(\cos\theta)\, d\phi = \delta_{ll'}\delta_{jj'}\delta_{mm'} \tag{6.11}$$

where we have used the orthogonality of the spin wave functions $[\chi^*_{1/2}(m_s), \chi_{1/2}(m_{s'})] = \delta_{m_s m_s'}$.

The remarkable thing about the structure (6.3) of the single-particle wave function for a single-particle potential is that its angular and spin-variable dependence is completely independent of the single-particle potential.

Problem. Show that the separation (6.3) holds also when $U(i)$ is a function of p^2; use a power expansion of $U(i)$ in p^2 to obtain the new radial equation for $R_{nlj}(r)$.

The eigenvalues E_{nlj}, which are determined by (6.2) through the requirement that $R_{nlj}(r) \to 0$ as $r \to \infty$, do not depend on the quantum number m. The states of a single particle in a three-dimensional scalar potential $U(i)$ are $(2j+1)$-fold degenerate. Occasionally the degeneracy may be even higher [it is $2(2l+1)$ if $U_s = 0$, for instance], but the $(2j+1)$-fold degeneracy of the eigenvalue E_{nlj} is a fundamental property of any scalar $U(i)$. Physically it is due to the fact that a scalar potential does not determine any preferred direction in space, and the energy of a particle bound in it cannot, therefore, depend along which axis we chose to quantize our angular momenta.

7. GROUPING OF LEVELS IN SINGLE-PARTICLE POTENTIALS

Although $R_{nlj}(r)$ and E_{nlj} depend on the detailed features of $U(i)$ there are some general statements that can be made about them. We shall confine ourselves now to central potentials $U(r)$, and assume, furthermore, that $U(r)$ is an attractive potential, monotonically nonincreasing in its absolute value.

$$U(r) < 0, \qquad |U(r_1)| \geq |U(r_2)| \quad \text{if} \quad r_1 < r_2 \tag{7.1}$$

and also $U(r) \to 0$ for $r \to \infty$, and $r^2 U(r) \to 0$ for $r \to 0$. (The last requirement is to assure the existence of a lowest bound state). Under these conditions we can show that of two bound states with the same orbital angular momenta, the one with more nodes is less bound. Or formally:

$$E_{n_2 l} > E_{n_1 l} \qquad \text{if} \qquad n_2 > n_1 \tag{7.2}$$

Similarly, of two orbits with the same number of nodes, the one with smaller l is more bound:

$$E_{n l_2} > E_{n l_1} \qquad \text{if} \qquad l_2 > l_1 \tag{7.3}$$

Both results become very plausible if we remember that the nodes of the radial wave function are all contained within the range of the potential, and that a big number of nodes implies more curvature of the radial wave function and, hence, more kinetic energy associated with that part of the wave function. If l is the same for the two orbits, (7.2) then follows. On the other hand, if n remains unchanged, increasing l leads to an increase in the angular kinetic energy and, hence, the relation (7.3). The formal proof of (7.2) and (7.3) is given in deShalit and Talmi (63), pp. 34–36.

The interplay between radial and orbital quantum numbers in affecting energies as indicated in (7.2) and (7.3) leads to the general phenomenon of the grouping of levels in central potentials. Roughly speaking an increase in l can be compensated by a decrease in n and vice versa. It is then possible to have $E_{n_1 l_1} \approx E_{n_2 l_2}$ when $(l_1 - l_2) = -\rho(n_1 - n_2)$, ρ being a positive number whose value depends on the form of the central potential ($\rho = 1$ for Coulomb potentials; $\rho = 2$ for harmonic oscillator and square well potentials). This grouping of levels is responsible for the formation of "shells" of particles in central potentials.

We shall now proceed to study some of the manifestations of this "shell structure" in nuclei. It will be instructive to mention also some aspects of the shell structure in atoms since in both cases there are additional effects that tend to enhance some phenomena associated with shell structure.

The most dramatic effects of shell structure are seen in the study of the variation of ionization energy with particle number. In Fig. 7.1 we show three shells, or clusters of levels in a central potential the lowest includes just one orbit, the next one two orbits and the third, three orbits. Each orbit is associated with a possible set of values for the independent quantum numbers. Assume that each orbit can accommodate just two particles. If there is no interaction among these particles, the ionization energy as a function of A will look like Fig. 7.2. Every time a new shell starts to fill there is a sharp decrease in the ionization energy of the last particle.

In real cases we encounter a different behavior. Thus in the atomic case, the transition from an atom with Z-electrons to one with $Z + 1$ electrons is connected also with an increase of the charge on the nucleus. The ionization energy, instead of staying constant, like in Fig. 7.2, as the shells fill, becomes larger because of the increase in the central potential. When a new shell starts to fill, the last electron not only occupies a higher, less bound, level in the Coulomb field but it is also screened from the attractive nuclear charge by the shell of electrons just filled; its binding energy is thus reduced even

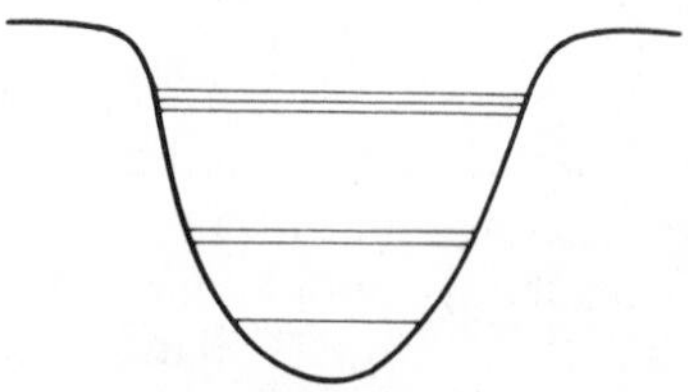

FIG. 7.1. Schematic clustering of levels in central potentials.

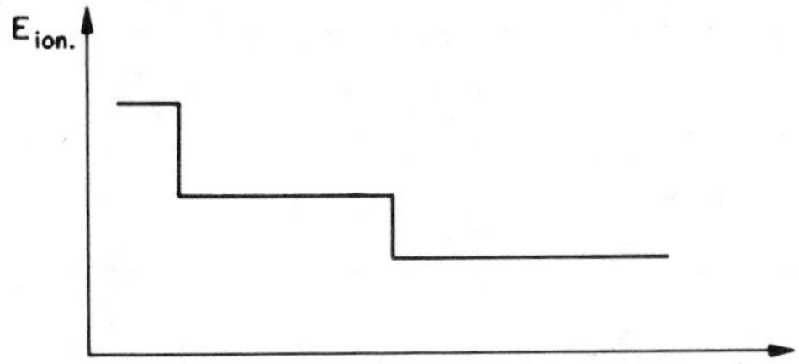

FIG. 7.2. Variation of ionization energy with size of system occupying the schematic potential shown in Fig. 7.1.

further. The ionization energy then shows large jumps at the closure of shells as shown in Fig. 7.3.

In nuclei the central potential does not get deeper as we keep increasing the mass number, since the nuclear density remains constant. Its dimensions change instead [compare, for instance, (2.11)]. This again makes each orbit more tightly bound as the nucleus increases in size although the effect is not

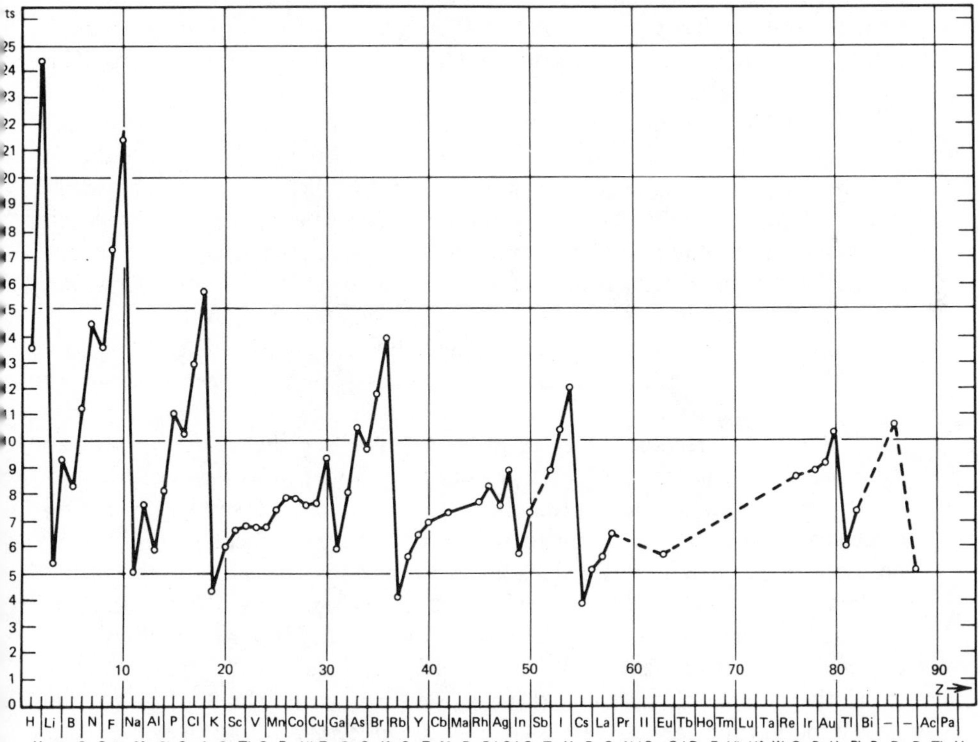

FIG. 7.3. Dependence of the ionization potential of the neutral atom on atomic number [taken from Herzberg (44)].

as dramatic as with the electrons in an atom. In addition, as we shall see later, the residual interaction and the possible momentum dependence of the central potential somewhat mask this effect. The jumps in the ionization energies everytime a new shell starts to fill are consequently less spectacular and amount to only 2 MeV, that is, $\sim 25\%$ of the ionization energy itself. They reflect predominantly the reduced binding of the new orbit more than the increase in binding of the particles in the filled shell. This can be seen in Figs. I.4.1 to I.4.3.

Discontinuities in the ionization energies are not the only indication for the existence of shells. Various other nuclear properties seem to undergo abrupt variations around the same region in the periodic table where discontinuities in ionization energies are observed. This was discussed in Chapter I.

Unlike the atomic case where all shells are filled by just one type of particle—the electron—the nuclear shells are simultaneously filled by two types of particles—neutrons and protons. Furthermore the basic two-body interactions $U(pp)$, $U(nn)$, and $U(pn)$ are all of the same order of magnitude. This provides us with an interesting possibility of testing the validity of the independent-particle approximation. We note that stable nuclei beyond ^{40}Ca have an increasing tendency toward an excess of neutrons over protons, thus indicating a somewhat different average potential for neutrons and protons. This is, of course, well understood in terms of the extra Coulomb potential that is effective for protons and not for neutrons. If the independent-particle picture is a valid zeroth-order approximation, the effects of the filling of shells should be recognized by the independent filling of proton shells and neutrons shells. In the absence of Coulomb effects, protons and neutrons will fill the same levels to produce stable nuclei, and we shall not be able to tell the filling of a neutron–proton shell from that of protons alone and neutrons alone. Since, however, the potential that the protons sense is slightly "elevated" by the Coulomb repulsion, stable heavy nuclei have appreciably more neutrons than protons (Fig. 7.4). We can then have a stable nucleus with a number of neutrons that corresponds to a filled shell without at the same time filling a proton shell. Whatever effects show up as the result of filling shells should show up *independently* for neutrons and for protons in the independent-particle picture. Furthermore, if the dominant features of nuclei are determined by the *nuclear* force rather than the Coulomb force, we should expect these shell effects to occur at the same numerical values for neutrons and protons, since the shapes of their potentials should be similar. Thus, if it is found that the 51st neutron and the 83rd neutron are particularly weakly bound, for various values of Z, and if we suspect that this is due to the closure of shells with 50 and 82 neutrons, then we should also find that the 51st proton and the 83rd proton are weakly bound for various values of N. This, indeed, is found to be the case, as was discussed in Chapter I.

The numbers at which neutrons or protons seem to fill a shell have been

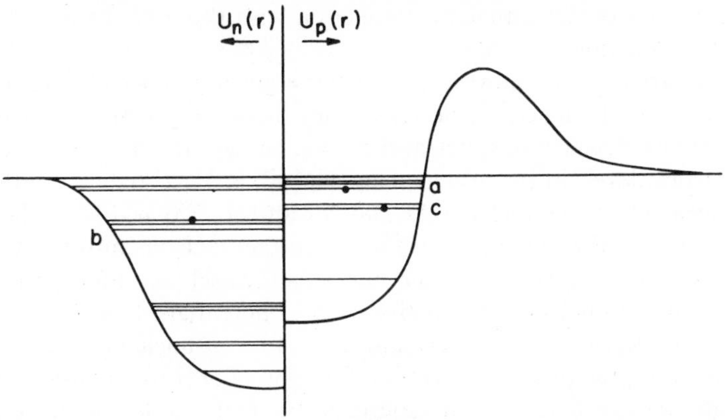

FIG. 7.4. Nuclear potentials for protons (right) and neutrons (left) for heavy nucleus. Because of the Coulomb repulsion, the average potential for a proton is less deep than that for neutron and exhibits the well-known Coulomb barrier outside the nucleus. This barrier is formed as a result of the concellation of the repulsive Coulomb force inside the nucleus by the stronger attractive nuclear potential. To obtain the lowest stable state of A-nucleons one has to fill the lowest A available states. For instance if $A = 26$ (in the unrealistic potential shown in this figure), the lowest state will be obtained by filling the 10 lowest neutron states with two neutrons each and the three lowest proton states with two protons each. If, instead, we were to put seven protons and 19 neutrons into this potential, then one of the protons will be in the state a (or higher) and there will be a neutron missing in the state b (or lower). This situation is unstable against β-decay, since by emitting a positron the proton in a could be converted into a neutron b and thereby reduce the energy of the A-nucleon system. Once this state is achieved the nucleus is stable; although the occupied proton orbit c is still higher than the neutron orbit b, no β-decay can take place since the neutron orbits b are all occupied.

called "*magic numbers*"; the fact that the same empirical magic numbers: 2, 8, 20, 28, 50, 82, and 126 have been found to be "magic" for neutrons and protons separately (except for 126; the heaviest nuclei known have $Z = 103$) is probably one of the strongest evidences in favor of the validity of the independent-particle picture as a zeroth-order approximation.

By now there are many unrelated phenomena that reflect the existence of shell structure in nuclei. In fact, we can even determine empirically the l, j, and m quantum numbers of the various single-particle orbits and their ordering. How exactly this is done we shall only learn later; at present we may satisfy ourselves with just the following observation that indicates the principle involved.

If the order of the various single-particle levels is fixed, then, given a number of nucleons the zeroth-order wave function is also fixed. All nuclear properties are thereby determined in zeroth order; to the extent that we concentrate on properties that are not too sensitive to further corrections in the wave function, the order of single-particle levels can be determined by associating the measured properties of the actual wave function with those of the model zeroth-order wave function.

Two remarks should, however, be made: a given number of nucleons for a given sequence of single-particle levels fixes the wave function uniquely only if there are no degenerate states at the same energy. In addition the order of single-particle levels in the nucleus may depend, at least slightly, on the number of nucleons in the nucleus. We should not forget that, like in the degenerate Fermi gas, the potential $U(i)$ in which the nucleons move nearly independently is formed by these same nucleons themselves, and a priori all we can say is that A-nucleons should give rise to a potential whose A lowest states would reproduce, roughly, the ground state of the A-nucleon system. In principle, therefore, the potential $U(i)$ is a function of A. From considerations of continuity we can expect no drastic changes in the order of single-particle levels as we go from one nucleus to its neighbor, but levels that are found to be close to each other in one nucleus may turn out to be in the reverse order in a neighboring nucleus.

8. THE NUCLEAR-SHELL STRUCTURE

The radial quantum numbers of the single-particle levels cannot be measured directly, but they can be inferred from (7.2) if we have good reasons to believe that we have not missed levels in between; if we concentrate on all the levels with the same values of l and j, the lowest one will have $n = 1$, the next one $n = 2$, etc.

Figure 8.1 shows the nuclear single-particle levels as determined experimentally. The absolute spacing between the levels in this figure is arbitrary since it anyhow decreases as the nucleus becomes larger. The relative spacing is also qualitative only, but it is meant to show which states lie closer to each other.

The notation we are using is the standard (nuclear) spectroscopic notation (nlj), where n stands for the radial quantum number, s, p, d, f, g, h, i stand for $l = 0, 1, 2, 3, 4, 5, 6$; and $j = l \pm 1/2$. (Note that in *atomic* spectroscopy another notation is used for the radial number n_a; the relation between n_a and our n is: $n = n_a - l$). The m quantum number is generally not shown because all states with the same value of nlj but different values of $m (m = -j, -j + 1, \ldots, j - 1, j)$ are degenerate. We shall refer to the set of the degenerate $2j + 1$ states ψ_{nljm} as the *level* (nlj).

In Fig. 8.1 we have also indicated the degeneracy $N(k)$ of each of the levels and the number of particles required to fill a shell, the magic numbers $\sum N(k)$. Although the clustering of levels into shells is obvious for the first four shells, it becomes less obvious for the higher shells. It is customary to talk of *major shells* that fill at the magic numbers 2, 8, 20, 28, 50, 82, and 126, and of *subshells* that fill at the submagic numbers 16, 38, and 40.

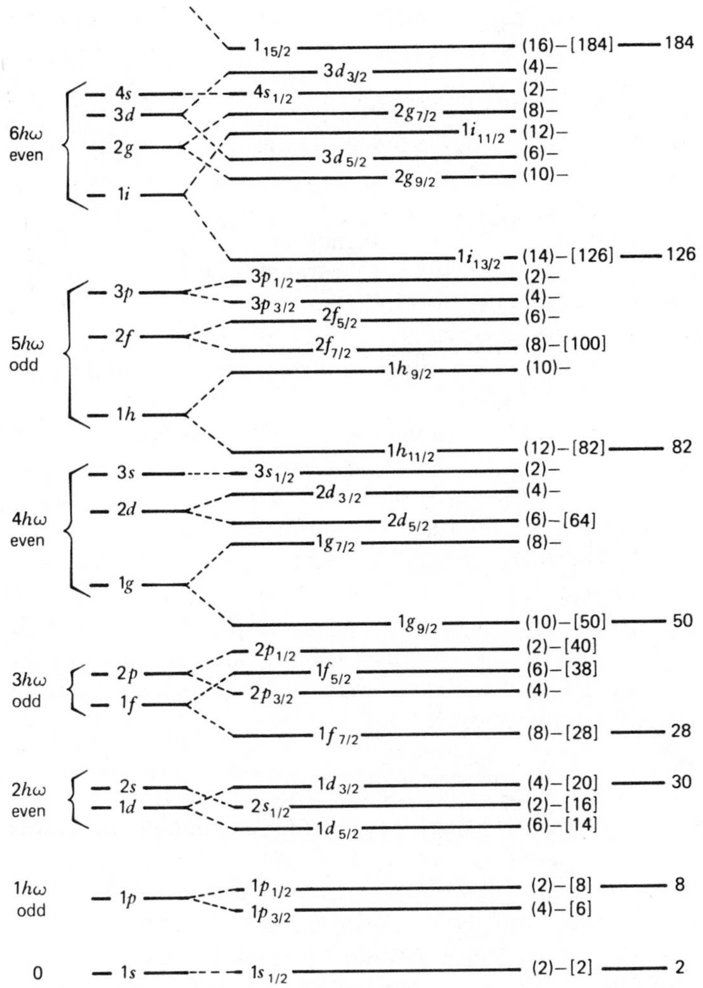

FIG. 8.1. Approximate level pattern for protons. The spin-orbit splitting is adjusted in such a way that the empirical level sequence is represented. For convenience the oscillator level grouping and the particles of these groups are indicated at the left side of Fig. 7.2. Round brackets (2), (4), etc. and square brackets [2], [6], etc. indicate the level degeneracies and the total occupation numbers. In the $6\hbar\omega$ oscillator group the splittings are not drawn in a proper scale, the $3d$ splitting is too large. For neutrons the level pattern is the same as that for protons up to $N = 50$. Beyond this value it is found that the $2d_{5/2}$ level is below the $1g_{7/2}$, $3s_{1/2}$ below the $2d_{3/2}$, and the $2f_{7/2}$ close to the $1g_{9/2}$ level. [From Mayer and Jensen (55)].

The empirical grouping of the levels into clusters is dominated by two factors. One is the effect referred to in Section IV.7, whereby an increase in n can be compensated energywise by a decrease in l. This effect is responsible for the clustering of the $2s$ and $1d$ states in the 3rd shell, the $2p$ and $1f$ in the 5th shell, the $1g$, $2d$, and $3s$ in the 6th shell, and the $1h$, $2f$, and $3p$ in the 7th shell. This clustering in itself, which is characteristic of potentials of a finite range (square well, etc.) or their equivalents (harmonic oscillator, etc.), is not sufficient to explain some of the data: indeed, the empirical fact that in the second shell the $p_{3/2}$ and $p_{1/2}$ levels are separated by about 30% of the separation between the two shells, shows that there must be an important spin-orbit component in the potential $U(i)$. The realization of this fact by Mayer and Jensen in 1949 was an important turning point in the study of nuclei.

The magnitude of the spin-orbit potential turns out to lead to spin-orbit splittings that are comparable to the energy differences between major shells. This empirical fact is, as far as we can tell now, accidental, although some of its trends can be understood. The distance in energy between major shells is roughly speaking $\hbar\omega$, where ω is the frequency of the harmonic oscillator potential that leads to the correct rms radius for the nucleus. In deriving (2.11) we showed that $\hbar\omega \sim 41$ MeV$/A^{1/3}$. In discussing (5.4) we concluded that the spin-orbit potential was a nuclear-surface effect; the splitting between the levels with $j = l + 1/2$ and $j = l - 1/2$ should therefore be proportional to $A^{-1/3}$ (the relative probability to find a nucleon at the surface). On the other hand this same splitting is proportional to

$$|\lambda_{l,j=l-1/2} - \lambda_{l,j=l+1/2}| = 2l + 1 \tag{8.1}$$

where the λ's were evaluated in (6.7). Thus the magnitude of the spin-orbit splitting, relative to $\hbar w$, is independent of A and proportional to $2l + 1$ and can exceed for large l the distance between major shells. This is why the two members of the spin-orbit doublets for $l = 1$ and $l = 2$ (i.e., $p_{3/2} - p_{1/2}$ and $d_{5/2} - d_{3/2}$) belong to the same shell. For $l = 3$ the situation is intermediate and we have the "anomaly" of a shell composed of a single level (the fourth, often called "the $f_{7/2}$ shell"); and for $l = 4, 5 \ldots$ the two members of the spin-orbit doublet are split so much that one of them has become a member of the lower shell. Since $\hbar\omega$ decreases with increasing A there will be some value for which $\hbar\omega$ and the spin-orbit splitting are comparable. For large values of A, the spin-orbit splitting will dominate.

9. PARITY AND ANGULAR MOMENTA OF NUCLEAR LEVELS

Given the empirical ordering of levels as in Fig. 8.1 we can go a long way in constructing nuclear wave functions and determining their properties. Such wave functions are referred to as shell-model wave functions; they are

A-particle wave functions constructed out of single-particle wave functions characterized by the scheme shown in Fig. 8.1. As has already been mentioned before, the angular and spin dependence of a single-particle wave function is completely determined by the quantum numbers l, j, and m. Physical quantities that are insensitive to details of the radial wave functions or better still, physical quantities that are rigorously independent of the radial wave functions, can already be computed at this stage and compared with experiment.

The simplest quantity of this type is the *parity* of an A-particle state. For a Slater determinant the parity is just the product of the parities of the single-particle wave functions that go into it. These are very simple to derive since the radial wave function $R_{nlj}(r)$ does not change under reflection, and the parity of $Y_{lm}(\theta, \phi)$ is $(-1)^l$. Thus the parity of a Slater determinant constructed out of states $n_i l_i j_i m_i$ $i = 1, \ldots, A$ is just $(-1)^{\Sigma l_i}$. We shall come to its experimental verification later.

Remark. We have disregarded the parity of the nucleon itself, that is, what happens to the spin and isospin functions under reflection. Since the baryonic quantum number is conserved in all known processes, the total intrinsic parity of the nucleons will remain the same in any process that does not change a nucleon into another baryon. For such processes the intrinsic parity of the nucleon is therefore irrelevant.

The next simplest physical quantity about which we shall be able to draw interesting conclusions, valid for any interaction $v(ij)$, is the total angular momentum

$$\mathbf{J} = \sum_{i=1}^{A} (\mathbf{l}_i + \mathbf{s}_i) \tag{9.1}$$

that commutes with a scalar Hamiltonian like (4.1) (with or without $\sum v_{ij}$). It is clear that if eigenstates of H are not already simultaneous eigenstates of $\mathbf{J}^2$ and of J_z, then they must be degenerate. It is then always possible to take a linear combination of these degenerate eigenstates of H in such a way as to produce a wave function that is also an eigenstate of $\mathbf{J}^2$ and J_z.

Problem. Prove it.

Let us examine the relation between the shell-model wave functions and the operator $\mathbf{J}$. We notice first that if the number of nucleons corresponds to a closed shell there is no degeneracy in the ground state and, therefore, such states must be automatically eigenstates also of $\mathbf{J}^2$ and J_z. We characterize the single-particle states by $nljm$ and construct a Slater determinant $\Phi_0(1, .., z_0)$

out of the states $n_i l_i j_i m_i$; $i = 1, \ldots, z_0$ (we consider for simplicity the filling of the proton states only; the generalization to protons and neutrons is straightforward). We see immediately that

$$J_z \Phi_0(1, \ldots, z_0) = \sum_{k=1}^{z_0} j_{kz} \Phi_0(1, \ldots, z_0)$$

$$= \sum_{i=1}^{z_0} [m_i \Phi_0(1, \ldots, z_0)] = M_0 \Phi_0(1, \ldots, z_0) \tag{9.2}$$

where

$$M_0 = \sum_{i=1}^{z_0} m_i \tag{9.3}$$

Physically this result is very simple; it says that if we quantize the single-particle angular momenta and the total angular momentum along the same axis, then the projection of the angular momentum of the state Φ on this axis is the sum of the projections of each of the particle's angular momentum on the same axis. It is a trivial thing to convince ourselves that

$$\sum_{\substack{\text{All states} \\ \text{of one level}}} m_i = \sum_{m_i=-j}^{+j} m_i = 0$$

Hence, if in (9.3) we have particles filling the various levels completely, there being no level that is partially filled, it follows that

$$M_0 = 0 \tag{9.4}$$

We shall now show that the value of $\mathbf{J}$ is also zero. We know that $\mathbf{J}$ commutes with H_0 (see eqs. 4.2 and 4.3); we also know that because of the Pauli principle there is no other state degenerate with the state Φ_0 that has all lowest levels filled; it follows that

$$\mathbf{J}\Phi_0 = \boldsymbol{\lambda}\Phi_0 \tag{9.5}$$

where $\boldsymbol{\lambda} = (\lambda_x, \lambda_y, \lambda_z)$ are three numbers. $J_x \pm iJ_y$ are the raising and lowering operators for M; since Φ_0 has a well-determined value of M, it follows from (9.5), by operating on Φ_0 with $J_x \pm iJ_y$, that $\lambda_x \pm i\lambda_y = 0$; from (9.4) we conclude that $\lambda_z = M_0 = 0$; hence $\boldsymbol{\lambda} = 0$ or

$$J\Phi_0 = 0 \qquad \text{and also} \qquad J^2\Phi_0 = 0 \tag{9.6}$$

Φ_0 is thus completely spherically symmetric state with no preferred direction.

We can also determine the parity of this state. Since each single-particle level nlj can accommodate $(2j + 1)$ particles, and since $2j + 1$ is always an even number, the parity of a filled level is $(-1)^{(2j+1)\cdot l} = +1$. Thus Φ_0 has positive parity too. Hence, a Slater determinant constructed out of completely filled levels is an eigenstate of J^2 and J_z corresponding to $J = 0^+$ and $J_z = 0$. (The notation $J = 0^+$ implies $J = 0$ and positive parity.)

Let us now add one particle to Φ_0 putting it in the level $n'l'j'$ that has hitherto had no particles in it, and let us put it there in the state m'. The corresponding Slater determinant Φ_1 has $z_0 + 1$ particles (where z_0 corresponds to a closed shell) and it can be shown immediately that

$$J_z\Phi_1(1, \ldots, z_0 + 1) = \sum_{i=1}^{z_0+1} m_i\Phi_1 = m'\Phi_1 \tag{9.7}$$

Φ_1 has states degenerate with it that are obtained by putting the last particle in a different m-state of the same level. However, there are generally no states degenerate with Φ_1 that also have the same eigenvalue of J_z. Since J^2 commutes with J_z, it follows that Φ_1 is also an eigenfunction of J^2 and it is easy to see that:

$$J^2\Phi_1(1, \ldots, z_0 + 1) = j'(j' + 1)\Phi_1(1, \ldots, z_0 + 1) \tag{9.8}$$

Problem. Prove (9.8) by breaking J into a part that operates on the filled levels and a part that operates on the added particle. Use (9.6) for the first part.

We can also convince ourselves that the parity of Φ_1 is $(-1)^{l'}$. Thus, a Slater determinant constructed out of a filled level plus one particle in an unfilled level, is an eigenstate of the parity operator, of $\mathbf{J}^2$ and J_z with eigenvalues equal to the parity, j, and m quantum numbers of the odd particle in the unfilled level.

Physically what happens is the following: the different m-states in a given single-particle level nlj correspond to different directions around which the nucleon rotates. Nucleons in states m and $-m$ rotate in opposite directions. It is generally *not* true that if we put one nucleon in a state m and another one in $-m$ their total angular momentum vanishes, because j_{1z} and j_{2z} do not commute separately with $(\mathbf{j}_1 + \mathbf{j}_2)^2$; however, by the time we fill all the possible m-states of a given level, all possible directions of rotation are equally present; there is thus no preferred direction in space, and the total angular momentum must vanish (otherwise its own direction will be a "preferred" direction). If we have one particle outside filled levels the total angular momentum is just due to this particle and, hence, the results (9.7) and (9.8).

It can be similarly shown that if we have $z_0 - 1$ particles, the resulting Slater determinant $\Phi_{-1}(1, \ldots, z_0 - 1)$ in which the level $n'l'j'$ has just one state m' empty, satisfies

$$J_z\Phi_{-1}(1, \ldots, z_0 - 1) = -m'\Phi_{-1}(1, \ldots, z_0 - 1) \tag{9.9}$$

$$J^2\Phi_{-1}(1, \ldots, z_0 - 1) = j'(j' + 1)\Phi_{-1}(1, \ldots, z_0 - 1) \tag{9.10}$$

It is instructive to derive (9.6) to (9.10) again in a slightly different way: We noted that H commutes with $\mathbf{J}$, and also $\mathbf{J}^2$ commutes with $\mathbf{J}$. We may thus choose $\Phi(J)$ to be a simultaneous eigenfunction of H and $\mathbf{J}^2$. It follows that $(J_x + iJ_y)\Phi(J)$ and $(J_x - iJ_y)\Phi(J)$ are also eigenstates of H and of $\mathbf{J}^2$. Moreover they belong to the same eigenvalues of H and $\mathbf{J}^2$ as the original function $\Phi(J)$. Operating again with $(J_x \pm iJ_y)$ will produce still another degenerate eigenfunction of H with the same eigenvalue $J(J + 1)$ of $\mathbf{J}^2$. It is well known that in this way we can produce exactly $2J + 1$ independent functions. Thus, if it is known that there is no degeneracy for a given eigenstate $\Phi_0(J_0)$ of H, it follows that it should satisfy $2J_0 + 1 = 1$, that is, its angular momentum must vanish. The closed-shell nuclei are, of course, a special case of this general theorem.

Similarly, if it is known that a state $\Phi_1(J_1)$ is one out of $2j' + 1$ degenerate states that can be obtained from each other through the use of some power of $(J_x + iJ_y)$, then it follows that J_1 should satisfy the relation $2J_1 + 1 = 2j' + 1$ or $J_1 = j'$. The important thing to note is that, given a level of degeneracy D, none of its states can have a total angular momentum bigger than J_D where $2J_D + 1 = D$. We also note that if we can construct an eigenstate of H that is a simultaneous eigenstate of J_z, with an eigenvalue M, then by repeated applications of $J_x - iJ_y$ we can produce at least $2M$ additional degenerate states of H. It follows that the lowest total angular momentum J associated with our original state satisfies $2J + 1 \geq 2M + 1$ or $J \geq M$. Physically this result says that an angular momentum is always larger than, or equal to, its projection on an arbitrary axis.

10. SYSTEMS WITH SEVERAL PARTICLES OUTSIDE FILLED LEVELS

If we pass now to systems that have more than one particle added to, or missing from, filled levels, the situation becomes more complicated. Let us consider a system in which there are $k > 1$ particles, occupying states outside of a closed shell. Let these states be $n^{(1)}l^{(1)}j^{(1)}m^{(1)}, n^{(2)}l^{(2)}j^{(2)}m^{(2)}, \ldots, n^{(k)}l^{(k)}j^{(k)}m^{(k)}$. Forming again a Slater determinant $\Phi_k(1, \ldots, z_0 + k)$ for the $z_0 + k$ particles we find that

$$J_z\Phi_k(1, \ldots, z_0 + k) = M_k\Phi_k(1, \ldots, z_0 + k) \tag{10.1}$$

where

$$M_k = \sum_{i=1}^{k} m^{(i)} \tag{10.2}$$

Thus again, the Slater determinant is an eigenstate of J_z as well as of H_0, and only the particles outside the filled levels contribute to the eigenvalue M_k of J_z. Unlike the previous cases, however, we *can* have now other $(z_0 + k)$

particle states, which are degenerate with Φ_k, and have the *same* eigenvalue M_k. In fact, if we leave the filled levels untouched but move the k-particles to other states in their levels, we have a new situation defined by

$$n^{(1)}l^{(1)}j^{(1)}m^{(\nu 1)}, \ldots, n^{(k)}l^{(k)}j^{(k)}m^{(\nu k)}$$

If we form a Slater determinant $\Phi_{\nu k}$ out of these new states (plus the previous untouched filled levels) we obtain:

$$J_z\Phi_{\nu k} = M_{\nu k}\Phi_{\nu k} \tag{10.3}$$

where

$$M_{\nu k} = \sum_{i=1}^{k} m^{(\nu i)} \tag{10.4}$$

$\Phi_{\nu k}$ is obviously degenerate with Φ_k since we have left all the particles in the same levels; we can also make it correspond to the same eigenvalue of J_z if we require that

$$\sum_{i=1}^{k} m^{(\nu i)} = \sum_{i=1}^{k} m^{(i)} \tag{10.5}$$

Unlike the previous cases, we cannot argue any longer that Φ_k should also be an eigenfunction of $\mathbf{J}^2$, and in fact this is generally not the case. In order to produce an eigenfunction of $\mathbf{J}^2$, we now have to take linear combinations of different $\Phi_{\nu k}$ that satisfy the condition (10.5).

The set of Slater determinants of $z = z_0 + k$ particles obtained from the single-particle states $n^{(1)}l^{(1)}j^{(1)}m^{(1)}, \ldots, n^{(z)}l^{(z)}j^{(z)}m^{(z)}$, forms a complete set of antisymmetric z-particle wave functions. It is referred to as the m-scheme, because, in addition to being characterized by the level quantum numbers $n^{(i)}l^{(i)}j^{(i)}$, it is characterized by the m-quantum number of the corresponding states. The Hamiltonian H_0 is diagonal in the m-scheme and there are generally quite high degeneracies in its spectrum. All the independent states obtained by keeping the quantum numbers $n^{(1)}l^{(1)}j^{(1)}, n^{(2)}l^{(2)}j^{(2)}, \ldots, n^{(z)}l^{(z)}j^{(z)}$ fixed and letting $m^{(1)}, m^{(2)}, \ldots, m^{(z)}$ take any permissible set of values, are said to form a *configuration*. That is, in specifying a configuration we need only give the (n, l, j) values involved. In $H_0 = \sum [T_i + U(i)]$ all the states of the same configuration are degenerate. This is obvious since for each one of the particles, keeping $n^{(i)}l^{(i)}j^{(i)}$ fixed and changing $m^{(i)}$ does not change its energy. Since in H_0 the particles do not interact with each other, it follows that all the states of the same configuration, obtained by taking all possible sets of $m^{(i)}$, are also degenerate.

In general, *different* configurations will *not* be degenerate with each other even in H_0, for if we choose to empty one of the occupied levels of whatever particles it has, and fill another level instead, this will generally lead to a different energy for H_0. An exception can occur if the two levels considered were accidentally degenerate (such as the $1d$ and $2s$ levels in the harmonic oscillator potential).

The importance of the concept of a configuration stems from the results we have just found. Indeed, we know from perturbation theory that if the zeroth-order Hamiltonian has degenerate eigenstates, the first-order effects of the perturbation are obtained by the diagonalization of the perturbation within the Hilbert subspace of the degenerate states. Thus, to the extent that in the shell-model Hamiltonian (4.1) the interaction $v(ij)$ can be considered as a perturbation, our first task will be to diagonalize $v(ij)$ within each set of degenerate states and, therefore, to consider just one configuration at a time. This drastically reduces the complexity of the problem.

A configuration is denoted by its occupied levels: $n^{(1)}l^{(1)}j^{(1)}n^{(2)}l^{(2)}j^{(2)}, \ldots, n^{(z)}l^{(z)}j^{(z)}$. A shorthand notation is very often used: if the same level $n^{(i)}l^{(i)}j^{(i)}$ is occupied by k particles, one writes $(n^{(i)}l^{(i)}j^{(i)})^k$ instead of repeating k times the same set of three quantum numbers. Very often the completely filled levels are not mentioned at all, since in many cases their effect can be disregarded. If it is clear which shell one is concerned with, a configuration is often denoted just by the j-values of the relevant unfilled levels. Thus, when we say that the lowest proton configuration of ${}^{51}_{23}V_{28}$ is $(7/2)^3$, what we really mean is the configuration

$$(1s_{1/2})^2(1p_{3/2})^4(1p_{1/2})^2(1d_{5/2})^6(2s_{1/2})^2(1d_{3/2})^4\underline{(1f_{7/2})^3}$$

Similarly if one says that certain levels in ${}^{40}_{20}Ca_{20}$ belong to the proton configuration $(d_{3/2}{}^{-1}f_{7/2})$, what one really means is the configuration

$$(1s_{1/2})^2(1p_{3/2})^4(1p_{1/2})^2(1d_{5/2})^6(2s_{1/2})^2\underline{(1d_{3/2})^3(1f_{7/2})}$$

(we have underlined that part which is referred to in the shorthand notation). Note the use of negative powers to denote "missing" particles in a would-be filled level, and also the use of $(l_j)^n$ rather than j^n when the latter may be confusing. Indeed, referring to some of the levels in ${}^{40}Ca$ as levels of the configuration $[(3/2)^{-1}, (7/2)]$ is not clear enough, since the missing particle may then be in the $1p_{3/2}$ level or the $1d_{3/2}$ level.

Before proceeding with the analysis of several particles outside close shells, let us summarize our findings thus far.

In an attempt to analyze the structure of a system described by the shell-model Hamiltonian

$$H = \sum_{i=1}^{A} T_i + U(i) + \sum_{i<j=1}^{A} v(ij) = H_0 + \sum_{i<j=1}^{A} v(ij)$$

we derive first the single-particle spectrum of H_0.

This is schematically shown in Fig. 10.1 stressing the difference between the proton and neutron single-particle spectra. Each level in this spectrum is characterized by the quantum numbers $nljm$ and is $(2j + 1)$-fold degenerate

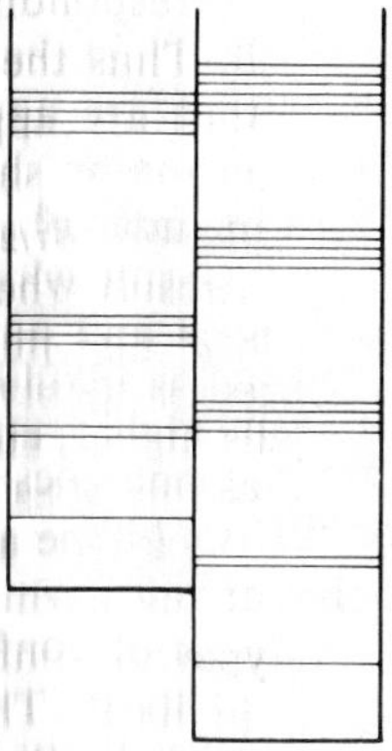

FIG. 10.1. Schematic diagram of energy levels of H_0.

(its energy is independent of m). If we wish we can add a fifth quantum number $m_\tau (= \pm 1)$ to tell us whether we are dealing with a proton state ($m_\tau = +1$) or a neutron state ($m_\tau = -1$). In the absence of Coulomb forces, and neglecting possible small charge dependence of nuclear forces, the levels will then be degenerate also with respect to m_τ.

To produce a state of Z-protons and N-neutrons we distribute these particles among the different levels of Fig. 10.1. For any given distribution of the particles among the levels—that is, a given configuration—we can have many states depending on the m-states in each level into which the particles go. The A-particle spectrum, still considering only H_0, will then look like Fig. 10.2. Each A-particle level will generally be highly degenerate; there will be group-

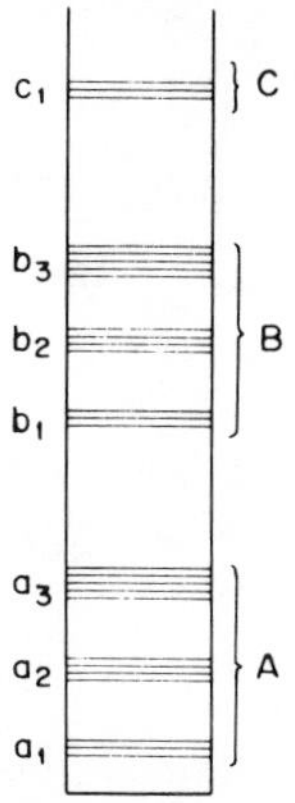

FIG. 10.2. Schematic diagrams of energy levels of a nucleus governed by H_0.

ings of such highly degenerate levels corresponding to different configurations that stay within the same major shell. Thus the levels a_1, a_2, a_3, . . . in Fig. 10.2 all result from configurations that are approximately degenerate with the lowest configuration to the same major shell. In $^{136}_{54}Xe_{82}$, for instance, these a-type configurations may include $g^4_{7/2}$, $g^2_{7/2}d^2_{5/2}$, $g^2_{7/2}d^3_{5/2}$, $g_{7/2}d^3_{5/2}$, $d^4_{5/2}g^2_{7/2}s^4_{1/2}$, etc. The levels b_1, b_2, . . . result when a particle in one of the a-configurations is removed from its level and put in a level one shell higher. The levels c result from the same process involving two particles, or moving one particle to a level two major shells higher, etc.

It is easy to convince oneself that as one goes higher in energy the possible variations become so numerous that the b-type and c-type configurations will begin to overlap each other. Only in the harmonic oscillator do we find a clean separation between different types of configurations and, indeed, each type is then completely degenerate in itself. The spacing between different types of configurations in the harmonic oscillator is $\hbar\omega$, as can be easily verified.

If we now want to take into account also the interaction $v(ij)$ in the shell-model Hamiltonian, the first thing is to diagonalize it within the subspace of degenerate states, that is, within each configuration. As a result of this diagonalization some of the degeneracy will be removed and the configurations of type a in Fig. 10.2 will look qualitatively as shown in Fig. 10.3. Each of the levels obtained from the diagonalization of $\sum v(ij)$ will still be degenerate; however, this particular degeneracy is simple. In fact, since we know that $\mathbf{J}^2$, the square of the total angular momentum, commutes with the shell-model Hamiltonian even when $v(ij)$ is included, J is a good quantum number, and each level characterized by J is $(2J + 1)$-fold degenerate. Apart from accidental degeneracies this is the only degeneracy that remains after the diagonalization of $v(ij)$ within each configuration, except of course for the de-

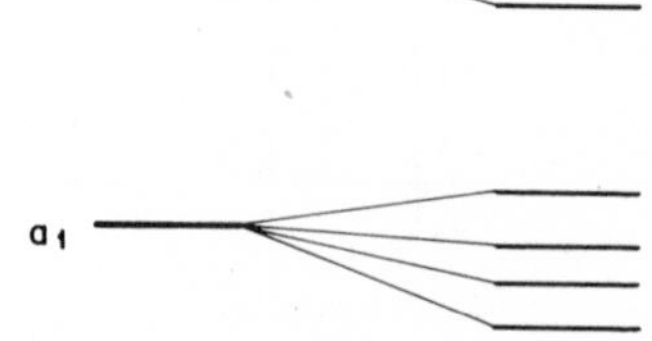

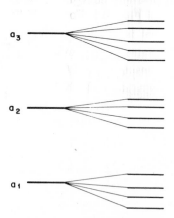

FIG. 10.3. Splitting of degenerate levels in H_0 under the influence of a perturbation $v(ij)$.

generacy with respect to m_τ the isospin "magnetic" quantum number that is removed only if $v(ij)$ includes the Coulomb potential.

We shall come later to the quantitative evaluation of the splittings between different levels after the diagonalization of the residual interaction $\sum v(ij)$, but it is immediately clear that this splitting reflects the strength of the residual interaction. If we look back at Fig. 10.3 we see that, provided the residual interaction is weak enough, the first few levels in a nucleus described by the shell model will be the levels of the lowest configuration. We cannot tell where these levels lie without knowing $v(ij)$, but we can still tell which are their possible values of J. Since these levels are obtained by the diagonalization of a matrix within a certain configuration, the new eigenfunctions are linear combinations of the various states of this configuration. To tell which values of the total angular momentum will be encountered, we have to determine which J-values can be obtained from that configuration.

11. EXAMPLES

The simplest examples are those of nuclei corresponding to closed shells or closed shells plus or minus one particle. In these cases, as we saw in (9.6) to (9.10), the lowest configurations have just one level with $J = 0^+$ for the closed shells and $J = j'$ for a closed-shell $\pm$one particle, where j' is the angular momentum of the additional or missing particle. Experimentally one finds indeed that all closed-shell nuclei have $J = 0^+$ for their ground state. As far as closed shells ± 1 nuclei are concerned, the angular momenta of their ground and low-lying states were used, as we shall see later, to determine the shell-model sequence of single-particle levels as shown in Fig. 8.1. We should not use them therefore as an "evidence" for the validity of our arguments.

There are, however, other simple cases that could illustrate the results obtained thus far. Consider, for instance, the nucleus $^{40}_{19}K_{21}$. From the sequence of levels in the shell model (Fig. 8.1) we would expect its lowest proton configuration to be $d_{3/2}^{-1}$, and its lowest neutron configuration to be $f_{7/2}$. The lowest configuration of ^{40}K is thus $(\pi d_{3/2}^{-1}, \nu f_{7/2})$ (where π stand for protons and ν for neutrons). In more detail, the proton configuration is

$$1s_{1/2}^2 1p_{3/2}^4 1p_{1/2}^2 1d_{5/2}^6 2s_{1/2}^2 1d_{3/2}^3$$

and the neutron configuration is

$$1s_{1/2}^2 1p_{3/2}^4 1p_{1/2}^2 1d_{5/2}^6 2s_{1/2}^2 1d_{3/2}^4 1f_{7/2}$$

The proton configuration by itself means one particle is missing at the top of the third major shell; the proton configuration therefore leads to a level of a definite angular momentum $J_p = 3/2^+$. Similarly the neutron configuration by itself corresponds to one particle at the beginning of the fourth major shell. This shell contains just one level $-1f_{7/2}$—hence the neutron configuration has $J_n = 7/2^-$. The total angular momentum J of the $(d_{3/2}^{-1}, f_{7/2})$ configuration can now be obtained by the simple rule of adding two angular momenta:

$$J = |J_p - J_n|,\ |J_p - J_n| + 1, \ldots, J_p + J_n \tag{11.1}$$

Since the neutron and the proton configurations have opposite parity, we expect all the lowest levels in ^{40}K to have negative parities.

The experimental levels of ^{40}K are shown in Fig. (11.1). The lowest levels have negative parities and total angular momenta 4, 3, 2, and 5 that just correspond to all possible integers between $|3/2 - 7/2|$ and $3/2 + 7/2$ as required by (11.1). We therefore feel rather certain that we observe here just all the levels of the configuration $(\pi d_{3/2}^{-1}, \nu f_{7/2})$. As one goes higher in energy no excited states are encountered until an excitation of about 2 MeV. By that time we probably encounter the next configuration that by inspection of Fig. 8.1 should involve the lowest neutron level in the fifth major shell that is, $(\pi d_{3/2}^{-1}, \nu p_{3/2})$. Four new levels are indeed found here, all of negative parity, and with angular momenta 3, 2, 1, and 0, which is again exactly what we would expect from the application of (11.1) to this configuration. The group of levels at 0, 29, 799, and 885 KeV form what we called a configuration of type *a* (Fig. 10.2). Those at 2.04 to 2.56 MeV are levels of a configuration type *b*. The fact that they do not lie at a considerably higher energy is probably due to the special nature of the $f_{7/2}$ shell. The splitting between the fourth and the fifth major shells is a "halfway" splitting: for the lower *l*-values the spin-orbit splitting is not sufficient to overcome the distance between shells, and for the higher *l*-values it does it completely. The spin-orbit splitting for the *f*-nucleon does not push the $1f_{7/2}$ level all the way down to the top of the third shell, but pushes it enough to split it from the bottom of the $2p - 1f$ group of levels. This is the reason for the occurrence of this $f_{7/2}$ shell with just one level in it.

There is also some indication of levels in ^{40}K, at about 2 MeV or even less, which may be the levels of the configuration $(\pi s_{1/2}^{-1}, \nu f_{7/2})$. This would be

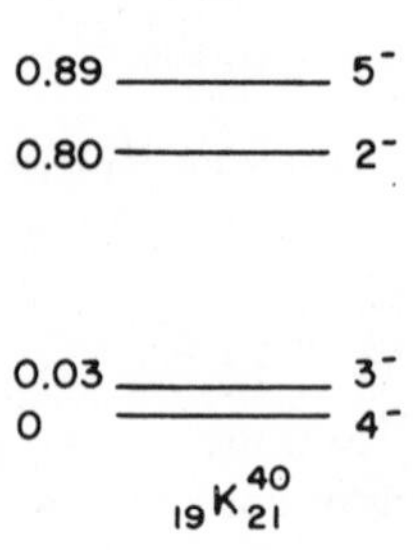

FIG. 11.1. Energy levels of ^{40}K.

another "type a" configuration. Since the evidence is, however, rather poor we shall not discuss it here.

The preceding discussion and the example of the levels in ^{40}K demonstrate how we can obtain specific information on levels of nuclei without a detailed knowledge of $v(ij)$. The fact that agreement between experimental findings and these simple-minded expectations is found over a wide range of nuclei gives further support to the basic assumptions underlying the independent-particle picture of the nucleus.

12. THE J-SCHEME

We can go further and deduce other important results, still having to do with the total angular momentum of the system, but demonstrating in a very convincing way the effects of antisymmetrization. We mentioned above that in order to obtain the possible angular momenta for the low-lying states of nuclei, all we have to do is to see which are the angular momenta that the lowest configurations can give rise to. These angular momenta were very simply determined when the number of particles was such that the lowest configurations produced exactly filled levels, or filled levels plus one particle in an unfilled level. In the case of ^{40}K we have one particle in an unfilled neutron level and one particle was missing in an otherwise full proton level; the angular momenta of this configuration could be still rather trivially deduced. But what if we have two or more particles in an unfilled level?

The wave functions in these cases can still be written rather easily in the m-scheme, where we specify the m-values of the states occupied by the various particles. Such wave functions are eigenfunctions of $j_{iz}, i = 1, \ldots, A$, and therefore also of

$$J_z = \sum_{i=1}^{A} j_{iz}.$$

What we are after, however, are wave functions, in the same configuration, which are still eigenfunctions of J_z, but are simultaneously eigenfunctions of $\mathbf{J}^2$. Since $[j_{iz}, \mathbf{J}^2] \neq 0$, the m-scheme wave functions do not generally satisfy this requirement. For an A-nucleon system there exist generally several wave functions that belong to the same value of J_z. By combining these wave functions with properly chosen coefficients, a wave function that is simultaneously an eigenfunction of J^2 and J_z can be obtained. This amounts to using a unitary transformation $\langle JM | m_1, \ldots, m_A \rangle$ such that the A-particle functions

$$\Phi(JM) = \sum_{m_1,\ldots,m_A} \langle JM | m_1, m_2, \ldots, m_A \rangle \Phi(m_1, m_2, \ldots, m_A) \quad (12.1)$$

are simultaneous eigenfunction of $\mathbf{J}^2$ with the eigenvalues $J(J+1)$, and of J_z with the eigenvalues M.

The coefficients of the transformation (12.1), written in full, actually should look as follows:

$$\langle j_1, j_2, \ldots, j_A; \alpha JM | j_1 m_1, j_2 m_2, \ldots, j_A m_A \rangle \tag{12.2}$$

They depend, of course, only on the quantum numbers of the states involved and not on the coordinates of the particles. The additional quantum number α is introduced to distinguish between orthogonal states with the same value of J and M, since for states with three or more particles it is generally possible to obtain a given total angular momentum in more than one way.

For $A = 2$, the coefficients (12.2) reduce to the well known Clebsch-Gordan coefficients (see Appendix A, A. 2.57). Also for $A > 2$ they effectively transform from the m-scheme to what is called the J-scheme, and can therefore be called generalized Clebsch–Gordan coefficients. They can be determined, like the Clebsch–Gordan coefficients, by operation on both sides of (12.1) with

$$J_z = \sum j_{kz},\ J_x + iJ_y = \sum_k (j_{kx} + ij_{ky}) \text{ and } J_x - iJ_y = \sum_k (j_{kx} - ij_{ky})$$

These operators affect only the m-quantum numbers and leave the quantum numbers n and j the same, thus not modifying the configuration. One obtains then the set of equations:

$$\langle JM | m_1, \ldots, m_A \rangle = 0 \quad \text{if} \quad m_1 + m_2 \ldots . m_A \neq M \tag{12.3}$$

$$\begin{aligned}
&\sqrt{(J-M)(J+M+1)}\langle J, M+1 | m_1, m_2, \ldots, m_A \rangle \\
&\quad = \langle JM | m_1 - 1, m_2, \ldots, m_A \rangle \sqrt{(j_1 - m_1 + 1)(j_1 + m_1)} \\
&\quad + \langle JM | m_1, m_2 - 1, \ldots, m_A \rangle \sqrt{(j_2 - m_2 + 1)(j_2 + m_2)} \\
&+ \ldots + \langle JM | m_1, m_2, \ldots, m_A - 1 \rangle \sqrt{(j_A - m_A + 1)(j_A + m_A)}
\end{aligned} \tag{12.4}$$

and

$$\begin{aligned}
&\sqrt{(J+M)(J-M+1)}\langle J, M-1 | m_1, m_2, \ldots, m_A \rangle \\
&\quad = \langle JM | m_1 + 1, m_2 \ldots, m_A \rangle \sqrt{(j_1 + m_1 + 1)(j_1 - m_1)} \\
&\quad + \langle JM | m_1, m_2 + 1, \ldots, m_A \rangle \sqrt{(j_2 + m_2 + 1)(j_2 - m_2)} \\
&+ \ldots + \langle JM | m_1, m_2, \ldots, m_A + 1 \rangle \sqrt{(J_A + m_A + 1)(j_A - m_A)}
\end{aligned} \tag{12.5}$$

We shall not investigate the question of the existence of solutions to these equations, but use them to derive some properties of the solutions assuming that they exist (see below for a method of actual construction of a solution).

(a) Since the coefficients of these equations are all real and the equations are homogeneous in $\langle JM|m_1, \ldots, m_A\rangle$, we can choose the solutions to be real as well, and normalize them so that

$$\sum_{m_1, \ldots m_A} |\langle JM|m_1, \ldots, m_A\rangle|^2 = 1 \tag{12.6}$$

(b) As was mentioned above (12.3) to (12.5) connect states of the *same configuration* only. The matrix $\langle JM|m_1, \ldots m_A\rangle$ therefore produces a state of angular momentum J *in the configuration* $(j_1, \ldots, j_A)$. This state is denoted by

$$|j_1, j_2, \ldots, j_A; JM\rangle$$

(c) If a solution to (12.3) to (12.5) exists for a given value of J and M, then the way we constructed our equations guarantees that solutions will also exist for the same J and all other possible values of M. Physically it means that if a certain configuration can give rise to a state of a given angular momentum J, and a given z-projection M, it will also give rise to all the other M-states of this level J. This conclusion follows basically from the fact that a configuration in which all the m-states are degenerate has no specific orientation in space, and this property should remain valid irrespective of how we choose to look at it: through the m-scheme, the J-scheme, or any other scheme.

(d) Equations 12.3 to 12.5 do not involve the wave functions $\Phi(m_1, \ldots, m_A)$ although we used these functions to derive the equations. We notice however, that all that has been used was the "response" of $\Phi(m_1, \ldots, m_A)$ to the operators j_z, and $j_x \pm ij_y$. These operators affect only the spin and angular variables of Φ; the matrix $\langle JM|m_1, \ldots, m_A\rangle$ is thus independent of the *radial* wave functions, and therefore also of the r^2 and p^2 dependence of the potential $U(i)$ that generates the radial wave functions. This result is also of great importance. It tells us that if we believe that a group of levels originates from one pure configuration, the possible angular momenta of these levels are independent of the rest of the dynamics of the system.

(e) The symmetry properties of $\Phi(m_1, m_2, \ldots, m_A)$ do not reflect themselves in (12.3) to (12.5). The matrix that transforms from the m-scheme to the J-scheme is independent of whether we transform symmetric states or antisymmetric states, or states of no definite symmetry at all. Of course, since $\Phi(j_1m_1, j_2m_2, \ldots, j_Am_A)$ is antisymmetric with respect to the exchange of the coordinates of its particles, the same will hold true for $\Phi(j_1, j_2, \ldots, j_A; JM)$; similarly a symmetric wave function in the m-scheme will give rise to a symmetric wave function also in the J-scheme.

In concrete cases the solution of (12.3) to (12.5) may not be the best way to obtain the coefficients $\langle JM|m_1, \ldots, m_A\rangle$. A more practical way of doing it, which has broader applications as we shall see later, is the following.

We observe that the m-scheme state

$$\Phi(m_1 = j_1, m_2 = j_2, \ldots, m_A = j_A) \tag{12.7}$$

is already an eigenstate of $\mathbf{J}^2$ and J_z with the eigenvalue $J_{\max}(J_{\max} + 1)$

Problem. Prove it by noting that

$$\mathbf{J}^2 = \mathbf{j}_1^2 + \mathbf{j}_2^2 + \ldots + \mathbf{J}_A^2 + \sum_{i<k} (j_{i+}j_{k-} + j_{i-}j_{k+} + 2j_{iz}j_{kz})$$

where

$$j_{k\pm} = j_{kx} \pm ij_{ky}$$

and $J_z = J_{\max}$, respectively, where

$$J_{\max} = j_1 + j_2 + \ldots + j_A \tag{12.8}$$

$J_{\max}$ will be recognized as the maximum possible angular momentum in the configuration $(j_1, \ldots, j_A)$, and the fact that (12.7) is simultaneously an eigenfunction in both the m-scheme and the J-scheme is because there are no other degenerate states in the m-scheme with the same value of $M = \sum m_k$. Thus we can write

$$\Phi(J_{\max}, M = J_{\max}) = \Phi(m_1 = j_1, \ldots, m_A = j_A) \tag{12.9}$$

and we conclude that there is only one level with $J = J_{\max}$ in the configuration $(j_1, \ldots, j_A)$.

We now operate on (12.9) with $J_x - iJ_y$ and obtain

$$\begin{aligned}\Phi(J_{\max}, M = J_{\max} - 1) = \frac{1}{\sqrt{2J_{\max}}} [\sqrt{2j_1}\Phi(m_1 = j_1 - 1,\\ m_2 = j_2, \ldots, m_A = j_A) + \sqrt{2j_2}\Phi(m_1 = j_1, m_2 = j_2 - 1, \ldots, m_A = j_A)\\ + \ldots + \sqrt{2j_A}\Phi(m_1 = j_1, m_2 = j_2, \ldots, m_A = j_A - 1]\end{aligned} \tag{12.10}$$

There are other independent m-scheme states with $M = J_{\max} - 1$. By choosing linear combinations that are orthogonal to (12.10) we obtain states with angular momenta $J = J_{\max} - 1$. Their total angular momentum J must satisfy $J \geq J_{\max} - 1$ (a vector cannot be shorter than its projection); but we said there was just *one* level with $J = J_{\max}$; our states, by construction, are orthogonal to the states with $J = J_{\max}$; hence $J = J_{\max} - 1$. Operating now with $J_x - iJ_y$ or $\Phi(J_{\max}, M = J_{\max} - 1)$ and on all the states $\Phi(J_{\max} - 1, M = J_{\max} - 1)$ we reach the states with $M = J_{\max} - 2$, and continue with the same procedure there. Thus, step by step, one can construct all the J-scheme states from those in the m-scheme and effectively solve (12.3) to (12.5).

13. CONFIGURATION WITH EQUIVALENT PARTICLES

Thus far we did not specify whether the levels of the A-particles $n_1l_1j_1, n_2l_2j_2, \ldots, n_Al_Aj_A$ are all different. Our considerations remain valid even if some of the levels are identical. However, additional new features then appear and we shall now proceed to discuss them, since they are of great importance in the study of many nuclear properties. The most general case would be the one characterized by the configuration

$$(n_1l_1j_1)^{N_1}(n_2l_2j_2)^{N_1}, \ldots, (n_kl_kj_k)^{N_k}, \qquad N_i \leq 2j_i + 1, \qquad \sum N_i = A$$

For our purpose it will be sufficient to consider just N-particles forming the configuration $(nlj)^N$. We shall find later that, in spite of the process of antisymmetrization, completely filled levels do not affect many nuclear properties; such properties depend then only on the particles in partially filled levels. In the extreme independent-particle picture there will be, for the ground state, at most two partially filled levels at a time: one for the protons and one for the neutrons. Protons and neutrons fill the same levels only for light nuclei; for heavier nuclei proton levels are pushed up by the Coulomb force and are consequently filled more slowly than neutron levels (see Section IV.10). We shall therefore find a good number of nuclei that can be described well enough by the configuration j^N.

Nucleons that occupy the same level are said to be *equivalent* to each other. This name derives from the property of the $2j + 1$ states of a level (nlj) that makes them transform among themselves under a rotation of the frame of reference. Since the radial and the orbital quantum numbers are often omitted, it is customary to denote by j^2, for example, a configuration of two equivalent nucleons, and by (j, j) one of two nonequivalent nucleons. Thus j^2 is short for $(nlj)^2$, whereas jj is short for $(nlj, n'l'j)$ where $(n, l) \neq (n', l')$.

The two configurations j^2 and (j, j) have different permissible values of J, although both are formed from two particles in states with the same j. In fact, in the m-scheme the wave functions of the different antisymmetric states of (j, j) have the form

$$\Phi(jj; mm') \equiv \Phi(nljm, n'l'jm') = \frac{1}{\sqrt{2}} \begin{vmatrix} \Phi_1(nljm)\Phi_1(n'l'jm') \\ \Phi_2(nljm)\Phi_2(n'l'jm') \end{vmatrix} \tag{13.1}$$

m and m' can take independently any of the $2j + 1$ values $m, m' = -j, -j + 1, \ldots, j$ and there are therefore $(2j + 1)^2$ different wave functions $\Phi(jj; mm')$. In the corresponding J-scheme, J can assume all integral values from $J = 0$ to $J = 2j$, so that the total number of independent states in the J-scheme is

$$\sum_{J=0}^{2j} (2J + 1) = (2j + 1)^2 \tag{13.2}$$

equal to that of the original number of independent states in the m-scheme.

When we pass now to the configuration j^2 we notice that the number of independent states in the m-scheme is reduced. Whereas $\Phi(nljm, n'l'jm')$ and $\Phi(nljm', n'l'jm)$ are independent states, we have for j^2:

$$\Phi(j^2, mm') \equiv \Phi(nljm, nljm') = -\Phi(nljm', nljm) = -\Phi(j^2, m'm) \quad (13.3)$$

Of course, as can be readily seen from (13.3), when $m = m'$, $\Phi = 0$. There are therefore only $j(2j + 1)$ independent antisymmetric m-scheme states in j^2. There should consequently be only $j(2j + 1)$ independent states also in the J-scheme.

This reduction in the number of J-scheme states of j^2 as compared to those of jj, comes out formally through the symmetry properties of the matrix for two particles, which transforms from the m-scheme to the J-scheme. This matrix is nothing but the matrix of the Clebsch–Gordan coefficients $\langle JM|jmjm'\rangle$. This matrix, as we have stressed already, depends only on j and m (and not on the radial and orbital quantum numbers) so that it is the same matrix for both the jj-configuration and the j^2 configuration. With the phase conventions adopted previously, it can be shown that the following symmetry of the Clebsch–Gordan coefficients holds (see Appendix A, A.2.65)

$$\langle JM|j_1m_1 j_2m_2\rangle = (-1)^{j_1+j_2-J}\langle JM|j_2m_2 j_1m_1\rangle \quad (13.4)$$

We now see that for $j_1 = j_2 = j$, and j half-integer (i.e., $(-1)^{2j} = -1$) the Clebsch–Gordan coefficient $\langle JM|jmjm'\rangle$ is symmetric or antisymmetric *with respect to the exchange of m and m'* if J is odd or even, respectively. Since by (13.3) $\Phi(j^2, mm')$ is antisymmetric with respect to the exchange of m and m', it follows that

$$\Phi(j^2JM) = \sum_{mm'} \langle JM|jmjm'\rangle\Phi(j^2, mm') = 0 \quad (13.5)$$

for j half-integer and odd J.

The configuration j^2 of equivalent particles can therefore give rise only to levels with even total angular momenta. The total number of independent states is therefore

$$\sum_{J\ \text{even}} (2J + 1) = 1 + 5 + \ldots + [2(2j - 1) + 1] = j(2j + 1) \quad (13.6)$$

in agreement with the number of independent states in the m-scheme for the same configuration.

14. EXAMPLES

We shall now apply these results to the study of a specific spectrum. There. are many examples of nuclear spectra that are believed to be associated with a j^2-configuration; one that may be particularly instructive is that of ${}^{90}_{40}Zr_{50}$ As was pointed out before, 38 is a semimagic number. Since the neutron number 50 is just magic, we expect the lowest states of ${}^{90}_{40}Zr_{50}$ to be determined by the proton configurations $(2p_{1/2})^2$, $(1g_{9/2})^2$ and $(2p_{1/2}1g_{9/2})$ (or in more detail $1s^2_{1/2}\ 1p^4_{3/2}\ 1p^2_{1/2}\ 1d^6_{5/2}\ 2s^2_{1/2}\ 1d^4_{3/2}\ 1f^8_{7/2}\ 1f^6_{5/2}\ 2p^4_{3/2}\ 2p^2_{1/2}$ etc., our shorthand notation specifies only the orbits occupied by the last two protons). In the $p^2_{1/2}$ configuration the total angular momentum J can take on just one value $J = 0^+$ (since it must be even, and since for the j^2 configuration $J \leq 2j - 1$); in the $g^2_{9/2}$ configuration we can have states with total angular momentum $J = 0^+, 2^+, 4^+, 6^+$, and 8^+. Finally, the configuration $(p_{1/2}\, g_{9/2})$ can lead to states with $J = 4^-$ and 5^- (note that p and g have opposite parities). Figure14.1 shows the observed spectrum of Zr^{90} with the assignment of spin and parity for the various levels [omitted is an uncertain level between 3.081 MeV and 3.455 MeV; the assignment 4^- to the level at 2.745 MeV is not certain—see Dickens et. al. (67)]. Between the ground state and up to an excitation of 4.5 MeV all one sees are just the levels expected on the basis of our discussion above and no others. There are no levels with even parity and odd values of J, and the only levels of odd parity are the ones that can be gotten from the $(p_{1/2}g_{9/2})$ configuration.

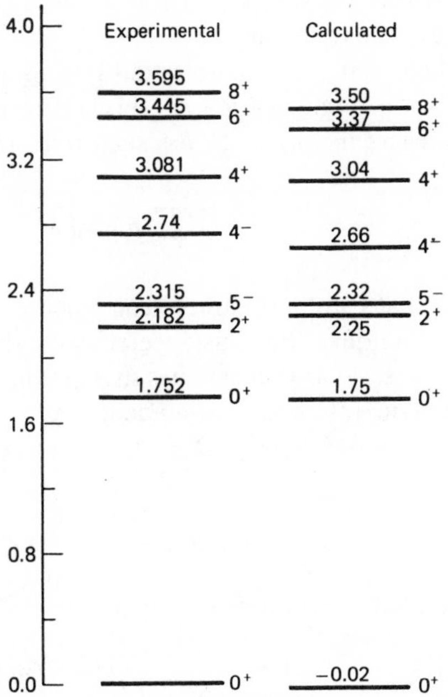

FIG. 14.1. Experimental and calculated energy levels of ⁹⁰Zr (in MeV). The total energies rather than energy differences, are plotted [taken from Auerbach and Talmi (65)].

Remark. Later on we shall see that in the region of 3 MeV excitation one finds systematically, in all even–even nuclei, a 3^- level whose nature is more complex. There is an indication that Zr^{90} has such a level as well, but we shall not discuss it here. Experimentally it is possible to distinguish these "collective 3^-" levels from other levels by their excitation cross section in some inelastic processes.

We notice in passing that the levels of the $g_{9/2}^2$ configuration are ordered according to J, and that they are more closely packed as J increases. This will turn out to be a rather general property of the levels of the j^n configuration, which will be discussed later (Section V.1).

15. THE ALLOWED J'S

The effects of antisymmetrization, which is responsible for the elimination of odd values of J from the permissible states of j^2, are felt also in the more complex configurations j^N with $N > 2$. The number of states in the configuration (7/2, 7/2, 7/2), for instance, is $8^3 = 512$; that of the three *equivalent* particles $(7/2)^3$ turns out to have only 56 states.

To be able to tell which are the permissible J's in a configuration j^N it is convenient to proceed in the following way.

It is easy to count how many permissible states of a given value of M there are in the m-scheme for the configuration j^N; this is determined by all the possible sets of N half-integers $m_1, m_2, \ldots, m_N$ such that no two m's are equal,

$$-j \leq m_i \leq j, \qquad \sum_{i=1}^{N} m_i = M,$$

and provided two sets that can be obtained from each other by permutations are considered to be identical. For instance, if we take the configuration $(7/2)^3$, the maximum possible M that satisfies these requirements is $M = 15/2$ and it can be obtained in just one independent way; $\sum m_i = 15/2$ if $(m_1, m_2, m_3) = (7/2, 5/2, 3/2)$. Also $M = 13/2$ can be obtained from just one set:

$$\sum m_i = 13/2 \quad \text{if} \quad (m_1, m_2, m_3) = (7/2, 5/2, 1/2)$$

$M = 11/2$ can be obtained in two nonequivalent ways:

$$\sum m_i = 11/2\colon \quad (m_1 m_2 m_3) = \begin{cases} (7/2, 5/2, -1/2) \\ (7/2, 3/2, 1/2) \end{cases}$$

$M = 3/2$ and $M = 1/2$ can be obtained with the following independent sets:

$$\sum m_i = 3/2: \quad (7/2, 3/2, -7/2), (7/2, 1/2, -5/2), (7/2, -1/2, -3/2)$$
$$(5/2, 3/2, -5/2), (5/2, 1/2, -3/2), (3/2, 1/2, -1/2) \tag{15.1}$$

$$\sum m_i = 1/2: \quad (7/2, 1/2, -7/2), (7/2, -1/2, -5/2), (5/2, 3/2, -7/2)$$
$$(5/2, 1/2, -5/2), (5/2, -1/2, -3/2), (3/2, 1/2, -3/2) \tag{15.2}$$

Now that we have counted the number $\gamma(M)$ of antisymmetric m-scheme states in j^N, with a definite value of M, we remark that this must equal the number of J-levels in the J-scheme of j^N with $J \geq M$. Indeed, every level with a given J in j^N gives rise to one state $|JM\rangle$ for each value of M satisfying $-J \leq M \leq J$. Thus if $\gamma(J)$ is the number of levels in j^N with a given value of J, we have

$$\gamma(M) = \sum_{J=|M|}^{J\max} \gamma(J) \tag{15.3}$$

and therefore

$$\gamma(J) = \gamma(|M| = J) - \gamma(|M| = J + 1) \tag{15.4}$$

Applying (15.4) to our example of $(7/2)^3$ we find that this configuration contains: no level with $J > 15/2$, one level with $J = 15/2$, no level with $J = 13/2$, one level each with $J = 11/2, 9/2, 7/2, 5/2$, and $3/2$, and no level with $J = 1/2$ [$\gamma(M = 1/2) = \gamma(M = 3/2) = 6$; see (15.1) and (15.2)].

An example pertinent to this situation is ^{51}V where the analysis of levels up to nearly 3 MeV shows the spectrum shown in Fig. 15.1. With 28 neutrons just filling the fourth shell, we expect the low-energy spectrum to be dominated by the protons in the unfilled shell, that is, the $(7/2)^3$ configuration. The ground state of $^{51}_{23}V_{28}$ has a measured spin $7/2^-$ and spins $5/2^-$, $3/2^-$, $11/2^-$, $9/2^-$, and $15/2^-$ have been assigned to levels at 0.320 MeV, 0.930 MeV, 1,601 MeV, 1.813 MeV, and 2.699 MeV, respectively.

The fact that the spins and parities of these levels coincide with those expected of an $f^3_{7/2}$ configuration lends great support to the identification of these levels as belonging to this configuration. It should, however, be mentioned that spins and parities alone are not sufficient for such an identification. As we shall see below, all the levels of a given configuration of particles of the same type (protons or neutrons) must also have the same gyromagnetic ratio; magnetic moments of the levels considered would therefore help considerably in ascertaining their identification as the levels of $f^3_{7/2}$. Other properties of these levels are also rather uniquely determined if they are identified as belonging to a definite configuration.

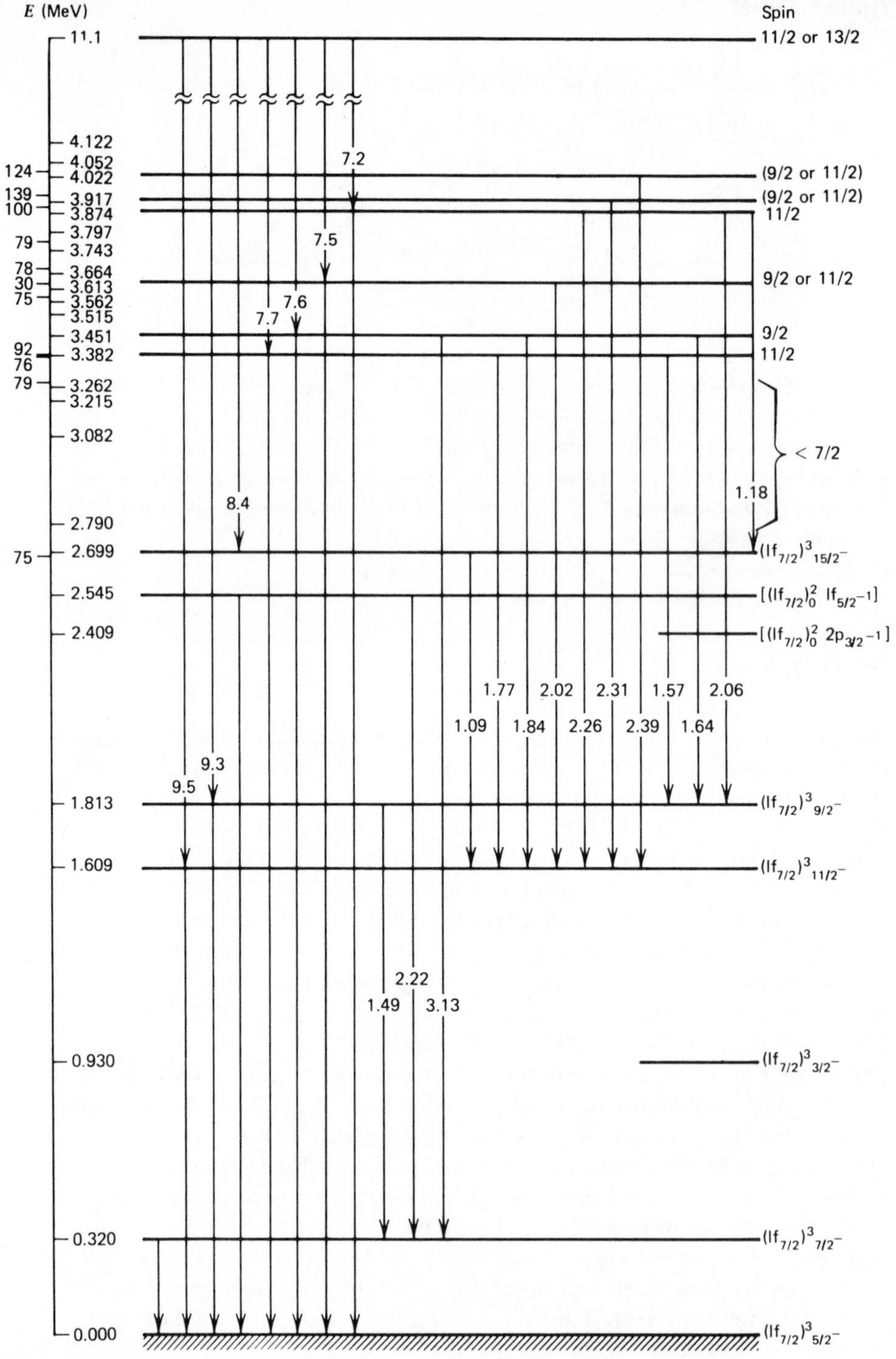

FIG. 15.1. Gamma-ray decay in $^{51}V_{28}$ [taken from Schwager (61)].

Two more levels in ^{51}V have been identified at 2.409 MeV ($3/2^-$) and 2.545 MeV ($5/2^-$). These are probably the lowest states of the proton configurations ($1f^2_{7/2}$, $2p_{3/2}$) and ($1f^2_{7/2}$, $1f_{5/2}$). It is to be noted that no level with $J = 13/2^-$ or with $J = 1/2^-$ has been observed in this nucleus below 2.7 MeV. Such levels are absent also in the $f^3_{7/2}$ configuration!

In our discussions and examples thus far we have deliberately excluded cases in which neutrons and protons are simultaneously in the same level, and referred only to the cases in which either the protons or the neutrons were just of the right number to fill completely a major shell. We shall tackle the more general problem later when we shall introduce the full use of isospin formalism in complex nuclei. But already with the cases studied thus far we see how and why it is possible to explain some features of low-lying nuclear levels without looking at the detailed dynamical affects of the interaction. Implicitly we do say something about the interaction, because our considerations were only valid if we were justified in describing these levels in terms of isolated configurations. This is probably not true for arbitrary interactions $v(ij)$. But once we accept the conjecture of the shell model that $v(ij)$ can be treated as a perturbation in lowest order, the rest of our discussion above follows without further dynamical specifications of $v(ij)$.

16. MAGNETIC MOMENTS IN THE NUCLEAR SHELL MODEL

The success we have encountered in explaining angular momenta of low-lying states in terms of simple configurations may be misleading. The total angular momentum operator is a rigorous constant of motion. Furthermore its eigenvalues are quantized so that its possible values are limited. How would our results have been affected, then, if the description of low-lying states in terms of simple configurations were not really as good an approximation as we have assumed it to be? Let us consider an idealized situation in which only two configurations are involved, $(5/2)^2$ and $(9/2)^2$. For the complete independent-particle picture each configuration is completely degenerate, as shown on the extreme left of Fig. 16.1. Then, to take $v(ij)$ into account in first order, we have to first diagonalize it separately within the subspaces of degenerate states. This gives rise to two groups of levels, as shown in the middle of Fig. 16.1, one originating from $(9/2)^2$ with $J = 0, 2, 4, 6$, and 8, and one originating from $(5/2)^2$ with $J = 0$, 2, and 4.

If $\sum v(ij)$ is not a small interaction, then, within our idealized situation, we have to diagonalize $\sum v(ij)$ in the *complete* Hilbert space of the configurations considered. Since $v(ij)$ is a scalar, and therefore conserves total angular momentum, it can have nonvanishing matrix elements only between states with the same J and M. The complete diagonalization of $\sum v(ij)$ can therefore be reduced to the diagonalization of considerably smaller matrices, each characterized by referring only to states of given J and M. States with $J = 6$

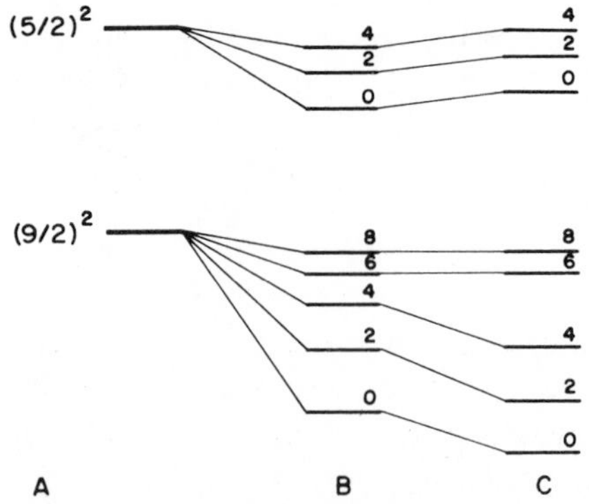

FIG. 16.1. Levels in the two-particle configuration which can populate either $(9/2)^2$ or $(5/2)^2$. *A–v(ij)* is switched off; configurations completely degenerate. *B—v(ij)* is on, but very weak; it is diagonalized only within each configuration separately. Levels split off but can still be assigned to well defined configurations. *C—v(ij)* is included in the complete diagonalization within these two configurations. The level ordering in *B* and *C* remain the same although in *C* the configurations are no longer pure.

and $J = 8$, *in our example*, appear only in the configuration $(9/2)^2$; they will therefore remain unaffected by taking into account the effects of the configuration $(5/2)^2$. For the states with $J = 0$, 2, and 4 we shall have to diagonalize a matrix of the form

$$\begin{pmatrix} E_1^{(0)}(J) & a(J) \\ a(J) & E_2^{(0)}(J) \end{pmatrix} \tag{16.1}$$

where $E_1{}^{(0)}(J)$ is the energy of the level J in the $(9/2)^2$ configuration taken by itself, and $E_2{}^{(0)}(J)$—the corresponding quantity for the $(5/2)^2$ configuration. $a(J)$ is the matrix element of $\sum v_{ij}$ between the $(9/2)^2$ and $(5/2)^2$ configurations. This is the term that is neglected in first-order perturbation theory. Thus

$$a(J) = \langle (9/2)^2 JM | \sum v(ij) | (5/2)^2 JM \rangle \tag{16.2}$$

It is well known that the eigenvalues $E_1(J)$ and $E_2(J)$ of (16.1) satisfy

$$E_1(J) - E_2(J) = [E_1^{(0)}(J) - E_2^{(0)}(J)] \sqrt{1 + \frac{4a^2(J)}{[E_1^{(0)}(J) - E_2^{(0)}(J)]^2}} \tag{16.3}$$

The net result of including all the effects of $v(ij)$ in our case is therefore to push the lowest states with $J = 0$, 2, and 4 even lower, leave $J = 6$ and $J = 8$ in their first-order position, and push the higher states with $J = 0$, 2, and 4 even higher. The exact solution therefore "magnifies" the first-order prediction: that of finding among the lowest levels just the J's allowed by the lowest configuration. This result can be shown to be valid also when more than two configurations are included. We therefore conclude that the prediction of spin values may not provide a very strong test of the shell-model picture.

For this reason, and for the reasons mentioned in our discussion of ^{51}V, we shall now look for additional physical quantities, which still depend only, or mostly, on the spin and angular coordinates, and try to explain them with shell-model wave functions. This way we shall have a more severe test for these wave functions. One such quantity is the magnetic moment of nuclei.

Formally, the magnetic moment operator for a given system is that operator that determines the response of the system to an outside homogeneous magnetic field. Given the Hamiltonian of the system $H(\mathbf{p}_i, \mathbf{x}_i)$ we know that its modification in the presence of an outside electromagnetic field characterized by a vector potential $\mathbf{A}$ is obtained by replacing $\mathbf{p}_i$ by $\mathbf{p}_i - (e/c)\mathbf{A}_i$, provided the Hamiltonian gives a complete description of the system, including the intrinsic structure of whatever particles it may have. Here we shall assume this to be the case.

Taking $\mathbf{A}$ to be the vector potential for a magnetic field $\mathbf{H}$, we can expand the Hamiltonian in the presence of $\mathbf{A}$ in powers of $\mathbf{H}$, to obtain

$$H(\mathbf{p}_i - \frac{e}{c}\mathbf{A}_i, \mathbf{x}_i, \boldsymbol{\sigma}_i, \boldsymbol{\tau}_i) = H(\mathbf{p}_i, \mathbf{x}_i, \boldsymbol{\sigma}_i, \boldsymbol{\tau}_i) - \hat{\boldsymbol{\mu}}(\mathbf{p}_i, \mathbf{x}_i, \boldsymbol{\sigma}_i, \boldsymbol{\tau}_i)\cdot\mathbf{H} + \ldots \quad (16.4)$$

The operator $\hat{\boldsymbol{\mu}}$ is, then, by definition, the magnetic moment operator of the system.

We see from (16.4) that to derive an expression for $\hat{\boldsymbol{\mu}}$ in terms of $\mathbf{x}_i$, $\mathbf{p}_i$, $\boldsymbol{\sigma}_i$, and $\boldsymbol{\tau}_i$ we must have an explicit expression for the Hamiltonian H. However, if the only $\mathbf{p}$-dependence of the Hamiltonian H is in the kinetic energy, that is, if the interaction $v(ij)$ in (16.4) is independent of the momenta of the particles, we can deduce $\hat{\boldsymbol{\mu}}$ without this knowledge of the complete Hamiltonian. As can be seen from (16.4), in this case the interaction $\sum v(ij)$ drops out and $\hat{\boldsymbol{\mu}}$ is determined solely by the kinetic energy.

It should be stressed that this conclusion applies only to the *functional dependence* of $\hat{\boldsymbol{\mu}}$, as an operator, on the operators $\mathbf{p}_i$, $\mathbf{x}_i$, $\boldsymbol{\sigma}_i$, and $\boldsymbol{\tau}_i$. The measured values of $\hat{\boldsymbol{\mu}}$, which are *expectation values* of the operator $\hat{\boldsymbol{\mu}}$, involve the eigenfunctions of H and therefore depend on $v(ij)$, even in the absence of momentum dependence for $v(ij)$.

The actual expression for the magnetic moment operator when only the kinetic energy is considered is derived in texts on quantum mechanics [see for example Gottfried (66)] with the result:

$$\hat{\boldsymbol{\mu}} = \sum_{i=1}^{A} \boldsymbol{\mu}_i \quad (16.5)$$

$$\hat{\boldsymbol{\mu}}_i = g_l^{(i)}\mathbf{l}_i + g_s^{(i)}\mathbf{s}_i \quad (16.6)$$

Equation 16.5 says that $\hat{\boldsymbol{\mu}}$ is a single-particle operator (Section I.8). Equation

16.6 identifies the two sources of magnetism. One is the current created by the orbital motion; this part of the magnetic moment is characterized by an orbital g-factor (or gyromagnetic ratio)

$$g_l^{(i)} = \begin{cases} 1\mu_0 & \text{for protons} \\ 0 & \text{for neutrons} \end{cases} \tag{16.7}$$

The second term represents the magnetic contributions of the spin characterized by a g-factor

$$g_s^{(i)} = \begin{cases} 5.5845\mu_0 & \text{for protons} \\ -3.8263\mu_0 & \text{for neutrons} \end{cases} \tag{16.8}$$

where

$$\mu_0 \equiv 1 \text{ nuclear magneton} = \frac{e\hbar}{2M_pc} = 5.049 \times 10^{-24} \frac{\text{erg}}{\text{gauss}} \tag{16.9}$$

It will be recognized immediately that (16.6) does not involve the radial coordinates at all ($\mathbf{l}_i$ can be expressed in terms of angular coordinates only). The expectation value of $\hat{\boldsymbol{\mu}}$ can therefore be calculated if the angular and spin dependence of $\Psi(1, \ldots, A)$ is known.

The number referred to as "the magnetic moment" of a system in a state of total angular momentum J is defined by

$$\mu \equiv \langle JM|\hat{\mu}_z|JM\rangle_{M=J} \tag{16.10}$$

that is, it is the expectation value of the z-component of the operator $\hat{\boldsymbol{\mu}}$ in the substate of maximum z-projection of $\mathbf{J}$.

Since μ_z is the z-component of a vector, we can use the Wigner–Eckart theorem (see Appendix A) to obtain

$$\mu = \begin{pmatrix} J & 1 & J \\ -J & 0 & J \end{pmatrix} (J||\hat{\boldsymbol{\mu}}||J) = \sqrt{\frac{J}{(J+1)(2J+1)}}\,(J||\hat{\boldsymbol{\mu}}||J) \tag{16.11}$$

For closed shells we have already found that $J = 0$; therefore the magnetic moment of nuclei composed of closed shells vanishes.

For the state $\Phi_1(JM)$ (see Eq. 9.7) of a configuration that includes just one particle in a level $(n'l'j')$ outside closed shells, we can use the fact that $\hat{\boldsymbol{\mu}}$ is a single-particle operator, (16.5), to obtain:

$$\langle\Phi_1(JM)|\hat{\boldsymbol{\mu}}|\Phi_1(JM)\rangle = \langle j'M|\hat{\boldsymbol{\mu}}|j'M\rangle \tag{16.12}$$

where $< j'M|$ is a single-particle wave function.

We have already seen, (9.8), that for such configurations $J = j'$. Equation 16.12 can therefore be interpreted in the following way: the magnetic moment of a configuration with one particle outside closed shells is equal to the magnetic moment of the odd nucleon taken by itself. The particles in the closed shells do not contribute to the magnetic moment of the system.

Physically this result is rather easy to understand. When a level is completely filled, then for every particle in that level going one way there is a particle, in the same level, going in the opposite way, and for every particle with its spin pointing up there is a particle, in the same level, with its spin pointing down. Thus the contributions of all the orbital currents to the magnetic moment cancel each other, and so do the spin contributions. We are left therefore with the contribution of the odd nucleon alone.

The magnetic moment of a single nucleon is easy to evaluate. From the Wigner–Eckart theorem we know that the matrix elements $(jm'|\mathbf{v}|jm)$ for all vectors $\mathbf{v}$ and a given fixed $\mathbf{j}$ are proportional to each other, the proportionality factor being independent of m and m'. We can therefore define g_{lj} through

$$\langle ljm'|\hat{\mathbf{\mu}}|ljm\rangle = g_{lj}\langle ljm'|\mathbf{j}|ljm\rangle \tag{16.13}$$

and g_{lj} will be independent of m and m'; g_{lj} is, of course, nothing but the gyromagnetic ratio for the particle in the level (nlj). To evaluate g_{lj} note that since the operator $\boldsymbol{j}$ cannot change the quantum numbers j, n, l, it follows from (16.13) that

$$\langle jm|\hat{\mathbf{\mu}}\cdot\mathbf{j})|jm\rangle = g_{lj}\langle jm|\mathbf{j}\cdot\mathbf{j}|jm\rangle \tag{16.14}$$

with the same g_{lj}.

We now introduce for $\hat{\mathbf{\mu}}$ the expression $\hat{\mathbf{\mu}} = g_l\mathbf{l} + g_s\mathbf{s}$ and use the operator identities, derived from squaring $\mathbf{j} = \mathbf{l} + \mathbf{s}$:

$$\begin{aligned} 2(\mathbf{l}\cdot\mathbf{j}) &= (\mathbf{j}\cdot\mathbf{j}) + (\mathbf{l}\cdot\mathbf{l}) - (\mathbf{s}\cdot\mathbf{s}) \\ 2(\mathbf{s}\cdot\mathbf{j}) &= (\mathbf{j}\cdot\mathbf{j}) + (\mathbf{s}\cdot\mathbf{s}) - (\mathbf{l}\cdot\mathbf{l}) \end{aligned} \tag{16.15}$$

Noticing that in the state $|ljm>$

$$\langle ljm|\mathbf{j}\cdot\mathbf{j}|ljm\rangle = j(j+1) \tag{16.16}$$

and

$$\langle ljm|\mathbf{l}\cdot\mathbf{l}|ljm\rangle = l(l+1)$$

and that $\mathbf{s}\cdot\mathbf{s} = 3/4$, we obtain finally

$$\mu = gj = \frac{j}{2}(g_l + g_s) + \frac{l(l+1) - 3/4}{2(j+1)}(g_l - g_s) \tag{16.17}$$

It is customary to write (16.17) separately for $j = l + 1/2$ and $j = l - 1/2$

$$\mu = \begin{cases} lg_l + \frac{1}{2}g_s & \text{for} \quad j = l + 1/2 \\ \dfrac{j}{j+1}\left[(l+1)g_l - \frac{1}{2}g_s\right] & \text{for} \quad j = l - 1/2 \end{cases} \tag{16.18}$$

Equation 16.17 and 16.18 are known as the *Schmidt moments* or the *single-particle moments.* Although they are defined only for half integral values of j it is common to refer to the diagram of μ versus j as the "*Schmidt diagram.*" Figure 16.2*a* shows the Schmidt diagram for protons ($g_l = \mu_0$, $g_s = 5.58\mu_0$) and Fig. 16.2*b* shows the same diagram for neutrons ($g_l = 0$, $g_s = -3.83\mu_0$). Each diagram has two lines in it—one corresponding to $j = l + 1/2$ and the other to $j = l - 1/2$. The dots in the diagram are measured moments of ground and excited states of odd-A nuclei (odd Z-even N-nuclei are plotted in the proton diagram and even Z-odd N are plotted in the neutron diagram). Many measured moments are shown, not just those corresponding to one particle outside closed shells. If the shell-model wave function were an exact wave function we would expect the moments of nuclei with one nucleon added to a

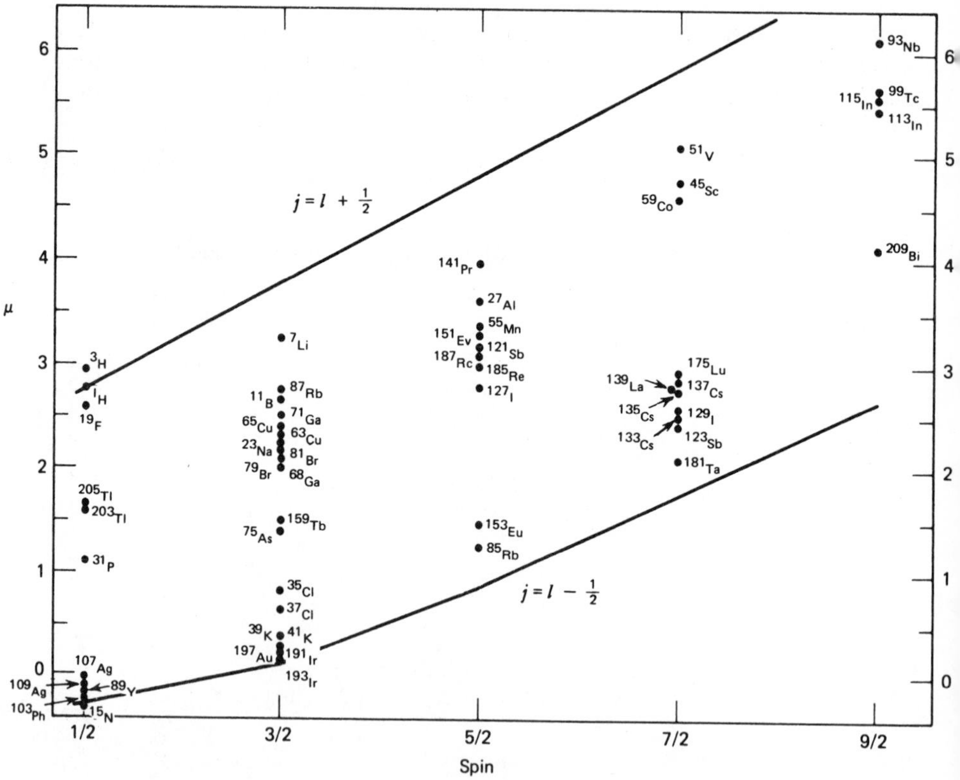

(a)

FIG. 16.2. (*a*) Schmidt diagram for odd-proton nuclei. The magnetic moment in nuclear magnetons is plotted against the spin [from R. J. Blin-Stoyle (56)].

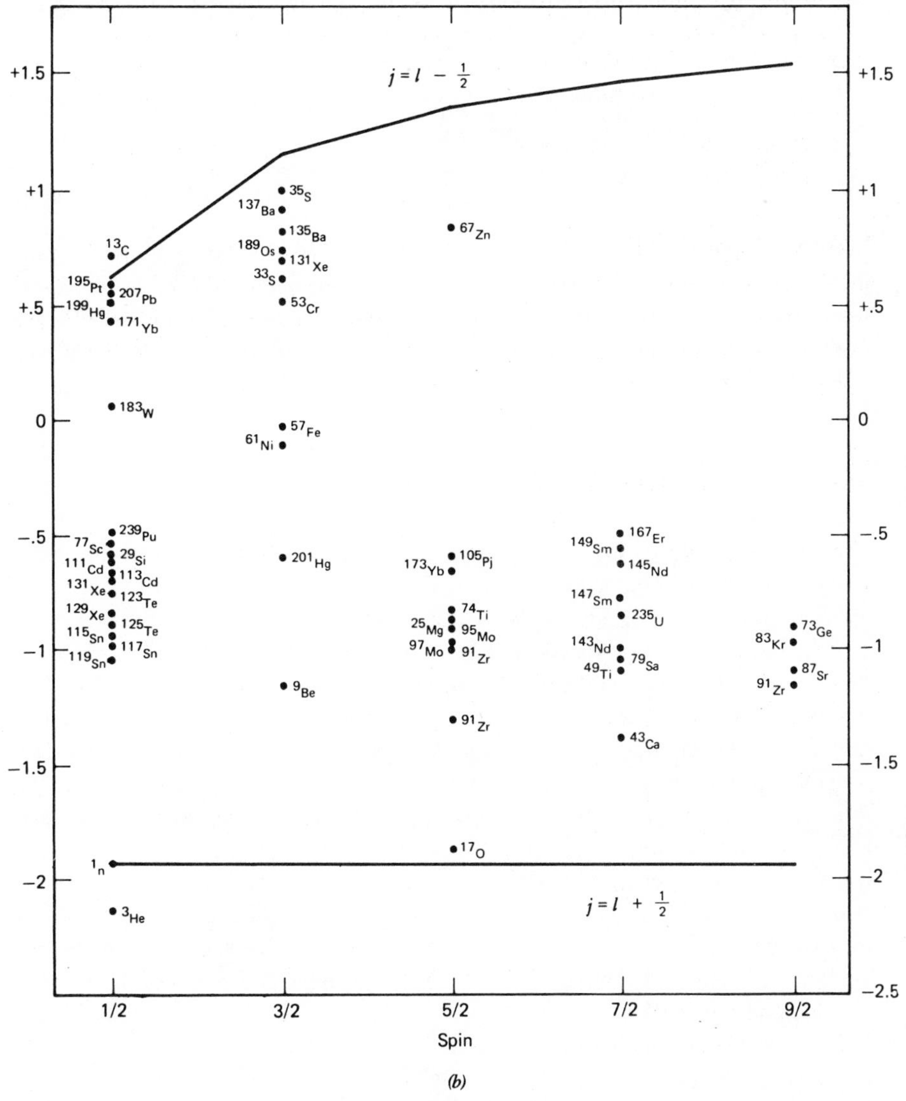

FIG. 16.2. (*b*) Schmidt diagram for odd-neutron nuclei. The magnetic moment in nuclear magnetons is plotted against the spin [from R. J. Blin–Stoyle, (56)].

closed shell to lie on the Schmidt lines. This is actually the content of (16.18) for a single particle outside closed shells.

Problem. Prove that the magnetic moment of a nucleus in a state with one particle missing from a closed shell [the function Φ_{-1} of (9.9) and (9.10)] is equal to the magnetic moment of the missing particle.

We see from Fig. 16.2 that the magnetic moment of ${}^{17}_{8}O_9$ lies indeed fairly close to the Schmidt line, and so do those of ${}^{3}_{1}H_2$, ${}^{3}_{2}He_1$, and ${}^{89}_{39}Y_{50}$. However other moments for nuclei containing one nucleon plus closed shells, notably those of ${}^{209}_{83}Bi_{126}$ and ${}^{41}_{20}Ca_{21}$ lie rather far from the Schmidt line, or at least do not come any closer to it than many of the other nuclei.

17. MAGNETIC MOMENTS—SEVERAL NUCLEONS

The empirical data on magnetic moments of nuclei shows a very striking regularity. It is remarkable that the magnetic moments of all nuclei fall between the Schmidt lines. The moments of the various nuclei seem to fall distinctly into two groups when plotted on a Schmidt diagram: one group lies closer to the "$j = l + 1/2$" line and the other lies closer to the "$j = l - 1/2$" line. This regularity has been used to determine parities semiempirically. If the nuclear angular momentum j and the magnetic moment μ of a level are known it is possible to determine by inspection to which Schmidt line this belongs and thereby assign to it the corresponding $l(l = j \pm 1/2)$. The parity of the level is then taken to be positive or negative according to whether l thus determined is even or odd. Independent studies of the parities of nuclear levels using nuclear reactions, or the emission of β- and γ-rays, confirm this empirical rule in most cases. It is interesting therefore to see whether the shell model can explain the observed regularities also for these more complex configurations. To this end we shall try now to obtain an estimate of these moments.

First, we shall calculate the magnetic moment in a configuration of N-protons or N-neutrons in unfilled shells. The contributions of the filled levels to the magnetic moment all vanish,

Problem. Prove it.

so that we need to take care only of the N-equivalent particles in the level nlj. In other words we want to evaluate

$$\langle j^N JM | \sum \hat{\mathfrak{u}}(i) | j^N JM \rangle \tag{17.1}$$

where the N-particles are either all protons or all neutrons. Transforming into the m-scheme we obtain

$$\langle j^N JM | \sum \hat{\boldsymbol{\mu}}(i) | j^N JM \rangle = \sum_{m_1 \ldots m_N m'_1 \ldots m'_N :} \langle JM | m_1, \ldots, m_N \rangle$$

$$\times \langle jm_1, jm_2, \ldots, jm_N | \sum \hat{\boldsymbol{\mu}}(i) | jm'_1 jm'_2, \ldots, jm'_N \rangle$$

$$\times \langle m'_1 m'_2, \ldots, m'_N | JM \rangle \quad (17.2)$$

Since $\hat{\boldsymbol{\mu}} = \sum \hat{\boldsymbol{\mu}}(i)$ is a one-particle operator the matrix element

$$\langle jm_1, jm_2, \ldots, jm_N | \sum \hat{\boldsymbol{\mu}}(i) | jm'_1, jm'_2, \ldots, jm'_N \rangle \quad (17.3)$$

will reduce to sum of matrix elements of the type $\langle jm | \hat{\boldsymbol{\mu}}(1) | jm' \rangle$. From (16.13) we know that

$$\langle jm | \hat{\boldsymbol{\mu}} | jm' \rangle = g_{lj} \langle jm | \mathbf{j} | jm' \rangle$$

where g_{lj}, the single-particle (Schmidt) g-factor, is independent of m and m'. We conclude, therefore, that as long as all the N-particles are of the same type (proton or neutron), so that g_{lj} is the same:

$$\langle jm_1, jm_2, \ldots, jm_N | \sum \hat{\boldsymbol{\mu}}(i) | jm'_1, jm'_2, \ldots, jm'_N \rangle$$

$$= g_{lj} \langle jm_1, jm_2, \ldots, jm_N | \sum \mathbf{j}(i) | jm'_1, jm'_2, \ldots, jm'_N \rangle \quad (17.4)$$

Introducing (17.4) into (17.3) and noting that $\sum \mathbf{j}(i) = \mathbf{J}$ we obtain finally:

$$\langle j^N JM | \sum \hat{\boldsymbol{\mu}}(i) | j^N JM \rangle = g_{lj} \langle j^N JM | \mathbf{J} | j^N JM \rangle \quad (17.5)$$

The magnetic moment is obtained by putting $M = J$ and taking the z-component of (17.5):

$$\mu(j^N, J) = g_{lj} J \quad (17.6)$$

Note the absence of any dependence of the final result on N: the magnetic moment of any number of particles in the level (nlj) is proportional to the total angular momentum of the level involved, with a proportionality factor—the gyromagnetic factor g_{lj}—which is independent of N. In other words particles that do not contribute to the angular momentum do not contribute to the magnetic moment.

At first glance these results seem to be of little relevance to real nuclei since most nuclei involve unfilled proton as well as neutron levels. But later when we shall discuss the energies of the different states of a given configuration, we shall find that very often, for odd values of N, the spin J of the lowest level of the configuration j^N of particles of the same type is $J = j$ (see Section V.11). Furthermore, if the lowest configuration involves an *even* number N_1 of particles of one type, say neutrons in the level j_1, and an odd number N_2 of particles of the other type in the level $j_2 \neq j_1$, then the ground state of the configuration $j_1^{N_1}j_2^{N_2}$ has again the total angular momentum determined by the odd configuration $J = j_2$. We shall also find that the group of even number of particles in the level j_1 is most of the time so correlated that its own total angular momentum vanishes. The even group does not contribute, therefore, to the magnetic moment. Thus the ground state of a configuration of the type $j_1^{N_1}j_2^{N_2}$ with N_1 even and N_2 odd looks like a single particle in the level j_2 both from the point of view of its total angular momentum and also from that of its magnetic moment. This is probably the reason why the measured moments of the ground states of odd-A nuclei group themselves into two distinct groups that form "bands" parallel, more or less, to the Schmidt single-particle lines, but not lying on them.

For the physics of magnetic moments, it is important to note that when the angular momenta of similar particles add up so as to cancel each other, their magnetic moment also vanishes; the whole resultant moment is due to the group of particles that cooperate in producing the finite total angular momentum. Since these are equivalent particles, all occupying the same level, their combined gyromagnetic ratio is the same as that of each one of them by itself.

Our discussion of magnetic moments has enabled us to obtain some general understanding of their basic features, still without considering the details of neither the single-particle field $U(i)$ nor the residual interaction $v(ij)$. However, at best we obtained only a qualitative understanding of the experimental data. At this present stage of our discussion we have not explained the fact that practically all the observed moments fall *between* the Schmidt lines rather than being scattered around them, as could have been expected. We also cannot, at this stage, explain why a nucleus such as $^{209}_{83}Bi_{126}$, which has one proton outside of both proton and neutron closed shells, deviates from its expected Schmidt single-particle moment more than most other observed nuclei. It seems that a more detailed study of nuclear structure is required to understand these, and other, phenomena. From the qualitative success we have had thus far, we can feel encouraged to adopt the shell-model wave functions as our starting point, but further refinements are required. Fortunately these refinements will turn out to have striking regularities; we shall therefore be able to obtain a considerably more detailed picture of the nucleus before resorting to elaborate numerical computations. The following sections will be devoted to the study of these further refinements.

18. THE SINGLE-PARTICLE POTENTIAL

The discussion of the magnetic moments of nuclei was based entirely on the l and j assignments to the single-particle levels, and did not involve the radial shape of the single-particle potential. Also the relative strength of the spin-dependent part did not enter into our considerations except when it affects the order of the single-particle levels. Although qualitatively we were able to account for some of the observed features of magnetic moments of nuclei, the fact that we failed to account for the more quantitative features may mean one (or both) of two things: either the wave function is too crude, or the operators (16.5) and (16.6) are not the appropriate operators for complex nuclei. There is some evidence, to which we shall come later (Chapter VIII), that to a good approximation the magnetic moment operator for nuclei is indeed of the form (16.5) and (16.6):

$$\hat{\mathbf{\mu}} = \sum_{i=1} (g_l^{(i)}\mathbf{l}_i + g_s^{(i)}\mathbf{s}_i)$$

We therefore conclude that it is the shell-model wave function that requires the refinement. It is obvious that this refinement should come about through the inclusion of the effects of the interaction $v(ij)$ that have thus far been neglected and that we shall now consider. This is no surprise. There are many phenomena that point to the importance of $v(ij)$. Historically the magnetic moment problem was among the first where it was most clearly indicated.

One way to procede is the following: we consider the single-particle Hamiltonian

$$H_0 = \sum_{i=1}^{A} T_i + \sum_{i=1}^{A} U(i) \tag{18.1}$$

as an operator that generates a complete orthonormal set of independent-particle antisymmetrized wave functions $\Phi_\alpha(1, \ldots, A)$ for the system of A-nucleons. Using these wave function we now construct the matrix

$$\langle\alpha|H|\beta\rangle = \langle\Phi_\alpha|\,[\sum T_i + \sum_{i<j} v(ij)]\Phi_\beta\rangle \tag{18.2}$$

and diagonalize it. Here $v(ij)$ is the full nucleon–nucleon potential. If we manage to diagonalize the complete, infinite, matrix $\langle\alpha|H|\beta\rangle$ we have solved the problem exactly, and the eigenvectors and eigenfunctions of H are then independent of the choice of H_0. H_0 in this case serves only as a convenient prescription for generating antisymmetrized A-particle wave functions. Generally, however, we shall have to satisfy ourselves with a partial diagonalization of H. Starting with a given state Φ_α we shall pick on a small set of states $\Phi_{\beta i}$, which connect with Φ_α via H particularly strongly, and diagonalize the matrix

$$\begin{vmatrix} \langle\alpha|H|\alpha\rangle\langle\alpha|H|\beta_1\rangle\langle\alpha|H|\beta_2\rangle \ldots \\ \langle\beta_1|H|\alpha\rangle\langle\beta_1|H|\beta_1\rangle\langle\beta_1|H|\beta_2\rangle \ldots \\ \langle\beta_2|H|\alpha\rangle\langle\beta_2|H|\beta_1\rangle\langle\beta_2|H|\beta_2\rangle \ldots \\ \vdots \qquad\qquad \vdots \qquad\qquad \vdots \end{vmatrix} \tag{18.3}$$

The validity of the diagonalization of (18.3) as an approximation to the complete diagonalization of (18.2) depends, obviously, on the choice of H_0.

The closer the eigenfunctions $\Phi_\alpha(1, \ldots, A)$ of H_0 come to the description of the real nucleus, the better is the approximation represented by (18.3). It is obvious that if we want to take into account the effects of $v(ij)$ on the independent-particle wave function, and involve only a few selected configurations, we shall do best by starting from the "best H_0" in the sense just discussed.

Let us make this statement more quantitative by using the variational principle. In other words let us look for the independent-particle wave function $\Phi_\alpha(1, \ldots, A)$ that satisfies

$$\delta\langle\Phi_\alpha|\,[\sum T_i + \sum_{i<j} v(ij)]\Phi_\alpha\rangle = 0 \qquad \text{subject to} \qquad \langle\Phi_\alpha|\Phi_\alpha\rangle = 1 \tag{18.4}$$

If the variation in $\Phi_\alpha(1, \ldots, A)$ is produced through a variation of the single-particle potential $U(i)$ in (18.1), the solution of (18.4) will give us the "best" single-particle potential $U(i)$.

Let us first solve (18.4) for the simpler case in which anti-symmetry is ignored. If $\alpha_1, \alpha_2, \ldots, \alpha_A$ are the A-lowest single-particle states of (18.1) then Φ_α will take the form

$$\Phi_\alpha(1, 2, \ldots, A) = \phi_{\alpha_1}(1)\phi_{\alpha_2}(2)\ldots\phi_{\alpha_A}(A) \tag{18.5}$$

A variation δ_0 of the single-particle potential $U(i)$ can be chosen so as to vary arbitrarily just one of the single-particle wave functions (see Remark below) say $\phi_{\alpha_A}(\mathbf{r})$. Hence in determining the best Φ_α using (18.4) we shall automatically obtain the best U.

Remark. Formally this can be seen in the following way: let $\phi_\alpha(r)$ be the eigenstates in the potential $U(r)$, where for simplicity we ignore spin and isospin variables:

$$[T + U_0(\mathbf{r})]\phi_\alpha(\mathbf{r}) = \epsilon_\alpha\phi_\alpha(\mathbf{r}) \tag{18.6}$$

Define now a new potential $U_1(r, r') \equiv U_0 + \delta U$ that will generally be a nonlocal potential (see Section III.5):

$$U_1(\mathbf{r}, \mathbf{r}') = U_0(\mathbf{r})\delta(\mathbf{r} - \mathbf{r}') + \left(\epsilon_\mu + \frac{\hbar^2}{2M}\nabla_r{}^2 - U_0(\mathbf{r})\right)\delta\phi_\mu(\mathbf{r})\int\phi_\mu^*(\mathbf{r}') \tag{18.7}$$

where ϕ_μ is one of the eigenstates ϕ_α of (18.6) and $\int\phi_\mu^*(\mathbf{r}')$ means that this is the *operator* of integration with $\phi_\mu^*(\mathbf{r}')$. Because of the orthogonality of the ϕ_α's, δU has been chosen so that $\delta U\phi_\mu$ is proportional to ϕ_μ while $\delta U\phi_\lambda$, $\lambda \neq \mu$ is zero. As a consequence

$$[T + U_1]\phi_\lambda(\mathbf{r}) = (T + U_0)\phi_\lambda(\mathbf{r}) + \left(\epsilon_\mu + \frac{\hbar^2}{2M}\nabla_r{}^2 - U_0(\mathbf{r})\right)\delta\phi_\mu(\mathbf{r}) \times \int\phi_\mu^*(\mathbf{r}')\phi_\lambda(\mathbf{r}')d\mathbf{r}' = \epsilon_\lambda\phi_\lambda(\mathbf{r}) \qquad \text{for} \quad \lambda \neq \mu$$

To first order in $\delta\phi_\mu$ we also have

$$[T + U_1][\phi_\mu(\mathbf{r}) + \delta\phi_\mu(\mathbf{r})] = (T + U_0)\phi_\mu(\mathbf{r}) + [T + U_0(\mathbf{r})]\delta\phi_\mu(\mathbf{r})$$

$$+ \left[\epsilon_\mu + \frac{\hbar^2}{2M}\nabla_r^2 - U_0(\mathbf{r})\right]\delta\phi_\mu(\mathbf{r})\int \phi_\mu^*(\mathbf{r}')\phi_\mu(\mathbf{r}')d\mathbf{r}' = \epsilon_\mu[\phi_\mu(\mathbf{r}) + \delta\phi_\mu(\mathbf{r})]$$

Hence $\delta U = U_1(\mathbf{r}, \mathbf{r}') - U_0(\mathbf{r})\delta(\mathbf{r} - \mathbf{r}')$ is the change in the potential $U(r)$ required to change just *one* of the eigenfunctions $\phi_\mu(\mathbf{r})$ by $\delta\phi_\mu(\mathbf{r})$.

The normalization of $\phi_{\alpha i}(\mathbf{r})$ can be taken care of through the addition of appropriate Lagrange multipliers λ_α in the well-known way. Equation 18.4 then reduces, for the simpler functions (18.5) to:

$$\delta\left\{\int \phi_{\alpha_1}^*(1)\phi_{\alpha_2}^*(2)\ldots\phi_{\alpha_A}^*(A)\left[\sum_{i=1}^{A} T_i + \sum_{i<j} v(ij)\right]\right.$$

$$\left.\times\, \phi_{\alpha_1}(1)\phi_{\alpha_2}(2)\ldots\phi_{\alpha_A}(A)d(1)\ldots d(A) - \sum_k \lambda_{\alpha_k}\int \phi_\alpha^*(\mathbf{r})\phi_{\alpha_k}(\mathbf{r})d\mathbf{r}\right\} = 0 \tag{18.8}$$

We vary the real and imaginary parts of ϕ_{α_A} independently, or as is more convenient, we can consider the variations of ϕ_{α_A} and of $\phi_{\alpha_A}^*$ as independent [see, for instance, Morse and Feshbach (53), p. 315]. Taking δ_0 to be $\delta\phi_{\alpha_A}^*(A)$, we then obtain:

$$\int \phi_{\alpha_1}^*(1)\phi_{\alpha_2}^*(2)\ldots\delta\phi_{\alpha_A}^*(A)\left[\sum_{i=1}^{A} T_i + \frac{1}{2}\sum_{i\neq j} v(ij)\right]$$

$$\times\, \phi_{\alpha_1}(1)\phi_{\alpha_2}(2)\ldots\phi_{\alpha_A}(A) - \lambda_{\alpha_A}\int \delta\phi_{\alpha_A}^*(\mathbf{r}_A)\phi_{\alpha_A}(\mathbf{r}_A) = 0 \tag{18.9}$$

For (18.9) to be valid for any variation $\delta\phi_{\alpha_A}^*(A)$, the expression multiplying $\delta\phi_{\alpha_A}^*(A)$ must vanish, that is,

$$\left[T_A + \int \phi_{\alpha_1}^*(1)\ldots\phi_{\alpha_{A-1}}^*(A-1)\left[\sum_{j\neq A}^{A-1} v(Aj)\right]\right.$$

$$\left.\times\, \phi_{\alpha_1}(1)\ldots\phi_{\alpha_{A-1}}(A-1) - \epsilon_{\alpha_A}\right]\phi_{\alpha_A}(A) = 0 \tag{18.10}$$

where

$$\epsilon_{\alpha_A} = \lambda_{\alpha_A} - \int \phi_{\alpha_1}^*(1)\ldots\phi_{\alpha_{A-1}}^*(A-1)\left[\sum_{i=1}^{A-1} T_i + \frac{1}{2}\sum_{i\neq j}^{A-1} v(ij)\right]$$

$$\times\, \phi_{\alpha_1}(1)\ldots\phi_{\alpha_{A-1}}(A-1) \tag{18.11}$$

The integral on the left-hand side of (18.10) is a function of the variables of the Ath particle. Putting

$$U_{\alpha_A}(A) = \int \phi^*_{\alpha_1}(1)\phi^*_{\alpha_2}(2)\ldots\phi^*_{\alpha_{A-1}}(A-1)\left[\sum_{j\neq A}^{A-1} v(Aj)\right] \times \phi_{\alpha_1}(1)\ldots\phi_{\alpha_{A-1}}(A-1) \quad (18.12)$$

we can write (18.10) in the more familiar form:

$$[T_A + U_{\alpha_A}(A)]\phi_{\alpha_A}(\mathbf{r}_A) = \epsilon_{\alpha_A}\phi_{\alpha_A}(\mathbf{r}_A) \quad (18.13)$$

Similar equations can be obtained for the other particles by using the variations $\delta\phi^*_{\alpha_k}(\mathbf{r}_k)$, with $k = 1, 2, \ldots, A-1$.

The potential $U_{\alpha_A}(A)$ in (18.13) has a very simple physical meaning: it is the potential felt by the particle A due to its interaction with all the rest of the particles when the latter are found in the orbits $\alpha_1, \alpha_2, \ldots, \alpha_{A-1}$. What (18.13) tells us, then, is that within the set of functions (18.5) the best Ath particle wave function is obtained when we choose the single-particle potential operating on particle k equal to the average potential created by all the rest of the $A - 1$ particles.

A potential $U(A)$ satisfying (18.12) is known as a *Hartree potential*; it was first suggested by Hartree for the handling of problems in atomic spectra. Implied in its application is a self-consistency procedure: one starts with an intelligent guess for $U(i)$ and obtains from it the single-particle wave functions $\phi_{\alpha_k}(i)$ corresponding to the A-lowest states. With these wave functions and the known two-body interactions, one then produces an improved $U(i)$ using (18.12). The new improved single-particle potential is used to generate new single-particle wave functions, which are in turn put into (18.12) to produce yet better $U(i)$, and so on. Since the process converges, one then winds up with a *self-consistent Hartree potential* $U(i)$; this potential has the unique property that its single-particle eigenfunctions reproduce the potential $U(i)$ when used in (18.12).

It will be recognized that the potential $U_{\alpha_A}(A)$ depends on the states occupied by the other $A - 1$ particles. Therefore the self-consistent potential will generally be different for each of the A-particles. This will generally destroy the orthogonality of some of the single-particle states. More precisely, two single-particle states $(nljm)$ and $(n'l'j'm')$, derived from a spherically symmetric potential will always be orthogonal to each other if $(ljm) \neq (l'j'm')$. But if $(ljm) = (l'j'm')$ and $n \neq n'$, the two states will be orthogonal to each other only if they are both states derived from the *same* potential $U(i)$.

We shall not discuss the handling of this difficulty here, since we are interested in deriving a Hartree potential when antisymmetric functions are used. In such cases, as we shall see, this problem takes care of itself.

19. THE HARTREE-FOCK SELF-CONSISTENT FIELD. NONDEGENERATE CASE

Hartree's method for the derivation of a best single particle potential in a system of interacting particles has been extended by Fock to systems of identical particles. The basic idea is again the same: the mutual interaction of the nucleons (or any set of identical particles for that matter) leads to an average potential felt by each one of the nucleon. It is conceivable that a good starting point for an approximate description of this system just takes into account this average potential and ignores all other effects. We form therefore a set of trial functions out of single-particle wave functions in an appropriate average single-particle potential, and ask for the best function within this set of trial functions. However, now the set of trial functions will consist of Slater determinants rather than simple products of the type (18.5).

To simplify notation we shall designate the A-lowest states in the potential $U(i)$ by the greek letters $\alpha, \beta, \ldots, \rho$; λ and μ will stand for any of the states $\alpha, \beta, \ldots, \rho$, so that

$$\sum_{\lambda} f(\lambda) = f(\alpha) + f(\beta) + \ldots + f(\rho), \text{ etc.} \tag{19.1}$$

ξ, η, and ζ will stand for states in $U(i)$ other than the lowest A-states (i.e., states unoccupied in the A-particle ground state). Each of the quantum numbers $\alpha, \ldots, \rho, \ldots, \zeta$ stands, of course, for the complex of quantum numbers of a particle in a single-particle potential; for spherically symmetric potentials we have

$$\alpha \equiv (n^{\alpha}, l^{\alpha}, j^{\alpha}, m^{\alpha}, m^{\alpha}{}_{\tau}) \tag{19.2}$$

We shall further assume that the number of particles A is such that the ground state in the single-particle potential consists of completely filled shells. The more general case leads to a number of degenerate "lowest" states, and will be treated later.

The variational wave function for the A-particle ground state in the Hartree–Fock self-consistent field approximation, is taken to be

$$\Phi_0(1, \ldots, A) = \frac{1}{\sqrt{A!}} \begin{vmatrix} \phi_\alpha(1) & \phi_\alpha(2) & \ldots & \phi_\alpha(A) \\ \phi_\beta(1) & \phi_\beta(2) & \ldots & \phi_\beta(A) \\ \cdot & & & \\ \cdot & & & \\ \cdot & & & \\ \phi_\rho(1) & \phi_\rho(2) & \ldots & \phi_\rho(A) \end{vmatrix} \tag{19.3}$$

The variation of Φ_0 will now be carried out by assuming independent variations for the $2A$ functions ϕ_λ and ϕ_λ^*, where ϕ_λ^* is understood to be the wave function describing the *state* λ irrespective of which particle is in that state. In other words a variation $\delta\phi_\alpha$ will change all the elements in the first row in (19.3) etc. $\phi_\lambda(\mathbf{r}, \sigma, \tau)$ satisfies, of course, the equations

$$(T + U)\phi_\lambda = \epsilon_\lambda\phi_\lambda \tag{19.4}$$

and

$$\langle\phi_\lambda | \phi_\mu\rangle = \delta(\lambda, \mu) \tag{19.5}$$

We notice first that there is a complete class of variations of ϕ_λ that leaves Φ_0 essentially unchanged. These are all the variations that carry a ϕ_λ into another state within the same Hilbert space spanned by $\phi_\alpha, \ldots, \phi_\rho$. Indeed, since the set $\phi_\alpha, \ldots, \phi_\rho, \ldots, \phi_\zeta$ is a complete set of functions, we can expand

$$\delta\phi_\lambda = \sum_\mu C_\lambda{}^\mu\phi_\mu + \sum_\xi C_\lambda{}^\xi\phi_\xi \tag{19.6}$$

where we use the convention (19.1) that μ runs over the A occupied states $\alpha, \ldots, \rho$ and ξ runs over all unoccupied states. If in (19.6) all the $C_\lambda{}^\xi$ vanish, then the Slater determinant taken with ϕ_λ or with $\phi_\lambda + \delta\phi_\lambda$ is the same. This follows from the well-known theorem that the addition of any row to another row in a determinant does not change the value of the determinant. We may therefore conclude that those variations of the ϕ_λ's that amount to a linear transformation within the Hilbert space of occupied states do not change Φ_0.

This result has the important consequence that one can disregard the orthogonality requirement (19.5) when determining the best ϕ_λ by means of the variational principle. Suppose that one obtained ϕ_λ's that were not orthogonal. Then by taking an appropriate linear combination of them a new set of single-particle wave functions could be obtained that were mutually orthogonal. These new functions could then be used in (19.3) for Φ_0; by the theorem derived in the preceding paragraph Φ_0 would not thereby be changed.

This result also follows directly from the symmetry of the determinant (19.3) for Φ_0. Since one can exchange any of the two rows in the determinant and only change the sign of Φ_0, it follows that the order α to ρ is arbitrary. It follows from this remark that the variational equations for ϕ_λ must be independent of λ so that the potential U will be independent of λ. Hence all the ϕ_λ's are eigenfunctions of the same effective Hamiltonian, $T + U$, and are mutually orthogonal. This will be shown explicitly below.

Thus the orthogonality problem, which is so vexing for the Hartree method, is not present in the Hartree–Fock method. The antisymmetrization results in a reduction in complexity although, as we shall see, there is an added difficulty in the more complicated U that is obtained.

Using again Lagrange multipliers to insure the *normalization* of ϕ_λ, we can carry out, as in the Hartree case, the variation

$$\delta \left\{ \int \Phi_0^*(1, \ldots, A) \left[\sum T_i + \frac{1}{2} \sum_{i \neq j} v_{ij} \right] \Phi_0(1, \ldots, A) - \sum \epsilon_\mu \int \phi_\mu^* \phi_\mu \right\} = 0 \quad (19.7)$$

A straightforward calculation then leads to the following equations that have to be satisfied by ϕ_λ to satisfy (19.7):

$$T_1 \phi_\lambda(1) + \left[\sum_\mu \int \phi_\mu^*(2) v(12) \phi_\mu(2) d(2) \right] \phi_\lambda(1) - \sum_\mu \left[\int \phi_\mu^*(2) v(12) \phi_\lambda(2) d(2) \right] \phi_\mu(1) = \epsilon_\lambda \phi_\lambda(1) \quad (19.8)$$

where the summation on μ is over *all* occupied states. Equation 19.8 replaces (18.10) for the non-antisymmetrized Hartree method. We see that the single-particle potential has now two parts. The first is referred to as the *direct potential* and the second as the *exchange potential.* This exchange potential results directly from the use of antisymmetrized wave functions. We can define a nonlocal potential that will exhibit the fact that now the potential is indeed the same one for all occupied states. We put

$$U(1, 1') = \sum_\mu \left[\int \phi_\mu^*(2) v(12) \phi_\mu(2) d(2) \right] \delta(1 - 1') - \sum_{\mu, \mu'} \int [\phi_\mu^*(2) v(12) \phi_{\mu'}(2)\, d(2)] \phi_{\mu'}^*(1') \phi_\mu(1) \quad (19.9)$$

Since

$$\int \phi_{\mu'}^*(1') \phi_\lambda(1')\, d(1') = \delta(\mu', \lambda)$$

it is readily seen that with $U(1, 1')$ defined by (19.9), (19.8) can be written in the form:

$$T_1 \phi_\lambda(1) + \int U(1, 1') \phi_\lambda(1')\, d(1') = \epsilon_\lambda \phi_\lambda(1) \quad (19.10)$$

Equation 19.10 is the Schrödinger equation for a single particle moving in a nonlocal single-particle potential, $U(1, 1')$. The potential $U(1, 1')$ in (19.9) is independent of the particular state λ since it involves summations over *all* occupied states μ. It is thus the same for all states of the system and, as we mentioned earlier, implies the mutual orthogonality of the set ϕ_λ.

We notice that in (19.8) the interaction $v(12)$ appears *without* the factor 1/2 that had to be used in the Hamiltonian in (19.7) to avoid double counting of the mutual interactions. This has the following important consequences:

From (19.8) we find that

$$\epsilon_\lambda = \int \phi_\lambda^*(1)T_1\phi_\lambda(1)\,d(1) + \sum_\mu \int \phi_\lambda^*(1)\phi_\lambda^*(2)v(12) \times [\phi_\lambda(1)\phi_\mu(2) - \phi_\lambda(2)\phi_\mu(1)]\,d(1)\,d(2) \tag{19.11}$$

On the other hand the expectation value of the Hamiltonian

$$H = \sum T_i + \frac{1}{2}\sum_{i\neq j} v_{ij}$$

taken with the wave function $\Phi_0(1, \ldots, A)$ defined by (19.3) leads to

$$E_0 = \int \Phi_0^* H \Phi_0 = \sum_\lambda \int \phi_\lambda^*(1)T_1\phi_\lambda(1)\,d(1) + \frac{1}{2}\sum_{\mu,\lambda} \int \phi_\lambda^*(1)\phi_\mu^*(2)v(12) \times [\phi_\lambda(1)\phi_\mu(2) - \phi_\lambda(2)\phi_\mu(1)]\,d(1)\,d(2) \tag{19.12}$$

Therefore

$$E_0 = \sum_\lambda \epsilon_\lambda - \frac{1}{2}\sum_{\lambda,\mu} \int \phi_\lambda^*(1)\phi_\mu^*(2)v(12) \times [\phi_\lambda(1)\phi_\mu(2) - \phi_\lambda(2)\phi_\mu(1)]\,d(1)\,d(2) \tag{19.13}$$

Since ϵ_λ, according to (19.10), is the single-particle energy associated with the state λ, $\sum \epsilon_\lambda$ is the energy of Φ_0 taken as a wave function of A-particles in the potential $U(1, 1')$. In fact it follows immediately from (19.10) that

$$\sum_{i=1}^{A} [T_i + U(i)]\Phi_0(1, \ldots, A) = (\sum \epsilon_\lambda)\Phi_0(1, \ldots, A) \tag{19.14}$$

where

$$U(i)\Phi_0 \equiv \int U(i, i')\Phi_0(1, 2, \ldots, i', \ldots, A)\,d(i')$$

Thus, although the Hartree–Fock self-consistent potential leads to the best wave function of the type (19.3) for the A-nucleon system, the nuclear energy that corresponds to this wave function *is not* the sum of the single-particle energies, the energy of Φ_0 in the self-consistent potential. This reflects once again the important point that if $U(1, 1')$ is determined in a self-consistent way, then it should be understood only as a prescription for the generation of a "best" nuclear wave function. Although Φ_0 is an eigenfunction of the self-consistent potential it will be erroneous to conclude that the eigenvalue that goes with it also gives the nuclear energy.

It is quite easy to understand the origin of this difference between E_0 and $\sum \epsilon_\lambda$. When we look for the best potential $U(i)$ to determine the behavior of a single particle in $\Phi_0(1, \ldots, A)$ we find that in some sense this potential is the average potential felt by the single particle due to its interaction with all the others. Since ϵ_λ is an eigenvalue of (19.10) that involves the potential $U(i)$, it too reflects this interaction. If we now add up all the ϵ_λ's we take into account each interaction v_{ij} twice. The Hartree–Fock energies $\sum \epsilon_\lambda$ therefore overestimate the nuclear binding and hence the difference between E_0 and $\sum \epsilon_\lambda$ in (19.13).

The nature of the nonlocality in $U(i)$ is also of some interest. As we see from (19.9) it is due entirely to the exchange term. This is even more clearly seen if we introduce an index σ that takes on all values—both of occupied and of nonoccupied states. Since the completeness of the set ϕ_σ implies:

$$\sum_\sigma \phi_\sigma^*(1')\phi_\sigma(1) = \delta(1 - 1')$$

and if we limit ourselves to $U(1, 1')$ operating on occupied states, then we see that (19.9) can be written in the form

$$U(1, 1') = \sum_{\mu,\sigma} \int \phi_\sigma^*(1')\phi_\mu^*(2)v(12)[\phi_\sigma(1)\phi_\mu(2) - \phi_\sigma(2)\phi_\mu(1)]\,d(2) \tag{19.15}$$

The exchange integral

$$\int \phi_\sigma^*(1')\phi_\mu^*(2)v(12)\phi_\sigma(2)\phi_\mu(1)\,d(2) \tag{19.16}$$

involves an integration over the coordinate (2) of a product of two different functions ϕ_μ^* and ϕ_σ. If the two states μ and σ are very different from each other the product $\phi_\mu^*\phi_\sigma$ will have many changes of sign within the range of $v(12)$ and (19.16) will be very small. This is to be contrasted with the direct integral that involves the product $\phi_\mu^*(2)\phi_\mu(2)$ and thus weighs all parts of $v(12)$ with the same phase.

We shall not solve the Hartree–Fock equations here but just describe how it is done. Equations 19.9 and 19.10 can form the basis for an iteration procedure to derive the best ground state nuclear wave function. Again one starts from an intelligent guess of $U(i)$, local or nonlocal, and derives its single-particle eigenfunctions ϕ_λ; these are then used to derive an improved single-particle potential according to (19.9), which in its turn is used in (19.10) to produce the next iteration of single-particle wave functions. The iterations are continued until they converge, thereby determining a *self-consistent Hartree–Fock potential* $U(i)$ with its eigenfunctions $\phi_\lambda(i)$, and *Hartree–Fock single-particle energies* ϵ_λ. The energy of the nuclear ground state can then be determined using (19.13).

It should be mentioned that this procedure works only if $v(ij)$ does not contain an infinite repulsive (or attractive) core. Since the trial functions $\Phi_0(1, \ldots, A)$ do not include in them any two-particle correlation beyond the one implied by the Pauli principle, integrals such as those appearing in (19.15) will diverge if $v(12)$ has a repulsive core. In Chapter VII we shall present some results of calculations using the Hartree–Fock method.

20. SYMMETRY OF THE HARTREE-FOCK POTENTIALS. CLOSED SHELLS*

Thus far we have only considered cases in which the number of particles A produced completely filled levels in $U(i)$. As soon as the number A of particles leaves some shells only partially filled, new problems come up. The origin of these problems lies in the fact that for a spherically symmetric potential it is then possible to form several wave functions $\Phi(1, \ldots, A)$, which are orthogonal to each other but correspond to the same energy. As we shall see, the Hartree–Fock self-consistent field approach then breaks down, unless we are prepared to introduce nonspherically symmetric single-particle potentials.

To see this let us first examine, in more detail, spherically symmetric potentials. For simplicity we shall assume that the interaction $v(12)$ depends on $|\mathbf{r}_1 - \mathbf{r}_2|$ only and does not involve the spin and isospin coordinate of the particles. In this case the single-particle potential will also be a function of $\mathbf{r}$ and $\mathbf{r}'$ only, and from (19.15) we obtain

$$U(\mathbf{r}_1, \mathbf{r}_1') = \sum_{\mu} \left[\int \phi_\mu^*(\mathbf{r}_2) v(|\mathbf{r}_1 - \mathbf{r}_2|) \phi_\mu(\mathbf{r}_2) r_2^2 \, dr_2 \, d\Omega_2 \right] \delta(\mathbf{r}_1 - \mathbf{r}_1)$$

$$- \sum_{\mu,\mu'} \left[\int \phi_\mu^*(\mathbf{r}_2) v(|\mathbf{r}_1 - \mathbf{r}_2|) \phi_{\mu'}(\mathbf{r}_2) r_2^2 \, dr_2 \, d\Omega_2 \right] \phi_{\mu'}^*(\mathbf{r}_1) \phi_\mu(\mathbf{r}_1) \qquad (20.1)$$

Consider the direct integral first. Since $|\mathbf{r}_1 - \mathbf{r}_2|^2 = r_1^2 + r_2^2 - 2r_1r_2 \cos\theta_{12}$, where θ_{12} is the angle between $\mathbf{r}_1$ and $\mathbf{r}_2$, we can expand $v(|\mathbf{r}_1 - \mathbf{r}_2|)$ in a series of Legendre polynomials of θ_{12}

$$v(|\mathbf{r}_1 - \mathbf{r}_2|) = \sum_l v_l(r_1, r_2) P_l(\cos\theta_{12}) \qquad (20.2)$$

where $v_l(r_1, r_2)$ depends only on the magnitude of $\mathbf{r}_1$ and $\mathbf{r}_2$.

Using the addition theorem for spherical harmonics we can write

$$P_l(\cos\theta_{12}) = \frac{4\pi}{2l+1} \sum_m Y_{lm}^*(\theta_1\phi_1) Y_{lm}(\theta_2\phi_2) \qquad (20.3)$$

---

*(See also Appendix p. 274).

The z-axis, which is required in order to define the angular coordinates of $\mathbf{r}_1$ and $\mathbf{r}_2$ can be chosen arbitrarily; from (20.2) we see that $v(|\mathbf{r}_1 - \mathbf{r}_2|)$ is independent of the choice of this axis, and this is reflected in (20.3′) through the summation over all values of m.

Introducing (20.3) into (20.2), the direct integral U_d in (20.1) takes the form

$$U_d(\mathbf{r}_1) = \sum_{l,m} \frac{4\pi}{2l+1} Y_{lm}^*(\theta_1\phi_1) \times \left[\sum_{\mu} \int \phi_\mu^*(\mathbf{r}_2) v_l(r_1, r_2) Y_{lm}(\theta_2\phi_2) \phi_\mu(\mathbf{r}_2) r_2^2 \, dr_2 \, d\Omega_2 \right] \tag{20.4}$$

Our aim is to check whether in the nondegenerate case a spherically symmetric Hartree–Fock potential is self-consistent. We assume, therefore, that ϕ_μ are eigenfunctions in such a potential and we shall check whether they give rise again to a spherically symmetric potential. ϕ_μ can be written in the form (spin is ignored):

$$\phi_\mu(\mathbf{r}) = \frac{1}{r} R_{n_\mu l_\mu}(r) Y_{l_\mu m_\mu}(\theta\phi) \tag{20.5}$$

Introducing (20.5) into (20.4) we obtain for the direct potential

$$U_d(\mathbf{r}_1) = \sum_{lm, n_\mu l_\mu} \frac{4\pi}{2l+1} F_l(n_\mu l_\mu; r_1) Y_{lm}^*(\theta_1\phi_1) \times \left[\sum_{m_\mu} \int Y_{l_\mu m_\mu}^*(\theta_2\phi_2) Y_{lm}(\theta_2\phi_2) Y_{l_\mu m_\mu}(\theta_2\phi_2) \, d\Omega_2 \right] \tag{20.6}$$

where

$$F_l(n_\mu l_\mu; r_1) = \int |R_{n_\mu l_\mu}(r_2)|^2 v_l(r_1, r_2) \, dr_2 \tag{20.7}$$

depends only on the magnitude of $\mathbf{r}_1$.

The angular integration $d\Omega_2$ can be carried out very easily. We notice, using the addition theorem for spherical harmonics, that

$$\sum_{m_\mu} Y_{l_\mu m_\mu}^*(\theta_2\phi_2) Y_{l_\mu m_\mu}(\theta_2\phi_2) = \frac{2l_\mu + 1}{4\pi} P_{l_\mu}(1) = \frac{2l_\mu + 1}{4\pi} \tag{20.8}$$

Introducing (20.8) into (20.6) we are left with an integral

$$\int Y_{lm}(\theta_2\phi_2) \, d\Omega = \sqrt{4\pi}\,\delta(l, 0)\delta(m, 0) \tag{20.9}$$

Hence

$$U_d(\mathbf{r}_1) = \sum_{n_\mu l_\mu} (2l_\mu + 1) F_0(n_\mu l_\mu; r_1) \tag{20.10}$$

For future use we want, however, to derive the same result in a more complicated way, whose intermediate steps will turn out to be useful. Thus, using Appendix A, A.2.36 for the integral over a product of three spherical harmonics, we obtain [11]

$$\int Y^*_{l_\mu m_\mu}(\theta\phi) Y_{lm}(\theta\phi) Y_{l_\mu m_\mu}(\theta\phi)\, d\Omega$$

$$= (-1)^{m_\mu} \begin{pmatrix} l_\mu & l & l_\mu \\ -m_\mu & m & m_\mu \end{pmatrix} (2l_\mu + 1) \sqrt{\frac{2l+1}{4\pi}} \begin{pmatrix} l_\mu & l & l_\mu \\ 0 & 0 & 0 \end{pmatrix} \tag{20.11}$$

where

$$\begin{pmatrix} l_1 & l_2 & l_3 \\ m_1 & m_2 & m_3 \end{pmatrix}$$

is a three-j symbol. Our assumption that Φ_0 is nondegenerate implies that all levels are completely occupied. For every value of l_μ the summation over m_μ in (20.6) therefore extends over all values of m_μ: $-l_\mu < m_\mu < +l_\mu$. Using the identity

$$(-1)^{l_\mu - m_\mu} = \sqrt{2l_\mu + 1} \begin{pmatrix} l_\mu & 0 & l_\mu \\ -m_\mu & 0 & m_\mu \end{pmatrix}$$

we can then use the orthogonality of the $3 - j$ symbols (see Appendix A, A.2.70) to obtain

$$\sum_{m_\mu} \int Y^*_{l_\mu m_\mu}(\theta, \phi) Y_{lm}(\theta, \phi) Y_{l_\mu m_\mu}(\theta, \phi)\, d\Omega$$

$$= (-1)^{l_\mu} (2l_\mu + 1)^{3/2} \sqrt{\frac{2l+1}{4\pi}} \begin{pmatrix} l_\mu & l & l_\mu \\ 0 & 0 & 0 \end{pmatrix} \sum_{m_\mu} \begin{pmatrix} l_\mu & 0 & l_\mu \\ -m_\mu & 0 & m_\mu \end{pmatrix} \times \begin{pmatrix} l_\mu & l & l_\mu \\ -m_\mu & m & m_\mu \end{pmatrix}$$

$$= (-1)^{l_\mu} (2l_\mu + 1)^{3/2} \sqrt{\frac{1}{4\pi}} \begin{pmatrix} l_\mu & l & l_\mu \\ 0 & 0 & 0 \end{pmatrix} \delta(l, 0) \delta(m, 0)$$

$$= \frac{2l_\mu + 1}{\sqrt{4\pi}} \delta(l, 0) \delta(m, 0) \tag{20.12}$$

Recalling that $Y_{00}(\theta, \phi) = 1/\sqrt{4\pi}$ we obtain finally from (20.6), by introducing (20.12) into it:

$$U_d(\mathbf{r}_1) = \sum_{n_\mu l_\mu} F_0(n_\mu l_\mu; r_1)(2l_\mu + 1) \tag{20.13}$$

The dependence on θ_1 and ϕ_1 has thus disappeared and we see that the direct part of the potential $U(\mathbf{r}_1)$ is a function only of the magnitude of r_1. As far as this part is concerned, and for nondegenerate states, a spherically symmetric potential can be self-consistent.

Our result (20.13) that $U_d(\mathbf{r}_1)$ is spherically symmetric and depends only on $|\mathbf{r}_1|$ could be obtained even without the detailed calculation given above: we recall (see Appendix A, A.2.26) that under an arbitrary rotation R of the frame of reference, a spherical harmonic $Y_{lm}(\theta, \phi)$ transforms into a linear combination of spherical harmonics of the same order:

$$Y_{lm}(\theta'\phi') = \sum_{m'} D^{(l)}_{m'm}(R) Y_{lm'}(\theta, \phi) \tag{20.14}$$

where (θ', ϕ') are the angular coordinates of the direction (θ, ϕ) with respect to the new axes, and $D^{(l)}_{m'm}(R)$ is Wigner's D-matrix. Since $U_d(\mathbf{r})$ in our case involves a sum over all the magnetic substates of an occupied level, each of them taken with the same weight, we can conclude from (20.14) that with respect to the rotated frame we shall also obtain the same structure for $U_d(\mathbf{r})$. Hence, $U_d(\mathbf{r})$ remains invariant under rotations, that is, for a given value of $|\mathbf{r}|$ the dependence of $U_d(\mathbf{r})$ on the angular variables of $\mathbf{r}$ is independent of the particular frame of reference. It follows that $U_d(\mathbf{r})$ is a scalar that does not depend on the angular variables, and is therefore a function of $|\mathbf{r}|$ only.

The detailed derivation of this result gives us additional interesting information. The expansion (20.2) of the interaction $v(|\mathbf{r}_1 - \mathbf{r}_2|)$ is a generalization of the familiar multipole expansion of the Coulomb interaction. Substituting (20.3) into (20.2) we obtain:

$$v(|\mathbf{r}_1 - \mathbf{r}_2|) = \sum_{lm} \frac{4\pi}{2l+1} v_l(r_1, r_2) Y^*_{lm}(\theta_1\phi_1) Y_{lm}(\theta_2\phi_2) \tag{20.15}$$

and for the special case $v(|\mathbf{r}_1 - \mathbf{r}_2|) = 1/|\mathbf{r}_1 - \mathbf{r}_2|$, (20.15) reduces to

$$\frac{1}{|\mathbf{r}_1 - \mathbf{r}_2|} = \sum_{l,m} \frac{4\pi}{2l+1} \left[\frac{1}{r_1^{l+1}} Y^*_{lm}(\theta_1, \phi_1)\right] [r_2{}^l Y_{lm}(\theta_2\phi_2)] \qquad \text{for } r_1 > r_2 \tag{20.16}$$

Equation (20.16) is the familiar result giving the interaction between two charge distributions in terms of their multipole moments. Although for a general interaction $v_l(r_1, r_2)$ does not separate into a product of functions of $\mathbf{r}_1$ and $\mathbf{r}_2$, we still refer to (20.15) as the *multipole expansion of the interaction* $v(|\mathbf{r}_1 - \mathbf{r}_2|)$, the term with $l = 0$ being the monopole–monopole part of the interaction, that with $l = 1$ being the dipole–dipole part, etc. These multipoles of the interaction should not be confused with the electromagnetic multipoles, which are generally unrelated to the nuclear interaction multipoles.

Our result (20.13) now tells us that the direct part of the self-consistent potential for closed-shell nuclei depends only on the monopole–monopole part of the interaction $v(|\mathbf{r}_1 - \mathbf{r}_2|)$. Two interactions v and v' that have the same monopole–monopole part but differ in the higher multipoles will lead to the same direct self-consistent potential. In a pictorial way we visualize the single-particle potential as taking care of some of the effects of the two-body interaction; (20.13) then tells us that the direct potential takes care only of the monopole part of $v(|\mathbf{r}_1 - \mathbf{r}_2|)$.

Our derivation of the spherical symmetry of $U_d(\mathbf{r})$ using the transformation properties of the spherical harmonics, suggests that this result is more general and will be valid for any self-consistent potential of closed-shell nuclei. Furthermore, it is valid for both the direct and the exchange part of the potential. It is merely a reflection of the fact that when all the magnetic substates of a level are occupied with equal probability, the system is truly spherical symmetric. We shall not go here into a detailed proof of the spherical symmetry of the general (spin-dependent) self-consistent potential of closed-shell nuclei, but only indicate the proof for the exchange part for spin-independent interactions.

Turning to the exchange energy from (20.1), using (20.2) and (20.3), we obtain for the exchange potential:

$$-U_{\text{ex}}(\mathbf{r}_1, \mathbf{r}_1) = \sum_{\mu,\mu',l,m} \frac{4\pi}{2l+1} Y^*_{lm}(\theta_1\phi_1)\phi^*_{\mu'}(\mathbf{r}_1')\phi_\mu(\mathbf{r}_1)$$

$$\times \int \phi^*_\mu(\mathbf{r}_2) v_l(r_1, r_2) Y_{lm}(\theta_2\phi_2)\phi_{\mu'}(\mathbf{r}_2) r_2^2\, dr_2\, d\Omega_2$$

$$= \sum_{\mu,\mu',l,m} \frac{4\pi}{2l+1} G_l(n_\mu l_\mu, n_{\mu'} l_{\mu'}; r_1, r_1') Y^*_{lm}(\Omega_1) Y^*_{l_{\mu'} m_{\mu'}}(\Omega_1') Y_{l_\mu m_\mu}(\Omega_1)$$

$$\times \int Y^*_{l_\mu m_\mu}(\Omega_2) Y_{lm}(\Omega_2) Y_{l_{\mu'} m_{\mu'}}(\Omega_2)\, d\Omega_2 \qquad (20.17)$$

where

$$G_l(n_\mu l_\mu, n_{\mu'} l_{\mu'}; r_1, r_1') = \frac{R^*_{n_{\mu'} l_{\mu'}}(r_1')}{r_1'} \cdot \frac{R_{n_\mu l_\mu}(r_1)}{r_1} \int R^*_{n_\mu l_\mu}(r_2) R_{n_{\mu'} l_{\mu'}}(r_2)$$

$$\times\, v_l(r_1, r_2)\, dr_2$$

We now use (20.8) and the identity (Appendix A, A.2.35)

$$Y^*_{lm}(\Omega_1)Y_{l_\mu m_\mu}(\Omega_1) = \sum_{l'm'} (-1)^{m_\mu} \begin{pmatrix} l_\mu & l & l' \\ -m_\mu & m & m' \end{pmatrix} \times \sqrt{\frac{(2l_\mu+1)(2l+1)(2l'+1)}{4\pi}} \begin{pmatrix} l_\mu & l & l' \\ 0 & 0 & 0 \end{pmatrix} Y_{l'm'}(\Omega_1) \quad (20.18)$$

to obtain

$$-U_{\text{ex}}(\mathbf{r}_1, \mathbf{r}_1') = \sum_{\mu\mu' l m} \frac{4\pi}{2l+1} G_l \frac{(2l_\mu+1)(2l+1)}{4\pi} \sqrt{(2l_{\mu'}+1)(2l'+1)} \times \begin{pmatrix} l_\mu & l & l_{\mu'} \\ 0 & 0 & 0 \end{pmatrix} \begin{pmatrix} l_\mu & l & l' \\ 0 & 0 & 0 \end{pmatrix} Y^*_{l_{\mu'} m_{\mu'}}(\Omega_1') Y_{l'm'}(\Omega_1) \times \begin{pmatrix} l_\mu & l & l_{\mu'} \\ -m_\mu & m & m_{\mu'} \end{pmatrix} \begin{pmatrix} l_\mu & l & l' \\ -m_\mu & m & m' \end{pmatrix} \quad (20.19)$$

We notice again that our assumption that we are dealing with closed-shell nuclei implies that the summation in (20.19) extends over all the magnetic quantum numbers m_μ and $m_{\mu'}$ in the occupied levels. We can therefore carry out the summation over m_μ and m, which appear now only in the $3-j$ symbols, and obtain, using the orthogonality of those symbols, a factor $\delta(l_{\mu'} l') \times \delta(m_{\mu'} m')$. We recall also the addition theorem for spherical harmonics (20.3), and introducing the angle $\theta_{11'}$ as the angle between the directions of $\mathbf{r}_1$ and $\mathbf{r}_1'$, we obtain finally:

$$-U_{\text{ex}}(\mathbf{r}_1, \mathbf{r}_1') = \sum_{\mu,\mu',l} \frac{(2l_\mu+1)(2l_{\mu'}+1)}{4\pi} \begin{pmatrix} l_\mu & l & l_{\mu'} \\ 0 & 0 & 0 \end{pmatrix}^2 G_l(n_\mu l_\mu, n_{\mu'} l_{\mu'}; r_1 r_1') \times P_{l_{\mu'}}(\cos\theta_{11'}) \quad (20.20)$$

Thus the nonlocal, exchange part of the self-consistent potential of closed-shell nuclei depends only on the magnitude of r_1 and r_1' and the angle $\theta_{11'}$ *between* them. It does not depend on the specific orientations of $\mathbf{r}_1$ or $\mathbf{r}_1'$ separately. Such a potential is spherically symmetric, since the angle between two vectors is obviously independent of the frame of reference. We see therefore that the spherical symmetry of a nonlocal exchange potential is slightly more complicated than that of the direct part, but at any rate, for closed-shell nuclei, the total self-consistent field is spherically symmetric.

21. HARTREE-FOCK POTENTIAL. NONCLOSED SHELLS

In obtaining the results (20.13) and (20.20) we used in an essential way the assumption that we were dealing with closed-shell nuclei. The sum over all occupied states that is prescribed in the self-consistent field equations (19.15) or (20.1) involves, among other things, a sum over the magnetic quantum numbers m of the occupied states. A closed-shell nucleus has all the m-states of a given level equally populated. This enabled us to carry out the m-summation explicitly and obtain the spherical symmetry of the potential.

Remark. Formally all we used was the requirement that there will be no partially filled levels. The closure of major *shells* has never been used in deriving (20.13) and (20.20). As we shall see later the residual interaction has the tendency to mix up the order of filling of close-lying levels. In practice, therefore, one rarely meets with filled levels in situations that do not involve also closed major shells. We shall thus continue to talk about "closed shells" etc., although formally we are referring to filled levels only.

When we have particles outside closed shells the situation is changed. To see how, let us consider the case in which there is just one particle outside closed shells, and assume $U(\mathbf{r})$ is spherically symmetric. The odd particle will then be in a state $\mu_0 = n_0 l_0 m_0$ (ignoring spin). The direct part of the new self-consistent potential $U_d'(\mathbf{r}_1)$ will again satisfy (20.4), except that the summation over the states μ should now include also the state μ_0. We obtain therefore

$$U_d'(\mathbf{r}_1) = U_d(r_1) + \sum_{lm} \frac{4\pi}{2l+1} Y^*_{lm}(\theta_1\phi_1) \int \phi^*_{\mu_0}(\mathbf{r}_2) v_l(r_1, r_2) \times Y_{lm}(\theta_2\phi_2)\phi_{\mu_0}(\mathbf{r}_2) r_2^2\, dr_2 d\Omega_2 \quad (21.1)$$

where $U_d(r_1)$ is the spherically symmetric self-consistent potential of the closed shells as given by (20.13). Application of (20.11) to the angular integral in (21.1) yields:

$$U_d'(\mathbf{r}_1) = U_d(r_1) + \sum_l (2l_{\mu_0}+1)(-1)^{m_{\mu_0}} \begin{pmatrix} l_{\mu_0} & l & l_{\mu_0} \\ -m_{\mu_0} & 0 & m_{\mu_0} \end{pmatrix} \begin{pmatrix} l_{\mu_0} & l & l_{\mu_0} \\ 0 & 0 & 0 \end{pmatrix} \times F_l(n_{\mu_0} l_{\mu_0}; r_1) P_l(\cos\theta_1) \quad (21.2)$$

where use has been made of that fact that

$$\begin{pmatrix} l_1 & l_2 & l_3 \\ m_1 & m_2 & m_3 \end{pmatrix}$$

vanishes if $m_1 + m_2 + m_3 \neq 0$, and also of the identity

$$P_l(\cos\theta) = \sqrt{\frac{4\pi}{2l+1}}\, Y_{l0}(\theta, \phi) \tag{21.3}$$

We see that $U_d'(\mathbf{r}_1)$ is not any longer a function of the magnitude of $\mathbf{r}_1$ only, as it involves a dependence on the direction of $\mathbf{r}_1$ as well. More precisely, $U_d'(\mathbf{r}_1)$ in (21.2), has a cylindrical rather than a spherical symmetry (it is still independent of ϕ_1). A spherically symmetric potential cannot, therefore, be a self-consistent potential, because starting from wave functions $\phi_{nlm}(\mathbf{r})$ in a spherically symmetric potential, the self-consistency equations lead to an *axially* symmetric potential.

The next natural question is whether an *axially* symmetric potential can be a self-consistent potential for nonclosed-shell nuclei. To answer this question we have to calculate the single-particle wave functions ϕ_μ in an axial potential, put them into (19.15), and see whether they give rise again to an axially symmetric potential. This problem is best treated in cylindrical coordinates (ρ, ϕ, z), (Fig. 21.1), since an axially symmetric potential in these coordinates is a potential that is independent of ϕ. An axial potential that is very commonly used is the antisotropic harmonic oscillator that in Cartesian coordinates is given by

$$V(\mathbf{r}) = \tfrac{1}{2}M[\omega_0^2(x^2 + y^2) + \omega_z^2 z^2] \tag{21.4}$$

where $\omega_z \neq \omega_0$. In cylindrical coordinates and, more generally, an axially symmetric potential is given by:

$$V(\rho, z) = \tfrac{1}{2}M[f(\rho) + g(z)] \tag{21.5}$$

The single-particle Schrödinger equation in an axially symmetric potential separates if we put [Morse and Feshbach (53), p. 1259]

$$\phi(\mathbf{r}) = \Psi(\rho, z)\Phi(\phi) \tag{21.6}$$

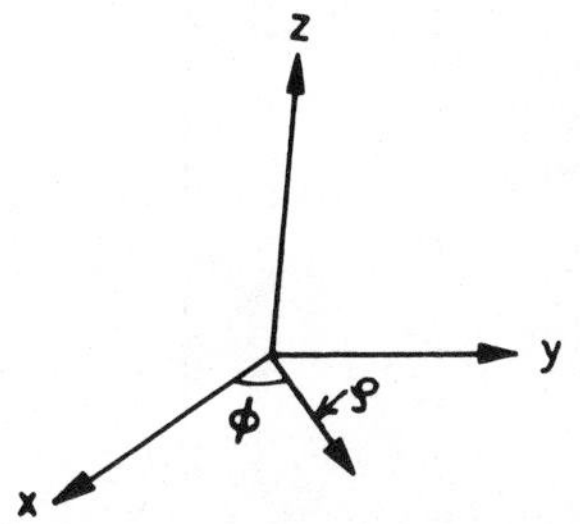

FIG. 21.1. Cylindrical coordinates for axially symmetric potentials.

leading to the equations:

$$\frac{d^2\Phi(\phi)}{d\phi^2} = -m^2\Phi(\phi) \tag{21.7}$$

and

$$-\frac{\hbar^2}{2M}\left[\frac{1}{\rho}\frac{\partial}{\partial\rho}\left(\rho\frac{\partial}{\partial\rho}\right) + \frac{\partial^2}{\partial z^2}\right]\Psi(\rho, z) + \left[V(\rho, z) - \frac{\hbar^2 m^2}{2M\rho^2}\right]\Psi(\rho, z) = E\Psi(\rho, z) \tag{21.8}$$

Equation 21.7 can be solved immediately:

$$\Phi_m(\phi) = \exp(im\phi) \tag{21.9}$$

In order for $\Phi(\phi)$ to be single valued, m must be an integer giving rise to the well-known quantization of the z-component of angular momentum for a particle moving in a cylindrical potential. Furthermore, since m appears squared in the eigenvalue equation (21.8), it is obvious that the states with $m = +m_0$ and $m' = -m_0$ are degenerate.

The single-particle energy levels in a cylindrical potential can be best visualized by considering small deviations from spherical symmetry. If in the spherical cases we have a grouping of levels, this grouping will remain also when we introduce small deviations from sphericity. Whereas in the spherical case each level with a given value of l was $2l + 1$-fold degenerate—in the cylindrical potential a good deal of this degeneracy is removed and each level is now at most doubly degenerate (it is nondegenerate for $m = 0$ and exactly doubly degenerate for $m \neq 0$).

The complete characterization of the single-particle eigenfunction in a cylindrical potential requires an explicit knowledge of $V(\rho, z)$ in (21.8). In the spherical-potential case, l and m are valid quantum numbers independent

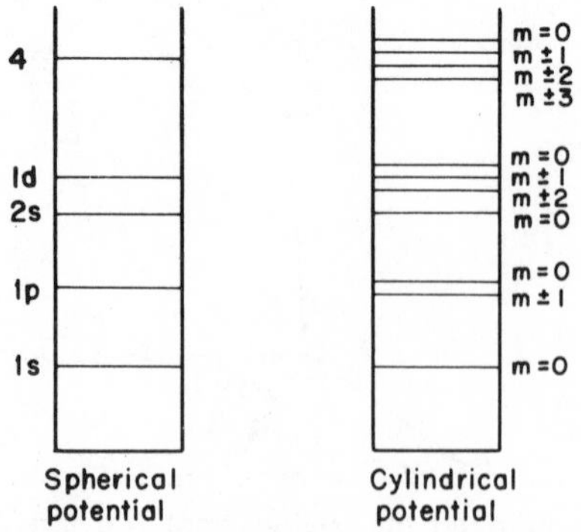

FIG. 21.2. Single-particle levels in spherical and cylindrical potentials.

of the shape of the potential. For the cylindrical potential only m is such a quantum number. Since, however, the spherical-potential wave functions form a complete set, we can expand the cylindrical-potential wave functions $\phi_{\alpha m'}(\mathbf{r})$ in them:

$$\phi_{\alpha m'}(\mathbf{r}) = \sum_{n',l'} a_{n'l'}^{\alpha m'} \frac{R_{n'l'}(r)}{r} Y_{l'm'}(\theta, \phi) \tag{21.10}$$

Here α are the additional quantum numbers for the cylindrical potential. Note that in (21.10) the summation extends over n and l only since m' is a good quantum number for both sets of wave functions. (For further discussion of single-particle wave functions in a cylindrical potential see Chapter VI).

The expression (21.10) for $\phi_{\alpha m}(\mathbf{r})$ can be used to give us an answer as to the self-consistency of cylindrical potentials. Limiting ourselves again to spin-independent interactions $v(|\mathbf{r}_1 - \mathbf{r}_2|)$, and considering for the moment the direct part of the self-consistent potential only, we introduce $\phi_{\alpha m}(\mathbf{r})$ for $\phi_\mu(\mathbf{r})$ in (20.4) to obtain

$$U_d(\mathbf{r}_1) = \sum_{l,m,\mu} \frac{4\pi}{2l+1} Y_{lm}^*(\theta_1\phi_1) \sum_{n',l',n'',l''} (a_{n'l'}^{\alpha_\mu m_\mu'})^* a_{n''l''}^{\alpha_\mu m_\mu'} F_l(n'l', n''l'', r_1)$$
$$\times \int Y_{l'm_\mu'}^*(\theta_2\phi_2) Y_{lm}(\theta_2\phi_2) Y_{l''m_\mu'}(\theta_2\phi_2)\, d\Omega_2 \tag{21.11}$$

where

$$F_l(n'l', n''l'', r_1) = \int R_{n'l'}^*(r_2) R_{n''l''}(r_2) v_l(r_1, r_2)\, dr_2 \tag{21.12}$$

Note that in the angular integral in (21.11) we have now different values of l' and l'', but the magnetic quantum numbers m_μ that go with them are all the same. This just reflects the fact that in the cylindrical potential l is no longer a good quantum number, whereas m still is.

The angular integration can again be carried out explicitly and the only contribution to (21.11) come from terms with $m = 0$. We notice that the dependence of $U_d(\mathbf{r}_1)$ on the direction of $\mathbf{r}_1$ comes entirely from $Y_{lm}(\theta_1\phi_1)$; if the summation includes only terms with $m = 0$ this reduces to $Y_{l0}(\theta_1\phi_1) = \sqrt{(2l + 1/4\pi)} P_l(\cos\theta_1)$. Thus $U_d(\mathbf{r}_1)$ will not depend on the angle ϕ_1, and an axially symmetric potential can be a self-consistent potential for any number of particles, in the sense that starting from such a potential and using the self-consistent field equations (19.15), we generate again a potential of the same symmetry.

To be precise we have demonstrated this result only for the direct part of the Hartree–Fock potential, and even this only for spin-independent interactions. It can be shown to be true also for the exchange part and for any scalar interaction $v(ij)$, but we shall not go into it here.

The emergence of an axially symmetric "deformed" potential as the more general self-consistent potential, rather than the intuitively expected spherical potential, raises a number of questions, two of which we shall proceed now to analyze:

(a) Would a cylindrical potential be a better potential also for closed-shell nuclei?

(b) What is the physics behind this deformation of the potential and how is a preferred direction determined by the nucleus?

22. CLOSED-SHELL NUCLEI WITH CYLINDRICAL POTENTIALS

Our treatment of the deformed (i.e., nonspherical) Hartree–Fock potential can be used for closed-shell nuclei as well, and an interesting question then arises: would such use automatically lead to a spherical potential as a special case of the deformed potential? And if not, which potential is better?

To obtain an answer to these questions we consider again the expansion (21.10) of the eigenfunction $\phi_{\alpha m}(\mathbf{r})$ in the cylindrical potential $V(\mathbf{r})$ in terms of the spherical-potential wave functions $\phi_{nlm}(\mathbf{r})$:

$$\phi_{\alpha m}(\mathbf{r}) = \sum_{n,l} a_{nl}^{\alpha m} \phi_{nlm}(\mathbf{r}) \tag{22.1}$$

In principle the summation in (22.1) includes an infinite number of terms, and the value of the coefficients $a_{nl}^{\alpha m}$ depends on the particular nondeformed potential $U(\mathbf{r})$ we choose to generate the functions $\phi_{nlm}(\mathbf{r})$. We can expect that if $U(\mathbf{r})$ is pretty close to the cylindrical potential $V(\mathbf{r})$ then the expansion of $\phi_{\alpha m}(\mathbf{r})$ will involve mostly wave functions $\phi_{nlm}(\mathbf{r})$ of more or less the same energy. This follows if we consider $V(\mathbf{r}) - U(r)$ as a perturbation on $U(r)$, and use the general results of perturbation theory.

A closed-shell nucleus is characterized by the special circumstance that all levels that are close to each other are filled completely, whereas the next available levels lie fairly far away and are empty. Thus in the potentials described by the levels in Fig. 22.1, if we put A-particles into the deformed potential we shall form a closed shell; similarly A'-particles in the nondeformed potential will form a closed shell for that case. If the difference between $U(r)$ and $V(\mathbf{r})$ is not too large we can expect to have $A' = A$, that is, the gaps in the single-particle energy level spectrum corresponding to the formation of major shells can be expected to lead to the same magic numbers for the potentials U and V. In addition we can expect that if $|\alpha m\rangle$ is one of these A-states in $V(\mathbf{r})$, then the states $|nlm\rangle$ that contribute most to its expansion (22.1) will all be included among the lowest $A' = A$ states of $U(r)$. The amplitude of other $|nlm\rangle$ states lie considerably higher in energy and their amplitude in (22.1) is expected to be small.

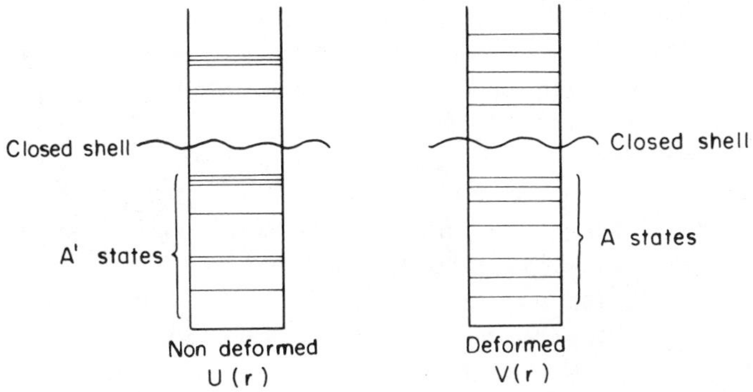

FIG. 22.1. The filling of levels in closed shell nuclei in deformed and nondeformed potentials.

If these expectations are valid, then for magic numbers A, (22.1) gives rise to a linear transformation between two sets of A-independent functions. An A-particle Slater determinant constructed out of the $\phi_{\alpha m}$'s is then identical to that constructed out of the ϕ_{nlm}'s. The latter has been shown to lead to a spherically symmetric potential; we conclude therefore that to the extent that the A-lowest $\phi_{\alpha m}$'s are expressible in terms of the A-lowest ϕ_{nlm} of a certain spherical potential, the self-consistent field will be spherically symmetric. If this condition is not satisfied, the self-consistent field will be deformed. Or, to state it differently: if the A-occupied states of a deformed potential are expressible in terms of exactly A-states in a spherical potential, and if the latter form a closed shell, then the self-consistent deformed potential has a special shape: a spherically symmetric one.

We understand now also why we confined our considerations of spherical potentials to closed-shell nuclei rather than to nuclei with any number of completely filled levels. In the latter case, the presence of empty levels close to the filled ones will generally make the deformed potential a better self-consistent potential than the spherical one.

The self-consistent potentials are derived from wave functions composed of a single Slater determinant. We see from (22.1) that a single determinant of $\phi_{\alpha m}$'s will generally be expressed in terms of a *sum* of Slater determinants of ϕ_{nlm}'s. Thus, in a certain sense, the use of deformed potentials is a particular way of using a *sum* of Slater determinants of spherical-potential wave functions, rather than limiting ourselves to just *one* such determinant. One should thus consider using single-particle deformed-potential wave functions as basis wave functions in the Slater determinant as a procedure for including the effects of other configurations that mix with the initial Slater determinant formed from spherical-potential single-particle wave functions. It is obvious that from the viewpoint of a variational principle this extra freedom can only improve the wave function. One can think of other, even more general, classes of potentials that will produce an even broader class of Slater determinants. For instance the condition on axial symmetry may be relaxed and potentials of any shape can be admitted. Furthermore the single-particle potential may include parity violating terms, etc. As one proceeds to more and more complicated potentials the self-consistent potential approximation becomes less and less practical. It is therefore preferable to limit ourself to potentials that follow more or less the shape of the nucleus.

In view of the fact that nuclear states are eigenstates of $\mathbf{J}^2$, the square of the total angular momentum, we can ask ourselves whether a linear combination of Slater determinants that has *this* property is not preferable to the one derived from a deformed potential. This will generally be a different combination than the one obtained from the deformed potential, since the latter does not lead to states of well-defined angular momentum. To answer this question we have to look deeper into the physical meaning of our results thus far.

23. THE VIOLATION OF CONSERVATION LAWS IN MANY-PARTICLE SYSTEMS

A deformed potential fixes a preferred direction in space; it is not invariant with respect to rotations of the frame of reference and, hence, leads to violation of the conservation of total angular momentum. If the potential has an axial symmetry around the z-axis then only J_z commutes with the single-particle Hamiltonian H_{sp} but $[J_x, H_{sp}] \neq 0$ and $[J_y, H_{sp}] \neq 0$ and, hence, $[\mathbf{J}^2, H_{sp}] \neq 0$.

How is it that starting from a Hamiltonian H, which is perfectly invariant under rotations and commutes with $\mathbf{J}^2$, we find that the best description of its ground state is generated by a Hamiltonian that violates this conservation law?

This paradoxical situation is very similar to that of the description of the center-of-mass motion. In that case too, the exact nuclear Hamiltonian commutes with the total linear momentum, yet we found it more appropriate to describe the system in terms of a shell-model Hamiltonian that does not conserve the total linear momentum.

Behind these paradoxes lies a fundamental conflict between two incompatible aims [see Lipkin (60)]. Our intuitive understanding of many particle systems is made possible only if we can visualize the motion of each particle as almost independent of the motion of all others. Conservation laws, on the other hand, tell us just the opposite: if one particle changes its momentum, then, to conserve total linear momentum, another particle must do so as well; or if the angular momentum of one particle in the system changes then, to conserve total angular momentum, the angular momentum of another particle has to change as well. The picture of a nearly independent-particle motion can only be compatible with conservation theorems if the quantity in question is exactly conserved for each particle separately. It is however not possible to conserve simultaneously for each particle both its linear momentum and its angular momentum and, hence, the conflict mentioned above.

To resolve this conflict we have to give up something. If we insist that our wave functions represent well-defined values of the conserved quantities, we have to give up the independent-particle picture and use wave functions that clearly manifest the correlations among the nucleons. Some useful results have been thus far obtained following this line and we shall discuss them later. The other possibility is to give up the conservation laws and buy in this way a simple wave function for the system. This approach has been found to be most fruitful, provided one is led by sound physical arguments as to which conservation laws to give up and under what circumstances. We discussed at length (Section IV.2) the reasons for sacrificing the conservation of total linear momentum. Let us now look into the question of sacrificing the conservation of angular momentum in favor of a simple Slater determinant generated from a deformed potential.

The nuclear shape is determined by several opposing tendencies. The dominant one is that of the Pauli principle, which by itself would lead to a spherical shape whenever the number of particles can exactly fill the lowest A-levels, leaving no particles in partially filled levels. On the other hand the mutual interaction among the nucleons favors maximum overlap between their mass distributions. We shall discuss this effect in more detail later, but a qualitative picture of it is appropriate here.

Consider a particle in a given spherical-potential state nlm. The angular dependence of its wave function is determined by l and m. For $m = l$ the wave function is concentrated in the plane perpendicular to $\mathbf{l}$ (i.e., to the z-axis), giving rise to a pancake like mass distribution. For $m = 0$ the distribution is concentrated along the direction of the z-axis giving rise to a cigar-shaped mass distribution. Although both distributions give rise to the same orbital angular momentum l, they are very different in shape. The fact that they correspond to the same energy in the central potential is just another manifestation of the symmetry of such potentials that can accommodate distributions of different shapes at the same average energy.

Suppose the central potential has a g-state and a d-state that lie very close to each other and are just beginning a new shell. Suppose, further, that the d-state is lower, and we start filling this level. The first nucleon will go to any of the d-states; if we include two, they will prefer to go to the $m = +2$ and $m = -2$ states of the d-level, since in this way their distributions will have the best overlap (see Fig. 23.1). The third particle, in the absence of the residual interaction would have gone again to the d-level; the Pauli principle prevents it from going into an $m = \pm 2$ state and it will therefore go, say to the $m = 1$ state. But the existence of the nearby g-level can make it energetically more favorable for the third particle to go into the $m = 4$ state of the g-level,

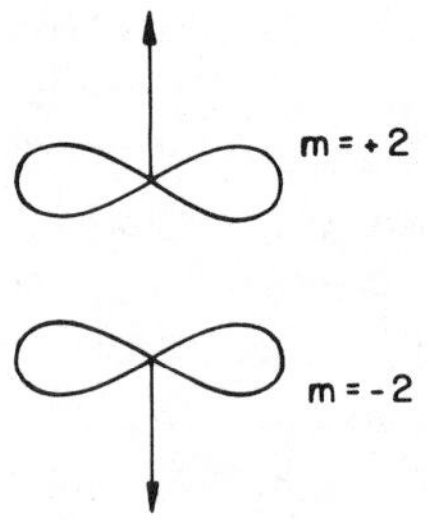

FIG. 23.1. Mass distributions of d-nucleons in states with $m = \pm 2$.

thereby having a better overlap with the two d-particles, since the $l = 4$ $m = 4$ state also has a pancake shape. The filling of this "shell," with its d- and g-levels, will therefore lead to pancake-shaped nuclei if the attractive interaction among the nucleons is large enough, and the distance in energy between the d- and g-levels is not too great.

Although the actual evaluation of nuclear shapes must be done much more carefully than the qualitative picture we have just given, we can see from it that under suitable conditions the nucleons in the nucleus may find it energetically more advantageous to be correlated in angle, thereby giving the nucleus a deformed rather than a spherical shape.

When we come now to describe such correlations in terms of a simple wave function we are led into difficulties similar to those involved in the motion of the center of mass. If we force the wave function into the simplest form—a single Slater determinant—it responds by making the self-consistent potential deformed. If we prefer to describe the nucleus in terms of wave functions in a spherical potential, we shall have to include a sum of several Slater determinants to produce the necessary intrinsic correlations among the particles.

Which particular approach to use depends greatly on the nature of the problem. Sometimes there are physical reasons to believe that only very few Slater determinants can be involved in the description of the state, or set of states considered. Such is the case, for instance, when the number of nucleons is close to a magic number so that only very few states are available at nearby energies. In these cases it may be preferable to work with wave functions generated from a spherical potential, with well-defined angular momenta, and produce the necessary correlations through the use of several Slater determinants, belonging even to different, but close enough, configurations. We shall see examples later on.

In other cases there may be physical evidence for a large deformation of the nucleus. This evidence may come for instance from measured electric quadrupole moments. The latter are proportional to

$$\int \rho(r)(3z^2 - r^2)\, d\mathbf{r} \tag{23.1}$$

where $\rho(r)$ is the charge distribution in the nucleus. Assuming a uniform density, the contributions to (23.1) that come from spherically symmetric parts, such as the shaded part in Fig. 23.2, vanish. The size and sign of the quadrupole moment is therefore a direct measure of the deviation of the nuclear shape from sphericity. If the measured quadrupole moment indicates large deformations, it is more appropriate to use a nonspherical potential as the starting point for the construction of the nuclear wave function.

Although we have stressed the similarity between the center-of-mass problem and that of deformed potentials, there is one important aspect in which they differ. The center-of-mass motion of a free nucleus is well understood and it has nothing to do with the dynamics of the interaction among the nucleons. When in Section IV.2 we chose to tie the center of mass to a point in space, we could prove that this did not affect the internal structure of the nucleus.

On the other hand a specific shape of the nucleus is already a dynamical consequence of the interactions among the nucleons. If we can describe a nucleus as a cigar rotating perpendicular to its symmetry axis, it is because of some complicated dynamical correlations between the nucleons. Such a description might lose its validity if nuclear forces were of a different character. Any attempt to fix the orientation of a nucleus in space, in a way similar to the fixing of its center-of-mass, will interfere with the internal dynamics of the nucleus and would therefore distort the resulting internal structure.

Mathematically the difference between the two cases stems from the following: it is possible to transform the coordinates of the A-particle system into a center-of-mass coordinate $\mathbf{R} = (1/A)\sum \mathbf{r}_i$ and intrinsic coordinates, which are functions of $\mathbf{r}_i - \mathbf{r}_j$ only; since the kinetic energy separates in these coordinates and the forces are also functions only of $\mathbf{r}_i - \mathbf{r}_j$, the dynamics of the system is separable from that of the center of mass.

We can similarly define an orientation of the nucleus by introducing the Eulerian angles $\theta = (\theta_1, \theta_2, \theta_3)$ describing the orientation of an intrinsic frame of reference; this frame can be defined to coincide with the principal axes of the system, for example, by requiring that the coordinates $\mathbf{r}_1'(\theta)$ referred to it satisfy

$$\sum_{i=1}^{A} x_i'(\theta)y_i'(\theta) = \sum_{i=1}^{A} y_i'(\theta)z_i'(\theta) = \sum_{i=1}^{A} z_i'(\theta)x_i'(\theta) = 0 \qquad (23.2)$$

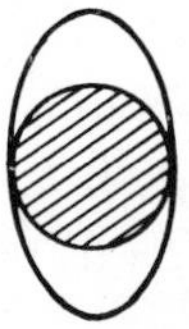

FIG. 23.2. The region that contributes to the quadrupole moment.

Equation 23.2 can be solved in principle to give the Euler angles (θ_1, θ_2, θ_3) in terms of the Cartesian coordinates $\mathbf{r}_i$ of the particles. However, unlike the case of the center-of-mass coordinate, the kinetic energy does not separate into a part depending on θ_i and a part that depends on the remaining independent coordinates.

In the shell model a motion of the potential as a whole sets, of course, the whole nucleus into a motion. This, however, will not concern us because it is essentially the motion of the center of mass of the nucleus. In a deformed potential, the rotation of the potential sets the whole nucleus into rotation. Since, as we have seen, the effect of such rotations cannot be separated from the internal dynamics of the system, they will be connected with possible intrinsic excitations and will concern us very greatly. Indeed these excitations can form the well-known rotational bands observed in highly deformed nuclei.

24. SUMMARY

Encouraged by the fair success of the independent-particle model in describing nuclear matter, and by our understanding of this success on the basis of the independent pair approximation, we discussed in this chapter independent-particle models for finite nuclei. We saw how a single-particle potential, for finite systems, may give rise to grouping of levels thereby producing the phenomenon of closed shells. The striking features associated with closed shells then enabled us to pin down some of the properties of the single-particle potential such as the existence of an important spin-orbit interaction.

Equipped with a sequence of single-particle levels determined essentially from experiment, we were then able to understand the parities and angular momenta of low-lying states of various nuclei. Both the presence of levels with some values of J and the absence of levels with other values of J could be accounted for if proper attention was paid to the Pauli principle.

A quick glance at another simple nuclear property—magnetic moments—revealed that their order of magnitude and general pattern was again nicely explained with the finite system independent-particle picture. However more detailed features of the observed moments seemed to show that one should go beyond the independent-particle approximation. In some way effects of the two-body interactions should be included, although the independent-particle wave function may serve as a good starting point.

We were thus led to ask ourselves which would be the "best" single-particle potential to start with. Using the variational approach we derived the Hartree–Fock self-consistent potential and discovered that, except for closed-shell nuclei, this potential is deformed and has an axial symmetry. (More precisely, we found that generally an axial-symmetric potential can serve as a self-consistent potential, whereas a spherically symmetric one generally cannot).

This naturally raised the question of the role of conservation laws in model wave functions and we have discussed the interplay between the simplicity of a wave function and the violation of conservation laws.

Now that we have clarified some aspects of the independent-particle wave function we are prepared to study the effects of introducing the residual interaction. There are different possibilities open before us: we may choose to work with the "best" independent-particle wave functions ignoring initially the conservation laws; or we may choose to work with "next-to-best" wave functions, giving more weight to some of the conservation laws. Moreover we may introduce into the Hartree–Fock approximation more flexibility by allowing the single-particle potential to vary in time—oscillate and change its shape slowly. If the characteristic frequencies of the potential oscillations are much smaller than those of the particles moving in it, we may use an adiabatic approximation to derive further interesting nuclear properties.

The richness of the possibilities before us is a mixed blessing. While providing alternative approaches it also requires that we study the particular features of each one of them on more general grounds, so that we shall be able to select the proper approach for each concrete case. We shall therefore start with a spherically symmetric single-particle potential, and study the effects of the residual interaction irrespective of whether the single-particle potential is self-consistent or not. We shall then see what further can be said if it is in fact self consistent. Subsequently we shall discuss deformed potentials in general, again irrespective of their being self-consistent or not. Finally we shall consider adiabatic oscillations of the single-particle potential and see what can be gained this way.

APPENDIX

SPHERICAL SYMMETRY OF SELF-CONSISTENT POTENTIALS FOR CLOSED-SHELL NUCLEI

A general interaction $\nu(12)$ can be decomposed into a sum of products of irreducible tensor operators (see Appendix A, 924)

$$\nu(12) = \sum \nu_l(r_1, r_2)T^{(l)}(1)\cdot T^{(l)}(2) \tag{A.1}$$

The direct potential is now given by (see Eq. 20.4):

$$U_d(i) = \sum_{l,m,\mu} T_m^{(l)*}(1) \sum_{\mu} \int \overset{*}{\phi}_\mu(2)\nu_l(r_1, r_2)T_m^{(l)}(2)\phi_\mu(2)\, d(2) \tag{A.2}$$

By the Wigner–Eckart theorem the integral in (A.2) is proportional to

$$(-1)^{j_\mu - m_\mu}\begin{pmatrix} j_\mu & l & j_\mu \\ -m_\mu & m & m_\mu \end{pmatrix} = \sqrt{2j_\mu + 1}\begin{pmatrix} j_\mu & 0 & j_\mu \\ -m & 0 & m_\mu \end{pmatrix}\begin{pmatrix} j_\mu & l & j_\mu \\ -m_\mu & m & m_\mu \end{pmatrix} \tag{A.3}$$

where j_μ is the angular momentum of the single-particle state μ. Summation over m_μ of (A.3) leads to a factor $\delta(l, 0)\delta(m, 0)$, so that finally $U_d(1)$ is proportional to $T_0^{(0)*}(1)$, which is by definition a scalar and hence spherically symmetric.

For the exchange integral one uses an obvious generalization of (20.17) to obtain

$$-U_{\text{ex}}(1, 1') = \sum T_m^{(l)*}(1)\overset{*}{\phi}_{\mu'}(1')\phi_\mu(1) \int \overset{*}{\phi}_\mu(2)\nu_l(r_1, r_2)T_m^{(l)}(2)\phi_{\mu'}(2)d(2) \tag{A.4}$$

The integral, again using Wigner–Eckart theorem, can be written as

$$(-1)^{j_\mu - m_\mu}\begin{pmatrix} j_\mu & l & j_{\mu'} \\ -m_\mu & m & m_{\mu'} \end{pmatrix} G(j_\mu j_{\mu'} l; r_1) \tag{A.5}$$

so that

$$-U_{\text{ex}}(1, 1') = \sum T_m^{(l)*}(1)\overset{*}{\phi}_{\mu'}(1')\phi_\mu(1)(-1)^{j_\mu - m_\mu} \times \begin{pmatrix} j_\mu & l & j_{\mu'} \\ -m_\mu & m & m_{\mu'} \end{pmatrix} G(j_\mu j_{\mu'} l; r_1) \tag{A.6}$$

The combination

$$\phi_{lm}^{j_\mu j_{\mu'}}(11') = \sum_{m_\mu m_{\mu'}} (-1)^{j_\mu - m_\mu} \begin{pmatrix} j_\mu & l & j_{\mu'} \\ -m_\mu & m & m_{\mu'} \end{pmatrix} \phi_{\mu'}^*(1')\phi_\mu(1) \qquad \text{(A.7)}$$

behaves, under rotations, like the mth component of an irreducible tensor of degree l. Therefore

$$Q_l^{j_\mu j_{\mu'}}(1, 1') = \sum_m T_m^{(l)*}(1)\phi_{lm}^{j_\mu j_{\mu'}}(1, 1') \qquad \text{(A.8)}$$

is a scalar.

Since

$$-U_{\text{ex}}(1, 1') = \sum_{l j_\mu j_{\mu'}} Q_l^{j_\mu j_{\mu'}}(1, 1')G_l(j_\mu j_\mu l; r_1)$$

it follows that U_{ex} is also a scalar and hence spherically symmetric.

The fact that we were dealing with closed-shell nuclei entered through the summation over *all* m_μ and $m_{\mu'}$ in (A.7), which is essential to deduce the transformation properties of $\phi_l^{j_\mu j_{\mu'}}(1, 1')$.

CHAPTER V
THE NUCLEAR SHELL MODEL

In the previous chapters we saw how a system of interacting fermions, close to its ground state, can show features of a system of independent, noninteracting, fermions with an appropriate density. The nuclear shell model starts from this description and attempts to refine it by taking into account, to some approximation, the fact that the motion of a nucleon in the nucleus is not entirely independent of the motions of the other nucleons. Even if the bulk of the interaction among the nucleons reflects itself in setting a certain equilibrium density, this is by no means the complete effect of the interaction and further important correlations are established among the nucleons by their mutual interaction.

The shell model limits itself to nuclei whose independent-particle approximation is spherically symmetric, that is, to nuclei, which in this approximation can be described by filling in the lowest levels of a spherically symmetric potential. We saw in Chapter IV that self-consistent potentials are very often not spherically symmetric, and we shall see later that important nuclear properties stem from these deviations from sphericity and their variation with time. Despite this, however, it turns out that the spherically symmetric approximation, even when it does not represent a self-consistent potential, produces a very good approximation for the lowest nuclear states. This is because the nucleus, as an isolated system, has a well-defined total angular momentum, a property that is nearly automatically incorporated into wave functions derived from spherically symmetric potentials. In many cases the correlations established among the nucleons by the requirement that the total angular momentum has a certain value are the crucial ones for the specific property we may be interested in. Wave functions derived from spherically symmetric potentials then become especially useful, as we shall see below.

The spherically symmetric potentials we are referring to are meant to be of the most general type; they include all potentials that remain invariant whenever a rotation is performed on the coordinates of any *single* nucleon at a time. Thus potentials of the type $\sum \zeta(r_i)\mathbf{l}_i \cdot \mathbf{s}_i$, or nonlocal* potentials

---

*Recall that a nonlocal potential acts on the wave function Ψ as follows:

$$\sum_i \int d\mathbf{r}_i' \, U(\mathbf{r}_i, \mathbf{r}_i') \, \Psi$$

$\sum U(\mathbf{r}_i, \mathbf{r}_i')$ that depend only on the angle *between* $\mathbf{r}_i$ and $\mathbf{r}_i'$ (in addition, of course, to $|\mathbf{r}_i|$ and $|\mathbf{r}_i'|$), are included along with the more obvious potentials of the type $\sum U(|\mathbf{r}_i|)$. All we shall be using in formulating the shell model is the fact that the single-particle wave functions have well-defined *single-particle* angular momentum.

The central problems considered in the shell model can be formulated as follows:

An A-particle system composed of Z-protons and $N = A - Z$ neutrons is put in a spherically symmetric (scalar) single-particle potential $U(i)$. $U(i)$ involves the ith particle's spatial, spin, and isospin coordinates, (i.e., there could be a different potential for protons and neutrons). The particles are furthermore under the influence of a two-body residual interaction $v(ij)$, where again i and j stands for space, spin, and isospin coordinates of the two particles. Three-body forces may be considered as well, but we shall limit our considerations to two-body forces only. As yet there is no evidence for important contributions of three-body forces to nuclear properties. The interaction $v(ij)$ is supposed to be weak in the sense of perturbation theory. Under these conditions:

(a) Can we devise experiments that will test our basic assumptions even before we specify $U(i)$ and $v(ij)$? In other words: are there experiments that will tell us that a certain set of nuclei can be described well enough by a model consisting of nucleons in a spherical potential and a weak residual interaction among them?

(b) From the analysis of the empirical data, and to the extent that a shell model can be shown to be applicable, what can we say about the dominant features of the empirical $U(i)$ and $v(ij)$?

(c) Using realistic potentials derived from the basic nucleon–nucleon forces for both $U(i)$ and $v(ij)$, how can we calculate energies of various nuclear states and the corresponding wave functions to the required approximation?

As we have stressed several times, we shall be able to go a long way in interpreting various nuclear phenomena before being forced to use concrete potentials for $U(i)$ and $v(ij)$. That this should happen in a shell model is of no surprise. Of the $5A$ degrees of freedom that an A-nucleon system possesses, $4A$: θ_i, ϕ_i, s_i, τ_i, are essentially determined by specifying the configuration to which a state belongs, the total angular momentum of the state and its isospin. Dealing with a shell-model wave function actually means that we work with a nearly completely known function even before we specify $U(i)$. The central potential determines only the radial wave functions, and for a wide class of smooth monotonic potentials, even these have similar properties such as those described in Section IV.7. Although in our calculations we shall try to avoid writing down explicitly the shell-model wave functions, we should always remember that for the most part they are universal, known functions. It is this fact that is reflected in the generality of the results that we shall obtain.

The fact that nuclei obey some of these general results does not prove their universal validity for any system of fermions. The neutral atom of ^{6}Li, for instance, is a fermion, and a collection of these atoms does not show the various characteristics of a shell structure. There is a deep physical significance to the fact that nuclei do show evidence of shell structure, and this was discussed at length in Chapter III. However, once the conditions for the validity of the shell-model approximation are there—the rest follows from the very specific and universal structure of shell-model wave functions. It is for this reason that problem (a) will occupy a good part of our discussion of the shell model.

The Appendix to this chapter lists some of the more common residual potentials that have been used in the analysis of nuclear structure.

1. THE CLASSIFICATION OF STATES. TWO-PARTICLE CONFIGURATIONS

We consider the shell-model Hamiltonian

$$H_{SM} = \sum [T_i + U(i)] + \frac{1}{2} \sum_{i \neq j} v(ij) \tag{1.1}$$

In order to treat $v(ij)$ as a perturbation we must first diagonalize it within the subspace of degenerate states of

$$H_0 = \sum [T_i + U(i)] \equiv \sum h(i) \tag{1.2}$$

Degeneracies in H_0 are due to two sources: they may result from a certain symmetry of H_0, or they may be accidental. Among the first type we can count:

(a) The degeneracy due to similar proton and neutron states if H_0 is charge-independent.

(b) The degeneracy of the states $j = l + 1/2$ and $j' = l - 1/2$ if H_0 is spin-independent.

(c) The degeneracy of the various m-states of any given single-particle level if H_0 is spherically symmetric.

The second type of degeneracies includes for our purpose, the well-known degeneracies of the three-dimensional harmonic-oscillator potential or the Coulomb potential (strictly speaking these are due to some special symmetries of the corresponding Hamiltonians, but we shall disregard these symmetries here since they are not connected with simple dynamical properties of the nuclear forces).

Whenever there are degeneracies in the zeroth-order Hamiltonian, the first-order corrections to the energy are obtained by the diagonalization of the perturbation within the subspace of degenerate states. This diagonalization can be greatly simplified if we use special linear combinations of the zeroth-order degenerate wave functions rather than an arbitrarily selected set. We shall illustrate this point now by studying the two identical particles system.

Let $\phi_\alpha(i)$ be the eigenstates of H_0:

$$H_0\phi_\alpha(i) = \epsilon_\alpha\phi_\alpha(i) \tag{1.3}$$

The quantum numbers α stand for $(n_\alpha l_\alpha j_\alpha m_\alpha)$, as discussed previously, and the $(2j_\alpha + 1)$ states of the single-particle configuration $(n_\alpha l_\alpha j_\alpha)$ are all degenerate. A two-particle configuration $(n_\alpha l_\alpha j_\alpha, n_\beta l_\beta j_\beta)$ will have a zeroth-order energy $\epsilon_\alpha + \epsilon_\beta$ and will be $(2j_\alpha + 1)(2j_\beta + 1)$-fold degenerate. A simple set of eigenfunctions for the two-particle configuration is the m-scheme set:

$$\Phi(m_\alpha, m_\beta) = \frac{1}{\sqrt{2}}\begin{vmatrix} \phi_\alpha(1) & \phi_\alpha(2) \\ \phi_\beta(1) & \phi_\beta(2) \end{vmatrix} \tag{1.4}$$

Since the residual interaction $v(12)$ commutes with J_z we find that

$$\langle\Phi(m_\alpha, m_\beta)|v(12)\Phi(m'_\alpha, m'_\beta)\rangle = 0 \quad \text{if} \quad m_\alpha + m_\beta \neq m'_\alpha + m'_\beta \tag{1.5}$$

In setting out to diagonalize $v(12)$ in the subspace of the $(2j_\alpha + 1)(2j_\beta + 1)$ functions $\Phi(m_\alpha, m_\beta)$ it will therefore be convenient to group together all the states with a given value for $m_\alpha + m_\beta$. The matrix for $v(12)$ will then look like Fig. 1.1. We have denoted the values of $M = m_\alpha + m_\beta$ that characterize each one of the rows (and columns) of the matrix. The square in the upper left-hand corner is of order 1×1 since there is just one state with $M = j_\alpha + j_\beta$ (we

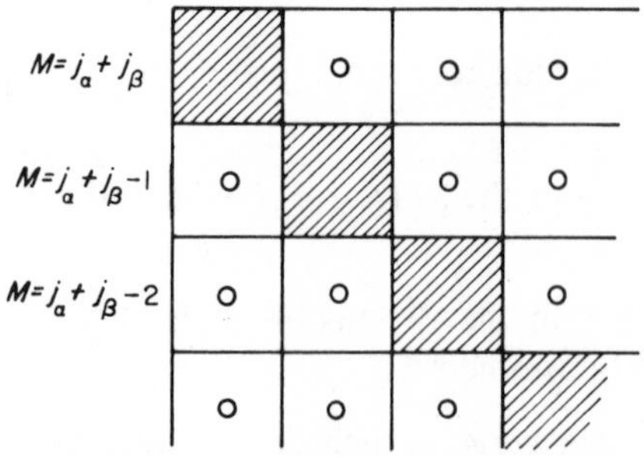

FIG. 1.1. The matrix elements of $v(12)$ in the M-scheme. Only matrix elements in the shaded areas can differ from zero.

assume $j_\alpha \neq j_\beta$ so that antisymmetry does not make this state vanish identically); the second square matrix is 2×2 since $M = j_\alpha + j_\beta - 1$ can be gotten in two independent ways: $m_\alpha = j_\alpha$, $m_\beta = j_\beta - 1$ and $m_\alpha' = j_\alpha - 1$, $m_\beta = j_\beta$; and similarly for the other squares along the diagonal.

We see that already by properly arranging the states $\Phi(m_\alpha, m_\beta)$ we reduce the task of diagonalizing the full matrix of order $[(2j_\alpha + 1)\cdot(2j_\beta + 1)] \times [(2j_\alpha + 1)\cdot(2j_\beta + 1)]$ into that of diagonalizing several, much smaller, matrices. However, we can carry out the diagonalization even further, without specifying $v(12)$, if we note that $v(12)$ commutes with $\mathbf{J} = \mathbf{j}_1 + \mathbf{j}_2$, that is, that the total angular momentum is conserved by the Hamiltonians (1.1) and (1.2). Indeed, if we make use of the transformation from the m-scheme to the J-scheme (see Section IV.12) using the Clebsch–Gordan coefficients, we arrive at another set, the J-scheme set of $(2j_\alpha + 1)(2j_\beta + 1)$ degenerate states of H_0 belonging to the eigenvalue $\epsilon_\alpha + \epsilon_\beta$:

$$|j_\alpha j_\beta JM\rangle = \sum_{m_\alpha m_\beta} (j_\alpha m_\alpha j_\beta m_\beta | JM)\Phi(m_\alpha, m_\beta) \tag{1.6}$$

$$J = |j_\alpha - j_\beta|, |j_\alpha - j_\beta| + 1, \ldots, j_\alpha + j_\beta$$

With this set of functions we find that:

$$\langle j_\alpha j_\beta JM | v(12) | j_\alpha j_\beta J'M'\rangle = 0 \quad \text{if} \quad J \neq J' \quad \text{and/or} \quad M \neq M' \tag{1.7}$$

Remark. In writing the wave function $|j_\alpha j_\beta JM\rangle$ in the form (1.6) special attention should be paid to its normalization. If $(n_\alpha l_\alpha j_\alpha) \neq (n_\beta l_\beta j_\beta)$ then it follows from (1.4) immediately that

$$\begin{aligned}\langle j_\alpha j_\beta JM | j_\alpha j_\beta J'M'\rangle &= \sum (j_\alpha m_\alpha j_\beta m_\beta | JM)(j_\alpha m_\alpha' j_\beta m_\beta' | J'M') \\ &\times \frac{1}{2}\int [\phi_{\alpha m_\alpha}(1)\phi_{\beta m_\beta}(2) - \phi_{\alpha m_\alpha}(2)\phi_{\beta m_\beta}(1)]^* \\ &\times [\phi_{\alpha m_\alpha'}(1)\phi_{\beta m_\beta'}(2) - \phi_{\alpha m_\alpha'}(2)\phi_{\beta m_\beta'}(1)]\, d(1)\, d(2) \\ &= \delta(J, J')\delta(M, M')\end{aligned} \tag{1.8}$$

We notice, in deriving (1.8) that the "cross integrals" like

$$\int \phi^*_{\alpha m_\alpha}(1)\phi^*_{\beta m_\beta}(2)\phi_{\alpha m_\alpha'}(2)\phi_{\beta m_\beta'}(1)\, d(1)\, d(2) \tag{1.9}$$

vanish on account of the assumption $(n_\alpha l_\alpha j_\alpha) \neq (n_\beta l_\beta j_\beta)$. If, however, $(n_\alpha l_\alpha j_\alpha) = (n_\beta l_\beta j_\beta)$, that is, if we have two equivalent particles, then the "cross integrals" (1.9) do not vanish:

$$\int \phi^*_{\alpha m_\alpha}(1)\phi^*_{\beta m_\beta}(2)\phi_{\alpha m_\alpha'}(2)\phi_{\beta m_\beta'}(1)\, d(1)\, d(2) = \delta(m_\alpha, m_\alpha')\delta(m_\beta, m_\beta') \quad \text{if} \quad (n_\alpha l_\alpha j_\alpha) = (n_\beta l_\beta j_\beta) \tag{1.10}$$

We then obtain for (1.8), using the symmetry properties of the Clebsch–Gordan coefficients (see Appendix A, A.2.69):

$$\langle j_\alpha j_\beta JM | j_\alpha j_\beta J'M' \rangle = [1 - (-1)^{2j+J}]\delta(J, J')\delta(M, M') \quad \text{if} \quad (n_\alpha l_\alpha j_\alpha) = (n_\beta l_\beta j_\beta) \tag{1.11}$$

where $j = j_\alpha = j_\beta$. Thus for equivalent nucleons $|j^2JM\rangle$ vanishes for odd values of J, while for even values of J it is properly normalized if we define

$$|j^2JM\rangle = \frac{1}{\sqrt{2}} \sum (jm_\alpha jm_\beta | JM)\Phi(m_\alpha, m_\beta) \tag{1.12}$$

Equation 1.12 supplements (1.6).

For the two-particle system, given j_α and j_β, there is only *one* state with a given value for J and M. Thus, if we take matrix elements of $v(12)$ with the wave functions (1.6), the matrix is automatically diagonal. We conclude that for two identical particles in the levels j_α and j_β, the first-order correction to the energy is given by

$$\Delta E(j_\alpha j_\beta JM) = \langle j_\alpha j_\beta JM | v(12) | j_\alpha j_\beta JM \rangle \tag{1.13}$$

Before we proceed with the study of some of the properties of $\Delta E(JM)$ in (1.13), we would like to remark on the simplicity of the result (1.13): the fact that we were able to write down an explicit expression for the eigenvalues of $v(12)$ in the configuration $j_\alpha j_\beta$, is typical only of the simplest two-particle configurations. If, for instance, there were a third level $j_{\beta'}$, accidentally degenerate with j_β, that is, $\epsilon_\beta = \epsilon_{\beta'}$, the situation would already be considerably more complex. We shall still find it useful to work in the J-scheme, but now our degenerate two-particle wave functions will include both $|j_\alpha j_\beta JM\rangle$ defined in (1.6), and the $(2j_\alpha + 1)(2j_{\beta'} + 1)$ states $|j_\alpha j_{\beta'} JM\rangle$ defined by replacing j_β by $j_{\beta'}$ in (1.6). The interaction $v(12)$ is still diagonal with respect to J and M, but if J happens to have values that satisfy simultaneously the two conditions:

$$|j_\alpha - j_\beta| < J < j_\alpha + j_\beta \quad \text{and} \quad |j_\alpha - j_{\beta'}| < J < j_\alpha + j_{\beta'} \tag{1.14}$$

there will be two degenerate states with such a value of J for any value of M satisfying $-J \leq M \leq J$. For these states the shifts in energy will then be given by the eigenvalues of the 2x2 matrices:

$$\begin{pmatrix} \langle j_\alpha j_\beta JM | v(12) | j_\alpha j_\beta JM \rangle & \langle j_\alpha j_\beta JM | v(12) | j_\alpha j_{\beta'} JM \rangle \\ \langle j_\alpha j_{\beta'} JM | v(12) | j_\alpha j_\beta JM \rangle & \langle j_\alpha j_{\beta'} JM | v(12) | j_\alpha j_{\beta'} JM \rangle \end{pmatrix} \tag{1.15}$$

Problems that involve a greater number of zeroth-order degenerate levels, or configurations with more particles, generally lead to still larger matrices even when we use the J-scheme. Nevertheless it remains, true that the matrices in the J-scheme are much smaller than those in the m-scheme. The transformation from the m-scheme to the J-scheme already amounts to a substantial, if even not complete, diagonalization of the residual interaction.

Let us come back now to (1.13). The interaction $v(12)$ is invariant under rotations. It follows then from the Wigner–Eckart theorem (see Appendix A, A.2.44) that

$$\langle j_\alpha j_\beta JM | v(12) | j_\alpha j_\beta JM \rangle = (-1)^{J-M} \begin{pmatrix} J & 0 & J \\ -M & 0 & M \end{pmatrix} (j_\alpha j_\beta J || v || j_\alpha j_\beta J) \quad (1.16)$$

where the double-barred reduced matrix element is independent of M. The special $3-j$ symbol in (1.16) is explicitly given in Appendix A, and it follows from (1.16) and (1.13) that

$$\Delta E(j_\alpha j_\beta JM) = \langle j_\alpha j_\beta JM | v(12) | j_\alpha j_\beta JM \rangle = \frac{1}{\sqrt{2J+1}} (j_\alpha j_\beta J || v || j_\alpha j_\beta J) \quad (1.17)$$

Thus the energy shift of a state with a given J and M is independent of M; the interaction $v(12)$ removes part of the degeneracy since it may shift states with different J's differently. But some degeneracy still remains: levels of a given J remain $(2J+1)$-fold degenerate.

This leftover degeneracy is of a similar origin as the $(2j_\alpha + 1)$-fold degeneracy of the single-particle levels in the central potential. It is due to the spherical symmetry of the Hamiltonian (1.1) that requires that there be no preferred axis in space determined by the solutions of (1.1).

To see how the interaction $v(12)$ removes part of the zeroth-order degeneracy it is instructive to study a special hypothetical interaction, the so-called *δ-force* (actually δ-potential)

$$v(12) = -V_0 \delta(\mathbf{r}_1 - \mathbf{r}_2) \quad (1.18)$$

where V_0 has the dimensions of (Energy) × (Volume).

The matrix elements of this potential measure directly the space overlap between the wave functions of particles 1 and 2. If the particle wave functions have a small overlap—say one is close to the center of the nucleus and the other at the surface—the δ-potential will make a very small contribution to the average energy. If, on the other hand, their density distribution is similar—this will show up in the greater magnitude of the matrix elements (1.13), with v given by (1.18). Nuclear forces are predominantly attractive and, due to their short range, are more effective when the overlap between $\phi(1)$ and $\phi(2)$ is largest. We have therefore chosen the negative sign in (1.18), so that states that come lowest with the δ-potential are states of best overlap, in accordance with what we expect for a more realistic $v(12)$. Although the δ-force does not resemble real nuclear forces, it does offer a useful approximation to matrix elements of real nuclear interactions. The latter depend crucially on the spatial overlap of the wave functions of the two interacting nucleons, and the δ-force measures this overlap directly.

To evaluate matrix elements of (1.18) (dropping V_0), that is,

$$\langle j_\alpha j_\beta JM | -\delta(\mathbf{r}_1 - \mathbf{r}_2) | j_\alpha j_\beta JM \rangle \tag{1.19}$$

we note that

$$\phi_\alpha(1) = \sum (1/2 m_{s\alpha} l_\alpha m_{l\alpha} | j_\alpha m_\alpha) \frac{R_{n_\alpha l_\alpha}(r_1)}{r_1} Y_{l_\alpha m_{l\alpha}}(\theta_1, \phi_1) \chi^{(1)}(m_{s\alpha}) \tag{1.20}$$

where $R_{nl}(r)$ is a radial function determined by the potential $U(i)$ and $\chi^{(i)}$ is the spin wave function for particle i. Hence, for identical particles, using (1.4), we can write the m-scheme wave function in the form:

$$\begin{aligned}\Phi(m_\alpha m_\beta) = & \sum_{m_{s\alpha} m_{s\beta}, m_{l\alpha} m_{l\beta}} (\tfrac{1}{2} m_{s\alpha} l_\alpha m_{l\alpha} | j_\alpha m_\alpha)(\tfrac{1}{2} m_{s\beta} l_\beta m_{l\beta} | j_\beta m_\beta) \\ & \times \frac{1}{2\sqrt{2}} \frac{1}{r_1 r_2} \{ [F_\alpha(1) F_\beta(2) - F_\alpha(2) F_\beta(1)] \\ & \times [\chi^{(1)}(m_{s\alpha})\chi^{(2)}(m_{s\beta}) + \chi^{(1)}(m_{s\beta})\chi^{(2)}(m_{s\alpha})] \\ & + [F_\alpha(1) F_\beta(2) + F_\alpha(2) F_\beta(1)] \\ & \times [\chi^{(1)}(m_{s\alpha})\chi^{(2)}(m_{s\beta}) - \chi^{(1)}(m_{s\beta})\chi^{(2)}(m_{s\alpha})]\}\end{aligned} \tag{1.21}$$

where we introduced the notation

$$F_\alpha(i) = R_{n_\alpha l_\alpha}(i) Y_{l_\alpha m_{l\alpha}}(\theta_i \phi_i) \tag{1.22}$$

It is obvious from (1.6) that

$$\begin{aligned}\langle j_\alpha j_\beta JM | -\delta(\mathbf{r}_1 - \mathbf{r}_2) | j_\alpha j_\beta JM \rangle = & \sum_{m_\alpha m_\beta, m_{\alpha'} m_{\beta'}} (j_\alpha m_\alpha j_\beta m_\beta | JM)(j_\alpha m'_\alpha j_\beta m'_\beta | JM) \\ & \times \langle m_\alpha m_\beta | -\delta(\mathbf{r}_1 - \mathbf{r}_2) | m'_\alpha m'_\beta \rangle\end{aligned} \tag{1.23}$$

To evaluate the matrix elements in (1.23) we use the $\phi(m_\alpha m_\beta)$ as given in (1.21) and note that, because of the $\delta(\mathbf{r}_1 - \mathbf{r}_2)$ function the term proportional to $F_\alpha(1)F_\beta(2) - F_\alpha(2)F_\beta(1)$ vanishes. It therefore follows that

$$\begin{aligned}\langle m_\alpha m_\beta | -\delta(\mathbf{r}_1 - \mathbf{r}_2) | m'_\alpha m'_\beta \rangle = & -\int \frac{1}{r^2} R^2_{n_\beta l_\beta}(r) R^2_{n_\alpha l_\alpha}(r)\, dr \\ & \sum_{\text{all } m_s, m_l} \int Y^*_{l_\alpha m_{l\alpha}}(\Omega) Y^*_{l_\beta m_{l\beta}}(\Omega) Y_{l_\alpha m'_{l\alpha}}(\Omega) Y_{l_\beta m'_{l\beta}}\, d\Omega\, (m_{s\alpha} m_{l\alpha} | m_\alpha) \\ & \times (m_{s\beta} m_{l\beta} | m_\beta) [(m_{s\alpha} m'_{l\alpha} | m'_\alpha)(m_{s\beta} m'_{l\beta} | m'_\beta) - (m_{s\beta} m'_{l\alpha} | m'_\alpha)(m_{s\alpha} m'_{l\beta} | m'_\beta)]\end{aligned} \tag{1.24}$$

In deriving (1.24) we used the expression for the δ-function in polar coordinates

$$\delta(\mathbf{r}_1 - \mathbf{r}_2) = \frac{1}{r_1 r_2} \delta(r_1 - r_2) \delta(\cos\theta_1 - \cos\theta_2) \delta(\phi_1 - \phi_2) \tag{1.25}$$

We have also introduced a shorthand notation for the Clebsch–Gordan coefficient:

$$(m_s m_l | m) \equiv (\tfrac{1}{2} m_s l m_l | jm) \tag{1.26}$$

and used the orthogonality of the spin wave function

$$\langle \chi^{(i)}(m_s) | \chi^{(i)}(m_s') \rangle = \delta(m_s, m_s')$$

Combining (1.23) and (1.24) we obtain finally:

$$\Delta E_\delta(J) = V_0 \langle j_\alpha j_\beta JM | -\delta(\mathbf{r}_1 - \mathbf{r}_2) | j_\alpha j_\beta JM \rangle = V_0 F^{(0)}(n_\alpha l_\alpha n_\beta l_\beta) A(j_\alpha j_\beta JM) \tag{1.27}$$

where the index δ was added explicitly to ΔE to remind us that it is derived from a δ-interaction, and

$$F^{(0)}(n_\alpha l_\alpha n_\beta l_\beta) = -\int_0^\infty \frac{1}{r^2} R^2_{n_\alpha l_\alpha}(r) R^2_{n_\beta l_\beta}(r)\, dr \tag{1.28}$$

Furthermore, the A's are universal functions defined by

$$A(j_\alpha j_\beta JM) = \sum_{\text{all } m\text{'s}} (m_\alpha m_\beta | M)(m_\alpha' m_\beta' | M)(m_{s\alpha} m_{l\alpha} | m_\alpha)(m_{s\beta} m_{l\beta} | m_\beta)$$
$$\times [(m_{s\alpha} m_{l\alpha}' | m_\alpha')(m_{s\beta} m_{l\beta}' | m_\beta') - (m_{s\beta} m_{l\alpha}' | m_\alpha)(m_{s\alpha} m_{l\alpha}' | m_\beta')]$$
$$\times \int Y_\alpha^*(\Omega) Y_\beta^*(\Omega) Y_{\alpha'}(\Omega) Y_{\beta'}(\Omega)\, d\Omega \tag{1.29}$$

Equation 1.27 has an interesting structure. It consists of a product of two factors: $F^{(0)}$ that implicitly depends on the potential $U(i)$ through the radial wave functions, and $A(j_\alpha j_\beta JM)$ that is a universal function of j_α, j_β, J, and M. Only the latter includes a dependence on J. Hence the important result that

$$\frac{\Delta E_\delta(J)}{\Delta E_\delta(J')} = \frac{A(j_\alpha j_\beta J)}{A(j_\alpha j_\beta J')} \quad \text{independent of} \quad U(i) \tag{1.30}$$

Equation 1.27 has to be understood as follows: the overlap of ϕ_α and ϕ_β, measured by the matrix element (1.19), is a product of two parts; one measures the overlap of the radial densities, $[F^{(0)}]$, and is independent of the total angular momentum; the other measures the overlap of the angular part of the wave functions and is entirely independent of the central potential $U(i)$. This separation is made possible for the matrix elements of a δ-interaction because as we can see from (1.25), the δ-interaction itself separates into a product of a part that depends on the magnitudes of $\mathbf{r}_1$ and $\mathbf{r}_2$ and a part that depends on the directions $\hat{\mathbf{r}}_1 = \mathbf{r}_1/r_1$ and $\hat{\mathbf{r}}_2 = \mathbf{r}_2/r_2$.

If we recall that the mass distribution of a particle is concentrated in the plane perpendicular to its angular momentum, then the contribution to $\Delta E_\delta(J)$ that depends on $\hat{\mathbf{r}}_1$ and $\hat{\mathbf{r}}_2$ can be pictured in a rather simple way: consider particle 1 with its angular momentum pointing in the direction $\mathbf{j}_1$. Its mass is concentrated in the cross-hatched region (Fig. 1.2); if the angular

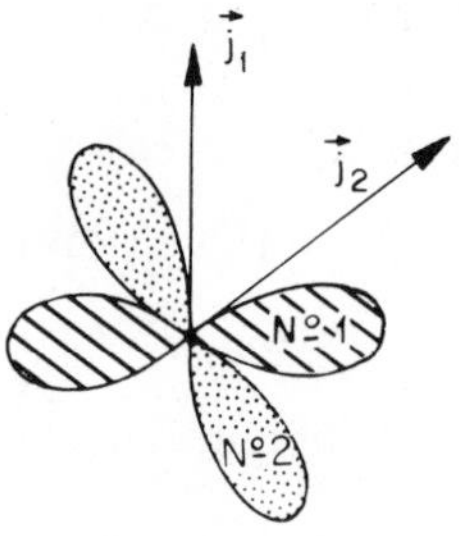

FIG. 1.2. The relative mass distributions of two nucleons.

momentum of particle 2 is pointing in the direction $\mathbf{j}_2$ its mass will be concentrated in the dotted region. Obviously the overlap between the two is going to depend on the angle between $\mathbf{j}_1$ and $\mathbf{j}_2$; that, however, is determined once we fix $\mathbf{J} = \mathbf{j}_1 + \mathbf{j}_2$. It is in this way that the overlap integral gets its dependence on the total angular momentum. Pictorially we can describe the situation as follows (Fig. 1.3): when the interaction $\nu(12) = -\delta(\mathbf{r}_1 - \mathbf{r}_2)$ is taken into account, a part of the $(2j_\alpha + 1) \times (2j_\beta + 1)$-fold degeneracy of the configuration $(j_\alpha j_\beta)$ is removed since states with different values of J get pushed by different amounts $\Delta E(J)$. A spectrum of levels, all belonging to the configuration $j_\alpha j_\beta$, is thereby created.

Equation 1.27 or 1.30 further tells us that if the residual nuclear interaction were of a very short range, then the spacings of the levels of a configuration $(j_\alpha j_\beta)$, apart from a common scale factor, would be the same for different nuclei, irrespective of their sizes and the particular shape of the single-particle potential appropriate for them. The latter will reflect itself only in the radial wave function $R_{nl}(r)$ and therefore through a multiplicative factor that can be absorbed in the scale factor. The ratios of spacings between the different

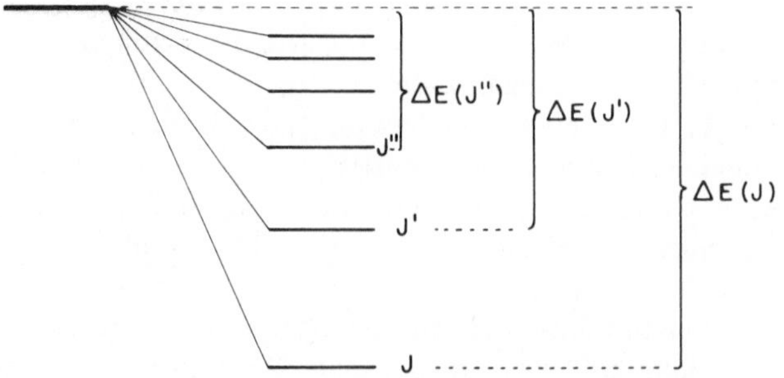

FIG. 1.3. Schematic description of the removal of the degeneracy in the configuration $j_\alpha j_\beta$ through a residual interaction.

levels are thus independent of radial quantum numbers, and are really determined just by the quantum numbers l_α, j_α, l_β, and j_β.

The coefficients $A(j_\alpha j_\beta JM)$ can be evaluated by direct substitution of the numerical values for the Clebsch–Gordan coefficients and using (Eq. IV.20.8) for the evaluation of the angular integrals. Later on we shall see an even simpler way of deriving more compact expressions for coefficients like $A(j_\alpha j_\beta JM)$. Here we shall just quote the result, which will be derived later (see Section V.4):

$$A(j_\alpha j_\beta JM) = \begin{cases} \dfrac{1}{4\pi}(2j_\alpha + 1)(2j_\beta + 1)\begin{pmatrix} j_\alpha & j_\beta & J \\ 1/2 & -1/2 & 0 \end{pmatrix}^2 & \text{if } l_\alpha + l_\beta - J \text{ is even} \\ 0 & \text{if } l_\alpha + l_\beta - J \text{ is odd} \end{cases} \tag{1.31*}$$

Equation (1.31) has several interesting features. It shows that for configurations of even parity (i.e., $l_\alpha + l_\beta$ even) only even values of J are affected by the δ-force, whereas for configurations of odd parity only odd values of J are affected. If we take as an example a configuration (5/2, 7/2), then if its parity is even its energy levels will look like Fig. 1.4*a*, whereas if its parity is odd they will look like Fig. 1.4*b*.

In both cases there are groups of levels that are not shifted at all and others that are shifted. The actual value of the energy shift depends on the numerical value of the $3-j$ symbol in (1.31). A table of these coefficients can be found in various places listed by Way and Hurley (66.) See also Rotenberg et al. (59).

The physics behind the phenomenon shown in Figs. 1.4*a* and 1.4*b* is of some interest. Take for instance the case of $J = 6^-$ and let us consider in more detail how such a state can be formed. If $j_\alpha = 5/2$ then l_α could be either 2 or 3 ($l = j \pm 1/2$); similarly if $j_\beta = 7/2$ then $l_\beta = 3$ or 4. If the parity of the configuration is negative we must have either the combination $(l_\alpha, l_\beta) =$ (2, 3) or $(l_\alpha, l_\beta) =$ (3, 4). Suppose we take $(l_\alpha, l_\beta) =$ (2, 3) so that we are considering the configuration $(d_{5/2}, f_{7/2})$. Since the orbital angular momenta can provide at most $2 + 3 = 5$ units of angular momentum, the state with $J = 6^-$ can only be obtained if the two spins are parallel and provide the extra unit of angular momentum missing (see Fig. 1.5). However, if the two spins are parallel, the two-particle wave function is symmetric with respect to the exchange of the spins; since it must be totally antisymmetric it follows that it is antisymmetric with respect to the exchange $(\mathbf{r}_1, \mathbf{r}_2) \to (\mathbf{r}_2, \mathbf{r}_1)$; the wave function for this particular state then vanishes, when $\mathbf{r}_1 = \mathbf{r}_2$. But a δ-interaction is effective only when $\mathbf{r}_1 = \mathbf{r}_2$. Hence the result that the δ-interaction

*The factor $1/4\pi$ should be changed to $1/8\pi$ if $(n_\alpha l_\alpha j_\alpha) = (n_\beta l_\beta j_\beta)$; see (1.11).

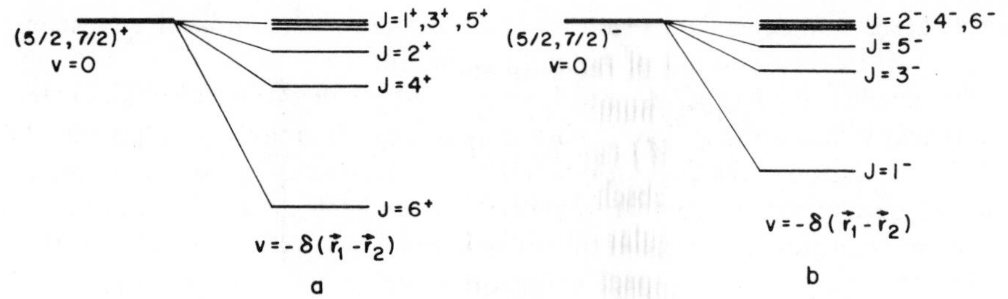

FIG. 1.4. The energy levels of the configuration (5/2, 7/2) even (*a*) and odd (*b*) parities, with a δ-interaction. [The scale of (*b*) is three times smaller than that of (*a*).]

does not affect the state with $J = 6^-$. Slightly more extensive consideration can be applied to the other states, and we shall return to them later. The important thing to realize is that, for the two-particle case, the even or odd values of J together with the parity of the state determines whether the probability of finding the two particles close together vanishes or not.

It is not very common to find two identical particles in configurations of odd parity among the low-lying states of actual nuclei. For two identical particles outside closed shells the lowest states are obtained when both particles go into the *same* lowest available level, and we thus obtain a configuration j^2 of two equivalent particles with even parity. For such configurations, as we saw in Eq. IV.13.5, the only allowed states are those with even values of J, so that the conclusions derived above for the δ-interaction do not lead to new information in this particular case. We can still use (1.31) to obtain the order of the different levels if we believe that the nuclear interaction is of a short range. An inspection of Table 1.1 shows that the state with $J = 0^+$ is then always the lowest, then comes $J = 2^+$, etc. This ascending order of J's for the excited states has been observed in most spectra of even–even nuclei. Examples taken from nuclei near closed shells are especially instructive in this respect since such nuclei are nearly spherical and assignment of orbits can be done with great reliability. Furthermore, as we shall see later on, the closed shells do not contribute to the splittings between the levels of the configuration of the valence nucleons; we can therefore compare empirical data

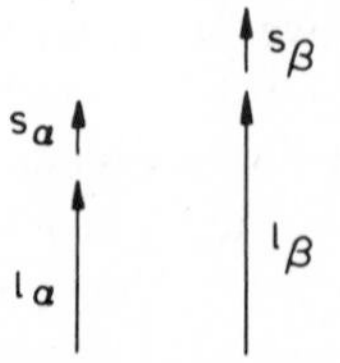

FIG. 1.5. The angular momenta in the state $J = 6^-$ of $(d_{5/2}, f_{7/2})$.

in such nuclei with our present calculations. The case of ^{90}Zr discussed in Section IV.14 is perhaps one of the nicest examples. Another one is that of ^{210}Po, shown in Fig. 1.6. Figure 1.7 shows the spectra of the configuration $(9/2)^2$ as calculated for a δ-interaction with the help of equations (1.27) and (1.31) on an arbitrary scale. We see that it reproduces qualitatively the clustering together of the levels of higher angular momenta observed both in ^{210}Po and ^{90}Zr (the configurations giving rise to the lowest excitations in these nuclei are, respectively, $1h_{9/2}^2$ and $1g_{9/2}^2$). Physically this means that the overlap between the two particles does not change appreciably as we keep decreasing the angle between their angular momenta beyond the angle corresponding to, say, $J = 4$.

TABLE 1.1 Values of $8\pi\ A(j_\alpha j_\beta JM)$ as Given by Equation 1.31 (see footnote below it) for $j_\alpha = j_\beta$

j_α	j_β	J	8π A
1/2	1/2	0	2
3/2	3/2	0	4
		2	4/5
5/2	5/2	0	6
		2	48/35
		4	4/7
7/2	7/2	0	8
		2	40/21
		4	200/77
		6	200/429
9/2	9/2	0	10
		2	80/33
		4	180/143
		6	320/429
		8	980/2439

As we go higher in excitation energy, we encounter configurations of the type (j_α, j_β) with $j_\alpha + j_\beta$ even for nuclei with two particles outside the closed shell. Usually one then still reaches configurations with positive parities because most of the levels in any given shell have the same parity (compare Fig. I.7.2). Our result (1.31) then tells us that the lowest states of these configuration should have even values of J. We obtain thus an explanation of the observed overwhelming predominance of levels with even J among the low-lying positive parity levels in even–even nuclei.

Occasionally an excitation of the two-particle system may lead also to

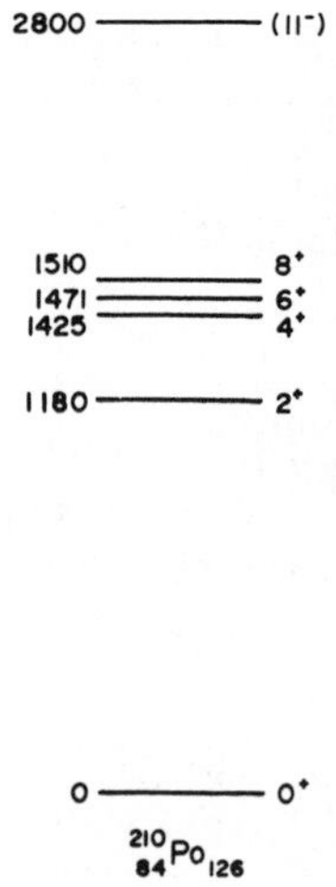

FIG. 1.6. Level scheme for ^{210}Po. The energy of each level is given in keV [taken from Yamazaki and Ewan (67)].

states of negative parity like the 5^- and 4^- states in ^{90}Zr, or the 5^- state in ^{210}Po, etc. From (1.31) we can expect that the lowest negative parity state in an even–even nucleus will have an odd value of J. This is indeed the case in both examples quoted, and in most other known cases (the *Talmi–Glaubman rule*).

2. THE EQUIVALENCE OF PROTONS AND NEUTRONS. ISOSPIN

Our discussion in the previous section was devoted to identical particles only. Let us now consider nonidentical particles as well, that is, a system composed of a proton and a neutron in the configuration $(j_p j_n)$. If the Hamiltonian is charge independent, that is, if we neglect for the moment electromagnetic contributions to H, we have an additional degeneracy; any state of the con-

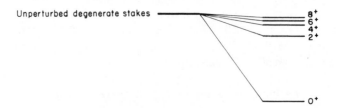

FIG. 1.7. The configuration $(9/2)^2$ with δ-interaction.

figuration (j_p, j_n) in which particle 1 is a proton and 2 is a neutron is degenerate with a state in which 1 is a neutron and 2 is a proton. Staying for the moment within a formalism without isospin, that is, considering the proton and the neutron as different particles, we conclude that the two states

$$|j_p j_n JM\rangle \tag{2.1}$$

and

$$|j_n j_p JM\rangle \tag{2.2}$$

are degenerate. Here we define a convenient shorthand notation

$$|j_p j_n JM\rangle = \sum_{m_\alpha m_\beta} (j_\alpha m_\alpha j_\beta m_\beta|JM)\phi_\alpha(1)\phi_\beta(2)p(1)n(2) \tag{2.3}$$

where $p(i)$ and $n(i)$ are proton and neutron wave functions, respectively, for the ith nucleon, ϕ_α gives the spatial and spin dependence. Similarly

$$|j_n j_p JM\rangle = \sum_{m_\alpha m_\beta} (j_\alpha m_\alpha j_\beta m_\beta|JM)\phi_\alpha(2)\phi_\beta(1)p(1)n(2) \tag{2.4}$$

Note that these wave functions are not antisymmetrized: in (2.3) the particle in state α is a proton and that in state β is a neutron, while in (2.4), the one in state α is a neutron and that in state β is a proton.

The two states (2.1) and (2.2), which are physically distinct, are degenerate in our approximation. To obtain the energy shifts of these two states it is therefore necessary to diagonalize a 2x2 matrix even if we are in the J-scheme. This matrix consists of the following elements

$$\begin{pmatrix} \langle j_p j_n J|v(12)|j_p j_n J\rangle & \langle j_p j_n J|v(12)|j_n j_p J\rangle \\ \langle j_n j_p J|v(12)|j_p j_n J\rangle & \langle j_n j_p J|v(12)|j_n j_p J\rangle \end{pmatrix} \tag{2.5}$$

where

$$\langle j_p j_n J|v(12)|j_p j_n J\rangle = \sum (m_\alpha m_\beta|M)(m'_\alpha m'_\beta|M)\langle\phi_\alpha(1)\phi_\beta(2)p(1)n(2)$$
$$|v(12)\phi_{\alpha'}(1)\phi_{\beta'}(2)p(1)n(2)\rangle \quad \text{and} \quad (m_\alpha m_\beta|M) \equiv (j_\alpha m_\alpha j_\beta m_\beta|JM) \tag{2.6}$$

If $v(12)$ is symmetric with respect to the exchange of the proton and the neutron then

$$v(pn) \equiv \langle p(1)n(2)|v(12)p(1)n(2)\rangle = \langle p(1)n(2)|v(21)p(1)n(2)\rangle \equiv v(np) \tag{2.7}$$

It then follows that

$$\langle j_p j_n J|v(12)|j_p j_n J\rangle = \langle j_n j_p J|v(12)|j_n j_p J\rangle \tag{2.8}$$

and

$$\langle j_p j_n J|v(12)|j_n j_p J\rangle = \langle j_n j_p J|v(12)|j_p j_n J\rangle \tag{2.9}$$

It follows now that the eigenvalues of (2.5) are

$$\Delta E^{(+)} = \langle j_p j_n J | v(12) | j_p j_n J \rangle + \langle j_p j_n J | v(12) | j_n j_p J \rangle \qquad (2.10)$$

and

$$\Delta E^{(-)} = \langle j_p j_n J | v(12) | j_p j_n J \rangle - \langle j_p j_n J | v(12) | j_n j_p J \rangle \qquad (2.11)$$

The corresponding normalized eigenstates are:

$$|j_p j_n J\rangle_\pm = \frac{1}{\sqrt{2}} [|j_p j_n J\rangle \pm |j_n j_p J\rangle] \qquad (2.12)$$

which, in view of (2.3) and (2.4), can be written as

$$|j_p j_n J\rangle_\pm = \frac{1}{\sqrt{2}} \sum_{m_\alpha m_\beta} (m_\alpha m_\beta | M) [\phi_\alpha(1)\phi_\beta(2) \pm \phi_\alpha(2)\phi_\beta(1)] p(1) n(2) \qquad (2.13)$$

The identification of the proton as particle 1 and the neutron as particle 2 has no physical meaning. We could have equally well taken as the wave functions of the two states the expression

$$|j_p j_n J\rangle'_\pm = \frac{1}{\sqrt{2}} \sum_{m_\alpha m_\beta} (m_\alpha m_\beta | M) [\phi_\alpha(1)\phi_\beta(2) \pm \phi_\alpha(2)\phi_\beta(1)] n(1) p(2) \qquad (2.14)$$

The two wave functions (2.13) and (2.14) are orthogonal to each other, since $\langle n(1) | p(1) \rangle = 0$. Furthermore, if the two particles p and n maintain their identities in the sense that the interaction $v(pn)$ cannot convert one into the other, then the matrix element of $v(pn)$ between the states (2.13) and (2.14) satisfies

$$_\pm\langle j_p j_n J | v(12) | j_p j_n J \rangle'_\pm = 0 \qquad (2.15)$$

We can therefore take as our wave functions, corresponding to the eigenvalues (2.10) and (2.11), any linear combinations of (2.13) and (2.14). This freedom is a reflection of the equivalence degeneracy, similar to the equivalence degeneracy in the two-electron system, and it need not concern us if the interactions in nature could not convert a proton into a neutron. However, we know that the real force between a proton and a neutron can lead to a charge exchange between them (see Section I.3). With such an interaction (2.15) will no longer hold. Since $|j_p j_n J\rangle_\pm$ and $|j_p j_n J\rangle'_+$ are degenerate in the absence of charge exchange potentials, the action of the charge exchange potential will lead to a combination of (2.13) and (2.14) that is either symmetric or antisymmetric with respect to the exchange of the charge quantum numbers (i.e., $p \to n$ and $n \to p$). We thus obtain four states that are either symmetric or antisymmetric with respect to the exchange of *all* the quantum numbers of particles 1 and 2. It is an empirical fact that nature has chosen the antisymmetric combination for the proton–neutron system. Since the two-proton system and the two-neutron systems are also antisymmetric, we see that the proton–neutron system can be considered as a special case of the general two-nucleon system (see Section I.6). We therefore adopt for the wave functions that go with $\Delta E^{(+)}$ and $\Delta E^{(-)}$ the following two linear combinations of (2.13) and (2.14):

$$|j_p j_n JM\rangle_0 = \frac{1}{2} \sum_{m_\alpha m_\beta} (m_\alpha m_\beta | M)[\phi_\alpha(1)\phi_\beta(2) + \phi_\alpha(2)\phi_\beta(1)] \times [p(1)n(2) - p(2)n(1)] \quad (2.16)$$

$$|j_p j_n JM\rangle_1 = \frac{1}{2} \sum_{m_\alpha m_\beta} (m_\alpha m_\beta | M)[\phi_\alpha(1)\phi_\beta(2) - \phi_\alpha(2)\phi_\beta(1)] \times [p(1)n(2) + p(2)n(1)] \quad (2.17)$$

It should be stressed that although we were led here to the totally antisymmetric wave function of the proton–neutron system starting from degenerate single-particle wave functions, this result is of a more general validity. The existence of an interaction that enables the neutron and the proton to exchange their charges makes the symmetry of their wave function with respect to the exchange of the two particles a meaningful quantity, and the antisymmetric wave function happens to be the one actually observed. The situation is very similar to that encountered with the electrons. As long as one neglects forces operating on the spins, it is possible to consider electrons with spin up as distinct from those with spin down, and then take a wave function of any total symmetry for a pair with one spin up, and the other down. However, as soon as one introduces spin-dependent forces between the electrons the symmetry acquires a physical meaning, and it then remains for nature to determine which symmetry to adopt. The fact that the two-electron wave function is antisymmetric also for one spin up and one down is what enables us to talk of "the electron" with its spin being considered as an intrinsic property. Similarly our results (2.16) and (2.17) enable us to talk of protons and neutrons as "nucleons" with their charge being an intrinsic property.

If we introduce now the isospin operators $\boldsymbol{\tau}(1)$ and $\boldsymbol{\tau}(2)$ the assumption of charge independence of the Hamiltonian implies that (see Section I.6)

$$[H, \mathbf{T}] = 0 \quad (2.18)$$

where

$$\mathbf{T} = \tfrac{1}{2}[\boldsymbol{\tau}(1) + \boldsymbol{\tau}(2)] \quad (2.19)$$

With the help of the total isospin vector $\mathbf{T}$ we can classify now the states in the *two-nucleon* configuration $j_\alpha j_\beta$ in the following way:

$$|j_\alpha j_\beta JMTM_T\rangle = \sum_{m_\alpha m_\beta m_{t\alpha} m_{t\beta}} (j_\alpha m_\alpha j_\beta m_\beta | JM)(\tfrac{1}{2}m_{t\alpha}\tfrac{1}{2}m_{t\beta}|TM_T) \times \Phi(m_\alpha m_{t\alpha}, m_\beta m_{t\beta}) \quad (2.20)$$

where Φ is the antisymmetric two-nucleon m-scheme function

$$\Phi(m_\alpha m_{t\alpha}, m_\beta m_{t\beta}) = \frac{1}{\sqrt{2}} \begin{vmatrix} \phi_\alpha(1) & \phi_\alpha(2) \\ \phi_\beta(1) & \phi_\beta(2) \end{vmatrix} \tag{2.21}$$

and the index α now stands for

$$\alpha = (n_\alpha l_\alpha j_\alpha m_\alpha m_{t\alpha}) \tag{2.22}$$

Remark on Notation. The operators $\boldsymbol{\tau}$, like $\boldsymbol{\sigma}$, have eigenvalues ± 1 ($\tau_x^2 = \tau_y^2 = \tau_z^2 = 1$). Because the operators τ_x, τ_y, and τ_z are 2x2 matrices, the vectors $\boldsymbol{\tau}$ add like spin-1/2 objects. It is convenient to introduce the notation $\mathbf{t} = (1/2)\boldsymbol{\tau}$ in analogy to $\mathbf{s} = (1/2)\boldsymbol{\sigma}$. It is for this reason that we take for the "z-projection" of $\mathbf{t}$ in (2.20) $m_t(= \pm 1/2)$; we shall reserve the notation $m_\tau(= \pm 1)$ for use in conjunction with $\boldsymbol{\tau}$ alone. The vector addition of two or more isospins always involves t's and not τ's. When $m_t = 1/2$ the particle is a proton, $m_t = -1/2$, a neutron.

Because of the assumed conservation of T by our Hamiltonian, Eq. 2.18, we obtain now, for two-nucleon configurations

$$\langle j_\alpha j_\beta JMTM_T | v(12) | j_\alpha j_\beta J'M'T'M_T' \rangle = \langle j_\alpha j_\beta JMTM_T | v(12) | j_\alpha j_\beta JMTM_T \rangle$$
$$\times\, \delta(J, J')\delta(M, M')\delta(T, T')\delta(M_T, M_T') \tag{2.23}$$

For a given value of the set $JMTM_T$ the two-particle state in the configuration $(j_\alpha j_\beta)$ is determined uniquely. T can assume just two values $T = 0$ and $T = 1$; for $T = 1$ and $M_T = 1$ or $M_T = -1$ we are back to the case of two protons or two neutrons, respectively, and then J and M define the state uniquely; for $T = 1$ $M_T = 0$ we have a state that is symmetric in isospin coordinates and hence antisymmetric with respect to the space-spin coordinates, that is, the state $|j_p j_n JM\rangle_1$ of (2.17); and finally for $T = 0$, $M_T = 0$ we obtain the state $|j_p j_n JM\rangle_0$ of (2.16), which is symmetric with respect to the space-spin coordinates.

We can now generalize our result (1.13), and state that for the two-nucleon configuration $(j_\alpha j_\beta)$ the first-order energy shifts caused by a charge-independent interaction $v(12)$ are given by

$$\Delta E(j_\alpha j_\beta JMTM_T) = \langle j_\alpha j_\beta JMTM_T | v(12) | j_\alpha j_\beta JMTM_T \rangle \tag{2.24}$$

where the state vectors are antisymmetric with respect to the exchange of the two nucleons. When we dealt with identical particles we showed that $\Delta E(j_\alpha j_\beta JM)$ is independent of M. This is still true in the more general case (2.24). Furthermore, since for all calculational purposes $\mathbf{T}$ acts like an angular momentum, we can conclude from the charge independence of $v(12)$, that is, from $[v(12), \mathbf{T}] = 0$, that $v(12)$ is a scalar with respect to isospin rotations, and therefore ΔE is independent also of M_T. Indeed, using the

Wigner–Eckart theorem twice, we can evaluate both the M and the M_T dependence of ΔE and obtain from (2.24)

$$\Delta E(j_\alpha j_\beta JMTM_T) = (-1)^{J-M}\begin{pmatrix} J & 0 & J \\ -M & 0 & M \end{pmatrix}(-1)^{T-M_T}\begin{pmatrix} T & 0 & T \\ -M_T & 0 & M_T \end{pmatrix}$$
$$\times\ (j_\alpha j_\beta JT\|v(12)\|j_\alpha j_\beta JT)$$
$$= \frac{1}{\sqrt{(2J+1)(2T+1)}}\,(j_\alpha j_\beta JT\|v(12)\|j_\alpha j_\beta JT) \qquad (2.25)$$

where now the double-barred element is independent of either M or M_T.

It is worthwhile to point out that although the interaction $v(12)$ is charge-independent and may not even involve the operators $\boldsymbol{\tau}_1$ and $\boldsymbol{\tau}_2$, the energy shift ΔE will in general depend on the isospin T of the state considered. This is because the value of T, for the two-particle system, determines whether the wave function is symmetric or antisymmetric in the remaining coordinates. Since $v(12)$ does depend on these coordinates, it will respond differently to the different states. The situation is similar to that encountered with electrons: In the two-electron system we find that triplets, $S = 1$, are appreciably lower than the singlets, $S = 0$, *not* because the electron–electron force is so strongly spin dependent, but because by fixing S we determine the spatial symmetry of the two-electron system, and thereby their average interaction.

In reality nuclear Hamiltonians are not charge independent. First the single-particle potential for protons is different from that for neutrons because of the presence of the average central Coulomb field. This by itself affects our results only slightly, as long as we ignore the Coulomb forces in $v(12)$ and assume that $[v(12), \mathbf{T}] = 0$. It is instructive to examine in more detail how this comes about.

If $(n_\alpha l_\alpha j_\alpha) = (n_\beta l_\beta j_\beta)$, the states $\phi_\alpha(p)\phi_\beta(n)$ and $\phi_\alpha(n)\phi_\beta(p)$ are still exactly degenerate even in the presence of a charge-dependent central potential, because in both cases we have one proton and one neutron in the same orbits. Here we denoted a proton in the state α by $\phi_\alpha(p)$, etc. Since $v(12)$ commutes with $\mathbf{T}$ we can again characterize the states by T and M_T and evaluate the eigenvalues of $v(12)$ within the configuration $j_\alpha j_\beta$ by using (2.24). Also the transition from (2.24) to (2.25) will still be valid because we assume that $v(12)$ commutes with $\mathbf{T}$. However, it will no longer be correct that $(j_\alpha{}^2JT\|v(12)\|j_\alpha{}^2JT)$ is independent of M_T; the single-particle wave function ϕ_α is an eigenfunction in a central field that now includes a Coulomb field; it therefore implicitly depends on $m_{t\alpha}$. The energy shifts $\Delta E(j_\alpha{}^2JMT = 1M_T)$ from the unperturbed zeroth energy of $j_\alpha{}^2$ may then be different for two protons in $j_\alpha(M_T = +1)$, two neutrons in this level $(M_T = -1)$ or a proton and a neutron in this level $(M_T = 0)$. The actual dependence of the empirical energy *shifts* on M_T is, nevertheless, not large for light nuclei. An example appropriate for our case is that of ${}^{18}_{8}O_{10}$, ${}^{18}_{9}F_9$, and ${}^{18}_{10}Ne_8$ where the 0^+, 2^+, and 4^+ $T = 1$ levels of the $d^2_{5/2}$ configuration have been identified. Although in ${}^{18}Ne$ the ground state and the 2^+ and 4^+ levels are due primarily to the excitation of two protons in the $d_{5/2}$ orbits and in ${}^{18}O$ they result from the excitation of two neutrons, the excitation energies differ only by roughly 5%. The energies in ${}^{18}Ne$ are slightly lower than in ${}^{18}O$, and this can be understood as follows: the centrifugal barrier for the two d-nucleons moving around ${}^{16}O$ keeps these nucleons relatively far from the center of the nucleus; at the nuclear surface the Coulomb potential is only about half as large as the centrifugal potential; its effect is therefore to push the charged protons in the d-orbit further out compared to the uncharged neutrons. The average distance between the two $d_{5/2}$ protons in ${}^{18}Ne$ is therefore larger than that of the two $d_{5/2}$ neutrons in ${}^{18}O$; the effects of the nuclear interactions are thus reduced, leading to an ex-

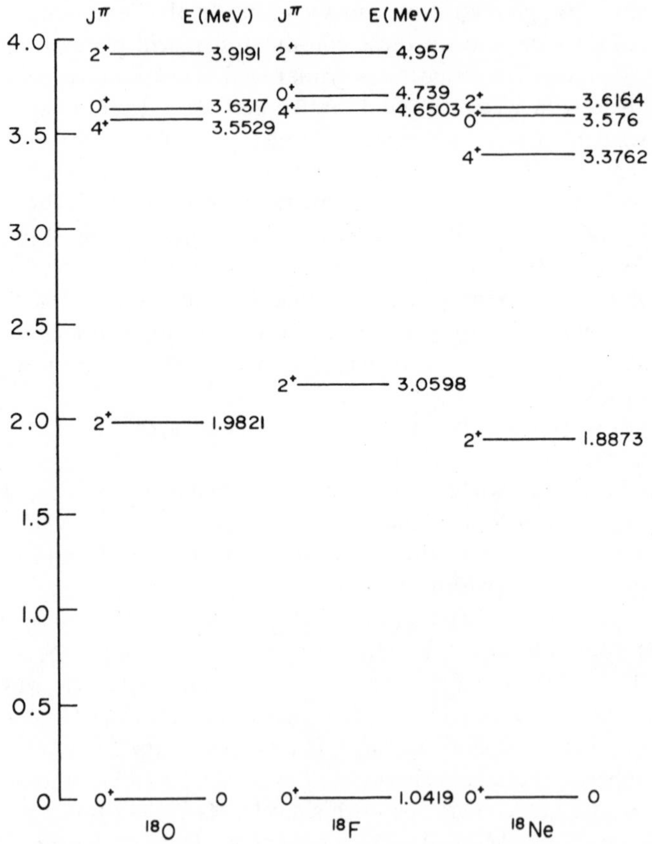

FIG. 2.1. $T = 1$ energy levels of $A = 18$ nuclei [data furnished by F. Ajzenberg-Selove (72)].

citation of 1.8873 MeV for the 2^+ state in ^{18}Ne, as compared with 1.9821 MeV for the corresponding state in ^{18}O.

The Coulomb effects can become more pronounced if the two nucleons are not equivalent, that is, if $(\alpha) \neq (\beta)$. In this case the zeroth-order states $\phi_\alpha(p)\phi_\beta(n)$ and $\phi_\alpha(n)\phi_\beta(p)$ are, strictly speaking, not even degenerate. First-order perturbation theory does not require now the diagonalization of the matrix $v(12)$ in (2.5). The strict first-order corrections are given by

$$\langle j_\alpha^{(p)} j_\beta^{(n)} JM | v(12) | j_\alpha^{(p)} j_\beta^{(n)} JM \rangle \tag{2.26a}$$

and

$$\langle j_\beta^{(p)} j_\alpha^{(n)} JM | v(12) | j_\beta^{(p)} j_\alpha^{(n)} JM \rangle \tag{2.26b}$$

Here $j_\alpha^{(p)}$ stands for a proton in the state α, etc. However, second-order corrections to the energies in (2.26a) and (2.26b) will be small only if

$$\Delta E_c = [\epsilon(j_\alpha^{(p)}) + \epsilon(j_\beta^{(n)})] - [\epsilon(j_\alpha^{(n)}) + (j_\beta^{(p)})] \gg \langle j_\alpha^{(p)} j_\beta^{(n)} JM | v(12) | j_\alpha^{(n)} j_\beta^{(p)} JM \rangle \tag{2.27}$$

In most cases condition (2.27) is not satisfied. The left-hand side is entirely due to the *difference* in Coulomb energies of the proton in the levels α and β. It is approximately given by

$$\Delta E_c \approx \int [|\phi_\alpha(\mathbf{r})|^2 - |\phi_\beta(\mathbf{r})|^2] V_c(r)\, d\mathbf{r} \tag{2.28}$$

where $V_c(r)$ is the Coulomb central potential. Inside the nucleus $V_c(r)$ does not vary much, and outside the nucleus $|\phi_\alpha(\mathbf{r})|^2 - |\phi_\beta(\mathbf{r})|^2$ is already quite small. The integral in (2.28) thus amounts nearly to the difference between two normalization integrals and is thus expected to be quite small. For $Z = 10$, ΔE_c could be $\sim$50 KeV if both ϕ_α and ϕ_β are reasonably bound states. On the other hand the matrix element on the right-hand side of (2.27), since it involves the *nuclear* interaction $v(12)$, is characteristically a couple of hundred KeV.

Thus even though the configurations $j_\alpha^{(p)} j_\beta^{(n)}$ and $j_\alpha^{(n)} j_\beta^{(p)}$ are not strictly degenerate when we take into account the Coulomb potential, in order to obtain a reliable estimate of the energies of the levels in the two-nucleon configuration $j_\alpha j_\beta$, we shall still do best to diagonalize the matrix (2.5). In fact one does not need to have strict degeneracy in order to use "degenerate perturbation theory"; it is enough that the difference between diagonal elements in the zeroth-order Hamiltonian be small compared with the off-diagonal matrix element that connects them in the perturbation, in order for a diagonalization to be preferable. The resulting normalized eigenstates will now no longer have the simple structure (2.13) and (2.14) but rather

$$|j_p j_n J\rangle_+ = a|j_p j_n J\rangle + b|j_n j_p J\rangle \tag{2.29}$$

$$|j_p j_n J\rangle_- = b|j_p j_n J\rangle - a|j_n j_p J\rangle \tag{2.30}$$

where $a^2 + b^2 = 1$. The states $|J\rangle_+$ and $|J\rangle_-$ are consequently no longer eigenstates of the isospin operator $\mathbf{T}^2$; one talks then of *isospin mixture* or *impurity* in the nuclear wave function. This impurity will be small if a $\simeq 1/\sqrt{2}$, or more precisely if $|a^2 - 1/2| \ll 1$. This, in fact, is true for most low-lying nuclear levels, which turn out to be rather close to pure eigenstates of $\mathbf{T}^2$ even for heavy nuclei, where the Coulomb field is much stronger. We shall come back to the reasons for this behavior later.

3. SPIN-INDEPENDENT HAMILTONIANS. COUPLING SCHEMES

Another important degeneracy occurs if in the Hamiltonian (1.1) the single-particle potential $U(i)$ is spin-independent. We know that for nuclei this is not really true since the presence of a spin-orbit term—$\zeta(r_i)\mathbf{l}_i \cdot \mathbf{s}_i$ in $U(i)$ has been established by many experiments. Still, our discussion of the Coulomb effects in the previous section showed that a situation in which there is a near degeneracy may require diagonalization of some submatrices of $v(12)$, so that we can expect that the use of degenerate perturbation theory will be similarly useful for relatively weak spin-orbit interactions.

In the absence of spin-dependent forces it makes no difference how the spin is coupled to the orbital angular momentum; it is then obvious that the single-particle levels $(n_\alpha l_\alpha j_\alpha = l + 1/2)$ and $(n_\alpha l_\alpha j_\alpha = l_\alpha - 1/2)$ are degenerate. To simplify notation we shall refer to the above degenerate spin doublet as composed of the levels $l_\alpha{}^+$ and $l_\alpha{}^-$. It is then clear that in the two-particle system the following four configurations are degenerate:

$$(l_\alpha^+, l_\beta^+)(l_\alpha^+, l_\beta^-)(l_\alpha^-, l_\beta^+) \text{ and } (l_\alpha^-, l_\beta^-) \tag{3.1}$$

Even if we form from each of these configuration eigenstates of J^2, J_z, in order to compute the first-order shifts in the energies of various levels, we shall still have to diagonalize, in the general case, a 4x4 matrix.

A great simplification occurs if in addition to $U(i)$, the interaction $v(12)$ is also spin-independent. In this case the whole Hamiltonian commutes with $\mathbf{S} = \mathbf{s}_1 + \mathbf{s}_2$, and like the case of the charge-independent Hamiltonian, we can diagonalize the interaction by constructing eigenstates of $\mathbf{S}^2$ from the configurations (3.1). We shall now show that this is possible.

We note that the states of any of the configurations in (3.1) can be obtained from the m-scheme states

$$\phi(n_\alpha l_\alpha m_{l\alpha} m_{s\alpha}, n_\beta l_\beta m_{l\beta} m_{s\beta}) \tag{3.2}$$

(for convenience we have dropped the m_t dependence; it does not add anything to our present discussion). Furthermore, the totality of all the independent states in the four configurations (3.1) is given by ($j = l + 1/2$, $j' = l - 1/2$):

$$(2j_\alpha + 1)(2j_\beta + 1) + (2j_\alpha + 1)(2j'_\beta + 1) + (2j'_\alpha + 1)(2j_\beta + 1) + (2j'_\alpha + 1)(2j'_\beta + 1) = [(2j_\alpha + 1) + (2j'_\alpha + 1)] \times [(2j_\beta + 1) + (2j'_\beta + 1)] = 4(2l_\alpha + 1)(2l_\beta + 1) \qquad (3.3)$$

But this is exactly the number of m-scheme states in (3.2). Since all the states in (3.1) are orthogonal to each other, and so are those of (3.2), there exists a unitary transformation which transforms from the set (3.1) to the set (3.2). The latter, however, can easily be combined into eigenstates of $\mathbf{S}^2$ by using proper Clebsch–Gordan coefficients. Furthermore, since $[\mathbf{L}, \mathbf{S}] = 0$ and $[\mathbf{J}^2, \mathbf{L}^2] = [\mathbf{J}^2, \mathbf{S}^2] = [\mathbf{J}_z, \mathbf{L}^2] = [\mathbf{J}_z, \mathbf{S}^2] = 0$, we can use the set of functions (3.2) to construct states that will be simultaneous eigenstates of the complete set of commuting observables $\mathbf{L}^2$, $\mathbf{S}^2$, $\mathbf{J}^2$, and J_z. They are given by:

$$|l_\alpha l_\beta SLJM\rangle = \sum_{M_S M_L m_{l\alpha} m_{l\beta} m_{s\alpha} m_{s\beta}} (SM_S LM_L | MJ)(\tfrac{1}{2}m_{s\alpha}\tfrac{1}{2}m_{s\beta} | SM_S) \times (l_\alpha m_{l\alpha} l_\beta m_{l\beta} | LM_L)\phi(m_{l\alpha}m_{s\alpha}, m_{l\beta}m_{s\beta}) \qquad (3.4)$$

Because of our previous remarks there exists also a unitary transformation that transforms from the set of functions (3.1), that is, $|l_\alpha j_\alpha l_\beta j_\beta JM\rangle$, to the set (3.4). This is a special case of the *change of coupling transformation* discussed in Appendix A, A.2.100 and is given by the appropriate $9 - j$ symbol through:

$$|l_\alpha l_\beta SLJM\rangle = \sum_{j_\alpha j_\beta} \sqrt{(2S+1)(2L+1)(2j_\alpha+1)(2j_\beta+1)} \begin{Bmatrix} 1/2 & 1/2 & S \\ l_\alpha & l_\beta & L \\ j_\alpha & j_\beta & J \end{Bmatrix} \times |l_\alpha j_\alpha l_\beta j_\beta JM\rangle \qquad (3.5)$$

The inverse transformation, which is again a change of coupling transformation, can be obtained, using the orthogonality properties of the $9 - j$ symbol. It reads:

$$|l_\alpha j_\alpha l_\beta j_\beta JM\rangle = \sum_{SL} \sqrt{(2j_\alpha+1)(2j_\beta+1)(2S+1)(2L+1)} \begin{Bmatrix} 1/2 & l_\alpha & j_\alpha \\ 1/2 & l_\beta & j_\beta \\ S & L & J \end{Bmatrix} \times |l_\alpha l_\beta SLJM\rangle \qquad (3.6)$$

The sum in (3.5) has four terms given by $j_\alpha = l_\alpha \pm 1/2$ and $j_\beta = l_\beta \pm 1/2$. The sum in (3.6) extends over all the allowed values of L: $|l_\alpha - l_\beta| < L < l_\alpha + l_\beta$, and over the values $S = 0, 1$ compatible with $|L - J| < S < L + J$.

If the interaction $v(12)$ is spin independent then $[\mathbf{L}, v(12)] = 0$. With the wave functions $|l_\alpha l_\beta SLJM\rangle$ we therefore find that

$$\langle l_\alpha l_\beta SLJM|v(12)|l_\alpha l_\beta S'L'J'M'\rangle = \langle l_\alpha l_\beta SLJM|v(12)|l_\alpha l_\beta SLJM\rangle \times \delta(S, S')\delta(L, L')\delta(J, J')\delta(M, M') \quad (3.7)$$

By fixing the values of SLJ and M we determine the two-particle states uniquely, even when all the four configurations in (3.1) are involved; it therefore follows from (3.7) that the set of functions (3.4) diagonalizes a spin-independent interaction $v(12)$ completely within the sub-Hilbert space of the four degenerate configurations (3.1).

The diagonalization (3.7) is valid also under slightly more general conditions. Actually what we used was not the spin-independence of $v(12)$, but rather the weaker requirement that $[\mathbf{L}, v(12)] = 0$, which is equivalent to $[\mathbf{S}, v(12)] = 0$. This requirement is satisfied also by interactions of the type $\boldsymbol{\sigma}_1 \cdot \boldsymbol{\sigma}_2 V(|\mathbf{r}_1 - \mathbf{r}_2|)$ which do depend on the spins of the particles involved.

Interactions $v(12)$ that commute with $\mathbf{S}$ necessarily commute also with $\mathbf{L} = \mathbf{l}_1 + \mathbf{l}_2$, since every interaction $v(12)$ commutes with $\mathbf{J} = \mathbf{j}_1 + \mathbf{j}_2$. They are therefore often called *central-interactions*. It is really more appropriate to call them scalar interactions, since they involve the spin operators $\boldsymbol{\sigma}_1$ and $\boldsymbol{\sigma}_2$ only in their scalar product. This will be in line with the names *vector-interaction* and *tensor-interaction* for interactions which involve the spin matrices via their vector product $\boldsymbol{\sigma}_1 \times \boldsymbol{\sigma}_2$ or their tensor combination $[\boldsymbol{\sigma}_1 \times \boldsymbol{\sigma}_2]^{(2)}$, respectively (see Section I.3).

Wave functions like those in (3.4), which are eigenstates of $\mathbf{S}^2$ and $\mathbf{L}^2$, in addition to $\mathbf{J}^2$ and $\mathbf{J}_z$, are called *LS-coupling wave functions*. The wave functions we have been considering until now, which are eigenstates of $\mathbf{j}_1^2$ and $\mathbf{j}_2^2$, in addition to $\mathbf{J}^2$ and J_z are called *jj-coupling wave functions*. Since the *LS*-coupling wave functions have well-defined values for the total orbital and total spin angular momenta separately, they are the appropriate wave functions to use whenever the dynamics of the system are such that orbital and spin angular momenta are separately conserved (at least approximately). A spin-independent single-particle potential with central two-body forces is such a case. In nature it is realized fairly well in atoms, where the dominant single-particle potential is the Coulomb potential of the nuclear charge, and the dominant two-body interaction is the Coulomb repulsion between the electrons. There is, in addition, an important spin-orbit interaction of each electron with its own orbit, but it is rather weak and in most atomic spectra it can be consistently handled as a small perturbation.

jj-coupling wave functions are more appropriate to use if the single-particle potential is strongly spin dependent; such is the case in most nuclei. However, even in this case it is sometimes convenient to write the *jj*-coupling wave functions in terms of *LS*-coupling wave functions, as we shall see later.

More precisely the situation is as follows: in the configuration $(l_\alpha l_\beta)$ (note that we specify only l, not j) there are generally four independent states for any given value of the total angular momentum J and its projection M.

Problem. Prove it.

The first-order shifts in the energies of these states due to a residual interaction $v(12)$ are obtained by the diagonalization of the matrix of $T(1) + T(2) + U(1) + U(2) + v(12)$ within the subspace of these four states (if these four states are degenerate in the zeroth order, then $T(1) + T(2) + U(1) + U(2)$ will have the same diagonal elements for all four states and we need diagonalize only $v(12)$; in general, however, this need not be the case). If we use the *jj*-coupling scheme our 4x4 matrix will have the general structure shown in Fig. 3.1. We have explicitly shown only one off-diagonal matrix element, but there

$$
\begin{array}{cccc}
\epsilon(j_\alpha) + \epsilon(j_\beta) + v(j_\alpha j_\beta) & \cdots & \cdots & \cdots \\
\cdots & \epsilon(j_\alpha) + \epsilon(j_\beta') + v(j_\alpha j_\beta') & \cdots & \langle j_\alpha j_\beta' J | v | j_\alpha' j_\beta' J \rangle \\
\cdots & \cdots & \epsilon(j_\alpha') + \epsilon(j_\beta) + v(j_\alpha' j_\beta) & \cdots \\
\cdots & \cdots & \cdots & \epsilon(j_\alpha') + \epsilon(j_\beta') + v(j_\alpha' j_\beta')
\end{array}
$$

Fig. 3.1. The 4×4 matrix of H in *jj*-coupling for the configuration $l_\alpha l_\beta$: $j_{\alpha,\beta} = l_{\alpha,\beta} + 1/2$, $j_{\alpha,\beta}' = l_{\alpha,\beta} - 1/2$. $v(j_\alpha j_\beta') = (j_\alpha j_\beta' J | v | j_\alpha j_\beta' J)$ etc. Dots stand for a matrix element that does not necessarily vanish.

will generally be a nonvanishing matrix element at each position in the matrix. The same values of (J, M), will lead to the matrix shown in Fig. 3.2 when written in terms of *LS*-coupling wave functions.

$$
\epsilon(l_\alpha) + \epsilon(l_\beta)\mathbf{1} +
\begin{array}{cccc}
V_{ls}(0, L_0; J) & \cdots & \cdots & \cdots \\
\cdots & \begin{array}{c} V_{ls}(1, L_0 - 1; J) \\ + v(1, L_0 - 1; J) \end{array} & \cdots & \cdots \\
\cdots & \cdots & \begin{array}{c} V_{ls}(1, L_0; J) \\ + v(1, L_0; J) \end{array} & \cdots \\
\begin{array}{c} \langle 1, L_0 + 1; J \mid V_{ls} \\ + v \mid 0, L_0; J \rangle \end{array} & \cdots & \cdots & \begin{array}{c} V_{ls}(1, L_0 + 1; J) \\ + v(1, L_0 + 1; J) \end{array}
\end{array}
$$

FIG. 3.2. The matrix Fig. 3.1, written in the *LS*-coupling scheme. The first quantum number is S $(=0, 1)$, the second the various values of L that for a fixed J, are $L = J, J + 1, J - 1, L_0 = J$.

Here we have explicitly noted the diagonal matrix elements that arise because of the single-particle spin-orbit interaction $V_{ls} = \zeta(r_1)\mathbf{l}_1\cdot\mathbf{s}_1 + \zeta(r_2)\mathbf{l}_2\cdot\mathbf{s}_2$, as well as those due to $v(12)$. Both generally have also off-diagonal matrix elements. This is to be contrasted with the *jj*-coupling scheme where only the residual interaction $v(12)$ contributes to the off-diagonal elements. In *jj*-coupling V_{ls} contributes only to the diagonal elements, and its contribution is included in the single-particle energy $E(j)$.

The most appropriate wave function is the one that gives the smallest value for the ratio of the off-diagonal matrix elements and the differences between the corresponding diagonal elements. Thus, in nuclei the *jj*-coupling scheme predominates, since the difference between diagonal elements in Fig. 3.1 includes the zeroth-order spin-orbit splitting for at least one of the particles. This makes differences between diagonal elements of the order of magnitude of 3 to 5 MeV (see Section I.12), whereas characteristic off-diagonal elements of $v(12)$ in medium weight and heavy nuclei are a couple of hundred KeV. In such cases, therefore, *jj*-coupling wave functions effectively diagonalize the interaction. In light nuclei, off-diagonal elements of $v(12)$ may be as large as 1 or 2 MeV. In such cases a complete diagonalization of the 4x4 matrix (Fig. 3.1 or Fig. 3.2) should be carried out for each configuration $(l_\alpha l_\beta)$. One then speaks of *intermediate coupling*. It is then often more convenient to write down the matrix of $v(12)$ using *LS*-coupling wave functions, and diagonalize it in this scheme. The results are, of course, independent of which way one chooses to write the matrix.

4. USES OF LS-COUPLING WAVE FUNCTIONS. δ-INTERACTION FOR IDENTICAL PARTICLES

Despite the fact that *jj*-coupling predominates in nuclei, *LS*-coupling wave functions offer a useful tool in some nuclear calculations. This is true if the single-particle Hamiltonian contains a strong spin-dependent force, like the spin-orbit potential in nuclei, but the residual interaction is spin-independent. To see how this comes about let us derive here some of the conclusions mentioned earlier for the δ-potential.

We assume therefore that the magnitude of the spin-orbit interaction is sufficient to justify the diagonalization of $v(12)$ within the partial Hilbert space of the *jj-coupling* configuration $(j_\alpha j_\beta)$. The first-order shifts in the energies of the levels with a given value of J is then given by (1.13).

$$\Delta E_\delta(j_\alpha j_\beta JM) = V_0\langle j_\alpha j_\beta JM| - \delta(|\mathbf{r}_1 - \mathbf{r}_2|)|j_\alpha j_\beta JM\rangle \tag{4.1}$$

Since the interaction here is spin independent, we shall find it useful to express $|j_\alpha j_\beta JM\rangle$ in terms of $|l_\alpha l_\beta SLJM\rangle$ using (3.6), and obtain

$$\Delta E_\delta(j_\alpha j_\beta JM) = V_0 \sum_{SLS'L'} (2j_\alpha + 1)(2j_\beta + 1) \times \sqrt{(2S+1)(2S'+1)(2L+1)(2L'+1)} \times \begin{Bmatrix} 1/2 & l_\alpha & j_\alpha \\ 1/2 & l_\beta & j_\beta \\ S & L & J \end{Bmatrix} \begin{Bmatrix} 1/2 & l_\alpha & j_\alpha \\ 1/2 & l_\beta & j_\beta \\ S' & L' & J \end{Bmatrix} \times \langle l_\alpha l_\beta SLJM| - \delta(|\mathbf{r}_1 - \mathbf{r}_2|)|l_\alpha l_\beta S'L'JM\rangle \quad (4.2)$$

Since $\delta(|\mathbf{r}_1 - \mathbf{r}_2|)$ is spin independent and commutes with $\mathbf{L} = \mathbf{l}_1 + \mathbf{l}_2$, we can use (3.7) and reduce (4.1) to the form

$$\Delta E_\delta(j_\alpha j_\beta JM) = V_0 \sum_{LS} (2j_\alpha + 1)(2j_\beta + 1)(2S+1)(2L+1) \times \begin{Bmatrix} 1/2 & l_\alpha & j_\alpha \\ 1/2 & l_\beta & j_\beta \\ S & L & J \end{Bmatrix}^2 \langle l_\alpha l_\beta LM_L| - \delta(|\mathbf{r}_1 - \mathbf{r}_2|)|l_\alpha l_\beta LM_L\rangle \quad (4.3)$$

where any value of $M_L(|M_L| \leq L)$ can be taken since the matrix element is independent of M_L.

Problem. Carry out the transition from (4.2) to (4.3).

Although the matrix element in (4.3) does not show an explicit dependence on S, there is a very important implicit dependence on it: since the total wave function $|j_\alpha j_\beta JM\rangle$ is assumed to be antisymmetric with respect to the exchange of particles 1 and 2, it follows that in (4.3) we have to take an *antisymmetric* spatial wave function $|l_\alpha l_\beta LM_L\rangle_a$ if $S = 1$ and a *symmetric* spatial wave function $|l_\alpha l_\beta LM_L\rangle_s$ if $S = 0$. Because of the $\delta(|\mathbf{r}_1 - \mathbf{r}_2|)$ function in the matrix element we must have $\mathbf{r}_1 = \mathbf{r}_2$, and only the latter will give a nonvanishing contribution to (4.3). Hence we conclude that in the summation in (4.3) only the $S = 0$ term will survive and we obtain:

$$\begin{aligned} \Delta E_\delta(j_\alpha j_\beta JM) &= V_0 \sum_{L} (2j_\alpha + 1)(2j_\beta + 1)(2L+1) \begin{Bmatrix} 1/2 & l_\alpha & j_\alpha \\ 1/2 & l_\beta & j_\beta \\ 0 & L & J \end{Bmatrix}^2 \\ &\quad \times \langle l_\alpha l_\beta LM_L| - \delta(|\mathbf{r}_1 - \mathbf{r}_2|)|l_\alpha l_\beta LM_L\rangle_s \\ &= \frac{(2j_\alpha + 1)(2j_\beta + 1)}{2} V_0 \begin{Bmatrix} j_\alpha & j_\beta & J \\ l_\beta & l_\alpha & 1/2 \end{Bmatrix}^2 \\ &\quad \times \langle l_\alpha l_\beta LM_L| - \delta(|\mathbf{r}_1 - \mathbf{r}_2|)|l_\alpha l_\beta LM_L\rangle_s \delta(L, J) \quad (4.4) \end{aligned}$$

where we have used Appendix A A.2.106 in reducing the $9-j$ symbol to a $6-j$ symbol.

To evaluate the matrix element in (4.4) we need the wave function only for $\mathbf{r}_1 = \mathbf{r}_2$. Let us evaluate therefore the single-variable function $\psi_{\alpha\beta}$, that is $|l_\alpha l_\beta L M_L\rangle$ with both variables $\mathbf{r}_1$ and $\mathbf{r}_2$ equated to $\mathbf{r}$

$$\psi_{\alpha\beta}(\mathbf{r}) = \frac{\sqrt{2}}{r^2} R_\alpha(r) R_\beta(r) \sum_{m_\alpha m_\beta} (l_\alpha m_\alpha l_\beta m_\beta | L M_L) Y_{l_\alpha m_\alpha}(\Omega) Y_{l_\beta m_\beta}(\Omega)$$

$$= \sqrt{2}\,\frac{(-1)^{l_\alpha - l_\beta}}{r^2} R_\alpha(r) R_\beta(r) \sqrt{\frac{(2l_\alpha+1)(2l_\beta+1)}{4\pi}} \times \begin{pmatrix} l_\alpha & l_\beta & L \\ 0 & 0 & 0 \end{pmatrix} Y_{LM_L}(\Omega) \qquad (4.5)^*$$

where Eq. IV.20.18 has been used to reduce the product over two spherical harmonics into one. Note that this became possible because of the δ-function.

We see immediately from (4.5) that because

$$\begin{pmatrix} l_\alpha & l_\beta & L \\ 0 & 0 & 0 \end{pmatrix}$$

vanishes (see Appendix A, A.2.78) when $l_\alpha + l_\beta + L$ is odd, $\psi_{\alpha\beta} = 0$ if $l_\alpha + l_\beta + L$ is odd. Formally this comes about because the parity of $Y_{lm}(\Omega)$ is $(-1)^l$, so that the product $Y_{l_\beta m_\beta}(\Omega) Y_{l_\alpha m_\alpha}(\Omega)$ can only lead to functions $Y_{LM_L}(\Omega)$ whose parity is $(-1)^{l_\alpha + l_\beta}$. Another way of looking at it is to notice that the two-particle space symmetric wave function $|l_\alpha l_\beta L M_L\rangle_s$ is symmetric separately in the radial coordinates and the angular coordinates if $l_\alpha + l_\beta + L$ is even; for odd values of $l_\alpha + l_\beta + L$ it is antisymmetric with respect to the separate exchange of the angular coordinates Ω_1 and Ω_2 or the radial coordinates r_1 and r_2. With a δ-potential the latter case leads to a vanishing $\psi_{\alpha\beta}(\mathbf{r})$ in (4.5).

Problem. Prove these statements by using the symmetry properties of the 3-j symbols.

---

*The $\sqrt{2}$ in front of (4.5) comes from the fact that in (4.4) we have to take a symmetric combination; for $\mathbf{r}_1 = \mathbf{r}_2 = \mathbf{r}$ both terms are identical, and together with the normalization factor $1/\sqrt{2}$, they yield the factor of $\sqrt{2}$.

Using the result (4.5), (4.4) can now be reduced to

$$\Delta E_\delta(j_\alpha j_\beta JM) = -V_0 \frac{(2j_\alpha + 1)(2j_\beta + 1)(2l_\alpha + 1)(2l_\beta + 1)}{4\pi}$$

$$\times \begin{pmatrix} l_\alpha & l_\beta & L \\ 0 & 0 & 0 \end{pmatrix}^2 \begin{Bmatrix} j_\alpha & j_\beta & J \\ l_\alpha & l_\beta & 1/2 \end{Bmatrix}^2 \delta(L, J) \int_0^\infty \frac{1}{r^2} R_{n_\alpha l_\alpha}(r) R_{n_\beta l_\beta}(r)\, dr \quad (4.6)$$

Finally we use the identity (see Appendix A, A.2.81)

$$(2l_\alpha + 1)(2l_\beta + 1) \begin{Bmatrix} l_\alpha & l_\beta & J \\ j_\beta & j_\alpha & 1/2 \end{Bmatrix}^2 \begin{pmatrix} l_\alpha & l_\beta & J \\ 0 & 0 & 0 \end{pmatrix}^2 = \begin{pmatrix} j_\alpha & j_\beta & J \\ 1/2 & -1/2 & 0 \end{pmatrix}^2$$

and obtain:

$$\Delta E_\delta(j_\alpha j_\beta JM) = -V_0 \frac{(2j_\alpha + 1)(2j_\beta + 1)}{4\pi} \begin{pmatrix} j_\alpha & j_\beta & J \\ 1/2 & -1/2 & 0 \end{pmatrix}^2$$

$$\times \int_0^\infty \frac{1}{r^2} R_{n_\alpha l_\alpha}(r) R_{n_\beta l_\beta}(r)\, dr \qquad \text{for} \qquad l_\alpha + l_\beta - J \qquad \text{even}$$

$$= 0 \qquad \text{for} \qquad l_\alpha + l_\beta - J \qquad \text{odd} \qquad (4.7)$$

Equation 4.7 will be recognized as identical to (1.27) with $A(j_\alpha j_\beta JM)$ given by (1.31). It constitutes thus a proof of (1.31).

The derivation of (4.7) demonstrates the usefulness of expressing the *jj*-coupling wave functions in terms of *LS*-coupling wave functions when $v(12)$ does not involve the spin coordinates. Later on we shall see other examples for which such transitions from one coupling scheme to another are of great convenience. In all these cases it is important to remember that the basic coupling scheme is determined by the dynamics of the system—like the *jj*-coupling scheme that formed the starting point for our calculations leading to (4.7). The transformations to other coupling schemes later in the calculation are just a matter of convenience.

5. NONIDENTICAL NUCLEONS IN HEAVY NUCLEI

Our considerations thus far have involved antisymmetrized wave functions for the two-nucleon system. Even when we considered a system composed of a neutron and a proton we were led to the introduction of the isospin formalism that requires that we handle these two particles as two states of the nucleon, antisymmetrizing the wave function with respect to their complete interchange.

As nuclei become heavier, the symmetry between protons and neutrons gradually diminishes. A neutron excess builds up, and around the middle of the periodic table we already find nuclei in which the protons and the neutrons are filling different major shells. The isospin formalism is of greatest value for light nuclei where the roughly equal number of protons and neutrons in nuclei of interest makes it very natural to treat them as two states of a single "elementary particle"—the nucleon. For heavier nuclei the requirement of complete antisymmetry remains strictly valid and the isospin formalism is applicable; we shall see later that its use even facilitates very much the understanding of several important nuclear properties such as the so-called "analog states" (see Section I.13). However, for many purposes we can use the fact that protons and neutrons are in different levels in the lowest states of heavy nuclei, and simplify their theoretical study by treating the proton and the neutron as distinguishable particles. An example may clarify the situation.

Consider the nucleus ${}^{38}_{17}Cl_{21}$. With 17 protons the shell-model proton configuration is $1d_{3/2}$ (or in greater detail: $1s_{1/2}^2\ 1p_{3/2}^4\ 1p_{1/2}^2\ 1d_{5/2}^6\ 2s_{1/2}^2\ 1d_{3/2}$). The neutron configuration is $1f_{7/2}$, or in slightly greater detail, $1d_{3/2}^4\ 1f_{7/2}$. Since all the levels up to the $1d_{3/2}$ level are occupied by both the protons and the neutrons, we shall not mention these 32 neutrons and proton states in our subsequent discussion and shall confine our attention just to the six remaining nucleons in the unfilled levels. Treating protons and neutrons as different particles we shall therefore write the configuration of ${}^{38}Cl$ simply as $(\pi d_{3/2}, \nu d_{3/2}^4 f_{7/2})$, where π and ν stand for protons and neutrons, respectively.

We can also treat the same nucleus assuming the neutrons and protons to be two states of the nucleon. Let us do it in some detail and compare various steps with the more "naive" approach of treating protons and neutrons as distinguishable. In the isospin formalism the configuration of ${}^{38}Cl$ is denoted by $(d_{3/2}^5, f_{7/2})$. Further, in order to make sure that we deal with ${}^{38}_{17}Cl_{21}$ we confine our consideration to states with $M_T = -2[M_T = (1/2)(Z - N)]$. The only way one can obtain this value of M_T is for the nucleon in the $f_{7/2}$ state to have $m_t = -1/2$, that is, be a neutron; in fact, if we let it have $m_t = +1/2$ the five nucleons in the $d_{3/2}$-level will have to add up to $M_T(d_{3/2}^5) = -5/2$ since $M_T = M_T(d_{3/2}^5) + m_t(f_{7/2})$. The $d_{3/2}$ level will then be occupied by five neutrons, violating the Pauli principle, there being only $2\cdot(3/2) + 1 = 4$ different states for the neutron in this level. It is therefore possible to specify our states, within the framework of the isospin formalism, even further, by stating that we are dealing with the configuration $(d_{3/2}^5(M_T = -3/2), f_{7/2}(m_t = -1/2)]$. We also notice that the states so constructed are eigenstates of the square of the total isotopic spin $\mathbf{T}^2$. Indeed, it is not possible to construct a state with $M_T = -3$ from the configuration $d_{3/2}^5 f_{7/2}$, since this will again lead to states $d_{3/2}^5(M_T = -5/2)$, which are not allowed by the Pauli principle. Thus $T = 2$ is the maximum allowed isospin for $(d_{3/2}^5 f_{7/2})$; since for a given J and M there is only one state with $M_T = -2$ in this configuration, such a state is automatically also an eigenstate of $\mathbf{T}^2$. A typical

Slater determinant in this configuration, expressed in terms of the single-particle wave functions $\phi_i(j, m, m_\tau)$ is:

$$\Phi(m, m') = \frac{1}{\sqrt{6!}}$$

$$\times \begin{vmatrix} \phi_1(3/2, -3/2, -1) & \phi_2(3/2, -3/2, -1) \ldots & \phi_6(3/2, -3/2, -1) \\ \phi_1(3/2, -1/2, -1) & \phi_2(3/2, -1/2, -1) \ldots & \cdot \\ \phi_1(3/2, +1/2, -1) & \cdot & \cdot \\ \phi_1(3/2, +3/2, -1) & \cdot & \cdot \\ \phi_1(3/2, m, +1) & \cdot & \phi_6(3/2, m, +1) \\ \phi_1(7/2, m', -1) & \cdot & \phi_6(7/2, m', -1) \end{vmatrix} \tag{5.1}$$

where the quantum numbers stand for j, m, and m_τ, respectively, and i denotes the particle number. Note that since no two particles can occupy the same state all the $d_{3/2}$ neutron states ($m_\tau = -1$) must be filled to obtain a nonvanishing Slater determinant. To compute energies in this configuration we have to take matrix elements of $\Sigma v(ik)$ between states Φ. It is immediately seen that a matrix element $\langle \Phi | \Sigma v(ik) \Phi \rangle$ taken with wave functions of the type (5.1) is reduced to a sum of diagonal matrix elements involving two-particle antisymmetrized wave functions. The latter will include terms of two types: one involves both particles in the $d_{3/2}$ level and the other involves one in the $d_{3/2}$ level, the second in the $f_{7/2}$ level. Diagonal matrix elements involving the first type correspond to the interaction of the $d_{3/2}$-neutrons ($m_\tau = -1$), in the closed-neutron shell, among themselves, or to the interaction of the $d_{3/2}$ proton ($m_\tau = +1$) with all the neutrons in the closed $d_{3/2}$ neutron shell. Such interactions, because they involve a closed shell (see discussion in Section V.7) do not contribute to the splitting between levels with different J's.

Only matrix elements of the second type will generally contribute to the splitting between states with different total angular momenta. They are typically of the form

$$\langle \phi_1(3/2, m, +1)\phi_2(7/2, m', -1) | v(12)[\phi_1(3/2, m, +1)\phi_2(7/2, m', -1) - \phi_1(7/2, m', -1)\phi_2(3/2, m, +1)] \rangle \tag{5.2}$$

The matrix element in (5.2) includes both a direct term and an exchange term. An important feature of this exchange term is that it involves not only the transformation of a proton into a neutron, but also its simultaneous transfer from a $d_{3/2}$ to an $f_{7/2}$ state.

To bring out the significance of this point better let us consider a two-nucleon state $d_{3/2}f_{7/2}$ with, say, $T = 0$. Such a state can occur as a negative parity excited state in ${}^{34}_{17}Cl_{17}$. A typical m-scheme wave function in this case will be

$$\Phi(mm') = \frac{1}{2}\left[\begin{vmatrix} \phi_1(3/2, m, +1) & \phi_2(3/2, m, +1) \\ \phi_1(7/2, m', -1) & \phi_2(7/2, m', -1) \end{vmatrix} - \begin{vmatrix} \phi_1(3/2, m, -1) & \phi_2(3/2, m, -1) \\ \phi_1(7/2, m', +1) & \phi_2(7/2, m', +1) \end{vmatrix}\right] \quad (5.3)$$

A difference of two determinants with different m_τ's was taken in (5.3) to assure that we are dealing here, like in (5.2) with an eigenstate of $\mathbf{T}^2 (T = 0)$ so that the comparison with (5.2) will be meaningful. Taking now matrix elements of $v(12)$ with $\Phi(mm')$ of (5.3), a typical element will be

$$\begin{aligned}\langle\phi_1(3/2, m, +1)\phi_2(7/2, m', -1)|\,v(12)\,[&\phi_1(3/2, m, +1)\phi_2(7/2, m', -1) \\ - \phi_1(7/2, m', -1)\phi_2(3/2, m, +1) &- \phi_1(3/2, m, -1)\phi_2(7/2, m', +1) \\ &+ \phi_1(7/2, m', +1)\phi_2(3/2, m, -1)]\rangle \quad (5.4)\end{aligned}$$

The first two terms yield a direct and an exchange term similar to the one in (5.2) but important additional terms appear as well—ones in which the *orbits* of particles 1 and 2 are unchanged and only their roles as protons and neutrons are interchanged. An exchange integral in which the particles change their orbits is usually small due to cancellations from positive and negative parts of the wave functions. On the other hand an exchange integral in which the particles just change their intrinsic charge but remain in the same orbits can be as large as a direct integral, if, of course, the interaction $v(12)$ provides for the charge exchange. Thus the exchange terms for the $(d_{3/2}f_{7/2})$ configuration are significantly different from, and larger than, those of the $(d^5_{3/2}f_{7/2})$ configuration.

The difference between the ${}^{38}_{17}Cl_{21}$ configuration $(d^5_{3/2}f_{7/2})$ and that of the negative parity excited states in ${}^{34}_{17}Cl_{17}$—coming from the configuration $(d_{3/2}\cdot f_{7/2})$—is easily understood also intuitively. In the latter case both the $d_{3/2}$ and the $f_{7/2}$ levels are occupied by just one nucleon so that exchanges of all types can take place as a result of the two-nucleon interaction, as shown in Fig. 5.1. In ${}^{38}_{17}Cl_{21}$, however, all the $d_{3/2}$-neutron states are occupied and only one type of exchange remains possible: that in which the two particles exchange both their charge and their orbits simultaneously as shown in Fig. 5.2.

In many calculations it is possible to neglect an exchange term like the one in ${}^{38}Cl$ in comparison with the direct term. This is especially true in heavier nuclei where the radial functions of the states involved may have a number of nodes. The radial integrals then have positive and negative contributions cancelling each other to a large extent and the exchange integrals that involve

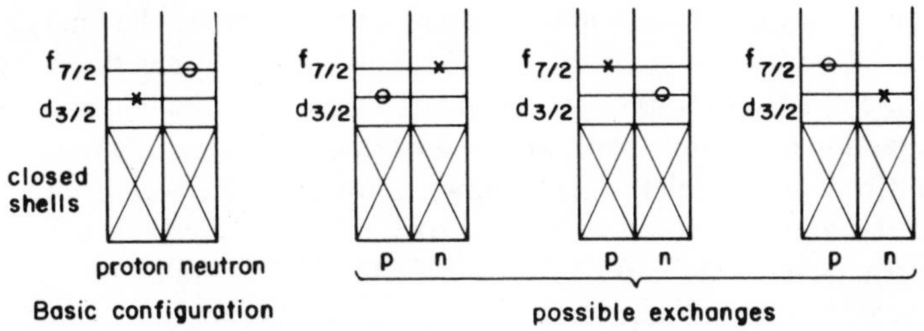

FIG. 5.1. The various ways for distributing two nucleons in the configuration $d_{3/2}\, f_{7/2}$ of $^{34}Cl_{17}$ over the two levels $d_{3/2}$ and $f_{7/2}$. x is nucleon no. 1 and 0 is nucleon no. 2.

two different orbits are characteristically less than 10% of the direct integrals involving the same orbits. If one makes the approximation of neglecting exchange integrals involving different orbits, then in cases like that of $^{38}_{17}Cl_{21}$ one is left just with the direct integrals, as is seen in (5.4) and recalling that when the charge is changed the orbits must also change. This then amounts to treating the neutron and the proton as distinct particles and not as two states of the nucleon, disallowing charge exchange between them. As we see, in order for this approximation to be valid the protons and the neutrons, whose interaction is considered, must be in different levels. But this in itself is not sufficient; in addition the neutron states corresponding to the proton's level must be completely filled so that the possibility of an exchange of a proton into a neutron without change of orbit will be blocked by the Pauli principle. (The same effect will show up if the proton states corresponding to the neutrons' level were all filled; in most actual cases, however, there are more neutrons than protons so that this is not a common situation.)

6. THE δ-INTERACTION FOR NONIDENTICAL NUCLEONS

The possibility of handling protons and neutrons in unfilled levels in heavy nuclei as distinct particles simplifies to some extent the theoretical analysis of

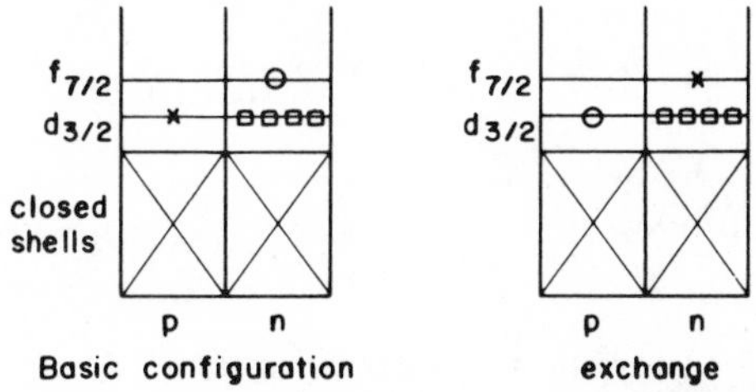

FIG. 5.2. In $^{38}_{17}Cl_{21}$, the states with $M_T = -2$ of the configuration $(d^5_{3/2} f_{7/2})$ must have four neutrons (□) filling the $d_{3/2}$ level. Only one possible exchange exists here.

such configurations. To see how it works let us calculate the energy shifts in a configuration (j_p, j_n) again taking a delta potential between the two particles. In the case of two protons or two neutrons, the Pauli principle assured us that a δ-interaction could take place only if the two spins of the interacting particles were antiparallel (singlet state). Now, however, since we propose to treat the neutron and the proton as distinct particles, there could be a δ-interaction both in the singlet $(S = 0)$ and the triplet $(S = 1)$ states, and these interactions can be of different strengths. We are therefore led to introduce the interaction

$$v(12) = \left[\frac{3 + \boldsymbol{\sigma}_1 \cdot \boldsymbol{\sigma}_2}{4} V_t + \frac{1 - \boldsymbol{\sigma}_1 \cdot \boldsymbol{\sigma}_2}{4} V_s\right] \delta(|\mathbf{r}_1 - \mathbf{r}_2|) \tag{6.1}$$

The operators $1/4(3 + \boldsymbol{\sigma}_1 \cdot \boldsymbol{\sigma}_2)$ and $1/4(1 - \boldsymbol{\sigma}_1 \cdot \boldsymbol{\sigma}_2)$ are projection operators on the triplet and singlet spin states, respectively, as can be easily verified. V_t and V_s measure the relative strengths of the two interactions. If we try to derive an estimate for V_t and V_s from the simplest $p - n$ system, we find that both V_t and V_s are attractive, and that V_t is more attractive than V_s. This reflects itself in the ground state of the deuteron, which is a ^3S state. The analysis of many proposed interactions shows that most of them are characterized by having [see Elliott and Lane (57)]

$$V_s \simeq 0.6 V_t \tag{6.2}$$

The evaluation of the energy shifts of the different states of the configuration (j_p, j_n) now proceeds in a straightforward way. We again work in the J-scheme, so that all we need calculate are expectation values of $v(12)$:

$$\Delta E(j_p j_n J) = \langle j_p j_n JM | v(12) | j_p j_n JM \rangle \tag{6.3}$$

We should remember however that now the states $|j_p j_n J\rangle$ are not antisymmetrized, that is,

$$|j_p j_n JM\rangle = \sum (j_p m_p j_n m_n | JM) \phi_p(j_p m_p) \phi_n(j_n m_n) \tag{6.4}$$

As in the derivation of (4.3) we find it useful to express the jj-coupling wave functions (6.4) in terms of LS-coupling wave functions since the projection operators in (6.1) can be very simply evaluated in this scheme. In fact we have

$$\begin{aligned} \left\langle s_1 s_2 S M_s \left| \frac{3 + \boldsymbol{\sigma}_1 \cdot \boldsymbol{\sigma}_2}{4} \right| s_1 s_2 S' M_s' \right\rangle &= \delta(S, 1) \delta(S, S') \delta(M_s, M_s') \\ \left\langle s_1 s_2 S M_s \left| \frac{1 - \boldsymbol{\sigma}_1 \cdot \boldsymbol{\sigma}_2}{4} \right| s_1 s_2 S' M_s' \right\rangle &= \delta(S, 0) \delta(S, S') \delta(M_s, M_s') \end{aligned} \tag{6.5}$$

We now use (3.6) for the transformation from the jj-coupling to the LS-coupling wave functions, and introduce (6.1) for $v(12)$ into (6.3). We notice that the change of coupling transformation (3.6) is independent of the magnetic quantum numbers, and obtain:

$$\Delta E(j_p j_n J)$$

$$= \langle j_p j_n JM| \left(\frac{3 + \boldsymbol{\sigma}_1\cdot\boldsymbol{\sigma}_2}{4} V_t + \frac{1 - \boldsymbol{\sigma}_1\cdot\boldsymbol{\sigma}_2}{4} V_s\right) \delta(|\mathbf{r}_1 - \mathbf{r}_2|)|j_p j_n JM\rangle$$

$$= \sum_L 3(2j_p + 1)(2j_n + 1)(2L + 1)$$

$$\times \begin{Bmatrix} 1/2 & l_p & j_p \\ 1/2 & l_n & j_n \\ 1 & L & J \end{Bmatrix}^2 V_t \langle l_p l_n LM_L|\delta(|\mathbf{r}_1 - \mathbf{r}_2|)|l_p l_n LM_L\rangle$$

$$+ (2j_p + 1)(2j_n + 1)(2L + 1) \begin{Bmatrix} 1/2 & l_p & j_p \\ 1/2 & l_n & j_n \\ 0 & L & J \end{Bmatrix}^2 V_s$$

$$\times \langle l_p l_n LM_L|\delta(|\mathbf{r}_1 - \mathbf{r}_2|)|l_p l_n LM_L\rangle \quad (6.6)$$

where (6.5) was used to carry out the S-summation. We can now use (4.5) for the evaluation of the matrix elements of the δ-interaction in (6.6). We need only remember that since the states in the matrix elements of (6.6) are not antisymmetric, the factor $\sqrt{2}$ in (4.5) will not appear here. (See remark leading to Eq. 4.5). Hence:

$$\langle l_p l_n LM_L|\delta(|\mathbf{r}_1 - \mathbf{r}_2|)|l_p l_n LM_L\rangle = \frac{(2l_p + 1)(2l_n + 1)}{4\pi} \begin{pmatrix} l_p & l_n & L \\ 0 & 0 & 0 \end{pmatrix}^2$$

$$\cdot \int_0^\infty \frac{1}{r^2} R^2_{n_p l_p}(r) R^2_{n_n l_n}(r)\, dr \quad (6.7)$$

We now use the identity (deShalit and Talmi (63) p. 517)

$$\begin{pmatrix} l_p & l_n & L \\ 0 & 0 & 0 \end{pmatrix} \begin{Bmatrix} 1/2 & l_p & j_p \\ 1/2 & l_n & j_n \\ S & L & J \end{Bmatrix} = \sum_{m_i} \begin{pmatrix} j_p & 1/2 & l_p \\ m_1 & m_2 & 0 \end{pmatrix} \begin{pmatrix} j_n & 1/2 & l_n \\ m_3 & m_4 & 0 \end{pmatrix}$$

$$\times \begin{pmatrix} J & S & L \\ m_5 & m_6 & 0 \end{pmatrix} \begin{pmatrix} 1/2 & 1/2 & S \\ m_2 & m_4 & m_6 \end{pmatrix} \begin{pmatrix} j_p & j_n & J \\ m_1 & m_3 & m_5 \end{pmatrix} \quad (6.8)$$

to obtain

$$\sum_L (2L+1)\begin{pmatrix} l_p & l_n & L \\ 0 & 0 & 0 \end{pmatrix}^2 \begin{Bmatrix} 1/2 & l_p & j_p \\ 1/2 & l_n & j_n \\ S & L & J \end{Bmatrix}^2 = \sum_{m_i m_i'} \begin{pmatrix} j_p & 1/2 & l_p \\ m_1 & m_2 & 0 \end{pmatrix}$$

$$\times \begin{pmatrix} j_p & 1/2 & l_p \\ m_1' & m_2' & 0 \end{pmatrix}\begin{pmatrix} j_n & 1/2 & l_n \\ m_3 & m_4 & 0 \end{pmatrix}\begin{pmatrix} j_n & 1/2 & l_n \\ m_3' & m_4' & 0 \end{pmatrix}\begin{pmatrix} 1/2 & 1/2 & S \\ m_2 & m_4 & m_6 \end{pmatrix}$$

$$\times \begin{pmatrix} 1/2 & 1/2 & S \\ m_2' & m_4' & m_6 \end{pmatrix}\begin{pmatrix} j_p & j_n & J \\ m_1 & m_3 & m_5 \end{pmatrix}\begin{pmatrix} j_p & j_n & J \\ m_1' & m_3' & m_5 \end{pmatrix} \quad (6.9)$$

We notice that $m_1 \ldots m_4$, $m_1' \ldots m_4'$ can take on only the values $\pm 1/2$. We also recall that (see Appendix A, A.2.82):

$$\begin{pmatrix} j & 1/2 & l \\ 1/2 & -1/2 & 0 \end{pmatrix} = (-1)^{j+l+1/2}\begin{pmatrix} j & 1/2 & l \\ -1/2 & 1/2 & 0 \end{pmatrix} = \frac{(-1)^{j-1/2}}{\sqrt{2(2l+1)}} \quad (6.10)$$

For $S = 0$ one then obtains from (6.9):

$$\sum (2L+1)\begin{pmatrix} l_p & l_n & L \\ 0 & 0 & 0 \end{pmatrix}^2 \begin{Bmatrix} 1/2 & l_p & j_p \\ 1/2 & l_n & j_n \\ 0 & L & J \end{Bmatrix}^2$$

$$= \frac{[1 + (-1)^{l_p + l_n + J}]}{4(2l_p+1)(2l_n+1)}\begin{pmatrix} j_p & j_n & J \\ 1/2 & -1/2 & 0 \end{pmatrix}^2 \quad (6.11)$$

For $S = 1$ the explicit summation of (6.9) is slightly more complex, and we shall not present it here [See deShalit (53)].

Problem. Carry out the summation over m_i and m_i' in (6.9) for $S = 1$ using the identity

$$\begin{pmatrix} j_p & j_n & J \\ 1/2 & 1/2 & -1 \end{pmatrix}^2 = \begin{pmatrix} j_p & j_n & J \\ 1/2 & -1/2 & 0 \end{pmatrix}^2$$

$$\times \frac{[(2j_p+1) + (-1)^{j_p+j_n+J}(2j_n+1)]^2}{4J(J+1)}$$

The final result takes the form:

$$\Delta E(j_p j_n J) = (2j_p + 1)(2j_n + 1)\frac{1}{8\pi}\int_0^\infty \frac{1}{r^2} R^2_{n_p l_p}(r) R^2_{n_n l_n}(r)\, dr \cdot F(J) \quad (6.12)$$

where

$$F(J) = \begin{pmatrix} j_p & j_n & J \\ 1/2 & -1/2 & 0 \end{pmatrix}^2 \Bigg\{ V_s \left[\frac{1 + (-1)^{l_p+l_n+J}}{2}\right]$$

$$+ V_t \left[\frac{1 - (-1)^{l_p+l_n+J}}{2}\right]\left[\frac{[(2j_p + 1) + (-1)^{j_p+j_n+J}(2j_n + 1)]^2}{4J(J + 1)}\right]\Bigg\} \quad (6.13)$$

Although (6.13) looks somewhat cumbersome, it is really rather simple in its structure. We notice first that $\Delta E(j_p j_n J)$ in (6.12) breaks up into a product of two factors: one that is independent of J and involves the integral of the radial wave functions; the other includes the whole dependence on J (see also Eq. 1.27). The relative spacing between the levels of the configuration (j_p, j_n), with a δ-interaction acting between the neutron and the proton, is therefore independent of the shape of the central potential and depends only on the values of l_p, j_p, l_n, j_n, and J. It depends, of course, also on the relative importance of the triplet and the singlet interactions. These results are true, of course, only to the extent that the interaction $v(12)$ can be considered as a perturbation. This, however, is a basic assumption that underlies all shell-model calculations.

Figure 6.1 shows the energy levels of the proton–neutron configuration $(3/2^+, 7/2^-)$ computed with the aid of (6.12) with $V_t/V_s = 2$ and $V_t < 0$. For comparison we also present the spectrum of the configuration $(3/2^-, 7/2^-)$. Both have been normalized to the same width of the configuration (i.e., the same distance between lowest and highest states in the configuration). We have also plotted the observed lowest energy levels of ^{38}Cl with their spin assignment. The spins of the ground states of $^{35}_{17}Cl_{18}$ and $^{37}_{17}Cl_{20}$ with 17 protons and an even number of neutrons are measured to be 3/2; that of $^{41}_{20}Ca_{21}$ with 21 neutrons and a closed proton shell is measured to be 7/2. Since ^{38}Cl has 17 protons and 21 neutrons we can expect the lowest states of this nucleus to belong to the configuration (3/2, 7/2). The shell model further tells us that the 3/2 state of the 17th proton is a $d_{3/2}$, and the 7/2 state of the 21st neutron is a $f_{7/2}$. Although the δ-interaction is at most a crude approximation to the real proton–neutron interaction we see from Fig. 6.1 that the spectrum calculated with it for the $(d_{3/2}, f_{7/2})$ configuration is indeed very close to that observed in $^{38}_{17}Cl_{21}$. The same interaction in the configuration $(p_{3/2}, f_{7/2})$ gives rise to a markedly different spectrum. Thus even the crude δ-force calculations

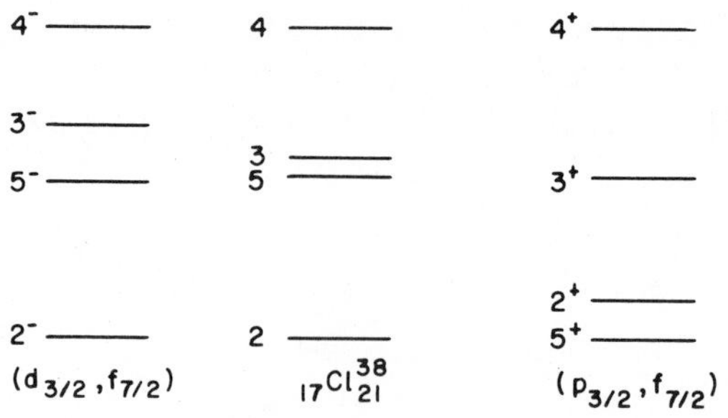

FIG. 6.1. Energy levels of a proton–neutron configuration.

can help determine whether the (3/2, 7/2) configuration of ^{38}Cl is ($3/2^+$, $7/2^-$) or ($3/2^-$, $7/2^-$).

Later on we shall see why a δ-interaction can be expected to simulate rather well the real proton–neutron interaction in nuclei. For the moment we shall just point out that if the real interaction has a short range, then the δ-potential is its natural limit. Calculations of nuclear spectra with δ-potential therefore may provide a first orientation for the assignment of nuclear excited states to their proper configurations.

In order for (6.12) to reproduce also the magnitude of the observed splitting between 2^- and 4^- levels in ^{38}Cl we have to take for V_t a value determined by:

$$V_t \int_0^\infty \frac{1}{r^2} R_{1d}{}^2(r) R_{1f}^2(r)\, dr = -11 \text{ MeV} \tag{6.14}$$

This value is in good agreement with values obtained from the study of other nuclei [see, for instance, Moinester, Schiffer, and Alford, (69)]. It also agrees fairly well with estimates of (6.14) based on the free nucleon–nucleon interaction. This is not a trivial point in view of the fact that the shell model takes for $v(12)$ only a residual interaction. We shall come to this question later (see below Section V.14).

Equation 6.12 was derived assuming just a two-particle configuration. There are very few nuclei, like $^{38}_{17}\mathrm{Cl}_{21}$, where we have just one proton and one neutron outside relatively well-established closed shells, and for which (6.12) can be expected to be valid. It is therefore hard to test its general validity, and to make further progress we shall have to generalize (6.12) to configurations involving several protons and several neutrons.

Among the few nuclei for which (6.12) can be tested, there is a particularly simple group—that in which either $j_p = 1/2$ or $j_n = 1/2$. Observing that the $3 - j$ symbol in (6.13) is then given by (6.10) it is a simple matter to see, recalling that V_s and V_t are negative and $|V_t| > |V_s|$, that if, say, $j_p = 1/2$, then of the two possible states of the configuration $(1/2, j_n)$, the lowest one is always the one that satisfies $(-1)^{l_p+l_n+J} = -1$. As an example we consider the nucleus $^{206}_{81}Tl_{125}$. The proton in the last unfilled level is known to be in the $3s_{1/2}$ level, while the neutron is in the $3p_{1/2}$ level. Actually both these levels can accommodate just two particle each, and the ground state of ^{206}Tl is indeed found to be $J = 0^-$. However, there is doublet in ^{206}Tl with a level $J = 2^-$ at an energy of 262 KeV and another with $J = 1^-$ at 301 KeV. (Fig. 6.2) This doublet can result from the coupling of the $3s_{1/2}$ proton to the $(3p_{3/2})^3$ neutron configuration [we shall see later that the configuration $(s_{1/2}, p_{3/2}{}^3)$ gives rise to a structure identical with that of $(s_{1/2}\, p_{3/2})$]. Again, the parity of these states is negative, so that the lowest member of the doublet must have an even J to satisfy $(-1)^{0+1+J} = -1$.

It is worthwhile to notice that the situation described here with proton–neutron configurations is, in a way, the opposite to that of two identical nucleons. There we found (see Section V.1) that configurations of odd parity had for their lowest state an *odd-value* of J. This difference can be easily traced to the predominance of the triplet state in the $p - n$ case (note that $|V_t| > |V_s|$), as against the predominance of the singlet state for short range inter-

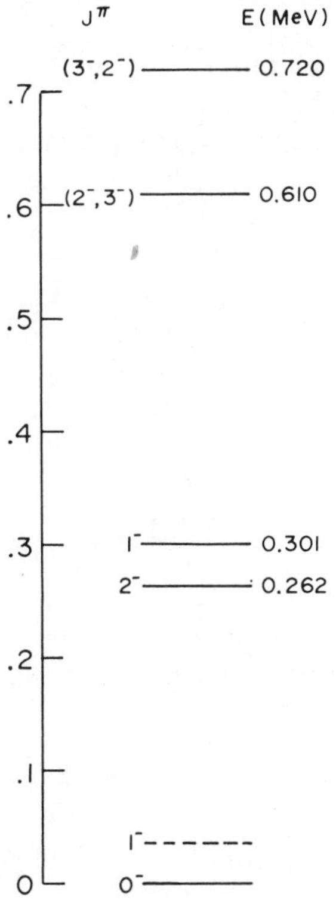

FIG. 6.2. Energy levels of ^{206}Tl, taken from Nuclear Data Sheet NRC-61-4-110.

actions of identical nucleons. In fact the total angular momentum of the lowest level is such as to lead to "parallel" intrinsic spins in a proton–neutron system and to "antiparallel" intrinsic spins in a proton–proton or a neutron–neutron system.

Although the study of two-particle configurations can be very interesting and suggestive, the real test of the shell model, as well as the understanding of its limitations and breakdowns, comes from the study of several-particle configurations. We shall proceed to do that now.

7. THE ROLE OF CLOSED SHELLS

Before embarking on the analysis of more-particle configurations it is advisable to clarify the role played by closed shells in various shell-model calculations. We have often noted in the past that in the calculation of relative spacing of energy levels the closed-shell contribution could be disregarded. This is also the case in atomic and molecular studies where one usually confines one's attention to the so-called valence electrons: those electrons that occupy unfilled levels.

It will be sufficient for our purpose to consider two levels—j filled with $n = (2j + 1)$ nucleons, and j' having $1 \leq n' < 2j' + 1$ nucleons. We are interested in the general question of relating a matrix element in the $n + n'$ particle configuration $j^{2j+1}j'^{n'}$ to a corresponding matrix element in the n' particle configuration $j'^{n'}$.

First let us establish a correspondence between the states of the two configurations. A state of $2j + 1$ fermions in the j-level is uniquely defined and has definite quantum numbers: its total angular momentum vanishes (see Eq. IV.9.6) and its parity is positive ($2j + 1$ is an even number, so that even if the parity of each one of the states in the level j is negative, that of the filled level is always positive). If we consider now a state of the $j'^{n'}$ configuration characterized by the quantum numbers $(\alpha J^\pi M)$, we can add to it the $2j + 1$ particles in the j-level and still characterize the state of the $n + n'$ by the same quantum numbers $(\alpha J^\pi M)$. This holds true even if we require that the resulting wave function be antisymmetrized with respect to the exchange of particles between the j and the j' levels. To prove this suppose that the state $|j'^{n'}\alpha J^\pi M\rangle$ is obtained from the m-scheme Slater determinants of this configuration by the unitary transformation $\langle \alpha J^\pi M | m_1' \ldots m_n{}' \rangle$:

$$|j'^{n'}\alpha J^\pi M\rangle = \sum_{m'} \langle \alpha J^\pi M | m_1', m_2', \ldots, m_{n'}' \rangle \Phi'(m_1', m_2', \ldots, m_{n'}') \tag{7.1}$$

The m-scheme states of the configuration $j^n j'^{n'}$ are given by

$$\Phi(m_1, m_2, \ldots, m_n, m_1', m_2', \ldots, m_{n'}') \tag{7.2}$$

Since the j-level is assumed to be completely filled we can specify the set $m_1, \ldots, m_n$ explicitly and write (7.2) in the form

$$\Phi(-j, -j+1, \ldots, j-1, j, m_1', m_2', \ldots, m_{n'}') \tag{7.3}$$

The number of independent m-scheme wave functions is therefore identical for both the $(j^{2j+1}j'^{n'})$ and the $j'^{n'}$ configurations and there is a one to one correspondence between them. We can therefore label also the m-scheme states of the configuration $j^n j'^{n'}$, using the set $m_1', m_2', \ldots, m_{n'}'$ alone. Using now the unitary transformation (7.1) on these m-scheme states of $j^n j'^{n'}$ we obtain a labeling of the states of the configuration $j^{2j+1}j'^{n'}$ which is identical to that of $j'^{n'}$. We have thus established a one-to-one correspondence between the two sets of states, and can proceed to compare their energies.

In order to see the contribution of closed shells to matrix elements, let us consider first a single-particle operator $\Omega = \Sigma\Omega(i)$. Let us further assume that $\Omega(i)$ is an irreducible tensor operator of degree k (see Appendix A, A.2.42), since any single-particle operator can be written as a sum of such operators. Thus we consider

$$\Omega_\kappa^{(k)} = \sum \Omega_\kappa^{(k)}(i) \tag{7.4}$$

Matrix elements of $\Omega_\kappa^{(k)}$ in the J-scheme can be obtained from those of the m-scheme by using the transformation (7.1). It is sufficient therefore to consider matrix elements of $\Omega^{(k)}$ in the m-scheme. Using the fact that $\Omega^{(k)}$ is a single-body operator, and the antisymmetry of the wave functions we have:

$$\begin{aligned}
&\langle m_1, \ldots, m_n, m_1', \ldots, m_{n'}' | \sum \Omega_\kappa^{(k)}(i) | \overline{m}_1, \ldots, \overline{m}_n, \overline{m}_1', \ldots, \overline{m}_{n'}' \rangle \\
&= (n+n')\langle m_1, \ldots, m_n, m_1', \ldots, m_{n'}' | \Omega_\kappa^{(k)}(1) | \overline{m}_1, \ldots, \overline{m}_n, \overline{m}_1', \ldots, \overline{m}_{n'}' \rangle \\
&= (n+n')\langle -j, -j+1, \ldots, j, m_1', \ldots, m_{n'}' | \Omega_\kappa^{(k)}(1) | -j, -j+1, \ldots, \\
&\qquad\qquad j, \overline{m}_1', \ldots, \overline{m}_{n'}' \rangle \\
&= \sum_{m=-j}^{j} \langle m | \Omega_\kappa^{(k)} | m \rangle \delta(m', \overline{m}') + \sum_{i=1}^{n'} \langle m_i' | \Omega_\kappa^{(k)} | \overline{m}_i' \rangle \delta^{(i)}(m', \overline{m}')
\end{aligned} \tag{7.5}$$

where the last step is obtained by expanding the Slater determinant along the row containing the single-particle wave functions of particle 1 and taking into account the orthogonality of single-particle wave functions with different quantum numbers. $\delta(m', \overline{m}')$ is a shorthand notation for

$$\delta(m', \overline{m}') = \prod_{k=1}^{n'} \delta(m_k', \overline{m}_k') \tag{7.6}$$

Similarly the symbol $\delta^{(i)}(m_k', \overline{m}_k')$ stands for the product of $n'-1$ δ-functions:

$$\delta^{(i)}(m', \overline{m}') = \prod_{k \neq i} \delta(m_k', \overline{m}_k') \qquad k = 1, \ldots, n' \tag{7.7}$$

Using the Wigner–Eckart theorem we can evaluate the first sum in (7.5):

$$\sum_{m=-j}^{j} \langle m|\Omega_{\kappa}^{(k)}|m\rangle = \sum_{m=-j}^{j} (-1)^{j-m}\begin{pmatrix} j & k & j \\ -m & \kappa & m \end{pmatrix}\langle j||\Omega^{(k)}||j\rangle$$

$$= \langle j||\Omega^{(k)}||j\rangle \sum_{m=-j}^{j} \sqrt{2j+1}\begin{pmatrix} j & 0 & j \\ -m & 0 & m \end{pmatrix}\begin{pmatrix} j & k & j \\ -m & \kappa & m \end{pmatrix}$$

$$= \sqrt{2j+1}\langle j||\Omega^{(k)}||j\rangle\delta(k,0)\delta(\kappa,0) \tag{7.8}$$

where we have used the relation

$$(-)^{j-m} = \sqrt{2j+1}\begin{pmatrix} j & 0 & j \\ -m & 0 & m \end{pmatrix}$$

There is therefore no contribution from closed shells to matrix elements of irreducible tensor operators except for scalars. In the latter case we have

$$\langle j||\Omega^{(0)}||j\rangle = \langle\Omega^{(0)}\rangle\langle j||1||j\rangle = \sqrt{2j+1}\ \langle\Omega^{(0)}\rangle \tag{7.9}$$

where $\langle\Omega^{(0)}\rangle$ is a number giving the expectation value of $\Omega^{(0)}$ in any of the j-states. Combining (7.9) with (7.8) we have the trivial result:

$$\sum_{m=-j}^{j} \langle m|\Omega^{(0)}|m\rangle = (2j+1)\langle\Omega^{(0)}\rangle \tag{7.10}$$

For nonscalar operators, the only terms that remain in (7.5) are those due to the particles in the unfilled level. We therefore conclude that for $k \neq 0$.

$$\langle j_0^{2j_0+1} j_1^{n_1}\alpha JM|\sum_{i=1}^{2j_0+1+n_1} \Omega_{\kappa}^{(k)}(i)|j_0^{2j_0+1} j_1^{n_1}\alpha' J' M'\rangle$$

$$= \langle j_1^{n_1}\alpha JM|\sum_{i=1} \Omega_{\kappa}^{(k)}(i)|j_1^{n_1}, \alpha' J' M'\rangle \qquad k \neq 0 \tag{7.11}$$

where a slight change in notation was introduced for convenience. Closed shells can therefore be ignored whenever one calculates matrix elements of a nonscalar, single-particle, irreducible tensor operator. The total angular momentum of a system or its magnetic moment are special examples that we have already discussed before. Other examples, which will be discussed later, are the operators of various multipole radiations, the β-decay operators, operators involved in some specific reactions, etc. In all these cases we can confine our attention to the valence nucleons only, ignoring the nucleons filling levels that remain fully occupied during the whole process.

The effects of closed shells on matrix elements of two-body operators is slightly more complicated. We shall treat here just two-body interactions

$v(12)$, that is, *scalar* two-body operators. More complex operators like, say, $\sum \mathbf{r}_i \times \mathbf{r}_j$ can be handled in a similar way.

We shall show now that the effect of a filled j-level on the matrix elements $v(ik)$ is such that the *spacings* between levels in the configuration $j^{2j+1}j'^{n'}$ are the same as those in the configuration $j'^{n'}$.

Taking matrix elements in the m-scheme, we find that

$$\left(m_1, \ldots, m_n, m_1', \ldots, m_{n'}' \left| \frac{1}{2} \sum_{i \neq k, 1}^{n+n'} v(ik) \right| \overline{m}_1, \ldots, \overline{m}_n, \overline{m}_1', \ldots, \overline{m}_{n'}' \right)$$

$$= \frac{(n+n')(n+n'-1)}{2}$$

$$\times \langle m_1, \ldots, m_n, m_1', \ldots, m_{n'}' | v(12) | \overline{m}_1, \ldots, \overline{m}_n, \overline{m}_1', \ldots, \overline{m}_{n'}' \rangle$$

$$= \sum_{i<k, 1}^{n} \langle m_i m_k | v(12) | m_i \overline{m}_k \rangle \delta(m' \overline{m}')$$

$$+ \sum_{i=1,\ldots,n, k=1,\ldots,n'} \langle m_i m_k' | v(12) | m_i \overline{m}_k' \rangle \delta^{(k)}(m', \overline{m}')$$

$$+ \sum_{m'_i < m'_k}^{n'} \langle m_i' m_k' | v(12) | \overline{m}_i' \overline{m}_k' \rangle \delta^{(ik)}(m', \overline{m}') \qquad (7.12)$$

where, in analogy with (7.7) we define

$$\delta^{(ik)}(m', \overline{m}') = \prod_{k, l \neq i} \delta(m_l', \overline{m}_l') \qquad (7.13)$$

and $|m_i m_k\rangle$ etc. are two-particle normalized Slater determinants.

There are three sums appearing on the right-hand side of (7.12). The first one represents the interaction energy within the filled level j. It is obvious from its structure that it is independent of the nature of the level j' or of the number of particles n' in this level. We notice that because of the factor $\delta(m'\overline{m}')$ this term appears only in the diagonal elements of the matrix of $(1/2)\sum v(ik)$. Hence in diagonalizing the matrix of $(1/2)\sum v(ik)$ to obtain the energies of the various states of a configuration, this term will represent the constant additional energy of the filled levels, no matter which state we consider.

The third sum in (7.12) is equally easy to handle: it involves only states in the unfilled levels and represents the interaction among the particles in this level.

The second sum involves states from both the filled and the unfilled levels and requires further study. We first notice that $\langle m_i m_k' | v(12) | m_i \overline{m}_k' \rangle$ vanishes unless $m_i + m_k' = m_i + \overline{m}_k'$. Hence

$$\langle m_i m_k' | v(12) | m_i \overline{m}_k' \rangle = \langle m_i m_k' | v(12) | m_i m_k' \rangle \delta(m_k' \overline{m}_k') \qquad (7.14)$$

Next, we would like to express the two-particle matrix element in (7.14) in terms of matrix elements in the J-scheme, since (7.14) is diagonal in this scheme. To this end we note that (see Appendix A, A.2.59)

$$|m_i m_k'\rangle = \sum_{J,M} (-1)^{j'-j-M}\sqrt{2J+1}\begin{pmatrix} j & j' & J \\ m_i & m_k' & -M \end{pmatrix}|jj'JM\rangle \qquad (7.15)$$

Introducing (7.15) into (7.14) we have

$$\langle m_i m_k'|v(12)|m_i\overline{m}_k'\rangle = \delta(m_k', \overline{m}_k')\sum_{JM}(2J+1) \times \begin{pmatrix} j & j' & J \\ m_i & m_k' & -M \end{pmatrix}^2 \langle jj'JM|v(12)|jj'JM\rangle \qquad (7.16)$$

We can now introduce (7.16) into the second sum in (7.12), carrying the summation over m_i only, and obtain

$$\sum_{m_i} \langle m_i m_k'|v(12)|m_i\overline{m}_k'\rangle\delta^{(k)}(m'\overline{m}') = \sum_{m_iJM}(2J+1)\begin{pmatrix} j & j' & J \\ m_i & m_k' & -M \end{pmatrix}^2 \times \langle jj'JM|v(12)|jj'JM\rangle\delta(m_k', \overline{m}_k')\delta^{(k)}(m', \overline{m}') \qquad (7.17)$$

In (7.17) the matrix element $\langle jj'JM|v(12)|jj'JM\rangle$ is independent of M. Since the level j is filled we have to sum over *all* values of m_i; we can therefore carry out explicitly the summation over m_i and M using the orthogonality and normalization of the $3-j$ symbol. We also notice that

$$\delta(m_k', \overline{m}_k')\delta^{(k)}(m', \overline{m}') = \delta(m'\overline{m}')$$

Hence we find finally

$$\sum_{m_i} \langle m_i m_k'|v(12)|m_i\overline{m}_k'\rangle\delta^{(k)}(m', \overline{m}') = \frac{1}{2j'+1}\sum_J (2J+1) \times \langle jj'JM_0|v(12)|jj'JM_0\rangle\delta(m', \overline{m}') \qquad (7.18)$$

where M_0 is any value consistent with $|M_0| \leq J$. Equation 7.18 tells us two things: the second sum in (7.12) contributes only to the *diagonal* elements of the matrix of $(1/2)\Sigma v(ik)$, and the contribution is the same for each m_k'. Note that in (7.18) we have not yet summed over m_k', but the result is independent of m_k'.

Since (7.18) is independent of m_k', the contribution of this second sum to (7.12) will depend on the number of particles in the level j' and, in fact, will be proportional to n'; for a given number of such particles it is therefore the same for all the diagonal elements of $(1/2)v(ik)$ in (7.12). Thus the interaction between the particles in the unfilled level j' and the particles in the filled level j amounts to a change in the single-particle energy of each one of the j' particles by an amount given by (7.18).

The off-diagonal elements in the matrix of $\Sigma v(ik)$, taken for the $A = n + n'$ particles, are therefore all determined by the third sum in (7.12), and are thus identical to those of the $j'^{n'}$ configuration. The existence of additional constant terms E_0 in the diagonal elements of (7.12) means that each of its eigenvalues can be obtained by adding E_0 to the eigenvalues of (7.12) with the constant terms omitted. The energy *differences* between states of the configuration $j^{2j+1}j'^{n'}$ are therefore identical to those in the configuration $j'^{n'}$.

We therefore conclude that *closed shells can be disregarded whenever one computes energy differences between states of the configuration* $j'^{n'}$.

The physical picture behind this result is rather simple to understand: the different levels of the configuration $j'^{n'}$ are obtained by letting the orbits of the n'-particles have different orientations in relation to each other. Different states are thus characterized by different probabilities of particle 1 to have a z-projection m_1' for its angular momentum, particle 2-m_2' etc. The overlap among the wave functions and, therefore, the total interaction energy of these n'-particles among themselves depends on this probability distribution. This is the third sum in (7.12). There is, however, an additional interaction which the n'-particles feel: their interaction with $2j + 1$ particles in the filled j-level. If this interaction turned out to depend on the orientation of a particle in j' relative to the particles in the j-level, it could have been different for different states of $j'^{n'}$ and thus contribute to energy differences in j'. But because the j-level is completely filled and the total angular momentum of the particles in it vanishes, the interaction of a particle in j' with the filled j-level cannot depend on the orientation of a j'-particle with respect to the filled level. Hence the m_k' independence of (7.18) and, as a result, the lack of any contribution to the splittings in the $j'^{n'}$ configuration from its interaction with filled levels.

Our results hold whenever a level, and not necessarily a whole shell, is filled. However in many cases, as we shall see later, several close-lying levels fill simultaneously, so that unless particular levels are well separated it is not safe to ignore them; the matrix elements connecting different close-lying levels may be large compared to their separation. We shall therefore ignore closed shells rather than filled levels.

8. PARTICLE-HOLE CONFIGURATIONS

We speak of a *hole* in a level whenever just one additional particle in it will make it completely filled. In ${}^{40}_{19}K^{21}$, for instance, we can say that the proton configuration has three particles in $1d_{3/2}$ (plus filled lower levels) or, equivalently, that it has one hole in the $1d_{3/2}$ level. The ${}^{40}K$ ground-state configuration will then be written as $(d_{3/2}^{-1}, f_{7/2})$, which is to be understood as completely equivalent to $(d_{3/2}^{3}, f_{7/2})$.

Since a filled level can be ignored in many calculations, we can expect a hole to behave like a particle in the same level with some "opposite" properties. We shall now derive some of the relations between a hole configuration and a particle configuration.

Let us first consider a single-particle operator $\Omega_\kappa^{(k)} = \Sigma\Omega_\kappa^{(k)}(i)$, and let us evaluate its diagonal matrix elements in the m-scheme first. As we can tell immediately from the Wigner–Eckart theorem only $\kappa = 0$ components will give nonzero diagonal matrix elements. We denote the configuration of n *holes* in the m-scheme by the unoccupied m-states, so that, for instance $|m_1^{-1}m_2^{-1}\rangle$ is a $(2j+1-2)$ particle Slater determinant in which all states except m_1 and m_2 are occupied. We obtain then

$$\langle m_1^{-1}m_2^{-1}\dots m_n^{-1}|\sum_{i=1}^{2j+1-n}\Omega_0^{(k)}(i)|m_1^{-1}m_2^{-1}\dots m_n^{-1}\rangle$$

$$= (2j+1-n)\langle m_1^{-1}\dots m_n^{-1}|\Omega_0^{(k)}(1)|m_1^{-1}\dots m_n^{-1}\rangle$$

$$= \sum_{m_i\neq m_1,\dots,m_n}\langle m_i|\Omega_0^{(k)}|m_i\rangle$$

$$= \sum_{m_i=-j}^{j}\langle m_i|\Omega_0^{(k)}|m_i\rangle - \sum_{m_i=m_1,\dots,m_n}\langle m_i|\Omega_0^{(k)}|m_i\rangle \tag{8.1}$$

From (7.8) we know that for $k \neq 0$ the first sum on the right-hand side of (8.1) vanishes. We also have obviously

$$\langle m_1m_2\dots m_n|\sum_{i=1}^{n}\Omega_0^{(k)}(i)|m_1m_2\dots m_n\rangle = \sum_{m=m_1m_2,\dots,m_n}\langle m_i|\Omega_0^{(k)}|m_i\rangle \qquad k\neq 0 \tag{8.2}$$

Hence, combining (8.1) with (7.8) and (8.2) we obtain

$$\langle m_1^{-1}m_2^{-1}\dots m_n^{-1}|\sum_{i=1}^{2j+1-n}\Omega_0^{(k)}(i)|m_1^{-1}m_2^{-1}\dots m_n^{-1}\rangle$$

$$= -\langle m_1m_2\dots m_n|\sum_{i=1}^{n}\Omega_0^{(k)}(i)|m_1m_2\dots m_n\rangle \qquad k\neq 0 \tag{8.3}$$

Equation 8.3, as expected, says that the expectation value, in the m-scheme, for any nonscalar single-particle tensor operator taken in a given hole configuration is minus the same quantity taken for the complementary-particle configuration.

As we mentioned earlier the $\kappa = 0$ matrix elements in (8.3) are the only nonvanishing diagonal elements of $\sum\Omega_\kappa^{(k)}(i)$ in the m-scheme. It follows therefore that (8.3) holds also in any other scheme, and in particular in the J-scheme. We must however be a little careful in identifying the J-scheme states that go into the two sides of (8.3). Later on we shall see that if the transformation matrix (Eq. IV.12.1) $\langle JM|m_1\dots m_n\rangle$ transforms the n-

particle m-scheme states into the J-scheme state $|j^nJM\rangle$, then the same transformation matrix will transform the n-hole m-scheme states $|m_1^{-1}m_2^{-1}\ldots m_n^{-1}\rangle$ into the state $|j^{-n}J,-M\rangle \equiv |j^{(2j+1-n)}J,-M\rangle$, that is,

$$\sum_{m_1\ldots m_n} |m_1^{-1}m_2^{-1}\ldots m_n^{-1}\rangle\langle m_1m_2\ldots m_n|JM\rangle = |j^{-n}J,-M\rangle \qquad (8.4)$$

Intuitively (8.4) is easy to understand, since in the hole configuration, $m_1\ldots m_n$ specify the *missing* particles, so that the state $|m_1^{-1}\ldots m_n^{-1}\rangle$ is a state whose total z-projection is

$$\sum_{m_i\neq m_1\ldots m_n} m_i = \sum_{-j}^{j} m_i - \sum_{m_i=m_1\ldots m_n} m_i = 0 - M = -M$$

The formal proof of (8.4) is, however, slightly more involved [deShalit and Talmi (63), p. 316] and we shall present it here just for the case $n = 1$, that is, a single-hole configuration.

We first notice that the total angular momentum J of a state of a single-hole configuration j^{2j} can only have one value: $J = j$. This follows directly from the fact that a filled level can have only $J = 0$; taking away one of the particles (jm) leaves us, therefore, with a state $|j^{2j}J = jM = -m\rangle$. More formally this result follows if we try to construct Slater determinants of $2j$ identical particles in the j-level. If we make the Slater determinant correspond to a given value of

$$M = \sum_{i=1}^{2j} m_i,$$

the only way to do it is to fill all m-states in the j-level except the state $m' = -M$. We then have

$$\sum_{i=1}^{2j} m_i = \sum_{i=1}^{2j+1} m_i - m' = 0 - m' = M \qquad (8.5)$$

as required.

Since $|m'| \leq j$ it follows that j^{2j} cannot have any state with $M > j$, and it has exactly one state for each value of M satisfying $|M| \leq j$. We have thus constructed in the configuration j^{2j} the $2j + 1$ states of total angular momentum j, and we see indeed that the hole state $|m_1^{-1}\rangle$ is a state of the same angular momentum as the particle state $|m_1\rangle$, but with the opposite value for the z-projection of its angular momentum.

Using (8.4) we can now take the appropriate linear combinations of (8.3) and obtain:

$$\langle j^{2j+1-n}J,-M|\sum_{i=1}^{2j+1-n}\Omega_0^{(k)}(i)|j^{2j+1-n}J,-M\rangle = -\langle j^nJM|\sum_{i=1}^{n}\Omega_0^{(k)}(i)|j^nJM\rangle$$

$$k \neq 0 \qquad (8.6)$$

We prefer to have a relation between matrix elements with the *same* value of M for the particles and holes. To do this we notice, using the Wigner–Eckart theorem, that

$$\begin{aligned}\langle J, -M|\Omega_0^{(k)}|J, -M\rangle &= (-1)^{J+M}\begin{pmatrix} J & k & J \\ M & 0 & -M\end{pmatrix}\langle J||\Omega^{(k)}||J\rangle \\ &= (-1)^{J+M+2J+k}\begin{pmatrix} J & k & J \\ -M & 0 & M\end{pmatrix}\langle J||\Omega^{(k)}||J\rangle \\ &= (-1)^{k+J-M}\begin{pmatrix} J & k & J \\ -M & 0 & M\end{pmatrix}\langle J||\Omega^{(k)}||J\rangle \\ &= (-1)^{k}\langle JM|\Omega_0^{(k)}|JM\rangle\end{aligned}$$

We therefore find finally from (8.6) that

$$\langle j^{2j+1-n}JM|\sum_{i=1}^{2j+1-n}\Omega_0{}^{(k)}(i)|j^{2j+1-n}JM\rangle$$

$$= (-1)^{k+1}\langle j^nJM|\sum_{i=1}^{n}\Omega_0^{(k)}(i)|j^nJM\rangle \qquad k \neq 0 \tag{8.7}$$

To be complete we also require the result for $k = 0$. One finds the trivial result:

$$\langle j^n\alpha JM|\Omega_0^{(0)}|j^n\alpha JM\rangle = n\langle jm|\Omega_0^{(0)}|jm\rangle$$

Hence we conclude that for $k = 0$

$$\langle j^{2j+1-n}JM|\sum_{i=1}^{2j+1-n}\Omega_0{}^{(0)}(i)|j^{2j+1-n}JM\rangle$$

$$= \frac{2j+1-n}{n}\langle j^nJM|\sum_{i=1}^{n}\Omega_0^{(0)}(i)|j^nJM\rangle \tag{8.8}$$

Equations 8.7 and 8.8 are very fundamental for many shell-model calculations. They show a qualitative difference between odd and even irreducible tensor operators and, as we shall see later, this difference reflects itself in many features of nuclear spectra, nuclear moments, etc. Odd tensors have the *same* expectation value for a particle state $|j^nJM\rangle$ and its conjugate hole state $|j^{-n}JM\rangle$, whereas even operators change their sign. Magnetic moments, being represented by an odd operator, have, therefore the same theoretical value for one-particle configuration and one-hole configuration, whereas quadrupole moments, for instance, that are represented by an even operator, change their sign at the middle of the shell. There are as yet not enough measurements of static moments to test these conclusions, since almost always one of the partners to be compared is a highly unstable nucleus. But one does find characteristically that quadrupole moments at the beginning of shells are negative [$Q({}^{17}_{8}O_9) = -0.0265$ barns, $Q({}^{45}_{21}Sc_{24}) = -0.22$ barns, $Q({}^{63}_{29}Cu_{34}) = -0.20$ barns, $Q({}^{121}_{51}Sb_{70}) = -0.20$ barns, $Q({}^{209}_{83}Bi_{126}) = -0.24$ barns, etc.] whereas quadrupole moments of nuclei near the end of a shell are positive [$Q({}^{25}_{12}Mg_{13}) = +0.22$ barns, $Q({}^{57}_{26}Fe_{31}) = +0.15$ barns, $Q({}^{67}_{31}Ga_{36}) = +0.217$ barns, $Q({}^{87}_{38}Sr_{49}) = +0.36$ barns, $Q({}^{197}_{79}Au_{118}) = +0.60$ barns, etc.].

It is sometimes more convenient to use (8.7) in a form that involves the reduced matrix elements only. Using the Wigner–Eckart theorem we find that

$$\langle j^{2j+1-n}J\| \sum_{i=1}^{2j+1-n} \Omega^{(k)}(i)\| j^{2j+1-n}J\rangle$$

$$= (-1)^{k+1}\langle j^nJ\| \sum_{i=1}^{n} \Omega^{(k)}(i)\| j^nJ\rangle \qquad k \neq 0 \tag{8.9}$$

Equation 8.7 can be used to derive an interesting relation between the energies in a particle-hole configuration and those of a particular particle-particle configuration. Let us consider, indeed, two nonidentical particles p and n, and compare the two configurations (j_p, j_n) and $(j_p{}^{2j_p}, j_n) \equiv (j_p{}^{-1}, j_n)$. As an example we can consider the two nuclei ${}^{38}_{17}Cl_{21}$ and ${}^{40}_{19}K_{21}$, with configurations $(d_{3/2}, f_{7/2})$ and $(d_{3/2}^{-1}, f_{7/2})$, respectively. We have noted already (Eq. IV.20.15) that an interaction $v(|\mathbf{r}_1 - \mathbf{r}_2|)$ can be written as a sum of products of multipole moments:

$$v(|\mathbf{r}_1 - \mathbf{r}_2|) = \sum_{lm} \frac{4\pi}{2l+1} v_l(r_1, r_2) Y^*_{lm}(\theta_1\phi_1) Y_{lm}(\theta_2\phi_2) \tag{8.10}$$

For more general interactions, involving spins as well, we can still decompose a rotationally invariant interaction $v(12)$ in the form (Chapter IV. App. A.1)

$$v(12) = \sum_{l} v_l(r_1, r_2) T^{(l)}(1) \cdot T^{(l)}(2) \tag{8.11}$$

where $T^{(l)}(i)$ is an irreducible tensor operator of rank l operating on the coordinates of the ith particle only, and the scalar product $T^{(l)}(1) \cdot T^{(l)}(2)$ is defined by

$$T^{(l)}(1) \cdot T^{(l)}(2) = \sum_{\lambda} (-1)^\lambda T^{(l)}_{-\lambda}(1) T^{(l)}_{\lambda}(2) \tag{8.12}$$

This decomposition of $v(12)$ will be useful in relating energies in (j_p, j_n) to those in $(j_p{}^{2j_p}, j_n)$.

We have already noted that the configuration $j_p{}^{-1}$ has just one level with $J = j_p$; it follows that the configuration $(j_p{}^{-1}, j_n)$ has as many levels as the configuration (j_p, j_n). Furthermore the splittings between the levels of the configuration $(j_p^{-1}j_n)$ are determined only by the interaction between the protons in j_p and the neutron in j_n; the interactions among the protons themselves give rise to the same shift in all the levels of the configuration (j_p^{-1}, j_n), since there is only one possible level for the j_p^{-1} protons. Thus, since we are interested in *energy differences* between states of the configuration $(j_p^{-1}j_n)$, all we need to calculate is

$$\Delta E(j_p^{-1}j_nJM) = \langle j_p^{-1}j_nJM|\sum_{i=1}^{2j} v(p_i, n)|j_p^{-1}j_nJM\rangle \tag{8.13}$$

where $v(p_i, n)$ is the interaction between the ith proton and the neutron. Introducing (8.11) into (8.13) we obtain

$$\Delta E(j_p^{-1}j_nJM) = \sum_l \langle j_p^{-1}j_nJM|\sum_i v_l(r_{pi}, r_n)T^{(l)}(p_i)\cdot T^{(l)}(n)|j_p^{-1}j_nJM\rangle \tag{8.14}$$

$v_l(r_{pi}, r_n)$ depends only on the magnitudes of r_{pi} and r_n; since the radial wave functions of the protons in j_p are independent of the magnetic quantum numbers, we see that in (8.14) the radial integral of $v_l(r_{pi}, r_n)$ can be factored out of the sum over the protons p_i. Introducing the notation

$$F^{(l)}(l_pl_n) = \int R^2_{n_pl_p}(r_p)R^2_{n_nl_n}(r_n)v_l(r_p, r_n)\, dr_p\, dr_n \tag{8.15}$$

we then find that

$$\Delta E(j_p^{-1}j_nJM) = \sum_l F^{(l)}(l_pl_n)\langle j_p^{-1}j_nJM|\left(\sum_{i=1}^{2j_p} T^{(l)}(p_i)\right) \cdot T^{(l)}(n)|j_p^{-1}j_nJM\rangle \tag{8.16}$$

We notice that in (8.16) we now have matrix elements of a scalar product of two tensors, $\sum_{i=1}^{2j_p} T^{(l)}(p_i)$ involving the protons only and $T^{(l)}(n)$ involving the neutron only, taken between states of a definite angular momentum j_p for the protons and j_n for the neutron. Using the relations developed in Appendix A we therefore obtain

$$\Delta E(j_p^{-1}j_nJM) = \sum_l F^{(l)}(l_pl_n)\begin{Bmatrix} j_p & j_n & J \\ j_n & j_p & l \end{Bmatrix}(-1)^{j_p+j_n+J} \times \langle j_p^{-1}, j_p||\sum_{i=1}^{2j_p} T^{(l)}(p_i)||j_p^{-1}, j_p\rangle\langle j_n||T^{(l)}(n)||j_n\rangle \tag{8.17}$$

In a similar way we obtain for the particle–particle configuration (j_nj_n)

$$\Delta E(j_p, j_n, JM) = \sum_l F^{(l)}(l_pl_n)\begin{Bmatrix} j_p & j_n & J \\ j_n & j_p & l \end{Bmatrix}(-1)^{j_p+j_n+J} \times \langle j_p||T^{(l)}(p)||j_p\rangle\langle j_n||T^{(l)}(n)||j_n\rangle \tag{8.18}$$

With the help of (8.9) we can now relate (8.17) and (8.18), which differ, as we notice, only in the reduced matrix elements of the proton operators $T^{(l)}(p)$. For our special case (8.9) reduces to

$$\langle j_p^{-1}, j_p|| \sum_{i=1}^{2j_p} T^{(l)}(p_i)|| j_p^{-1}, j_p\rangle = (-1)^{l+1}\langle j_p||T^{(l)}(p)||j_p\rangle \qquad l \neq 0 \quad (8.19)$$

To obtain an explicit connection between the energies of the two configurations [Pandya (56)] we use the identity (see Appendix A, 2.2.90)

$$(-1)^l \begin{Bmatrix} j_p & j_n & J \\ j_n & j_p & l \end{Bmatrix} = \sum_{J'} (-1)^{J+J'}(2J'+1) \begin{Bmatrix} j_p & j_n & J \\ j_p & j_n & J' \end{Bmatrix} \begin{Bmatrix} j_p & j_n & J' \\ j_n & j_p & l \end{Bmatrix}$$

and (8.19):

$$\begin{aligned} \Delta E(j_p^{-1}j_nJM) &= \sum_l F^{(l)}(l_pl_n) \begin{Bmatrix} j_p & j_n & J \\ j_n & j_p & l \end{Bmatrix} (-1)^{j_p+j_n+J+l+1} \\ &\quad \times \langle j_p||T^{(l)}(p)||j_p\rangle\langle j_n||T^{(l)}(n)||j_n\rangle \\ &= \sum_{l,J'} F^{(l)}(l_pl_n)\langle j_p||T^{(l)}(p)||j_p\rangle\langle j_n||T^{(l)}(n)||j_n\rangle \\ &\quad \times (-1)^{j_p+j_n+2J+J'+1}(2J'+1) \begin{Bmatrix} j_p & j_n & J \\ j_p & j_n & J' \end{Bmatrix} \begin{Bmatrix} j_p & j_n & J' \\ j_n & j_p & l \end{Bmatrix} \\ &= (-1)^{2J+1} \sum_{J'} (2J'+1) \begin{Bmatrix} j_p & j_n & J \\ j_p & j_n & J' \end{Bmatrix} \Delta E(j_pj_n, J'M) \end{aligned} \quad (8.20)$$

Or, since $2J$ is even,

$$\Delta E(j_p^{-1}j_nJM) = -\sum_{J'} (2J'+1) \begin{Bmatrix} j_p & j_n & J \\ j_p & j_n & J' \end{Bmatrix} \Delta E(j_pj_nJ'M) \qquad (8.21)$$

In deriving (8.21) we disregard the term with $l = 0$. As seen from (8.9) and (8.10) this term will add a constant (J-independent) energy to both the configurations (j_p^{-1}, j_n) and (j_p, j_n). Although this constant term may be different for the two configurations, it does not contribute to the splittings between levels in any one of them (because of its J-independence); thus it does not affect (8.21). We remind ourselves that (8.21) is valid only as long as we are interested in spacings between the levels of the configuration $(j_p^{-1}j_n)$; if we want the absolute position of these levels we should add an appropriate constant (J-independent) energy to the right-hand side of (8.21).

Equation 8.21 can be inverted to give $\Delta E(j_p j_n J)$ in terms of $\Delta E(j_p^{-1} j_n J)$ by using the orthogonality of the $6 - j$ (Racah) coefficients

$$\Delta E(j_p j_n JM) = -\sum_{J'} (2J'+1) \begin{Bmatrix} j_p & j_n & J \\ j_p & j_n & J' \end{Bmatrix} \Delta E(j_p^{-1} j_n J'M) \tag{8.22}$$

We thus see that the splittings in a particle-hole and a particle–particle configurations are the negative Racah-transforms of each other, independent of the nature of the interaction $v(p, n)$.

The spectra of the two configurations (j_p, j_n) and (j_p^{-1}, j_n) generally look very different from each other, like the spectra of K^{40} and ^{38}Cl shown in Fig. 8.1. Yet the connection between the two spectra implied by (8.21) or (8.22) is quite good, as is seen from Table 8.1.

TABLE 8.1 Energy Differences of Levels in ^{38}Cl Calculated from the Observed Energies in ^{40}K Using (8.22). All Differences Were Computed Relative to the Ground State ($J = 2^-$) of ^{38}Cl.

J	Calculated	Experiment
2	0 Mev	0 Mev
3	0.75	0.762
4	1.32	1.312
5	0.70	0.672

Relations of the type (8.21) or (8.22) are examples of the possible tests of the shell model for selected nuclei. The assumptions that went into the derivations of these relations are the following:

(a) A specific configuration assignment for the levels of the two nuclei to be compared and the validity of first-order perturbation theory.

(b) A general two-body residual interaction $v(p, n)$, which is assumed to be invariant under rotations.

(c) The constancy of the radial matrix elements $F^{(l)}(l_p l_n)$ as we go from $(j_p j_n)$ to $(j_p^{-1} j_n)$. This assumption is implicit in the final step in (8.20), and can be expected to hold fairly well.

Because of assumption (a), tests similar to the one described here are limited to states that are expected to be described relatively well in terms of a single configuration. This is generally true when the proton and the neutron numbers are close to magic numbers. In other cases there may be several close-lying levels that can be occupied by the nucleons, and our previous discussions have shown that we may have to diagonalize the residual interaction

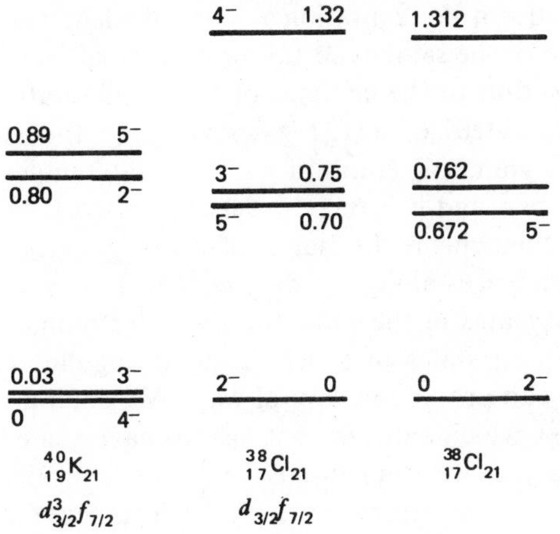

FIG. 8.1. Energy levels of ^{38}Cl and ^{40}K in MeV. Also shown are the ^{38}Cl levels calculated from the observed levels in ^{40}K [from Talmi and Unna (60)].

in a larger sub-Hilbert space, consisting of a few configurations. The resulting states will then involve components in several configurations at a time, and the possible relations between the spectra of different nuclei will depend on the details of the interaction $v(12)$. The limitation to one configuration determines the dependence of the wave function on $4A$ out of $5A$ degrees of freedom that it has (three coordinates, spin, and isospin for each of the nucleons), and it is this very limited remaining freedom that allows us such generalities in formulating our tests. Later on we shall see how similar tests of the shell model can be formulated employing quantities other than energies. Similar tests will also be developed for other models.

9. CONFIGURATIONS WITH MORE THAN TWO PARTICLES

Our study of the relation between the structure of particle hole and particle-particle configurations has shown how a connection can be established for particular nuclei, between spectra that look very different. Ultimately, of course, we hope to obtain an even broader connection among the spectra of all nuclei through their common interpretation in terms of a Schrödinger equation with a specific nuclear interaction. This stage, however, still lies ahead of us, and for now we must satisfy ourselves with a partial accomplishment of this aim by establishing connections between the properties of only a few, more closely related, nuclei. We shall therefore proceed now to analyze the energy spectra of more complicated configurations to see how far such connections among different nuclei can lead us.

Let us first consider a configuration of n equivalent nucleons j^n. All the n-nucleons are to be of the same type (all protons or all neutrons). To obtain the first-order correction to the energies of this configuration we again have to diagonalize the matrix of $(1/2(\sum_{i\neq k}^{n} v(ik)$ in the sub-Hilbert space spanned by the antisymmetric states of j^n. The total angular momentum is a good quantum number, and it is therefore useful to work from the beginning in the J-scheme; in this scheme the matrix of $(1/2)\sum v(ik)$ will break down into many smaller matrices along the diagonal, each having off diagonal elements only between states of the same total angular momentum.

To actually construct states of a definite total angular momentum in the j^n-configuration we can proceed in several ways. We shall demonstrate here a very convenient way, which will turn out also to have some formal uses later on. To simplify matters we shall discuss only configurations j^3 and j^4; the generalization to j^n is straightforward [see, for instance, deShalit and Talmi (63)].

In the configuration j^2 the construction of a wave function in the J-scheme is achieved through the use of the Clebsch–Gordan coefficients:

$$|j^2JM\rangle = \sum_{mm'} (jmjm'|JM)\phi_1(jm)\phi_2(jm') \tag{9.1}$$

Although the state $|j^2JM\rangle$, when written in this form, does not look antisymmetric, we know that it follows from the properties of the Clebsch–Gordan coefficients that (9.1) is indeed antisymmetric in particles 1 and 2 if J is even (it is symmetric for odd values of J; see Eq. IV.13.5).

With three particles we can easily construct a wave function that corresponds to a given total angular momentum by defining:

$$|j^2(J_{12}), j; JM\rangle = \sum_{M_{12}m_3} (J_{12}M_{12}jm|JM)|j^2J_{12}M_{12}\rangle\phi_3(jm_3) \tag{9.2}$$

In (9.2) we have taken a particular antisymmetric two-particle state $|j^2J_{12}M_{12}\rangle$ and added to it a third particle using the Clebsch–Gordan coefficients for the addition of the vector $\mathbf{J}_{12}$ to the vector $\mathbf{j}$ yielding a state of total angular momentum $\mathbf{J}$ and a z-projection M. Since the total angular operator $\mathbf{J}$ can be written as

$$\mathbf{J} = \mathbf{j}_1 + \mathbf{j}_2 + \mathbf{j}_3 = \mathbf{J}_{12} + \mathbf{j}_3 \tag{9.3}$$

it is obvious that the state defined in (9.2) is an eigenstate of $\mathbf{J}^2$ and of J_z.

However, unlike the case of the two-particle configuration it is not generally true that (9.2) will be automatically antisymmetric for some particular values of J. By construction, (9.2) is antisymmetric with respect to the exchange of particles 1 and 2, if we require that J_{12} be even; it is not antisymmetric with respect to the exchange of particles 1 and 3 or 2 and 3. To achieve totally antisymmetric states we antisymmetrize (9.2) with respect to 1 and 3, and 2 and 3, and define:

$$|j^3[J_{12}]JM\rangle = \mathcal{A}_{123}\sum_{m_3M_{12}}(J_{12}M_{12}\,jm_3|JM)|j^2J_{12}M_{12}\rangle\phi_3(jm_3)$$

$$= C\sum_{m_1m_2m_3M_{12}}(J_{12}M_{12}\,jm_3|JM)(jm_1\,jm_2|J_{12}M_{12})$$

$$\times\ [\phi_1(m_1)\phi_2(m_2)\phi_3(m_3) - \phi_1(m_3)\phi_2(m_2)\phi_3(m_1) - \phi_1(m_1)\phi_2(m_3)\phi_3(m_2)] \tag{9.4}$$

Since we want $|j^3JM\rangle$ to be normalized we have introduced a coefficient C that will be evaluated later. The expression's, (9.4), antisymmetry with respect to the exchange of 1 and 3 or 2 and 3 follows from its antisymmetry with respect to the exchange of 1 and 2. This follows from the relation:

$$(jm_1\,jm_2|J_{12}M_{12}) = -(jm_2\,jm_1|J_{12}M_{12}) \qquad \text{for even } J_{12} \tag{9.5}$$

Problem. Prove the antisymmetry of (9.4) with respect to 1 and 2 by noting that m_1 and m_2 in (9.4) are dummy indices.

The notation $|j^3[J_{12}]JM\rangle$ is meant to imply that we are dealing with an antisymmetric state of three particles that was constructed by antisymmetrizing a state in which particles 1 and 2 were coupled to J_{12}. (Compare with notation used in Eq. 9.2). It is to be noted that although the state (9.2) is an eigenstate of $\mathbf{J}_{12}^2 = (\mathbf{j}_1 + \mathbf{j}_2)^2$, this is no longer true of the antisymmetrized state $|j^3[J_{12}]JM\rangle$. The antisymmetrization has completely destroyed the special role played by particles 1 and 2 in (9.2) and in (9.4) all particles appear on an exactly equal footing.

We notice further that an antisymmetrization of (9.2) may lead to a vanishing result as well. Thus we know (see Section IV.15) that the highest allowed value for J in an antisymmetric state in the configuration j^3 is $J = 3j - 3$; in (9.2) we can construct a state with $J = 3j - 1$ by taking $J_{12} = 2j - 1$ (we recall that J_{12} must be even, and, of course, $J_{12} \leq 2j - 1$). An antisymmetrization of this particular state must therefore lead to an expression that vanishes identically.

Problem. Show by actually working out (9.4) that

$$|j^3[J_{12}]J = 3j - 1, M\rangle = 0$$

The states $|j^2(J_{12})jJM\rangle$, for different values of J_{12}, are orthogonal to each other, even if (JM) remain the same. A straightforward calculation shows that

$$\langle j^2(J_{12})jJM|j^2(J'_{12})jJM\rangle = \delta(J'_{12}, J_{12}) \tag{9.6}$$

This property, too, is lost after antisymmetrization. In fact we saw in Section IV.15 that in the $(7/2)^3$ configuration the allowed values of J (for antisymmetric states) were $J = 15/2, 11/2, 9/2, 7/2, 5/2$, and $3/2$, there being *one* allowed state for each value of J and M. If we take a state $|j^3 J = 9/2 M = 9/2\rangle$, for instance, it can be obtained, through the antisymmetrization procedure outlined above, from three *different* states (9.2) with $J_{12} = 2, 4$, and 6. Since we know that there is just one antisymmetric state $|j^3 J = 9/2 M = 9/2\rangle$, they are all bound to give rise to the *same* state after antisymmetrization. Indeed using (9.4) we find that

$$\begin{aligned}
\langle j^3 [J_{12}]JM | j^3 [J'_{12}]JM\rangle &= C^* \cdot C' \sum (J_{12}M_{12}\, jm_3 | JM)(jm_1\, jm_2 | J_{12}M_{12}) \\
&\quad \times (J'_{12}M'_{12}\, jm'_3 | JM)(jm'_1\, jm'_2 | J'_{12}M'_{12}) \\
&\times \int [\phi_1(m_1)\phi_2(m_2)\phi_3(m_3) - \phi_1(m_3)\phi_2(m_2)\phi_3(m_1) - \phi_1(m_1)\phi_2(m_3)\phi_3(m_2)]^* \\
&\times [\phi_1(m'_1)\phi_2(m'_2)\phi_3(m'_3) - \phi_1(m'_3)\phi_2(m'_2)\phi_3(m'_1) - \phi_1(m'_1)\phi_2(m'_3)\phi_3(m'_2)] \\
&\quad \times d\tau_1\, d\tau_2\, d\tau_3 \\
&= 3C^* \cdot C' \sum (M_{12}m_3 | M)(m_1 m_2 | M_{12}) \\
&\times (M'_{12}m'_3 | M)(m'_1 m'_2 | M'_{12})[\delta(m_1 m'_1)\delta(m_2 m'_2)\delta(m_3 m'_3) - \delta(m_1 m'_3)\delta(m_2 m'_2) \\
&\quad \times \delta(m_3 m'_1) - \delta(m_1 m'_1)\delta(m_2 m_3)\delta(m_3 m'_2)] \\
&= 3C^* C' \Bigg[\delta(J_{12}, J'_{12}) + 2\sqrt{(2J_{12}+1)(2J'_{12}+1)} \\
&\quad \times \begin{Bmatrix} j & j & J_{12} \\ j & J & J'_{12} \end{Bmatrix} \Bigg]
\end{aligned} \tag{9.7}$$

where the shorthand notation (1.26) has been introduced for the Clebsch-Gordan coefficients, and we have made use of the explicit expression of the Racah $6-j$ coefficients in terms of the $3j$-coefficient (see Appendix A, A.2.94). Since

$$\begin{Bmatrix} j & j & J_{12} \\ j & J & J'_{12} \end{Bmatrix}$$

does not generally vanish for $J_{12} \neq J'_{12}$, we see that the two states $|j^3[J_{12}]JM\rangle$ and $|j^3[J'_{12}]JM\rangle$ are not generally orthogonal to each other.

By taking all possible values of J_{12} in (9.4) we generally produce a redundant set of antisymmetric wave functions of the configuration j^3. Our proof, (9.7), that they need not be orthogonal to each other explains how such a situation may arise. In most practical cases one can produce all the allowed states of j^3 with two or three J_{12}'s. The number J_{12} is sometimes referred to as the *prin-*

cipal parent of the state $|j^3[J_{12}]JM\rangle$; the same state can have, then, several principal parents.

We can use (9.7) to derive the normalization constant in (9.4). In fact, putting $J_{12} = J'_{12}$ and requiring that $\langle j^3[J_{12}]JM|j^3[J_{12}]JM\rangle = 1$, we obtain

$$C = \left[3 + 6(2J_{12} + 1)\begin{Bmatrix} j & j & J_{12} \\ j & J & J_{12} \end{Bmatrix}\right]^{-1/2} \tag{9.8}$$

To construct antisymmetric wave functions of four particles in j, that is, for the configuration j^4, we can proceed in a similar way by constructing

$$|j^3(J_{123})jJM\rangle = \sum (J_{123}M_{123}jm_4|JM)|j^3J_{123}M_{123}\rangle\cdot\phi_4(jm_4) \tag{9.9}$$

and antisymmetrizing it. Here $|j^3J_{123}M_{123}\rangle$ is an antisymmetric three particle wave function. It is often more convenient, however, to start from another wave function: the one in which particles 1 and 2 and 3 and 4 are paired separately:

$$|j^2(J_{12})j^2(J_{34});JM\rangle = \sum (J_{12}M_{12}J_{34}M_{34}|JM)|j^2J_{12}M_{12}\rangle|j^2J_{34}M_{34}\rangle \tag{9.10}$$

Again the actual construction of the wave function guarantees that it is an eigenfunction of $\mathbf{J}^2 = (\mathbf{J}_{12} + \mathbf{J}_{34})^2 = (\mathbf{j}_1 + \mathbf{j}_2 + \mathbf{j}_3 + \mathbf{j}_4)^2$. It is also antisymmetric with respect to the exchange of particles 1 and 2 and of 3 and 4, if we choose J_{12} and J_{34} to be even. We now antisymmetrize it with respect to other exchanges (1 and 3, 2 and 4, etc.) and obtain an antisymmetrized state of the configuration j^n that is also an eigenstate of J^2 and of J_z:

$$|j^4[J_{12}, J_{34}], JM\rangle = C\mathcal{A}|j^2(J_{12})j^2(J_{34})JM\rangle \tag{9.11}$$

The same considerations with respect to orthogonality, etc. hold in this case as well, but we shall not discuss them. It is important to notice is that an explicit expression for properly antisymmetrized n particle wave functions can be written in a rather compact form if one makes use of the Clebsch–Gordan coefficients. This way of writing the wave functions makes it possible to use the Racah algebra and greatly facilitates actual computations.

10. COEFFICIENTS OF FRACTIONAL PARENTAGE

There is another expression for the antisymmetrized wave functions in the configuration j^n that, very often, is even more convenient to use than the expression (9.4) or (9.11). To understand the motivation behind this way of constructing the wave functions, consider a matrix element of $(1/2)\cdot\Sigma v(ik)$ in the configuration j^n:

$$V_J \equiv \langle j^n\alpha JM|\frac{1}{2}\sum_{i\neq k}^{n} v(ik)|j^n\alpha' JM\rangle \tag{10.1}$$

In (10.1) we have denoted by α any additional quantum numbers that are required to describe the particular state in question if J and M do not specify it uniquely. We shall see later what these quantum numbers are; for the moment they are irrelevant.

Because of the assumed antisymmetry of the wave functions $|j^n\alpha JM\rangle$ we have

$$V_J = \frac{n(n-1)}{2}\langle j^n\alpha JM|v(12)|j^n\alpha' JM\rangle \tag{10.2}$$

Any pair of coordinates could have been singled out, since each $v(ik)$ contributes an equal amount to (10.1). In the matrix element (10.2), the operator involves only the coordinates of particles 1 and 2; if we know the wave functions $|j^n\alpha JM\rangle$ we can, therefore, integrate out the coordinates of particles 3, 4, . . . , n. It will therefore be very convenient to write the state $|j^n\alpha JM\rangle$ in a way that singles out particles 1 and 2 from the rest of the particles. This can be done in the following way.

Consider all the antisymmetric two-particle states $|j^2J_2M_2\rangle$ for particles 1 and 2, and all the antisymmetric $(n-2)$ particle states $|j^{n-2}\beta J_{n-2}M_{n-2}\rangle$ for particles 3, 4, . . . , n. Any n-particle state in the j^n configuration that is antisymmetric in particles 1 and 2, and in particles 3, 4, . . . , n can be expanded in terms of the products $|j^2J_2M_2\rangle\cdot|j^{n-2}\beta J_{n-2}M_{n-2}\rangle$. The states that are antisymmetric in *all* particles, being a subclass of those antisymmetric separately in 1 and 2 and 3, 4, . . . , n, can therefore also be expanded in these products:

$$|j^n\alpha JM\rangle = \sum_{J_2M_2\beta J_{n-2}M_{n-2}} \langle\alpha JM|J_2M_2,\beta J_{n-2}M_{n-2}\rangle|j^2J_2M_2\rangle|j^{n-2}\beta J_{n-2}M_{n-2}\rangle \tag{10.3}$$

Since the right-hand side should represent a state of total (n-particle) angular momentum J and z-projection M, we shall find it useful to introduce the wave functions:

$$|j^2(J_2)j^{n-2}(\beta J_{n-2});JM\rangle = \sum_{M_2M_{n-2}}(J_2M_2J_{n-2}M_{n-2}|JM)\cdot|j^2J_2M_2\rangle|j^{n-2}\beta J_{n-2},M_{n-2}\rangle \tag{10.4}$$

Note that (10.4) is antisymmetric *separately* in 1 and 2 and in 3, 4, . . . , n, it being implied by the notation that particles 1 and 2 are coupled to total angular momentum J_2M_2, and the $n-2$ particles 3, . . . , n are coupled to form the state $|\beta J_{n-2}M_{n-2}\rangle$. The n-particle wave function $|j^n\alpha JM\rangle$ can now be expanded in terms of the wave functions in (10.4)

$$|j^n\alpha JM\rangle = \sum_{J_2,\beta,J_{n-2}}(\alpha J\{|j^2(J_2)j^{n-2}(\beta J_{n-2}),J)|j^2(J_2)j^{n-2}(\beta J_{n-2});JM\rangle \tag{10.5}$$

Although each of the functions on the right-hand side of (10.5) is not completely antisymmetric, their combination *is* antisymmetric in all n-particles.

This, in fact, is the purpose of taking a linear combination of such functions. The situation is similar to that encountered in the three-particle configuration as detailed in (9.4).

In (10.5) the coefficients $(\alpha J\{|j^2(J_2)j^{n-2}\beta J_{n-2}), J)$ are to be chosen so that $|j^n\alpha JM\rangle$ is normalized to unity. We note that they are independent of M.

Problem. Prove that the coefficients in (10.5) are independent of M by studying the behavior of both sides of the equation under rotations.

The coefficients $(\alpha J\{|j^2(J_2)j^{n-2}(\beta J_{n-2})J)$ in (10.5) are called *coefficients of fractional parentage*, because they determine the extent to which each of the *parent states* $|j^2(J_2)j^{n-2}(\beta J_{n-2});JM\rangle$ participates in building up the antisymmetric state $|j^n\alpha JM\rangle$. They are characteristic of the configuration and the angular momenta involved, and *do not* depend on the dynamics of the system. The special notation $(\{|)$ is used for these coefficients to stress the fact that they do not form a square matrix. The states $|\alpha JM\rangle$ are totally antisymmetric, whereas the states $|j^2(J_2)j^{n-2}(\beta J_{n-2});JM\rangle$ are only partially antisymmetric. There are therefore fewer states $|\alpha JM\rangle$ and the transformation is defined only in the one direction. Actually it is a projection of the states (10.4) onto a subspace of totally antisymmetric states.

To assure the normalization of $|j^n\alpha JM\rangle$ in (10.5) the coefficients of fractional parentage are normalized as well, so that

$$\sum_{J_2J_{n-2}\beta} |(\alpha J\{|j^2(J_2)j^{n-2}(\beta(J_{n-2});J)|^2 = 1 \tag{10.6}$$

With the help of these coefficients it is now possible to achieve our objective of integrating out the $n-2$ coordinates in (10.2) that are not involved in the operator $v(12)$. Substituting (10.5) into the matrix element of (10.2) we find that

$$\begin{aligned}\langle j^n\alpha J|v(12)|j^n\alpha' J\rangle = &\sum_{J_2J_{n-2}\beta J_2'J'_{n-2}\beta'} (\alpha' J\{|j^2(J_2)j^{n-2}(\beta J_{n-2});J)\\ &\times (j^2(J_2')j^{n-2}(\beta' J_{n-2}');J|\}\alpha J)\\ &\times \langle j^2(J_2')j^{n-2}(\beta' J_{n-2}');J|v(12)|j^2(J_2)j^{n-2}(\beta J_{n-2});J\rangle\end{aligned} \tag{10.7}$$

where we denote

$$(\alpha J\{|j^2(J_2)j^{n-2}(\beta J_{n-2})J)^* \equiv (j^2(J_2)j^{n-2}(\beta J_{n-2})J|\}\alpha J) \tag{10.8}$$

and we have omitted M from the matrix elements.

The operator $v(12)$ commutes with $\mathbf{J}_2 = \mathbf{j}_1 + \mathbf{j}_2$; hence it connects only states with $J_2 = J_2'$. Due to the orthonormality of the functions $|j^{n-2}\beta J_{n-2}\rangle$, (10.7) for $\alpha = \alpha'$, reduces to

$$\langle j^n\alpha J|v(12)|j^n\alpha J\rangle = \sum_{J_2} W(j^n\alpha J;J_2)\langle j^2J_2|v(12)|j^2J_2\rangle \tag{10.9}$$

where

$$W(j^n\alpha J; J_2) = \sum_{\beta, J_{n-2}} |(\alpha J\{|j^2(J_2)j^{n-2}(\beta J_{n-2}), J)|^2 \tag{10.10}$$

A similar expression is obtained when $\alpha \neq \alpha'$.

Problem. Show that in the general case

$$\langle j^n\alpha J|v(12)|j^n\alpha' J\rangle = \sum_{J_2} W(j^n\alpha\alpha' J; J_2)\langle j^2J_2|v(12)|j^2J_2\rangle \tag{10.11}$$

where

$$W(j^n\alpha\alpha' J; J_2) = \sum_{\beta J_{n-2}} (\alpha' J\{|j^2(J_2)j^{n-2}(\beta J_{n-2})J)(j^2(J_2)j^{n-2}(\beta J_{n-2})J|\}\alpha J) \tag{10.12}$$

The physical interpretation of (10.9) is straightforward: $W(j^n\alpha J, J_2)$ is the probability to find, in the state $|\alpha JM\rangle$ of the configuration j^n, a pair whose total angular momentum is J_2; the average interaction energy of a pair in the state $|j^n\alpha JM\rangle$ is then obtained by summing the interaction energies of that pair in the *pair states* $|j^2J_2M_2\rangle$, weighing each of them with its probability $W(j^n\alpha J; J_2)$.

Since the coefficients of fractional parentage are universal and independent of the particular dynamics of the system, it is possible to use (10.9) to test the configuration assignment of a group of levels when there are more than two particles in the configuration. In fact, for $2 < n < 2j - 1$ the configuration j^n always has more levels than the configuration j^2. If in (10.9) we consider now all the possible n-particle states obtained by taking the different values of α, α', and J, we see that they are all given in terms of the fewer matrix elements $\langle j^2J_2|v(12)|j^2J_2\rangle$. Thus some linear relations, independent of the nature of either the central field $U(i)$ or the residual interaction $v(12)$ must exist between the energies of the various levels of the configuration j^n. These relations can in fact be used to test the assignment of a group of levels to the configuration j^n. We shall explore such relations in the next section, and shall conclude this section by establishing a relation between the representation (10.5) of $|j^n\alpha JM\rangle$ and the method proposed in the previous section for the construction of the wave function—(9.4) or (9.11).

Let us consider, for simplicity, the case $n = 3$, so that on the one hand we have the representation (9.3)

$$|j^3[J_{12}]JM\rangle = C\mathcal{A}|j^2(J_{12})j; JM\rangle \tag{10.13}$$

and on the other hand we have the representation in terms of the coefficients of fractional parentage

$$|j^3\alpha JM\rangle = \sum_{J_2} (\alpha J\{|j^2(J_2)j; J)|j^2(J_2)j; JM\rangle \tag{10.14}$$

The similarity between the two expressions is quite transparent: both express the totally antisymmetric three-particle wave function in terms of wave functions in which the first two particles are singled out. However one method does it by directly antisymmetrizing $|j^2(J_{12})j; JM\rangle$, whereas the other achieves the same thing by summing over different values of J_2.

When there is only one state of the configuration j^3 with a given value of J and M, then (10.13) and (10.14) must be identical. Thus we can write

$$|j^3[J_{12}]; JM\rangle = \sum_{J_2} (J\{|j^2(J_2)j; J)|j^2(J_2)j; JM\rangle \tag{10.15}$$

Multiplying both sides of (10.15) by $\langle j^2(J_2')j; JM|$ and integrating over the coordinates of the three particles we obtain

$$\langle j^2(J_2')j; JM|j^3[J_{12}]; JM\rangle = (J\{|j^2(J_2')j; J) \tag{10.16}$$

The scalar production the left-hand side of (10.16) can be evaluated in a straightforward way. However it is more conveniently evaluated if we replace the antisymmetric function $|j^3[J_{12}]; JM\rangle$ by $(\frac{1}{3})\mathcal{A}_3|j^3[J_{12}]; JM\rangle$ where $\mathcal{A}_3$ is the antisymmetrizer of (10.13) (i.e., antisymmetrization only with respect to exchange with particle 3). Using the hermiticity of the operator $\mathcal{A}_3$ we then obtain:

$$\begin{aligned}(\alpha J\{|j^2(J_2')j; J) &= \frac{1}{3}\langle \mathcal{A}_3 j^2(J_2')i; JM|j^3[J_{12}]JM\rangle \\ &= \frac{1}{3C}\langle j^3[J_2']JM|j^3[J_{12}]JM\rangle\end{aligned}$$

We now use (9.7) and (9.8) to obtain finally

$$(j^3[J_{12}]J\{|j^2(J_2)j; J) = \frac{\delta(J_2, J_{12}) + 2\sqrt{(2J_2+1)(2J_{12}+1)}\begin{Bmatrix} j & j & J_2 \\ j & J & J_{12}\end{Bmatrix}}{\sqrt{3 + 6(2J_{12}+1)\begin{Bmatrix} j & j & J_{12} \\ j & J & J_{12}\end{Bmatrix}}} \quad \text{for } J_2 \text{ even}$$

$$= 0 \quad \text{for } J_2 \text{ odd} \tag{10.17}$$

In (10.17) we have denoted explicitly that the particular coefficient of fractional parentage is that of a state whose "godfather" is the even angular momentum J_{12}.

Similar expressions can be developed for configuration with more particles. For the j^4 one finds that [Schwartz and deShalit (54)]

$$(j^4[J_{12}, J_{34}]J\{|j^2(J_2)j^2(J_4); J) = C\Big[\delta(J_{12}, J_2)\delta(J_{34}, J_4)$$
$$+ (-1)^J\delta(J_{12}, J_4)\delta(J_{34}, J_2) - [1 + (-1)^{J_{12}}][1 + (-1)^{J_{34}}]$$
$$\times [(2J_{12} + 1)(2J_{34} + 1)(2J_2 + 1)(2J_4 + 1)]^{1/2} \begin{Bmatrix} j & j & J_{12} \\ j & j & J_{34} \\ J_2 & J_4 & J \end{Bmatrix}\Big] \tag{10.18}$$

Here $|j^4[J_{12}, J_{34}]JM\rangle$ is an antisymmetric four-particle state obtained by the antisymmetrization of the state in which particles 1 and 2 are antisymmetrically coupled to J_{12}, and particles 3 and 4 are similarly coupled J_{34}.

Coefficients of fractional parentage can be constructed in a variety of ways depending on the particular use one wants to put them to. In the four-particle configuration, for instance, we can define the coefficients $(\alpha J\{|j^3(J_3)j; J)$ that allow the expansion of the antisymmetric state $|j^4\alpha JM\rangle$ in terms of the states $|j^3(J_3)j; JM\rangle$ defined in (9.9). Again, these states, which are antisymmetric with respect to the first three nucleons only, build up a Hilbert space that includes as a subspace that of the fully antisymmetric states $|\alpha j^4JM\rangle$. Projection operators P can then be defined that project out of the states $|j^3(J_3)j; JM\rangle$ their antisymmetric part, and in terms of the coefficients of fractional parentage we have

$$P = \sum_{J_3J} |j^4\alpha J\rangle(\alpha J\{|j^3(J_3)j; J)\langle j^3(J_3)j; J| \tag{10.19}$$

Further discussions of the properties of coefficients of fractional parentage can be found in deShalit and Talmi (63).

11. AN EXAMPLE

We shall see now how (10.9) can be used to test some of the assumptions of the nuclear shell model and obtain further features of it.

The matrix element $\langle j^n\alpha J|(1/2)\Sigma v(ik)|j^n\alpha J\rangle$ gives the first-order shift in energy of the state $|JM\rangle$ from its unperturbed position. The energy difference between two levels αJ and $\alpha' J'$ in the j^n configuration is then given by

$$\begin{aligned} E(j^n\alpha J) - E(j^n\alpha' J') &= \frac{n(n-1)}{2}[\langle j^n\alpha J|v(12)|j^n\alpha J\rangle \\ &\quad - \langle j^n\alpha' J'|v(12)|j^n\alpha' J'\rangle] \\ &= \frac{n(n-1)}{2}\sum_{J_2}[W(j^n\alpha J; J_2) - W(j^n\alpha' J'; J_2)] \\ &\quad \times \langle j^2J_2|v(12)|j^2J_2\rangle \end{aligned} \tag{11.1}$$

Equations 10.2 and 10.9 were used in deriving (11.1). We note that from (10.6) it follows that

$$\sum_{J_2} W(j^n\alpha J;\, J_2) = 1$$

Hence in (11.1) we can subtract from $\langle j^2J_2|\,v(12)\,|j^2J_2\rangle$ a constant, independent of J_2 without affecting the equality. It is convenient to choose as this constant the shift of the lowest state in the configuration j^2, which, for attractive interactions turns out to be the state $J_2 = 0$. We obtain therefore

$$E(j^n\alpha J) - E(j^n\alpha' J') = \frac{n(n-1)}{2} \sum_{J_2\ \text{even}} [W(j^n\alpha J;\, J_2) - W(j^n\alpha' J';\, J_2)]$$
$$\times\ [\langle j^2J_2|\,v(12)\,|j^2J_2\rangle - \langle j^2J_{12} = 0|\,v(12)\,|j^2J_{12} = 0\rangle] \quad (11.2)$$

Equation 11.2 thus gives energy differences in the j^n-configuration in terms of energy differences in a two-body configuration j^2. Note that this two-body configuration refers to energies of pairs of particles in the n-nucleon state, and *not* to energies in another, lighter, nucleus, with just two nucleons in the levels j.

The simplest nontrivial example of the use of (11.2) is for $j = 7/2$ and $n = 3$; for $j = 3/2$ and $n = 3$ there is just one allowed level with $J = 3/2$ and we cannot use (11.2); in the $(5/2)^3$ configuration there are three allowed levels with $J = 9/2$, $J = 5/2$, and $J = 3/2$ but there are also three allowed levels in $(5/2)^2$ with $J = 4$, $J = 2$, and $J = 0$, so that (11.2) reduces to the expression of two energy differences in $(5/2)^3$ in terms of two energy differences in $(5/2)^2$. However in $(7/2)^3$ there are six allowed levels (see Section IV.15) with $J = 15/2$, $11/2$, $9/2$, $7/2$, $5/2$ and $3/2$, whereas $(7/2)^2$ has only four allowed levels with $J = 0$, 2, 4, and 6. In this case (11.2) gives us five independent energy differences in $(7/2)^3$ in terms of three energy differences in $(7/2)^2$, so that we have at our disposal a test for the validity of the underlying model. Table 11.1 contains the coefficients of fractional parentage for $(7/2)^3$ calculated with the help of (10.17).

Using Table 11.1 we find that

$$\begin{aligned}
E(3/2) - E(7/2) &= \frac{1}{84}\,[19V(2) + 135V(4) - 91V(6)] \\
E(5/2) - E(7/2) &= \frac{1}{132}\,[187V(2) - 75V(4) - 13V(6)] \\
E(9/2) - E(7/2) &= \frac{3}{308}\,[-11V(2) + 123V(4) - 35V(6)] \qquad (11.3) \\
E(11/2) - E(7/2) &= \frac{1}{132}\,[55V(2) - 21V(4) + 65V(6)] \\
E(15/2) - E(7/2) &= \frac{1}{132}\,[-55V(2) - 9V(4) + 163V(6)]
\end{aligned}$$

TABLE 11.1 The coefficients of Fractional Parentage for $(7/2)^3$

J_2 \ J	15/2	11/2	9/2	7/2	5/2	3/2
	$((7/2)^2\,(J_2)7/2;J\,\|\}\,(7/2)^3J)$					
0	0	0	0	$\frac{1}{2}$	0	0
2	0	$-\frac{1}{3}\sqrt{\frac{5}{2}}$	$\frac{1}{3}\sqrt{\frac{13}{14}}$	$-\frac{\sqrt{5}}{6}$	$\frac{1}{3}\sqrt{\frac{11}{2}}$	$\sqrt{\frac{3}{14}}$
4	$\sqrt{\frac{5}{22}}$	$\sqrt{\frac{13}{66}}$	$-5\sqrt{\frac{2}{77}}$	$-\frac{1}{2}$	$\sqrt{\frac{2}{33}}$	$-\sqrt{\frac{11}{14}}$
6	$\sqrt{\frac{17}{22}}$	$\frac{2}{3}\sqrt{\frac{13}{11}}$	$\frac{7}{3\sqrt{22}}$	$-\frac{\sqrt{13}}{6}$	$-\frac{1}{3}\sqrt{\frac{65}{22}}$	0

where we have introduced the notation

$$V(J) = [\langle j^2J|v(12)|j^2J\rangle - \langle j^2J = 0|v(12)|j^2J = 0\rangle] \qquad (11.4)$$

it being understood that the matrix elements in (11.4) are taken with wave functions appropriate to the nucleus whose j^3 configuration we want to study.

An example of a spectrum that has been analyzed using (11.3) is provided by ${}^{51}_{23}V_{28}$. This is shown in Fig. 11.1*a*. A somewhat less successful prediction from (11.3) is shown in Fig. 11.1*b*.

12. CONFIGURATION MIXING IN MANY-PARTICLE SYSTEMS

In the previous section we saw how the basic assumptions of the shell model can be tested. More precisely, we started out from the assumption that the dominant nuclear interaction in nuclei is a two-body potential $v(ij)$, which can be treated as a first-order perturbation, once a central field approximation is used to derive zeroth-order wave functions. We were then led to the derivation of certain linear relations (11.2) between the energy levels of any configuration j^n with $2 < n < 2j - 1$. The linear dependence between the energies of the levels of the configuration j^n is a reflection of the basic assumption of the shell model: that the interaction among the nucleons is due to a two-body force, which can be treated as a perturbation and is entirely independent of any further assumption about the detailed nature of that force, or, for that matter, of the exact form of the central potential $U(i)$. If a set of levels in a given nucleus, which are all ascribed to a given configuration j^n, fails to satisfy these linear relations, we must conclude one of two things: either we were wrong in assigning these levels to the pure configuration j^n, or two-body forces

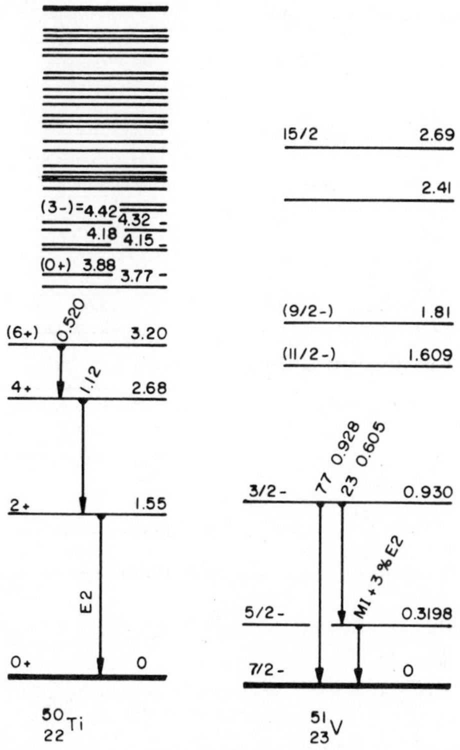

FIG. 11.1a. Shows the observed levels in $^{50}_{22}Ti_{28}$ and $^{51}_{23}V_{28}$. Taking the values of $V(J)$ in (11.3) from the levels of ^{50}Ti [$V(2) = 1.55$ MeV, $V(4) = 2.68$ MeV, and $V(6) = 3.20$ MeV] we obtain for the calculated energies in ^{51}V the values listed below. (Data from Lederer et. al. (67)).

Level	3/2	5/2	7/2	9/2	11/2	15/2
Calculated E	1.19	0.36	0	1.95	1.79	3.13
Experimental E	0.930	0.32	0	1.81	1.61	2.70

are not sufficient for the complete description of the dynamics of that nucleus. In principle one can apply further tests to the levels in question to try and decide between these two possibilities. We saw in Section IV.17, for instance, that the magnetic moment in a state $|j^nJ\rangle$ of n equivalent nucleons is given by

$$\langle j^nJ|\left(\sum_i \hat{\mu}_{iz}\right)|j^nJ\rangle = \frac{J}{j}\langle j|\hat{\mu}_z|j\rangle \tag{12.1}$$

where $\langle j|\hat{\mu}_z|j\rangle$ is the magnetic moment of a single particle in the j-level. Thus, if all the levels in question belong to the same configuration their magnetic moments should all lead to the *same* g-factor, where $g_J = \langle j^nJ|\,\Sigma\hat{\mu}_{iz}|j^nJ\rangle/J$. Other similar tests can be thought of, involving other moments, transition

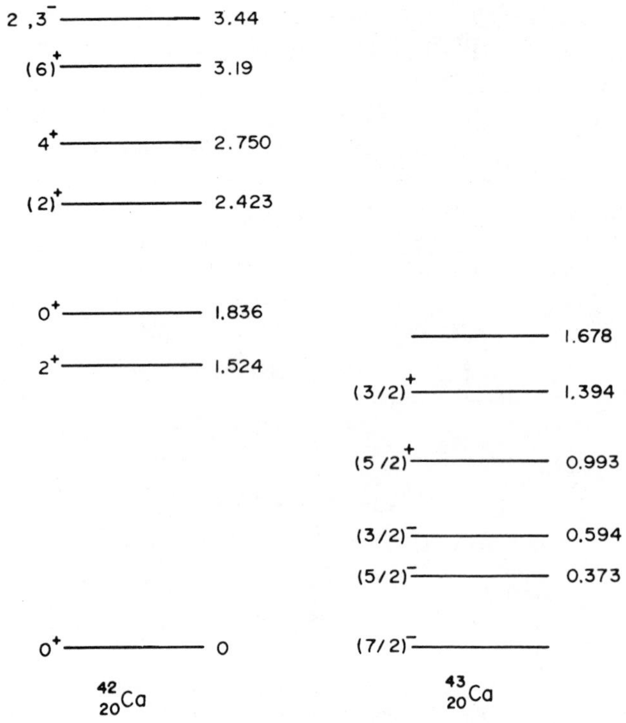

FIG. 11.1b. Shows the observed levels in ${}^{42}_{20}Ca_{22}$ and ${}^{43}_{20}Ca_{23}$. These are again supposed to be $(f_{7/2})^2$ and $(f_{7/2})^3$ nuclei, respectively. $V(J)$ is taken from ${}^{42}Ca$ [$V(2) = 1.524$ MeV, $V(4) = 2.750$ MeV, $V(6) = 3.19$ MeV] leading to ${}^{43}Ca$, through (11.3), to $E(3/2) = 1.32$ MeV (experimental 0.59 MeV) and $E(5/2) = 0.360$ MeV (experimental 0.373 MeV). Equation 11.3 describes rather well the connection between the spectra of ${}^{51}V$ and ${}^{50}Ti$, and it does less well for the connection between ${}^{42}Ca$ and ${}^{43}Ca$. Data was taken from Lederer et al. (67).

probabilities, or reaction cross sections. Today there is still no case where such complete data is available on all the levels of a given configuration. We shall therefore not go here into the detailed study of the interrelations of such possible tests.

Another question may, however, be raised: suppose a group of levels that are assigned to a configuration j^n do satisfy the linear relations implied by (11.2). Can we conclude then that these levels really represent the levels of the pure configuration j^n? This is a rather important question, for if the answer will turn out to be in the affirmative, then (11.2) will enable us to obtain the matrix elements $\langle j^2 J_2 | v(12) | j^2 J_2 \rangle$ of the nuclear interaction $v(12)$ in nuclear matter, and to compare it then to matrix elements of the free two-nucleon interaction. We shall then know to what extent $v(12)$ is modified when nucleons 1 and 2 are immersed in the dense environment of other nucleons.

Unfortunately the answer to our question turns out to be negative, and we shall now proceed to see why this is so. This analysis will give us also a deeper understanding of the complexities involved in the interpretation of a limited set of data of a multinucleon system.

Suppose that an observed level of angular momentum J cannot be described as a level of the pure configuration j^n, but should involve another configuration for its description. For simplicity we shall assume that the proper description of the n-particle state $|nJ\rangle$ can be written in the form

$$|nJ\rangle = |j^nJ\rangle + \sum_\alpha a_\alpha |j'^2 j^{n-2}\alpha J\rangle \qquad |a_\alpha| \ll 1 \tag{12.2}$$

In other words we assume, for reasons to be explained later, that the correct description of the n-particle state $|nJ\rangle$ is obtained by adding to $|j^nJ\rangle$, with small amplitudes a_α, a set of states in which two of the nucleons in the j-level actually move to another level j'. The energy shift of the level $|J\rangle$ is then given by 2nd order perturbation theory through

$$E^{(2)}(nJ) = \langle j^nJ|\Sigma v(ik)|j^nJ\rangle + \sum_\alpha \frac{|\langle j'^2 J^{n-2}\alpha J|\Sigma v(ik)|j^nJ\rangle|^2}{\epsilon(jj')} \tag{12.3}$$

In (12.3) $\epsilon(jj')$ is the zeroth-order energy difference between the configuration j^n and $j^{n-2}j'^2$.

Let us compute the n-dependence of the second term in (12.3) to see whether it is similar to that of the first term. The wave function $|j^nJ\rangle$ is antisymmetric in all n-particles; $\Sigma v(ik)$ is a symmetric operator; the product $\Sigma v(ik)|j^nJ\rangle$ can therefore connect only with an antisymmetric wave function. We therefore need not worry about the symmetry of the states $|j'^2 j^{n-2}\alpha J\rangle$ as long as we are sure to include all states of this configuration and have them properly normalized; the matrix element in (12.3) will automatically pick only the antisymmetric part of $|j'^2 j^{n-2}\alpha J\rangle$. We shall therefore take for the states $|j'^2 j^{n-2}\alpha J\rangle$ the set of states defined by

$$|j'^2(J'_{ik})j^{n-2}(\beta J_{n-2}); J\rangle \qquad (i < k) \tag{12.4}$$

In (12.4) the notation implies that the particles i and k are in the level j' coupled to angular momentum J'_{ik} and the rest of the particles are in the j-level forming the state $|j^{n-2}\beta J_{n-2}\rangle$. The states (12.4) are *not* totally antisymmetric. The second term in (12.3) can then be written as

$$\Delta E_2(nJ) = \sum_\alpha \frac{|\langle j'^2 j^{n-2}\alpha J|\Sigma v(ik)|j^nJ\rangle|^2}{\epsilon(jj')}$$

$$= \frac{1}{\epsilon(jj')} \sum_{i<k,\, i'<k',\, J'_{i'k'}\beta,\, J_{n-2}} |\langle j'^2(J'_{i'k'})j^{n-2}(\beta J_{n-2}); J|v(ik)|j^nJ\rangle|^2 \tag{12.5}$$

In (12.5) if $(ik) \neq (i'k')$ then the corresponding term in the sum vanishes. This is easily seen if we notice that $v(ik)$ can change the orbit of particles i and k only, and if they are not identical with i' and k', respectively, then the matrix element will vanish due to the orthogonality of the states j and j'. Since, in addition,

$$|\langle j'^2(J'_{ik})j^{n-2}(\beta J_{n-2}); J|v(ik)|j^nJ\rangle|^2$$

is the same for all pairs (ik), we obtain from (12.5) that:

$$\Delta E_2(nJ) = \frac{n(n-1)}{2}\frac{1}{\epsilon(jj')}\sum_{J'_{12}\beta' J'_{n-2}} |\langle j'^2(J'_{12})j^{n-2}(\beta' J'_{n-2})J|v(12)|j^nJ\rangle|^2 \tag{12.6}$$

We now use the fractional coefficient expansion for $|j^nJ\rangle$, and obtain

$$\Delta E_2(nJ) = \frac{n(n-1)}{2}\frac{1}{\epsilon(jj')}\sum_{J'_{12}\beta' J'_{n-2}, J_2\beta J_{n-2}} |(J\{|j^2(J_2)j^{n-2}(\beta J_{n-2}); J)|^2$$
$$\times |\langle j'^2(J'_{12})j^{n-2}(\beta' J'_{n-2})J|v(12)|j^2(J_2)j^{n-2}(\beta J_{n-2})J\rangle|^2 \tag{12.7}$$

Since $v(12)$ can only connect states of the same total angular momentum for particles 1 and 2 we obtain

$$\langle j'^2(J'_{12})j^{n-2}(\beta' J'_{n-2})J|v(12)|j^2(J_2)j^{n-2}(\beta J_{n-2})J\rangle$$
$$= \langle j'^2J_2|v(12)|j^2J_2\rangle\delta(J'_{12}J_2)\delta(\beta'\beta)\delta(J'_{n-2}J_{n-2}) \tag{12.8}$$

Using the definition of $W(j^nJ; J_2)$ in (10.10) we obtain finally for the second-order correction

$$\Delta E_2(nJ) = \left[\sum_{J_2}\frac{|\langle j'^2J_2|v(12)|j^2J_2\rangle|^2}{\epsilon(jj')}W(j^nJ; J_2)\right]\frac{n(n-1)}{2} \tag{12.9}$$

The first-order correction is given, as in (11.1) by

$$\Delta E_1(nJ) = \langle j^nJ|\Sigma v(ik)|j^nJ\rangle$$
$$= [\sum_{J_2}\langle j^2J_2|v(12)|j^2J_2\rangle W(j^nJ; J_2)]\frac{n(n-1)}{2} \tag{12.10}$$

We define now the two-body matrix of an *effective interaction* $\tilde{v}(12)$ by

$$\langle j^2J_2|\tilde{v}(12)|j^2J'_2\rangle = \left\{\left[\langle j^2J_2|v(12)|j^2J_2\rangle + \frac{|\langle j'^2J_2|v(12)|j^2J_2\rangle|^2}{\epsilon(jj')}\right]\right\}\delta(J_2, J'_2) \tag{12.11}$$

We notice that although the matrix elements of $\tilde{v}(12)$ may look rather simple, $\tilde{v}(12)$ itself may be a very complicated, nonlocal interaction. However it is still a two-body interaction, and a scalar one because of the factor $\delta(J_2, J'_2)$

on the right-hand side of (2.11). In terms of the effective interaction $\tilde{v}(12)$ we can write the energy shifts $E^{(2)}(nJ)$, correct to second order in $\Sigma v(ik)$, as

$$E^{(2)}(nJ) - E^{(2)}(nJ') = \frac{n(n-1)}{2} \sum_{J_2} [W(j^nJ; J_2) - W(j^nJ'J_2)] \times \langle j^2J_2|\tilde{v}(12)|j^2J_2\rangle \quad (12.12)$$

Comparison of (12.12) and (11.1) shows that the energies of the n-nucleon levels satisfy *the same linear relations*, whether we consider them to be the levels of the pure configuration j^n as in (11.1), or improve the wave functions by the inclusion of the small corrections $|j'^2j^{n-2}\alpha J\rangle$ as in (12.2). The interpretation of the resulting two-body matrix elements will be different in the two cases: if the pure configuration condition applies we obtain from (11.1) the matrix elements $\langle j^2J_2|v(12)|j^2J_2\rangle$ of the actual interaction $v(12)$. If, on the other hand, a configuration impurity is introduced, as in (12.2), the resulting two-body matrix element will include a correction to the actual interaction as is evident from (12.11).

In Chapter VII we shall show that this result, (12.12), holds true for an important class of corrections to the pure configuration j^n—the so-called *ladder-diagram corrections.* Our present calculation already suffices, however, to draw an important conclusion: if a set of levels in a nucleus is ascribed to a certain configuration j^n, and if these levels do satisfy the relations implied by (11.1), then all we can conclude is the existence of an *effective* two-body interaction $\tilde{v}(12)$, with its matrix elements given by (12.12), which explains the data. The relation of this interaction to the actual two-body nuclear forces in the nucleus may be rather complex, as is seen from (12.11).

When the description of a nuclear state requires the addition of small amplitudes of other configurations, like in (12.2), we usually talk of *configuration mixing* (as opposed to pure configurations). What we have just shown is that *some configuration mixing can be simulated,* as far as the energies are concerned, by pure configurations applied to a renormalized, or effective, interaction $\tilde{v}(12)$. Similar results can be shown to be valid for other nuclear properties. The apparent success of the shell model in interrelating various nuclear properties such as energies, magnetic moments, and magnetic dipole transitions should therefore be understood against this background. It shows at the same time both the power and the limitations of using a definite model in our analysis of nuclear structure: the results implied by a given model can have a range of validity that is broader than that of the model itself.

In view of the results obtained thus far it is evident that nuclei for which the shell model can be successfully applied can provide us, at most, with matrix elements of an effective two-body interaction. In order to be able to interpret these matrix elements in terms of some physical properties of the two-body interaction we need to understand better the connection between such matrix elements and the range and exchange character of $\nu(12)$. This is beyond the scope of our discussion and some remarks on this question are given in the appendix to this chapter. We shall devote the rest of this chapter to a very commonly used special case of the shell model, that derived from the harmonic oscillator potential.

13. NUCLEAR CONFIGURATIONS IN HARMONIC-OSCILLATOR POTENTIALS

The evaluation of matrix elements of the type

$$E(jj'J) = \langle jj'J|\nu(12)|jj'J\rangle \tag{13.1}$$

leads to complex expressions because $\nu(12)$ and the two-particle wave function $|jj'J\rangle$ have different natural coordinates: the interaction $\nu(12)$ is naturally expressed in terms of the relative coordinate $|\mathbf{r}_1 - \mathbf{r}_2|$; while the wave function $|jj'\rangle$, which describes two particles moving independently in a spherical wave, is naturally expressed in terms of the coordinates $\mathbf{r}_1$ and $\mathbf{r}_2$. The multipole expansion described in Section IV.20 can be conceived as a way of breaking up the $|\mathbf{r}_1 - \mathbf{r}_2|$ dependence of $\nu(12)$ into a separate $\mathbf{r}_1$ and $\mathbf{r}_2$ dependence, to match that of the wave function $|jj'J\rangle$ and even this can generally be done only with respect to the angular dependence Ω_1 and Ω_2. Once this is achieved the integrations proceed relatively simply.

Another conceivable approach to the evaluation of $E(jj'J)$ would be to write the wave function $|jj'J\rangle$ in terms of the relative coordinate $\mathbf{r} = \mathbf{r}_1 - \mathbf{r}_2$ and the center-of-mass coordinate $\mathbf{R} = (1/2)(\mathbf{r}_1 + \mathbf{r}_2)$. Since $\nu(12)$ does not depend on $\mathbf{R}$, the center-of-mass integration in (13.1) could then be carried out explicitly, and we shall be left only with the integration over the relative coordinate $\mathbf{r}$. Knowing some of the properties of $\nu(12)$ we may be able to approximate such integrals rather well.

The general wave function of two particles in a central field does not, generally, separate into a product of functions of the relative coordinate of the two particles and their center-of-mass coordinate. There is, however, one notable exception: that of particles moving in a harmonic-oscillator potential (see Section IV.3). Since this single-particle potential is used in many nuclear calculations, (it is the only potential that allows a clear separation of the center-of-mass motion of the whole nucleus), we shall devote this section to its study.

The potential energy of two particles of the same mass M moving in a harmonic oscillator potential can be written as

$$\frac{1}{2} M\omega^2(r_1^2 + r_2^2) = \frac{1}{2}\left(\frac{M}{2}\right)\omega^2 r^2 + \frac{1}{2}(2M)\omega^2 R^2 \tag{13.2}$$

where

$$\mathbf{r} = \mathbf{r}_1 - \mathbf{r}_2 \qquad \text{and} \qquad \mathbf{R} = \tfrac{1}{2}(\mathbf{r}_1 + \mathbf{r}_2) \tag{13.3}$$

The kinetic energy $T_1 + T_2$ of the two particles is also separable into relative and center-of-mass kinetic energies, with the masses $(M/2)$ and $2M$, respectively. It follows that an eigenstate of

$$H_0 = \frac{\hbar^2}{2M}(\nabla_1^2 + \nabla_2^2) + \frac{1}{2}M\omega^2(r_1^2 + r_2^2)$$

$$= \frac{\hbar^2}{2(M/2)}\nabla_r^2 + \frac{\hbar^2}{2(2M)}\nabla_R^2 + \frac{1}{2}\left(\frac{M}{2}\right)\omega^2 r^2 + \frac{1}{2}(2M)\omega^2 R^2 \qquad (13.4)$$

can be written either as a product $\psi_1(\mathbf{r}_1)\psi_2(\mathbf{r}_2)$ or as a product $\Psi_1(\mathbf{R})\phi_2(\mathbf{r})$. Furthermore, since the eigenvalues of H_0 are degenerate, we can always expand

$$\psi_1(\mathbf{r}_1)\psi_2(\mathbf{r}_2) = \sum_{i,k} a_{ik}\Psi_i(\mathbf{R})\phi_k(\mathbf{r}) \qquad (13.5)$$

where $\Psi_i(\mathbf{R})\phi_k(\mathbf{r})$ and $\psi_1(\mathbf{r}_1)\psi_2(\mathbf{r}_2)$ belong to the *same eigenvalue* E_0 [Talmi (52)].

The shell-model wave functions, ignoring spin for the moment, are certain linear combinations of products of the type $\psi_1(\mathbf{r}_1)\psi_2(\mathbf{r}_2)$, which also have a definite total orbital angular momentum LM. We shall denote them by

$$\psi_{12}(n_1l_1n_2l_2, LM) \qquad (13.6)$$

and notice that they are constructed from a state in which one of the particles has n_1 radial nodes and orbital angular momentum l_1, while the other has n_2 radial nodes and angular momentum l_2. ψ_{12} may be either symmetric or antisymmetric in 1 and 2. The state (13.6) is an eigenstate of H_0 that belongs to the eigenvalue

$$E_0 = (2n_1 + l_1 - 1/2)\hbar\omega + (2n_2 + l_2 - 1/2)\hbar\omega \qquad (13.7)$$

Using the expression of H_0 in terms of $\mathbf{R}$ and $\mathbf{r}$ we can construct wave functions $\Psi_{N\Lambda M_\Lambda}(\mathbf{R})$ and $\phi_{nlm}(\mathbf{r})$ that have the indicated numbers of nodes N and n, respectively, and correspond to the indicated angular momenta of the center of mass (ΛM_Λ) and the relative coordinate (lm). The energies associated with them are $(2N + \Lambda - 1/2)\hbar\omega$ and $(2n + l - 1/2)\hbar\omega$, respectively. Note that (ΛM_Λ) is the orbital angular momentum of the center of mass of the two particles in their motion in the central potential.

By using proper Clebsch–Gordan coefficients we can now construct from $\Psi_{N\Lambda M_\Lambda}(\mathbf{R})$ and $\phi_{nlm}(\mathbf{r})$ wave functions of well-defined angular momentum. It now follows from (13.5) that we can always find proper coefficients such that

$$\psi_{12}(n_1l_1n_2l_2LM) = \sum_{N\Lambda nl,mM_\Lambda} a_{N\Lambda nl}^{n_1l_1n_2l_2,L}(lm\Lambda M_\Lambda|LM)\phi_{nlm}(\mathbf{r})\Psi_{N\Lambda M_\Lambda}(\mathbf{R}) \qquad (13.8)$$

Furthermore, since the right-hand side of (13.8) should contain only wave functions that correspond to the same energy as that of ψ_{12}, the summation in (13.8) is limited to values of N, Λ, n, and l that satisfy

$$2N + \Lambda + 2n + l = 2n_1 + l_1 + 2n_2 + l_2 \qquad (13.9)$$

In addition the values of Λ, l, and L must also satisfy the vector-addition inequalities

$$|l - \Lambda| < L < l + \Lambda \qquad |l_1 - l_2| < L < l_1 + l_2 \tag{13.10}$$

We can now use (13.8) in order to evaluate the matrix element (13.1), or better still the matrix element in the l-scheme alone:

$$E(l_1l_2L) = \langle l_1l_2LM|\,v(|\mathbf{r}_1 - \mathbf{r}_2|)\,|l_1l_2LM\rangle$$

It follows then from (13.8), noting that $v(|\mathbf{r}_1 - \mathbf{r}_2|)$ does not depend on $\mathbf{R}$, or the direction of $\mathbf{r}$, that

$$E(l_1l_2L) = \sum_{nlN\Lambda} |a_{N\Lambda nl}^{n_1l_1n_2l_2,L}|^2 \int |R_{nl}(r)|^2 v(r)\,dr \tag{13.11}$$

Here $R_{nl}(r)$ is the normalized radial part of $\phi_{nlm}(\mathbf{r})$

$$\phi_{nlm}(\mathbf{r}) = \frac{1}{r} R_{nl}(r) Y_{lm}(\theta, \phi) \tag{13.12}$$

The a-coefficients in (13.8) are independent of $v(r)$ and of ω. The summation over N and Λ can therefore be carried out independent of the interaction $v(r)$. We define therefore new universal coefficients $b_{nl}{}^L$ (for convenience we do not indicate their dependence on $n_1l_1n_2$ and l_2)

$$b_{nl}{}^L = \sum_{N\Lambda} |a_{N\Lambda nl}^{n_1l_1n_2l_2,L}|^2 \tag{13.13}$$

and rewrite (13.11) in the form:

$$E(l_1l_2L) = \sum_{nl} b_{nl}{}^L \int |R_{nl}(r)|^2 v(r)\,dr \tag{13.14}$$

This expression can be further simplified if we notice that [see Morse and Feshbach (53), p. 1662; deShalit and Talmi, (63), p. 238] the square of the radial wave function $|R_{nl}(r)|^2$ in a harmonic oscillator potential can be written as

$$|R_{nl}(r)|^2 = P_{(2n+l-1)}(2\gamma r^2) \exp(-2\gamma r^2) \tag{13.15}$$

where $P_m(x)$ is a polynomial of degree m in x and

$$\gamma = \frac{M}{2}\frac{\omega}{2\hbar} \tag{13.16}$$

We see from (13.15) that $|R_{nl}(r)|^2 \cdot \exp(2\gamma r^2)$ is a polynomial of degree $(2n + l - 1)$ in $x = 2\gamma r^2$. In particular $|R_{1l'}(r)|^2 \exp(2\gamma r^2)$ is a polynomial of degree $l' + 1$ in $x = 2\gamma r^2$. We can express then $|R_{nl}(r)|^2 \exp(2\gamma r^2)$ as a linear combination of the more restricted set of functions $|R_{1l'}(r)|^2 \exp(2\gamma r^2)$.

It follows, therefore, that we can find coefficients $C_{l'}{}^{(nl)}$, independent of ω and $v(r)$, such that

$$|R_{nl}(r)|^2 = \sum_{0<l'<l+2n-2} C_{l'}^{(nl)} |R_{1l'}(r)|^2 \tag{13.17}$$

We therefore define new universal coefficients, constructed out of $b_{nl}{}^L$ and $C_{l'}{}^{(nl)}$:

$$\alpha_l{}^L = \sum_{n'l'} b_{n'l'}^L C_l{}^{(n'l')} \tag{13.18}$$

In terms of $\alpha_l{}^L$ we can write (13.14) in the form

$$E(l_1 l_2 L) = \sum_l \alpha_l{}^L I_l \tag{13.19}$$

where the *Talmi integrals* I_l are defined by

$$I_l = \int |R_{1l}(r)|^2 v(r)\, dr = \frac{(2\gamma)^{l+3/2} 2^{l+2}}{\sqrt{\pi}(2l+1)!!} \int_0^\infty r^{2l+2} \exp(-2\gamma r^2) v(r)\, dr \tag{13.20}$$

The Talmi integral I_l [see deShalit and Talmi (63), loc. cit.] is therefore an average value of the interaction $v(r)$ taken for two particles in a state of relative angular momentum l with respect to each other. In other words it is the average of the interaction in a relative l-state, weighted with harmonic-oscillator radial wave functions.

We thus see that when $U(i)$ is the harmonic oscillator potential, the energies of the configuration $(l_1 l_2)$ can be expressed in terms of some simple integrals over the interaction $v(r)$.

The coefficients $\alpha_l{}^L$ have been calculated once and for all [Brody and Moshinsky (60)], so that now it is a simple matter to write down explicitly expressions for (13.19). For instance the configuration ($1s1d$) can have two states with $L = 2$—the symmetric (^1D) and the antisymmetric (^3D). Their energies in terms of the Talmi-integrals turn out to be

$$\left.\begin{aligned} E(^1D) &= \frac{1}{2}(I_0 + I_2) \\ E(^3D) &= I_1 \end{aligned}\right\} \quad \text{for} \quad (1s, 1d)$$

The configuration ($2s$, $1d$) has five Talmi integrals contributing to the energies of its levels

$$\left.\begin{aligned} E(^1D) &= \frac{1}{48}[15I_0 - 2I_1 + 70I_2 - 98I_3 + 63I_4] \\ E(^3D) &= \frac{1}{8}[7I_1 - 6I_2 + 7I_3] \end{aligned}\right\} \quad \text{for} \quad (2s, 1d)$$

The configuration $(1p)^2$ takes on the following form:

$$\left.\begin{aligned} E(^1S) &= \frac{5}{4}(I_0 + I_2) - \frac{3}{2}I_1 \\ E(^3P) &= I_1 \\ E(^1D) &= \frac{1}{2}(I_0 + I_2) \end{aligned}\right\} \qquad \text{for} \qquad (1p)^2$$

We see from (13.20) that if $v(r)$ has a very long range and is constant within this range, I_l is approximately equal to that constant for all l's. Such an interaction, which is effectively a constant, should give rise to no splittings among the different levels. Indeed, we find in all the above examples that setting $I_l \approx$ constant leads always to a degeneracy of all the levels of the same configuration. Note that it can be shown in general that $\Sigma_l \alpha_l{}^L = 1$.

For very short ranges, I_0 is the dominating integral and we can often neglect the Talmi integrals I_l with $l > 0$. In the limit of a δ-interaction

$$v(r) = \frac{1}{4\pi r^2}\,\delta(r)$$

for which we have exactly

$$I_l(\delta) = 0 \qquad \text{for} \qquad l \neq 0. \tag{13.21}$$

The representation (13.14) for the energies of the configuration $l_1 l_2$ is thus particularly convenient for short range interactions.

The method described above for the evaluation of matrix elements using harmonic oscillator wave functions can be easily generalized to off-diagonal elements as well. It can also be applied to *jj*-coupling and to interactions $v(12)$ that depend on the spins as well, provided both particles move in zeroth order, in the same harmonic oscillator potential. For computational purposes the method has proved to be very convenient to use, especially since tables of the coefficients $a_{N\Lambda nl}^{n_1 l_1 n_2 l_2, L}$ in (13.11) have become available [see Brody and Moshinsky (60)].

We have described a method for the evaluation of two-particle matrix elements of the interaction $v(12)$. If the exact form of $v(12)$ and the single-particle wave functions are known one can use any method for the evaluation of the matrix elements—either a straightforward integration or one of the methods described above. But, if on the other hand, we have only partial information concerning either $v(12)$ or the single-particle wave functions, then the method of the multipole expansion, or that using the harmonics oscillator wave functions, may provide a very convenient scheme for carrying out some specific approximations. In the next section we shall see an example of how it works in detail.

14. THE EXCITED STATES OF THE α-PARTICLE—A DETAILED EXAMPLE

The α-particle is, in many respects, the simplest complex nucleus. It has enough particles in it to exhibit structure and, at the same time, their small number permits the use of some interesting approximations as we shall presently see.

Experimentally it is known that the α-particle has no bound state except the ground state. A number of resonances have been observed in ^{4}He, and the existence of others has been inferred indirectly from resonances in ^{4}H and ^{4}Li. Neglecting Coulomb forces, which are rather unimportant in these nuclei, all three mass-4 nuclei should show similar structure in parts of their spectra (see Figs. 14.1*a* and 14.1*b*). The observed widths of the resonances in ^{4}He with respect to various particle emissions are a few MeV. The corresponding excited states of ^{4}He have therefore a lifetime of about 2 x 10^{-22} sec. The potential energy of a nucleon in ^{4}He is about -40 MeV; its binding energy is about –7 MeV. Hence the average velocity of a nucleon in ^{4}He is about $(1/4)c$, where c is the velocity of light. During the lifetime of the excited states of ^{4}He a nucleon traverses, therefore, a distance of some 15 fermi, which is 4 or 5 times the nuclear diameter of ^{4}He. The nucleons are therefore somewhat "trapped" in ^{4}He even in these very short-lived states. We can try then to approximate the energies of these unbound levels by treating them as if they were bound states. The calculation of corrections to this zero width approximation will be discussed in Vol. II.

In calculating excitation energies in light nuclei special care has to be given to center-of-mass motion. Its omission can lead to sizeable errors in the calculated spectrum. If $H = \Sigma T_i + (1/2)\Sigma v_{ij}$ is the Hamiltonian describing the free α-particle, we shall now treat it directly by transforming, as in Eq. IV.2.7, to the modified Hamiltonian

$$H' = H + \frac{1}{2} AM\omega^2 \left(\frac{\Sigma \mathbf{r}_i}{A}\right)^2 \tag{14.1}$$

In H' the α-particle's center of mass, $\mathbf{R} = \Sigma \mathbf{r}_i/A$, is bound by a harmonic force to the point $\mathbf{R} = 0$. This does not affect the intrinsic structure of the α, but, of course, makes all its states bound (via the binding of the center of mass). The eigenvalues E' of H' consist, therefore, of intrinsic excitation energies E_i to which are added all possible excitations of the center of mass in the harmonic potential by which it is bound. Thus E' will have the general structure

$$E'_{in} = E_i + (n + 3/2)\hbar\omega \tag{14.2}$$

We are obviously interested only in E_i, and we shall see below how to distinguish between a real intrinsic excitation and a center-of-mass excitation that may lead to the same energy.

Using the transformation, Eq. IV.2.6, H' can be written in the form

$$H' = \sum_i \left[T_i + \frac{1}{2} M\omega^2 r_i^2 \right] + \frac{1}{2} \sum_{i \neq j} \left[v(ij) - \frac{M\omega^2}{2A} (\mathbf{r}_i - \mathbf{r}_j)^2 \right] \tag{14.3}$$

An exact diagonalization of H' would have led to the spectrum (14.2). This cannot be carried out, however, because of the complexity of $v(ij)$ and we have to use some approximation to obtain the eigenvalues of H'. In the spirit of the shell model we then assume that if ω is chosen properly (see Section IV.2) we can take

$$H_0 = \sum \left(T_i + \frac{1}{2} M\omega^2 r_i^2 \right) \tag{14.4}$$

as the zeroth-order shell-model Hamiltonian, and treat

$$\frac{1}{2} \sum_{i \neq j} v'(ij) = \frac{1}{2} \sum_{i \neq j} \left[v(ij) - \frac{M\omega^2}{2A} (\mathbf{r}_i - \mathbf{r}_j)^2 \right] \tag{14.5}$$

as a perturbation. We notice that H_0 is the Hamiltonian for independent particles moving in a central field, and $v'(ij)$ remains a two-body interaction even after the correction imposed on it by the binding of the center of mass of the nucleus.

Starting from this approximation we expect the lowest four-particle state to be that in which all four particles are in the lowest level of the harmonic single-particle potential $(1/2)M\omega^2 r^2$. Thus the ground state of ^{4}He is $(1s)^4$. Since an s-level can accommodate at most two neutrons and two protons, we conclude immediately that this state must have $J = 0$. Furthermore the state must have $S = 0$ as well, since to accommodate two protons in an s-level their spins must be in an antisymmetric state with $S_p = 0$; the same holds true for the pair of neutrons and hence $\mathbf{S} = \mathbf{S}_p + \mathbf{S}_n = 0$. Finally, the isotopic spin T of this state must also vanish. Indeed, had it been $T = 1$ or $T = 2$ we could have had states with the same spin and space dependence but with, say, $M_T = 1$. These would correspond to states of three protons and one neutron, all four in the $1s$-level. But we know from the Pauli principle that at most two protons can be accommodated in an s-level; hence the conclusion:

The ground state of ^{4}He is $(1s)^4$ with $J = 0$, $S = 0$, and $T = 0$ (14.6)

The energy of this ground state of ^{4}He, treating $v'(ij)$ as a perturbation, is

$$E'_{00} = 4 \cdot \left(\frac{3}{2} \hbar\omega \right) + \langle (1s)^4 J = 0\ T = 0 | \sum_{i<j} v'(ij) | (1s)^4 J = 0\ T = 0 \rangle \tag{14.7}$$

where the first term comes from H_0 and represents the energy of four particles in the lowest $1s$-state of the harmonic-oscillator potential.

It follows from (14.2) that the intrinsic energy of the lowest state of ^{4}He is given by

$$E_0 = 3\cdot\left(\frac{3}{2}\hbar\omega\right) + \langle (1s)^4 J = 0\, T = 0 |\sum_{i<j} v'(ij)| (1s)^4 J = 0\, T = 0\rangle \tag{14.8}$$

If we now look for excited states of the Hamiltonian H', we should expect two types of excitations: an intrinsic excitation and a center-of-mass excitation. According to (14.2) we should find the lowest center-of-mass excited state at an energy

$$E'_{01} = E_0 + \frac{5}{2}\hbar\omega = 7\hbar\omega$$
$$+ \langle (1s)^4 J = 0\, T = 0 |\sum_{i<j} v'(ij)| (1s)^4 J = 0\, T = 0\rangle = E'_{00} + \hbar\omega \tag{14.9}$$

Let us see what should be the quantum numbers of this state. Since it represents a center-of-mass motion of ^{4}He whose intrinsic structure is that of the ground state (14.6), the state at E'_{01} should have the same values for S and T as the ground state of ^{4}He. Center-of-mass motion cannot affect the spin or isospin dependence of the nuclear wave function. Furthermore, we know that the first excited state in a harmonic-oscillator potential is a p-state, and is thus associated with a negative parity and the addition of one unit of angular momentum. Since the ground state of ^{4}He has $J = 0^+$ we conclude that

the first center-of-mass excited state of ^{4}He at an energy E'_{01} has $S = 0$, $T = 0$, and $J = 1^-$ (14.10)

The result (14.10) makes it possible for us to separate center-of-mass excitation from intrinsic excitation, as we shall now see. The lowest excited configuration of H_0 is the one in which one particle is lifted from the $1s$ level into the next level—the $1p$—at an energy $\hbar\omega$ higher. Thus we expect the lowest excited states to belong to the configuration $(1s)^3 1p$. Let us see what are the allowed quantum numbers of the various levels in this configuration:

The $(1s)^3$ configuration must have even parity and $\mathbf{L} = \Sigma \mathbf{l}_i = 0$; furthermore we can convince ourselves easily that it must have $S = 1/2$ and $T = 1/2$. The nucleon in the p-level has, of course, $l = 1^-$, $S = 1/2$, and $T = 1/2$. We

Problem. Prove that the allowed states of $(1s)^3$ are ^{22}S, where the notation stands for $^{2S+1,2T+1}L$. How would you further characterize the nuclei ^{3}He and ^{3}H, whose ground states both belong to the configuration $(1s)^3$?

therefore conclude that the states of the configuration $(1s)^3 1p$ must all have orbital angular momentum and parity $L = 1^-$, and the spins and isospins associated with them could be $S = 0$ and $S = 1$, and $T = 0$ and $T = 1$. We thus expect to find the following eight levels comprising the excited configuration $(1s)^3 1p$:

$$^{11}P_1,\ ^{13}P_1,\ ^{31}P_0,\ ^{31}P_1,\ ^{31}P_2,\ ^{33}P_0,\ ^{33}P_1,\ ^{33}P_2 \qquad (14.11)$$

where the notation stands for $^{2S+1,2T+1}L_J$, and P stands for $L = 1$ [$^{13}P_1$ is pronounced singlet–triplet-P-one, meaning the state that is singlet in spin ($S = 0$), triplet in isospin ($T = 1$), has orbital angular momentum $1\hbar(P)$, and total angular momentum $1\hbar$]. In $^{2S+1,2T+1}L_J$ we must always satisfy, of course, $|L - S| \leq J \leq L + S$. Hence if $2S + 1 = 1$, it follows that $J = L$.

The eight levels (14.11) of the configuration $(1s)^3 1p$ lie at a zeroth-order excitation energy of $1\hbar\omega$ above the ground state of ^{4}He. Also the first center-of-mass excited state, (14.9), lies at that energy. Since the eight levels (14.11) cover all the excited states of (14.4) at this excitation, we conclude that one of the levels in (14.11), or one linear combination of them, corresponds to a center-of-mass excitation. From (14.10) we know that the latter has $S = 0$, $T = 0$, and $J = 1^-$: There is only one such level among the levels in (14.11); the level $^{11}P_1$. All other center-of-mass excitations lie at an energy of $2\hbar\omega$ or higher above the ground state of ^{4}He. We conclude therefore that all the remaining seven levels in (14.11) correspond to *intrinsic* excitations of ^{4}He. If $v'(ij)$ can really be treated as a perturbation we expect therefore to find *seven* negative parity excited levels at an excitation energy of about $1\hbar\omega$ in ^{4}He. They should have the following quantum numbers:

$$\text{Levels with } T = 0: \quad J = 0^-, J = 1^-, \text{ and } J = 2^-$$
$$\text{Levels with } T = 1: \quad J = 0^-, J = 1^-, J = 1^-, \text{ and } J = 2^- \qquad (14.12)$$

We notice that among the levels with $T = 1$ there are two levels with $J = 1^-$. In LS-coupling they are the levels $^{13}P_1$ and $^{33}P_1$ of (14.11). In other coupling schemes they could be any two independent linear combinations of these levels.

The levels with $T = 0$ must have $M_T = 0$ and could therefore show up only in ^{4}He ($Z = N$). The levels with $T = 1$ can have $M_T = -1, 0$, or $+1$. They should show up, therefore, in ^{4}H, ^{4}He, and ^{4}Li (see Section I.6). In ^{4}H and ^{4}Li they are the lowest group of levels, since the configuration $(1s)^4$ has $T = 0$ and therefore does not show up in either ^{4}H ($M_T = -1$) or ($M_T = +1$). The known energy levels of ^{4}Li, ^{4}He, and ^{4}H are shown in Fig. 14.1 together with the spins and isospins assigned to them. We see that ^{4}Li with $M_T = +1$ does have exactly four negative parity states at an average excitation energy of $\backsim$25 MeV above the ground state of ^{4}He, with total angular momenta as indicated in (14.12) for $T = 1$. Only two states have thus far been reported in ^{4}H($M_T = -1$) and their energies agree well with

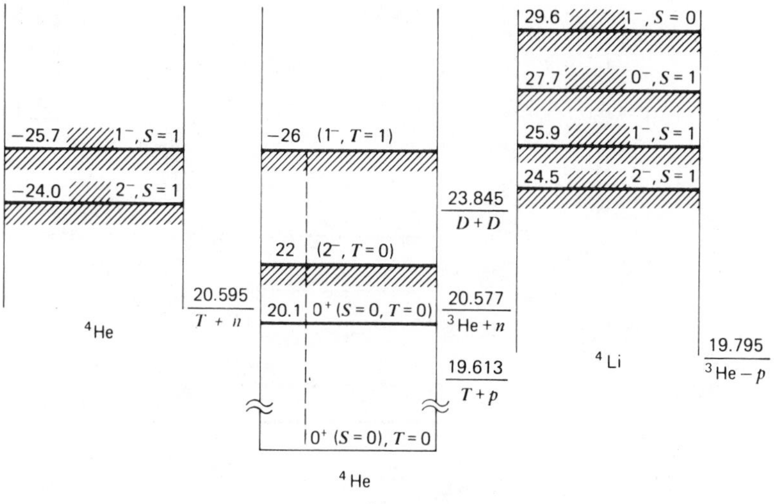

FIG. 14.1. (a) The energy levels in mass-4 nuclei [taken from deShalit and Walecka (66)].

those of the corresponding levels in ^{4}Li. In ^{4}He seven negative parity excited levels have been reported (Fig. 14.1b) [Werntz and Meyerhof (68)]. Four of them have $T = 1$, and agree well with the corresponding levels in both ^{4}H and ^{4}Li. Of the others, the one with $J = 2^-$, lies substantially lower than the $J = 2^-$ level in either ^{4}H or ^{4}Li. This is not surprising since it has $T = 0$ whereas the ^{4}H and ^{4}Li levels have $T = 1$. The other two negative parity states also have $T = 0$.

Before we proceed to study the detailed structure of the spectrum of the negative parity states in the mass-4 nuclei, let us see whether their excitation energy is reasonable. In zeroth-order the excitation energy of the configuration $(1s)^3 1p$ is $1\hbar\omega$. The value of ω, as we have stated before, should be chosen so as to make the zeroth-order wave function approximate the real wave function as well as we can. As was argued before (see Eq. IV.2.11) this can be achieved if the zeroth-order wave function reproduces the observed radius of the nucleus in question. The charge radius of ^{4}He was measured by electron scattering, and it was found that a harmonic oscillator potential that leads to this radius for the $(1s)^4$ configuration must have $\hbar\omega = 21.8$ MeV [B. Goulard, C. Goulard, and H. Primakoff (64)]. Thus the negative parity states in the mass-4 nuclei do appear at an average energy that is consistent with the size of ^{4}He and with our interpretation that these states belong to the configuration $(1s)^3 1p$.

A slightly lower value for $\hbar\omega$, 18 MeV, [Carlson and Talmi (54)] is obtained from the study of Coulomb energies. This can be easily understood if we

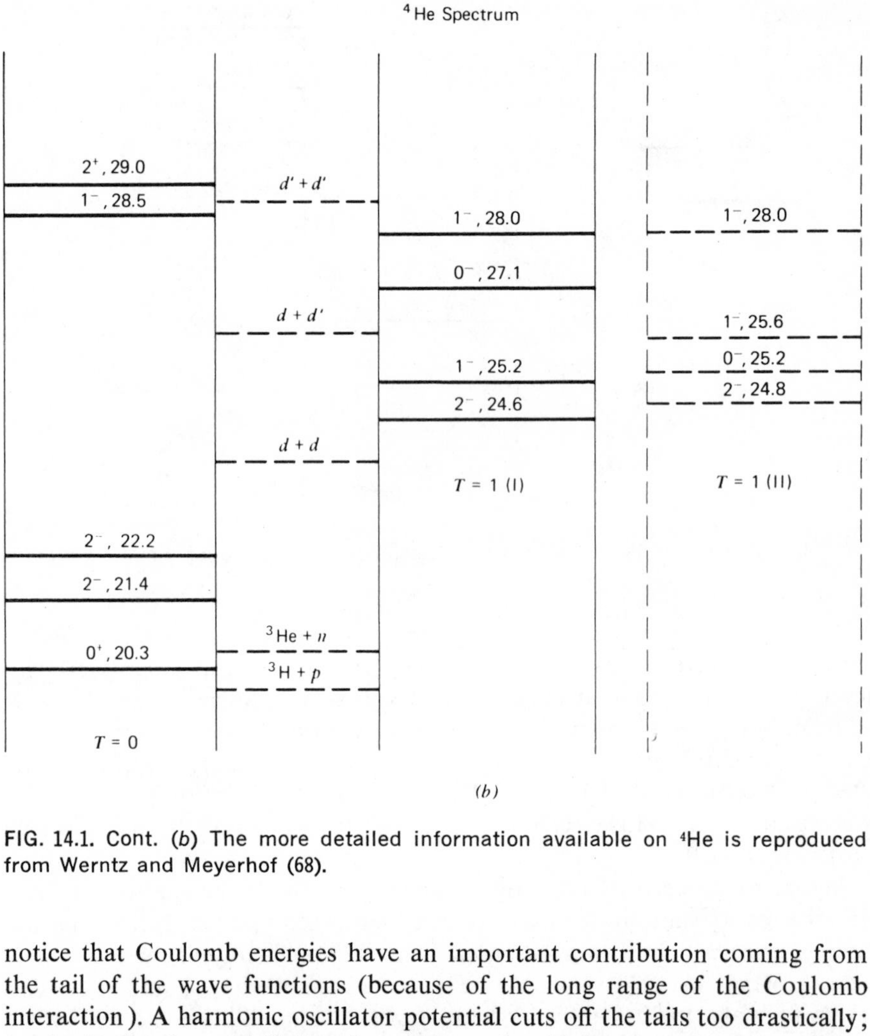

(*b*)

FIG. 14.1. Cont. (*b*) The more detailed information available on ^{4}He is reproduced from Werntz and Meyerhof (68).

notice that Coulomb energies have an important contribution coming from the tail of the wave functions (because of the long range of the Coulomb interaction). A harmonic oscillator potential cuts off the tails too drastically; to obtain the observed Coulomb energy we should therefore use a wave function of somewhat larger dimensions, that is, one which corresponds to a lower value of ω. For our purposes the part of the wave function inside the nucleus plays the dominant role, so that it seems more appropriate to use the value of $\hbar\omega = 21.8$ MeV as determined from electron scattering. This also agrees with the single-particle splitting between the observed $1p$ and $1s$ levels of 22.6 MeV as determined from the masses of ^{5}He, ^{4}He, and ^{3}He [deShalit and Walecka (66)].

Next, we shall evaluate the splittings among the seven levels of the configuration $(1s)^3 1p$. To this end we have to diagonalize the matrix

$$\langle (s^3p)^{2S+1,2T+1}L, J | \sum_{i<k} v'(ik) | (s^3p)^{2S'+1,2T'+1}L', J \rangle \tag{14.13}$$

We already know that all levels of $(1s)^3 1p$ have the same value for the total orbital angular momentum $L = 1$. We can also assume that T is a good quantum number, since nuclear interactions commute with T and the Coulomb interaction is rather unimportant for these nuclei. The matrix (14.13) is therefore reduced to several smaller matrices, along the diagonal, most of them consisting actually of just one row and column. The only exception is for $T = 1\ J = 1^-$ where the matrix is 2x2 as shown in (14.14)

$$\begin{pmatrix} \langle {}^{33}P_1 | v' | {}^{33}P_1 \rangle & \langle {}^{33}P_1 | v' | {}^{13}P_1 \rangle \\ \langle {}^{13}P_1 | v' | {}^{33}P_1 \rangle & \langle {}^{13}P_1 | v' | {}^{13}P_1 \rangle \end{pmatrix} \tag{14.14}$$

To evaluate the matrix elements (14.14) we notice that they contain a contribution from the interaction of the three $1s$-nucleons among themselves, and a contribution from the interaction of the p-nucleon with each one of the $1s$ nucleons. The interactions within the $(1s)^3$ configuration do not contribute to the *splittings* within the $(1s)^3 1p$ configuration. For the evaluation of the interaction of the $1p$ nucleon with the three $1s$ nucleons, we use fractional parentage coefficients to write

$$|(1s)^3 JT\rangle = \sum_{J'T'} (JT\{|(1s)^2 J'T'(1s)JT)|(1s)^2 J'T'(1s)JT\rangle \tag{14.15}$$

where J' and T' can range over the values $(J'T') = (0, 1)$ or $(J', T') = (1, 0)$, these being the only allowed antisymmetric states of the configuration $(1s)^2$. As in (10.9) we can now reduce the interaction of the $1p$-nucleon with the three $1s$ nucleons to a sum over interactions of the $1p$ nucleon with just a single $1s$-nucleon. In other words we can find coefficients B derived from the fractional parentage coefficients in (14.15), by means of which a matrix element of (14.13) can be written as follows:

$$\langle (s\,{}^3p)^{2S+1,2T+1}P, J | \sum_{i<k} v'(i,k) | (s\,{}^3p)^{2S'+1,2T+1}P, J\rangle = \epsilon\cdot\delta(S, S')$$
$$+ \sum_{S_2 S_2' T_2 J_2} B^{SS',J}_{S_2 S_2', T_2 J_2} \langle (sp)^{2S_2+1,2T_2+1}P, J_2 | v'(12) | (sp)^{2S_2'+1,2T_2+1}P, J_2\rangle \tag{14.16}$$

We shall not go here into the derivation of the detailed relation between the B-coefficients in (14.16) and the coefficients of fractional parentage (14.15). It is important to notice that the numerical values of the B's are independent of the interaction $v'(12)$ or of the central potential $U(i)$. They are entirely determined by the fact that we are dealing with the configuration $(1s)^3 1p$.

The first term in (14.16), $\epsilon\cdot\delta(S, S')$, represents the interaction among the three s-nucleons and is thus independent of T or J. It does not contribute to the splittings between the various levels of s^3p.

Evaluating the coefficients B, it is possible to express energy differences in

$(s\,{}^3p)$ in terms of two-body interactions [see deShalit and Walecka (66)]. As an example we give below such relations:

$$E(J=0^-,T=0)=\frac{1}{2}V({}^{11}P_1)+V({}^{31}P_1)+\frac{3}{2}V({}^{33}P_0)$$
$$-\frac{1}{\sqrt{2}}\langle{}^{11}P,1|v'(12)|\,{}^{31}P,1\rangle$$
$$E(J=0^-,T=1)=\frac{1}{2}V({}^{31}P_0)+V({}^{33}P_0)+\frac{1}{2}V({}^{13}P_1)+V({}^{33}P_1)$$
$$-\frac{1}{\sqrt{2}}\langle{}^{13}P,1|v'(12)|\,{}^{33}P,1\rangle$$
$$E(J=2^-,T=0)=\frac{1}{2}V({}^{11}P_1)+\frac{1}{4}V({}^{31}P_1)+\frac{3}{4}V({}^{31}P_2)+\frac{3}{2}V({}^{33}P_2)$$
$$+\frac{1}{2\sqrt{2}}\langle{}^{11}P,1|v'(12)|\,{}^{31}P,1\rangle$$
$$E(J=2^-,T=1)=\frac{1}{2}V({}^{31}P_2)+\frac{1}{2}V({}^{13}P_1)+\frac{1}{4}V({}^{33}P_1)+\frac{7}{4}V({}^{33}P_2)$$
$$+\frac{1}{2\sqrt{2}}\langle{}^{13}P,1|v'(12)|\,{}^{33}P,1\rangle \quad (14.17)$$

Here we used the notation

$$E(J,T)=\langle(s\,{}^3p)\,{}^{3,2T+1}P,J|\sum_{i<k}v'(ik)|\,(s\,{}^3p)\,{}^{3,2T+1}P,J\rangle$$

and

$$V({}^{ST}P_J)=\langle(1s1p)\,{}^{2S+1,2T+1}P,J|\,v'(ik)|\,(1s\,1p)\,{}^{2S+1,2T+1}P,J\rangle$$

We notice several rather trivial properties of the relations (14.17).

If the two-body interaction $v'(12)$ is independent of isospin, that is, if $V({}^{2S+1,1}P_J)=V({}^{2S+1,3}P_J)$, then it also follows, as can be expected, that the energies of the four-particle configuration are independent of T: $E(J,T=0)=E(J,T=1)$.

If we have a two-body interaction $v(12)$, which is spin-independent (but may be isospin dependent), it then follows that $E(J=0^-,T)=E(J=2^-,T)$. This result can also be easily understood. If $v'(12)$ is spin independent then there is no splitting between the various levels of the two-particle configuration. The *two*-particle configuration acts therefore, as if there were no interaction between the nucleons. It follows that the same holds true for the splittings in the four-particle configuration.

The fact that experimentally the levels of the configuration $(1s)^3 1p$ are split, again points at the spin dependence of $v'(12)$ [and therefore also of $v(12)$, see (14.5)]. Furthermore the fact that the $J=2^-$ level lies lower for

$T = 0$ than it does for $T = 1$ is a direct indication of the T-dependence of $v'(12)$ [and therefore again of $v(12)$].

We noticed above that a spin-independent part of $v'(12)$ cannot give rise to a splitting between the levels of $(1s)^2 1p$. $v'(12)$ differs from the real nucleon–nucleon interaction $v(12)$ by a spin-independent term $-(1/2)\cdot(M\omega^2/2A)(\mathbf{r}_1 - \mathbf{r}_2)^2$ [see (14.5)]. It follows that since we are interested in the *splittings* within the configuration $(1s)\ ^{31}p$ we can use the real nuclear interaction $v(12)$ in (14.17), rather than the modified interaction $v'(12)$ that reflected the binding of the center of mass of ^{4}He. Formally, the calculation will then look like a straightforward shell-model calculation with harmonic oscillator wave functions using the full nuclear interaction $v(12)$. Our derivation, however, shows that this is correct just for the splittings within this particular configuration, and not for the absolute position of these levels.

The matrix elements of the two-particle configuration $1s1p$ required for (14.13) can be evaluated using the Talmi integrals, which were introduced in the previous section. From (13.9) we conclude immediately that in our case only two Talmi integrals show up: I_0 and I_1, corresponding to averages over $v'(12)$ taken in relative s and relative p-states, respectively. Furthermore, as mentioned above, for the purpose of evaluating the energy splittings in $(1s)\ ^{31}p$ we can employ the nuclear interaction $v(12)$ instead of $v'(12)$. From nucleon–nucleon scattering data we know that there is very little interaction in a relative p-state between two nucleons. Thus only I_0 will be significantly different from zero. The whole evaluation of the structure of the configuration $(1s)\ ^{31}p$ therefore boils down to the evaluation of one Talmi integral—I_0—of the nuclear interaction $v(12)$.

In an actual calculation [deShalit and Walecka (66)] $v(12)$ was taken to be a Serber force whose parameters were made to fit the low energy nucleon–nucleon scattering data (up to $\backsim$90 MeV) [Hulthen and Sugawara (57), pp. 52, 62]:

$$V(r) = -\left[46.9\,\frac{1 - \boldsymbol{\sigma}_1\cdot\boldsymbol{\sigma}_2}{4}\,\frac{\exp(-\mu_1 r)}{\mu_1 r} + 52.1\,\frac{3 + \boldsymbol{\sigma}_1\cdot\boldsymbol{\sigma}_2}{4}\,\frac{\exp(-\mu_3 r)}{\mu_3 r}\right] \times \frac{1 + P_x}{2}\ \text{MeV} \qquad (14.18)$$

where $\mu_1 = 0.855 fm^{-1}$, $\mu_3 = 0.726 fm^{-1}$, and P_x is the space exchange operator (see Appendix B, p. 372).

The effect of tensor forces can be evaluated as well. For a Serber force, however, we are left, in our case, with an interaction only in the relative s-state. In this case no effects of the tensor forces can be felt since a second rank tensor vanishes when averaged in an s-state. We therefore expect the contribution of tensor forces to be at any rate small, and in the present calculation it has been neglected. The effect of mutual spin-orbit interaction was simulated by artificially splitting the $p_{3/2}$ and the $p_{1/2}$ single particle levels by 3.2 MeV as derived from the data on ^5He The various levels then come out and their energies are indicated in Fig. 14.2 (the number in parenthesis are the experimental values when available). The energy of the lowest level $(J = 2^- \; T = 0)$ was normalized to fit with the experimental data.

The fit between theory and experiment, at least for the known levels, is seen to be remarkably good. [An even better fit was obtained by Barrett (67) using more realistic interactions. See Fig. 14.3.] This seems to support our basic approximation—the shell-model approach—and also the feeling that the observed nucleon–nucleon force is adequate to describe the structure of the α-particle as well. It should be emphasized that the splittings in the $(1s)^3 1p$ configuration are sensitive only to the spin- and isospin-dependent parts of $\nu(12)$, so that they respond to a "weighting" of the nuclear interaction that is different from that of the free nucleon–nucleon scattering. It is in this way that spectra in complex nuclei can test different parts of the nuclear interaction with different sensitivities.

We have described the analysis of the spectrum ^4He at some length since it shows many of the features that one encounters in actual calculations of nuclear spectra. In spectra of more complex nuclei one encounters in general a substantially more complicated situation. Still, with careful handling of selected nuclei and levels it is possible to probe deeper and deeper into the details of nuclear interaction inside nuclei.

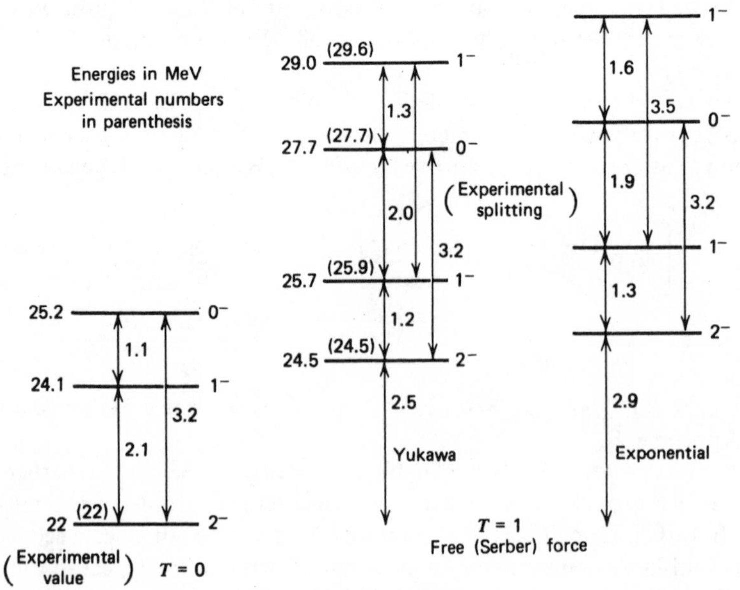

FIG. 14.2. Calculated levels for ^4He using the Serber force taken from deShalit and Walecka (66).

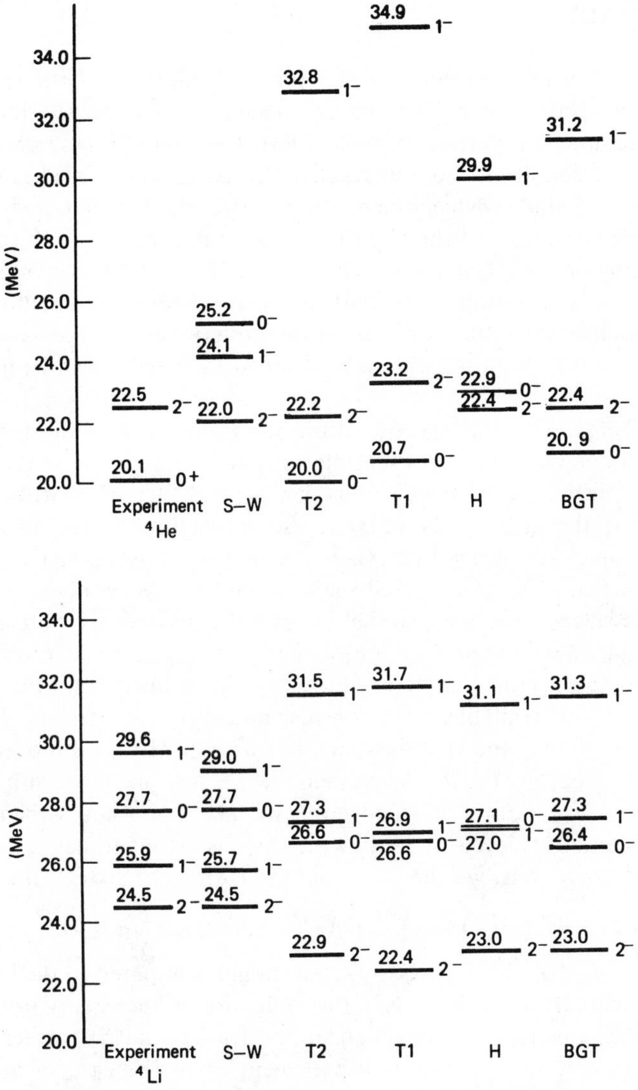

FIG. 14.3. The levels calculated for ^4He by Barrett (67) with various more realistic forces labeled *T2*, *T1*, *H*, and *BGT*. Note in particular the big difference in the $T = 0$ states between this calculation and the one using Serber force only ($S - W$). The recent experimental evidence on the $T = 0$ levels in ^4He (Fig. 14.1*b*) (also Meyerhof and Tombrello (68)] shows how such data can distinguish between the realistic forces and the simplified Serber force.

15. SUMMARY

We have seen in this chapter how the basic assumptions of the shell model enable us to draw rather far-reaching conclusions on the relations among various measured properties of nuclei. Two fundamental assumptions went into the derivation of most of our results. The assumed spherical symmetry of the average potential in which the nucleons move enabled us to assign definite orbits to the nucleons and thereby fix uniquely the dependence of the nuclear wave function on the angular and spin variables. The assumption of the zeroth-order independent motion of the particles then led to the possibility of interrelating the first-order effects of the residual interaction on the various levels of any given configuration, without a detailed knowledge of this interaction itself.

The analysis of nuclear spectra within the framework of the shell model, using equations like (11.2), is feasible only for relatively few nuclei: those which are fairly close to magic numbers. Several factors contribute to the complexity of the situation as we try to use the shell model for nuclei further removed from closed shells. First, there is a simple computational complexity: as we get further away from closed shells, more levels generally become available for the valence nucleons, so that instead of considering a pure configuration j^n we may have to spread the n valence particles over two or more levels jj', etc. For nuclei immediately beyond ^{16}O, for instance, both the $1d_{5/2}$ and the $2s_{1/2}$ levels are available. If we consider now ^{19}O, with three nucleons outside the core of ^{16}O, and if we assume, in line with the shell model, that all three neutrons occupy the $d_{5/2}$ level, then there is just one level with $J = 5/2$: that given by $(d_{5/2})^3\, J = 5/2$. If, however, we take into account the proximity of the $1d_{5/2}$ and $2s_{1/2}$ levels and allow the three neutrons to occupy either the $1d_{5/2}$ or the $2s_{1/2}$ levels, we can construct three different states with $J = 5/2$:

$$(d_{5/2})^3_{J=5/2},\ (d_{5/2}\, s^2_{1/2})_{J=5/2} \qquad \text{and} \qquad [(d^2_{5/2})_{J_2=2}\, s_{1/2}]_{J=5/2}$$

Thus, if the $1d_{5/2}$ and $2s_{1/2}$ levels are close enough compared to the interaction matrix elements between them, then the evaluation of the energy of the lowest $J = 5/2$ state involves the diagonalization of a 3x3 matrix, rather than the computation of one diagonal matrix element as in the case of a pure $d^3_{5/2}$ configuration.

In heavier nuclei the number of levels available between closed shells can be rather large. We are then very often confronted with a situation in which several close-lying levels are available for the valence nucleons. This is especially true when we consider nuclei sufficiently removed from closed shells like $^{176}_{70}Yb_{106}$ or $^{178}_{72}Hf_{106}$. In such nuclei one often finds that a level with a given value of J^π can be formed in a dozen different ways involving close-lying states. The calculation of the energies of such levels therefore requires the diagonalization of matrices of order $n \times n$ where n may be as large as 30 or more.

The meaning of the shell model as an approximate theory then gradually disappears. It is indeed obvious that, by the time we diagonalize large enough matrices, any single-particle potential that generates a complete set of orthonormal A-particle wave functions will be equally good. The whole point of a shell model is the attempt to select a particular single-particle potential with whose wave functions most of the off-diagonal elements of the residual interaction become small enough, so that the diagonalization problem reduces to a trivial, or at least manageable, one. To be consistent with its role a shell model should therefore lead to well-separated zeroth-order configurations, and cannot be conveniently used when it fails to do so.

Second, the diagonalization of large matrices is not only a complex numerical work that obscures the connection between the input and the output, but also makes it increasingly difficult to compare different nuclei without employing a definite residual interaction. Relations like (11.2) are easily obtained only for pure configurations, and they can be generalized also to slightly more complex situations involving two competing configurations. If we know the residual interaction in sufficient detail, then nothing is lost if relations like (11.2) cannot be conveniently used. But if our main aim is to test how far the nucleon–nucleon interaction is modified in nuclei, if we want to know whether it is still a two-body force, whether it is spin-dependent, attractive or repulsive for $T = 0$, etc.—then relations like (11.2) become essential. For then we want to derive empirical matrix elements of the residual interaction from the experimental data, rather than derive a theoretical spectrum from an empirical interaction.

It is obvious from the foregoing that the shell model has a limited range of useful applicability, covering nuclei in the vicinity of closed shells only. Among the light nuclei the magic numbers are very close to each other so that the shell model can actually be applied to most nuclei up to ${}^{40}_{20}\mathrm{Ca}_{20}$. In many cases pure configurations can be safely assumed, while in others, configurations involving two levels have to be invoked. For nuclei with Z and N between 20 and 28 we are dealing with a complete shell that involves just one level—$1f_{7/2}$. It is often called the *$f_{7/2}$-shell*, and corresponding nuclei have been intensively analyzed in terms of the shell model. Then again in the region of Sr, Y, and Zr, with proton numbers 38, 39, and 40, all levels up to the $2p_{1/2}$ and $1g_{9/2}$ are filled; the fact that these two levels have different parities, and that one of them can accommodate only two nucleons makes shell-model analysis very appropriate. It is a fortunate coincidence that for these nuclei the stable isotopes have $N \approx 50$—another magic number; the shell-model analysis is then simplified even further.

Another region in the periodic table in which proton and neutron closed shells coincide is found around ${}^{208}_{82}\mathrm{Pb}_{126}$. ${}^{208}\mathrm{Pb}$ itself is a double magic nucleus. Shell-model analysis can therefore be carried out for nuclei in this region of the periodic table, and many of them have actually been studied.

The most important conclusion to be drawn from the analysis of those cases in which the shell model can be expected to work is that it actually does work. In other words, the evidence that nucleons in the nucleus move rather independently of each other is substantiated beyond the mere observation of magic numbers, islands of isomerism, and the other evidence on which the shell model was first founded. Also when studied in greater detail, the shell model, with its few basic assumptions, explains relations between different experimentally measured quantities, such as magnetic moments of some levels and rates of magnetic dipole radiations between them, the relationship among the energies of levels in one nucleus, say ^{38}Cl and those in another one—^{40}K, etc. Without an underlying structure it would have been very difficult to understand such relations, so that the success of the shell model in interpreting and interrelating these data should be taken as a strong evidence in support of the validity of its underlying assumptions.

APPENDIX

A. TWO-PARTICLE CONFIGURATIONS—ENERGIES

Let us consider the configuration (jj'). The energy shifts of its levels, to first order in v, are given by

$$E(jj', J) = \langle jj'J|v(12)|jj'J\rangle$$

We shall consider first an interaction $v(12)$ that depends only on the relative coordinate $|\mathbf{r}_1 - \mathbf{r}_2|$ of the two particles and not on their spins or isospins and see which restrictions it imposes on nuclear spectra. It is then convenient to expand $v(|\mathbf{r}_1 - \mathbf{r}_2|)$ as in (8.10)

$$v(|\mathbf{r}_1 - \mathbf{r}_2|) = \sum_{k\kappa} \frac{4\pi}{2k+1} v_k(r_1, r_2) Y^*_{k\kappa}(\boldsymbol{\Omega}_1) Y_{k\kappa}(\boldsymbol{\Omega}_2) \tag{A.1}$$

where $\boldsymbol{\Omega}_1$ and $\boldsymbol{\Omega}_2$ are the directions of $\mathbf{r}_1$ and $\mathbf{r}_2$, respectively, taken with respect to an arbitrary z-axis. The expansion (A.1) is a generalization of the well-known multipole expansion of the Coulomb interaction, for which

$$v_k(r_1, r_2) = \left(\frac{r_<}{r_>}\right)^k \frac{1}{r} \qquad \text{for} \qquad v(|\mathbf{r}_1 - \mathbf{r}_2|) = \frac{1}{|\mathbf{r}_1 - \mathbf{r}_2|}$$

We shall refer to the average value of $v_k(r_1, r_2)$ taken with the appropriate radial wave functions, as the strength of the kth multipole interaction.

Using (A.1), the energy shifts $E(jj'J)$ can be written as

$$E(jj'J) = \sum_k f_k(J) F^{(k)} \tag{A.2}$$

where the *Slater integrals* $F^{(k)}$ are defined by

$$F^{(k)} \equiv F^{(k)}(nl, n'l') = \int |R_{nl}(r_1)|^2 |R_{n'l'}(r_2)|^2 v_k(r_1, r_2)\, dr_1\, dr_2 \tag{A.3}$$

and f_k are universal functions, independent of either $v(|\mathbf{r}_1 - \mathbf{r}_2|)$ or the central potential $U(i)$:

$$f_k(J) \equiv f_k(lj, l'j'; J) = \langle jj'J| \frac{4\pi}{2k+1} \sum_\kappa Y'_{k\kappa}(\boldsymbol{\Omega}_1) Y_{k\kappa}(\boldsymbol{\Omega}_2)|jj'J\rangle \tag{A.4}$$

Problem. Derive (A.2) from (A.1) and the expression for the wave function $|jj'J\rangle$ in terms of its radial and angular parts.

Using Appendix A (A.2.54) we can evaluate the matrix element in (A.4) in terms of the reduced matrix elements of the spherical harmonics. We then obtain

$$f_k(J) = (-1)^{j+j'+J} \left(lj|| \sqrt{\frac{4\pi}{2k+1}}\, Y_k||lj\right) \left(l'j'|| \sqrt{\frac{4\pi}{2k+1}}\, Y_k||l'j'\right) \times \begin{Bmatrix} j & j' & J \\ j' & j & k \end{Bmatrix} \tag{A.4}$$

The sum over k in (A.2) extends, in principle, over all values of $k = 0, 1, 2, \ldots, \infty$; however, because of (A.4) it is actually limited to only a very small number of terms. Indeed, from the $6-j$ symbol in (A.4) we conclude that $f_k(J)$ vanishes unless $k \leq \text{Min}\,(2j, 2j')$. Furthermore, the reduced matrix elements of the spherical harmonics vanish for odd values of k (see Appendix, A.2.49), whereas for even values of k they vanish unless $k \leq \text{Min}\,(2l, 2l', 2j-1, 2j'-1)$ (see Appendix, A.2.56). Hence we conclude that the sum over k in (A.2) extends only over the values

$$k \text{ even} \qquad k \leq \text{Min}\,(2l, 2l', 2j-1, 2j'-1) \tag{A.5}$$

This way we now obtain from (A.2) the finite sum

$$E(jj'J) = \sum_{k \leqslant \text{Min}(2l, 2l', 2j-1, 2j'-1)} f_k(J) F^{(k)} \quad \text{k even} \tag{A.6}$$

For ${}^{38}_{17}\text{Cl}_{21}$, for instance, where the relevant configuration is $(d_{3/2}, f_{7/2})$, the allowed values of k are according to (A.5), just $k = 0$ and $k = 2$. In this case, therefore, if the proton-neutron interaction depends only on the relative distance between the two particles, the energies of the four states of the configuration $(d_{3/2}, f_{7/2})$ are all given in terms of the *two* Slater integrals $F^{(0)}$ and $F^{(2)}$ by

$$E(d_{3/2}, f_{7/2}; J) = f_0(J)F^{(0)} + f_2(J)F^{(2)} \tag{A.7}$$

where $f_k(J)$ is determined by (A.4).

The two Slater Integrals $F^{(0)}$ and $F^{(2)}$ depend, of course, on the nature of the interaction $v(12)$, or more specifically on the multipole moments $v_0(12)$ and $v_2(12)$. They also depend on the detailed form of the radial wave functions $R_{nl}(r)$ and $R_{n'l'}(r)$ that according to (A.3) are required for the evaluation of $F^{(k)}$. $F^{(k)}$ therefore depends on the properties of the central potential $U(i)$. But from the fact that (A.7) gives us the energies of the *four* levels of the configuration $d_{3/2}, f_{7/2}$ in terms of the *two* Slater integrals $F^{(0)}$ and $F^{(2)}$, we can conclude that the energies of these four levels should satisfy two independent linear relations valid for any $v(|\mathbf{r}_1 - \mathbf{r}_2|)$ and $U(i)$. In fact a straightforward substitution reveals that we must have

$$[E(5) - E(2)] : [E(3) - E(2)] : [E(4) - E(2)] = 2 : 5 : 7 \tag{A.8}$$

The experimental ratios of these three energy differences in ^{38}Cl are

$$0.672 : 0.762 : 1.312 = 2 : 3.3 : 3.9 \tag{A.9}$$

We can therefore conclude from the discrepancy between (A.8) and (A.9) that the spectrum of ^{38}Cl cannot be explained under any circumstances in terms of the configuration $(d_{3/2}, f_{7/2})$ and an interaction $v(12)$ that depends on $|\mathbf{r}_1 - \mathbf{r}_2|$ only. Within the framework of the shell model we must have a spin dependence in $v(12)$ to be able to explain the observed spectrum of ^{38}Cl in terms of the configuration $d_{3/2}, f_{7/2}$. This result should have been expected if $v(12)$ is the real nuclear interaction, since this interaction is known to be spin dependent. We should not be surprised that, as derived here, it also holds for the effective interaction in ^{38}Cl.

Coming back to (A.2) and (A.6), it is easy to understand also the physics behind the severe limitations on the values of k that contribute to $E(jj'J)$: Although the interaction $v(|\mathbf{r}_1 - \mathbf{r}_2|)$ may have in its expansion all possible moments, we know (see Section I.8) that a particle in a level nlj cannot contribute to moments higher than Min $(2l, 2j)$. Furthermore, if we are talking of spatial moments only (i.e., not involving the spin of the particle), only the even moments can have nonvanishing expectation values; the odd ones vanish because of parity conservation (see Section I.8). Thus the condition (A.5) becomes a natural one to expect.

Although as we have stressed before, $F^{(k)}$ depends on a detailed knowledge of $v(|\mathbf{r}_1 - \mathbf{r}_2|)$ and $U(i)$ we can still relate its k-dependence to some general properties of the interaction $v(|\mathbf{r}_1 - \mathbf{r}_2|)$. To this end we note that from (A.1) it follows that:

$$v_k(r_1, r_2) = \frac{2k+1}{2} \int v(|\mathbf{r}_1 - \mathbf{r}_2|) P_k(\cos\omega_{12})\, d(\cos\omega_{12}) \tag{A.10}$$

where $\cos\omega_{12}$ is the angle between $\mathbf{r}_1$ and $\mathbf{r}_2$. The region of integration that contributes significantly to (A.10) is limited by the range b of $v(|\mathbf{r}_1 - \mathbf{r}_2|)$: for given values of r_1 and r_2, the angles ω_{12} in that region must satisfy $r_1 + r_2^2 - 2r_1r_2 \cos\omega_{12} < b^2$. Thus for given values of r_1 and r_2, $v_k(r_1, r_2)$ will become very small if $P_k(\cos\omega_{12})$ oscillates many times in that significant region of integration. As we see from Fig. A.1 the angles ω_{12} are effectively confined to a range of $|\omega_{12}| < \omega_0$ where $\omega_0 \sim (b/r)$, r being an average of r_1 and r_2. For $P_k(\cos\omega_{12})$ to have many oscillations within this region, k must satisfy

$$\frac{\pi}{k} \ll \omega_0 \approx \frac{b}{r}$$

We conclude, therefore, that

$$v_k(r_1, r_2) \text{ is small for } k \gg \frac{(\pi/2)(r_1 + r_2)}{b} \approx \frac{r_1 + r_2}{b} \tag{A.11}$$

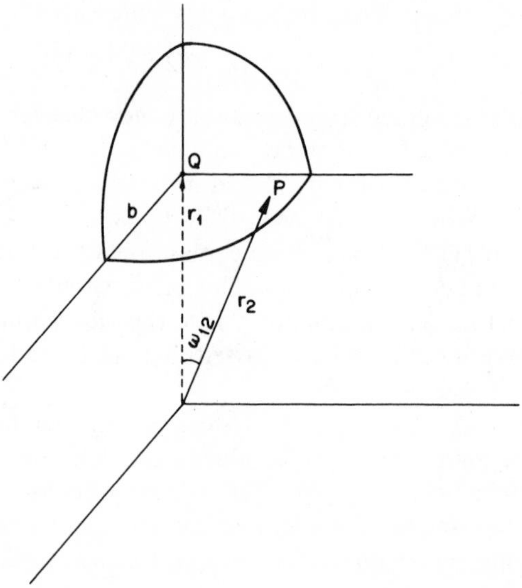

FIG. A.1. For fixed r_1 and r_2, ω_{12} is limited by the requirement that P and Q can be separated by no greater distance than b.

Although the derivation of (A.11) is very qualitative, we can draw from it some interesting conclusions.

For shorter ranges, high multipoles of the interaction become increasingly important. In fact, for a δ-potential we obtain, using (1.25),

$$v_k(r_1,r_2) = \frac{2k+1}{2}\frac{\delta(r_1 - r_2)}{r_1 r_2} \qquad \text{for} \qquad v = \delta(\mathbf{r}_1 - \mathbf{r}_2) \qquad \text{(A.12)}$$

and the multipole strength increases with k.

As the range of the interaction increases its high multipoles become less and less important. Also, as we see from (A.11) $v_k(r_1; r_2)$ when considered as a function of r_1 and r_2, starts to be of importance only for $r_1 + r_2 > kb$. We therefore conclude that the contribution of the higher multipoles tends to be concentrated closer to the nuclear surface; at the surface their relative importance increases as the range of the interaction becomes shorter. The interior of the nucleus contributes only to the lowest multipole moments of the interaction.

In some cases it is convenient to use simple potentials for the derivation of some approximate features of nuclear spectra. To emphasize the separate role of the short range and the long range parts of the interaction, an interaction of the following form is often used:

$$v(12) = V_{s.r.}\delta(|\mathbf{r}_1 - \mathbf{r}_2|) + V_{l.r.}r_1^2 r_2^2 P_2(\cos\omega_{12}) \qquad \text{(A.13)}$$

where $V_{s.r.}$ and $V_{l.r.}$ are constants. The short range part—$V_{s.r.}\delta(|\mathbf{r}_1 - \mathbf{r}_2|)$ is naturally simulated by a δ-function; the long range nature of the remaining part is assured by limiting its Legendre polynomial expansion to just $P_2(\cos \omega_{12})$. Furthermore, to accentuate the nuclear surface contribution to this part, $v_2(r_1, r_2)$ is artificially set equal to $V_{l.r.}r_1^2 \cdot r_2^2$. The interaction $v(12)$ in (A.13) is not a function of $|\mathbf{r}_1 - \mathbf{r}_2|$, since $r_1^2 r_2^2 P_2(\cos \omega_{12})$ cannot be expressed as a function of $|\mathbf{r}_1 - \mathbf{r}_2|$. However, because it depends only on the angle ω_{12} *between* $\mathbf{r}_1$ and $\mathbf{r}_2$, and on the magnitude of $\mathbf{r}_1$ and $\mathbf{r}_2$, it is still rotationally invariant and, as such, conserves angular momentum.

An even more drastic approximation is very often employed to simulate the effects of $v(12)$—the so-called *pairing plus quadrupole interaction.* We recall (see discussion following (1.30)] that in a configuration j^2 of two equivalent nucleons with a short range attractive interaction between them, the state with $J = 0$ comes out always to be the lowest one. This was explained in terms of the exceptionally good overlap of the wave functions of the two particles when their angular momenta are antiparallel. To simulate the short range interaction one introduces therefore a pairing interaction $v_p(12)$, defined through its matrix elements by

$$\langle jj'J | v_p(12) | jj'J \rangle = -G\delta(jj')\delta(J, 0) \tag{A.14}$$

This interaction is thus effective only if the particles are coupled to zero total angular momentum. The constant G may be taken to be a function of j and the mass number A of the nucleus considered. We shall discuss later (Chapter VI) in greater detail some generalizations of (A.14), but for now we just notice that a nuclear interaction may be approximately simulated by the pairing plus quadrupole interaction:

$$v_{pQ}(12) = v_p(12) + V_Q r_1^2 r_2^2 P_2(\cos \omega_{12}) \tag{A.15}$$

Again, $v_p(12)$ takes account, in some sense, of the short range part of the interaction, whereas $V_Q r_1^2 r_2^2 P_2(\cos \omega_{12})$ represents the long range effects.

The pairing interaction (A.14) does not have a short range in the strict sense. It is a nonlocal interaction and does not vanish even if r_1 or r_2 become very large. However, the short range nature of it shows up in another important way: long range interactions give rise to scattering primarily in the forward direction. As their range becomes shorter the scattering becomes more uniform. The pairing interaction gives rise to a completely uniform scattering, and as such it is sometimes referred to as "having a range even shorter than that of a δ-function." It should be emphasized however, that this is just a way of talking and does not refer to the strict definition of a range being the distance beyond which $v(r)$ vanishes.

Equation A.6 emphasizes the very limited number of multipoles that contribute to the first-order energy shifts of the levels of any given configuration jj'. It is not surprising, therefore, that interactions that look very different can give equally good fit to the empirical data. If two interactions lead to similar multipoles of low-order k, they will also give rise to similar nuclear spectra, even though they may differ wildly in their higher multipoles. It is important to bear this point in mind when analyzing the structure of various nuclear spectra.

The limitation to even values of k in the multipole expansion (A.6) is valid if $v(12)$ is a function of the relative coordinate $|\mathbf{r}_1 - \mathbf{r}_2|$ only. A similar expansion can be written also for spin-dependent interactions but this time odd and even values of k may occur. As a matter of fact the situation is even slightly more complicated: A general interaction $v(12)$, which is spin and space dependent, can be expanded in the following manner [see deShalit and Talmi (63), p. 213]

$$v(12) = \sum_{sks'k'r\rho} v_{sk,s'k';r}(r_1, r_2)(-1)^{\rho} T_{-\rho}{}^{(sk)r}(1) T_{\rho}{}^{(s'k')r}(2) \qquad \text{(A.16)}$$

$T^{(sk)r}$ is a "mixed tensor" involving in general both spin and space coordinates. s can take on the values 0 and 1 and T is defined by

$$T_{\rho}{}^{(sk)r}(i) = \begin{cases} \sqrt{\dfrac{4\pi}{2r+1}}\, Y_{r\rho}(i)\, \delta(k, r) & \text{if} \quad s = 0 \\ \\ \sum (1\lambda k\kappa | r\rho) \sigma_{\lambda}{}^{i} Y_{k\kappa}(i) & \text{if} \quad s = 1 \end{cases} \qquad \text{(A.17)}$$

In (A.17) $Y_{r\rho}$ are the usual spherical harmonics, and $\sigma_{\lambda}{}^{i}$ are the three components of the i-particle spin vector operator

$$\sigma_1 = -\frac{1}{\sqrt{2}}(\sigma_x + i\sigma_y),\ \sigma_0 = \sigma_z,\ \sigma_{-1} = \frac{1}{\sqrt{2}}(\sigma_x - i\sigma_y) \qquad \text{(A.18)}$$

Using the expansion (A.16) we can still write $E(jj'J)$ in a form similar to (A.6), but now the expansion will include many more terms. Their number, however, is still finite:

$$E(jj'J) = \sum_{sks'k'r} f_{sk,s'k';r} F^{(sk,s'k';r)} \qquad \text{(A.19)}$$

where the Slater integral is given again by (A.3) with $v_{sk,s'k';r}(r_1, r_2)$ replacing $v_k(r_1, r_2)$, and

$$f_{sk,s'k';r} = \langle jj'J | \sum_{\rho} (-1)^{\rho} T_{-\rho}{}^{(sk)r}(1) T^{(sk)r}(2) | jj'J \rangle$$

$$= (-1)^{j+j'+J} (j \| T^{(sk)r} \| j)(j' \| T^{(s'k')r'} \| j') \begin{Bmatrix} j & j' & J \\ j' & j & r \end{Bmatrix} \qquad \text{(A.20)}$$

It follows from (A.20) that r in (A.19) is limited by

$$0 \leq r \leq \text{Min}(2j, 2j') \qquad \text{(A.21)}$$

Since, from (A.17) it follows that $(k', k) \leq r + 1$ we see that the summation over k and k' is also limited to a finite number of terms. The considerably greater complexity of (A.19) in comparison with (A.6) is due to the larger variety of moments that can be constructed when the spin degrees of freedom are used as well. One is then faced with the task of constructing multipole moments involving not only the matter distribution but also spin distribution and their correlation (more "spin up" in some directions and more "spin down" in others, etc.).

Due to the complexity of the expansion (A.19) it has not been used for the derivation of some general properties of $E(jj'J)$. There are generally more independent Slater integrals in (A.19) than there are levels in the configuration jj' and, as a result, no general restrictions are imposed on the spacings $\Delta E(jj'J)$ when a spin-dependent interaction is allowed as well.

We did not exhibit explicitly the antisymmetrization in this section. For configurations j^2 antisymmetrization is automatically taken into account if J is allowed to assume only even values. For configurations jj' of two identical particles the multipole expansion (A.1) or (A.16) can be used to derive properly antisymmetrized generalizations of (A.2) or (A.19). We shall not go into these details here [the reader is referred to deShalit and Talmi (63)] It is enough for our purposes just to realize that general knowledge of the relative strengths of the various nuclear interaction multipoles can be used to impose limitations on the possible spectra of various nuclei. This may sometimes be a convenient way of studying the effects of variations in the range, or in the spin dependence, on nuclear spectra.

We end this section with an important theorem that follows easily from the expression (A.19) for the energies of the levels of the configuration jj'. Suppose the interaction between the two nucleons is such that only *one* multipole dominates, that is, we have

$$F^{(sk,s'k';r)} = 0 \qquad \text{if} \qquad r \neq r_0 \tag{A.22}$$

It then follows from (A.19) and (A.20) that the energy shifts $E(jj'J)$ can be written as

$$E(jj'J) = (-1)^{j+j'+J} \begin{Bmatrix} j & j' & J \\ j' & j & r_0 \end{Bmatrix} E_0 \tag{A.23}$$

where E_0 is independent of J. The relative shifts of the different levels are then uniquely determined by the Racah coefficient

$$(-1)^{j+J'+J} \begin{Bmatrix} j & j' & J \\ j' & j & r_0 \end{Bmatrix}$$

In particular, if $r_0 = 0$ we find that

$$E(jj'J) = \frac{E_0}{\sqrt{(2j+1)(2j'+1)}} \tag{A.24}$$

Thus, a monopole interaction ($r_0 = 0$) cannot give rise to a splitting among the levels of the configuration jj'. Splittings can only occur if the interaction has nonvanishing multipoles of a finite order $r_0 > 0$.

A configuration in which one of the particles is in $j = 1/2$, the other in j', is a case of special interest. It can have two levels, one with $J_+ = j' + 1/2$ and one with $J_- = j' - 1/2$. If the interaction between the particles depends on $|\mathbf{r}_1 - \mathbf{r}_2|$, and is spin independent, then according to (A.5) only a term with $k = 0$ can appear in the expansion (A.2). However, according to the theorem that we have just stated, such a monopole interaction can lead to no splitting between the levels $J_+ = j' + 1/2$ and $J_- = j' - 1/2$. Experimentally many cases are known in which such doublets, involving a proton–neutron configuration of the type $(1/2, j')$, are split by about 40 kev to 100 kev (see, for instance, the example of ^{206}Tl, Fig. 6.2). We are thus led to conclude once again that the proton–neutron interaction (or at any rate, the effective interaction) in nuclei, even as heavy as ^{206}Tl, must be spin dependent.

B. RESIDUAL TWO-BODY INTERACTIONS

In this appendix the residual two-body interactions that have been often used in shell-model calculations will be listed. These are given in the literature in a variety of forms:

(a) In terms of exchange operators: these involve the

Spin exchange	$P_B = \frac{1}{2}(1 + \boldsymbol{\sigma}_1 \cdot \boldsymbol{\sigma}_2)$	(Bartlett)	(B.1)
Isospin exchange	$P_H = \frac{1}{2}(1 + \boldsymbol{\tau}_1 \cdot \boldsymbol{\tau}_2)$	(Heisenberg)	(B.2)
Space exchange	$P_M = -P_B P_H$	(Majorana)	(B.3)
Ordinary	$P_W = 1$	(Wigner)	(B.4)

The spin-exchange operator P_B has the property

$$P_B \chi_{1/2}(1)\chi_{-1/2}(2) = \chi_{1/2}(2)\chi_{-1/2}(1) \tag{B.5}$$

where χ_m are the state vectors for a particle of spin 1/2. The isospin exchange operator has the identical effect when applied to the two-body isospin state vectors. The space exchange operator P_M has the property

$$P_M \psi(\mathbf{r}_1, \mathbf{r}_2) = \psi(\mathbf{r}_2, \mathbf{r}_1) \tag{B.6}$$

Because of the Pauli principle P_M can be given the form (B.3).

(b) In terms of the projection operators for a given spin and isospin of the two-body system: since these are 0 (singlet) and 1 (triplet) there are four possibilities corresponding to a linear combination of (B.1) to (B.4). Let

P^{TS} be the projection on to a two-body isospin and spin state with total isospin T and spin S, respectively. Then

$$(T = 0, S = 0): P^{00} = \frac{1}{16}(1 - \boldsymbol{\tau}_1 \cdot \boldsymbol{\tau}_2)(1 - \boldsymbol{\sigma}_1 \cdot \boldsymbol{\sigma}_2) \tag{B.7}$$

$$(T = 0, S = 1): P^{01} = \frac{1}{16}(1 - \boldsymbol{\tau}_1 \cdot \boldsymbol{\tau}_2)(3 + \boldsymbol{\sigma}_1 \cdot \boldsymbol{\sigma}_2) \tag{B.8}$$

$$(T = 1, S = 0): P^{10} = \frac{1}{16}(3 + \boldsymbol{\tau}_1 \cdot \boldsymbol{\tau}_2)(1 - \boldsymbol{\sigma}_1 \cdot \boldsymbol{\sigma}_2) \tag{B.9}$$

$$(T = 1, S = 1): P^{11} = \frac{1}{16}(3 + \boldsymbol{\tau}_1 \cdot \boldsymbol{\tau}_2)(3 + \boldsymbol{\sigma}_1 \cdot \boldsymbol{\sigma}_2) \tag{B.10}$$

In terms of P_B, P_N, P_M, P_W

$$\begin{aligned} P^{00} &= \tfrac{1}{4}(1 - P_H - P_B - P_M), \qquad & P^{01} &= \tfrac{1}{4}(1 - P_H + P_B + P_M) \\ P^{10} &= \tfrac{1}{4}(1 + P_H - P_B + P_M), \qquad & P^{11} &= \tfrac{1}{4}(1 + P_H + P_B - P_M) \end{aligned} \tag{B.11}$$

(c) As indicated in (B.1) to (B.4) and (B.7) to (B.10) any of these forms may be directly expressed in terms of $\boldsymbol{\sigma}_1$, $\boldsymbol{\sigma}_2$, $\boldsymbol{\tau}_1$ and $\boldsymbol{\tau}_2$. With this notation in mind some examples are given below:

Rosenfeld

$$V(12) = \tfrac{1}{3}\boldsymbol{\tau}_1 \cdot \boldsymbol{\tau}_2(a + b\boldsymbol{\sigma}_1 \cdot \boldsymbol{\sigma}_2)V_0 \exp(-\mu r)/\mu r \tag{B.12}$$

$V_0 = 45$ MeV, $a = 0.3$, $b = 0.7$, $\mu = (1/1.37)\ \text{fm}^{-1}$ [Elliott and Flowers (57)]

Kurath

$$V = -[P_M + \tfrac{1}{4}P_B]V_0 \exp(-\mu r/)\mu r \tag{B.13}$$

$V_0 = 36$ MeV $\quad \mu = 0.714\ \text{fm}^{-1}$ [Kurath (56); deShalit and Walecka (66)]

Serber-Yukawa

$$V = \left[{}^1V(r)\frac{1 - \boldsymbol{\sigma}_1 \cdot \boldsymbol{\sigma}_2}{4} + {}^3V(r)\frac{3 + \boldsymbol{\sigma}_1 \cdot \boldsymbol{\sigma}_2}{4} \right]\left(\frac{1 + P_M}{2}\right) \tag{B.14}$$

${}^aV(r) = -{}^aV_0 \exp(-\mu r)/\mu r$

${}^1V_0 = 46.9$ MeV $\quad {}^1\mu = 0.855\ \text{fm}^{-1}$

${}^3V_0 = 52.1$ MeV $\quad {}^3\mu = 0.726\ \text{fm}^{-1}$ [Hulthen and Sugawara (57)]

Serber-Exponential

The same form as (B.14) but

$$V(r) = -V_0 \exp(-\mu r) \tag{B.15}$$

$$^1V_0 = 108 \text{ MeV} \qquad ^1\mu = 1.409 \text{ fm}^{-1}$$

$$^3V_0 = 193 \text{ MeV} \qquad ^3\mu = 1.506 \text{ fm}^{-1}$$ [Hulthen and Sugawara (57)]

GGS Potential

$$V = \exp[-(r/\mu)^2][V_{01}P^{01} + V_{10}P^{10} + V_{00}P^{00} + V_{11}P^{11}] \tag{B.16}$$

where $\mu = 1.68$ fm and

GGS I: $V_{01} = -40$ MeV, $V_{10} = -20$ MeV, $V_{00} = 26$ MeV, $V_{11} = 6$ MeV

GGS II: $V_{01} = -40$ MeV, $V_{10} = -24$ MeV, $V_{00} = 24$ MeV, $V_{11} = 25$ MeV

[Gillet, Green, and Sanderson (66)]

Kallio-Kòltveit

$$V = P^{01}V_t + P^{10}V_s \qquad S \text{ state only} \tag{B.17}$$

where

$$V = \begin{cases} -V_k \exp[-\alpha_k(r - c) & r > c \\ \infty & r < c \end{cases}$$

$$c = 0.4 \text{ fm}$$

$$V_t = 475 \text{ MeV} \qquad V_s = 330.8 \text{ MeV}$$

$$\alpha_t = 2.521 \text{ fm}^{-1} \qquad \alpha_s = 2.402 \text{ fm}^{-1}$$

[A. Kallio and K. Kollveit (64)]

Clark and Elliott

$$V(12) = [V_{01}P^{01} + V_{10}P^{10} + V_{00}P^{00} + V_{11}P^{11} + (W_{01}P^{01} + W_{11}P^{11})S_{12} + (X_{01}P^{01} + X_{11}P^{11})\mathbf{S}\cdot\mathbf{L}]\exp(-r^2/\mu^2) \tag{B.18}$$

where

$$S_{12} = 3(\boldsymbol{\sigma}_1\cdot\hat{\mathbf{r}})(\boldsymbol{\sigma}_2\cdot\hat{\mathbf{r}}) - \boldsymbol{\sigma}_1\cdot\boldsymbol{\sigma}_2$$

$$\mathbf{S} = \tfrac{1}{2}(\boldsymbol{\sigma}_1 + \boldsymbol{\sigma}_2), \qquad \mathbf{L} = (\mathbf{r}_1 - \mathbf{r}_2) \times \tfrac{1}{2}(\mathbf{p}_1 - \mathbf{p}_2)$$

$$\mu = 1.8 \text{ fm}$$

$$V_{01} = 41.5\ (\pm 13.5) \text{ MeV} \qquad W_{01} = -95.1\ (\pm 38.8) \text{ MeV}$$

$$V_{10} = -38.0\ (\pm 7.8) \text{ MeV} \qquad W_{11} = -12.1\ (\pm 11.6) \text{ MeV}$$

$$V_{00} = 97.3\ (\pm 35.4) \text{ MeV} \qquad X_{01} = 5.8\ (\pm 31.2) \text{ MeV}$$

$$V_{11} = 14.8\ (\pm 9.3) \text{ MeV} \qquad X_{11} = 53.3\ (\pm 17.5) \text{ MeV}$$

Values obtained from a variety of nuclei; the uncertainty reflects the variation in values present in the data [Clark and Elliott (65)].

The exchange mixtures often referred to are concerned with the relative values of the coefficients B, H, M, and W in the following expression for V_{12}:

$$V_{12} = -[W + MP_M + HP_H + BP_B]J(r_{12}) \tag{B.19}$$

where $J(r_{12})$ gives the spatial dependence of V_{12} and

$$W + M + H + B = 1 \tag{B.20}$$

	Rosenfeld	Kurath	Soper	Serber I	Serber II
W	−.087	0	.30	.5	.5
M	.609	0.8	.43	.5	.5
B	.304	0.2	.27	0	0.1
H	.174	0	0	0	−0.1

We have not listed the separable potential devised by Tabakin (64) which is often used.

Another method for describing the residual interaction is to give its matrix elements between two-particle shell-model states. This procedure that has been more frequently employed in recent years is exemplified by the work of Halbert, McGrory, Wildenthal, and Pandya (71) wherein lists of the matrix elements employed are given.

CHAPTER VI

ROTATIONAL STATES IN DEFORMED NUCLEI AND OTHER COLLECTIVE MODES

1. EVIDENCE FOR COLLECTIVE ROTATIONS

Our studies of the shell model in the previous two chapters have indicated that the complexity of nuclear spectra increases very rapidly as we go farther and farther away from closed shell. In the configuration j^2 we expect only the levels with $J = 0,2,4 \ldots, 2j - 1$. But for $j^3 (j > 5/2)$ many more levels are allowed by the Pauli principle. If, in addition, there are two or more close-lying levels, the expected complexity of the spectrum of excited states increases with the number of particles even faster. This expectation is borne out by the empirical data quite dramatically as is shown in Fig. 1.1.

However, as we go farther away from closed shells some new, very simple and systematic, features start to show up for some nuclei. This is true for nuclei in the range $155 < A < 185$, for $A > 225$, for nuclei in the *s-d* shell $19 \leqslant A \leqslant 25$, and for *p* shell nuclei $9 \leqslant A \leqslant 14$. Odd A nuclei in these regions are characterized by exceptionally large positive quadrupole moments, Q, as shown in Fig. 1.2; even-even nuclei in the same region all have a rather low-lying first excited state with $J = 2^+$ (Fig. I.7.1) and electric quadrupole radiations are strongly enhanced (Fig. I.8.1).

Furthermore, when several low-lying excited states are known in these nuclei they show additional remarkable regularities. In Fig. 1.3 we show the few excited states of several nuclei in the mass region 176-178: Their spectra are seen to be exceedingly simple, with the angular momenta of the levels

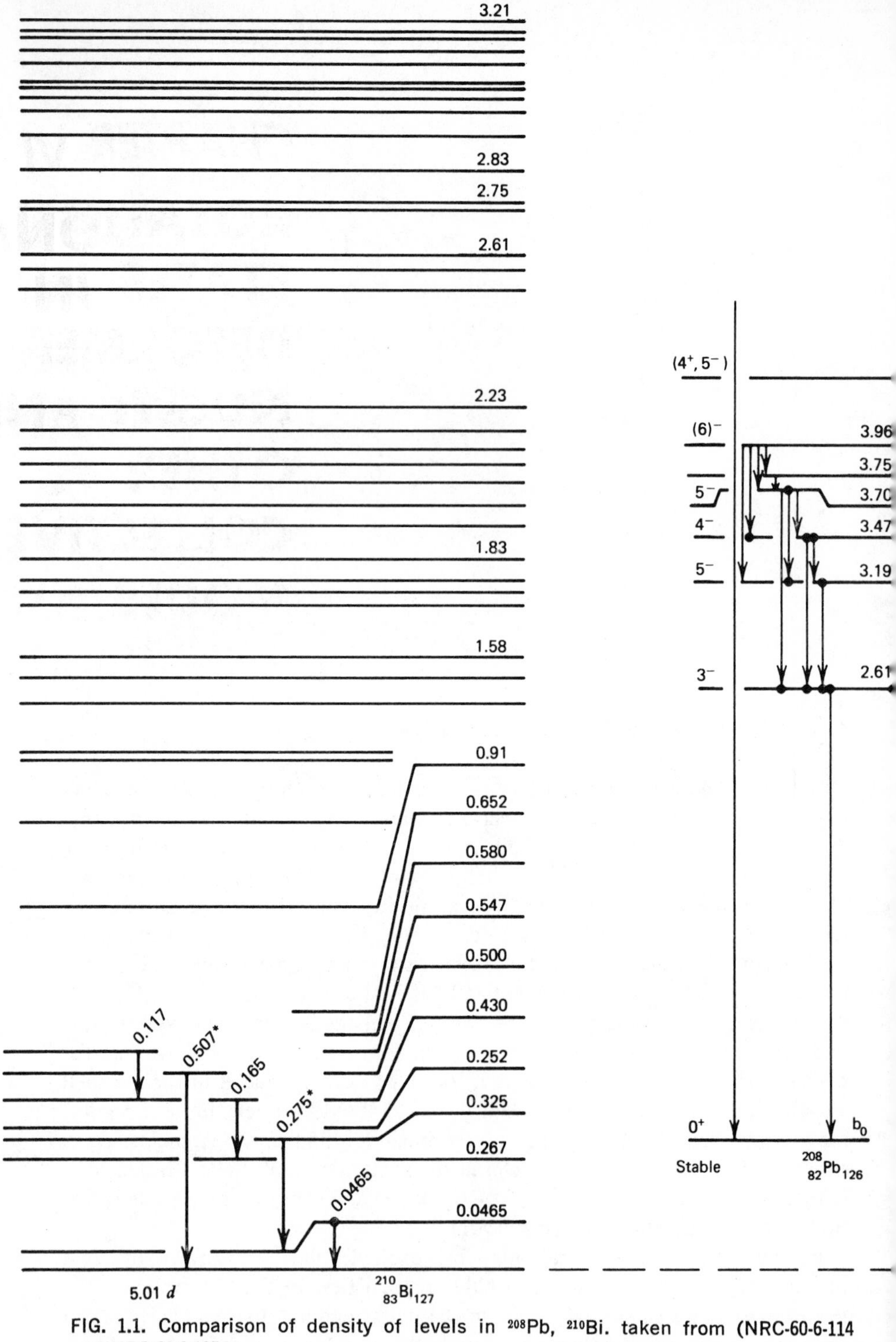

FIG. 1.1. Comparison of density of levels in ^{208}Pb, ^{210}Bi. taken from (NRC-60-6-114 and NRC-61-3-125.)

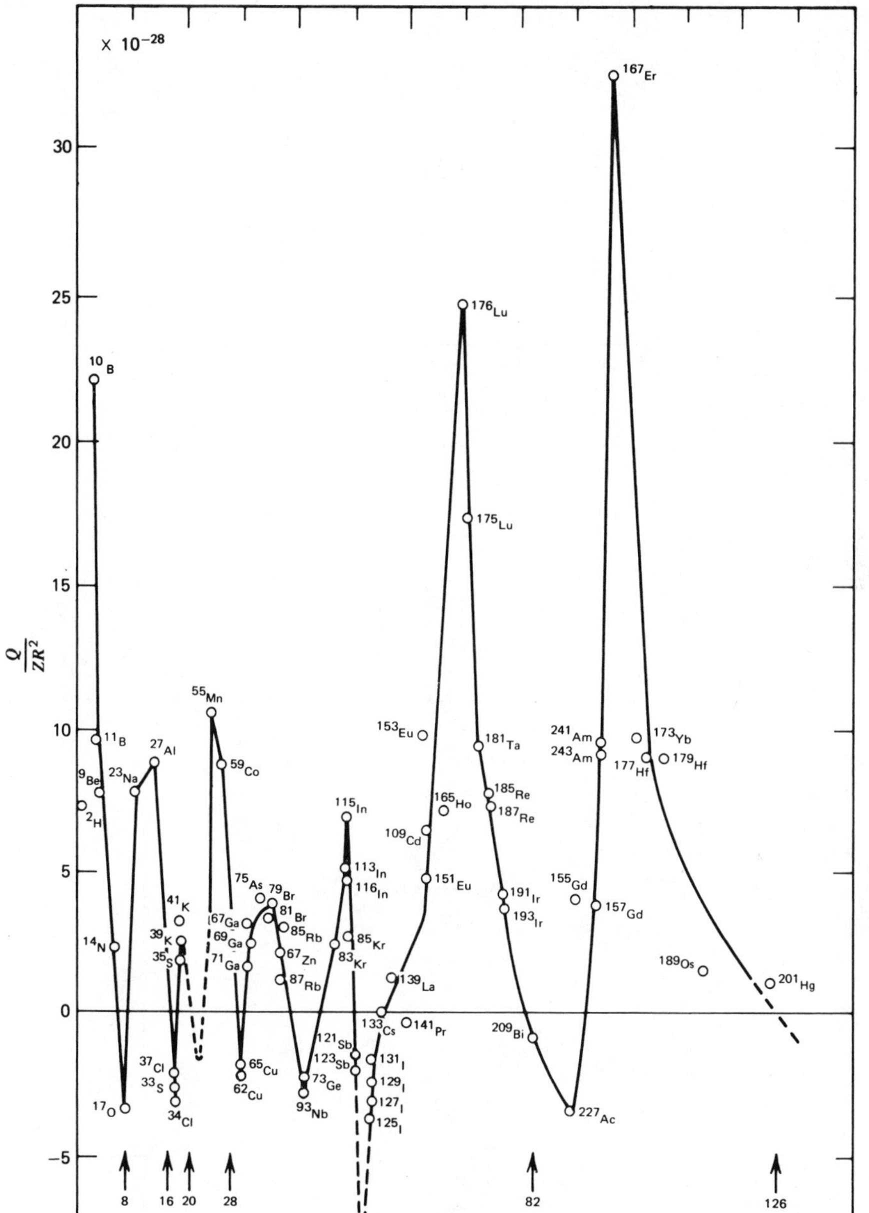

FIG. 1.2. Reduced nuclear quadrupole moments as a function of the number of odd nucleons. The quantity (Q/ZR^2) gives a measure of the nuclear deformation independent of the size of the nucleus [taken from Segre (64)].

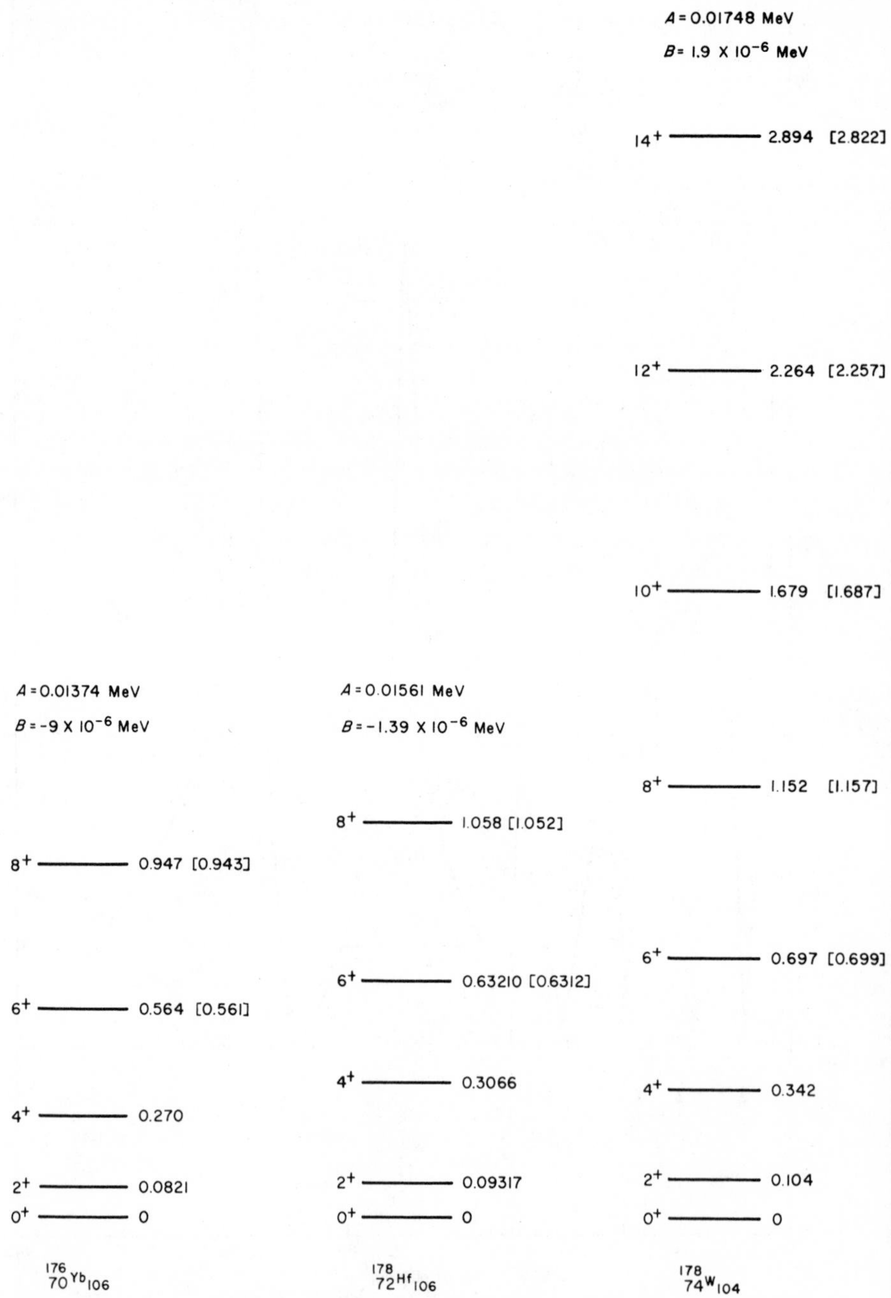

FIG. 1.3. Excited states of some deformed nuclei. Energies are in MeV. The numbers in brackets give the energies of the expected states as calculated with (1.1) using indicated constants A and B in MeV.

increasing monotonically with E. Furthermore the energies of various levels follow very closely the simple formula

$$E(J) = AJ(J+1) + BJ^2(J+1)^2 \tag{1.1}$$

(In Fig. 1.3 the coefficients A and B (in KeV) are given for each nucleus as determined from the first two excited states; the calculated values of $E(J)$ are then given in square brackets for the higher levels, next to the measured value of the excitation energy). In fact, it is indicated by the observed value of B that the energies can be approximated well enough, especially for the lower values of J, by the simpler formula

$$E(J) = A'J(J+1) \tag{1.2}$$

A spectrum like the one shown in Fig. 1.3 could have been easily obtained from a shell model. A configuration $(9/2)^2$, for instance, has, as its allowed values of J, exactly the angular momenta observed for the five lowest states in ^{176}Yb or ^{178}Hf. Furthermore, it is very easy to construct an interaction $v(ik)$ that will lead to the spectrum (1.2); all we have to do is put

$$v(ik) = 2A'(\mathbf{j}_i \cdot \mathbf{j}_k) \tag{1.3}$$

where $\mathbf{j}_i$ is the operator of total angular momentum of the ith particle. Indeed we have from (1.3)

$$\langle j^2J|v(12)|j^2J\rangle = 2A'\langle j^2J|\mathbf{j}_1\cdot\mathbf{j}_2|j^2J\rangle$$
$$= A'\langle j^2J|\mathbf{J}^2 - \mathbf{j}_1{}^2 - \mathbf{j}_2{}^2|j^2J\rangle = A'(J(J+1) - 2j(j+1))$$

The excitation energies will then follow the rule (1.2).

However, if this were the correct interpretation of the spectrum of $^{176}_{70}\text{Yb}_{106}$, then in going to $^{177}_{71}\text{Lu}_{106}$ we should have found the corresponding spectrum of the configuration j^3, with its energies given by,

$$E(j^3J) = \langle j^3J|2A'\sum_{i<k}((\mathbf{j}_i\cdot\mathbf{j}_k)|j^3J\rangle = A'\langle j^3J|\mathbf{J}^2 - \mathbf{j}_1{}^2 - \mathbf{j}_2{}^2 - \mathbf{j}_3{}^2|j^3J\rangle$$
$$= A'[J(J+1) - 3j(j+1)] \tag{1.4}$$

(We should stress that 1.4 follows directly from (Eq. V.11.1) in the previous chapter, if we put $n = 3$ and take for the two body energies $\langle j^2J_2|v(12)|j^2J_2\rangle$ the empirical values given by (1.2). The introduction of the explicit form (1.3) of $v(12)$, is made just for convenience). Now, in the $(9/2)^3$ configuration the spins allowed by the Pauli principle are (see Appendix A, Sec. 4).

$$J = 21/2,\ 17/2,\ 15/2,\ 13/2,\ 11/2,\ 9/2,\ 7/2,\ 5/2, \text{ and } 3/2.$$

It follows from (1.4) that the lowest state of ^{177}Lu should have $J = 3/2$, and the energy should again increase with J according to the same law (1.2). Instead, we find the situation presented in Fig. 1.4. ^{177} Lu has at least three series of excited states and in each one of them the levels show increasing angular momenta as the excitation energy goes up. Furthermore the observed

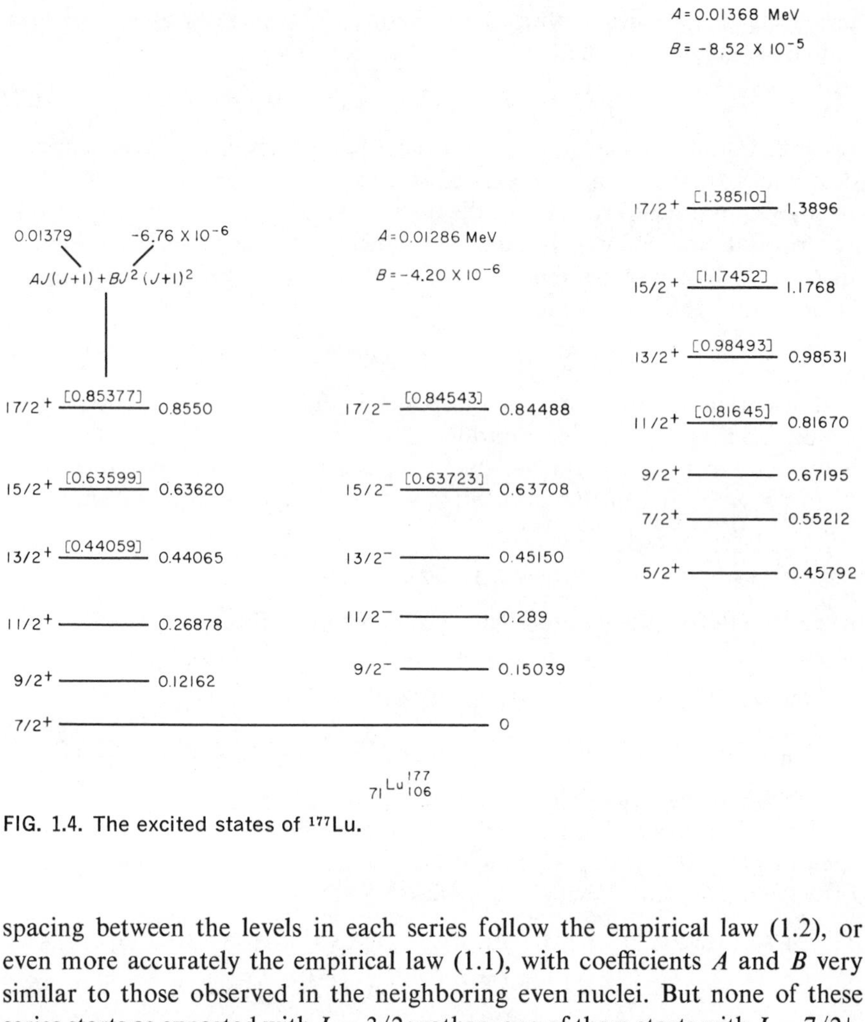

FIG. 1.4. The excited states of ^{177}Lu.

spacing between the levels in each series follow the empirical law (1.2), or even more accurately the empirical law (1.1), with coefficients A and B very similar to those observed in the neighboring even nuclei. But none of these series starts as expected with $J = 3/2$; rather, one of them starts with $J = 7/2^+$, the other with $J = 9/2^-$, and the third with $J = 5/2^+$.

To summarize then if ^{176}Yb were to be interpreted as a pure j^2 configuration, it follows that the spectrum of the j^3 configuration must follow (1.4). This result holds, according to (Eq. V.11.1) of the previous chapter, both if $v(ik)$ in (1.3) is the "true" interaction or just simulates it sufficiently well for ^{176}Yb. Thus the absence of $J = 3/2$ and $J = 5/2$ as the lowest states in ^{177}Lu absolutely rules out any attempt to explain the spectra of ^{176}Yb and ^{177}Lu on the basis of a pure shell-model configuration. It is important to realize the full strength of this conclusion, because it tells us that we have to look for an entirely different interpretation of the simple spectra of these nuclei, as exhibited by the empirical law (1.1).

There is, of course, additional important evidence to indicate that the nuclei discussed above are far from being describable in terms of a pure shell-model wave function. The static quadrupole moment of ^{177} Lu is measured to be $\sim 5.5 \times 10^{-24}$ cm^2, which is about 25 times larger than anything a reasonable shell-model wave function can give (see Section I.8); the observed quadrupole transition probabilities (see Chapter VIII) in ^{178}Hf are about 250 times faster than anything that could be reasonably expected from a shell-model wave function, etc.

A description of these nuclei and many other deformed nuclei has been given by a model proposed and developed by A. Bohr and B. Mottelson (53). Its essential point is the assumption, strongly suggested by the high positive quadrupole moment observed for these nuclei, that we are dealing here with nuclei with a permanent prolate deformation. The situation is assumed to be analogous to that of a diatomic molecule where rotational spectra with energies given by equations identical to (1.2) are known to exist. There the rotational spectrum results from the rotation of the molecule about an axis perpendicular to its symmetry axis, which, for a diatomic molecule, is the line joining the two constituent nuclei. In the Bohr-Mottelson model the nucleus, which resembles a short, thick cigar, is again assumed to rotate slowly around an axis perpendicular to its symmetry axis. It thus gives rise to a spectrum of a rotor, which is identical with (1.2). Unlike the diatomic molecule, where the two atomic nuclei provide a natural axis of symmetry for the system, there is no such natural axis in the nucleus. Rather, it is the dynamics of the many-nucleon system that is believed to lead to such an arrangement of the nucleons' orbits that an "instantaneous picture" taken of the nucleus at any given moment will reveal a mass distribution of a prolate spheroidal shape. At different instants the symmetry axis of the spheroid may point in different directions, and we should interpret this change in the orientation of the deformed nucleus as being due to the correlated motion of the nucleons themselves. In other words, there is a very strong correlation among all the nucleons in the nucleus, which manifests itself in a relatively slow rotation of the deformed overall pattern into which the nucleons shape their combined mass distribution.

We are, of course familiar with similar correlations. A system of interacting nucleons in free space always produces nucleon motion so that their center of mass moves in a straight line with a constant momentum. Viewed from the center of mass, the motion of the nucleons may look very erratic, with no apparent correlations among them. Yet in reality we know that there is a very strong A-particle correlation that gives rise to the very regular motion of their center of mass; furthermore this correlation is due to a specific property of the mutual interaction: its dependence only on the *mutual* distances among the particles, and its independence of their absolute positions in space.

It is however important to realize that there is a difference between the collective motion of the center of mass and the collective rotations. While the

A-particle correlation that gives rise to the collective uniform motion of the center of mass is exact and valid for any system of interacting particles, the same cannot be said of the collective overall *rotations* of systems of interacting nucleons in the sense to be described below.

We have already had some indication that nuclei are deformed from our discussion of the Hartree-Fock results. In Chapter IV we saw that self consistent fields are only axially symmetric for all but closed-shell nuclei. The appearance of a nonspherical potential as a self-consistent field originates in the fact that even a spherical field leads to independent-particle Slater determinants with an axial, rather than spherical, symmetry. If we insist on a single Slater determinant as our Hartree-Fock trial wave function we are then automatically led to an axially, rather than spherically symmetric potential as a possible self-consistent potential.

A single Slater determinant is not an eigenstate of the total angular momentum of the system (except for closed shells; see Section IV.9). For some nuclear properties, such as the average density of nuclei or the symmetry energy, the error that results when working with an approximate wave function, which is not an eigenfunction of the total angular momentum, may not be so bad. But for many properties, like various multipole moments, or splittings between some related levels, the use of a wave function that is not an eigenfunction of $\mathbf{J}^2$ may lead to large errors. We are thus faced with the problem of constructing a wave function that is an eigenfunction of $\mathbf{J}^2$ and yet represents a deformed nucleus. These two conditions, constant J and axial, rather than spherical, symmetry can be simultaneously satisfied only if the spheroidal self-consistent distribution, into which the particles may fit themselves as a result of their mutual interactions, rotates in space. For if it were stationary (or oscillating around an equilibrium direction) a preferred direction in space would have resulted from the mutual interaction of the particles, contrary to the assumption of the isotropy of space; it is the isotropy of space that leads, as is well known, to the constancy of the total angular momentum of the system.

Still, this does not mean that the collective overall rotation of the deformed nucleus is an exact concept like the collective motion of the center of mass. A general Hamiltonian cannot be exactly separable into a rotational part and an "intrinsic" part, similar to the separation into a center of mass motion and internal motion. The rotation of a system of particles leads, as is well known, to the appearance of Coriolis and centrifugal forces in the rotating system. Thus the "intrinsic" motion of the particles is subject to different forces depending on the rate of the "external" rotation, and the two motions cannot be exactly separated. The basic difference between translation and rotation is that the latter involves accelerations whereas the former does not, and this is the origin of the exact separability of the center-of-mass motion, as compared with the at-best approximate separability of a collective rotation that may occur only under special circumstances.

A nuclear shape can have an approximate meaning only if the motion of the nucleons in the nucleus is fast compared to the rotational velocity ω. With typical nucleon velocities of about $(1/4)c$ where c is the light velocity, a "shape" has therefore a meaning only if its frequency of rotation ω satisfies

$$\omega \ll \frac{c}{4} \cdot \frac{1}{2r_0 A^{1/3}} \tag{1.5}$$

where $r_0 A^{1/3}$ is the nuclear radius. Or, in other words, if the time $1/\omega$ for one rotation is long compared with the time for a nucleon to traverse the nuclear diameter. If this condition is not satisfied, the nucleus may have different shapes at different stages of the rotation. The energy associated with the rotation should therefore satisfy

$$\hbar\omega \ll \frac{\hbar c}{8r_0 A^{1/3}} \approx \frac{20}{A^{1/3}} \text{ MeV (with } r_0 \approx 1.2 \times 10^{-13} \text{ cm)} \tag{1.6}$$

Thus we expect the approximation of a rotor to be applicable only as long as the highest level ascribed to the rotor's excitation satisfies (1.6). For ^{178}W, for instance, (1.6) gives $\hbar\omega \ll 3.5$ MeV; indeed, the spectrum of a rotor, (1.2), is found to agree with experiment to about 10% accuracy up to energies of $\sim$1.1 MeV (see Fig. 1.3); the much better fit up to 2.894 MeV is obtained only with the improved formula (1.1).

The analogy between a rotating deformed nuclei and a rotating rigid body should not be pushed too far. Whereas in the rigid body all volume elements are at rest with respect to the body-fixed frame, this need not be the case for the deformed rotating nucleus. In the body-fixed frame the nucleons can still possess a very meaningful average flow pattern, to which we shall come later. Nevertheless the constant A' in (1.2) is often expressed in terms of an effective moment of inertia $\mathcal{J}$ defined by

$$A' = \frac{\hbar^2}{2\mathcal{J}} \tag{1.7}$$

Nuclei that do manifest rotational motion must have a very high A-particle correlation that accounts for the fact that, despite the fast motion of each of the nucleons, the totality of the nucleons, that is, the nucleus as a whole, displays this slow rotation of its shape. Since the nucleus is not spherically symmetric, we can define a "body-fixed" frame of reference whose z'-axis is taken along the axis of symmetry of the nucleus. The analogy to diatomic molecules, or to a rigid axially symmetric rotator, is obvious. The body-fixed frame itself slowly rotates, of course, with the nucleus. But viewed from the body-fixed frame the motion of the nucleons may seem to be "erratic" and free of the A-particle correlation that gives rise to the collective rotation. Thus when we shall later want to exploit the results found in spherical nuclei, that nucleons move in the nucleus nearly independently of each other, we shall find it better to use this approximation relative to the body-fixed frame of reference rather than relative to the lab-fixed system. This way one important A-particle correlation is automatically taken into account. The total motion of the nucleons is thus composed of two parts: an internal motion with respect to the body-fixed reference frame, described by an internal wave function, and the motion of the body-fixed reference frame itself.

The three series of excited states in ^{177}Lu, or "rotational bands" as they are called in Fig. 1.4, would then represent series of rotational levels of a deformed nucleus based on three different internal states in the body-fixed frame of reference. The three rotational bands observed in ^{177}Lu represent three possible rotations for nuclei with 177 nucleons but with three different intrinsic structures, which can, and do, give rise to different deformations.

The model we are about to develop unifies two basic features of nuclei: the nearly independent motion of the nucleons in some common potential, and the collective, coherent motions of the nucleus as a whole. It is therefore referred to as the *unified model*, and in the next section we shall proceed to study some of its basic features.

In Section IV.23 we pointed out why any attempt to describe a system of interacting particles by means of an independent-particle wave function is connected with a violation of a conservation law. We see here another example of it. In the body-fixed frame of reference the nucleus has a static spheroidal shape. If we want to describe it there in terms of a Slater determinant of single-particle wave functions, we should use a deformed axially symmetric potential to generate these functions. Such functions, however, do not have a well-defined total angular momentum. Thus the part of the nuclear wave function that is expressible in terms of an independent-particle wave function violates the conservation of angular momentum. The total wave function, which takes the rotation of the body-fixed frame of reference into account, will, of course, have a well-defined total angular momentum.

2. THE UNIFIED MODEL. BASIC ASSUMPTIONS

As we have indicated earlier a study of the collective motion of the center of mass may help us to understand some of the problems that arise in the study of rotational collective motion. A problem common to both is that of *redundant variables*. A nucleus with A-interacting nucleons is, basically, a system with $3A$ degrees of freedom (we ignore for the moment the intrinsic spin degrees of freedom, since they are irrelevant to our discussion). Three of these degrees of freedom, those describing the motion of the center of mass, can be trivially handled. Thus the problem of the intrinsic structure of a system with A-interacting nucleons involves the study of the dynamical behavior of only $3A$-3 degrees of freedom. In the shell model we use for the same purpose A-independent wave functions with $3A$-independent coordinates. These are just three coordinates too many. These extra coordinates are referred to as redundant variables, and actually each shell-model wave function should have been multiplied by $\delta(\mathbf{r}_1 + \ldots + \mathbf{r}_A)$ indicating that the origin of the potential should be at the center of mass. In Section IV.1 we saw how to handle these extra coordinates: if we introduce artificially an external potential in which the center of mass of the nucleons moves we convert the problem into that of the dynamical behavior of the full $3A$ degrees of freedom; the center of mass is no longer moving freely and its dynamical behavior is determined by the external potential.

An A-particle wave function derived from a model in which the particles are moving in a single-particle potential presents therefore a very distorted picture of the center of mass motion, even if it does approximate relatively well the *internal* structure of the nucleus. In other words, if $H = \sum T_i + \sum v_{ij}$ is the nuclear Hamiltonian and $\psi(\text{s.p.})$ is a wave function derived from a single-particle potential, the energy

$$\langle \psi(\text{s.p.}) | H | \psi(\text{s.p.}) \rangle$$

includes an artificial energy connected with the motion of the center of mass, and cannot, therefore, be directly compared with empirical nuclear energies.

Still, in the kinetic energy $\sum T_i$ the center-of-mass contribution can be separated out exactly. If $\mathbf{P}$ is the operator of the total linear momentum of the system then $\sum T_i - (1/2AM)\mathbf{P}^2$ contains only the contribution of the internal $3A$-3 degrees of freedom to the kinetic energy. The interaction $\sum v_{ij}$ depends, by its own structure, only on the internal degrees of freedom. If we evaluate, therefore,

$$\langle \psi(\text{s.p.}) | \left(H - \frac{1}{2AM} \mathbf{P}^2 \right) \psi(\text{s.p.}) \rangle \tag{2.1}$$

we calculate only internal energies, and if $\psi(\text{s.p.})$ reproduces well the internal correlations among the nucleons in a given state, then expressions of the type (2.1) should be close to the actual measured energies of the corresponding nuclear states. Although $\psi(\text{s.p.})$ describes the center-of-mass motion erroneously, the fact that $H' = (1/2AM)\ \mathbf{P}^2$ is expressible entirely in terms of the $3A$-3 intrinsic variables, makes the expression (2.1) less sensitive to the special center of mass motion incorporated in $\psi(\text{s.p.})$. Thus $\psi(\text{s.p.})$ can be used as if it were a wave function of intrinsic coordinates only, although in reality it is a function of the $3A$-3 intrinsic coordinates and the 3 redundant center-of-mass coordinates.

For wave functions generated from deformed potentials the problem of redundant variables, which now include also the potential's orientations in space becomes considerably more acute. To see how this complication is important, let us discuss first a two-dimensional system of A-particles. In such a system we can transform from the laboratory frame of reference to another frame $\sum'$ by means of the transformation

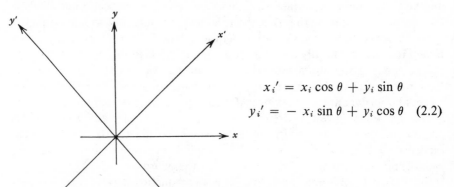

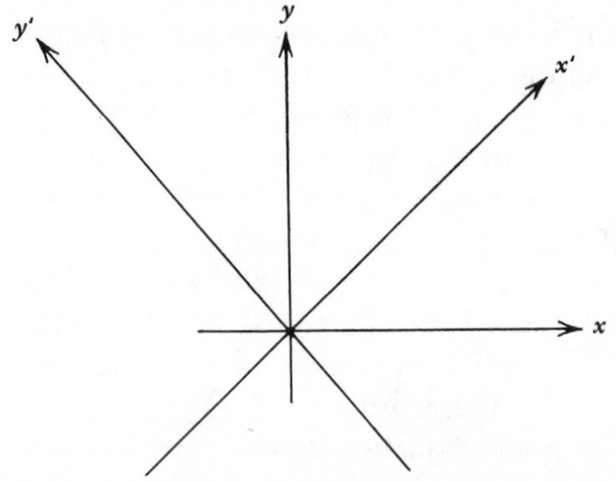

$$x_i' = x_i \cos\theta + y_i \sin\theta$$
$$y_i' = -x_i \sin\theta + y_i \cos\theta \qquad (2.2)$$

If we want the frame $\sum'$ to be a "body-fixed" frame of reference pointing along the principal axes of the system's tensor of inertia, then the angle θ should be such that we have identically,

$$\sum_i x_i' y_i' = 0 \qquad (2.3)$$

From (2.2) we obtain therefore

$$\left(\sum y_i^2 - \sum x_i^2\right) \sin\theta \cos\theta + \sum x_i y_i (\cos^2\theta - \sin^2\theta) = 0$$

or

$$(\cos 2\theta) \sum x_i y_i = \left[\frac{1}{2} \sin 2\theta\right] \sum (x_i^2 - y_i^2)$$

and hence

$$\theta = \frac{1}{2} \arctan \frac{2 \sum x_i y_i}{\sum (x_i^2 - y_i^2)} \qquad (2.4)$$

Equation 2.4 is the transformation that expresses the "collective coordinate" θ in terms of the particle coordinates $(x_i y_i)$, in the same sense that

$$X = \frac{1}{A} \sum_i x_i \qquad Y = \frac{1}{A} \sum_i y_i \qquad (2.5)$$

give the collective coordinate of the center of mass in terms of the coordinates of the particles.

The transformation (2.5), involving the center of mass, can be completed with the introduction of $2A - 2$ internal variables $\xi_\alpha = \xi_\alpha(x_i y_i)$ so that the Hamiltonian $H(x_i y_i)$, when expressed in terms of the new coordinates X, Y, and ξ_α, separates into two terms

$$H(x_i y_i) = H_1(X, Y) + H_2(\xi_\alpha)$$

The same thing *cannot* be done with the transformation (2.4). We can indeed introduce $2A - 1$ "internal" coordinates $\eta_\alpha(x_i y_i)$ so that, together with θ defined in (2.4), they will constitute a canonical set of variables for our system. However, a general Hamiltonian will *not* separate into a sum of $H_1(\theta)$ and $H_2(\eta_\alpha)$. Instead, the Hamiltonian written in terms of the new $2A$ variables θ, η_α, will take the form

$$H = H_1(\theta) + H_2(\eta_\alpha) + H_3(\theta,\eta_\alpha) \tag{2.6}$$

where, as a rule, $H_3(\theta,\eta_\alpha)$ cannot be made to vanish for physically meaningful systems; it involves the coordinates θ and η_α in a nonseparable way.

Hamiltonians of the type (2.6) are known also in molecular physics. For instance in the case of a diatomic molecule, once the center-of-mass coordinates and the motion of rotation about the axis of symmetry are eliminated, it is natural to introduce the direction of the axis connecting the two atoms as one coordinate θ and take the electron coordinates relative to this axis to be the internal coordinates η_α. The term $H_3(\theta,\eta_\alpha)$ then represents the centrifugal and Coriolis forces acting on the electrons in the rotating frame of the two atoms.

Under some circumstances it may be possible to treat the Hamiltonian (2.6) in an approximation known as the *Born-Oppenheimer approximation*, which makes it essentially separable. If the frequencies associated with the coordinate θ are very small, that is, if the angular motion is slow compared with the motion associated with the coordinates η_α, it is possible to consider first the variable θ as a numerical parameter and solve for the eigenvalues of the Hamiltonian

$$[H_2(\eta_\alpha) + H_3(\theta,\eta_\alpha)]\chi = E(\theta)\chi \tag{2.7}$$

in which the η_α are the *only* dynamical variables. The eigenvalues E in (2.7) will depend, generally, on the particular value of the parameter θ. They can then be taken as an additional "potential energy" that affects the θ-motion, in the following way:

$$[H_1(\theta) + E(\theta)]\phi(\theta) = E\phi(\theta) \tag{2.8}$$

The eigenvalue E of (2.8) is, in this approximation, the energy of the system in a state characterized by an internal wave function χ superimposed on the slow adiabatic motion described by $\phi(\theta)$.

The extent to which this approximation is valid also for deformed nuclei cannot at present be inferred from the properties of the nuclear interaction. However, assuming the approximate validity of the shell model, we find that for nuclei far from closed shells the onset of some adiabatic slow collective motion is to be expected. The presence of many close-lying levels of the same J and parity in the shell model description of such nuclei, makes it possible for the nucleons to respond to the internucleon forces and establish further correlations over and above the Pauli correlations that underlie the shell model. Such further correlations tend to give to the nucleus a certain shape, undergoing changes that are slow compared with the frequency with which the particles move across the nucleus.

However these arguments do not establish the existence of a deformed shape quantitatively. We must proceed on an empirical basis. Experimentally we know of the existence of rotational spectra in many nuclei whose proton and neutron numbers place them far from closed shells. We can therefore assume that the Hamiltonian for these nuclei does separate effectively into a sum of the Hamiltonians of rotation and intrinsic motion:

$$H = H_{\text{int}} + T_{\text{rot}} \tag{2.9}$$

At least we can argue that this separation is empirically valid for the low-lying states of these nuclei. In other words we imply in (2.9) that in principle we can transform, in some approximation, from the 3A coordinates $\mathbf{r}_i$ into a new set of coordinates θ_k similar to (2.4), plus internal coordinates ξ_α, and thereby separate the Hamiltonian H. The angle variables θ_k, $k = 1,2,3$ define the orientation of the three-dimensional deformed potential relative to a fixed laboratory frame of reference, and ξ_α are the remaining $3A$-3 internal coordinates of the system. (We ignore the separation of the center-of-mass coordinate just for simplicity.)

As in the two-dimensional case, θ_k can be implicitly defined in terms of $\mathbf{r}_i$ in the following way: let $A(\theta_k)$ be the 3×3 matrix that transforms $\mathbf{r}_i$ from the laboratory frame to a frame defined by the angles θ_k (these can be the Euler angles defining the new frame with respect to the laboratory fixed frame). If $\mathbf{r}_i'$ are the coordinates of particle i with respect to the new frame then

$$\mathbf{r}_i' = A(\theta_k)\mathbf{r}_i \tag{2.10}$$

Notice that (2.10) is a shorthand notation for the product of the 3×3 matrix A with the vector $\mathbf{r}_i$ of the ith particle. Thus $(A\,\mathbf{r}_i)_{x'} = A_{x'x}x_i + A_{x'y}y_i + A_{x'z}z_i$, etc. If the axes of the new frame are required to be along the principal axes of the nucleus, we obtain the following implicit equations for θ_k in analogy to (2.3):

$$\begin{aligned}
0 &= \sum_i x_i'y_i' = \sum_i [A(\theta_k)\mathbf{r}_i]_{x'}\,[A(\theta_k)\mathbf{r}_i]_{y'} \\
0 &= \sum_i y_i'z_i' = \sum_i [A(\theta_k)\mathbf{r}_i]_{y'}[A(\theta_k)\mathbf{r}_i]_{z'} \\
0 &= \sum_i z_i'x_i' = \sum_i [A(\theta_k)\mathbf{r}_i]_{z'}[A(\theta_k)\mathbf{r}_i]_{x'}
\end{aligned} \tag{2.11}$$

T_{rot} in (2.9) is the kinetic energy associated with the variation of the angles θ_k with time. If $\mathbf{R}$ is the angular momentum conjugate to the angles θ_k then T_{rot} is given by:

$$T_{\text{rot}} = \frac{\hbar^2}{2\mathcal{J}_{x'}} R_{x'}^2 + \frac{\hbar^2}{2\mathcal{J}_{y'}} R_{y'}^2 + \frac{\hbar^2}{2\mathcal{J}_{z'}} R_{z'}^2 \tag{2.12}$$

where $\mathcal{J}_{x'}$ $\mathcal{J}_{y'}$ and $\mathcal{J}_{z'}$ are the moments of inertia around the three principal axes.

The assertion of (2.9) (which, as we said, has thus far only an empirical basis) is that, provided proper numerical values are taken for the moments of inertia in (2.12), the operator H_{int} will be to a good approximation a function only of the $3A - 3$ intrinsic coordinates ξ_α. However, if $\chi(\xi_\alpha)$ is an eigenfunction of the intrinsic Hamiltonian H_{int}, we can express it also in terms of the coordinates $\mathbf{r}_i'$ of the particles relative to the body-fixed frame. These $3A$ coordinates are not independent, since they satisfy the three equations (2.11). Nevertheless, as was pointed out in Section 1, we shall adopt the approximation that, viewed from the body-fixed frame of reference, the nucleons move independently in a deformed potential whose axes coincide with those of the body-fixed frame. This is not the same as assuming the exact validity of (2.11), but it does assure us of the validity of (2.11) *on the average*. That is, if $\chi(\mathbf{r}_i')$ is a Slater determinant of single-particle wave functions in a deformed potential with an axial symmetry around the z'-axis it can be shown that

$$<\chi|(\textstyle\sum z_i'x_i')\chi> \;=\; <\chi|(\textstyle\sum x_i'y_i')\chi> \;=\; <\chi|(\textstyle\sum y_i'z_i')\chi> \;=\; 0 \quad (2.13)$$

Transformation (2.10) does not lead to the separable form of the Hamiltonian (2.9). Instead, as we pointed out in our discussion of the two-dimensional case (see Eq. 2.6), a term H_{coupl}, that couples the internal degrees of freedom with the rotational degrees of freedom, is also present. The correct Hamiltonian is therefore

$$H = H_{\text{int}} + H_{\text{coupl}} + T_{\text{rot}} \quad (2.14)$$

However the occurence of rotational spectra leads us to assume that H_{coupl} can be treated as a perturbation. We therefore take the approximate Hamiltonian of (2.9) as the starting point of our theory. The first step is to determine its eigenvalues and eigenfunctions.

3. THE WAVE FUNCTIONS OF THE UNIFIED MODEL

The eigenfunctions of the Hamiltonian (2.9), that is, the zero-order eigenfunctions of (2.14) in which H_{coupl} is neglected, have the separable form

$$\Psi \simeq \chi(\mathbf{r}_i')\ \Phi(\theta_k) \quad (3.1)$$

where the functions $\chi(\mathbf{r}_i')$ satisfy the conditions (2.13). Let us now see what can be said about eigenfunctions of this type, in which the internal and collective coordinates are separated.

First let us study the symmetry properties of the wavefunction Ψ defined by (3.1). In other words we are interested in questions like: what is the total angular momentum that corresponds to a given Ψ; what is its parity, etc.

The angular momentum of a given function Ψ can be extracted from its behavior under rotations. If $\mathbf{I}$ is the total angular momentum operator for the system described by Ψ, then a rotation by an angle $\delta\alpha$ around the unit vector $\mathbf{n}$ produces in Ψ a change given by (see Appendix A, A.2.2)

$$\delta\Psi = (e^{-i\delta\alpha\mathbf{n}\cdot\mathbf{I}} - 1)\Psi \tag{3.2}$$

A rotation of the system as a whole does not change the numerical value of any of the internal coordinates. If $\chi(\mathbf{r}_i')$ in (3.1) is a function of these internal coordinates, then it follows that under the above rotation $\delta\chi = 0$. We conclude therefore that

$$\mathbf{I}\ \chi(\mathbf{r}_i') = 0 \tag{3.3}$$

The internal coordinates $\mathbf{r}_i'$, because of (2.11), are not independent of each other; for given values of the independent coordinates $\mathbf{r}_i$, the resulting values for $\mathbf{r}_i'$ depend on the particular values of θ_k. However, once we *define* a function $\chi(\mathbf{r}_i')$ as a function of the A-vectors $\mathbf{r}_i'$, we can ask the question as to what is the change in χ induced by a rotation of the A-vectors $\mathbf{r}_i'$ relative to the (body-fixed) frame $\sum'$. This rotation, by an angle $\delta\alpha$, can be carried out again around the direction $\mathbf{n}$ fixed in space, and the change in χ is then given by an appropriate operator $\mathbf{J}$ through

$$\delta\chi(\mathbf{r}_i) = [e^{-i\delta\alpha\mathbf{n}\cdot\mathbf{J}} - 1]\ \chi(\mathbf{r}_i') \tag{3.4}$$

The *functional dependence* of $\chi(\mathbf{r}_i')$ on $\mathbf{r}_i'$ does not reflect the fact that the A-vectors $\mathbf{r}_i'$ are not independent. This latter fact affects only their *allowed range of variation* in a physical wave function Ψ of the type (3.1). Thus the operator $\mathbf{J}$ has all the features of an angular momentum operator for our system, and it is often referred to as the operator of *internal angular momentum.*

It is worthwhile mentioning that although the operator $\mathbf{J}$ rotates the vectors $\mathbf{r}_i'$ relative to the body-fixed frame of reference $\sum'$we can of course, take its components either relative to $\sum'$, in which case we designate them by J_A, J_B, and J_C, or relative to the laboratory frame $\sum$, when we designate them as J_j, J_k. and J_l.

In all cases we have

$$[J_A, J_B] = i \sum_C \epsilon_{ABC}\, J_C \qquad \text{or} \qquad [J_k, J_l] = i \sum_j \epsilon_{klj}\, J_j \tag{3.5}$$

where ϵ_{klj} is the antisymmetric tensor. ($\epsilon_{klj} = 0$ if any two indices are equal, $\epsilon_{123} = 1$, $\epsilon_{klj} = 1$ for an even permutation of (123), -1 for an odd permutation.)

It is also obvious that, by definition, $\mathbf{J}$ does not operate on the coordinates θ_k, so that

$$\mathbf{J}[\chi(\mathbf{r}_i')\ \Phi(\theta_k)] = [\mathbf{J}\ \chi(\mathbf{r}_i')]\ \Phi(\theta_k) \tag{3.6}$$

The operator of the total angular momentum $\mathbf{I}$ can also be decomposed into components I_j, I_k, I_l in the laboratory frame $\sum$ or into components I_A, I_B, I_C in the body-fixed frame of reference $\sum'$. The commutation relations of I_A,

I_B, and I_C are, however, different from those of I_j, I_k, and I_l, since I_A, for instance, is a component of **I** along an axis that itself undergoes a rotation by the operation I_B, etc.

To derive the commutation relations of I_A, I_B, and I_C let us introduce the three unit vectors in the directions of the three axes in $\sum'$, $\mathbf{e}_A$, $\mathbf{e}_B$, and $\mathbf{e}_C$, which satisfy:

$$\mathbf{e}_A \cdot \mathbf{e}_B = \sum_k e_{Ak}\, e_{Bk} = \delta_{AB};\ \sum_A e_{Ak}\, e_{Al} = \delta_{kl} \tag{3.7}$$

Since $\mathbf{e}_A$ is a vector quantity built from the coordinates of the particles it satisfies the following commutation relations with I_k

$$[I_k, e_{Al}] = i\sum_j \epsilon_{klj}\, e_{Aj} \tag{3.8}$$

The components of **I** with respect to the body-fixed frame of reference are given by

$$I_A = \sum I_k\, e_{Ak} = \sum e_{Ak}\, I_k \tag{3.9}$$

where we used the fact that, because of (3.8),

$$[I_k, e_{Ak}] = 0 \tag{3.10}$$

It follows from (3.9) and (3.8) that

$$\begin{aligned}[I_A, e_{Bk}] &= \sum_l [I_l\, e_{Al}, e_{Bk}] = \sum_l e_{Al}\, [I_l, e_{Bk}] \\ &= \sum_{lm} i\epsilon_{lkm}\, e_{Al}\, e_{Bm} = -\sum_C i\epsilon_{ABC}\, e_{Ck}\end{aligned} \tag{3.11}$$

In deriving (3.11) *we have used the fact that the body-fixed unit vectors* $\mathbf{e}_A$, $\mathbf{e}_B$ *and* $\mathbf{e}_C$ *commute among themselves*; we have also used the identity

$$\sum_{kl} \epsilon_{klm}\, e_{Ak}\, e_{Bl} = \sum_C \epsilon_{ABC}\, e_{Cm} \tag{3.12}$$

The component I_A is a component along an axis in the body-fixed frame of reference; it is therefore a scalar with respect to rotations of the total system. Hence

$$[I_k, I_A] = 0 \qquad \text{for any } k \text{ and } A \tag{3.13}$$

Combining (3.13) and (3.11) and using (3.9) we obtain finally

$$[I_A, I_B] = -\,i\sum_C \epsilon_{ABC}\, I_C \tag{3.14}$$

In the laboratory frame $[I_k, I_l] = i\sum \epsilon_{klm}\, I_m$. Thus the components of **I** along the body-fixed frame of reference satisfy commutation relations similar to those of the components of **I** with respect to the fixed laboratory frame, *except for a change in sign.*

Problem. Use the fact that the internal angular momentum **J** commutes with the vectors $\mathbf{e}_A$, and going through a similar derivation to that outlined above prove (3.5). Note that (3.13) is *not* valid when **I** is replaced by **J**.

The operator **I** describes, as we have indicated, the physical rotation of the system as a whole; **J**, on the other hand, describes the rotation of the particles relative to the body-fixed frame of reference. This rotation is not physically realizable since the body-fixed frame is carried with the particles whenever they are moved; yet when one expresses the wave function Ψ in the form (3.1): $\Psi = \chi(\mathbf{r}_i') \, \Phi(\theta_k)$, it is possible to investigate the effects of such nonphysical rotations by subjecting the coordinates $\mathbf{r}_i'$ of the particles to a transformation that takes them out of the physical region of variation allowed by their interdependence as implied by (2.11).

There is another nonphysical rotation **R**, which is the rotation of the body-fixed frame leaving the particles untouched. It is obvious that for any function $\chi(\mathbf{r}_i')$ of the internal coordinates defined with respect to the body-fixed frame, the result of a rotation **R** of the frame by an angle $\delta\alpha$ around **n** is equal to a rotation **J** of the particles by an opposite angle $-\delta\alpha$ around the same direction **n**. Thus

$$[e^{-i\delta\alpha \mathbf{n}\cdot\mathbf{R}} - 1] \, \chi(\mathbf{r}_i') = [e^{+i\delta\alpha \mathbf{n}\cdot\mathbf{J}} - 1] \, \chi(\mathbf{r}_i')$$

and hence

$$(\mathbf{R} + \mathbf{J}) \, \chi(\mathbf{r}_i') = 0 \tag{3.15}$$

Comparison of (3.15) and (3.3) suggests that

$$\mathbf{I} = \mathbf{R} + \mathbf{J} \tag{3.16}$$

In fact, if we fix a direction **n** in space and rotate first the frame $\sum'$ by an angle $\delta\alpha$ around **n**, and then the particles with respect to $\sum'$ by the same angle, we obtain the combined change in Ψ

$$\delta\Psi = [e^{-i\delta\alpha \mathbf{n}\cdot\mathbf{J}} e^{-i\delta\alpha \mathbf{n}\cdot\mathbf{R}} - 1]\Psi \tag{3.17}$$

For small values of $\delta\alpha$ this should be the same as the effect of a total rotation of Ψ by an angle $\delta\alpha$ around **n**, that is, we also have

$$\delta\Psi = [e^{-i\delta\alpha \mathbf{n}\cdot\mathbf{I}} - 1]\Psi \tag{3.18}$$

Comparison of (3.18) with (3.17) then leads to (3.16).

The operators **R** and **J** operate on different degrees of freedom of the system and they therefore commute with each other

$$[R_k, J_l] = 0 \tag{3.19}$$

Since **J** commutes also with the vectors $\mathbf{e}_A$. we obtain from (3.16) and (3.8) that

$$[R_k, e_{Al}] = i\sum_j \epsilon_{klj} \, e_{Aj} \tag{3.20}$$

From (3.16), (3.19), and (3.20) it now follows that **R** satisfies commutation relations similar to those of **I**:

$$[R_k, R_l] = i\sum_m \epsilon_{klm} R_m \qquad [R_A, R_B] = -i\sum_C \epsilon_{ABC} R_C \tag{3.21}$$

It also follows from (3.19) and (3.20) that

$$[R_A, J_B] = -i\sum_C \epsilon_{ABC} J_C \tag{3.22}$$

Thus, although the components I_A, etc. and R_A, etc. satisfy the "abnormal" commutation relations with $-i\epsilon_{ABC}$, while the components J_A, etc. satisfy the "normal" commutation relations with $+i\epsilon_{ABC}$, the relation $\mathbf{I} = \mathbf{R} + \mathbf{J}$ is still consistent. The consistency is maintained because when taken with respect to $\sum'$, the components of **R** and **J** no longer commute as in (3.19) but rather satisfy (3.22).

The decomposition of the total angular momentum **I** into an *internal angular momentum* **J** and a *collective angular momentum* **R**, associated with the rotation of the body-fixed frame $\sum'$, makes sense only in a certain approximation. The physical wave function $\Psi = \chi(\mathbf{r}_i') \, \Phi(\theta_k)$ is not really defined for values of $\mathbf{r}_i'$ and θ_k that are inconsistent with (2.11). It is therefore impossible, *in principle*, to study the behavior of a realistic Ψ under transformations like **J** and **R** that effect separately $\mathbf{r}_i'$ and θ_k and disregard their interdependence via (2.11). The introduction of the operators **R** and **J** makes sense only in the approximation in which we construct the functions $\chi(\mathbf{r}_i')$ and $\Phi(\theta_k)$ through a prescription that disregards their interdependence. As we shall see later, this is possible, since $\chi(\mathbf{r}_i')$ is taken to be a Slater determinant of single-particle wave functions in a deformed potential whose orientation is given by θ_k, and the actual Ψ has redundant coordinates in it. (Recall a similar prescription for the center-of-mass problem). In this case (2.11) is satisfied only *on the average*, and it becomes possible to investigate changes in $\chi(\mathbf{r}_i')$ as a result of rotations of the particles relative to the axes of the deformed potential.

In view of this nature of the operators **J** and **R** it is a priori not obvious whether the rotational kinetic energy T_{rot} should be given as in (2.12) in terms of **R** or in terms, say, of a similar expression where **R** is replaced by **I**. To resolve this question one should really carry out the transformation from $\mathbf{r}_i$ to the coordinates $\mathbf{r}_i'$ and θ_k, introduce explicitly the assumed approximations, and derive the structure of the kinetic energy in terms of the new set of redundant coordindates. This program has been carried out by Villars and Cooper (69) who showed that the form (2.12) for T_{rot} in terms of **R**, rather than **I**, is indeed plausible. We shall not try to reproduce their arguments here, but will just note that also empirically **R** seems to be more appropriate than **I** for T_{rot} in (2.12). As we shall see below $T_{\text{rot}} = 1/2\mathcal{J}\,\mathbf{I}^2$ would have led to rotational spectra, with a characteristic $I(I+1)$ spacing between levels, for all deformed nuclei. On the other hand $T_{\text{rot}} = (1/2\mathcal{J})\,\mathbf{R}^2$ does so only approximately and in particular, it deviates badly from the $I(I + 1)$ law for nuclei whose ground state has $I = 1/2$. This is indeed found to be the case also empirically, so that the form (2.12) for the rotational energy is preferred also from this point of view.

We are now in a position to write down the wave function (3.1) in a more explicit form. To this end we note that the total angular momentum $\mathbf{I}$ is the real constant of motion of the system. It is therefore desirable to express $T_{\rm rot}$ as far as possible in terms of it. We shall confine ourselves for now to axially symmetric deformations, so that two of the moments of inertia are equal to each other. Taking z' as the axis of symmetry we have

$$\mathcal{J}_{x'} = \mathcal{J}_{y'} = \mathcal{J} \tag{3.23}$$

We then obtain from (2.12):

$$\begin{aligned} T_{\rm rot} &= \frac{\hbar^2}{2\mathcal{J}}(R_{x'}{}^2 + R_{y'}{}^2) + \frac{\hbar^2}{2\mathcal{J}_z{}'} R_{z'}{}^2 = \frac{\hbar^2}{2\mathcal{J}}\mathbf{R}^2 + \left(\frac{\hbar^2}{2\mathcal{J}_{z'}} - \frac{\hbar^2}{2\mathcal{J}}\right) R_{z'}{}^2 \\ &= \frac{\hbar^2}{2\mathcal{J}}[\mathbf{I}^2 + \mathbf{J}^2 - 2\mathbf{I}\cdot\mathbf{J}] + \left(\frac{\hbar^2}{2\mathcal{J}_{z'}} - \frac{\hbar^2}{2\mathcal{J}}\right)(I_{z'} - J_{z'})^2 \end{aligned} \tag{3.24}$$

In deriving (3.24) we have used (3.16) to put $\mathbf{R} = \mathbf{I} - \mathbf{J}$.

Since we assume the nucleus to have an axial symmetry, the eigenfunctions $\chi(\mathbf{r}_i')$ of the internal motion can be chosen to be eigenfunctions of J_z'. Let us characterize these functions by the quantum number Ω giving the projection of the particles' angular momentum on the body-fixed z'-axis:

$$J_{z'}\chi_\Omega(\mathbf{r}_i') = \Omega\chi_\Omega(\mathbf{r}_i') \tag{3.25}$$

The function $\Phi(\theta_k)$ in (3.1) should be taken to be an eigenfunction of $T_{\rm rot}$. Since $T_{\rm rot}$ in (2.12) has the form of the Hamiltonian of a rigid rotator, $\Phi(\theta_k)$ should be the eigenfunctions of the symmetric top [see van Winter (54)]. As we know these can be taken as simultaneous eigenfunctions of $\mathbf{R}^2$, R_z, and $R_{z'}$. Since $\mathbf{J}$ does not operate on the angle variables θ_k, it follows from (3.16) that we can choose the functions Φ to be simultaneous eigenfunctions of $\mathbf{I}^2$, I_z, and $I_{z'}$; denoting them by the conventional notation for Wigner's D-matrices (see Appendix A, p. 918) we have:

$$\begin{aligned} \mathbf{I}^2 D_{MK}^{I*}(\theta_k) &= I(I+1)\, D_{MK}^{I*}(\theta_k) \\ I_z D_{MK}^{I*}(\theta_k) &= M D_{MK}^{I*}(\theta_k) \\ I_z D_{MK}^{I*}(\theta_k) &= K D_{MK}^{I*}(\theta_k) \end{aligned} \tag{3.26}$$

where, of course, $-I \leq (K, M) \leq I$. It is important to note that because of the phase in the commutation relation (3.14) the combination $I_{x'} + iI_{y'}$ (primes indicate components with respect to the body-fixed frame) is a lowering operator. When it acts on $D_{M,K}^{I*}$ it yields $D_{M,K-1}^{I*}$ in contrast with $I_x + iI_y$ that raises the value of M by unity.

With the help of $\chi_\Omega(\mathbf{r}_i')$ of (3.25) and $D^{I*}_{MK}(\theta_k)$ of (3.26) we can now construct the zeroth eigenfunctions of H, (2.14). More precisely we rewrite H, (2.14) taking into account (3.24), and cast it in the form:

$$H = \left[H_{\text{int}} + \frac{\hbar^2}{2\mathcal{J}}\mathbf{J}^2\right] + \left[\frac{\hbar^2}{2\mathcal{J}}(\mathbf{I}^2 - 2I_{z'}J_{z'}) + \left(\frac{\hbar^2}{2\mathcal{J}_{z'}} - \frac{\hbar^2}{2\mathcal{J}}\right)(I_{z'} - J_{z'})^2\right]$$
$$+ \left[H_{\text{coupl}} - \frac{\hbar^2}{\mathcal{J}}(I_+'J_-' + I_-'J_+')\right] \tag{3.27}$$

$$\text{where } I_\pm' = \mp\frac{1}{\sqrt{2}}(I_{x'} \pm iI_{y'}),\ J_\pm' = \mp\frac{1}{\sqrt{2}}(J_{x'} \pm iJ_{y'})$$

We shall see later that in many cases the

$$\textit{Coriolis term}, -(\hbar^2/\mathcal{J})(I_+'J_-' + I_-'J_+')$$

in (3.27), which gives rise to a particle-rotation coupling, can be neglected because it leads to a small perturbation on the rest of the Hamiltonian. If we also adopt the empirical conclusion that H_{coupl} may be neglected for low-lying levels, we arrive at a zeroth-order Hamiltonian

$$H_0 = H_{\text{int}}' + \frac{\hbar^2}{2\mathcal{J}}\left[\mathbf{I}^2 - 2I_{z'}J_{z'} + \left(\frac{\hbar^2}{2\mathcal{J}_{z'}} - \frac{\hbar^2}{2\mathcal{J}}\right)(I_{z'} - J_{z'})^2\right] \tag{3.28}$$

with

$$H_{\text{int}}' = H_{\text{int}} + \frac{\hbar^2}{2\mathcal{J}}\mathbf{J}^2 \tag{3.29}$$

The new intrinsic Hamiltonian H' still has axial symmetry. If we take $\chi_\Omega(\mathbf{r}_i)$ in (3.25) to be the normalized eigenfunctions of H_{int}' we obtain for the normalized eigenfunctions of H_0:

$$\Psi = \sqrt{\frac{2I+1}{8\pi^2}}\, D^{I*}_{MK}(\theta_k)\, \chi_\Omega(\mathbf{r}_i') \tag{3.30}$$

The coupling of the angular momenta involved in (3.30) is shown in Fig. 3.1. **J** and **R** need not be, and generally are not, separately constants of the motion since we are dealing with a system with an axial symmetry.

Actually we shall see that due to some degeneracies, (3.30) is not the final form of Ψ, but it can serve to give us a first feeling for the wave functions of the unified model.

4. SYMMETRIES OF THE ROTATIONAL-MODEL WAVE FUNCTIONS

The condition of axial symmetry that we have imposed on H_{int} enables us to limit somewhat the class of allowed eigenfunctions (3.30). If the nucleus has an

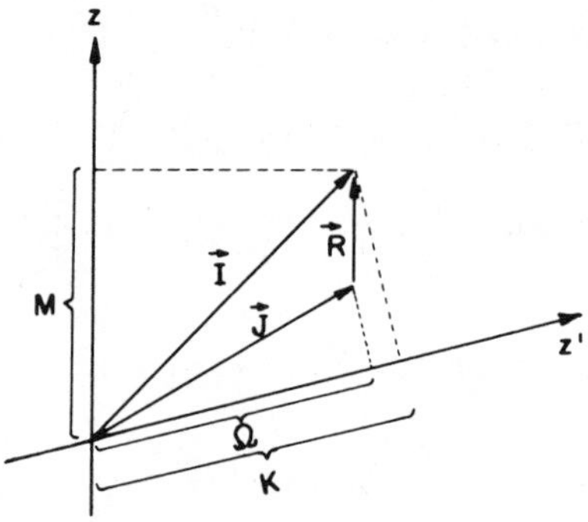

FIG. 3.1. The coupling of **J**, **R**, and **I** in (3.30).

axial symmetry and we choose the z'-axis to be in the direction of this axis of symmetry, then it is obvious that the solution of (2.11) cannot give us a unique direction for the x'- and y'-axes. Only the z'-direction can be fixed by (2.11).

If we could solve (2.11) we would get the angles θ_k, which determine the orientation of the body-fixed axes, as a function of the particle coordinates $\mathbf{r}_i$. As the particles move around, θ_k and, consequently, the body-fixed axes will change as well; $\dot{\theta}_k$ is in principle calculable from the motion of the particles through the dependence of θ_k on $\mathbf{r}_i$ as given by (2.11). If the motion of the particles is such that axial symmetry is retained all throughout the motion, then, obviously, this particle motion will lead to no rotation of any choice of the x'- and y'-axes. This is actually the content of the statement that the particles in their motion retain the axial symmetry of the nucleus. The collective part of the rotation in this case can only be in a direction perpendicular to z'. Whatever angular momentum there is to the system in the z'-direction must come from the internal angular momentum and is not to be ascribed to the collective angular momentum. Thus we reach the conclusion that in the wave function (3.30) we must have $K = \Omega$.

It is worthwhile to stress that the axial symmetry of a *Hamiltonian* around the z'-axis enables us only to conclude that the wave function must have a definite dependence on the azimuthal angle: $e^{iM\phi}$. However, here we have *assumed* an axial symmetry of the *wave function* with respect to the body-fixed z'-axis. This is a much stronger requirement that says, formally, that the total wave function remains *unchanged* under rotations around the z'-axis.

$$e^{i\alpha R_{z'}}\Psi = \Psi \qquad \text{for any } \alpha$$

It follows, therefore, that

$$R_{z'}\Psi = 0 \qquad \text{or} \qquad (I_{z'} - J_{z'})\Psi = 0 \tag{4.1}$$

Since $I_{z'}$, operates only on $\Phi(\theta_k)$ and $J_{z'}$ operates only on $\chi(\mathbf{r}_i')$ (see Eqs. 3.3 and 3.6) we obtain from (4.1) and (3.30).

$$\chi_\Omega(\mathbf{r})\,[I_{z'}\,D^{I*}_{MK}(\theta_k)] = D^{I*}_{MK}(\theta_k)\,[J_{z'}\chi_\Omega(\mathbf{r}_i')]$$

Using (3.25) and (3.26) it then follows that for axially symmetric nuclei we must have

$$K = \Omega \tag{4.2}$$

in agreement with our conclusion above, which followed from the assertion that there is no collective rotation around the z'-axis. As can be seen from Fig. 3.1, 4.2 means that the rotation $\mathbf{R}$ of the body-fixed frame of reference can only be perpendicular to the axis of the symmetry. No quantum-mechanical rotation is possible about an axis of symmetry of the system.

Another requirement on the wave function Ψ follows from the fact that (2.11) cannot fix a *sense of direction* for the z'-axis either; only the line along which it should lie is determined. In fact, if the tensor of inertia is diagonal with respect to one choice of the axes $x'y'z'$, it will remain diagonal also if we choose any of the axes along the opposite direction. If we rotate the body-fixed frame of reference by 180° around the y'-axis, for instance, we obtain a new frame whose z'- and x'-axes coincide with the negative directions of the previous z'- and x'-directions, respectively. Since (2.11) to begin with could have not distinguished between these two choices of the body-fixed frame of reference, the wave function Ψ should remain invariant under this particular rotation of the body-fixed frame. We encounter again an invariance of the wave function with respect to a certain transformation, which is due to the fact that, for an axially symmetric nucleus, (2.11) cannot provide a solution that violates this symmetry. We can therefore write

$$e^{-i\pi R_{y'}}\,\Psi = \Psi \tag{4.3}$$

The rotation (4.3) affects both the D-function and the χ- function, and let us study each effect separately.

Since $\mathbf{J}$ does not affect the D-functions, we obtain from (3.16) that

$$e^{-i\pi R_{y'}}\,D^{I*}_{MK}(\theta_k) = e^{-i\pi I_{y'}}\,D^{I*}_{MK}(\theta_k) = (-1)^{I+K}\,D^{I*}_{M,-K}(\theta_k) \tag{4.4}$$

where use has been made of the properties of the D-functions as summarized in Appendix A Sect 2.

To study the effect of $R_{y'}$ on $\chi_\Omega(\mathbf{r}_i')$ we note that as far as $\chi_\Omega(\mathbf{r}_i')$ is concerned, a rotation of the body-fixed frame by an angle α around $\mathbf{n}$ has the same effect as that of rotating $\chi_\Omega(\mathbf{r}_i')$ by an angle $(-\alpha)$ around the same direction $\mathbf{n}$ (see also Eq. 3.15). Therefore

$$e^{-i\pi R_{y'}}\,\chi_\Omega(\mathbf{r}_i') = e^{+i\pi J_{y'}}\,\chi_\Omega(\mathbf{r}_i')$$

We shall further find it useful to expand $\chi_\Omega(\mathbf{r}_i')$ in a series of functions of $\mathbf{r}_i'$ that are also eigenfunctions of $\mathbf{J}^2$:

$$\chi_\Omega(\mathbf{r}_i') = \sum a_J\chi_\Omega^{(J)}\,(\mathbf{r}_i') \tag{4.5}$$

where $\mathbf{J}^2\,\chi_\Omega^{(J)}\,(\mathbf{r}_i') = J(J+1)\,\chi_\Omega^{(J)}\,(\mathbf{r}_i')$.

We obtain then (see Appendix A, A.2.2, for finite rotations of angular momentum eigenfunctions)

$$e^{-i\pi R_{y'}} \chi_\Omega(\mathbf{r}_i') = e^{i\pi J_{y'}} \chi_\Omega(\mathbf{r}_i') = \sum a_J \, e^{i\pi J_{y'}} \chi_\Omega{}^{(J)}(\mathbf{r}_i')$$

$$= \sum a_J(-1)^{\Omega+J} \chi_{-\Omega}{}^{(J)}(\mathbf{r}_i') \quad (4.6)$$

For convenience it is customary to introduce the notation

$$\widehat{(-1)^J} \chi_{-\Omega}(\mathbf{r}_i') \equiv \sum a_J(-1)^J \chi_{-\Omega}{}^J(\mathbf{r}_i') \quad (4.7)$$

It has to be realized, however, that since $\chi_\Omega(\mathbf{r}_i')$ is not generally an eigenstate of $\mathbf{J}^2$, $\widehat{(-1)^J}$ is an *operator* defined by (4.7).

Using (4.7) and (4.4) and noting that $K = \Omega$, we obtain finally

$$e^{-i\pi R_{y'}} \Psi = e^{-i\pi R_{y'}} \sqrt{\frac{2I+1}{8\pi^2}} D^{I*}_{MK}(\theta_k) \chi_K(\mathbf{r}_i')$$

$$= \widehat{(-1)^{I-J}} \sqrt{\frac{2I+1}{8\pi^2}} D^{I*}_{M,-K}(\theta_k) \chi_{-K}(\mathbf{r}_i') \quad (4.8)$$

It can be shown that if the intrinsic Hamiltonian is invariant under time reversal, which indeed we shall assume to be the case, then for $\Omega \neq 0$ the states $\chi_{+\Omega}$ and $e^{-i\pi R_{y'}}\chi_{+\Omega} = \widehat{(-1)^{\Omega+J}} \chi_{-\Omega}$ derived in (4.6) are degenerate. Hence the corresponding energy is an even function of Ω, and because of (4.2), (3.26), and (3.27) also the states $D^I{}_{MK}(\theta_k)$ and $D^I{}_{M,-K}(\theta_k)$ correspond to the same energy of the rigid rotator. It follows therefore that we can obtain eigenfunctions of H_0, (3.28), which have the required symmetries (4.1) and (4.3), by taking the linear combination

$$\Psi(IKM) = \sqrt{\frac{2I+1}{16\pi^2}} \{D^{I*}_{MK}(\theta_k) \chi_K(\mathbf{r}_i') + \widehat{(-1)^{I-J}} D^{I*}_{M,-K}(\theta_k) \chi_{-K}(\mathbf{r}_i')\}$$

$$\text{for} \quad K > 0 \quad (4.9)$$

It is to be noted that only *one* such state can be built combining K and $-K$. We adopt, therefore, the convention that K is always positive. We also notice that as a result of the symmetry requirement (4.3), the total wave function is no longer separable into a product of an intrinsic wave function and a collective wave function. Rather, it is a sum of two such products. This implies a coupling of the intrinsic motion to the collective motion, whose origin has to do with the quantum-mechanical symmetry properties of the wave function.

The case $K = \Omega = 0$ deserves special attention because in this case $\chi_\Omega = \chi_{-\Omega} \equiv \chi_0$. If the intrinsic Hamiltonian has a *reflection symmetry* in the $x'y'$-, plane in addition to the axial symmetry, then a rotation of the body-fixed frame of reference by an angle π around the y'-axis commutes with the Hamiltonian; it follows that χ_0 is an eigenstate also of $e^{-i\pi R_{y'}}$:

$$e^{-i\pi R_{y'}} \chi_0 = \lambda \chi_0 \quad (4.10)$$

The expansion of χ_0 in terms of eigenstates of $\mathbf{J}^2$ contains integral values of J only since $\Omega = 0$ implies an even number of particles; since from (4.6) it follows that $e^{-i\pi R_{y'}}\,\chi_0 = e^{i\pi J_{y'}}\,\chi_0$, we conclude that

$$e^{-2i\pi R_{y'}}\,\chi_0 = e^{2i\pi J_{y'}}\,\chi_0 = +\,\chi_0 \tag{4.11}$$

Hence λ in (4.10) takes on the values

$$\lambda = \pm 1 \tag{4.12}$$

We can also use (4.6) to evaluate $e^{-i\pi R_{y'}}\,\chi_0$:

$$e^{-i\pi R_{y'}}\,\chi_0 = e^{i\pi J_{y'}}\,\chi_0 = (-1)^J\,\chi_0 \tag{4.13}$$

Comparing (4.13) with (4.10), and noting (4.12) we conclude that intrinsic wave functions with $\Omega = 0$ can have in them components of either even J or odd J but not a mixture of both. If χ_0 belongs to the eigenvalue $\lambda = +1$ of $e^{-i\pi R_{y'}}$ it will have only even J's in its expansion (4.5), and only odd J's will show up if it belongs to $\lambda = -1$. In other words

$$(-1)^J\,\chi_0 = \lambda\,\chi_0 \qquad \lambda = \pm 1 \tag{4.14}$$

The derivation of (4.9) can now be repeated for the case $K = 0$. One finds then, using (4.14):

$$\begin{aligned}\Psi(I, K = 0, M) &= N\{D^{I*}_{MO}(\theta_k)\,\chi_0(\mathbf{r}_i') + (-1)^{I-J}\,D^{I*}_{MO}(\theta_k)\,\chi_0(\mathbf{r}_i')\}\\ &= N\{D^{I*}_{MO}(\theta_k)\,\chi_0(\mathbf{r}_i') + \lambda\,(-1)^I\,D^{I*}_{MO}(\theta_k)\,\chi_0(\mathbf{r}_i')\}\\ &= N\,[1 + \lambda\,(-1)^I]\,D^{I*}_{MO}(\theta_k)\,\chi_0(\mathbf{r}_i')\end{aligned} \tag{4.15}$$

where N is a normalization constant. One sees, therefore, that for $\lambda = +1$, $\Psi(I, K = 0, M)$ vanishes for odd values of I, whereas for $\lambda = -1$ it vanishes for even values of I. Thus we expect rotational bands built on intrinsic states with $K = 0$ to have either even I's only or odd I's only unlike rotational bands built on other values of K, which can have all values of I satisfying, of course, $I \geqslant K$.

The wave function $\Psi(IKM)$, (4.9) or (4.15), are eigenstates of H_0, (3.28), that satisfy the symmetry requirements resulting from the ambiguity in the definition of θ_k from (2.11) when the nucleus has an intrinsic axial symmetry. It is easy to evaluate the eigenvalues E_0 of H_0 that go with these wave functions:

$$H_0\,\Psi(IKM) = \left\{\epsilon(K) + \frac{\hbar^2}{2\mathcal{J}}\,[I(I+1) - 2K^2]\right\}\,\psi(IKM) \tag{4.16}$$

Here $\epsilon(K)$ is the eigenvalue of the intrinsic Hamiltonian H'_{int}, (3.29), to which χ_K belongs:

$$H'_{\text{int}}\,\chi_K = \left[H_{\text{int}} + \frac{\hbar^2}{2\mathcal{J}}\,\mathbf{J}^2\right]\chi_K = \epsilon(K)\,\chi_K \tag{4.17}$$

We have noted before, in deriving (4.9), that $\epsilon(K)$ depends actually only on $|K|$, so that χ_{-K} satisfies (4.17) with the *same* value for $\epsilon(K)$. Since, as indicated by (3.26) $|K| < I$, the levels I that involve the same internal structure, and therefore the same value of $K \neq 0$, must satisfy

$$I = |K|, |K| + 1, |K| + 2, \ldots \tag{4.18}$$

These are the levels that constitute a rotational band constructed on the intrinsic state χ_K.

The Hamiltonian H_0 is, of course, not the complete Hamiltonian of the rotating nucleus. The full Hamiltonian (3.27) contains a coupling term H_{coupl} that is believed to be small, and the Coriolis term $-(\hbar^2/\mathcal{J})(I_+'J_-' + I_-'J_+')$ that has thus far been neglected. Because of the occurence of $J_\pm'$ in the Coriolis term it can couple an intrinsic state χ_K only with another intrinsic state χ_{K+1}. Normally, for deformed nuclei, we find that the empirical moment of inertia $\mathcal{J}$ satisfies

$$\frac{\hbar^2}{2\mathcal{J}} \ll |\epsilon(K) - \epsilon(K')| \qquad K \neq K' \tag{4.19}$$

That is, the spacing of the intrinsic levels is large compared to the spacing of the rotational levels. Since $J_\pm'$ has no diagonal elements in K, the Coriolis potential can affect the spectrum of H only in second order, and because of (4.19) its effects are expected to be small for small values of I.

There is, however, one case that deserves special care because of the appearance of both $+K$ and $-K$ on $\Psi(IKM)$. If K is such that

$$K \pm 1 = -K, \qquad \text{that is,} \qquad K = +\frac{1}{2} \tag{4.20}$$

then the Coriolis term does have *diagonal* terms in $\Psi(IKM)$ and therefore does lead to first-order corrections to the eigenvalues of H. Equation 4.16 should then be corrected to include first-order contributions of the Coriolis potential, and we obtain

$$E(IKM) = \langle\Psi(IKM)|H\Psi(IKM)\rangle = \epsilon(K) + \frac{\hbar^2}{2\mathcal{J}}[I(I+1) - 2K^2] + \delta\epsilon\left(\frac{1}{2}\right)\delta\left(|K|, \frac{1}{2}\right) \tag{4.21}$$

where a straightforward calculation gives, using the expansion (4.5):

$$\delta\epsilon\left(\frac{1}{2}\right) = \frac{\hbar^2}{2\mathcal{J}}\sum_J (-1)^{I+J}\frac{(2I+1)(2J+1)}{4}|a_J|^2 \tag{4.22}$$

It is customary to introduce the decoupling parameter* a defined by:

$$a = \sum_J (-1)^{J-1/2}\left(J + \frac{1}{2}\right)|a_J|^2 \qquad (4.23)$$

and to lump together all the K-dependent parts of $E(IKM)$. Equation 4.21 then takes the following form, valid for any value of K,

$$E(IKM) = E(K) + \frac{\hbar^2}{2\mathcal{J}}\left[I(I+1) + a(-1)^{I+1/2}\left(I+\frac{1}{2}\right)\delta\left(|K|,\frac{1}{2}\right)\right] \qquad (4.24)$$

where $E(K)$ is the I-independent part of $E(IKM)$.

It is instructive to summarize what went into the derivation of (4.24) since, as we shall see, its agreement with experiment is so spectacular.

We asserted that the true nuclear Hamiltonian is approximately separable in the Euler angles that define via (2.11) the orientation of a body-fixed frame of reference, and the coordinates $\mathbf{r}_i'$ relative to this frame. We further assumed that the intrinsic wave function obtained from this separation has an axial symmetry around z' and a reflection symmetry in the $x'y'$-plane. Then, for the case in which the empirical moment of inertia $\mathcal{J}$ required to carry out the separation satisfies (4.19), we obtained (4.24) correct to the first order in the Coriolis potential. Nothing further had to be assumed about the nature of the intrinsic Hamiltonian H_{int}. For a given band K, (4.24) has in it at most three parameters: $E(K)$, $\mathcal{J}$, and a. These can be functions of K but are expected to be independent of I within the band. Since a rotational band may be expected to have appreciably more than three levels in it, we can use the experimental data not just to determine the internal energies $E(K)$, the moment of inertia $\mathcal{J}$, and the decoupling parameter a (for $|K| = 1/2$), but also to check the consistency of the whole underlying model that gave us (4.24).

5. COMPARISON WITH EXPERIMENT

We saw in Section VI.1 a few examples of rotational spectra in the region of $A \sim 176$ to 178. The deformed even-even nuclei have rotational spectra with $I = 0^+$ for their ground state. We note that the levels of any rotational band must satisfy $I \geqslant K$ ($D^I{}_{MK}$ vanishes if $I < K$; the projection of an angular momentum cannot be bigger than the angular momentum itself). We conclude therefore that in even-even nuclei, the lowest rotational bands have

---

*The origin of this name comes from the observation that if **R** is perpendicular to the nuclear axis, as implied by (4.2), and if **J** lies along this axis, then $\mathbf{I}\cdot\mathbf{J} \approx \mathbf{J}^2$ and the Coriolis potential introduces no I-dependent correction to the energy. When the I-dependence is important it measures therefore the decoupling of the vector of **J** of the intrinsic motion from the nuclear axis of symmetry.

$K = 0$. According to (4.15) they should have only even values of I for the angular momenta of their excited levels. This expectation is very convincingly borne out by the actual experimental data in all known cases.

Consider now an example of an odd-A nuclei, ^{177}Lu. We observe at least three rotational bands in ^{177}Lu (see Fig. 1.4). Again, using the relation $I > K$ for the various levels of the same rotational band, we associate these bands with $K = 7/2^+$ (ground-state band), $K = 9/2^-$ (band starting at 0.1504 MeV), and $K = 5/2^+$ (band starting at 0.458 MeV). We note that the moments of inertia associated with these three bands are rather close to each other:

$$\frac{\hbar^2}{2\mathcal{J}} = 0.01379 \text{ MeV for the } K = \frac{7^+}{2} \text{ band}$$

$$= 0.01286 \text{ MeV for the } K = \frac{9^-}{2} \text{ band}$$

$$= 0.01368 \text{ MeV for the } K = \frac{5^+}{2} \text{ band}$$

They are on the whole larger than the moments of inertia of the neighboring even-even nuclei ($\hbar^2/2\mathcal{J} = 0.01374$ MeV for ^{176}Yb, $(\hbar^2/2\mathcal{J}) = 0.01482$ MeV for ^{176}Hf, and $(\hbar^2/2\mathcal{J}) = 0.01561$ MeV for ^{178}Hf). This seems to be true for most deformed odd-A nuclei.

$K = 1/2$ bands are known in the different deformed regions of the periodic table; two of them are shown in Fig. 5.1 with the calculated energies shown in square brackets (lower levels were used to derive the parameters). Again the agreement of the empirical date with (4.24) for $K = 1/2$ is striking. We notice the large deviation of the spectra, in these cases, from the simple $I(I + 1)$ rule, clearly indicating the special role played by the decoupling parameter for $K = 1/2$ bands.

Several questions pose themselves as one studies the phenomenal success of the simple formula (4.24) in reproducing so well and systematically the low-lying energy levels of practically all even-even and odd even nuclei in certain regions of the periodic table (the situation with odd-odd nuclei is slightly more complex):

(a) Are there other tests that could be used to confirm the simple structure of the rotational wave function $\Psi(IKM)$ given by (4.9)? What we should look for are tests that make, if possible, no explicit use of the wave functions χ_Ω beyond the fact that they involve internal coordinates only.

(b) Is the empirical value of the moment of inertia reasonable and does it tell us anything about nuclear dynamics?

(c) Can we understand the systematics encountered in the empirical moments of inertia? We mentioned above that moments of inertia of odd-A

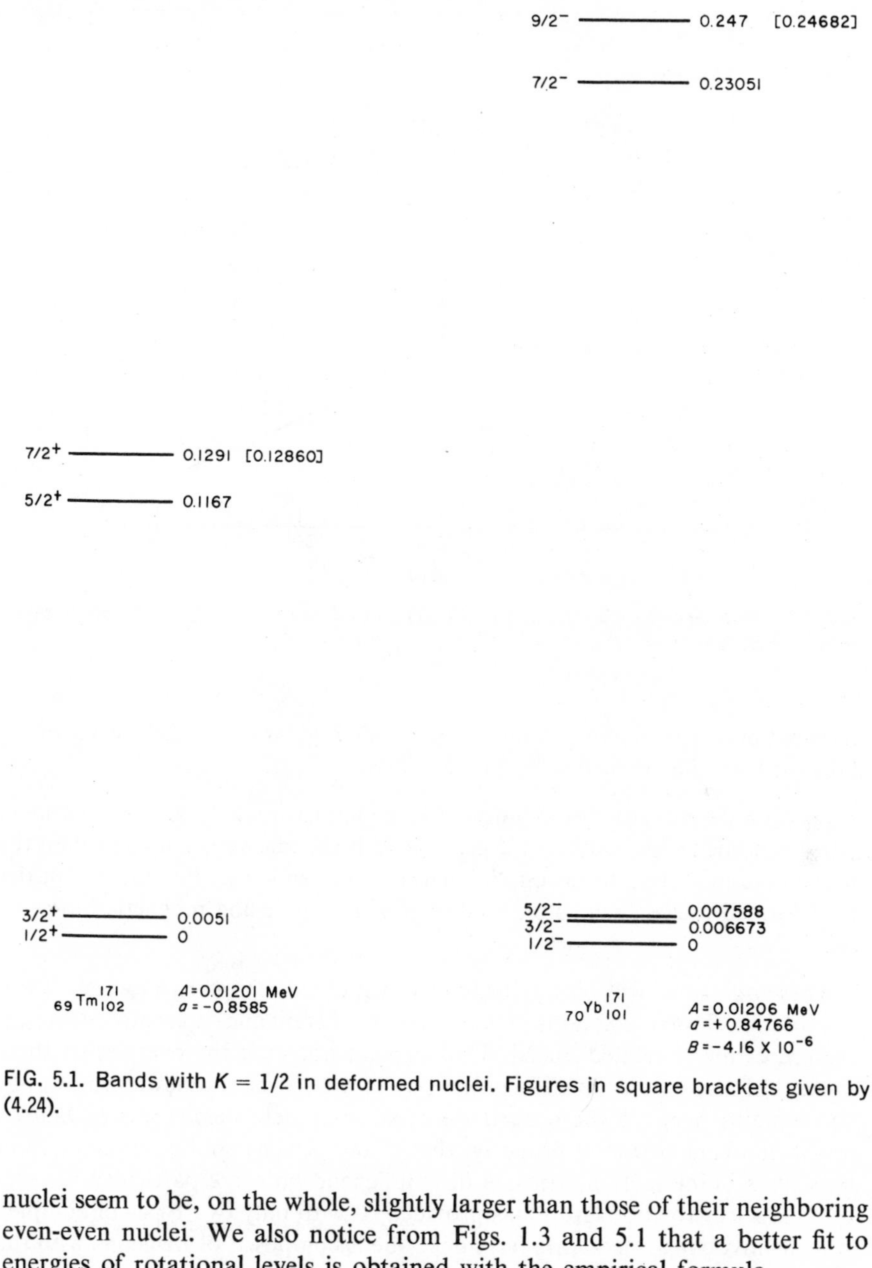

FIG. 5.1. Bands with $K = 1/2$ in deformed nuclei. Figures in square brackets given by (4.24).

nuclei seem to be, on the whole, slightly larger than those of their neighboring even-even nuclei. We also notice from Figs. 1.3 and 5.1 that a better fit to energies of rotational levels is obtained with the empirical formula

$$E(I) = E_0 + AI(I + 1) + BI^2 (I + 1)^2 \tag{5.1}$$

Although B comes out always to about 10^3 to 10^4 times smaller than A, its sign is always negative. This can be interpreted as a slight I-dependence of the

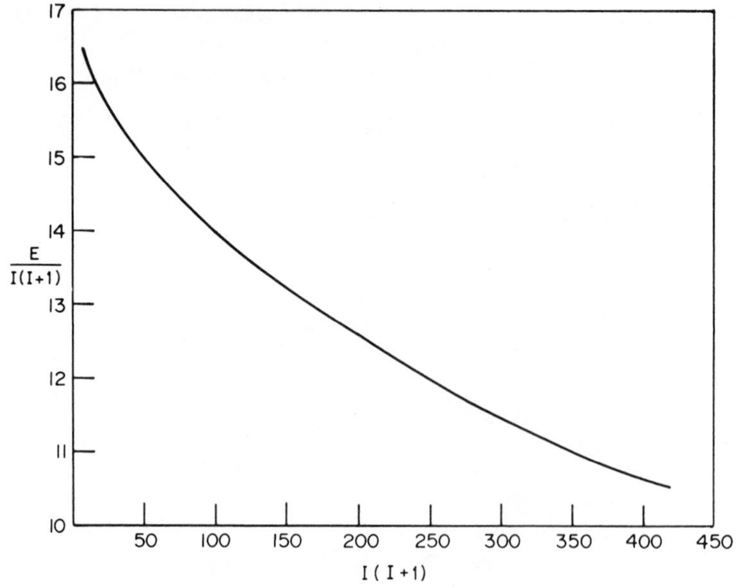

FIG. 5.2. The energies of Gd divided by $I(I+1)$ plotted against $I(I+1)$. The data is taken from Theiberger et al. (71).

moment of inertia, making $\mathcal{J}$ bigger for higher angular momenta. This formula fails for large I as indicated by Fig. 5.2.

(d) Can the intrinsic wave function χ_Ω be approximated by an independent particle model using a deformed field? And if so, can we use it to explain the moment of inertia $\mathcal{J}$, the decoupling parameter a, and other intrinsic quantities of deformed nuclei such as their intrinsic quadrupole and magnetic moments (see below)?

These questions will occupy us for the rest of this chapter. As we shall see, the rotational wave functions present a most coherent and internally consistent picture of many related nuclei. They explain not only the energies of these nuclei but many other nuclear properties as well. A basic question, however, still remains open: is the occurrence of strongly deformed nuclei in betwen magic numbers a general property, that is, one which could be expected for a large class of interactions $v(ik)$, or does it depend on some particular features of the nuclear force? Nature has provided us with only two types of complex systems involving interacting fermions; one is composed of nucleons with the nuclear interaction among them, and the other of atoms with electromagnetic interaction among their electrons. One, the nuclear case, shows in its ground state under some circumstances a rotational structure; the other, the atomic case shows a crystaline structure. It is hard, therefore, to judge, on the basis of experimental evidence alone, which behavior is the rule and which results from some special features of the interaction. In Chapter VII we shall try to study this question more closely by studying several model Hamiltonians and their approximate solutions. We shall find several arguments that make the existence of deformed nuclei in between closed shells quite plausible, but a definitive answer to our question cannot, at this stage, be produced.

6. OTHER TESTS OF ROTATIONAL WAVE FUNCTIONS. ELECTRIC-TRANSITION PROBABILITIES

We recall that the validity of adopting shell-model wave functions for a set of levels could be tested with very few additional assumptions about the dynamics of the system. This was due to the fact that, through the assumptions of the shell model, physical properties of various levels became interrelated, so that the expression of one in terms of the other became a test of the particular shell-model wave function assignment.

For the rotational model such tests become even more natural. The wave function (4.9) is supposed to describe a whole band of levels, all with the same intrinsic wave functions χ_Ω and $\chi_{-\Omega}$. Since the D-functions are known, universal functions, we can expect that the values of a physical quantity taken for all levels of the same band will be simply related to each other.

Most nuclear quantities of interest, like moments of various kinds, transition amplitudes, etc., can be expressed in terms of irreducible tensor operators (Appendix A, A.2.39). Let $\hat{T}_\kappa^{(k)}$ be a component of such an operator referred to the laboratory frame of reference. We shall be generally interested in the matrix element

$$\langle IKM|\hat{T}_\kappa^{(k)}|I'K'M'\rangle = \int \Psi^* (IKM)\, \hat{T}_\kappa^{(k)}\, \Psi(I'K'M')\, d(1) \ldots d(A) \tag{6.1}$$

since $\hat{T}_\kappa^{(k)}$ is a component of an irreducible tensor operator we can easily express it relative to the body-fixed frame of reference. Using the relation in Appendix A, A.2.39 and A.2.19 we have

$$\hat{T}_\kappa^{(k)} = \sum_{\kappa'} D_{\kappa\kappa'}^{k*}(\theta_k)\, \hat{T}_{\kappa'}^{(k)} \tag{6.2}$$

In (6.2) $T_{\kappa'}^{(k)}$ are the components of the tensor $\mathbf{T}^{(k)}$ referred to the body-fixed frame. They will generally depend on both the non-independent coordinates $\mathbf{r}_i'$ and the orientation θ_k of the body-fixed frame. In many special cases, however, this dependence on $\mathbf{r}_i'$ and θ_k can be separated. For one important class of tensors $\mathbf{T}^{(k)}$, $T_{\kappa'}^{(k)}$ depends only on $\mathbf{r}_i'$ and does not depend at all on θ_k. This happens when $\mathbf{T}^{(k)}$ $(\mathbf{r}_i')$ is constructed out of the coordinates $\mathbf{r}_i$ of the particles but involves no momenta, or the operators $\partial/\partial x_i$, etc. By definition, irreducible tensor operators (see Appendix A, A.2.39), when referred to another frame, involve only the coordinates with respect to that new frame, and do not depend explicitly on the angles between the two frames. In this case we can write (6.2) in the form:

$$\hat{T}_\kappa{}^{(k)}(\mathbf{r}_i) = \sum_{\kappa'} D^{k*}_{\kappa\kappa'}(\theta_k)\,\hat{T}_{\kappa'}{}^{(k)}(\mathbf{r}_i') \tag{6.3}$$

We note that (6.3) gives us the operator $\hat{T}_\kappa{}^{(k)}(\mathbf{r}_i)$ essentially in a separated form, as a product of a function of the coordinates $\mathbf{r}_i'$ relative to the body-fixed frame, and a function of the collective coordinates θ_k. Our basic assumptions, (4.9), say that $\Psi(IKM)$ can be similarly separated. We can therefore carry out explicitly all the θ_k integrations—they involve only known functions of θ_k—and thereby obtain the complete I- and I'-dependence of (6.1). Introducing (6.3) and (4.9) into (6.1) we obtain, in fact:

$$\begin{aligned}
&\langle IKM|\hat{T}_\kappa{}^{(k)}(\mathbf{r}_i)|I'K'M'\rangle \\
&= \frac{1}{16\pi^2}\sqrt{(2I+1)(2I'+1)}\sum_{\kappa'}\Big\{\int D^I_{MK}(\theta_k)\,D^{k*}_{\kappa\kappa'}\,D^{I'*}_{M'K'}(\theta_k)\,\langle K|T_{\kappa'}{}^{(k)}|K'\rangle \\
&\quad + (-1)^{I-I'}\int D^I_{M,-K}(\theta_k)\,D^{k*}_{\kappa\kappa'}(\theta_k)\,D^{I'*}_{M',-K'}(\theta_k)\,\langle -K|T_{\kappa'}{}^{(k)}|-K'\rangle \\
&\quad + (-1)^{I'}\int D^I_{MK}(\theta_k)\,D^{k*}_{\kappa\kappa'}(\theta_k)\,D^{I'*}_{M',-K'}(\theta_k)\,\langle K|T_{\kappa'}{}^{(k)}|-K'\rangle \\
&\quad + (-1)^{I}\int D^I_{M,-K}(\theta_k)\,D^{k*}_{\kappa\kappa'}(\theta_k)\,D^{I'*}_{M',K'}(\theta_k)\,\langle -K|T_\kappa{}^{(k)}|K'\rangle\Big\}
\end{aligned} \tag{6.4}$$

where we used the notation

$$\begin{aligned}
\langle K|T_\kappa{}^{(k)}|K'\rangle &= \int \chi^*_K(\xi_\alpha)T_\kappa{}^{(k)}(\xi_\alpha)\,\chi_{K'}(\xi_\alpha)d\xi_\alpha \\
\langle -K|T_\kappa{}^{(k)}|K'\rangle &= \int [(\widehat{-1})^{-J}\,\chi_{-K}(\xi_\alpha)]^*\,T_\kappa{}^{(k)}(\xi_\alpha)\,\chi_{K'}(\xi_\alpha)d\xi_\alpha, \text{ etc,}
\end{aligned} \tag{6.5}$$

We can now use the relation (see Appendix A, A.2.73)

$$\frac{1}{8\pi^2}\int D^I_{MK}(\theta_k)\,D^{k*}_{\kappa\kappa'}(\theta_k)\,D^{I'*}_{M'K'}(\theta_k)\,dR = \begin{pmatrix} I & k & I' \\ -M & \kappa & M' \end{pmatrix}\begin{pmatrix} I & k & I' \\ -K & \kappa' & K' \end{pmatrix}(-1)^{M-K} \tag{6.6}$$

where dR stands, symbolically, for the integration over the whole angular range of θ_k with the proper volume element in the space of these angles. Equation 6.4 can then be reduced to the form

$$\langle IKM|\hat{T}_\kappa^{(k)}(\mathbf{r}_i)|I'K'M'\rangle = \frac{(-1)^{M-K}}{2}\sqrt{(2I+1)(2I'+1)}\begin{pmatrix} I & k & I' \\ -M & \kappa & M' \end{pmatrix}$$

$$\times \sum_{\kappa'} \left[\begin{pmatrix} I & k & I' \\ -K & \kappa' & K' \end{pmatrix} \langle K|T_{\kappa'}^{(k)}|K'\rangle \right.$$

$$+ (-1)^{I+I'} \begin{pmatrix} I & k & I' \\ K & \kappa' & -K' \end{pmatrix} \langle -K|T_{\kappa'}^{(k)}|-K'\rangle$$

$$+ (-1)^{I'} \begin{pmatrix} I & k & I' \\ -K & \kappa' & -K' \end{pmatrix} \langle K|T_{\kappa'}^{(k)}|-K'\rangle$$

$$\left. + (-1)^{-I} \begin{pmatrix} I & k & I' \\ K & \kappa' & K' \end{pmatrix} \langle -K|T_{\kappa'}^{(k)}|K'\rangle \right] \tag{6.7}$$

Equation 6.7 can be further reduced if we notice the following relation between $\langle -K|T_{-\kappa'}^{(k)}|-K'\rangle$ and $\langle K|T_{\kappa'}^{(k)}|K'\rangle$: By (6.5) and (4.7) and using the Wigner-Eckart theorem (see Appendix A, A.2.44):

$$\langle -K|\hat{T}_{-\kappa'}^{(k)}|-K'\rangle = \sum_{JJ'} [(-1)^J a^J]^*[(-1)^{J'} a_{J'}] \langle \chi_{-K}{}^J|T_{-\kappa'}^{(k)}|\chi_{-K'}{}^{J'}\rangle$$

$$= \sum_{JJ'} (-1)^{-J+J'} a_J^* a_{J'} (-1)^{J+K} \begin{pmatrix} J & k & J' \\ K & -\kappa' & -K' \end{pmatrix} \langle \chi^J||T^{(k)}||\chi^{J'}\rangle$$

$$= \sum_{JJ'} (-1)^{K-k-J} a_J^* a_{J'} \begin{pmatrix} J & k & J' \\ -K & \kappa' & K' \end{pmatrix} \langle \chi^J||T^{(k)}||\chi^{J'}\rangle$$

$$= (-1)^k \sum_{JJ'} a_J{}^* a_{J'} \langle \chi_K{}^J|T_{\kappa'}^{(k)}|\chi_{K'}{}^{J'}\rangle = (-1)^k \langle K|T_{\kappa'}^{(k)}|K'\rangle \tag{6.8}$$

Note that $[(-1)^J]^* = (-1)^{-J}$ since J can be half integer. We obtain then finally

$$\langle IKM|\hat{T}_{\kappa'}^{(k)}(\mathbf{r}_i)|I'K'M'\rangle = (-1)^{M-K}\sqrt{(2I+1)(2I'+1)}\begin{pmatrix} I & k & I' \\ -M & \kappa & M' \end{pmatrix}$$

$$\times \sum_{\kappa'} \left[\begin{pmatrix} I & k & I' \\ -K & \kappa' & K' \end{pmatrix} \langle K|T_{\kappa'}^{(k)}|K'\rangle \right.$$

$$\left. + (-1)^{I'} \begin{pmatrix} I & k & I' \\ -K & \kappa' & -K' \end{pmatrix} \langle K|T_{\kappa'}^{(k)}|-K'\rangle \right] \tag{6.9}$$

Equation 6.9 thus expresses any matrix element of $\widehat{T}^{(k)}(\mathbf{r}_i)$ between a level of the band K and a level of the band K' in terms of the internal matrix elements

$$\langle K|T_{\kappa'}{}^{(k)}|K'\rangle \qquad \text{and} \qquad \langle K|T_{\kappa'}{}^{(k)}|-K'\rangle \tag{6.10}$$

We notice that although (6.9) involves formally, a summation over κ', the sum reduces to one term in (6.9) if the value of k is sufficiently small. The first term contributes only if $\kappa' = K - K'$ and the second only if $\kappa' = K + K'$. In many applications of (6.9) one considers an irreducible tensor of a rather low order—$k = 1$ or $k = 2$. In that event these conditions on κ' can be satisfied for a few (often only one) value of κ'. Indeed, if the bands K and K' are such that $K + K' > k$, we cannot satisfy $\kappa' = K + K'$ since by definition $\kappa' \leqslant k$. Only the first term in (6.9) contributes then to the matrix element in question.

As an example of the possible uses of (6.9) let us consider the quadrupole moment operator (see Section 1.8)

$$Q_\mu{}^{(2)} = \sqrt{\frac{16\pi}{5}} \sum r_i{}^2\, Y_{2\mu}(\theta_i\phi_i)\, \frac{1+\tau_{iz}}{2} \tag{6.11}$$

which is a special irreducible tensor of rank 2 and of the type that satisfies (6.3). The quadrupole moments of the various levels of the same K-band are then given by:

$$Q(I) = \langle IK\, M = I|\, Q_0{}^{(2)}\, |I\, K\, M = I\rangle \tag{6.12}$$

If $K \geq 3/2$, the second term in (6.9) vanishes and we obtain from (6.12)

$$Q(I) = (-1)^{I-K}(2I+1)\begin{pmatrix} I & 2 & I \\ -I & 0 & I \end{pmatrix}\begin{pmatrix} I & 2 & I \\ -K & 0 & K \end{pmatrix} Q_0 \qquad K \geq 3/2 \tag{6.13}$$

where the intrinsic quadrupole moment Q_0 is given by

$$Q_0 = \sqrt{\frac{16\pi}{5}}\, \langle \chi_K|\sum_i r_i'^2\, Y_{20}(\theta_i'\phi_i')\, \frac{1+\tau_{iz}}{2}|\chi_K\rangle \tag{6.14}$$

A direct substitution of the expression (see Appendix A, A.2.80) for the special 3-j symbol involved in (6.13) yields (also, incidentally, for $K = 1/2$ and $K = 0$):

$$Q(I) = \frac{3K^2 - I(I+1)}{(I+1)(2I+3)}\, Q_0 \qquad K \neq 1 \tag{6.15}$$

For the lowest state of a band we have $I = K$, and (6.15) reduces to

$$Q(I = K) = \frac{I(2I-1)}{(I+1)(2I+3)}\, Q_0 \qquad K \neq 1 \tag{6.16}$$

A careful analysis shows that (6.15) and (6.16) are valid also for $K = 0$ and $K = 1/2$; a slightly different formula results for $K = 1$ due to the contribution of the second term in (6.9).

Problem. Verify (6.15) and (6.16) for $K = 1/2$ by using (6.9), and for $K = 0$ by using the wave function (4.15) to derive the appropriate modification of (6.9). Derive the expression of $Q(I)$ for $K = 1$.

Experimental information about quadrupole moments of excited nuclear states have just begun to accumulate so that it is hard to test (6.15). From a measurement of the ground state quadrupole moment in a rotational band we can derive a value for Q_0, which is a measure of the intrinsic shape of the deformed nucleus, but again, we cannot compare it to other quadrupole moments in the same rotational band. However, this value of Q_0 can be compared with the moment of inertia of the same band, which is also related in a definite way to the nuclear deformation, and some important conclusions regarding the dynamics of nuclear flow can then be drawn (see below).

The value derived from the intrinsic quadrupole moment Q_0 from (6.16) can also serve to estimate transition probabilities between levels of the same rotational band when the radiation emitted is known to be an electric quadrupole one. The transition probability, per unit time, for the emission of a photon of energy $\hbar\omega = \hbar ck$ and of multipolarity λ, if the nucleus undergoes a transition from the initial state i to the final state f is given by (see Chapter VIII)

$$T(\lambda) = \frac{8\pi(\lambda + 1)}{\lambda[(2\lambda + 1)!!]^2} \frac{k^{2\lambda+1}}{\hbar} B(\lambda, I_i \rightarrow I_f) \tag{6.17}$$

where

$$B(\lambda, I_i \rightarrow I_f) = \frac{1}{2I_i + 1} \sum_{M_i, M_f, \mu} |\langle I_f M_f | T_\mu^{(\lambda)}(\mathbf{r}_i) | I_i M_i \rangle|^2 \tag{6.18}$$

For electric quadrupole radiations the transition operator $T_\mu^{(2)}(\mathbf{r}_i)$ in (6.18) is very simply related to Q_μ:

$$T_\mu^{(2)} = e\sqrt{\frac{5}{16\pi}}\, Q_\mu \tag{6.19}$$

We therefore obtain from (6.9), (6.14), and (6.18), that for electric quadrupole transitions with a given band

$$B(E2, I_i \rightarrow I_f) = \frac{e^2}{2I_i + 1} \sum_{M_i M_f \mu} \frac{5}{16\pi} |\langle I_f K M_f | Q_\mu | I_i K M_i \rangle|^2$$

$$= \frac{e^2}{2I_i + 1} \frac{5}{16\pi} (2I_f + 1)(2I_i + 1) \sum_{M_i M_f \mu} \begin{pmatrix} I_f & 2 & I_i \\ -M_f & \mu & M_i \end{pmatrix}^2 \begin{pmatrix} I_f & 2 & I_i \\ -K & \kappa & K \end{pmatrix}^2$$

$$\times |\langle K | Q_\kappa | K \rangle|^2 = \frac{5}{16\pi} e^2 Q_0^2 (2I_f + 1) \begin{pmatrix} I_f & 2 & I_i \\ -K & 0 & K \end{pmatrix}^2 \qquad \text{for } K \neq \frac{1}{2}, 1 \tag{6.20}$$

Equation 6.20 tells us that within a given band there is essentially one number, Q_0, which determines all electric quadrupole properties—both the static moments as given by (6.15) as well as the transition moments, as indicated by (6.20).

The best experimental tests of (6.20) are obtained from the so-called Coulomb excitations of deformed odd-A nuclei. These are experiments in which a nucleus in its ground state is struck by a charged projectile whose energy is less than the Coulomb barrier between it and the target nucleus. The resulting variation with time of the electric field of the projectile as seen by the target, can then induce electric multipole transitions in the target. Since the role of initial and final states is reversed, compared to that encountered in γ decay, we see from (6.18) that the reduced matrix element $B(E2)$ that determines the Coulomb quadrupole excitation rate, is related to the one that determines the γ decay, by

$$\frac{B(E2;\, I_i \to I_f)}{B(E2;\, I_f \to I_i)} = \frac{2I_f + 1}{2I_i + 1} \tag{6.21}$$

γ-decay, $B(E2, I_i \to I_f)$, Coulomb excitation $B(E2, I_f \to I_i)$, (See Figure (6.1)).

Equation 6.20 has been tested for a number of odd-A deformed nuclei where one generally observes the quadrupole Coulomb excitations from the $I = K$ ground state to the levels $I' = K + 1$ and $I'' = K + 2$. It follows from (6.20) that, for these cases,

$$R = \frac{B(E2, I = K \to I'' = K + 2)}{B(E2, I = K \to I' = K + 1)} = \frac{2(K + 1)}{K(2K + 3)} \qquad (K > 3/2) \tag{6.22}$$

Some selected results for R are shown in Table 6.1, and it is seen that at least for nuclei well within the deformed region, like ^{175}Lu and ^{177}Hf, the agreement between theory and experiment is very good.

An even more interesting test of (6.20) is the comparison of Q_0 as derived from an absolute measurement of the lifetime of a level decaying by emitting an electric quadrupole, with that obtained from an absolute measurement of the static quadrupole moment of the ground state of the same band, using (6.16). Table 6.2 gives some such comparisons.

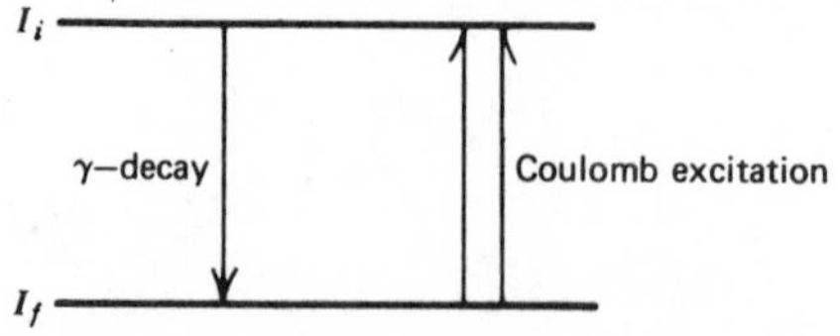

FIG. 6.1. γ-decay and Coulomb excitation.

TABLE 6.1 Theoretical and Experimental Values of R (Eq. 6.22) for Deformed Nuclei. The Theoretical Values are shown in Parentheses. [taken from Mottelson (60)]

Nuclei	$I_0 = K$	$\dfrac{B(E2;\, K \to K+2)}{B(E2;\, K \to K+1)}$
^{153}Eu	5/2	0.31(0.35)
^{155}Gd	3/2	0.52(0.56)
^{157}Gd	3/2	0.55(0.56)
^{159}Tb	3/2	0.51(0.56)
^{161}Dy	5/2	0.27(0.35)
^{163}Dy	5/2	0.27(0.35)
^{165}Ho	7/2	0.26(0.26)
^{167}Er	7/2	0.23(0.26)
^{171}Yb	1/2	1.5 (1.5)
^{173}Yb	5/2	0.31(0.35)
^{175}Lu	7/2	0.24(0.26)
^{177}Hf	7/2	0.31(0.26)
^{179}Hf	9/2	0.25(0.20)
^{181}Ta	7/2	0.29(0.26)

Although the agreement between the two independent determinations of Q_0 is not perfect, it is rather satisfactory if one considers that one is determined from nuclear reactions and the other from atomic spectra. The extraction of absolute quantities from such measurements requires a good theoretical analysis of the experiments; whereas several uncertainties may cancel in the comparison of two similar measurements, like the comparison in Table 6.1. They will generally not drop out from comparisons of the type shown in Table 6.2.

Numerous additional tests have been suggested for the structure of the wave functions (4.9) utilizing various electric multipole radiations in conjunction with (6.9). One of the most impressive among these concerns the *K-selection rules* that follow from (6.9). It is obvious from the 3-j symbols that appear in (6.9) that in order to have nonvanishing results for the transition amplitude $\langle IKM|T_{-\kappa'}{}^{(k)}|I'K'M'\rangle$ we must at least satisfy either

$$K + K' = \kappa' \leqslant k$$

$$\text{or} \qquad |K - K'| = \kappa' \leqslant k \tag{6.23}$$

In other words, a kth multipole radiation cannot take place between two bands with Min $(|K - K'|, K + K') > k$. This is known as the K-selection rule for multipole radiation from deformed nuclei, and is seen to be strongly

TABLE 6.2 The Intrinsic Quadrupole Moments of Heavy Odd Nuclei. The Table Gives the Values Obtained by Coulomb Excitation and by Hyperfine Structure Measurements. [taken from Kerman (59)]

Nucleus	I_0	$\lvert Q_0 \rvert$ (Coulomb excitation)	Q	Q_0 (hfs)
$^{153}_{63}$Eu	5/2	7.7	2.5	7.0
$^{155}_{64}$Gd	3/2	8.0	1.1	5.5
^{157}Gd	3/2	7.7	1.0	5.0
$^{159}_{65}$Tb	3/2	6.9		
$^{165}_{67}$Ho	7/2	7.8	≈ 2	≈ 4
$^{167}_{68}$Er	7/2		≈ 10	≈ 20
$^{169}_{69}$Tm	1/2	8.0	0	
$^{173}_{70}$Yb	5/2		3.9	11
$^{175}_{71}$Lu	7/2	8.2	5.7	12
$^{177}_{72}$Hf	7/2	7.5		
^{179}Hf	9/2	≈ 7		
$^{181}_{73}$Ta	7/2	6.8	4.3	9.2
$^{185}_{75}$Re	5/2	5.4	2.8	7.8
^{187}Re	5/2	5.0	2.6	7.3
$^{227}_{89}$Ac	3/2		-1.7	-8
$^{233}_{92}$U	5/2	14	≈ 6	≈ 17
^{235}U	7/2	9	≈ 8	≈ 17
$^{237}_{93}$Np	5/2	9		
$^{239}_{94}$Pu	1/2	8.3	0	

dependent on the validity of the approximation (4.9) for the nuclear wave function.

Several cases are known in which this K-selection rule seems to be the only way to explain an unusually large retardation in an electromagnetic radiation. The most spectacular is the 8^- state at 1.1429 MeV in ^{180}Hf that decays by emitting an electric dipole radiation to the 8^+ state at 1.0853 MeV. The measured lifetime of this 8^- level is 5.5 hours, which is about a factor 10^{13} to 10^{14} times longer than normal lifetimes of nuclear electric dipole radiations of comparable energies. The $I = 8^+$ state at 1.0853 MeV of ^{180}Hf fits nicely into the $K = 0$ band constructed on the ground $I = 0^+$ state as its lowest member. No other levels, except members of this rotational band, are known in ^{180}Hf below this energy of 1.085 MeV. On the other hand the $I = 8^-$ level, being the lowest negative parity state in ^{180}Hf, is probably the lowest member of a new band. This band is therefore characterized by $K = 8^-$. Normally we expect an electric dipole radiation between a level with $I = 8^-$ and another with $I = 8^+$. But in rotational nuclei an electric dipole radiation to a band with $K = 0$ can take place according to (6.23), only from a band with $K = 1$. The high forbideness of the $E1$ transition in ^{180}Hf is thus an indication of an extremely pure description of the levels involved in terms of the rotational-model wave function (4.9). From the overall point of view, ^{180}Hf in its $I = 8^-$ state, differs only by zero units of angular momentum from its state of motion in the $I = 8^+$ state, yet the internal structure that goes to produce these two states is so different that an $E1$ operator can hardly connect them.

There are additional tests of the separability assertion for the rotational-model wave functions. To some of them—those involving magnetic moments and magnetic dipole radiation—we shall come later. At this stage we would just like to stress that thus far we have not found it necessary to be too explicit above the internal wave functions χ_Ω. All we used was their axial symmetry and the existence of the quantum number Ω that characterizes them (in addition to other quantum numbers, if and when necessary). The validity of the approximation (4.9) tells us nothing further about χ_Ω; it is the interpretation of parameters like the moment of inertia $\mathcal{J}$, or Q_0 that would teach us more about the internal state χ_Ω. These we shall now proceed to discuss.

7. NUCLEAR DEFORMATIONS, QUADRUPOLE MOMENTS, AND MOMENTS OF INERTIA

Having seen that experimental evidence supports the assertion (4.9) for the wave functions of rotational levels in deformed nuclei, our next natural question concerns the parameters obtained from the analysis of experimental data. Do they make sense?

As a first orientation let us describe a deformed nucleus as having a uniform nucleon density within a deformed box. To make our discussion more concrete we shall consider a particular type of deformation, the quadrupole deformation, so that the distance of the surface of the box from its center, measured in the direction $(\theta\phi)$ is given by

$$R = R_0 \left[1 + \sum_{\mu} \beta_{2\mu} \, Y_{2\mu}(\theta,\phi)\right] \tag{7.1}$$

Equation 7.1 is the equation for the surface of a deformed nucleus in spherical coordinates.

If we assume that the box is axially symmetric and take the z-axis along the axis of symmetry of the box, then R cannot depend on ϕ and we are left with

$$R = R_0 \left[1 + \beta Y_{20}(\theta)\right] \tag{7.2}$$

The axes of this spheroid are given by

$$R(\theta = 0) = R_0 \left[1 + \beta \sqrt{\frac{5}{4\pi}}\right] \quad \text{(a single axis)}$$

$$\text{and} \qquad R\left(\theta = \frac{\pi}{2}\right) = R_0 \left[1 - \beta \sqrt{\frac{5}{16\pi}}\right] \quad \text{(a double axis)} \tag{7.3}$$

For $\beta > 0$ the single axis is larger than the double axis and the nucleus is a prolate spheroid, that is cigarshaped. For $\beta < 0$ the single axis is smaller than the double axis and the nucleus is an oblate spheroid; that is, diskshaped.

Another parameter often used to describe nuclear deformations is the difference in length between the single and the double axis measured in units of R_0:

$$\delta = \frac{\Delta R}{R_0} = \frac{3}{2}\beta\sqrt{\frac{5}{4\pi}} \approx 0.946\,\beta \tag{7.4}$$

As is seen δ and β are practically equal to each other.

Characteristic values of β (or δ) for deformed nuclei [obtained from (7.5)] range between about $\beta = .2$ and $\beta = .5$. All known deformed nuclei with large deformations have positive that is, cigarshaped, deformations.

The moment of inertia about the z'-axis for a uniform rigid body of mass AM with a deformation β can be easily calculated. To lowest order in β one obtains then by straightforward integration:

$$\mathcal{J}_{\text{rig}} = \frac{2}{5} AM\, R_0^2\, [1 + 0.31\beta] + 0(\beta^2) \tag{7.5}$$

The same spheroid, when uniformly charged, with a total charge Ze leads to an intrinsic quadrupole moment given by [Bohr and Mottelson (55)]

$$Q_0 = \frac{3}{\sqrt{5\pi}} Z\, R_0^2\, \beta(1 + 0.16\beta) + 0(\beta^3) \tag{7.6}$$

Thus Q_0 and $\mathcal{J}_{\text{rig}}$ are connected via the single parameter β. (It is worthwhile to note that Q_0 is related to the *deviation* from sphericity and thus vanishes when $\beta \to 0$. The rigid-body moment of inertia includes, however, also the contribution of the spherical, central part of the deformed rotator. As $\beta \to 0$, $\mathcal{J}_{\text{rig}}$ therefore approaches a finite value). From the measured $E2$ transition probability in ^{178}Hf, for instance, we obtain, using (6.20), $Q_0 \approx 8.1 \times 10^{-24}$ cm^2. This leads, using (7.6) with $R_0 = 1.2 \times A^{1/3}$ fm, to $\beta \sim 0.3$. With this value of β the moment of inertia $\mathcal{J}_{\text{rig}}$ turns out to be such that

$$\frac{\hbar^2}{2\mathcal{J}_{\text{rig}}} \approx 3 \text{ KeV} \tag{7.7}$$

The observed value for this quantity in ^{178}Hf is

$$\left(\frac{\hbar^2}{2\mathcal{J}}\right)_{\text{exp}} = 15.5 \text{ KeV} \tag{7.8}$$

Thus, although the observed moment of inertia does come out to be of the right order of magnitude, the rigid moment $\mathcal{J}_{\text{rig}}$ seems to be too large by a factor of five. Furthermore, an inspection of (7.5) shows that in the region of deformations deduced from the measured values of Q_0, that is, $0.1 < \beta < 0.4$, the dependence of $\mathcal{J}_{\mathrm{rig}}$ on β is very slight, contrary to experimental evidence, which shows a considerably higher sensitivity of $\mathcal{J}_{\mathrm{exp}}$ to β, as shown in Fig. 7.1 taken from Bohr and Mottelson (55).

Taking another extreme view we can consider the nucleus as a frictionless fluid filling a deformed vessel. When such a vessel is set in rotation its central part essentially does not rotate, and an irrotational flow sets in at the "deformed regions" as shown in Fig. 7.2. The effective radius of the rotating motion is reduced and it can be shown that the moment of inertia is given by [Gustafson (55) and Moszkowski (57)]:

$$\mathcal{J}_{\mathrm{irrot}} = \frac{9}{8\pi} AM\, R_0^2\, \beta^2 \tag{7.9}$$

The irrotational-flow moment of inertia vanishes for vanishing deformations, and as such is a priori more suited for our estimates, because as we shall see later the moment of inertia of a nucleus, composed as it is from nucleons, should vanish for the spherically symmetric case, that is, for $\beta = 0$.

Equation 7.9, despite its appropriate behavior for $\beta \to 0$, still does not reproduce the observed moments of inertia of highly deformed nuclei. In fact, if we introduce into (7.9) the deformation $\beta = 0.3$ found above for ^{178}Hf from its Q_0, we obtain

$$\frac{\hbar^2}{2\mathcal{J}_{\mathrm{irrot}}} \approx 60 \text{ KeV} \tag{7.10}$$

which is about four times larger than the observed values (7.8). This result suggests that the experimental value of the moment of inertia $\mathcal{J}$ satisfies

$$\mathcal{J}_{\mathrm{irrot}} < \mathcal{J}_{\mathrm{exp}} < \mathcal{J}_{\mathrm{rig}} \tag{7.11}$$

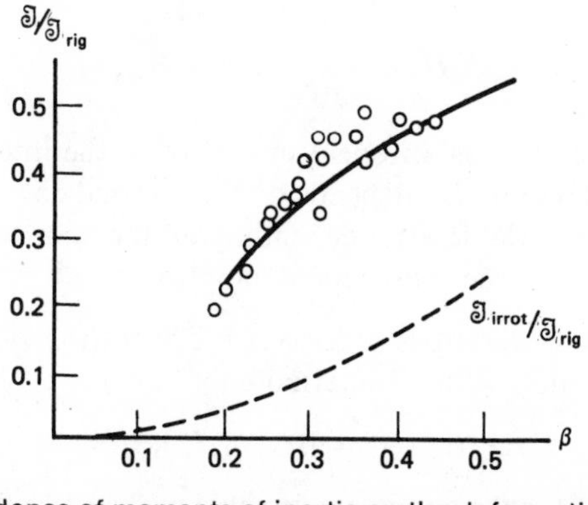

FIG. 7.1. The dependence of moments of inertia on the deformation. The experimental points lie between one-fifth and one-half of the rigid value for nuclei in the rare earth region. The moments of inertia corresponding to the irrotational flow of a liquid is given by the dotted curve. (Taken from Bohr and Mottleson (55))

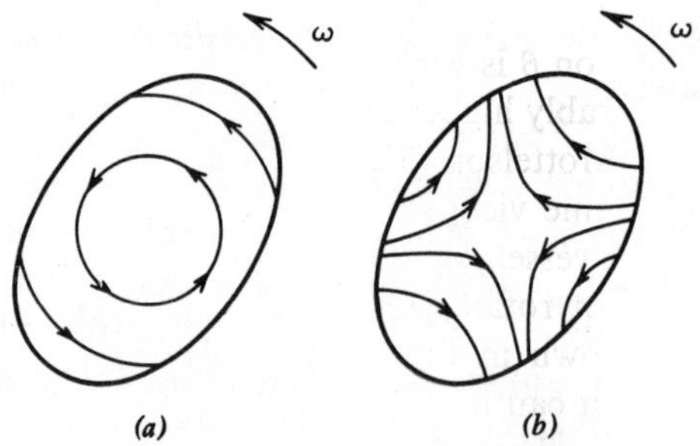

FIG. 7.2. Velocity fields for collective rotations. (a) illustrates the rotation of a rigid body, while in (b) the velocity field for the wavelike rotation of an irrotational fluid is shown. While the former motion refers to a system in which each particle is located near an equilibrium position, the latter type of rotation is characteristic of a shell structure, like the nucleus, in which the individual particles freely traverse the entire volume of the system [taken from Bohr and Mottelson (55)].

This relation is satisfied by all measured moments of inertia of deformed nuclei for which the deformation can be independently obtained from quadrupole moments. We therefore see from (7.11) that important information on the pattern of nuclear currents in rotating deformed nuclei can probably be obtained from the observed moments of inertia. The difference between $\mathcal{J}_{\text{irrot}}$ and $\mathcal{J}_{\text{rig}}$ for a given value of β is big enough, so that the exact location of $\mathcal{J}_{\text{exp}}$ in the interval (7.11) can tell us how much the actual pattern of nuclear currents resembles that of a rigid rotator or of an irrotational flow.

Let us now consider the connection between the interval wave functions χ_Ω and the moment of inertia. Suppose the nuclear Hamiltonian can be transformed into collective and internal coordinates so that to a good approximation it can be written as:

$$H \approx \frac{\hbar^2}{2\mathcal{J}(\mathbf{r}_i')} \mathbf{R}^2 + H_{\text{int}} \tag{7.12}$$

where at this stage $\mathcal{J}(\mathbf{r}_i')$ is an operator involving the internal coordinates of the various particles and H_{int} depends on the internal coordinates only. Let us introduce an angle variable $\phi(\mathbf{r}_i)$ depending on the coordinates of all the particles, and let ϕ satisfy the following assumptions:

(1) The angle ϕ describes a collective orientation of the nucleus in the sense that it commutes with all internal coordinates.

$$[\mathcal{J}(\mathbf{r}_i'),\phi] = 0 \qquad [H_{\text{int}},\phi] = 0 \tag{7.13}$$

(2) The angle ϕ is the conjugate variable to the collective angular momentum R in the sense that

$$[R,f(\phi)] = -\,i f'(\phi) \tag{7.14}$$

Strictly speaking (7.14) makes sense only for two-dimensional systems, where the angular momentum **R** has just one component.

To simulate the situation in rotating deformed nuclei, where we find empirically that the moment of inertia operator can be replaced by its expectation value, we also assume that

$$[\mathbf{R}, \mathcal{J}(\mathbf{r}_i')] = 0 \tag{7.15}$$

and

$$[H_{\text{int}}, \mathcal{J}(\mathbf{r}_i')] = 0 \tag{7.16}$$

We now observe the identity, valid for any operator A, and for any eigenstate $|0>$ of H,

$$<0|[[H,A], A]|0> = \sum_k <0|[H,A]|k><k|A|0>$$

$$- <0|A|k> <k|[H,A]|0> = 2 \sum_{E_k \neq E_0} \frac{|<0|[H,A]|k>|^2}{E_k - E_0} \tag{7.17}$$

Here $|k>$ is the complete set of eigenstates of the Hamiltonian H. The last step in (7.17) follows from the identity

$$<a|A|b> = \frac{<a|[H,A]|b>}{E_a - E_b} \quad \text{for} \quad E_a \neq E_b \tag{7.18}$$

and provided $[H,A]$ is a Hermitian operator.
Introducing

$$A = i\mathcal{J} \sin \phi \tag{7.19}$$

and using (7.13) to (7.16) we obtain

$$[H,A] = \frac{i\hbar^2}{2\mathcal{J}} [\mathbf{R}^2, \mathcal{J} \sin \phi] = \frac{1}{2} \hbar^2 (R \cos \phi + \cos \phi\, R) \tag{7.20}$$

so that $[H,A]$ is hermitian. We can now obtain further

$$[[H,A], A] = \frac{1}{2} i\hbar^2\mathcal{J} \{[R \cos \phi, \sin \phi] + [\cos \phi\, R, \sin \phi]\} = \hbar^2\mathcal{J} \cos^2\phi \tag{7.21}$$

Introducing (7.20) into the right-hand side of (7.17) and (7.21) into its left-hand side we obtain finally:

$$<0|\mathcal{J} \cos^2 \phi|0> = \frac{\hbar^2}{2} \sum_{E_k \neq E_0} \frac{|<0|(\cos \phi)\, R + R \cos \phi|k>|^2}{(E_k - E_0)} \tag{7.22}$$

Equation 7.22 should be interpreted as follows: if we assume that our Hamiltonian H has a rotational structure, and if we know the spectrum of excited energies E_k, then guessing the angle variable ϕ, we can use (7.22) to calculate the expectation value of $\mathcal{J}(\mathbf{r}_i')$ in any given state $|0>$. It is obvious that as such, (7.22) is not very useful. For one thing, if the spectrum E_k is known we can extract $<0|\mathcal{J}|0>$ from it directly by trying to fit it into rotational bands with energies $E(K) + [\hbar^2 I(I+1)]/2<0|\mathcal{J}|0>]$. However, a somewhat modified use of (7.22) can cast it into a very useful form, as we shall now see.

The Hamiltonian (7.12) describes a freely rotating nucleus. If it really separates into a collective rotation and an intrinsic part, then we can force the nucleus to oscillate in an external potential $V(\phi)$, instead of have it rotating freely, *without changing its internal structure.* The Hamiltonian

$$H' = \frac{\hbar^2}{2\mathcal{J}(\mathbf{r}_i')}\mathbf{R}^2 + H_{\text{int}} + V(\phi) \tag{7.23}$$

will therefore lead to the same value of the moment of inertia as the Hamiltonian (7.12). Physically it is obvious that we can measure the moment of inertia of a system either by letting it rotate as in (7.12) or oscillate in as (7.23), using (7.22) in both cases.

If we make now the potential $V(\phi)$ in (7.23) strong and deep enough, we can practically freeze the orientation of the nucleus so that its axis of symmetry lies along the direction determined by $V(\phi)$. In other words the energies corresponding to collective excitations of the nucleus— its oscillations in $V(\phi)$— will become very high. On the right hand side of (7.22) we can replace then $\cos\phi$ by its expectation value in the ground state $|0>$, and consider in the sum only *internal* excitations $E_{k'}$. We then obtain

$$\langle 0|\mathcal{J}|0\rangle = 2\hbar^2 \sum_{E_{k'}\neq E_0} \frac{|\langle 0|R|k'\rangle|^2}{E_{k'} - E_0} \tag{7.24}$$

Equation 7.24 differs from (7.22) not only in the disappearance of the variable ϕ, but also in the meaning of the states $<0|$ and $|k'>$: these are now to be taken as the states of the system when placed in a strong outside potential $V(\phi)$ that keeps the nucleus well oriented in space. In practice these states can therefore be approximated by the wave functions of particles moving in a deformed potential with a residual interaction among them. These functions then become internal states χ_0 and $\chi_{k'}$. We can use (3.15) and replace $\mathbf{R}$ by $\mathbf{J}$, thereby obtaining

$$\langle 0|\mathcal{J}|0\rangle = 2\hbar^2 \sum_{E_{k'}\neq E_0} \frac{|\langle 0|J_{x'}|k'\rangle|^2}{E_{k'} - E_0} \tag{7.25}$$

We have indicated the x'-component of $\mathbf{J}$ in (7.25) to stress that it is a rotation around the direction of $\mathbf{R}$ that is perpendicular to the nuclear axis of symmetry.

Equation 7.25 was first obtained by Inglis (54) starting from somewhat different assumptions: if we take the internal structure of the nucleus to be approximated by the motion of the nucleons in a deformed potential $U(x,y,z)$, then we may try to estimate the nucleus moment of inertia by calculating how much energy is required to set this potential rotating with a frequency ω. We envisage therefore a classical crank rotating the deformed nucleus with a frequency ω, say around the x-direction, The additional energy will then have the form $A\omega^2$, and by putting $A = \mathcal{J}/2$ we shall obtain an expression for the moment of inertia. If x', y', and z' are the coordinates of a particle relative to the body-fixed frame, and x, y, z are the laboratory coordinates, then:

$$\begin{aligned} x' &= x \\ y' &= y \cos \omega t + z \sin \omega t \\ z' &= -\, y \sin \omega t + z \cos \omega t \end{aligned} \tag{7.26}$$

Let $\psi(x,y,z,t)$ be the particle's wave function. Then by substitution of (7.26) we can express it in terms of the internal coordinates

$$\psi(x,y,z,t) = \chi(x',y',z',t)$$

The Schrödinger equation then becomes, using (7.26),

$$H\psi = i\hbar \frac{\partial \psi}{\partial t} = i\hbar \left[\frac{\partial \chi}{\partial t} + \frac{\partial \chi}{\partial x'}\frac{\partial x'}{\partial t} + \frac{\partial \chi}{\partial y'}\frac{\partial y'}{\partial t} + \frac{\partial \chi}{\partial z'}\frac{\partial z'}{\partial t}\right] = i\hbar \frac{\partial \chi}{\partial t} + \hbar\omega l_x \chi \tag{7.27}$$

H includes, of course, the rotating deformed potential $U(x,y,z,t)$ which, when transformed into the body-fixed frame of reference becomes stationary $U(x,y,z,t) = U[x'(t), y'(t), z'(t), 0]$. Thus the internal wave function χ satisfies a Schrödinger equation with a stationary potential and a Coriolis correction

$$[T + U(x',y',z') - \hbar\omega\, l_x]\,\chi = i\hbar \frac{\partial \chi}{\partial t} \tag{7.28}$$

Equation 7.28 determines the wave function $\chi(x',y',z')$ if we assume it to be stationary in the rotating frame.

$$H'\chi \equiv (H - \hbar\omega l_x)\chi = E'\chi \tag{7.29}$$

However it *does not* determine the energy of the system. To obtain the energy one has to take the expectation value of H with χ, either in the laboratory frame or in the rotating frame:

$$E = \frac{\langle \chi | H\chi \rangle}{\langle \chi | \chi \rangle} \approx \frac{\langle \chi | H'\chi \rangle}{\langle \chi | \chi \rangle} + \frac{\langle \chi | \omega\hbar l_x \chi \rangle}{\langle \chi | \chi \rangle} \tag{7.30}$$

To evaluate E assume that $\hbar\omega l_x$ is a small perturbation and let χ_n be the eigenfunctions in the *static* deformed potential with no Coriolis forces, that is, in

H' for $\omega = 0$, χ_0 being the lowest state. Then, to the first order in $\hbar\omega$ (7.29) has the solution:

$$\chi = \chi_0 + \sum_{n \neq 0} \frac{\langle n | \hbar\omega l_x | 0 \rangle}{E_n - E_0} \chi_n \tag{7.31}$$

where E_n are the eigenvalues in the static-deformed potential. Introducing (7.31) into (7.30), and noting that $H = H' + \hbar\omega l_x$, we obtain, to second order in $\hbar\omega$

$$E = E' + 2\hbar^2\omega^2 \sum_n \frac{|\langle 0 | l_x | n \rangle|^2}{E_n - E_0} \tag{7.32}$$

where E' is expectation value of H' with respect to χ (Eq. 7.30). E' as we see from (7.29) is also a function of $\hbar\omega$. Considering $-\hbar\omega l_x$ as a perturbation we notice that it contributes to the energy E' only in the second order;

$$E' = E_0 - \hbar^2\,\omega^2 \sum_n \frac{|\langle 0 | l_x | n \rangle|^2}{E_n - E_0} \tag{7.33}$$

Introducing (7.33) into (7.32) we obtain finally

$$E = E_0 + \hbar^2\,\omega^2 \sum_n \frac{|\langle 0 | l_x | n \rangle|^2}{E_n - E_0} \tag{7.34}$$

If we identify the last term in (7.34) with $(1/2)\,\mathcal{J}\omega^2$, we obtain immediately Inglis' formula (7.25) for the moment of inertia $\mathcal{J}$.

The use of (7.25) becomes rather simple if we take for the intrinsic wave function $|0\rangle$ and $|k\rangle$ Slater determinants in a deformed harmonic oscillator potential and ignore the residual interaction among the nucleons. Employing the potential energy

$$U(x,y,z) = \frac{1}{2} M\,[\omega_x^2\,(x^2 + y^2) + \omega_z^2\,z^2] \tag{7.35}$$

we can characterize the single-particle states by the number of oscillator quanta along the x-, y-, and z-directions: n_x, n_y, and n_z.* The energies of the various levels are then given by:

$$E(n_x, n_y, n_z) = \hbar\omega_x\,(n_x + n_y + 1) + \hbar\omega_z \left(n_z + \frac{1}{2}\right) \tag{7.36}$$

The matrix elements of the operator l_x between two states with a different value of n_x, the number of oscillator quanta along the x-direction, vanishes.

*The parameters ω_x and ω_z are as yet undetermined. We shall fix this ratio later by using the variational principle for the energy.

However, matrix elements in which n_y and n_z change are finite. These break up into two types:

(*i*) Matrix elements between states within the same major shell, that is such that $n_y + n_z$ is not changed. For instance

$$\langle n_y - 1, n_z + 1 | l_x | n_y n_z \rangle = \frac{i}{2} \sqrt{n_y(n_z + 1)} \frac{\omega_x + \omega_z}{\sqrt{\omega_x \omega_z}} \tag{7.37}$$

(ii) Matrix elements between states in different major shells. Matrix elements of l_x of this latter type can connect only states two shells apart, in order for the parity to remain the same; these matrix elements should also vanish for the spherical case $\omega_x = \omega_z$, since then the states *within* each major shell build up states of well defined $\mathbf{l}^2$. For example, we have

$$\langle n_y + 1, n_z + 1 | l_x | n_y n_z \rangle = \frac{i}{2} \sqrt{(n_y + 1)(n_z + 1)} \frac{\omega_z - \omega_x}{\sqrt{\omega_x \omega_z}} \tag{7.38}$$

Using (7.37) and (7.38), and noting that $\mathbf{l} = \sum_i \mathbf{l}_i$ is a one-body operator, we obtain for (7.25):

$$\langle 0 | \mathcal{J} | 0 \rangle = 2\hbar^2 \sum_{\substack{\text{all occupied } n \\ \text{all } n'}} \frac{|\langle n_x n_y n_z | l_x | n_x' n_y' n_z' \rangle|^2}{E_{n'} - E_{n0}}$$

$$= 2\hbar^2 \sum_{\text{occ } n} \Bigg[\frac{|\langle n_x n_y n_z | l_x | n_x n_y + 1 \; n_z - 1 \rangle|^2}{\hbar\omega_x - \hbar\omega_z}$$

$$+ \frac{|\langle n_x n_y n_z | l_x | n_x n_y - 1 \; n_z + 1 \rangle|^2}{\hbar\omega_x - \hbar\omega_z} + \frac{|\langle n_x n_y n_z | l_x | n_x n_y + 1 \; n_z + 1 \rangle|^2}{\hbar\omega_x + \hbar\omega_z}$$

$$+ \frac{|\langle n_x n_y n_z | l_x | n_x n_y - 1 \; n_z - 1 \rangle|^2}{-\hbar\omega_x - \hbar\omega_z} \Bigg]$$

$$= \frac{\hbar}{2\omega_x \omega_z} \left[\frac{(\omega_z - \omega_x)^2}{\omega_z + \omega_x} \sum_{\text{occ}} (n_y + n_z + 1) - \frac{(\omega_z + \omega_x)^2}{\omega_z - \omega_x} \sum_{\text{occ}} (n_z - n_y) \right] \tag{7.39}$$

It is interesting to compare (7.39) with the expression for the moment of inertia of a rigid body. (In this case we have

$$\mathcal{J}_{\text{rig}} = M \sum_{\text{occ}} \langle n_x n_y n_z | y_i^2 + z_i^2 | n_x n_y n_z \rangle$$

$$= \frac{\hbar}{2\omega_x \omega_z} \left[(\omega_z + \omega_x) \sum_{\text{occ}} (n_y + n_z + 1) - (\omega_z - \omega_x) \sum_{\text{occ}} (n_z - n_y) \right] \tag{7.40}$$

where we have used the virial theorem

$$\frac{1}{2} M \omega_z^2 \langle z_i^2 \rangle = \frac{1}{2} \hbar\omega_z \left(n_z^{(i)} + \frac{1}{2}\right) \tag{7.41}$$

Although (7.39) and (7.40) look different, it turns out that for the *stable* deformations they are practically equal.

To show this we should determine the stable configuration; in other words, for a given occupation of states we ask for the values of ω_x and ω_z that will minimize the energy. The total internal energy in our case of particles moving independently in a deformed oscillator is given by:

$$E = \sum_{\text{occ}} \left[\hbar\omega_x (n_x + n_y + 1) + \hbar\omega_z \left(n_z + \frac{1}{2}\right)\right] \tag{7.42}$$

To obtain the most stable deformation we shall minimize E under the condition that the volume of the nucleus does not change*, that is, we allow *shape* variations without a change in density; we must therefore require that the variation of ω_x and ω_z will be constrained by:

$$\omega_x\omega_y\omega_z = \omega_x^2\omega_z = \omega_0^3 \tag{7.43}$$

where ω_0 is a constant. Thus we obtain from (7.42)

$$E(\omega_x) = \hbar\omega_x \left[\sum_{\text{occ}} (n_x + n_y + 1) + \frac{\omega_0^3}{\omega_x^3} \sum_{\text{occ}} \left(n_z + \frac{1}{2}\right)\right] \tag{7.44}$$

The equilibrium deformation $\bar{\omega}_x$ is obtained by setting $\partial E/\partial\omega_x = 0$ and hence satisfies

$$\bar{\omega}_x \sum_{\text{occ}} (n_x + n_y + 1) = \bar{\omega}_z \sum_{\text{occ}} (2n_z + 1) \tag{7.45}$$

or

$$[\bar{\omega}_z - \bar{\omega}_x]\left[\sum_{\text{occ}} (n_x + n_y + 1) + \sum_{\text{occ}} (2n_z + 1)\right]$$
$$= [\bar{\omega}_z + \bar{\omega}_x]\left[\sum_{\text{occ}} (n_x + n_y + 1) - \sum_{\text{occ}} (2n_z + 1)\right] \tag{7.46}$$

For an axially symmetric potential we expect

$$\sum_{\text{occ}} n_x = \sum_{\text{occ}} n_y \tag{7.47}$$

*The condition of constant volume has to be imposed to assure a given density, since the single-particle potential cannot lead to an equilibrium density all by itself.

We obtain then for the equilibrium deformation

$$\frac{\bar{\omega}_z - \bar{\omega}_x}{\bar{\omega}_z + \bar{\omega}_x} \sum_{\text{occ}} (n_y + n_z + 1) = - \sum_{\text{occ}} (n_z - n_y)$$

or

$$\frac{(\bar{\omega}_z - \bar{\omega}_x)^2}{(\bar{\omega}_z + \bar{\omega}_z)} \sum_{\text{occ}} (n_y + n_z + 1) = - (\bar{\omega}_z - \bar{\omega}_x) \sum_{\text{occ}} (n_z - n_y) \tag{7.48}$$

and

$$- \frac{(\bar{\omega}_z + \bar{\omega}_x)^2}{(\bar{\omega}_z - \bar{\omega}_x)} \sum_{\text{occ}} (n_z - n_y) = (\bar{\omega}_z + \bar{\omega}_x) \sum (n_y + n_z + 1)$$

Comparing (7.48) with (7.39) and (7.40), we obtain the general result that at equilibrium deformation

$$\langle 0 | \mathcal{J} | 0 \rangle = \mathcal{J}_{\text{rig}} \qquad \text{at} \qquad \omega_x = \bar{\omega}_x \tag{7.49}$$

The result (7.49) is very surprising at first glance. Intuitively one may expect that an independent-particle motion inside a deformed box would lead to an irrotational moment of inertia that, as we have seen, is considerably smaller than $\mathcal{J}_{\text{rig}}$. Instead we find that the moment of inertia comes out to be exactly equal to that of a rigid rotator with the same deformation.

We can obtain a better insight into this result if we consider the velocity distribution of the particles in the deformed potential in the body-fixed frame of reference. From the virial theorem we have

$$\frac{1}{2} M \langle v_{ix}^2 \rangle = \frac{1}{2} \hbar \, \omega_x \left[n_x^{(i)} + \frac{1}{2} \right]$$

$$\frac{1}{2} M \langle v_{iy}^2 \rangle = \frac{1}{2} \hbar \, \omega_y \left[n_y^{(i)} + \frac{1}{2} \right]$$

$$\frac{1}{2} M \langle v_{iz}^2 \rangle = \frac{1}{2} \hbar \, \omega_z \left[n_z^{(i)} + \frac{1}{2} \right]$$

For the equilibrium deformation it follows from (7.45) and (7.47) that

$$\bar{\omega}_x \sum_{\text{occ}, i} \left(n_x^{(i)} + \frac{1}{2} \right) = \bar{\omega}_y \sum_{\text{occ}, i} \left(n_y^{(i)} + \frac{1}{2} \right) = \bar{\omega}_z \sum_{\text{occ}, i} \left(n_z^{(i)} + \frac{1}{2} \right) \tag{7.50}$$

a result that can be simple interpreted: at equilibrium, energy is equally distributed in every direction. We thus conclude that at the equilibrium

$$\langle \overline{v_{ix}^2} \rangle = \langle \overline{v_{iy}^2} \rangle = \langle \overline{v_{iz}^2} \rangle \tag{7.51}$$

where the bar indicates the average over all the nucleons. Thus the velocity field of the nucleons in the body-fixed frame of reference is isotropic. At a given point there is no net current in the body-fixed frame of reference. In the *laboratory* frame of reference the velocity field will therefore be that of a rigid rotator, leading, thereby, to the rigid moment of inertia.

The role played by the equilibrium deformation in deriving this result should be noticed. Without the condition (7.50) there is no reason for the velocity field to be isotropic in the body-fixed frame, and with preferred velocities in this frame, the velocity field in the laboratory frame will not be that of a rigid rotator. Thus we expect this result to hold not only for deformed harmonic oscillator potential, but for any deformed potential that for the equilibrium deformation leads to an isotropic velocity field in the body-fixed frame of reference.

It is worthwhile to note the values taken on by the moment of inertia $\mathcal{J}$, given by (7.39), in some special cases for which the values of ω_x and ω_z do not necessarily satisfy conditions (7.48). If the system consists of just one nucleon in its ground state, then the only occupied state has $n_x = n_y = n_z = 0$ and we obtain

$$\mathcal{J} = \frac{\hbar}{2\omega_x\omega_z}\frac{(\omega_z - \omega_x)^2}{\omega_z + \omega_x} \tag{7.52}$$

From (7.41) we derive, for our particular case,

$$\omega_z = \frac{\hbar}{2M\,\overline{z^2}} \qquad \text{and similarly} \qquad \omega_x = \frac{\hbar}{2M\,\overline{x^2}} = \frac{\hbar}{2M\,\overline{y^2}} \tag{7.53}$$

To be able to compare $\mathcal{J}$ in (7.53) with previous expressions for the moment of inertia, let us introduce, in line with (7.3) two constants R_0 and β such that

$$\sqrt{\overline{z^2}} = R_0\left(1 + \beta\sqrt{\frac{5}{4\pi}}\right) \qquad \sqrt{\overline{x^2}} = R_0\left(1 - \frac{1}{2}\beta\sqrt{\frac{5}{4\pi}}\right) \tag{7.54}$$

We then obtain from (7.52) and (7.53), to lowest order in β,

$$\mathcal{J} = \frac{9}{8\pi}(5MR_0)^2\,\beta^2 \tag{7.55}$$

Equation 7.55 is the moment of inertia characteristic of an irrotational flow (see Eq. 7.9). The occurrence of $M' = 5M$ in (7.55) is understood if we recall that (7.9) stands for the moment of inertia of a uniform mass distribution in a box of average radius R_0. To obtain the equivalence between the two masses we should have

$$M\langle r^2\rangle = M'\frac{\int_0^{R_0} r^2\, r^2\, dr}{\int_0^{R_0} r^2\, dr} = \frac{3}{5}M'R_0^2$$

But from (7.54)

$$M\langle r^2\rangle = M(\overline{x^2} + \overline{y^2} + \overline{z^2}) = 3MR_0^2$$

Hence $M' = 5M$.

It is quite natural to expect the irrotational moment of inertia for a single particle, since its mass-current density is given by

$$\mathbf{j} = \frac{\hbar}{2im}[\psi^* \nabla \psi - (\nabla\psi^*)\psi] \tag{7.56}$$

A straightforward calculation gives us then:

$$\text{curl } \mathbf{j} = \frac{\hbar}{2im}\nabla \times [\psi^*\nabla\psi - (\nabla\psi^*)\psi] = 0 \tag{7.57}$$

The flow pattern of a single particle corresponds, therefore, to an irrotational flow.

Equation 7.39 leads to the irrotational moment of inertia also for closed-shell configurations. In this case, in fact, we have $\sum_{\text{occ}} n_x = \sum_{\text{occ}} n_y = \sum_{\text{occ}} n_z$, so that in (7.39) we are left again with the term proportional to $(\omega_z - \omega_x)^2$ and therefore to β^2. However, as can be seen from (7.46), for closed-shell configuration $\bar{\omega}_z - \bar{\omega}_x = 0$ so that there is no finite equilibrium deformation, and consequently

$$\mathcal{J} = 0 \qquad \text{for closed shells} \tag{7.58}$$

In between closed shells the Pauli principle forces the nucleons to occupy successively higher levels, and the second term in (7.39), with its small energy denominator $(\omega_z - \omega_x)$, makes important contributions to $\mathcal{J}$ (see 7.48) leading to the large value of $\mathcal{J}_{\text{rig}}$.

It was shown by Niels Bohr in 1911 that no diamagnetic effects are present in a classical electron gas confined to a given volume. Although a magnetic field does induce electron currents, the reflection of these currents from the walls of the container exactly cancel the induced current so that the diamagnetic effect vanishes. It is well known that the effects of a homogeneous magnetic field on a system of charged particles can be simulated by rotating the system with the appropriate Larmor frequency around the direction of the magnetic field. The absence of diamagnetic effects in the electron gas is then another way of saying that the moment of inertia of the system has not changed by the rotation. In both cases the crucial point is that in the body-fixed frame no additional currents are induced by the rotation of the system.

The observed moments of inertia are different from $\mathcal{J}_{\text{rig}}$ and, in fact, we have always,

$$\mathcal{J}_{\text{exp}} < \mathcal{J}_{\text{rig}}$$

We therefore suspect, because of the foregoing considerations, that an independent-particle approximation is too crude for the estimate of moments of inertia and that the residual interaction among the nucleons must be taken into account as well. Qualitatively we can see from (7.24) that taking this interaction into account will change $\mathcal{J}$ in the right direction: we mentioned before (see Chapter V) that a pair of interacting nucleons makes best use of the attractive interaction if their total angular momentum vanishes. Since pairs coupled to zero total angular momentum do not contribute to (7.24), most of the contributions to the sum in (7.24) will come from "broken pairs," that is, pairs of particles that in the ground state $|0\rangle$ are coupled to $J = 0$ and in the excited state $|k'\rangle$ are coupled to $J = 1$. For such pairs the zeroth-order energy difference $E_{k'} - E_0$ is an *underestimate* of the actual energy difference, because in addition to changing the levels of the particles we have to overcome a part of their strong mutual interaction. The introduction of a more realistic energy denominator in (7.24) will therefore *reduce* the value of $\langle 0|\mathcal{J}|0\rangle$ in comparison with the independent-particle value [Bohr and Mottelson(55)].

Furthermore, we can expect this effect of the introduction of the residual interaction to be more important for even-even nuclei than for odd-even nuclei. In odd-A nuclei there is at least one nucleon that is not paired off in the ground state, and *its* excitation energies may therefore be closer to the zeroth-order excitations $E_{k'} - E_0$. We therefore expect, as a rule, that moments of inertia of odd-A nuclei will be larger than those of neighboring even-even nuclei. This is indeed true, as was mentioned above (Section VI.5), thus supporting our conjecture that the difference between the observed moments of inertia and the independent-particle estimate can be traced to the residual interaction.

The importance of considering the residual interactions and coupling term in the Hamiltonian (2.14) can also be inferred from the correction term $BI^2(I + 1)^2$ to the energies in the rotational bands, as discussed in (5.1). From the physical viewpoint we expect that a nonrigid structure with a positive Q will become more elongated as it spins faster because of the centrifugal forces. (We remember that the axis of rotation is perpendicular to the symmetry axis). This leads to an increase in the moment of inertia with increasing angular momentum I. The fact that the coefficient B in (5.1) is found always to be negative supports this general expectation. It also means that within the framework of our model of the rotating nucleus, where the deformation is taken to be constant, we have to consider mixing of different internal structures in the levels of a given rotational band. By allowing the extent of this mixing of different internal wavefunctions to depend on I, we can obtain an I-dependent variation of the moment of inertia, but we shall not go into these questions here.

8. MAGNETIC DIPOLE MOMENTS AND CURRENTS IN DEFORMED NUCLEI

The discussions in the previous section have indicated the importance of a better knowledge of the velocity distributions in deformed nuclei for the understanding of their moments of inertia. We also indicated why the residual interaction among the nucleons will have to be taken into account, at least approximately, if we want to understand the values of the observed moments of inertia.

A certain velocity pattern for the mass distribution in the nucleus implies also a certain electric current distribution. Strictly speaking, the presence of charge-exchange interactions among the nucleons makes the connection between mass currents and electric currents rather complicated. However, our experience with nondeformed nuclei indicates that such exchange effects, if detectable at all, make a rather small contribution when we consider magnetic moments and magnetic dipole transition probabilities. Formally this conclusion reflects itself through the fact that the quantum mechanical operators representing these quantities are, to a good approximation, single-particle operators

$$\hat{\boldsymbol{\mu}} = \sum \hat{\boldsymbol{\mu}}_i \tag{8.1}$$

A contribution to $\hat{\boldsymbol{\mu}}$ from the exchange of charge between protons and neutrons will have to depend on the coordinates of both particles, and will thus have the form of a two-particle operator:

$$\hat{\boldsymbol{\mu}}_{\text{exch}} = \frac{1}{2} \sum_{i \neq j} \hat{\boldsymbol{\mu}}_{ij} \tag{8.2}$$

If we maintain the approximation of neglecting $\hat{\boldsymbol{\mu}}_{\text{exch}}$ also for deformed nuclei, then we expect that a particular *mass*-current pattern that gives rise to the observed moment of inertia should also produce corresponding observed effects in magnetic properties through the associated *charge* currents. Before embarking on the study of this question, however, we have to ascertain that an operator of the type given in (8.1) actually describes the data consistently. We shall again follow the technique, developed in Section VI.6 for the study of the quadrupole operator, and employed previously to test the consistency of the shell-model description of nondeformed nuclei.

To the extent that we can talk about internal motion as separated from the collective rotation we can always write a single-particle magnetic moment operator as

$$\begin{aligned} \hat{\boldsymbol{\mu}} &= g_R \mathbf{R} + g_\Omega \mathbf{J} \\ &= g_R \mathbf{I} + (g_\Omega - g_R)\, \mathbf{J} \end{aligned} \tag{8.3}$$

Here g_R and g_Ω are the g-factors, in nuclear magnetons, for the collective rotation and internal motions, respectively; both may, and generally will, depend on the detailed features of the internal structure and the current distributions that it implies. The description of $\hat{\mathbf{u}}$ in terms of (8.3) is valid under slightly more general conditions than (8.1), since it can still accommodate exchange-current contributions to the internal g-factor g_Ω.

The structure (8.3) of the magnetic dipole operator can be tested by comparing magnetic dipole moments and magnetic dipole transition probabilities involving different states of the same rotational bands. If (8.3) is verified, the resulting empirical values of g_R and g_Ω then have to be explained in terms of a specific internal structure of the deformed nucleus.

To evaluate matrix elements of (8.3) we express the components of $\mathbf{J}$ in the laboratory frame of reference through their components in the fixed frame of reference. We note also the occurrence in (8.3) of the operator $g_R I_\kappa$ that cannot be described in terms of internal variables alone. No corresponding term appears in (6.9) since there we assumed that $T_\kappa^{(k)}$ $(\mathbf{r}_i)$, when referred to the body-fixed frame, involved only coordinates relative to that frame. Using the wave functions (4.9) and (6.6), we obtain for $\hat{\mathbf{u}}$ just as in (6.9),

$$\langle IKM|\hat{\mu}_\kappa|I'K'M'\rangle = \langle IKM|g_R I_\kappa|I'K'M'\rangle$$

$$+ (-1)^{M-K}\sqrt{(2I+1)(2I'+1)}\begin{pmatrix} I & 1 & I' \\ -M & \kappa & M' \end{pmatrix}$$

$$\times (g_\Omega - g_R)\sum_{\kappa'}\left[\begin{pmatrix} I & 1 & I' \\ -K & \kappa' & K' \end{pmatrix}\langle K|J_{\kappa'}|K'\rangle\right.$$

$$\left. + (-1)^{I'}\begin{pmatrix} I & 1 & I' \\ -K & \kappa' & -K' \end{pmatrix}\langle K|J_{\kappa'}|-K'\rangle\right] \tag{8.4}$$

The first term in (8.4) can be easily evaluated with the aid of the Wigner-Eckart theorem, and considering the fact that the wave functions we are working with are assumed to be eigenfunctions of I^2 and $I_0 = I_z$. We obtain (see Appendix A, A.2.44)

$$\langle IKM|g_R I_\kappa|I'K'M'\rangle$$

$$= \delta(I,I')\,\delta(K,K')\,g_R(-1)^{I-M}\begin{pmatrix} I & 1 & I \\ -M & \kappa & M' \end{pmatrix}\sqrt{I(I+1)(2I+1)} \tag{8.5}$$

where used has been made of the relation

$$\langle I||\mathbf{I}||I\rangle = \sqrt{I(I+1)(2I+1)}$$

Thus this term contributes only to the static moments of the various levels of a rotational band (and also to transistion moments when band mixing occurs; see below).

$M1$ transition probabilities in purely deformed nuclei are completely determined by the second term in (8.4). The transition probability per unit time is given by (see (6.17))

$$T(M1, IK \rightarrow I'K') = \frac{16\pi}{9}\frac{k^3}{\hbar} B(M1, IK \rightarrow I'K') \tag{8.6}$$

where the tensor operator giving rise to magnetic dipole radiation is (Chapter VIII)

$$T_\kappa^{(1)} = \sqrt{\frac{3}{4\pi}}\frac{e\hbar}{2Mc}(g_R - g_\Omega)\, J_\kappa \tag{8.7}$$

Using the wave function given by (4.9) we obtain therefore:

$$B(MI, IK \rightarrow I'K') = \frac{3}{4\pi}\left(\frac{e\hbar}{2Mc}\right)^2 \frac{(g_R - g_\Omega)^2}{2I + 1} \sum_{MM'\kappa} |\langle IKM|J_\kappa|I'K'M'\rangle|^2$$

$$= \frac{3}{4\pi}\left(\frac{e\hbar}{2Mc}\right)^2 (g_R - g_\Omega)^2 (2I' + 1)\left\{\sum_{\kappa'} \begin{pmatrix} I & 1 & I' \\ -K & \kappa' & K' \end{pmatrix} \langle K|J_{\kappa'}|K'\rangle\right.$$

$$\left. + (-1)^{I'}\begin{pmatrix} I & 1 & I' \\ -K & \kappa' & -K' \end{pmatrix} \langle K|J_{\kappa'}|-K'\rangle\right\}^2 \tag{8.8}$$

Equation 8.8 simplifies further if we deal with transitions within the same band $K = K' \neq 1/2$. In this case the only internal matrix element required is

$$\langle K|J_0|K\rangle = K$$

and the equation reduces to:

$$B(M1, IK \rightarrow I'K) = \frac{3}{4\pi}\left(\frac{e\hbar}{2Mc}\right)^2 (g_R - g_\Omega)^2 K^2 (2I' + 1) \begin{pmatrix} I & 1 & I' \\ -K & 0 & K \end{pmatrix}^2$$

$$K \neq \frac{1}{2} \tag{8.9}$$

For the same band, (8.4) gives for the static magnetic moment

$$\mu = \langle IKM = I|\mu_0|IKM = I\rangle = g_R I + (g_\Omega - g_R)\frac{K^2}{I + 1}$$

$$= \left[g_R + (g_\Omega - g_R)\frac{K^2}{I(I + 1)}\right] I \qquad K \neq \frac{1}{2} \tag{8.10}$$

For $K = 1/2$, the second term in (8.4), that proportional to $\langle K|J_{\kappa'}|-K\rangle$, gives an extra contribution, similar to the decoupling term in the $K = 1/2$ rotational band. We shall not discuss its derivation here but refer the reader to Davidson (65). The final result for this case is given in (8.11) below.

Equation 8.9 and 8.10 can serve to check the consistency of our assumption about the structure of the magnetic moment operator, (8.3), and the description of the rotational wave functions in terms of (4.9). In a typical rotational band in an odd-A nucleus there may be five or six levels known; each one can decay by an $M1$ radiation to the level below it, and each has a static magnetic moment. We can have, then, 10 pieces of data, all of which should be determined in terms of the two parameters g_R and g_Ω in (8.9) and (8.10). The internal consistency of this whole approach can therefore be tested rather thoroughly.

Figure 8.1 shows the ratios of $(g_\Omega - g_R)_1^2/(g_\Omega - g_R)_2^2$ derived from $M1$ rates between two different pairs of levels within the same rotational band ($K \neq 1/2$). According to (8.9) this ratio should be unity. The deviations from unity are indeed found to be rather small, and nearly always within the experimental uncertainty of the measurement.

In one case, that of ^{169}Tm, there are three static moments and three transition moments known within the same $K = 1/2$ band. Enough data is available in this case to derive g_R and g_Ω to check the consistency of the model. Actually three parameters have to be determined in this case, the third one, b_0, being similar in its nature to the decoupling parameter a in (4.24). One finds, for $K = 1/2$ [see Davidson (65), p. 139)]

$$\mu = g_R I + (g_\Omega - g_R)\frac{K^2}{I+1}[1 + (2I+1)(-1)^{I+1/2} b_0] \qquad \text{for } K = \frac{1}{2}$$

$$B(M1, I+1, I) = \frac{3}{64\pi}\left(\frac{e\hbar}{2Mc}\right)^2 \frac{2I+1}{I+1}(g_\Omega - g_R)^2 [1 + (-1)^{I-1/2} b_0]^2 \tag{8.11}$$

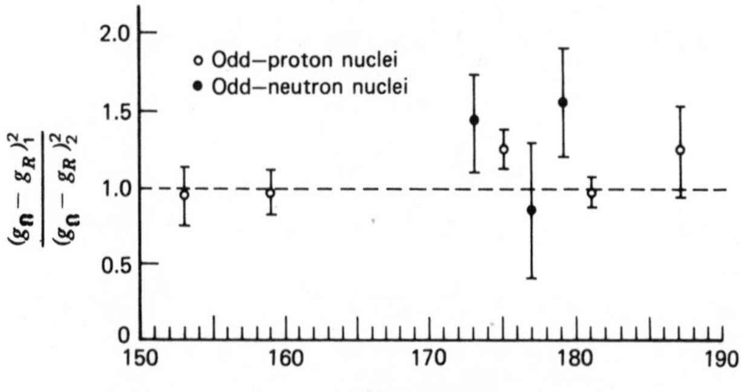

FIG. 8.1. Ratio of the parameters $(g_\Omega - g_R)_1^2$ and $(g_\Omega - g_R)_2^2$ derived, respectively, from the $M1$ transition probabilities between first excited and ground states, and between second and first excited states (taken from Bodenstadt and Rogers (64)].

Using these equations the values of g_R, g_Ω, and b_0 were determined for the ground-state rotational band of ^{169}Tm, from the measured moments of the ground state and the first excited level $I = 3/2$ at 8 KeV, and from the $M1$ transition from the $I = 5/2$ to the $I = 3/2$ levels. The values obtained this way for the three parameters are:

$$g_\Omega = -1.57 \pm 0.11 \qquad g_R = 0.406 \pm 0.019 \qquad b_0 = -0.16 \pm 0.03$$

These parameters were then used to derive theoretical values for $B(M1, I = 3/2 \rightarrow I = 1/2)$, for $B(M1, I = 7/2 \rightarrow I = 5/2)$, and for $\mu(I = 5/2)$. These theoretical values agree rather well with the experimental values as shown in Fig. 8.2.

In many nuclei there is not sufficient data to check the consistency of (8.9) and (8.10), but still enough data to derive empirical values for g_R and g_Ω. The values derived this way from g_R in even-even nuclei are shown in Fig. 8.3, while those derived from odd-A nuclei are shown in Fig. 8.4 separately for odd-Z and odd-N nuclei.

The most naive estimate of g_R can be made as follows: we recall that a g-factor measures the ratio between the magnetization produced by a current and the angular momentum carried by this same current. It is therefore proportional to the charge and inversely proportional to the mass, of the entity whose motion gives rise to the magnetic moment. In the collective rotation **R** it is the whole mass AM that contributes to the material current, but it carries only a charge of Ze units. Expressed in terms of nuclear magnetons we therefore expect to have

$$g_R = \frac{Z}{A} \tag{8.12}$$

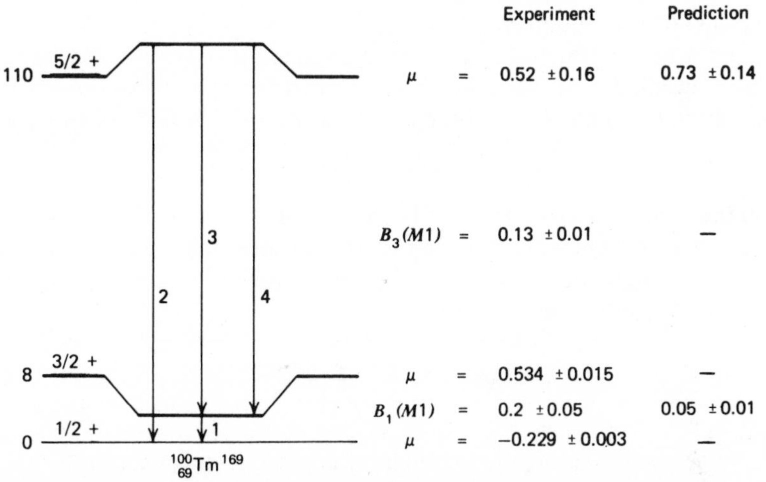

FIG. 8.2. The ground state rotational band of ^{169}Tm showing predicted and experimental values of the magnetic parameters [taken from Bodenstadt and Rogers (64)].

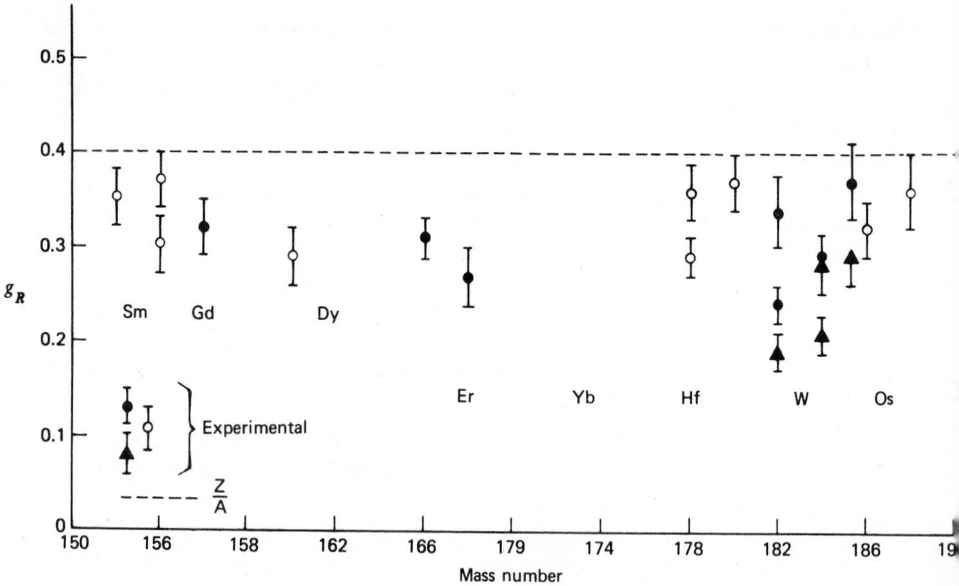

FIG. 8.3. The collective g_R values for even–even deformed nuclei in the rare earth region. With exception of Hf, W, and Os averaged values are shown [taken from Bodenstadt and Rogers (64)].

It is remarkable that the observed values of g_R all lie in a rather close neighborhood of Z/A. They do show systematic deviations from Z/A, and we shall see later how these can be understood; but the order of magnitude of g_R, including its sign, does testify in favor of the collective interpretation of the angular momentum **R**.

A slightly better estimate of g_R can be obtained. Consider a state $|0\rangle$ in a deformed, even-even nucleus, which belongs to the $K = 0$ band ($|0\rangle$ is not necessarily the ground state.) It then follows from (8.10) that for this state

$$\mu(I) = g_R I \tag{8.13}$$

On the other hand we can put $\hat{\boldsymbol{\mu}} = \sum \hat{\boldsymbol{\mu}}_i$ and try to compute its expectation value using a detailed wave function of the nucleus. Such an internal wave-function, correct to first order in the Coriolis potential $-\hbar\omega l_x$, is given by (see the discussion following 7.28):

$$\chi = \chi_0 + \hbar\omega \sum_{k\neq 0} \frac{\langle k|J_x|0\rangle}{E_k - E_0} \chi_k \tag{8.14}$$

where we replaced l_x by J_x for the internal angular momentum.

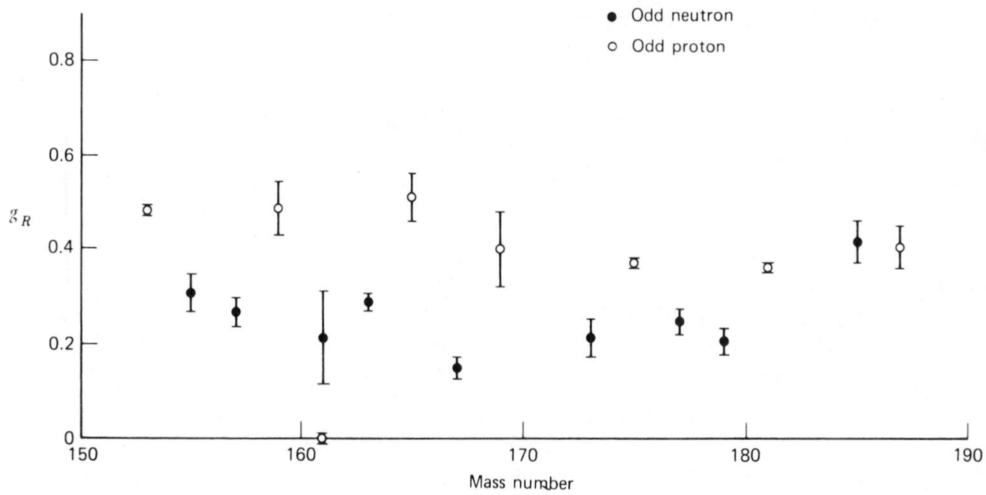

FIG. 8.4. The collective g_R values for odd-A deformed nuclei [taken from Bodenstadt and Rogers (64)].

The collective rotation **R** is in the x-direction and, hence, the magnetic moment that goes with it has a nonvanishing component only in this direction. Since by hypothesis it is determined by the internal coordinates only, we have

$$g_R I = \langle \Psi(I) | \hat{\mu}_x | \Psi(I) \rangle = \langle \chi(I) | \hat{\mu}_x | \chi(I) \rangle \tag{8.15}$$

Introducing now $\chi(I)$ from (8.14) into (8.15), and recalling that in our approximation $\omega = \hbar I/\mathcal{J}$, we can write (8.15), to the lowest order, in the form

$$g_R I = \frac{\hbar^2 I}{\mathcal{J}} \sum_{k \neq 0} \frac{\langle 0 | J_x | k \rangle \langle k | \hat{\mu}_x | 0 \rangle}{E_k - E_0} + \text{c.c.} \tag{8.16}$$

We shall see later that in $\hat{\mu}_x = \sum \hat{\mu}_{ix}$ the intrinsic spins of the particles largely cancel each other's contribution, so that to some approximation we can put

$$\hat{\mu}_x \approx \sum g_{li}\, l_{xi} \tag{8.17}$$

with $g_l = 1$ for protons and $g_l = 0$ for neutrons. The states k in (8.16) involve, therefore, only proton excitations. We can define a moment of inertia $\mathcal{J}_p$ for the "proton fluid" in the rotating nucleus which, in analogy with (7.24), will be given by

$$\mathcal{J}_p = 2\hbar^2 \sum_{k \neq 0} \frac{|\langle 0 | J_{xp} | k \rangle|^2}{E_k - E_0} \tag{8.18}$$

where $\mathbf{J}_p$ is the proton orbital angular momentum operator. From (8.16), (8.17), and (8.18) we now obtain

$$g_R = \frac{\mathcal{J}_p}{\mathcal{J}} = \frac{\mathcal{J}_p}{\mathcal{J}_p + \mathcal{J}_n} \tag{8.19}$$

where we have introduced the "neutron fluid" moment of inertia $\mathcal{J}_n$ through $\mathcal{J} = \mathcal{J}_p + \mathcal{J}_n$, which follows immediately from (7.24).

It is obvious from (8.19) that if the protons and the neutrons have the same spatial distribution in the nucleus, then $g_R = Z/(Z + N) = Z/A$, which was our first guess, (8.12). It can also be shown that an independent-particle model in a deformed potential, if used for the computation of $\mathcal{J}_p$ and $\mathcal{J}_n$, $\mathcal{J}_n$, leads to $g_R = Z/A$ in much the same way that it leads to $\mathcal{J} = \mathcal{J}_{\text{rig}}$ for the total moment of inertia. Thus the deviations of the experimental values of g_R from Z/A should be accounted for by the deviations of the actual internal wave function from that of an independent-particle model. The same conclusion was arrived at when we discussed the moments of inertia.

Furthermore, we have seen before (Section VI.5) that moments of inertia of odd-A nuclei are systematically larger than those of even-even nuclei. If this effect is due, as was mentioned above, to the relatively lower energy required to excite the odd-particle configuration, we can expect this effect to show up in $\mathcal{J}_p$ or in $\mathcal{J}_n$ according to whether the odd-A nucleus is an odd-Z or and odd-N nucleus. Thus for an odd-Z nucleus $\mathcal{J}_p$ will be larger than the corresponding moment of inertia $\mathcal{J}_p$ in the neighboring even-A nucleus. It then follows from (8.19) that for odd-Z nuclei $g_R^{\text{odd}} > g_R^{\text{even}}$ where g_R^{even} refers to the neighboring even-Z even-N nucleus.

Similarly for an odd-N nucleus, $\mathcal{J}_n$ will be larger than the corresponding moment of inertia $\mathcal{J}_{n'}$ in the neighboring even-A nucleus. Hence the use of (8.19) shows that for odd-N nuclei, $g_R^{\text{odd}} < g_R^{\text{even}}$

Figure 8.5 shows the measured values of g_R for some odd-A nuclei. We see that the odd-Z values are invariably higher than the odd-N values, and that the even-A values for g_R fall between. We can therefore expect that a theory that will account for the detailed features of moments of inertia will also reproduce the detailed systematics found for g_R.

9. FIRST-ORDER CORRECTIONS IN THE COLLECTIVE HAMILTONIAN

In attempting to improve the estimates of mass and electric current distributions in deformed nuclei we should concentrate on two main points. One of them has to do with the intrinsic Hamiltonian H_{int} that as we have stressed several times, should include at least some part of the mutual interaction among the nucleons, in addition to the average deformed potential in which

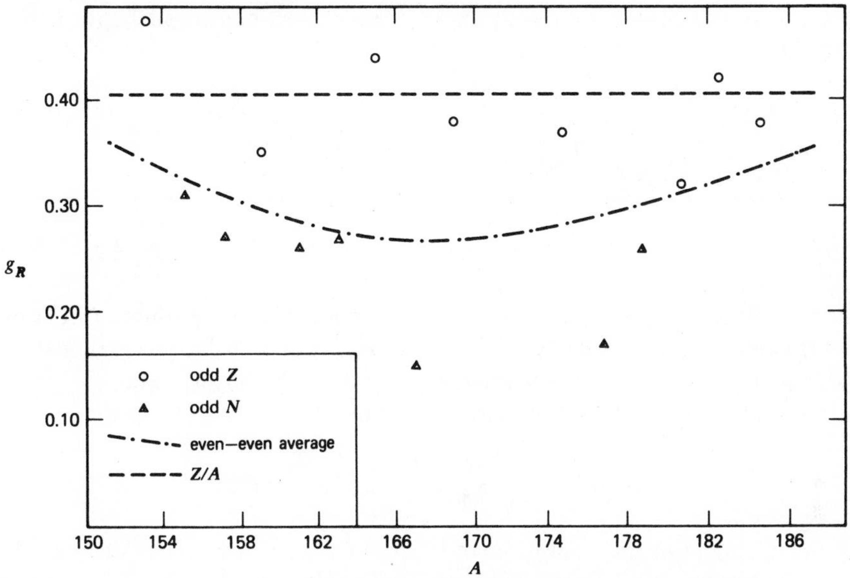

FIG. 8.5. Measured values of g_R for odd-A nuclei [taken from Nilsson (64)].

they are assumed to move independently. There is, however, another term in the complete Hamiltonian (3.27) whose effects we have thus far disregarded:

$$H_{PRC} = H_{\text{coupl}} - \frac{\hbar^2}{\mathcal{J}}(I_+' J_-' + I_-' J_+') \tag{9.1}$$

This part of the Hamiltonian, which is often referred to as the *particle-rotation-coupling* term, consists of two parts:

(i) H_{coupl} that is a true coupling Hamiltonian (2.14) between the collective rotation **R** and the intrinsic structure described by H_{int}. The empirical occurrence of rotational spectra of such high purity allows us to conclude that actual nuclear dynamics make H_{coupl} a very small perturbation in the highly deformed nuclei.

(*ii*) The Coriolis term $-(\hbar^2/\mathcal{J})(I_-' J_+' + I_-' J_+')$ is not really a particle-rotation coupling, since it couples **J** to **I** and not to the collective angular momentum **R**. It results as we recall from our insistence on describing rotational motion in terms of $(1/2\mathcal{J})\,\mathbf{I}^2$ instead of $1/2\mathcal{J}\,\mathbf{R}^2$. From the empirical rotational structure of spectra of deformed nuclei we cannot therefore automatically conclude that the Coriolis term can be neglected. We did notice that, except for $K = 1/2$, this term would mix in states of different internal structure, and if (4.19) is valid, its effects are probably small. But in view of our intention to study later the effects on nuclear properties of the mutual interactions in H_{int}, we should check first the corrections to such properties implied by the Coriolis potential.

Let us therefore consider the Hamiltonian

$$H = \frac{\hbar^2}{2\mathcal{J}} (\mathbf{I}^2 - 2I_{z'}J_{z'}) + \left[H_{\text{int}} + \frac{\hbar^2}{2\mathcal{J}} \mathbf{J}^2 \right] - \frac{\hbar^2}{\mathcal{J}} (I_+'J_-' + I_-'J_+') = \frac{\hbar^2}{2\mathcal{J}} \mathbf{R}^2 + H_{\text{int}} \tag{9.2}$$

which will be recognized to differ from the complete Hamiltonian (3.27) or (2.14) essentially through the omission of H_{coupl} [The term proportional to $(I_{z'} - J_{z'})^2$ in (3.27) is omitted since we are only considering states for which $\Omega = K$]. The cylindrically symmetric wave functions

$$\Psi(IMK) = \sqrt{\frac{2I + 1}{8\pi^2}}\, D^{I*}_{MK}(\theta_k)\, \chi_K(\mathbf{r}_i') \tag{9.3}$$

do not diagonalize the Hamiltonian (9.2), since for given values of I and M the Coriolis term in (9.2) will mix in different values of K.

Suppose, for simplicity, that the intrinsic Hamiltonian H_{int} in (9.2) is such that only the band $K' = K_0 + 1$ be considered together with the band $K = K_0$. All other bands are assumed to involve much higher internal excitations, so that they can be neglected. In this case the eigenfunctions and eigenvalues of H corresponding to a given value of I are obtained by the diagonalization of the matrix

$$\begin{pmatrix} \epsilon(K_0) + \dfrac{\hbar^2}{2\mathcal{J}} I(I + 1) & V' \\ V' & \epsilon(K_0 + 1) + \dfrac{\hbar^2}{2\mathcal{J}} I(I + 1) \end{pmatrix} \tag{9.4}$$

where $\epsilon(K)$ is given by

$$\left\{ H_{\text{int}} + \frac{\hbar^2}{2\mathcal{J}} [\mathbf{J}^2 - 2K^2] \right\} \chi_K = \epsilon(K)\, \chi_K \tag{9.5}$$

and the off-diagonal element V' is given by

$$\begin{aligned} V' &= \frac{2I + 1}{8\pi^2} \langle D^{I*}_{M,K_0+1}\, \chi_{K_0+1} | -\frac{\hbar^2}{\mathcal{J}} (I_+'J_-' + I_-'J_+') | D^{I*}_{M,K_0}\chi_{K_0} \rangle \\ &= \frac{2I + 1}{8\pi^2} \langle D^{I*}_{M,K_0+1}\, \chi_{K_0+1} | -\frac{\hbar^2}{\mathcal{J}} I_-'J_+' \,|\, D^{I*}_{M,K_0}\, \chi_{K_0} \rangle \\ &= \sqrt{\frac{I(I + 1) - K_0(K_0 + 1)}{2}}\, v'(K_0) \end{aligned} \tag{9.6}$$

where

$$v'(K) = \frac{\hbar^2}{\mathcal{J}} \langle \chi_{K+1} | J_+' | \chi_K \rangle \tag{9.7}$$

In deriving (9.6) we used the fact that $J_- \chi_{K_0}$ gives χ_{K_0-1} and therefore does not contribute to the matrix element in question, and also the relation (see remark following 2.36.)

$$I_-' D^{I*}_{M,K} = - \sqrt{\frac{I(I+1) - K(K+1)}{2}}\, D^{I*}_{M,K+1} \tag{9.8}$$

The diagonalization of (9.4) leads to the eigenvalues

$$\Delta E(I) = \frac{1}{2}[\epsilon(K_0) + \epsilon(K_0+1)] + \frac{\hbar^2}{2\mathcal{J}} I(I+1)$$
$$\pm \sqrt{\left[\frac{\epsilon(K_0) - \epsilon(K_0+1)}{2}\right]^2 + \frac{[I(I+1) - K_0(K_0+1)]}{2} |v'(K_0)|^2} \tag{9.9}$$

If the intrinsic wave functions χ_{K_0} are such that

$$[I(I+1) - K_0(K_0+1)]\, |v'(K_0)|^2 < [\epsilon(K_0) - \epsilon(K_0+1)]^2 \tag{9.10}$$

that is, if the mixing of the bands K_0 and $K_0 + 1$ is small, then we obtain, for that solution $\Delta E(I)$ of (9.9) that goes over to $E(IKM)$ of (4.24) when $v'(K_0) \to 0$:

$$\Delta E(I) = \epsilon(K_0) + \frac{1}{2} \frac{I(I+1) - K_0(K_0+1)}{\epsilon(K_0+1) - \epsilon(K_0)} |v'(K_0)|^2 + \frac{\hbar^2}{2\mathcal{J}} I(I+1) \tag{9.11}$$

We thus get, in this approximation, for the excitation energies inside the band K_0:

$$E(IK_0M) - E(K_0K_0M) = \Delta E(I) - \Delta E(K_0)$$
$$= \left[\frac{\hbar^2}{2\mathcal{J}} + \frac{1}{2} \frac{|v'(K_0)|^2}{\epsilon(K_0+1) - \epsilon(K_0)}\right] [I(I+1) - K_0(K_0+1)] \tag{9.12}$$

(For $K_0 = 1/2$ we should replace in (9.12) $[E(IK_0M) - E\,(K_0K_0M)]$ by $[E(IK_0M) - E(K_0K_0M) + (-1)^{I-K_0} a(I - K_0)\, \delta(K_0, 1/2)]$).

Equation 9.12 shows that, to lowest order, the Coriolis term does not change the I-dependence of the band structure; its only effect is to change the effective moment of inertia from $\mathcal{J}$ to $\mathcal{J}'$, where $\hbar^2/2\mathcal{J}'$ is the new coefficient of $(I(I+1) - K_0(K_0+1)]$:

$$\frac{\hbar^2}{2\mathcal{J}'} = \frac{\hbar^2}{2\mathcal{J}} + \frac{1}{2} \frac{|v'(k_0)|^2}{\epsilon(K_0+1) - \epsilon(K_0)} \tag{9.13}$$

Thus, to the extent that a moment of inertia is deduced from the empirical data, *it already includes the lowest-order effects of the Coriolis potential.* Therefore from the fact the energy given by the rotational formula (4.21) are in good agreement with experiment we *cannot* conclude that the wave functions are given by the rotational wave functions (4.9). In the approximation for which (9.12) is valid we have, in fact,

$$\Psi(IK_0M) = \sqrt{\frac{2I+1}{8\pi^2}}\Big[D^{I*}_{MK_0}\,\chi_{K_0} + \sqrt{\frac{I(I+1)-K_0(K_0+1)}{2}}\,\frac{v'(K_0)}{\epsilon(K_0+1)-\epsilon(K_0)}\,D^{I*}_{M,K_0+1}\,\chi_{K_0+1}\Big] \tag{9.14}$$

or, more precisely, to take into account the symmetry requirements that led to (4.9):

$$\Psi(IK_0M) = \sqrt{\frac{2I+1}{16\pi^2}}\Big\{D^{I*}_{MK_0}\,\chi_{K_0} + (\widehat{-1})^{I-J}\,D^{I*}_{M,-K_0}\,\chi_{-K_0} + \sqrt{\frac{I(I+1)-K_0(K_0+1)}{2}}\,\frac{v'(K_0)}{\epsilon(K_0+1)-\epsilon(K_0)}\Big[D^{I*}_{M,K_0+1}\,\chi_{K_0+1} + (\widehat{-1})^{I-J}\,D^{I*}_{M,-(K_0+1)}\,\chi_{-(K_0+1)}\Big]\Big\} \tag{9.15}$$

Whereas the use of the pure rotational wave function (4.9) has led us to the formulation of K-selection rules like (6.23), the presence of small components in $\Psi(IK_0M)$ with $K = K_0 + 1$, as in (9.14) or (9.15), may lead to small violations of these selection rules; in fact even if K_0 does not satisfy $|K_0 - K'| \leq k$, this relation may be satisfied by $K_0 + 1$ and thus lead to a k-multipole transition between the small admixed component of one level and the main component of the other. As a rule we can expect, therefore, that predictions concerning transition probabilities will tend to be less well satisfied than those concerning the energies of the corresponding levels. For the energies, the mixing of bands has in lowest order, the effect of renormalizing the moment of inertia, and only in higher order does it affect also the I-dependence of the energies. With transition probabilities, however, the mixing of bands can cause more drastic changes than just renormalization of charges or moments, even in the lowest order.

Rotational levels in several nuclei have been analyzed in terms of a possible existence of band mixtures. Both energies and transition probabilities have been included in the analysis, with a substantial improvement in the comparison of theory and experiment for both. An example is offered by ^{183}W [Kerman (56)].

Two negative parity rotational bands are known in ^{183}W, one with $K = 1/2^-$ and the other with $K = 3/2^-$. Trying to fit the data on the basis of pure rotational bands, the best one can do for the $K = 3/2^-$ band, for instance, is to take $\hbar^2/2\mathcal{J} = 17$ KeV and obtain the following predictions: $5/2^-$ level at 293.8 KeV (experimentally at 291.7 KeV), $7/2^-$ level at 412.8 KeV (experimentally at 412.1 KeV), and $9/2^-$ level at 565.8 KeV (experimentally at 554.2 KeV). However with band mixture taken into account, a much better overall agreement is obtained for both the levels of the $K = 1/2^-$ band and the $K = 3/2^-$ band, and this at the cost of introducing essentially just one additional parameter $\nu'(K_0)$. The improvement is most significant in $B(E2)$ values for transitions between different bands. Using the notation $B(E2, IK \to I'K')$ one finds experimentally

$$R = \frac{B(E2,\, 1/2\ 1/2 \to 5/2\ 3/2)}{B(E2,\, 1/2\ 1/2 \to 5/2\ 1/2)} = 0.09$$

Theoretically, without mixing of bands, the ratio R should be characteristic of the ratio of single-particle transitions (between two bands) to collective transitions (within the band). Depending on details of the model various values can be obtained for this theoretical ratio, but they are all in the neighborhood of $R_{\text{th}} \approx 0.001$ nearly 90 times smaller than the observed ratio. With the mixing implied by the parameters of Fig. 9.1 interband transitions include a component that looks like an intraband transition, and the theoretical ratio becomes $R_{\text{th}} \sim 0.14$, in substantial agreement with the measured value.

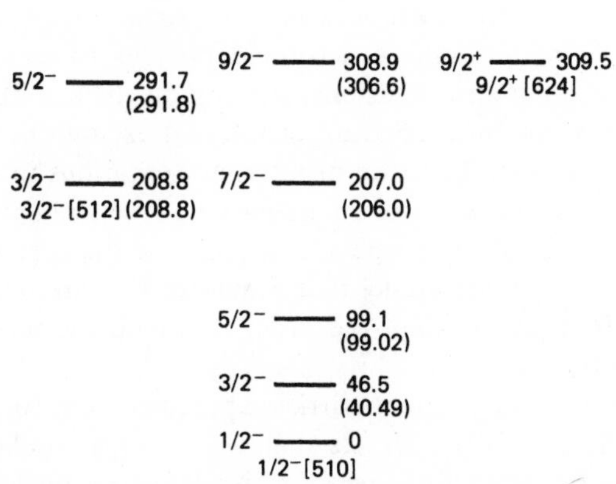

FIG. 9.1. Experimental energy levels for ^{183}W. The numbers in parentheses are calculated from (9.9). [See also note below (9.12). Taken from Davidson (65)].

It is obvious from this example that transition probabilities offer a much more sensitive test for the "purity" of a rotational level than their energies alone, and they can be most effectively used to pin down the importance of various mechanisms for band admixtures. These are not the only tests available. The variety of tests that can be used to establish the existence and importance of such admixtures are to be found in Davidson (65). We shall not discuss them here.

10. MODELS FOR INTRINSIC STRUCTURE IN DEFORMED NUCLEI

Our considerations thus far have been centered on the particular structure of the rotational wave function $\Psi(IKM)$ as given by (4.9), without going into the details of the internal wave functions χ_Ω. The properties of these internal wave functions that enter into the description of the various phenomena discussed above are contained in a number of parameters such as the decoupling parameter a defined in (4.23), the band-mixing parameter $\nu'(K)$ defined by (9.7), the moment of inertia $\mathcal{J}$, the collective g-factor, g_R, the internal quadrupole moment Q_0, etc.

An attempt to go one step beyond this was made in Section VI.7 when we use the Inglis formula in conjunction with independent-particle wave functions derived from a deformed harmonic oscillator potential, and obtained the rigid moment of inertia $\mathcal{J}_{\text{rig}}$, as given by (7.49). We notice there that a more detailed knowledge of the intrinsic wave function χ_Ω is required to have a better estimate of the moment of inertia, and that such an improved χ_Ω can also give us better information of g_R, as well as "theoretical" values for the decoupling parameter a, the band-mixing parameter $\nu'(K_0)$, and other intrinsic properties.

There is another important set of facts that a theory for χ_Ω ought to give us. As we proceed from one odd-A deformed nucleus to another we observe different bands at different excitations. Picking on the lowest $I = K$, level of each band we thus obtain a spectrum of internal excitations as a function of the mass number A. This structure of internal excitations in deformed nuclei plays the role of the spectrum of single-particle levels in spherical nuclei, in that both teach us about the shape and strength of the corresponding single-particle potential. A proper model that generates the internal wave functions χ_Ω should therefore also explain the order in which the internal excitations appear in actual nuclei.

The success of the independent-particle approximation for spherical nuclei near closed shells naturally suggests that we adopt a similar procedure for deformed nuclei. Thus as a first guess for the deformed nucleus internal wave function we want to take an independent-particle wave function, generated from a deformed potential. This wave function will have to be improved upon for the derivation of properties like the moment of inertia, but for now we shall use this lowest approximation. Limiting ourselves to harmonic-oscillator potentials, we thus take as our zeroth-order Hamiltonian for the intrinsic wave function χ_Ω:

$$H_{\text{int}} = -\sum \frac{\hbar^2}{2M} \nabla_i^2 + \frac{1}{2} M \sum_i [\omega_x^2(x_i^2 + y_i^2) + \omega_z^2 z_i^2] \qquad (10.1)$$

The single-particle eigenfunctions of H_{int} can be characterized by the three harmonic oscillator quantum numbers n_x, n_y, and n_z, so that in terms of $U_n(x)$, the one-dimensional harmonic-oscillator eigenfunctions, we obtain

$$\chi(n_x n_y n_z) = U_{nx}(x)\, U_{ny}(y)\, U_{nz}(z) \qquad (10.2)$$

The energy corresponding to this state is

$$E(n_x n_y n_z) = \hbar\omega_x(n_x + n_y + 1) + \hbar\omega_z\left(n_z + \frac{1}{2}\right) = \hbar\omega_x(N - n_z + 1)$$
$$+ \hbar\omega_z\left(n_z + \frac{1}{2}\right) \qquad N \equiv n_x + n_y + n_z \qquad (10.3)$$

Clearly all the states with the same value for $n_x + n_y$ and a fixed value for n_z are degenerate. We shall therefore find it useful to go over to cylindrical coordinates ρ, ϕ, and z

$$x = \rho \cos\phi \qquad y = \rho \sin\phi \qquad z = z \qquad (10.4)$$

In these coordinates the potential energy in (10.1) is independent of ϕ, so that the eigenfunctions take on the form

$$\chi(N, n_z, \Lambda) = R_{N-n_z}(\rho)\, e^{i\Lambda\phi}\, U_{n_z}(z) \qquad (10.5)$$

$$\Lambda = \pm n, \pm(n-2), \pm(n-4), \ldots, \pm 1 \text{ or } 0 \qquad n = N - n_z \qquad (10.6)$$

Here U_n is again the eigenfunction of a one-dimensional harmonic oscillator, and $R_n(\rho)$ is the radial part of the eigenfunction of a two-dimensional harmonic oscillator. The energies that go with $\chi(N,n_z,\Lambda)$ are independent of Λ:

$$E(N,n_z,\Lambda) = \hbar\omega_x(N - n_z + 1) + \hbar\omega_z\left(n_z + \frac{1}{2}\right) \qquad (10.7)$$

For any given value of N and n_z, the z-projection of the two-dimensional orbital angular momentum that is, Λ, can take on the values indicated by (10.6), since all the states in one major harmonic-oscillator shell have the same parity. In any given major shell the angular momentum Λ can therefore take on either odd values only, or only even values.

The quantum numbers N, n_z, and Λ characterize the single-particle orbits of the Hamiltonian (10.1). Each level is doubly degenerate, with Λ and $-\Lambda$ belonging to the same eigenvalue. As we fill the levels with an odd number of particles we obtain wave functions characterized, among other things, by the set Λ_i for all the occupied levels. If we are dealing with the lowest state of the system, or more generally if each level Λ_i is filled in pairs as much as possible, we find immediately that $\Lambda = \Sigma\, \Lambda_i$ will equal the Λ-value of the last odd particle. It is therefore commonly found that the notation for the Λ-value of the odd single particle is the same as that for $\Lambda_{\text{total}} = \Sigma\, \Lambda_i$.

The Hamiltonian (10.1) is spin independent; since it is also invariant under rotations around the z-axis, the spin projection of a single particle on this axis, which can take on the values $\Sigma = \pm 1/2$, and the total projection of the angular momentum on the z-axis $\Omega = \Lambda + \Sigma$ are both good quantum numbers.

Finally, H_{int} is also invariant under reflections and hence its eigenfunctions are characterized by a definite parity. It is easy to see that the parity is even or odd according to whether N is even or odd.

The eigenstates of (10.1) are thus characterized by the set:

$$\Omega, \pi, [N, n_z, \Lambda] \tag{10.8}$$

where Ω, the z-projection of the total angular momentum of the particle satisfies

$$\Omega = \Lambda \pm \frac{1}{2} \tag{10.9}$$

The parity π satisfies

$$\pi = (-1)^{\text{N}} \tag{10.10}$$

The z-oscillator quantum number satisfies, of course,

$$n_z \leqslant N \tag{10.11}$$

and the Λ-quantum number satisfies

$$|\Lambda| = N - n_z,\ N - n_z - 2,\ N - n_z - 4, \ldots, \qquad 1 \text{ or } 0 \tag{10.12}$$

In labeling a level it is customary to use only positive values for Λ and Ω (10.8) since the two states $\pm\Omega$ are degenerate. The quantum numbers (10.8) are sometimes referred to as the *asymptotic quantum numbers*. This refers to the situation prevailing at very large deformations, when various additional terms that are usually added to H_{int}, and that do not commute with Λ or N (see below), can be neglected. The ground state of ^{183}W which is measured to be $I = 1/2^-$, is, for instance, assigned the asymptotic quantum numbers $1/2^-$ [5, 1, 0]. The lowest state of the $K = 3/2^-$ band at 208.8 KeV is assigned the asymptotic quantum numbers $3/2^-$ [5, 1, 2]. Within the approximation of the intrinsic Hamiltonian (10.1) these two states are degenerate.

We know from the study of spherical nuclei that a spin-independent central potential results in a very poor approximation to the actual nuclear wave functions, and that a rather strong spin-orbit interaction has to be included in the zeroth-order single-particle Hamiltonian. Furthermore, for heavier nuclei the harmonic-oscillator potential is too rounded and the real potential comes closer to a square well potential. To introduce these two effects also into the deformed potential it was found useful to add two more terms to (10.1): a spin-orbit term $C\mathbf{l}\cdot\mathbf{s}$ and a term $D\mathbf{l}^2$ where C and D are constants; the second term artificially produces a splitting between levels with different values of l. This simulates the transition from an oscillator potential, in which all the l-values within a major shell are degenerate, to a square well potential in which higher l-levels in a major shell are more bound than the lower l-levels. The exact values of C and D are fixed to reproduce the observed level order and splittings near magic numbers at zero deformatıons, and may sometimes vary in going from one nucleus to another. The resulting Hamiltonian is known as the *Nilsson Hamiltonian* [Nilsson (55)]:

$$H = \frac{\hbar^2}{2M} \Sigma \nabla_i^2 + \frac{1}{2} M \sum_i [\omega_x^2(x_i^2 + y_i^2) + \omega_z^2 z_i^2] + C \Sigma \mathbf{l}_i\cdot\mathbf{s}_i + D \Sigma \mathbf{l}_i^2 \qquad (10.13)$$

It is very widely used in generating zeroth-order internal wave functions for deformed nuclei.

The single-particle energies and eigenfunctions of the Nilsson Hamiltonian are available in the literature [see, for instance, Davidson (68)]. They were computed in the following approximation.

First, one introduces a deformation parameter δ (see Eq. 7.4) and a frequency ω_0 by:

$$\omega_z^2 = \omega_0^2\left(1 - \frac{4}{3}\delta\right) \qquad \omega_x^2 = \omega_y^2 = \omega_0^2\left(1 + \frac{2}{3}\delta\right) \qquad (10.14)$$

To keep the volume of the nucleus constant we must have $\omega_x\omega_y\omega_z$ = constant and hence ω_0 and δ are related through

$$\omega_0\left(1 - \frac{4}{3}\delta^2 - \frac{16}{27}\delta^3\right)^{1/6} = \text{constant} = \omega_{00} \qquad (10.15)$$

With the help of the deformation parameter δ the Nilsson single-particle Hamiltonian (10.13) can be written as

$$H = -\frac{\hbar^2}{2M}\nabla^2 + \frac{1}{2}M\omega_0^2r^2 - M\omega_0^2r^2\,\delta\,\frac{4}{3}\sqrt{\frac{\pi}{5}}\,Y_{20}(\theta,\phi) + C\mathbf{l}\cdot\mathbf{s} + D\mathbf{l}^2 \qquad (10.16)$$

Eigenfunctions of the central-field Hamiltonian

$$H_0 = -\frac{\hbar^2}{2M}\nabla^2 + \frac{1}{2}M\omega_0^2r^2 \qquad (10.17)$$

are now computed and characterized by the quantum numbers N, l, Λ, and Σ so that:

$$H_0|Nl\Lambda\ \Sigma> = \left(N + \frac{3}{2}\right)\hbar\omega_0|Nl\Lambda\ \Sigma> \qquad l_z|Nl\Lambda\ \Sigma> = \Lambda|Nl\Lambda\ \Sigma>$$
$$\mathbf{l}^2|Nl\Lambda\ \Sigma> = l(l+1)|Nl\Lambda\ \Sigma> \qquad s_z|Nl\Lambda\ \Sigma> = \Sigma|Nl\Lambda\ \Sigma> \tag{10.18}$$

The rest of the Hamiltonian (10.16) does not commute with $\mathbf{l}^2$, l_z, or s_z, and it also connects states of different values of N. One adopts the approximation of neglecting the terms nondiagonal in N, since they involve coupling to states with excitation energy $2\hbar\omega$. The operator $j_z = l_z + s_z$ does commute with (10.16); one therefore diagonalizes (10.16) in the subspace of all wave functions (10.18) with fixed values of N and Ω. Thus the approximate eigenstates of (10.16) are given by

$$\chi_{N\Omega} = \sum_{l\,\Lambda} a_{l\,\Lambda}{}^{\Omega}|Nl\Lambda\ \Sigma = (\Omega - \Lambda)> \tag{10.19}$$

The coefficients $a_{l\Lambda}{}^{\Omega}$, which depend of course also on the deformation δ, were calculated by Nilsson and tabulated by him [Nilsson (55)]. Also tabulated are the eigenvalues of (10.16) computed in the approximation of neglecting off-diagonal elements in N, and given for different values of the deformation δ.

For large deformations δ, the terms $C\mathbf{l}\cdot\mathbf{s}$ and $D\mathbf{l}^2$ in (10.16) can be neglected in comparison with $M\omega_0{}^2\ \delta\ (4/3)\ \sqrt{\pi/5}\ Y_{20}(\theta\phi)$. At this limit the asymptotic quantum numbers (10.8) become good quantum numbers. It is therefore customary to label the states (10.19) by the asymptotic quantum numbers of the state that $\chi_{N\Omega}$ approaches for $\delta \gg 1$.

Figure 10.1 shows the eigenvalues belonging to the different Nilsson wave functions. The energy in units of $\hbar\omega_0(\delta)$ is plotted as a function of δ. Another parameter η is often is introduced; it is related to δ through

$$\eta = \frac{2\hbar\omega_0(\delta)}{C}\,\delta \tag{10.20}$$

Positive values of δ correspond to prolate deformations and negative values to oblate deformations. For zero deformation Fig. 10.1 shows just the spherical shell-model sequence of single-particle levels with the gaps between major shells determining the magic numbers. For small deformations each shell-model j is split into $(1/2)\ (2j + 1)$-levels characterized by the different values of $|m| \leq j$. The value of $|m|$ then becomes the quantum number Ω that remains unchanged as the deformation increases.

For small positive deformations δ the states with the smallest values of m becomes more bound, whereas for small negative deformations the states with $m = j$ are the most bound. This is easy to understand if we recall that the mass distribution of a particle is concentrated in a plane perpendicular to its angular momentum. Thus if the attractive central potential has a prolate

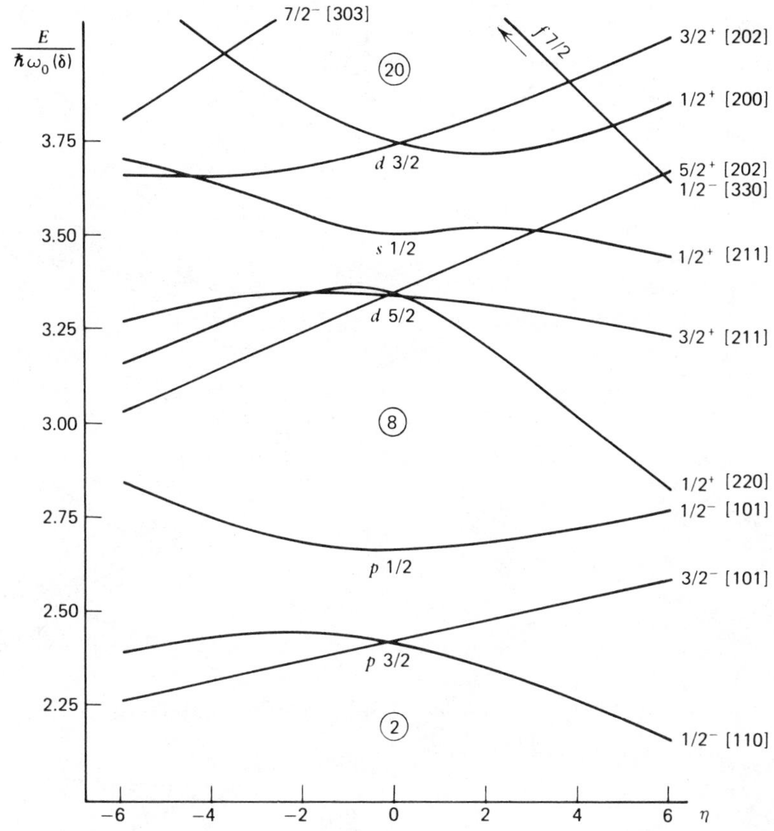

FIG. 10.1. (a) Single-particle levels in the regions $8 < Z < 20$ and $8 < N < 20$ [taken from Nathan and Nilsson (65)].

deformation the particles that benefit from it most are those whose angular momentum is perpendicular to the z-axis. This requires that the j-projection be minimal. On the other hand an oblate attractive potential is more advantageous for particles whose angular momentum is parallel to the z-axis. More formally we can derive the same results if in (10.16) we consider

$$H' = -M\omega_0^2 r^2 \, \delta \, \frac{4}{3}\sqrt{\frac{\pi}{5}} \, Y_{20}(\theta,\phi) \tag{10.21}$$

as a perturbation on the spherically symmetric Hamiltonian

$$H_1 = H - H' = -\frac{\hbar^2}{2M}\nabla^2 + \frac{1}{2}M\omega_0^2 r^2 + C\mathbf{l}\cdot\mathbf{s} + D\mathbf{l}^2 \tag{10.22}$$

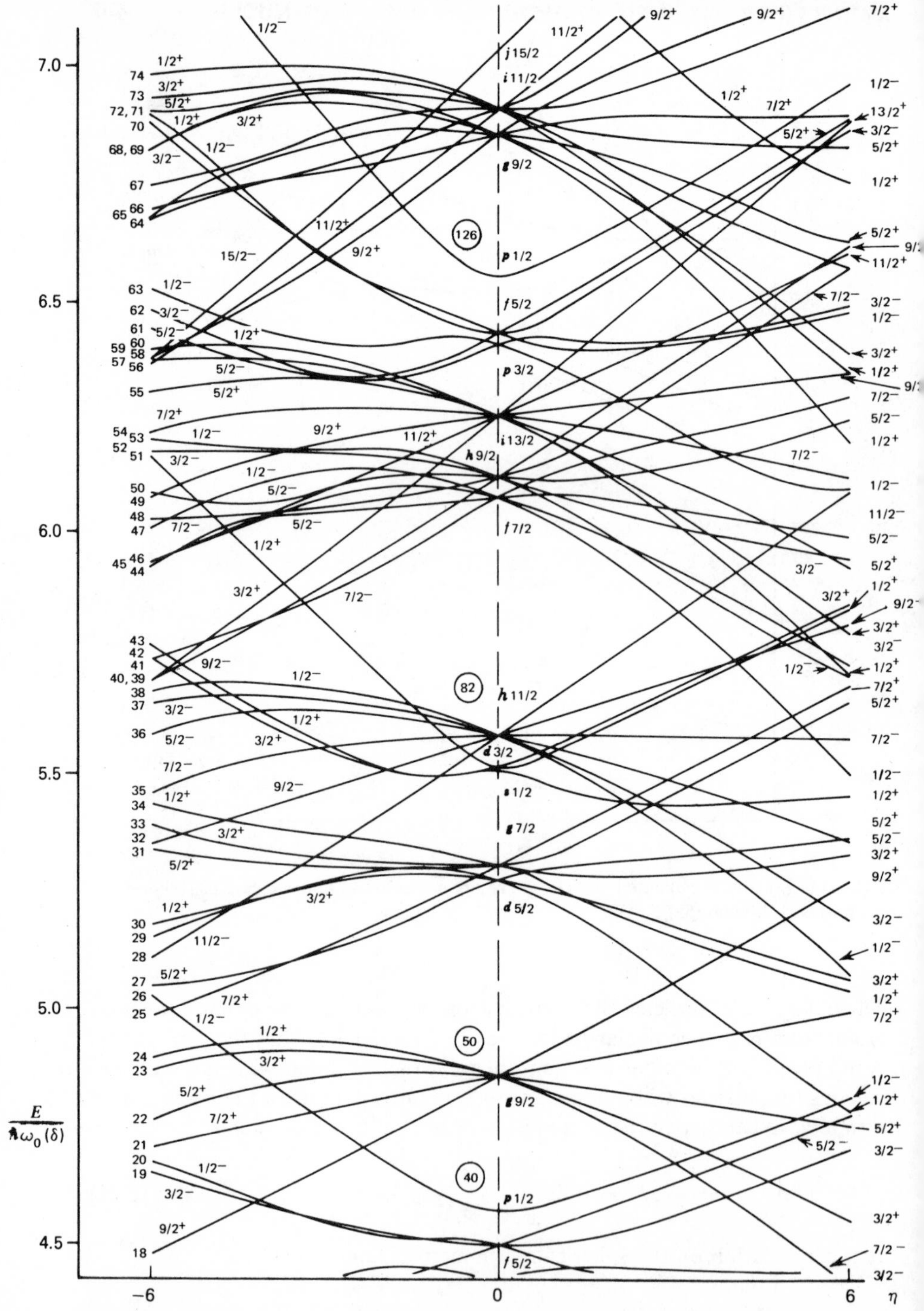

FIG. 10.1. Cont. (*b*) Single-particle Levels for the Nilsson Hamiltonian [taken from Kerman (59)].

The eigenfunctions of H_1 are characterized by the quantum numbers $(Nljm)$ and the eigenvalues are the energies shown on the Nilsson diagram for zero deformation. The shift in energy for the state $|Nljm>$ as the perturbation (10.21) is switched on is then given by

$$\Delta E(Nljm) = -\frac{4}{3}\sqrt{\frac{\pi}{5}}\, M\omega_0^2\, \delta\, \langle Nljm|r^2 Y_{20}(\theta,\phi)|Nljm\rangle$$

$$= -\frac{4}{3}\sqrt{\frac{\pi}{5}}\,\hbar\left(N+\frac{3}{2}\right)\omega_0\delta\,(-1)^{j-m}\begin{pmatrix} j & 2 & j \\ -m & 0 & m\end{pmatrix}(lj||Y_2||lj) \tag{10.23}$$

Here we have used the fact that (Eq. 7.41)

$$\frac{1}{2}M\omega_0^2\,\langle Nljm|r^2|Nljm\rangle = \frac{1}{2}\hbar\omega_0\left(N+\frac{3}{2}\right) \tag{10.24}$$

Introducing the explicit expressions for the $3-j$ symbol (Appendix A, A.2.79) and the reduced matrix element of Y_2 (Appendix A, A.2.49), (10.23) can be written in the form

$$\Delta E(Nljm) = -\frac{2}{3}\hbar\omega_0\left(N+\frac{3}{2}\right)\delta\,\frac{[3m^2-j(j+1)][3/4-j(j+1)]}{(2j-1)\,j(j+1)\,(2j+3)} \tag{10.25}$$

Several interesting features can be seen from (10.25). First we notice that

$$\text{For}\quad \delta > 0 \quad \Delta E(Nljm) < 0 \quad \text{if} \quad 3m^2 < j(j+1)$$

$$\Delta E(Nljm) > 0 \quad \text{if} \quad 3m^2 > j(j+1) \tag{10.26}$$

The opposite relations hold for $\delta < 0$. This confirms our previous qualitative considerations.

Furthermore we notice that

$$\sum_m \Delta E\,(Nljm) = 0 \tag{10.27}$$

Thus for small deformations, of either sign, the "center of gravity"of the various levels is not shifted. There are always some m-levels that are pushed up and others that are pushed down. We also see from (10.27) that the sum of single-particle energies of closed shells remains unchanged by small deformations δ. In fact the lowest-order contribution to such sums turns out to be proportional to δ^2.

$\Delta E(Nljm)$ is linear in δ and the slope is proportional to N. Thus for the higher-oscillator shells relatively small deformations produce large variations in the energy. Heavy nuclei are therefore more susceptible to gain in energy by passing from a spherical to a deformed shape. The dependence of the shape of the Nilsson curves near $\delta = 0$ on the quantum number N is clearly seen in Fig. 10.1. Its physical interpretation is also straightforward. As we see from (10.16) a given deformation δ has an effect that is proportional to r^2. Particles that extend to larger distances are therefore more strongly affected by the same deformation δ.

As the deformation δ increases it is no longer permissible to treat H' as a perturbation and a diagonalization procedure has to be undertaken. This has the effect of "mixing" into a state $|ljm\rangle$ other states $|l'j'm\rangle$ with the same value of m but different values of $l'j'$. If we take the shell between 8 and 20, for instance, it has, for $\delta = 0$ three levels: $1d_{5/2}$ $2s_{1/2}$, and $1d_{3/2}$. As a positive deformation sets in the state $|d_{5/2}\, m = 1/2\rangle$ is pushed down with a slope proportional to δ. However as δ increases this level is pushed down more strongly due to its interaction with the $|s_{1/2}\, m = 1/2\rangle$ state above it. This latter level, according to (10.25) is unaffected, to first order by the deformation but due to its interaction with the $|d_{5/2}\, m = 1/2\rangle$ level on the one hand and the $|d_{3/2}\, m = 1/2\rangle$ level on the other, its position does change with δ as δ increases. This is seen more dramatically for $\delta < 0$. For such negative values of δ the level $|d_{5/2}\, m = 1/2\rangle$ is pushed up; it comes therefore closer to the state $|s_{1/2}\, m = 1/2\rangle$ as δ becomes more negative. The effect of their interaction is that the $|d_{5/2}\, m = 1/2\rangle$ state reverses its slope and goes *down* in energy with the decrease of δ (increase of $|\delta|$!), while pushing the $|s_{1/2}\, m = 1/2\rangle$ level up.

Since the Nilsson calculations confine the diagonalization of H to the levels in one major shell only, there are always some levels for which (10.25) is valid also for arbitrarily large values of $|\delta|$. In the shell we have just been considering this is true for the level $|d_{5/2}\, m = 5/2\rangle$. In fact in this shell there is no other level with $m = 5/2$, so that first-order perturbation theory becomes identical with complete diagonalization within the shell. It is seen that the energy of this level does indeed change linearly with δ over the complete range of variation of δ.

Many features of the Nilsson diagrams can be understood in terms of considerations similar to the ones we have described above. We shall not go into them here, but refer the reader to Nilsson's original paper [Nilsson (55)].

11. NUCLEAR CONFIGURATIONS IN THE NILSSON POTENTIAL

As we mentioned above the quantum number N actually tells us in which major shell the state in question originates. The neutron states in ${}^{183}_{74}W_{109}$, for instance, fall in the major shell between 82 and 126. All its negative parity states (p, f, h) states therefore belong to $N = 5$; the positive parity states, which stem from the $i_{13/2}$ single-particle shell-model state, actually belong to the next harmonic-oscillator shell and were pushed down by the spin-orbit interaction. They therefore have $N = 6$.

With the single-particle wave functions in the deformed potential available through (10.19), it is now a relatively simple to figure out the nuclear configuration for a given number N and Z of neutrons and protons. All one has to do is to choose a deformation δ, look at the Nilsson diagram, and fill the lowest

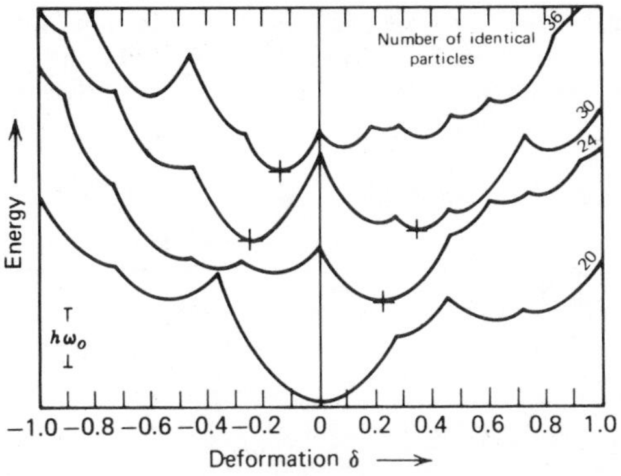

FIG. 11.1. Deformation energies for various stages of shell filling-spheroidal harmonic-oscillator binding potential [taken from Moszkowski (57)].

available levels at this deformation with neutrons and protons, remembering that each Nilsson orbit can accommodate two protons and two neutrons (because of the $\pm\Omega$ degeneracy). Having done that, we add up the single-particle energies and obtain the total energy of the lowest independent particle state corresponding to this deformation δ.* This procedure is now repeated, with the same values of Z and N, for different deformations δ. Since as δ changes, different Nilsson orbits cross each other, the structure of the lowest independent particle state may change with δ. The total energy plotted as a function of δ then has the form shown in Fig. 11.1. The lowest energy therefore gives us both the expected deformation δ_0 and the configuration of Nilsson orbits that goes with it. Table 11.1 shows the expected ground state spins and parities of certain light nuclei based on the successive filling of the levels of the deformed harmonic-oscillator potential (10.16). Only nuclei removed from the closed shells (^{12}C and ^{16}O) are listed. Perfect agreement with experiment is obtained.

For odd-odd nuclei the states with $\Omega = \Omega_p + \Omega_n$ and $\Omega = |\Omega_p - \Omega_n|$ are degenerate (see below) for the Hamiltonian (10.16). Since the free proton-neutron interaction favors parallel intrinsic spins (in the deuteron the ground state has $S = 1$), it has been assumed that the addition of a residual two-body interaction to (10.16) will lower that combination $\Omega_p \pm \Omega_n$ that makes the intrinsic spins parallel. From the asymptotic quantum numbers of each level we find that the spin projections of the proton and the neutron are given by

$$\Sigma_p = \Omega_p - \Lambda_p \qquad \Sigma_n = \Omega_n - \Lambda_n$$

*If the Nilsson potential is "considered in the spirit" of the Hartree-Fock self-consistent field, we are counting the two-body interactions twice. We shall not discuss this question here; the reader is referred to Bassichis and Wilets (69).

TABLE 11.1 The Expected Ground State Spins of Certain Light Nuclei Based on the Filling of the Levels of a Deformed Well. For Odd-Odd Nuclei the Theory is Ambiguous. The Highest Ω Value Seems to be the Correct One. [taken from Kerman (59)]

Nucleus	Configuration: Protons	Configuration: Neutrons	Ω_{theo}	I_{exp}
$^{6}_{3}$Li	1/2−	1/2−	1	1
$^{7}_{3}$Li	1/2−	$(1/2-)^2$	1/2	1/2
$^{8}_{3}$Li	1/2−	$(1/2-)^2 3/2-$	2	2
$^{9}_{4}$Be	$(1/2-)^2$	$(1/2-)^2 3/2-$	3/2	3/2
$^{10}_{5}$B	$(1/2-)^2 3/2-$	$(1/2-)^2 3/2-$	3	3
$^{19}_{8}$O		$(1/2+)^2 3/2+$	3/2	
$^{18}_{9}$F	1/2+	1/2+	1	1
$^{19}_{9}$F	1/2+	1/2+	1/2	1/2
$^{20}_{9}$F	1/2+	$(1/2+)^2 3/2+$	2	2
$^{19}_{10}$Ne	$(1/2+)^2$	1/2+	1/2	1/2
$^{21}_{10}$Ne	$(1/2+)^2$	$(1/2+)^2 3/2+$	3/2	3/2
$^{23}_{10}$Ne	$(1/2+)^2$	$(1/2+)^2(3/2+)^2(5/2+)$	5/2	5/2
$^{21}_{11}$Na	$(1/2+)^2 3/2+$	$(1/2+)^2$	3/2	3/2
$^{22}_{11}$Na	$(1/2+)^2 3/2+$	$(1/2+)^2 3/2+$	3	3
$^{23}_{11}$Na	$(1/2+)^2 3/2+$	$(1/2+)^2(3/2+)^2$	3/2	3/2
$^{24}_{11}$Na	$(1/2+)^2 3/2+$	$(1/2+)^2(3/2+)^2 5/2+$	4	4
$^{25}_{11}$Na	$(1/2+)^2 3/2+$	$(1/2+)^2(3/2+)^2(5/2+)^2$	3/2	5/2
$^{23}_{12}$Mg	$(1/2+)^2(3/2+)^2$	$(1/2+)^2 3/2+$	3/2	3/2
$^{25}_{12}$Mg	$(1/2+)^2(3/2+)^2$	$(1/2+)^2(3/2+)^2 5/2+$	5/2	5/2
$^{27}_{12}$Mg	$(1/2+)^2(3/2+)^2$	$(1/2+)^2(3/2+)^2(5/2+)^2 1/2+$	1/2	1/2
$^{25}_{13}$Al	$(1/2+)^2(3/2+)^2 5/2+$	$(1/2+)^2(3/2+)^2$	5/2	5/2
$^{26}_{13}$Al	$(1/2+)^2(3/2+)^2 5/2+$	$(1/2+)^2(3/2+)^2 5/2+$	5	5
$^{27}_{13}$Al	$(1/2+)^2(3/2+)^2 5/2+$	$(1/2+)^2(3/2+)^2(5/2+)^2$	5/2	5/2
$^{28}_{13}$Al	$(1/2+)^2(3/2+)^2 5/2+$	$(1/2+)^2(3/2+)^2(5/2+)^2 1/2+$	3	3
$^{27}_{14}$Si	$(1/2+)^2(3/2+)^2(5/2+)^2$	$(1/2+)^2(3/2+)^2 5/2+$	5/2	5/2

If both Σ_p and Σ_n are parallel or antiparallel to Λ_p and Λ_n, respectively, Σ_p and Σ_n will be parallel to each other in the state $\Omega = \Omega_p + \Omega_n$ (Fig. 11.2). If on the other hand Σ_p is parallel to Λ_p while Σ_n is antiparallel to Λ_n, or vice versa, Σ_p and Σ_n will be parallel to each other in the state $\Omega = |\Omega_p - \Omega_n|$ (Fig. 11.3).

$$4(\Omega_p - \Lambda_p)(\Omega_n - \Lambda_n) = \begin{cases} +1 \text{ if } \Sigma_p \text{ parallel to } \Lambda_p \text{ and } \Sigma_n \text{ parallel to } \Lambda_n, \text{ or} \\ \quad \text{if } \Sigma_p \text{ antiparallel to } \Lambda_p \text{ and } \Sigma_n \text{ antiparallel to } \Lambda_n \\ -1 \text{ if } \Sigma_p \text{ parallel to } \Lambda_p \text{ and } \Sigma_n \text{ antiparallel to } \Lambda_n, \text{ or} \\ \quad \text{if } \Sigma_p \text{ antiparallel to } \Lambda_p \text{ and } \Sigma_n \text{ parallel to } \Lambda_n \end{cases}$$

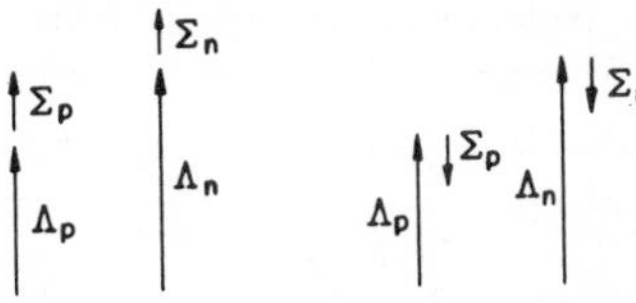

FIG. 11.2. Coupling of Ω's for odd-odd nuclei; $\Omega_p = \Lambda_p \pm \Sigma_p$, $\Omega_n = \Lambda_n \pm \Sigma_n$.

It then follows that the lowest configuration of a deformed odd-odd nucleus whose odd proton and odd neutron are in the levels Ω_p and Ω_n, respectively, will be given by

$$\Omega = |\Omega_p + 4(\Omega_p - \Lambda_p)(\Omega_n - \Lambda_N)\Omega_n| \tag{11.1}$$

Equation 11.1 was used to determine the K-value of the lowest configuration in the odd-odd nuclei listed in Table 11.2. With the exception of ^{28}Al the observed spin of the ground state agrees with the one expected from the Nilsson diagram. In that case the expected spin of 2^+ is very close at 31.2 KeV. A detailed study of deformed odd-odd nuclei was carried out by Gallagher and Mozkowski (58).

For heavier nuclei the Nilsson orbits come very close to each other, and the above procedure can only be used to give two or three alternatives for the intrinsic state of lowest energy. Generally one of them does turn out to be the lowest state and the others form the "heads" of low-lying excited bands. Table 11.3 [Kerman, (59)] shows the expected Ω-values for the lowest intrinsic states of some heavy nuclei with the corresponding deformation δ. The deformation, in these cases, has been determined from the measured quadrupole moment of the ground state of the nucleus in question. Again the observed spin of the nuclear ground state always agrees with one of the two or three lowest expected values of Ω.

Finally, Table 11.4 [Harmatz, Handley, and Mihelich (62)] gives the various intrinsic excitations identified in odd-A nuclei from ^{151}Sm to ^{189}Os together with their excitation energy. We notice that the 5/2 + [642] state, for instance, first appears as an excited state in ^{151}Sm (209 KeV), then in ^{153}Gd (129 KeV), ^{155}Gd (105 KeV), and ^{157}Gd (64 KeV). In ^{159}Gd it becomes the ground state and in ^{163}Er it is an excited state again (22 KeV), and again in ^{165}Er (47 KeV)

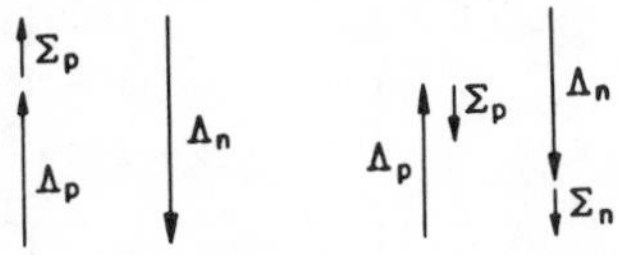

FIG. 11.3. Coupling of 's for odd-odd nuclei: $\Omega_p = \Lambda_p \pm \Sigma_p$, $\Omega_n = \Lambda_p \pm \Sigma_p$.

TABLE 11.2 Observed Ground-State Spins as well as the Nilsson Odd-Neutron and Odd-Proton States for Some Deformed Odd-Odd Nuclei. Where Known, Similar Information for Excited Bands is Given along with the Band-Head Energy. [taken from Davidson (65)]

Nucleus	Band Head (keV)	$I\pi$	Odd-Particle Configuration: Proton	Odd-Particle Configuration: Neutron
^{20}F	0	2+	$\frac{1}{2}$ + [220]	$\frac{3}{2}$ + [211]
^{22}Na	0	3+	$\frac{3}{2}$ + [211]	$\frac{3}{2}$ + [211]
^{21}Na	0	4+	$\frac{3}{2}$ + [211]	$\frac{5}{2}$ + [202]
	472	1+	$\frac{3}{2}$ + [211]	$\frac{5}{2}$ + [202]
^{26}Al	0	5+	$\frac{5}{2}$ + [202]	$\frac{5}{2}$ + [202]
	229	0+	$\frac{5}{2}$ + [202]	$\frac{5}{2}$ + [202]
^{28}Al	0	3+	$\frac{5}{2}$ + [202]	$\frac{1}{2}$ + [211]
	31.2	2+	$\frac{5}{2}$ + [202]	$\frac{1}{2}$ + [211]
^{30}P	0	1+	$\frac{1}{2}$ + [211]	$\frac{1}{2}$ + [211]
^{32}P	0	1+	$\frac{1}{2}$ + [211]	$\frac{1}{2}$ + [200]
^{34}Cl	0	0+	$\frac{3}{2}$ + [202]	$\frac{3}{2}$ + [202]
	143	3+	$\frac{3}{2}$ + [202]	$\frac{3}{2}$ + [202]
^{36}Cl	0	2+	$\frac{3}{2}$ + [202]	$\frac{1}{2}$ + [200]
^{152}Eu	0	3−	$\frac{3}{2}$ + [411]	$\frac{3}{2}$ − [521]
	55	0−	$\frac{5}{2}$ − [532]	$\frac{5}{2}$ + [642]
^{154}Eu	0	3−	$\frac{3}{2}$ + [411]	$\frac{3}{2}$ − [521]
^{156}Tb	0	3−	$\frac{3}{2}$ + [411]	$\frac{3}{2}$ − [521]
^{160}Tb	0	3−	$\frac{3}{2}$ + [411]	$\frac{3}{2}$ − [521]
^{160}Ho	0	5+	$\frac{7}{2}$ − [523]	$\frac{3}{2}$ − [521]
^{162}Ho	0	1+	$\frac{7}{2}$ − [523]	$\frac{5}{2}$ − [523]
	90	6−	$\frac{7}{2}$ − [523]	$\frac{5}{2}$ + [642]
^{164}Ho	0	1+	$\frac{7}{2}$ − [523]	$\frac{5}{2}$ − [523]
	139	6−	$\frac{7}{2}$ − [523]	$\frac{5}{2}$ + [642]
^{166}Ho	0	0−	$\frac{7}{2}$ − [523]	$\frac{7}{2}$ + [633]
	12	7−	$\frac{7}{2}$ − [523]	$\frac{7}{2}$ + [633]
	190	3+	$\frac{7}{2}$ − [523]	$\frac{1}{2}$ − [521]
	419	1+	$\frac{7}{2}$ − [523]	$\frac{5}{2}$ − [523]
	499	4+	$\frac{7}{2}$ − [523]	$\frac{1}{2}$ − [521]
^{164}Tm	0	1+	$\frac{7}{2}$ − [523]	$\frac{5}{2}$ − [523]

TABLE 11.2—Continued

Nucleus	Band Head (keV)	$I\pi$	Odd-Particle Configuration: Proton	Odd-Particle Configuration: Neutron
^{166}Tm	0	2+	$\frac{1}{2}+[411]$	$\frac{5}{2}+[642]$
^{168}Tm	0	3+	$\frac{1}{2}+[411]$	$\frac{7}{2}+[633]$
^{170}Tm	0	1+	$\frac{1}{2}+[411]$	$\frac{1}{2}-[521]$
^{172}Tm	0	2−	$\frac{1}{2}+[411]$	$\frac{5}{2}+[512]$
^{172}Lu	0	4−	$\frac{7}{2}+[404]$	$\frac{1}{2}-[521]$
^{174}Lu	0	1−	$\frac{7}{2}+[404]$	$\frac{5}{2}-[512]$
	171	6−	$\frac{7}{2}+[404]$	$\frac{5}{2}-[512]$
^{176}Lu	0	7−	$\frac{7}{2}+[404]$	$\frac{7}{2}-[514]$
	~200	1−	$\frac{7}{2}+[404]$	$\frac{7}{2}-[514]$
^{178}Ta	0	7−	$\frac{7}{2}+[404]$	$\frac{7}{2}-[514]$
	?	1+	$\frac{9}{2}-[514]$	$\frac{7}{2}-[514]$
^{180}Ta	0	9−	$\frac{9}{2}-[514]$	$\frac{9}{2}+[624]$
	212	1+	$\frac{7}{2}+[404]$	$\frac{9}{2}+[624]$
^{182}Ta	0	3−	$\frac{7}{2}+[404]$	$\frac{1}{2}-[510]$
^{182}Re	0	7+	$\frac{5}{2}+[402]$	$\frac{9}{2}+[624]$
	~200	2+	$\frac{5}{2}+[402]$	$\frac{9}{2}+[624]$
^{184}Re	0	3−	$\frac{5}{2}+[402]$	$\frac{1}{2}-[510]$
^{186}Re	0	1−	$\frac{5}{2}+[402]$	$\frac{3}{2}-[512]$
^{188}Re	0	1−	$\frac{5}{2}+[402]$	$\frac{3}{2}-[512]$
	169	4+	$\frac{1}{2}-[505]$	$\frac{3}{2}-[512]$
^{188}Ir	0	2−	$\frac{1}{2}+[411]$	$\frac{3}{2}-[512]$
^{192}Ir	0	4−	$\frac{5}{2}+[402]$	$\frac{3}{2}-[512]$
^{194}Ir	0	1−	$\frac{5}{2}+[402]$	$\frac{3}{2}-[512]$
^{242}Am	0	1−	$\frac{5}{2}-[523]$	$\frac{5}{2}+[622]$
	48.6	5−	$\frac{5}{2}-[523]$	$\frac{5}{2}+[622]$
^{250}Bk	0	2−	$\frac{3}{2}-[521]$	$\frac{1}{2}+[620]$
	85.2	7+	$\frac{7}{2}+[633]$	$\frac{7}{2}+[613]$
	99	5−	$\frac{3}{2}-[521]$	$\frac{7}{2}+[613]$
	382	6+	$\frac{5}{2}+[642]$	$\frac{7}{2}+[613]$

TABLE 11.3 The Ground State Spins for Heavy Nuclei as Deduced from a Deformed Well. In this Region it is Necessary to Know the Deformations and These are Given in Column 2 as Obtained from Measured Intrinsic Quadrupole Moments. As has been Noted in the Text the Predicted Values are Usually Ambiguous But the Experimental Value Always Agrees with One of Them. [taken from Kerman (59)].

Nucleus	Assumed Deformations	$\Omega_{\text{theo}}\pi$	I_{exp}
$^{151}_{63}$Eu	0.16	3/2±, 5/2±	5/2
$^{153}_{63}$Eu	0.30	5/2+, 3/2+	5/2
$^{159}_{65}$Tb	0.31	3/2+, 5/2+	3/2
$^{165}_{67}$Ho	0.30	7/2−, 1/2+	7/2
$^{169}_{69}$Tm	0.28	1/2+, 7/2−	1/2
$^{175}_{71}$Lu	0.28	7/2+, 5/2+	7/2
$^{181}_{73}$Ta	0.23	5/2+, 7/2+	7/2
$^{185}_{75}$Re	0.19	9/2−, (5/2+)	5/2
$^{187}_{75}$Re	0.19	9/2−, (5/2+)	5/2
$^{191}_{77}$Ir	0.14	3/2+, 1/2+, 11/2−	3/2
$^{193}_{77}$Ir	0.12	3/2+, 1/2+, 11/2−	3/2
$^{155}_{64}$Gd	0.31	5/2+, 3/2−	3/2
$^{157}_{64}$Gd	0.31	3/2−, 5/2+	3/2
$^{161}_{66}$Dy	0.31	5/2−	
$^{167}_{68}$Er	0.29	1/2−, 7/2+	7/2
$^{171}_{70}$Yb	0.29	7/2+, 1/2−	1/2
$^{173}_{70}$Yb	0.29	5/2−	5/2
$^{177}_{72}$Hf	0.26	7/2−	
$^{179}_{72}$Hf	0.27	9/2+	
$^{183}_{74}$W	0.21	1/2−, 7/2−, 3/2−	1/2
$^{187}_{76}$Os	0.18	1/2−, 3/2−, 9/2+	
$^{189}_{76}$Os	0.15	1/2−, 3/2−, 11/2+	3/2

and in ^{167}Yb (29 KeV). Until it becomes the ground state in ^{159}Gd, an excitation to the 5/2 + [642] state probably takes place via the promotion of a particle from a lower level into this level. After ^{159}Gd the $5/2^+$ [642] level is filled in the nuclear ground state. It can appear again as an excited state if a *hole* is created in this level and the particle promoted to pair off with the odd particle in a higher orbit. It is obvious that since each level can accommoderate only two particles of the same type, a hole is characterized exactly by the same quantum numbers as a particle.

Thus we obtain the systematic appearance of levels first as excited states with their excitation decreasing as A increases, then as a ground state, and then again as excited states with excitation increasing with A. An independent-particle picture explains this trend very naturally, and the fact that it is observed so nicely, as shown in Table 11.4, strongly supports our independent-particle assumption for the intrinsic wave functions χ_Ω.

The fact that the Nilsson model gives us usually two or three alternatives for the lowest state of a given nucleus points to its limit of validity. We saw in the previous paragraphs that it takes very little energy to go from one intrinsic wave function to another—the heads of different bands sometimes lie only 50 KeV to 100 KeV apart. Obviously we cannot expect a simple approximation for the internal wave function like Nilsson's Hamiltonian (10.16) to be able to reproduce such fine details of the intrinsic structure. It does, however, go as far as indicating the reason for the existence of so many close-lying internal excitations in some regions of the periodic table.

The availability of intrinsic wave functions makes it possible to try to understand also some of the other parameters of the collective nuclear rotations. The decoupling parameter a is given by (4.23) with the coefficients a_J determined by (4.5). The Nilsson calculation provides an expansion of χ_Ω in terms of l and Λ. It is very easy to transform it into an expansion in terms of J. In fact we notice that

$$|Nl\Lambda\,\Sigma> = \sum_j (l\Lambda 1/2\;\Sigma|j\Omega)|Nlj\Omega> \tag{11.2}$$

where $(l\Lambda 1/2\;\Sigma|j\Omega)$ is a Clebsch-Gordon coefficient and

$$\Omega = \Lambda + \Sigma \tag{11.3}$$

It then follows from (10.19) that

$$\chi_{N\Omega} = \sum_{lj}\left[\sum_\Lambda a_{l\Lambda}{}^{\Omega}\,(l\Lambda 1/2\;\Sigma|j\Omega)\right]|Nlj\Omega> \tag{11.4}$$

We note that the sum over l can actually be omitted. Since the wave function χ_Ω is assumed to have a definite parity its expansion (10.19) will include only even l's or only odd l's. Since, in addition, $j = l \pm 1/2$, we see that j and the parity of χ_Ω fix which one of the two possible values of l will really contribute to (11.4). By comparing (11.4) with (4.5) and (4.23), we obtain for the decoupling parameter a the expression

$$a = \sum_{j\Lambda\Lambda'\Sigma\Sigma'} (-1)^{j-1/2}\,(j+1/2)\,a_{l\Lambda}{}^{\Omega}\,a_{l\Lambda'}{}^{\Omega*}\,(l\Lambda 1/2\Sigma|j\Omega)\,(l\Lambda' 1/2\Sigma'|j\Omega) \tag{11.5}$$

A comparison of theoretical values of a with results obtained from the analysis of $K = 1/2$ rotational bands is given by Mottelson and Nilsson (59).

Other parameters can be calculated in a similar way. We shall not enter here into their evaluation since it turns out, as we saw for the moment of inertia, that the effects of the residual interaction cannot be neglected if a more detailed comparison with experiment is desired. We only give here an evaluation of the equilibrium deformation computed, as outlined before, with the Nilsson eigenvalues, and its comparison with the observed deformation as deduced from nuclear quadrupole moments. (See Fig. 11.4) The agreement is really remarkable in view of the crude assumptions that go into the Hamiltonian (10.13).

TABLE 11.4 Summary of Intrinsic Excitations (in keV) Listed According to the Nilsson Orbital Description for Odd-A Nuclei in the Region of Odd-Neutron Numbers 89 to 113. Parentheses are Used to Designate Decreasing Reliability of the Level Assignments; Double Parentheses Designate Least Certainty [from Harmatz et al. (62)]

N	Nucleus	3/2− [532]	3/2+ [402]	1/2+ [660]	3/2+ [651]	3/2− [521]	5/2+ [642]	5/2− [523]	7/2+ [633]	1/2− [521]	5/2− [512]	7/2− [514]	9/2+ [624]	1/2− [510]	3/2− [512]	7/2− [503]
89	^{151}Sm	(344)	((104))	(4)	(0)	((167))	((209))									
89	^{153}Gd	(212)	(109)		(0)	((303))	(129)									
91	^{153}Sm					0										
91	^{155}Gd	((326))	((367))	(247)	(86)	0	(105)	((286))								
93	^{157}Gd					0	(64)									
95	^{159}Gd					0										
95	^{161}Dy					74	0	25								
95	^{163}Er				((415))	(104)	((22))	(0)		(345)						
97	^{163}Dy							0								
97	^{165}Er		((1427))		((853))	(242)	(47)	0	((117))	(297)	(608)					
97	^{167}Yb					((239))	(29)	(0)								
99	^{165}Dy								(0)	(108)						
99	^{167}Er							((585))	0	207	((347))					
99	^{169}Yb							(570)	0	(24)	(191)	((962))				((1465))
101	^{171}Yb								(95)	0	122	(835)	((936))			
101	^{173}Hf									0	((107))					
103	^{173}Yb								351		0	636				
103	^{175}Hf								(207)	125	0	(348)				((1045))
105	^{175}Yb											(0)		((455))		

105	^{175}Yb				(0)		((455))		
105	^{177}Hf	(746)		(508)	0	321			(1058)
105	^{179}W				(0)		((221))		
107	^{177}Yb				((104))	(0)	((326))		
107	^{179}Hf				(215)	0	(375)		
107	^{181}W	((953))	((746))	365	((408))	0	((385))	((560))	((807))
107	^{183}Os					0	(170)		
109	^{183}W					((309))	0	208	453
109	^{185}Os						0	127	((356))
111	^{185}W							(0)	
111	^{187}Os						0	(74)	
113	^{189}Os						36	0	((217))

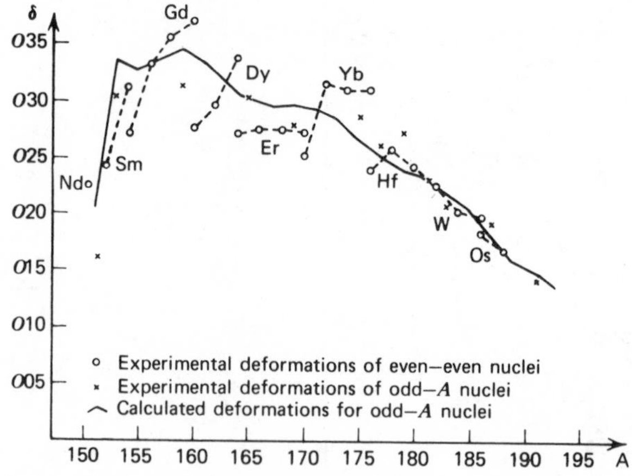

FIG. 11.4. The calculated ground state equilibrium deformations are compared with those deduced from the observed intrinsic quadrupole moments [taken from Mottelson and Nilsson (55)].

In summarizing this section we can say that the rotational-model wave function (4.9) has proved to be useful not only in determining general relations among various properties of rotational band, but also in requiring for the internal wave function χ_Ω a rather simple structure. The treatment of χ_Ω as a wave function for independent particles moving in a deformed field helps us understand the sequence of the observed intrinsic excitations as single-particle excitations in a deformed field. At the same time it also gives us at least the correct order of magnitude for the observed nuclear deformation and other parameters pertinent to deformed nuclei.

It should be realized that in deriving these properties we made no direct use of the properties of the underlying nucleon-nucleon force, Indirectly it came in through the three parameters of (10.13); The average harmonic-oscillator frequency ω_0 (or more precisely the constant ω_{00} in Eq. 10.15) that is determined by the observed nuclear density; and the constant C and D that in (10.13) are made to fit the zero-deformation order and spacings of single-particle levels. It is obvious that the actual values adopted for ω_{00}, C and D represent some average properties of the underlying nucleon forces, but it is remarkable that so much could be achieved without further specification of the nuclear force. Here, as in the case of the shell model, once the density and shape of the nucleus are taken from experiment, many of the other properties of nuclei are determined within rather narrow margins. To obtain even narrower limits requires a more proper handling of the nuclear forces. We have neglected thus far both the coupling of the collective rotation $\mathbf{R}$ to the intrinsic structure (H_{coup} in Eq. 2.14), and the residual interaction among the articles moving in the deformed potential. Both terms can be included at least in some approximation, leading to the understanding of further details in the experimental data.

12. PROJECTED WAVE FUNCTIONS

The success of the unified model in describing the behavior of deformed nuclei suggests also a way of constructing wave functions for nuclei in other regions of the periodic table. Basically this prescription is motivated by the following considerations.

The simplest wave function to construct and to operate with is a Slater determinant. This wave function is composed of single-particle wave functions generated by an arbitrary potential. One can exploit this arbitrariness in the single-particle potential to obtain the best possible Slater determination for the problem at hand. Alternatively one can start with a fixed potential U_0, derive from it the single-particle wave functions $\phi_\alpha^{(0)}(i)$, construct Slater determinants $\Psi^{(0)}(\alpha_1, \ldots, \alpha_A)$ and proceed to improve the approximation to the real wave function by taking linear combinations of $\Psi^{(0)}(\alpha_1, \ldots, \alpha_A)$.

Suppose, for the sake of argument, it is found that given the single-particle potential U_0, a good description of a certain nuclear state is provided by the following simple combinations of two Slater determinants:

$$\Psi = \Psi^{(0)}(\alpha_1, \alpha_2, \ldots, \alpha_A) + c\,\Psi^{(0)}(\beta_1, \alpha_2, \ldots, \alpha_A) \tag{12.1}$$

Equation 12 can be written as a single Slater determinant in which the first state has been charged from $\phi_{\alpha 1}$ into

$$\phi_{\alpha 1}{}^{(0)} \rightarrow \phi_{\alpha 1}{}^{(0)} + c\phi_{\beta 1}{}^{(0)} \equiv \phi_{\alpha 1}{}^{(1)} \tag{12.2}$$

However this same change can be induced also by an appropriate change in U_0. We saw in Section IV.18 that it is possible to change a single-particle potential so that only *one* of its eigenfunctions changes. If we change U_0 to U_1 so that

$$\phi_{\alpha 1}{}^{(1)} = \phi_{\alpha 1}{}^{(0)} + c\,\phi_{\beta 1}{}^{(0)}, \qquad \phi_{\alpha_i}{}^{(1)} = \phi_{\alpha_i}{}^{(0)} \qquad i \neq 1 \tag{12.3}$$

the new Slater determinant, formed from the functions $\phi_{\alpha_i}{}^{(1)}$, will give us the desired good description of the nuclear state considered.

The success of the description of many nuclei by means of a deformed potential can be taken as an indication that by distorting a spherical potential in this manner we *automatically* obtain the right combination of spherical eigenfunctions that makes the corresponding Slater determinant a better approximation to the real nuclear wave function. From this point of view the *deformed* potential is a definite prescription for a convenient mixing of various configurations of the *spherical* potential. We need hardly remind ourselves that a single-particle wave function derived from a spherical potential does not itself possess spherical symmetry and, depending on its m-value, may represent a mass distribution concentrated in one direction or another. The deformed potential is thus equivalent to a prescription of mixing together spherical wave functions that help each other build up the required deformation.

Wave functions in a deformed potential are not eigenfunctions of the total angular momentum. Still we can use the prescription dictated by the deformed potential to generate wave functions that are eigenstates of $\mathbf{J}^2$ and J_z. To this end let us take a wave function $\chi_K'(\mathbf{r}_i', \ldots, \mathbf{r}_A')$ that is defined with respect to the deformed potential whose axes Σ' are given by the Euler angles θ_i and assume χ_K' has a z'-component of the angular momentum equal to K. To find out how χ_K would look in the laboratory frame Σ all we have to do is to replace in it each $\mathbf{r}_k'$ by its appropriate expression in terms of $\mathbf{r}_k$ and θ_i; we shall be thus led to a function

$$\chi_K(\mathbf{r}_1, \ldots, \mathbf{r}_A; \theta_i) = \chi_K'(\mathbf{r}_1'(\mathbf{r}_1, \theta_i), \ldots, \mathbf{r}_A'(\mathbf{r}_A, \theta_i)) \tag{12.4}$$

However, there is also another way to obtain $\chi(\mathbf{r}_1, \ldots, \mathbf{r}_A; \theta_i)$: we expand first χ_K in a series of eigenfunctions of $\mathbf{J}^2$ and J_z:

$$\chi_K' = \sum_J a_J \, \phi'_{\Sigma'}(JK) \tag{12.5}$$

The wave functions $\phi(JK)$, being eigenfunctions of $\mathbf{J}^2$ and J_z, transform under rotations with Wigner's D-matrix (Appendix A, A.2.1). Hence, in terms of the wave functions $\phi_\Sigma(JM)$ in the laboratory frame, they are given by

$$\phi'_{\Sigma'}(JK) = \sum_M \phi_\Sigma(JM) \, D_{MK}^{(J)}(\theta_i) \tag{12.6}$$

Introducing (12.6) into (12.5) we obtain χ_K' expressed in terms of the laboratory coordinates contained in $\phi_\Sigma(JM)$ and the direction θ_i of the frame Σ' of the deformed potential contained in the D's:

$$\chi_K'(\mathbf{r}_1', \ldots, \mathbf{r}_A') = \chi_K(\mathbf{r}_1, \ldots \mathbf{r}_A; \theta_i) = \sum_J a_J \phi_\Sigma(JM) \, D_{MK}^{(J)}(\theta_i) \tag{12.7}$$

Using the orthogonality relations of Wigner's D-matrices (appendix A, A.2.11) we can now project out of the known function $\chi(\mathbf{r}_1, \ldots, \mathbf{r}_A; \theta_i)$, as given by (12.4), a function of a well-defined angular momentum, that is,

$$\phi_\Sigma(JM) \approx N \int D_{MK}^{(J)*}(\theta_i) \, \chi_K(\mathbf{r}_1, \ldots, \mathbf{r}_A; \theta_i) \, dR \tag{12.8}$$

where N is a normalization factor, and $\int \ldots dR$ stands for the integration over all orientations of the deformed potential [Peierls and Yoccoz (57)]. Note that as a consequence of the projection ϕ_Σ depends only upon intrinsic coordinates. There are no extra, that is redundant, variables.

As we present it here (12.8) seems to be just a convenient way to extract an eigenfunction of $\mathbf{J}^2$ and J_z out of the deformed intrinsic state $\chi_K(\mathbf{r}_1', \ldots, \mathbf{r}_A')$.

The following remarks will explain why the funtions $\phi_\Sigma(JM)$ obtained this way may be of physical interest.

Suppose it is found empirically that an A-particle system has a rotational spectrum with an angular momentum J that extends from $J = K$ up to $J = K + n - 1$. If $H(A)$ is the exact Hamiltonian of the A-particle system then the Hamiltonian

$$H' = H(A) - \frac{1}{2\mathcal{J}} J^2 \tag{12.9}$$

has a degenerate ground state with angular momenta $J = K, \ldots, J = K + n - 1$. Generally, the eigenfunctions of an exact Hamiltonian like $H(A)$ can be chosen to be eigenfunctions of $\mathbf{J}^2$, since $\mathbf{J}^2$ commutes with $H(A)$. $\mathbf{J}^2$ commutes, of course, with H' as defined in (12.9); furthermore the eigenfunction $\phi\ (JM)$ of $H(A)$ are also eigenfunctions of H'. But because H' has an n-fold degenerate ground state, we can take for the n-ground-state wavefunctions any n-independent combinations of $\phi(J = K, M)$, $\phi(J = K + 1, M)$, $\ldots\ \phi(J = K + n - 1, M)$. These combinations *need not* be eigenstates of $\mathbf{J}^2$ and may represent a state deformed in a special direction. If the functions $\phi(JM)$ of this rotational band really describe the *same* intrinsic structure, rotating at different speeds, then a combination of the type (12.5) can describe fairly well the deformed intrinsic structure [the description will be better the larger the number of states $\phi(JK)$ that are included on the right-hand side of (12.5); because of the uncertainty relations a wave function can describe an orientation with a precision $\Delta\theta$ only if it includes a range of ΔJ of angular momenta satisfying $\Delta J \geq (2\pi/\Delta\theta)$]. If we now reverse the argument and start from a wave function $\chi_K\ (\mathbf{r}', \ldots, \mathbf{r}_A')$ in a deformed potential we can expect the wave function $\phi(JM)$ projected out of it via (12.8) to be a good description of the appropriate nuclear state with angular momentum quantum number (JM).

This method of generating wave functions of a definite angular momentum from Slater determinants in deformed potentials was first suggested by Redlich (58). It has since then found many uses, particularly in conjunction with self-consistent deformed Hartree-Fock potentials. Wave functions projected from Slater determinants of single-particle wave functions in a deformed Hartree-Fock potential are briefly referred to as projected-Hartree-Fock wave functions. They incorporate many of the advantages of both deformed Hartree-Fock wave functions and wave functions of a definite total angular momentum. Table 12.1, taken from Bassichis et al. (65), shows energies of various nuclei calculated with projected Hartree-Fock wave functions using a Rosenfeld mixture for the two-body interaction (see Appendix B, p. 375). These energies represent expectation values of the Hamiltonian $\sum T_i + \sum_{i<j} v_{ij}$ taken with the appropriate projected wave function for each level. Table 12.2 shows a similar calculation carried out by Gunye (68), where quadrupole and dipole moments as well as transition probabilities, were

TABLE 12.1 Calculated and Experimental Energies of Levels Arising from Various Configurations in A = 18, 19, and 20 Nuclei. [taken from Bassichis et al. (65)]

Nucleus	J	Experimental	Calculated	Configuration
^{20}Ne	0^+	−33.02	−33.7	$4p$
	2^+	−31.39	−32.5	$4p$
	4^+	−28.77	−29.9	$4p$
	6^+	−25.42	−26.3	$4p$
	8^+		−22.5	$4p$
^{19}F	$\frac{1}{2}^+$	−20.18	−20.5	$3p$
	$\frac{3}{2}^+$	−18.63	−18.5	$3p$
	$\frac{5}{2}^+$	−19.99	−19.9	$3p$
	$\frac{7}{2}^+$		−15.2	$3p$
	$\frac{9}{2}^+$	−17.39	−18.0	$3p$
	$\frac{11}{2}^+$		−11.9	$3p$
	$\frac{13}{2}^+$		−15.8	$3p$
^{19}F	$\frac{1}{2}^-$	−20.07	−19.1	$4p$–$1h$
	$\frac{3}{2}^-$	−18.73	−17.5	$4p$–$1h$
	$\frac{5}{2}^-$	−18.84	−18.0	$4p$–$1h$
	$\frac{7}{2}^-$		−14.6	$4p$–$1h$
	$\frac{9}{2}^-$		−15.5	$4p$–$1h$
	$\frac{11}{2}^-$		−10.7	$4p$–$1h$
^{18}O	0^+	−12.19	−11.5	$2p$
	0^+	− 8.56	− 8.7	$2p$
	0^+	− 6.86	− 6.0	$4p$–$2p$
	0^+		1.5	$2p$
	2^+	−10.21	− 9.8	$2p$
	2^+	− 8.27	− 8.2	$2p$
	2^+	− 6.94	− 4.6	$4p$–$2h$
	2^+		− 2.7	$2p$
	3^+	− 6.82	− 6.6	$2p$
	4^+	− 8.64	− 8.7	$2p$
	4^+		− 1.7	$4p$–$2h$
^{18}F	1^+	− 9.74	− 9.9	$2p$
	1^+	− 8.04	− 6.2	$2p$
	1^+		− 5.4	$4p$–$2h$
	1^+		− 4.4	$2p$
	2^+		− 6.6	$2p$
	2^+		− 4.5	$4p$–$2h$
	3^+	− 8.80	− 9.3	$2p$
	3^+		− 5.5	$2p$
	3^+		− 3.8	$4p$–$2h$
	3^+		− 0.8	$2p$
	4^+		− 3.8	$2p$
	4^+		− 1.8	$4p$–$2h$
	5^+	− 8.61	− 8.9	$2p$
	5^+		− 1.1	$4p$–$2h$

TABLE 12.2 The Type of Electromagnetic Transition between the Initial State (Spin J_i) and Final State (Spin J_f) is Shown in Column 5; the γ-ray Energy (in MeV) is Shown in Column 4. The Number ($-n$) in the Last Three Columns Indicates the Multiplying Factor 10^{-n}. [taken from Gunye (68)]

Table A

Quadrupole moment (in units of e $\times$ 10^{-24} cm^2) and magnetic moment (in units of nuclear magnetons) for some nuclear states

Nucleus	J	Q(proj.)	Q (S. M.)	Q (expt.)	μ (proj.)	μ (S. M.)	μ (expt.)
^{19}F	1/2				2.57	2.67	2.63
	5/2	−0.11	−0.11	±0.11	3.18	3.28	3.69
	3/2				−1.88	—	—
^{21}Ne	3/2	0.10	—	0.09	−0.36	—	−0.66
	5/2	−0.04	—	—			
^{22}Na	3	0.20			1.67	—	1.75
^{23}Na	3/2	0.11	—	0.10	1.88	—	2.22
	5/2	−0.03	—	—	2.12	—	—

Table B

Mean lives (in seconds) of the nuclear states

Nucleus	J_i	J_f	E	Transition	τ (proj.)	τ (S. M.)	τ (expt.)
^{18}O	2	0	1.98	E2	5.10(−12)	5.09(−12)	(3.7 ± 0.7)(−12)
	4	2	1.57	E2	2.12(−11)	2.14(−11)	>3 (−12>
^{18}F	3	1	0.94	E2	6.80(−11)	6.67(−11)	(6.8 ± 0.7)(−11)
^{19}F	5/2	1/2	0.20	E2	1.14(−7)	1.25(−7)	1.25(−7)
	3/2	1/2	1.56	E2	4.33(−12)	4.60(−12)	(3 ±1)(−12)
^{20}Ne	2	0	1.63	E2	1.39(−12)	1.45(−12)	1.23(−12)
	4	2	2.62	E2	1.06(−13)	1.13(−13)	1.34(−14)
^{21}Ne	5/2	3/2	0.35	M1	1.66(−12)		
	7/2	5/2	1.40	M1	2.04(−14)		
^{22}Ne	2	0	1.27	E2	4.65(−12)		(4 ± 2)(−12)
	4	2	2.06	E2	3.27(−12)		4.0 (−13)
^{22}Na	5	3	1.53	E2	5.66(−12)		(3.2 ± 1.0)(−12)
	4	3	0.89	E2	1.66(−11)		(1.8 ± 0.7)(−11)
^{23}Na	5/2	3/2	0.44	M1	5.50(−13)		1.50(−12)
	7/2	5/2	1.63	M1	9.10(−15)		

calculated with the projected Hartree-Fock wave-functions and the results compared with rather elaborate shell-model calculations of Arima et al (68). It is seen that the projected wave functions describe the actual states rather well; their usefulness becomes apparent if we take into account the relative ease with which these wave functions are generated. [For a recent discussion of the projection procedure, see Villars and Rogerson (71)].

These examples of the use of the projected Hartree-Fock wave functions are taken from calculations on the properties of light nuclei. It is thus suggested that it would be useful to consider these as well as the rare earth and actinide nuclei as deformed, and thus leads to an approximation to nuclear wave functions that might be referred to as projected deformed shell-model wave functions. This method involves applying the simple Nilsson shell-model orbitals to develop a nuclear wave function in a deformed basis and then projecting according to (12.8) to obtain the "true" nuclear wave function whose properties are to be compared with experiment. Such a procedure was applied by Kurath and Picman (59) to the p-shell nuclei. These authors found that the wave function generated in this way had a strong overlap with wave functions formed using a spherical shell-model potential and taking into account a central two-body residual interaction. These results are shown in Table 12.3 where the amplitude of each type of wave function is expanded in terms of spherical shell-model wave functions $\phi_{IM}^{(s)}$ with spin quantum numbers I and M while s denotes the particular configuration involved. If the wave function is Ψ_{IM}, then

$$\Psi_{IM} = \sum_{s} d_s \, \phi_{IM}^{(s)}$$

TABLE 12.3 Comparison of Projected Deformed Shell-Model Wave Functions with Spherical Shell-Model Wave Functions. The Latter Includes Configuration Mixing Induced by the Residual Interactions. [from Kurath and Pitman (59)].

	η \ s	(1)	(2)	(3)	(4)	(5)	(6)	(7)	(8)
Generated	$-\infty$	0.157	0.416	−0.644	−0.172	0.100	−0.214	0.360	0.416
	−8	0.081	0.212	−0.485	−0.130	0.111	−0.237	0.400	0.682
	−4	0.032	0.084	−0.283	−0.076	0.095	−0.204	0.343	0.859
	γ \ s	(1)	(2)	(3)	(4)	(5)	(7)	(7)	(8)
Computed	0.	0.165	0.410	−0.642	−0.181	0.105	−0.224	0.363	0.410
	1.5	0.133	0.335	−0.603	−0.166	0.112	−0.247	0.385	0.507
	3.0	0.102	0.264	−0.543	−0.150	0.117	−0.259	0.388	0.609
	4.5	0.075	0.201	−0.471	−0.133	0.118	−0.257	0.372	0.704
	6.0	0.055	0.152	−0.402	−0.188	0.115	−0.244	0.342	0.780
	7.5	0.041	0.115	−0.340	−0.105	0.109	−0.226	0.306	0.838

It is the coefficient d_s that is given in Table 12.3 for various values of η and and γ. The parameter η is proportional to the ratio of the deformation parameter to the spin-orbit coupling in the Nilsson Hamiltonian (Eq. 10.20). The parameter γ gives the ratio of the spin-orbit strength in the shell-model potential to the two-particle interaction strength as measured by the exchange integral over this interaction [Inglis (53)]. The comparison in the table is given for the lowest $I = 2$, $T = 0$ state of ^{12}C. It is clear that a very close one-to-one relation exists between the two types of wave function when roughly $\eta\gamma \approx -30$. From this we learn that for these nuclei the effect of a residual interaction in the spherical shell model, which of course mixes in many configurations to the simple first-order configuration, can be simulated by assuming these particles move in a nonspherical potential with an appropriately chosen deformation but with no residual interaction. In other words the effect of the residual interaction is to deform the nucleus.

A corollary of this discussion is that the energy spectra of these nuclei should have a rotational character [Clegg (61)]. However, band mixing is substantial and usually must be included before a good match with the data is achieved [Clegg (62)].

Considerable evidence has also accumulated for rotational structure for the *s-d* shell nuclei. A review of this phenomenon at an early stage of its understanding has been given by Gove (60). [See also Litherland, Paul, Bartholomew, and Gove (56), Bromley, Gove, and Litherland (57), and Litherland, McManus, Paul, Bromley, and Gove (58).]

An example of a rotational spectrum is shown in Fig. 12.1. We see that ^{24}Mg has two bands, $K = 0$ and $K = 2$. The values of A and B in the expression

$$E = AI(I + 1) + B[I(I + 1)]^2 \tag{5.1}$$

that fit this spectrum and the spectra of a number of nuclei in the *s–d* shell are given in Table 12.4. These particular nuclei were chosen because the deviation from the simple formula (5.1) is small either because the bands that might mix with the ground state via the rotation–vibration coupling or the particle–rotation coupling are distant or ΔK is large. However in many cases in these light nuclei, for example, ^{19}F, ^{23}Na, ^{25}Na, the bands do overlap considerably and lead to considerable mixing, an effect that has been discussed for these nuclei by several authors [see, for example, Paul (57), Rakavy (57), Kerman (59), and Brink and Kerman (59)]. As examples where these effects are large, the ground-state bands of those nuclei with N or $Z = 11$ are shown in Fig. 12.2. In Fig. 12.3 we see the comparatively (compare with the rare earth nuclei) irregular growth of the energy with $I(I + 1)$ for ^{23}Na. For these nuclei the spacing of the levels does not appear to be rotational. It is only if perturbations involving other bands are taken into account that the rotational model can be applied to the data. This analysis is borne out by comparison with β-decay and γ-decay transition probabilities. To obtain a fit, the effect of the particle-rotation coupling must be calculated

TABLE 12.4 A and B Coefficients for Light Nuclei. [from Gove (60)].

Nucleus	A (keV)	B (keV)	$\mathcal{J}/\mathcal{J}_{\text{rig}}$
	Ground State Band		
^{24}Mg	237	−1.56	0.54
^{23}Ne	292	−1.29	0.44
^{25}Mg	279	−1.99	0.46
^{25}Al	271	−1.67	0.47
	First Excited Band		
^{23}Ne	271	−0.54	0.47
^{25}Mg	177	−1.32	0.72
^{25}Al	182	−1.62	0.70

to at least third order (see Section 9 for the second-order calculation). Another explanation suggests that the rotational Hamiltonian $(\hbar^2/2I)\mathbf{R}^2$ is only the first term in a power-series expansion in even powers of **R** [Bohr and Mottelson (69)]. The higher-order $\mathbf{R}^2$ terms with $\mathbf{R} = \mathbf{I} - \mathbf{J}$ give results similar to those obtained with the third-order particle-rotation interactions. Explanation based on the shell model [Wildenthal, McGrory, Halbert, and Glaudemans (68)], the SU(3) model of Elliott (58) [see review by Harvey (68)] which is

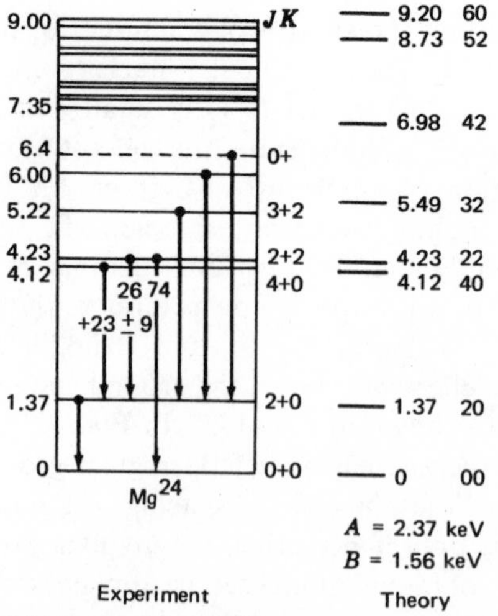

FIG. 12.1. Spectrum of ^{24}Mg [from Gove (60)].

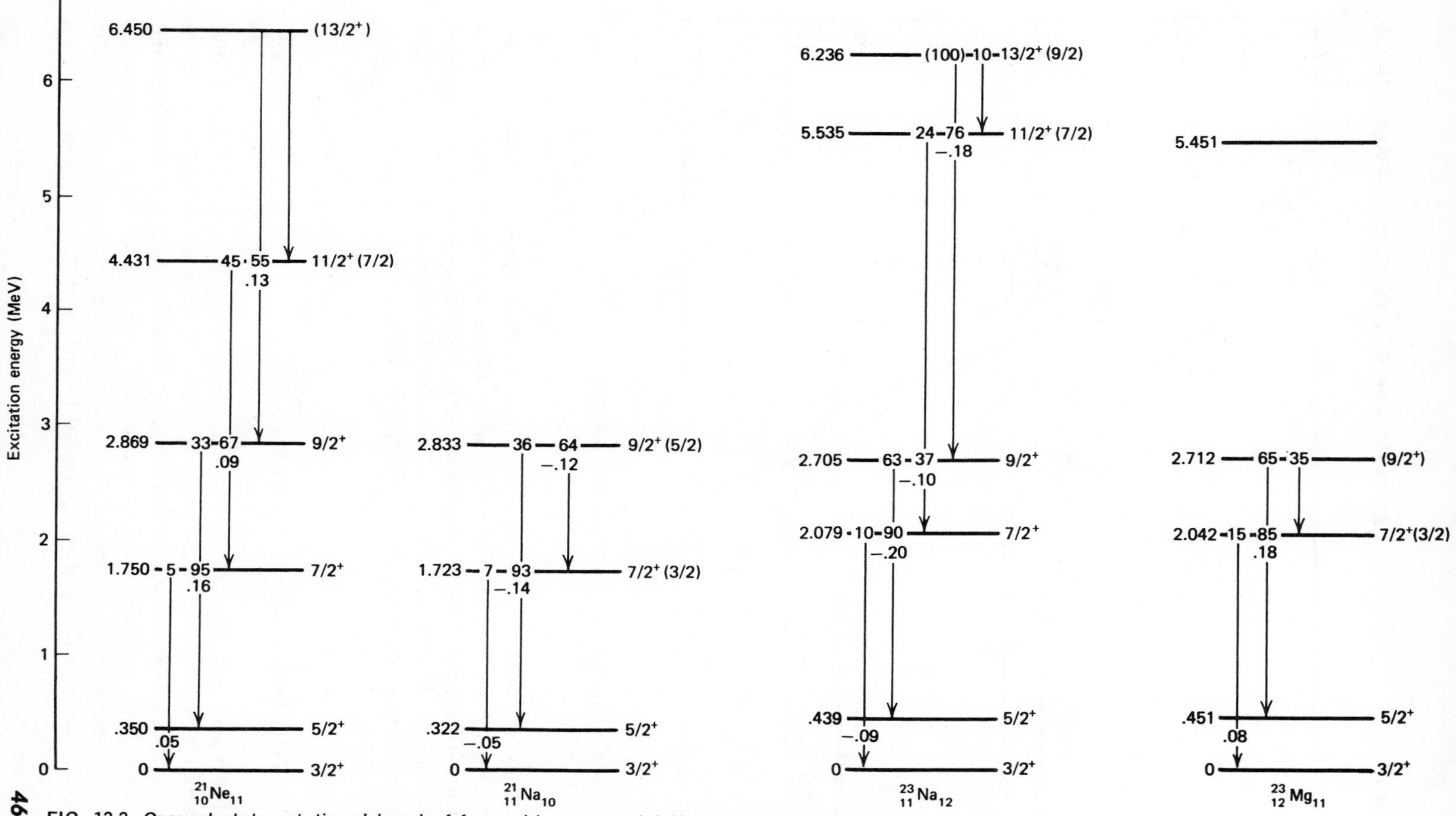

FIG. 12.2. Ground state rotational band of four odd–even nuclei [from Lindgren (69)].

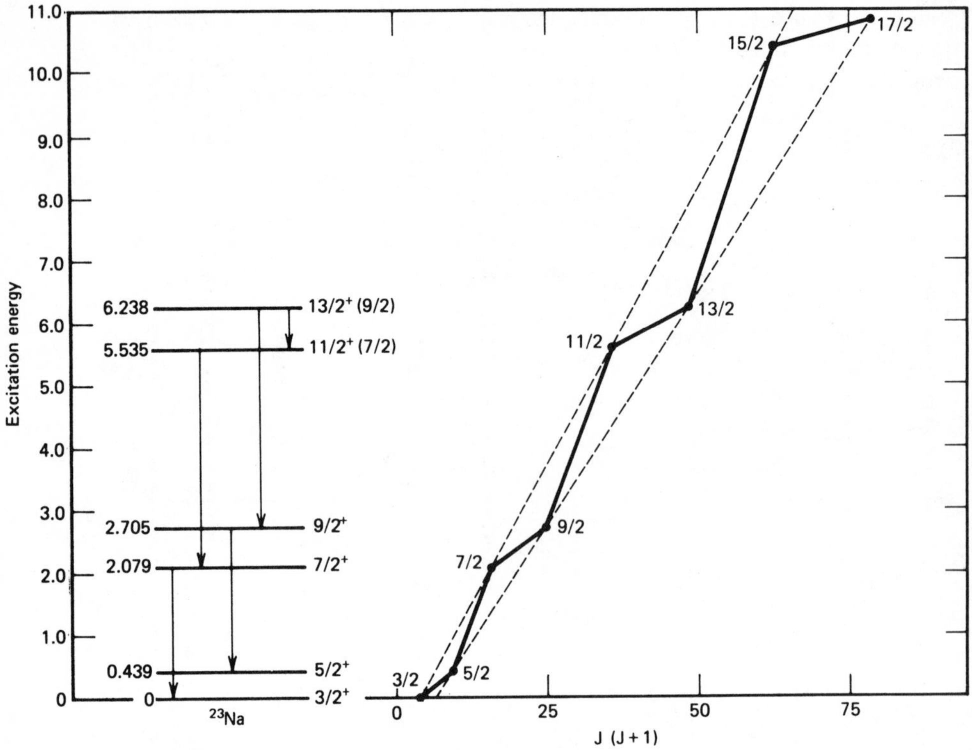

FIG. 12.3. Spectrum of ^{23}Na [from Lindgren (69)].

briefly described in Chapter VII. The nonaxially symmetric model is discussed by Chi and Davidson (63) while Malik and Scholz (66,67) in their Coriolis coupling model diagonalize the total Hamiltonian including the rotational, Nilsson-deformed shell-model Hamiltonian and the particle-rotation coupling using the deformed *s-d* shell-wave functions as basis wave functions.

13. THE BOHR HAMILTONIAN

The success of the description of deformed nuclei in terms of collective rotations of the nucleus as a whole raises the question of the existence of other possible modes of collective motion. Since the average nuclear potential in which the particles move independently is, in the last analysis, generated by the particles themselves, we should expect the parameters determining the shape, size, and depth of this potential to undergo time-dependent variations as well. To be sure, these variations are not independent of the particle motion, and they represent just another way of looking at some average functions of the particles' coordinates. But our experience with deformed nuclei suggest that treating the particle coordinates and the collective coordinates as independent variables may actually represent a very good approximation. Behind this approximation there is, of course, the expectation that the parameters of the average potential, such as its shape or size, undergo *slow* variations around an equilibrium value. The frequencies characteristic of the motion of the particles in the potential are supposed to be much higher than those characterizing the variations in the average potential. One can then use the Born-Oppenheimer approximation and treat the motion of the particles for *fixed* values of the parameters; the resulting total energy of the particles, which is a function of these parameters, can then be used as the potential energy for their slower oscillations.

A procedure for introducing collective variables was suggested by Bohr (52). Let us consider the case of small oscillations about an equilibrium shape. As a first example suppose that shape is spherical. Then the corresponding energy for the equilibrium configuration will involve the nuclear radius R_0 as a parameter. If we are dealing with a shell-model description, the single-particle shell-model potential would involve a radius parameter explicitly, as is true for the Wood-Saxon potential that is commonly employed:

$$V(r) = -V_0 \frac{1}{1 + e^{(r-R0)/a}}$$

This Wood-Saxon R_0 would then be a parameter in the expression for the energy of the nucleus. The small oscillations that occur because of a change in shape can be described by replacing R_0 by $R_0(1 + \delta)$, and allowing δ to oscillate. More specifically for quadrupole shaped oscillations*

$$R = R_0 [1 + \Sigma \alpha_\mu Y_{2\mu} (\theta,\phi)] \tag{13.1}$$

One condition is usually imposed: that the nuclear volume is unchanged by this deformation. This gives an overall condition on the α_μ's that we shall not write out explicitly. Introducing this expression for R in the place of R_0 in the single-particle potential would mean that the potential is no longer spherical. The energy of the system will now involve not only R_0 but also the

*More generally $R = R_0 [1 + \Sigma \alpha_{\lambda\mu} Y_{\lambda\mu}]$. In this expression the $\lambda = 1$ value is omitted since it corresponds to a motion of the nucleus as whole. However, if these deformations are given isospin dependence: that is, if the neutron oscillations differ from that of the protons, it should be retained. This is discussed in Section 14. The $\lambda = 0$, the monopole, is expected to occur at relatively high excitation energies. The $\lambda = 2$ and $\lambda = 3$ oscillations are the ones of greatest experimental interest at present.

parameters α_μ and their time derivatives. For small oscillations about R_0, one can expect the energy to consist additively of a kinetic-energy term $(1/2)\ B\sum_\mu|\dot{\alpha}_\mu|^2$ and a potential energy term $(1/2)\ C\sum_\mu|\alpha_\mu|^2$. We are thus led to write for the collective Hamiltonian the expression

$$H_{\text{coll}} = \frac{1}{2}B\sum|\dot{\alpha}_\mu|^2 + \frac{1}{2}C\sum|\alpha_\mu|^2 \tag{13.2}$$

The more complex analysis required when the equilibrium shape is a deformed one will be given later. Before going to discuss some of the consequences of (13.1) and (13.2) it should be emphasized that the procedure described here can be employed for any macroscopic variable. The important and difficult problem is to know for which variables this is a useful method. The discovery of the significant collective variables is of course a central problem of nuclear physics.

Returning to (13.1) and (13.2) we note that in order to make the former invariant against rotation, the new dynamical variables α_μ must behave, under rotations, like $Y^*_{2\mu}\ (\theta,\ \phi)$. Then $\Sigma\alpha_\mu Y_{2\mu}\ (\theta,\ \phi)$ behaves under rotations like $\Sigma\ Y^*_{2\mu}\ (\theta',\phi')\ Y_{2\mu}\ (\theta,\ \phi)$, and we know from the addition theorem for spherical harmonics (Appendix A, A.2.34) that the latter expression is a scalar. With these transformation properties for α_μ, (13.2) will also be a scalar and the total Hamiltonian will have eigenstates with well-defined $\mathbf{J}^2$ and J_z. Since R is invariant under an inversion, it follows that under that operator α_μ will not be changed by an inversion [$\alpha_{\lambda\mu}$ under inversion becomes $(-)^\lambda\alpha_{\lambda\mu}$]. Another condition on the α_μ comes from the reality of the nuclear radius. Since $Y^*_{\lambda\mu} = (-)^\mu Y_{\lambda,-\mu}$ it follows that

$$\alpha_\mu{}^* = (-)^\mu\ \alpha_{-\mu} \tag{13.3}$$

It follows that the description of the quadrupole oscillations about a spherical shape requires five independent real parameters.

Secondly, we notice that the Hamiltonian (13.2) has a structure of that of a harmonic oscillator. The operators α_μ, upon quantization, will therefore become creation and annihilation operators of the quantum of oscillation—the phonon— associated with the α-degrees of freedom. The term $\alpha_\mu Y_{2\mu}\ (\theta,\phi)$ in (13.1) will therefore provide a mechanism for the particles, whose coordinates appear in $Y_{2\mu}(\theta,\phi)$, to interact with the α-degrees of freedom and cause the excitation or deexcitation of the phonons.

To describe these conclusions more completely it is necessary to carry out the quantization of H_{coll}. It is now assumed that α_μ is a dynamical variable and therefore an operator. The procedure is well known since (13.2) simply describes five harmonic oscillators each with "mass" B, "force constant" C and frequency

$$\omega = \sqrt{\frac{C}{B}} \tag{13.4}$$

The results are most conveniently expressed in terms of two new operators a_μ and $a_\mu^\dagger$ that are related to α_μ and $\dot{\alpha}_\mu$ through the following equations

$$\alpha_\mu = \sqrt{\frac{\hbar}{2B\omega}}\,[a_\mu + (-)^\mu a_{-\mu}^\dagger]$$

$$\dot{\alpha}_\mu{}^* = i\sqrt{\frac{\hbar}{2B\omega}}\,[a_\mu^\dagger - (-)^\mu a_{-\mu}] \tag{13.5}$$

The operators a_μ satisfy the commutation relations

$$[\alpha_\mu, a_\nu^\dagger] = \delta_{\mu\nu} \tag{13.6}$$

$$[a_\mu, a_\nu] = [a_\mu^\dagger, a_\nu^\dagger] = 0$$

the commutation rules of a boson field, in this case a scalar phonon. The states of such a field will be characterized by the value of λ, in this case 2, the value of μ and n_μ (more generally $n_{\lambda\mu}$), the number of such phonons. The number operator $\hat{n}_\mu$ is

$$\hat{n}_\mu \equiv a_\mu^\dagger a_\mu \tag{13.7}$$

In view of the transformation properties of $\alpha_{\lambda\mu}$ it is no surprise and easy to show that the phonons have a total spin of λ with a "z" component of μ and a parity $(-)^\lambda$.

The operators a_μ are "destruction" operators because when they act on a state $|n_\mu, 2, \mu>$ the number n_μ is reduced by one. Similarly the $a_\mu^\dagger$ are "creation" operators:

$$a_\mu|n_\mu, 2, \mu> = \sqrt{n_\mu}|n_\mu - 1, 2, \mu> \tag{13.8}$$

$$a_\mu^\dagger|n_\mu, 2, \mu> = \sqrt{n_\mu + 1}\,|n_\mu + 1, 2, \mu> \tag{13.9}$$

The no-phonon state referred to briefly as the "vacuum" satisfies

$$a_\mu|0, 2, \mu> = 0$$

Normalized one-phonon states according to (13.9) are constructed from the vacuum state as follows:

$$|1, 2, \mu> = a_\mu^\dagger|0, 2, \mu>$$

Two-phonon states can have angular momentum 0^+, 2^+, and 4^+. The odd angular momenta do not occur because of the boson character of the phonons that require their state vectors to be symmetric. This can be immediately seen as follows. A two-boson state will involve

$$a_\mu^\dagger a_{\mu'}^\dagger\,|0, 2, \mu>|0, 2, \mu'>$$

Because of commutation rule (13.6), μ and μ' can be interchanged without any change proving the point.

A two-boson state with a given angular momentum I and "z"-component M is constructed as follows:*

$$|2; IM\,22\rangle = \sum \langle 2\mu_1, 2\mu_2|IM\rangle\, a^\dagger_{\mu 1}\, a^\dagger_{\mu 2}\, |0\,2\,\mu_1\rangle\, |0\,2\,\mu_2\rangle \qquad (13.10)$$

From the properties of the Clebsch-Gordon coefficients it is easy to prove (we leave it as an exercise for the reader) that the state (13.10) is symmetric under interchange of the two phonons as long as I is not odd, and vanishes when it is.

The energies of these states are the eigenvalues of the Hamiltonian (13.2) which in terms of a_μ and $a^\dagger_\mu$ is

$$H_{\text{coll}} = \hbar\omega \sum_\mu \left(\hat{n}_\mu + \frac{1}{2}\right) = \hbar\omega\left(\frac{5}{2} + n_\mu\right) \qquad (13.11)$$

The spectrum thus consists of a set of equally spaced levels. We see that the two-phonon state is triply degenerate since there are three two-phonon states. The three 2^+ phonon states can have total I of 0^+, 2^+, 3^+, 4^+, and 6^+. Similar results are obtained with the 3^- phonon. The two 3^- phonon state can have total angular momentum of 0^+, 2^+, 4^+, and 6^+. Two-phonon states of which one phonon is of the 2^+ variety the other of 3^- can also in principle exist. These will give rise to odd-parity states extending in angular momentum from 1^- to 5^- with an energy equal to the sum of the 2^+ and 3^- boson.

We note once again that this model has been developed for the vibration of spherical undeformed nuclei. The simplest test of its validity is by comparison with the observed energy spectrum. Of course the degeneracies mentioned above are not exactly maintained. The Hamiltonian (13.2) is "correct" only for small oscillations. It is thus the first term in an expansion in α_μ. The next term, $0(\alpha_\mu{}^3)$, will give rise to anharmonic forces and will split the degeneracies. Thus in comparing with experiment, the two-phonon states are not expected to be exactly degenerate but should be split. The model makes sense only if the splitting is small compared to $\hbar\omega$.

Tests of the model can be made by observing the excitation of one- and two-phonon levels by projectiles or by observing the electromagnetic and β-decay of these levels. As a first example suppose the interaction of a particle (e.g., an α-particle) with a spherical nucleus is given by a potential that is a function of $\mathbf{r}$, $\mathbf{p}$, and R_0:

$$V = V(\mathbf{r},\mathbf{p},R_0) \qquad (13.12)$$

where $\mathbf{r}$ and $\mathbf{p}$ are relative target–α-particle distance and momentum, respectively. If the incident projectile has spin then spin operator $\mathbf{s}$ would also be

---

*The notation is based on the more general case for which the phonons have a total angular momentum λ_1 and λ_2, respectively. Then the two-phonon state is described by $|2; IM\,\lambda_1\,\lambda_2\rangle$.

involved in V. However let us consider the simplest case (13.12) which is appropriate for α-particles. If the nucleus can vibrate, R_0 is replaced by R of (13.1). In the limit of small oscillations, that is, to first order in α_μ

$$V(\mathbf{r},\mathbf{p},R) = V(\mathbf{r},\mathbf{p},R_0) + \sqrt{\frac{\hbar}{2B\omega}}\, R_0 \frac{\partial V}{\partial R_0} \sum_\mu [a_\mu + (-)^\mu a^\dagger_{-\mu}]\, Y_{2\mu} + \ldots \quad (13.13)$$

We see that the interaction now involves the possibility of exciting a vibrational state. Two-phonon states can either be excited by a two-step process in which each phonon state is created in turn or through the effect of the quadratic term in the expansion of V, the first omitted term in expansion (13.13). The calculation of the inelastic cross sections is postponed to the second volume of this work.* However it is clear that comparing the excitation of the single-phonon state with the various two-phonon states or with the excitation (or decay) of these levels by electromagnetic means will provide a sensitive test of the model.

We also note that the particle-phonon interaction as given by (13.13) is proportional to $\partial V/\partial R_0$. For potentials that are a function of $r - R_0$ this derivative equals $-\partial V/\partial r$. Since V is generally flat within the nuclear interior decreasing rapidly at the nuclear surface, we see that $\partial V/\partial r$ will be peaked at the surface. Thus the particle-phonon interaction is a surface interaction and will be induced most effectively when the projectile wave function is maximal at the nuclear surface. This is the case for heavy composite projectiles such as the α-particle or heavier ions.

The transition probability for the electromagnetic decay of a nuclear level is given in eq. VIII.5.35. It is proportional to the B-coefficients. If we consider the decay of the one-phonon state, where the phonon is a 2^+, to the zero-phonon 0^+ state the quadrupole multipole will be involved so that in this discussion we are concerned with $B(E2)$:

$$B(E2; i \to f) = \frac{1}{5}\,(0\,||\,\hat{Q}_2^{(E)}\,||\,2)^2 \quad (13.14)$$

$$= \langle 00|\,\hat{Q}_{2\mu}^{(E)}\,|2\mu\rangle^2 \quad (13.15)$$

For the relation of $B(E2)$ with the lifetime of a state, see (6.17). In this expression the operator $\hat{Q}_{2\mu}^{(E)}$ is

$$\hat{Q}_{2\mu}^{(E)} = \int d\mathbf{r}\, r^2\, Y_{2\mu}^*\, \rho_p \quad (13.16)$$

*For a review of recent work with proton projectiles, see Satchler (70); with alpha particle projectiles, see J. S. Blair (66). See also Drozdov (60).

In this expression ρ_p, the proton-charge density is a function of the nuclear radius R. Hence

$$\rho_p(\mathbf{r}, R) = \rho_p(\mathbf{r}, R_0) + \sqrt{\frac{\hbar}{2\omega B}}\, R_0 \frac{\partial \rho_p}{\partial R_0} \sum_{\mu} [a_\mu + (-)^\mu a^\dagger_{-\mu}]\, Y_{2\mu} + \ldots \tag{13.17}$$

Inserting this expression into (13.16) and assuming that $\rho(\mathbf{r}, R_0)$ is spherical, that is, depends only on r, one obtains

$$\hat{Q}_{2\mu}{}^{(E)} = \left[\sqrt{\frac{\hbar}{2\omega B}}\, R_0 \int_0^\infty dr\, r^4 \frac{\partial \rho_p}{\partial R_0}\right] [a_\mu + (-)^\mu a^\dagger_{-\mu}] \tag{13.18}$$

As expected $Q_{2\mu}{}^{(E)}$ can either destroy a phonon, the nucleus radiating a photon, or create a one-phonon state the nucleus absorbing a photon. The value of $BE(2)$ is thus

$$B(E2, 2 \rightarrow 0) = \frac{\hbar}{2\omega B} R_0{}^2 \left[\int_0^\infty dr\, r^4 \frac{\partial \rho_p}{\partial R_0}\right]^2 \tag{13.19}$$

In the experimental analysis it has been customary to take ρ_p to be uniform up to the radius R_0:

$$\rho_p = \frac{Z}{\Omega}\, \theta\, (R_0 - r) \tag{13.20}$$

where Ω is the nuclear volume and $\theta\,(x)$ is the unit function

$$\begin{aligned} \theta\,(x) &= 1 \qquad x > 0 \\ &= 0 \qquad x < 0 \end{aligned}$$

Differentiating with respect to R_0 but keeping Ω constant (see footnote following Eq. 13.1). then

$$\frac{\partial \rho_p}{\partial R_0} = \frac{Z}{\Omega}\, \delta\, (R_0 - r)$$

Thus the electromagnetic interaction probes the nuclear surface. Finally

$$B(E2, 2 \rightarrow 0) = \frac{\hbar}{2\sqrt{BC}} \left(\frac{3}{4\pi} Ze\, R_0{}^2\right)^2 \tag{13.21}$$

Compare this result with that for a single particle. $B_{sp}(E2) = (\tfrac{1}{4}\pi)[(\tfrac{3}{5})eR_0]^2$ (see VIII.5.48). The vibration shows a collective enhancement as indicated by the Z^2 factor. It occurs because an oscillation of the radius given by (13.1) and (13.2) leads to a coherent radiation by all the protons.

A second conclusion that follows from the quadrupole operator (13.17) is that the quadrupole moment of any of the phonon states is zero. This is because $Q_{2\mu}$ is linear in a_μ and $a^\dagger_\mu$. Physically this means only that the vibration involves oblate and prolate deformations with equal weight.*

*Nonzero values can be obtained if second-order terms in (13.17) are kept.

We can now proceed to make some comparison with experiment. By observing transition probabilities and energy spectra it becomes possible (see Eqs. 13.26, 13.11, and 13.4) to determine the parameters B and C. The values thus obtained are shown in Figs. 13.1 and 13.2 taken from Temmer and Heydenburg (56). [For more recent data see Nuclear Data Group (66).] In each of these, the empirical values are compared with the predictions if the nucleus is assumed to be a vibrating charged liquid drop. According to Bohr (52) and assuming irrotational flow, [see derivations in Davidson (68) Eisenberg and Greiner (70)], these are

$$(B_\lambda)_{\text{irrot}} = \frac{1}{\lambda}\frac{3}{4\pi} AMR_0^2 \tag{13.22}$$

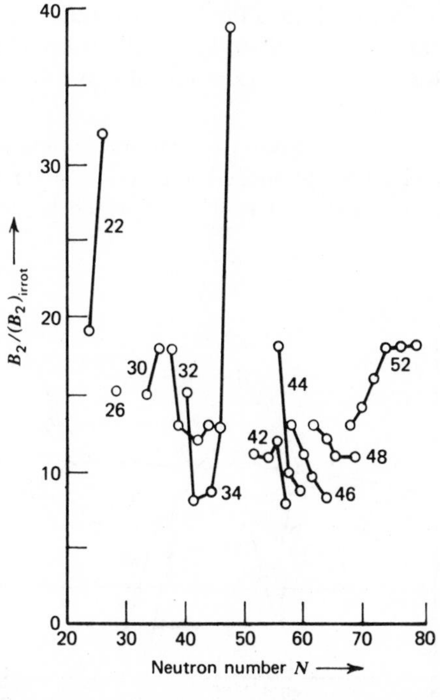

FIG. 13.1. Empirical mass parameters for quadrupole vibrations of even–even nuclei with $22 \leq Z \leq 52$. The irrotational value is used as the unit [taken from Temmer and Heydenburg (56)].

and

$$C_\lambda = (\lambda - 1)(\lambda + 2) R_0^2 S - \frac{3}{2\pi} \frac{(\lambda + 1)}{2\lambda + 1} \frac{Z^2 e^2}{R_0} \tag{13.23}$$

Here S is the surface energy per unit area. In the semiempirical mass formula, $4\pi R_0^2 S$ is the coefficient of the $A^{2/3}$ term and has the value of 18.56 MeV. For the quadrupole case $\lambda = 2$.

If we turn now to the energy spectrum, as mentioned above, there should be groups of levels with spins 0^+, 2^+, and 4^+ at energies of $2\hbar\omega$. Figure 13.3 taken from Stelson and McGowan (61) shows a systematic occurrence of these levels in qualitative agreement with expectation.

In even-even nuclei, the *second* 2^+ state thus represents a two-phonon excitation. Its decay straight to the ground state is thus forbidden and we should find $B(E2, 2' \to 2) \gg B(E2, 2' \to 0)$ where $2'$ denotes the second 2^+ excited state, and 2, the first one. Figure 13.4, taken from Van Patter (59), confirms this expectation very dramatically for many nuclei. The group of nuclei that seem to violate this expectation are all in the region of highly deformed nuclei, where the Hamiltonian (13.2), whose oscillations center around $\alpha_\mu = 0$, is not expected to be applicable. This shows up also in a plot of E_2/E_1 — the ratio of the energy of the second excited state to the first excited state, for the various nuclei, given in Fig. 13.5. For deformed nuclei one expects this ratio to be $J_2(J_2 + 1)/J_1(J_1 + 1) = 20/6$. For oscillations of the type considered here it should be closer to 2. This expectation is borne out by experiment as Fig. 13.5 demonstrates.

The above discussion is limited to even mass-number nuclei. A simple extension of the model to odd mass number $(A + 1)$ would consider the states of such a nucleus to consist of a *core* formed by a neighboring even-A

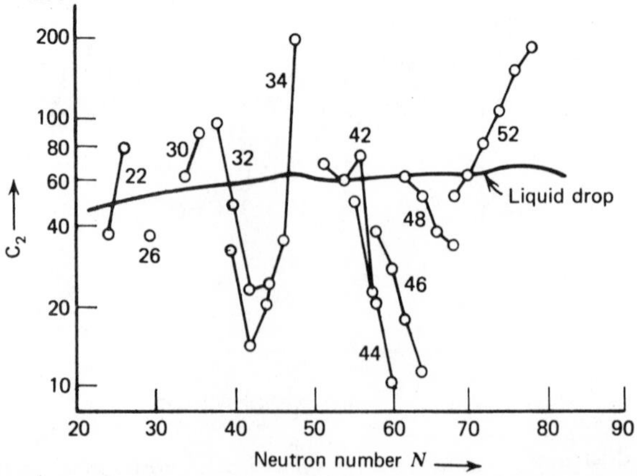

FIG. 13.2. Empirical restoring force parameters for quadrupole vibrations of even-even nuclei with $22 \leqq Z \leqq 52$ [taken from Temmer and Heydenburg [56)].

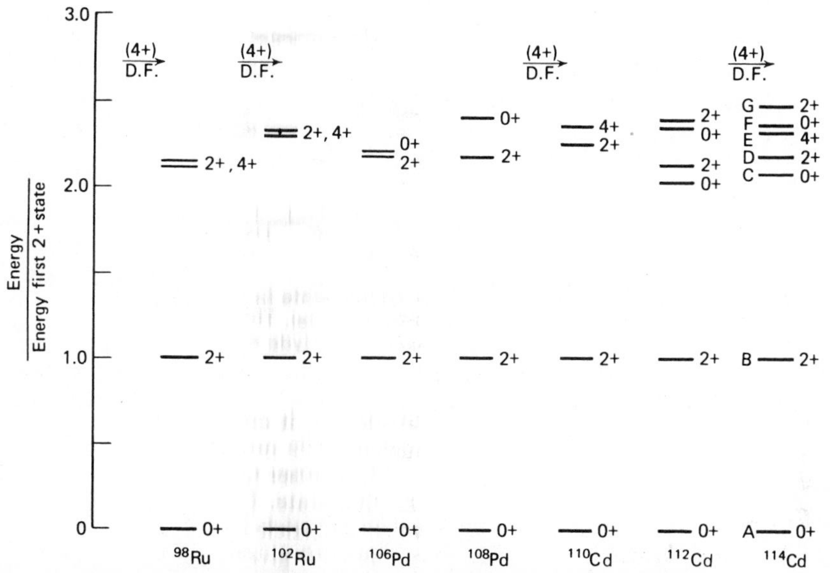

FIG. 13.3. Systematic occurrence of levels in medium weight nuclei showing characteristics of vibrational excitation [taken from Stelson (61)].

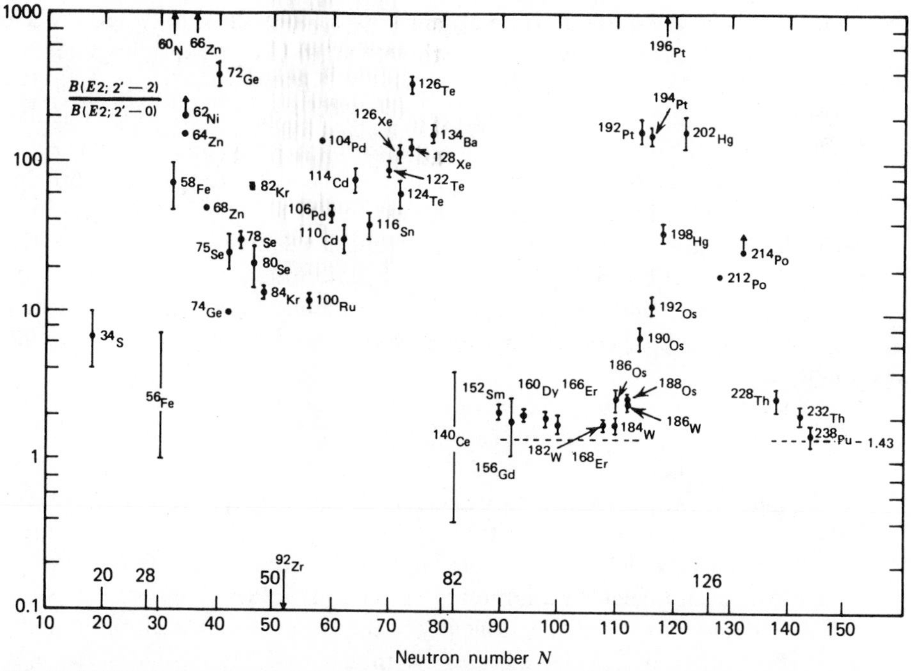

FIG. 13.4. Experimental values of $B(E2;\ 2' \rightarrow 2)B(E2;\ 2' \rightarrow 0)$ for even–even nuclei, as a function of the neutron number [taken from Van Patter (59)].

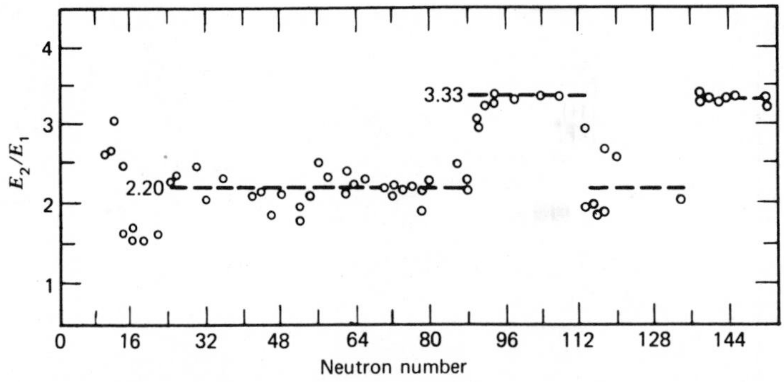

FIG. 13.5. Ratio of the energy of the second excited state to the first excited state as a function of the number of neutrons in even–even nuclei. The ratio 3.33 corresponds to rotational excitation of deformed nuclei [taken from Hyde et al. (64)].

nucleus plus a valence particle. Or equivalently, it could be thought of as a hole in a core formed by the $A + 2$ nucleus. The model to be chosen is determined by whichever nucleus, A or $A + 2$, is closer to closed shell and therefore more likely to be spherical in its ground state. Let us consider the case for which the core is nucleus A and the valence particle has a spin of j. Coupling this spin to that of the ground state of the core gives one state with spin j. Coupling to the one-phonon state gives rise to states with spin lying between $|j - 2|$ and $j + 2$, and so on for multiphonon states. The perturbing potential is given by (13.13). As can be readily verified in this case, the first term in this potential does not give any splitting between the various values of the angular momentum of the combined phonon plus particle. The second term has zero-diagonal value so that the splitting must be second order in the particle-phonon coupling. To be consistent with expansion (13.13) the splitting must be small compared to $\hbar\omega$. If it is, the coupling is genuinely weak; if not, the coupling is said to be strong, and the simple description given above must be revised to include higher-order terms. If in a given nucleus these are sufficiently significant, the whole concept of vibrational states for that nucleus loses its validity.

The situation for a number of odd-mass nuclei and ground-state spin of 1/2 is shown in Fig. 13.6. In this case the spins of the first excited state, if that has a 2^+ phonon core, are 3/2 and 5/2. The figure also shows the energy of the 2^+ state of the adjacent even nuclei. In all cases except for ^{195}Pt only two levels are seen. However the splitting is often appreciable being of the order of 100-200 KeV, which is to be compared with energies $\hbar\omega$ given by the 2^+ state of the even nuclei. In addition ^{195}Pt has four rather than two states.

This is by no means the only difficulty of the vibrational model. In the same spirit (first order in α_μ) the quadrupole moment of the excited states was

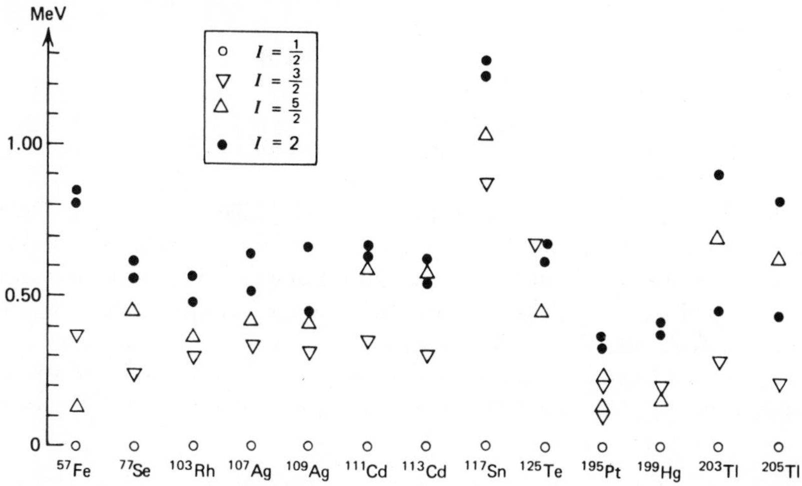

FIG. 13.6. Collective quadrupole states in odd mass nuclei with spherical equilibrium shape and ground state spin 1/2. The 2^+ states are those of adjacent even nuclei. From Nathan (64).

shown above to be zero. Coulomb excitation experiments indicate that these moments are far from zero. [For a review of this subject see de Boer (68).] Again we see that the coupling is not weak in many nuclei.

This last result must be interpreted to mean that in the vibration the prolate and oblate deformations do not have equal probabilities. A theory that yields such a result has been developed by Kumar and Baranger (68). This theory will be discussed in Chapter VII. It is based on the "Bohr Hamiltonian" Bohr (52) that we shall now describe.

We have alluded briefly to another "vibration," the 3^- phonon. This will be discussed in Section 14 of this chapter. For the present it will suffice to point out that for many odd nuclei the splitting analogous to that just discussed for the 2^+ phonon is small and the coupling does seem to be weak.

The parameters α_μ in (13.1) represent a deformation in the potential, and can therefore serve to define a body-fixed frame of reference. To transform to coordinates in frame Σ' we have to use Wigner's matrix $D^{(2)}$ (Appendix A, and write

$$\alpha'_{\mu'} = \sum \alpha_\mu \, [D^{(2)}_{\mu\mu'}(\theta_i)]^* \tag{13.24}$$

(We remember here that the α_μ's transform like $Y_{2\mu}{}^*(\theta\,\phi)$, in order to make $\sum \alpha_\mu Y_{2\mu}(\theta_i\,\phi_i)$ a scalar: hence the use of $D^{(2)*}$ in (13.24)). In order to make the frame Σ' body fixed, representing a deformed system, with a reflection symmetry in the $x'y'$-plane, we have to put

$$\alpha_1' = \alpha_{-1}' = 0 \qquad \alpha_2' = \alpha_{-2}' \equiv a_2 \qquad \alpha_0' \equiv a_0 \tag{13.25}$$

This will make $\sum \alpha_\mu Y_{2\mu}(\theta,\phi)$ transform into

$$\alpha_0' Y_{20}(\theta', \phi') + \alpha_2' [Y_{2,2}(\theta', \phi') + Y_{2,-2}(\theta', \phi')]$$

$$= \sqrt{\frac{5}{16\pi}} [a_0 (3 \cos^2 \theta' - 1) + a_2 \sqrt{6} \sin^2 \theta' \cos 2\phi']$$

which is seen to be symmetric with respect to reflection in the $x'y'$-plane ($\theta' \to \pi - \theta'$) and to allow nonaxially symmetric shapes as well. The condition $\alpha_1' = \alpha_{-1}' = 0$ assures that the z'-axis is along one of the principle axes of the ellipsoid, and $\alpha_2' = \alpha_{-2}'$ assures that the whole expression is real. We now transform from the five variables α_μ to the two deformation variables a_0 and a_2 and the three Euler angles θ_i that represent the orientation of the axes in space. Furthermore it is convenient to replace a_0 and a_2 by β and γ defined by

$$a_0 = \beta \cos \gamma \qquad a_2 = \frac{1}{\sqrt{2}} \beta \sin \gamma \tag{13.26}$$

β then gives the nuclear deformation

$$\beta^2 = a_0^2 + 2a_2^2 = \sum |\alpha_{\mu'}'|^2 = \sum |\alpha_\mu|^2 \tag{13.27}$$

while γ is a shape parameter of the nucleus. Using (13.1) one sees that in the body-fixed frame of reference the radii in the directions of the three principle axes are:

$$R_k = R_0 \left[1 + \frac{5}{4\pi} \beta \cos \left(\gamma - \frac{2\pi k}{3}\right)\right] \tag{13.28}$$

One sees from (13.28) that going from γ to $\gamma + n(2\pi/3)$, where n is an integer, only changes the role of the various axes. Thus to describe a given (quadrupole) shape it is enough to let γ vary in the range $0 < \gamma < \pi/3$.

In terms of the variables β, γ, and θ_i, the kinetic energy $1/2\ \beta \sum |\dot{\alpha}_\mu|^2$ can be written in the form known as the *Bohr Hamiltonian* [Bohr (52)]:

$$T = \frac{1}{2} B(\dot{\beta}^2 + \beta^2\dot{\gamma}^2) + \frac{1}{2} \sum \mathcal{J}_{k'} \omega_{k'}^2 \tag{13.29}$$

where $\omega_{k'}$ is the angular velocity of the k'-axis with respect to the laboratory frame and $\mathcal{J}_{k'}$, the moments of inertia, are given by

$$\mathcal{J}_{k'} = 4B\beta^2 \sin^2 \left(\gamma - \frac{2\pi k'}{3}\right) \qquad k' = 1,2,3 \tag{13.30}$$

Taking the expression (13.22), B_{irrot}, for B and recalling that (see Eq. 7.5) for a sphere of radius R_0 the rigid body moment of inertia is

$$\mathcal{J}_{\text{rig}} = \frac{2}{5} AMR_0^2$$

we can write the moments of inertia in the form

$$\mathcal{J}_{k'} = \frac{15}{4\pi} \mathcal{J}_{\text{rig}}\, \beta^2 \sin^2\left(\gamma - \frac{2\pi k'}{3}\right) \tag{13.31}$$

To quantize the Bohr Hamiltonian with the kinetic energy (13.29) we have to take into account the possible noncommutativity of the various variables. As is well known there is no unique prescription to go from a classical Hamiltonian to a quantum mechanical one, except for variables that can be transformed to Cartesian coordinates. [see, for instance, W. Pauli (33)]. Our collective coordinates are expressible, in principle, in terms of the particles coordinates, and through such expressions their quantization should have been determined. This program has not, however, been carried out [see Kumar and Baranger (67)] and one uses the Pauli prescription for the quantization: if

$$T = \frac{1}{2} \sum g_{mn}(q)\, \dot{q}_m \dot{q}_n$$

then its quantized counterpart is

$$T = -\frac{\hbar^2}{2} \sum \frac{1}{\sqrt{|g|}} \frac{\partial}{\partial q_m} \sqrt{|g|}\, g^{mn}(q) \frac{\partial}{\partial q_n}$$

where g^{mn} is the inverse of g_{mn}, and $|g|$ is the determinant of g_{mn}. Applying this procedure to the Bohr Hamiltonian one obtains

$$T = -\frac{\hbar^2}{2B}\left\{\frac{1}{\beta^4}\frac{\partial}{\partial\beta}\left(\beta^4\frac{\partial}{\partial\beta}\right) + \frac{1}{\beta^2 \sin 3\gamma}\frac{\partial}{\partial\gamma}\left(\sin 3\gamma\frac{\partial}{\partial\gamma}\right)\right\} + T_{\text{rot}} \tag{13.32}$$

where

$$T_{\text{rot}} = \sum \frac{\hbar^2 R_k'^2}{2\mathcal{J}_{k'}} \tag{13.33}$$

and **R** is the angular momentum operator for the rotation of the system around its body-fixed axes.

If the nucleus is highly deformed, if the fluctuations of this deformation are small so that we can put $\dot{\beta} = \dot{\gamma} = 0$, then the Bohr Hamiltonian (13.29), or (13.32), reduces to just the rotational kinetic energy (2.12). However (13.32) allows us now, in principle, to solve the complete problem of oscillations plus rotations, including the effects of one on the other. All we have to do for that purpose is to specify the potential energy for the new collective variables that we have introduced. For physical reasons we want this potential energy to depend only on β and γ, since the rotation of the axes, represented by $\dot{\theta}_i$, should proceed freely. Thus we assume that

$$V_{\text{coll}} = V(\beta,\gamma) \tag{13.34}$$

A possible way to proceed now is to use the Born-Oppenheimer approximation; solve the problem of A-independent particles moving in a potential characterized by the deformation parameters β and γ via (13.1); determine the energy $E(\beta,\gamma)$ of the A-particles in this potential, and use it as the potential energy for the collective Hamiltonian. This program has not been carried out yet. Instead one tries to expand $V(\beta,\gamma)$ around equilibrium positions of the two variables β_0 and γ_0, and limit oneself to the harmonic terms for oscillations around β_0 or γ_0, possibly also considering some unharmonic corrections. Tamura and Komai (59), for instance, take

$$V(\beta,\gamma) = \frac{1}{2} C(\beta - \beta_0)^2 + K(\beta - \beta_1) \cos 3\gamma$$

and obtain a fairly good fit to a broad set of data by properly choosing C, β_0, β_1, and K.

Kumar and Baranger (67) carried out an extensive numerical solution of the Bohr Hamiltonian with a parameterized potential and were able to test thereby various approximations to that Hamiltonian. They have been able to obtain a complete description of the transition from the nondeformed to the deformed region with all the peculiar features of nuclei in the transition region. The reader will find in their paper also an extensive discussion of the symmetry properties of the solutions to the Bohr Hamiltonian, etc.

Davydov and Filippov (58) carried out extensive calculations in which γ was assumed to take a constant numerical value and the collective Hamiltonian was then solved for the rotations and the oscillations in the variable β, the so-called *β-vibrations.* Figure 13.7 shows the expected position of the various collective states as a function of γ on the Davydov-Filippov model, while Fig. 13.8, taken from McGowan and Stelson (61), shows the actual trend of excited states in even-even nuclei in a transition region from the deformed nucleus ^{182}W to the spherical nucleus ^{196}Pt.

We see from Fig. 13.7 that, as expected from (13.29) and (13.30), small values of γ lead to the rotational spectrum of an axially symmetric top. In fact, since, according to (13.13) $\mathcal{J}_3 = 0$ for $\gamma = 0$, there can be no rotation around this axis [$(\hbar^2 R_3{}^2/2\mathcal{J}_3)$ is infinite for $R_3 \neq 0$), and consequently the rotations take place around an axis perpendicular to the symmetry axis of the nucleus, as in the pure rotational model. As γ increases and the nucleus ceases to be spherically symmetric, new levels come down and a more complex spectrum shows up. One notices in particular the coming down of a level with $J = 2^+$, which shows up characteristically also in Fig. 13.8, as one moves from the deformed nucleus ^{182}W towards nuclei of a more spherical shape.

Figure 13.9 and 13.10 illustrate the variations in the shapes of nuclei as we change β and γ. *β-vibrations* for $\gamma = 0$ nuclei are vibrations in which the nucleus oscillates and changes its deformation around a given equilibrium deformation, retaining always its axial symmetry. *γ-vibrations,* on the other hand, are oscillations in which the nucleus may retain its deformation param-

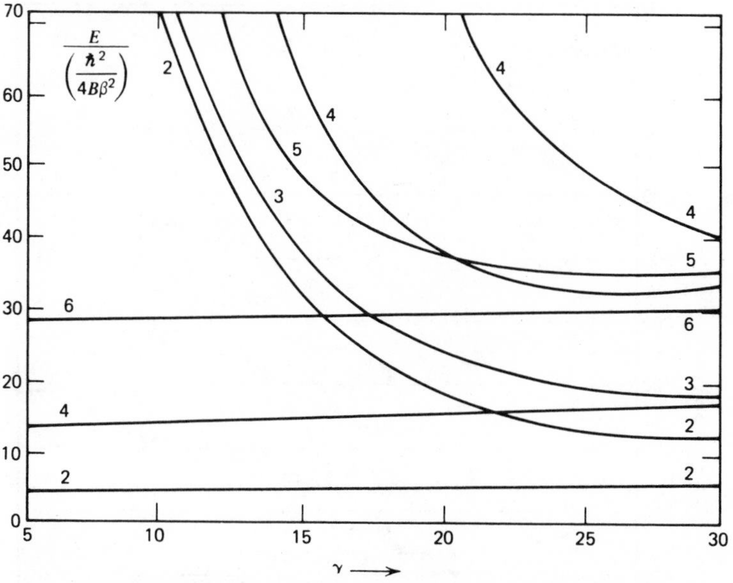

FIG. 13.7. The computed energy levels for various values of the deformation parameter γ [Davydov and Philippov (58)].

eter β, but changes from being nonaxially symmetric with $R_{x'} > R_{y'}$, via an axially symmetric shape, to nonaxially symmetric shape with $R_{y'} > R_{x'}$, and back to axial symmetry, etc.

It is obvious from the derivation of the Bohr Hamiltonian that each vibrational mode can still have, superimposed on it, a rotational band, so that one can obtain a rather complex spectrum consisting of ground-state rotational band, rotational bands based on β-vibrations and γ-vibrations, and rotational bands based on different intrinsic excitations, as discussed in previous sections. We shall not go further into these modes of excitation here but rather refer the reader to the more specialized literature on this subject. In the next chapter we shall discuss the microscopic structure of some of these modes.

14. OTHER COLLECTIVE MODES

In the last section we described two types of collective quadrupole vibrations and how they relate to the rotation of deformed nuclei. It is, however, quite obvious that there is no reason a priori to limit possible collective features of nuclei to quadrupole modes alone. Higher multipoles could show up as well, and more complex types of collective motion involving the spin and isospin degrees of freedom could be envisaged.

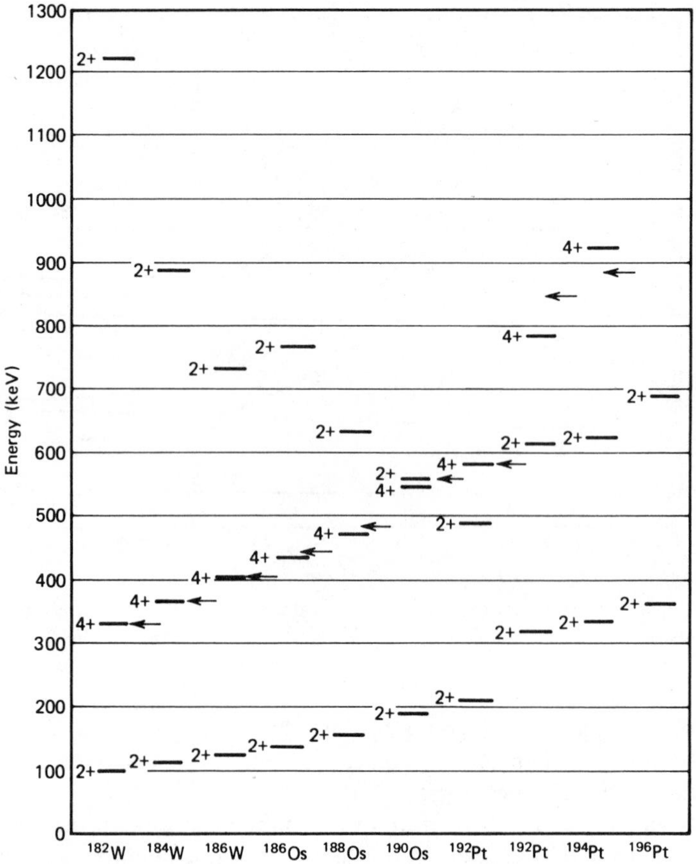

FIG. 13.8. Low-lying levels for even–even nuclei in the transition region from spheroidal to spherical shape [taken from McGowan and Stelson (61)].

Empirically there are two general indications for collective states. Because these states are associated with some bulk properties of nuclei, and because nuclei have a constant density, collective states of a given type tend to show up in all nuclei, or at least in a wide range of neighboring nuclei, and their excitation energy tends to vary smoothly from one nucleus to another. This is of course is only a necessary condition since noncollective properties can also vary smoothly with A and Z.

Secondly, a collective state, from the point of view of its microscopic structure, displays strong correlations among the nucleons. The trivial collective motion of the center of mass, for instance, reveals itself in a very strong correlation of the average momenta of the nucleons and establishes a corre-

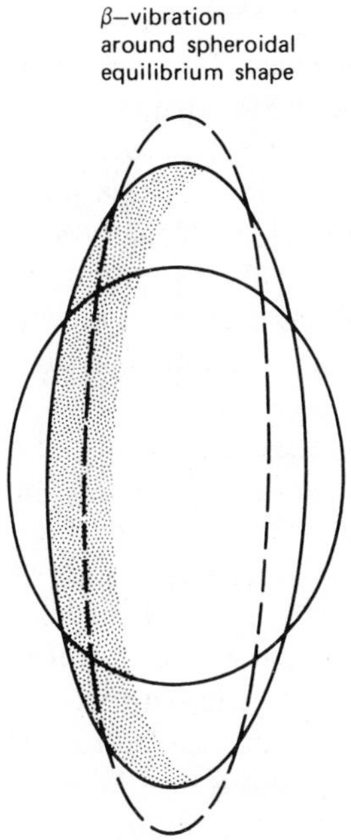

FIG. 13.9. β-vibrations for $\gamma = 0$ [taken from Hyde et al. (64)].

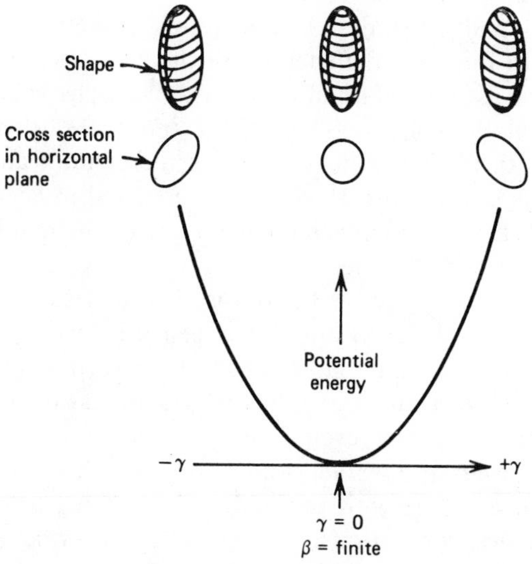

FIG. 13.10. γ-vibration around a spheroidal equilibrium shape [taken from Hyde et al. (64)].

lated phase, determined by $e^{i(\mathbf{p}\cdot\mathbf{x})/\hbar A}$, for each of the nucleons. Here **p** is the momentum of the center of mass. The collective rotation, when viewed from the laboratory frame of reference, also manifests itself in a well-defined correlation among the various nucleons that assures that the overall shape of the nucleus undergoes "slow" rotations. These phase relations among the wave functions of the individual nucleons usually lead also to an enhancement of corresponding transition probabilities. Associated with the quadrupole oscillations are then enhanced electric quadrupole radiations. In fact, enhanced transitions of one sort or another is the second characteristic feature of collective states.

Two well-established collective excitations, in addition to the ones we have already mentioned, are the *octupole excitation* and the *giant dipole excitation.* The former shows up as a $J = 3^-$ state in even-even nuclei, at an excitation of about 6 MeV for the lighter nuclei, going down to around 2 MeV for the heavier ones. It also shows up in odd-A nuclei where, however, its structure is slightly more complex (see below). The collective octupole excitations are associated with an enhanced electric octupole radiation, and they can therefore also be easily excited in inelastic electron scattering or via other electro magnetic interactions [see de Forest and Walecka (66)]. They are clearly observed, of course, also in reaction like (p,p') or (α,α'). For reviews of this problem see Nathan (64) and Bernstein (70).

The collective octupole states can be described in terms of octupole deformations of the nuclear potential, that is, deformations that do not conserve the reflection symmetry in the $x'y'$-plane and lead to pearshaped nuclei. They have also been equally well described in terms of their microscopic structure (Chapter VII). In Fig. 14.1 the energies of the 3^- collective vibrational states in even nuclei as given by Nathan (64) are shown. The values predicted from the hydrodynamic model ($\lambda = 3$ in 13.22 and 13.23) also shown gives the general trend with A. We shall not discuss these states any further as their analysis parallels that for the 2^+ vibration discussed in Section 13. However, there is one substantial difference connected with the manifestation of these collective states in odd-A-nuclei.

We assume that the collective excitations can be treated as independent of the particle motion. We also adopt the general conclusion of the various independent-particle models that nucleons tend to pair off in time-reversed conjugate states. Odd-A nuclei can then be pictured as having an odd-nucleon loosely attached to the even-even *core*. The collective excitations, involving as they do the nucleus as a whole, can be considered to a good approximation as excitations of the core with the odd-nucleon being a "spectator" having little to do with these excitations. This model is known as the *core-excitation* model. The core excitation involves the angular momentum J_c (such as $J_c = 3^-$ for the octupole excitations); the odd particle's orbit on the other hand is characterized by an angular momentum j. In the approximation

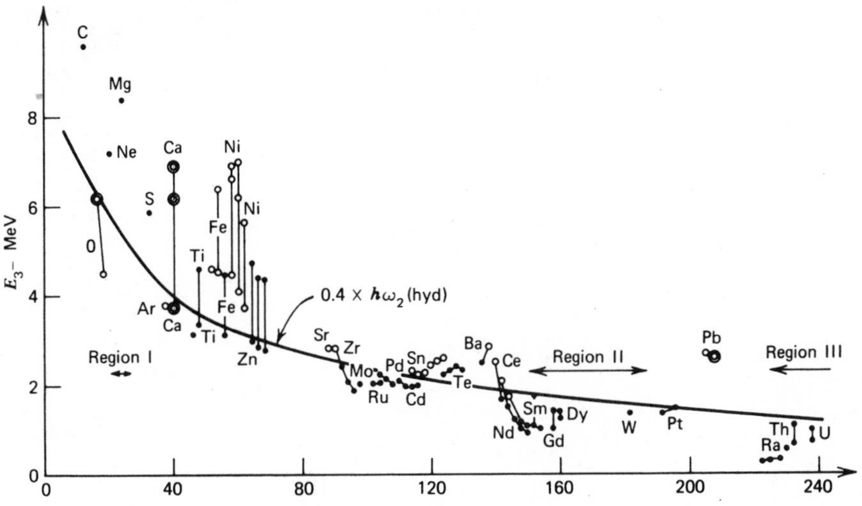

FIG. 14.1. The energies of 3⁻ collective vibrational states in even nuclei. The open circles represent closed shell nuclei [taken from Nathan (64)].

in which we neglect the interaction between the odd-particle and the collective-surface wave the collective excitation of the odd-*A* nucleus will consist then of a set of degenerate levels with $J = |J_c - j|, |J_c - j| + 1, \ldots, J_c + j$. A weak interaction between the odd particle and the surface wave will split these degenerate levels and lead to a *core multiplet* at approximately the excitation energy of the collective state J_c in the neighboring even-even nucleus. Figure 14.2 taken from Hafele and Woods (66) shows a typical situation of this kind. ^{206}Pb and ^{208}Pb both possess $J = 3^-$ collective states at 2,649 MeV and 2.615 MeV, respectively; these are clearly seen, for instance, in the inelastic proton scattering shown in Fig. 14.2*a*. The neighboring odd-*A* nuclei ^{207}Pb and ^{209}Bi have very different spins in their ground states, that of $^{107}_{82}Pb_{125}$ being $1/2^-$ while that of $^{209}_{83}Bi_{126}$ is $9/2^-$. The inelastic proton scattering on ^{207}Pb shows a doublet of levels around 2.6 MeV associated with an octupole excitation, with spins $J = 5/2^+$ and $J = 7/2^+$, whereas inelastic proton scattering on ^{209}Bi shows at least six, and possibly seven levels around 2.6 MeV, all associated with an octupole excitation, their angular momenta taking all the allowed values between $|J_c - j| = |3 - 9/2| = 3/2^+$, to $J_c + j = 15/2^+$ (with the levels $11/2^+$ and $13/2^+$ probably unresolved). These are precisely the multiplets we would expect if the odd particle was weakly coupled to the collective state. They are referred to as *weak-coupling* multiplets or *core multiplets.*

The identification of multiplets, like the doublet in ^{270}Pb and the septuplet in ^{209}Bi, as core multiplets can be checked in a variety of ways [see Lawson

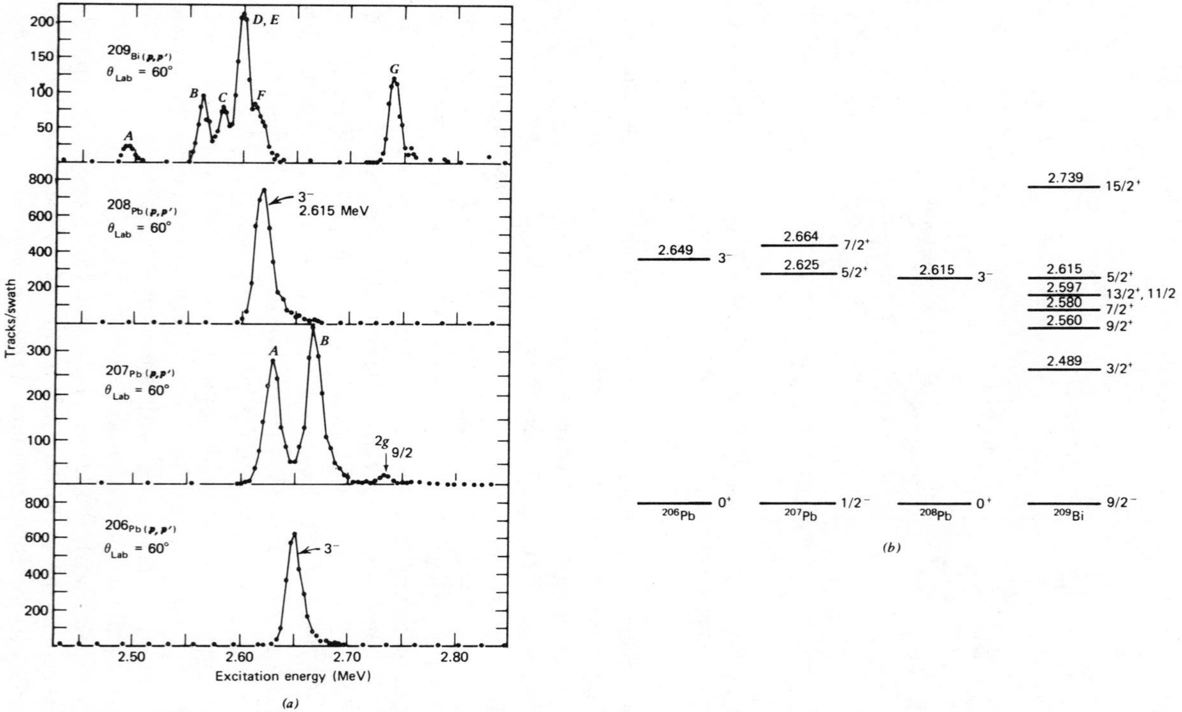

FIG. 14.2. (*a*) Typical inelastic proton spectra for excitation energies in the region around the 2.6 MeV octupole state. The incident proton energy was 21 MeV [taken from Hafele and Woods (66)]. (*b*) Energy diagram for the 2.6 MeV octupole levels observed by Hafele and Woods (66). Other known states for these nuclei are not shown. The spacing to the ground state is not to scale.

and Uretsky (57) and deShalit (61)], and many examples are known of groups of levels that lend themselves naturally to this interpretation. One method that will be described in Chapter VIII involves comparison of the radiative transition rates of the members of the core multiplet. These should be related directly to the radiative transition rates of the core implying relations among these rates as well as to the rates in neighboring nuclei. The fact that such multiplets involving collective states occur so often is another experimental verification of the validity of the approximation that treats collective excitations independently of the particle degrees of freedom.

That this is a general phenomenon is demonstrated by Fig. 14.3 where the 3^- core excited states for odd-A nuclei are shown. The j of the valence nucleon is that of the ground state given at the bottom of the figure. At the time these data were collected not all the states that could be formed by the coupling of the valence nucleon to the ground state had been observed. This has been remedied in experiments done since that time [e.g., the review of Bromley and Weneser (68) of the Pb region]. Good resolution is essential so that for example in ^{209}Bi all seven levels can be seen. Included in the figure are the 3^- levels from neighboring even-even nuclei. It is clear that for the heavier nuclei the effect of the particle-vibration coupling induces a small splitting. Hence in this region a weak coupling regime prevails in contrast with the 2^+ core excitation discussed earlier.

The *giant-dipole* states form another mode of collective excitation. Their experimental manifestation is seen in both photo-absorption experiments [i.e., (γ,n) or (γ,p) reactions] and in the inverse radiative capture reactions.

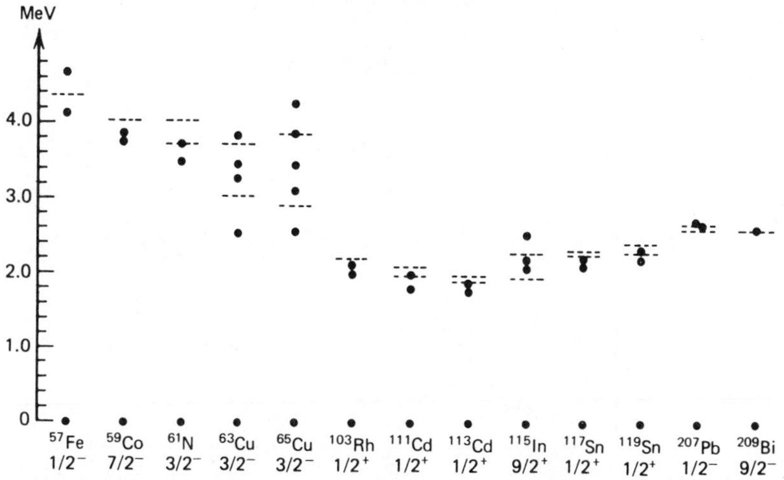

FIG. 14.3. Collective octupole states in odd mass nuclei with spherical equilibrium shape. The spin and parity of the ground state is given. The energies of the 3^- states in adjacent even nuclei are represented by the dashed lines [from Nathan (64)].

Figure 14.4, taken from Danos and Fuller (65), shows a typical example of the greatly enhanced absorption of photons on ^{208}Pb at an energy of about 114 MeV. Such resonances, which can be shown experimentally to involve an electric-dipole radiation, are known all over the periodic table. For $A \gtrsim 40$ they occur at an energy given by

$$E_{\text{giant dipole}} = E_{\text{g.d.}} \approx 80 \times A^{-1/3}\ \text{MeV} \tag{14.1}$$

A collection of some of the data on the energy of the giant dipoles is shown in Fig. 14.5, reproduced from Fuller (73). It is quite obvious from this figure that we are dealing here with a phenomenon that repeats itself regularly across the periodic table.

An early understanding of the phenomenon was achieved by Goldhaber and Teller (48) and Steinwedel and Jensen (50). These authors use a classical hydrodynamic model in which the neutrons are supposed to form one liquid, the protons another. The effect of the incident electromagnetic wave is to displace the proton liquid with respect to the neutron liquid. Nuclear interactions provide a restoring force, and oscillations develop. Goldhaber and Teller developed rough estimates of the effect of the nuclear interaction. The following description will use the Steinwedel and Jensen procedure in which the restoring force is related to the symmetry energy in the semi-empirical mass formula, which is itself a manifestation of the force between neutrons and protons. The oscillation of this system will be calculated in the absence of an external force. The resonance frequency should be close to this free field frequency.

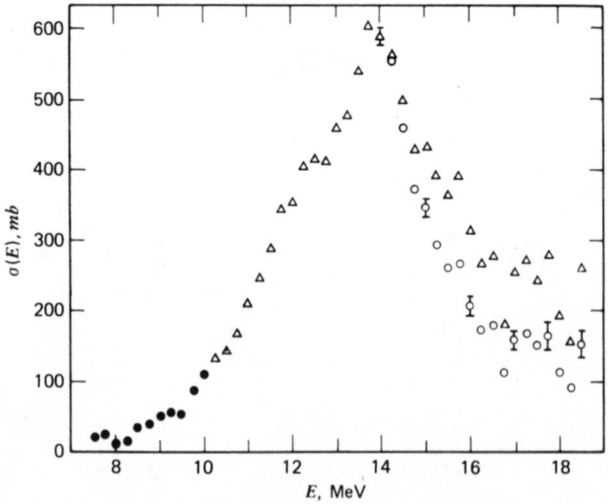

FIG. 14.4. Photon absorption in ^{208}Pb [taken from Danos and Fuller (65)].

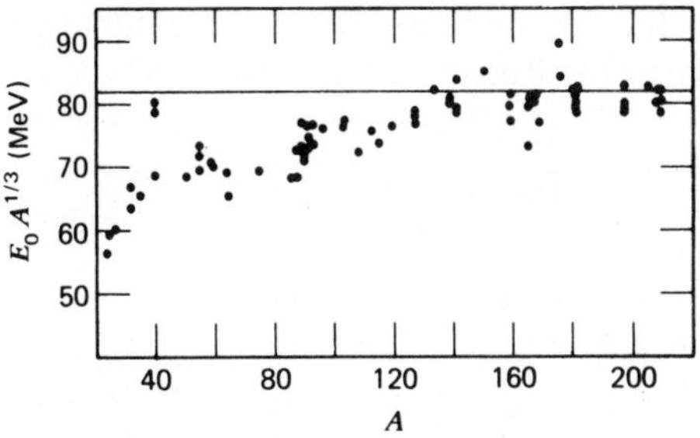

FIG. 14.5. Giant resonance energy times $A^{1/3}$ as a function of A [taken from Fuller and Hayward (62)].

In a hydrodynamic model one ascribes a density ρ_p to the proton fluid, a velocity $\mathbf{v}_p$ of the fluid in a volume element, that velocity being related to the displacement $\mathbf{s}_p$ by $\mathbf{v}_p = \partial \mathbf{s}_p/\partial t$. (This relation holds only for small velocities. More accurately $\mathbf{v}_p = d\mathbf{s}_p/dt = \partial \mathbf{s}_p/\partial t + (\mathbf{v}_p \cdot \nabla)\mathbf{v}_p$, Morse and Feshbach (52), Chapter II. The discussion below will be limited to small velocities so that the correction terms that are non linear will be omitted.)

Since we are interested only in the relative motion of the two fluids, it will be useful to introduce the relative and center-of-mass velocities:

$$\mathbf{v} \equiv \mathbf{v}_p - \mathbf{v}_n$$
$$\mathbf{V} = \frac{\rho_p \mathbf{v}_p + \rho_n \mathbf{v}_n}{\rho_0} \tag{14.2}$$
$$\rho_0 \equiv \rho_p + \rho_n$$

In terms of these quantities the kinetic energy T of the two fluids

$$T = \frac{1}{2} M \int (\rho_p \mathbf{v}_p^2 + \rho_n \mathbf{v}_n^2)\, d\mathbf{r} \tag{14.3}$$

becomes

$$T = \frac{1}{2} M \left(\int \frac{\rho_p \rho_n}{\rho_0} \mathbf{v}^2\, d\mathbf{r} + \int \rho_0 \mathbf{V}^2\, d\mathbf{r} \right) \tag{14.4}$$

where M is the nucleon mass.

The potential energy is based on the symmetry term in the semiempirical mass formula (Eq. II.3.1)

$$U_S = a_4 \frac{(N - Z)^2}{A} \tag{14.5}$$

which measures in an average way the effectiveness of the neutron-proton force that provides the restoring force acting between the neutron and proton liquids. In terms of the variables ρ_n and ρ_p and assuming that the density of each liquid is uniform (i.e., disregarding surface effects) U_S can be written

$$U_S = a_4 \int \frac{(\rho_n - \rho_p)^2}{\rho_0} d\mathbf{r} \tag{14.6}$$

To determine the restoring force we ask for the change in energy that occurs when there is a change $\delta\mathbf{s}$ in the relative position of volume elements of the two types of liquid. If $\delta\mathbf{s}_p$ and $\delta\mathbf{s}_n$ are the changes in the position of the proton and neutron volume elements, respectively, then

$$\delta\mathbf{s} = \delta\mathbf{s}_p - \delta\mathbf{s}_n \tag{14.7}$$

while

$$\delta\mathbf{S} = \frac{1}{\rho_0} (\rho_p \delta\mathbf{s}_p + \rho_n \, \delta\mathbf{s}_n) \tag{14.8}$$

gives the change in their center of mass.

The change in U_S following from changes in the densities $\delta\rho_p$ and $\delta\rho_n$ is

$$\delta U_S = \frac{2a_4}{\rho_0} \int d\mathbf{r} \, (\rho_p - \rho_n)(\delta\rho_p - \delta\rho_n) \tag{14.9}$$

To relate $\delta\rho_p$ and $\delta\rho_n$ we make use of the continuity equations

$$\frac{\partial\rho_p}{\partial t} + \operatorname{div} \rho_p \mathbf{v}_p = 0 \tag{14.10}$$

or approximately

$$\frac{\partial\rho_p}{\partial t} + \rho_p^{(0)} \operatorname{div} \mathbf{v}_p = 0 \tag{14.11}$$

where $\rho_p = \rho_p^{(0)} + \delta\rho_p$.
The quantity $\rho_p^{(0)}$ is the equilibrium value of ρ_p. Therefore

$$\delta\rho_p + \rho_p^{(0)} \operatorname{div} (\delta\mathbf{s}_p) = 0 \tag{14.12}$$

Using a similar relation for $\delta\rho_n$, δU_S becomes

$$\delta U_S = -\frac{2a_4}{\rho_0} \int d\mathbf{r} \, (\rho_p - \rho_n)(\rho_p{}^0 \operatorname{div} \delta\mathbf{s}_p - \rho_n{}^0 \operatorname{div} \delta\mathbf{s}_n)$$

Replacing $\delta\mathbf{s}_p$ and $\delta\mathbf{s}_n$ from (14.7) and (14.8), and dropping the term in $\delta\mathbf{S}$ to simplify the formulas (this is not an approximation), δU_S becomes

$$\delta U_S = -\frac{2}{\rho_0} \int d\mathbf{r} \, (\rho_n - \rho_n) \frac{2\rho_p{}^0\rho_n{}^0}{\rho_0} \nabla\cdot(\delta\mathbf{s}) \tag{14.13}$$

Integrating by parts and assuming the surface contribution to vanish (14.13) becomes

$$\delta U_S = \frac{8a_4\, \rho_p{}^0\, \rho_n{}^0}{\rho_0{}^2} \int d\mathbf{r}\; \nabla\rho_p \cdot \delta\mathbf{s} \tag{14.14}$$

Hence the restoring force in the relative coordinate system is

$$\mathbf{F} = -\frac{8a_4\, \rho_p{}^0\, \rho_n{}^0}{\rho_0{}^2} \nabla\, \rho_p \tag{14.15}$$

From the kinetic energy (14.5) and this force the Newton equation of motion is readily obtained:

$$M \frac{\rho_p{}^{(0)}\, \rho_n{}^{(0)}}{\rho_0} \frac{\partial \mathbf{v}}{\partial t} + \frac{8a_4\, \rho_p{}^0\, \rho_n{}^0}{\rho_0{}^2} \nabla\rho_p = 0$$

or

$$\frac{\partial \mathbf{v}}{\partial t} + \frac{8a_4}{M\rho_0} \nabla\rho_p = 0 \tag{14.16}$$

This should be combined with the difference of the current conservation equations for ρ_p and ρ_n:

$$\frac{\partial \rho_p}{\partial t} + \frac{\rho_p{}^0\, \rho_n{}^0}{\rho_0} \operatorname{div} \mathbf{v} = 0 \tag{14.17}$$

Taking the divergence of (14.16) and replacing div $\mathbf{v}$ from the above equation yields the wave equation for ρ_p:

$$\nabla^2\, \rho_p = \frac{M\rho_0{}^2}{8a_4\, \rho_p{}^0\, \rho_n{}^0} \frac{\partial^2 \rho_p}{\partial t^2} \tag{14.18}$$

The velocity of propagation of the wave, u, is given by

$$u^2 = \frac{8a_4\, \rho_p{}^0\, \rho_n{}^0}{M\rho_0{}^2} = \frac{8a_4}{M} \frac{ZN}{A^2} \tag{14.19}$$

Thus the velocity with which proton-density fluctuations propagate through nuclear matter in virtue of the neutron-proton interaction is given by u. Inserting a rough value of $NZ/A^2 \sim 1/4$, $a_4 \sim 20$ MeV, u is of the order of $0.21c$.

We now look for simple harmonic solutions for ρ_p

$$\rho_p = \rho_0 + \eta e^{-i\omega t} \tag{14.20}$$

The value of ω will give the resonant frequencies. The wave equation becomes

$$(\nabla^2 + k^2)\, \eta = 0 \tag{14.21}$$

where

$$k \equiv \frac{\omega}{u} \tag{14.22}$$

We now look for a solution η that is of the class excited by an oscillating electric dipole field $\mathbf{E}$. This immediately tells us that

$$\eta = Cj_1(kr)\, P_1(\cos\theta) \tag{14.23}$$

where C is a constant, j_1 is the spherical Bessel function of order 1, and θ is the angle between $\mathbf{E}$ and $\mathbf{r}$. The boundary condition on η follows from the requirement that the surface contribution vanishes in going from (14.13) to (14.14). This is accomplished if $\mathbf{r}\cdot\mathbf{v}$ is zero at the surface taken to be of the order of the nuclear radius. Physically this condition requires that there be zero outgoing current at this radius. From form (14.23) for η this boundary condition is

$$\frac{dj_1(kr)}{dr} = 0 \qquad \text{at} \qquad r = R \tag{14.24}$$

The roots of this equation are well known (see Morse and Feshbach (52), Appendix to Chapter XI). The first two are

$$kR = 2.08,\ 5.95$$

Taking the first one, the corresponding resonant frequency is

$$\omega = \frac{2.08\ u}{R} \tag{14.25}$$

Since u is essentially a constant for much nuclei we see that the resonant frequency and thus the resonant energy, $\hbar\omega$, decreases with increasing A like $A^{-1/3}$, a result that is in agreement with experiment. A rough evaluation of the multiplicative constant yields $\hbar\omega \sim 70\ A^{-1/3}$ that is to be compared with the experimental value of 80 $A^{-1/3}$.

Of course a spherical nucleus has been assumed. For deformed nuclei it becomes necessary to solve (14.21) with boundary condition $\mathbf{n}\cdot\nabla\eta = 0$ where $\mathbf{n}$ is normal to the boundary to be applied to, say, prolate (or oblate) spheroidal surfaces. Such solutions can be readily obtained using prolate (or oblate) spheroidal coordinates (see Morse and Feshbach (52), Chapter XI). These calculations have been performed by Danos (58) and Okamoto (58). As may be expected whereas the spherical nucleus has only one such frequency the spheroidal nucleus will have two. This result first discovered by Fuller, Weiss (58) and by Spicer et al. (58) has been observed for several nuclei. [See Spicer (69) for a review of this subject.]

These calculations have omitted many effects. The sharp nuclear surface, the simplified density dependence of the potential energy given by (14.6) are surely inaccurate. In addition there are the limitations of hydrodynamic model of the nucleus that does not take into account its particle structure. The model is probably applicable to medium and large mass-number nuclei. For light nuclei a more detailed model is needed as is evident from the complex structure of the cross section for photon absorption since there are many peaks not just the single one shown for Pb in Fig. 14.4. Such a model is discussed below.

We shall give a more detailed study of the giant dipole states in Chapter VIII. They will also be discussed in Volume II since these states are in the continuum and their analysis involves techniques of reaction theory. Here we wish only to make some remarks about the giant dipole states in light nuclei, ignoring the fact that they are unbound and treating them as if they were bound states (compare the study of the excited states of the α-particle in Section V.14). We shall use the shell-model analysis of the giant dipole states in order to see how a detailed microscopic structure can lead to collective effects and, thus, build at least a conceptual bridge between the collective states and their underlying multinucleon structure.

To this end let us consider an even-even nucleus with its ground state $J = 0^+$, and for the sake of simplicity let it be a closed shell nucleus. Dipole excitations involve the addition of one unit of angular momentum and a negative parity. Therefore we should look for excited shell-model states with $J = 1^-$, a linear combination of which will make up the giant dipole state. These can be obtained by taking the ground-state shell-model wave function Φ_0 ($J = 0$) and lifting a particle from an occupied level $j_a^\pm$ to an unoccupied $j_k^\pm$ such that $|j_a - 1| \leq j_k \leq j_a + 1$, or

$$|j_a - j_k| \leq 1 \qquad P(a)\cdot P(k) = -1 \tag{14.26}$$

where $P(a)$ is the parity of level j_a, etc. and forming the appropriate linear combination of the various particle-hole wave functions. (14.26) can be satisfied by taking a particle from the last filled major shell in Φ_0 and putting it in an appropriate level in the next shell. As is well known, successive shells in the nuclear shell model have states with alternating parities, so that the negative parity of the resulting state is guaranteed (Fig. I.7.2). As an example suppose Φ_0 represents the shell-model ground state of O^{16}; the proton configuration is then $(1s_{1/2})^2\,(1p_{3/2})^4\,(1p_{1/2})^2$, and we can choose j_a to be $(1p_{1/2})$ or $(1p_{3/2})$, and lift the particle from this level into an appropriate level in the next shell $1d_{5/2}$, $2s_{1/2}$ or $1d_{3/2}$. For $j_a = (1p_{1/2})$ we take $j_k = (2s_{1/2})$ or $j_k = (1d_{3/2})$, while if we choose to vacate a particle from a $j_a = (1p_{3/2})$ level we can take j_k to be $j_k = (2s_{1/2})$, $j_k = (1d_{3/2})$ or $j_k = (1d_{5/2})$.

A state formed by the promotion of one particle from the shell-model ground state is called a *one-particle one-hole state*, or briefly $1p1h$-state. In O^{16} we have just described five possible $1p1h$-states that can lead to a state of $J = 1^-$. There are, of course, many more such states: we can lift one particle from the ground state of O^{16} three shells high into the $3s_{1/2}$ or $2d_{3/2}$ states, or even further up if we wish. We can also create $3p3h$-states with $J = 1^-$ by lifting three particles from the p-level into the sd shell, etc.

The $1p1h$-states are often written as $|j_a^{-1}\,j_k;\,JM\rangle$ indicating that a particle was vacated from j_a and promoted to j_k. We obtain the lowest excitation for a $1p1h$, $J = 1^-$ states by exciting nucleons from the last filled shell in Φ_0 to the next (unfilled) shell such as the state $|1p_{3/2}^{-1}\,2s_{1/2};\,1^-\rangle$ in O^{16}. The energy involved in doing so is about $\hbar\omega$, where ω is the frequency of the harmonic-oscillator potential that is closest to the shell-model potential used to describe Φ_0. Equation IV.2.11 gives $\hbar\omega$ as a function of A:

$$\hbar\omega = (41 \times A^{-1/3})\text{ MeV} \tag{14.27}$$

If we compare (14.27) with (14.1) we see that the A-dependence of $E_{\text{g.d.}}$ is correctly reproduced, but the lowest order shell-model estimate of $E_{\text{g.d.}}$ misses the empirical value nearly by a factor of two!

This can be improved upon if we consider the residual interaction among the nucleons. As we have seen in the case of O^{16} there are five different ph-states for the protons that give us a $J = 1^-$ state, and there are five similar states for the neutrons. In a harmonic-oscillator potential all these states are degenerate in lowest order; in potentials of a more realistic shape they lie fairly close to each other. In order to obtain a better estimate of the actual position of the $J = 1^-$ levels we should therefore diagonalize the Hamiltonian, including the interaction, in the subspace of all these levels.

Before we carry out this diagonalization, however, we have to pay attention also to the isospin of the levels whose position we want to compute. The electric dipole operator is given by

$$\mathbf{D} = \sum_i e\left(\frac{1 + \tau_{i3}}{2}\right)\mathbf{r}_i = \frac{e}{2}\sum \mathbf{r}_i + \frac{e}{2}\sum \tau_{i3}\,\mathbf{r}_i \tag{14.28}$$

The isoscalar part of $\mathbf{D}$ is just proportional to the center-of-mass coordinate; it cannot therefore contribute to matrix elements of $\mathbf{D}$ between two states of different *intrinsic* structure. Thus an electric dipole radiation can proceed only via the second, isovector, part. An electric dipole radiation therefore carries with it not only one unit of angular momentum and parity but also one unit of isospin. If Φ_0 has $T = 0$, like in O^{16}, we are only interested in excited states with $J = 1^-$ and $T = 1$ for the description of the giant dipole state. (This, incidentally, also assures us that we do not mix in center-of-mass excitations; see discussion in Section V.14. In O^{16} there are five such $1p1h$-states whose zeroth-order energy is $\hbar\omega$ (the proper combinations of proton ph-excitation and a similar neutron ph-excitation).

The matrix elements required for the diagonalization in the space of the $1p1h$-configurations can be computed using the techniques developed in Section V.8; the approximation of confining oneself to these matrix elements alone is known as the Tamm-Dancoff approximation (or sometimes as the $1\hbar\omega$ TD approximation). The inclusion of all the ($J = 1^-$, $T = 1$) states at a zeroth-order excitation of $3\hbar\omega$ is referred to as *Second-Order Tamm-Dancoff (or $3\hbar\omega$ TD) approximation.* To be consistent such $3\hbar\omega$ TDA calculations for the excited states must be accompanied by a $2\hbar\omega$ TDA calculation for the ground state of the system (i.e., deriving the ground state energy and wave function from the diagonalization of the Hamiltonian in the subspace of all states involving no excitations and $2p2h$ excitations).

Such Tamm-Dancoff calculations were carried out for a number of nuclei [see, for instance, Elliott and Flowers (57), Lane (64), Brown (64), Danos and Fuller (65)]. They yield both the corrected energies of the various states and their structure in terms of the configurations that span the space of the particular TDA. The inclusion of the interaction among the particles pushes the energy levels from their zeroth-order position given by (14.27) up in the direction actually required by the empirical data. This is due to the fact that the interaction in $T = 1$ ph-states is repulsive. One or two states are pushed up particularly strongly, and their wave functions assume a structure that enhances strongly the dipole transitions between them and the ground state. We shall not analyze here these results in detail, but present a much over simplified model—the schematic model of Brown and Bolsterli (59)—that will explain why these particular features show up in these calculations:

Assume that there are N-degenerate, zeroth-order, $J = 1^-$ states $\phi_1, \ldots, \phi_N$ (each of them is an A-particle wave function) and assume further that the interaction between the particles can be written in the form

$$v(12) = \sum_k g_k \mathbf{T}^{(k)}(1) \cdot \mathbf{T}^{(k)}(2) \tag{14.29}$$

where g_k are constants and $\mathbf{T}^{(k)}(i)$ are irreducible tensors of degree k operating on the coordinates of the ith particle. A component of $\mathbf{T}^{(k)}$ is written $\mathbf{T}_\kappa^{(k)}$. Let us now consider a matrix element

$$\begin{aligned} \langle n|v(12)|m\rangle &= \sum g_k \langle n|\mathbf{T}^{(k)}(1) \cdot \mathbf{T}^{(k)}(2)|m\rangle \\ &= \sum_k g_k \sum_s \langle n|\mathbf{T}^{(k)}(1)|s\rangle \cdot \langle s|\mathbf{T}^{(k)}(2)|m\rangle \end{aligned} \tag{14.30}$$

In (14.30) m and n are two of the N-states ϕ_i, and s is a complete set of states. The state m differs from Φ_0 in that one of the particles is lifted to an unoccupied level; in the state n it is *another* particle that is thus lifted. Noting that $\mathbf{T}^{(k)}(i)$ can change only the state of particle i we see right away that (14.30) can be different from zero only if in ϕ_n it is particle 1 that is lifted and in ϕ_m it is particle 2. Furthermore only $|s\rangle = \Phi_0$ will contribute to the sum over s; indeed if particle 2, for instance is not back in its assigned level in $|s\rangle$, then $\langle n|\mathbf{T}^{(k)}(1)|s\rangle = 0$, since in ϕ_n it is particle 1 that is lifted to an excited state, and $T^{(k)}(1)$ cannot change the state of particle 2. We thus obtain

$$\langle n|v(12)|m\rangle = \sum g_k \langle n|\mathbf{T}^{(k)}(1)|0\rangle \cdot \langle 0|\mathbf{T}^{(k)}(2)|m\rangle \tag{14.31}$$

We now note that $\Phi_0 = |0\rangle$ has $J = 0^+$, while ϕ_n and ϕ_m have $J = 1^-$ since we are confining ourselves to diagonalization in the $J = 1^-$ subspace. The matrix element

$$\langle n\, J = 1^-|\mathbf{T}^{(k)}(1)|0\, J = 0^+\rangle$$

therefore does not vanish only if $k = 1$ and $\mathbf{T}^{(1)}$ has a negative parity (see the Wigner-Eckart theorem, Appendix A, A.2.44).

Moreover, if the eigenvalue of J_z for the state $|n, J = 1^-\rangle$ is M, then the only component of $\mathbf{T}^{(k)}$ that contributes is the $\kappa = k$ component. Finally because of the scalar property of $v(12)$, the J_z value for the state ϕ_m is also M.

Effectively, as far as the matrix element is concerned,

$$\mathbf{T}^{(1)}(1) = \lambda(r)\,\mathbf{r}_1 \tag{14.32}$$

We can write the matrix element (14.31) in the form

$$\langle n|v(12)|m\rangle = \frac{1}{3} g_1 V_n V_m \tag{14.33}$$

where V_n is the reduced matrix element (Appendix A, Sect. 2) of $\lambda(r_1)\,\mathbf{r}_1$ between the ground state Φ_0 and the state Φ_n:

$$V_n = (0||\lambda(r_1)\,\mathbf{r}_1||n) \tag{14.34}$$

Note that V_n is independent of M. A simplified model that is often used asserts that

$$V_n = \lambda D_n \qquad D_n \equiv (0||\mathbf{r}_1||n) \tag{14.35}$$

where λ is independent of n. This can only be exact if the various states Φ_n involve the same orbitals. Inserting (14.35) into (14.33) yields

$$\langle n|v(12)|m\rangle = GD_nD_m \tag{14.36}$$

with $G = (1/2)\,\lambda^2 g_1$. This is the result we shall use below. However, as it should be clear, (14.33) and its simplified form (14.36) follow from the fact that within the set of $1p1h$-states only one multipole in the expansion (14.29) is operative: the one with $k = 1$.

We can now write down the complete matrix $\langle n|\Sigma v(ik)|m\rangle$. It takes the form*

$$G\begin{pmatrix} D_1D_1, & D_1D_2, & \ldots, & D_1D_N \\ D_2D_1, & D_2D_2, & \ldots, & D_2D_N \\ \cdot & \cdot & & \cdot \\ \cdot & \cdot & & \cdot \\ \cdot & \cdot & & \cdot \end{pmatrix} \tag{14.37}$$

Its diagonalization can be carried out easily if we note that its rank is 1; (the mth row can be obtained from the first by multiplying the latter by D_m/D_1); all its eigenvalues therefore vanish except one, and since the trace of (14.37)

*Note that the following analysis does not require (14.35). Only form (14.33) is essential.

is invariant under diagonalization we find that the one nonvanishing eigenvalue is

$$\Delta E = G \sum D_i^2 \tag{14.38}$$

The eigenfunctions are also easy to write down; those which belong to the eigenvalues $\Delta E = 0$ are, for instance

$$\psi = \frac{D_2 \phi_1 - D_1 \phi_2}{\sqrt{D_1^2 + D_2^2}} \tag{14.39}$$

as can be verified by direct substitution. The one that belongs to the only nonvanishing eigenvalue shift ΔE of (14.38) is an admixture of all the states

$$\phi = (D_1^2 + D_2^2 + \ldots + D_N^2)^{-1/2} (D_1\phi_1 + D_2\phi_2 + \ldots + D_N\phi_N) \tag{14.40}$$

We now see a very interesting feature of this calculation: The states that were not shifted in energy, like ψ in (14.39), have a vanishing transition probability to the ground state via the dipole operator:

$$|(\psi|\,|\mathbf{D}|\,|0)|^2 = \frac{1}{(D_1^2 + D_2^2)} [D_2 (\phi_1|\,|\mathbf{D}|\,|0) - D_1 (\phi_2|\,|\mathbf{D}|\,|0)]^2 = 0 \tag{14.41}$$

where we have used (14.35). On the other hand the dipole transition probability from the state ϕ is enhanced:

$$\begin{aligned} |(\phi|\,|\mathbf{D}|\,|0)|^2 &= \frac{1}{D_1^2 + D_2^2 + , \ldots , + D_N^2} \left|\sum D_i (\phi_i|\,|\mathbf{D}|\,|0)\right|^2 \\ &= \frac{1}{\sum D_i^2} \cdot (\textstyle\sum D_i^2)^2 = \sum D_i^2 = N \langle D_i^2 \rangle \end{aligned} \tag{14.42}$$

where N equals the number of ph-states used in the TDA.

The result of switching on the interaction among all the degenerate $J = 1^-$ $1p1h$-states has thus been twofold: it shifted one of the levels by an amount ΔE, so as to decrease the discrepancy between (14.27 and (14.1), and at the same time enhanced the dipole transition probability to this state by a factor N at the cost of depleting the other $J = 1^-$ states of their dipole transition probabilities.

This fact is made more obvious by noting that state (14.40) can be written (multiply D_m by the 3j factor $\begin{pmatrix} -1 & 1 & 0 \\ M & M & 0 \end{pmatrix}$ so that the reduced matrix D_m can be replaced by the matrix element $\langle 0|\mathbf{r}_i|m, J = 1^-, M\rangle$)

$$\Phi_M = F D_M \Phi \tag{14.43}$$

where F is a normalization factor:

$$F^{-2} = \langle \Phi \,|D_M^2|\, \Phi \rangle$$

and D_M is the isovector part of the dipole operator:

$$D_M = \sum_i \tau_3(i)\, Y_{1M}(\hat{\mathbf{r}}_i)\, r_i \tag{14.44}$$

If Φ_0 has zero spin, then the Φ_M given by (14.43) has unit spin and with projection, M. From (14.43) and (14.44) Φ_M is just the state developed by the interaction of the dipole component of an incident electromagnetic wave with the target nucleus. Clearly the transition induced can only go to Φ_M; the transition probabilities to all other states is zero.

The collective nature of the giant-dipole state in this schematic model comes therefore through the special mixing of configurations implied by (14.40). All the zeroth-order states are included in (14.40), each of them with a coefficient that has the right phase to assure a constructively interfering contribution of each ϕ_n in (14.40) to the dipole transition probability.

It now becomes apparent why it was so relevant for the $J = 1^-$ $1p1h$-states to have picked out of the interaction $v(12)$ in (14.29) exactly the dipole-dipole term, $k = 1$. In general the kth multipole in the nucleon-nucleon interaction is specifically suited to endow a corresponding core state having $J = k$ with collective features for the emission of k-pole radiation in analogy with (14.43). This can be seen immediately if we inspect the expression (14.31) for the matrix element $\langle n|v(12)|m\rangle$: for a given value of $k = k_0$ the matrix element $\langle n|T^{(k)}|0\rangle$ will be different from zero only if the angular momentum J_n of the state $|n\rangle$ satisfies $J_n = k_0$. This relation is sometimes referred to as the *specificity* of the particular multipole in the nuclear interaction.

An estimate of the energy at which these other multipoles will resonate can be obtained from the hydrodynamic model that would require $j_\lambda'(kR)$ to be zero where λ is the multipole order. The lowest resonant energy for each multipole are shown in the following table.

TABLE 14.1

λ	E (MeV)
0	$151A^{-1/3}$
1	$70A^{-1/3}$
2	$112A^{-1/3}$
3	$152A^{-1/3}$
4	$190A^{-1/3}$

The specificity extends not only to multipole order but also to the isospin character of the excitation. The giant dipole resonance is not only a 1^- state, it also has a well-defined isospin. As can be seen from (14.44), the isospin can be $T_0 \pm 1$ where T_0 is the isospin of the ground state. In an even-even nucleus, for which the ground state has zero spin, the isospin of the giant dipole resonance has an isospin of 1. For other nuclei there can be two differing states that will resonate [see Fallieros, Goulard, and Venter (65)]. The $T = 1$ dipole resonance that the hydrodynamic model emphasizes involves the two kinds of nucleons moving in opposite directions. Resonances with $T = 0$ in which these nucleons move in phase are also possible. The hydrodynamic model also suggests other collective modes involving spin as well as spin-isospin oscillations [Glassgold, Heckrotte, and Watson (59)].

The observability of these resonances will depend very much upon their width; as was noted earlier these states can be unbound. They acquire their width in virtue of coupling with other nuclear states via the residual interaction. For example the $1p1h$-states just discussed above will couple to $3p3h$ states [Shakin and Wang (71)]. If the coupling is strong the width will be correspondingly large with the consequence that it will be difficult to distinguish the resonance from the background produced by other nonresonant mechanisms. The state generated by the interaction specificity, as in the case of (14.43), is called a *doorway state*. Equation 14.43 is an example of a more general mechanism leading to doorway state resonances [Feshbach (66), Block and Feshbach (63)].

15. SUMMARY

The last four chapters dealt with various manifestations of nuclear structure, stressing both the underlying fundamental feature of a nearly independent-particle motion, and the important role played by the average interaction and the residual interaction. The picture that emerges now is the following.

The nucleons in the nucleus are held together by their mutual interaction. Because of the hard repulsive core in this interaction they are prevented from coming too close to each other, and a universal average density is obtained for all nuclei. The Pauli principle limits the effects of the nucleon-nucleon interaction to relatively small internucleon separations, thus providing the ground of an approximately independent-particle motion for the nucleons in the nucleus.

Using the Hartree-Fock method it then becomes reasonable to generate self-consistent levels for the individual nucleons. The self-consistent potential is generally ellipsoidal and represents a balance between the requirements of the kinetic energy, which prefers equal extension of the single-particle wave function in all directions (i.e., a spherical potential), and the tendency of the attractive nuclear interaction to maximize the overlap between the single-particle wave functions, thus leading to a deformed potential. The Pauli principle again plays a decisive role in establishing the final equilibrium shape of the potential.

It is a general feature of potentials of spherical symmetry that they lead to a "bunching" of single-particle levels. Depending on the number of nucleons in the system one may therefore have one shape of the nucleus or another. Magic numbers, corresponding to closed shells, have particular stability associated with them. The actual numerical values of the magic numbers turn out to be crucially determined by the spin-dependent part of the nucleon-nucleon interaction, thus indicating how an important gross feature of nuclei is related to an "accidental" feature of the fundamental two body interaction.

The average self-consistent potential cannot completely replace the nucleon-nucleon interaction for two reasons: first, there are strong variations in the nucleon-nucleon force when the two nucleons come close together, which an average potential, acting on only one nucleon at a time, cannot reproduce. And second, since the overall potential is produced by the particles themselves, it undergoes variations in time that reflect the motion of the particles.

Both corrections to the average potential can in principle be accounted for by properly taking into account the residual interaction. If the residual interaction is treated exactly, one has, of course, an exact solution of the nuclear problem. It turns out, however, that in some cases it is enough to take the residual interaction only to the first order in perturbation theory to account for the major part of the phenomena observed, and one witnesses in these cases a most remarkable success of the description of nuclei in terms of the shell model, with its many underlying symmetries. In other cases, first-order perturbation theory may not be enough for the proper handling of the residual interaction. Its treatment to higher orders in perturbation theory is then called for. Fortunately there exists in many of these cases an alternative way to tackle the problem: it is possible to convert some of the parameters determining the average potential into dynamical variables and handle them as slowly varying degrees of freedom in an adiabatic approximation. Various types of collective motions can be invoked this way and there is empirical evidence for the extreme usefulness of this approach in some regions of the periodic table. In particular the collective rotations of deformed nuclei seem to persist to the fairly high excitations, as can be seen in Figs. 5.2, 15.1, and 15.2. Spins as high as 20 have been observed (Thieberger et al. (71)).

The parameters of the collective motions in nuclei are related to the detailed structure of the nucleus. In the spirit of the adiabatic approximation, for instance, the potential energy for the "slow" motion (such as nuclear vibrations of various types) is the instantaneous energy of the "fast" intrinsic motion, and this establishes a relation between the two. Many of these parameters can now be derived this way even quantitatively, in nice agreement with experiment.

Empirical evidence seems to indicate that although the collective variables

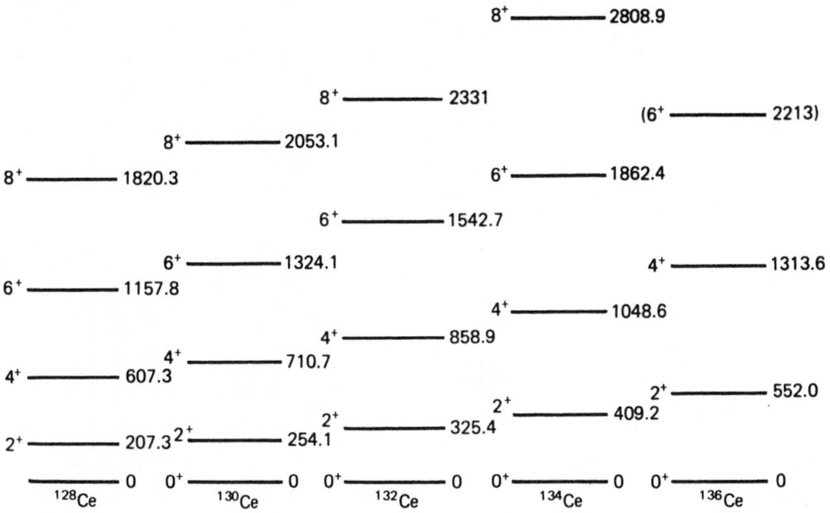

FIG. 15.1. Ground band energy levels in the light even Cerium nuclei [taken from Ward et al. (68)].

are fundamentally functions of the particle variables, in many cases the two sets can be treated independently as if the system had extra degrees of freedom. This raises immediately the question of the dynamical interplay between these extra degrees of freedom and those of the particles and also between one collective mode and another. This will be most probably an area of fruitful future development in the theory of nuclear structure.

There are some modes of motion, like the giant dipole states, that can be equally well treated with present-day techniques both from the point of view of their microscopic structure and also in terms of collective variables. They offer therefore a good opportunity to learn how particular features of one approach are reflected in the other. The general question of expressing the collective variables in terms of the particle variables remains very difficult. Some formal studies of this problem have been carried out [Klein and Kerman (65) or Peierls and Thouless (62)], and in the case of collective rotations it was possible to justify the Bohr Hamiltonian with a substantial degree or rigor [Villars and Cooper (69)]. We are, however, not yet in a position to show formally that a definite part of the interaction is responsible for the onset of collective motion of a certain type in nuclei of a given mass. This will have to await further developments in the many-body theory. The closest one has come to it are the calculations of Kumar and Baranger (68), which are based on a special simple model interaction and will be described in Section VII.18.

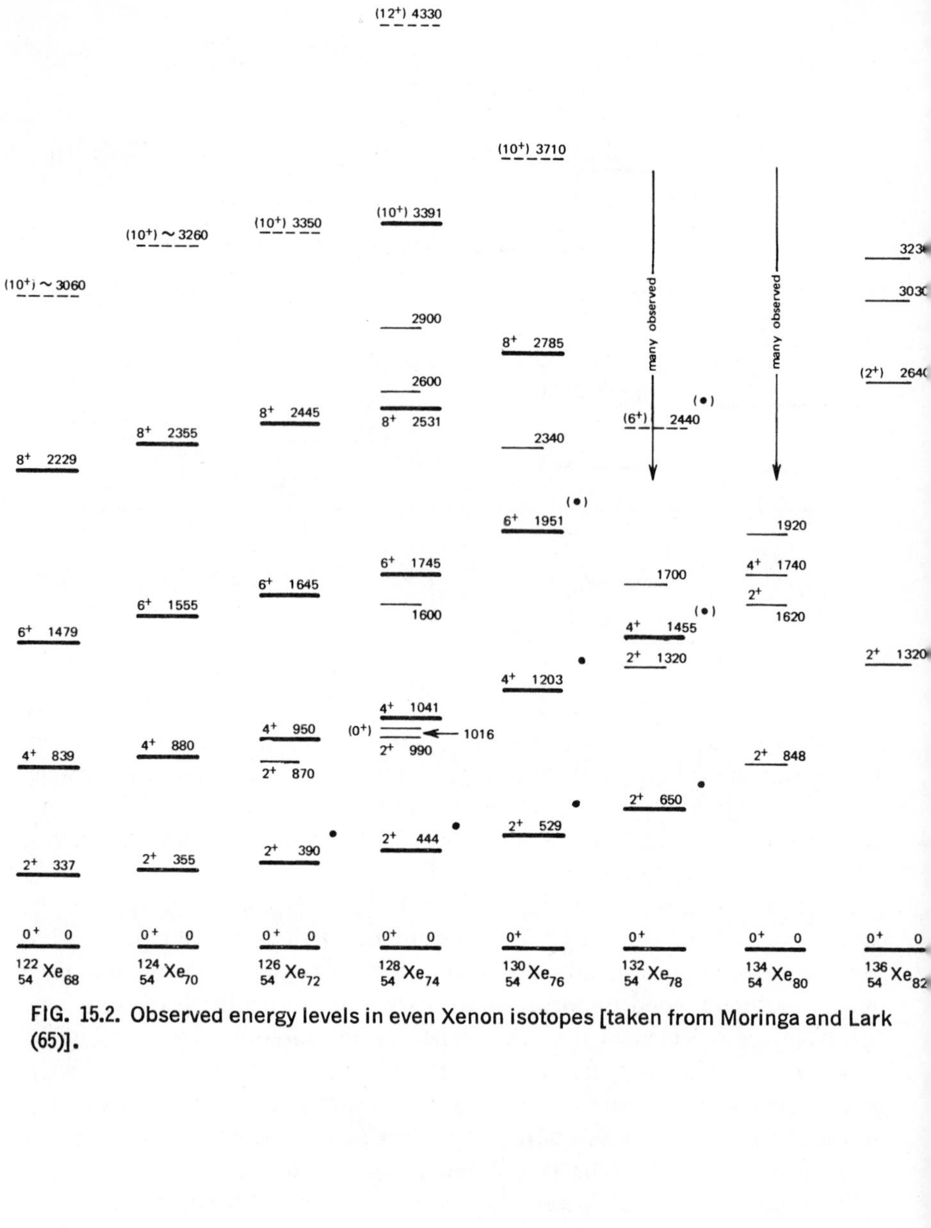

FIG. 15.2. Observed energy levels in even Xenon isotopes [taken from Moringa and Lark (65)].

CHAPTER VII
MULTINUCLEON SYSTEMS

We discussed in the last two chapters two different, but relatively simple, aspects of the behavior of multiparticle systems. In the framework of the shell model we dealt with particles moving in a self-produced average spherical field. The approach toward the more realistic situation in this framework was obtained by the introduction of an effective mutual interaction among the particles as a perturbation, to be considered usually only in first order. In another extreme situation we realized that a system of nucleons can perform also slow rotations and vibrations in which all the nucleons take part in a coherent way. We then introduced a set of essentially phenomenological parameters to describe these motions, and found that with the use of only a few it was possible to account for a variety of data. We also found it possible to integrate the description of such collective motions and the outstanding features of independent-particle motion, arriving in this way at Bohr and Mottelson's unified model.

Despite these remarkable successes in interpreting and interrelating nuclear data of various sorts, the fundamental questions in nuclear structure have yet to be faced. We still have to understand how a Hamiltonian of the general type $\Sigma T_i + (1/2)\Sigma v(ij)$ can be approximated by a shell-model Hamiltonian in some cases, and leads to collective motions in others. The two-body nuclear force $v(ij)$ whose complete nature is still not fully understood is most probably responsible for the systematics observed in nuclei and for the validity of some of the simple approximations that we have been using. We would therefore like to derive the relation between some features of $v(ij)$ and the main parameters of the shell-model, or the unified-model, Hamiltonians. We want to be able to relate the electromagnetic properties of nuclei to the properties of free nucleons, and to determine to what extent the packing of nucleons in the nucleus does or does not modify significantly the intrinsic properties of the individual nucleons.

These, and other problems, require a considerably more detailed study of the multinucleon system. This will be the subject of this chapter. We stress from the outset that at present we are still far from having a complete answer to most of the questions that were raised above. Many aspects of the multinucleon system are still understood only qualitatively and their quantitative understanding requires further developments in the theory of such systems. However, the progress made thus far in such theories, and the availability of faster and larger computers, do enable us to make some meaningful statements about the origin of some of the regularities observed in the properties of nuclei.

The theory of multinucleon systems should predict the energy spectra of such systems, and the corresponding wave functions, expressed in terms of the coordinates of all the particles of the system. A less ambitious program is concerned only with the ground states and low-lying levels of the multinucleon systems, and is satisfied with the one- and two-nucleon density matrices for these states instead of with their complete wave functions.

With present-day computing facilities this less ambitious program can be accomplished for selected nuclei by a straightforward diagonalization of large-enough matrices. This is true especially for light nuclei near closed shells, where the use of relatively few configurations may constitute a valid approximation. For most nuclei, however, more elaborate approximation methods have to be used.

An important feature of the multinucleon system is the required antisymmetry of its wave function. When many particles are involved, the exact bookkeeping of the effects of antisymmetrization becomes very complex if one uses wave functions in ordinary space. A method has therefore been developed to take antisymmetry automatically into account, and to relieve us, to a large extent, from the necessity of worrying about phases and normalizations. It is referred to as *the method of second quantization*, and has wide uses in field theory, where it was originally introduced. We shall proceed now to describe some of its main features.

1. THE METHOD OF SECOND QUANTIZATION

To facilitate the analysis of systems with many identical particles it is convenient to work with wave functions defined in *Fock space*. State vectors in Fock space are characterized by giving:

(i) the possible single-particle states $|\alpha\rangle$ that any of the particles can occupy, and

(ii) the set of occupation numbers for these states.

The occupation number for a given particle state $|\alpha\rangle$ describes how many particles actually occupy that state. For fermions the Pauli principle tells us that this occupation number can be either 0 or 1. For bosons it can, of course, be any nonnegative integer.

The single-particle states $|\alpha\rangle$ are taken to be a complete set of mutually orthogonal states, such as the set of eigenstates of a single-particle Hamiltonian

$$H_0|k_\alpha\rangle = \epsilon_\alpha|k_\alpha\rangle \tag{1.1}$$

H_0 might be the Hamiltonian $(1/2M)p^2$ for free particles in a box, in which case the states $|k_\alpha\rangle$ are plane waves multiplied by the appropriate spin and isospin wave functions. H_0 might, however, be a Hamiltonian describing the motion of a single particle in a hypothetical spin and isospin dependent central potential $U(r, \mathbf{s}, \mathbf{t})$, in which case the states $|k_\alpha\rangle$ will be considerably more complicated. The formalism we are going to develop in this section is independent of the nature of H_0. This freedom in the choice of the Hamiltonian generating the single-particle states $|k_\alpha\rangle$ will be made use of in later sections.

A single-particle state in Fock space is denoted by $|\alpha\rangle$. It describes a particle in the state $|k_\alpha\rangle$ of H_0. A two-particle state is denoted by $|\alpha\beta\rangle$ and describes a state in which the states $|k_\alpha\rangle$ and $|k_\beta\rangle$ of H_0 are occupied by two identical particles. Fock space contains all the states with an arbitrary number of particles as well as the state in which there is no particle present. This latter state is referred to as the *vacuum* and is usually denoted by $|0\rangle$. It is usually assumed that this state is unique.

To the linear vector space formed by the "kets" $|\alpha_1, \ldots, \alpha_n\rangle$ there is a dual vector space formed by the "bras" $\langle\alpha_1, \alpha_2, \ldots, \alpha_n|$, and the scalar product of two states $\langle\alpha_1, \alpha_2, \ldots, \alpha_n|$ and $|\beta_1, \beta_2, \ldots, \beta_n\rangle$ is denoted by

$$\langle\alpha_1, \alpha_2, \ldots, \alpha_n|\beta_1, \beta_2, \ldots, \beta_m\rangle \tag{1.2}$$

If $n \neq m$, or if the set $(\alpha_1, \ldots, \alpha_n) \neq (\beta_1, \ldots, \beta_m)$ then

$$\langle\alpha_1, \alpha_2, \ldots, \alpha_n|\beta_1, \beta_2, \ldots, \beta_m\rangle = \langle\beta_1, \beta_2, \ldots, \beta_m|\alpha_1, \alpha_2, \ldots, \alpha_n\rangle = 0 \tag{1.3}$$

defines the nondiagonal elements of the metric tensor in the Fock space; we have still left open the diagonal elements of the metric tensor, that is, the normalization of the various states. This will be described in (1.16) and (1.17).

It is convenient to introduce in the Fock space operators that connect various states to each other. The simplest operators of that kind are the *creation operators* $a_k^\dagger$ and the *annihilation operators* a_k. They are defined as follows: let $|\alpha_1, \ldots, \alpha_n\rangle$ be an n-particle state in Fock space and let $|\beta_1\alpha_1, \ldots, \alpha_n\rangle$ be an $n+1$ particle state. Then we can define a creation operator $a_\beta^\dagger$ by

$$a_\beta^\dagger \equiv \sum_{n=0}^{\infty} |\beta, \alpha_1, \ldots, \alpha_n\rangle\langle\alpha_1, \ldots, \alpha_n| \tag{1.4}$$

so that

$$a_\beta^\dagger|\alpha_1, \alpha_2, \ldots, \alpha_n\rangle = |\beta, \alpha_1, \alpha_2, \ldots, \alpha_n\rangle \tag{1.5}$$

We have used normalization condition (1.16).

Similarly

$$a_{\alpha_1} \equiv \sum_{n=1}^{\infty} |\alpha_2, \ldots, \alpha_n\rangle\langle\alpha_1\alpha_2, \ldots, \alpha_n| \tag{1.6}$$

so that

$$a_{\alpha_1}|\alpha_1, \alpha_2, \ldots, \alpha_n\rangle = |\alpha_2, \ldots, \alpha_n\rangle \tag{1.7}$$

and

$$a_\beta|\alpha_1, \alpha_2, \ldots, \alpha_n\rangle = 0 \quad \text{if} \quad \beta \neq \alpha_1, \alpha_2, \ldots, \alpha_n \tag{1.8}$$

We shall use the notation

$$a_{k_\alpha}^\dagger \equiv a_\alpha^\dagger \qquad a_{k_\alpha} \equiv a_\alpha \tag{1.9}$$

Note that the operator $a_\beta^\dagger$ converts a n-particle state into an $n + 1$ particle state, and puts the additional state in a specific place relative to the others. Since we are dealing with identical particles, the phase of a state can depend on the order that the single-particle states in it are specified and, thus, the order of the indices is important.

It follows from (1.5) that

$$a_\alpha^\dagger|0\rangle = |\alpha\rangle \tag{1.10}$$

The definition of the vacuum $|0\rangle$ is made complete by requiring also that, in agreement with (1.8),

$$a_\alpha|0\rangle = 0 \tag{1.11}$$

In other words the annihilation of any particle from the vacuum leads to a vanishing result.

We shall be dealing primarily with fermions. In this case a certain convention has to be adopted with regard to the phase of a multiparticle state $|\alpha_1, \alpha_2, \ldots, \alpha_n\rangle$. We want the state vectors in the Fock space to be in a one-to-one correspondence with the usual Slater determinants. There, too, one has to introduce a certain convention to determine the phases of the wave functions. The way this is done is to arrange the single-particle wave functions $\phi_{k_i}(\mathbf{r})$ according to a certain order, say $k_1 < k_2 < k_3, \ldots$, and then write a "standard" Slater determinant of particles in the orbits $k, l, m \ldots$ as

$$\begin{vmatrix} \phi_k(1) & \phi_l(1) & \phi_m(1) \ldots \\ \phi_k(2) & \phi_l(2) & \phi_m(2) \ldots \\ \vdots & & \end{vmatrix} \equiv \Phi_{klm\cdots}(1, 2, \ldots) \tag{1.12}$$

where it is understood that for a standard determinant $k < l < m \ldots$. It is obvious from (1.12) that

$$\Phi_{klm\cdots} = - \Phi_{lkm\cdots} \text{ etc.} \tag{1.13}$$

A similar procedure is adopted in the second quantization formalism. The states $|k_1\rangle$, $|k_2\rangle$, $\ldots$ generated by H_0 are arranged according to an arbitrary standard order, say, the order

$$k_1 < k_2 < k_3 \tag{1.14}$$

An n-particle states $|\alpha_1, \alpha_2, \ldots, \alpha_n\rangle$ will then be written in the standard form if

$$k_{\alpha_1} < k_{\alpha_2} < \ldots < k_{\alpha_n}$$

To assure the antisymmetry we require in analogy with (1.13), that,

$$|\alpha_1, \alpha_2, \alpha_3, \ldots, \alpha_n\rangle = -|\alpha_2, \alpha_1, \alpha_3, \ldots, \alpha_n\rangle \tag{1.15}$$

We can now complete the definition of the metric in the Fock space by defining

$$\langle \alpha_1, \alpha_2, \ldots, \alpha_n | \alpha_1, \alpha_2, \ldots, \alpha_n \rangle = 1 \tag{1.16}$$

or more generally

$$\langle \alpha_1, \alpha_2, \ldots, \alpha_n | \beta_1, \beta_2, \ldots, \beta_n \rangle = (-1)^{P_\alpha + P_\beta} \tag{1.17}$$

(provided the set $\{\alpha\}$ is a permutation of the set $\{\beta\}$). P_α is the number of permutations required to bring the set $\{\alpha\}$ into the standard order, and similarly for P_β.

From (1.5) we conclude that

$$|\alpha_1, \alpha_2, \alpha_3, \ldots, \alpha_n\rangle = a_{\alpha_1}^\dagger a_{\alpha_2}^\dagger |\alpha_3, \ldots, \alpha_n\rangle \tag{1.18}$$

and

$$|\alpha_2, \alpha_1, \alpha_3, \ldots, \alpha_n\rangle = a_{\alpha_2}^\dagger a_{\alpha_1}^\dagger |\alpha_3, \ldots, \alpha_n\rangle$$

Comparing (1.18) and (1.17) we see that to satisfy them both we must require that the operators $a_\beta^\dagger$ satisfy

$$a_\beta^\dagger a_\gamma^\dagger = - a_\gamma^\dagger a_\beta^\dagger \tag{1.19}$$

Using (1.8) we can conclude in a similar way that

$$a_\beta a_\gamma = - a_\gamma a_\beta \tag{1.20}$$

Thus, for fermions, two different creation operators or two annihilation operators always anticommute among themselves.

From (1.15) we conclude that if the set $(\alpha_1, \alpha_2, \ldots, \alpha_n)$ contains two identical labels, say if $\alpha_1 = \alpha_2$, then $|\alpha_1, \alpha_2, \ldots, \alpha_n\rangle = 0$. This is another way of saying that two fermions cannot occupy the same state at the same time. Note the difference between the vacuum state $|0\rangle$ and the statement $|\alpha_1\alpha_1\rangle = 0$: The vacuum state $|0\rangle$ is a legitimate state in the Fock space that happens not to have any particle in it. On the other hand the statement $|\alpha_1\alpha_1\rangle = 0$ means that the state $|\alpha_1\alpha_1\rangle$ does not belong to the Fock space we are considering.

Consider (1.5) again. In view of what we have just said it can be rewritten in the form

$$a_\beta^\dagger|\alpha_1, \alpha_2, \ldots, \alpha_n\rangle = \begin{cases} |\beta, \alpha_1, \alpha_2, \ldots, \alpha_n\rangle & \text{if} \quad \beta \neq \alpha_1, \alpha_2, \ldots, \alpha_n \\ 0 & \text{otherwise} \end{cases} \tag{1.21}$$

Suppose now $\beta \neq \alpha_1, \alpha_2, \ldots, \alpha_n$. From (1.21) and (1.7) we obtain

$$a_\beta a_\beta^\dagger|\alpha_1, \alpha_2, \ldots, \alpha_n\rangle = a_\beta|\beta, \alpha_1, \alpha_2, \ldots, \alpha_n\rangle = |\alpha_1, \ldots, \alpha_n\rangle \tag{1.22}$$

whereas from (1.11) we obtain

$$a_\beta^\dagger a_\beta|\alpha_1, \alpha_2, \ldots, \alpha_n\rangle = a_\beta^\dagger \cdot 0 = 0 \tag{1.23}$$

Hence

$$(a_\beta a_\beta^\dagger + a_\beta^\dagger a_\beta)|\alpha_1, \alpha_2, \ldots \alpha_n\rangle = |\alpha_1, \alpha_2, \ldots, \alpha_n\rangle \quad \text{if} \quad \beta \neq \alpha_1, \alpha_2, \ldots, \alpha_n \tag{1.24}$$

In case β does equal one of the α's, say, $\beta = \alpha_2$, we obtain from (1.21)

$$a_\beta a_\beta^\dagger|\alpha_1, \alpha_2, \ldots, \alpha_n\rangle = 0 \quad \text{if} \quad \beta = \alpha_2$$

and from (1.7) and (1.21), again for $\beta = \alpha_2$

$$\begin{aligned} a_\beta^\dagger a_\beta|\alpha_1, \alpha_2, \ldots, \alpha_n\rangle &= -a_\beta^\dagger a_\beta|\alpha_2, \alpha_1, \alpha_3, \ldots, \alpha_n\rangle \\ &= -a_\beta^\dagger|\alpha_1, \alpha_3, \ldots, \alpha_n\rangle \\ &= -|\beta, \alpha_1, \alpha_3, \ldots, \alpha_n\rangle \\ &= -|\alpha_2, \alpha_1, \alpha_3, \ldots, \alpha_n\rangle \\ &= |\alpha_1, \alpha_2, \alpha_3, \ldots, \alpha_n\rangle. \end{aligned}$$

Thus, in this case again we find that (1.24) is valid. Hence (1.24) is valid for every value of β and we can conclude that

$$a_\beta a_\beta^\dagger + a_\beta^\dagger a_\beta = 1 \tag{1.25}$$

where 1 is the unit operator.

Finally we can show in a similar way that for any β and γ, $\beta \neq \gamma$

$$a_\beta^\dagger a_\gamma|\alpha_1, \ldots, \alpha_n\rangle = -a_\gamma a_\beta^\dagger|\alpha_1, \ldots, \alpha_n\rangle \tag{1.26}$$

and therefore

$$a_\beta^\dagger a_\gamma = -a_\gamma a_\beta^\dagger \quad \text{for} \quad \beta \neq \gamma \tag{1.27}$$

Equations (1.19), (1.20), (1.25), and (1.27) can be summarized in the equations:

$$\{a_\alpha^\dagger, a_\beta^\dagger\} = \{a_\alpha, a_\beta\} = 0$$
$$\{a_\alpha^\dagger, a_\beta\} = 1\delta_{\alpha\beta} \tag{1.28}$$

where we have introduced the notation of the anticommutator

$$\{AB\} \equiv AB + BA \tag{1.29}$$

Furthermore it is easy to see that (1.5) and (1.7), with their extension (1.21), and the metric defined in the Fock space, lead to the conclusion that the operator $a_\alpha^\dagger$ is the Hermitian conjugate of a_α, so that

$$(a_\alpha)^\dagger \equiv a_\alpha^\dagger \tag{1.30}$$

The creation and annihilation operators connect states with n-particles with states with $n \pm 1$ particles. In most applications to nuclear physics, however, we are interested in operators that connect states of n-particles with other states with the *same* number of particles. All known interactions conserve the number of baryons. Such operators are easily constructed with the help of the annihilation and creation operators. Consider the operator

$$a_\beta^\dagger a_{\alpha_1} \tag{1.31}$$

and assume it operates on the state $|\alpha_1, \ldots, \alpha_n\rangle$. If $\beta \neq \alpha_2, \ldots, \alpha_n$ we obtain then

$$a_\beta^\dagger a_{\alpha_1}|\alpha_1, \alpha_2, \ldots, \alpha_n\rangle = a_\beta^\dagger|\alpha_2, \ldots, \alpha_n\rangle = |\beta, \alpha_2, \ldots, \alpha_n\rangle \tag{1.32}$$

Hence

$$\langle\beta, \alpha_2, \ldots, \alpha_n|a_\beta^\dagger a_{\alpha_1}|\alpha_1, \alpha_2, \ldots, \alpha_n\rangle = 1 \qquad \text{if} \qquad \beta \neq \alpha_2, \ldots, \alpha_n \tag{1.33}$$

Thus, the operator $a_\beta^\dagger a_{\alpha_1}$ connects the n-particle state to the n-particle state $|\beta, \alpha_2 \ldots, \alpha_n\rangle$. We can say that the operator $a_\beta^\dagger a_{\alpha_1}$ shifts a particle from the state $|\alpha_1\rangle$ to the state $|\beta\rangle$. More precisely, it does so provided it acts on a state that, to begin with, contains a particle in the state $|\alpha_1\rangle$ and no particles in the state $|\beta\rangle$. If these conditions are not fulfilled, then the application of $a_\beta^\dagger a_{\alpha_1}$ to such states leads to a vanishing result.

In particular the operator $\hat{N}_\beta = a_\beta^\dagger a_\beta$ leads from the state $|\alpha_1, \alpha_2, \ldots, \alpha_n\rangle$ to itself if β is identical to one of the α_i's, and to a vanishing result otherwise:

$$a_\beta^\dagger a_\beta|\alpha_1, \ldots, \alpha_n\rangle = \begin{cases} |\alpha_1, \ldots, \alpha_n\rangle & \text{if} \quad \{\alpha\} \supset \beta \\ 0 & \text{if} \quad \{\alpha\} \not\supset \beta \end{cases} \tag{1.34}$$

It follows from (1.34) that

$$\langle\alpha_1, \alpha_2, \ldots, \alpha_n|\sum_{k=1}^{\infty} a_k^\dagger a_k|\alpha_1, \alpha_2, \ldots, \alpha_n\rangle = n \tag{1.35}$$

where the summation is extended over *all* the eigenstates $|\alpha_k\rangle$ of H_0. The operator

$$\hat{N} = \sum_k a_k^\dagger a_k \equiv \sum \hat{N}_k \tag{1.36}$$

is furthermore diagonal in the representation $|\alpha_1, \alpha_2, \ldots\rangle$, and because of (1.35) it is referred to as *the number operator*. Its eigenvalues, as can be seen from (1.35), are the integers 0, 1, 2, . . . , n, . . . , each eigenvalue, except zero, being infinitely degenerate if the Hamiltonian H_0 has an infinite spectrum of eigenstates. It follows from the commutation relations (1.28) that

$$a_\alpha^\dagger a_\alpha a_\alpha^\dagger a_\alpha = a_\alpha^\dagger a_\alpha \tag{1.37}$$

Thus the number operator $\hat{N}_\alpha = a_\alpha^\dagger a_\alpha$ satisfies the equation

$$\hat{N}_\alpha^2 = \hat{N}_\alpha \tag{1.38}$$

and consequently its eigenvalues, which must satisfy the same equation, are 1 and 0. This, of course, just reflects our requirement that the operators $a_\alpha^\dagger$ and a_α be connected with the creation and annihilation of *fermions*: the number of such particles in any state α can be either 1 or 0.

Using (1.28) it is easy to prove that

$$[a_\alpha, \hat{N}] = a_\alpha \qquad [a_\alpha^\dagger, \hat{N}] = -a_\alpha^\dagger \tag{1.39}$$

It follows from (1.39) that for any α and β

$$[a_\alpha^\dagger a_\beta, \hat{N}] = [a_\beta a_\alpha^\dagger, \hat{N}] = 0 \tag{1.40}$$

In fact, (1.40) can be generalized:

$$[a_{\alpha_1}^\dagger a_{\alpha_2}^\dagger \ldots a_{\alpha_n}^\dagger a_{\beta_1} a_{\beta_2} \ldots a_{\beta_n}, \hat{N}] = 0 \tag{1.41}$$

Equation 1.41 is easy to understand: the operator $a_{\alpha_1}^\dagger a_{\alpha_2}^\dagger \ldots a_{\alpha_n}^\dagger a_{\beta_1} a_{\beta_2} \ldots a_{\beta_n}$ when operating on any state of m-particles ($m \geq n$) annihilates the n-particles in the states $\beta_1, \beta_2, \ldots, \beta_n$ and replaces them by n-particles in the states $\alpha_1, \alpha_2, \ldots, \alpha_n$. To the extent that in so doing the operator is not called on to annihilate nonexisting particles or to create particles in occupied states, it will reproduce a state with the same number m of particles (otherwise it gives 0). Thus this operator does not change the number of particles in the states on which it operates, and hence its commutativity with the number operator $\hat{N}$. Such an operator is called a "number-conserving operator" and is characterized by having a product of a certain number of annihilation operators with an equal number of creation operators. It is obvious that a sum of number-conserving operators also commutes with the number operator, $\hat{N}$, so that for instance the operator

$$\sum_{\alpha,\beta} \epsilon_{\alpha\beta} a_\alpha^\dagger a_\beta + \frac{1}{4} \sum_{\mu\lambda\rho\sigma} v_{\mu\lambda\rho\sigma} a_\mu^\dagger a_\lambda^\dagger a_\sigma a_\rho \tag{1.42}$$

also commutes with $\hat{N}$.

The eigenvalues of an operator of the type (1.42) can thus be characterized by the quantum numbers of $\hat{N}$, that is, (1.42) has eigenvalues that correspond to systems with no particles, with 1 particle, with 2 particles, . . . , n particles, . . . , etc. ad infinitum. This property does not hold for operators of the type

$$\sum U_{\lambda\mu\rho}a_\lambda^\dagger a_\mu a_\rho$$

Since physical properties of nuclei are always related to systems of a definite number of particles (the number of nucleons is conserved in all known processes), it is clear that we expect physical observables to be represented by operators of the type (1.42), or their generalizations, including other number-conserving operators.

To obtain the connection between the formalism of operations in the Fock space and the physical problems we are interested in, we make the following observation. Let $\hat{\Omega} = \sum_1^n \hat{\omega}_k$ be a one-body operator in coordinate space, and let $\Phi(\alpha_1, \ldots, \alpha_n)$ be an n-particle Slater determinant formed out of the single-particle states $\phi(\alpha_1)$, $\phi(\alpha_2)$, . . . taken in the standard order. Then we have

$$\langle\phi(\alpha_1, \ldots, \alpha_n)|\hat{\Omega}\phi(\alpha_1, \ldots, \alpha_n)\rangle = \sum_{i=1}^{n} \langle\phi(\alpha_i)|\hat{\omega}\phi(\alpha_i)\rangle \tag{1.43}$$

and, for $\alpha_1 \neq \alpha_1'$

$$\langle\phi(\alpha_1, \alpha_2, \ldots, \alpha_n)|\hat{\Omega}\phi(\alpha_1', \alpha_2, \ldots, \alpha_n\rangle = \langle\phi(\alpha_1)|\hat{\omega}\phi(\alpha_1')\rangle \tag{1.44}$$

Furthermore, if $(\alpha_1, \alpha_2) \neq (\alpha_1'\alpha_2')$ then

$$\langle\phi(\alpha_1, \alpha_2, \alpha_3, \ldots, \alpha_n)|\hat{\Omega}\phi(\alpha_1', \alpha_2', \alpha_3, \ldots, \alpha_n)\rangle = 0 \tag{1.45}$$

Let us construct now, in Fock space, the operator

$$\hat{\Omega} = \sum_{k,k'} \omega_{kk'}a_k^\dagger a_{k'} \tag{1.46}$$

where the summation extends over all the states k (and *not* just the states $\alpha_1, \alpha_2, \ldots, \alpha_n$ occupied in $\Phi(\alpha_1, \ldots, \alpha_n)$) and where $\omega_{kk'}$ are numerical coefficients given by

$$\omega_{kk'} = \langle\phi(k)|\hat{\omega}\phi(k')\rangle \tag{1.47}$$

Construct now the n-particle state in Fock space corresponding to $\Phi(\alpha_1, \ldots, \alpha_n)$:

$$|\alpha_1, \alpha_2, \ldots, \alpha_n\rangle = a_{\alpha_1}^\dagger a_{\alpha_2}^\dagger \ldots a_{\alpha_n}^\dagger|0\rangle \tag{1.48}$$

Then the diagonal matrix element of $\hat{\Omega}$ is

$$\langle\alpha_1, \alpha_2, \ldots, \alpha_n|\hat{\Omega}|\alpha_1, \alpha_2, \ldots, \alpha_n\rangle$$
$$= \sum_{k,k'} \omega_{kk'}\langle 0|a_n \ldots a_{\alpha_1}a_k^\dagger a_{k'}a_{\alpha_1}^\dagger a_{\alpha_2}^\dagger \ldots a_{\alpha_n}^\dagger|0\rangle \tag{1.49}$$

The sum on the right-hand side of (1.49) contains an infinite number of terms. Most of them, however, vanish. Indeed, in order for the coefficient of $\omega_{kk'}$ not to vanish, k' must equal one of the α_i's $i = 1, \ldots, n$, and the same should hold for k. Furthermore in the particular matrix element in (1.49), since the creation operators $a_{\alpha_i}^\dagger$ and the annihilation operators a_{α_1} refer to the *same* n states $\alpha_1, \alpha_2, \ldots, \alpha_n$, we have nonvanishing contributions only from terms with $k = k'$. We thus obtain:

$$\langle \alpha_1, \alpha_2, \ldots, \alpha_n | \hat{\Omega} | \alpha_1, \alpha_2, \ldots, \alpha_n \rangle = \sum_{i=1}^{n} \omega_{\alpha_i \alpha_i} = \sum_{i=1}^{n} \langle \alpha_i | \hat{\omega} | \alpha_i \rangle \qquad (1.50)$$

Comparison of (1.50) with (1.43) shows that the diagonal elements of $\hat{\Omega}$ defined by (1.46), taken with the n-particle Fock-space functions (1.48), are identical with the diagonal elements of $\hat{\Omega}$ taken with the configuration-space wave functions $\Phi(\alpha_1, \ldots, \alpha_n)$. It is easy to show that the same holds true for the matrix elements (1.44) and (1.45). Thus, the operator $\hat{\Omega}$ defined by (1.46) is equivalent to the configuration space operator $\Omega = \sum_{k=1}^{n} \omega_k$. The two operators are, of course, not identical; they operate in different spaces and cannot be compared directly. In configuration space we have different operators $\Omega(n) = \sum_{k=1}^{n} \omega_k$ for systems with a different number of particles, whereas the operator $\hat{\Omega}$, (1.46), is *independent* of the number of particles in the system. When the matrix of $\hat{\Omega}$ is constructed with the help of the states (1.48), taking into account all possible values of n, we obtain a series of submatrices along the diagonal, each corresponding to a different number of particles. This follows because we are dealing generally with operators that conserve the number of particles. Hence there will be no matrix elements of $\hat{\Omega}$ connecting states with a different number of particles. Each one of these smaller matrices with a given n is identical with the matrix of $\Omega(n)$ taken in the configuration space of the corresponding number of particles.

The fact that $\hat{\Omega}$ corresponds to a one-body operator is reflected in its form through the presence of *one* annihilation operator and *one* creation operator in each one of its terms. When $\hat{\Omega}$ operates on any state in the Fock space, it changes it into another state that at most can differ from the first one through the state occupied by *one* of its particles.

To construct an operator in Fock space that corresponds to a two-body operator

$$V = \frac{1}{2} \sum_{i \neq j} v(ij) \qquad (1.51)$$

where $v(ij)$ is symmetric in i and j, we proceed in a similar way. Let the two-body matrix elements of $v(ij)$ be given by:

$$\begin{aligned} v_{\alpha\beta,\alpha'\beta'} &= \langle\alpha\beta|\,v(ij)\,|\alpha'\beta'\rangle \\ &= \langle\phi_\alpha(i)\phi_\beta(j)|\,v(ij)\,|\,\phi_{\alpha'}(i)\phi_{\beta'}(j)\rangle \\ &\qquad - \langle\phi_\alpha(i)\phi_\beta(j)|\,v(ij)\,|\,\phi_{\alpha'}(j)\phi_{\beta'}(i)\rangle \end{aligned} \tag{1.52}$$

Note that $v_{\alpha\beta,\alpha'\beta'}$ is defined in terms of *antisymmetric*, *normalized*, two-particle wave functions

$$\frac{1}{\sqrt{2}}\begin{vmatrix} \phi_\alpha(i) & \phi_\beta(i) \\ \phi_\alpha(j) & \phi_\beta(j) \end{vmatrix}.$$

It includes both the direct and the exchange terms, so that

$$v_{\alpha\beta,\alpha'\beta'} = -v_{\beta\alpha,\alpha'\beta'} = v_{\alpha\beta,\beta'\alpha'} = v_{\beta\alpha,\beta'\alpha'} \tag{1.53}$$

We construct now the Fock-space operator

$$\hat{V} = \frac{1}{4}\sum_{k_1k'_2k'_1k_2} v_{k_1k_2,k'_1k'_2}a^\dagger_{k_1}a^\dagger_{k_2}a_{k'_2}a_{k'_1} \tag{1.54}$$

where the summation extends again over the complete, infinite, set of states k. [The expression (1.54) for the potential energy is sometimes written without the factor 1/4. It is then implied that the sum extends over *different* pairs (k_1k_2) and $(k'_1k'_2)$, so that if k_1k_2 is included, k_2k_1 is not. We shall adhere to the convention of summing over all indices, thus including for given four numbers $k_1k_2k'_1k'_2$ the four terms $k_1k_2,k'_1k'_2$; $k_1k_2,k'_2k'_1$; $k_2k_1,k'_1k'_2$; and k_2k_1, $k'_2k'_1$. $\hat{V}$ is obviously a number-conserving operator, and it can be shown, in complete analogy to our discussion of $\hat{\Omega}$, that its matrix elements in the sub-Fock space of n-particles are identical with the matrix elements of the configuration space operator V defined by (1.51) in the Hilbert space of n-particle wave functions.

The multinucleon problem can now be translated from a problem in configuration space to a problem in Fock space.

Given the Hamiltonian

$$H = \sum T_i + \frac{1}{2}\sum_{i\neq j} v(ij) \tag{1.55}$$

in configuration space. Generate an arbitrary complete set of single-particle wave functions $|k_\alpha\rangle$ and evaluate the numbers

$$\epsilon_{kk'} = \langle k|T|k'\rangle \tag{1.56}$$

$$v_{k_1k_2,k'_1k'_2} = \langle k_1k_2|\,v(12)\,|k'_1k'_2\rangle \tag{1.57}$$

Construct now the Fock-space Hamiltonian

$$\hat{H} = \sum \epsilon_{kk'} a_k^\dagger a_{k'} + \frac{1}{4} \sum v_{k_1 k_2, k'_1 k'_2} a_{k_1}^\dagger a_{k_2}^\dagger a_{k'_2} a_{k'_1} \tag{1.58}$$

The Hamiltonian $\hat{H}$ commutes with the number operator $\hat{N} = \sum a_k^\dagger a_k$; hence its matrix breaks down to submatrices along the diagonal, each one corresponding to a different number of particles. Each of these submatrices is identical with a corresponding matrix of H. Hence if we manage to diagonalize $\hat{H}$ it will be equivalent to the *simultaneous* diagonalization of H for systems of 0, 1, 2, 3, . . . , n, . . . particles.

At first glance it seems as if we have gained nothing by going over from configuration space to Fock space, and if anything we might have even complicated the problem: in configuration space we have to diagonalize H for a system of a given number of particles at a time, whereas in Fock space all systems, with an arbitrary number of particles in them, get diagonalized simultaneously.

It is, however, just this last feature that makes Fock-space calculations more powerful. Once we have expressed our *physical* problem in terms of the *mathematical* problem of diagonalization of a certain matrix (1.58), we can resort to any approximation method that is mathematically justifiable for the derivation of the eigenvalues of $\hat{H}$. Since we are now dealing with a larger matrix, which includes systems of all sizes, there is more flexibility in the use of mathematical approximations. In fact, one of the most powerful methods of approximating the eigenvalues of $\hat{H}$ involves a transformation to a scheme that *mixes* states of different numbers of particles. We already mentioned before in Section IV.23 that by violating conservation laws we can arrive at simpler and better approximations to a given system. The transition to Fock space allows us to violate the conservation of the number of particles in order to obtain improved approximations to some nuclear properties.

Remark. Since the number operators $\hat{N}_\alpha = a_\alpha^\dagger a_\alpha$ for different states commute with each other

$$[\hat{N}_\alpha, \hat{N}_\beta] = 0 \tag{1.59}$$

a diagonalization of the Hamiltonian $\hat{H}$ often consists of transforming $\hat{H}$, as closely as possible, into a function of the operators $\hat{N}_\alpha$. These are then chosen to be diagonal, leading to the approximate diagonalization of $\hat{H}$. Thus one may introduce a unitary transformation leading from the operators $a_k^\dagger$ and a_k to new operators $b_\mu^\dagger$ and b_μ satisfying again the anticommutation relations (1.28). The transformation is to be defined so as to make the Hamiltonian $\hat{H}$, expressed in terms of the b's, take on the form

$$\hat{H} = \sum E_\mu b_\mu^\dagger b_\mu + \hat{H}' \tag{1.60}$$

where $\hat{H}'$ is small. The approximate eigenvalues of $\hat{H}$ are then expressible in terms of sums of E_μ, whose structure depends on which particular b_μ orbits are occupied.

The following trivial example illustrates the procedure: consider a Hamiltonian

$$\hat{H} = \sum \epsilon_{kk'} a_k^\dagger a_{k'} \tag{1.61}$$

Define new creation and annihilation operators $b_\mu^\dagger$ and b_μ through

$$b_\mu = \sum U_{\mu k} a_k \qquad b_\mu^\dagger = \sum U_{\mu k}^* a_k^\dagger \tag{1.62}$$

We have

$$\{b_\mu, b_{\mu'}^\dagger\} = \sum U_{\mu k} U_{\mu' k'}^* \{a_k, a_{k'}^\dagger\} = \sum U_{\mu k} U_{\mu' k}^* = (UU)^\dagger_{\mu\mu'} \tag{1.63}$$

Thus if $U_{\mu k}$ is a unitary matrix, the operators b_μ satisfy the anticommutation relation (1.28). Expressed in terms of the b's, (1.61) becomes:

$$\hat{H} = \sum U_{\mu k} \epsilon_{kk'} U_{k'\mu'}^\dagger b_\mu^\dagger b_{\mu'} \tag{1.64}$$

If we now define the unitary transformation so that it diagonalizes the matrix $\epsilon_{kk'}$ then $\hat{H}$ assumes the form

$$\hat{H} = \sum E_\mu b_\mu^\dagger b_\mu$$

where

$$\sum_{k,k'} U_{\mu k} \epsilon_{kk'} U_{k'\mu'}^\dagger = E_\mu \delta_{\mu\mu'}$$

The transformation (1.62) also yields

$$\sum a_k^\dagger a_k = \sum b_\mu^\dagger b_\mu$$

Thus if the eigenvalues E_μ are arranged in ascending order $E_1 < E_2 < \ldots$, the ground state (g.s.) of $\hat{H}$ for a system of n particles is obtained by putting one particle in each of the levels $\mu = 1, \mu = 2, \ldots, \mu = n$, and no particles in any other state. In other words,

$$|\text{g.s.}\rangle = b_n^\dagger \ldots b_2^\dagger b_1^\dagger |0\rangle \tag{1.65}$$

and

$$E_{\text{g.s.}} = \sum_{i=1}^{n} E_\mu \tag{1.66}$$

We realize, of course, that all we have done in this example was to go from the basis of single-particle state vectors $a_k^\dagger|0\rangle$ that were derived from an arbitrary single-particle Hamiltonian H_0, to the single-particle states $b_\mu^\dagger|0\rangle$ of the Hamiltonian $\hat{H}$. Such a transformation could be carried out trivially because $\hat{H}$, (1.61), is a single-particle Hamiltonian. In the more general case a transformation to a Hamiltonian that looks like a single-particle Hamiltonian can be generally accomplished only to a certain approximation, and the "particles" described by $b_\mu^\dagger|0\rangle$ may not bear then a simple relation to the physical particles described by the single-particle states $a_k^\dagger|0\rangle$.

Before we close this section we want to introduce another representation of the creation and annihilation operators that is frequently very useful.

Let the states $\phi_k(\mathbf{r})$ be the complete set of eigenstates of the single-particle Hamiltonian H_0, so that they satisfy the orthonormality conditions:

$$\langle\phi_k(\mathbf{r})|\phi_{k'}(\mathbf{r})\rangle = \delta_{kk'} \tag{1.67}$$

$$\sum_k \phi_k(\mathbf{r})\phi_k^*(\mathbf{r}') = \delta(\mathbf{r} - \mathbf{r}') \tag{1.68}$$

Let the operators a_k and $a_k^\dagger$ be labeled by the same indices k. [The operational meaning of the index k is specified, of course, only when we start building up physical operators involving the operators a_k and $a_k^\dagger$. The Hamiltonian $\hat{H}$, (1.58), has the coefficients $\epsilon_{kk'}$ and $v_{k_1k_2,k_1'k_2'}$ that are derived from T and $v(ij)$ through the use of definite functions like $\phi_k(\mathbf{r})$]. We construct now the *operator*

$$\hat{\psi}(\mathbf{r}) = \sum_k \phi_k(\mathbf{r})a_k \tag{1.69}$$

$$\hat{\psi}^\dagger(\mathbf{r}) = \sum_k \phi_k^*(\mathbf{r})a_k^\dagger \tag{1.70}$$

We see immediately that

$$\{\hat{\psi}(\mathbf{r}), \hat{\psi}^\dagger(\mathbf{r}')\} = \sum_{k,k'} \phi_k(\mathbf{r})\phi_{k'}^*(\mathbf{r}')\{a_k, a_{k'}^\dagger\} = \sum_k \phi_k(\mathbf{r})\phi_k^*(\mathbf{r}') = \delta(\mathbf{r} - \mathbf{r}') \tag{1.71}$$

Similarly

$$\{\hat{\psi}(\mathbf{r}), \hat{\psi}(\mathbf{r}')\} = \{\hat{\psi}^\dagger(\mathbf{r}), \hat{\psi}^\dagger(\mathbf{r}')\} = 0 \tag{1.72}$$

It is of course not necessary to define the operators $\hat{\psi}(\mathbf{r})$ in terms of the operators a_k and the wave functions ϕ_k. One may instead use the state vectors $|\mathbf{r}_1\rangle$, $|\mathbf{r}_1\mathbf{r}_2\rangle$, etc., describing states in which a particle is known to be at $\mathbf{r}_1$, or one at $\mathbf{r}_1$ and another at $\mathbf{r}_2$, etc. The normalization of these states is:

$$\langle\mathbf{r}_1\mathbf{r}_2 \ldots |\mathbf{r}_1'\mathbf{r}_2' \ldots\rangle = \delta(\mathbf{r}_1 - \mathbf{r}_1')\,\delta(\mathbf{r}_2 - \mathbf{r}_2') \ldots$$

In terms of these states [note the analogy with (1.4)]:

$$\hat{\psi}^\dagger(\mathbf{r}) = |\mathbf{r}\rangle\langle 0| + \int |\mathbf{r}, \mathbf{r}_1\rangle\langle\mathbf{r}_1|\, d\mathbf{r}_1 + \int |\mathbf{r}, \mathbf{r}_1, \mathbf{r}_2\rangle\langle\mathbf{r}_1,\mathbf{r}_2|\, d\mathbf{r}_1\, d\mathbf{r}_2 + \ldots$$

while

$$\hat{\psi}(\mathbf{r}) = |0\rangle\langle\mathbf{r}| + \int |\mathbf{r}_1\rangle\langle\mathbf{r}, \mathbf{r}_1|\, d\mathbf{r}_1 + \int |\mathbf{r}_1, \mathbf{r}_2\rangle\langle\mathbf{r}, \mathbf{r}_1, \mathbf{r}_2|\, d\mathbf{r}_1\, d\mathbf{r}_2 + \ldots$$

For example

$$\hat{\psi}^\dagger(\mathbf{r})|\mathbf{r}_1, \mathbf{r}_2\rangle = \int |\mathbf{r}, \mathbf{r}_1', \mathbf{r}_2'\rangle\langle\mathbf{r}_1', \mathbf{r}_2'|\mathbf{r}_1, \mathbf{r}_2\rangle\, d\mathbf{r}_1'\, d\mathbf{r}_2' = |\mathbf{r}, \mathbf{r}_1, \mathbf{r}_2\rangle$$

where we have used

$$\langle\mathbf{r}_1, \mathbf{r}_2 \ldots \mathbf{r}_n|\mathbf{r}_1', \mathbf{r}_2', \ldots, \mathbf{r}_m'\rangle = 0 \qquad \text{if} \qquad n \neq m$$

Using the antisymmetry of the states $|\mathbf{r}_1 \ldots\rangle$:

$$|\mathbf{r}_1, \mathbf{r}_2 \ldots\rangle = -|\mathbf{r}_2, \mathbf{r}_1 \ldots\rangle$$

it becomes possible to directly verify (1.72), and by using methods analogous to that employed in the derivation of (1.58), one can obtain (1.76). Thus the operators $\hat{\psi}^\dagger(\mathbf{r})$ and $\hat{\psi}(\mathbf{r})$ have the same anticommutation relations as the creation and annihilation operators $a_k^\dagger$ and a_k, if we consider $\mathbf{r}$ as the (continuous) label of the states $\hat{\psi}(\mathbf{r})$. $\psi^\dagger(\mathbf{r})|0\rangle$ represents therefore a particle created at the point $\mathbf{r}$, and similarly $\hat{\psi}^\dagger(\mathbf{r}')\psi(\mathbf{r})|G\rangle$ represents a situation where a particle was removed from $|G\rangle$ at $\mathbf{r}$ and another particle was then added at $\mathbf{r}'$.

It is obvious from (1.69) and (1.70) that

$$a_k = \int \phi_k^*(\mathbf{r})\hat{\psi}(\mathbf{r})\, d\mathbf{r} \qquad a_k^\dagger = \int \phi_k(\mathbf{r})\hat{\psi}^\dagger(\mathbf{r})\, d\mathbf{r} \tag{1.73}$$

We can therefore express an operator, say, like

$$\hat{T} = \sum \langle k|T|k'\rangle a_k^\dagger a_{k'}$$

also in terms of the $\hat{\psi}$ operators. We obtain:

$$\begin{aligned}
\hat{T} &= \sum \int d\mathbf{r}\, d\mathbf{r}' \langle k|T|k'\rangle \phi_k(\mathbf{r})\phi_{k'}^*(\mathbf{r}')\hat{\psi}^\dagger(\mathbf{r})\hat{\psi}(\mathbf{r}') \\
&= \sum \int d\mathbf{r}\, d\mathbf{r}'\, d\varrho\, \phi_k^*(\varrho)T(\varrho)\phi_{k'}(\varrho)\phi_k(\mathbf{r})\phi_{k'}^*(\mathbf{r}')\hat{\psi}^\dagger(\mathbf{r})\hat{\psi}(\mathbf{r}') \\
&= \int \hat{\psi}(\mathbf{r})T(\mathbf{r})\hat{\psi}(\mathbf{r})\, d\mathbf{r}
\end{aligned} \tag{1.74}$$

Similarly for the potential (1.51) with the Fock-space operator (1.54) we obtain

$$\hat{V} = \frac{1}{2}\int d\mathbf{r} \int d\mathbf{r}' \hat{\psi}^\dagger(\mathbf{r}')\hat{\psi}^\dagger(\mathbf{r})v(\mathbf{r}, \mathbf{r}')\hat{\psi}(\mathbf{r})\hat{\psi}(\mathbf{r}') \tag{1.75}$$

Thus the Hamiltonian (1.58) can also be written as

$$\hat{H} = -\hbar^2/2m \int \hat{\psi}^\dagger(\mathbf{r})\nabla^2\hat{\psi}(\mathbf{r})\, d^3\mathbf{r} + \hat{V} \tag{1.76}$$

where $\hat{V}$ is given by (1.75). The number operator (1.36) becomes

$$\hat{N} = \int \hat{\psi}^\dagger(\mathbf{r})\hat{\psi}(\mathbf{r})\, d\mathbf{r} \tag{1.77}$$

In the classical limit obtained by replacing the field operators $\hat{\psi}$ by a Schrödinger wave function ψ, $\hat{V}$ becomes simply the interaction energy of a classical fluid of density $\rho(\mathbf{r}) = \psi^*(\mathbf{r})\psi(\mathbf{r})$ in which the potential between a volume element at $\mathbf{r}$ and another at $\mathbf{r}'$ is $v(\mathbf{r}, \mathbf{r}')$. The first term in (1.76) becomes the "kinetic energy" of such a fluid. The quantum mechanical Hamiltonian (1.76) is the consequence of quantizing the classical field ψ.

2. THE HARTREE-FOCK POTENTIAL

Before we proceed to use the method of second quantization for the derivation of some special approximations to the many-body problem, let us apply it to a problem that we have already tackled using other methods: to that of the Hartree-Fock potential. We formulate the problem in the following manner.

A Hamiltonian $H_0 = \Sigma T(i) + U(i)$ is used to generate single-particle states $a_k^\dagger|0\rangle$, and from these the N-particle states $a_{k_1}^\dagger a_{k_2}^\dagger \ldots a_{k_N}^\dagger|0\rangle$. We now allow variations in the potential $U(i)$ that would lead to variations in the single-particle wave function $\phi_k(\mathbf{r})$, and, through the single-particle state vector $a_k^\dagger|0\rangle$, to corresponding variations in the N-particle states. We are asked to find those N-particle wave functions that minimize the expectation value of

$$\hat{H} = \sum \epsilon_{kk'} a_k^\dagger a_{k'} + \frac{1}{4}\sum v_{k_1k_2,k_1'k_2'} a_{k_1}^\dagger a_{k_2}^\dagger a_{k_2'} a_{k_1'} \tag{2.1}$$

where $v_{k_1k_2,k_1'k_2'}$ are the antisymmetric matrix elements of the interaction $v(ij)$ given by (1.52) and (1.53).

To solve this problem let us first derive a general expression for the variation of an N-particle wave function. Given the state vector

$$|G\rangle = a_{k_1}^\dagger a_{k_2}^\dagger \ldots a_{k_N}^\dagger|0\rangle \tag{2.2}$$

we can obtain from it another vector by operating on $|G\rangle$ with $a_{k_m}^\dagger a_{k_i}$, where $1 \le i \le N$, $m > N$; indeed

$$a_{k_m}^\dagger a_{k_i}|G\rangle = a_{k_m}^\dagger a_{k_i} a_{k_1}^\dagger a_{k_2}^\dagger \ldots a_{k_N}^\dagger|0\rangle = \pm a_{k_m}^\dagger a_{k_1}^\dagger \ldots a_{k_{i-1}}^\dagger a_{k_{i+1}}^\dagger \ldots a_{k_N}^\dagger|0\rangle$$
$$i \le N \qquad m > N \tag{2.3}$$

is a state in which k_i is no longer occupied, and k_m is occupied instead. (We shall use the indices i, j to denote occupied states in $|G\rangle$, and the indices m, n to denote unoccupied states in $|G\rangle$. $|G\rangle$ is assumed to have the lowest

N-particle states occupied, so that the state $a_N^\dagger|0\rangle$ can be considered the highest in the Fermi distribution. k_N is then called the Fermi momentum even if k_i does not stand for a momentum eigenstate).

We shall be able to show that the most general variation on the state $|G\rangle$ that is not orthogonal to $|G\rangle$ and that involves the variation of single-particle states can be obtained through the operation [Thouless (61)]

$$|G'\rangle = \left[\exp \sum_{m=N+1}^{\infty} \sum_{i=1}^{N} (C_{mi}a_m^\dagger a_i)\right] |G\rangle \tag{2.4}$$

where C_{mi} are arbitrary constants and the exponential is defined in terms of its power expansion. $|G'\rangle$ is known as the *Thouless variational wave function.* We note that for $r > N$

$$(\sum_{m,i} C_{mi}a_m^\dagger a_i)^r = 0 \qquad r > N \tag{2.5}$$

since there are at most N different a_i's in the sum, and a product of $r > N$ a_i's must involve at least one of the a_i's in the combination a_i^2; from (1.28) we have $a_i^2 = 0$ and hence (2.5). Thus in (2.4) only the first $N + 1$ terms in the power expansion of the exponential may have nonvanishing contributions.

It is easy to see that (2.4) is indeed the most general variation that corresponds to a change in the single-particle states. We shall now show that by defining

$$b_i^\dagger = a_i^\dagger + \sum_{m=N+1}^{\infty} C_{mi}a_m^\dagger \tag{2.6}$$

we can write (2.4) in the form

$$|G'\rangle = b_{k1}^\dagger b_{k2}^\dagger \ldots b_{kN}^\dagger |0\rangle \tag{2.7}$$

Since (2.6) represents the most general transformation of a single-particle state $a_i^\dagger|0\rangle$ that is not orthogonal to $a_i^\dagger|0\rangle$ itself, the proof of (2.7) will have demonstrated our point.

To carry out the proof we note that the N-operators

$$\sum_{m=N+1}^{\infty} C_{mi}a_m^\dagger a_i \qquad i = 1, 2, \ldots, N$$

commute among themselves

$$\left[\sum_{m=N+1}^{\infty} C_{mi}a_m^\dagger a_i, \sum_{m'=N+1}^{\infty} C_{m'j}a_{m'}^\dagger a_j\right] = 0 \tag{2.8}$$

Hence we have, from (2.4),

$$|G'\rangle = \exp\left(\sum_{i=1}^{N} \sum_{m=N+1}^{\infty} C_{mi}a_m^\dagger a_i\right) |G\rangle = \prod_{i=1}^{N} \exp\left(\sum_{m=N+1}^{\infty} C_{mi}a_m^\dagger a_i\right) |G\rangle \tag{2.9}$$

But because $a_i^2 = 0$,

$$\exp\left(\sum_{m=N+1}^{\infty} C_{mi}a_m^\dagger a_i\right) = \left(1 + \sum_{m=N+1}^{\infty} C_{mi}a_m^\dagger a_i\right)$$

and hence

$$\begin{aligned}
|G'\rangle &= \prod_{i=1}^{N}\left(1 + \sum_{m=N+1}^{\infty} C_{mi}a_m^\dagger a_i\right)|G\rangle \\
&= \prod_{i=1}^{N}\left(1 + \sum_{m=N+1}^{\infty} C_{mi}a_m^\dagger a_i\right)a_1^\dagger a_2^\dagger \ldots a_N^\dagger|0\rangle \\
&= \left(a_1^\dagger + \sum_{m=N+1}^{\infty} C_{m1}a_m^\dagger a_1 a_1^\dagger\right)\left(a_2^\dagger + \sum_{m=N+1}^{\infty} C_{m2}a_m^\dagger a_2 a_2^\dagger\right)\ldots \\
&\qquad \cdot\left(a_N^\dagger + \sum_{m=N+1}^{\infty} C_{mN}a_m^\dagger a_N a_N^\dagger\right)|0\rangle \\
&= \left(a_1^\dagger + \sum_{m=N+1}^{\infty} C_{m1}a_m^\dagger\right)\ldots\left(a_N^\dagger + \sum_{m=N+1}^{\infty} C_{mN}a_m\right)|0\rangle \\
&= b_1^\dagger b_2^\dagger \ldots b_N^\dagger|0\rangle
\end{aligned} \tag{2.10}$$

which is just the result (2.7).

Actually to make (2.6) the most general expression, the sum over m should have been extended from $m = 1$ to $m = \infty$; but it is obvious that in the N-particle wave function (2.7), the terms

$$\sum_{m=1}^{N} C_{mi}a_m^\dagger$$

will make no contribution.

We note, incidentally that the Thouless wave function $|G'\rangle$ is not normalized to unity. Instead, its normalization is given by

$$\langle G|G'\rangle = 1 \tag{2.11}$$

This normalization assures us that $|G'\rangle$ is obtained from a variation of $|G\rangle$ and is not orthogonal to it.

To return to our original problem, we want to define an arbitrary infinitesimal variation on the state $|G\rangle$, which corresponds to the most general change in the single-particle potential $U(i)$. This can be achieved by taking C_{mi} in (2.4) to be infinitesimal quantities. Thus we put

$$|\delta G\rangle = [\exp(\sum_{im} \delta C_{mi}a_m^\dagger a_i) - 1]|G\rangle = \sum \delta C_{mi}a_m^\dagger a_i|G\rangle \tag{2.12}$$

and

$$\langle\delta G| = \langle G|\sum \delta C_{mi}^* a_i^\dagger a_m \tag{2.13}$$

The operator $a_m^\dagger a_i$ is often referred to as a "particle-hole producing" operator; implied in this name is the assumption that this operator operates on a state $|G\rangle$ in which the single-particle state $a_i^\dagger|0\rangle$ is occupied while the states $a_m^\dagger|0\rangle$ are unoccupied. $a_m^\dagger a_i|G\rangle$ then produces a hole in $|G\rangle$ corresponding to the annihilation of the particle in $a_i^\dagger|0\rangle$, and in return adds a particle in the state $a_m^\dagger|0\rangle$ to $a_i|G\rangle$. We shall be dealing often with such *particle-hole excitations* since they form a very common and important subclass of nuclear excitations. The variational principle [note normalization condition (2.11)],

$$\delta\langle G|\hat{H}|G\rangle = 0$$

in which all the variations δC_{mi} are independent of each other, now reads

$$\langle G|a_i^\dagger a_m \hat{H}|G\rangle = 0 \qquad \text{for} \qquad m > N, i \leq N \tag{2.14}$$

In (2.14) we have applied the variation to the state $\langle G|$ to derive an equation for $|G\rangle$ (the variations of $|G\rangle$ and $\langle G|$ can be considered independently).

To solve (2.14) we note that because in $|G\rangle$ only the states $|k_1\rangle, \ldots, |k_N\rangle$ are occupied, we have

$$\begin{aligned} \langle G|a_i &= 0 = a_i^\dagger|G\rangle \qquad i \leq N \\ \langle G|a_m^\dagger &= 0 = a_m|G\rangle \qquad m > N \end{aligned} \tag{2.15}$$

It therefore follows from (2.14) and (2.15) that

$$\langle G|\,[a_i^\dagger a_m, \hat{H}\,]|G\rangle = \langle G|a_i^\dagger a_m \hat{H} - \hat{H} a_i^\dagger a_m|G\rangle = 0 \tag{2.16}$$

Using the definition (2.1) of $\hat{H}$ we find that

$$\begin{aligned} [\hat{H}, a_k^\dagger] &= \sum_{k'} \epsilon_{k'k} a_{k'}^\dagger + \frac{1}{2} \sum_{k_1,k_2,k'} v_{k_1k_2,k'k} a_{k_1}^\dagger a_{k_2}^\dagger a_{k'} \\ [\hat{H}, a_k] &= -\left[\sum \epsilon_{kk'} a_{k'} + \frac{1}{2} \sum v_{kk',k_1k_2} a_{k'}^\dagger a_{k_2} a_{k_1}\right] \end{aligned} \tag{2.17}$$

where we have used (1.53). Hence

$$\begin{aligned} [a_i^\dagger a_m, \hat{H}\,] &= a_i^\dagger [a_m, \hat{H}\,] + [a_i^\dagger, \hat{H}\,] a_m \\ &= \sum_{k'} \epsilon_{mk'} a_i^\dagger a_{k'} - \sum_{k'} \epsilon_{k'i} a_{k'}^\dagger a_m + \frac{1}{2} \sum v_{mk',k_1k_2} a_i^\dagger a_{k'}^\dagger a_{k_2} a_{k_1} \\ &\qquad - \frac{1}{2} \sum v_{k_1k_2,k'i} a_{k_1} a_{k_2} a_{k'} a_m \end{aligned} \tag{2.18}$$

Introducing (2.18) into (2.16) we obtain as our variational condition

$$\langle G|\epsilon_{mi} + \sum_{k,k'} v_{mk,ik'} a_k^\dagger a_{k'}|G\rangle = 0 \tag{2.19}$$

We notice that in (2.19) there is no contribution from the terms with either $k > N$ or $k' > N$, because of (2.15). Furthermore, we have, again using (2.15), that for $k, k' \leq N$

$$\langle G|a_k^\dagger a_{k'}|G\rangle = \langle G|\delta(k, k') - a_{k'}a_k^\dagger|G\rangle = \delta(k, k')$$

Equation 2.19 is therefore satisfied if

$$\eta_{mi} \equiv \epsilon_{mi} + \sum_{j=1}^{N} v_{mj,ij} = 0 \qquad \text{for} \qquad i \leq N \qquad m > N \tag{2.20}$$

Equation 2.20 is equivalent to Eq. IV.19.8. Its interpretation is simple: the matrix elements η_{mi} in (2.20) contain two terms; the first, ϵ_{mi}, is just the matrix element of the kinetic energy; the second, $U_{m,i} \equiv \sum_j v_{mj,ij}$, can be considered as the matrix element of a single-body operator U, which we can call the Hartree-Fock single-particle potential; together they then form a single-particle Hartree-Fock Hamiltonian. The best single-particle states from the variational point of view are then those that guarantee that the Hartree-Fock Hamiltonian defined by the single-particle matrix elements

$$\eta_{kk'} \equiv \epsilon_{kk'} + \sum_j v_{kj,k'j}$$

does not connect occupied states $k' = i$ with nonoccupied states $k = m$. We note that (2.20) does not say anything about the elements $\eta_{ii'}$ connecting two occupied states, nor about $\eta_{mm'}$. This is only natural because a linear transformation of the occupied states among themselves does not change the antisymmetric states $|G\rangle$.

We can therefore conveniently choose the single-particle states so that they will diagonalize separately the submatrices $\eta_{ii'}$ and $\eta_{mm'}$. Equation 2.20 will then read

$$\epsilon_{kk'} + \sum_{j=1}^{N} v_{kj,k'j} = \eta_k \delta_{kk'} \tag{2.21}$$

Note that the Hartree-Fock single-particle Hamiltonian (2.21) depends on the states we assume to be occupied. This dependence comes into (2.21) through the sum $\sum_{j=1}^{N} v_{kj,k'j}$ in which the states $|j\rangle$ are limited to the assumed set of occupied single-particle states.

Equation 2.21 gives the conditions for a *self-consistent* solution. In fact it says that if the states $a_i^\dagger|0\rangle$ are the "best" single-particle wave functions, then they have the following special property: starting from these states and the interaction

$$\frac{1}{4}\sum v_{k_1k_2,k_1'k_2'} a_{k_1}^\dagger a_{k_2}^\dagger a_{k_2'} a_{k_1'}$$

we pick on the lowest N-states $|k_1\rangle, \ldots, |k_N\rangle$ and construct the single-particle potential

$$U_{k,k'} = \sum_{j \leq N} v_{kj,k'j} \tag{2.22}$$

Then, the N lowest eigenstates of

$$\sum_{kk'} (\epsilon_{kk'} + U_{k,k'})a_k^\dagger a_{k'}$$

will give us back the states $a_i^\dagger|0\rangle$, $i = 1, \ldots, N$, identical to the states we started with.

The potential $\sum U_{k,k'}a_k^\dagger a_{k'}$, satisfying these conditions, is therefore called the *self-consistent potential* and the states $a_i^\dagger|0\rangle$ are referred to as the *Hartree–Fock self-consistent wave functions.* We see from (2.21) and (2.22) that their characteristic feature is summarized by (2.20).

The self-consistent field approximation can be used to derive the best estimate of the ground-state energy obtainable within the set of functions generated by the Thouless transformation (2.4), that is, within the set of N-particle functions that can be written as a Slater determinant of some N-single-particle wave functions. But this class of trial functions may not be adequate. It is possible, for example, that the ground state of a given system has such strong correlations that a description of it in terms of a Slater determinant of independent particles is not possible even as an approximation. We shall later see how these correlations can be included (see Section 3).

In order for the self-consistent solution to be of physical interest, the energy it leads to should be a *minimum* with respect to small variations in the coefficients C_{mi}. Thus far we have only shown that it is an *extremum.* To derive the condition for it to be a minimum we have to evaluate the ground state energy and the normalization of $|G'\rangle$ to second order in C_{mi}. This we now proceed to do.

From (2.4) and (2.15) we obtain:

$$\langle G'|G'\rangle = \langle G| \left(1 + \sum C_{mi}^* a_i^\dagger a_m + \frac{1}{2} \sum C_{mi}^* C_{m'i'}^* a_i^\dagger a_m a_{i'}^\dagger a_{m'}\right)$$
$$\left(1 + \sum C_{nj} a_n^\dagger a_j + \frac{1}{2} \sum C_{nj} C_{n'j'} a_n^\dagger a_j a_{n'}^\dagger a_{j'}\right) |G\rangle$$

To second order in the C's we can neglect the terms with $C_{mi}^* C_{m'i'}^*$ since they involve $a_i^\dagger a_{i'}^\dagger|G\rangle$, which, because of (2.15), vanish. Similar considerations hold for the terms with $C_{nj}C_{n'j'}$. Remembering that $i, j \leq N$ and $m,n > N$ we are therefore left only with

$$\langle G'|G'\rangle = \langle G|1 + \sum (C_{mi}^* a_i^\dagger a_m)(C_{nj} a_n^\dagger a_j)|G\rangle = 1 + \sum_{i \leq N} \sum_{m=N+1}^{\infty} |C_{mi}|^2 \tag{2.23}$$

To obtain the energy to second order in the C's we have to evaluate

$$\langle G'|\hat{H}|G'\rangle = \langle G|\left(1 + \sum C^*_{mi}a^\dagger_i a_m + \frac{1}{2}\sum C^*_{mi}C^*_{m'i'}a^\dagger_i a_m a^\dagger_{i'}a_{m'}\right)\hat{H}$$
$$\cdot\left(1 + \sum C_{nj}a^\dagger_n a_j + \frac{1}{2}\sum C_{nj}C_{n'j'}a^\dagger_n a_j a^\dagger_{n'}a_{j'}\right)|G\rangle \quad (2.24)$$

In (2.24) there is one term with no C's:

$$\langle G|\hat{H}|G\rangle \quad (2.25)$$

There are terms involving C or C^* linearly; these vanish identically because of the self-consistency requirement (2.14). There are terms involving the product C^*C; these have the general structure

$$\langle G|a^\dagger_i a_m \hat{H} a^\dagger_n a_j|G\rangle = \langle G|a^\dagger_i a_m[\hat{H}, a^\dagger_n a_j]|G\rangle + \langle G|a^\dagger_i a_m a^\dagger_n a_j \hat{H}|G\rangle \quad (2.26)$$

Because of (2.15), the second term in (2.26) yields

$$\langle G|a^\dagger_i a_m a^\dagger_n a_j \hat{H}|G\rangle = \delta(m,n)\delta(i,j)\langle G|\hat{H}|G\rangle \quad (2.27)$$

The first term in (2.26) can be evaluated, using (2.18):

$$\langle G|a^\dagger_i a_m[\hat{H}, a^\dagger_n a_j]|G\rangle = 2v_{jm,ni} + \delta_{ij}[\epsilon_{mn} + \sum_{i'} v_{mi',i'n}]$$
$$- \delta_{mn}[\epsilon_{ji} + \sum_{i'} v_{ji',ii'}] \quad (2.28)$$

Using (2.21) we obtain finally for the terms involving C^*C

$$\langle G|a^\dagger_i a_m \hat{H} a^\dagger_n a_j|G\rangle = \delta(m,n)\delta(i,j)[\langle G|\hat{H}|G\rangle + \eta_m - \eta_i] + 2v_{jm,ni} \quad (2.29)$$

Using similar techniques, one can evaluate also the terms involving CC and C^*C^*. Combining all terms together we obtain:

$$\langle G'|\hat{H}|G'\rangle = \langle G|\hat{H}|G\rangle(1 + \sum |C_{mi}|^2) + \sum_{m>N,i\le N}(\eta_m - \eta_i)|C_{mi}|^2$$
$$+ 2\sum_{i,j\le N;m,n>N} v_{jm,ni}C^*_{mi}C_{nj} + \sum_{i,j\le N;m,n>N} v_{ij,mn}C_{mi}C_{nj}$$
$$+ \sum_{i,j\le N;m,n>N} v_{mn,ij}C^*_{mi}C^*_{nj} \quad (2.30)$$

The first term in (2.30) gives the Hartree-Fock energy to first order in the C's

$$E_0 = \langle G|\hat{H}|G\rangle$$

In order for this value to represent a minimum, the remaining part of (2.30) must be a positive definite function of the C's. We notice that (2.30) is a quadratic form in the "vector" (C_{mi}, C^*_{nj}). In order for it to be positive definite the eigenvalues λ of the corresponding matrix, formed from the coefficients of the quadratic form, must all be positive. The equation determining the principal axis of the quadratic form are:

$$(\eta_m - \eta_i)C_{mi} + 2\sum_{jn} v_{jm,ni}C_{nj} + 2\sum_{jn} v_{mn,ij}C^*_{nj} = \lambda C_{mi}$$

$$(\eta_m - \eta_i)C^*_{mi} + 2\sum_{j,n} v_{ni,jm}C^*_{nj} + 2\sum_{n,j} v_{ij,mn}C_{nj} = \lambda C^*_{mi} \qquad (2.31)$$

Because of the variational principle associated with the principal axes of quadratic forms, the left-hand side of (2.31) can be obtained from (2.30) by differentiating with respect to C^*_{mi} and C_{mi}, omitting the terms proportional to $E_0 = \langle G|H|G\rangle$. Thus equations (2.31) are given by

$$\delta\{\langle G'|H|G'\rangle - (E_0 + \lambda)\langle G'|G'\rangle\} = 0 \qquad (2.32)$$

We see therefore right away that λ represents the additional energy acquired by the system when single-particle states with $m > N$ are allowed in the variational wave function. In order for the Hartree-Fock solution to be the state of lowest energy, we must have

$$\lambda > 0 \qquad (2.33)$$

It is obvious that (2.32) and (2.33) need not be satisfied for an arbitrary interaction even if the states $a^\dagger_k|0\rangle$ satisfy the Hartree-Fock equation (2.21). If (2.33) for λ is not satisfied, then the Hartree-Fock solution represents an unstable equilibrium and cannot be expected to be a valid description of the nuclear ground state.

3. TIME-DEPENDENT HARTREE-FOCK SOLUTION

An interesting question now presents itself in connection with *stable* Hartree-Fock solutions that do satisfy (2.31) and (2.33): since a small variation in the C's increases the total energy, the system behaves as if it has a "restoring force" around its Hartree-Fock self-consistent state. Classically we can therefore expect it to develop harmonic oscillations around this equilibrium point. Quantum mechanically these oscillations will lead to an excited state that involves a coherent excitation of the various particles. Such states usually exhibit various collective features, and we see that their presence is deeply connected with the stability of the Hartree-Fock solution.

To study these collective states quantitatively it is convenient to work with the time-dependent Schrödinger equation. Given a solution $\Phi(x, t)$ of the time-dependent Schrödinger equation

$$\hat{H}\Phi(x, t) = i\hbar\frac{\partial\Phi(x, t)}{\partial t} \qquad (3.1)$$

it is possible to obtain the allowed energies of the system by Fourier analyzing $\Phi(x, t)$ with respect to the time t. In fact if

$$\Phi(x, t) = \int \exp(-i\omega t)\Phi(x, \omega)\, d\omega$$

then it follows that any ω for which $\Phi(x, \omega) \neq 0$, satisfies

$$\hat{H}\Phi(x, \omega) = \hbar\omega\Phi(x, \omega)$$

and is thus an eigenvalue of $(1/\hbar)\hat{H}$.

The time-dependent Schrödinger equation can be obtained from a variational principle

$$\delta\langle\Phi(t)|\hat{H} - i\hbar\frac{\partial}{\partial t}|\Phi(t)\rangle = 0 \tag{3.2}$$

where $\Phi(t)$ is a time-dependent variational wave function.

In analogy with (2.4) we choose as our variational wave function

$$|\Phi(t)\rangle = \exp(-iE_0t/\hbar)\exp[\Sigma C_{mi}(t)a_m^\dagger a_i]|G\rangle \tag{3.3}$$

where now the coefficients $C_{mi}(t)$ are time dependent. When all $C_{mi} = 0$, $\Phi(t)$ reduces to the stationary state $|G\rangle$ with its appropriate time dependence $\exp(-iE_0t/\hbar)$. However, using the more general variational wave function (3.3), together with the variational principle (3.2), we can expect to obtain information also about the excited states of our system near the ground state.

We again evaluate $\langle\Phi(t)|\hat{H}|\Phi(t)\rangle$ to second order in the C's obtaining an expression similar to (2.30). In fact as we can see from (2.32), the variational result for (3.2) will be identical with (2.31) except that the parameter λ will be replaced by $i\hbar\partial/\partial t$. We thus obtain

$$(\eta_m - \eta_i)C_{mi}(t) + 2\sum_{n,j} v_{jm,ni}C_{nj}(t) + 2\sum_{n,j} v_{mn,ij}C_{nj}^*(t) = i\hbar\dot{C}_{mi}(t) \tag{3.4}$$

where again i, $j \leq N$ and $m, n > N$, N being the number of particles in the system. The left-hand sides of (3.4) and (2.31) are of course identical in form. This similarity is responsible for the connection that we are going to establish between the stability condition of the Hartree-Fock solution and the existence of collective modes of excitation. To show this connection notice that (3.4) involves both C and C^* so that we must look for a solution involving both frequencies ω and ω^*:

$$C_{mi}(t) = X_{mi}\exp(-i\omega t) + Y_{mi}^*\exp(i\omega^* t) \tag{3.5}$$

Substituting (3.5) into (3.4) we obtain two equations for the X's and the Y's:

$$\begin{aligned}(\eta_m - \eta_i)X_{mi} + 2\sum v_{jm,ni}X_{nj} + 2\sum v_{mn,ij}Y_{nj}^* &= \hbar\omega X_{mi}\\ (\eta_m - \eta_i)Y_{mi} + 2\sum v_{ni,jm}Y_{nj} + 2\sum v_{ij,mn}X_{nj}^* &= -\hbar\omega Y_{mi}\end{aligned} \tag{3.6}$$

where use has been made of the relation

$$v^*_{k_1k_2,k_1'k_2'} = v_{k_1'k_2',k_1k_2} \tag{3.7}$$

It is convenient now to introduce two (infinite) matrices

$$\begin{aligned}\langle mi|A|nj\rangle &\equiv (\eta_m - \eta_i)\delta(m,n)\delta(i,j) + 2v_{jm,ni}\\ \langle mi|B|nj\rangle &\equiv 2v_{mn,ij}\end{aligned} \tag{3.8}$$

We notice that A is hermitian

$$\langle mi|A|nj\rangle^* = \langle nj|A|mi\rangle \tag{3.9}$$

and B is symmetric

$$\langle mi|B|nj\rangle = \langle nj|B|mi\rangle \tag{3.10}$$

Equation 2.31 can now be written in the form

$$\begin{pmatrix} A & B \\ B^* & A^* \end{pmatrix}\begin{pmatrix} C \\ C^* \end{pmatrix} = \lambda \begin{pmatrix} C \\ C^* \end{pmatrix} \tag{3.11}$$

while (3.6) take the form

$$\begin{pmatrix} A & B \\ -B^* & -A^* \end{pmatrix}\begin{pmatrix} X \\ Y \end{pmatrix} = \hbar\omega \begin{pmatrix} X \\ Y \end{pmatrix} \tag{3.12}$$

$\hbar\omega$ is thus an eigenvalue of a matrix that can be constructed from the self-consistent energies η_k and matrix elements of the interaction v.

Because of (3.4) and (3.10) the matrix M formed out of the matrix elements of A and B:

$$M \equiv \begin{pmatrix} A & B \\ B^* & A^* \end{pmatrix} \tag{3.13}$$

is hermitian and its eigenvalues λ, (3.11), are real. The matrix

$$N = \begin{pmatrix} A & B \\ -B^* & -A^* \end{pmatrix} \tag{3.14}$$

is, however, not necessarily hermitian and its eigenvalues $\hbar\omega$ need not therefore be real. The conditions for the existence of real eigenvalues $\hbar\omega$ of the matrix N can be obtained as follows. We have, from (3.12)

$$\begin{aligned}(X^*Y^*)\begin{pmatrix} A & B \\ B^* & A^* \end{pmatrix}\begin{pmatrix} X \\ Y \end{pmatrix} &= (X^*Y^*)\begin{pmatrix} 1 & 0 \\ 0 & -1 \end{pmatrix}\begin{pmatrix} A & B \\ -B^* & -A^* \end{pmatrix}\begin{pmatrix} X \\ Y \end{pmatrix}\\ &= (X^*Y^*)\begin{pmatrix} 1 & 0 \\ 0 & -1 \end{pmatrix}\hbar\omega\begin{pmatrix} X \\ Y \end{pmatrix}\\ &= \hbar\omega(X^*X - Y^*Y)\end{aligned} \tag{3.15}$$

If M is positive definite in addition to being Hermitian, that is, if its eigenvalues are real and positive, then the left-hand side of (3.15) is real and positive. It follows therefore that

$$\hbar\omega(X^*X - Y^*Y) \qquad \text{is real and positive} \tag{3.16}$$

Since $X^*X - Y^*Y$ is always real we conclude from (3.16) that, under these conditions, $\hbar\omega$ is also real. Furthermore

$$\hbar\omega > 0 \qquad \text{if} \qquad X^*X - Y^*Y > 0$$

and (3.17)

$$\hbar\omega < 0 \qquad \text{if} \qquad X^*X - Y^*Y < 0$$

Thus, when the Hartree–Fock solution is stable and the bilinear form, (2.30), or the matrix M, (3.13), is positive definite, the expression (3.5) for the C's that solves the time-dependent Schrödinger equation allows real solutions for ω. If M is not positive definite we cannot draw this conclusion about ω, and since $\hbar\omega$ is an eigenvalue of the nonhermitian matrix N, it will generally be complex. It is now seen from (3.5) that a complex solution for ω leads to a solution for $C_{mi}(t)$ that diverges in time, so that the coefficients $C_{mi}(t)$ do not stay small. Physically it means that the Hartree-Fock solution is not stable, and if we let the system develop in time it will "slide over" to other states rather than oscillate periodically around the Hartree-Fock solution.

To interpret our results we note that the frequencies appearing in $\Phi(t)$, according to the discussion following (3.1), should correspond to excited states of our system. Actually this would have been a precise statement had $\Phi(t)$ been an exact solution of the complete time-dependent Schrödinger equation. As it stands, $\Phi(t)$ is at best a good approximation to the exact solution. This can show itself in two ways:

1. The frequencies that show up in the Fourier decomposition of $\Phi(t)$ will be only approximately equal to the exact eigenfrequencies of the system.

2. $\Phi(t)$ may include small components that oscillate with frequencies which do not at all belong to the system.

It is not hard to identify the latter components and discard the energies they predict. In fact we know the spectrum of excited states of our system in the limit $v \to 0$; it corresponds to the excitations of individual particles to the different single-particle states of the Hartree-Fock Hamiltonian. Hence we can take the limit of $\Phi(t)$ as $v \to 0$, and observe which Fourier components in $\Phi(t)$ persist in this limit and lead to the known unperturbed excitation energies.

To proceed with this program we use (2.10) and (3.5) in order to bring $\Phi(t)$ into a form whose Fourier analysis is easy to carry out. In doing so we shall use the form (3.5) for the coefficients $C_{mi}(t)$, with ω being determined as an eigenvalue of (3.12) and X_{mi} and Y_{mi} being determined by (3.6). $\Phi(t)$ can then be written as:

$$\Phi(t) = \exp(-iE_0t/\hbar)\left(a_1^\dagger + \sum_{m=N+1}^{\infty} C_{m1}(t)a_m^\dagger\right)\ldots\left(a_N^\dagger + \sum_{m=N+1}^{\infty} C_{mN}(t)a_m^\dagger\right) \cdot |0\rangle \quad (3.18)$$

Introducing (3.5) for the C_{mi}'s in (3.18) we obtain

$$\begin{aligned}\Phi(t) &= \exp(-iE_0t/\hbar)a_1^\dagger a_2^\dagger \ldots a_N^\dagger|0\rangle + \exp[-i(E_0 + \hbar\omega)t/\hbar]\\ &\times\left\{\sum_{m=N+1}^{\infty} X_{m1}a_m^\dagger a_2^\dagger \ldots a_N^\dagger + \sum_{m=N+1}^{\infty} X_{m2}a_1^\dagger a_m^\dagger a_3^\dagger \ldots a_N^\dagger + \ldots\right\}|0\rangle\\ &\quad + \exp[-i(E_0 - \hbar\omega^*)t/\hbar]\\ &\times\left\{\sum_{m=N+1}^{\infty} Y_{m1}^* a_m^\dagger a_2^\dagger, \ldots a_N^\dagger + \sum_{m=N+1}^{\infty} Y_{m2}^* a_1^\dagger a_m^\dagger a_3^\dagger \ldots a_N^\dagger + \ldots\right\}|0\rangle\\ &+ \exp[-i(E_0 + 2\hbar\omega)t/\hbar]\{\sum_{m,m'=N+1} X_{m1}X_{m'2}a_m^\dagger a_{m'}^\dagger a_3^\dagger \ldots a_N^\dagger + \ldots\}|0\rangle\\ &\quad + \ldots\\ &= \exp(-iE_0t/\hbar)|G\rangle + \exp[-i(E_0 + \hbar\omega)t/\hbar]\sum_{i=1}^{N}\sum_{m=N+1}^{\infty} X_{mi}a_m^\dagger a_i|G\rangle\\ &\quad + \exp[-i(E_0 - \hbar\omega^*)t/\hbar]\sum_{i=1}^{N}\sum_{m=N+1}^{\infty} Y_{mi}^* a_m^\dagger a_i|G\rangle\\ &+ \exp[-i(E_0 + 2\hbar\omega)t/\hbar]\sum_{i,j=1}^{N}\sum_{m,m'=N+1}^{\infty} X_{mi}X_{m'j}a_m^\dagger a_{m'}^\dagger a_i a_j|G\rangle + \ldots \quad (3.19)\end{aligned}$$

where $|G\rangle$ is the Hartree-Fock ground state $|G\rangle = a_1^\dagger a_2^\dagger \ldots a_N^\dagger|0\rangle$.

Thus, from our discussion following (3.1), we conclude that when (3.12) allows real solutions for $\hbar\omega$, that is, when the Hartree-Fock solution is stable, the system has excited states at energies $E_0 + \hbar\omega$, $E_0 + 2\hbar\omega, \ldots$, these states being given respectively by $Q^\dagger|G\rangle$, $(Q^\dagger)^2|G\rangle$ etc. where

$$Q^\dagger = \sum_{i=1}^{N}\sum_{m=N+1}^{\infty} X_{mi}a_m^\dagger a_i \quad (3.20)$$

The occurrence of spectra with equal spacings $\hbar\omega$ is characteristic of harmonic vibrations, and we interpret, therefore, the states with energies $E_0 + \hbar\omega$, $E_0 + 2\hbar\omega$, etc. as *collective vibrational states* of the system, which can be gotten from the Hartree-Fock ground state $|G\rangle$ through the excitations of the "particle-hole" states $a_m a_i^\dagger$, each with the amplitude X_{mi}. This excitation is represented by the operator $Q^\dagger$ in (3.20).* Note that the index m goes to infinity so that an infinite number of particle-hole excitations are involved.

Equation 3.19 also gives us states at an excitation $E_0 - \hbar\omega^*$, etc. To test the relevance of these solutions we should take the limit $v \to 0$. We notice that in (3.6), $\eta_m - \eta_i$ is always positive if $|G\rangle$ is the ground state of the Hartree-Fock Hamiltonian. Hence in the limit $v \to 0$, (3.6) leads to two solutions:

If $\hbar\omega > 0$ then, it can take on one of the values $\eta_m - \eta_i$, say, $\hbar\omega = \eta_{m0} - \eta_{i0}$, and we have

$$Y_{mi} = 0 \qquad \text{for all } m, i;\ X_{mi} = 0 \quad \text{for } (m, i) \neq (m_0, i_0) \qquad \hbar\omega > 0 \tag{3.21}$$

If $\hbar\omega < 0$, then it can take on one of the values $-(\eta_m - \eta_i)$, say $\hbar\omega = -(\eta_{m_0} - \eta_{i_0})$, and we have

$$X_{mi} = 0 \qquad \text{for all } m, i;\ Y_{mi} = 0 \quad \text{for } (m, i) \neq (m_0, i_0) \qquad \hbar\omega < 0 \tag{3.22}$$

For $\hbar\omega > 0$ the component in (3.19) oscillating with the frequency $(1/\hbar)(E_0 - \hbar\omega)$, would have corresponded to an "excited" state below the ground state. Because of (3.21), however, we see that in the limit $v \to 0$ this state has a zero norm. Similarly, for $\hbar\omega < 0$ the component in (3.19) with $E_0 + \hbar\omega$ would have led to an "excited" state below the ground state. Equation 3.22 tells us that *this* state has a zero norm in the limit $v \to 0$.

These remarks make it clear that the appearance of the negative frequency solutions is associated with the extension of our Hilbert space, (i.e., the Hilbert space dimension is increased) and that these solutions "disappear" when, at the limit $v \to 0$, this extension is not utilized. In the more general case, $v \neq 0$, but still

$$|v_{ni,jm}|, \ |v_{ij,mn}| \ll |\eta_m - \eta_n| \tag{3.23}$$

so that the perturbation approach may be valid, we expect again two solutions: one with $X_{mi} \gg Y_{mi}$ and the other with $Y_{mi} \gg X_{mi}$. We see from (3.17) that the first one goes with positive frequencies whereas the second goes with negative frequencies. Actually it is easy to convince oneself that the solution with $Y_{mi} \gg X_{mi}$ is obtained from that with $X_{mi} \gg Y_{mi}$ by the substitution

$$X_{mi} \to Y_{mi}^*, \qquad Y_{mi} \to X_{mi}^*$$

$$\omega \to -\omega \qquad \text{(assuming } \omega \text{ is real)} \tag{3.24}$$

---

*The operator $Q^\dagger$ in (3.20) is often said to lead to "coherent excitation of particle-hole states." One means, then, to say that in the representation employing the independent-particle states $|k\rangle$, $Q^\dagger|G\rangle$ is a state involving many particle-hole states with *comparable amplitudes*. The coherence as such is, of course, trivial.

Thus in (3.19), states with energies $E_0 - \hbar\omega > E_0$ are always associated with small norms, going to zero as $v \to 0$, and appear as a result of the approximate nature of our procedure. We shall therefore disregard them. The states with the finite norms correspond to real excitations above the ground state.

The full expansion of $\Phi(t)$ in (3.18) actually gives rise to more elaborate wave functions than the ones we wrote down. Thus, for instance, picking up the term with $X_{m1} \exp(-i\omega t) a_m^\dagger$ from the first bracket in (3.18), and the term with $Y_{m'2}^* \exp(i\omega t) a_{m'}^\dagger$ from the second, we obtain a correction term to the ground-state wave function. With this term included, the component in $\Phi(t)$ that oscillates with the frequency $E_0/\hbar$ will now look as follows:

$$\exp(-iE_0t/\hbar)\{a_1^\dagger a_2^\dagger \ldots a_N^\dagger + \sum_{m,m'} X_{m1} Y_{m'2}^* a_m^\dagger a_{m'}^\dagger a_3^\dagger \ldots a_N^\dagger\} |0\rangle$$

$$= \exp(-iE_0t/\hbar)\{1 + \sum_{m,m'} X_{m1} Y_{m'2}^* a_m^\dagger a_{m'}^\dagger a_2 a_1\} |G\rangle \qquad (3.25)$$

A correction to the ground-state wave function of the type described by the second term in the braces takes into account the virtual scattering of the pair of particles in the states k_1 and k_2 into the states k_m and $k_{m'}$. It thus makes it possible to account better for the correlations still present among the nucleons in the real ground state even after choosing the "best" Hartree-Fock single-particle potential. The excitation energies $\hbar\omega$, which are associated with solutions of the N-body problem that do take into account these correlations in the ground and excited states can therefore be expected to represent a better approximation to the true excitation energies.

We can obtain further insight into the nature of the solutions to (3.6) by choosing v to be a very simple interaction:

$$2v_{jm,ni} = 2v_{mn,ij} = \begin{cases} V & \text{if} \quad 1 \leqslant i,j \leqslant N \quad \text{and} \quad N+1 \leqslant m,n \leqslant M \\ 0 & \text{otherwise} \end{cases} \qquad (3.26)$$

(V is real). Physically this amounts to assuming that the interaction v is effective in scattering particles only among states in the vicinity of the highest occupied state N, and that for these states $v(ij)$ has the same constant amplitude for all relevant scatterings. Equations 3.6 then reduce to

$$(\eta_m - \eta_i - \hbar\omega)X_{mi} = -V \sum_{nj} (X_{nj} + Y_{nj})$$

$$(\eta_m - \eta_i + \hbar\omega)Y_{mi} = -V \sum_{nj} (X_{nj} + Y_{nj}) \qquad (3.27)$$

Dividing the first equation by $(\eta_m - \eta_i - \hbar\omega)$, the second by $(\eta_m - \eta_i + \hbar\omega)$, then adding both sides we find

$$(\sum_{mi} X_{mi} + Y_{mi}) = -V\left(\sum_{mi} \frac{2(\eta_m - \eta_i)}{(\eta_m - \eta_i)^2 - (\hbar\omega)^2}\right)(\sum X_{nj} + Y_{nj})$$

The eigenvalues $\hbar\omega$ must therefore satisfy the relation

$$f(\omega) \equiv \sum_{m,i} \frac{2(\eta_m - \eta_i)}{(\eta_m - \eta_i)^2 - (\hbar\omega)^2} = -\frac{1}{V} \tag{3.28}$$

Figure 3.1 shows $f(\omega)$ as a function of $\hbar\omega$. $f(\omega)$ has poles at $\hbar\omega = \eta_m - \eta_i$. These will be recognized as being the energies required to lift a particle from the occupied state $|i\rangle$ to the unoccupied state $|m\rangle$ in the Hartree-Fock single-particle approximation. To obtain the eigenvalues $\hbar\omega$ we have to look for the intersection of $f(\omega)$ with the horizontal line $\phi(\omega) = -(1/V)$. We see that if $|V|$ is very small, so that $|1/V|$ is large, these intersections will take place at energies very close to the poles of $f(\omega)$. In other words, very weak interactions V, attractive or repulsive, reproduce the "particle-hole excited state" $a_m^\dagger a_i|G\rangle$ as the "collective" state, with its energy being very close to the Hartree-Fock energy $\eta_m - \eta_i$. If V is positive, large or small, there will still be one root for $\hbar\omega$ above each Hartree-Fock energy $\eta_m - \eta_i$, and rather close to it.

For strongly attractive interactions a new interesting situation may occur. We note that the minimum of the curve centered about $\omega = 0$ occurs at $f(\omega) = \Sigma 2/(\eta_m - \eta_i)$. Therefore, if V is such that

$$-\frac{1}{V} > \sum \frac{2}{|\eta_m - \eta_i|}$$

there will not only be solutions $\hbar\omega$ close to the Hartree-Fock energies $\eta_m - \eta_i$ (points B, C, etc), but also one solution, the lowest one (point A), which may be quite further removed from its corresponding Hartree-Fock solution. It will thus represent a state that is significantly affected by the interaction. Its collective nature, as we shall see later, will be especially outstanding.

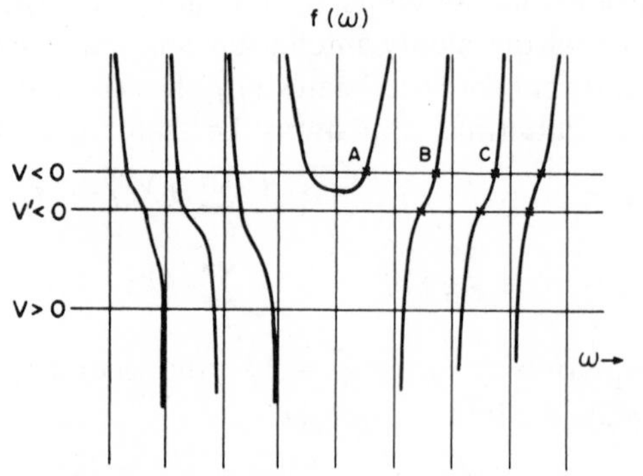

FIG. 3.1. The function $f(\omega)$ of (3.28).

If $V' < 0$ and the interaction is so strong that

$$-\frac{1}{V'} < \sum_{m,i} \frac{2}{|\eta_m - \eta_i|}$$

then the intersection point A will go over into a complex solution for $\hbar\omega$. The Hartree-Fock solution in this case becomes unstable and the system "decays" into a state of a different nature. We notice that this happens for strong attractive ($V' < 0$) interactions when, indeed, the independent-particle approximation may cease to be a good approximation.

4. SPURIOUS STATES

Equation 2.31, defining the stability of the Hartree–Fock solution, can also have an eigenvalue $\lambda = 0$. We can see immediately that if (2.31) is satisfied with $\lambda = 0$, then (3.6), which determines the frequency of the collective excitation, allows a solution with $\hbar\omega = 0$. Such a solution is neither decaying (complex $\hbar\omega$) nor oscillating (real $\hbar\omega$). It implies that if the system is set in motion with the help of the operator $\exp [\Sigma C_{mi}(t) a_m^\dagger a_i]$ corresponding to this solution, it will continue to move without either getting back or changing its structure.

Such a phenomenon is expected as a result of the conservation theorems. If P is a conserved single-particle quantity such as the total momentum, then the operator

$$\hat{P} = \sum P_{kk'} a_k^\dagger a_{k'} \tag{4.1}$$

commutes with the Hamiltonian. The state

$$\exp \left(i\alpha \sum P_{kk'} a_k^\dagger a_{k'}\right) |G\rangle \tag{4.2}$$

is a state in which the center of mass is shifted by a distance α and has the same intrinsic energy as the state $|G\rangle$. The Hartree–Fock solution has a neutral equilibrium with respect to modifications in the wave function of the type (4.2). Thus the existence of solutions to (3.6) with a zero energy $\hbar\omega$ is a necessary consequence of the existence of operators that exactly commute with the Hamiltonian. Such operators lead to modifications in the wave function of the system that do not change its intrinsic structure. The solutions that belong to $\hbar\omega = 0$ are usually referred to as "spurious solutions," and (3.6) indeed offers a convenient way of recognizing them. Of course, the eigenvalue $\hbar\omega = 0$ is only expected to come out if both the representation $a_k|0\rangle$ and the energies η_k have been obtained by a Hartree–Fock method. In actual calculations one often "guesses" approximate Hartree–Fock wave functions and energies. In such cases $\hbar\omega = 0$ may not automatically be a solution of (3.6), and the degree to which the actual solution deviates from $\hbar\omega = 0$ determines the extent to which the particular representation and energies are inconsistent with the conservation law considered.

5. THE MOMENT OF INERTIA IN TIME-DEPENDENT HARTREE-FOCK POTENTIALS

The time-dependent Hartree–Fock equations offer a convenient method for handling the unified model, which attempts to combine single-particle aspects of a system of nucleons with its collective motions. We noted before (see Eq. 2.6) that the transition from the state $|G\rangle$ to the state $\exp(\Sigma C_{mi}a_m^\dagger a_i)|G\rangle$ is equivalent to a transformation from the single-particle states $a_i^\dagger|0\rangle$ to the single-particle states $b_i^\dagger|0\rangle$ where $b_i^\dagger = a_i^\dagger + \sum_{m=N+1}^{\infty} C_{mi}a_m^\dagger$. If the coefficients C_{mi} are now made to be time dependent, that means that the physical interpretation of the states $b_i^\dagger|0\rangle$ is also time dependent. These single-particle states can be visualized as being in a time-dependent potential. The particle state $b_{k1}^\dagger b_{k2}^\dagger \ldots b_{kn}^\dagger|0\rangle$ is then a state of n-particles that changes with time. If the frequencies associated with this change are small compared with the characteristic frequencies of the particles in the Hartree–Fock potential well, then we can visualize the state $b_{k1}^\dagger \ldots b_{kn}^\dagger|0\rangle$ as representing the adiabatic adaption of the n-particle system to the oscillations of the potential. The latter represents the collective motion, whereas the structure of the wave function as a Slater determinant of n-particles represents the independent-particle aspect of this motion.

As an example of the use of the time-dependent Hartree Fock formalism we consider the derivation of an expression for the moment of inertia of a rotating system: assume that the Hamiltonian $H = \Sigma T(i) + (1/2)\Sigma v(ij)$ gives rise to a rotational spectrum, corresponding to the rotation around the x-axis of a deformed nucleus with a symmetry axis along the z-axis. If $|G\rangle$ is the ground state of H, then we can expect that an ω exists such that $|G'\rangle = \exp(-i\omega J_x t)|G\rangle$ is a nonrotating deformed state where J_x is the x-component of the angular momentum. Indeed we know that $\exp(i\alpha J_x)\Psi$ gives a wave function rotated by an angle α with respect to Ψ, but otherwise identical with Ψ. Since we assume that $|G\rangle$ represents a rotating deformed nucleus, by rotating it "backwards" with the appropriate frequency ω, we can make the deformation stay static in space.

To the wave function

$$|G(t)\rangle = \exp(-i\omega t J_x)|G\rangle = \exp\left(-i\omega t \sum \langle k'|J_x|k\rangle a_{k'}^\dagger a_k\right)|G\rangle \tag{5.1}$$

we now apply the Thouless transformation

$$\exp\left(\sum C_{mi}a_m^\dagger a_i\right) \tag{5.2}$$

where the states $a_k^\dagger|0\rangle$ are single-particle states in a deformed potential. Since

$$|G'\rangle = \exp\left(\sum C_{mi}a_m^\dagger a_i\right)\exp\left(-i\omega t \sum \langle k'|J_x|k\rangle a_{k'}^\dagger a_k\right)|G\rangle \quad (5.3)$$

depends explicitly on the time, t, we have to use the time-dependent Hartree–Fock method to derive the best solution for $a_k^\dagger|0\rangle$. We see from (5.3) that if the C's are small and $\omega t \ll 1$ we can use the formalism developed above (See 3.3) if we put

$$\dot{C}_{mi}(t) = -i\omega\langle m|J_x|i\rangle \quad (5.4)$$

Introducing (5.4) into (3.4) we obtain for the best C's:

$$(\eta_m - \eta_i)C_{mi} + 2\sum_{n,j} v_{jm,ni}C_{nj} + 2\sum_{n,j} v_{mn,ij}C_{nj}^* = \hbar\omega\langle m|J_x|i\rangle \quad (5.5)$$

The energy that corresponds to the state (5.3) can be obtained from (2.30)

$$\begin{aligned} E = \frac{\langle G'|\hat{H}|G'\rangle}{\langle G'|G'\rangle} &= E_0 + \sum (\eta_m - \eta_i)|C_{mi}|^2 + 2\sum v_{jm,ni}C_{mi}^*C_{nj} \\ &\quad + \sum v_{ij,mn}C_{mi}C_{nj} + \sum v_{mn,ij}C_{mi}^*C_{nj}^* \\ &= E_0 + \frac{1}{2}\hbar\omega\sum\left[\left(\langle m|J_x|i\rangle C_{mi}^* + \langle i|J_x|m\rangle C_{mi}\right)\right] \end{aligned} \quad (5.6)$$

where E_0 is the Hartree–Fock energy taken with the deformed nonrotating, Hartree–Fock wave function $|G\rangle$, (15.1)

$$E_0 = \langle G|H|G\rangle$$

and we have used (5.5) for the last step in (5.6).

From (5.5) it follows that the C's are proportional to ω. We can therefore define an ω-independent quantity

$$\mathcal{J} = \frac{\hbar}{\omega}\sum_{mi}\left(\langle m|J_x|i\rangle C_{mi}^* + \langle i|J_x|m\rangle C_{mi}\right) \quad (5.7)$$

The energy E, as given by (5.6), now takes the form

$$E = E_0 + \frac{1}{2}\mathcal{J}\omega^2 \quad (5.8)$$

To the extent that the nucleus under consideration actually exhibits a rotational spectrum, (5.8) shows that $\mathcal{J}$, as defined by (5.7), is the effective moment of inertia.

If in (5.5) we ignore the effects of the interaction and set $v = 0$, we obtain

$$C_{mi} = \frac{\hbar\omega\langle m|J_x|i\rangle}{\eta_m - \eta_i} \quad (5.9)$$

In this approximation the expression (5.7) for the moment of inertia takes on the form

$$\mathcal{J} = 2\hbar^2 \sum \frac{|\langle m|J_x|i\rangle|^2}{\eta_m - \eta_i} \tag{5.10}$$

Equation 5.10 will be recognized as the Inglis formula in the cranking model (Eq. VI.7.25). As derived here it is a valid approximation to $\mathcal{J}$ provided the states $a_k^\dagger|0\rangle$ are the "best" deformed single-particle wave functions that reproduce the energy of the nucleus in the body-fixed frame of reference, and provided we ignore the interaction effects on the moment of inertia. The latter can be included if we use (5.7) rather than (5.10) for the evaluation of $\mathcal{J}$. The procedure will then be as follows: start from a deformed potential $\Sigma U_{kk'}^{(0)} a_k^\dagger a_{k'}$, solve for the single-particle states $a_k^\dagger|0\rangle$; fill the n-lowest levels, and calculate $U_{kk'}^{(1)}$ using (2.22); rederive the single-particle states in $\Sigma U_{kk'}^{(1)} a_k^\dagger a_{k'}$ and repeat the procedure until self-consistency is obtained. With these self-consistent deformed states go into (5.5) (and its complex conjugate) and solve for C_{mi} (or rather for C_{mi}/ω). Introduce the values so obtained for $(1/\omega)C_{mi}$ into (5.7) and get a value for the moment of inertia.

It should be stressed that although in principle (5.5) represents an infinite set of equations in an infinite number of unknowns C_{mi}, in practice one limits the C's to states m and i close to the last filled level. The reason for doing so is that C_{mi} is of the order of magnitude of $1/(\eta_m - \eta_i)$ (as can be seen from Eq. 5.5 by ignoring v). Thus the C's that correspond to particle-hole excitations of high energy are relatively small. Or, stated in another way, a small rotation of a deformed wave function, when expressed in terms of particle-hole excitations of the unrotated wave function, involves mostly states near the Fermi surface. The deeper states are unaffected by a small rotation, nor are the highly excited states populated this way.

The treatment of the time-dependent Hartree–Fock approximation is a natural extension of the time-independent Hartree–Fock method. Our search for the time dependence of solutions close to the stable Hartree–Fock solution is motivated by the existence of classical oscillations around points of stability. But it should be emphasized that the actual manipulation of the equations is carried out completely within the framework of quantum mechanics.

Many approaches have been developed for the approximate handling of many interacting fermions. Some of them are more appropriate for the description of collective states, while others may be more useful in the description of excitations involving few nucleons. Unfortunately a common feature of almost all these methods is the lack of a rigorous estimate of the error involved in their application. We have to resort to semiquantitative, basically intuitive, arguments for the justification of these approximations. It is therefore useful to look at related, or even identical, approximations from different points of view, with the hope that we shall be gaining, thereby, a better feeling for their range of validity.

We shall discuss only few of these approximations in this book, their choice being largely a question of taste. The Hartree–Fock method is one such approximation, and since it is derived from a variational principle, it gives the "best" account of the nuclear binding energy within the framework of the independent-particle approximation. The time-dependent Hartree–Fock method then leads to a vibrationlike excitation spectrum, and can be expected to present a good approximation for systems that actually possess vibrationlike spectrum for their excitations.

We shall describe now another approach to the approximate description of generalized vibrational states, which leads to results very similar to the time-dependent Hartree–Fock method. It will indicate to us more clearly other aspects involved in the approximate description of these collective vibrations. The method is known by the name of "Random-Phase Approximation" (RPA) and is used extensively, in various modifications, also in other branches of physics dealing with many particle systems.

6. THE RANDOM-PHASE APPROXIMATION

Let us assume that a system of nucleons has excited states at energies $\hbar\omega$ $2\hbar\omega, \ldots$ and that they show the characteristic features of harmonic vibrations. This situation is not met with in reality with high precision, but many nuclei show an approximate vibrational spectrum manifesting itself through the equal spacings of the vibrational states, their spins, and their related transition probabilities (see Section VI.13). Given the ground state $|E_0\rangle$ of the system and the Hamiltonian $\hat{H}$, it is possible to obtain an excited vibrational state of the system if a solution exists to the operator equation for $\Omega^\dagger$

$$[\hat{H}, \Omega^\dagger] = \hbar\omega\Omega^\dagger \tag{6.1}$$

In fact if an operator $\Omega^\dagger$ satisfying (6.1) can be found then the state

$$|E\rangle = \Omega^\dagger |E_0\rangle \tag{6.2}$$

has the property that

$$\begin{aligned} \hat{H}|E\rangle &= \hat{H}\Omega^\dagger|E_0\rangle = (\hbar\omega\Omega^\dagger + \Omega^\dagger\hat{H})|E_0\rangle \\ &= (E_0 + \hbar\omega)\Omega^\dagger|E_0\rangle = (E_0 + \hbar\omega)|E\rangle \end{aligned}$$

Thus $|E\rangle$ given by (6.2) is an eigenstate of the Hamiltonian $\hat{H}$ that belongs to the eigenvalue $E_0 + \hbar\omega$, where E_0 is the ground-state energy.

We notice further that if a solution to (6.1) exists then $(\Omega^\dagger)^2|E_0\rangle$ is also an eigenstate of $\hat{H}$, and since from (6.1) it follows that

$$[\hat{H}, (\Omega^\dagger)^2] = 2\hbar\omega(\Omega^\dagger)^2$$

we obtain

$$\hat{H}(\Omega^\dagger)^2|E_0\rangle = (E_0 + 2\hbar\omega)(\Omega^\dagger)^2|E_0\rangle$$

Thus we expect to find solutions to (6.1) whenever the spectrum of the Hamiltonian includes a band of equally spaced "vibrational" states.

Finally we notice that it follows from (6.1) that the state $\Omega|E_0\rangle$ is also an eigenstate of $\hat{H}$ with energy $E_0 - \hbar\omega$. If we want $|E_0\rangle$ to be the ground state it must therefore be the "vacuum" for the Ω-excitations, that is, it must satisfy

$$\Omega|E_0\rangle = 0 \tag{6.3}$$

The exact solutions to (6.1) are in such cases completely equivalent to the Schrödinger equation

$$\hat{H}\psi = E\psi$$

for the corresponding states, and the replacement of the wave function ψ by the operator $\Omega^\dagger$ that generates ψ from the ground state does not basically modify the complexity of the problem.

We can try, however, to solve (6.1) in an approximate form, in which case its structure may be easier to handle, though completely equivalent to that of the Schrödinger equation. In particular we shall limit ourselves now to operators $\Omega^\dagger$ that can be expressed in terms of particle-hole operators:

$$\Omega^\dagger = \sum_{m,i} (X_{mi}a_m^\dagger a_i + Y_{mi}^* a_m a_i^\dagger) \tag{6.4}$$

where the X's and the Y's will turn out to be the same as those introduced in (3.5). Here again the indices i, j stand for occupied states in the single-particle Slater determinant that is dominant in the ground state, and m, n are unoccupied states.

We notice the similarity between $\Omega^\dagger$ as defined by (6.4) and the operator $Q^\dagger$ defined by (3.20). We shall find later on that the X_{mi}'s in (6.4) satisfy the same equation satisfied by the X_{mi}'s in (3.6) (and hence the use of the same notation). Since $a_i^\dagger|G\rangle = 0$ we see (see Eq. 3.20) that

$$\Omega^\dagger|G\rangle = Q^\dagger|G\rangle$$

However the form adopted for the operator $\Omega^\dagger$ in (6.4) anticipates the possibility that the *actual* ground state of our system has states $|i\rangle$ that are unoccupied part of the time, so that an excitation from the *real* physical ground state can be achieved also with the operators $a_m a_i^\dagger$. This will be discussed in more detail later on. Since the operators $a_m^\dagger a_i$ represent independent excitations, we can satisfy (6.1) only to the extent that we can satisfy the separate equations

$$[\hat{H}, a_m^\dagger a_i] = \hbar\omega a_m^\dagger a_i$$

$$[\hat{H}, a_m a_i^\dagger] = \hbar\omega a_m a_i^\dagger \tag{6.5}$$

with the *same* value of ω for all pairs i and m for which X_{mi} and Y_{mi} are different from zero.

We further notice that the form (6.4) for the operator $\Omega^\dagger$ anticipates the result that the physical ground state $|E_0\rangle$ of the system is not a single Slater determinant of the states $|k_i\rangle$. In fact if $|E_0\rangle$ were given by the Hartree–Fock ground state,

$$|E_0\rangle = |G\rangle = a_1^\dagger a_2^\dagger \ldots a_N^\dagger|0\rangle$$

then contrary to (6.3) $\Omega|E_0\rangle$ will not be zero if any of the Y^*_{mi} differ from zero. Only if $|E_0\rangle$ is a *sum* of Slater determinants involving N-particles, spread over different states, can we expect to have $\Omega|E_0\rangle = 0$ with $\Omega^\dagger$ given by (6.4) and some nonvanishing Y^*_{mi}.

We still find it convenient to distinguish between the states $|i\rangle$ occupied in the Hartree–Fock ground state $|G\rangle$, and the states $|m\rangle$ that are unoccupied in $|G\rangle$, in the sense that we have no particle-hole pairs of the type $a_i^\dagger a_{i'}$, or $a_m^\dagger a_{m'}$ in (6.4); but as we shall see, (6.4) does allow for the most important correlations in the ground state to be included in our solution.

Inserting (6.4) for $\Omega^\dagger$ into (6.1) we shall obtain on the left-hand side terms involving products of two a-operators and terms involving the products of four such operators; since the right-hand side of (6.1) includes only products of two a's, (6.1) cannot generally be satisfied by an $\Omega^\dagger$ of the form (6.4). This was to be expected since a transition from an exact ground state to an exact excited state of a system of n-interacting fermions involves generally all k-particle-hole excitations with $k = 1, 2, \ldots, n$.

To satisfy (6.1) with an Ω of the type (6.4) we must therefore invoke further approximations. We mentioned that the commutator $[\hat{H}, \Omega^\dagger]$ generally includes terms of the type $a_p^\dagger a_q^\dagger a_r a_s$. To reduce such a product into products of the type $a_{p'}^\dagger a_{q'}$ we make the following approximation (see the derivation of Eqs 6.8 and 6.9).

$$a_p^\dagger a_q^\dagger a_r a_s \rightarrow \langle a_q^\dagger a_r\rangle a_p^\dagger a_s - \langle a_q^\dagger a_s\rangle a_p^\dagger a_r - \langle a_p^\dagger a_r\rangle a_q^\dagger a_s + \langle a_p^\dagger a_s\rangle a_q^\dagger a_r \tag{6.6}$$

Here $\langle a_q^\dagger a_r\rangle$ is the expectation value of $a_q^\dagger a_r$ taken in a state that most conveniently approximates the ground state $|E_0\rangle$. This, sometimes, is chosen to be the dominant single-Slater determinant in $|E_0\rangle$, that is, the Hartree–Fock ground state $|G\rangle$, which is often referred to as the *bare ground state*; ($|E_0\rangle$ is then referred to as the *physical ground state*; the bare ground state includes in it only those correlations among the nucleons that can be simulated by the single-particle Hartree–Fock potential; the physical ground state has in addition also those further correlations induced by the interaction to the extent that $\Omega^\dagger$ can represent them). If $a_p^\dagger a_q^\dagger a_r a_s$ is looked upon as a bilinear form in the operators $a_p^\dagger a_q$, the approximation (6.6) can be considered as a *linearization approximation*, in which one power of this operator is replaced by its expectation value. This approximation is also referred to as a *random-phase approximation.*

The essence of this approximation lies in neglecting simultaneous two-particle two-hole excitations and in assuming that whenever we are faced with such an operator, say $a_p^\dagger a_q^\dagger a_r a_s$, we can consider only *one*-particle-hole pair to be effective, whereas the other is "inert" and takes on its expectation value. Furthermore it is assumed that the Hartree–Fock uncorrelated ground state is good enough for estimating the effects of the "inert" part of $a_p^\dagger a_q^\dagger a_r a_s$. The four terms in (6.6) represent, of course, all four possible ways for selecting this inert part.

We notice that in our approximation

$$\langle a_q^\dagger a_r \rangle = \begin{cases} \delta_{qr} & \text{if} \quad r \leqslant N \\ 0 & \text{if} \quad r > N \end{cases} \tag{6.7}$$

Using this result we obtain then from (2.18) and its complex conjugate, together with (2.21) and (6.5), that

$$\hbar\omega a_m^\dagger a_i = [\hat{H}, a_m^\dagger a_i] = (\eta_m - \eta_i) a_m^\dagger a_i + 2 \sum v_{in,mj} a_n^\dagger a_j + 2 \sum v_{ij,mn} a_j^\dagger a_n \tag{6.8}$$

$$\begin{aligned} \hbar\omega a_m a_i^\dagger &= [\hat{H}, a_m a_i^\dagger] \\ &= -(\eta_m - \eta_i) a_m a_i^\dagger - 2 \sum v_{mj,in} a_n a_j^\dagger - 2 \sum v_{mn,ij} a_n a_j^\dagger \end{aligned} \tag{6.9}$$

Equations 6.8 and 6.9 were derived under the random-phase assumption that replaces some operators by their expectation values for a given state $|E_0\rangle$ or $|G\rangle$. They are therefore expected to be valid only when operating on this state; furthermore, since they were derived from (6.4), they are limited to pairs of indices (i, m) for which either $X_{mi} \neq 0$ or $Y_{mi} \neq 0$.

We note the similarity between the right-hand side of (6.8) and (6.9) and the left-hand side of (3.6) deduced from the time-dependent Hartree–Fock approximation. To establish the connection between the two we "invert" (6.4) to give X_{mi} and Y_{mi} in terms of Ω and $a_m^\dagger a_i$. In fact, it follows from (6.4) that

$$\begin{aligned} [\Omega, a_n^\dagger a_j] &= \sum X_{mi}^* [a_i^\dagger a_m, a_n^\dagger a_j] + Y_{mi} [a_i a_m^\dagger, a_n^\dagger a_j] \\ &= \sum X_{mi}^* [a_i^\dagger a_m, a_n^\dagger a_j] \\ &= \sum X_{mi}^* (\delta_{mn}\delta_{ij} - \delta_{mn} a_j a_i^\dagger - \delta_{ij} a_n^\dagger a_m) \end{aligned} \tag{6.10}$$

Taking expectation values of (6.10) with the correlated (physical) ground state $|E_0\rangle$, but still using the approximation

$$a_i^\dagger |E_0\rangle \approx 0 \qquad a_m |E_0\rangle \approx 0 \tag{6.11}$$

we obtain

$$\langle E_0 | [\Omega, a_n^\dagger a_j] | E_0 \rangle = \langle E | a_n^\dagger a_j | E_0 \rangle \approx X_{nj}^* \tag{6.12}$$

In deriving (6.12) we used (6.2) in the form $\langle E_0|\Omega = \langle E|$ as well as (6.3): $\Omega|E_0\rangle = 0$. By similar arguments we can obtain also

$$\langle E|a_j^\dagger a_n|E_0\rangle \approx Y_{nj} \tag{6.13}$$

Taking now matrix elements of (6.8) and (6.9) between $|E_0\rangle$ and $\langle E|$, and using (6.12) and (6.13), we obtain that the X's and Y's introduced in (6.4) should satisfy

$$\begin{aligned}(\eta_m - \eta_i - \hbar\omega)X_{mi}^* + 2\sum v_{in,mj}X_{nj}^* + 2\sum v_{ij,mn}Y_{nj} &= 0\\ (\eta_m - \eta_i + \hbar\omega)Y_{mi} + 2\sum v_{mj,in}Y_{nj} + 2\sum v_{mn,ij}X_{nj}^* &= 0\end{aligned} \tag{6.14}$$

These equations are the complex conjugate of (3.6). Equation 6.14 is a weaker form of (6.8) and (6.9); although (6.14) follows from these two equations, the inverse is not correct. Actually what is implied in going over from (6.8) and (6.9) to (6.14) is the assertion that we do not need the validity of the *operator* equations (6.8) and (6.9) in general, but only when operating between the states $|E_0\rangle$ and $|E\rangle$. The excitation energy $\hbar\omega$ appears as an eigenvalue problem of a nonhermitian matrix (compare Eq. 3.12), and the conditions for the existence of a real, positive, solution for $\hbar\omega$ is also the condition for the stability of the Hartree–Fock solution for the ground state. The reader is referred to Rowe (66) for further analysis of the relation between the RPA and the time dependent HF approximation.

7. EXCITATIONS OF DEFINITE MULTIPOLES

Our treatment of the structure of collective excited states has been, thus far, quite general. We did not have to specify the states $|k\rangle$ except in saying that they are the single-particle eigenstates in a self-consistent Hartree–Fock potential. We did not even have to specify whether this potential is spherically symmetric, axially symmetric, or has any other symmetry.

If the Hartree–Fock potential is spherically symmetric, then, on account of the invariance of the interaction $v(ij)$ under rotations, the eigenstates of the system are also eigenstates of the total angular momentum squared $\mathbf{J}^2$ and its z-component J_z. The evaluation of the energies of collective excited states through the solution of the eigenvalue equations (6.14) or (3.6) can then be greatly simplified if we start from certain linear combinations of X_{mi} and Y_{mi} so that the excited states generated by $\Omega^\dagger$ have a well-defined angular momentum.

We recall, from (6.2) and (6.4) that an excited state of the system at an energy $E = E_0 + \hbar\omega$ is given by

$$|E\rangle = \Omega^\dagger|E_0\rangle = (\sum X_{mi}a_n^\dagger a_i + Y_{mi}^* a_n a_i^\dagger)|E_0\rangle \tag{7.1}$$

Here X_{mi} and Y_{mi} are the components of the "eigenvector" (X, Y) that belongs to the eigenvalue $\hbar\omega$ of (6.14). Suppose now that the ground state $|E_0\rangle$ has $J = 0$, and that we are interested in excited states $|E\rangle$ of a definite value J of the total angular momentum. It is obvious that this can be achieved only if we destroy a particle of angular momentum j_i and create a particle in j_n such that j_i and j_n can combine to give the desired total J. These considerations are of great practical importance for the applications of this method in actual calculations, and we therefore develop them below in some detail.

Assume that the states $|k\rangle$ are characterized by their angular momenta $|j_k m_k\rangle$. Then in the ground state $|E_0\rangle$, with $J = 0$, we can single out the ith orbit and write it as

$$|E_0\rangle = \sum_{m_i j_i' m_i'} (j_i m_i j_i' m_i'|00) a^\dagger_{j_i m_i} |E_0'(j_i' m_i')\rangle \tag{7.2}$$

where $|E_0'(j_i' m_i')\rangle$ are $(N-1)$ particle states characterized by the total angular momentum quantum numbers j_i' and m_i' and possibly other quantum numbers as well. The Clebsch–Gordan coefficient $(j_i m_i j_i' m_i'|00)$ was introduced to assure that $|E_0\rangle$ has the appropriate total angular momentum $J = 0$. The application of the term

$$X_{j_n m_n, j_i m_i} a^\dagger_{j_n m_n} a_{j_i m_i}$$

in $\Omega^\dagger$ to the state $|E_0\rangle$ will yield

$$\begin{aligned} X_{j_n m_n, j_i m_i} a^\dagger_{j_n m_n} a_{j_i m_i} |E_0\rangle &= \sum_{j_i' m_i'} (j_i m_i j_i' m_i'|00) X_{j_n m_n, j_i m_i} a^\dagger_{j_n m_n} |E_0(j_i' m_i')\rangle \\ &= (2j_i+1)^{-1/2} (-1)^{j_i - m_i} X_{j_n m_n, j_i m_i} a^\dagger_{j_n m_n} |E_0'(j_i, -m_i)\rangle \end{aligned} \tag{7.3}$$

where we have used the relation (Appendix A, A.2.76)

$$(j_i m_i, j_i' m_i'|00) = (-1)^{j_i - m_i} \delta(j_i, j_i')\, \delta(m_i, -m_i')(2j_i+1)^{-1/2}$$

Then from (7.1) the X-term on its right-hand side becomes

$$\frac{1}{\sqrt{2j_i+1}} \sum (-1)^{j_i - m_i} X_{j_n m_n, j_i m_i}\, a^\dagger_{j_n m_n} |E_0'(j_i, -m_i)\rangle$$

In order for this term to represent a state of a given J and M we must have

$$(-1)^{j_i - m_i} X_{j_n m_n, j_i m_i} = (j_n m_n, j_i - m_i|JM) X_{j_n j_i}$$

or

$$X_{j_n m_n, j_i m_i} = X_{j_n j_i} (-1)^{j_i - m_i} (j_n m_n, j_i - m_i|JM) \tag{7,4}$$

where $X_{j_n j_i}$ is independent of the magnetic quantum numbers m_i and m_n.

In a similar way we can show that Y^*_{ni} in (7.1) should be such that

$$\begin{aligned} Y^*_{j_n m_n j_i m_i} &= -Y^*_{j_n j_i} (-1)^{j_n - m_n} (j_i m_i, j_n - m_n|JM) \\ &= -Y^*_{j_n j_i} (-1)^{-j_i + J - m_n} (j_n - m_n, j_i m_i|JM) \end{aligned} \tag{7.5}$$

where we have used the properties of the $3-j$ symbols to change the order of j's in the Clebsch–Gordan coefficient. It is easy to trace the origin of the minus sign in m_i; it results from the fact that we are coupling a *particle* in the state $j_n m_n$ to a *hole* in the state $j_i m_i$. The minus sign in front of $Y^*_{j_n j_i}$ in (7.5) is a matter of convenience as (7.5) really serves to define $Y^*_{j_n j_i}$.

It is sometimes convenient to define a particle-hole creation operator of a definite angular momentum. In view of (7.4) this is defined as follows

$$A^\dagger(j_n j_i, JM) = \sum_{m_i m_n} (-1)^{j_i - m_i}(j_n m_n, j_i - m_i|JM) a^\dagger_{j_n m_n} a_{j_i m_i} \quad (7.6)$$

It follows from (7.6) that

$$\begin{aligned} A(j_n j_i, JM) &= \sum_{m_i m_n} (-1)^{j_i - m_i}(j_n m_n, j_i - m_i|JM) a^\dagger_{j_i m_i} a_{j_n m_n} \\ &= -(-1)^{j_i - j_n + M} \sum (-1)^{-j_i + J - m_i}(j_n - m_n, j_i m_i|J - M) \\ &\qquad a_{j_n m_n} a^\dagger_{j_i m_i} \end{aligned} \quad (7.7)$$

where use has been made of the relation

$$(j_1 m_1, j_2 m_2|JM) = (-1)^{j_1 + j_2 - J}(j_1 - m_1, j_2 - m_2|J - M)$$

With the help of (7.6), (7.7), (7.4), and (7.5) we can now write the general expression for the operator $\Omega^\dagger$ appropriate for the creation of a state $|EJM\rangle$ out of the state $|E_0\, J = 0\rangle$:

$$\Omega^\dagger(JM) = \sum_{j_n j_i} [X_{j_n j_i} A^\dagger(j_n j_i, JM) + (-1)^{j_i - j_n - M} Y^*_{j_n j_i} A(j_n j_i, J - M)] \quad (7.8)$$

If one starts with the operators (7.8) rather than (6.4), then Equations 6.14 for the eigenvalues $\hbar\omega$ separate automatically into independent equations for each value of J and M. In actual calculations this may result in a significant simplification.

The operator $\Omega^\dagger(JM)$ in (7.8) is sometimes referred to as the operator for the excitation of the Jth multipole. The origin of the name is obvious, since it is, by construction an operator that adds to the system the angular momentum J with z-component M. For a further discussion of the RPA see Gillet (66) and Section 19 of this chapter.

8. THE BOGOLJUBOV-VALATIN TRANSFORMATION

A number of approximations had to be made in the quantum mechanical derivation of (6.14), the most important of them being the linearization approximation, (6.6), and the approximate identity between the correlated ground state $|E_0\rangle$ and the uncorrelated Hartree–Fock ground state $|G\rangle$ to the extent required by (6.11) (and its equivalent, Eq. 6.7). Obviously the validity of these approximations ultimately depends on the interaction $v(ij)$. It seems that a decisive role in determining whether the approximation works is played by the Hartree–Fock energies η_i and η_m, which represent, of course, some averaged properties of the interaction $v(ij)$ in a system of a given number of particles. If for a particular problem the Hartree–Fock energies turn out to be such that $\eta_m - \eta_i \gg |v|$ for any pair of particle-hole states (m,i), where v represents a typical interaction matrix element, then, as we can see from (2.31), the Hartree–Fock solution can be expected to be stable, and the rest follows. Physically this means that the main effect of the interaction $v(ij)$ is to establish a correlation among the nucleons, which keeps them together, and which can be well simulated by the effects of a single-particle potential properly chosen. This situation may be true for one number of nucleons, but need not necessarily be true for another. Indeed, if for the particular number of particles we are considering it turns out that there are enough particle-hole pairs for which $\eta_m - \eta_i \approx |v|$, then the preceding analysis must be modified. Physically the system may find it energetically advantageous to have its particles distributed over a wide range of levels near the highest Hartree–Fock occupied state; although the system loses an excitation energy $\eta_m - \eta_i$ by promoting one of its particles from $|k_i\rangle$ to $|k_m\rangle$, it may gain an extra mutual interaction energy v, and we have assumed that there are enough pairs for which $\eta_m - \eta_i \approx |v|$. Mathematically the stability condition may then become barely satisfied, making our various approximations less valid.

The conclusion that the validity of our results depends primarily on the Hartree–Fock values of $\eta_m - \eta_i$ is, in a sense, very encouraging. The whole structure of the Hartree–Fock approximation actually makes the differences $\eta_m - \eta_i$ as large as possible. In fact this approximation consists of minimizing the total energy of the first N-occupied states, and making the matrix elements of $\hat{H}$ between an occupied and unoccupied state vanish (see Eq. 2.20). It thus aims at giving the single-particle potential such a shape that it will make occupied states have energies as low as possible, and unoccupied states as high as possible, leading to an energy "gap" between occupied and unoccupied states [Kelson and Levinson (64)]. This gap is shown in Fig. 8.1. This increases the differences $\eta_m - \eta_i$, and thus works in the direction of making our approximation more valid. Nevertheless this gap between occupied and unoccupied states remains sometimes quite small. We noticed already in Section IV.7 that a single-particle central potential tends to lead to a grouping of levels. If the number of particles is such that a group of such levels is only partially filled, that is, if we are far from a magic number, we may find ourselves with a group of nearly degenerate levels near the top of the Fermi distribution only some of which are occupied. In that case another procedure should be followed to obtain the ground state of the system. We shall now develop such a method utilizing to its full power the method of second quantization.

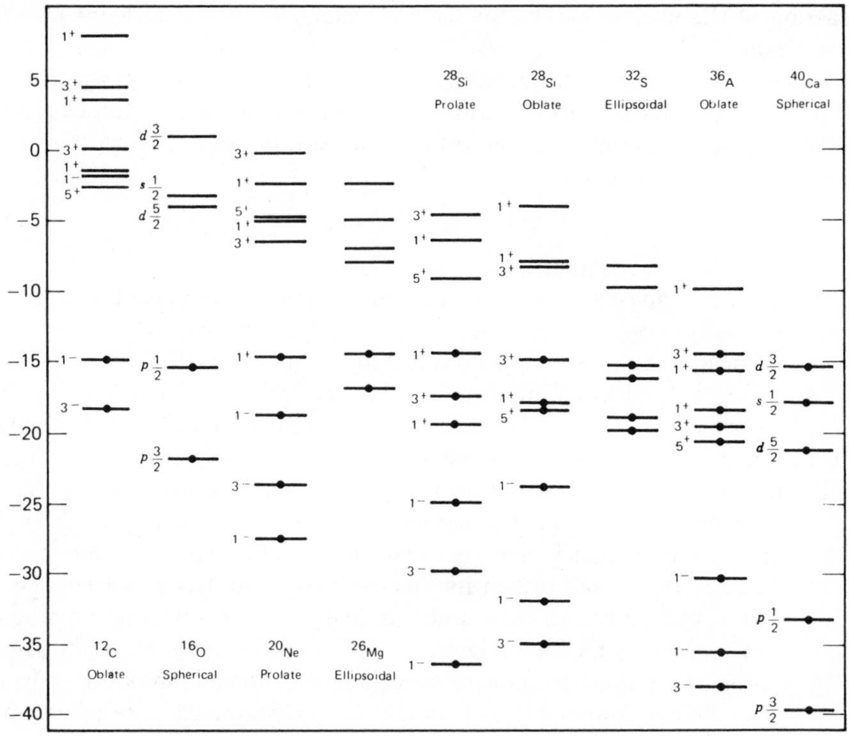

FIG. 8.1. Spectrum of the Hartree–Fock orbits of lowest energy solutions of even–even $N = Z$ nuclei between ^{12}C and ^{40}Ca. The levels are fourfold degenerate and the occupied orbits are marked with a dot. Note the energy gap between occupied and unoccupied states. For spherical and axially symmetric cases, the $1p$ and $2s$–$1d$ shell orbits are shown. For the ellipsoidal solutions only the $2s$–$1d$ shell orbits are given [from Ripka (68)].

In the Hartree–Fock self-consistent potential method the Hamiltonian

$$\hat{H} = \sum \epsilon_{kk'} a_k^\dagger a_{k'} + \frac{1}{4} \sum v_{k_1 k_2, k_1' k_2'} a_{k_1}^\dagger a_{k_2}^\dagger a_{k_2'} a_{k_1'} \tag{8.1}$$

expressed in terms of a set of single-particle states $|k\rangle$, was made to operate on a new set of single-particle states produced from the vacuum by $b_k^\dagger|0\rangle$ where (see Eq. 2.6)

$$b_i^\dagger = a_i^\dagger + \sum_{m=N+1} C_{mi} a_m^\dagger \tag{8.2}$$

The choice of these new single-particle states satisfies the requirement that the new single-particle energy $\Sigma \eta_k b_k^\dagger b_k$ will contain the maximum possible fraction of the nuclear energy for the particular number of nucleons we are considering.

Equation 8.2 can be considered as a transformation from the set $\{a^\dagger\}$ to $\{b^\dagger\}$. It is a canonical transformation to first order in the C's. It can be made fully canonical by proper normalization, and can, of course, be written in a more general form as

$$b_k^\dagger = \sum U_{kk'} a_{k'}^\dagger \tag{8.3}$$

where $U_{kk'}$ is a unitary matrix.

It is possible, however, to introduce a more general canonical transformation, originally suggested by Bogoljubov and Valatin. We notice that if a single-particle Hamiltonian is invariant under time reversal* then a particle in the state $|k\rangle$ and a hole in the time-reversed state $|k\rangle$ have very similar physical properties. Thus if $|k\rangle$ is a state of linear momentum $\mathbf{k}$, then the time-reversed state will have a momentum $-\mathbf{k}$, and the existence of a particle with momentum $\mathbf{k}$, or the absence of a particle with momentum $-\mathbf{k}$, have the same effect on the total momentum of the system. Similarly if $|k\rangle$ is a state with a z-component k of the angular momentum, that is, the ϕ-dependence of $|k\rangle$ is $\exp(-ik\phi)$, then its time-reversed state has a ϕ-dependence $\exp(+ik\phi)$ and represents a rotation in the opposite direction. Again a particle in $\exp(-ik\phi)$, or the absence of one in $\exp(+ik\phi)$, make the same contribution to the total component of the angular momentum of the system.

We shall find it convenient to describe the time-reversed state to $|k\rangle$ by $|-k\rangle$, and define for $k > 0$ the Bogoljubov–Valatin transformation from a_k to the operators α_k through

$$\alpha_k = U_k a_k - V_k a_{-k}^\dagger \qquad \alpha_{-k} = U_k a_{-k} + V_k a_k^\dagger \qquad k > 0 \tag{8.4}$$

Here U_k and V_k are numbers *defined only for positive values of* k; it turns out, as the reader can easily verify in the derivation of the subsequent relations, that they can be consistently chosen to be real and positive. The physical motivation that suggests this transformation will be described in Section 9: For the moment, we shall simply see what some of the consequences of (8.4) are. It follows then immediately from (8.4) that

$$\alpha_k^\dagger = U_k a_k^\dagger - V_k a_{-k} \qquad \alpha_{-k}^\dagger = U_k a_{-k}^\dagger + V_k a_k \qquad k > 0 \tag{8.5}$$

We further require that the α_k operators satisfy the anticommutation relations

$$\{\alpha_k, \alpha_{k'}\} = \{\alpha_k^\dagger, \alpha_{k'}^\dagger\} = 0 \qquad \{\alpha_k^\dagger, \alpha_{k'}\} = \delta_{kk'} \tag{8.6}$$

It is seen from (8.4) and (8.5), that (8.6) will be automatically satisfied if U_k and V_k satisfy

$$U_k^2 + V_k^2 = 1 \tag{8.7}$$

*For a discussion of time reversal see Appendix A (Sec 3).

In what follows we shall impose this condition on U_k and V_k. We observe that since the α_k's obey fermion anticommution rules the numbers operator in "α space," $\alpha_k^\dagger \alpha_k$, will have eigenvalue 1 or 0. The meaning of the vacuum state $|0\rangle_\alpha$ and of the single *quasi-particle state* $\alpha_k^\dagger |0\rangle_\alpha$ will be discussed below.

Equations 8.4 and 8.5 can be inverted to express the a_k's in terms of the α_k's:

$$\begin{aligned} a_k &= U_k \alpha_k + V_k \alpha_{-k}^\dagger \\ & \qquad\qquad\qquad k > 0 \\ a_{-k} &= U_k \alpha_{-k} - V_k \alpha_k^\dagger \end{aligned} \tag{8.8}$$

The Bogoljubov transformation has some very peculiar features, which we shall now proceed to study. Consider for example the zero quasi-particle state $|0\rangle_\alpha$. This state satisfies the equations

$$\alpha_k |0\rangle_\alpha = 0 \qquad \alpha_{-k}|0\rangle_\alpha = 0 \tag{8.9}$$

In terms of the a_k-type real-particle states, $|0\rangle_\alpha$ is given by the linear combination

$$|0\rangle_\alpha = \lambda_0 |0\rangle + \sum \lambda_{k'} a_{k'}^\dagger |0\rangle + \sum \lambda_{-k'} a_{-k'}^\dagger |0\rangle + \sum \mu_{k'} a_{k'}^\dagger a_{-k'}^\dagger |0\rangle + \ldots \tag{8.10}$$

The coefficients $\lambda_0, \lambda_{k'} \ldots$ are determined by the condition (8.9). Rather than show, step by step, how they can be obtained we shall prove that the solution

$$|0\rangle_\alpha = N \prod_{k'} \left(1 + \frac{V_{k'}}{U_{k'}} a_{k'}^\dagger a_{-k'}^\dagger \right) |0\rangle \tag{8.11}$$

which is of the general type given by (8.10), and satisfies the conditions (8.9). Notice that

$$\begin{aligned} \alpha_k \left(1 + \frac{V_k}{U_k} a_k^\dagger a_{-k}^\dagger \right) &= (U_k a_k - V_k a_{-k}^\dagger) \left(1 + \frac{V_k}{U_k} a_k^\dagger a_{-k}^\dagger \right) \\ &= U_k a_k + V_k a_k a_k^\dagger a_{-k}^\dagger - V_k a_{-k}^\dagger \end{aligned} \tag{8.12}$$

where we used the fact that $a_\alpha^\dagger a_\alpha^\dagger = 0$. Note that any specific index k appears in only one factor in (8.11), and thus in the form considered in (8.12). The various factors in (8.11) commute with each other. Thus we can consider the expression in (8.12) as operating directly on $|0\rangle$. We can therefore replace, in (8.12), $a_k a_k^\dagger$ by 1 and $U_k a_k$ by zero. We find

$$\alpha_k \left(1 + \frac{V_k}{U_k} a_k^\dagger a_{-k}^\dagger \right) |0\rangle = 0$$

and similarly

$$\alpha_{-k} \left(1 + \frac{V_k}{U_k} a_k^\dagger a_{-k}^\dagger \right) |0\rangle = 0$$

It follows that (8.11) satisfies the conditions (8.9).

It is easy to work out the value of the normalization constant N. Straightforward calculation shows that

$$N = \prod_{k'} U_{k'}$$

so that (8.11) can be written in the form

$$|0\rangle_\alpha = \prod (U_{k'} + V_{k'} a_{k'}^\dagger a_{-k'}^\dagger)|0\rangle \tag{8.13}$$

which is sometime more useful.

We can expand the product in (8.11) and obtain

$$|0\rangle_\alpha = N\left\{1 + \sum \frac{V_{k'}}{U_{k'}} a_{k'}^\dagger a_{-k'}^\dagger + \sum \frac{V_{k'}V_{k''}}{U_{k'}U_{k''}} a_{k'}^\dagger a_{-k'}^\dagger a_{k''}^\dagger a_{-k''}^\dagger + \ldots\right\}|0\rangle \tag{8.14}$$

It is therefore seen that the state that corresponds to no α-type quasi particles, when translated into the a-type states, becomes a mixture of the state $|0\rangle$ that has no a-type particles, states $a_k^\dagger a_{-k}^\dagger|0\rangle$ that have a pair of a-type particles in the states k and $-k$, states like $a_{k'}^\dagger a_{-k'}^\dagger a_{k''}^\dagger a_{-k''}^\dagger|0\rangle$ that have two pairs of a-type particles, etc.

It seems, therefore, that in performing the Bogoljubov transformation we have been led to introduce new entities that occupy the states α_k, such that the state of no α-type entities corresponds to a mixture of states with no a-type particles, with two such particles, and four such particles etc. Since the number of nucleons is absolutely conserved in any nuclear process, this looks like a very unphysical transformation. Yet we recall that by violating conservation theorems we were able in the past to come forth with simpler mathematical approximations to the problem of solving the Schrödinger equation. We violated conservation of linear momentum when we introduced a fixed average potential in the shell model; we violated conservation of angular momentum when we introduced deformed potential to derive the unified model; and we are about to violate now the conservation of the number of particles in order to achieve a substantial improvement in an independent-particlelike treatment of the Schrödinger equation.*

To make a clear distinction between the entities that occupy the α-type states and the particles that occupy the a-type states, the former are usually referred to as *quasi particles.*

9. THE BCS GROUND STATE

The numbers U_k and V_k in the Bogoljubov transformation have not been specified thus far. The freedom in choosing them, subject to the condition

---

*It is indeed possible to demonstrate far-reaching analogies among these various approximationes [see Broglia (73)].

(8.7), can be used now to produce better approximations for the ground states of the many-nucleon system.

To see how it works let us first consider a simple case: that of noninteracting nucleons in a single-particle potential. The Hamiltonian is then given by

$$H = \sum \epsilon_k a_k^\dagger a_k = \sum_{k>0} \epsilon_k (a_k^\dagger a_k + a_{-k}^\dagger a_{-k}) \tag{9.1}$$

where ϵ_k are the single-particle energies. Since the general Bogoljubov transformation does not conserve the number of particles, we want to solve for the energy of the system described by (9.1) subject to the condition that the *average* number of particles is fixed. This, as is well known, [Morse and Feshbach (52)] can be done through the introduction of a Lagrange multiplier λ multiplying the number operator $\hat{N} = \Sigma\, a_k^\dagger a_k = \sum_{k>0} (a_k^\dagger a_k + a_{-k}^\dagger a_{-k})$. We are thus looking for the eigenvalues and eigenfunctions of

$$H - \lambda \hat{N} = \sum_{k>0} (\epsilon_k - \lambda)(a_k^\dagger a_k + a_{-k}^\dagger a_{-k}) \tag{9.2}$$

where λ is to be fixed by the prescribed expectation value of $\hat{N}$;

$$N = \langle \hat{N} \rangle \tag{9.3}$$

Introducing the transformation (8.8) into (9.2) we obtain:

$$\begin{aligned} H - \lambda \hat{N} &= \sum_{k>0} (\epsilon_k - \lambda)[U_k^2(\alpha_k^\dagger \alpha_k + \alpha_{-k}^\dagger \alpha_{-k}) + V_k^2(\alpha_k \alpha_k^\dagger + \alpha_{-k}\alpha_{-k}^\dagger) \\ &\qquad + 2U_kV_k(\alpha_k^\dagger \alpha_{-k}^\dagger + \alpha_{-k}\alpha_k)] \\ &= \sum_{k>0} 2(\epsilon_k - \lambda)V_k^2 + \sum_{k>0} (\epsilon_k - \lambda)(U_k^2 - V_k^2)(\alpha_k^\dagger \alpha_k + \alpha_{-k}^\dagger \alpha_{-k}) \\ &\qquad + \sum_{k>0} 2(\epsilon_k - \lambda)U_kV_k(\alpha_k^\dagger \alpha_{-k}^\dagger + \alpha_{-k}\alpha_k) \end{aligned} \tag{9.4}$$

The operators $\alpha_k^\dagger \alpha_k$ commute with the quasi-particle number operator $\hat{N}_\alpha = \sum_{k>0} (\alpha_k^\dagger \alpha_k + \alpha_{-k}^\dagger \alpha_{-k})$; they can therefore be brought to a diagonal form simultaneously with $\hat{N}_\alpha$. The operators $\alpha_k^\dagger \alpha_{-k}^\dagger$ and $\alpha_{-k}\alpha_k$, however, do not commute with $\hat{N}_\alpha$. Thus to diagonalize $H - \lambda N$ we want to chose U_k and V_k in such a way that the last term in (9.4) vanishes. This leads immediately to the result

$$U_kV_k = 0 \qquad \text{for all } k\text{'s} \tag{9.5}$$

which, together with (8.7), leads to one of two possibilities:

$$U_k = 0 \qquad V_k = 1 \tag{9.6a}$$

or

$$U_k = 1 \qquad V_k = 0 \tag{9.6b}$$

The lowest eigenstate of (9.4) is now the quasi-particle vacuum $|0\rangle_\alpha$ which satisfies, in view of $\alpha_k|0\rangle_\alpha = 0$,

$$(H - \lambda\hat{N})|0\rangle_\alpha = [\sum_{k>0} 2(\epsilon_k - \lambda)V_k^2]|0\rangle_\alpha \tag{9.7}$$

In view of (9.6) we see now that the lowest eigenvalue of $H - \lambda\hat{N}$ is obtained by setting

$$\begin{aligned} V_k &= 1 \qquad U_k = 0 \qquad \text{if} \qquad \epsilon_k < \lambda \\ V_k &= 0 \qquad U_k = 1 \qquad \text{if} \qquad \epsilon_k > \lambda \end{aligned} \tag{9.8}$$

The value of λ should be determined by the condition

$$_\alpha\langle 0|\hat{N}|0\rangle_\alpha = {}_\alpha\langle 0|\sum_{k>0} a_k^\dagger a_k + a_{-k}^\dagger a_{-k}|0\rangle_\alpha = N \tag{9.9}$$

which can be easily shown to reduce to

$$\sum_{k>0} 2V_k^2 = N \tag{9.10}$$

Combining (9.8) and (9.10) we see that the lowest state of the system is obtained, as expected, by filling the $N/2$ lowest pairs of states k and $-k$; furthermore we see that λ, as determined by the condition (9.9) is the energy of the highest occupied state. It is, in other words, the Fermi energy of the system.

It is instructive to introduce the solution (9.8) back into the Bogoljubov transformations (8.4) and (8.5):

$$\alpha_k^\dagger = \begin{cases} -a_{-k} & \text{if} \quad \epsilon_k < \lambda \\ a_k^\dagger & \text{if} \quad \epsilon_k > \lambda \end{cases} \tag{9.11}$$

Thus the quasi-particle creation operator in this case coincides with the creation operator for real particles, when referring to the unoccupied states above the top of the Fermi sea; it is however equal to a real-particle destruction operator for the occupied states. This is, of course, the natural way to define creation and annihilation operators when we want our ground state to be the lowest reference state: that is, the vacuum. It leads to a symmetric handling of particles and holes, *both* of which represent an excitation from the ground state.

In the general case of interacting nucleons, when the coefficients V_k and U_k turn out to take on values that are different from 0 or 1, the quasi-particle vacuum will not correspond to a real-particle state of a definite number of particles. Rather, as can be seen from (8.13), V_k^2 is the probability to find in the quasi-particle vacuum a pair of real particles in the states k, and $-k$ and $U_k^2 = 1 - V_k^2$ is, therefore, the probability that this state is empty. The expectation value of the number operator $\hat{N} = \sum_{k>0} (a_k^\dagger a_k + a_{-k}^\dagger a_{-k})$ in the quasi-particle vacuum is consequently given by (9.10).

It is instructive to see also the structure of the single quasi-particle state $\alpha_k^\dagger|0\rangle_\alpha$ in terms of the real-particle states. We find, using (8.13) and (8.5):

$$\begin{aligned}\alpha_k^\dagger|0\rangle_\alpha &= (U_k a_k^\dagger - V_k a_{-k}) \prod (U_{k'} + V_{k'} a_{k'}^\dagger a_{-k'}^\dagger)|0\rangle \\ &= (U_k^2 a_k^\dagger + V_k^2 a_k^\dagger) \prod_{k \neq k'} (U_{k'} + V_{k'} a_{k'}^\dagger a_{-k'}^\dagger)|0\rangle \\ &= a_k^\dagger \prod_{k \neq k'} (U_{k'} + V_{k'} a_{k'}^\dagger a_{-k'}^\dagger)|0\rangle \end{aligned} \tag{9.12}$$

Thus, to obtain the state in which one quasi-particle occupies the quasi-particle state k, we should put one real particle in the corresponding particle state and fill all the *other* states with *pairs* of particles and with amplitudes identical to those with which they appear in the quasi-particle vacuum.

A pair of quasi particles in the states k and $-k$ give rise to a different type of wave function:

$$\begin{aligned}\alpha_k^\dagger \alpha_{-k}^\dagger|0\rangle_\alpha &= (U_k a_{-k}^\dagger - V_k a_{-k}) a_{-k}^\dagger \prod_{k \neq k'} (U_{k'} + V_{k'} a_{k'}^\dagger a_{-k'}^\dagger)|0\rangle \\ &= (U_k a_k^\dagger a_{-k}^\dagger - V_k) \prod_{k \neq k'} (U_{k'} + V_{k'} a_{k'}^\dagger a_{-k'}^\dagger)|0\rangle \end{aligned} \tag{9.13}$$

Thus we see that a single quasi-particle wave function (9.12) is very closely related to that of a single real particle; a pair of quasi particles, on the other hand, involves the corresponding pair of real particles in a more complex fashion.

To go now over to the general case of a Hamiltonian with interaction, we shall limit ourselves, for simplicity and for reasons that will become clear later, to a subclass of interactions—the so-called *pairing interactions*, which can be written in the form

$$\frac{1}{4} \sum v_{k_1 k_2, k_1' k_2'} a_{k_1}^\dagger a_{k_2}^\dagger a_{k_2'} a_{k_1'} \, \delta(k_1 + k_2)\, \delta(k_1' + k_2') = \sum_{k, k' > 0} v_{k-k, k'-k'} a_k^\dagger a_{-k}^\dagger a_{-k'} a_{k'} \tag{9.14}$$

These interactions are characterized by the fact that they are operative only between particles occupying a pair of mutually time-reversed state, scattering a pair of particles from one such pair of states to another. In view of the structure of the Bogoljubov transformation we expect it to be most appropriate for these interactions. We shall see in the next section why we can expect an important part of the nuclear interaction to show up in this form.

We shall also assume that the single-particle part of the Hamiltonian has been brought to a diagonal form by a proper choice of the single-particle states $a_k^\dagger|0\rangle$, so that our total Hamiltonian reads

$$\hat{H} = \sum_{k>0} \eta_k (a_k^\dagger a_k + a_{-k}^\dagger a_{-k}) + \sum_{k, k' > 0} v_{k-k, k'-k'} a_k^\dagger a_{-k}^\dagger a_{-k'} a_{k'} \tag{9.15}$$

Introducing now the transformation (8.8), $\hat{H}$ can be written in terms of the α-operators. To make it, however, more useful for our purpose we want to use also the commutation relations of the α's in order to bring each term to a *normal form* in which the annihilation operators are put to the right of the creation operators. Thus, for instance, we have

$$\sum_{k>0} \eta_k a_k^\dagger a_k = \sum \eta_k (U_k\alpha_k^\dagger + V_k\alpha_{-k})(U_k\alpha_k + V_k\alpha_{-k}^\dagger)$$

$$= \sum \eta_k [U_k{}^2\alpha_k^\dagger\alpha_k + V_k^2\alpha_{-k}\alpha_{-k}^\dagger + U_kV_k(\alpha_k^\dagger\alpha_{-k}^\dagger + \alpha_{-k}\alpha_k)]$$

$$= \sum \eta_k V_k^2 + \eta_k [U_k^2\alpha_k^\dagger\alpha_k - V_k^2\alpha_{-k}^\dagger\alpha_{-k} + U_kV_k(\alpha_k^\dagger\alpha_{-k}^\dagger + \alpha_{-k}\alpha_k)] \tag{9.16}$$

Notice the appearance in (9.16) of the term $\Sigma\eta_k V_k{}^2$ free of any α-operators. It resulted from transforming $V_k{}^2\alpha_{-k}\alpha_{-k}^\dagger$ to its normal form. In carrying out the complete transformation it is convenient to group together all the terms that have no α''s in them, and call it U. We also want to group together the terms that are proportional to $\alpha_k^\dagger\alpha_k$ or $\alpha_{-k}^\dagger\alpha_k$, which we call H_{11}, and all terms proportional to $\alpha_k^\dagger\alpha_{-k}^\dagger$ or $\alpha_k\alpha_{-k}$, which we call H_{20}; finally all other terms proportional to products of four α's such as $\alpha_k^\dagger\alpha_{-k}^\dagger\alpha_{k'}\alpha_{-k'}$ or $\alpha_k^\dagger\alpha_{-k}^\dagger\alpha_{k'}^\dagger\alpha_{-k'}^\dagger$, or $\alpha_k^\dagger\alpha_{-k}^\dagger\alpha_{-k'}^\dagger\alpha_{-k'}$ etc., we group together under H'. Thus we obtain

$$\hat{H} = U + H_{11} + H_{20} + H' \tag{9.17}$$

where

$$U = \sum_{k>0} [(2\eta_k + V_k^2 v_{k-k,k-k})V_k^2 + \sum_{k'>0} v_{k-k,k'-k'}U_kV_kU_{k'}V_{k'}] \tag{9.18a}$$

$$H_{11} = \sum_{k>0} [(\eta_k + v_{k-k,k-k}V_k^2)(U_k^2 - V_k^2) - 2\sum_{k'>0} v_{k-k,k'-k'}U_kV_kU_{k'}V_{k'}] \times [\alpha_k^\dagger\alpha_k + \alpha_{-k}^\dagger\alpha_{-k}] \tag{9.18b}$$

$$H_{20} = \sum_{k>0} [(2\eta_k + 2v_{k-k,k-k}V_k^2)U_kV_k + \sum_{k'} v_{k-k,k'-k'}U_{k'}V_{k'}(U_k^2 - V_k^2)] \times [\alpha_k^\dagger\alpha_{-k}^\dagger + \alpha_{-k}\alpha_k] \tag{9.18c}$$

Like the simpler case of $v = 0$ discussed previously we want now to choose U_k and V_k in such a way as to make $H_{20} \equiv 0$. This will leave us with U and H_{11} that commute with the number operator in the α-scheme and can therefore be brought to diagonal form. However in the present case we shall be left also with the term H' that does not commute with the number operator in the α-scheme, and whose diagonalization in that scheme is rather complicated.

The operator H', it will be recalled, has only terms that are products of four α-operators. We see from (9.16) that $\Sigma\eta_k(a_k^\dagger a_k + a_{-k}^\dagger a_{-k})$ does not contribute to H'. Hence H' is proportional to the strength of the interaction v. Furthermore, since we have shifted all the annihilation operators to the right and the creation operators to the left, every term in H' has got either an α at its extreme right or an $\alpha^\dagger$ at its extreme left (or both). Since $\alpha_k|0\rangle_\alpha = {}_\alpha\langle 0|\alpha_k^\dagger = 0$, we conclude that

$$ {}_\alpha\langle 0|H'|0\rangle_\alpha = 0 \tag{9.19}$$

Thus H' is an operator that does not commute with N_α; it is proportional to the strength of the interaction v and its ground-state expectation value in the α-scheme vanishes. We shall therefore neglect it now and assume that it can be taken up later as a perturbation.

Two points should be made in this connection:

(i) We have not really proved that H' is small in the sense of perturbation theory; at most we have given some vague plausibility arguments in this direction. It is conceivable that H' may be small for some interactions, and nonnegligible for others.

(ii) The full Hamiltonian commutes with the real-particle number operator $\hat{N} = \sum_{k>0} a_k^\dagger a_k + a_{-k}^\dagger a_{-k}$. On the other hand $U + H_{11}$ commutes with the quasi-particle number operator $\hat{N}_\alpha = \Sigma\alpha_k^\dagger\alpha_k + \alpha_{-k}^\dagger\alpha_{-k}$. Since in general $[\hat{N}, \hat{N}_\alpha] \neq 0$, we see that in dropping H' we go from the physical situation in which the number of real particles is fixed, to the nonphysical situation with a fixed number of quasi particles, and an undetermined number of real particles.

To reduce the impact of the second difficulty we want to make sure that our choice of U_k and V_k also leads at least to a given *average* number of real particles in the quasi-particle ground state. Thus we want to impose on U_k and V_k the further condition that

$$\begin{aligned}\langle\hat{N}\rangle = {}_\alpha\langle 0|2\sum V_k^2 + (U_k^2 - V_k^2)(\alpha_k^\dagger\alpha_k + \alpha_{-k}^\dagger\alpha_{-k}) \\ + 2U_kV_k(\alpha_k^\dagger\alpha_{-k}^\dagger + \alpha_{-k}\alpha_k)|0\rangle_\alpha = N_0\end{aligned} \tag{9.20}$$

This can be achieved, as is well known by diagonalizing $\hat{H} - \lambda\hat{N}$ rather than $\hat{H}$, and choosing the Lagrange multiplier λ so that (9.20) is satisfied. We are thus led to the problem of diagonalizing the operator

$$\hat{H} - \lambda\hat{N} = U' + H_{11}' + H_{20}' + H' \tag{9.21}$$

where

$$\begin{aligned}U' = U - 2\lambda\sum V_k^2 = \sum_{k>0}\{[2(\eta_k - \lambda) + v(k,k)V_k^2]V_k^2 \\ +(\sum_{k'>0} v(k,k')U_{k'}V_{k'})U_kV_k\}\end{aligned} \tag{9.22}$$

$$H_{11}' = H_{11} - \lambda \sum (U_k^2 - V_k^2)(\alpha_k^\dagger \alpha_k + \alpha_{-k}^\dagger \alpha_{-k})$$
$$= \sum_{k>0} \{ [(\eta_k - \lambda) + v(k,k)V_k^2](U_k^2 - V_k^2)$$
$$- 2[\sum_{k>0} v(k,k')U_{k'}V_{k'}]U_kV_k\} [\alpha_k^\dagger \alpha_k + \alpha_{-k}^\dagger \alpha_{-k}] \quad (9.23)$$

$$H_{20}' = H_{20} - 2\lambda \sum U_kV_k(\alpha_k^\dagger \alpha_{-k}^\dagger + \alpha_{-k}\alpha_k) = \sum_{k>0} \{ [2(\eta_k - \lambda)$$
$$+ 2v(k,k)V_k^2]U_kV_k + [\sum_{k'>0} v(k,k')U_{k'}V_{k'}](U_k^2 - V_k^2)\}$$
$$\times [\alpha_k^\dagger \alpha_{-k}^\dagger + \alpha_{-k}\alpha_k] \quad (9.24)$$

and we used the notation

$$v(k,k') = v_{k-k,k'-k'} \quad (9.25)$$

The diagonalization of $H_0' = U' + H_{11}' + H_{20}'$ amounts to choosing U_k and V_k such that $H_{20}' = 0$. Introducing the notation

$$\xi_k = (\eta_k - \lambda) + v(k,k)V_k^2 \quad (9.26)$$

$$\Delta_k = -\sum_{k'>0} v(k,k')U_{k'}V_{k'} \quad (9.27)$$

and noting the mutual independence of the operators $(\alpha_k^\dagger \alpha_{-k}^\dagger + \alpha_{-k}\alpha_k)$ for different values of k, we see that $H_{20}' = 0$ leads to the equations

$$2\xi_kU_kV_k - \Delta_k(U_k^2 - V_k^2) = 0 \quad (9.28)$$

These equations have the trivial solution

$$U_kV_k = 0 \qquad \text{for all } k\text{'s} \quad (9.29)$$

because then, according to (9.27), $\Delta_k = 0$ and thus (9.28) is automatically satisfied. Noting that $U_k^2 + V_k^2 = 1$, the trivial solution (9.29) is actually equivalent to taking into account only the diagonal elements $v(k,k)$ of v and considering states that are either fully occupied or completely unoccupied by real particles. We shall therefore assume now that at least for some values of k, the product $U_kV_k \neq 0$ and proceed to see whether we obtain thereby lower eigenvalues for our Hamiltonian. From (9.28) and the condition $U_k^2 + V_k^2 = 1$ we obtain then:

$$U_k^2 = \frac{1}{2}\left(1 + \frac{\xi_k}{\sqrt{\xi_k^2 + \Delta_k^2}}\right) \quad (9.30a)$$

$$V_k^2 = \frac{1}{2}\left(1 - \frac{\xi_k}{\sqrt{\xi_k^2 + \Delta_k^2}}\right) \quad (9.30b)$$

Equation 9.30 constitute only an implicit solution of (9.26) and (9.27) since ξ_k and Δ_k are themselves functions of U_k and V_k. However we can use the solutions (9.30) to write an integral equation for Δ_k. Multiplying (9.30a) by (9.30b) we obtain

$$U_k V_k = \frac{\Delta_k}{2\sqrt{\xi_k^2 + \Delta_k^2}} \tag{9.31}$$

Hence, from (9.27)

$$\Delta_k = -\frac{1}{2}\sum_{k'} \frac{v(k, k')}{\sqrt{\xi_{k'}^2 + \Delta_{k'}^2}} \Delta_{k'} \tag{9.32}$$

Assuming that the amplitudes U_k and V_k were determined from (9.32) and (9.30), the remaining part of the Hamiltonian $U' + H'_{11}$ is then diagonal in the α-scheme. The quasi-particle vacuum $|0\rangle_\alpha$ is the lowest state for $U' + H'_{11}$, and its energy is given by

$$\begin{aligned} E_0 &= {}_\alpha\langle 0 | U' + H'_{11} | 0 \rangle_\alpha \\ &= 2\sum_{k>0} \left[(\eta_k - \lambda)V_k^2 + \frac{1}{2} v(k, k)V_k^4 - \frac{1}{2}\Delta_k U_k V_k \right] \end{aligned} \tag{9.33}$$

To complete our calculation we still have to determine the Lagrange multiplier λ (see the definition, Eq. 9.26, of ξ_k). This can be done using (9.20). Introducing (9.30b) for V_k^2 we obtain

$$N_0 = 2\sum V_k^2 = \sum_k \left(1 - \frac{\xi_k}{\sqrt{\xi_k^2 + \Delta_k^2}}\right) \tag{9.34}$$

Equations 9.26, 9.27, 9.30, and 9.34 constitute the so called *superconducting* or *BCS solution** to our problem, leading to the ground-state energy (9.33). We shall now study some properties of this solution to see the conditions for its existence, and when it actually leads to values for the ground-state energy that are better than those obtained by the trivial, or as it is often called the *normal solution* $U_k V_k = 0$, given by (9.29).

10. PROPERTIES OF THE SUPERCONDUCTING SOLUTION

In the limit of $v \to 0$ we see, from (9.26), that

$$\frac{\xi_k}{|\eta_k - \lambda|} = \begin{cases} +1 & \text{for} \quad \eta_k > \lambda \\ -1 & \text{for} \quad \eta_k < \lambda \end{cases} \qquad v = 0 \tag{10.1}$$

---

*The origin of the name has to do with the theory of superconductivity, where the Bogoljubov transformation was first introduced, and where a solution of the many-electron problem, similar to the one we have been discussing here, turned out to be necessary to explain the phenomenon of superconductivity. See, for instance, Bardeen (63). The same solution was also suggested earlier by Bardeen, Cooper, and Schrieffer (57) (BCS) using a slightly different method.

In the same limit, (9.27) yields $\Delta_k \to 0$. Thus, in this limit from (9.30b)

$$V_k^2 = \frac{1}{2}(1 - \xi_k/|\xi_k|)$$

or

$$V_k^2 = \begin{cases} 0 & \text{for} \quad \eta_k > \lambda \\ 1 & \text{for} \quad \eta_k < \lambda \end{cases} \tag{10.2}$$

We therefore see that λ plays the role of the Fermi energy for noninteracting fermions: states with $\eta_k < \lambda$ are fully occupied while states with $\eta_k > \lambda$ are completely empty.

For interacting fermions we continue to call λ the Fermi energy (or, more precisely, *the chemical potential*) and we expect V_k^2 for this case to be a smooth function of k that comes closer to (10.2) as the interaction gets weaker. Figure 10.1 shows V_k, U_k, and $2U_kV_k$ as functions of k. As we see V_k—the probability amplitude for the occupation of the state k—attains the values $V_k = 1$ for the low states in the Fermi sea, and $V_k = 0$ for states far above the Fermi momentum. Around the Fermi momentum k_F (or the corresponding Fermi energy λ) V_k goes gradually from 1 to 0, indicating a certain diffuseness in the surface of the Fermi distribution. U_k, satisfying $U_k^2 = 1 - V_k^2$ behaves, of course, in a complementary manner, and U_kV_k is a Gaussian-like function, whose width measures the width of the diffuse surface of the Fermi distribution.

To obtain a numerical solution for the superconducting state one can proceed as follows: starting with the single-particle potential as determined by the energies η_k in (9.15) and the number of particles N_0 in the system, derive the Fermi energy λ (i.e., the energy of the highest occupied state). Then approximate ξ_k by

$$\xi_k = \begin{cases} (\eta_k - \lambda) + v(k, k) & \text{if} \quad \eta_k < \lambda \\ (\eta_k - \lambda) & \text{if} \quad \eta_k > \lambda \end{cases} \tag{10.3}$$

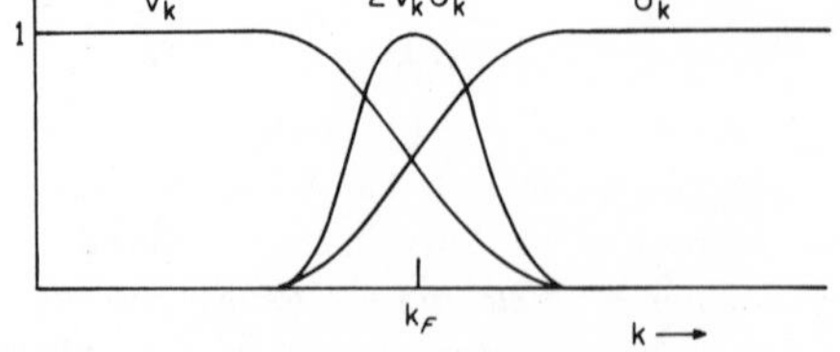

FIG. 10.1. Properties of V_k^2, $2U_kV_k$, and U_k^2 for interacting fermions.

With these values of ξ_k solve simultaneously (9.32) and (9.34) for Δ_k and λ. Generally it will be sufficient to carry out the summation in (9.32) assuming that $\Delta_k \neq 0$ only for a finite range of states around the Fermi energy, (provided a solution to Eqs. 9.32 and 9.34 exists). Once values for Δ_k and λ are obtained, U_k and V_k can be obtained from (9.30) and used to iterate the procedure in order to improve the approximation. That is, a new value of ξ_k can be calculated from (9.26), a new set of ξ_k^2's and λ's determined, and so on.

Necessary conditions for the existence of solutions to (9.32) and (9.34) are easy to formulate. In fact, we shall show that for attractive interactions, $v(k, k') < 0$; we must have therefore

$$1 < -\frac{1}{2}\sum \frac{v(k, k')}{|\xi_{k'}|} = \frac{1}{2}\sum_{k'>0} \frac{-v(k, k')}{|(\eta_{k'} - \lambda) + v(k', k')V_{k'}^2|} \tag{10.4}$$

for a solution to (9.32) to be found. As a first step in the proof let Δ_{k_0} be the largest of all the Δ_k's. Then it follows from (9.32) that

$$1 = -\frac{1}{2}\sum_{k'} \frac{v(k_0, k')}{\sqrt{\xi_{k'}^2 + \Delta_{k'}^2}} \frac{\Delta_{k'}}{\Delta_{k_0}} < -\frac{1}{2}\sum \frac{v(k_0, k')}{\sqrt{\xi_{k'}^2 + \Delta_{k'}^2}} < -\frac{1}{2}\sum \frac{v(k_0, k')}{|\xi_{k'}|}$$

There will therefore be no solution to (9.32) except the trivial one $\Delta_k = 0$ if $v(k, k')$ is so weak that

$$\frac{1}{2}\sum_{k'>0} \frac{|v(k, k')|}{|(\eta_{k'} - \lambda) + v(k', k')V_{k'}^2|} < 1 \rightarrow \text{no BCS solution} \tag{10.5}$$

Next we want to see whether the energy of the BCS ground-state is lower than that of the normal state. It can be easily shown [Lane (64)] that the BCS solution can be obtained from a variational principle using the trial wave function (8.13) with the Hamiltonian $\hat{H} - \lambda\hat{N}$ given by (9.21); therefore, if the energy E_0 in (9.33) turned out to be lower than that of the normal state, we can conclude that the BCS state constitutes a better approximation to the actual ground state. To evaluate the desired difference

$$E_0 - 2\sum_{k \leq k_F}\left[(\eta_k - \lambda_0) + \frac{1}{2}v(k, k)\right]$$

we note that given the single-particle states $a_k^\dagger|0\rangle$ we can separate, in the original Hamiltonian (9.15), the terms in the interaction with $k = k'$ from the rest of the terms. Within the limits of the random-phase approximation we can put (see Eq. 6.6)

$$v_{k-k,k-k}a_k^\dagger a_{-k}^\dagger a_{-k}a_k \approx v(k, k)(\langle a_{-k}^\dagger a_{-k}\rangle a_k^\dagger a_k - \langle a_{-k}^\dagger a_k\rangle a_k^\dagger a_{-k} - \langle a_k^\dagger a_{-k}\rangle a_{-k}^\dagger a_k + \langle a_k^\dagger a_k\rangle a_{-k}^\dagger a_{-k}) \tag{10.6}$$

Taking expectation values in the normal state we obtain therefore

$$v_{k-k,k-k}a_k^\dagger a_{-k}^\dagger a_{-k}a_k \approx \begin{cases} v(k,k)(a_k^\dagger a_k + a_{-k}^\dagger a_{-k}) & \text{if } k \text{ is occupied} \\ 0 & \text{if } k \text{ is unoccupied} \end{cases} \tag{10.7}$$

Within the approximation (10.6) we can therefore renormalize our single-particle energies η_k:

$$\tilde{\eta}_k = \begin{cases} \eta_k + v(k,k) & \text{if} \quad k \text{ is occupied} \\ \eta_k & \text{if} \quad k \text{ is unoccupied} \end{cases} \tag{10.8}$$

and put $v(k, k) = 0$ in the rest of our calculation. Assume that this has been done originally and that the U_k's and V_k's were determined from the single-particle energies $\tilde{\eta}_k$ rather than η_k; we then obtain for the energy shift δE of the superconducting BCS state relative to the normal state (see p. 559):

$$\begin{aligned} \delta E &= E_0 - 2\sum_{k \le k_F} (\eta_k - \lambda_0) \\ &= 2\sum_{k>0} \left[(\eta_k - \lambda)V_k^2 - \frac{1}{2}\Delta_k U_k V_k \right] - 2\sum_{k \le k_F} (\eta_k - \lambda_0) \\ &= \sum_{k>0} \left\{ (\eta_k - \lambda)\left[1 - \frac{\eta_k - \lambda}{\sqrt{(\eta_k - \lambda)^2 + \Delta_k^2}} \right] - \frac{1}{2}\frac{\Delta_k^2}{\sqrt{(\eta_k - \lambda)^2 + \Delta_k^2}} \right\} \\ &\qquad - 2\sum_{k \le k_F} (\eta_k - \lambda_0) \end{aligned} \tag{10.9}$$

To estimate δE we now make the following approximations:

(i) We assume the single-particle energy levels $|k\rangle$ to be equally spaced a distance D apart where $D \ll \lambda$.

(ii) We then replace summations by integrations

$$\sum_k \rightarrow \frac{1}{D}\int d\eta_k$$

(iii) We take the Fermi energies λ_0 and λ to be equal to each other.

(iv) We replace Δ_k by an average value Δ_0 (Δ_k turns out to be a very slowly varying function of k. This can be expected also from its definition (9.27). The contribution to $\sum_{k'>0} v(k, k')U_{k'}V_{k'}$ comes mostly from the region near the top of the Fermi distribution since $U_{k'}V_{k'}$ vanishes elsewhere. The value of Δ_k is therefore a measure of the effectiveness of the interaction $v(ij)$ in scattering a pair of particles from the state $(k_F, -k_F)$ at the top of the Fermi distribution to the state $(k, -k)$. Limiting ourselves to values of k such that $|k - k_F| < 1/b$, where b is range of the interaction v, Δ_k will not be very k-dependent).

Introducing $\epsilon = \eta_k - \lambda$ we then obtain, integrating from $-\epsilon_0$ up to $+\epsilon_0$ with $|\epsilon_0| \gg \Delta_0$

$$\delta E = \frac{1}{D}\int_{-\epsilon_0}^{\epsilon_0} d\epsilon\, \epsilon\left(1 - \frac{\epsilon}{\sqrt{\epsilon^2 + \Delta_0^2}}\right) - \frac{\Delta_0^2}{2D}\int_{-\epsilon_0}^{\epsilon_0} \frac{d\epsilon}{\sqrt{\epsilon^2 + \Delta_0^2}} - \frac{2}{D}\int_{-\epsilon_0}^{0} \epsilon\, d\epsilon$$

$$= -\frac{1}{2D}\int_{-\epsilon_0}^{\epsilon_0} \frac{2\epsilon^2 + \Delta_0^2}{\sqrt{\epsilon^2 + \Delta_0^2}}\, d\epsilon - \frac{2}{D}\int_{-\epsilon_0}^{0} \epsilon\, d\epsilon$$

$$= \frac{1}{D}\left[\epsilon_0^2 - \epsilon_0\sqrt{\epsilon_0^2 + \Delta_0^2}\right] \approx -\frac{\Delta_0^2}{2D} \tag{10.10}$$

We thus see that in going from the normal to the BCS state there is, indeed, a gain in energy by an amount $\Delta_0^2/2D$.

To estimate the value of δE for actual nuclei we need an estimate for Δ_0 and D. We can obtain a crude estimate for Δ_0 from the integral equation (9.32). Taking for $v(k, k')$ an average matrix element $v(k, k') = -G$, (9.32), with the above approximations, becomes:

$$\Delta_0 = \frac{G\Delta_0}{2D}\int_{-\epsilon_0}^{\epsilon_0} \frac{d\epsilon}{\sqrt{\epsilon_k^2 + \Delta_0^2}} = \frac{G}{2}\frac{\Delta_0}{D}\left(2 \text{ arc sinh } \frac{\epsilon_0}{\Delta_0}\right)$$

or

$$\text{arc sinh } \frac{\epsilon_0}{\Delta_0} = \frac{D}{G}$$

that is,

$$\Delta_0 = \frac{\epsilon_0}{\sinh D/G} \tag{10.11*}$$

G, being an average matrix element $v(k, k')$, is of the order of magnitude of

$$G \approx -\langle k - k|v|k' - k'\rangle \approx -\frac{1}{\Omega^2}\left\{\int \exp\,[-i\mathbf{k}\cdot(\mathbf{r}_1 - \mathbf{r}_2)]v(|\mathbf{r}_1 - \mathbf{r}_2|)\right.$$

$$\times \exp\,[i\mathbf{k}'\cdot(\mathbf{r}_1 - \mathbf{r}_2)] - \int \exp\,[i\mathbf{k}\cdot(\mathbf{r}_1 - \mathbf{r}_2)]v(|\mathbf{r}_1 - \mathbf{r}_2|)$$

$$\left.\times \exp\,[i\mathbf{k}'\cdot(\mathbf{r}_2 - \mathbf{r}_1)]\right\} \tag{10.12}$$

---

*It is worth noting that Δ_0 is not an analytic function of G at $G = 0$; in fact near $G = 0$, Δ_0 is given by $\Delta_0 \approx 2\xi e^{-D/G}$. Δ_0 cannot be expanded in a power series of G near $G = 0$, and it is apparent therefore that we could have not reached the result (10.11) by perturbation theory. The superconducting state does *not* represent a small perturbation on the normal state even for very weak interactions G. Note also that there is a lower bound (10.5) for G below which there is no superconducting solution.

where Ω is the nuclear volume. In (10.12) either the exchange or the direct integrals are small (depending on whether $\mathbf{k}'$ is more-or-less in the direction of $\mathbf{k}$ or of $-\mathbf{k}$). It is enough therefore to take only the direct integral for cases in which $\mathbf{k}$ is in the direction of $\mathbf{k}'$. Taking for $v(|\mathbf{r}_1 - \mathbf{r}_2|)$ a square well of depth $V_0 \approx -30$ MeV and range $b = 1.8$fm we obtain from (10.12) that, for $A \approx 100$,

$$G \approx \begin{cases} 30\left(\dfrac{1.8}{1.2A^{1/3}}\right)^3 \approx 1 \text{ MeV} & \text{if} \quad |\mathbf{k} - \mathbf{k}'| < \dfrac{1}{b} \\ 0 & \text{if} \quad |\mathbf{k} - \mathbf{k}'| > \dfrac{1}{b} \end{cases} \tag{10.13}$$

More realistic calculations yield values for G of about 0.4 to 0.5 MeV for matrix elements that involve changes in momentum smaller than $\sim 1/b$.

The value of D—the average spacing between single-particle levels—can also be estimated. For a nucleus with $A = 100$ there are $A/4 = 25$ occupied levels with different values of $|\mathbf{k}|$; taking the Fermi energy to be about 40 MeV we obtain $D \approx 1.5$ MeV. We then obtain sinh $(D/G) \approx 10$. If, to comply with (10.13), we let the range of integration in (10.11) extend from $\epsilon_0 = -(\epsilon_F/2)$ to $\epsilon_0 = (\epsilon_F/2)$, where ϵ_F is the Fermi energy, we obtain finally

$$\Delta_0 \approx 2 \text{ MeV} \tag{10.14}$$

Thus for a nucleus of mass 100 with 50 protons and 50 neutrons, the gain in energy of the superconducting state over the normal state is roughly $2\delta E = -(\Delta_0/D) \approx 2.5$ MeV, where we took one δE each for the protons and the neutrons.

It should be stressed that this is a very rough estimate of the quantity $2\delta E$, but it does show that δE is certainly very small compared to the *total* binding energy of the nucleus. For the calculation of *binding energies*, the Bogoljubov transformation is therefore a complication that yields, relatively speaking, very little improvement. Its real value in nuclear physics is connected, as we shall see in the next section, with the study of excitation energies near the ground state. Here it may lead even to important qualitative improvements over the approximation using the normal states.

11. EXCITED STATES IN THE FREE QUASI-PARTICLE SYSTEM

The Bogoljubov transformation was constructed so that the vacuum of the quasi particles $|0\rangle_\alpha$ should correspond to the ground state of a system with N-particles. To create an excited state from the ground state with real particles we have to operate on it with an operator of the type $a_k^\dagger a_{k'}$ where $k \neq k'$. Such an operator removes a particle from the state k' and puts a particle into the state k.

Using the transformation (8.8) we see that (for $k \neq k'$)

$$a_k^\dagger a_{k'} = U_k U_{k'} \alpha_k^\dagger \alpha_{k'} + U_{k'} V_k \alpha_{-k} \alpha_{k'} - V_k V_{k'} \alpha_{-k'}^\dagger \alpha_{-k} + U_k V_{k'} \alpha_k^\dagger \alpha_{-k'}^\dagger \quad (11.1)$$

If we operate with (11.1) on the quasi-particle vacuum, only the last term: $U_k V_{k'} \alpha_k^\dagger \alpha_{-k'}^\dagger$ will give nonvanishing results, since $\alpha_k |0\rangle_\alpha = 0$. We see, therefore, that to obtain results equivalent to those of operating with $a_k^\dagger a_{k'}$ on the real-particle ground state, we have to operate generally with $\alpha_{k1}^\dagger \alpha_{k2}^\dagger$ on the *quasi-particle* vacuum. A two quasi-particle excitation is thus equivalent to an excitation within the same nucleus. The fact that in one representation the number of quasi particles grows by two as one goes from the ground state to the excited state, whereas in the other the number of real particles remains unchanged, is connected, of course, with the fact that a state of a well-defined number of quasi particles corresponds to a state of an undetermined number of real particles.

We need not stress that in constructing the excited states of the quasi-particle system one is to use the same values of U_k and V_k as determined by (9.30) and (9.32) for the ground state of that system.

Since the Bogoljubov transformation is designed so as to convert the Hamiltonian into an independent quasi-particle Hamiltonian (i.e., one that commutes with the quasi-particle number operator $\Sigma \alpha_k^\dagger \alpha_k + \alpha_{-k}^\dagger \alpha_{-k}$), it follows that the two quasi-particle excited state $\alpha_{k1}^\dagger \alpha_{k2}^\dagger |0\rangle$ is automatically an eigenstate of the Hamiltonian

$$\hat{H} = U' + H_{11}' + H_{20}'$$

which, in view of the choice of U_k and V_k, reduces to

$$\hat{H} = U' + H_{11}' \quad (11.2)$$

Here U' and H_{11}' are given by (9.22) and (9.23) so that

$$\begin{aligned} \hat{H} = \sum_{k>0} [2(\eta_k - \lambda) + v(k,k)V_k^2]V_k^2 + \sum_{k,k'>0} v(k,k')U_k V_k U_{k'} V_{k'} \\ + \sum_{k>0} \{[(\eta_k - \lambda) + v(k,k)V_k^2](U_k^2 - V_k^2) \\ -2\sum_{k'>0} v(k,k')U_k V_k U_{k'} V_{k'}\}(\alpha_k^\dagger \alpha_k + \alpha_{-k}^\dagger \alpha_{-k}) \quad (11.3) \end{aligned}$$

The excitation energy (above the vacuum energy) of the two quasi-particle state

$$|k_1 k_2\rangle_\alpha = \alpha_{k1}^\dagger \alpha_{k2}^\dagger |0\rangle_\alpha$$

will then be

$$E(k_1, k_2) = E(k_1) + E(k_2) \quad (11.4)$$

where

$$E(k) = [(\eta_k - \lambda) + v(k,k)V_k^2](U_k^2 - V_k^2) - 2\sum_{k'>0} v(k,k')U_k V_k U_{k'} V_{k'}$$

Using (9.26), (9.27), (9.30), and (9.31) to express the U_k's and V_k's in terms of ξ_k and Δ_k we obtain

$$E(k) = \sqrt{\xi_k^2 + \Delta_k^2} \tag{11.5}$$

so that the two quasi-particle excitation energy is

$$E(k_1, k_2) = \sqrt{\xi_{k_1}^2 + \Delta_{k_1}^2} + \sqrt{\xi_{k_2}^2 + \Delta_{k_2}^2} \tag{11.6}$$

Thus, even if the states k_1 and k_2 lie right above the Fermi energy λ so that $\xi_k \approx 0$, the two quasi-particle excitation requires at least an energy $E(k_1, k_2) \geq \Delta_1 + \Delta_2 \approx 2\Delta_0$. There is a *gap* of at least $2\Delta_0$ between the quasi-particle ground state and the lowest two quasi-particle excited state of the system.

This phenomenon of a gap between the ground state of a system of an even number of fermions and its excitations plays a very important role in determining the properties of such systems. In the theory of superconductivity it is responsible for the persistence of the superconducting current. In nuclei it gives rise to exceptionally large energy denominators associated with excitations from the ground states of even–even nuclei, leading to important modifications of the predictions based on the various independent-particle approximations.

An odd-A system represents a different situation. We saw in (9.12) that a state of an odd number of real particles corresponds to a state of one quasi particle $\alpha_k^\dagger|0\rangle_\alpha$. Thus, if $\alpha_{k0}^\dagger|0\rangle_\alpha$ corresponds to the ground state of the odd-A system, and $\alpha_{k1}^\dagger|0\rangle_\alpha$ to an excited state of the system, the excitation energy above the ground state will be given by:

$$\Delta E = E(k_1) - E(k_0) = \sqrt{\xi_{k_1}^2 + \Delta_{k_1}^2} - \sqrt{\xi_{k0}^2 + \Delta_{k0}^2} \tag{11.7}$$

We see that here, unlike the case of even-A systems, if the single-particle states k_1 and k_0 are rather close to each other, the energy required to go from one state to the other in the quasi-particle representation will also be small. The phenomenon of the gap is connected with the *even group of particles*, filling in pairs the states k and $-k$, and does not affect appreciably the odd particle's spectrum.

Figure 11.1 shows a characteristic excitation spectrum of an even–even nucleus, and that of a neighboring odd-A nucleus. It is seen that in the odd-A nucleus the spectrum of excited states becomes denser with excitation energy right from the ground state. The corresponding spectrum in the even–even nucleus remains "dilute" up to an excitation of 2 to 3 MeV, and only then starts to grow dense at rate comparable to that of the neighboring odd-A nucleus. The especially high stability of the ground state of a system of an even number of fermions is thus clearly reflected in these spectra.

We shall not go here into the derivation of various nuclear properties in the quasi-particle representation. The calculations, although rather tedious, are in most cases straightforward. We shall only quote the result for the moment of inertia, which is characteristic of other results as well [see G. Brown

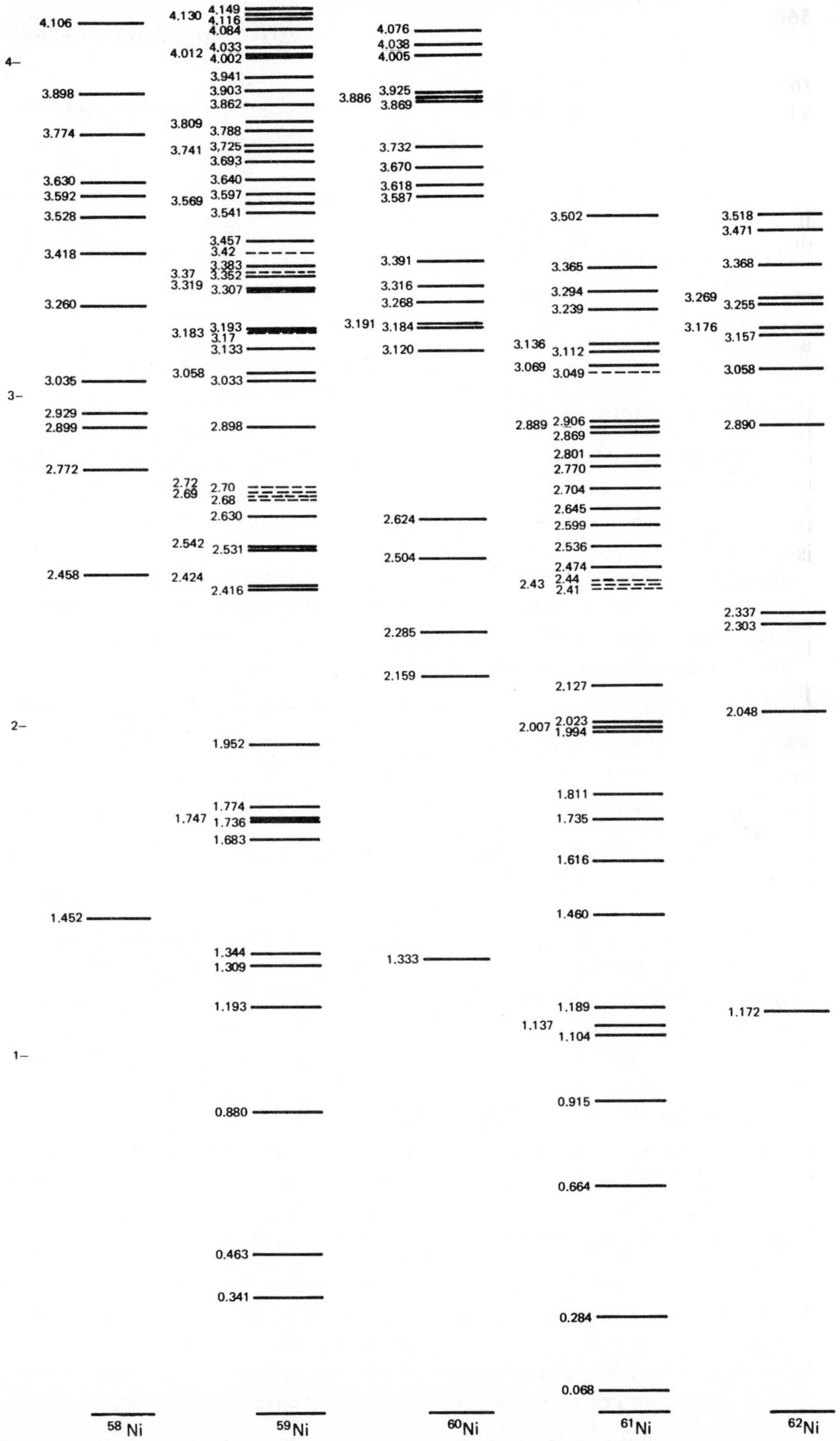

FIG. 11.1. Energy Levels of the nickel isotopes [from Buechner (58)].

(64), p. 109]. For the analogue of the Inglis cranking formula (see Eq. VI.7.25) one obtains in the quasi-particle representation

$$\mathcal{J} = \hbar^2 \sum_{k,k'>0} \frac{|\langle k|j_x|k'\rangle|^2}{E(k) + E(k')} (U_k V_{k'} - U_{k'} V_k)^2 \tag{11.8}$$

If we compare it with the cranking formula (VI.7.25) obtained with the independent-particle model

$$\mathcal{J}_{ipm} = 2\hbar^2 \sum_n \frac{|\langle n|j_x|0\rangle|^2}{E_n - E_0} \tag{11.9}$$

we notice two important differences:

(i) Equation 11.9 allows only for the excitation via j_x from the fully occupied state $|0\rangle$ to a fully unoccupied state $|n\rangle$ whereas the quasi-particle formula (11.8) allows also for the excitation from *partially* filled states to *partially* empty states. The corresponding matrix element is then naturally multiplied by a factor $U_k V_{k'}$ which is the amplitude for k' to be occupied and k to be empty. (The occurrence of a coherent factor $U_k V_{k'} - U_{k'} V_k$ multiplying the matrix element $\langle k|j_x|k'\rangle$ has to do with the fact that the operator j_x is odd under time reversal so that $\langle -k'|j_x|-k\rangle = -\langle k|j_x|k'\rangle$).

(ii) The energy denominator in (11.8), since it involves quasi-particle excitation, is generally larger than the corresponding denominator in (11.9) by $2\Delta_0$.

Problem.
Show that for the trivial solution (9.29), expression (11.8) goes over to (11.9).

Both effects tend to make the quasi-particle moment of inertia $\mathcal{J}$ smaller than the independent-particle model $\mathcal{J}_{ipm}$. Since $\mathcal{J}_{ipm}$ was shown to be essentially the rigid-body moment of inertia, we expect (11.8) to give a moment of inertia that comes closer to the actual experimental moments of inertia. The results of some numerical calculations are shown in Fig. 11.2 together with the experimental results.

In discussing the effects of the residual interaction on $\mathcal{J}_{ipm}$, we anticipated that this interaction will reduce the value of $\mathcal{J}_{ipm}$. The result (11.8) is, therefore, a realization of this expectation since the quasi-particle representation is just a way of including in the energies of independent quasi particles at least a part of the residual interaction of the real particles.

12. PAIRING INTERACTIONS VS. REALISTIC INTERACTIONS

Our whole discussion of the quasi-particle representation was based on the special pairing interaction, (9.14), which is effective only between pairs in

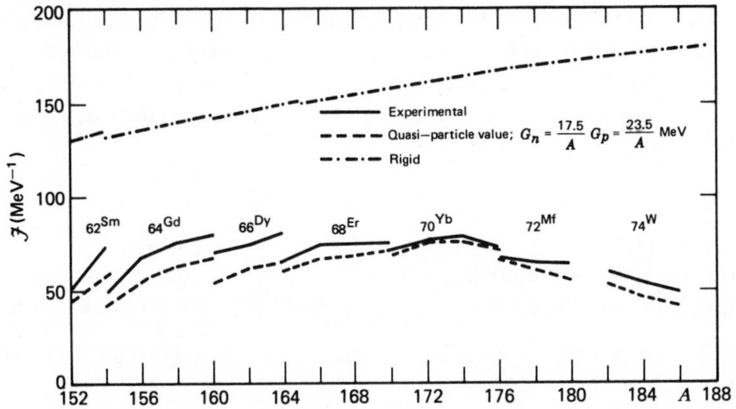

FIG. 11.2. Moments of inertia for even–even deformed nuclei in the rare earth region [from Bodenstedt and Rogers (64)].

the states $(k, -k)$. The Bogoljubov transformation can be carried out also for more general interactions. In fact, starting from the general two-body interaction $(1/4)\Sigma v_{k_1k_2,k_1'k_2'} a^\dagger_{k_1} a^\dagger_{k_2} a_{k_2'} a_{k_1'}$, and using the transformation (8.8) we can cast the Hamiltonian $\hat{H} - \lambda\hat{N}$ again into the form

$$\hat{H} - \lambda\hat{N} = U' + H_{11}' + H_{20}' + H'$$

where now

$$U' = \sum_{k>0} [2(\eta_k - \lambda) + \sum_{k'>0} (v_{k-k',k-k'} + v_{kk',kk'})V_{k'}^2 + \sum v_{k-k,k'-k'} U_k V_k U_{k'} V_{k'}] \quad (12.1)$$

$$H_{11}' = \sum_{k''k>0} \{[(\eta_k - \lambda) + \sum_{k'>0} (v_{kk',k''k'} + v_{k-k',k''-k'})V_{k'}^2] \times (U_k U_{k''} - V_k V_{k''}) - \sum_{k'>0} v_{k-k'',k'-k'} U_{k'} V_{k'} (U_k V_{k''} + V_k U_{k''})\} \times (\alpha_k^\dagger \alpha_{k''} + \alpha_{-k}^\dagger \alpha_{-k''}) \quad (12.2)$$

$$H_{20}' = \sum_{k,k''>0} \{[(\eta_k - \lambda) + \sum_{k'>0} (v_{kk',k''k'} + v_{k-k',k''-k'})V_{k'}^2] \times (U_k V_{k''} + V_k U_{k''}) + \sum_{k'>0} v_{k-k'',k'-k'} U_{k'} V_{k'} (U_k U_{k''} - V_k V_{k''})\} \times (\alpha_k^\dagger \alpha_{-k''}^\dagger + \alpha_k \alpha_{-k''}) \quad (12.3)$$

H' consists again of all terms involving four α-operators arranged in the normal order.

In the pairing interaction the terms in H' are neglected in first approximation, and the remaining part of the Hamiltonian is brought to a diagonal form by choosing V_k and U_k so that H_{20}' vanishes identically. We shall now make H_{20}' vanish also for the more general case, again by properly choosing V_k and U_k. In fact we can generalize the formalism even further and relax the condition that the single-particle part of the original Hamiltonian be already diagonal in the scheme of the states $a_k^\dagger|0\rangle$. This has the effect of replacing, in (12.2) and (12.3), $\eta_k - \lambda$ by

$$\eta_k - \lambda \rightarrow \epsilon_{kk''} - \lambda\delta_{kk''} \tag{12.4}$$

whereas in (12.1) $\eta_k - \lambda$ is replaced by $\epsilon_{kk} - \lambda$. Here the matrix elements $\epsilon_{kk'}$ are the appropriate coefficients in the one-body part of the Hamiltonian $\Sigma\epsilon_{kk'}a_k^\dagger a_{k'}$. The condition $H_{20}' = 0$ in the general case then reads

$$\{[(\epsilon_{kk''} - \lambda\delta_{kk''}) + \sum_{k'>0} (v_{kk',k''k'} + v_{k-k',k''-k'})V_{k'}^2](U_kV_{k''} + U_{k''}V_k) + \sum_{k'>0} v_{k-k'',k'-k'}U_{k'}V_{k'}(U_kU_{k''} - V_kV_{k''})\} = 0 \tag{12.5}$$

The solution of (12.5) can be now made to incorporate both the self-consistency requirements and the maximum effects of the correlations accounted for in the Bogoljubov transformation. Thus the trivial solution of (12.5)

$$\begin{aligned} V_k &= 1 \qquad U_k = 0 \qquad \text{for occupied states} \\ V_k &= 0 \qquad U_k = 1 \qquad \text{for unoccupied states} \end{aligned} \tag{12.6}$$

yields immediately

$$\epsilon_{kk''} + \sum_{k' \text{ occupied } (k' \lessgtr 0)} v_{kk',k''k'} = 0 \qquad \text{for} \qquad \begin{cases} k'' \text{ occupied} \\ k \;\; \text{unoccupied} \end{cases} \tag{12.7}$$

This will be recognized as the Hartree–Fock condition, (2.20). However, we can now do better than the trivial solution in looking for a nontrivial solution to (12.5). In principle one can start with a good guess for the single-particle states and a *given* set of values of U_k and V_k, and consider (12.5) for $k \neq k''$ as a self-consistency requirement on the single real-particle states $|k\rangle$. Then with these states fixed, one can go to the $k = k''$ terms in (12.5) and consider them, together with the condition $\langle \hat{N} \rangle = N_0$, as equations for λ, V_k's, and U_k's. When we solve these equations, the new values of U_k and V_k can be introduced again into the $k \neq k''$ terms of (12.5) to determine self-consistent single-particle states, and the whole procedure can be iterated until it converges on a self-consistent solution that is also a nontrivial simultaneous solution of the BCS problem.

This program has not been carried through yet, but one can come close to it by making the following observation: the last term in (12.5) is very small if there is a high probability for k'' to be occupied and for k to be empty, since then $U_{k''} \ll 1$ and $V_k \ll 1$, and therefore $(U_kU_{k''} - V_kV_{k''}) \ll 1$.

Furthermore, the product $U_{k'}V_{k'}$ is different from zero only near the top of the Fermi distribution, and for $k \neq k''$ the matrix element $v_{k-k'',k'-k'}$ is also small for reasonable interactions. We thus neglect the third term in (12.5) in deriving the self-consistency relations and obtain them in the form

$$\epsilon_{kk''} + \sum_{\text{occupied } k' \lessgtr 0} v_{kk',k''k'} V_{k'}^2 = 0 \quad \text{for} \quad \begin{cases} k'' \text{ mostly occupied} \\ k \text{ mostly unoccupied} \end{cases} \tag{12.8}$$

The proper self-consistent single-particle states therefore come out as the states in the single-particle potential

$$U_{kk''} = \sum_{k'} v_{kk',k''k'} V_{k'}^2 \tag{12.9}$$

This is, in fact, the natural generalization of the conventional Hartree–Fock potential since $V_{k'}^2$ is the probability for the state k' to be occupied and $v_{kk',k''k'}$ is the potential induced by a particle that occupies k'. For the self-consistent potential we therefore have

$$\epsilon_{kk''} + \sum_{k'} v_{kk',k''k'} V_{k'}^2 = \eta_k \, \delta_{kk''} \tag{12.10}$$

and (12.5), for $k = k''$, takes on the form

$$2(\eta_k - \lambda)U_k V_k + [\sum_{k'>0} v(k, k')U_{k'}V_{k'}](U_k^2 - V_k^2) = 0 \tag{12.11}$$

Equation 12.11 will be recognized as identical in form (although not in content!) to the corresponding equation derived for the pairing interaction from (9.24), so that all the properties of the BCS solution derived earlier hold for the present case as well.

We thus find that a general interaction leads to results similar to those of a pairing interaction if the single-particle states are determined in a self-consistent manner in the sense of (12.9) or (12.10), provided the terms in H' can be treated as a small perturbation, and if the resulting values of U_k and V_k are such that $|U_k U_{k''} - V_k V_{k''}| \ll 1$.

Our discussion of the BCS solution has been very general. We did not specify the symmetry of the single-particle potential (except for being invariant under time reversal), nor did we limit ourselves to any particular interaction. Thus it holds equally well for systems with spherical symmetry or axial symmetry. Most treatments of many-particle systems with mutual interactions aim at reducing the problem to that of some equivalent noninteracting entities, since then we know how to solve the problem. The Hartree–Fock approach does it by inducing the bulk of the correlations among the interacting particles through an outside effective single-particle potential that holds the noninteracting particles together. The BCS-method goes further than that by shifting most of the pairing part of the interaction into the single-particle parameters of the problem, at the cost, however, of giving up the conservation of the number of particles. The great value of taking this pairing part of the interaction so seriously stems from the fact that the nonpairing part is taken care of, to a large extent, by the self-consistency requirement as was stressed in the discussion above, and the generality of this result makes it all the more attractive.

The pairing part of the interaction $v_{k-k,k'-k'}$, when $\mathbf{k}$ now stands for the linear momentum of plane waves, turns out not to be too sensitive to the relative direction of $\mathbf{k}$ and $\mathbf{k}'$; what is contributed by the direct integral when $\mathbf{k}$ is parallel to $\mathbf{k}'$ is taken over by the exchange integral when $\mathbf{k}$ is antiparallel to $\mathbf{k}'$. In fact it is not uncommon to approximate the pairing part of the interaction by the expression

$$\sum v_{k-k,k'-k'} a_k^\dagger a_{-k}^\dagger a_{k'} a_{-k'} \approx -G \sum_{k'k'>0} a_k^\dagger a_{-k}^\dagger a_{k'} a_{-k'} \tag{12.12}$$

In this case the scattering of a pair $(\mathbf{k}', -\mathbf{k}')$ is entirely isotropic [equal amplitudes to all final directions $(\mathbf{k}, -\mathbf{k})$]. This is a feature of a *short* range potential, and one therefore often refers to the pairing part of the interaction as the one responsible for the short range properties of the full interaction. Consequently, the nonpairing part is the part that is responsible for the long range properties of the full interaction. Since it is the nonpairing part that is taken care of by the self-consistent Hartree–Fock potential, it is often referred to as the *field-producing part of the interaction.* The pairing part, on the other hand, is the *correlation part of the interaction.*

The field producing, or long range, part of the interaction, in determining the features of the Hartree–Fock potential also determines the collective oscillations of the system as discussed previously. The pairing part is then of critical importance in determining the inertial parameters of these collective oscillations, since the flow pattern of the particles during these oscillations is greatly affected by their short range correlations. In this way the various parts of the interparticle interaction are seen to affect different properties of the many-nucleon system. In the next section we shall see another useful way of looking at these results.

A very direct consequence of the pairing appears in (p, t) and (t, p) reactions in which pairs of neutrons primarily in a relative ${}^1S_0$ state are transferred. These reactions are enhanced when the transitions are between superconducting states as exist, for example, among the tin isotopes. This phenomenon will be discussed more completely in Vol. II.

13. PERTURBATION TREATMENT OF THE MANY-FERMION SYSTEM

The Hartree–Fock–Bogoljubov method for the handling of the many-fermion system is nonperturbative in its nature. Even if the interaction among the particles is small compared to characteristic frequencies of the system, that part of the interaction which is taken into account in deriving the BCS ground state does not allow a power-series expansion of the ground-state energy in powers of the interaction. This was clearly demonstrated in (10.11), where the energy gap Δ_0 was shown to be proportional to (sinh D/G) with G being the strength of the two-body interaction.

However the Hartree–Fock–Bogoljubov method does not take account of the whole of the two-body interaction in the Hamiltonian. There is a remaining term H' in the Hamiltonian that cannot be brought to a form which commutes with either the particle or the quasi-particle number operator, and that is neglected in the independent quasi-particle approximation. Such terms, generally referred to as *residual interactions*, can be taken into account using perturbation theory. We described some simple applications of residual-interaction perturbation theory in the chapters on the shell model; here we want to expand on the theory somewhat, introducing some of the more general features that derive from it.

The problem can be formulated in the following way: given a Hamiltonian $H = H_0 + V$ in which H_0 is a single-particle (or single quasi-particle) operator and V is a two-body operator, we want to write down expressions for the ground-state wave function and energy in terms of a power series in V. Furthermore, in this power series we want to identify special terms that can be formally summed up without specifying the interaction, so that the actual task of carrying out explicit numerical summations for specific interactions can be limited to a smaller number of terms.

Formally a power-series expansion of the eigenfunctions and eigenvalues is easy to write down. Let

$$H\Psi_0 = E\Psi_0 \tag{13.1}$$

and

$$H_0\Phi_0 = \epsilon_0\Phi_0 \tag{13.2}$$

We shall assume that the solution Ψ_0 "belongs" to Φ_0 in the sense that

$$\Psi_0 \to \Phi_0 \quad \text{as} \quad V \to 0 \tag{13.3}$$

and $\langle\Psi_0|\Phi_0\rangle \neq 0$

From (13.1) we obtain

$$E\langle\Phi_0|\Psi_0\rangle = \langle\Phi_0|H\Psi_0\rangle = \langle\Phi_0|(H_0 + V)\Psi_0\rangle = \epsilon_0\langle\Phi_0|\Psi_0\rangle + \langle\Phi_0|V\Psi_0\rangle$$

Hence, since $\langle\Phi_0|\Psi_0\rangle \neq 0$, the energy shift due to the interaction V is given by:

$$\Delta E = E - \epsilon_0 = \frac{\langle\Phi_0|V\Psi_0\rangle}{\langle\Phi_0|\Psi_0\rangle} \tag{13.4}$$

To obtain Ψ_0, we note that because of (13.3) we can normalize it so that $\langle\Phi_0|\Psi_0\rangle = 1$. It can then be verified directly that

$$\Psi_0 = \Phi_0 + \frac{Q_0}{E - H_0} V\Psi_0 \tag{13.5}$$

satisfies (13.1). Here Q_0 is a projection operator that projects on the space orthogonal to Φ_0, that is, for any function f

$$Q_0 f = f - \Phi_0\langle\Phi_0|f\rangle \tag{13.6}$$

Equation 13.5 can be iterated to yield

$$\Psi_0 = \left[1 + \sum_{n=1}^{\infty}\left(\frac{Q_0}{E - H_0} V\right)^n\right]\Phi_0 = \left[\sum_{n=0}\left(\frac{Q_0 V}{E - H_0}\right)^n\right]\Phi_0 \tag{13.7}$$

Provided the sum in (13.7) converges, we have thereby obtained an expansion of Ψ_0 in terms of powers of V. Similarly, (13.4) yields, using (13.7):

$$\Delta E = \langle\Phi_0|\left[\sum_{n=0} V\left(\frac{Q_0 V}{E - H_0}\right)^n\right]|\Phi_0\rangle \tag{13.8}$$

Equations 13.7 and 13.8, known as the *Wigner–Brillouin expansion*, do not constitute an explicit solution of our problem since the *exact* energy E appears in the power expansion. Yet, if the series is summed to any finite order it is possible, in principle, to extract E from (13.8) and obtain thereby also an approximate expression for Ψ_0 to the same order.

Despite its simple structure, the Wigner–Brillouin series is not always useful for large systems. To see the origin of some of the difficulties we shall estimate the order of magnitude of its various terms. We shall consider infinitely large systems of volume Ω with a constant nucleon density $\rho \equiv (A/\Omega)$. For these systems the appropriate individual-particle orbitals are plane waves. The state $|\Phi_0\rangle$ is an antisymmetrized product of plane waves (see Chapter III). Thus in the term of lowest order

$$\Delta E_1 = \langle\Phi_0|V|\Phi_0\rangle = \frac{1}{2}\sum \langle ij|v|ij\rangle$$

$|ij\rangle$ is a two-particle state in which the first particle is in plane-wave state with momentum $\mathbf{k}_i$ the second in a state with momentum $\mathbf{k}_j$. Here

$$V = (1/2)\sum_{i\neq j} v(ij)$$

and the matrix element includes the exchange term. The summation is extended over all occupied states i and j. If the strength of the two-body potential is denoted by v_0 and its range by b, ΔE_1 is given by

$$\Delta E_1 \approx \frac{A(A-1)}{2} v_0 \frac{b^3}{\Omega} \tag{13.9}$$

Here A is the number of particles in the system. Introducing the density of nucleons ρ we see that

$$\Delta E_1 = v_0 b^3 0(\rho A) \tag{13.10}$$

where $0(\rho A)$ is a quantity of order ρA.

If we go now to the next term in (13.8) we obtain

$$\Delta E_2 = \langle \Phi_0 | V \frac{Q_0}{E - H_0} V | \Phi_0 \rangle = \frac{1}{4} \sum_{ij,mn} \frac{\langle ij | v | mn \rangle \langle mn | v | ij \rangle}{E - \epsilon(m, n)} \tag{13.11}$$

where the summation over (m, n) is restricted to states orthogonal to Φ_0. Because v is invariant with respect to translations, the state (m, n) is further restricted by the requirement that $\mathbf{k}_m + \mathbf{k}_n = \mathbf{k}_i + \mathbf{k}_j$. For given momenta $\mathbf{k}_i$ and $\mathbf{k}_j$ the summation over $\mathbf{k}_m$ and $\mathbf{k}_n$ therefore represents only a single sum over an independent variable; since summation over $\mathbf{k}$ can be replaced by the integral $(\Omega/(2\pi)^3 \int_0^{k_0} d\mathbf{k}$, we see that there will be roughly $(4\pi/3)\cdot [\Omega k_0^3/(2\pi)^3]$ terms in (13.11) for each pair (ij). Here k_0 is some effective upper limit on the momentum integral. If we set $k_0 = k_F$, then $(4\pi/3)\Omega k_0^3 \approx \Omega\rho = A$. Normally k_0 will be larger than k_F and, at any rate, it depends on the interaction, whereas k_F depends on the density of the system. The energy denominator is of order of magnitude

$$E - \epsilon(m, n) = [\epsilon(i) + \epsilon(j)] - [\epsilon(m) + \epsilon(n)] + \Delta E \approx 0(\rho A) v_0 b^3$$

since already to first order $\Delta E_1 \approx v_0 b^3 0(\rho A)$. Thus we find that the order of magnitude of ΔE_2 is

$$\Delta E_2 = 0\left[\frac{A(A-1)}{2}\left(v_0 \frac{b^3}{\Omega}\right)^2 k_0^3 \frac{\Omega}{\rho A v_0 b^3}\right] = v_0 b^3 0(k_0^3) \tag{13.12}$$

It can be verified that this is also the order of magnitude of each one of the next terms. We have to add up corrections ΔE_n up to the $n = A$ order to obtain a total correction of order A, comparable to ΔE_1. For large values of A the Wigner–Brillouin series, therefore, converges very slowly, and is consequently of little practical value.

Another expansion in powers of V is offered by the *Rayleigh-Schrödinger series*. To make this expansion more transparent we call the interaction now λV where λ is a small number and we expand both E and ψ in powers of λ:

$$E = \epsilon^{(0)} + \lambda\epsilon^{(1)} + \lambda^2\epsilon^{(2)} + \dots \tag{13.13}$$

$$\Psi_0 = (1 + \lambda U_1 + \lambda^2 U_2 + \dots)\Phi_0 \tag{13.14}$$

where the U's are operators operating on Φ_0. In the equation $(H_0 + \lambda V)\Psi_0 = E\Psi_0$ one now equates equal powers of λ, obtaining the following relations

$$\epsilon^{(0)} = \langle \Phi_0 | H_0 \Phi_0 \rangle$$

$$\epsilon^{(1)} = \langle \Phi_0 | V \Phi_0 \rangle$$

$$\epsilon^{(2)} = \langle \Phi_0 | V \frac{Q_0}{E_0 - H_0} V \Phi_0 \rangle$$

$$\epsilon^{(3)} = \langle \Phi_0 | V \frac{Q_0}{E_0 - H_0} (V - \langle \Phi_0 | V \Phi_0 \rangle) \frac{Q_0}{E_0 - H_0} V \Phi_0 \rangle \tag{13.15}$$

Equations 13.15 differ in two important ways from the corresponding terms in the Wigner–Brillouin series:

(i) The energy denominators involve now only zeroth-order energy differences, and are therefore of order ϵ_F, which, being the order of magnitude of the average single-particle energy, is very closely equal to $\rho v_0 b^3$.

(ii) The structure of the terms of higher order becomes very complicated. Thus the third-order term consists of a part that looks like the product of first-order and second-order corrections (with an extra energy denominator). Because of the first point we now find that each term is at least of order of magnitude ρA, but because of the second point we find, for instance, that

$$\epsilon^{(3)} = 0(\rho A) + 0(\rho^2 A^2) \tag{13.16}$$

A term in the energy of order $\rho^2 A^2$ leads to an energy per particle proportional to $\rho^2 A$, which increases indefinitely with the number of particles. For a system of a constant density and with finite-range interactions this result cannot be possibly true. A deeper inspection of the various terms in the Rayleigh–Schrödinger series does, indeed, reveal that terms in $\epsilon^{(n)}$ that are not of order ρA are cancelled by similar terms in $\epsilon^{(m)}$, so that the *total* energy remains of order ρA as expected. However the fact that some terms in the nth approximation are cancelled by terms of the mth order approximation makes it hard to stop the series at any one place. We shall therefore explore now methods that enable us to overcome this difficulty.

14. THE LINKED-CLUSTER EXPANSION*

Let us introduce the operator $U(\beta)$ defined by

$$U(\beta) = \exp(\beta H_0) \exp(-\beta H) \tag{14.1}$$

If Ψ_n are the eigenstates of the full Hamiltonian and Φ_0 is the ground state of H_0, we have, using (13.1) and (13.2):

$$\begin{aligned} \langle \Phi_0 | U(\beta) \Phi_0 \rangle &= \sum_n \langle \Phi_0 | \exp(\beta H_0) \Psi_n \rangle \langle \Psi_n | \exp(-\beta H) \Phi_0 \rangle \\ &= \sum_n \exp[\beta(\epsilon_0 - E_n)] |\langle \Phi_0 | \Psi_n \rangle|^2 \end{aligned} \tag{14.2}$$

where ϵ_0 is the lowest eigenvalue of H_0, and E_n are the eigenvalues of H, with

*Goldstone (57), Hugenholtz (57), Hubbard (57), Coester (58), Bloch (58), and Bloch and Horowitz (58).

E_0 the lowest among them. It follows from (14.2) that for large values of β one term will dominate the sum:

$$\lim_{\beta\to\infty} \langle \Phi_0 | U(\beta)\Phi_0 \rangle = \exp\,[\beta(\epsilon_0 - E_0)]\,|\langle \Phi_0 | \Psi_0 \rangle|^2 \tag{14.3}$$

Normalizing $\langle \Phi_0 | \Psi_0 \rangle = 1$, it follows that

$$\Delta E = E_0 - \epsilon_0 = -\lim_{\beta\to\infty} \frac{1}{\beta} \ln \langle \Phi_0 | U(\beta)\Phi_0 \rangle \tag{14.4}$$

To evaluate $\langle \Phi_0 | U(\beta)\Phi_0 \rangle$ we use Dyson's expansion* to write

$$\begin{aligned} U(\beta) &= \exp(\beta H_0)\exp[-\beta(H_0 + V)] \\ &= 1 - \int_0^\beta \exp(\beta_1 H_0) V \exp(-\beta_1 H_0)\, d\beta_1 + \int_0^\beta d\beta_2 \int_0^{\beta_2} d\beta_1 \\ &\quad \times [\exp(\beta_2 H_0) V \exp(-\beta_2 H_0)][\exp(\beta_1 H_0) V \exp(-\beta_1 H_0)] - + \ldots \\ &= \sum_{n=0}^{\infty} \frac{(-1)^n}{n!} \int_0^\beta \ldots \int_0^\beta T[V(\beta_n)V(\beta_{n-1}) \ldots V(\beta_1)]\, d\beta_1 \ldots d\beta_n \end{aligned} \tag{14.5}$$

---

*Dyson's expansion is obtained by examining the differential equation for $U(\beta)$. Differentiating (14.1) with respect to β yields

$$\frac{dU}{d\beta} = e^{\beta H_0}(-V)e^{-\beta H} = -e^{\beta H_0} V e^{-\beta H_0} U$$

or

$$\frac{dU}{d\beta} = -V(\beta)U$$

where

$$V(\beta) \equiv e^{\beta H_0} V e^{-\beta H_0}$$

Integrating the differential equation yields an integral equation for $U(\beta)$:

$$U(\beta) = 1 - \int_0^\beta V(\beta_1) U(\beta_1)\, d\beta_1$$

The Dyson expansion is obtained by solving this equation by iteration. After two iterations

$$U(\beta) \simeq 1 - \int_0^\beta V(\beta_1)\, d\beta_1 + \int_0^\beta d\beta_1\, V(\beta_1) \int_0^{\beta_1} d\beta_2\, V(\beta_2)$$

$$= 1 - \int_0^\beta V(\beta_1)\, d\beta_1 + \frac{1}{2} \int_0^\beta d\beta_1 \int_0^\beta d\beta_2\, \{V(\beta_1)\, V(\beta_2)\, \theta(\beta_1 - \beta_2) + V(\beta_2)\, V(\beta_1)\, \theta(\beta_2 - \beta_1)\} \cdots$$

where

$$\theta(\beta) = \begin{cases} 1 & \beta > 0 \\ 0 & \beta < 0 \end{cases}$$

This is just (14.7).

where

$$V(\beta) = \exp(\beta H_0) V \exp(-\beta H_0) \tag{14.6}$$

and the symbol $T[\quad]$ implies that the operators within the square brackets should always be taken in descending order of β's: $\beta_n > \beta_{n-1} > \ldots > \beta_1$, (*time-ordered product*). Thus, the term with $n = 2$, for instance, will be

$$\frac{1}{2}\left[\int_0^\beta d\beta_2 \int_0^{\beta_2} d\beta_1 V(\beta_2)V(\beta_1) + \int_0^\beta d\beta_1 \int_0^{\beta_1} d\beta_2 V(\beta_1)V(\beta_2)\right] \text{etc.} \tag{14.7}$$

In taking the ground-state matrix element of $U(\beta)$ as required in (14.4) the different terms in the sum (14.5) will make contributions of different powers of V. The second-order term (14.7), for instance, will have one contribution of the form

$$\langle \Phi_0 | V(\beta_2)V(\beta_1)\Phi_0 \rangle = \sum_n \langle \Phi_0 | V(\beta_2)\Phi_n \rangle \langle \Phi_n | V(\beta_1)\Phi_0 \rangle \tag{14.8}$$

We shall now assume V to be a sum of two-body interactions and put it equal to

$$V = \frac{1}{2} \sum w_{k_1k_2,k_1'k_2'} a_{k_1}^\dagger a_{k_2}^\dagger a_{k_2'} a_{k_1'} \tag{14.9}$$

In (14.9) the matrix element $w_{k_1k_2,k_1'k_2'}$ *does not* include the exchange term, that is, it is given by

$$w_{k_1k_2,k_1'k_2'} = \int \overset{*}{\phi}_{k_1}(1)\overset{*}{\phi}_{k_2}(2) v(12) \phi_{k_1'}(1)\phi_{k_2'}(2)\, d(1)\, d(2) \tag{14.10}$$

The reason for this choice will become clear as the discussion below proceeds. It is, of course, clear that the total sum in (14.9) includes, effectively, also the exchange terms because of the anticommutation relations satisfied by the a's. With (14.9) and the convention that $ij \ldots$ are labels of states occupied in Φ_0 while $m, n \ldots$ are unoccupied, we see that the matrix element

$$\langle \Phi_n | V(\beta)\Phi_0 \rangle$$

can give rise to the following matrix elements of v:

If $\Phi_n = \Phi_0$ we have elements of the type $w_{ij,ij}$ and $w_{ij,ji}$.

If $\Phi_n \neq \Phi_0$ we have elements of the type $w_{mn,ij}$, $w_{in,ij}$, $w_{mi,ij}$.

We further note that because of the definition (14.6) of $V(\beta)$, every matrix element $w_{k_1k_2,k_1'k_2'}$ will be accompanied by the factor $\exp \beta[(\epsilon_{k1} + \epsilon_{k2}) - (\epsilon_{k_1'} + \epsilon_{k_2'})]$. The second-order term (14.8) will then lead to the following contributions from the $\langle \Phi_0 | V\Phi_0 \rangle \langle \Phi_0 | V\Phi_0 \rangle$ term:

$$\frac{1}{4}\sum w_{i'j',i'j'}w_{ij,ij} - \frac{1}{4}\sum w_{i'j',i'j'}w_{ji,ij} + \frac{1}{4}\sum w_{i'j',j'i'}w_{ji,ij} - \frac{1}{4}\sum w_{i'j',j'i'}w_{ij,ij} \tag{14.11}$$

The other terms in (14.8) will contribute

$$\frac{1}{4}\exp [(\beta_2 - \beta_1)(\epsilon_j - \epsilon_m)](\sum w_{i'j,i'm}w_{im,ij} - \sum w_{i'j,i'm}w_{mi,ij} - \sum w_{ji,im}w_{i'm,i'j} + \sum w_{ji,im}w_{mi,ij}) \tag{14.12}$$

and

$$\frac{1}{4}\sum (w_{ij,mn}w_{mn,ij} - w_{ij,mn}w_{nm,ij} + w_{ji,nm}w_{mn,ij} - w_{ji,mn}w_{mn,ij}) \cdot \exp [(\beta_2 - \beta_1)(\epsilon_i + \epsilon_j - \epsilon_m - \epsilon_n)] \tag{14.13}$$

It is convenient to describe such matrix elements graphically using the Goldstone diagrams. To this end we denote an interaction by a broken line ------. Two solid lines, representing the initial and final state of a nucleon, meet at each of the end points of a broken line. We further adopt the convention that an arrow pointing away from the point of interaction represents a particle in a given state, whereas an arrow pointing toward the point of interaction represents a hole in a given occupied state. Thus the graph

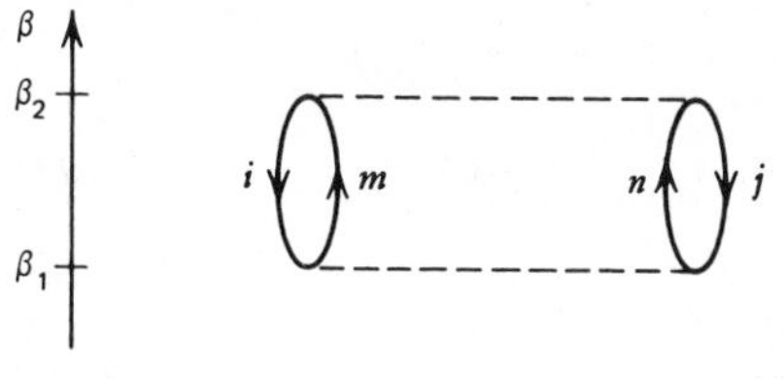

stands for

$$\tfrac{1}{4}\{w_{ij,mn}w_{mn,ij} \exp [(\beta_2 - \beta_1)(\epsilon_i + \epsilon_j - \epsilon_m - \epsilon_n)]$$

since at the lower interaction line (for $\beta = \beta_1$) it indicates that a hole was created in the state i, the particle being excited to the state m and another hole was created in the state j, the particle being excited to the state n. At the upper interaction line, that is, for $\beta = \beta_2$ a hole is created in the state m refilling the state i, and similarly the particle in n jumps back to j. The exponential factor that goes with each interaction is determined as follows: it involves the β-factor appropriate to this interaction ($\beta_1, \beta_2, \ldots,$ in ascending order) a factor exp $(\beta\epsilon_n)$ for each particle line and exp $(-\beta\epsilon_k)$ for each hole line that meet at this interaction.

The term $w_{i'j,i'm}w_{im,ij}$ exp $[(\beta_2 - \beta_1)(\epsilon_j - \epsilon_m)]$ is given by the graph.*

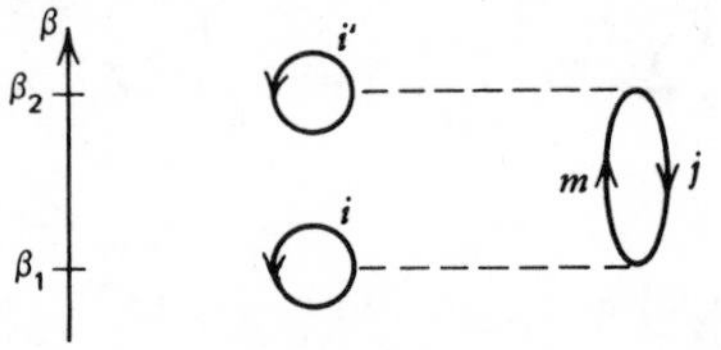

At $\beta = \beta_1$ a hole is created in j (line coming down from above the dotted line) and in i (line coming up from below the dotted line), and a particle is created in m and in i (lines going up above the dotted line). etc.

The term $w_{ij,ij}$ looks like

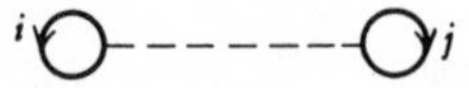

whereas the exchange term $w_{ij,ji}$ looks like

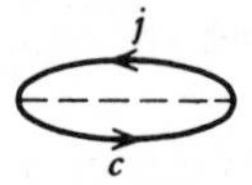

The second order term $\frac{1}{4}\, w_{i'j',i'j'}w_{ij,ij}$ looks like

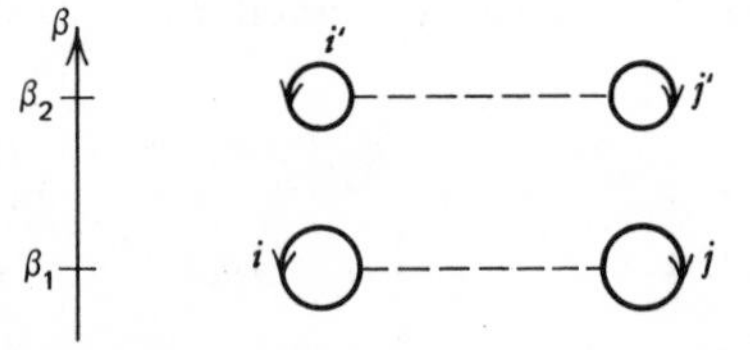

and is an example of an *unlinked* diagram.

The rules for establishing the connection between a graph and its corresponding matrix element should be augmented by giving its sign. We shall not prove here which sign goes with which graph [see Thouless (61)] but just summarize the rules for reading these graphs:

1. For each interaction line write down a factor $(1/2)w_{k_1k_2,k_1'k_2'}$ where k_1 and k_2 are the lines leaving the interaction line from left and right, respectively,

---

*Note that a particle that does not change its state during the interaction is represented by a circle: it leaves the interaction line and comes back to it in the same state.

while k_1' and k_2' are the lines entering the interaction line from left and right, respectively.

2. Write the exponential factors for each interaction line with its appropriate β multiplying the energies of the particles and holes that meet at this interaction line; lines leaving the interaction line come with a positive sign, those entering the interaction line come with a negative sign. For instance the diagram

will have associated with it the factor

$$\tfrac{1}{2}w_{n'm,ni} \exp\left[\beta(\epsilon_{n'} + \epsilon_m - \epsilon_n - \epsilon_i)\right]$$

the diagram

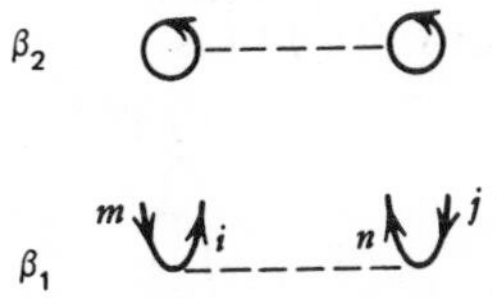

will be associated with

$$\{\tfrac{1}{2}w_{in,mj} \exp\left[\beta_1(\epsilon_i + \epsilon_n - \epsilon_m - \epsilon_j)\right]\} \cdot \{\tfrac{1}{2}w_{i'j',i'j'} \exp\left[\beta_2(\epsilon_{i'} + \epsilon_{j'} - \epsilon_{i'} - \epsilon_{j'})\right]\}$$

etc.

3. Count the number of closed fermion loops L and the number of hole lines H in the graph, and multiply the result of the previous step by $(-1)^{L+H}$.

4. In evaluating $U(\beta)$ integrate over the β's keeping their order in the diagram and sum over all particle and hole indices.

We shall now prove an important theorem: if a graph has two unlinked parts, then the sum of all the relative orientations of these two graphs is given by the product of the two linked graphs taken each one for itself.

To prove this theorem we shall look first in detail into a concrete example—that of a fourth-order graph consisting of two linked graphs, unlinked to each other—

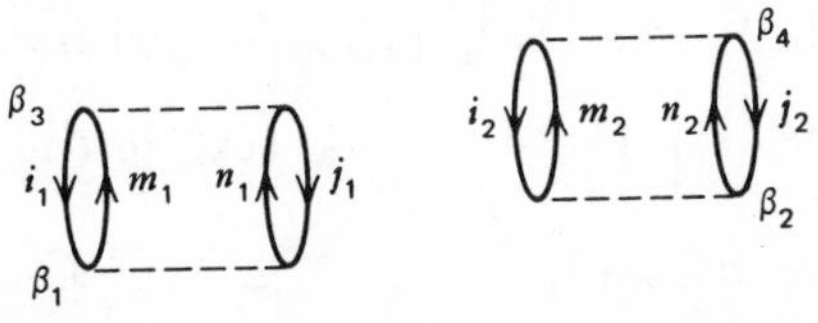

(14.14)

The contribution from this graph to $\langle \Phi_0 | U(\beta)\Phi_0 \rangle$ is:

$$\Delta U(\text{I}) = \int_0^{\beta} d\beta_4 \int_0^{\beta_4} d\beta_3 \int_0^{\beta_3} d\beta_2 \int_0^{\beta_2} d\beta_1 \sum_{i_1 j_1 m_1 n_1, i_2 j_2 m_2 n_2} \frac{1}{16}$$

$$\times \{\exp(\beta_4 \epsilon_0) w_{i_2 j_2, m_2 n_2} \exp(-\beta_4 \Delta_2) \exp(\beta_3 \Delta_2) w_{i_1 j_1, m_1 n_1}$$

$$\times \exp[-\beta_3(\Delta_2 + \Delta_1)] \exp[\beta_2(\Delta_2 + \Delta_1)] w_{m_2 n_2, i_2 j_2} \exp(-\beta_2 \Delta_1)$$

$$\times \exp(\beta_1 \Delta_1) w_{m_1 n_1, i_1 j_1} \exp(-\beta_1 \epsilon_0)\} \quad (14.15)$$

where

$$\Delta_1 = [\epsilon(m_1) + \epsilon(n_1) - \epsilon(i_1) - \epsilon(j_1)],$$

$$\Delta_2 = [\epsilon(m_2) + \epsilon(n_2) - \epsilon(i_2) - \epsilon(j_2)] \quad (14.16)$$

Another graph that can be obtained by changing the relative orientation of the two unlinked graphs in (14.14) is

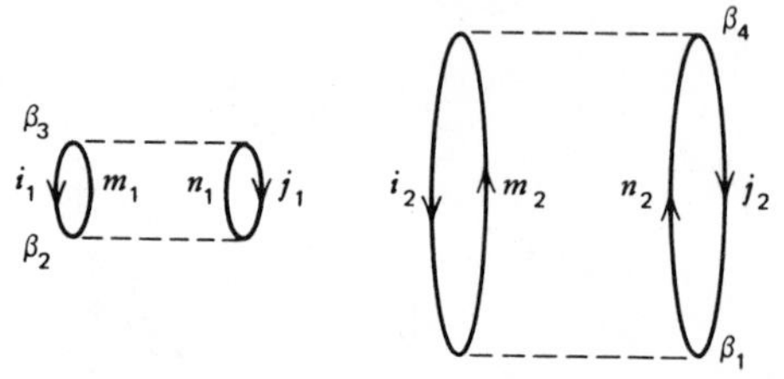

(14.17)

Its contribution to ΔU is given by

$$\Delta U(\text{II}) = \int_0^{\beta} d\beta_4 \int_0^{\beta_4} d\beta_3 \int_0^{\beta_3} d\beta_2 \int_0^{\beta_2} d\beta_1 \sum_{i_1 j_1 m_1 n_1, i_2 j_2 m_2 n_2} \frac{1}{16}$$

$$\times \exp(\beta_4 \epsilon_0) w_{i_2 j_2, m_2 n_2} \exp(-\beta_4 \Delta_2) \exp(\beta_3 \Delta_2) w_{i_1 j_1, m_1 n_1}$$

$$\times \exp[-\beta_3(\Delta_2 + \Delta_1)] \exp[\beta_2(\Delta_2 + \Delta_1)] w_{m_1 n_1, i_1 j_1} \exp(-\beta_2 \Delta_2)$$

$$\times \exp(\beta_1 \Delta_2) w_{m_2 n_2, i_2 j_2} \exp(-\beta_1 \epsilon_0) \quad (14.18)$$

We see that the matrix elements that are involved in (I) and (II) are the same, and the only difference between the contribution of these two graphs comes from the range of the integration of the exponential factors. For graph (I) these factors can be written as

$$\Delta U(\text{I}): \int \{\exp[-\beta_4(\Delta_2 - \epsilon_0)] \exp[\beta_2(\Delta_2 - \epsilon_0)] \exp(+\beta_2 \epsilon_0)\}$$

$$\times \{\exp[-\beta_3(\Delta_1 - \epsilon_0)] \exp[\beta_1(\Delta_1 - \epsilon_0)] \exp(-\beta_3 \epsilon_0)\}$$

$$= \int_{\beta > \beta_4 > \beta_3 > \beta_2 > \beta_1 > 0} d\beta_4\, d\beta_3\, d\beta_2\, d\beta_1 \exp[-(\beta_4 - \beta_2)\Delta_2] \exp[-(\beta_3 - \beta_1)\Delta_1]$$

where we have put $\epsilon_0 = 0$.

Similarly for graph (II) we obtain

$$\Delta U(\mathrm{II}): \int\limits_{\beta>\beta_4>\beta_3>\beta_2>\beta_1>0} d\beta_4\, d\beta_3\, d\beta_2\, d\beta_1 \exp\,[-(\beta_4-\beta_1)\Delta_2]\exp\,[-(\beta_3-\beta_2)\Delta_1]$$

$$= \int\limits_{\beta>\beta_4>\beta_3>\beta_1>\beta_2>0} d\beta_4\, d\beta_3\, d\beta_2\, d\beta_1 \exp\,[-(\beta_4-\beta_2)\Delta_2]\exp\,[-(\beta_3-\beta_1)\Delta_1]$$

(note change of integration variables in second step).
Thus the sum $\Delta U(\mathrm{I}) + \Delta U(\mathrm{II})$ involves the β-integrations subject only to the restrictions $\beta > \beta_4 > \beta_3 > (\beta_1, \beta_2) > 0$ where β_1 and β_2 are no longer restricted to be larger or smaller with respect to each other.

If one takes all possible relative orientation of these two parts of the graph, we shall always have the same matrix elements showing up, and the sum of all the β-integrations would yield:

$$\sum_{i=\mathrm{I,\,II}} \Delta U(i) = \int\limits_{\substack{\beta>\beta_4>\beta_2>0\\ \beta>\beta_3>\beta_1>0}} d\beta_4\, d\beta_3\, d\beta_2\, d\beta_1 \exp\,[-(\beta_4-\beta_2)\Delta_2]\exp\,[-(\beta_3-\beta_1)\Delta_1]$$

Since the sum over the indices $i_1 j_1 m_1 n_1$ is independent of that over the indices $i_2 j_2 m_2 n_2$, we see that the contribution to U from adding up all the contributions from the relative positions of the two unlinked parts is just the product of the contributions from each one of the linked parts of the graph. There is just one additional factor to be taken into account in the particular case we are considering. If, for example, we take the product of the two linked graphs, we shall be counting the two contributions

as different. Since the two linked parts of the total graph are topologically identical, we must divide their product by 2 in order to obtain the contribution of the 4th-order factorizable graph to ΔE. We thus obtain for the contribution of the 4th-order graphs

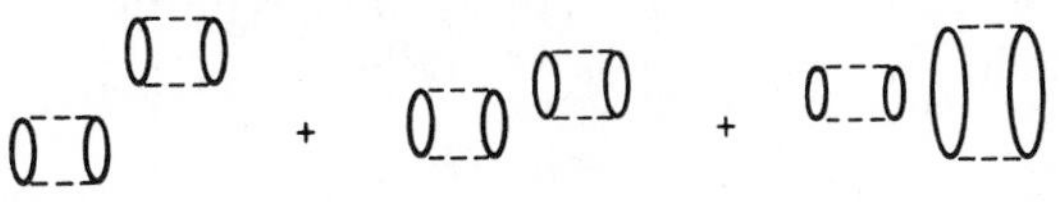

the expressions

$$\Delta U = \frac{1}{2}\left[\sum_{ijmn}\int_0^\beta d\beta_2 \int_0^{\beta_2} d\beta_1 \exp\,[-(\beta_2-\beta_1)\Delta]\frac{1}{4}\,|w_{ij,mn}|^2\right]^2$$

It is now easy to convince ourselves that if we have a graph consisting of n_1 unlinked parts of topological type Γ_1, each one of which by itself being linked, n_2 of type $\Gamma_2 \ldots$, n_k of type $\rightarrow\Gamma_k$, then the total contribution to ΔU, from all their relative positions, will be given by

$$\Delta U = \frac{1}{n_1!n_2! \ldots n_k!} \Gamma_1^{n_1}\Gamma_2^{n_2} \ldots \Gamma_k^{n_k} \tag{14.19}$$

where Γ_i refers to, the integrals and sums that are implied by the linked graph of topological type Γ_i.

A remark should be added here about the choice of $w_{k_1k_2,k_1'k_2'}$ to be the non-antisymmetric matrix elements of $v(ij)$. If we take, for instance graph II, (14.17), and if the indices j_1 and i_2 are summed independently (which is essential if we are to write this term as a product of two terms), then they include also the terms in which $j_1 = i_2 = i$.

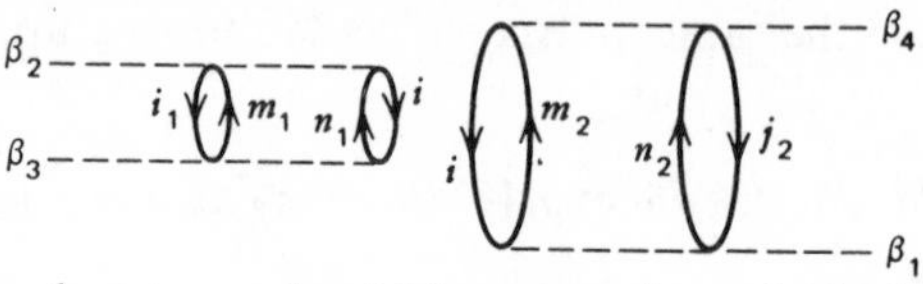

(14.20)

Strictly speaking such a term should have not been included since it violates the Pauli principle; in fact it says that the particle in the state i was annihilated by the interaction $V(\beta_1)$ and again by $V(\beta_2)$. Our prescription for reading graphs does not, however, make this graph vanish! This apparent shortcoming is rectified when we take *all* graphs into account, because among them we shall also find the graph

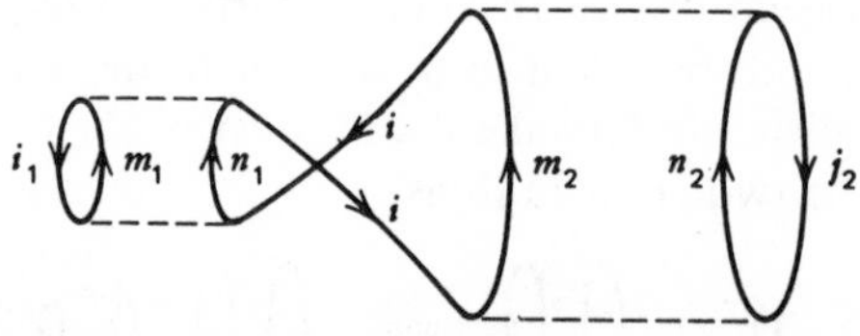

(14.21)

Graph (14.21) contributes exactly the same matrix elements and exponential factors to ΔE as the graph (14.20); since, however, it has only three fermion loops, as against the four loops in graph (14.20) its *sign* will be reversed. Counting both of them, therefore, brings our result into agreement with the Pauli principle. It is worthwhile to note that while graph (14.20) is unlinked, (14.21) is linked.

If we introduce now (14.5) into (14.4), we obtain the energy shift ΔE in terms of an infinite sum of integrals of products of $V(\beta)$'s:

$$\Delta E = -\lim_{\beta\to\infty} \frac{1}{\beta} \ln \langle \Phi_0 | \sum \frac{(-1)^n}{n!} \int_0^\beta \cdots \int_0^\beta T[V(\beta_n) \ldots V(\beta_1)]\, d\beta_1 \ldots d\beta_n | \Phi_0 \rangle \tag{14.22}$$

When we assign graphs to the various terms, we see that every conceivable graph is realized by one of the terms. Taking into account (14.19) we obtain then

$$\Delta E = -\lim_{\beta\to\infty}\frac{1}{\beta}\ln\left\{\sum \frac{1}{n_1!n_2!\ldots n_k!}\Gamma_1^{n_1}\Gamma_2^{n_2}\ldots\Gamma_k^{n_k}\right\}$$

$$= -\lim_{\beta\to\infty}\frac{1}{\beta}\ln\{\exp(\Gamma_1+\Gamma_2+\ldots+\Gamma_k+\ldots)\}$$

$$= -\lim_{\beta\to\infty}\frac{1}{\beta}\langle\Phi_0|\sum\frac{(-1)^n}{n!}\int_0^\beta\ldots\int_0^\beta T[V(\beta_n)\ldots V(\beta_1)]\,d\beta_1\ldots d\beta_n|\Phi_0\rangle_L \tag{14.23}$$

Here the index L at the end of the last matrix element in (14.23) indicates that in evaluating this matrix element *only linked graphs* are to be taken.

Although superficially (14.22) and (14.23) resemble each other, we should realize that they are very different. In (14.22) we have the *logarithm* of an expression involving all possible topological types of graph, whereas in (14.23) *there is no logarithm*, and the class of graphs over which this expression is to be summed is greatly reduced.

We can now write (14.23) in a slightly more familiar form. To this end we note that:

$$\langle\Phi_0|\frac{(-1)^n}{n!}\int_0^\beta\ldots\int_0^\beta T[V(\beta_n)\ldots V(\beta_1)]\,d\beta_1\ldots d\beta_n|\Phi_0\rangle_L$$

$$= (-1)^n\int_0^\beta d\beta_n\int_0^{\beta_n}d\beta_{n-1}\ldots\int_0^{\beta_2}d\beta_1\langle\Phi_0|\exp(\beta_n\epsilon_0)V$$

$$\times\exp[-(\beta_n-\beta_{n-1})H_0]V\exp[-(\beta_{n-1}-\beta_{n-2})H_0]$$

$$\times\ldots\exp[-(\beta_2-\beta_1)H_0]V\exp(-\beta_1\epsilon_0)|\Phi_0\rangle_L$$

$$= (-1)^n\int_0^\beta d\beta_n\ldots\int_0^{\beta_2}d\beta_1\langle\Phi_0|V\exp[-\zeta_n(H_0-\epsilon_0)]V$$

$$\times\exp[-\zeta_{n-1}(H_0-\epsilon_0)]\ldots\exp[-\zeta_2(H_0-\epsilon_0)]V|\phi_0\rangle_L \tag{14.24}$$

where we have introduced $\zeta_k = \beta_k - \beta_{k-1}$. Since in each integration $d\beta_k$, β_k changes from $0 < \beta_k < \beta_{k+1}$, and since we are interested in the limit $\beta \to \infty$, we see that we can change our variables from $\beta_1\beta_2\ldots\beta_n$ to $\beta_1, \zeta_2, \ldots, \zeta_n$ and let each one of them go from 0 to ∞ as β goes to infinity. Furthermore we note that since Φ_0 is the ground state, and since we should consider only linked graphs in (14.24), all intermediate states will be above Φ_0. It follows that exp $[-\zeta_k(H_0 - \epsilon_0)]$ in (14.24) has effectively a negative, nonvanishing exponential. We thus obtain for (14.24), carrying out the integrations on $\beta_1(0 < \beta_1 < \beta)$ and $\zeta_k(0 < \zeta_k < \infty)$:

$$(-1)^n\beta\langle\Phi_0|V\frac{1}{H_0-\epsilon_0}V\frac{1}{H_0-\epsilon_0}\cdots\frac{1}{H_0-\epsilon_0}V|\Phi_0\rangle_L$$

$$= -\beta\langle\Phi_0|V\frac{1}{\epsilon_0-H_0}V\frac{1}{\epsilon_0-H_0}\cdots\frac{1}{\epsilon_0-H_0}V|\Phi_0\rangle_L \quad (14.25)$$

where the factor β comes from the integration $d\beta_1$ (for which there is no exponential factor). Introducing (14.25) into (14.23) we obtain finally

$$\Delta E = \langle\Phi_0|V\sum_{n=0}^{\infty}\left(\frac{1}{\epsilon_0-H_0}V\right)^n|\Phi_0\rangle_L \quad (14.26)$$

Equation 14.26, which is an exact expression, is known as *Goldstone*'s linked cluster expansion [Goldstone (57)]. It resembles the Wigner–Brillouin series, except for the fact that it has the *unperturbed* ground-state energy ϵ_0 in the denominator, rather than the exact energy E_0; in addition it represents a very important simplification over the Wigner–Brillouin series in that (14.26) involves the summation over linked graphs only.

It is easy to show that a linked graph is always of order A, either ρA, or $\rho^2 A$ etc. In going over to the linked graphs we have thus also automatically gotten rid of the difficulty in the Rayleigh–Schrödinger series, which, it will be recalled, has terms of order A^2, A^3, etc. cancelling each other in different orders of perturbation theory.

We can now look for approximations to ΔE by limiting ourselves to a subset of linked graphs that can be summed explicitly. Rather than do this we shall now show how some of the approximations we studied before are formulated in terms of the linked-cluster expansion (14.26).

15. THE BETHE-GOLDSTONE EQUATION

Consider the series of linked graphs given by a sequence of interactions between ascending particle lines:

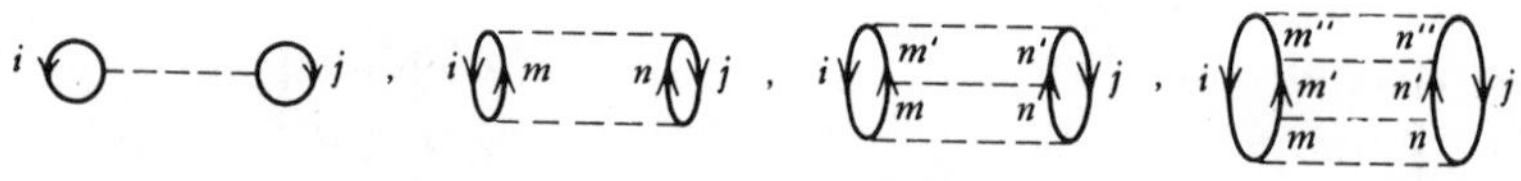

(15.1)

They correspond to a situation in which a pair of particles ij are excited from the ground state into the unoccupied state (m, n), then scatter again into the unoccupied state (m', n'), and then again to $(m''n'')$ etc., until after several such scatterings they fall back into the original state ij. To the graphs in (15.1) we can add the antisymmetric part in ij, that is, graphs like

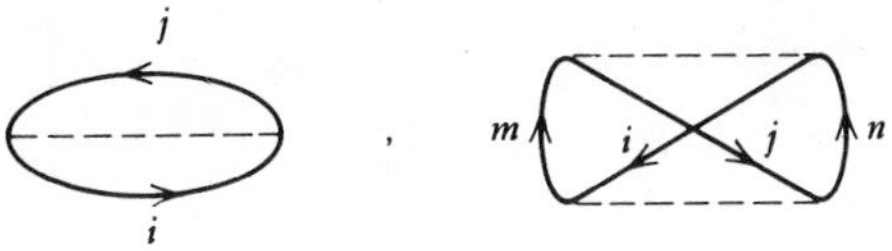

, , etc. (15.2)

We can easily write down the contribution to ΔE of the graph (15.1) and (15.2), which for obvious reasons are called *ladder graphs*:

$$\Delta E_{\text{lad}} = \frac{1}{2}\sum_{ij} v(ij, ij) + \frac{1}{4}\sum_{ij,mn} |v(ij, mn)|^2 \frac{1}{\epsilon_i + \epsilon_j - \epsilon_m - \epsilon_n} + \frac{1}{8}$$

$$\times \sum_{ij,mn,m'n'} v(ij, m'n') \frac{1}{\epsilon_i + \epsilon_j - \epsilon_{m'} - \epsilon_{n'}} v(m'n', mn) \frac{1}{\epsilon_i + \epsilon_j - \epsilon_m - \epsilon_n}$$

$$\times v(mn, ij) + \dots \quad (15.3)$$

where the rules for the structure of the energy denominators are obviously the same as those for the exponentials (see preceding section). In (15.3) $v(k_1k_2, k_3k_4)$ are the antisymmetric matrix elements

$$v(k_1k_2, k_3k_4) = w_{k_1k_2,k_3k_4} - w_{k_1k_2,k_4k_3} \quad (15.4)$$

We can define now a matrix G, also called *the reaction matrix*, by the following integral equation for its matrix elements:

$$\langle kl|G|ij\rangle = v(kl, ij) + \frac{1}{2}\sum_{mn} v(kl, mn) \frac{1}{\epsilon_i + \epsilon_j - \epsilon_m - \epsilon_n} \langle mn|G|ij\rangle \quad (15.5)$$

Here the states kl can be either occupied or unoccupied, whereas (ij) are limited to occupied states and (mn) to unoccupied states. An iteration of (15.5) shows immediately that with the help of this G-matrix it is possible to write ΔE_{lad}, (15.3), as the expectation value of G:

$$\Delta E_{\text{lad}} = \frac{1}{2}\sum_{i,j} \langle ij|G|ij\rangle \quad (15.6)$$

where in (15.6) the summation extends over all occupied states.

Equation 15.5 is a "two-body" equation, and its solution, as we have just shown, is equivalent to a summation over all ladder graphs in the many-body system.

Actually we can write (15.5) in a more familiar form. Introducing the projection operator Q_F that projects out of the Fermi sea the states occupied in Φ_0, we see that (15.5) is equivalent to the operator equation

$$G = v + v \frac{Q_F}{\epsilon_0 - H_0} G \quad (15.7)$$

[Note that in (15.5) we have $(1/2)\sum_{m,n}$ since the summation over m and n is independent; in the operator equation (15.7) repetition of mn and nm does not occur and, hence, the omission of the factor (1/2)].

If $\phi(ij)$ satisfies the equation

$$(\epsilon_0 - H_0)\phi(ij) = 0$$

then

$$\psi(ij) \equiv v^{-1}G\phi(ij) = \left(1 + \frac{Q_F}{\epsilon_0 - H_0} G\right)\phi(ij) \tag{15.8}$$

satisfies the equation

$$(\epsilon_0 - H_0)\psi(ij) = [(\epsilon_0 - H_0) + Q_F G]\phi(ij) = Q_F G\phi(ij) = Q_F v\psi(ij)$$

or

$$(H_0 + Q_F v)\psi(ij) = \epsilon_0\psi(ij) \tag{15.9}$$

Furthermore from (15.8) it follows that

$$\langle\phi|\psi\rangle = 1 \tag{15.10}$$

We can show that

$$\psi(ij) \to \phi(ij) \qquad \text{when} \qquad |\mathbf{r}_i - \mathbf{r}_j| \to \infty \tag{15.11}$$

Equation 15.11 results from the observation that all components of $\psi(ij)$ other than $\phi(ij)$ belong to energies different from ϵ_0; $\psi(ij)$ cannot therefore manifest any phase shift at infinity and must coincide with $\phi(ij)$.

Equation 15.9, which is completely equivalent to (15.7), has the form of a Schrödinger equation. We thus see that the solution of the Bethe–Goldstone equation as formulated in Section III.6 is equivalent to the summation of the ladder graphs in the linked-cluster expansion for the energy (14.26). Rather than carry out the infinite sums of the ladder graphs we can, therefore, solve the differential equation (15.9) subject to the normalization (15.10) and, having found the solutions $\psi(ij)$ evaluate the corresponding corrections to the energy through (15.6), that is,

$$\Delta E_{\text{lad}} = \frac{1}{2}\sum_{i,j} \langle\phi_{ij}|G\phi_{ij}\rangle = \frac{1}{2}\sum \langle\phi_{ij}|v\psi_{ij}\rangle \tag{15.12}$$

16. THE HARTREE-FOCK POTENTIAL; THE SELF-CONSISTENT METHOD FOR THE G-MATRIX

Let us consider the second-order graph

(16.1)

Together with it we shall now consider the series of graphs

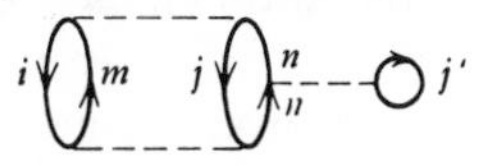

(a)

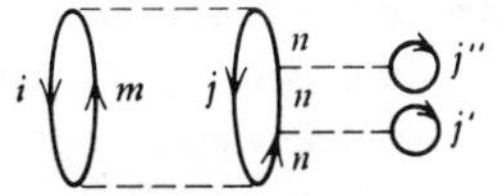

(b)

etc. (16.2)

as well as the corresponding exchange graphs

(16.3)

We note that these graphs are characterized by the particle line n being interrupted by successive interaction with one of the particles in the occupied states, *without*, however, exciting these other particles. Also the particle in the n-state is not scattered into another state during its interaction with the j'-particle, so that the energy denominators in the linked-cluster expansion become particularly simple. In fact, we see that for each interaction line with a j'-particle for example, (16.2a) contributes to the basic graph (16.1) a factor

$$\frac{1}{\epsilon_i + \epsilon_j - \epsilon_m - \epsilon_n} \sum_{j'} v(nj', nj') = A(ij, nm) \tag{16.4}$$

where the summation over j' affects the interaction v only. Since successive graphs with more interactions with j'-particles involve higher powers of $A(ij, nm)$ as given by (16.4), we see that we can sum up this geometrical series, obtaining for the graph (16.1) and its "satellites" (16.2) and (16.3):

$$\begin{aligned} \Delta E' &= \sum |v(mn, ij)|^2 \frac{1}{\epsilon_i + \epsilon_j - \epsilon_m - \epsilon_n} \sum_{k=0} \left(\frac{\sum_{j'} v(nj', nj')}{\epsilon_i + \epsilon_j - \epsilon_m - \epsilon_n} \right)^k \\ &= \sum |v(mn, ij)|^2 \frac{1}{\epsilon_i + \epsilon_j - \epsilon_m - [\epsilon_n + \sum_{j'} v(nj', nj')]} \end{aligned} \tag{16.5}$$

The sum of all these particular diagrams has thus the effect of redefining single-particle energies: $\epsilon_n \to \epsilon_n + \sum_{j'} v(nj', nj')$, and it is easy to see that each one of the ϵ's in (16.5) will turn into an $\epsilon_k \to \epsilon_k + \sum_{j'} v(kj', kj')$ if we allow its corresponding line to interact with j'-particles.

The sum of the geometrical series in (16.5) converges to its expression on the right-hand side of this equation only if $\sum_{j'} v(nj', nj') < [\epsilon_i + \epsilon_j - \epsilon_m - \epsilon_n]$. If this relation is not satisfied, it is still possible for the complete linked-cluster expansion, (14.26), to converge. In any event it is clear from (16.5) that it will pay to redefine the single-particle energies by introducing an additional central potential $U = \sum U_{kk'} a_k^\dagger a_{k'}$, such that the single-particle energies η_k in this central potential satisfy

$$\eta_k = \epsilon_k + \sum_{j'} v(kj', kj') \tag{16.6}$$

Comparison with (2.21) now shows that if the single-particle states $|k\rangle$ are chosen to be the states in the potential U, this potential automatically becomes the Hartree–Fock self-consistent potential.

The situation can now be described formally as follows: we go from the Hamiltonian $H_0 + V$ to the Hamiltonian $(H_0 + U) + (V - U)$ where U is a single-body operator. We take our single-particle states to be those of

$$H_0' = H_0 + U \tag{16.7}$$

With these states, and with

$$V' = V - U \tag{16.8}$$

as our new perturbation, we can associate with each diagram of the type (16.2) or (16.3) a corresponding diagram of the type

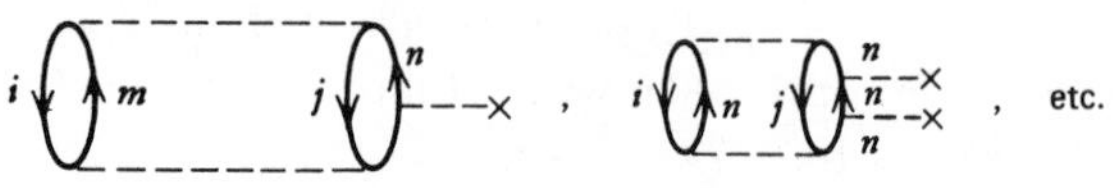

(16.9)

where 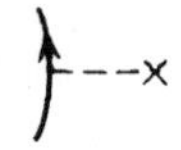 stands for a single-particle scattering through the potential U in (16.8). If the potential U is chosen self-consistently, that is, if the eigenvalues η_k of H_0' in (16.7) satisfy (2.21):

$$\epsilon_{kk'} + \sum_{j'} v(kj', k'j') = \eta_k\, \delta_{kk'}$$

then the graphs (16.9) will just cancel the graphs (16.2) and (16.3). Effectively, therefore, if we choose our states self-consistently we can forget about all the graphs (16.2), (16.3), and (16.9).

We see here again a case in which a whole class of graphs can be taken into account by a proper choice of the single-particle states. The convenience of working with graphs in the identification of various terms in the perturbation series is thus clearly manifested. But it should be stressed that these considerations are only valid when the convergence properties of the linked-cluster expansion is not affected by changes in the order of summation. If this is not true, special care must be exercised in performing partial infinite sums.

The concept of self-consistent field can now be extended to include also the graphs represented by the G-matrix. Thus we consider the series of graphs in which a particle line n appears with the following ladder "satellites":

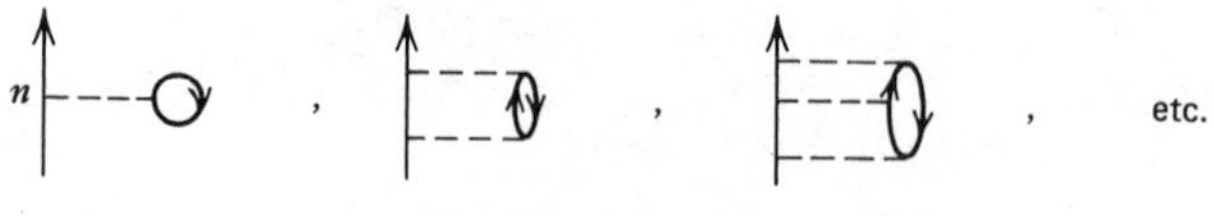

(16.10)

It is obviously possible to introduce a single-particle potential U whose matrix elements are just the sum of all the graphs in (16.10). Furthermore we can extend this potential also to occupied states by requiring that its matrix elements in these states be given by the sum of

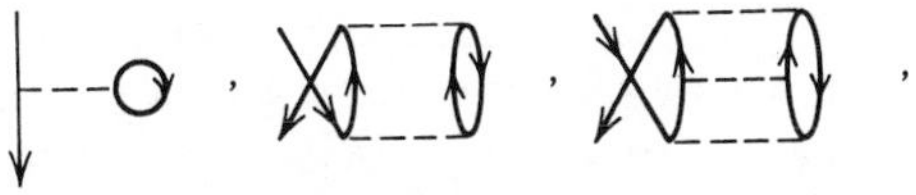 etc. (16.11)

We notice that the ladders of interactions are between up-going lines, both in (16.10) and in (16.11), like in the graphs that led to the G-matrix. In fact it is not difficult to convince ourselves that the potential that should be used to replace these graphs is given by

$$U_{kk'} = \sum_i \langle ki|G|k'i\rangle \tag{16.12}$$

The derivation of the self-consistent potential is now slightly more complex: one starts with a set of single-particle wave functions (i.e., an H_0) and solves the G-matrix equation (15.7)

$$G = v + v\frac{Q_F}{\epsilon_0 - H_0}G \tag{16.13}$$

Having solved this equation one constructs the potential $U_{kk'}$ and solves the Schrödinger equation for the Hamiltonian

$$H_0' = H_0 + U$$

With the new single-particle wave functions one enters again (15.7) to obtain new matrix elements for G, and so on until the whole thing converges. The single-particle energies in this G-self-consistent potential include then corrections to the ordinary ladder approximation that look like

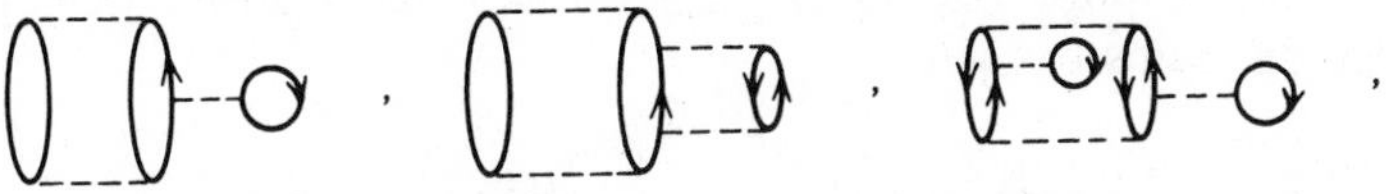 etc.

Actually the description given here of the G-self consistent potential is misleading in its apparent simplicity. If we look back at (15.5) we notice that the definition of the reaction matrix G is intimately connected with a specific choice of two states i and j, since the energy denominator contains $\omega = \epsilon_i + \epsilon_j$ explicitly. Thus, a G-matrix is really a function of the energy ω and we should write it as $G(\omega)$. When we want to replace a series of diagrams by a single-particle potential, like in (16.12), the question is immediately raised: which value of ω do we have to take for $G(\omega)$ in (16.12)?

Consider, for instance, the following set of diagrams

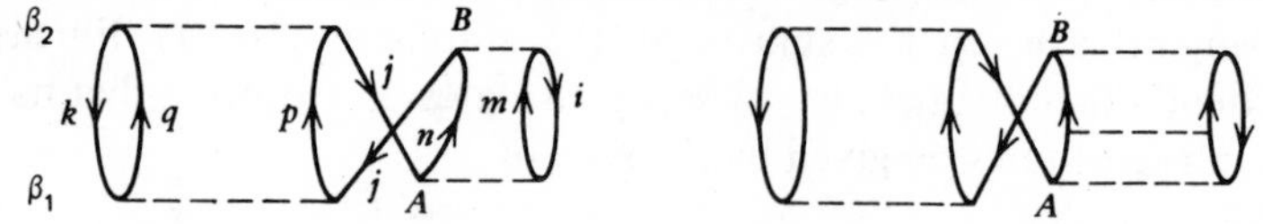

(a) (b) (16.14)

All these diagrams consist of a pair of particle-hole states (p, j) and (q, k) created at β_1 and annihilated at β_2. However, in between another pair of particle-hole states, (nj) and (mi) are created at $\beta_A > \beta_1$, and the two particles n and m are allowed to interact any number of times via the ascending ladders, until they annihilate the holes in j and i at $\beta_B < \beta_2$. (Note, incidentally, that two holes exist in j simultaneously; this is due to our disregard for the Pauli principle).

We may want to replace those interactions by modifying the single-particle potential using (16.12). However if we take diagram (a), for instance, the exponential β-factor or equivalently, the energy denominator after the second interaction is

$$-\epsilon_q - \epsilon_p - \epsilon_n - \epsilon_m + \epsilon_k + 2\epsilon_j + \epsilon_i$$
$$= [(\epsilon_i + \epsilon_j) + (\epsilon_k + \epsilon_j - \epsilon_p - \epsilon_q)] - \epsilon_m - \epsilon_n \quad (16.15)$$

Normally, if we were to compute the G-matrix from (16.13) disregarding the other particles, the energy denominator would have been (see Eq. 16.4)

$$\epsilon_i + \epsilon_j - \epsilon_m - \epsilon_n \quad (16.16)$$

The G-matrix computed with just the initial or final energies as in (16.16) is said to be computed *on the energy shell.* In the graphs (16.14) we need to compute $G(\omega')$ where $\omega' = \epsilon_i + \epsilon_j + (\epsilon_k + \epsilon_j - \epsilon_p - \epsilon_q)$, and $G(\omega')$ is then said to be taken *off the energy shell.* Which value of ω one has to take for $G(\omega)$ in (16.12) depends therefore on the set of diagrams for which the ladders are to be replaced by a self-consistent potential.

An important simplification takes place if one adds to all the diagrams in (16.14) also diagrams of the type

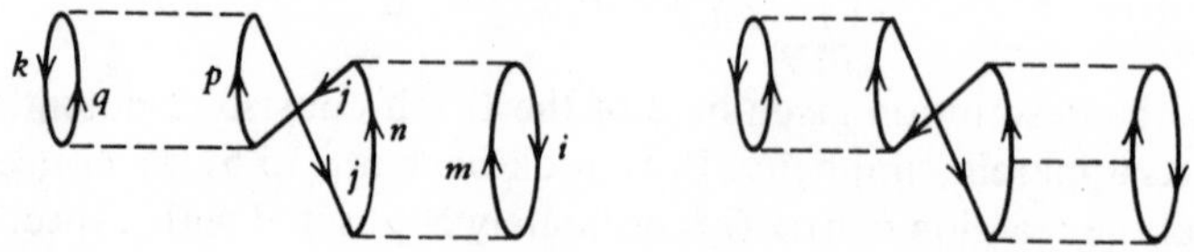

(16.17)

It was shown by Bethe, Brandow and Petschek (63) that the sum of *all* these diagrams can be replaced by a reaction matrix taken *on* the energy shell. Thus the sum of all the diagrams in (16.14) and (16.17) can be replaced by

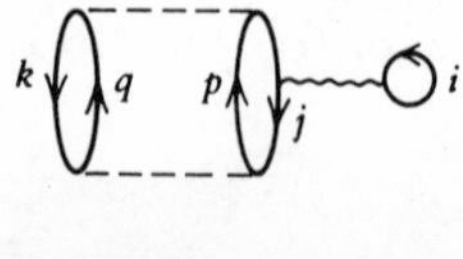

(16.18)

where the wiggly line indicates that a v has to be replaced by a G in evaluating this diagram, and that the G-matrix is to be evaluated on the energy shell.

17. BRUECKNER THEORY

The replacement of $v(ij)$ by a reaction matrix $G(ij)$ as defined in (15.7) is not only a convenient way for the partial summation of many diagrams. Actually it was first introduced by Brueckner (54) and (55) in order to be able to do calculations with singular nuclear potentials such as hard core potentials.

Simple perturbation theory as well as the Hartree–Fock method fail for singular potentials since the matrix elements of the nucleon–nucleon potential with respect to uncorrelated wave functions

$$\phi(ij) \sim \phi_\alpha(i)\phi_\beta(j)$$

are infinite. Even if the core of the nucleon–nucleon potential is not infinitely repulsive but comparatively large, perturbation theory and the Hartree–Fock method do not yield good first approximations.* The Brueckner method consists in replacing the potential by the reaction matrix G and the uncorrelated wave function $\phi(ij)$ by the correlated one $\psi(ij)$ according to (15.8)

$$v(ij)\psi(ij) = G(ij)\phi(ij) \tag{17.1}$$

A hard core potential is effective in scattering a pair of nucleons to high momentum states; what the transition from v to G does is to sum up all these scatterings for a given pair, and it thus provides us with an effective interaction that takes *these* scatterings into account. The ladder diagrams we have been considering involve a summation over a sequence of interactions between ascending particle lines; it is possible to use similar techniques to sum up descending ladder diagrams with a sequence of interactions between hole-lines. A hole line can take on any of the N occupied states labels i; a particle line can take on any of the unoccupied labels m. A sum over the ascending ladders will generally involve a more important correction to the energy than that of the descending ladders.

The theory of nuclear structure that uses the reaction matrix to sum up all short range interaction effects is known as the *Brueckner theory* [(Brueckner (54) and (55)]. It has found many important applications in the calculation of properties of both nuclear matter and finite nuclei. In combining this theory with the Goldstone diagrams, special care should be taken to account for the fact that $G(ij)$ is not really an interaction in the Hamiltonian sense. Thus, whereas it makes perfect sense to add up ladder diagrams like those shown in

---

*Actually, as we shall discuss later, the Hartee-Fock method has considerable difficulty with the tensor-force component of the nuclear force [Riihimaki (70)].

(15.1) when one deals with a real interaction, to do the same for the effective interaction $G(ij)$ would be wrong. A diagram like (17.2):

(17.2)

where the wiggly lines stand for matrix elements of $G(ij)$, *should not* be included in the theory; it just duplicates contributions of the real interaction $v(ij)$ that are already included in the diagram (17.3).

(17.3)

Thus the transition from $v(ij)$ to $G(ij)$ is accompanied also by a large reduction in the number of diagrams that have to be included for the evaluation of the energy of the A-particle systems. They consist of diagrams of the type shown in (17.4), in addition to (17.3).

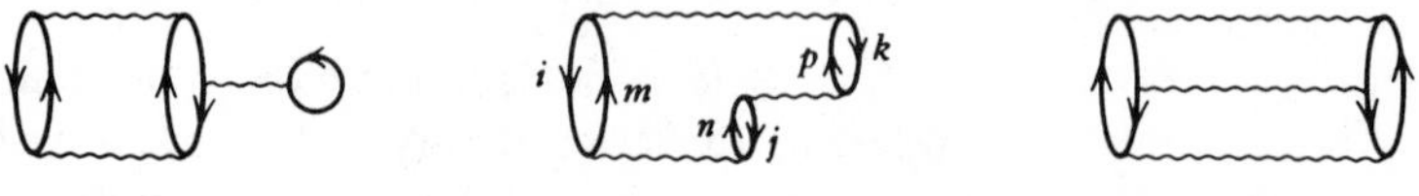

(17.4)

(a) (b) (c)

Note that (17.4) does include a ladderlike diagram connecting downgoing lines (diagram c). The reason for it becomes obvious if we note that the reaction matrix $G(ij)$ was defined to take care of all the ladder diagrams with v-type interactions between ascending lines only (see Eq. 15.1). Thus an example of a Goldstone diagram that is included in (17.4c) and is not included in other diagrams is given in (17.5)

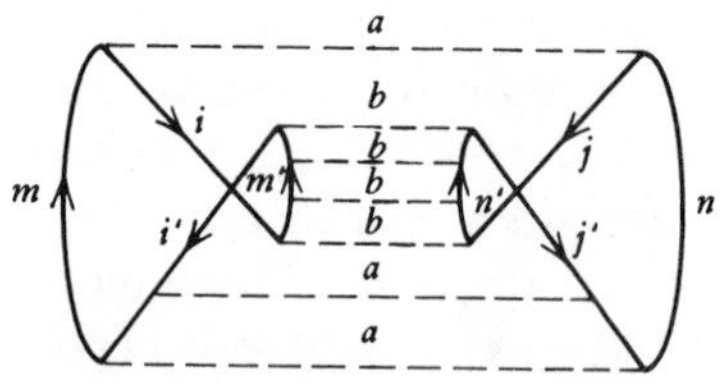

(17.5)

in which we see a mixture of descending (a) and ascending (b) ladders. It is a typical diagram that comes up when we are dealing with two particle two-hole and four-particle four-hole states.

Diagrams of the type (17.4b) generate three-particle correlations. In the first interaction two particles in level i and j are promoted to unfilled levels m and n. In the second interaction the j-level is filled again but a third particle is promoted from level k to level p. In the final interaction the particles in the m and p levels return to the i and k levels. It is clear that the contribution to such diagrams comes from regions in configuration space in which three particles are close together and hence their association with three-particle correlations.

Diagrams of the type (17.4a) can be taken care of by properly choosing the central potential (self-consistent field). This was discussed in the previous section, and it was also pointed out that, as was shown by Bethe, Brandow and Petschek (63), each G-matrix is to be calculated on the energy shell if one sums up ladders of all relative positions [for a more detailed discussion, see Day (67) and Baranger (67)].

In actual calculations the aim is of course to try and account for nuclear properties with an interaction $v(ij)$ derived from the two-nucleon problem. Since such interactions are strongly repulsive at short distances, it is most convenient to resort to the Brueckner–Goldstone theory that expresses the energy in terms of special linked diagrams with a G-matrix taken for the interaction.

However, the evaluation of a G-matrix from the interaction $v(ij)$ is generally a rather complicated task because of two factors: the projection operator Q_F is a nonlocal operator, when expressed in coordinate space, and the energy denominator may become rather complicated when realistic potentials U are introduced (see Eq. 16.7). Several approximations have been suggested to overcome these difficulties. Some of them have been utilized to carry out actual numerical calculations.

The approximations involved separating the low-lying and the high-lying excitations. The former are generated essentially by the long range parts of the nuclear interaction, the high-lying ones by the strong short range components. Because the latter states lie so high (the average excitation energy is a few hundred MeV) important simplifications can be made. First, it becomes a reasonable approximation to use noninteracting wave functions for the two particles in the excited state. Second, it is a good approximation to drop the Pauli projection operator Q_F. This is because the probability of exciting the low-lying states is relatively small for this short range force and it is these states for which Q_F is most important. A second approximation that is made is in the single-particle spectrum. Because we are dealing with only the high energy excitations, it is possible to represent the single-particle spectrum (and therefore the single-particle potential) by a simple formula. In the case of nuclear matter [Bethe, Brandow, and Petschek (63)] the single-particle spectrum is represented by a simple quadratic function. The comparison between this approximation and the actual spectrum is shown in Fig. 17.1 taken from Day (67). A corresponding approximation for finite nuclei is given by Kuo and Brown (66).

Remark: If the two particles have momenta $\hbar\mathbf{k}_1$ and $\hbar\mathbf{k}_2$ the Bethe, Brandow, Petschek (63) representation for the energy is:

$$\left[\frac{-\hbar^2}{2m}(\nabla_1^2+\nabla_2^2)+U(1)+U(2)\right]|\mathbf{k}_1,\mathbf{k}_2\rangle=\left[\frac{\hbar^2}{2m^*}(k_1^2+k_2^2)+2A_2\right]|\mathbf{k}_1,\mathbf{k}_2\rangle$$

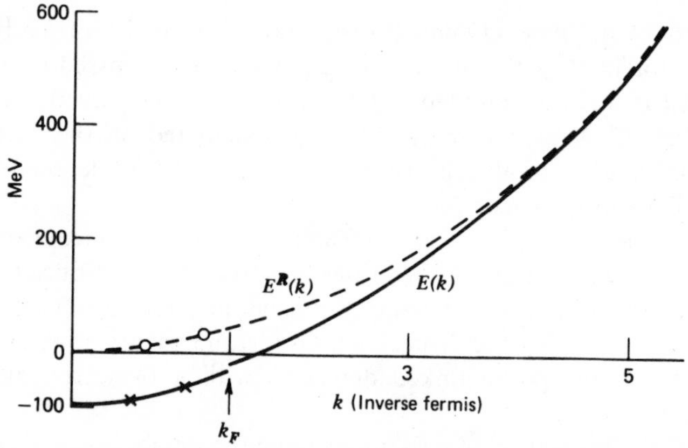

FIG. 17.1. The solid line curve gives a typical plot of the energy spectrum $E(k)$ used in nuclear matter. The dashed curve gives the reference spectrum that is chosen so as to approximate the actual spectrum for k between 3 fm^{-1} and 5 fm^{-1} [from Day (67)].

where m^* is an effective mass. Hence the single-particle energy and the corresponding one-body operator are

$$\epsilon(k) = \frac{\hbar^2}{2m^*} k^2 + A_2$$

and

$$h(1) = -\frac{\hbar^2}{2m^*} \nabla_1^2 + A_2$$

Choosing the single-particle potential involves a generalization of the method described in Section 16. In that case U was chosen so as to cancel out the diagrams (16.2) and (16.3). The generalization involves choosing U so that a wider class of three-particle diagrams can be cancelled. This involves a detailed consideration of the three-particle system. A discussion of this would involve too large a digression. The reader is referred to Bethe (71), Brown (67), and Rajaraman and Bethe (67), for a discussion.

Once the short range potential is treated, the effect of the long range components of the nucleon–nucleon force can perhaps be taken into account by perturbative methods. In the following we outline such a perturbation formalism and discuss one method [Moszkowski and Scott (60), (61)], for separating the short range and long range parts of the nuclear force.

To discuss the perturbative method let us introduce the notation:

$$P \equiv \frac{Q_F}{\epsilon_0 - H_0} \tag{17.6}$$

so that the G-matrix equation (15.7) can be written as

$$G = v + vPG \tag{17.7}$$

We define now a new set of operators $\bar{G}, \bar{P}$ that are obtained from an interaction $\bar{v}$ by a relation similar to (17.7):

$$\bar{G} = \bar{v} + \bar{v}\bar{P}\bar{G} \tag{17.8}$$

Equation 17.7 can be formally solved for v to give

$$v = G\frac{1}{1 + PG} = \frac{1}{1 + GP}G \tag{17.9}$$

where the second step in (17.9) can be most easily obtained from the first by expanding both inverses in powers of GP and PG, respectively. We have similarly

$$\bar{v} = \bar{G}\frac{1}{1 + \bar{P}\bar{G}} = \frac{1}{1 + \bar{G}\bar{P}}\bar{G} \tag{17.10}$$

By subtracting (17.10) from (17.9) we have

$$v - \bar{v} = G\frac{1}{1 + PG} - \frac{1}{1 + \bar{G}\bar{P}}\bar{G}$$

and hence

$$\begin{aligned}(1 + \bar{G}\bar{P})(v - \bar{v})(1 + PG) &= (1 + \bar{G}\bar{P})\,G - \bar{G}(1 + PG)\\ &= G - \bar{G} + \bar{G}(\bar{P} - P)G\end{aligned}$$

or

$$G = \bar{G} + (1 + \bar{G}\bar{P})(v - \bar{v})(1 + PG) + \bar{G}(P - \bar{P})G \tag{17.11}$$

If the operator $\bar{v}$ and $\bar{P}$ are chosen so that (17.8) is relatively easy to solve, then it is possible to calculate G using the solution $\bar{G}$ and (17.11). We notice that the "correction terms" in (17.11) are proportional to $v - \bar{v}$ and $P - \bar{P}$, and can thus be rather small if $\bar{v}$ and $\bar{P}$ are properly chosen. Equation 17.11 can then be solved by iteration, building up essentially a perturbation series in can $(v - \bar{v})$ and $(P - \bar{P})$.

This method for dealing with the Brueckner–Goldstone theory of nuclear structure has been used in combination with a method of Moszkowski and Scott (60) for decomposing the actual interaction v into a short range part v_s and a long range part v_l by writing

$$v(r) = v_s(r)\theta(d - r) + v_l(r)\theta(r - d) \tag{17.12}$$

where $\theta(x)$ is the step function

$$\theta(x) = \begin{cases} 1 & \text{if} \quad x > 0 \\ 0 & \text{if} \quad x > 0 \end{cases} \tag{17.13}$$

(see Fig. 17.2 (*a*) for v_s). In (17.12) the *separation distance d* is a parameter yet to be determined; the main requirement on d is that the resulting $v_s(r)$ will not affect nucleons in low-lying nuclear states. $v_s(r)$ will then be responsible only for the scattering to high-lying states and in *its* evaluation we can approximate the propagator $P = Q_F/(\epsilon_0 - H_0)$ by $\bar{P} = 1/(\epsilon_0 - \overline{H}_0)$ where $\overline{H}_0$ has been discussed earlier, as well as the replacement of Q_F by unity. In terms of the notation of (17.6) to (17.11), $\bar{v}$ becomes just $v_s(r)\theta(d - r)$ and $v - \bar{v}$, to be treated perturbatively becomes $v_l(r)\theta(r - d)$.

It turns out that in order to achieve the desired property for $v_s(r)$, the separation distance d should be chosen so as to have the hard core repulsive effects exactly balanced for the low momentum states by enough attraction from the potential outside the core as is shown in Fig. 17.2*b*. The remaining potential $v_l(r)$ is then weak and nonsingular. One can handle that part of the potential by simple perturbation theory without using Brueckner's theory. The projection operator Q_F is very important for $v_l(r)$.In fact, since the long range part of the interaction scatters pairs of particles to levels that lie near the Fermi surface, the Pauli principle, coming in through the projection operator Q_F, reduces the effects of $v_l(r)$ even further. The use of perturbation theory to handle that part of the interaction is thus especially justified, and the incorporation of Q_F presents no special problems.

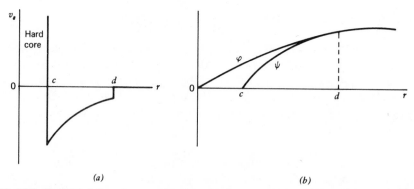

FIG. 17.2. Sketch of the short range potential v_s and corresponding wave function ψ compared to the noninteracting wave function ϕ. The separation distance is d [from Baranger (67)].

Applications of the Brueckner theory to the problem of nuclear matter have been described in Chapter III while its application to the theory of finite nuclei will be discussed in Sections 18, 19, 20 of this chapter. A recent review has been given by Bethe (71). His results, summarized in Chapter III indicate that we have achieved a qualitative understanding of the properties of nuclear matter. The excellent quantitative agreement must however be viewed with a healthy scepticism. We have alluded to the convergence questions associated with the evaluation of the linked-cluster expansions. The values of the first few terms may suggest, but, of course, do not prove convergence. Even if one were certain of its convergence there is the question of the best single-particle potential U to be used in H_0. Presumably there is a best U in the sense that it leads to the most rapid convergence. At even a more fundamental level it is not even known whether the various self-consistent calculations necessarily converge to the ground state [Calogero (70), Baker et. al. (65), (69), (70)]. What is missing of course is an appropriate variational principle.

Phase Equivalent Potentials

But in addition to these questions that, it is to be emphasized, are not entirely mathematical in nature, there is an uncertainty that originates in the uncertain nature of the nuclear force. As we mentioned in Chapter I, there are several phenomenological potentials that predict the properties of the deuteron and and the nucleon–nucleon scattering with equal accuracy. These potentials are not by any means identical (Lomon and Feshbach (68)] particularly at short range. In momentum space the corresponding t-matrices differ off the energy shell* although they agree on the energy shell. The question arises whether these differing potentials give substantially different results, say, for nuclear matter. (Identical results would suggest that the properties of nuclear matter depend only upon the phase shifts.) And, in fact, according to Haftel and Tabakin (70) variations up to 20 MeV per particle in the binding energy of nuclear matter can be obtained from differing potentials that fit the two-body data. However they do not give identical fits.

This problem has recently been attacked through the use of "phase-equiva-

*The scattering amplitude is proportional to $\langle \mathbf{k} | t(E) | \mathbf{k}' \rangle$ where $\mathbf{k}$ and $\mathbf{k}'$ are the incident and final relative momenta of the two-body system and the corresponding energies are equal to E. Under these circumstances the t-matrix is said to be on the energy shell. When $E(k) \neq E(k')$ and/or $E(k) \neq E$, the t-matrix is said to be off the energy shell. These matters will be discussed in more detail in Volume II.

lent" interactions [Ekstein (60), Baker (62), Mittelstaedt (65)]. If $\psi(\mathbf{r})$ is the scattering solution to a two-nucleon Schrödinger equation then $U\psi(\mathbf{r})$ will have the same asymptotic dependence if U is a unitary operator such that

$$\langle \mathbf{r}|U|\mathbf{r}'\rangle \underset{r\to\infty}{\longrightarrow} \delta(\mathbf{r}-\mathbf{r}') \tag{17.14}$$

The wave function

$$\bar{\psi} = U\psi \tag{17.15}$$

satisfies the equation

$$\bar{H}\bar{\psi} = E\bar{\psi} \tag{17.16}$$

where

$$\bar{H} = UHU^{\dagger} \qquad H = T + V \tag{17.17}$$

Writing

$$\bar{H} \equiv T + \bar{V} \tag{17.18}$$

where T is the kinetic energy operator, we see that $\bar{V}$ is a phase-equivalent potential, that is, a potential that gives the same asymptotic amplitude or phase shifts as V. $\bar{V}$ is of course a nonlocal potential. Thus by choosing various forms for U satisfying (17.14) a set of interactions that give identical phase shifts are generated. Employing these interactions in a nuclear matter calculation will permit a test of the sensitivity of the properties of nuclear matter to the nature of the interactions.*

Another procedure involves a change in the radial variable of the two-body system from r to $R(r)$ [see Coester, Cohen, Day, Vincent (70)], where

$$R(r) \to r \qquad \text{as} \qquad r \to \infty \tag{17.19}$$

If ψ satisfies the differential equation

$$\frac{d^2\psi}{dR^2} + \left[k^2 - \frac{l(l+1)}{R^2} - \frac{2m}{\hbar^2}V(R)\right]\psi = 0$$

Then

$$\phi = \sqrt{R'}\,\psi \tag{17.20}$$

*Note that the many-body Hamiltonian is written: $\bar{H}_N = \Sigma T_i + \Sigma_{i\neq j} V(ij)$. If a suitable generalization of U for the many-body system is used to transform $\Sigma T_i + \Sigma_{i\neq j} V(ij)$, the resultant interaction will contain not only the two-body potential $\bar{V}(ij)$ but also many-body potentials. Of course, if these are included there would be no change in the results for a many-body system from those obtained with $\Sigma T_i + \Sigma_{i\neq j} V_{ij}$. These many-body terms are not included in $\bar{H}_N$; the unitary transformation is used to obtain equivalence for two-body systems only.

satisfies

$$\phi'' + \left\{k^2 - \frac{l(l+1)}{R^2(r)} - \frac{2m}{\hbar^2} V(R(r)) + \frac{1}{R'^2}\left[\frac{1}{2}\frac{R'''}{R'} - \frac{7}{4}\left(\frac{R''}{R'}\right)^2\right] + \frac{1}{2}\left[\left(\frac{1}{R'^2} - 1\right)\frac{d^2}{dr^2} + \frac{d^2}{dr^2}\left(\frac{1}{R'^2} - 1\right)\right]\right\}\phi = 0 \quad (17.21)$$

From 17.19 and 17.20 we see that

$$\phi \to \psi \qquad \text{as} \qquad r \to \infty$$

so that the solution of (17.21) will have the same phase shift as ψ. The phase-equivalent potential can then be obtained from the coefficient of ϕ in (17.21). We see that this potential is l and momentum dependent, that is, nonlocal as expected.

Haftel and Tabakin (71) have investigated the dependence of the energy per particle of nuclear matter for a set of potentials that are phase equivalent to the Reid soft core potential that is used in Bethe's nuclear matter calculation. Haftel and Tabakin find variations of up to 9.5 MeV in the binding energy per particle and 0.33 fm^{-1} in the value of k_F. These should be compared with the empirical values of approximately 16 MeV and 1.36 fm^{-1} respectively. These results do not include the effects of third-order graphs but these authors estimate that these effects are small.

Other authors [e.g., Coester, Cohen, Day, and Vincent (70)] obtain even larger variations.

A similar result is obtained by Lomon (72) using the boundary condition model of nuclear forces [Lomon and Feshbach (68)]. This interaction accurately predicts the scattering, the bound state, and the electromagnetic interactions of the two-nucleon system. At medium and long range beyond about 0.7 fm the interaction is given by a potential derived perturbatively* from a field theoretical calculation involving the exchange of one and two pions, single η, ρ, and ω mesons by the nucleons. The particle masses and coupling constants agree with the experimental values where they are known and otherwise with theoretical expectations. The interaction inside 0.7 fm is represented by an independent-boundary condition for which some justification has been given [Feshbach and Lomon (64)]. The full statement of the boundary conditions will be given in Volume II. Here the nature of the boundary condition will be indicated for a single-channel case, as prevails for example for the two nucleons in a relative $^1S_0$ state. The wave function

*The perturbative results for these potentials can be improved [M. H. Partovi and Lomon (70) and F. Partovi and Lomon (72)].

then satisfies a one-dimensional Schrödinger equation for $r > r_0$ (=0.7 fm). At $r = r_0$, ψ satisfies the boundary condition

$$r_0\left(\frac{d\psi}{dr}\right) = f\psi(r_{0+}) \qquad \text{as} \qquad r \to r_{0+} \tag{17.22}$$

where $\psi(r_{0+})$ means $\psi(r)$ evaluated as r approaches r_0 from the $r > r_0$ side. The boundary condition parameters are fixed by comparison with experiment.

It is to be noted that no information about the interaction is needed for the spatial region $r < r_0$. However the results for nuclear matter will depend upon this interior interaction. Thus we can, in this model, obtain a set of phase-equivalent interactions simply by varying the interior interaction. Lomon has chosen to specify the interior interaction again by a boundary condition. Again for the relative $^1S_0$ state it has the form:

$$r_0\left(\frac{d\psi}{dr}\right) = b\psi \qquad \text{as} \qquad r \to r_{0-} \tag{17.23}$$

$$V(r) = 0 \qquad r < r_0$$

where now r_0 is approached from the $r < r_0$ side. This is the simplest calculational model. It can be readily generalized to include nonzero interior potentials and/or $b = f$.

With these assumptions the binding energy and k_F for nuclear matter can be calculated as a function of the b's. Changing their value does not change the fit of the boundary-condition model to the two-nucleon data. Lomon (72) can obtain variations in the binding energy as large as 30 MeV per particle. Arguments are presented to indicate that three- and higher-body graphs will not change this result appreciably.

These results indicate that the calculated properties of nuclear matter are quite sensitive to the properties of the two-nucleon interaction. It is thus not possible to predict the properties of nuclear matter from only the on the energy-shell nucleon–nucleon data. Rather it may be possible to choose among the various proposed potentials by requiring it to fit not only the two-nucleon data but the properties of nuclear matter as well. That this is possible is indicated by the fact that Lomon can find a set of b's that give the empirical binding energy and saturation density for nuclear matter.

18. APPLICATIONS; THE MANY-BODY THEORY FOR FINITE NUCLEI

The general methods developed in this chapter have been applied to the study of finite nuclei, the resultant developments being referred to as the *microscopic* theory of finite nuclei. This has been an area of intense activity and it will not be possible to describe these developments in detail. However we hope to give the flavor of these attempts by describing a few examples.

Under attack is the most fundamental problem in the study of nuclear structure. The central problem in dealing with bound states of nuclei is to determine the energy levels and corresponding wave functions of the nuclear Hamiltonian

$$H = \sum T_i + \sum_{i<j} v_{ij} \tag{18.1}$$

The solution of this problem will reveal the characteristic modes of motion of nuclei and the manner in which they develop under the action of the nucleon–nucleon forces. These forces, to the extent to which they are known from the two-body systems, are very complex. Although they are on the average only of moderate strength, there are regions, for example, small interparticle distances, in which their interaction strength is very large. In some phenomenological potentials the nuclear potential is represented by an infinitely strong hard core. The nucleon–nucleon potential even in the intermediate range is still of considerable strength but here one has the additional complication of a strong tensor force. It is the effect of these tensor forces that has been a principal stumbling block in the development of a microscopic theory of nuclear structure.

In order to be able to treat this complex force, new methods, some of which have been described earlier in this chapter, have been devised. Approximations that have proven useful for the understanding of many-body problems in other fields of physics have been adapted. These adaptations are by no means trivial, on the one hand, because of the nature of nuclear forces and, on the other hand, because nuclei are composed of only a relatively few particles. As a compensation for these difficulties, nuclear systems exhibit a rich variety of phenomena and form a fertile field for the investigation of many-body quantal effects and their theoretical explication.

Two general methods for the treatment of the very strong short range interaction potential have been tried. In one the attempt is made to replace the hard core by a sufficiently soft core and still match the nucleon–nucleon data. This possibility exists because the evidence for the hard core is based upon the energy dependence of the nucleon–nucleon phase shifts; particularly that of the $^1S_0$ phase shift that goes through a zero at about 200 MeV laboratory energy. The hard core is the simplest way to describe this observation. But it is also possible to obtain it by shaping the short range potential appropriately [Bressel, Kerman, and Rouben (69)] or through the use of a combination of short range local and nonlocal potentials, Riihimaki (70). Once such a soft core (sometimes referred to as a "*supersoft*" core) is obtained the Hartree–Fock method might be applicable. It should be realized that the physical situation that would prevail in nuclei if this approach were to prove satisfactory is, of course, strikingly different from that which would prevail if the nuclear potential had a hard core. In the latter case it is necessary to at least introduce two-particle correlations, correlations that are generated by the G-matrix (or have been described in the independent pair approximation in Chapter III). If the supersoft potential description is valid and the Hartree–Fock approximation can be used, a Slater determinant will suffice in that approximation to describe the nuclear wave function. In that event the only correlations present in the wave functions are those arising from the Pauli exclusion principle. Correlations then would be introduced through corrections to the Hartree–Fock wave function.

The Hartree-Fock Method

We have discussed the theory of the Hartree–Fock method several times in this text. Our aim this time is to describe its application to the problem of the finite nucleus. We remind the reader of some of the fundamental formulas. The nuclear Hamiltonian is given by (18.1). The Hartree–Fock method attempts a variational solution of the corresponding Schrödinger equation, the trial function Φ being written as a single Slater determinant of single-particle wave functions:

$$\Phi = \frac{1}{\sqrt{A!}} \begin{vmatrix} \phi_{\alpha 1}(1) & \phi_{\alpha 2}(1) & \cdots & \phi_{\alpha A}(1) \\ \phi_{\alpha 1}(2) & \phi_{\alpha 2}(2) & \cdots & \phi_{\alpha A}(2) \\ \cdot & \cdot & & \cdot \\ \cdot & \cdot & & \cdot \\ \cdot & \cdot & & \cdot \\ \cdot & \cdot & & \cdot \\ \phi_{\alpha 1}(A) & \phi_{\alpha 2}(A) & & \phi_{\alpha A}(A) \end{vmatrix} \tag{18.2}$$

The nonlinear equations determining ϕ_α are obtained by varying the expression for the energy

$$E = \langle \Phi | H | \Phi \rangle / \langle \Phi | \Phi \rangle \tag{18.3}$$

with respect to ϕ_α^*. These can then be solved by an iterative procedure referred to as the self-consistent method. In terms of the single-particle states $|\alpha\rangle = \phi_\alpha$

$$E = \sum_\alpha \langle \alpha | t | \alpha \rangle + \frac{1}{2} \sum_{\alpha\beta} \langle \alpha\beta | v | [\alpha\beta] \rangle \tag{18.4}$$

where

$$|\alpha\beta\rangle = |\alpha\rangle|\beta\rangle \qquad \text{and} \qquad |[\alpha\beta]\rangle = |\alpha\beta\rangle - |\beta\alpha\rangle \tag{18.5}$$

It is assumed here that Φ is normalized to unity. Many authors have adopted a somewhat different scheme for the determination of the single-particle wave functions. Instead of determining the equations for ϕ_α they have expanded ϕ_α in a set of a known orthonormal set:

$$\phi_\alpha = \sum C_n{}^\alpha \chi_n \tag{18.6}$$

The energy in terms of the C's is:

$$E = \sum_{\alpha, nn'} (C_n{}^\alpha)^* C_{n'}{}^\alpha \langle n|t|n'\rangle + \frac{1}{2} \sum_{\alpha\beta, nn'qq'} (C_n{}^\alpha)^* (C_q{}^\beta)^* C_{n'}{}^\alpha C_{q'}{}^\beta \langle nq|v|[n'q']\rangle \tag{18.7}$$

The unknowns are now the coefficients $C_n{}^\alpha$ so that the variational principle becomes

$$\frac{\partial}{\partial C^*} \{E - \sum_a [\sum_n (C_n{}^\alpha)^* C_n{}^\alpha - 1]\epsilon_\alpha\} = 0 \tag{18.8}$$

where C is one member of the set $C_n{}^\alpha$. The constants ϵ_α are Lagrange multipliers and the solutions must satisfy the condition

$$\sum_n (C_n{}^\alpha)^* C_n{}^\alpha = 1 \qquad \text{for all } \alpha$$

Inserting (18.7) into (18.8) yields

$$\sum_{n'} C_{n'}{}^\alpha [\langle n|t|n'\rangle + \sum_{\beta q q'} (C_q{}^\beta)^* C_{q'}{}^\beta \langle nq|v|[n'q']\rangle] = \epsilon_\alpha C_n{}^\alpha \tag{18.9}$$

The matrix elements of the Hartree–Fock Hamiltonian, $\langle n|h|n'\rangle$ are placed equal to the square bracket in (18.9) so that (18.9) becomes

$$\sum_{n'} C_{n'}{}^\alpha \langle n|h|n'\rangle = \epsilon_\alpha C_n{}^\alpha \tag{18.10}$$

so that the Lagrange multipliers are just the Hartree–Fock single-particle energies. In terms of ϵ_α, (18.7) becomes

$$E = \frac{1}{2} \sum \epsilon_\alpha + \frac{1}{2} \sum C_n{}^{\alpha *} C_{n'}{}^\alpha \langle n|t|n'\rangle \tag{18.11}$$

The solution of (18.9) involves choosing an appropriate set χ_n. It must be such that the series for ϕ_α (18.6) converges rapidly—indeed so that one can cut it off at $n = N$ where N is not too large. The convergence can be tested by observing the consequences of changing N. Once χ_n is chosen, the solution proceeds by iteration. One guesses a set of values for $C_n{}^\alpha$ from which $\langle n|h|n'\rangle$ is determined. Equation 18.10 is then solved for a new set of $C_n{}^\alpha$ from which a new $\langle n|h|n'\rangle$ is calculated. This procedure is continued until a self-consistent solution is obtained.

Calculations of this type have been performed using spherical harmonic-oscillator wave functions for χ_n. The "soft" nucleon–nucleon potentials have included the Tabakin (64) potential, the Feshbach–Lomon (68) interaction, as well as potentials specially designed for this purpose by Bressel, Kerman, and Rouben (69) and by Riihimaki (70).

The use of the harmonic oscillator set for χ_n has been tested by Kerman, Svenne, and Villars (66), by Davies, Krieger, and Baranger (66) and most recently by Bassichis, Pohl, and Kerman (68). Two parameters are involved, the cutoff N to the number of terms in the series (5) and the oscillator parameter $\gamma(\equiv\hbar/M\omega)$. In principle, that is, if N is infinite, the results are exact and independent of γ. This feature is illustrated by Fig. 18.1, which shows that the Hartree–Fock energy for ^{16}O using the Tabakin potential becomes insensitive to γ as N increases. Moreover the sensitivity of E_0 to the value of N is not great. By increasing N from 1 to 3 yields an increase in $|E_0|$ of about 15% while increasing it from 3 to 4 yields no appreciable change. We also note that a great improvement in E_0 is obtained in going from $N = 1$ to $N = 3$, whereas the transition from $N = 1$ to $N = 2$ does not lead to big changes in E_0. This is due to the fact that neighboring shells in the harmonic-oscillator potential have alternating parities, and $N = 1$ wave functions can be improved by adding states from the next shell only if two particles are moved simul-

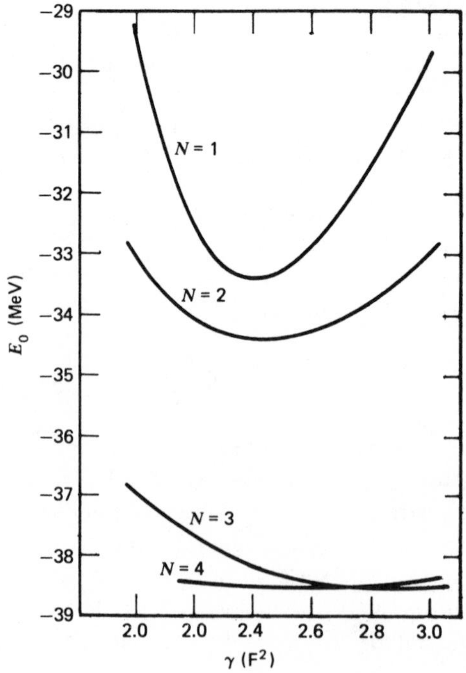

FIG. 18.1. Behavior of the Hartree–Fock energy with γ and N for ^{16}O [from Kerman, Svenne and Villars (66)].

taneously from one shell to the other. Similar results are reported by Davies et al (66) in that the single-particle wave function for occupied states for $N = 1$ and $N = 5$ had an average overlap of 99.9%. A more extensive test on N has been made by Bassichis et al. (68), who tested a variety of nuclei including deformed nuclei. In the original calculation of Kerman et al (66) the space had been truncated at the $2p1f$ shell. In the new calculation single-particle orbitals up to $1i_{13/2}$ were included in the complete set χ_n. A constant value of γ was used. The results, Fig. 18.2, show that the binding energy is insensitive to this extension of the set χ_n. However this is not true for the quadrupole moments, Fig. 18.3, where the ratio of the calculated Q for the large space is compared to that for the small space. It is found that comparatively small admixtures (a few percent) from the higher-lying high spin orbitals had very large effect on Q.

Hartree–Fock calculations cannot be compared directly with experiment because of the important effect of tensor forces. These give no contribution to the Hartree–Fock potential. Thus a correction to the Hartree–Fock energy must be calculated. Bassichis, Kerman, and Svenne (67) have made a second-order estimate of this effect, while Kerman and Pal (67) have done a more

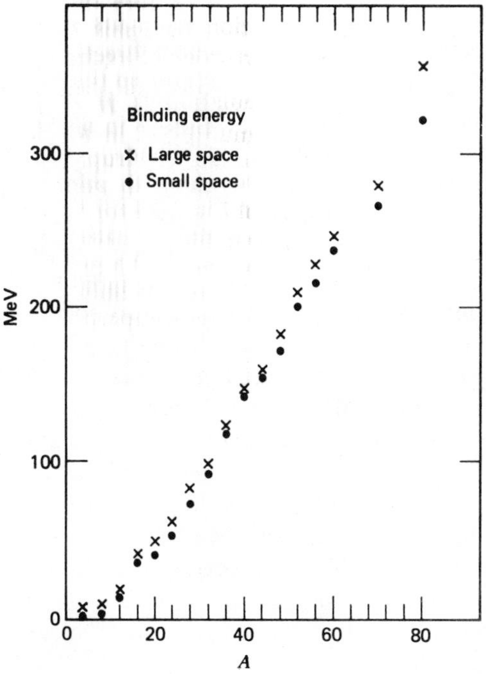

FIG. 18.2. The Hartree-Fock energies as computed in small (●) and large (X) spaces [from Bassichis et al (68)].

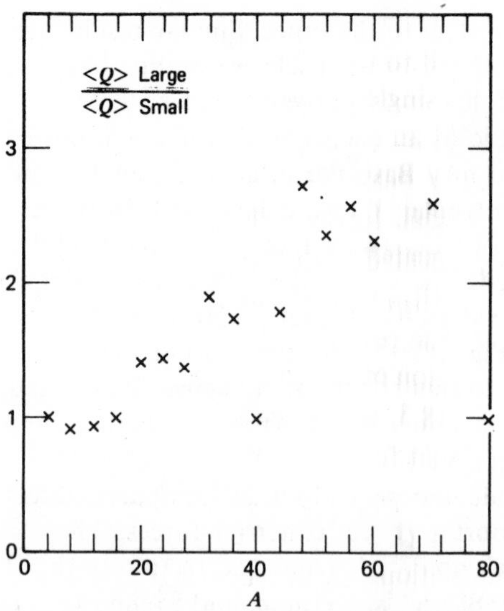

FIG. 18.3. The ratio of the quadrupole moment as calculated in the large space to that calculated in the small space [From Bassichis et al. (68)].

accurate calculation, using approximations that are similar to those employed by Kuo and Brown (66) discussed later in this chapter.

The Hartree–Fock method will for closed subshell nuclei automatically give a spherical undeformed nucleus. To explore the manner in which the nuclear energy changes with deformation Bassichis (67) et al introduced a forcing term in the Hamiltonian that provides a direction in space. The forcing term is $\mu\hat{Q}$ where $\hat{Q}$ is the quadrupole operator so that the Hamiltonian used in the *constrained* Hartree–Fock calculation is $H + \mu\hat{Q}$. This is an application of the method of Lagrange multipliers in which one would require a specific expectation value for the nuclear quadrupole moment. By varying μ one will generate nuclear wave functions with differing values of Q. An example of these calculations is given in Fig. 18.4 for ^{12}C. Although the energy has a stationary value at $Q = 0$, a deep minimum exists for an oblate deformation ($Q < 0$) and another not quite as deep with a prolate deformation. Both minima lie below the $Q = 0$ value. These results indicate that the ^{12}C nucleus in its ground state is oblate in agreement with experiment. The wave function developed in this way is no longer an eigenstate of J^2 (it has a good J_z value but because of the term $\mu\hat{Q}$ will be a superposition of different J-states). Projection to an eigenstate of J^2 was not made, rather a "naive" semiempirical correction was evaluated, the true energy $E_{J=0}$ was obtained from the Hartree–Fock energy E_{HF} as follows:

$$E_{J=0} = E_{HF} - \frac{\hbar^2}{2\mathcal{J}} \langle J^2 \rangle \tag{18.12}$$

The empirical value of $\mathcal{J}$ rather than a calculated one was used.

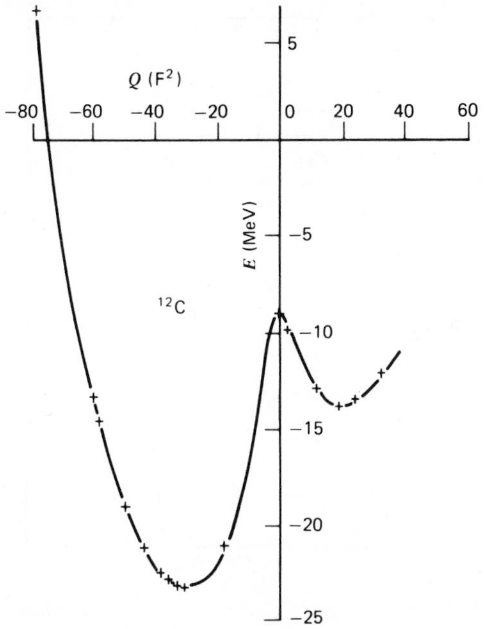

FIG. 18.4. The Hartree–Fock energy of ^{12}C as a function of the deformation. A similar behavior is obtained for all closed subshell nuclei [from Bassichis et al. (67)].

The results are shown in Fig. 18.5 and Table 18.1. The resulting energies including the second-order correction were found to be in good agreement with "experiment." Note that in making this comparison, the Coulomb energy was subtracted from the experimental value for the binding energies (B.E.). It is this difference that is labeled "experiment" in Fig. 18.5 and Table 18.1. However the value for the nuclear radius obtained this way [see also Bassichis et al. (68)] are much smaller than the experimental value as determined from electron scattering, indicating that the predicted nuclear densities are much greater than the experimental values. The large size of the second-order contribution was thought not to be an indication of a lack of convergence of the perturbation theory since the second-order term was small compared to the first-order potential energy.

Of course the Tabakin potential is not realistic and one can only regard the above calculations as indicative. [For other examples of Hartree–Fock calculations see Nester, Davies, Krieger, and Baranger (68), Nemeth and Vautherin (70) as well as the earlier work of Kelson and Levinson (64).] More realistic potentials have been used by Shao, Bassichis, and Lomon (72) and Riihimaki (70). The first of these use the Feshbach–Lomon potential that has the merit of being based on the boson-exchange theory of nuclear forces,

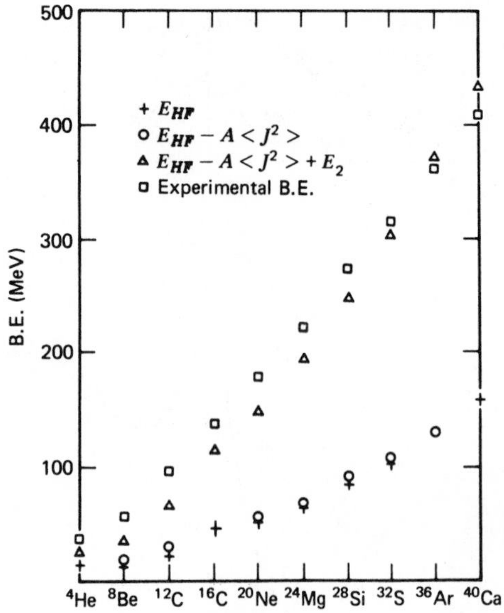

FIG. 18.5. The Hartree–Fock energy (+), with angular momentum correction (○) and second-order correction (△) compared to observed energies (after Coulomb subtraction) [from Bassichis et al. (67)].

for $r > r_c$. Instead of describing the interaction for $r > r_c$, it is parametrized by giving boundary conditions on the two-particle wave function at r_c^+ (where $r_c^+ \equiv r_c + 0^+$). Another set of conditions at $r_c^{(-)}$ is fixed by the properties of nuclear matter. The agreement with two-particle data is excellent. In spite of the singular nature of this interaction, Hartree–Fock calculations can be performed. The methods used are essentially those of Bassichis et al (67). The results are very encouraging for not only are binding energies similar in quality to those obtained with the Tabakin potential but the nuclear radii are now much more in agreement with experiment as shown in Table 18.2. However it should be remembered that these radii are calculated without the modifications in the wave functions that are introduced by the second-order corrections discussed earlier. Be that as it may the success of these results indicates that the Hartree–Fock wave functions are a suitable starting point for a Brueckner, Hartree–Fock calculation.* The results obtained with this approach are discussed in the next sections.

Riihimaki's thesis (70) proceeds by first constructing a "soft" phenomenological nucleon–nucleon potential. For $r > r_c$ the potential is taken to be that of Hamada and Johnston (62). For $r < r_c$ it is a mix of local and nonlocal

---

* The Brueckner, Hartree-Fock method is described below, p. 612.

TABLE 18.1 The Hartee-Fock Energy with Second-Order Corrections for the Tabakin Potential[a]

Nucleus	ΔE_2	$E^{J=0}$	E_{tot}	E_{obs}	E_{tot}/A	E_{obs}/A
^{4}He	− 7.04	− 12.78	− 19.82	− 28.5	− 4.96	− 7.13
^{8}Be	− 17.57	− 17.57	− 35.14	− 59.0	− 4.29	− 7.38
^{12}C	− 37.57	− 29.70	− 67.27	− 98.6	− 5.61	− 8.21
^{16}O	− 68.46	− 46.99	−115.45	−139.5	− 7.22	− 8.72
^{20}Ne	− 94.43	− 56.05	−150.48	−179.2	− 7.52	− 8.96
^{24}Mg	−126.47	− 68.31	−694.78	−224.7	− 8.12	− 9.36
^{28}Si	−159.11	− 91.23	−250.34	−272.0	− 8.94	− 9.71
^{32}S	−197.33	−109.27	−306.60	−317.3	− 9.58	− 9.92
^{36}Ar	−239.05	−131.92	−370.97	−363.4	−10.35	−10.10
^{40}Ca	−282.99	−154.70	−437.69	−410.8	−10.94	−10.27

[a] $E^{J=0}$ ≡ Hartree-Fock energy corrected for rotational energy; ΔE_2 = second-order contributions; E_{obs} = experimental energy with Coulomb energy subtraction. Energies are in MeV [from Bassichis et al (67)].

potentials. Non-local potentials were used only for the $^1S_0$ and $^3S_1 + {}^3D_1$ two-nucleon states. A different r_c was used for each spin and isospin state and for each component (central, tensor, etc.) of the interaction, but generally they were all close to one pion Compton wavelength. Moderately good agreement, $\chi^2 \backsim 4$ for p–p scattering and 2.1 for np data, was obtained.

In the present context perhaps the most significant result of Riihimaki's work, a result that is corroborated by the calculations of Bassichis and Strayer (72) that examine third-order contributions is that the perturbation series (using in the case of the latter authors, a set of Hartree–Fock wave functions) is not converging rapidly. Moreover the origin of this problem is not in the short range region but in the intermediate range $r \backsim 1$ fm where the tensor force seems to be the main culprit. As we shall see the tensor force causes convergence difficulties even for the Brueckner–Hartree–Fock method.

TABLE 18.2 Hartee-Fock Binding Energy and Root-Mean Square Radius for Feshbach-Lomon Potential.[a]

Nucleus	Binding Energy (MeV)	Binding Energy (MeV) (Exp.; Coulomb Energy subtracted)	R_{rms} (fm)	R_{exp}^{rms} (fm)
^{16}O	−132.7	−140.0	2.70	2.73
^{40}Ca	−338.4	−410.—	3.58	3.50
^{48}Ca	−373.2	−481.—	3.7	3.49

[a] Taken from Shao, Bassichis, Lomon (72).

We leave the Hartree–Fock method at this point. The need for substantial higher-order corrections suggests that the simple determinantal wave function is too naive a starting point; that correlations need to be built in from the beginning. The Brueckner–Hartree–Fock method that we are about to discuss inserts short range correlations. But as mentioned above this does not seem to suffice. Perhaps long range correlations such as those connected with collective states, for example, vibrations, need to be included as well.

Another variation on the conventional Hartree–Fock approximation is to perform first a Bogoljubov transformation on the Hamiltonian (18.1), thus taking into account in a better way the strong pairing part of the interaction $v(ij)$, and then to carry out the Hartree–Fock prescription in the space of the quasi particles with the remaining part of the interaction. Since, however, the Bogoljubov transformation starts from particles in a central field, the self-consistency problem becomes somewhat more complicated for this Hartree–Fock–Bogoljubov approximation. [See Baranger and Kumar (68), where the time-dependent Hartree–Bogoljubov approximation theory is developed for pairing-plus-quadrupole interaction.]

The reader may find detailed descriptions of the various approximations, as well as references to the extensive work that has been done using the Hartree–Fock approach, in review papers by Ripka (68), by Kerman (69), and in the earlier work of Villars (66).

Brueckner-Hartree-Fock

The second procedure involves the use of the G-matrix, adapting the methods developed for nuclear matter (see Section 16 and 17) to finite nuclei. In one such method, the solutions of the G-matrix equation

$$G = v + v \frac{Q_F}{\epsilon_0 - H_0} G \tag{18.13}$$

are used in the expression (see Eq. III.4.11) for the energy

$$E = \sum_i \langle i|T|i\rangle + \frac{1}{2}\sum \langle ij|G(\epsilon_i + \epsilon_j)|ij\rangle \tag{18.14}$$

Antisymmetrized G's are used in these and the following expressions. As in the case of nuclear matter, the critical elements in this procedure are the choice of ϵ_0 and the single-particle U appearing in H_0. As indicated in (18.3) the value of ϵ_0 that is best employed is on the energy shell [Baranger (69)], the same prescription as that given by Bethe, Brandow, and Petschek (63). The same discussion leads to a definition of U that is fixed so as to cancel important second-order G-matrix diagrams. Baranger (69) finds

$$\langle j|U|i\rangle = \frac{1}{2}\sum \langle jk|G(\epsilon_j + \epsilon_k) + G(\epsilon_i + \epsilon_k)|ik\rangle$$

$$\langle j|U|m\rangle = \sum_k \langle jk|G(\epsilon_j + \epsilon_k)|mk\rangle$$

$$\langle m|U|i\rangle = \sum_k \langle mk|G(\epsilon_k + \epsilon_i)|ik\rangle \tag{18.15}$$

where $|m\rangle$ is an unoccupied state.

To complete the definition of U the matrix elements between particles in states that are originally unoccupied need to be determined. As in the case of nuclear matter these are chosen in order to cancel a suitable set of third-order graphs. This is still a controversial matter. The reader is referred to Davies and McCarthy (71) for a recent discussion and additional references. Bethe and Rajaraman (67) and Negele (70) place $U = 0$ for these excited state matrix elements.

Once the relation between U and G is fixed a self-consistent procedure becomes possible. Starting from a "reasonable" set of single-particle wave functions and the corresponding energies one can compute G from the Bethe–Goldstone equation (18.13). These energies are needed for both the energy ϵ_0 as well as for the energies ϵ_m of the intermediate states. Once G is determined U can be determined from (18.14) and (18.15). New single-particle energies are then obtained by finding the eigenvalues of $H_0 = T + U$. These new energies, and wave functions form the start of a new cycle in which G and then U are once more determined.

This iterative scheme should be compared with the ordinary Hartree–Fock in which the single-particle potential is obtained from the potential directly. In the *Brueckner Hartree-Fock* (BHF) method, as it is called, the G-matrix takes the place of the potential. Although in the Hartree–Fock method the potential is fixed, in the BHF method the effective potential is computed self-consistently.

Several authors have made direct attacks on this difficult problem. These are based upon the use of harmonic-oscillator wave functions to describe the solutions of H_0 in (18.13). This has the advantage that the two-particle wave function can be written either as a product of single-particle wave functions or as a product of a wave function depending only on the center of mass $\mathbf{R}$ of the two particles and a wave function depending only upon their relative coordinate $\mathbf{r}$. This Talmi transformation (52) has been discussed earlier (Section 1.3). If the single-particle wave functions are functions of $\mathbf{r}_1$ and $\mathbf{r}_2$, respectively, with principle quantum numbers n_1 and n_2 and orbital angular momenta l_1 and l_2, and these are coupled together to give a total orbital angular momentum Λ, the resultant wave function can be expressed in terms of the two-particle center-of-mass wave function with quantum numbers N and L, and the relative co-ordinate with quantum numbers n and l as follows:

$$|n_1l_1, n_2l_2, \Lambda\rangle = \sum_{NL,nl} |NL, nl, \Lambda\rangle\langle n_1l_1, n_2l_2, \Lambda|NL, nl, \Lambda\rangle \tag{18.16}$$

where the brackets $\langle n_1l_1, n_2l_2, \Lambda|NL, l, \Lambda\rangle$ have been tabulated by Brody and Moshinsky (67). Once this separation has been made it is then proposed to modify the relative coordinate wave function $\phi_{nl}(\mathbf{r})$ by means of the Bethe–Goldstone equation (18.13). There are a number of approximations and guesses that are usually made. [For a review of this subject, see Baranger (69), Bethe (71), and Davies and McCarthy (71).] First the solutions of H_0 are not harmonic-oscillator wave functions. However this does not seem to be critical since it appears that the solutions can be approximately represented by a linear combination of such functions, and in light nuclei even one term will suffice [Kerman, Svenne, and Villars (66) and Davies, Krieger, and Baranger (66)]. Second and of somewhat more critical importance is the question of the proper approximation for the projection operator Q_F. To illustrate the difficulties involved, suppose that the harmonic-oscillator wave functions are good descriptions of the single-particle wave functions. Then

$$Q_F = 1 - \sum |n_1l_1, n_2l_2, \Lambda\rangle\langle n_1l_1, n_2l_2, \Lambda| \tag{18.17}$$

where the sum goes over all states for which *either* oscillator state (n_1l_1) or (n_2l_2) is occupied. This operator, when expressed in terms of relative and center-of-mass wave functions, will not only be quite complicated but will also thoroughly complicate the Bethe–Goldstone equation. More explicitly

$$|n_1l_1, n_2l_2, \Lambda\rangle\langle n_1l_1, n_2l_2, \Lambda| = \sum_{NL,N'L',nl,n'l'} |NL, nl, \Lambda\rangle\langle N'L', n'l', \Lambda|$$

$$\cdot \langle n_1l_1, n_2l_2, \Lambda|NL, nl, \Lambda\rangle\langle n_1l_1, n_2l_2, \Lambda|N'L', n'l', \Lambda\rangle \tag{18.18}$$

We see that in the center-of-mass relative coordinate system Q_F is no longer diagonal. This means that the equations for the relative coordinate wave functions are coupled so that many coupled equations would need to be solved in a rigorous treatment. A similar problem occurs in the nuclear-matter problem where it is "solved" by averaging over the momentum of the center of mass. We shall not go into the necessarily very detailed discussion involved in the various choices made for the approximation for finite nuclei. Baranger (69) and Bethe (71) discuss them thoroughly and give the references to the original papers [Eden and Emery (58), Eden, Emery, and Sampanthar (59), Becker, Mackellar, and Morris (68), Kallio and Day (69), McCarthy (69), Wong (67a), (67b), and Kohler and McCarthy (67a), (67b)]. The investigations by Baranger and his collaborators use the method of Wong and of McCarthy.

A final problem concerns the choice of the particle spectrum for the two-particle unperturbed Hamiltonian that appears in the Bethe–Goldstone equation once the center-of-mass effects are taken into account. Although the wave functions are well represented by the harmonic-oscillator wave functions, the energies are not those of a harmonic oscillator. Davies and McCarthy (71) attack this problem by parametrizing the spectrum. A solution is said to have been obtained to the extent that the final result is insensitive to these parameters. The point here is that the final solution does not depend upon the choice of the intermediate energy spectrum. Their calculations, that are the most recent of this genre, give good agreement with the experimental binding energies but do not fit the charge radii at all well. The charge density does not have the flat region that is needed for agreement with the electron-scattering experiments and drops off much too precipitously from its central value. In addition, tremendous sensitivity to the intermediate energy spectrum was found. Davies and McCarthy found that the binding energy per particle could be shifted by as much as 7 or 8 MeV, although their preferred solutions showed a somewhat smaller sensitivity of about 1.5 MeV in ^{16}O and ^{40}Ca. Corresponding large changes in the charge radius were obtained. In ^{40}Ca, for example, the binding energy per particle can in the extreme be changed from 2.3 to 10.7 MeV while the charge radius decreases from 3.33 fm to 2.85 fm.

The discussion just given describes methods in which the intermediate states are also harmonic-oscillator functions. The intermediate states have also been described by plane waves [Wong (67a), (67b)]. The separation method of Moskowski and Scott (60, 61) as applied to finite nuclei has been discussed by Baranger (69). But little calculation seems to have been done using this procedure.

Investigations that make greater use of the results of nuclear-matter calculations have been recently made by Shao and Lomon (72), by Negele (70), and Sprung and Banerjee (71). The latter use a local density approximation that will be described below. Each of these attempts exploit the fact that the healing distance is small compared with nuclear dimensions except for the lightest nuclei. Hence the wave function for a pair of nucleons both of which are inside the nuclear interior should be very similar to the wave functions for nuclear matter at the same density. There is one important difference. For nuclear matter the two-particle wave function heals to a plane wave solution. In finite nuclei this is of course not true. Shao and Lomon assume that they heal to the Hartree–Fock solutions obtained with the identical two-particle interaction. They employ the FL boundary-condition interaction [Lomon and Feshbach (68)] that has the merit of not only providing a good fit to the two-nucleon data but also by virtue of the ability to easily adjust the off-energy shell behavior or the potential, to obtain a good description of nuclear matter [Lomon (72)]. The Hartree–Fock method has been applied to the FL interaction [Shao, Bassichis, and Lomon (72)], and the results obtained are in somewhat better agreement with the experimental data than Hartree–Fock calculations are with other interactions (see discussion p. 610).

Shao and Lomon then require that the two-particle solutions to the Bethe–Goldstone equation heal to the Hartree–Fock two-particle wave functions of the same energy. More precisely the latter are expressed in terms of a center-of-mass wave function and a relative coordinate wave function. The Bethe–Goldstone wave function is required to heal to the relative coordinate wave function. The following important result is obtained. The nuclear-matter wave functions and the Hartree–Fock relative coordinate wave functions are identical over a range 1 fm $\leq r \leq$ 1.5 fm (see Fig. 18.6). It is thus possible to join the two smoothly, with very little dependence on exactly where the join is made. It now becomes possible to write the two-particle wave function in the deShalit–Weisskopf form (see Chapter III):

$$\psi_{\alpha\beta}(i, j) = \phi_\alpha(i)\phi_\beta(j)[1 + f_{\alpha\beta}(i, j)] \tag{18.19}$$

where α and β denote the quantum numbers and i and j the coordinates. From (18.19) the deShalit–Weisskopf wave function can be formed:

$$\Psi = \frac{1}{N} \det |\phi_{\alpha_1}(1) \ldots \phi_{\alpha_A}(A)(1 + \sum_{i<j} f_{\alpha_i\alpha_j})| \tag{18.20}$$

where N is the normalization. The energy of the nucleus is then evaluated with this wave function. This may be regarded as a use of the variational principle for the energy. This represents an important advantage of the Shao–Lomon calculation over calculations that employ effective Hamiltonians such as the local-density approximation to be discussed below or the harmonic-oscillator method discussed above where the expression for the energy is based on the inclusion of a subset of graphs. The results obtained by Shao and Lomon are encouraging as indicated in Table 18.3.

TABLE 18.3 Shao-Lomon Binding Energy for Finite Nuclei Including Effects of Bethe-Goldstone Correlations. [from Shao and Lomon (72)]

Nucleus	^{16}O	^{40}Ca	^{48}Ca
Hartree–Fock Energy	− 86 MeV	−224.5 MeV	−269.8 MeV
Correlation Energy	− 39.4	−176.9	−206.1
Total	−125.4	−401.4	−475.9
Experimental	−140.0	−410.0	−490.0

The important effect of the correlations should be noted.

This promising approach needs to be extended to heavier nuclei and the charge radius computed and compared with experiment.

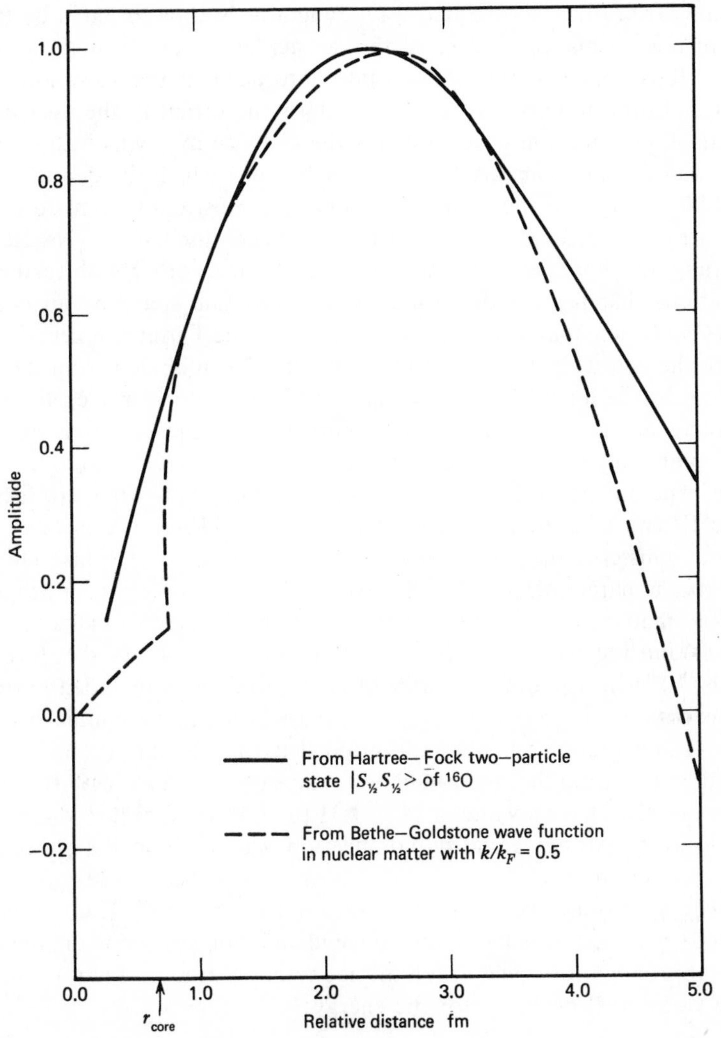

FIG. 18.6. Comparison of $^1S_0$ relative wave functions obtained from the Hartree-Fock wave function for ^{16}O and the Bethe-Goldstone wave function in nuclear matter with $(k/k_F) = 0.5$ [from Shao and Lomon (72)].

Local-Density Approximation

Another approach to this problem is referred to as the *local-density approximation* (LDA) originally introduced by Brueckner, Gammel, and Weitzner (58), and Brueckner, Lockett, and Rotenberg (61). In this approximation the value of the G matrix for a finite nucleus in a small region around a given point in which the Bethe–Goldstone correlated wave functions differ appreciably from the uncorrelated wave functions is approximated by the G-matrix in infinite nuclear matter of the density that prevails at the point in question. It is hoped in this way to take into account the variation in the neutron and proton density that is particularly important in the nuclear surface. This approximation is suggested if the distance over which the correlations prevail is small compared to the distance over which the density has an appreciable change. In these considerations it is most convenient to express the G-matrix in coordinate space. In the most recent discussions [Negele (70) and Sprung and Banerjee (71)], this is carried out together with further approximations that replace the G-matrix in coordinate space by an *effective local two-body potential* that is a function of k_F, the Fermi momentum, and therefore the density ρ. At this point the potential is modified from the strict LDA form by the introduction of a number of parameters; in Negele's calculation there are finally three. This is motivated by the fact that nucleon–nucleon potentials that are used in the calculation of nuclear matter do not give the experimental volume energy. Various corrections that are listed in Chapter III must be added to the value obtained. There are effects of the three-body clusters, three-body forces, and relativistic corrections. The data used to fix the parameters of the effective potential are taken from properties of nuclear matter; that is, the effective potential is used to calculate these properties and the parameters fixed so as to yield the observed values. The properties of nuclear matter that were fitted in this manner include the volume energy as obtained from the semiempirical mass formula, the symmetry energy from the same source, and the saturation density. The latter [Bethe (71)] was fixed by requiring that the calculated radius of ^{40}Ca agree with the experimental value. This gave a value of $k_F = 1.31\ \text{fm}^{-1}$. With respect to the effective two-body potential, the greatest modification was made in the short range ($r < 1$ fm) component.

In this way an effective two-body potential is obtained. The consequent many-body potential, which is density dependent is constructed using the local-density approximation. The resultant many-body Hamiltonian is inserted into the variational principle for the energy:

$$\delta\{\langle\Psi|H_{\text{eff}}(\mathbf{r}_1 \ldots \mathbf{r}_A, \rho)|\Psi\rangle - E\langle\Psi|\Psi\rangle\} = 0 \tag{18.21}$$

The Hartree–Fock type equations are obtained if Ψ is a Slater determinant. Since ρ depends on Ψ, the variation must also be applied to the ρ appearing in H_{eff}. As a consequence the Hartree–Fock potential will contain a contribution from $\delta H_{\text{eff}}/\delta\rho$. This term plays an important role in the calculation of the saturation density, giving rise to a two-body potential term in the Hartree–Fock equations depending on $\partial v_{\text{eff}}/\partial\rho$ where v_{eff} is the two-body density-dependent potential. These terms are referred to as the "saturation potential" by Brandow.

It is to be emphasized that there is *no* justification for the application of the variational principle to the effective Hamiltonian. That is, it is not possible to relate variational principle (18.21) to the variational principle satisfied by the actual Hamiltonian.

Negele has applied this analysis to the calculation of the binding energy, the single-particle energies, and charge density of nuclei throughout the periodic table. The original nucleon–nucleon force is the soft core Reid potential (68). We give a few examples of his results. Table 18.4 [Bethe (71)]

TABLE 18.4 Total Binding Energies per Particle in MeV and Nuclear Radii in fm According to the Theory of Negele [from Bethe (71)]

Nucleus	^{16}O	^{40}Ca	^{48}Ca	^{90}Zr	^{208}Pb
Theoretical binding energy	7.59	7.99	7.96	8.33	7.83
Experimental binding energy	7.98	8.55	8.67	8.71	7.87
Proton rms radius (theoretical)	2.71	3.41	3.45	4.18	5.44
Proton rms (experimental)	2.64	3.43	3.42	—	5.44
Neutron rms radius (theoretical)	2.69	3.37	3.60	4.30	5.67

gives the binding energy per particle, the proton rms-nucleon radius as well as the neutron radius. The agreement with experiment is excellent. These results indicate that once the nuclear forces are adjusted to give the correct binding energy and symmetry energy for nuclear matter as well as the ^{40}Ca radius, the binding energy and charge radius are correctly predicted for a variety of nuclei. The saturation potential plays an important part in these calculations. Because of its presence it is possible to simultaneously obtain the single-particle energies and the total binding energy. That portion of the energy that depends upon the saturation potential is negative so that the nucleus is more strongly bound than the single-particle energies indicate [(Bethe (71), see also Eq. 18.38)].

The large effect of the density dependence and the adjustment of the nuclear forces mentioned above on the charge density of the ^{40}Ca nucleus is shown in Figure 18.7. The solid line gives the empirical charge density as determined from high energy electron scattering [Bellicard et al (67) and Frosch et al (68)]. The broken line with the long dashes is calculated with a density-independent two-body potential. The dot-dash curve employs Negele's density-dependent theory, but without any adjustment of the interaction while the curve with short dashes includes that adjustment. The great importance of the density dependence is clearly seen.

The comparison of the full calculation with empirical charge distribution is shown in Fig. 18.8. It will be noticed that the theoretical distributions have more oscillations than the empirical distribution. These oscillations are a manifestation of shell structure. A more critical comparison of theory and experiment is obtained by comparing the experimentally observed electron scattering with that predicted by the calculated charge distribution. Such

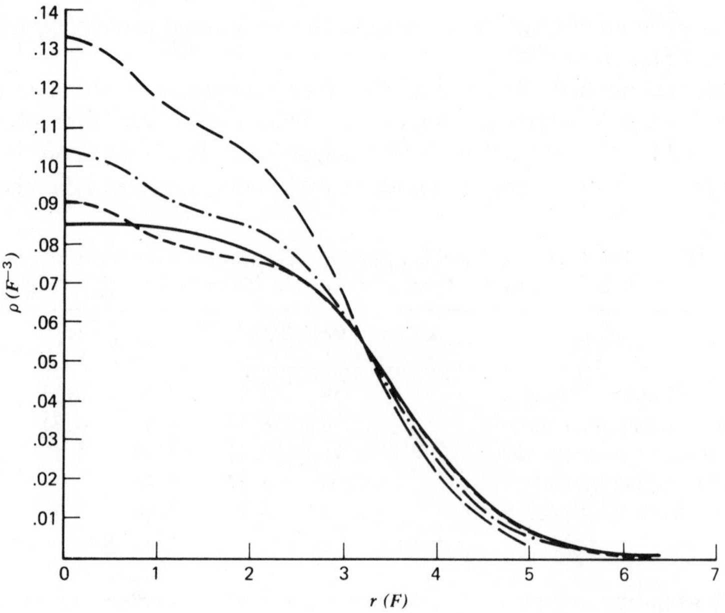

FIG. 18.7. ^{40}Ca charge density. See text for description [from Negele (70)].

a comparison is shown in Fig. 18.9. Again the agreement is excellent. However as is shown by Fig. 18.10, not a very good prediction of the difference between electron scattering by ^{40}Ca and that by ^{48}Ca is obtained, although substantial agreement is obtained for scattering by each nucleus.*

The wave functions obtained in this theory are Slater determinants. The entire effect of the correlations are contained in the effective density-dependent interaction. The calculation of the correlated wave functions has not been discussed. Of course configuration interaction will also in some cases be of importance.

The question naturally arises how much of the success of this formulation is a consequence of the nuclear force employed, (the Reid soft core) and how much can be ascribed to the relatively smooth behavior of nuclear binding energy as a function of A and Z. Because of the latter it may be entirely possible that the adjustments to the nuclear potential, involving in Negele's case three empirical parameters, are sufficient to insure the agreement with experiment that has been obtained.

*A recent calculation Bertozzi et. al. (72) indicates that this difference is due to the interaction of the electron with the neutron that is commonly omitted in the analysis of these experiments.

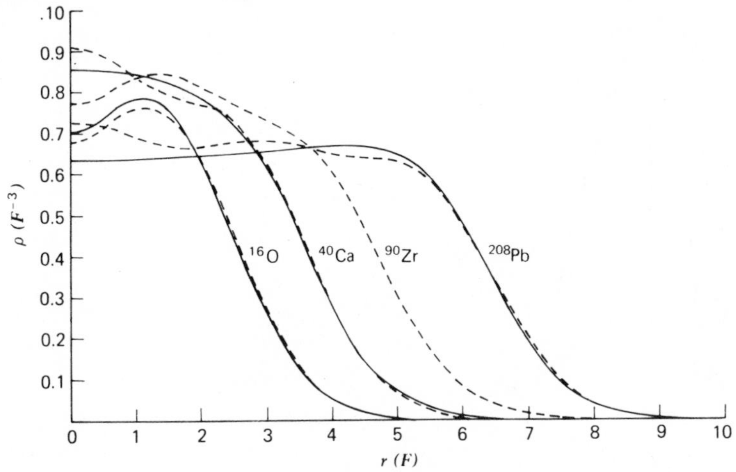

FIG. 18.8. Negele's charge distributions (dashed lines) and empirical (solid lines) results obtained from electron scattering according to Bellicard et al. (67), Frosch et al. (68), Ehrenberg et al. (59), and Bellicard and van Oosbrun (67) [from Negele (70)].

That this is to some extent true is indicated by the results obtained by Vautherin and Brink (72) and Vautherin, Veneroni, and Brink (70) who employ a very simplified model proposed by Skyrme (56, 59) [see also Moskowski (70)] that leads to a density-dependent Hamiltonian. Because of the simplicity of the model some insights into the relevant issues can be obtained.

Skyrme used a simplified two-body force and in addition a three-body force

$$V = \sum_{i<j} v_{ij} + \sum_{i<j<k} v_{ijk} \tag{18.22}$$

The effect of the three-body force, which is taken to be of the form

$$v_{123} = t_3 \delta(\mathbf{r}_1 - \mathbf{r}_2)\delta(\mathbf{r}_2 - \mathbf{r}_3) \tag{18.23}$$

where t_3 is a parameter, is equivalent to the introduction of a two-body density-dependent potential as we shall indicate below. The two-body force is of the form

$$v_{12} = t_0(1 + x_0 P_\sigma)\delta(\mathbf{r}_1 - \mathbf{r}_2) + \frac{1}{2} t_1 [\delta(\mathbf{r}_1 - \mathbf{r}_2)k^2 + k'^2 \delta(\mathbf{r}_1 - \mathbf{r}_2)]$$

$$+ t_2 \mathbf{k}' \cdot \delta(\mathbf{r}_1 - \mathbf{r}_2)\mathbf{k} + iW_0(\boldsymbol{\sigma}_1 + \boldsymbol{\sigma}_2) \cdot \mathbf{k}' \times [\delta(\mathbf{r}_1 - \mathbf{r}_2)\mathbf{k}] \tag{18.24}$$

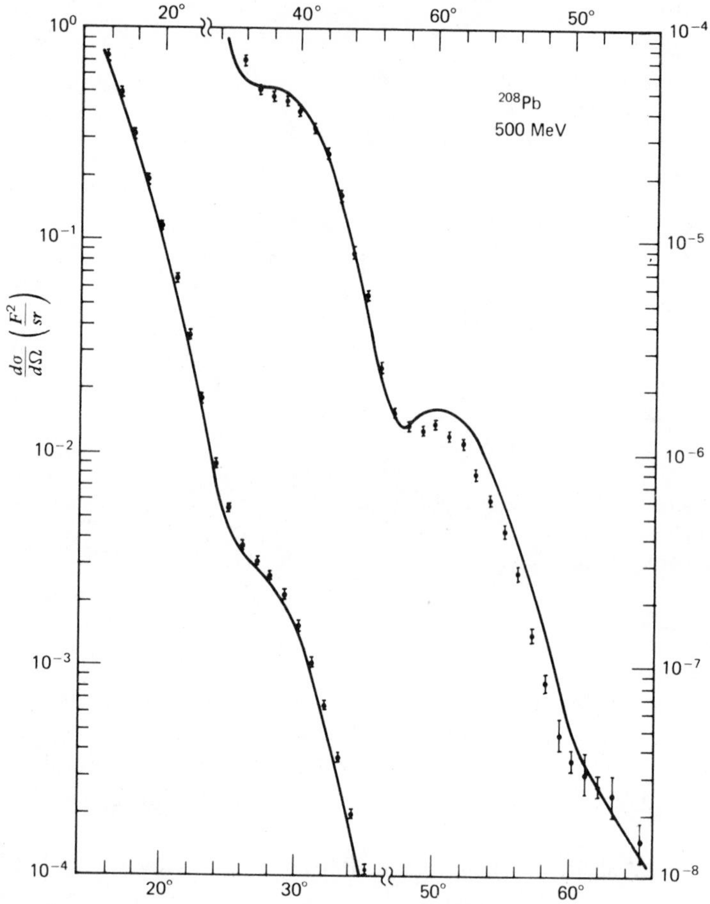

FIG. 18.9. Scattering of 500 MeV electrons by ^{208}Pb. Experimental points, Stanford data courtesy of Heisenberg; theoretical curve calculated from Negele's charge distributions [quoted by Bethe (71)]

In this expression t_0, t_1, t_2, W_0, and x_0 are parameters to be adjusted by comparison with experiment, P_σ is the spin-exchange operator $(1 + \boldsymbol{\sigma}_1 \cdot \boldsymbol{\sigma}_2)/2$ and k^2 and k'^2 are differential operators defined by

$$\mathbf{k} \equiv \frac{1}{2i} (\overrightarrow{\nabla}_1 - \overrightarrow{\nabla}_2) \tag{18.25}$$

$$\mathbf{k}' \equiv -\frac{1}{2i} (\overleftarrow{\nabla}_1 - \overleftarrow{\nabla}_2)$$

where the direction of the arrow indicate the direction (right or left) of the action of the differential operator.

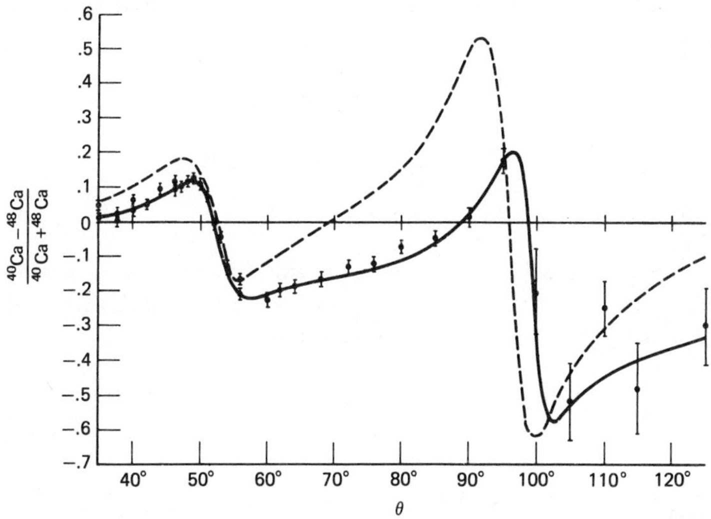

FIG. 18.10. The ratio of the difference of the scattering cross sections by ^{40}Ca and ^{42}Ca over the sum. The solid line is obtained from the empirical charge distributions of Bellicard et al. (67), Frosch et al. (68). The experimental points were obtained by these authors. The dashed line is given by Negele's theory (see footnote p. 620) [from Negele (70)].

Problem: Prove that other types of exchange forces do not give contributions differing from (18.24) in form if the spatial dependence of the potential is given by a delta function.

Note that v_{12} is momentum dependent. The first two terms correspond to interactions in only the relative *S*-state of the two particles while the last two terms correspond to *P* wave interactions.

Problem: Prove these relationships by taking matrix elements of the potential between relative wave functions of the form $R(r)Y_{lm}(\hat{\Omega})$.

Form (18.24) also has a simple interpretation in momentum space. There the plane wave matrix element of (18.9) $\langle \mathbf{k} | v_{12} | \mathbf{k}' \rangle$ has the form

$$\langle \mathbf{k} | v_{12} | \mathbf{k}' \rangle = t_0(1 + x_0 P_\sigma) + \frac{1}{2} t_1(k^2 + k'^2) + t_2 \mathbf{k}\cdot\mathbf{k}' + iW_0(\boldsymbol{\sigma}_1 + \boldsymbol{\sigma}_2) \cdot (\mathbf{k} \times \mathbf{k}') \quad (18.26)$$

It is evident that (18.26) includes the first few terms of the expansion of $\langle \mathbf{k} | v_{12} | \mathbf{k}' \rangle$ in a power series in $\mathbf{k}$ and $\mathbf{k}'$.

It is simple to apply the Hartree–Fock method. It is only necessary to insert interaction (18.26) into the variational principle for the energy using a Slater determinant for the variational wave function. The matrix element for the energy can be written in the following form:

$$E = \langle\Psi|(T+V)|\Psi\rangle \equiv \int H(\mathbf{r})\,d\mathbf{r} \tag{18.27}$$

where

$$\Psi = \frac{1}{\sqrt{A!}} \det \phi_i(\mathbf{r}_j, m_s, m_t) \tag{18.28}$$

where $\mathbf{r}_j$, m_s, m_t give the spatial, spin, and isospin coordinates of each particles. Inserting (18.28) for Ψ into (18.27) yields the following expression for the energy density $H(\mathbf{r})$, according to Vautherin and Brink:

$$H(\mathbf{r}) = \frac{\hbar^2}{2m} T(\mathbf{r}) + \frac{1}{2} t_0 \left[\left(1 + \frac{1}{2} x_0\right)\rho^2 - \left(\frac{1}{2} + x_0\right)(\rho_n^2 + \rho_p^2)\right]$$

$$+ \frac{1}{4}(t_1 + t_2)\rho T + \frac{1}{8}(t_2 - t_1)(\rho_n T_n + \rho_p T_p) + \frac{1}{16}(t_2 - 3t_1)\rho\nabla^2\rho$$

$$+ \frac{1}{32}(3t_1 + t_2)(\rho_n\nabla^2\rho_n + \rho_p\nabla^2\rho_p) + \frac{1}{16}(t_1 - t_2)(\mathbf{J}_n^2 + \mathbf{J}_p^2)$$

$$+ \frac{1}{4} t_3 \rho_n \rho_p \rho - \frac{1}{2} W_0(\rho\nabla\cdot\mathbf{J} + \rho_n\nabla\cdot\mathbf{J}_n + \rho_p\nabla\cdot\mathbf{J}_p) + H_c \tag{18.29}$$

where ρ is the nucleon density:

$$\rho = \rho_n + \rho_p$$

and where

$$\rho_p = \sum_{i,m_s} |\phi_i(\mathbf{r}, m_s, m_t = 1/2)|^2 \tag{18.30}$$

$T(\mathbf{r})$ is the kinetic-energy density. For protons it is

$$T_p = \sum_{i,m_s} |\nabla\phi_i(\mathbf{r}, m_s, m_t = 1/2)|^2 \tag{18.31}$$

and $T = T_p + T_n$.

Finally $\mathbf{J}$ is related to the spin density:

$$\mathbf{J}_p = i \sum_{i,m_s,m'_s} \phi_i^*(\mathbf{r}, m_s, 1/2)[\langle m_s|\boldsymbol{\sigma}|m'_s\rangle \times \nabla\phi_i(\mathbf{r}, m'_s, 1/2)]$$

$$\mathbf{J} = \mathbf{J}_p + \mathbf{J}_n \tag{18.32}$$

The cubic term proportional to t_3 in (18.29) is a consequence of the three-body force demonstrating that it is equivalent to a density-dependent interaction. All the other terms are bilinear in the quantities ρ, T, and $\mathbf{J}$ while the kinetic-energy term is linear. H_c is the Coulomb term.

The Hartree–Fock equations for closed-shell nuclei are:

$$\left[-\boldsymbol{\nabla}\cdot\frac{\hbar^2}{2m^*}\boldsymbol{\nabla}+U(m_t,\mathbf{r})+i\mathbf{W}(m_t,\mathbf{r})\cdot(\boldsymbol{\sigma}\times\boldsymbol{\nabla})\right]\phi_i=\epsilon_i\phi_i \tag{18.33}$$

where ϵ_i is the single-particle energy. The effect of the velocity-dependent terms in (18.24) that are proportional to t_1 and t_2, is, among other things, to replace the nucleon mass by an effective mass $m^*\ (m_t, \mathbf{r})$

$$\frac{\hbar^2}{2m^*}=\frac{\hbar^2}{2m}+\frac{1}{4}(t_1+t_2)\rho+\frac{1}{8}(t_2-t_1)\rho(m_t,\mathbf{r}) \tag{18.34}$$

The potential $U(m_t, \mathbf{r})$ is momentum dependent and because of the three-body force bilinear in the density:

$$\begin{aligned}U(m_t,\mathbf{r}) &= t_0\left[\left(1+\frac{1}{2}x_0\right)\rho-\left(\frac{1}{2}+x_0\right)\rho(m_t,\mathbf{r})\right]+\frac{1}{4}t_3[\rho^2-\rho^2(m_t,\mathbf{r})]\\ &\quad -\frac{1}{8}(3t_1-t_2)\nabla^2\rho+\frac{1}{16}(3t_1+t_2)\nabla^2\rho(m_t,\mathbf{r})+\frac{1}{4}(t_1+t_2)T\\ &\quad +\frac{1}{8}(t_2-t_1)T(m_t,\mathbf{r})-\frac{1}{2}W_0[\boldsymbol{\nabla}\cdot\mathbf{J}+\boldsymbol{\nabla}\cdot\mathbf{J}(m_t,\mathbf{r})]+H_c\end{aligned} \tag{18.35}$$

The terms involving the gradients of the density will be sensitive to its spatial dependence particularly near the nuclear surface. For $N = Z$ nuclei the parameter modifying $\nabla^2\rho$ is proportional to $(5t_2 - 9t_1)$. It is this parameter that will be important for surface effects. Under the same circumstances the change in $\hbar^2/2m$ depends upon $(3t_1 + 5t_2)$ and, therefore, that parameter is important for the determination of the single-particle level energies.

The factor $\mathbf{W}(m_t, \mathbf{r})$ is given by

$$\mathbf{W}(m_t,\mathbf{r})=\frac{1}{2}W_0[\boldsymbol{\nabla}\rho+\boldsymbol{\nabla}\rho(m_t,\mathbf{r})]+\frac{1}{8}(t_1-t_2)\mathbf{J}(m_t,\mathbf{r}) \tag{18.36}$$

The dependence on W_0 is expected (see Eqs. 18.24 and 18.26) but there is an additional dependence on $\mathbf{J}$ whose origin is in the momentum-dependent terms in the nuclear forces.

One final formal point is of great importance. In the Hartree–Fock approximation with density-independent forces the total energy of the nucleus is given by

$$E=\frac{1}{2}\sum(T_i+\epsilon_i) \tag{18.37}$$

where T_i denotes the single-particle kinetic energy and ϵ_i the single-particle energy. Because of this relation (we have alluded to this point before) it is not possible to obtain the binding energy, radius, and single-particle energies of ^{16}O and ^{40}Ca using a density-independent interaction [Kerman (69)]. The essential point is that there are three experimental quantities and essentially only two quantities in the theoretical expression (18.37) so that it would be very significant if all the experimental quantities were predicted correctly. In the presence of a density-dependent force of the Skyrme-type, relation (18.37) is modified to

$$E = \frac{1}{2} \sum (T_i + \epsilon_i) - \frac{1}{8} t_3 \int \rho_n \rho_p \rho \, d\mathbf{r} \tag{18.38}$$

This formula shows that for this model the binding energy of a nucleus is greater (assuming $t_3 > 0$) than would be expected from the single-particle energies and (18.37). It was found by Negele that this was critical in obtaining the experimental binding energy. Vautherin and Brink obtain a similar result as will be described below.

The parameters, six in all were chosen so as to give the properties of ^{16}O and ^{208}Pb; binding energies and radii. The splitting of the $1p$ levels in ^{16}O were used to select the value of W_0. To illustrate the results we choose one set tabulated in Table 18.5 determined in this fashion (their set II),

TABLE 18.5 Parameters of the Density-Dependent Hamiltonian of Vautherin and Brink (72)

t_0 (MeV fm^3)	t_1 (MeV fm^5)	t_2 (MeV fm^5)	t_3 (MeV fm^6)	W_0 (MeV fm^5)	x_0
−1169.9	585.6	−27.1	9333.1	105	0.34

With these values the results obtained for a number of closed-shell nuclei are tabulated in Table 18.6. In this table the rms radii of the neutrons is denoted by r_n, of the proton r_p and of the charge r_c. Here ρ_c and thereby r_c is obtained from ρ_p by folding in the finite extent of the proton charge distribution

$$\rho_c(\mathbf{r}) = \int f_p(\mathbf{r} - \mathbf{s}) \rho_p(\mathbf{s}) \, d\mathbf{s}$$

where $f_p(\mathbf{r})$ is the proton charge density.
This model gives agreement with experiment that is as good as that obtained with Negele's theory.

Comparison with electron scattering is shown in Fig. 18.11. Again excellent agreement with experiment is obtained.

It is seen from this discussion that a simple model (but with the important component of a density-dependent Hamiltonian generated by a simple three-body interaction and with six parameters to be determined by comparison

TABLE 18.6 Nuclear Binding and Radii from the Theory of Vautherin and Brink (72) Compared to Experiment

Nucleus		^{16}O	^{40}Ca	^{48}Ca	^{90}Zr	^{208}Pb
Experimental	r_c	2.73 fm	3.49	3.48	4.27	5.50
	E/A	−7.98 MeV	−8.55	−8.67	−8.71	−7.87
Theory	r_n	2.61	3.35	3.63	4.32	5.69
	r_p	2.63	3.40	3.45	4.24	5.49
	r_c	2.75	3.49	3.54	4.31	5.55
	E/A	−7.89	−8.41	−8.39	−8.43	−7.54

with experiment) is capable of explaining the A-dependence of the binding energy and electron-scattering experiments. It must be concluded that the excellence of the results obtained by Negele with a much more elaborate theory and complex effective interaction based directly upon phenomenological two-nucleon potentials is in considerable part due to the presence of three parameters that are determined empirically. These parameters play an important role as can be seen from Fig. 18.7 where the adjustment of the density-dependent interaction is essential to obtain agreement with the empirical charge distribution. It also appears that a common element that is of critical importance is the density dependence of the effective Hamiltonian. It can be seen from Fig. 18.7 that it plays an essential role in flattening the nuclear charge density. It is of importance in producing saturation, the additional

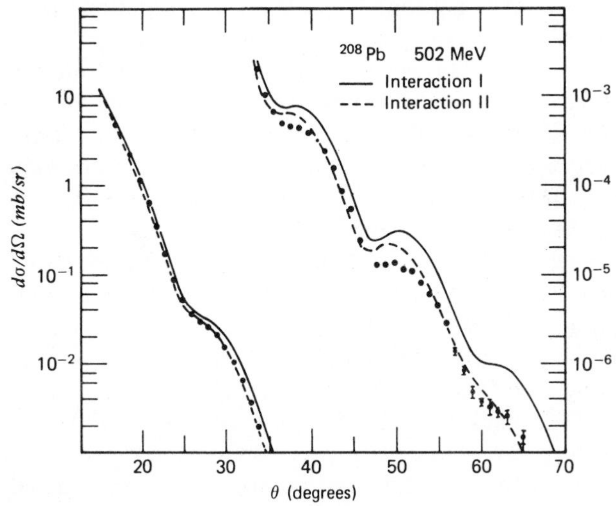

FIG. 18.11. Elastic electron scattering of 502 Mev electrons by ^{208}Pb compared with theory [Vautherin and Brink (72)].

term in the effective Hamiltonian being referred to as the saturation potential. Finally it resolves the problem of obtaining simultaneously the observed single-particle energies and the correct binding energy for nuclei. Since the very schematic form of Vautherin and Brink seems to yield excellent results one must conclude that the detailed nature of the density-dependent interaction is not essential. [Recently Vautherin (72) has extended this theory to deformed nuclei.]

Of course the problem remains of relating their empirical parameters to the nucleon–nucleon interaction. One very recent and promising attempt has been made by Negele and Vautherin (72). These authors attempt to derive the Vautherin–Brink interaction essentially by expanding the G-matrix in powers of $\mathbf{k}$ and $\mathbf{k}'$ (actually they do not use a power series but their approach is close enough to a power-series expansion so that in a qualitative sense this description is not misleading), eventually arriving at an effective Hamiltonian. This procedure then relates the nucleon–nucleon amplitude to an effective interaction resembling the one used by Vautherin and Brink.

19. EFFECTIVE INTERACTIONS

The last sections have been primarily devoted to the relation of the nuclear ground state, its energy as well as its charge and particle density, to the underlying nuclear forces. But of course there are many other properties of nuclei we wish to understand—in particular the nature of the excited states, their energy, their electromagnetic properties, their electromagnetic and β-ray transition rates.

In the shell model, deformed or spherical, excitations are described by moving particles from occupied to unoccupied single-particle states. These elementary excitations do not form a solution of the nuclear Schrödinger equation. Linear combinations must be used in the attempt to diagonalize the nuclear Hamiltonian. In principle an infinite set of excitations must be used. In practice, the Hilbert space is truncated and/or a subset of excitations is considered. The first procedure has been discussed in Chapter V, for the shell model, the second is exemplified by the RPA method.

In either event use of the nucleon–nucleon potential as the residual interaction is not correct. Because of the truncation, for example, excitation to states outside the space have been omitted. To account for these states the nucleon–nucleon interaction should be modified. The two-body potential $v(ij)$ should be replaced by an effective interaction $v_0(ij)$. It is also worthwhile to remind ourselves that for many nuclear properties it is possible to limit ourselves to a smaller subspace by considering effective, or renormalized, operators. In Section V.12 we saw how configuration mixing can be simulated, for the purpose of calculating energies, by a renormalization of the interaction, and in Section VI.9 we saw how, to lowest order, the Coriolis mixture of rotational bands can be simulated by a modification of the moment of inertia. Similar renormalization effects will be shown in Chapter VIII to lead to effective charges of nucleons in the nucleus, etc.

We shall discuss how these effective interactions and operators may be obtained from first principles in the next section. However a large number of calculations have been performed with simple semiempirical effective interactions. The justification for doing so comes from the fact that we are dealing with a very limited set of matrix elements, and for such *limited sets* the matrix elements $\langle\alpha|v(ij)|\beta\rangle$ of the real interaction can be reproduced by corresponding matrix elements $\langle\alpha|v_0(ij)|\beta\rangle$ of a model, or effective interaction $v_0(ij)$ (see also Sections V.12 and VI.9).

The size of the *model space*, as the truncated Hilbert space is referred to, depends upon the nature of the states under discussion. We have already seen in the discussion of the Hartree–Fock method using oscillator wave functions how a rather small subspace sufficed in that case.

Figure 19.1, reproduced from Gillet, Green, and Sanderson (66), shows the effect on the energies of two 3^- states in ^{208}Pb, of increasing the dimensions of the subspace in which the nuclear Hamiltonian is diagonalized. It is seen that the higher 3^- state (upper solid line) does not change its energy appreciably as the diagonalization space increases from 5 configurations whose maximum zeroth-order excitation lies at 5 MeV to 45 configurations with a

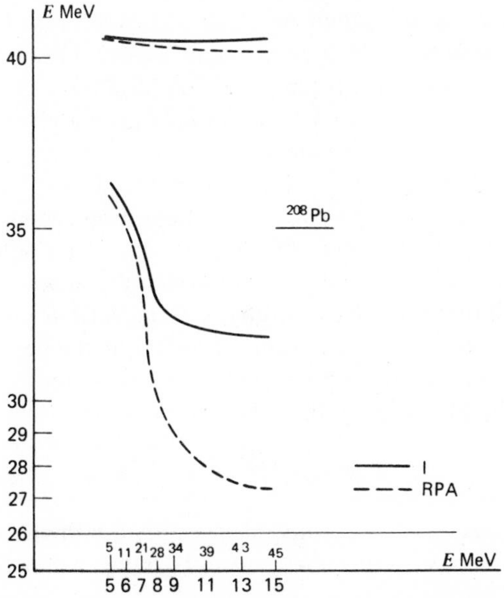

FIG. 19.1. The position of the 3^- states of ^{208}Pb as a function of the dimension of the Hilbert space. The dashed line curve is obtained by using the RPA [from Gillet et al. (66)].

maximum excitation of 15 MeV. The lowest 3^- state that is the collective octupole state, (lower solid line) is more strongly affected by the increase in the dimensionality of the diagonalization subspace, but again seems to reach a "plateau" at about 40 configurations involving zeroth-order excitation of about 10 MeV.

Tests similar to the one shown in Fig. 19.1 can be applied also to other quantities calculated with the help of wave functions derived from the partial diagonalization of the Hamiltonian. They amount to a semiempirical test of the convergence of the whole method. As a rule one finds that, provided the initial choice of single particle wave functions is reasonable, one does not have to go to very large subspaces to obtain the "saturation" value for the energies of regular states, like the higher 3^- state in Fig. 19.1. States that show collective features, like the lowest 3^- state in Fig. 19.1, require larger subspaces to obtain a good approximation for their energies. Furthermore, if we want the wave function of a given state for purposes other than the evaluation of its energy, it is generally required to go to larger subspaces. Energies, being eigenvalues of the Hamiltonian, are less sensitive to small variations in the wave functions. Other quantities, such as moments or transition probabilities, do not have this property. The dimensions of the subspace in which the nuclear Hamiltonian is to be diagonalized thus depend on the quantities that are to be calculated with the resulting wave functions.

In choosing model interactions for use in a truncated Hilbert space one is led by both intuition and arguments of simplicity. A very popular interaction is the δ-interaction discussed in Sections V.1, V.4, and V.6. The extent to which it can be made to fit the data on some low-lying excited states of various nuclei can be judged from a recent work of Moinester, Schiffer, and Alford (69). It will be recalled (Section V.A-2) that within a given configuration the energies of the various states can be expressed in terms of pure multipole–multipole interactions between pairs of nucleons. Conversely, if the energies of all the states of a given two-particle configuration are known, it is possible to derive from them the strength of the individual multipole–multipole interactions α_k. Moinester et al (loc. cit.) find from the analysis of various nuclei from ^{38}Cl to ^{210}Bi that the *relative* strength of the various pure multipole interactions remains more or less the same throughout the periodic table. In fact one can conveniently define a pure multipole interaction of order k to be the one for which the Slater integrals in Section V.A-19 reduce to

$$F^{(r)} = \sqrt{2k+1}\,\delta(r, k) \tag{19.1}$$

($\sqrt{2k+1}$ is introduced for convenience only), and the irreducible tensor operators in Section V.A-20 reduce to unity independent of k, so that

$$(j||T^{(k)}||j) = (j||1||j) = \sqrt{2j+1} \tag{19.2}$$

The spectrum of energy levels in the configuration $(j_1 j_2)$ for a pure multipole is given, using Section V.A-19, by:

$$E^{(k)}(j_1 j_2 J) = (-1)^{j_1+j_2+J}\sqrt{(2j_1+1)(2j_2+1)(2k+1)}\begin{Bmatrix} j_1 & j_2 & J \\ j_2 & j_1 & k \end{Bmatrix} \tag{19.3}$$

The strength α_k of the kth multipole is then defined through the requirement that the actual experimental spectrum in the configuration $(j_1 j_2)$ be given by

$$E(j_1 j_2 J) = \sum_k \alpha_k E^{(k)}(j_1 j_2 J) \tag{19.4}$$

It follows from (19.3) and (19.4), using the orthogonality of the 6–j symbols, that

$$\alpha_k = \sqrt{\frac{(2k+1)}{(2j_1+1)(2j_2+1)}}\sum_J (-1)^{j_1+j_2+J}(2J+1)\begin{Bmatrix} j_1 & j_2 & J \\ j_2 & j_1 & k \end{Bmatrix} E(j_1 j_2 J) \tag{19.5}$$

Thus from the measured energies $E(j_1 j_2 J)$ one can deduce α_k in a straightforward manner.

One point has to be clarified, however, with respect to (19.5): what is the reference energy from which $E(j_1 j_2 J)$ is measured. It is within the spirit of all shell-model calculations that this reference energy should be the unperturbed zeroth-order energy and one therefore takes it from the observed single particle (or single-hole energies) in neighboring nuclei.

Defining

$$\beta_k = \left(\frac{\alpha_k}{\alpha_0}\right)(-1)^{k(l_1+j_1+l_2+j_2+\sigma)}$$

$$\sigma = \begin{cases} 0 \text{ for } pp \text{ or } hh \text{ states} \\ 1 \text{ for } ph \text{ states} \end{cases}$$

Moinester et al obtain the values shown in Table 19.1 for the relative multipole moments β_k in ^{208}Bi; the table shows also average values of β_k for various odd–odd nuclei all over the periodic table, as well as the theoretical values for a δ-interaction.

It is therefore seen that the δ-interaction seems to reproduce features of nuclear spectra fairly accurately, at least for simple configurations. This probably indicates that the main effects of the actual residual interaction produce substantially extra binding when the wave functions of two nucleons overlap each other well.

TABLE 19.1 Comparison of Multipole Coefficients $100\langle\beta_k\rangle$ from δ-Function Potential with Experiment [from Moinester et al. (69)]

Multipolarity k	^{208}Bi only Experiment	^{208}Bi only δ function	All data Experiment	All data δ function
2	58	53	52	52
4	38	36	32	36
6	22	27	20	25
8	22	20	15	19
1	−20	−17	−18	−18
3	− 7	− 7	− 7	− 7
5	2	− 7	− 4	− 7
7	11	− 3	2	− 3
9	8	− 2	− 2	− 2

The δ-interaction can be used for $v(ij)$ also when diagonalizing the effective nuclear Hamiltonian in truncated spaces containing more than one shell-model configuration. A closer agreement with experiment is obtained, however, with a slight modification of this interaction, proposed by Green and Moszkowski (65) and often called the *surface δ-interaction*. It amounts to ascribing to the Slater integral $F^{(k)}(nl, n'l')$ (see Section V.A.3) a value that is independent of n, l, n', and l':

$$F^{(k)}(nl, n'l') = (2k + 1)F^{(0)} \tag{19.6}$$

The argument behind this approximation is that the residual interaction among nucleons seems to be most effective when both are at the nuclear surface. The one-particle radial wave functions at the nuclear surface turn out to have nearly the same value for all states and, hence, the assumption (19.6). Figure 19.2 shows the energy levels of the two-nucleon system when the truncated space includes the $2s$ and $1d$ states. It is seen that the surface–δ-interaction gives a ratio for $E(4^+)/E(2^+)$ that is close to the observed ratio in nondeformed nuclei; the ordinary δ-interaction does not do so well for this ratio, and the pairing interaction Section V.A.14 is, of course, even poorer.

There are other "popular" effective interactions used in the framework of the shell model. Some of them are mentioned in Appendix B, Chapter V. Since, however, the justification for the use of these simplified interactions is connected with the use of limited subspaces for the diagonalization of the effective nuclear Hamiltonian, one can adopt also another point of view, largely developed by Racah for atoms and by Talmi and coworkers for nuclei [see Talmi (65)]. In this approach one derives directly the matrix elements of the effective interaction from the analysis of simple configurations and uses

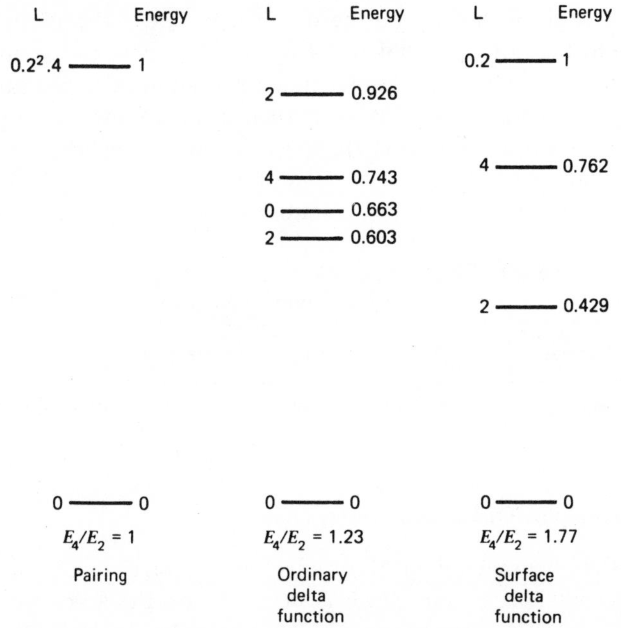

FIG. 19.2. Energy levels of the $(s,d)^2$ configuration for pairing, ordinary delta function, and surface delta function interactions; the ground state is taken to be zero, and the unperturbed energy is at unity [from Green and Moskowski (65)].

these semiempirical matrix elements in calculations involving more complex configurations. As an example, consider the isotopes of K, from ${}^{39}_{19}K_{20}$ to ${}^{43}_{19}K_{24}$. The shell-model configuration of the protons in these isotopes is $d_{3/2}^{-1}$, and that of the neutrons is $f_{7/2}^n$ with $n = 0, 1, 2, 3$, and 4. Thus in the diagonalization of the Hamiltonian (18.1) in the subspace of the configurations $(d_{3/2}^{-1}, f_{7/2}^n)$ only the following matrix elements of $v(ij)$ will appear (see also Section V.8):

$$\langle d_{3/2}^{-1} f_{7/2}\, J | v(12) | d_{3/2}^{-1} f_{7/2}\, J\rangle \qquad J = 2^-, 3^-, 4^- \quad \text{and} \quad 5^- \tag{19.7}$$

$$\langle f_{7/2}^2\, J | v(12) | f_{7/2}^2\, J\rangle \qquad J = 0^+, 2^+, 4^+ \quad \text{and} \quad 6^+ \tag{19.8}$$

These eight matrix elements can be deduced from the spectra and energies of ${}^{40}_{19}K_{21}$ [for the matrix elements in (19.7)] and of ${}^{42}_{20}Ca_{22}$ [for those in (19.8)]. They can then be fed into the matrices for ${}^{41}K$, ${}^{42}K$, and ${}^{43}K$, giving both the spectra and energies of these nuclei, as well as the detailed structure of their wave functions within the configurations $(d_{3/2}^{-1}, f_{7/2}^n)$. The latter can be used to compute also other properties of various levels in these nuclei. In particular the magnetic moments of all levels in these configurations are uniquely determined, once the wave function is known, by the two magnetic moments

$$\mu(d_{3/2}^{-1}) \qquad \text{and} \qquad \mu(f_{7/2}) \tag{19.9}$$

The former is given directly by the moment of ${}^{39}_{19}K_{20}$ whose configuration is pure $d_{3/2}^{-1}$ and the latter by that of ${}^{41}_{20}Ca_{21}$ which is a pure $f_{7/2}$ nucleus. Table 19.2, reproduced from Talmi and Unna (60), shows the fit that can be obtained to the experimental data with such calculations. The significance of these studies lies in the fact that they test directly the validity of the assumption that solutions of the nuclear Hamiltonian can be approximated by its diagonalization in a small truncated subspace.

TABLE 19.2 Magnetic Moments of the K-Isotopes [from Talmi and Unna (60)]

Nucleus	^{39}K	^{40}K	^{41}K	^{42}K	^{43}K
Experimental	0.391	−1.30	0.215	−1.14	0.163
Calculated	0.335	−1.256	0.208	−1.199	0.187

The Pairing-plus-Quadrupole Interaction

Among the phenomenological, or effective, interactions that are commonly used, a specially prominent place is occupied by the so-called *pairing plus quadrupole* interaction, one form of which is given by Eq. V.A.15. As explained in that section this interaction is taken to represent an approximate separation of the two-nucleon interaction into a "short range" part (the pairing interaction) and a "long range" part (the quadrupole–quadrupole interaction). The pairing interaction is often taken in a slightly more general form than that adopted in Eq. V.A.14:

$$\langle j_a j_a' J | v_p(12) | j_b j_b' J\rangle = \sqrt{(2j_a+1)(2j_b+1)}\,G\,\delta(j_a j_a')\,\delta(j_b j_b')\,\delta(J,0) \tag{19.10}$$

where j_a and j_b are allowed to take on all the j-values in the last filled major shell. The physical idea behind (19.10) is that the short range interaction allows the scattering of pairs coupled to $J = 0$ from $|j_a^2\, J = 0\rangle$ to $|j_b^2 J = 0\rangle$ with equal amplitudes for all pairs of states that are close to the Fermi surface.

Although the pairing-plus-quadrupole interaction can be handled, in principle, by diagonalization in a large enough subspace, in many cases it is found to be considerably easier to make use of the Bogoljubov transformation and absorb thereby the main effects of the pairing interaction into the single-particle properties of the quasi particles. One can then express the remaining quadrupole–quadrupole interaction in terms of the quasi-particle coordinates and deal with it as a perturbation in that scheme. Extensive calculations for both even–even and odd–even nuclei have been carried out using this approach by Kisslinger and Sorensen (63) who applied the RPA to the quasi-particle quadrupole interaction. This quasi-particle random-phase approximation (QRPA) resembles very much the one developed in Section VII.6, except that now one wants to include also pairs like $\alpha_k^\dagger\alpha_{k'}^\dagger$ and $\alpha_k\alpha_{k'}$. We shall not go into its description here and the reader is referred to the paper of Kisslinger and Sorensen (63) for further details. Figure 19.3 shows an example of their results for the odd-A Xe isotopes. The open points are the experimental values for the energies of the levels indicated, whereas the solid lines indicate the calculated positions of these levels. We see that the calculations account for the "crossing" of the $3/2^+$ and the $1/2^+$ levels between ^{129}Xe and ^{131}Xe, but other features of the empirical spectra are not reproduced that well. The approximate nature of this calculation is quite evident from Fig. 19.3, but its relative transparency may compensate for this. A broader program of more complete calculations for the pairing plus quadrupole interaction has been undertaken by Kumar and Baranger (68), and references given there. We shall come back to this program toward the end of this section.

It is, of course, possible to include also other parts of the full nuclear interaction $v(ij)$ in calculations of this type, and Kisslinger and Sorensen (63) tried to improve their wave functions even further by mixing in higher configurations using a δ-interaction between pairs of nucleons. This calculation was motivated by the success of earlier calculations of Blin–Stoyle (53), Arima and Horie (54), and Arima, Horie, and Sano (57), who introduced configuration mixing in the shell model to explain the deviations of magnetic moments from the Schmidt lines (see Chapter VIII). It has, however, to be

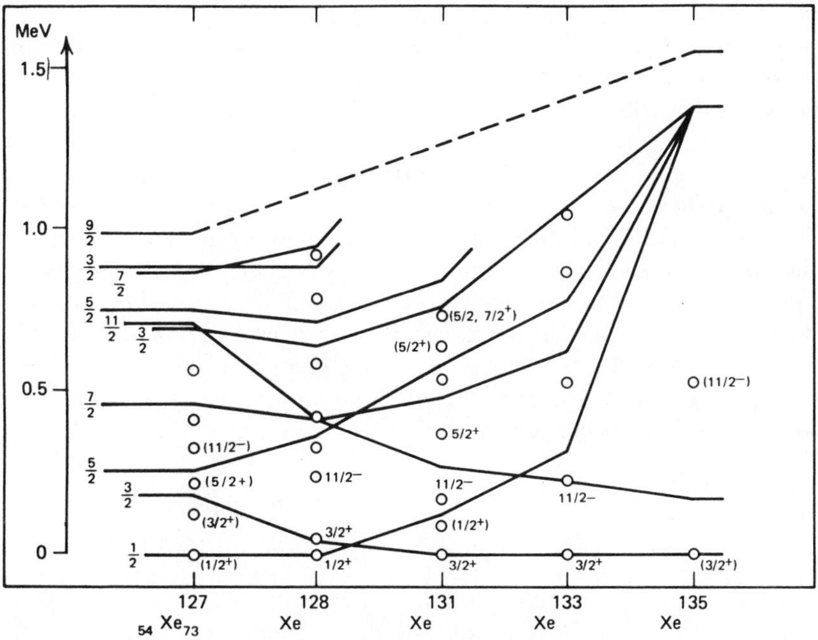

FIG. 19.3. Energy levels of the odd-mass Xe isotopes. Experimental values are indicated by small circles [from Kisslinger and Sorensen (63)].

borne in mind when doing such calculations, that the inclusion of various bits and pieces of the nuclear interaction involves a danger of including some parts of the interaction more than once while still omitting others. It therefore seems advisable either to consider the whole interaction, or simulate it by an effective interaction that is simple to handle within a limited subspace of configurations close to the chosen configuration.

Collective Excitations

The approximations reviewed up to this point can be considered as the first few steps in the complete diagonalization of (18.1) using as a basis the single-particle wave functions generated by a single-particle potential. By increasing the subspace of functions in which the diagonalization is carried out, we can in principle approach the solution of (18.1) as closely as we wish.

However there is another possible procedure by which we can improve upon the description of the single-particle potential by realizing that the combined effect of the nucleons on a particular nucleon has fluctuations about its average value. The average value is given by the Hartree–Fock method. The fluctuations about the average can be conveniently treated by the converting some of the parameters of the potential into dynamical variables (Section VI.13). Or equivalently we may use the time-dependent Hartree–Fock method. In terms of the direct diagonalization method, this procedure corresponds to incorporating the effects of selected higher configurations. Because of this the residual interaction must be modified. It is not correct to employ the residual interaction of the direct diagonalization method since clearly we would be counting some effects "twice."

The single-particle potential is in the last analysis produced by the nucleons themselves. It is therefore conceivable that *collective* motions of the nucleons, especially if their frequencies are small compared to the frequencies of the individual-particle motion, can be best described via some time variation of the parameters that determine the single-particle potential. In the framework of the Hartree–Fock approximation, this can be achieved by going over to the time-dependent Hartree–Fock approximation in which the variational parameters are allowed to be functions of t. As we have seen in Sections VII.5 and VII.6, this leads to the same results as the linearized equations of motion (RPA) for the collective particle-hole excitation operator (6.4). In terms of Goldstone diagrams it includes a whole set of diagrams of the type shown in (19.11).

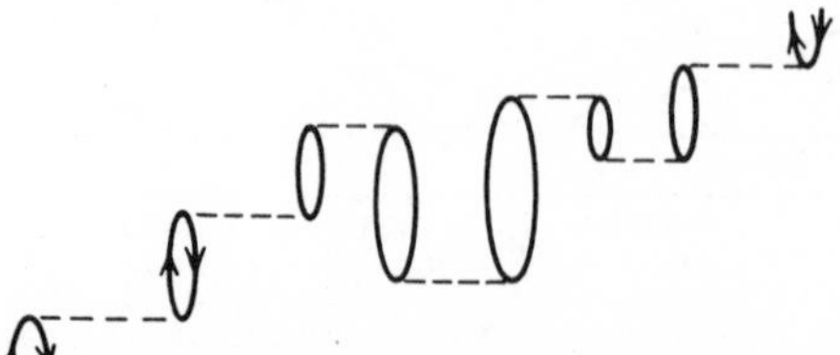

(19.11)

Remark: The diagram in (19.11) is characteristic of particle-hole diagrams and a few comments may help read it more easily:

Consider the diagram:

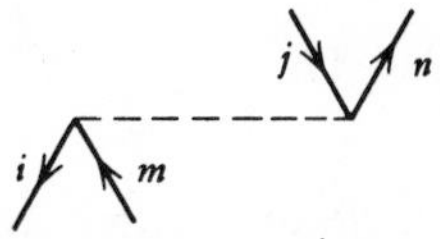

The rules for reading graphs (Section VII.14) associate with this diagram the factor

$$\tfrac{1}{2}v_{in,mj} \exp\,[\beta(\epsilon_i + \epsilon_n - \epsilon_m - \epsilon_j)] \tag{b}$$

Since we are now concerned with states of well-defined angular momentum for the *ph*-pair, it will be more appropriate to write the diagram (a) in terms of a particle-hole interaction rather than the particle–particle interaction v. This can be done in a way similar to the one that led to Eq. V.8.21. If we decompose the interaction $v(k, r)$ into a sum of products of irreducible tensor operators

$$v(k, r) = \sum_s \mathbf{T}^{(s)}(k)\cdot\mathbf{T}^{(s)}(r)$$

then we have, for the *ph*-configurations,

$$\langle j_1^{-1}j_2J|\sum_k v(k, r)|j_1'^{-1}j_2'J\rangle = \sum (-1)^{j_2+j_1'+J}\begin{Bmatrix} j_1 & j_2 & J \\ j_2' & j_1' & s \end{Bmatrix}$$

$$\times\, (j_1^{-1}, j_1||\Sigma T^{(s)}(k)||j_1'^{-1}, j_1')(j_2||T^{(s)}(r)||j_2') \tag{c}$$

where k runs over the particles in all occupied states. Using arguments similar to that which led to Eq. V.8.19, and assuming that the occupied states correspond to closed j-shells, we have

$$(j_1^{-1}, j_1||\sum_k T^{(s)}(k)||j_1'^{-1}, j_1') = (-1)^{s+1}(j_1||T^{(s)}||j_1') \qquad s \neq 0 \tag{d}$$

Introducing (d) into (c) and using the relation

$$\begin{Bmatrix} j_1 & j_2 & J \\ j_2' & j_1' & s \end{Bmatrix} = \sum_l (-1)^{l+s+J}(2l + 1)\begin{Bmatrix} j_1 & j_2 & J \\ j_1' & j_2' & l \end{Bmatrix}\begin{Bmatrix} j_1 & j_1' & s \\ j_2 & j_2' & l \end{Bmatrix}$$

we obtain

$$\langle j_1^{-1}j_2J|\sum_k v(k, r)|j_1'^{-1}j_2'J\rangle = \sum_l (-1)^{j_2-j_2'+1}(2l + 1)\begin{Bmatrix} j_1 & j_2 & J \\ j_1' & j_2' & l \end{Bmatrix}$$

$$\times \sum_s (-1)^{j_2'+j_1'+l}\begin{Bmatrix} j_1 & j_2' & l \\ j_2 & j_1' & s \end{Bmatrix}(j_1||T^{(s)}||j_1')(j_2||T^{(s)}||j_2') \tag{e}$$

The sum over s in (e) will be recognized as $\langle j_1 j_2' l | v | j_1' j_2 l \rangle$ and thus

$$\langle j_1^{-1} j_2 J | \sum_k v(k,r) | j_1'^{-1} j_2' J \rangle \equiv \langle j_1 j_2 J | v_{ph} | j_1' j_2' J \rangle$$

$$= -\sum (-1)^{j_2 - j_2'} (2l+1) \begin{Bmatrix} j_1 & j_2 & J \\ j_1' & j_2' & l \end{Bmatrix} \langle j_1 j_2' l | v_{pp} | j_1' j_2 l \rangle \quad \text{(f)}$$

where v_{pp} is the particle–particle interaction and v_{ph} is the ph-interaction defined by (f). If we now draw the diagram (a) for the ph-states in the left-hand side of (f) we obtain

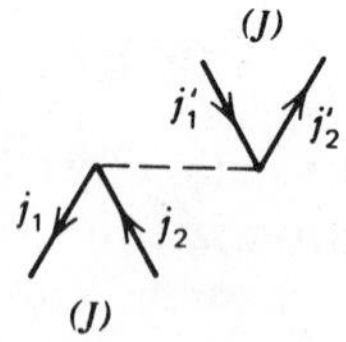

(g)

We see that the matrix elements that appear on the right-hand side of (f) have the expected structure (b); but if the ph-states have a definite angular momentum, it is more convenient to define a ph-interaction by means of (f) and assocate diagrams like (g) in an obvious way with matrix elements of this ph-interaction. We can visualize (g) as indicating that the particle j_2 annihilates the hole j_1^{-1}, sending their combined angular momentum J via the interaction to be picked up by the particle-hole pair $(j_1'^{-1}, j_2')$.

In a diagram like (19.11), every "ph-bubble" carries the same angular momentum J determined by the initial pair. This may no longer be the case if we have a situation like that shown in (h):

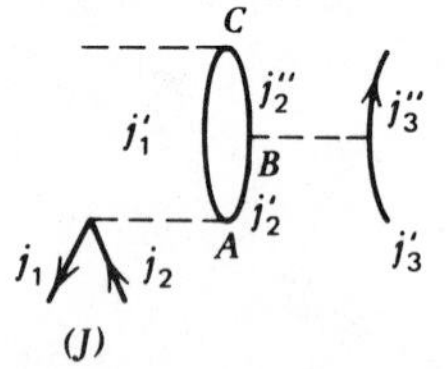

(h)

Here the ph-pair $j_1'^{-1} j_2'$ is still formed at A with an angular momentum J, but at B particle j_2' interacts with another particle j_3'. The interaction conserves the total angular momentum of particles 2 and 3, but j_2'' can differ from j_2'. Thus when j_2'' annihilates j_1' at C, they can transfer via the interaction line an angular momentum differing from J.

In these diagrams the particles and holes can be in any of the corresponding states that are included in the time-dependent HF approximation or in the definition (6.4) of the collective particle-hole operator $\Omega^\dagger$. Each particle-hole pair is coupled to the same angular momentum J (Section VII.7), which is also the angular momentum of the collective state considered.

If we want to express this approximation in terms of a diagonalization of the Hamiltonian in a truncated space, then we see from (19.11) that we have to choose this space in the following way: construct all possible particle-hole excitations with total angular momentum J; these will be states of the type $|j_i^{-1}j_m; J\rangle$. Now construct states of the same parity with three such *ph*-excitations, five, seven, . . . , coupling each pair to the same J, and coupling all the J's again to J:

$$|j_i^{-1}j_m(J), j_i'^{-1}j_m'(J), j_i''^{-1}j_m''(J), \ldots; J\rangle \tag{19.12}$$

Diagonalize the Hamiltonian in the space of these special 1, 3, 5, . . . *ph*-excitations. Do the same to the ground state in the space of the similar special 0, 2, 4, . . . *ph*-excitations. The energy difference between the two (collective) states is the excitation energy given by the RPA.

We notice that a shell-model diagonalization procedure is normally taken in a different way. Thus if one *ph*-excitation requires an energy $1\hbar\omega$, we would normally tend to diagonalize the Hamiltonian in the subspace of all states that have about the same excitation energy (first-order perturbation theory calls for the diagonalization of H in the subspace of degenerate states). This is then called the $1\hbar\omega$ Tamm–Dancoff approximation and in terms of Goldstone diagrams it corresponds to adding all diagrams of the type shown in (19.13).

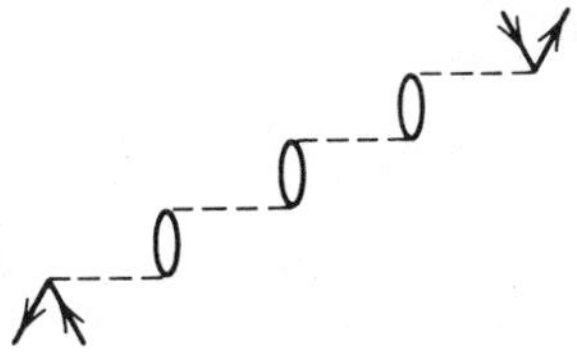

(19.13)

In going to higher-order Tamm–Dancoff approximation, say that of $3\hbar\omega$, the shell model does *not* restrict each pair to have the same angular momentum J as in (19.12); indeed it can have states like

$$|j_i^{-1}j_m(J_0), j_i'^{-1}j_m'(J_0'), j_m''^{-1}j_m''(J_0''), \ldots; J\rangle \tag{19.14}$$

where the only requirement is that $\mathbf{J}_0 + \mathbf{J}_0' + \mathbf{J}'' + \ldots = \mathbf{J}$. In terms of diagrams it will then have, in addition to (19.11), also diagrams like (19.15):

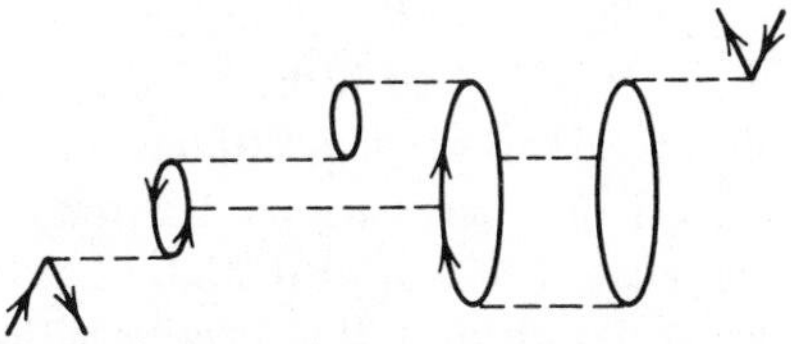

(19.15)

It seems that for the computation of properties of collective states the RPA is better suited than the complete diagonalization in truncated spaces of the type generated by (19.14) or by the corresponding diagrams (19.15). Figure

19.1 shows the 3^- state in ^{208}Pb as computed also in the RPA (broken lines). We see that for the upper state, RPA and partial diagonalization give more or less the same result. However, for the lowest, collective, state RPA gives a lower energy closer to the experimental value. The effect is seen even more dramatically in computing the electric octupole transition rate as shown in Table 19.3. COP and CAL are two different two-particle forces that were

TABLE 19.3 Comparison of the Energy and Octupole Transition Rate in Weisskopf Units (G (3)) for the Lowest Octupole States in ^{208}Pb Using Either Diagonalization within a Subspace or RPA with the Same Levels. Energies are in MeV [from Gillet (66)]

Force	Approximation	$V_0 =$	-40	-50
COP	I	Energy	3.15	2.64
		$G(3)$	6.01	7.04
	RPA	Energy	2.74	1.62
		$G(3)$	12.94	25.55
CAL	I	Energy	3.27	2.77
		$G(3)$	6.10	7.04
	RPA	Energy	2.75	1.45
		$G(3)$	13.91	30.59

used in this calculation [see Gillet (66)] and V_0 is the strength of the interaction. Approximation I consists of complete diagonalization of the interaction within a certain truncated subspace, that is, it contains all states, including those given by (19.14). The RPA calculation was confined to the same subspace. $G(3)$ is the octupole transition rate corrected for the energy dependence and given in Weisskopf units (Chapter VIII). We see from this table that the transition rate from the lowest 3^- state is enhanced by roughly a factor of 4 in going from the description in terms of Approximation I to the RPA. This enhancement is required to bring the calculated rate into agreement with experiment, and thus shows the usefulness of the RPA. It is possible that the RPA turns out to be as successful as it is because of its equivalence to the time-dependent Hartree–Fock approximation. The latter provides an intuitively easily acceptable description of collective vibrations. But it should be emphasized that the exact nature of the approximation involved in the RPA is not fully understood. Thus it has not been possible to give a reliable estimate of the terms neglected in the RPA, and in some cases [see, for instance, Gillet and Sanderson (67) and Barrett and Kirson (72)], the RPA tends to overemphasize the collectivity of some states. The same ideas, that is, of TDA and RPA have been applied to the heavier nuclei where the quasi-particle description of Bogoljubov and Valatin has been used [see Sawicki (69) and Klein (72)].

The Bohr Hamiltonian

The association of independent degrees of freedom with the single-particle potential is of particular importance for the parameters that describe the quadrupole deformations of this potential. Higher multipole deformations are necessarily connected with higher kinetic energy for the corresponding surface waves. We can therefore expect quadrupole deformations to be the most abundant ones. This, indeed, is found to be true, and they have therefore attracted special attention. The Bohr Hamiltonian, which deals with the quadrupole deformations of the single-particle potential, was studied at great length in Chapter VI. We shall not go therefore into its detailed review here except for a brief outline of the recent work of Kumar and Baranger (68) that is the most extensive and consistent treatment presently available of the Bohr Hamiltonian.

Kumar and Baranger consider a system of A-particles interacting via the pairing-plus quadrupole interaction (actually the pairing interaction is assumed to pair off only nucleons of the same kind: p–p and n–n pairs; furthermore, in the quadrupole–quadrupole interaction the exchange terms are neglected). They then develop the time-dependent Hartree–Fock–Bogoljubov theory (actually a Hartree–Bogoljubov approximation, since the exchange terms in the interaction are neglected), and because of the quadrupole–quadrupole nature of the two-body interaction it is evident that the time-dependent average potential displays quadrupole oscillations. Using the adiabatic approximation they calculate the total energy of the A-particles in the slowly oscillating potential and use it as the potential energy for the oscillations. Transforming then to intrinsic axes (through the requirement that they coincide at any moment with the principal axes of the ellipsoidal potential) they obtain explicit expressions for the mass parameters for the rotational and vibrational kinetic energies and for the potential energy for the β- and γ-vibrations. They are thus equipped to solve numerically the equations of motion of the Bohr Hamiltonian and relate the results to the fundamental features of the interaction.

Actually another simplification is introduced that is consistent with the whole approximation of using the pairing-plus-quadrupole Hamiltonian. These simplified forces make sense, if at all, only if they are limited to particles near the Fermi surface. A pairing interaction that is valid *between all states* can be shown to lead to a force of an infinite range, and a quadrupole–quadrupole interaction, if it becomes effective between all pairs of nucleons, would have led to a needlelike nucleus [see Baranger and Kumar (68)]. Thus the whole picture makes sense only if the calculations are limited to the "valence" nucleons, and proper approximate methods are developed to deal with the core, its deformation, and possible polarization [see Baranger and Kumar (68)].

Figure 19.4 shows the calculated results for the excited states of the even–even isotopes of $_{74}W$, $_{76}Os$, and $_{78}Pt$ together with the experimental values for these states. It is noteworthy to point out that these results were obtained

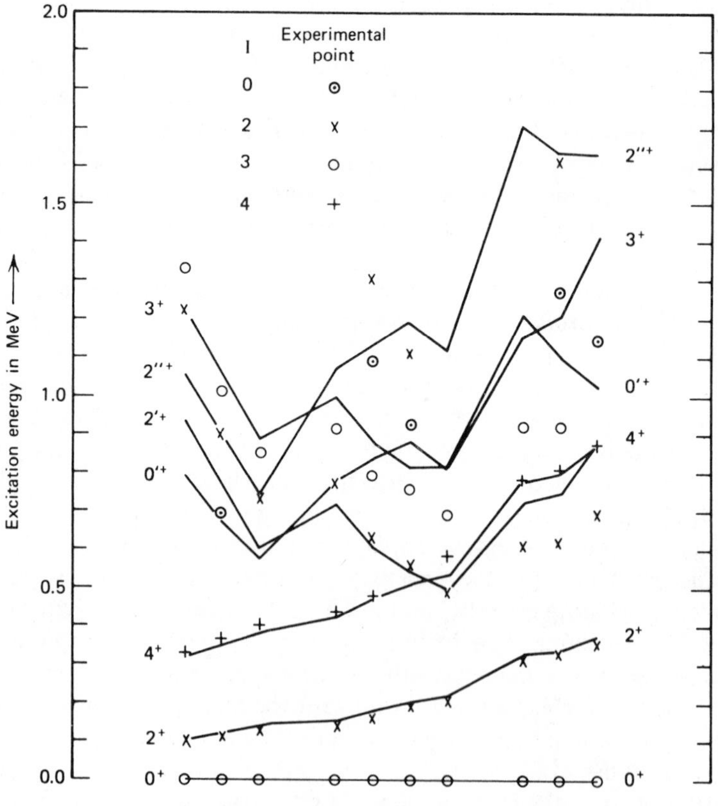

FIG. 19.4. Energy levels of isotopes of W, Os, and Pt. The calculated values are connected by straight lines. The unconnected symbols are the experimental values.

using the single-particle states in two major shells, with a pairing potential whose strength was taken to be (see Eq. 19.10)

$$G = \begin{cases} \dfrac{27}{A} \text{ MeV for protons} \\ \\ \dfrac{22}{A} \text{ MeV for neutrons} \end{cases}$$

and a quadrupole–quadrupole interaction whose strength varies slightly from nucleus to nucleus ($\pm 4\%$) to fit the energy of the first 2^+ state, and is given by

$$g = (75 \pm 3)A^{-1.4} \text{ MeV}$$

The validity of the Bohr Hamiltonian as a dynamical theory for these nuclei is clearly indicated by Fig. 19.4. For the light W isotopes we find ourselves close to the rotational limit with large equilibrium deformations, whereas for the heavier Pt isotopes we are closer to a pure vibrational model. The transition in between shows how the nucleus gradually develops its full deformation as one goes further away from closed shells.

Actually these calculations have been used also to derive electro-magnetic properties of these nuclei. In analogy with other calculations it is assumed that the polarizability of the core endows the valence nucleons with an extra effective charge. This charge does not really reside on the nucleon; instead it is associated directly with the polarization of the core. Since, however, the polarization moves with the polarizing nucleon this effective charge can be ascribed to the nucleon in the computations of moments and transition probabilities that match the polarization of the core. Thus for the computation of quadrupole moments and quadrupole transition probabilities the quadrupole deformation of the core induces an *effective charge for the quadrupole operator*. In the calculations of Kumar and Baranger the proton's effective charge is taken as $e_p = [1 + 1.7(Z/A)]e$, and that of the neutron is $e_n = 1.7(Z/A)$. With these values for e_p and e_n Kumar and Baranger computed the $B(E2)$ values for quadrupole radiation between different levels as shown in Table 19.4. They also computed nuclear magnetic moments as shown in Table 19.5. One sees that in these more sensitive tests their nuclear wave functions seem to reproduce the data very well. One cannot but feel that despite the crude approximation involved in taking the pairing-plus-quadrupole interaction to replace the actual nuclear interaction, there is enough in it to mimic an important part of the real interaction. On the other hand we should also be aware of the fact that a quadrupole–quadrupole interaction is particularly suited to describe quadrupole deformations in nuclei, and it is not impossible that the

TABLE 19-4

B ($E2$; $i \to f$) values in $e^2 \times 10^{-48}$ cm^4 [from Kumar and Baranger (68)]

i	f	^{182}W	^{184}W	^{186}W	^{186}Os	^{188}Os	^{190}Os	^{192}Os	^{192}Pt	^{194}Pt	^{196}Pt
0^+	2^+	4.019 (4.15)	3.745 (3.66)	3.501 (3.55)	2.950 (3.11)	2.731 (2.75)	2.595 (2.55)	2.576 (2.15)	1.812 (2.09)	1.704 (1.93)	1.428 (1.63)
0^+	$2'^+$	0.027	0.085	0.154 (0.171)	0.190 (0.209)	0.184 (0.261)	0.143 (0.250)	0.035 (0.204)	0.005	0.005 (0.008)	0.022
0^+	$2''^+$	0.173 (0.125)	0.135 (0.141)	0.083	0.042	0.022	0.011	0.007	0.001	0.001	0.001
$0'^+$	2^+	0.170	0.231	0.264	0.158	0.112	0.079	0.041	0.065	0.141	0.215
$0'^+$	$2'^+$	3.000	2.501	1.857	0.770	0.621	0.527	0.714	0.420	0.307	0.244
$0'^+$	$2''^+$	0.844	1.748	3.125	2.475	2.564	2.521	2.340	1.398	1.263	1.161
2^+	$2'^+$	0.104	0.176	0.302 (0.072)	0.256 (0.100)	0.403 (0.157)	0.539 (0.293)	0.743 (0.354)	0.555	0.451 (0.228)	0.306

2^+	$2''^+$	0.016	0.007	0.002	0.002	0.003	0.003	0.000	0.000	0.000	0.001
		(0.045)	(0.050)								
2^+	3^+	0.084	0.098	0.104	0.104	0.092	0.069	0.018	0.003	0.002	0.009
2^+	4^+	2.169	2.070	1.972	1.632	1.509	1.429	1.405	1.004	0.963	0.853
		(2.20)	(1.98)	(1.62)	(1.52)	(1.45)	(1.19)	(1.00)		(0.85)	(0.68)
$2'^+$	$2''^+$	0.989	0.737	0.401	0.035	0.001	0.003	0.010	0.029	0.025	0.043
$2'^+$	3^+	0.356	0.561	0.804	0.978	1.046	1.077	1.030	0.774	0.694	0.543
$2'^+$	4^+	0.042	0.032	0.020	0.000	0.002	0.005	0.000	0.001	0.001	0.004
$2''^+$	3^+	1.044	0.743	0.471	0.257	0.233	0.250	0.282	0.219	0.260	0.138
$2''^+$	4^+	0.067	0.120	0.185	0.104	0.076	0.046	0.044	0.042	0.052	0.094
3^+	4^+	0.054	0.078	0.121	0.119	0.194	0.261	0.297	0.226	0.181	0.116

Experimental values in parenthesis.

TABLE 19.5. Magnetic Moments in nm [from Kumar and Baranger (68)]

State	^{182}W	^{184}W	^{186}W	^{186}Os	^{188}Os	^{190}Os	^{192}Os	^{192}Pt	^{194}Pt	^{196}Pt
2^+	0.522 (0.48)	0.557 (0.55)	0.570 (0.60)	0.609 (0.63)	0.613 (0.5)	0.593 (0.56)	0.590 (0.60)	0.425 (0.5)	0.440 (0.5)	0.458 (0.54)
$2'^+$	0.509	0.519	0.521	0.535	0.539	0.536	0.568	0.441	0.448 (0.3)	0.469
$2''^+$	0.520	0.527	0.521	0.554	0.561	0.561	0.567	0.416	0.431	0.467
3^+	0.796	0.822	0.820	0.876	0.861	0.839	0.871	0.643	0.668	0.701
4^+	1.024	1.092	1.115	1.210	1.219	1.181	1.176	0.826	0.854	0.884

Experimental values in parentheses are taken from the 1967 Table of Isotopes.

very impressive success of the calculations of Kumar and Baranger is related to this fact.

20. MICROSCOPIC THEORY OF VALENCE FORCES

The question now arises: can one derive the effective potentials among the valence particles from fundamental principles, that is, relate them to the nucleon–nucleon interaction? Such a theory would proceed by dividing the system into a core, and valence particles and describe the interaction between the core and the valence particles, and among the valence particles. It would describe the effect of the interaction of the valence particles with the core on the interaction of the valence particles. This effect is of great importance as we shall see. The interaction between a valence particle and the core will excite the core. The latter can be deexcited by a further interaction with another valence particle. This process, known as *core excitation*, will then give rise to an effective force between the valence particles since two of them will have shifted their state, as a consequence of their interaction with the core, while the core has returned to its original state. More picturesquely, this process can be described as a polarization of the core by one of the valence particles, the polarization affecting the motion of the other valence particles.

Not only is it necessary to justify the use of the effective potential but also the use of a truncated Hilbert space referred to as a model space (see Section 19). In a shell-model calculation, for example, it is not possible to include all the levels to which the valence particles can be virtually excited. A few are selected, those in which the particles are placed according to the independent-particle approximation (Chapter IV), as well as those levels that are nearby. As we have repeatedly emphasized, at best only a few such levels can be considered as the calculation becomes rapidly impractical with an increasing number of levels.

The fundamental equation for the effective interaction acting between valence particles, analogous to the Goldstone linked-cluster result has been given by Brandow (67). The procedure used owes a great deal to Eden and Francis (55) and Bloch and Horowitz (58). Suppose that the model space includes a given set of states. Suppose moreover that these states are eigenstates of an unperturbed Hamiltonian H_0. In practice H_0 is the independent-particle model operator $T + U$ where U is a one-body operator, that is, the shell-model potential. The eigenstates are denoted by Φ_i. The model space consists of a subset of these states. Let P be a projection operator that when acting on an arbitrary state vector Ψ projects out that component of Ψ that is in the model space. P is given by

$$P = \sum_i |\Phi_i\rangle\langle\Phi_i| \qquad \Phi_i \text{ in the model space} \tag{20.1}$$

The complementary operator Q projects on to the rest of the Hilbert space:

$$Q \equiv 1 - P \tag{20.2}$$

Obviously

$$\begin{aligned} P^2 &= P \\ PQ &= QP = 0 \\ Q^2 &= Q \end{aligned} \tag{20.3}$$

We now seek the effective interaction in the model space by eliminating the complementary Q-space from the equation

$$H\Psi = E\Psi \tag{20.4}$$

satisfied by the exact state vector Ψ. The Hamiltonian H is written $H_0 + V$ so that (20.4) becomes

$$(E - H_0 - V)\Psi = 0 \tag{20.5}$$

Note that P and Q commute with H_0. To eliminate the Q-space, that is, to find the equation for $P\Psi$ [Feshbach (62)] operate with P on the left-hand side:

$$(E - H_0)P\Psi - PV\Psi = 0$$

or using (20.2)

$$(E - H_0 - V_{PP})P\Psi = V_{PQ}Q\Psi \tag{20.6}$$

where

$$V_{PP} \equiv PVP \qquad \text{and} \qquad V_{PQ} \equiv PVQ \tag{20.7}$$

By operating with Q on the left side of (20.5) one similarly obtains

$$(E - H_0 - V_{QQ})(Q\Psi) = V_{QP}(P\Psi)$$

Solving this equation formally for $Q\Psi$

$$Q\Psi = \frac{1}{E - H_0 - V_{QQ}} V_{QP}(P\Psi)$$

and substituting the result in (20.6) yields

$$\left(E - H_0 - V_{PP} - V_{PQ}\frac{1}{E - H_0 - V_{QQ}}V_{QP}\right)(P\Psi) = 0 \tag{20.8}$$

We see that the effective interaction $PV_{\text{eff}}P$ is given by

$$V_{\text{eff}} = V + VQ\frac{1}{E - H_0 - V_{QQ}}QV \tag{20.9}$$

This equation can be rewritten as an integral equation for V_{eff} by the following algebraic manipulation:

$$V_{\text{eff}} = V + VQ\frac{1}{E - H_0}QV + VQ\left(\frac{1}{E - H_0 - V_{QQ}} - \frac{1}{E - H_0}\right)QV$$

$$= V + VQ\frac{1}{E - H_0}QV + VQ\frac{1}{E - H_0}QVQ\frac{1}{E - H_0 - V_{QQ}}QV$$

In the last equation replace

$$VQ\frac{1}{E - H_0 - V_{QQ}}QV \quad \text{by} \quad (V_{\text{eff}} - V)$$

according to (20.9). Hence finally

$$V_{\text{eff}} = V + VQ\frac{1}{E - H_0}QV_{\text{eff}} \tag{20.10}$$

Equations 20.9 and 20.10 show that V_{eff} is energy dependent, that is, the iterated solution of (20.10) has the character of the Brillouin–Wigner perturbation theory. It is desirable to obtain an energy-independent V_{eff}. It would also be useful to separate the core and valence energies since shell-model theory or any of the more sophisticated theories described earlier in this chapter permit $E - E_c$ to be calculated where E, the energy in (20.10), is the total energy of the system and E_c is the ground-state energy of the core. Fortunately both of these goals can be simultaneously achieved. We shall not describe this process in detail but shall refer the reader to Brandow (67) as well as to review articles by MacFarlane (69) and Barrett and Kirson (72). Diagrammatic considerations very similar to those that were employed to obtain the Goldstone linked-cluster expansion for nuclear-matter energy are used to obtain an expression for the core energy. Elimination of the core energy replaces (20.10) by

$$\mathcal{V}_{\text{eff}} = V + V\frac{Q}{E_v - H_0^{(v)}}\mathcal{V}_{\text{eff}} \tag{20.11}$$

where $H_0^{(v)}$ is just the original Hamiltonian H_0 minus the unperturbed core energy E_{0c}, that is the core energy determined with H_0. The energy E_v is likewise just $E - E_c$. It is important to remember that matrix elements of $\mathcal{V}_{\text{eff}}$ are taken only in the model space. It is actually possible to eliminate the dependence on E_v and use $(E_0)_v$ in (20.11). This involves the use of "folded diagrams." The description of this concept and its consequences would require too great a digression. The reader is referred to Brandow (67), Johnson and Baranger (70), and the review article of Barrett and Kirson. Our further discussion requires only (20.11).

To make further progress it is useful to have a specific situation in mind. Following McFarlane (69) let us consider nuclei with two-valence nucleons with closed-shell cores such as ^{18}O, ^{42}Sc, and ^{90}Zn. We shall therefore be concerned with two-particle matrix elements of V_{eff}. Intermediate excited states that occur in such terms as

$$V + V \frac{Q}{E_v - H_0(v)} V$$

can involve two-particle excited states in which both valence nucleons are excited into empty levels, the core remaining unexcited, or one-particle excited states the core still quiescent, and finally states in which the core is excited, with nucleons promoted from the core to empty or partially filled orbitals (core excitation). In the first approximation it is customary to include as intermediate states the two-particle excited states with an undisturbed core and add in the one-particle and core-excitation contributions at a later time. For this case the matrix element of $\mathcal{V}_{\text{eff}}$ is

$$\langle ij|\mathcal{V}_{\text{eff}}|kl\rangle = \langle ij|V|kl\rangle + \sum_{ab,\text{unocc}} \frac{\langle ij|V|ab\rangle\langle ab|\mathcal{V}_{\text{eff}}|kl\rangle}{E_v - \epsilon_a - \epsilon_b} \tag{20.12}$$

This equation is similar to the one satisfied by G_0, the Bethe–Goldstone equation with Q_F replaced by Q_{2p} the projection operator into two-particle unfilled orbits, and the "starting energy" ϵ_0 being given by E_v. However for the case being considered here, $Q_F = Q_{2p}$ so that

$$\langle ij|\mathcal{V}_{\text{eff}}|kl\rangle = \langle ij|G_0(E_v)|kl\rangle \tag{20.13}$$

Note the difference from the preceding section in which G was determined self-consistently. In the present discussion the eigenstates defining the model space are given. Equation 20.13 states the $\mathcal{V}_{\text{eff}}$ and $G_0(E_v)$ are the same within the model space. This is an important result stating that within the model space and to the extent that core and one-particle excitations can be neglected the G_0-matrix and $\mathcal{V}_{\text{eff}}$ are identical.

To now determine the spectra of the nucleus with the two-valence particle we need only diagonalize the effective Hamiltonian with the matrix element

$$\langle ij|T_1 + T_2|kl\rangle + \langle ij|G_0(E_v)|kl\rangle$$

It is however necessary to estimate the G_0-matrix at the energy E_v, which of course is to be determined by diagonalizing the effective Hamiltonian. Kuo and Brown (66) have introduced the approximation

$$E_v \simeq \tfrac{1}{2}\{[\epsilon(i) + \epsilon(j)] + [\epsilon(k) + \epsilon(l)\} \tag{20.14}$$

To evaluate the matrix element of G_0, Kuo and Brown (66) and Kuo (67) [see also Brown (67)] use the harmonic-oscillator method described in the previous section. Separating out the center-of-mass wave functions, the matrix element of G_0 is evaluated by assuming that G_0 is a function only of the relative coordinate. The separation method is used so that G_0 is written as a sum of G_s, the short range G-matrix, and a long range component. The approximations that the Pauli principle can be disregarded and that the plane-wave description of the intermediate states is adequate for the calculation of G_s is made while the effect of the long range component is calculated perturbatively using the long range part of V, V_l, as determined for example by the separation method (see also Eq. 17.6 et seq.). If

$$V = V_s + V_l \tag{20.15}$$

and

$$G_s(\omega) = V_s + V_s \frac{Q}{\omega - H_0} G_s$$

Then by eliminating V_s from the equation for G_0 we obtain

$$G_0(\omega) = G_s(\omega) + \left(1 + G_s \frac{Q}{\omega - H_0}\right) V_l \left(1 + \frac{Q}{\omega - H_0} G_0\right) \tag{20.16}$$

and

$$G_0(\omega) = G_s(\omega) + \left(1 + G_0 \frac{Q}{\omega - H_0}\right) V_l \left(1 + \frac{Q}{\omega - H_0} G_s\right) \tag{20.17}$$

By substituting the second of these equations into the first one obtains a second approximation for $G(\omega)$

$$\begin{aligned} G_0(\omega) \simeq G_s(\omega) &+ \left(1 + G_s \frac{Q}{\omega - H_0}\right) V_l \left(1 + \frac{Q}{\omega - H_0} G_s\right) \\ &+ \left(1 + G_s \frac{Q}{\omega - H_0}\right) V_l \left(\frac{Q}{\omega - H_0} + \frac{Q}{\omega - H_0} G_s \frac{Q}{\omega - H_0}\right) V_l \\ &\times \left(1 + \frac{Q}{\omega - H_0} G_s\right) \end{aligned} \tag{20.18}$$

According to (20.13) we are interested in the matrix element of G and therefore on the effect of G acting on an uncorrelated wavefunction $\phi_{ij} \equiv |ij\rangle$. Defining the correlated wave function $\psi_{ij}^{(s)}(\omega)$ by the equation

$$\psi_{ij}^{(s)}(\omega) = \phi_{ij} + \frac{Q}{\omega - H_0} G_s \psi_{ij}^{(s)}(\omega) \tag{20.19}$$

The superscript s indicates that only short range correlations have been incorporated in $\psi^{(s)}$. Employing (20.18) and (20.19) the matrix element of $G_0(\omega)$ becomes

$$\begin{aligned} \langle ij | G_0(\omega) | kl \rangle \simeq \langle ij | G_s(\omega) | kl \rangle &+ \langle \psi_{ij}^{(s)}(\omega) | V_l | \psi_{kl}^{(s)}(\omega) \rangle \\ &+ \langle \psi_{ij}^{(s)}(\omega) | V_l \frac{Q}{\omega - H_0} V_l | \psi_{kl}^{(s)}(\omega) \rangle \end{aligned} \tag{20.20}$$

where a term involving V_l twice and G_s is assumed to be third order.

The last term is of particular importance for treating the tensor-force component of V_l since, as we have discussed earlier, it is necessary to take the tensor interaction, particularly in the intermediate range, into account to at least second order. However just for that order Kuo and Brown assert that the evaluation of the last term of (20.20) can be simplified. The tensor force, (they consider the Hamada–Johnston potential) connects the wave function $\psi_{kl}^{(s)}(\omega)$ for the energies ω of interest with intermediate states lying in a rather narrow energy range, with a median energy of about 200 MeV (see Fig. 20.1). Under these circumstances

$$V_{Tl}\frac{Q}{\omega - H_0}V_{Tl} \simeq -\frac{1}{\overline{E}}(V_{Tl})^2 = -\frac{1}{\overline{E}}(S_{12})^2 v_{Tl} \qquad (20.21)$$

where $\overline{E}$ is approximately given by the median energy described above and where V_{Tl} is the "long" range part of the tensor force. The form of the tensor force is

$$V_{Tl} = S_{12} v_{Tl}(r)$$

where $S_{12} \equiv 3\boldsymbol{\sigma}_1\cdot\hat{\mathbf{r}}\,\boldsymbol{\sigma}_2\cdot\hat{\mathbf{r}} - \boldsymbol{\sigma}_1\cdot\boldsymbol{\sigma}_2$ and v_{Tl} gives the radial part of V_{Tl}. From

$$S_{12}^2 = 6 - 2S_{12} + 2\boldsymbol{\sigma}_1\cdot\boldsymbol{\sigma}_2$$

and placing $\boldsymbol{\sigma}_1\cdot\boldsymbol{\sigma}_2$ equal to one since S_{12} vanishes for singlet states we obtain

$$V_{Tl}\frac{Q}{\omega - H_0}V_{Tl} \simeq \left(\frac{-8 + 2S_{12}}{\overline{E}}\right) v_{Tl}^2 \qquad (20.22)$$

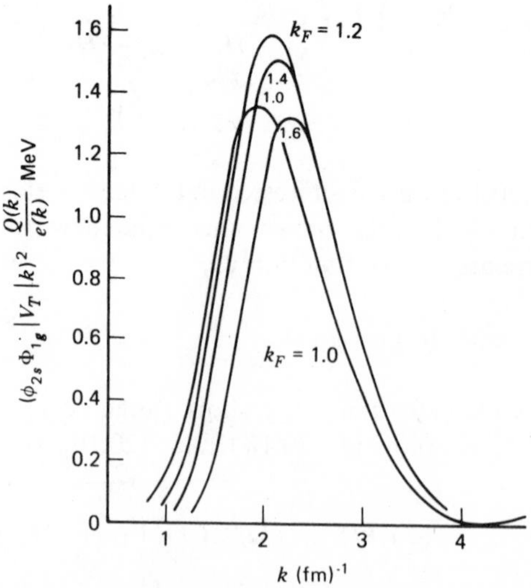

FIG. 20.1. The second-order contribution for the (2s, 1g) state as a function of the momentum k of the intermediate state, using the long range part of the triplet even Hamada-Johnston potential [from Brown (67)].

Thus in second order one obtains the familiar result that the tensor force contributes a strong attractive central force plus a weaker tensor term. The surmise (20.21) is verified by direct calculation [Kuo (67)]. Direct calculation is in any event superior but approximation (20.22) is very useful.

Another term that also contributes to the matrix element of G in an important way is the core excitation term in which the intermediate states involve the core in an excited state. Recall that G_0 does not contain any core excitation. For a two-valence particle system a nuclear state will not only contain the two-particle state but also three-particle, one-hole states (abbreviated $3p1h$), $4p2h$ states, etc. These last involve core excitations in which one particle (in the $3p1h$ case) or two particles (in the $4p2h$ case) are excited out of the core and placed in unoccupied states. Diagrammatically the core excitation process is shown in Fig. 20.2. The simplest (and therefore the first one to try) calculation of the effect of core excitation takes into account the $3p1h$ states only. However we need not specify the nature of the core excitations in developing an appropriate perturbation theory. Let us decompose Q into two parts

$$Q = Q_{2p} + \delta Q \tag{20.23}$$

where Q_{2p} projects on to that part of the Hilbert space for which the core is quiescent and both valence particles change levels. δQ projects onto the remainder of the excited state space, most importantly, the core-excited states. The equation for $G(\omega)$ reads

$$G(\omega) = V + V\frac{Q_{2p} + \delta Q}{\omega - H_0}G(\omega) \tag{20.24}$$

Let $G_0(\omega)$ be the G-matrix we have been dealing with up to this point (20.13)

$$G_0(\omega) = V + V\frac{Q_{2p}}{\omega - H_0}G_0(\omega)$$

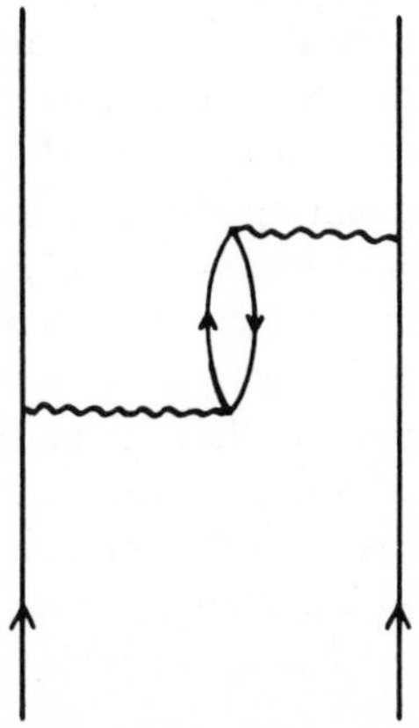

FIG. 20.2. A core excitation diagram.

It is a simple exercise to eliminate V between these two equations to obtain

$$G = G_0 + G_0 \frac{\delta Q}{\omega - H_0} G \tag{20.25}$$

The first approximation yields

$$G \simeq G_0 + G_0 \frac{\delta Q}{\omega - H_0} G_0 \tag{20.26}$$

where δQ

$$\delta Q = \sum_{3p1h} \phi_{3p1h}\rangle\langle\phi_{3p1h} + \sum \phi_{4p2h}\rangle\langle\phi_{4p2h} + \ldots \tag{20.27}$$

Inserting (20.26) into (20.13) to obtain the final matrix elements gives results that are similar to those given by ordinary perturbation theory except that the perturbing potential is not the bare interaction V but G_0

With the development of the approximate formulas for G_0 (20.20) and G (20.26) an approximate expression for the effective potential acting between valence particles becomes available. The physical consequences that result will be developed in course of the comparison with experiment that will now be presented.

As a first example, we consider the calculation of Kuo and Brown (66) of the spectrum of ^{18}O. They used the Hamada–Johnston (62) potential for the nucleon–nucleon potential. The results are shown in Fig. 20.3. The levels labeled a were calculated with an inert core, that is, using G_0 of (20.20). The calculation of the levels labeled b included the particle hole core excitation leading to the $3p1h$ intermediate states. Thus the effect of the first term of (20.27) was calculated. We see that the effect of the core excitation is to bring the low-lying levels down substantially, the resultant agreement with experiment is very good indeed. The only empirical data employed are the single-particle level energies that were taken from the observed levels in ^{17}O. Similarly in the Ni isotopes the inclusion of the core-excitation term improves the comparison with experiment substantially (Fig. 20.4).

Bertsch (65) has pointed out the close relationship between the core-excitation mechanism and the pairing-plus quadrupole force discussed in the preceding section. The matrix elements of the G-matrix calculated between particle pairs that are coupled to give $J = 0$, can be compared with the matrix elements of the pairing force used by Kisslinger and Sorensen. This comparison is shown in Fig. 20.5, indicating that the matrix elements of the effective interaction as given by G are of the correct order of magnitude. That the match with the data is good has already been demonstrated in Fig. 20.4.

The formal reason for the connection between core excitation and the expansion in multipoles (pairing plus quadrupole are the first terms) is quite

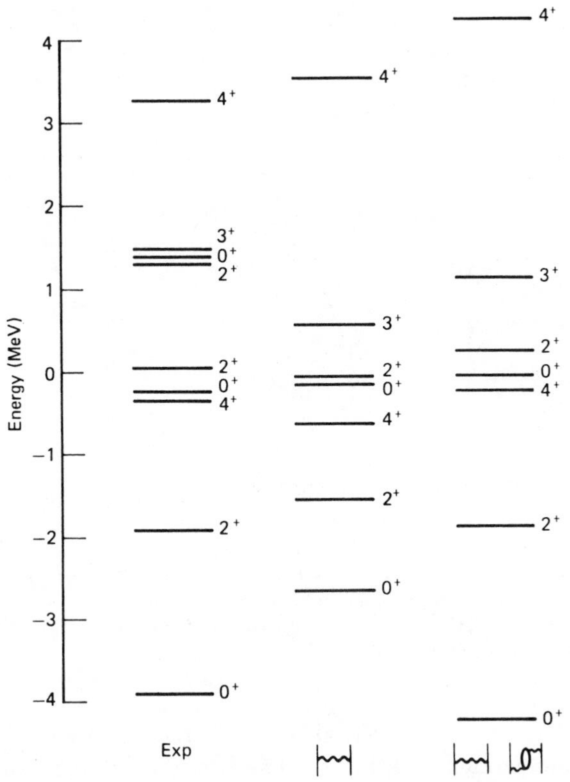

FIG. 20.3. The spectrum of ^{18}O (*a*) calculated using only G_0 (*b*) including core excitation [from Kuo and Brown (66)].

simple. (See also Remark p. 637) In (20.26), both G and G_0 are scalars. Thus in a typical matrix element (omitting the energy denominator) corresponding to Fig. 20.6:

$$\langle j_{a'}|g_{c\gamma}|j_a\rangle\langle j_{b'}|g_{\gamma c}|j_b\rangle \tag{20.28}$$

where

$$g_{c\gamma} \equiv \langle j_c|G_0|j_\gamma\rangle$$

it is clear that $g_{c\gamma}$ must transform as a spherical tensor of order J'' where $(j_c + j_\gamma) > J'' > |j_c - j_\gamma|$. But since G itself must transform as a scalar, the actual matrix elements of G must involve a sum of products like (20.28) each of which involves a product of tensors with identical value of J''. This means that the matrix element and, hence, the particle-hole pair in Fig. 20.6 carry a specific angular momentum corresponding to the multipole order.

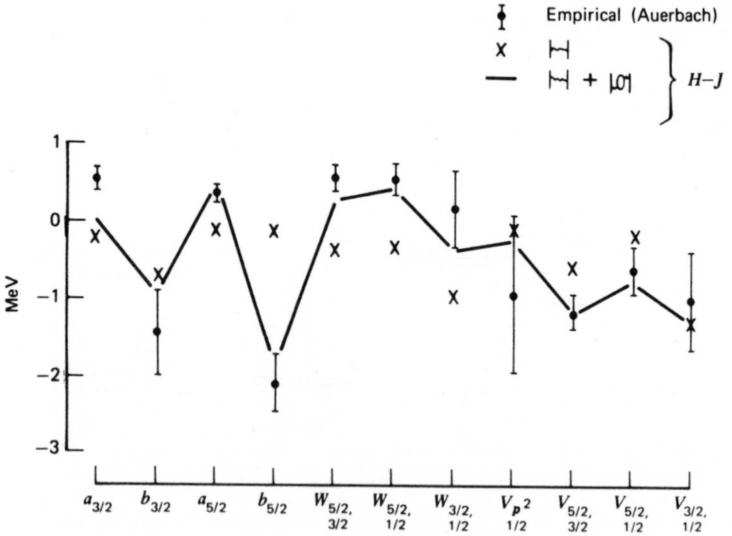

FIG. 20.4. Effective interaction in the Ni isotopes. Various parameters $a_{3/2}$, etc. along the abscissa are combinations of two-body matrix elements obtained by Auerbach (66) from the Ni spectrum. The X give the results without the core-excitation effect; the solid line includes it [from Brown (67)].

Thus the particle-hole pair that gives rise to the quadrupole force must have an angular momentum of 2. Bertsch has asserted and direct calculations have substantiated that the even multipoles make the biggest contribution to the core-excitation process. Figure 20.7 shows a comparison with the empirical values of Kisslinger and Sorensen. Again excellent results are obtained.

The question arises as to the dependence of this agreement on the nature of the nucleon–nucleon potential. D. Clement and E. Baranger (68) have tested this dependence by carrying out the calculation of the effective potential and the level structure of ^{18}O for the soft Tabakin (64) potential. Their results are shown in Fig. 20.8 and compared to those of Kuo (67) based on the Hamada–Johnston potential. It is clear that there is no substantial difference in the calculated spectra.

These results would seem to indicate that a real understanding of the interaction between valence particles has been achieved by these considerations. However before this conclusion is secure it is necessary to understand the errors involved in the various approximations. Agreement with experiment of a badly understood approximation is not enough for a sound theory. The recent review of Barrett and Kirson (72) arrives at the pessimistic conclusion that the agreement with experiment is not comprehensible at the present time.

Actually the agreement is not as good as the above discussion would seem to indicate. For example, Kuo and Brown (66) obtained good agreement for

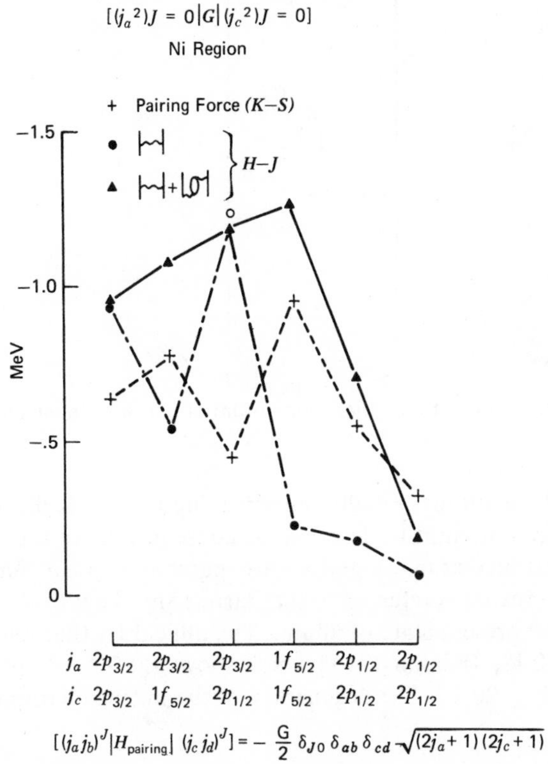

FIG. 20.5. Pairing component of the effective interaction. Computation is compared with pair force used by Kisslinger and Sorensen (60) [from Brown (67)].

$A = 18$ nuclei using only the $3p1h$ terms in (20.27). But when a better calculation of G was made by Kuo (67) the agreement was lost. It could only be resolved if in (20.27) the $4p2h$ terms and also the $3p1h$ terms were retained. If this were generally true the simple relationship between the second-order core excitation and the pairing plus quadrupole would not be strictly valid.

But beyond this a systematic investigation of the higher-order effects has been carried out by several authors, for example, Barrett and Kirson (72) and Kirson and Zamick (70). They have classified the higher-order effects into three types: (1) "vertex," (2) "propagator," (3) "screening." If we consider the simple second-order core-excitation process, vertex renormalization refers to the introduction of higher-order processes between the particle-hole diagram and the particle line as illustrated in Fig. 20.9. Of course there are many other possibilities including the exchange of upward-going lines. The "propagator" correction corresponds to the introduction of additional bubble diagrams as in Fig. 20.10. The effect of these can be calculated using

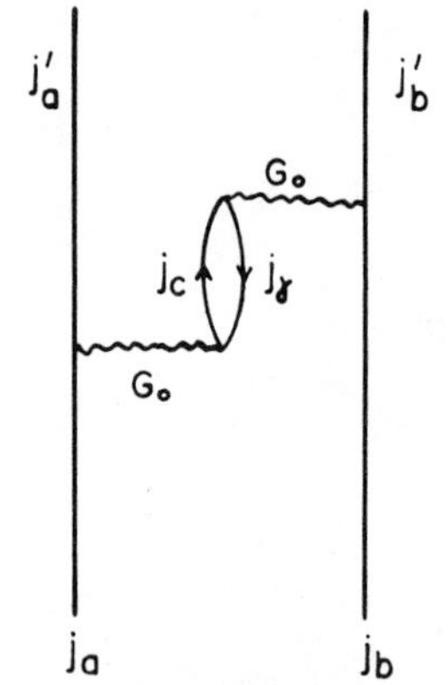

FIG. 20.6. Core excitation with angular momentum of particles involved indicated.

the TDA or RPA methods. Finally the screening process is illustrated in Fig. 20.11. We shall not attempt to describe the consequences of these effects since it seems to us that further investigations are required before a complete picture will emerge. The reader is referred to the Barrett and Kirson (72) review for a description of the present state of affairs. The difficulties that remain are illustrated by Fig. 20.12, [Kirson (71)] which shows the $T = 1$ spectrum of ^{18}O (also shown in Fig. 20.3). The degeneracy of the independent-particle picture

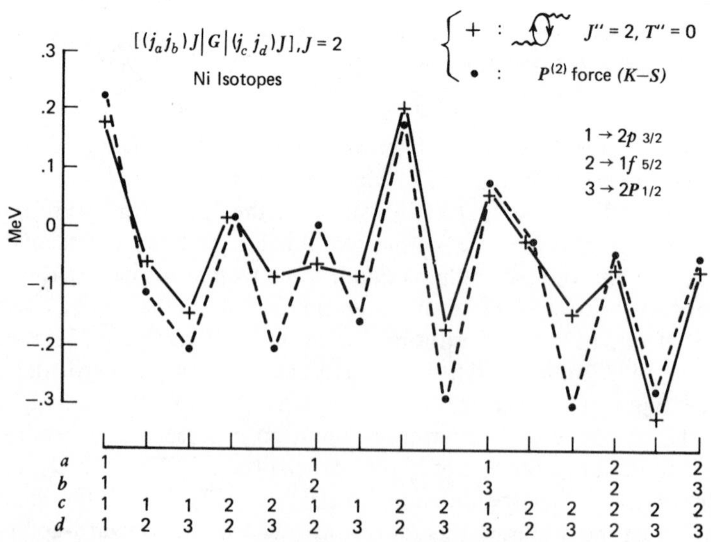

FIG. 20.7. The $J''= 2$, $T = 0$ part of the core-excitation term compared with the quadrupole-quadrupole force used by Kisslinger and Sorensen (60) [from Brown (67)].

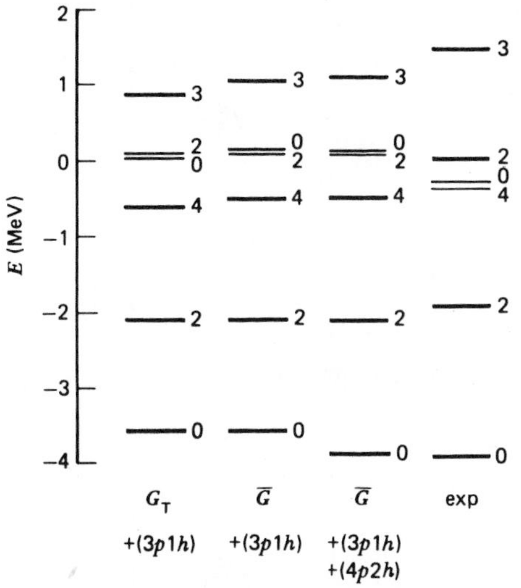

FIG. 20.8. Calculated and observed spectrum of ^{18}O. The $3p,1h$ label corresponds to the inclusion of $3p1h$ intermediate states, while $4p2h$ indicates the inclusion $4p2h$ intermediate states. G_T refers to the use of the Tabakin potential [Clement and E. Baranger (68)] while $\overline{G}$ has been evaluated using the Hamada-Johnston potential [Kuo (67)] [from McFarlane (69)].

is lifted when the interaction potential G_0 is used. Inclusion of the second-order core excitation gives rise to spectrum labeled $3p1h$. The propagator can be corrected for by using forward-going particle-hole pairs as in Fig. 20.10. This is labeled TDA and gives quite good agreement with experiment. If, however, both forward and backward graphs are included, the RPA approximation is obtained. Too much binding is obtained. However, when screening effects (Fig. 20.12) are included, the spectrum labeled nRPA is obtained. The vertex effects in Fig. 20.9 give rise to the bbRPA spectrum while the last spectrum bbnRPA includes both vertex, screening and RPA processes summed to all orders in G_0. The bbRPA spectrum is almost identical with the G_0 spectrum while the bbnRA is even further away from agreement with experiment. It is clear that the series of approximations involves several large terms of opposite sign so that the answer obtained is dependent upon where the approximation stops.* This is hardly a satisfactory situation. Barrett and Kirson (72) point to the effect of collective states as one of the sources of this slow convergence. A related problem originates in the tensor force. Even after the tensor force is "smoothed" by the replacement of V by G, the resultant G remains too

*The TDA approximation appears to achieve the best agreement with experiment!

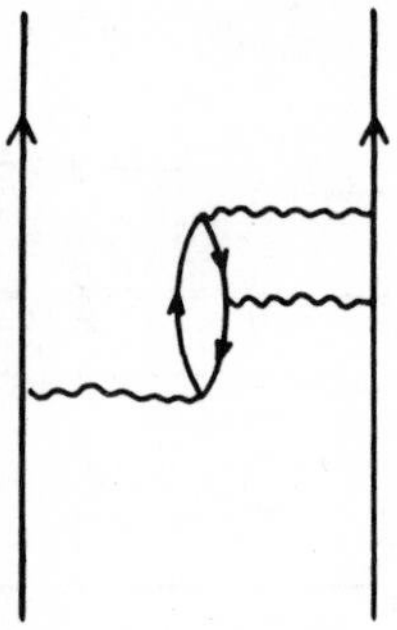

FIG. 20.9. Vertex-correction diagram.

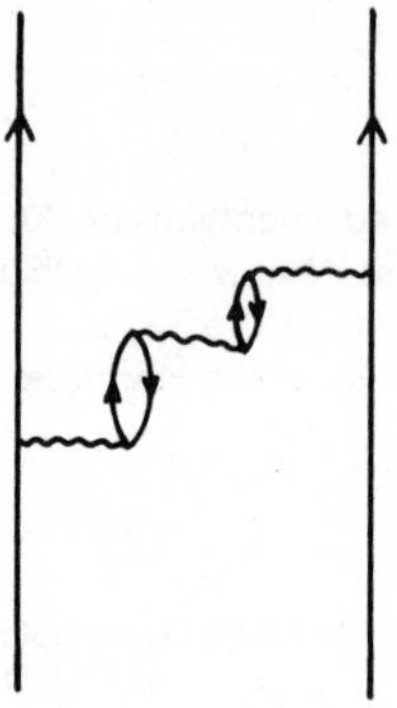

FIG. 20.10. Propagator-correction diagram.

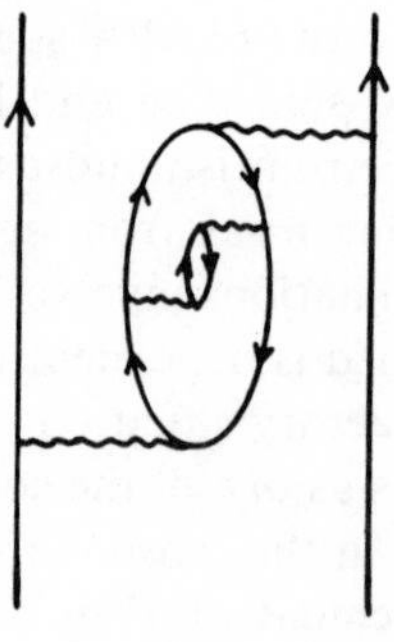

FIG. 20.11. Screening.

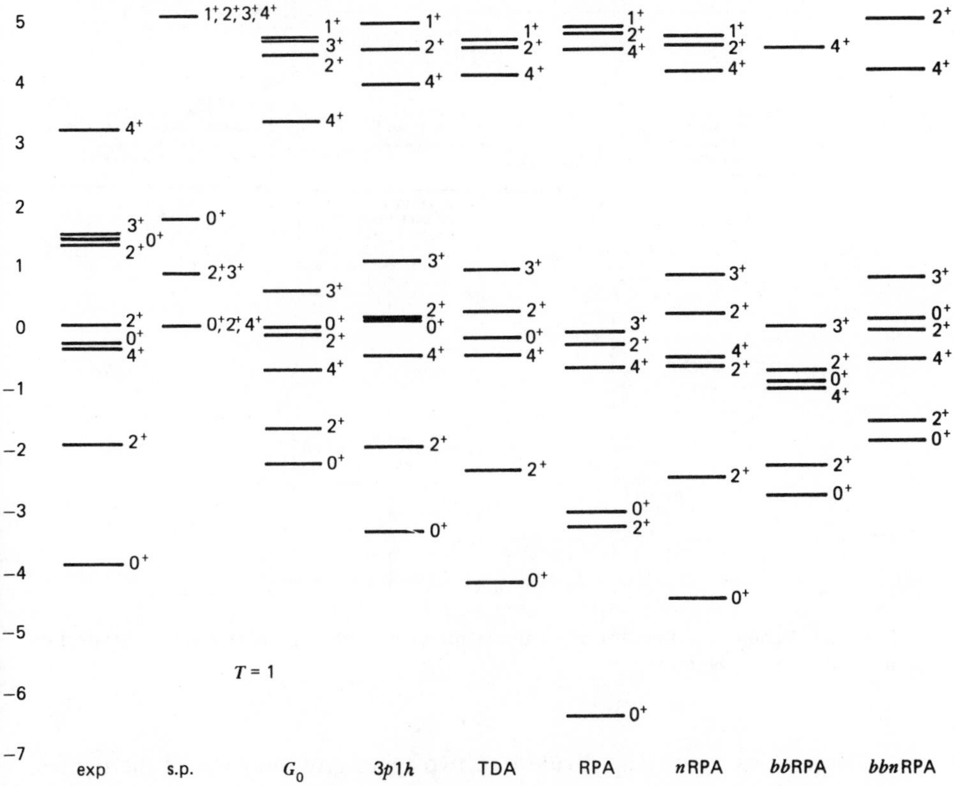

FIG. 20.12. Spectrum of ^{18}O calculated by Kirson (71) under different approximations in the calculation of the effective interaction [from Barrett and Kirson (72)].

strong for perturbation treatment [Barrett (71)]. Although we have gained great qualitative insights into the nature of the low-lying states of nuclei, their quantitative description still remains an unsolved problem.

21. DESCRIPTIVE SUMMARY

In Figs. 21.1 and 21.2 we have tried to put together the various approaches to the solution of the finite many-body problem with their interrelations. The central problem is that of solving the many-body Schrödinger equation either through the diagonalization of the nuclear Hamiltonian $H = \Sigma Ti + [v(ij)]$ or, equivalently, through the summation of all linked diagrams in the Goldstone expansion.

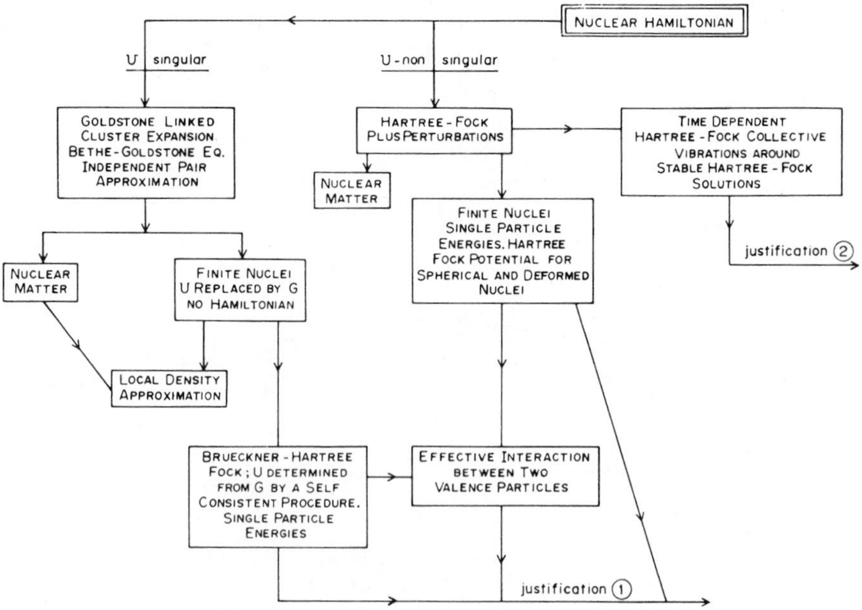

FIG. 21.1. Schematic diagram of methods for direct calculation of nuclear properties from the nuclear Hamiltonian.

The problems naturally divide into two types given in each of the figures, although there is a great deal of overlap. In Fig. 21.1 the emphasis is on the direct solution of the nuclear problem starting from the nucleon–nucleon forces. In Fig. 21.2 the main emphasis is on the understanding of the properties of the nucleus and its excited states. The generalized shell model, comprising a spherical or deformed single-particle field for all the nucleons, as well as interactions between the particles leading to configuration mixing is the main concept employed in these discussions. The single-particle potential and mutual interactions are often described semiempirically. Recently it has become possible to use "realistic" potentials obtained from the effective interactions derived from the *G*-matrix. These are generally presented as matrix elements within a limited subspace [see E. Halbert et al (71) and E. Baranger (71) for examples]. Another portion of Fig. 21.2 deals with collective motion whose description was also originally semiempirical, but by now contains "microscopic" elements. The introduction of dynamics into the single-particle potential is probably the simplest way to account for collective features of the nucleus. This may be achieved by considering the possible oscillations of the Hartree–Fock potential about its equilibrium value (see Fig. 21.1). This turns out to be equivalent to the RPA. Another procedure involves the Bohr Hamiltonian (Fig. 21.2) that is an effective Hamiltonian describing collective

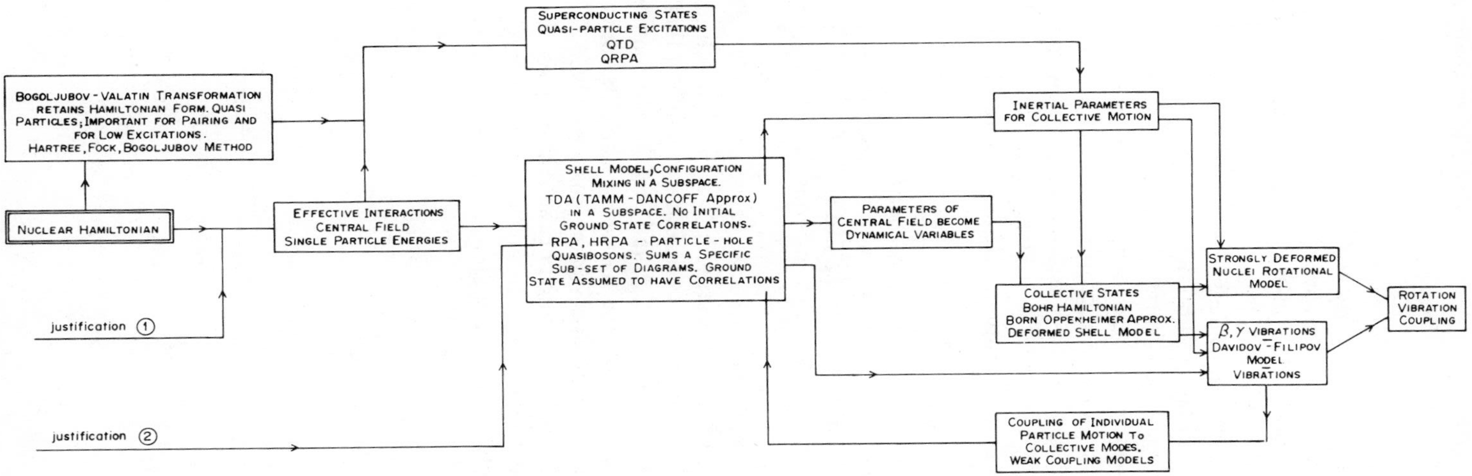

FIG. 21.2. Schematic diagram of methods and relationships among them for the calculation of nuclear properties through effective interactions that are either semiempirical or justified by calculations indicated in Fig. 21.1.

motions. The inertial parameters of this model are obtained from a microscopic description of the associated particle motion, for example, the independent particle or the quasi-particle models.

Single-particle motion, vibrations, and rotations are examples of fundamental modes of motion of nuclei. But none of these exist in a "pure" form. To some extent these deviations can be described in terms of coupling between these modes, e.g. between rotations and vibrations. When the pure forms provide relatively accurate descriptions, the effect of these couplings will naturally be weak (a "circular" statement!) and weak-coupling models such as the core-excitation model (see Chapter VIII) will be useful. The nature of the coupling between fundamental modes as well as the discovery of new modes by, for example, using new projectiles, broader energy, and angular ranges but better energy and angle resolution, polarizations, etc., are at the frontiers of our subject.

Another frontier area is shown in Fig. 21.1. The passage from the nucleon–nucleon potential to the effective interactions between particles and to the single-particle potential and energies is still not understood, as emphasized in the last section. The calculations described in Fig. 21.2 remain to some extent ad hoc until quantitative justifications in terms of the nucleon–nucleon potentials are achieved.

The extent to which nuclear structure can be completely understood in terms of the free nucleon–nucleon interaction in conjunction with the Schrödinger equation is thus still an open question. But it does seem that with the various approaches summarized in Fig. 21.1 and 21.2 we are coming close to answering this question. Thus far all the indications are that quantum mechanics as formulated for the atom works for the nucleus as well. It is a great tribute to human ingenuity that on the basis of the relatively scanty information on atoms and their radiations a logically consistent theory was developed that turns out to be equally applicable under conditions so vastly different from those encountered at the atomic level.

APPENDIX

HARMONIC-OSCILLATOR WAVE FUNCTIONS; SU(3) AND THE QUADRUPOLE-QUADRUPOLE INTERACTION

If we work with *HO* wave functions we can use a further symmetry of the single-particle Hamiltonian to generate convenient wave functions, as we shall explain below:

Writing the HO Hamiltonian in the form

$$H_0 = \mathbf{r}^2 + \alpha^2\mathbf{p}^2 = (\mathbf{r} + i\alpha\mathbf{p})\cdot(\mathbf{r} - i\alpha\mathbf{p})$$

we see that $x - i\alpha p_x$, etc. are creation operators for the quanta of the *HO*, whereas $x + i\alpha p_x$, etc. are annihilation operators for these quanta. Out of the three creation operators and three annihilation operators it is possible to construct nine bilinear operators that leave the number of oscillator quanta unchanged and therefore commute with H_0. It is convenient to group them in three groups:

(i) The scalar product of $\mathbf{r} + i\alpha\mathbf{p}$ and $\mathbf{r} - i\alpha\mathbf{p}$ that gives back H_0.

$$(\mathbf{r} + i\alpha\mathbf{p}) \cdot (\mathbf{r} - i\alpha\mathbf{p}) \tag{A.1}$$

(ii) The antisymmetric tensor constructed from these two vectors, that is, their vector product that gives the orbital angular momentum

$$\mathbf{L} = (\mathbf{r} \times \mathbf{p}) \tag{A.2}$$

(iii) The symmetric tensor, which has five components, conveniently written as

$$\mathrm{Q}_q = \sqrt{\frac{4\pi}{5}}\,[r^2 Y_{2q}(\mathbf{\Omega}_r) + \alpha^2 p^2 Y_{2q}(\mathbf{\Omega}_p)]$$

where $\mathbf{\Omega}_r$ and $\mathbf{\Omega}_p$ are the directions of $\mathbf{r}$ and $\mathbf{p}$ respectively.

$$(A.3)$$

The nine operators (A.2) and (A.3) are the generators of the group of unitary transformations in three dimensions, and H_0 is invariant under all the transformations of these groups. (We note that these include as a subgroup the rotations in three dimensions as given by $\mathbf{L}$.) In the same way that an invariance of a Hamiltonian under rotations leads to a classification of its eigenstates according to the eigenvalues of $\mathbf{J}$ and $\mathbf{J}_z$, the invariance of H_0 under the larger group $U(3)$ suggests a corresponding classification of the eigenstates of H_0. Actually $U(3)$ is slightly too general for physical purposes, since an overall change in phase of the wave function has no physical significance; one can then limit oneself to unimodular transformations only, and dropping the identity operator, be left with the eight generators of the group $SU(3)$.

We shall not go into the details of this classification of *HO* wave functions. The reader is referred to the original work of Elliott (58) on this subject, and to a simplified, two-dimensional model discussed by Lipkin (61) [see also Brown (64), p. 79]. We only want to point out that the particular quadrupole–quadrupole two-body interaction

$$v_{QQ}(12) = -gr_1^2r_2^2 \sum_m Y_{2m}^*(\theta_1\phi_1)Y_{2m}(\theta_2\phi_2) \tag{A.4}$$

also commutes with the operators (A.3) and (A.4). It is therefore possible to achieve a more complete diagonalization of the A-particle Hamiltonian

$$H = \sum_{i=1}^{A} \left(\frac{1}{2M}p_i^2 + \frac{1}{2}M\omega^2r_i^2\right) + \frac{1}{2}\sum_{i,j} v_{QQ}(i,j) \tag{A.5}$$

if we take its matrix elements with the states that belong to irreducible representations of $SU(3)$ [see Gasiorowicz (67)]. The Hamiltonian (A.5) is, of course, not a realistic one. It turns out, however, to reproduce several nuclear properties, such as deformations and rotational states, rather well. It is therefore believed that even with realistic interactions the use of *HO* wave functions that belong to the irreducible representations of $SU(3)$ may lead to particularly small off-diagonal elements in the matrix of the Hamiltonian and thus make approximate diagonalization easier. This method has been applied extensively for nuclei in the $2s1d$ shell ($9 < Z$, or $N \leqslant 20$) which is the only shell-model major shell that coincides completely with a *HO* shell (the other shells are either trivial, like the first and the second shells, or contain one level from the next *HO* shell because of the strong spin-orbit interaction). For an extensive study of the use of *HO* wave functions see Moshinsky (68).

CHAPTER VIII

ELECTROMAGNETIC TRANSITIONS

1. INTRODUCTION

Qualitatively it is clear that the nuclear forces are stronger than electromagnetic forces, which are much larger than gravitational forces. In this intuitive scale the forces that determine the internal structure of the elementary particles (defined to have baryon number less than or equal to one in magnitude) appear to be considerably stronger than nuclear forces while the interaction responsible for β-decay is considerably weaker than electromagnetic forces but still much stronger than gravitational forces. These qualitative remarks are often made quantitative by giving a dimensionless "coupling constant" describing the interaction between the particle and the field that transmits the force in the simplest possible diagram shown in Fig. 1.1. The solid lines represent the interacting particles while the broken line represents the transmitted field. In the electromagnetic case the broken line represents a photon. The particle-photon coupling strength at A is of the order of $e/\sqrt{\hbar c}$. Since this factor occurs a second time at B, the electromagnetic coupling is proportional to $e^2/\hbar c \sim 1/137$. If the exchanged particle is a π-meson we obtain the long range weak part of the nucleon-nucleon force, the so called OPEP potential. Using the so called pseudovector coupling (see Volume II) the coupling constant $g^2/\hbar c$ turns out to be about .081 (see Eq. I.3.7) about 10-times larger than the electromagnetic coupling constant. Of course this does not mean that this component of the nuclear force is stronger than the electromagnetic force for all interparticle distances since the Coulomb force is of infinite range while the nuclear force is of short range. The nuclear force is much stronger for these interparticle distances that are less than exchanged particle's Compton wavelength which for a π-meson is 1.4 fm. The force carried by the π-meson is only one (and a weak one at that) component of the nuclear force so that the Coulomb force is more than one order of magnitude weaker than the nuclear force.

The electron-neutrino field associated with β-decay can also give rise to forces between nucleons. At a distance equal to the π-meson Compton wavelength this force is the order of 10^{-14} times the OPEP potential at that point. The gravitational force is still weaker. The ratio of the gravitational force to

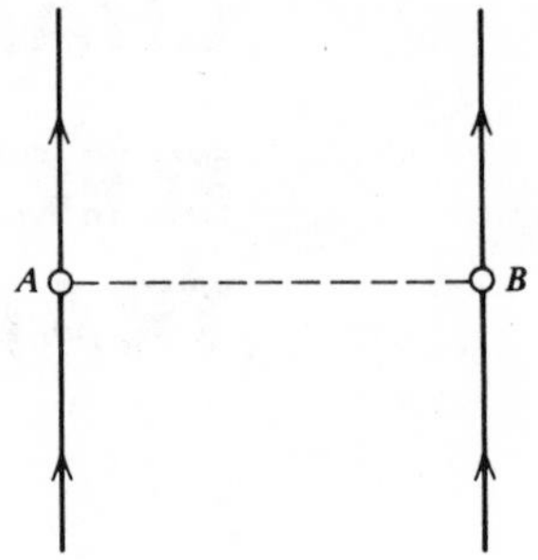

FIG. 1.1. Exchange diagram for interacting particles.

the Coulomb force is Gm^2/e^2 where G is the gravitational constant. This ratio is of the order of 10^{-25}.

Thus far in this volume our primary interest has been in the strong nuclear forces and in the associated nuclear structures that they dominate. In this chapter and the next we shall be concerned with the weak forces associated with the electromagnetic and the electron-neutrino fields. We shall be particularly interested in the use of these interactions as probes of nuclear structure.

The use of light to examine a structure should of course occasion no surprise. Radiative transitions form the basis upon which our understanding of atomic structure is founded and as expected play an important role in our investigations of the nucleus. The new feature that enters the scene in the nuclear case is the presence of a second weakly coupled field, the isospin- and spin-dependent electron-neutrino field. This field can also be used, with some important practical differences, to obtain "pictures" of the nucleus. To the new degree of freedom the isospin, which distinguishes nuclear from atom physics, nature has provided as a sort of compensation, an additional weak field. This process is continued in the elementary-particle area where one finds additional fields involving the new degree of freedom of hypercharge.

Let us be somewhat more precise about the word "weak." Its operational meaning is that the effect of weak fields on nuclei can be calculated with sufficient accuracy with first-order perturbation theory. Stated in another way the change in the nuclear wave function and the field wave function that occur because of the weak interaction can be neglected in calculating the effects of the corresponding weak interactions. Speculations have been made that assert that for energy and momentum changes well above the ranges that are of interest for nuclear physics, the weak interactions become strong in some sense. An example exists for the electromagnetic field in which a photon converts via the weak-coupling strength to the strongly interacting ρ-meson [Gottfried and Yennie (69)]. The possibility of a similar transformation occurring for the electron-neutrino field has been suggested but the particle corresponding to the ρ has not been found. In any event these possibilities give rise to effects that are unobservable in typical nuclear-physics experiments at least with present experimental techniques.*

Let us be somewhat more explicit as to how the weakness of the coupling make it possible to use these fields to probe nuclear structure. Take the electromagnetic field as an example. In that case the interaction Hamiltonian responsible for radiative transitions is given by

$$H_{\text{int}} = -\frac{1}{c}\int \mathbf{j} \cdot \mathbf{A}(\mathbf{r})\, d\mathbf{r} \tag{1.1}$$

where $\mathbf{j}$ is the nuclear-current density and $\mathbf{A}$ is the vector potential in the radiation gauge

$$\operatorname{div} \mathbf{A} = 0 \tag{1.2}$$

Both $\mathbf{j}$ and $\mathbf{A}$ are quantum mechanical field operators. Equation 1.1 assumes that the corresponding field quanta, for example, the photons for $\mathbf{A}$ and the particles for $\mathbf{j}$ are elementary, that is, have no internal structure. The matrix element for the transition from a state consisting of a nucleus in its initial state $|i\rangle$ and zero photons to a state consisting of one photon of frequency ω (energy $\hbar\omega$) and a corresponding final nuclear state $|f\rangle$ is given by

$$M = \langle f, 1_\omega | -\frac{1}{c}\int \mathbf{j} \cdot \mathbf{A}\, d\mathbf{r} | i, 0_\omega \rangle \tag{1.3}$$

As we shall show below, to first order in $\mathbf{A}$, it is possible to factor M into a matrix element of the electromagnetic field that is readily calculable and a nuclear matrix element:

$$-\frac{1}{c}\,\boldsymbol{\varepsilon}\cdot\int d\mathbf{r}\, \langle f|\mathbf{j}\, e^{i\mathbf{k}\cdot\mathbf{r}}|\, i\rangle \qquad k \equiv \frac{\omega}{c} \tag{1.4}$$

where $\boldsymbol{\varepsilon}$ is the polarization vector of the emitted wave and $\mathbf{k}$ is its momentum in units of $\hbar$. Note that $(\boldsymbol{\varepsilon}\cdot\mathbf{k}) = 0$. We see that measurements of M (actually is it often $|M|^2$ that is measured) will provide us with information on the nature of the nuclear electromagnetic current for energy transfers equal to $\hbar\omega$ and for corresponding momentum transfers $\hbar\omega/c$. As one can readily imagine, the

*Of course nuclear wave functions are certainly modified by the Coulomb interaction between protons and for some transitions these modifications play an important role. There are other second-order effects that are observable because of the nuclear charge such as the reorientation effect in heavy ion scattering.

investigation of the dependence of (1.4) on $\mathbf{k}$ and $\boldsymbol{\varepsilon}$ will yield sensitive measures of the components of the nuclear wave functions and thus test any nuclear model that purports to develop such wave functions.

A decisive feature is the long wavelength of characteristic nuclear radiation compared to the nuclear radius. If E is the energy of the gamma ray in MeV and R the nuclear radius in fermis, we obtain

$$kR = \frac{ER}{197 \text{ MeV fermi}} \tag{1.5}$$

Thus for gamma-ray energies of the order of several MeV and for even large nuclei $kR \ll 1$. As a consequence in the expansion $\mathbf{j}\, e^{i\mathbf{k}\cdot\mathbf{r}}$ in (1.4) in a power series in $\mathbf{k}\cdot\mathbf{r}$ (see Appendix VIII.A) the first nonvanishing term usually (but not always) dominates. Actually as we shall see in Section 4 it is more convenient to expand in terms of quantities that behave like spherical tensors under spatial rotations. This is the *multipole expansion.* A multipole of order L transfers angular momentum L in radiative transitions. In order to conserve angular momentum such transitions must then satisfy the selection rules:

$$|J_f - J_i| \le L \le J_f + J_i \qquad J_i = 0 \rightarrow J_f = 0 \quad \text{forbidden} \tag{1.6}$$

where J_i and J_f are the spins of the initial and final nucleus, respectively. Since the spin of the photon is 1, $J_i = 0$ to $J_f = 0$ transitions are forbidden. A second selection rule follows if the transition conserves parity. The multipoles divide into two types, *magnetic* and *electric.* The parity associated with a magnetic multipole of order L is $(-)^{L+1}$; the electric multipole of order L has the parity $(-)^L$, so that the odd-L magnetic multipoles and the even-L electric multipoles have the same parity. If the parity of the initial and final state are Π_i and Π_f, respectively, conservation of parity implies transitions will occur only for

$$\left.\begin{array}{ll}\text{Even-}L & \text{electric multipoles} \\ \text{Odd-}L & \text{magnetic multipoles}\end{array}\right\} \quad \text{if} \quad \Pi_i\,\Pi_f = 1$$

$$\left.\begin{array}{ll}\text{Odd-}L & \text{electric multipoles} \\ \text{Even-}L & \text{magnetic multipoles}\end{array}\right\} \quad \text{if} \quad \Pi_i\,\Pi_f = -1$$

The effect of $kR \ll 1$ can be stated as follows: the angular momentum barrier for photon emission selects the two lowest possible multipole orders of the same parity. A transition with $|J_f - J_i| = 1$ and no parity change will be dominated by the electric quadrupole ($L = 2$) or the magnetic dipole ($L = 1$). From this example we can see how determining the nature of a radiative transition, its polarity and electric and magnetic character will lead to information on the nuclear states involved in the transition. We shall derive all these results later in this chapter but they should be well known to anyone familiar with elementary quantum mechanics.

Nuclear models can involve additional characteristic selection rules. Thus determining their validity experimentally is a severe test of the model. See, for example, the discussion of the transition in ^{180}Hf from a $J^\pi = 8^+$ to an 8^- state (Section VI.6).

The information extracted from photon reactions is limited because the energy difference between the nuclear states, $\hbar\omega$, is directly proportional to the momentum carried off by the recoiling nucleus ($\hbar\omega/c$). This direct relationship is avoided if the reaction is induced by a charged particle. In such an inelastic scattering it becomes possible to transfer a momentum $\mathbf{q}$ (in units of $\hbar$) and an energy ω (in units of $\hbar$) where $q \neq \omega/c$. Elastic and inelastic scattering by charged particles are discussed in Volume II.

The electron-neutrino field also acts as a weak probe. Qualitatively speaking (for the more complete description see Chapter 9) the electromagnetic field in (1.1) is replaced by the electron-neutrino field while the nuclear electromagnetic current is replaced by a weak nuclear current. There are in fact two such currents, the *vector* and *axial vector* currents that we may provisionally characterize as leading to β-decay without (vector) or with (axial vector) nucleon-spin flip. Measurements of β-decay of nuclei provides information on these currents and thus once more on nuclear wave functions and nuclear models. Because the isospin and spin character of the weak interaction differs from that of the electromagnetic interaction, the components of the nuclear wave function under scrutiny in both cases will differ and therefore the information obtained via the two interactions will supplement each other.

From a practical point of view the experiments that can be conducted with the electron-neutrino field are limited. Because of the weakness of the coupling, enormous neutrino sources are required before induced reactions are observable. Some experiments of this sort have been done using the neutrinos produced in reactions [F. Reines and C. L. Cowan, Jr. (59)]. But for the most part, the electron-neutrino field has been observed in the spontaneous decay of radioactive nuclei. This is in contrast with the electromagnetic case for which the interaction and sources are sufficiently strong as to permit the ready observation of photon-induced reactions.

We shall begin the discussion with the electromagnetic interaction. It is assumed that the reader is familiar with the treatment of the quantized electromagnetic field as contained in such texts as Gottfried's (66).

2. THE NUCLEAR ELECTROMAGNETIC CURRENT

In the first approximation the nuclear current and charge are simply the sum of the currents and charges of the individual nucleons making up the nucleus, each of which are assumed to be point particles, that is, to have no internal structure. In the nonrelativistic limit we obtain

$$\rho(\mathbf{r}) = \sum_i \frac{1}{2} e\,[1 + \tau_3(i)]\,\delta(\mathbf{r} - \mathbf{r}_i) \tag{2.1}$$

and

$$\mathbf{j}(\mathbf{r}) = \mathbf{j}_c + \mathbf{j}_m \tag{2.2}$$

where

$$\mathbf{j}_c(\mathbf{r}) = \sum_i \frac{1}{2} e\,[1 + \tau_3(i)]\,\frac{1}{2}\,[\mathbf{v}_i\,\delta(\mathbf{r} - \mathbf{r}_i) + \delta(\mathbf{r} - \mathbf{r}_i)\,\mathbf{v}_i] \tag{2.3}$$

and

$$\mathbf{j}_m(\mathbf{r}) = \frac{e\hbar}{4m} \sum_i \frac{1}{2}\,[(g_p + g_n) + \tau_3(i)\,(g_p - g_n)]\,\text{curl}_r\,[\boldsymbol{\sigma}_i\delta(\mathbf{r} - \mathbf{r}_i)] \tag{2.4}$$

The current $\mathbf{j}_c$ is the convection current arising from the motion of the protons. The quantity $\mathbf{v}_i$ is the velocity operator for the ith particle. It is defined by

$$\mathbf{v}_i = \frac{i}{\hbar}\,[H, \mathbf{r}_i] = \frac{\partial H}{\partial \mathbf{p}_i} \tag{2.5}$$

In this equation H is the full Hamiltonian including the electromagnetic terms. H therefore contains the additional dependence on the vector potential $\mathbf{A}$ that is required for gauge invariance. The required gauge invariance for a nuclear Hamiltonian that in the absence of radiation has the functional dependence $H_0(\mathbf{p}_1 \ldots \mathbf{p}_A, \mathbf{r}_1 \ldots \mathbf{r}_A)$ is obtained by the substitution

$$\mathbf{p}_i \rightarrow \mathbf{p}_i - \frac{e}{2c}\,[1 + \tau_3(i)]\,\mathbf{A}(\mathbf{r}_i) \equiv \mathbf{D}_i \tag{2.6}$$

For example, if the dependence of H_0 on $\mathbf{p}_i$ is only via the kinetic energy, that is,

$$H_0 = \frac{1}{2m} \sum p_i^2 + V$$

then

$$H = \frac{1}{2m} \sum_i D_i^2 + V$$

From (2.5) we find the familiar result

$$\mathbf{v}_i = \frac{1}{m}\,\mathbf{D}_i$$

The vector $\mathbf{j}_m$ is the current associated with the spin of the nucleons. The quantity

$$\frac{e\hbar}{8m}\left[(g_p + g_n) + \tau_3(i)(g_p - g_n)\right] \boldsymbol{\sigma}_i\, \delta(\mathbf{r} - \mathbf{r}_i)$$

is just the volume magnetization of the ith particle.

The nuclear current and charge density are related by the continuity equation:

$$\operatorname{div} \mathbf{j} + \frac{\partial \rho}{\partial t} = 0 \tag{2.7}$$

The presence of momentum dependent terms in V will change the relation between $\mathbf{v}_i$ and $\mathbf{p}_i$. For example, suppose V contains spin-orbit forces [Jensen and Mayer (52)]*

$$V_{so} = \sum_k f(r_k)\, \boldsymbol{\sigma}_k \cdot \mathbf{l}_k$$

Since $\mathbf{l} = (1/\hbar)(\mathbf{r} \times \mathbf{p})$, gauge invariance requires that in the presence of an electromagnetic field, V_{so} becomes

$$V_{so} = \frac{1}{\hbar} \sum_k f(r_k)\, \boldsymbol{\sigma}_k \cdot \left[\mathbf{r}_k \times \left\{\mathbf{p}_k - \frac{e}{2c}(1 + \tau_3(k))\, \mathbf{A}(\mathbf{r}_k)\right]\right\}$$

The contribution of this term to the velocity $\mathbf{v}_i$ is readily computed from (2.5). We find

$$\mathbf{v}_k = \frac{1}{m}\mathbf{D}_k + \frac{1}{\hbar} f(r_k)(\boldsymbol{\sigma}_k \times \mathbf{r}_k) \tag{2.8}$$

The modification in the definition of $\mathbf{v}$ that occurs with potentials with more involved dependence on $\mathbf{p}$, such as nonlocal potentials, (see Volume II) can also be obtained using (2.5).

From (1.1) and (2.3) we find the following as the contribution of the convection current to the electro-magnetic interaction:

$$H_{\text{int}} = -\frac{e}{4c} \sum_i \left[1 + \tau_3(i)\right]\left[\mathbf{v}_i \cdot \mathbf{A}(i) + \mathbf{A}(i) \cdot \mathbf{v}_i\right] \tag{2.9}$$

---

*This case is given primarily as an example of how a velocity-dependent force can affect the definition of the current [see N. Austern and R. G. Sachs (51)]. However one must be certain that one is dealing with a true momentum dependence rather than an apparent dependence generated by an approximation. When dealing with an effective interaction, it is safer to apply the criterion of gauge invariance to the original Hamiltonian from which it is obtained and then approximate that resultant Hamiltonian.

3. EXCHANGE CURRENTS* AND THE EFFECTS OF FINITE NUCLEON SIZE

Equations 2.3 and 2.4 are in fact approximations. It will be recalled (see Chapter I, p. 17 and Volume II) that the nucleons are able to emit and absorb mesons such as pions, ρ-mesons, etc; the exchange of these particles giving rise to nuclear forces. But there is another consequence of the particle exchange: the presence of currents of these particles flowing between nucleons. These currents are called *exchange currents*. Since some of the bosons are charged, the exchange currents can interact with an external electro-magnetic field. It is thus apparent that (2.3) and (2.4) are incomplete. To the nucleon current one must add the currents of the various mesons that may be present in nuclei. The electromagnetic interaction is thus of the form $-1/c(\mathbf{j}_N + \mathbf{j}_B)\cdot\mathbf{A}$ where $\mathbf{j}_N$ is the nucleon current and $\mathbf{j}_B$ is the boson current.

For many of the problems of interest to us the mesons are virtual. (This is of course not true when the mesons are produced, absorbed or scattered by nuclei). In principle it is therefore possible to eliminate the meson coordinates and to express the entire Hamiltonian in terms of nucleon coordinates. A familiar example is provided by nuclear forces. This elimination can also be made for the electromagnetic terms, but with the effect that form (1.1) will no longer be correct. This conclusion can be arrived at intuitively with the aid of the diagrams, Figure 3.1 that show the interaction of a pair of nucleons that exchange a π^+ with the electromagnetic field, indicated by a wave line. Form (2.1) is adequate for the description of Fig. 3.1*a* and 3.1*b* but not for Figure 3.1*c*, the exchange current diagram. There is one exception to this general rule; in the infinite wavelength limit form, (1.1) is correct [(Estrada and Feshbach (63)]. This result is comparatively obvious since in this limit the electromagnetic field has no spatial dependence. As a consequence the interaction Hamiltonian (after elimination of the meson coordinates) contains the electromagnetic field as a simple factor. More formally this can be viewed as a consequence of Ward's identity [Bjorken and Drell (65) p. 299] which is a consequence of charge conservation. We shall see how this works out in the discussion that now follows.

We begin with the interaction Hamiltonian given by (1.1) where the *current* **j** *includes not only the nucleon current but also any boson currents that may be present*. When a photon of polarization $\boldsymbol{\varepsilon}$ is either emitted or absorbed **A** is proportional to $\boldsymbol{\varepsilon}\, e^{i\mathbf{k}\cdot\mathbf{r}}$ so that

$$H_{\text{int}} \sim \int (\mathbf{j}\cdot\boldsymbol{\varepsilon})\, e^{i\mathbf{k}\cdot\mathbf{r}}\, d\mathbf{r} \tag{3.1}$$

*For a recent discussion see Heller (71).

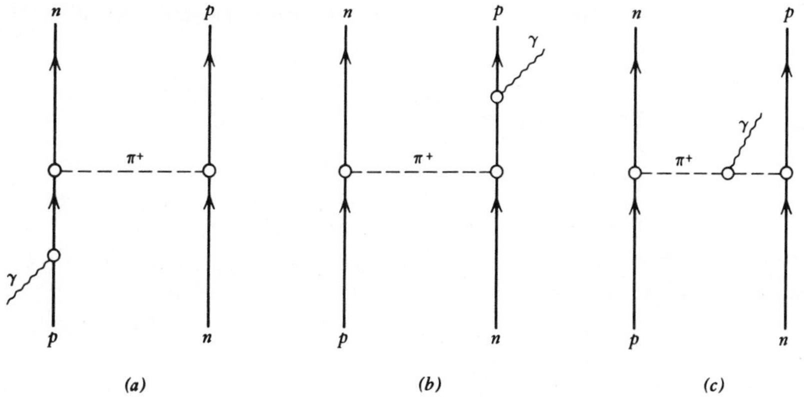

FIG. 3.1. Particle exchange and exchange currents.

We now make use of the vector identity

$$\boldsymbol{\varepsilon}\, e^{i\mathbf{k}\cdot\mathbf{r}} = \text{grad}\,(\boldsymbol{\varepsilon}\cdot\mathbf{r}\, e^{i\mathbf{k}\cdot\mathbf{r}}) - i\mathbf{k}\,(\boldsymbol{\varepsilon}\cdot\mathbf{r})\, e^{i\mathbf{k}\cdot\mathbf{r}}$$

In the long wavelength limit

$$\boldsymbol{\varepsilon}\, e^{i\mathbf{k}\cdot\mathbf{r}} \simeq \text{grad}\,(\boldsymbol{\varepsilon}\cdot\mathbf{r}\, e^{i\mathbf{k}\cdot\mathbf{r}}) + 0\,(kR) \tag{3.2}$$

where R is the size of the radiating system. We recall (Eq. 1.5) that for nuclear transitions, $kR \ll 1$. The factor $e^{i\mathbf{k}\cdot\mathbf{r}}$ is retained in (3.2) because we wish to retain the contribution of one term or possibly two (one electric, one magnetic) in the expansion of $\mathbf{j}e^{i\mathbf{k}\cdot\mathbf{r}}$ to multipoles that dominate. In (3.2) we have retained only the electric multipole terms as we shall see and our error term is kR times the dominant electric multipole term. Our conclusions will thus apply only to electric multipole transitions. Inserting (3.2) and (3.1) yields

$$H_{\text{int}} \sim \int \mathbf{j}\cdot\text{grad}\,(\boldsymbol{\varepsilon}\cdot\mathbf{r}\, e^{i\mathbf{k}\cdot\mathbf{r}})\, d\mathbf{r}$$

The assumption that $\mathbf{j}(\mathbf{r})$ goes to zero for large r with sufficient speed allows us to rewrite the above as

$$H_{\text{int}} \sim -\int (\text{div}\,\mathbf{j})\,(\boldsymbol{\varepsilon}\cdot\mathbf{r})\, e^{i\mathbf{k}\cdot\mathbf{r}}\, d\mathbf{r}$$

Or using the equation of continuity (2.7) yields

$$H_{\text{int}} \sim \int \frac{\partial\rho}{\partial t}\,(\boldsymbol{\varepsilon}\cdot\mathbf{r})\, e^{i\mathbf{k}\cdot\mathbf{r}}\, d\mathbf{r}$$

We make further progress if we now make use of the Heisenberg equation of motion:

$$\frac{\partial \rho}{\partial t} = \frac{i}{\hbar} [H, \rho] \tag{3.3}$$

where H is the total Hamiltonian including not only the nucleon terms but the boson terms as well. The density ρ is the charge density including the contributions of both the boson and nucleon fields. Now take the matrix element of H_{int} between the initial and final states and use (3.3). The initial state consists of a nucleus in a given energy state $|i\rangle$ plus zero bosons:

$$\text{Initial state} = |i, \text{o}\rangle$$

while the final state involves the final nucleus $|f\rangle$ and again zero bosons. It follows that

$$\langle f, \text{o}| H_{\text{int}} | i, \text{o}\rangle \sim \frac{i}{\hbar} (E_f - E_i) \langle f | \rho_0 \, \boldsymbol{\varepsilon} \cdot \mathbf{r} \, e^{i\mathbf{k}\cdot\mathbf{r}} | i \rangle \tag{3.4}$$

where ρ_0 is the expectation value of ρ, the total charge density, for the boson vacuum.

Comparing this result with (3.1) we obtain the theorem that in the matrix element of H_{int} for a radiative transition, the quantity $\mathbf{j}\cdot\boldsymbol{\varepsilon}$ in (3.1) may be replaced by

$$\mathbf{j}\cdot\boldsymbol{\varepsilon} \rightarrow \frac{i}{\hbar} (E_f - E_i) \rho_0 \, \boldsymbol{\varepsilon}\cdot\mathbf{r} \tag{3.5}$$

This is a generalization of Siegert's theorem [A. Siegert (37), N. Austern and R. G. Sachs (51)]. We note that the theorem is exact only in the limit $k \rightarrow 0$. In that limit (3.4) shows that the matrix element of H_{int} is proportional to the matrix element of the electric dipole moment $\rho_0 \mathbf{r}$. If this should vanish the next nonvanishing term involves the matrix element of $(\boldsymbol{\varepsilon}\cdot\mathbf{r})(\mathbf{k}\cdot\mathbf{r})$ which since $\boldsymbol{\varepsilon}\cdot\mathbf{k}$ is zero is a pure electric quadrupole. And indeed it can be verified that (3.4) involves only *electric* multipole moments and it is to the transitions induced by these that Siegert's theorem applies.

As can readily be seen from the derivation, the magnetization currents do not contribute to the approximate form (3.4) The spin magnetization current $\mathbf{j}_m$ (2.4) does not contribute since its divergence is zero corresponding to the fact that this current does not result in a net motion of charge. The orbital angular momentum and its associated current do not contribute for the same reason. Hence substitution (3.5) is valid only for electric multipole transition. The error is kR times the first nonvanishing term of (3.4).

The usefulness of (3.4) lies in the fact that the structure of ρ_0 is simpler than that of $\mathbf{j}$ in the original expression (1.1). A simple example will illustrate this point. Suppose ρ_0 is given by (2.1). Then

$$\frac{\partial \rho_0}{\partial t} = \frac{i}{\hbar} [H, \rho_0] = \frac{ie}{2\hbar} \sum_i [H, (1 + \tau_3(i)) \, \delta(\mathbf{r} - \mathbf{r}_i)] \tag{3.6}$$

Inserting the kinetic-energy terms in H, and using the continuity equation (2.7) lead to a $\mathbf{j}$ corresponding to $\partial\rho_0/\partial t$ given by $\mathbf{j}_c$, (2.3). What interests us in the present context is the isospin-dependent part of the potential. If we assume charge independence, one term in V is of the form (see Eq. I.3.5)

$$V_\tau = \frac{1}{2}\sum_{j\neq k} \boldsymbol{\tau}(j)\cdot\boldsymbol{\tau}(k)\, v(\boldsymbol{\sigma}_j, \boldsymbol{\sigma}_k, \mathbf{r}_{jk}), \qquad \mathbf{r}_{jk} = \mathbf{r}_j - \mathbf{r}_k$$

where v is symmetric against a permutation of j and k. The isospin dependence is present because of the exchange of various bosons by the interacting nucleons. Because of the presence of V_τ and the isospin dependence of the charge operator, there will be an isospin contribution to $(\partial\rho_0/\partial t)$ that we shall call the exchange contribution

$$\left(\frac{\partial\rho_0}{\partial t}\right)_{\text{exch}} = -\frac{e}{\hbar}\sum_{i\neq j} v(\boldsymbol{\sigma}_i, \boldsymbol{\sigma}_j, \mathbf{r}_{ij})\, \delta(\mathbf{r} - \mathbf{r}_i)\, [\boldsymbol{\tau}(i) \times \boldsymbol{\tau}(j)]_3 \tag{3.7}$$

It follows from the continuity equation that there must be a corresponding exchange current proportional to $[\boldsymbol{\tau}(i) \times \boldsymbol{\tau}(j)]_3$ but it otherwise is not uniquely determined. To determine it one must go back to the boson-exchange origin of

Problem. Prove that a possible $\mathbf{j}$ is

$$\mathbf{j} = -\frac{e}{2\hbar}\sum_{i\neq j} v(\boldsymbol{\sigma}_i, \boldsymbol{\sigma}_j, \mathbf{r}_{ij})\, [\boldsymbol{\tau}(i) \times \boldsymbol{\tau}(j)]_3\, (\mathbf{r}_i - \mathbf{r}_j)\cdot \int_0^1 d\alpha\, \delta[(\mathbf{r} - \mathbf{r}_i) + \alpha(\mathbf{r}_i - \mathbf{r}_j)]$$

Why is it not unique? (Sachs 53)

For a more general prescription for $\mathbf{j}$ see Osborn and Foldy (50); also Heller (71) for a recent discussion.

the potential V_τ and include the electromagnetic interactions from the beginning. If one is interested only in leading terms of the matrix element of a particular electric multipole moment, one need not actually calculate this current. We need only to know the charge density operator, which in this example is the simple (2.1) form and substitute it into Siegert's theorem (3.4). If $\langle f|$ and $|i\rangle$ are exact, (3.4) includes all exchange effects for the electric multipole moments to within the limits of validity of the approximations leading to (3.4). The exchange effects are now contained in $\langle f|$ and $|i\rangle$ with the important compensation that the complicated operator $\mathbf{j}$ is replaced by ρ_0.

Of course ρ_0 does not have the point-particle form given by (2.1). The very processes, which give rise to nuclear forces and exchange currents, also mean that the nucleons have a finite size. This finite size is a consequence of the "bare" nucleon emitting and absorbing bosons. Experiments in which high energy electrons [Hofstadter (63)] were elastically scattered from nucleons have been interpreted in terms of nucleon magnetization and charge form factors, which, as we shall see shortly, do inform us about the current and charge distributions inside a nucleon. A convenient pair of form factors has been proposed by Sachs (Sachs (62)]. Their relation to the nucleon magnetization current and charge density is:

$$\mathbf{j}_m(\mathbf{r}) = \frac{ie}{(2\pi)^3} \int d\mathbf{q}\, (\boldsymbol{\sigma} \times \mathbf{q})\, G_M(t)\, e^{i\mathbf{q}\cdot\mathbf{r}}$$

$$\rho(\mathbf{r}) = \frac{e}{(2\pi)^3} \int d\mathbf{q}\, G_E(t)\, e^{i\mathbf{q}\cdot\mathbf{r}} \tag{3.8}$$

G_M and G_E are the magnetic and electric form factors; there being separate ones for the neutron and proton (or for differing isospins of the nucleon.) The vector $\mathbf{q}$ is the momentum transferred by the electron $(\mathbf{p} - \mathbf{p}')/\hbar$ to the nucleon in the scattering. This quantity will depend upon the reference frame. The quantity t is however invariant. It is

$$t = q^2 - \omega^2 \qquad \omega \equiv \frac{(E - E')}{\hbar c}$$

For elastic collisons, in the center of mass of the colliding electron-nucleon systems $\omega = 0$, so that $t = q^2$

Empirically it is found that [Feld (69)]

$$G_E^{(p)} \simeq \frac{G_M^{(p)}}{\mu_p} \simeq \frac{G_M^{(n)}}{\mu_n} \simeq \frac{4M^2c^2}{\mu_n t} G_E^{(n)} \simeq \left[\frac{1}{1 + \dfrac{t}{.71\left(\dfrac{\text{GeV}}{c}\right)^2}}\right]^2 \tag{3.9}$$

The magnetic moments μ_p and μ_n are given in units of the nuclear magneton. These results imply that the current distribution within the neutron and proton are quite similar.

It is easy to show by inverting (3.8) that the mean-square-charge radius defined by

$$\langle r_c^{\,2} \rangle^{1/2} = \frac{1}{e} \int r^2 \rho(\mathbf{r})\, d\mathbf{r}$$

is given by

$$\langle r_c^{\,2} \rangle^{1/2} = -\frac{1}{6}\left(\frac{d\,G_E}{dt}\right)_{t=0} \tag{3.10}$$

The radius for the proton is found to be $0.813 \pm .008$ fm. with nearly identical results for the corresponding magnetization radius. It should be borne in mind that the neutron form factors are experimentally not as well determined as those for the protons, the quoted error for the neutron magnetization radius being of the order of 20%.

This discussion suggests that (2.1) for point particles be replaced by

$$e \sum \frac{1}{2} [1 + \tau_3(i)] \, \rho \, (\mathbf{r} - \mathbf{r}_i) \tag{3.11}$$

where ρ is given by (3.8). Such a description is incomplete unless one adds the corresponding value of the current. This expression is not available. It is of course not needed if we are calculating electric multipoles, for then we may use Siegert's theorem. The details of nucleon structure may also be neglected if the emitted (or absorbed) radiation has a wavelength much larger than the nucleon size. As we can see from (1.5) the gamma-ray energy would need to be of the order of 200 MeV before nucleon structure would play a significant role in radiative transitions.

The quantity ρ_0 in (3.4) differs from the point-particle expression (2.1) not only because the nucleons have a finite size. Expression (2.1) and (3.11), which includes finite size effects, is a sum of one-body operators. Again because of the exchange of charged particles between nucleons, it is possible that a more precise ρ_0 would contain two- and more-body operators. Not much is known about these terms but it is expected that they will be of importance only when the nucleons are relatively close and again this region would be of interest only if the radiation involved has high energy. The energies involved are generally much larger than the energy of the gamma rays generated by nuclear radiative transitions.

We note that in our comments about radiative transtions in this section we have not been able to make very definitive statements about the effect of exchange currents on the divergenceless component of the current operator. This has the advantage that such related quantities as the magnetic moment or the magnetic multipole operators are more sensitive to the nature of the exchange current. We shall return to this problem later in this chapter.

4. THE QUANTIZED ELECTROMAGNETIC FIELD

In this volume we are concerned primarily with the matrix elements of the nuclear electromagnetic current $\mathbf{j}$ since these provide information with respect to the nuclear wave function. However we shall also need to know the matrix elements of the probing electromagnetic field $\mathbf{A}$ in order to arrive at values for the matrix elements of $\mathbf{j}$. This section will be devoted to a discussion of $\mathbf{A}$ and its matrix elements. Only a brief review is given since this material is covered in many quantum mechanics texts [see Gottfried (66) p. 406].

The free-field vector potential $\mathbf{A}$ in the *radiation* gauge satisfies the vector wave equation:

$$\nabla^2 \mathbf{A} - \frac{1}{c} \frac{\partial^2 \mathbf{A}}{\partial t^2} = 0 \tag{4.1}$$

$$\operatorname{div} \mathbf{A} = 0$$

The electric and magnetic fields expressed in terms of **A** are:

$$\mathbf{E} = -\frac{1}{c}\frac{\partial \mathbf{A}}{\partial t} \tag{4.2}$$

$$\mathbf{H} = \text{curl } \mathbf{A}$$

The space-time dependence of the solutions of (4.1) is given by

$$\mathbf{A}_k = \boldsymbol{\varepsilon}_{k\alpha}\, e^{i(\mathbf{k}\cdot\mathbf{r}-\omega_k t)} \qquad \omega_k \equiv ck \tag{4.3}$$

where $\boldsymbol{\varepsilon}_k$ is the polarization vector with the properties:

$$\boldsymbol{\varepsilon}_{k\alpha}\cdot\mathbf{k} = 0 \qquad \boldsymbol{\varepsilon}_{k\alpha}\cdot\boldsymbol{\varepsilon}_{k\alpha'} = \delta_{\alpha\alpha'} \tag{4.4}$$

That is the polarization is perpendicular to the propagation vector **k**; and there are two mutually orthogonal polarizations denoted by α.

Solutions of eq. (4.1) in the spherical coordinate system are also known Morse and Feshbach (53) Blatt and Weisskopf (52). There are two types, one describing the magnetic $2l$ pole radiation, labeled with a superscript M, the other describing electric $2l$ pole radiation, hence, the superscript E. The magnetic multipole vector potential is

$$\mathbf{A}_{lm}{}^{(M)} = \frac{1}{\sqrt{l(l+1)}}\,\mathbf{L}\,[j_l\,(kr)\;Y_{lm}(\hat{\mathbf{r}})] \tag{4.4a}$$

where the time dependence is exp $(-i\omega_k t)$. Here $\mathbf{L} = (\mathbf{r} \times \text{grad})$, j_l is the spherical Bessel function. Replacement of j_l by h_l, the spherical Hankel function, also leads to a solution. A second solution, the electric multipole case, is obtained by taking the curl of (4.4a). Choosing a particular phase and normalization this second solution is

$$\mathbf{A}_{lm}{}^{(E)} = \frac{1}{k\sqrt{l(l+1)}}\,\text{curl curl }\mathbf{r}\,j_l\,(kr)\;Y_{lm}\,(\hat{\mathbf{r}}) \tag{4.4b}$$

Evaluating the curl operation gives the explicit expression

$$\mathbf{A}_{lm}{}^{(E)} = \frac{1}{\sqrt{l(l+1)}}\left\{k\mathbf{r}\,j_l Y_{lm} + \frac{1}{k}\,\text{grad}\left[Y_{lm}\frac{d}{dr}(r j_l)\right]\right\} \tag{4.4c}$$

Using the small kr behavior of j_l:

$$j_l(kr) \to \frac{(kr)^l}{(2l+1)!!} \qquad kr \to 0 \tag{4.4d}$$

yields

$$\mathbf{A}_{lm}{}^{(E)} \to \frac{k^{l-1}}{(2l+1)!!}\sqrt{\frac{l+1}{l}}\,\text{grad}\,(r^l\;Y_{lm}) \tag{4.4e}$$

In the discussion that follows it will be useful to express these vectors in Cartesian coordinates. Suppose the unit vectors in the x, y, z directions are $\hat{\mathbf{x}}$, $\hat{\mathbf{y}}$, and $\hat{\mathbf{z}}$, respectively. Let the corresponding spherical tensors of first rank be

$$\mathbf{u}_1 = -\frac{\hat{\mathbf{x}} + i\hat{\mathbf{y}}}{\sqrt{2}}$$
$$\mathbf{u}_0 = \hat{\mathbf{z}} \tag{4.4f}$$
$$\mathbf{u}_{-1} = \frac{\hat{\mathbf{x}} - i\hat{\mathbf{y}}}{\sqrt{2}}$$

Then it can be shown [Rose (55)] that:

$$\mathbf{A}_{lm}^{(M)*} = (-)^l \sqrt{(2l+1)} \sum_{\mu,\nu} \mathbf{u}_\nu \begin{pmatrix} l & 1 & l \\ m & \nu & \mu \end{pmatrix} Y_{l\mu}\, j_l\,(kr) \tag{4.4g}$$

and

$$\mathbf{A}_{lm}^{(E)*} = (-)^l [\sqrt{l+1} \sum_{\mu\nu} \mathbf{u}_\nu \begin{pmatrix} l & 1 & l-1 \\ m & \nu & \mu \end{pmatrix} Y_{l-1,\mu}\, j_{l-1}$$
$$- \sqrt{l} \sum \mathbf{u}_\nu \begin{pmatrix} l & 1 & l+1 \\ m & \nu & \mu \end{pmatrix} Y_{l+1,\mu}\, j_{l+1}] \tag{4.4h}$$

We note that the angular momentum carried by a photon described by $\mathbf{A}_{lm}^{(E)}$ or $\mathbf{A}_{lm}^{(M)}$ is l in units of $\hbar$, the z projection is m. The parity of the electric multipole is $(-)^l$, the magnetic $(-)^{l+1}$.

We shall find it convenient to work in a *helicity* representation of the photon. Because of the vanishing mass of the photon its intrinsic spin ($S = 1$ for a photon) can have only two orientations with respect to the momentum $\hbar\mathbf{k}$ of the photon. It can be either parallel or antiparallel. The photon state in which the photon spin is parallel to $\mathbf{k}$ is called a state of positive helicity, ($\lambda = 1$) while when the spin is antiparallel it is a state of negative helicity ($\lambda = -1$).

Photons of a definite helicity are in fact just circularly polarized waves, the sign of λ being related directly to the sense of the circular polarization. This is indicated by the relation between the helicity polarization vectors $\mathbf{e}_{k,1}$, $\mathbf{e}_{k,-1}$, which we now introduce, and the plane polarizations $\boldsymbol{\varepsilon}_{k\alpha}$. Suppose $\boldsymbol{\varepsilon}_{k1}$, $\boldsymbol{\varepsilon}_{k2}$, and $\hat{\mathbf{k}}$ form a right-handed coordinate system, $\hat{\mathbf{x}}$, $\hat{\mathbf{y}}$, $\hat{\mathbf{z}}$, respectively. Then

$$\mathbf{e}_{k,1} = -\frac{1}{\sqrt{2}}(\boldsymbol{\varepsilon}_{k1} + i\boldsymbol{\varepsilon}_{k2})$$
$$\tag{4.5}$$
$$\mathbf{e}_{k,-1} = \frac{1}{\sqrt{2}}(\boldsymbol{\varepsilon}_{k1} - i\boldsymbol{\varepsilon}_{k2})$$

These helicity vectors satisfy

$$\mathbf{e}_{k\lambda}^* \cdot \mathbf{e}_{k\lambda'} = \delta_{\lambda\lambda'} \tag{4.6}$$
$$\mathbf{e}_{k\lambda}^* \times \mathbf{e}_{k\lambda'} = i\lambda\hat{\mathbf{k}}\,\delta_{\lambda\lambda'} \tag{4.7}$$
$$\mathbf{e}_{k,-\lambda} = -\mathbf{e}_{-k,\lambda} \tag{4.7'}$$

Note that the definition of helicity vectors conforms with the phase convention for irreducible tensors of rank 1 as it should be in order to represent a spin one system. (See Appendix A, A.2.40, A.2.41).

These vectors satisfy the eigenvalue equation

$$(\mathbf{S}\cdot\hat{\mathbf{k}})\mathbf{u}_\lambda = \lambda\mathbf{u}_\lambda$$

where **S** is the spin operator for particles of unit spin. If we take the z-direction along $\hat{\mathbf{k}}$, $S_z \equiv \mathbf{S}\cdot\hat{\mathbf{k}}$ is diagonal with

$$\mathbf{u}_1 = \mathbf{e}_1$$

$$\mathbf{u}_{-1} = -\mathbf{e}_{-1}$$

where we have dropped the k subscript. To complete the eigenvalue spectrum 1, 0, -1, of S_z we add

$$\mathbf{u}_0 \equiv \hat{\mathbf{k}}$$

with eigenvalue zero. This eigenvector does not appear in the description of the electromagnetic field because of the radiation condition div $\mathbf{A} = 0$. The fact that the electromagnetic field has only two degrees of freedom is connected with the zero mass of the photon, and is a common feature for all massless particles with finite spin. In a matrix representation

$$\mathbf{u}_1 \equiv \begin{pmatrix}1\\0\\0\end{pmatrix} \qquad \mathbf{u}_0 = \begin{pmatrix}0\\1\\0\end{pmatrix} \qquad \mathbf{u}_{-1} = \begin{pmatrix}0\\0\\1\end{pmatrix} \tag{4.7$''$}$$

$$S_z = \begin{pmatrix}1 & 0 & 0\\0 & 0 & 0\\0 & 0 & -1\end{pmatrix}$$

It can be verified that the other members of the spin operator that satisfy

$$[S_x, S_y] = iS_z$$

are

$$S_x = \frac{1}{\sqrt{2}}\begin{pmatrix}0 & 1 & 0\\1 & 0 & 1\\0 & 1 & 0\end{pmatrix} \cdot \quad S_y = \frac{1}{\sqrt{2}}\begin{pmatrix}0 & -i & 0\\i & 0 & -i\\0 & i & 0\end{pmatrix}$$

Since S is the infinitesimal rotation operator, under an infinitesimal rotation $\boldsymbol{\theta}$ of the coordinate system, the change in an arbitrary state **u** is

$$\delta\mathbf{u} = -i\mathbf{S}\cdot\boldsymbol{\theta}\mathbf{u}$$

To illustrate let **u** be $\mathbf{u}_0$. Then

$$\delta\mathbf{u}_0 = -i\mathbf{S}\cdot\boldsymbol{\theta}\mathbf{u}_0 = -i\frac{\theta_x}{\sqrt{2}}(\mathbf{u}_1 + \mathbf{u}_{-1}) + \frac{\theta_y}{\sqrt{2}}(-\mathbf{u}_1 + \mathbf{u}_{-1}) = \theta_y\boldsymbol{\varepsilon}_{k1} - \theta_x\boldsymbol{\varepsilon}_{k2}$$

This result can be obtained directly from the vector properties of $\mathbf{u}_0 \equiv z$. The agreement obtained establishes the isomorphism between the spin 1 Hilbert space defined by (4.7$''$) and the vectors $\mathbf{e}_{k,\lambda}$.

The quantization of the electromagnetic field involves the introduction of operators that create and destroy photons of definite helicity. Let $b_\lambda(\mathbf{k})$ be the destruction operator and $b_{\lambda'}^\dagger(\mathbf{k})$ the creation operator. Since $\mathbf{k}$ is a con-continuous variable these operators satisfy the commutation rules

$$[b_\lambda(\mathbf{k}), b_{\lambda'}(\mathbf{k}')] = 0 = [b_{\lambda'}^\dagger(\mathbf{k}), b_\lambda^\dagger(\mathbf{k}')]$$
$$[b_\lambda(\mathbf{k}), b_{\lambda'}^\dagger(\mathbf{k}')] = \delta(\mathbf{k} - \mathbf{k}')\, \delta_{\lambda\lambda'} \tag{4.8}$$

The vector field operator $\mathbf{A}$ can now be written in terms of these operators as follows [Gottfried (66)]:

$$\mathbf{A}(\mathbf{r},t) = \sqrt{\frac{\hbar c^2}{2\pi^2}} \sum_\lambda \int \frac{d\mathbf{k}}{\sqrt{2\omega_k}} [\mathbf{e}_{k\lambda}^*\, b_\lambda^\dagger(\mathbf{k})\, e^{-i(\mathbf{k}\cdot\mathbf{r}-\omega_k t)} + \mathbf{e}_{k\lambda}\, b_\lambda(\mathbf{k})\, e^{i(\mathbf{k}\cdot\mathbf{r}-\omega_k t)}] \tag{4.9}$$

Thus the operator $\mathbf{A}$ can create and destroy photons of all momenta and helicities. If we substitute (4.9) into (1.1) we obtain an interaction Hamiltonian that can describe the emission and absorption of single photons.

The interaction Hamiltonian density

$$\mathcal{H}_{\text{int}} = -\frac{1}{c}\, \mathbf{j}(\mathbf{r}, t)\cdot \mathbf{A}(\mathbf{r}, t) \tag{4.10}$$

is a scalar with respect to spatial rotations. Thus angular momentum is conserved in all nuclear electro-magnetic transitions. In order to exhibit this important property and for future applications it is convenient to expand $\mathbf{A}$ in a sum of creation and destruction operators for states of a definite *angular* rather than *linear* momentum as in (4.9).

Toward this end [K. Gottfried (66), M. Jacob and G. C. Wick (59)] let us consider a photon moving in the z-direction, that is $\mathbf{z} = k\hat{\mathbf{z}}$, where $\hat{\mathbf{z}}$ is a unit vector in the $\mathbf{z}$-direction. Since a particle moving in a given direction can have no orbital angular momentum component in that direction, its angular momentum in the z-direction is entirely due to its intrinsic spin. Hence the z-component of the angular momentum for such a photon is given by its helicity λ .Denoting the state of the photon by $|k\hat{\mathbf{z}};\lambda\rangle$

$$S_z|k\hat{\mathbf{z}};\lambda\rangle = \lambda|k\hat{\mathbf{z}};\lambda\rangle \tag{4.10$'$}$$

we can immediately write:

$$|k\hat{\mathbf{z}};\lambda\rangle = \sum_{j=1}^{\infty} |k; j\lambda\rangle\, \langle k; j\lambda|k\hat{\mathbf{z}};\lambda\rangle \tag{4.11}$$

where $|k; jm\rangle$ is the photon state where the photon moving in the z-direction has angular momentum j and z-projection m. The summation starts with $j = 1$ since a state with $j = 0$ cannot have $\lambda = \pm 1$.

From (4.11) (we shall obtain the coefficients $\langle k;j\lambda|k\hat{\mathbf{z}};\lambda\rangle$ later) it is possible to obtain the representation of a plane wave propagating in an arbitrary direction **k** and of course with the same helicity by rotating the state. The effect of rotation as we know from the transformation properties of states with a definite angular momentum is to replace

$$|k; j\lambda\rangle \quad \text{by} \quad \sum_m |k; jm\lambda\rangle\, D_{m\lambda}{}^j(R) \tag{4.12}$$

where R is the operator which rotates from $k\hat{\mathbf{z}}$ to $\mathbf{k}$. The photon state $|k;jm\lambda\rangle$ describes a photon of energy $\hbar ck$, angular momentum j, z-component m and helicity λ.

Introducing (4.12) into (4.11) yields

$$|\mathbf{k};\lambda\rangle = \sum_{mj} |k; jm\lambda\rangle\, D_{m\lambda}{}^j(R)\, \langle k; j\lambda | k\hat{\mathbf{z}};\lambda\rangle \tag{4.13}$$

Note since R describes the rotation of a single direction and not that of a whole rigid body, only the two angles α,β specifying the direction $\hat{\mathbf{k}}$ relative to $\hat{\mathbf{z}}$ are required to describe R. Hence

$$D_{m\lambda}{}^j(R) = D_{m\lambda}{}^j(\alpha,\beta,0) = D_{m\lambda}{}^j(\hat{\mathbf{k}})$$

The desired rotation is thus obtained by rotating the coordinate axes about the z-axis through the angle α until the y-axis becomes perpendicular to the plane formed by $\hat{\mathbf{z}}$ and $\hat{\mathbf{k}}$ and then about the new y axis through an angle β equal to the angle between $\hat{\mathbf{z}}$ and $\hat{\mathbf{k}}$. Since only two angles are involved, the D's satisfy a simpler orthogonality relation:

$$\int [D_{m\lambda}{}^j(\hat{\mathbf{k}})]^*\, D_{m'\lambda}{}^{j'}(\hat{\mathbf{k}})\, d\hat{\mathbf{k}} = \frac{4\pi}{2j+1}\, \delta_{jj'}\, \delta_{mm'} \tag{4.14}$$

(This equation can be obtained from the more general orthogonality relations given in Appendix A (A.2.11) by noting that since λ is common to both D functions the integration $d\gamma$ is trivial yielding just 2π). Employing this orthogonality relation, we can invert (4.13) to obtain

$$|k; jm\lambda\rangle\, \langle k;j\lambda | k\hat{\mathbf{z}};\lambda\rangle = \frac{2j+1}{4\pi} \int [D_{m\lambda}{}^j(\hat{\mathbf{k}})]^* |\mathbf{k},\lambda\rangle\, d\hat{\mathbf{k}} \tag{4.15}$$

The D-functions project out of the plane wave the angular momentum state characterized by j and m.

To complete this discussion we shall need the amplitudes $\langle k; j\lambda | k\hat{\mathbf{z}};\lambda\rangle$. We first note the normalization of the state vectors involved in (4.15):

$$\langle \mathbf{k},\lambda | \mathbf{k}',\lambda'\rangle = \delta(\mathbf{k} - \mathbf{k}')\, \delta_{\lambda\lambda'} \tag{4.16}$$

and

$$\langle k,jm\lambda | k'; j'm'\lambda'\rangle = \frac{\delta(k-k')}{kk'}\, \delta_{jj'}\, \delta_{mm'}\, \delta_{\lambda\lambda'} \tag{4.17}$$

Inserting representation (4.15) into this last orthogonality relation and using (4.16) and (4.14) as well as

$$\delta(\mathbf{k} - \mathbf{k}') = \frac{\delta(k - k')}{kk'} \delta(\hat{\mathbf{k}} - \hat{\mathbf{k}}')$$

yields

$$\langle k; j\lambda | k\hat{\mathbf{z}};\lambda \rangle = \sqrt{\frac{2j+1}{4\pi}} \tag{4.18}$$

where a choice has been made for the phase of this amplitude. Inserting this result into (4.15) and (4.13) yields

$$|k;jm\lambda\rangle = \sqrt{\frac{2j+1}{4\pi}} \int [D_{m\lambda}{}^{j}(\hat{\mathbf{k}})]^* \, |\mathbf{k},\lambda\rangle \, d\hat{\mathbf{k}} \tag{4.19}$$

and

$$|\mathbf{k},\lambda\rangle = \sum_{jm} \sqrt{\frac{2j+1}{4\pi}} \, |k;jm\lambda\rangle \, D_{m\lambda}{}^{j}(\hat{\mathbf{k}}) \tag{4.20}$$

This expansion suggests a corresponding expansion for the creation and destruction operators $b_\lambda^\dagger(\mathbf{k})$:

$$b_\lambda^\dagger(\mathbf{k}) = \sum_{jm} \sqrt{\frac{2j+1}{4\pi}} \, D_{m\lambda}{}^{j}(\hat{\mathbf{k}}) \, b_{jm\lambda}^\dagger(k) \tag{4.21}$$

where

$$b_{jm\lambda}^\dagger = \sqrt{\frac{2j+1}{4\pi}} \int [D_{m\lambda}{}^{j}(\hat{\mathbf{k}})]^* b_\lambda^\dagger(\hat{\mathbf{k}}) \, d\hat{\mathbf{k}} \tag{4.22}$$

It is easy to verify that the only nonvanishing commutator is

$$[b_{jm\lambda}(k), b_{j'm'\lambda'}^\dagger(k')] = \frac{\delta(k - k')}{kk'} \delta_{jj'} \, \delta_{mm'} \, \delta_{\lambda\lambda'} \tag{4.23}$$

These new operators $b_{jm\lambda}^\dagger(k)$ create a photon with the quantum number k, j, m, and λ. From (4.22) it can be verified that b_{jm} behaves under rotations like the mth component of a spherical tensor of rank j.

To obtain an expansion of $\mathbf{A}(\mathbf{r},t)$ in terms of these new operators, substitute (4.21) into (4.9):

$$\mathbf{A}(\mathbf{r},t) = \sqrt{\frac{\hbar c^2}{8\pi^3}} \sum_{j,m,\lambda} \sqrt{2j+1} \int_0^\infty \frac{k^2 dk}{\sqrt{2\omega_k}} [b_{jm\lambda}^\dagger \, \mathbf{f}_{jm\lambda}(k,\mathbf{r}) e^{i\omega_k t} + b_{jm\lambda} \mathbf{f}_{jm\lambda}{}^*(k,\mathbf{r}) e^{-i\omega_k t}] \tag{4.24}$$

where

$$\mathbf{f}_{jm\lambda}(k,\mathbf{r}) \equiv \int \mathbf{e}_{k\lambda}^* \, e^{-i\mathbf{k}\cdot\mathbf{r}} \, D_{m\lambda}{}^{(j)}(\hat{\mathbf{k}}) \, d\hat{\mathbf{k}} \tag{4.25}$$

Since λ takes on only the values ± 1, j cannot take on the value 0. This is a consequence of the fact that the photon has an intrinsic spin of one. This is, as it turns out, as far as we need to go in the decomposition of **A** into partial waves of well-defined helicity.

In the long run we shall be concerned with the emission or absorption of electromagnetic radiation by states having well-defined parities. Assuming conservation of parity in the transition,* has the consequence that only electromagnetic radiation of a given parity will be emitted. Since the helicity is the component of the spin in the direction of motion, it is odd under a reflection (see Eq. 4.10′) so that radiation of a definite helicity does not have a well-defined parity.

To determine the components that have a specific parity we consider the behavior of the state vectors $|\mathbf{k},\lambda\rangle$ and then $|k; jm\lambda\rangle$ under reflection. Since both the helicity and momentum are odd under a reflection we have (P = reflection operator)

$$P|\mathbf{k};\lambda\rangle = |-\mathbf{k}; -\lambda\rangle \tag{4.26}$$

The constant of proportionality (unity in Eq. 4.26) is obtained by comparing the behavior of the transformation function of both sides of this equation under reflection. Note that

$$\langle \mathbf{r}|-\mathbf{k}; -\lambda\rangle = \mathbf{e}^*_{-k,-\lambda}\, e^{i\mathbf{k}\cdot\mathbf{r}} = -\,\mathbf{e}^*_{k,\lambda}\, e^{i\mathbf{k}\cdot\mathbf{r}}$$

On the other hand

$$\langle \mathbf{r}|P|\mathbf{k},\lambda\rangle = \langle P\mathbf{r}|\mathbf{k},\lambda\rangle = -\mathbf{e}^*_{k,\lambda}\, e^{i\mathbf{k}\cdot\mathbf{r}}$$

proving (4.26).

Applying (4.26) to expansion (4.13) yields

$$D_{m\lambda}{}^{(j)}(\hat{\mathbf{k}})\, P|k; jm\lambda\rangle = D_{m,-\lambda}{}^{(j)}(-\hat{\mathbf{k}})\, |k; jm-\lambda\rangle$$

Using (see Appendix A, A.2.19).

$$D^{(j)}_{mn}(-\hat{\mathbf{k}}) = D^{(j)}_{mn}(\alpha+\pi, \pi-\beta, 0) = (-)^j D^{(j)}_{m,-n}(\alpha,\beta,0) = (-)^j D^{(j)}_{m,-n}(\hat{\mathbf{k}})$$

we obtain

$$P|k; jm\lambda\rangle = (-)^j\, |k; jm-\lambda\rangle \tag{4.27}$$

It immediately follows that the creation and destruction operators satisfy

$$Pb_{jm\lambda}P^{-1} = (-)^j\, b_{jm-\lambda} \tag{4.28}$$

*There is some evidence for the violation of the conservation of parity in the nucleon-nucleon interaction. This violation arises from the addition to nuclear forces coming from the weak interactions that do not conserve parity. The resulting effects are very weak and thus need not be considered at this point. We shall return to this discussion after we describe the theory of weak interactions in Chapter IX.

We can now define combinations of $b^\dagger_{jm\lambda}$ and $b^\dagger_{jm-\lambda}$ that will remain invariant under reflection:

$$b^{(\sigma)\dagger}_{jm} = \frac{1}{\sqrt{2}}(b^\dagger_{jm,1} + (-)^\sigma\, b^\dagger_{jm,-1}) \tag{4.29}$$

where σ can take on the values zero or one and

$$[b^{(\sigma)}_{jm}(k), b^{(\tau)\dagger}_{j'm'}(k')] = \frac{\delta(k-k')}{kk'}\,\delta_{jj'}\,\delta_{mm'}\,\delta_{\sigma\tau} \tag{4.30$'$}$$

The operator $b^{(o)\dagger}_{jm}$ is the creation operator for *electric* multipole radiation of multiple order 2^j while $b^{(1)\dagger}_{jm}$ is the corresponding operator for *magnetic* multipole radiation. Under reflection

$$P\,[b^{(\sigma)\dagger}_{jm}]\,P^{-1} = (-)^{j+\sigma}\, b^{(\sigma)\dagger}_{jm} \tag{4.30$''$}$$

so that electric multipoles ($\sigma = 0$) of multipole order 2^j carry the parity $(-)^j$ while magnetic multipoles ($\sigma = 1$) of order 2^j have the opposite parity $(-)^{j+1}$.

The states these new operators create have a well-defined parity. They are

$$|k\sigma; jm\rangle \equiv \frac{1}{\sqrt{2}}\,[|k; jm1\rangle + (-)^\sigma|k;jm-1\rangle] \tag{4.30$'''$}$$

where according to (4.27), $\sigma = 0$ state has the parity $(-)^j$, the $\sigma = 1$ state the parity $(-)^{j+1}$. Finally one can construct plane wave linearly polarized states that for $\sigma = 0$ is a linear superposition of electric multipole states, and for $\sigma = 1$ decomposes into magnetic multipoles:

$$|\mathbf{k}\sigma\rangle = \frac{1}{\sqrt{2}}\,[|\mathbf{k},1\rangle + (-)^\sigma\,|\mathbf{k},-1\rangle] \tag{4.30^{iv}}$$

We can now rewrite expansion (4.24) in terms of these new operators with the result:

$$\mathbf{A}(\mathbf{r},t) = \sqrt{\frac{\hbar c^2}{8\pi^3}}\sum_{jm\sigma}\sqrt{2j+1}\int_0^\infty \frac{k^2dk}{\sqrt{2\omega_k}}\,(b^{(\sigma)\dagger}_{jm}\,\mathbf{f}^{(\sigma)}_{jm}\,e^{-i\omega_k t} + \text{c.c.}) \tag{4.31}$$

where

$$\mathbf{f}^{(\sigma)}_{jm} = \frac{1}{\sqrt{2}}(\mathbf{f}_{jm1} + (-)^\sigma\,\mathbf{f}_{jm-1})$$

Using (4.25) defining $f_{jm\lambda}$ and the relations between $D_{m\lambda}{}^{(j)}(-\hat{\mathbf{k}})$ and $D_{m,-\lambda}{}^{(j)}(-\hat{\mathbf{k}})$ (see Appendix A, A.2.19)

$$\mathbf{f}^{(\sigma)}_{jm} = \frac{1}{\sqrt{2}}\int d\hat{\mathbf{k}}\,\overset{*}{\mathbf{e}}_{k1}\,D_{m1}{}^{(j)}(\hat{\mathbf{k}})\,[e^{-i\mathbf{k}\cdot\mathbf{r}} - (-)^{j+\sigma}\,e^{i\mathbf{k}\cdot\mathbf{r}}] \tag{4.32}$$

When we need to distinguish states of a given parity, it is more convenient to use (4.31) while states of a well-defined helicity are most easily considered with expansion (4.24).

5. EMISSION OF ELECTROMAGNETIC RADIATION

In this volume we shall restrict our considerations to the case of emission (or absorption) of radiation accompanying the transition of a nucleus from a state $|i\rangle$ to a state $|f\rangle$. The quantum numbers of these nuclear states include the total angular momentum J, its "z"-component M and the parity Π. The isospin quantum numbers T and $M_{(T)}$ will not be explicitly given until we consider isospin-selection rules. We shall assume that parity is conserved in the transition. The effects of the weak parity nonconserving interactions are outlined after the discussion of the weak interactions in Chapter IX. It will therefore be most convenient to let the emitted photon have a polarization σ (see Eq. 4.30^{iv}) and to use expansion (4.31) to describe its emission.

The probability for the transition is:

$$dw\,(i \rightarrow f\mathbf{k}\sigma) = \frac{2\pi}{\hbar}\frac{k^2}{\hbar c}\,|M_{fi}|^2\,d\Omega \tag{5.1}$$

where $d\Omega$ is the differential solid angle and σ the polarization of the plane wave. The factor $k^2/\hbar c$ is the density of final states appropriate for a plane-wave normalization $(1/2\pi)^{3/2}\,e^{i\mathbf{k}\cdot\mathbf{r}}$. We find $\rho_F = k^2dk/dE$ where $E = \hbar ck$. The matrix element

$$M_{fi}\,(\mathbf{k},\sigma) \equiv \langle f;\,\mathbf{k}\sigma\,|\,H_{\text{int}}\,|\,i;\,00\rangle \tag{5.2}$$

where H_{int} is given by (1.1). The matrix element is evaluated at

$$E_f = E_i + \hbar\omega \tag{5.3}$$

where E_f and E_i are the energies of the nuclear states involved in the transition. The dependence of M_{fi} on $\mathbf{k}$ and σ can be made explicit. We note that

$$M_{fi}\,(\mathbf{k},\sigma) = \sum_{jm\sigma'} \langle \mathbf{k},\sigma\,|\,k\sigma';\,jm\rangle\,\langle f;\,\sigma' jm\,\,|\,H_{\text{int}}\,|\,i;\,000\rangle$$

$$= \sum_{jm\sigma'} \langle \mathbf{k},\sigma\,|\,k\sigma';\,jm\rangle\,\langle f;\,000\,|\,[b_{jm}^{(\sigma)'},\,H_{\text{int}}]\,|\,i;\,000\rangle$$

or

$$M_{fi}\,(\mathbf{k},\sigma) = \sum_{jm\sigma'} \langle \mathbf{k},\sigma\,|\,k\sigma';\,jm\rangle \int d\mathbf{r}\,\langle f|\,\mathbf{j}(\mathbf{r})\,|i\rangle\cdot\langle 000|\,[b_{jm}^{(\sigma)'},\,\mathbf{A}]\,|000\rangle$$

Inserting expansion (4.31) and using (4.30′) yields

$$M_{fi}\,(\mathbf{k},\sigma) = -\frac{\hbar}{16\pi^3\omega_k}\sum_{jm\sigma'} \langle \mathbf{k}\sigma\,|\,k;\,jm\sigma'\rangle\,\sqrt{2j+1}\,\langle f|\,M_{jm}^{(\sigma)'}\,\,|i\rangle \tag{5.4}$$

where

$$M_{jm}^{(\sigma)} \equiv \int \mathbf{f}_{jm}^{(\sigma)} \cdot \mathbf{j}(\mathbf{r})\, d\mathbf{r} \tag{5.5}$$

Using definitions (4.30‴) and (4.30^{iv}) and relation (4.19) yields

$$M_{fi}(\mathbf{k},\sigma) = -\frac{1}{16\pi^2}\sqrt{\frac{\hbar}{\omega_k}}\sum_{jm\sigma'}(2j+1)\,[D_{m1}{}^{(j)*}(\mathbf{k}) + (-)^{\sigma+\sigma'} D_{m,-1}{}^{(j)*}(\hat{\mathbf{k}})] \cdot \langle f|M_{jm}^{(\sigma)'}|i\rangle \tag{5.6}$$

The dependence of M_{fi} on the direction of emission from which the angular distribution will eventually be obtained is given explicitly in the square bracket. The dependence on nuclear structure is contained in the matrix element of the operator $M_{jm}^{(\sigma)'}$·(See Eqs. 2.3 and 2.4). Since $M_{jm}^{(\sigma)'}$ is the matrix

$$\langle 000|b_{jm}^{(\sigma)'}\, H_{\text{int}}|000\rangle$$

where H_{int} is invariant against a rotation, it follows that $M_{jm}^{(\sigma')}$ has the same behavior under rotation as $b_{jm}^{(\sigma')}$, that is, it is the mth component of a tensor operator of rank j. It immediately follows that $\langle f|M_{jm}^{(\sigma)'}|i\rangle$ vanishes unless

$$|J_f - J_i| \le j \le (J_f + J_i) \tag{5.7}$$

and

$$M_f = M_i + m \tag{5.8}$$

Moreover since the minimum value of j is one (see Eq. 4.25) transitions by the emission of a *single photon from $J_i = 0$ to $J_f = 0$ are forbidden.* Finally since H_{int} is reflection invariant the parity carried by the photon, that is, by $b_{jm}^{(\sigma)'}$ and therefore by $M_{jm}^{(\sigma)'}$ is $(-)^j$ if σ' is 0, that is, electric and $(-)^{j+1}$ if σ' is magnetic. Hence in a transition

$$\begin{aligned} \Pi_f &= (-)^j\,\Pi_i \qquad \text{electric multipole} \\ \Pi_f &= (-)^{j+1}\,\Pi_i \qquad \text{magnetic multipole} \end{aligned} \tag{5.9}$$

Thus an even parity change will involve electric multipoles of even order and magnetic multipoles of odd order. Magnetic dipole radiation ($j = 1$) and electric quadrupole radiation ($j = 2$) are both possible if the appropriate angular momentum selection rules (5.7) and (5.8) are satisfied.

By using the Wigner-Eckart theorem we can exhibit the angular momentum selection rules explicitly:

$$\langle f|M_{jm}^{(\sigma)}|i\rangle = \langle J_f M_f|M_{jm}^{(\sigma)}|J_i M_i\rangle = (-)^{J_f - M_f}\begin{pmatrix} J_f & j & J_i \\ -M_f & m & M_i \end{pmatrix}(J_f||M_j{}^{(\sigma)}||J_i) \tag{5.10}$$

Substituting (5.10) and (5.6) into (5.1) gives:

$$\frac{dw(J_iM_i\Pi_i \to J_fM_f\Pi_f;\, \mathbf{k}\sigma)}{d\Omega}$$

$$= \frac{k}{(128\,\pi^3)\,\hbar c^2} \sum_{\substack{j\sigma' \\ j'\sigma''}} (2j'+1)(2j+1) \begin{pmatrix} J_f & j & J_i \\ -M_f & m & M_i \end{pmatrix} \begin{pmatrix} J_f & j' & J_i \\ -M_f & m & M_i \end{pmatrix}$$

$$\times\, (J_f\Pi_f\,||\,M_j^{(\sigma')}\,||\,J_i\Pi_i)\,(J_f\Pi_f\,||\,M_{j'}^{(\sigma'')}\,||\,J_i\Pi_i)$$

$$\times\, [D_{m1}^{(j)*}(\hat{\mathbf{k}}) + (-)^{\sigma+\sigma'}\, D_{m,-1}^{(j^*)}(\hat{\mathbf{k}})]\,[D_{m1}^{(j)'}(\hat{\mathbf{k}}) + (-)^{\sigma+\sigma''}\, D_{m,-1}^{(j')}(\hat{\mathbf{k}})] \quad (5.11)$$

Only a few terms of this series, that is, one or two multipole orders are needed for the accurate prediction of experiments. As a first step let us consider the case often met in practice in which only one multipole contributes importantly.

A. Single Multipole

In that event j must equal j' in (5.11). Because the parity change is determined by $j + \sigma'$, parity conservation requires, $\sigma' = \sigma''$ where σ' is fixed by the parity change and j. Under these circumstances (5.11) becomes

$$\frac{dw(J_iM_i\Pi_i \to J_fM_f\Pi_f;\, \mathbf{k}\sigma)}{d\Omega}$$

$$= \frac{k}{128\pi^3\hbar c^2}(2j+1)^2 \begin{pmatrix} J_f & j & J_i \\ -M_f & m & M_i \end{pmatrix}^2 |(J_f\Pi_f\,||\,M_j^{(\sigma)'}\,||\,J_i\Pi_i)|^2$$

$$\cdot(|D_{m1}^{(j)}(\hat{\mathbf{k}})|^2 + |D_{m,-1}^{(j)}(\hat{\mathbf{k}})|^2 + 2\,(-)^{\sigma+\sigma'}\, D_{m,-1}^{(j)*}(\hat{\mathbf{k}})\, D_{m,1}^{(j)}(\hat{\mathbf{k}})) \quad (5.12)$$

We observe that the angular distribution is a function of β only, where $\hat{\mathbf{k}}$ is specified by the spherical angles β and α. Because of the last term in the round bracket the angular distribution depends on σ', that is, on the multipole type, electric $\sigma' = 0$, magnetic $\sigma' = 1$ involved in the transition. This dependence on σ' disappears if a sum over σ, that is, over the polarization of the emergent wave is performed. Thus one method for determining the character of the radiated multipole requires the measurement of the angular distribution for a given polarization of the radiated wave.

If σ is summed then

$$\frac{dw(J_iM_i\Pi_i \to J_fM_f\Pi_f;\, \mathbf{k})}{d\Omega}$$

$$= \frac{k(2j+1)^2}{64\pi^3\hbar c^2} \begin{pmatrix} J_f & j & J_i \\ -M_f & m & M_i \end{pmatrix}^2 |(J_f\Pi_f\,||\,M_j^{(\sigma')}\,||\,J_i\Pi_i)|^2$$

$$\cdot([d_{m1}^{(j)}(\beta)]^2 + [d_{m1}^{(j)}(\pi - \beta)]^2) \quad (5.13)$$

where we have employed the relation from Appendix A (A.2.18′), $|d_{m1}^{(j)}(\pi - \beta)| = |d_{m,-1}^{(j)}(\beta)|$. The round bracket* gives the angular distribution of the emitted radiation of a given multipole order in which the detector does not distinguish between the two possible polarizations of the emitted radiation. We observe that this angular distribution does not depend on σ'. In other words the angular distribution depends only upon the multipole order and not upon its electric or magnetic character; that is not upon the parity of the emitted radiation.

One result of interest follows from the fact that $d_{m1}^{(j)}(\beta)$ vanishes for $\beta = 0$ unless $m = 1$. Similarly $d_{m1}{}^{j}(\pi - \beta)$ vanishes at $\beta = 0$ unless $m = -1$. This means that the intensity in the direction of propagation of the emitted photon is zero for all m except $|m| = 1$. This result is to be expected on the following intuitive grounds. We note that the magnitude of the component of the angular momentum in the direction of motion of a particle equals its intrinsic spin, in this case unity. Thus the component of the angular momentum change of the nucleus $m = (M_i - M_f)$ in the direction of the emission of the gamma ray is unity.

B. Averaged Cross Sections

If the various initial magnetic substates, M_i, are equally populated and if the final M_f's are not detected, we may then average (5.11) over M_i and sum over M_f to obtain the observed transition rate

$$\frac{dw(J_i\Pi_i;\, J_f\Pi_f;\, \mathbf{k}\sigma)}{d\Omega} = \frac{1}{2J_i + 1}\sum_{M_iM_f}\frac{dw(J_iM_i\Pi_i;\, J_fM_f\Pi_f;\, \mathbf{k}\sigma)}{d\Omega}$$

Summing over M_i yields the partial transition rate

$$\frac{dw(J_i\Pi_i \to J_f\Pi_f;\, m\mathbf{k}\sigma)}{d\Omega} = \frac{k}{128\pi^3\hbar c^2}\sum_j \frac{(2j+1)}{2J_i+1}(J_f\Pi_f\,||\,M_j{}^{(\sigma')}\,||\,J_i\Pi_i)^2 \cdot |D_{m1}^{(j)} + (-)^{\sigma+\sigma'} D_{m,-1}^{(j)}|^2 \tag{5.15}$$

where $m = M_f - M_i$. Summing over final polarizations, that is over σ, yields

$$\frac{dw(J_i\Pi_i \to J_f\Pi_f;\, m\mathbf{k})}{d\Omega} = \frac{k}{64\pi^3\hbar c^2}\sum_j \frac{(2j+1)}{2J_i+1}(J_f\Pi_f\,||\,M_j^{(\sigma')}\,||\,J_i\Pi_i)^2 \cdot [|D_{m1}^{(j)}|^2 + |D_{m,-1}^{(j)}|^2] \tag{5.16}$$

---

*In other derivations [see Blatt and Weisskopf (52) p. 594] this result is given in terms of the spherical harmonics as follows:

$$\frac{(j+m)(j-m+1)}{2j(j+1)}|Y_{j,m-1}(\beta,\alpha)|^2 + \frac{m^2}{j(j+1)}|Y_{j,m}(\beta,\alpha)|^2 \cdot + \frac{(j-m)(j+m+1)}{2j(j+1)}|Y_{j,m+1}(\beta,\alpha)|^2 \tag{5.14}$$

Because the absolute values of the spherical harmonics are required this expression is a function only of β.

Finally summing over m yields

$$\frac{dw(J_i\Pi_i \to J_f\Pi_f, \mathbf{k})}{d\Omega} = \frac{k}{32\pi^3\hbar c^2} \sum \frac{(2j+1)}{2J_i+1} (J_f\Pi_f || M_j^{(\sigma')} || J_i\Pi_i)^2 \tag{5.17}$$

It will be useful to have the results in inverse order; first summing over polarization, that is, over σ and then over M_i and M_f. We obtain

$$\frac{dw(J_iM_i\Pi_i \to J_fM_f\Pi_f; \mathbf{k})}{d\Omega}$$

$$= \frac{k}{64\pi^3\hbar c^2} \sum_{j,j'} (2j'+1)(2j+1) \begin{pmatrix} J_f & j & J_i \\ -M_f & m & M_i \end{pmatrix} \begin{pmatrix} J_f & j' & J_i \\ -M_f & m & M_i \end{pmatrix}$$

$$\cdot (J_f\Pi_f || M_j^{(\sigma')} || J_i\Pi_i)(J_f\Pi_f || M_{j'}^{(\sigma'')} || J_i\Pi_i)$$

$$\cdot [D_{m1}^{(j)*} D_{m1}^{(j')} + D_{m,-1}^{(j)*} D_{m,-1}^{(j')}] \tag{5.18}$$

Averaging over M_i and summing over M_f yields (5.17).

The angular distribution given by (5.17) is isotropic. This is to be expected since we have averaged over the possible orientations of the emitting nucleus (i.e., over the various values of M_i) and summed over the orientations of the final nucleus. There is thus no "preferred" spatial direction.

An anisotropic angular distribution can be obtained only if there is a special direction selected by the experimental arrangement. For example if the magnetic substates of the initial state were not uniformly populated, then (5.18) would be correct, but (5.16) and therefore (5.17) would not be. The angular distribution would then be given by the squarebracket terms, the "z" axis being defined by the direction of quantization for the magnetic substates M_i. If only one value of j contributes importantly the, multipole order can be determined from the angular distribution. The electric or magnetic character can also be obtained if the reduced matrix elements of $M_{jm}{}^{(\sigma)}$ for differing j's are comparable in magnitude. Then interference between these differing j's will contribute. The angular distribution would then yield the multipole order, the parity of the radiation, and finally the ratio of the corresponding reduced matrix elements.

One procedure that has been used to weight the various M_i's differently involves placing the emitting nucleus in an external magnetic field. The direction of the field is then the preferred direction. The field splits the magnetic substates of the emitting nucleus. If thermal equilibrium is established, the population in each level will differ thus achieving the necessary condition for the observation of an angular distribution of the emitted gamma rays.

In another class of experiments of interest in this connection usually referred to as *angular correlation* experiments, the gamma ray can be the second leg of a sequential process. Examples are the cascades (γ,γ) and (β,γ) illustrated in Fig. 5.1a and Fig. 5.1b. Another general class are the reactions $(x,y\gamma)$ in which the emission of a particle y leads to an excited state of the residual system

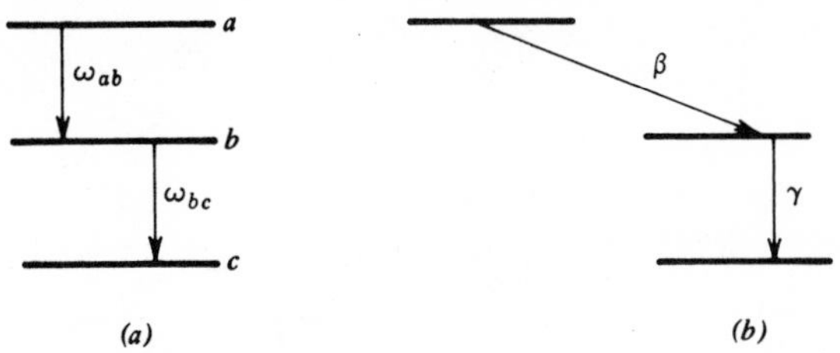

FIG. 5.1. Example of angular correlation experiments.

that then decays via γ-emission. The direction of the first particle, a gamma ray in Fig. 5.1*a*, a β-particle in Fig. 5.1*b*, particle y in the reaction provides a preferred spatial direction. For example, taking the direction of the initial gamma ray in Fig. 5.1*a* as the z-direction means that $M_a - M_b$ can only equal ± 1, leading to a nonuniform population of the magnetic substates of state $|b\rangle$ as is true, for example, when the spin $J_a < J_b$. It is for this reason that the final gamma ray ω_{bc} can have an anisotropic intensity distribution. For further details we refer the reader to monographs in which the theory of angular correlation experiments is considered in detail.

In heavy ion reactions a preferred spatial direction is naturally selected by the collision of the ion with the nucleus. Because of the large momentum of the heavy ion and the grazing nature of the ion-nucleus collision, one would expect that the residual nucleus after the collision would spin about an axis perpendicular to the scattering plane. The angular distribution of the gamma rays that are emitted after the collision would then be anisotropic. This phenomenon is illustrated by the angular distribution of gamma rays emitted by excited states of ^{110}Cd that is formed by the reaction ^{96}Zr(^{18}O,4n) ^{110}Cd. These are shown in Figs. 5.2 and 5.3. In Fig. 5.2 the gamma rays are all quadrupole while in Fig. 5.3 they are dipole. The 998 and 401 keV lines are magnetic dipole; the 178 keV is a mixture. The characteristic nature of the angular distribution is unmistakeable in each group, and consequently a unique determination of the multipole order involved in each transition can be obtained.

We turn now to the calculation of the reduced matrix element $(J_f \| M_j^{(\sigma)} \| J_i)$. The operator $M_{jm}^{(\sigma)}$ is given by (5.5). To evaluate it we first expand $\mathbf{f}_{jm}^{(\sigma)}$, (Eq. 4.32) employing the familiar expansion of the plane wave

$$e^{i\mathbf{k}\cdot\mathbf{r}} = 4\pi \sum_{l\mu} i^l \, Y_{l\mu}(\hat{\mathbf{r}}) \, Y^*_{l\mu}(\hat{\mathbf{k}}) \, j_l(kr)$$

This yields:

$$\mathbf{f}_{jm}^{(\sigma)} = \frac{4\pi}{\sqrt{2}} \sum_{l\mu} i^l \, j_l(kr) \, Y_{l\mu}(\hat{\mathbf{r}}) \, [(-)^l - (-)^{j+\sigma}]$$

$$\cdot \int d\hat{\mathbf{k}} \, \mathbf{e}^*_{kl} \, D_{m1}^{(j)}(\hat{\mathbf{k}}) \, Y^*_{l\mu}(\hat{\mathbf{k}})$$

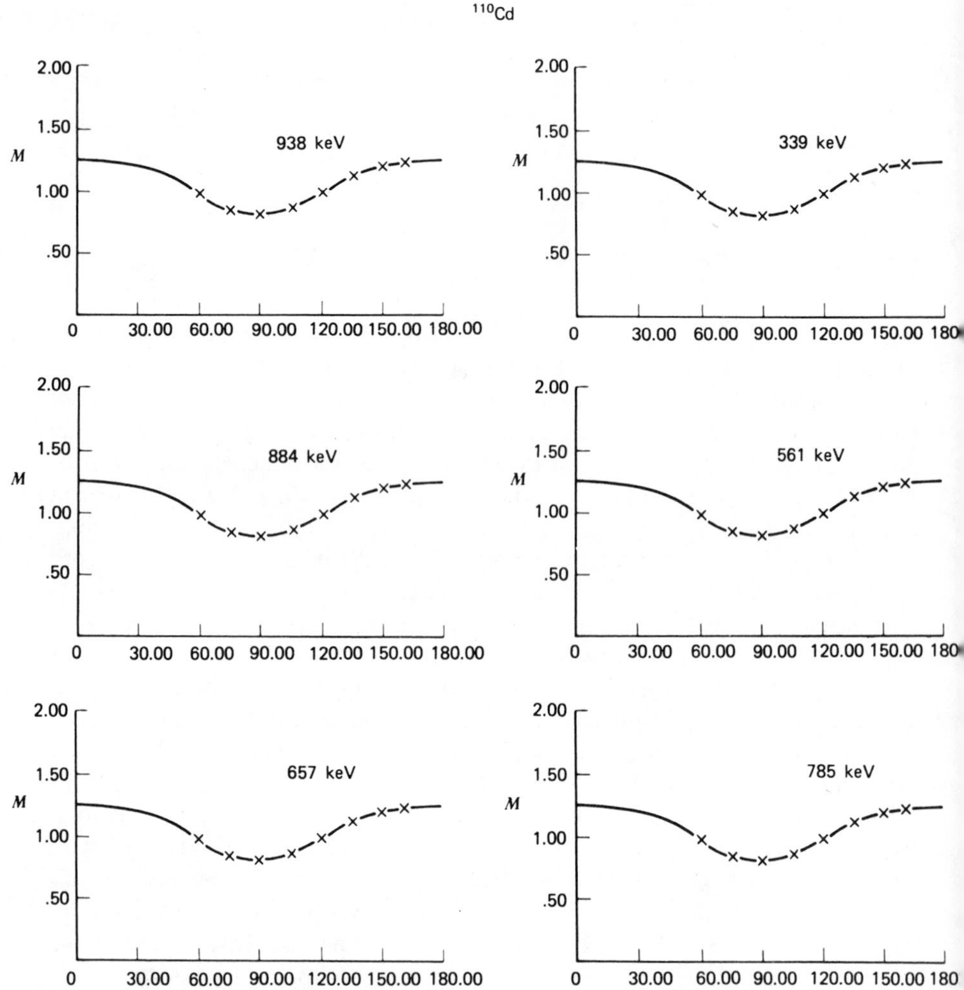

FIG. 5.2. Angular distribution of electric quadrupole gamma rays emitted in various transitions between excited states of ^{110}Cd found in the heavy ion reaction ^{96}Zr(^{18}0,4n) ^{110}Cd.| [from Sunyar [(73)].

In order to perform the indicated integration it is necessary to make the dependence of $\mathbf{e}_{k1}^*$ on $\mathbf{k}$ explicit. The vector $\mathbf{e}_{k1}^*$ can be obtained from the fixed set of coordinate directions given by $\mathbf{u}_\nu$ (4.4f) by a rotation:

$$\mathbf{e}_{k1}^* = \sum_\nu \mathbf{u}_\nu^* \, D_{\nu 1}^{(1)*}(\hat{\mathbf{k}}) \tag{5.19}$$

Using

$$\mathbf{u}_\nu^* = (-)^\nu \, \mathbf{u}_{-\nu}$$

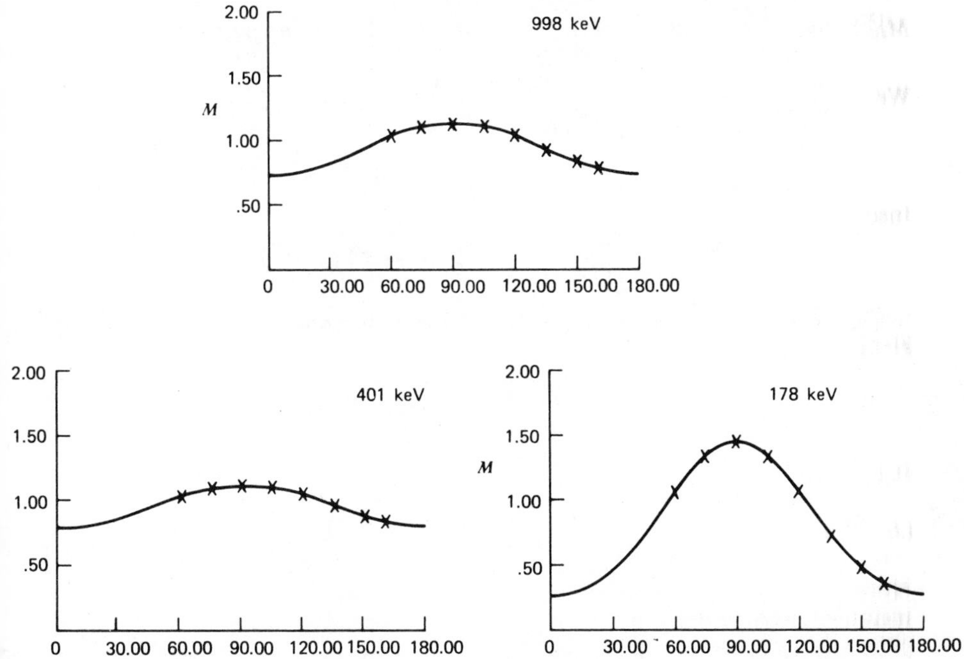

FIG. 5.3. Same as Fig. 5.2 except that the gamma rays are magnetic dipole radiation or in the case of the 178 keV line mixed magnetic dipole and electric quadrupole radiation [from Sunyar (73)].

and a similar relation for the D-function, (5.19) can also be written

$$\mathbf{e}_{k1}^{*} = -\sum \mathbf{u}_\nu \, D^{(1)}_{\nu,-1}(\hat{\mathbf{k}}) \tag{5.20}$$

The $\hat{\mathbf{k}}$-integration may now be performed using the results given in Appendix A (A.2.11) to obtain for $M_{jm}^{(\sigma)}$ the following

$$M_{jm}^{(\sigma)} = -16\pi^2 \sum_{l\mu\nu} \sqrt{\frac{2l+1}{8\pi}} \begin{pmatrix} j & 1 & l \\ m & \nu & \mu \end{pmatrix} \begin{pmatrix} j & 1 & l \\ 1 & -1 & 0 \end{pmatrix} i^l \left[(-)^l - (-)^{j+\sigma}\right] \cdot \int d\mathbf{r} \; Y_{l\mu}(\mathbf{r}) \, j_l(kr) \, \mathbf{u}_\nu \cdot \mathbf{j} \tag{5.21}$$

We shall consider first magnetic multipoles. The matrix element for magnetic multipole radiation, $\sigma = 1$, differs from zero only when $l = j$. Then

$$M_{jm}^{(M)} = -32\pi^2 (-i)^j \sqrt{\frac{2j+1}{8\pi}} \sum_{\mu\nu} \begin{pmatrix} j & 1 & j \\ m & \nu & \mu \end{pmatrix} \begin{pmatrix} j & 1 & j \\ 1 & -1 & 0 \end{pmatrix} \cdot \int d\mathbf{r} \; Y_{j\mu}(\hat{\mathbf{r}}) \, j_j(kr) \, \mathbf{u}_\nu \cdot \mathbf{j}$$

Employing (4.4g) and (4.4a) the sum over μ and ν can be performed:

$$M_{jm}^{(M)} = -i^j \frac{32\pi^2}{\sqrt{8\pi}} \frac{1}{\sqrt{j(j+1)}} \begin{pmatrix} j & 1 & j \\ 1 & -1 & 0 \end{pmatrix} \int d\mathbf{r}\, \mathbf{j}\cdot \mathbf{L}(j_j(kr)\, Y_{jm}^*) \tag{5.22}$$

We introduce the operator $\mathfrak{M}_{jm}^{(M)}$:

$$\mathfrak{M}_{jm}^{(M)} \equiv -\frac{1}{c(j+1)} \int d\mathbf{r}\, \mathbf{j}\cdot (\mathbf{r} \times \boldsymbol{\nabla}) [j_j(kr)\, Y_{jm}^*] \tag{5.23}$$

Inserting this definition and the value of the $3j$ symbol:

$$\begin{pmatrix} j & 1 & j \\ 1 & -1 & 0 \end{pmatrix} = \frac{(-)^{j+1}}{\sqrt{2(2j+1)}}$$

yields

$$M_{jm}^{(M)} = (-i)^{j-1}\, 8\pi^2 c \sqrt{\frac{j+1}{\pi(2j+1)j}}\, \mathfrak{M}_{jm}^{(M)} \tag{5.24}$$

It immediately follows that

$$(J_f||M_j^{(M)}||J_i) = (-i)^{j-1}\, 8\pi^2 c \sqrt{\frac{j+1}{\pi j(2j+1)}}\, (J_f||\mathfrak{M}_j^{(M)}||J_i) \tag{5.25}$$

Note that no long wavelength approximations were made in deriving this result. Before discussing it further we will find it convenient to derive the corresponding result for electric multipole transitions.

Problem. Show that $\mathfrak{M}_{jm}^{(M)}$ can also be written as follows:

$$\mathfrak{M}_{jm}^{(M)} = -\frac{1}{c(j+1)} \int d\mathbf{r}\, j_j(kr) Y_{jm}^* \operatorname{div}(\mathbf{r} \times \mathbf{j})$$

Recalling that the magnetization is $(1/2c)\,(\mathbf{r} \times \mathbf{j})$ and that the divergence of the magnetization is $\rho^{(M)}$ where $\rho^{(M)}$ is the magnetic pole density [Stratton (41) p. 228] we obtain

$$\mathfrak{M}_{jm}^{(M)} = -\frac{2}{j+1} \int d\mathbf{r}\, j_j(kr)\, Y_{jm}^*\, \rho^{(M)} \tag{5.26}$$

For long wavelength

$$\mathfrak{M}_{jm}^{(M)} \simeq -\frac{2}{j+1} \frac{k^j}{(2j+1)!!} \int d\mathbf{r}\, r^j\, Y_{jm}^*\, \rho^{(M)}$$

We observe the close connection of $\mathfrak{M}_{jm}^{(M)}$ with the magnetic multipole given in the integral in terms of the magnetic pole density.

We turn to the electric multipole case. In (5.21) we now place σ equal to zero and combine the $l = j + 1$ and $l = j - 1$ terms. Note also that

$$\begin{pmatrix} j & 1 & j-1 \\ 1 & -1 & 0 \end{pmatrix} = (-)^{j-1} \sqrt{\frac{(j+1)}{2(2j-1)(2j+1)}}$$

and

$$\begin{pmatrix} j & 1 & j+1 \\ 1 & -1 & 0 \end{pmatrix} = (-)^{j+1} \sqrt{\frac{j}{2(2j+3)(2j+1)}}$$

Comparing with (4.4h) it follows that

$$M_{jm}^{(E)} = \frac{(-i)^{j-1}\, 8\pi^2}{\sqrt{(2j+1)\pi}} \int d\mathbf{r}\, \mathbf{j}\cdot \mathbf{A}_{jm}^{(E)\,*} \tag{5.27}$$

An exact expression for $\mathbf{A}_{jm}^{(E)}$ is given in (4.4*c*); the long wavelength approximation in (4.4e). Defining

$$\mathfrak{M}_{jm}^{(E)} \equiv \frac{1}{c}\sqrt{\frac{j}{j+1}} \int d\mathbf{r}\, \mathbf{j}\cdot \mathbf{A}_{jm}^{(E)\,*} \tag{5.28}$$

yields

$$M_{jm}^{(E)} = (-i)^{j-1}\, 8\pi^2 c \sqrt{\frac{j+1}{\pi j(2j+1)}}\, \mathfrak{M}_{jm}^{(E)}$$

and

$$(J_f\,||\,M_j^{(E)}\,||\,J_i) = (-i)^{j-1}\, 8\pi^2 c \sqrt{\frac{j+1}{\pi j(2j+1)}}\, (J_f\,||\,\mathfrak{M}_j^{(E)}\,||\,J_i) \tag{5.29}$$

We see that (5.29) and (5.25), the corresponding expression for the magnetic multipole case, differ only in the operator given by (5.23) for the magnetic and (5.28) for the electric case. We shall now discuss each case in the long wavelength limit for which

$$j_l\,(kr) \to \frac{(kr)^l}{(2l+1)!!}$$

and

$$\mathbf{A}_{lm}^{(E)} \to \frac{k^{l-1}}{(2l+1)!!}\sqrt{\frac{l+1}{l}}\ \text{grad}\ (r^l\ Y_{lm})$$

Then

$$\mathfrak{M}_{jm}^{(E)} \simeq \frac{1}{c}\frac{k^{j-1}}{(2j+1)!!}\int d\mathbf{r}\, \mathbf{j}\cdot \text{grad}\ r^j\ Y_{jm}^*$$

or

$$\mathfrak{M}_{jm}^{(E)} \simeq -\frac{1}{c}\frac{k^{j-1}}{(2j+1)!!}\int d\mathbf{r}\, r^l\, Y_{lm}^*\, \operatorname{div}\mathbf{j}$$

Introducing current conservation and

$$\frac{\partial\rho}{\partial t} = \frac{i}{\hbar}[H,\rho]$$

so that

$$\langle f|\operatorname{div}\mathbf{j}|i\rangle = -\langle f|\frac{\partial\rho}{\partial t}|i\rangle = -\frac{i}{\hbar}(E_f - E_i)\langle f|\rho|i\rangle = ikc\,\langle f|\rho|i\rangle$$

Hence

$$\mathfrak{M}_{jm}^{(E)} \simeq -\frac{ik^j}{(2j+1)!!}\int d\mathbf{r}\, r^j\, Y_{jm}^*\,\rho \tag{5.30}$$

and

$$(J_f||M_j^{(E)}||J_i) \sim \frac{8\pi^2 c(-ik)^j}{(2j+1)!!}\sqrt{\frac{j+1}{\pi j(2j+1)}}(J_f||Q_j^{(E)}||J_i) \tag{5.31}$$

where

$$Q_{jm}^{(E)} \equiv \int d\mathbf{r}\, r^j\, Y_{jm}^*\,\rho \tag{5.32}$$

are the electric multipole moments.

The long wavelength approximation when applied to $\mathfrak{M}_{jm}^{(M)}$ yields an expression identical to (5.31) except that $Q_{jm}^{(E)}$ is replaced by $Q_{jm}^{(M)}$ where from (5.23)

$$Q_{jm}^{(M)} \equiv -\frac{1}{c(j+1)}\int d\mathbf{r}\, r^j\, \mathbf{j}\cdot(\mathbf{r}\times\nabla)\, Y_{jm}^* \tag{5.33}$$

The total transition probability can now be readily evaluated by multiplying (5.17) by 4π and inserting (5.31). The result is

$$w(i \to f\mathbf{k}) \equiv \sum_j w(\sigma j;\, i \to f) \tag{5.34}$$

where

$$w(\sigma j;\, i\to f) = 8\pi c\,\frac{e^2}{\hbar c}\,\frac{j+1}{j[(2j+1)!!]^2}\,k^{2j+1}\,B(\sigma j;\, i\to f) \tag{5.35}$$

The B coefficients are then

$$B(\sigma j;\, i\to f) = \frac{1}{[2J_i+1]e^2}(J_f||Q_j^{(\sigma)}||J_i)^2 \tag{5.36}$$

where σ can be either E or M for electric and magnetic multipoles, respectively.

Generally only one multipole of a given type will contribute significantly to w. The ratio of $w(\sigma j + 1; i \rightarrow f)$ to $w(\sigma j; i \rightarrow f)$ is of the order $(kR)^2/(2j + 3)^2$ where R is of the order of the nuclear radius. For nuclear gamma rays this ratio is very small, as can be seen from (1.5), and decreases very rapidly with increasing multipole order. States that can decay only by emitting low energy gamma rays of high multipole order will have long lifetimes and are thus easily observable. As discussed in Chapter 1 (see Fig. I.7.3) these long-lived states that are called *isomers* occur more plentifully as shells close. The existence of these isomeric islands was one of the observations that pointed to the shell model.

One word of caution: the reader should bear in mind that (5.35) is a long wavelength approximation obtained by replacing $j_l\,(kr)$ with the first term in its power-series expansion. It is possible (usually because of the validity of some approximate selection rule) that this first term is not in fact dominant, and that an accurate answer must include the next term. We shall see some specific cases below.

As the next part of this discussion, we shall evaluate $\mathfrak{M}_{jm}^{(\sigma)}$ employing the point charge and point magnetic dipole expression (2.1) to (2.4), which are commonly used. The error following from the neglect of the finite size of the nucleons and from exchange current has been presumed to be small. The details involve only some straightforward applications of vector-operator identities.

We obtain

$$
\begin{aligned}
\mathfrak{M}_{jm}^{(E)} = & -\frac{i}{j+1}\left\{\frac{e}{2}\sum_i\,[1 + \tau_3(i)]\; Y^*_{jm}\,(\hat{\mathbf{r}}_i)\,\frac{d}{dr_i}\,[r_i j_j\,(kr_i)]\right.\\
& + i\,\frac{e\hbar}{2m_pc}\,k\sum_i \frac{1}{4}\,[(g_p + g_n)\\
& + \tau_3(i)\,(g_p - g_n)]\,j_j\,(kr_i)\;\boldsymbol{\sigma}_i\cdot(\mathbf{r}_i \times \nabla_i)\;Y^*_{jm}\,(\hat{\mathbf{r}}_i)\\
& \left. + \frac{ek^2}{4}\sum_i\,[1 + \tau_3(i)]\;r_i^2\,j_j\,(kr_i)\;Y^*_{jm}\,(\hat{\mathbf{r}}_i)\right\}
\end{aligned}
\tag{5.37}
$$

The dominant term in the long wavelength limit is given by the first sum yielding (5.30) in that limit.

The result for the corresponding magnetic multipole radiation is

$$
\begin{aligned}
\mathfrak{M}_{jm}^{(M)} = & \frac{e\hbar}{2m_pc}\,\frac{2}{(j+1)}\left\{\sum_i \frac{1}{2}\,[1 + \tau_3(i)]\,[\boldsymbol{\nabla}_i\,(Y^*_{jm}(\hat{\mathbf{r}}_i) j_j)]\cdot\mathbf{l}_i\right.\\
& + \frac{1}{8}\sum_i\,[(g_p + g_n) + \tau_3(i)\,(g_p - g_n)]\,[k^2\boldsymbol{\sigma}_i\cdot\mathbf{r}_i\;Y^*_{jm}\,(\hat{\mathbf{r}}_i)\,j_j\,(kr_i)\\
& \left. + [\boldsymbol{\nabla}_i\,(Y^*_{jm}\,(\hat{\mathbf{r}}_i)\,\frac{d}{dr_i}(r_i j_j)]\cdot\boldsymbol{\sigma}_i]\right\}
\end{aligned}
\tag{5.38}
$$

In the long wavelength limit

$$\mathfrak{M}_{jm}^{(M)} \simeq \frac{e\hbar}{2m_p c}\frac{k^j}{(2j+1)!!}\sum_i \left\{\frac{1}{j+1}[1+\tau_3(i)]\mathbf{l}_i + \frac{1}{4}[g_p + g_n + \tau_3(i)(g_p - g_n]\,\boldsymbol{\sigma}_i\right\}\cdot\boldsymbol{\nabla}_i\,[r_i^j\, Y_{jm}^*(\hat{\mathbf{r}}_i)]$$

The resulting value of $Q_{jm}^{(M)}$ is

$$Q_{jm}^{(M)} \simeq \frac{ie\hbar}{2m_p c}\sum_i \left\{\frac{1}{j+1}[1+\tau_3(i)]\,\mathbf{l}_i + \frac{1}{4}[g_p + g_n + \tau_3(i)(g_p - g_n)]\,\boldsymbol{\sigma}_i\right\} \cdot \nabla_i\,(r_i^j\, Y_{jm}^*(\hat{\mathbf{r}}_i)) \tag{5.39}$$

The corresponding result for $Q_{jm}^{(E)}$ can be readily obtained from (5.37) or from (5.32) combined with (2.1), the expression for the charge density. It is

$$Q_{jm}^{(E)} = e\sum_i \frac{1}{2}[1+\tau_3(i)]\, r_i^j\, Y_{jm}^*(\hat{\mathbf{r}}_i) \tag{5.40}$$

This is as far as the formalism can be carried without explicit consideration of the initial and final nuclear states. As we have remarked earlier, the value of the reduced matrix elements of $Q_{jm}^{(\sigma)}$ is sensitive to the nature of these states. An outstanding example is furnished by the collective states. When the nuclear states are collective, electromagnetic transitions generally involve a change in motion of many nucleons. As a consequence the resulting transition probabilities are generally much larger than the probabilities for these transitions in which only one particle changes its state. Thus the collectivity of the states involved in a transition is measured by the ratio of an observed transition probability to the corresponding single-particle transition probability.

The single-particle transition probability for an electric multipole transition, the particle involved being a proton, is proportional to (see Eq. 5.36):

$$B_{\text{s.p.}}(Ej;\, i\to f) = \frac{1}{(2j_i+1)}\left(n_f\,\tfrac{1}{2}\,l_f\,j_f\,||\,r^j\, Y_j(\hat{\mathbf{r}})\,||\,n_i\,\tfrac{1}{2}\,l_i j_i\right)^2$$

The single-particle states are characterized by the orbital angular momentum l, total angular momentum j, spin 1/2, and radial quantum number n, the subscript i denoting initial and f final state, respectively.

Using the results tabulated in Appendix A, (A.2.49) we obtain

$$B_{s\cdot p\cdot}(Ej,\, i\to f) = \begin{cases} \begin{pmatrix} j_f & j & j_i \\ -1/2 & 0 & 1/2 \end{pmatrix}^2 \dfrac{(2j_f+1)(2j+1)}{4\pi}(l_f j_f||r^j||l_i j_i)^2 & \text{if } (l_f + l_i + j) \text{ is even} \\ 0 & \text{If } l_f + l_i + j \text{ is odd or if } j \text{ does not satisfy the inequality } |l_i - l_f| \le j \le (l_i + l_f) \end{cases} \tag{5.41}$$

The calculation of B for the magnetic case is more involved because of the presence of the gradient operator in (5.39). We refer the reader to deShalit and Talmi's book (see their equation 17.26) for the details. Their results can be further simplified with the aid of some identities among the nj coefficients given in Appendix A (Sect 2). The result is

$$B_{\text{s.p.}}(Mj; i \to f) = \left(\frac{\hbar}{2m_p c}\right)^2 \frac{(2j_f + 1)(2j + 1)}{4\pi} (l_f j_f \| r^{j-1} \| l_i j_i)^2$$
$$\cdot \left\{ \left(\mu_p - \frac{1}{j+1}\right)^2 [(2j_f + 1)(j_f - l_f)\right.$$
$$+ (2j_i + 1)(j_i - l_i) - j] \begin{pmatrix} j_f & j_i & j \\ 1/2 & -1/2 & 0 \end{pmatrix}$$
$$+ (-)^{j_f - l_f - (j_i - l_i)} \frac{1}{j+1} \begin{pmatrix} j_f & j_i & j-1 \\ -1/2 & 1/2 & 0 \end{pmatrix}$$
$$\left. \cdot \sqrt{(j_i + j_f + j + 1)(j_i + j - j_f)(j_f + j - j_i)(j_i + j_f - j + 1)} \right\} \tag{5.42}$$

Equation 5.42 is valid when $l_f + l_i + j$ is odd. In addition the triad l_f, l_i, and j must satisfy the vector addition condition $|l_i - l_f| \leq j - 1 \leq l_i + l_f$. If $l_f + l_i + j$ is even, $B_{\text{s.p.}}(Mj; i \to f)$ is zero. The quantity μ_p is the proton magnetic moment in nuclear magnetons.

Equation 5.42 simplifies if j takes on its minimum value consistent with the angular momentum change in the transition: $j = |j_i - j_f|$. The second term in (5.42) is zero since the $3j$ coefficient $\begin{pmatrix} j_f & j_i & j-1 \\ -1/2 & 1/2 & 0 \end{pmatrix}$ is then zero. Hence

$$B_{\text{s.p.}}(Mj; i \to f) = \left(\frac{\hbar}{2m_p c}\right)^2 \frac{(2j_f + 1)(2j + 1)}{4\pi} (l_f j_f | r^{j-1} | l_i j_i)^2 \begin{pmatrix} j_f & j & j_i \\ -1/2 & 0 & 1/2 \end{pmatrix}^2$$
$$\cdot \left(\mu_p - \frac{1}{j+1}\right)^2 [(2j_f + 1)(j_f - l_f) + (2j_i + 1)(j_i - l_i) - |j_i - j_f|]^2$$
$$\text{when} \quad j = |j_i - j_f| \tag{5.43}$$

In the event that in addition $|l_i - l_f|$ is equal to $j - 1$,

$$[(2j_f + 1)(j_f - l_f) + (2j_i + 1)(j_i - l_i) - |j_i - j_f|]^2 = 4j^2 \tag{5.44}$$

These results hold if the radiating particle is a proton. If the particle is a neutron, the second term in (5.42) is no longer present. In the first term and in (5.43) $\mu_p - (1/j + 1)$ is replaced by μ_n, the neutron magnetic moment.

Rough estimates of the single-particle transition probabilities that are based on (5.41), (5.43), and (5.44), have been given by Weisskopf and are known as the *Weisskopf units*. The correction to the Weisskopf unit that can be obtained from the more precise results (5.41) and (5.43) are of the order of unity. The radial integral in (5.41) have been estimated to be $3R^j/(j+3)$, a result that assumes the radial wavefunction is a constant up to a radius R. The Weisskopf result for electric multipole radiation is:

$$w_{\text{s.p.}}(Ej; i \rightarrow f) \simeq \frac{2(j+1)}{j[(2j+1)!!]^2}\left(\frac{3}{l+3}\right)^2 \frac{e^2}{\hbar c}(kR)^{2j}\, kc$$

$$\simeq \frac{4.4\,(j+1)}{j[(2j+1)!!]^2}\left(\frac{3}{l+3}\right)^2\left(\frac{E_\gamma}{197\text{ MeV}}\right)^{2j+1}\left[\frac{R}{10^{-13}\text{ cm}}\right]^{2j} 10^{21}\text{ sec}^{-1} \qquad (5.45)$$

In calculating the magnetic multipole, the large size of the anomalous magnetic moments must be taken into account. Weisskopf takes the factor μ^2 to be about 10 so that

$$w_{\text{s.p.}}(Mj; i \rightarrow f) \simeq 10\left(\frac{\hbar}{m_p cR}\right) w(Ej; i \rightarrow f) \qquad (5.46)$$

$$w_{\text{s.p.}}(Mj; i \rightarrow f) \simeq \frac{1.9\,(j+1)}{j[(2j+1)!!]^2}\left(\frac{3}{j+3}\right)^2\left(\frac{E_\gamma}{197\text{ MeV}}\right)^{2j+1}\left(\frac{R}{10^{-13}\text{ cm}}\right)^{2j-2} \times 10^{21}\text{ sec}^{-1} \qquad (5.47)$$

The gamma ray width Γ_γ may be calculated directly from the transition probability

$$\Gamma_\gamma = \hbar w$$

Some numerical values calculated by Wilkinson (60) are given in Table 5.1, where

$$R = r_0 A^{1/3} \qquad r_0 = 1.2\text{ fm}$$

E_γ is in MeV, Γ_j in eV.

TABLE 5.1 Single-Particle Widths (Weisskopf Units)

Multipole Order	Electric	Magnetic
1	$(6.8\ 10^{-2})A^{2/3}E_\gamma^3$	$(2.1\ 10^{-2})E_\gamma^3$
2	$(4.9\ 10^{-8})A^{4/3}E_\gamma^5$	$(1.5\ 10^{-8})A^{2/3}E_\gamma^5$
3	$(2.3\ 10^{-14})A^2E_\gamma^7$	$(6.8\ 10^{-15})A^{4/3}E_\gamma^7$
4	$(6.8\ 10^{-21})A^{8/3}E_\gamma^9$	$(2.1\ 10^{-21})A^2E_\gamma^9$
5	$(1.6\ 10^{-27})A^{10/3}E_\gamma^{11}$	$(4.9\ 10^{-28})A^{8/3}E_\gamma^{11}$

The $B(Ej)$ and $B(Mj)$ to be obtained from (5.45) and (5.47) are

$$B_{\text{s.p}}(Ej) = \frac{1}{4\pi}\left(\frac{3}{j+3}\right)^2 (1.2\ A^{1/3})^{2j}\,(\text{fm})^{2j} \qquad (5.48)$$

and

$$B_{\text{s.p.}}(Mj) = \frac{10}{\pi}\left(\frac{3}{j+3}\right)^2 (1.2\,A^{1/3})^{(2j-2)} \left(\frac{\hbar}{2m_p c}\right)^2 (\text{fm})^{2j-2}$$

$$B_{\text{s.p.}}(M1) = 1.79\left(\frac{\hbar}{2m_p c}\right)^2 \tag{5.49}$$

6. ISOSPIN SELECTION RULES [WARBURTON AND WENESER (69)]

Nuclear levels are not only characterized by spin, energy, parity, but also by the considerably more approximate quantum number of isotopic spin T. The usual radiative selection rules on spin, energy, and parity changes must be expanded to include selection rules on isospin changes. An obvious rule that follows because the photon is uncharged is

$$\Delta T_3 = 0$$

An even stricter condition applies since radiative transitions cannot change the number of neutrons or protons:

$$\Delta N = 0 \qquad \Delta Z = 0$$

The isospin selection rules can be obtained from the matrix element $\langle f|M_{jm}^{(\sigma)}|i\rangle$ of (5.4) and (5.6). We must bear in mind that the two symmetries of interest, charge symmetry and charge independence, are broken by the electromagnetic interaction since protons are charged and neutrons are not, the masses of the neutron and proton are not equal, and they have unequal magnetic moments. In addition the charge independence of nuclear forces is approximate since the masses of the charged mesons and the masses of the neutral mesons responsible for these forces are not equal. We shall discuss these matters more thoroughly in Volume II [see also Henley (69)]. Their existence means that rigorously the total isospin of the initial and final states involved in the radiative transition under discussion are only approximate quantum numbers. Consider, for example, the Coulomb interaction:

$$\frac{e^2}{4}\sum_{i<j} [1 + \tau_3(i)]\,[1 + \tau_3(j)]\frac{1}{r_{ij}}$$

This interaction involves an isoscalar, an isovector, and an isotensor of second rank. Hence a state that is approximately an eigenstate of T^2 with eigenvalue $T(T + 1)$ will in first-order perturbation theory have isospin impurities coming from states with isospin $T + 1$, and $T + 2$. We shall not discuss the amplitudes of these impurities here [see Soper (69)]. Our major purpose here is to point out that the selection rules to be derived below are approximate, and that the extent of their failure is a measure of the isospin impurities.

The isospin character of the nuclear current is the decisive element for our discussion. Recalling expressions (2.2) to (2.4) we see that the current and therefore the operator $M_{jm}^{(\sigma)}$ can be resolved into two parts

$$\mathbf{j} = \mathbf{j}_0 + \mathbf{j}_1 \tag{6.1}$$

where $\mathbf{j}_0$ is an isoscalar and

$$\mathbf{j}_1 = \sum_i \tau_3(i)\, \mathbf{j}_1(i) \tag{6.2}$$

transforms under a rotation in isospace like an isovector. Correspondingly

$$M_{jm}^{(\sigma)} = M_{jm0}^{(\sigma)} + M_{jm1}^{(\sigma)} \tag{6.3}$$

We now ask: disregarding the isospin symmetry breaking interactions discussed earlier what are the selection rules for the electromagnetic transitions? Let us consider first the weaker condition of charge symmetry. Charge symmetry relates the states of *conjugate nuclei.* A nucleus is said to be conjugate to a nucleus with Z-protons and N-neutrons if it contains N-protons and Z-neutrons. Charge symmetry implies that corresponding states of conjugate nuclei are "essentially identical." More precisely, if Ψ is the state vector for a particular state in a given nucleus, the vector for the corresponding state in the conjugate nucleus $\Psi^{(c)}$ is obtained as follows:

$$\Psi^{(c)} = \tau_2(1)\, \tau_2(2) \ldots \tau_2(A)\Psi \tag{6.4}$$

Using this result, we can now relate the matrix element for electromagnetic transitions between corresponding states in conjugate nuclei as follows:

$$\langle f^{(c)} | M_{jm}^{(\sigma)} | i^{(c)} \rangle = \langle f | \tau_2(A) \ldots \tau_2(1)\, M_{jm}^{(\sigma)}\, \tau_2(1) \ldots \tau_2(A) | i \rangle$$

Introducing (6.3) and recalling (6.2) we obtain

$$\langle f^{(c)} | M_{jm}^{(\sigma)} | i^{(c)} \rangle = \langle f | M_{jm0}^{(\sigma)} | i \rangle - \langle f | M_{jm1}^{(\sigma)} | i \rangle \tag{6.5}$$

whereas between the corresponding levels

$$\langle f | M_{jm}^{(\sigma)} | i \rangle = \langle f | M_{jm0}^{(\sigma)} | i \rangle + \langle f | M_{jm1}^{(\sigma)} | i \rangle$$

Thus a simple relation between these two differing transition amplitudes will occur only if one of the two amplitudes, the isoscalar or isovector amplitudes, vanish. (Alternatively, by measuring both transition amplitudes, we can determine the isoscalar and isovector components.) Since the isoscalar term $M_{jm0}^{(\sigma)}$ is in fact not dependent upon isospin, its matrix element will vanish if the states $|i\rangle$ and $|f\rangle$ are orthogonal as far as their isospin dependence is concerned. This will occur if $i\rangle$ and $|f\rangle$ have opposite symmetry with respect to a proton $\leftrightarrow$ neutron exchange. We thus obtain the result: corresponding transitions in conjugate nuclei between states with opposite symmetry under neutron-proton exchange have identical widths. The same result will hold approximately if

$$|\langle f|M_{jm0}^{(\sigma)}|i\rangle| \ll |\langle f|M_{jm1}^{(\sigma)}|i\rangle|$$

As we shall see in the discussion to follow, this inequality holds for electric dipole ($E1$) and for magnetic multipole (Mj) radiation.

These results as well as others also follow from the stronger condition of charge independence. Then both initial and final states have good isospin T_f and T_i as well as "z" components giving the value of $(Z - N)/2$, denoted by M_{T_f} and M_{T_i}. The dependence of the radiative matrix element on these quantum numbers is given explicitly by

$$\begin{aligned}\langle J_f\, M_f;\, T_f\, M_{Tf}|M_{jm}^{(\sigma)}|J_iM_i;\, T_iM_{Ti}\rangle\\ = \langle J_f\, M_f;\, T_fM_{Tf}|M_{jm0}^{(\sigma)} + M_{jm1}^{(\sigma)}|J_iM_i;\, T_iM_{Ti}\rangle\end{aligned}$$

Using the Wigner-Eckart theorem with respect to the isospin dependence gives:

$$\begin{aligned}&\langle J_f\, M_f;\, T_f\, M_{Tf}|M_{jm}^{(\sigma)}|J_i\, M_i;\, T_i\, M_{Ti}\rangle\\ &= (-)^{T_f - M_{Tf}}\left\{\begin{pmatrix} T_f & 0 & T_i \\ -M_{T_f} & 0 & M_{T_i}\end{pmatrix}\langle J_f\, M_f;\, T_f||M_{jm0}^{(\sigma)}||J_iM_i;\, T_i\rangle\right.\\ &\left.+\begin{pmatrix} T_f & 1 & T_i \\ -M_{T_f} & 0 & M_{T_i}\end{pmatrix}\langle J_fM_f;\, T_f||M_{jm1}^{(\sigma)}||J_i\, M_i;\, T_i\rangle\right\}\end{aligned} \tag{6.6}$$

The reduced matrix elements have been "reduced" with respect to the isospin dependence and thus depend only on the isospins T_f and T_i. The dependence on $J_{f,i}, M_{f,i}$ and other properties of these states remain. We observe at once *the selection rule* $\Delta T = 0, \pm 1$ as is expected because of the mixed isoscalar and isovector character of $M_{jm}^{(\sigma)}$. [Radicati (52), Gell-Mann and Telegdi (53)]. In addition $M_{T_i} = M_{T_f}$ which simply says that $Z - N$ is not changed in the transition; again an obvious result. Inserting explicit values for the $3j$ coefficients gives:

$$\langle J_fM_f;\, T\, M_T|M_{jm0}^{(\sigma)}|J_iM_i;\, TM_T\rangle = \langle J_fM_f|M_{jm0}^{(\sigma)}|J_iM_i\rangle \tag{6.7}$$

and

$$\begin{aligned}\langle J_fM_f;\, T+1\, M_T|M_{jm1}^{(\sigma)}|J_iM_i;\, TM_T\rangle &= \left(\frac{(T+M_T+1)\,(T-M_T+1)}{(2T+3)\,(T+1)\,(2T+1)}\right)^{1/2}\\ &\cdot\langle J_fM_f;\, T+1||M_{jm1}^{(\sigma)}||J_iM_i;\, T\rangle\end{aligned} \tag{6.8a}$$

$$\begin{aligned}\langle J_fM_f;\, T\, M_T|M_{jm1}^{(\sigma)}|J_iM_i;\, T\, M_T\rangle &= \frac{M_T}{[(2T+1)\,(T+1)T]^{1/2}}\\ &\cdot\langle J_fM_f;\, T||M_{jm1}^{(\sigma)}||J_iM_i;\, T\rangle\end{aligned} \tag{6.8b}$$

$$\langle J_f M_f; TM_T | M^{(\sigma)}_{jm1} | J_i M_i; T+1\ M_T \rangle = -\left(\frac{(T+M_T+1)(T-M_T+1)}{(2T+3)(T)(2T+1)}\right)^{1/2} \langle J_f M_f; T || M^{(\sigma)}_{jm\,1} || J_i M_i; T+1 \rangle \tag{6.8c}$$

Observe that the transitions $\Delta T = \pm 1$ given by (6.8a) and (6.8c) are invariant against the substitution $M_T \to -M_T$. This is just the transformation from a nucleus to its conjugate. Thus in the language of isospin, we find that the transition probabilities between corresponding levels in conjugate nuclei are identical if $\Delta T = \pm 1$ [Morpurgo (59)]. We obtained earlier a similar result based only on charge symmetry. A similar rule does not hold for $\Delta T = 0$ since the isovector transition (6.8b) amplitude changes sign when M_T changes sign while the isoscalar component (6.7) is unchanged. Relations between corresponding transitions in conjugate nuclei for $\Delta T = 0$ become possible only if the isoscalar matrix element is much smaller (or zero) than the isovector one, as is generally true, or vice versa. The resulting approximate rules are derived and listed by Warburton and Weneser (69).

Experimental probing of the $\Delta T = \pm 1, 0$ rule has consisted in looking for $\Delta T = \pm 2$ transitions. In recent years a number of $T = 2$ levels have been located in *light nuclei*, and one can investigate the selection rules by searching for radiative transitions to the $T = 0$ ground state. Upper bounds of only a few percent of the allowed transitions have been established for these isospin forbidden transitions [Hanna (69)]. An example of the conjugate nuclei rule is obtained by comparing the $M1$ decay of levels in C^{13} and N^{13} at about 15 MeV. The widths for the C^{13} decay is 25 ± 7 eV [Peterson (68)] and for N^{13}, 27 ± 5 eV [Adelberger, Cocke, Davids, McDonald (69)]. For a critical evaluation of this last result, see Warburton and Weneser (69).

7. NUCLEAR RECOIL AND RADIATIVE TRANSITIONS

Momentum is of course conserved in a radiative transition. If, for example, a proton changes its state by radiation, the rest of the nucleus must recoil so that the nucleus as a whole has a momentum equal and opposite to that of the radiated photon. This gives rise to an additional current that must be included in the calculation of the transition matrix element. This effect is automatically included in our results as can be seen if we make the conservation of momentum explicit.

Toward this end we introduce the center-of-mass coordinate **R**

$$\mathbf{R} \equiv \frac{1}{A}\sum \mathbf{r}_i$$

and the *intrinsic* coordinates $\boldsymbol{\xi}_i$

$$\boldsymbol{\xi}_i = \mathbf{r}_i - \mathbf{R} \tag{7.1}$$

where

$$\sum \xi_i = 0 \tag{7.2}$$

Taking the initial nuclear momentum to be 0, the initial nuclear wave function ψ_i will be a function of the intrinsic coordinates only while the final wave function will be given by

$$|f\rangle = e^{i\mathbf{P}\cdot\mathbf{R}}\,\psi_f(\xi_1 \ldots \xi_A)$$

where $\mathbf{P}$ is the momentum of the nucleus as a whole. We turn now to the matrix element for the transition. We need consider only the convective current (2.3). (Why?) Thus

$$M_{fi} \sim \sum_i \epsilon\cdot\langle f|e^{i\mathbf{k}\cdot\mathbf{r}_i}\,[H, \mathbf{r}_i]|i\rangle \tag{7.3}$$

$$\sim \sum_i \epsilon\cdot\langle\psi_f|e^{i(\mathbf{k}-\mathbf{P})\cdot\mathbf{R}}\,e^{i\mathbf{k}\cdot\xi_i}\,[H, \mathbf{r}_i]|\psi_i\rangle$$

Now H can be decomposed into two parts, the intrinsic Hamiltonian H_ξ involving only intrinsic coordinates and the kinetic energy T_R of the center of mass. Hence

$$[H, \mathbf{r}_i] = [H_\xi, \xi_i] + [T_R, \mathbf{R}]$$

The second term is proportional to the momentum of the center of mass and thus gives rise to a term in M_{fi} of the form

$$\epsilon\cdot(\mathbf{k} - \mathbf{P})\,e^{i(\mathbf{k}-\mathbf{P})\cdot\mathbf{R}}$$

which vanishes upon integration over $\mathbf{R}$. We are therefore left with

$$M_{fi} \sim \sum_i \epsilon\cdot\langle\psi_f|e^{i\mathbf{k}\cdot\xi_i}\,[H_\xi, \xi_i]|\psi_i\rangle \tag{7.4}$$

This is identical in form with (7.3) except that the coordinates $\mathbf{r}_i$ are replaced by the intrinsic coordinates ξ_i. Equation 5.40 remains valid if this substitution is made. The change is not so easily managed for the corresponding magnetic multipoles expression (5.39) because of the need to interpret the operator $\mathbf{l}_i$ in virtue of (7.2). In this case it is best to simply begin with (5.39) and transform to the center-of-mass and intrinsic coordinates dropping any final dependence on the former variable.

One qualitative change should be commented on. Because of (7.2)

$$[H_\xi, \xi_A] = -\sum_{i\neq A} [H_\xi, \xi_i]$$

Hence the Ath term in (7.4) has the following form

$$\epsilon\cdot\langle\psi_f|\exp\Big(-i\mathbf{k}\cdot\sum_{j\neq A}\xi_j\Big)\sum_{i\neq A}[H_\xi, \xi_i]|\psi_i\rangle \tag{7.5}$$

This is clearly a many-body term. It is thus no longer possible, once the transformation to intrinsic coordinates is introduced, to say that the current consists of a sum of single-body operators. This statement is correct as long as $k \neq 0$. In the dipole approximation the single-body character remains but many-body terms enter for the higher multipoles.

Since in many models intrinsic coordinates are not used, it is useful to express the multipole operators in terms of $\mathbf{r}_i$. Let us begin with the dipole moment ($j = 1$) (5.40) which we shall rewrite in vector form:

$$\mathbf{D} = e \sum \frac{1}{2} [1 + \tau_3 (i)] \boldsymbol{\xi}_i \tag{7.6}$$

From (7.2) and (7.1) we obtain

$$\mathbf{D} = \frac{e}{2} \sum \tau_3(i)\, \boldsymbol{\xi}_i = \frac{e}{2} [\sum_i \tau_3(i)\, \mathbf{r}_i - \mathbf{R} \sum_i \tau_3(i)] \tag{7.7}$$

The sum over $(1/2)\, \tau_3(i)$ is just T_3 and hence may be replaced by $(Z - N)/2$. Finally inserting the expression for $\mathbf{R}$ in terms of $\mathbf{r}_i$ yields:

$$\mathbf{D} = \frac{e}{2} \sum_i \left[\tau_3(i) - \frac{Z - N}{A} \right] \mathbf{r}_i$$

or

$$\mathbf{D} = e \left[\sum_{\text{protons}} \mathbf{r}_i \left(\frac{N}{A} \right) - \sum_{\text{neutrons}} \mathbf{r}_i \left(\frac{Z}{A} \right) \right] \tag{7.8}$$

We thus see that for the dipole operator *each proton may be taken to have an effective charge of* $(N/A)\, e$, *each neutron an effective charge of* $-\ (Z/A)\, e$. The neutrons do not really radiate; it is the protons that move with the neutrons in the recoil that do.

In the course of this discussion we have also derived another selection rule. From (7.7) it follows that the dipole moment transforms like an isovector, the isoscalar component is zero. Hence the matrix element (6.7) vanishes and it follows:

(a) Corresponding $E1$ transitions in conjugate nuclei have equal strengths

(b) $\Delta T = 0$, $E1$ transitions in self conjugate nuclei ($M_T = 0$) are forbidden.

It is instructive to develop the corresponding formulae for the quadrupole moment that we write in dyadic form:

$$Q_2 = \frac{1}{2} \sum [1 + \tau_3(i)]\, \boldsymbol{\xi}_i \boldsymbol{\xi}_i \tag{7.9}$$

Inserting $\boldsymbol{\xi}_i = \mathbf{r}_i - \mathbf{R}$ yields

$$Q_2 = \frac{1}{2} \sum (\mathbf{r}_i - \mathbf{R})\mathbf{r}_i + \frac{1}{2} \sum \tau_3(i)\, [\mathbf{r}_i \mathbf{r}_i - \mathbf{R}\, \mathbf{r}_i - \mathbf{r}_i \mathbf{R}] + \frac{Z - N}{2} \mathbf{R}\, \mathbf{R}$$

In each term replace **R** by its expansion in terms of $\mathbf{r}_i$. *Finally keep only single-particle terms.* One obtains

$$Q_2 \simeq \frac{1}{2}\left(1 - \frac{2}{A} + \frac{2Z}{A^2}\right) \sum \mathbf{r}_i\mathbf{r}_i + \frac{1}{2}\left(1 - \frac{2}{A}\right) \sum \tau_3(i)\, \mathbf{r}_i\mathbf{r}_i$$

Hence the effective charge for protons $[\tau_3(i) = 1]$ is $[1 - (2/A) + (Z/A^2)]$, while the effective charge for neutrons is Z/A^2. Thus except for the light nuclei the recoil effect on the one-body part of the quadrupole operator is not of importance for the $E2$ transition. The many-body terms are generally neglected although accurate evaluations of these have not been made. The effective charge for the one-body part of the higher multipoles can be readily determined with result

$$\text{Effective proton charge} = e\left[\left(1 - \frac{1}{A}\right)^j + (-)^j \frac{(Z - 1)}{A^j}\right] \tag{7.10}$$

$$\text{Effective neutron charge} = eZ\left(-\frac{1}{A}\right)^j$$

Similar results can be obtained for the orbital part of the magnetic multipoles. In both cases it is clear that effects of quantitative importance are present for the protonic component of the multipole moments for light nuclei for the lower-order multipoles.

8. SUM RULES

When the sum of all the radiative-transition probabilities from a given level weighted by some power of each transition energy can be given in a closed form, generally as the expectation value of some operator, a sum rule is said to exist. Clearly such a rule bounds the size of any given matrix element and as such provides a useful benchmark for the distribution in energy of any given multipole strength.

Of considerable importance is the direct connection of the sums with measureable quantities such as the photoabsorption cross sections. In these experiments the attenuation of an incident beam of gamma rays is measured and the photoabsorption cross section is deduced. To obtain that cross section from the results of Section 5 for radiative transitions, we shall make use of the principle of detailed balance discussed in Volume II. In radiative decay the photon is emitted, the nucleus making a transition from state f to state i. In absorption, the photon is absorbed, the nucleus going from state i to state f. The principle of detailed balance states that the square magnitude of the matrix element for the transition is the same in both cases. To apply it here, we shall first strip off from (5.35), the emission-transition probability, those factors that modify the matrix element and then multiply by those needed to obtain the absorption. Thus (1) divide (5.35) by the density of final states for photon emission, $k^2/\hbar c$, (2) divide by 4π to correct for the angular integration over an isotropic angular distribution performed in obtaining (5.35), and (3) remove the average over the initial spin states of the nucleus. Turning to the absorption experiment we must (1) average over the initial photon spin degeneracy $2(2j + 1)$ and (2), to get the absorption cross section per photon, divide by the incident photon-current density $c/(2\pi)^2$ for the normalization used in obtaining (5.35). It is finally necessary to multiply by the density of final states. Since in this case the transition is to a definite final state, the density is a delta function. Rather than carry the δ-function along on the calculation, the cross-section is integrated over the possible final energies, a contribution occurring only at a final energy corresponding to the definite final state. The result is designated by I. Collecting all the factors

$$\frac{1}{4\pi}\frac{(2\pi)^3}{c}\frac{\hbar c}{k^2}\frac{1}{2(2j+1)}$$

and multiplying (5.35) by this result yields

$$\mathrm{I}(i \to f) = (2\pi)^3 \frac{e^2}{\hbar c}\frac{j+1}{j(2j+1)\,[(2j+1)!!]^2}(E_f - E_i)\,k^{2j-2}\,B(\sigma j;\, i \to f) \qquad (8.1)$$

The sum rules are concerned with the sum of this result weighted by powers of $(E_f - E_i)$, over all possible final energy conserving states, connected to the initial state by a radiative multipole transition. An example will suffice to show how they are derived and their possible exploitation.

Consider first of all

$$S_j^{(1)} \equiv \frac{8\pi^3 e^2}{\hbar c}\frac{j+1}{j(2j+1)\,[(2j+1)!!]^2}\sum_f (E_f - E_i)\,B(\sigma j;\, i \to f) \qquad (8.2)$$

The superscript on S indicates the first power of $(E_f - E_i)$ in the sum. Comparing with (8.1), we see that

$$S_j^{(1)} = \sum_f \frac{1}{k^{2j-2}}\, I\,(i \to f)$$

so that $S_1^{(1)}$ does in fact yield the total absorption cross section that, summed over all possible final states, is a measurable quantity. $S_j^{(1)}$ may be rewritten (See Eqs. 5.32 and 5.36) as

$$S_j^{(1)} = \frac{8\pi^3 e^2}{\hbar c}\frac{(j+1)}{j[(2j+1)!!]^2\,(2J_i+1)}\sum_{m,f,M_i}(E_f - E_i)$$

$$\times\, |\langle f|\sum_\alpha e_\alpha r_\alpha{}^j\, Y^*_{jm}\,(\hat{\mathbf{r}}_\alpha)|i\rangle|^2 \qquad (8.2')$$

Here i (and correspondingly f) denote the quantum numbers J_iM_i; $T_iM_{T_i}$. The quantity e_α is $[1 + \tau_3(\alpha)]/2$ for $j \geq 2$. For $j = 1$, that is, the dipole case, it is according to (7.8) given by $(1/2)\,[\tau_3(\alpha) - (Z - N)/A]$.

The sum rule will be obtained mainly through closure. We note

$$(E_f - E_i)|\langle f|\,Q_{jm}\,|i\rangle|^2 = \frac{1}{2}\,\langle i|\;[Q^\dagger_{jm}, H]\;|f\rangle\langle f|\,Q_{jm}\,|i\rangle$$

$$+\frac{1}{2}\,\langle i|\,Q^\dagger_{jm}\,|f\rangle\;\langle f|\;[H, Q_{jm}]\;|i\rangle$$

where H is the Hamiltonian for the initial and final nuclear wave function and $Q_{jm} = \Sigma\, e_i r_i^j\, Y^*_{jm}(\hat{\mathbf{r}}_i)$. If both sides are summed over f we obtain

$$\sum_f (E_f - E_i)\;|\langle f|\,Q_{jm}\,|i\rangle|^2 = \frac{1}{2}\,\langle i|\;[Q^\dagger_{jm}, H]\;Q_{jm} + Q^\dagger_{jm}\;[H, Q_{jm}]|\,i\rangle$$

Recalling that $Y^*_{jm} = (-)^j Y_{j,-m}$, we see that for every term of the type in the above equation, there is another in which $Q^\dagger_{jm} \leftrightarrow Q_{jm}$. Hence

$$\sum_{fm} (E_f - E_i)|\,\langle f|\,Q_{jm}\,|i\rangle|^2 = \frac{1}{2}\sum_m \langle i|\;[[Q^\dagger_{jm}, H], Q_{jm}]\;|i\rangle \tag{8.3}$$

Let us now work out the double commutator, starting with the kinetic-energy term:

$$\sum_a \left[\left[Q^\dagger_{jm}, -\frac{\hbar^2}{2m}\,\nabla_a^2\right], Q_{jm}\right]$$

$$= \sum_{a,i,k} e_k e_i \left[\left[r_i^j\, Y^*_{jm}(\hat{\mathbf{r}}_i), -\frac{\hbar^2}{2m}\,\nabla_a^2\right], r_k^j\, Y^*_{jm}\,(\hat{\mathbf{r}}_k)\right]$$

$$= -\frac{\hbar^2}{2m}\sum_a e_a^2 \left[[r_a^j\; Y_{jm}\,(\hat{\mathbf{r}}_a), \nabla_a^2], r_a^j\; Y^*_{jm}\,(\hat{\mathbf{r}}_a)\right]$$

$$= \frac{\hbar^2}{m}\sum_a e_a^2\; \nabla_a\,(r_a^j\; Y_{jm}(\hat{\mathbf{r}}_a))\cdot\nabla_a\,(r_a^j\; Y^*_{jm}\,(\hat{\mathbf{r}}_a))$$

$$= \frac{\hbar^2}{2m}\sum_a e_a^2\; \nabla_a^2\;[r_a^{2j}\; Y_{jm}\,(\hat{\mathbf{r}}_a)\; Y^*_{jm}\,(\hat{\mathbf{r}}_a)] \tag{8.4}$$

In this calculation we have made repeated use of the result

$$\nabla_a^2\,(r_a^j\; Y_{jm}\,(\hat{\mathbf{r}}_a)) = 0$$

We sum the last result over m using

$$\sum_m Y_{jm}(\hat{\mathbf{r}})\, Y^*_{jm}(\hat{\mathbf{r}}) = \frac{2j+1}{4\pi}$$

Therefore

$$\frac{1}{2}\sum_{a,m}\left[\left[Q^\dagger_{jm},\, -\frac{\hbar^2}{2m}\nabla_a{}^2\right],\, Q_{jm}\right] = \frac{\hbar^2}{2m}\frac{j(2j+1)^2}{4\pi}\sum_a e_a{}^2\, r_a{}^{2j-2} \tag{8.5}$$

We note that the matrix elements of this operator are independent of M_i.

The potential energy contained in H will also contribute to the double commutator if it is dependent on the isospin. From our present-day understanding of nuclear forces (see Chapter I) it is reasonable to assume that in the zeroth-order approximation these forces are charge independent so that the isospin-dependent terms describing the interaction between particles a and b is of the form $\boldsymbol{\tau}(a)\cdot\boldsymbol{\tau}(b)$. The isospin-dependent part of H is thus

$$\mathcal{V}^{(\tau)} = \sum_{a<b}\boldsymbol{\tau}(a)\cdot\boldsymbol{\tau}(b)\, V_{ab}{}^{(\tau)}$$

where $V^{(\tau)}_{ab}$ depends on the spin and space coordinates of particles a and b. These terms do not commute with Q_{jm} because of the latter's dependence on $\tau_3(i)$. Thus the potential-energy contribution to the double commutator, (8.3), is

$$\frac{1}{8}\frac{2j+1}{4\pi}\sum_{i,k}[[\tau_3(i)\, r_i{}^j,\, \mathcal{V}^{(\tau)}],\, \tau_3(k)\, r_k{}^j]$$

Evaluation of this double commutator is a simple exercise. We obtain for (8.3),

$$\frac{1}{2}\sum_m[[Q^\dagger_{jm},\, \mathcal{V}^{(\tau)}],\, Q_{jm}] = -\frac{1}{2}\frac{2j+1}{4\pi}\sum_{a<b}V^{(\tau)}_{ab}\,(r_a{}^j - r_b{}^j)^2\,[\boldsymbol{\tau}(a)\cdot\boldsymbol{\tau}(b) - \tau_3(a)\tau_3(b)] \tag{8.6}$$

Again note the independence of the matrix elements of this result with respect to M_i.

In the case of a pure Majorana exchange force

$$V^{(\tau)}_{ab} = -\frac{1}{2}(1 + \boldsymbol{\sigma}_a\cdot\boldsymbol{\sigma}_b)\, V^{(M)}$$

The expectation value of $V^{(\tau)}_{ab}\,[\boldsymbol{\tau}(a)\cdot\boldsymbol{\tau}(b) - \tau_3(a)\,\tau_3(b)]$ for spin singlet, isospin triplet is $V^{(M)}$ multiplied by 0 and 2 for the two-proton and the neutron-proton system, respectively, while for a spin triplet, isospin singlet the multiplication factor is 2. Hence the sign of this term in (8.6) will be determined by the sign of $V^{(M)}$. If, therefore, $\mathcal{V}^{(\tau)}$ is a pure Majorana and $V^{(M)}$ is negative (as would be required in such a theory for say a bound deuteron), *its contri-*

bution to the sum is positive. The identical result holds for the OPEP (the one pion exchange potential).

The final result for $S_j^{(1)}$ is obtained by substituting (8.5) and (8.6) into (8.2′):

$$S_j^{(1)} = \left(\frac{e^2}{\hbar c}\right)\frac{(j+1)\pi^2}{[(2j-1)!!]^2}\left\{\frac{\hbar^2}{m}\langle i|\sum_a e_a^2\, r_a^{2j-2}\,|i\rangle\right.$$

$$\left.-\frac{1}{j(2j+1)}\sum_{a<b}\langle i|\, V_{ab}^{(\tau)}\,(r_a^j - r_b^j)^2\cdot[\boldsymbol{\tau}(a)\cdot\boldsymbol{\tau}(b) - \tau_3(a)\,\tau_3(b)]\,|i\rangle\right\} \tag{8.7}$$

The one-body operators coming from the kinetic energy are either model independent ($j = 1$) or related to the $2j - 2$ moment of the charge density of the target nucleus ($j \geqslant 2$). The latter in principle can be measured by observation of the radiative transitions in atoms formed by these nuclei with negatively charge particles such as electrons, muons, and more recently pions, kaons, and antiprotons. The energy of the x-rays emitted by these particles depends directly upon the charge moments. Related information is also obtained from electron scattering. The two-body terms originating in $\mathcal{V}^\tau$ require more detailed knowledge of the target wave function, particularly of the neutron-proton correlations. A *very rough* estimate of the term is obtained as follows. First factor $r_a^j - r_b^j$ into $(r_a - r_b)\,(r_a^{j-1} + \ldots + r_b^{j-1})$. The spatial expectation value is then approximated by $[jr_0R^{j-1}]^2\, \bar{V}$ where r_0 is the range of $V_{ab}^{(\tau)}$, R is the nuclear radius and $\bar{V}$ is some average value of the potential. The isospin term can be readily evaluated to give $-(A + N - Z) = -2Z$. Similarly the kinetic term is roughly $(\hbar^2/m)\, ZR^{2(j-1)}, j \neq 1$. Taking the ratio of the two-body to one-body terms yields about $(\hbar^2/2m\, r_0^2\, \bar{V})$ a number that does not differ greatly from unity. It is thus not possible to neglect the potential-energy term.

It has become customary to write (8.7) as if the potential-energy term were a correction V_j:

$$j > 2 \qquad S_j^{(1)} = \frac{e^2}{\hbar c}\frac{(j+1)\pi^2}{[(2j-1)!!]^2}\frac{\hbar^2}{m} Z\,\langle r^{2j-2}\rangle\,[1 + V_j] \tag{8.8}$$

where

$$\langle r^{2j-2}\rangle \equiv \frac{1}{Z}\langle i|\sum_a e_a^2\, r_a^{2j-2}|i\rangle$$

For $j = 1$ (see Eq. 7.8) it is easy to evaluate the kinetic-energy term. We obtain

$$S_1^{(1)} = 2\pi^2\frac{e^2}{\hbar c}\frac{NZ}{A}\frac{\hbar^2}{m}(1 + V_1) = 60\,\frac{NZ}{A}(1 + V_1)\text{ MeV } mb \tag{8.9}$$

An estimate of V_1 has been obtained by Levinger and Bethe (50, 51, 60) who found it to be between 0.4 and 0.5 for the nuclear-force models in fashion at that time. However recent theoretical estimates as well as the empirical data (see Brown (73) for a review) gives a value that is considerably larger than one. Much of this enhancement originates in the tensor force.

Similar sum rules can be generated by replacing the factor $(E_f - E_i)$ in (8.2) by $(E_f - E_i)^n$ where n can be any positive integer including zero. The value $n = 1$ provides a sum rule that is directly connected to the integrated absorption cross section for dipole radiation, $n = 3$ for quadrupole, etc. The dipole-sum rule has proven to be especially interesting because of the existence of the giant dipole resonance (see Section 11.B) that exhausts a considerable fraction of the sum (8.9). Although "giant" resonances have been discovered for other multipoles, they do not have the prominence of the electric dipole.

Problem. Derive the sum rule for the form factor

$$\sum_f (E_f - E_i)\,\{\langle i|\sum_a e^{i\mathbf{q}\cdot\mathbf{r}_a}|f\rangle\,\langle f|\sum_b e^{i\mathbf{q}'\cdot\mathbf{r}_b}|i\rangle$$

$$+\ \langle i|\sum_b e^{i\mathbf{q}'\cdot\mathbf{r}_b}|f\rangle\langle f|\sum_a {}^{i\mathbf{q}\cdot\mathbf{r}_a}|i\rangle\}$$

$$= -\frac{\hbar^2}{m}\,\mathbf{q}\cdot\mathbf{q}'\,\langle i|\sum_{a,b} e^{i\mathbf{q}\cdot\mathbf{r}_a}\,e^{i\mathbf{q}'\cdot\mathbf{r}_b}|i\rangle$$

[Kao and Fallieros (70) and Noble (71)]

Another important type of sum rule is obtained by application of dispersion theory. As pointed out by Gell-Mann, Goldberger, and Thirring (54) the Kramers-Kronig dispersion relation (see Volume II for discussion) can be exploited for this purpose. The relation is

$$\operatorname{Re}\,[f(E) - f(0)] = \frac{E^2}{2\pi^2\hbar c}\,\mathcal{P}\int_0^\infty \frac{\sigma(E')}{E'^2 - E^2}\,dE' \tag{8.10}$$

In this equation $f(E)$ is the amplitude for the scattering at 0° of a photon, while $\sigma(E)$ is the total photoabsorption cross section including all multipoles. The integral is a principal value integral as denoted by $\mathcal{P}$. An important point is that we know the value of $f(0)$ for any system. Since zero-frequency photons correspond to a static electric field, the amplitude can only depend upon the charge. Its value for a system of charge Z and mass Am is therefore given by the zero-frequency limit of the Compton cross section, that is, by the Thomson limit

$$f(0) = -\frac{(Ze)^2}{Amc^2} \tag{8.11}$$

Similarly for a proton and neutron the result is

$$f_p(0) = -\frac{e^2}{mc^2} \qquad f_n(0) = 0 \tag{8.12}$$

We note that the photon-absorption process for the neutron and proton has a threshold: that of the pion mass, μ. At energies below the pion mass the photon-nucleon interaction results only in scattering. Hence the Kronig-Kramers relation for neutron and proton becomes

$$\operatorname{Re} f_p(E) + \frac{e^2}{mc^2} = \frac{E^2}{2\pi^2\hbar c}\,\mathcal{P}\int_{\mu c^2}^{\infty} \frac{\sigma_p(E')}{E'^2 - E^2}\,dE' \tag{8.13}$$

$$\operatorname{Re} f_n(E) = \frac{E^2}{2\pi^2\hbar c}\,\mathcal{P}\int_{\mu c^2}^{\infty} \frac{\sigma_n(E')}{E'^2 - E^2}\,dE' \tag{8.14}$$

A sum rule results if we make the reasonably physical assumption that, at infinite photon energy, the forward- scattering amplitude by a nucleus of Z-proton and N-neutrons is just the additive sum of the scattering by each of the nucleons:

$$f(\infty) = Zf_p(\infty) + Nf_n(\infty) \tag{8.15}$$

To derive the sum rule we note that

$$\operatorname{Re}\,[f(E) - Zf_p(E) - N f_n(E)] = -\left[\frac{(Ze)^2}{Amc^2} - Z\frac{e^2}{mc^2}\right]$$

$$+\frac{E^2}{2\pi^2\hbar c}\,\mathcal{P}\left[\int_0^{\mu c^2} \frac{\sigma(E')}{E'^2 - E^2}\,dE' + \int_{\mu c^2}^{\infty} \frac{\sigma(E') - Z\sigma_p(E') - N\sigma_n(E')}{E'^2 - E^2}\,dE'\right]$$

Evaluate this equation at infinite E, assuming that the second integral converges with sufficient rapidity. Then

$$\frac{NZ}{A}\frac{e^2}{mc^2} = \frac{1}{2\pi^2\hbar c}\left[\int_0^{\mu c^2} \sigma(E)dE + \int_{\mu c^2}^{\infty} [\sigma(E) - Z\sigma_p(E) - N\sigma_n(E)]\,dE\right]$$

or

$$\int_0^{\mu c^2} \sigma(E)dE = \frac{NZ}{A}\frac{2\pi^2 e^2}{\hbar c}\frac{\hbar^2}{m} + \int_{\mu c^2}^{\infty} [Z\sigma_p(E) + N\sigma_p(E) - \sigma(E)]\,dE \tag{8.16}$$

This sum rule differs from the sum rule (8.9) since there is no multipole decomposition. Nevertheless there is a close correspondence, the first term on the right-hand side of (8.16) being identical to the first term on the right-hand side of (8.9). We shall postpone the comparison of these results with experiment until Section 11.B.*

---

*It turns out that the "reasonably" physical assumption 8.15 is in fact not correct. The high frequency limit is modified because of the possibility that a photon can convert into a vector meson. (See Weise (73) for a recent review). As a consequence (8.16) is not correct.

9. EFFECTIVE CHARGE

In a simple shell-model description, the electric multipole radiation emitted in a transition in which a neutron changes its state would be small. It is not exactly zero because of the second term in (5.37). This conclusion is in violent disagreement with experiment. For electric dipole radiation, we have a ready explanation for as we have seen in Section 7 (Eq. 7.8) the neutron has an effective charge of $-(Z/A)\,e$ and the proton charge should be reduced to $(N/A)\,e$. When the neutron changes its state the rest of the nucleus recoils. However this mechanism (Eq. 7.10) will not suffice for the higher multipoles. The answer lies in the inadequacy of the simple shell-model description. The nuclear states are more complex. Although the principal component of the wave function may indeed be a quiescent core plus an excited neutron, there will also be components in which the protons in the core are excited. For example in the case of ^{17}0 the excited state, in which the neutron is in an $s_{1/2}$ level (radiating to a $d_{5/2}$ level by $E2$), the wave function will also contain a two particle–one hole ($2p1h$) component in which a proton in the ^{16}0 core is excited from a $p_{1/2}$ orbit into a p or f orbit forming a ($1p1h$) proton state that is then combined with the neutron in its final $d_{5/2}$ state (Fig. 9.1). It is the excited proton that does the radiating by making a transition to the hole, the core returning to its ground state. More complex components such as three particle–two hole states ($3p2h$) will also be present in the exact ^{17}0 wave function. Because the electromagnetic interaction is a one-body operator the radiative transitions of the $3p2h$ states can only be to $3p2h$ or $2p1h$ states. When the amplitudes of these components are small, their contributions can be neglected as a second-order contribution so that the $2p1h$ states are then largely responsible for the enhanced radiative-transition probability.

Another interpretation of effective charge is given by Bohr and Mottelson (69). In their description the core is deformed by the nonspherical field generated by the valence nucleon. As a consequence the core acquires a quadrupole moment. A very rough estimate of the induced quadrupole

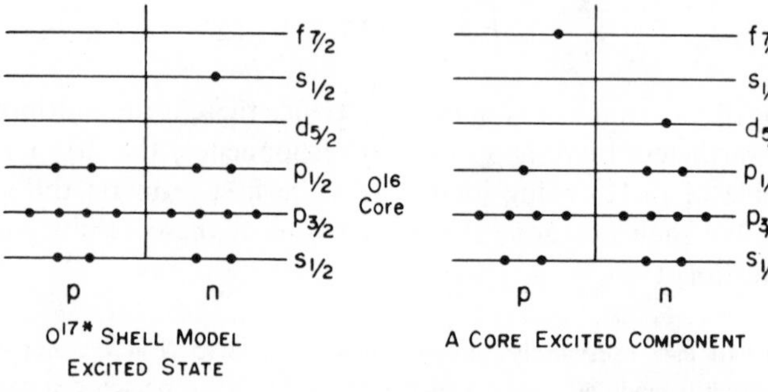

FIG. 9.1. Core excitation.

moment according to Bohr and Mottelson (19) is $(Z/A)\,Q_{\text{s.p.}}$, where $Q_{\text{s.p.}}$ is the single-particle quadrupole moment. This argument does not differ qualitatively from the preceding discussion since the multiparticle-hole components that the exact wave function contains can correspond to a deformation.

The core excitations may also be collective. A vibrational state, which in first approximation is a constructively interfering linear combination of $1p1h$ states, is an obvious candidate. The radiating component would then consist of the core in an excited vibrational state, plus the neutron in its final level, the radiation being emitted in the deexcitation. If the core is deformed, the core excited state can be an excited rotational state of the rotational band based on the core.

The general phenomenon being described is referred to as core excitation [deShalit (61); see also Section 12 and Section VII.19]. The valence nucleon polarizes the core by exciting it. As a consequence the core can contribute to the radiative-transition probability. Although this effect is most dramatic for the "radiating" neutron, it is of course also present if the valence nucleon is a proton. A similar phenomenon occurs if there are several valence nucleons. The first excited state of ^{18}O behaves for radiative transitions very much as if the valence particles were protons rather than neutrons.

The consequent enhanced radiative-transition probability can often be approximately described by giving the valence neutrons and protons an effective charge:

$$e_p = (1 + \alpha_p)e \tag{9.1}$$

$$e_n = \alpha_n e$$

with the matrix element calculated with the simple initial and final wave functions omitting core excitations. This is of course an empty statement if it refers to a single transition since there is then only one experimental number, the lifetime, assuming for the moment that we are dealing with a pure electric multipole. It becomes a significant assumption if we assert that the *same* values of e_n and e_p can be used for other transitions in the same nucleus or for similar transitions for nuclei in the same shell.

This concept of the effective charge can be extended to more complex model wave functions than the simple shell-model case with which we started this discussion. Of course if a complete set of wave functions are used to describe the initial and final state, and these are chosen so as to diagonalize the Hamiltonian, there is no need for the effective charge concept. However, if the Hilbert space in which the Hamiltonian is diagonalized is truncated (ie., if diagonalization is performed in only part of the space), there will then generally be components outside of the truncated space that can contribute to the transition probability, and then the concept of effective charge may prove valuable.

When the shell-model theory described in Chapters 4 and 5 is employed, the core is considered inert, and the residual interaction is diagonalized usually with the use of perturbation theory, the wave functions for the valence nucleons being restricted to a particular shell, except for special cases where configurations involving nearby orbitals turn out to be important. The effective charge then arises from core excitations in which the nucleons in the core are excited to unfilled orbits. It would appear possible that excitation of the valence nucleons to orbits not considered in the shell-model diagonalization would also contribute. However as shown in deShalit and Talmi (Sec. 37) such excitation of a *single* particle is equivalent to the use of an improved shell-model potential that may be deformed. Only when multiparticle excitations of the valence nucleons are important components of the exact initial and final wave functions will it be necessary to include valence-nucleon excitations, if, as we emphasize, an appropriate single-particle potential is used.

For a given shell-model theory the matrix element for a radiative transition is linearly dependent on the effective charges e_n and e_p. By comparing nuclei in the same shell one can then check the concept of effective charge and also determine their values [Wilkinson (67)]. An example of such a fit is shown in Fig. 9.2 based on the $E2$ transitions in ^{17}O, ^{18}O, ^{18}F, ^{19}F, and ^{19}Ne, nuclei with valence nucleons in the s-d shell and with an ^{16}O core. The shell-model calculations involved were performed by Elliott and Wilsdon (67), the experiments by Becker, Olness, and Wilkinson [Becker (64), Becker (67)]. Except for ^{18}F, the $E2$ transitions are the lowest in the nucleus. A given nucleus is represented by a straight line. If the concept of effective charge were exact, the lines for the various nuclei should intersect in a single point. The rather small region defined by the crossing of the lines indicate that the notion of effective charge is a good one but not an exact one for these nuclei, and for the $E2$ transition. Moreover, the additional charge $\alpha_n e$ and $\alpha_p e$ are about the same and equal to about $0.6e$. This equality would be expected because of charge independence of nuclear forces and the charge symmetry of the core nucleus.

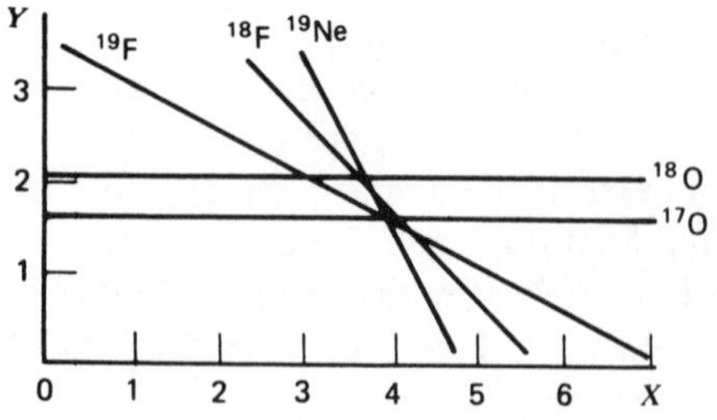

FIG. 9.2. $X \sim (1 + \alpha_p)$, $Y \sim \alpha_n$; $\alpha_p e$ and $\alpha_n e$ are the effective charges to be attached to the protons and neutrons, respectively, outside the closed ^{16}O core to gain agreement between the predictions with shell-model wave functions and the low-lying $E2$ transition rates in the nuclei shown [taken from Wilkinson (67)].

Effective charges have also been used to describe transitions in the p-shell nuclei. In this case the core is ^{4}He. Because there are only two protons in ^{4}He and because of the large energies needed for excitation of ^{4}He, the main contribution to the effective charge comes from valence-nucleon excitations. For these nuclei it is found empirically that $\alpha_n \simeq \alpha_p \sim 0.5$, [Warburton (63)], as Table 9.1 makes clear. Shell-model calculations using the Nilsson deformed orbitals have been made [Cohen and Kurath (65), Kurath (65), Poletti, Warburton, and Kurath (67)] and as expected an explanation of the value of the effective charge for the $1p$-shell nuclei has been developed.

TABLE 9.1 The Energies of the States are Shown in MeV. x^a_{theo}: is the Theoretical E2/M1 Amplitude Ratio without Using Effective Charges; x^b_{theo}: is That Ratio Using $\alpha = 0.5$.

Nucleus	Transition	x^a_{theo}	x^b_{theo}	x_{exp}
^{11}B	$4.46 \to 0$	-0.123	-0.224	$-(0.20 \pm 0.02)$
^{11}C	$4.32 \to 0$	$+0.090$	$+0.204$	$+(0.17 \pm 0.02)$
^{13}C	$3.68 \to 0$	-0.054	-0.100	$-(0.096 \pm 0.025)$
^{13}N	$3.51 \to 0$	$+0.030$	$+0.068$	$+(0.09 \pm 0.02)$
^{14}N	$7.03 \to 0$	$+0.361$	$+0.722$	$+(0.60 \pm 0.10)$
^{15}N	$6.32 \to 0$	$+0.081$	$+0.122$	$+(0.12 \pm 0.015)$
^{15}O	$6.18 \to 0$	0	-0.048	$-(0.16 \pm 0.016)$

The ratios x^a are the theoretical values obtained with $\alpha_n = \alpha_p = 0$. The impressive improvement (see column x^b) obtained using the effective charge $\alpha_n \simeq \alpha_p = 1/2$ is apparent. The sign is given correctly in all cases, while the magnitude of the $E2/M1$ amplitude ratio is remarkably close for most cases.

Analysis of the $f_{7/2}$ shell $E2$ transitions yields $\alpha_p = 0.97$ and $\alpha_n = 1.87$ [Zamick and McCullen (65)]. In this case the effective charge of the neutron and proton are nearly identical so that the isovector component of the charge

$$\frac{1}{2}(e_p + e_n) + \frac{1}{2}(e_p - e_n)\tau_3 \tag{9.2}$$

is almost completely quenched. A similar effect, but not as large, occurs for the lighter nuclei where $\alpha_p = \alpha_n = 1/2$.

One word of caution! These values of e_p and e_n are dependent on the size of the Hilbert space included in the shell-model calculation. As we mentioned earlier, the larger this space, the smaller α_n and α_p. This phenomenon is seen in the calculations of Federman and Zamick (69) on the effective charge for ($E2$) transitions in ^{58}Ni. Using a ^{40}Ca core the average α_n and α_p are 0.47 and 0.17, respectively. When ^{56}Ni serves as a core the values of α_n and α_p are much larger: 1.02 and 0.37, respectively.

We conclude this discussion by describing the calculation of the effective charge for the case of a single-valence nucleon. A similar calculation that is more detailed is given in Section 12 for the effective magnetic moment operator. For the single-valence nucleon case, the model wave function for the excited state, which is to decay electromagnetically, is

$$\psi_i^{(\mathrm{Mod})} = \frac{1}{\sqrt{A+1}}\, \mathcal{A}\, (\psi_c \phi_i) \tag{9.3}$$

where ψ_c is the normalized core wave function, ϕ_i the single particle wave function for the valence nucleon, A is the number of nucleons in the core, and $\mathcal{A}$ is the antisymmetrization operator. Similarly the wave function for the final state is

$$\psi_f^{(\mathrm{Mod})} = \frac{1}{\sqrt{A+1}}\, \mathcal{A}\, (\psi_c \phi_f) \tag{9.4}$$

These two functions must be orthogonal. This is achieved by requiring the orthogonality of ϕ_f and ϕ_i, and by insisting that ϕ_i and ψ_c are orthogonal when the coordinate of ϕ_i are identical with one of the coordinates of ψ_c. These requirements are satisfied if ψ_c and therefore $\psi_{i,f}$ are Slater determinants.

Under these circumstances the matrix element of the multipole Q_{jm} is

$$M_{fi}^{(\mathrm{Mod})} = \langle \psi_f^{(\mathrm{Mod})} | \sum_k Q_{jm}(\mathbf{r}_k) | \psi_i^{(\mathrm{Mod})} \rangle = \langle \phi_f(\mathbf{r}_i) | Q_{jm}(\mathbf{r}_1) | \phi_i(\mathbf{r}_1) \rangle \tag{9.5}$$

where

$$Q_{jm}(\mathbf{r}_1) = e_1\, r_1^{\,j}\, Y_{jm}(\hat{\mathbf{r}}_1)$$

The effect of the residual interaction, V_R, on the wave function can be represented as follows:

$$\Psi_i = N_i\, [1 + R_i]\, \psi_i^{(\mathrm{Mod})} \tag{9.6}$$

where Ψ_i is the exact wave function. The operator R_i is

$$R_i = \frac{q_i}{E_i - H_0}\, V_R \tag{9.7}$$

Here H_0 is the model Hamiltonian, E_i the energy for Ψ_i, q_i is a projection operator that annihilates $\psi_i^{(\mathrm{Mod})}$, and N_i, the normalization factor. A similar description applies to Ψ_f. In terms of R_i and R_f the exact matrix element for the radiative transition is

$$M_{fi} = \langle \psi_f^{(\mathrm{Mod})} | (1 + R_f^\dagger)\, Q_{jm}\, (1 + R_i)\, \psi_i^{(\mathrm{Mod})} \rangle \tag{9.8}$$

where

$$Q_{jm} = \sum_k Q_{jm}(\mathbf{r}_k)$$

Assuming that first-order perturbation theory is sufficiently accurate $N_i = 1$, E_i is the energy for $\psi_i^{(\text{Mod})}$ and (9.8) can be written as the sum of three terms, $M_{fi}^{(0)}$, $M_{fi}^{(1)}$ and $M_{fi}^{(2)}$. $M_{fi}^{(0)}$ is given by (9.5) while

$$M_{fi}^{(1)} \equiv \langle R_f\, \psi_f^{(\text{Mod})} \,|\, Q_{jm}\, \psi_i^{(\text{Mod})} \rangle \tag{9.9}$$

$$M_{fi}^{(2)} \equiv \langle \psi_f^{(\text{Mod})} \,|\, Q_{jm}\, R_i\, \psi_i^{(\text{Mod})} \rangle \tag{9.10}$$

Let us now consider $M_{fi}^{(1)}$ is some detail. Note that Q_{jm} operating on $\psi_i^{(\text{Mod})}$ (Eq. 9.3) will generally contain terms in which the core or the valence particle changes state. We restrict the discussion to core excitations only by replacing $Q_{jm}\, \psi_i^{(\text{Mod})}$ as follows:

$$Q_{jm}\, \psi_i^{(\text{Mod})} \rightarrow \frac{1}{(A+1)^{3/2}} \sum_{c'} \mathcal{A}\, (\psi_{c'}\, \phi_i)\, \langle \mathcal{A}\, \psi_{c'}\, \phi_i |\; Q_{jm} | \mathcal{A}\, \psi_c \phi_i \rangle \tag{9.11}$$

This makes sense only if $\{\mathcal{A}\psi_{c'}\, \phi_i\}$ form an orthogonal set. This is possible only if ϕ_i is orthogonal to each $\psi_{c'}$ when its coordinate and one of $\psi_{c'}$ are identical. This condition can be insured only if the set $\{\psi_{c'}\}$ of interest here is not complete. If it were the condition would imply $\phi_i \equiv 0$! The condition is easily met when the wave functions are Slater determinants where it simply means that the valence nucleon and a nucleon of the core cannot occupy the same orbit. This condition is no longer needed if the antisymmetrization is dropped. See the discussion in Section 12 for an example of that approximation.

Turn now to the other factor in (9.9), $R_f \psi_f^{(\text{Mod})} = 1/\sqrt{A+1}\; R_f\, \mathcal{A}\, [\psi_c \phi_f]$. Since R_f is a symmetric operator, it follows that

$$R_f\, \mathcal{A}\, [\psi_c \phi_f] = \mathcal{A}\, [R_f \psi_c \phi_f]$$

Expanding $R_f \psi_c$ in core states yields

$$R_f\, \mathcal{A}\, [\psi_c \phi_f] = \mathcal{A}\, [\textstyle\sum \psi_{c'} \phi_f\, \langle \psi_{c'} | R_f | \psi_c \rangle] \tag{9.12}$$

Note that the coefficients $\langle \psi_{c'} | R_f | \psi_c \rangle$ are functions of the coordinates of the valence particle. Since R_f is a scalar, the tensor character of these functions are determined by the angular momenta of $\psi_{c'}$ and ψ_c. From the sum (9.12) we select only those $\psi_{c'}$ that appear in the sum (9.11); that is, only those $\psi_{c'}$ that can be generated from ψ_c by Q_{jm} are of interest since only they can contribute to the multipole moment. Thus $\langle \psi_{c'} | R_f | \psi_c \rangle$ will contain components that transform like Q_{jm}. But these need not be the only components; though in the frequently encountered case where the angular momentum J_c of the original core state is zero, the behavior of $\langle \psi_{c'} | R_f\, \psi_c \rangle$ under a rotation is unique and given by the tensor Q_{jm}. In any event a necessary condition for the validity of the concept of effective charge is that the important components of $\langle \psi_{c'} | R_f | \psi_c \rangle$ transform like Q_{jm}. Assuming this condition to hold, we may rewrite (9.12) as follows:

$$R_f\, \mathcal{A}\, [\psi_c \phi_f] \rightarrow \mathcal{A}\, [\textstyle\sum \psi_{c'} \phi_f\, R_{jm;f}^{cc'}] \tag{9.13}$$

where

$$R^{c'c}_{jm;f} = \langle \psi_{c'} | R_f \, \psi_c \rangle_{jm} \tag{9.14}$$

that is, the jm component of $\langle \psi_{c'} | R_f \, \psi_c \rangle$. Note that $R^{c'c}_{jm;f}$ is a function of the coordinates which occur in ϕ_f. Inserting (9.13) and (9.11) into (9.9) for $M^{(1)}_{fi}$ yields

$$M^{(1)}_{fi} = \frac{1}{(A+1)^2} \sum_{c',c''} \langle \mathcal{A} \, [\psi_{c''} \phi_f \, R^{c''c}_{jm;f}] | \mathcal{A} \, [\psi_{c'} \, \phi_i] \rangle \, \langle \mathcal{A} \, [\psi_{c'} \, \phi_i] | \, Q_{jm} \, \mathcal{A} \, [\psi_c \, \phi_i] \rangle$$

or

$$M^{(1)}_{fi} = \frac{1}{(A+1)} \sum_{c'} \langle \phi_f | (R^{c'c}_{jm;f})^\dagger \, \phi_i \rangle \, \langle \mathcal{A} \, [\psi_{c'} \, \phi_i] | \, Q_{jm} | \, \mathcal{A} \, [\psi_c \, \phi_i] \rangle \tag{9.15}$$

The quantity $M^{(2)}_{fi}$ can be evaluated in a similar fashion to yield

$$M^{(2)}_{fi} = \frac{1}{(A+1)} \sum_{c'} \langle \phi_f | \, R^{cc'}_{jm;i} \, \phi_i \rangle \, \langle \mathcal{A} \, [\psi_c \phi_f] | \, Q_{jm} | \, \mathcal{A} \, [\psi_c' \, \phi_f] \rangle \tag{9.16}$$

The factors $(1/A+1) \, \langle \phi_f | \, R^{cc'}_{jm;i} \, \phi_i \rangle$ and the similar factor in (9.15) are proportional to the model-matrix element (9.5), the proportionality factor being a ratio of radial integrals. Let these ratios be denoted by $\gamma^{c'c}_j \, (E_f)$ and $\gamma^{c'c}_j \, (E_i)$. Then the addition αe to be added to the original charge to yield the effective charge is:

$$\alpha e = \frac{1}{A+1} \sum_{c'} \Big(\langle \gamma_j{}^{c'c} \, (E_i) \, \langle \mathcal{A} \, [\psi_c \phi_f] | \, Q_{jm} | \, \mathcal{A} \, [\psi_{c'} \phi_f] \rangle$$

$$+ \, \gamma_j{}^{c'c} \, (E_f) \, \langle \mathcal{A} \, [\psi_{c'} \phi_i] | \, Q_{jm} | \, \mathcal{A} \, [\psi_c \phi_i] \rangle \Big) \tag{9.17}$$

It is also possible to define an effective multipole operator $\hat{Q}^{\text{eff}}_{jm}$

$$\hat{Q}^{\text{eff}}_{jm} = Q_{jm} + \frac{1}{A+1} \sum_{c'} \{ R^{cc'}_{jm;i} \, \langle \mathcal{A} \, [\psi_c \phi_f] | \, Q_{jm} | \, \mathcal{A} \, [\psi_{c'} \phi_f | \, \rangle$$

$$+ \, R^{c'c}_{jm;f} \, \langle \mathcal{A} \, [\psi_{c'} \phi_i] | \, Q_{jm} | \, \mathcal{A} \, [\psi_c \phi_i] \rangle \}$$

From these results we can draw the following conclusions, presuming the validity of perturbation theory. First, note that the effective charge is a function of the multipole order of the transition. This means that the effective charge for say electric quadrupole transitions will not necessarily be identical with the effective charge for say ($E3$) transitions. Second, note that the effective charge rigorously depends also upon the initial and final states involved in the transition. For the effective charge to be a useful concept it is necessary that it be approximately state independent for a class of transitions. A necessary condition following from (9.15) and (9.16) is that the core states involved in these transitions be identical. This could be the case for analogous transitions in a set of nuclei in the same shell (i.e., having the same core) or in a given nucleus involving transitions from differing excited states if the core excitation, is the same for each. Another condition for state independence has to do with the factors $\gamma_j^{c'c}$ (E_f) and $\gamma_j^{cc'}$ (E_i). From (9.14) and 9.7) we see that the state dependence of $\gamma_j^{cc'}$ (E_f) arises from the matrix element $\langle\psi_{c'}|(q_f/E_f - H_0)|V_R\psi_c\rangle$. Let us investigate in the case of the shell model, the circumstances under which this matrix element is not sensitive to the final state. In that case core excitation occurs by lifting a particle in the core to an excited state. Since the parity of the shells, at least for the light nuclei, alternates with shells, the excited core particle cannot go to the same shell as the valence particle but rather to the next higher shell. Therefore the average value of H_0 for the states involved will be considerably larger than E_f and, therefore, the above mentioned matrix element is not sensitive to which of the low lying levels the transition occurs. A similar reasoning applies to $\gamma_j^{cc'}(E_i)$, that is, the value of this quantity is not sensitive to E_i, the value of the energy of the initial state as long as we are dealing with a low-lying state. A final source of state dependence comes from the radial integrals involving the single particle wave functions. This dependence will not result in a state dependence if we restrict the transitions considered to those involving appropriately the single-particle wave functions from the same shell. To all these conditions we must add the important qualification mentioned earlier [see below (9.12)] that the most significant component in the matrix element $\langle\psi_{c'}|R\psi_c\rangle$ transform like a tensor of rank j, order m, that is, like Q_{jm}, a condition that is satisfied if ψ_c (or ψ_c') has zero spin. From all these requirements it is clear that one must carefully choose the class of transitions to which the concept of effective charge is applied.

But these considerations should not becloud the important remark: restricting the Hilbert space employed in describing ψ_f and ψ_i will generally result in incorrect predictions for electromagnetic transition rates. Some mechanism like that of core excitation, which draws upon the excluded portion of Hilbert space, must be invoked to obtain agreement with experiment. For the class of transitions for which the effective charge concept is valid, the magnitude of the correction can be obtained semiempirically. But again recall that the correction depends on the size of the Hilbert space. We leave it to the reader to apply the qualitative considerations developed above to the $E2$ transitions in the light nuclei (Fig. 9.2) to the p-shell nuclei (Table 9.1), and to the $f_{7/2}$ nuclei. Would one expect the notion of effective charge to be valid for these nuclei? For a more detailed discussion of the various elements entering into the shell model description of effective charge, see deShalit and Talmi (63) p. 503 ff.

The core excited state could of course be a collective state, for example, a vibration such as the giant electric dipole resonance (Section 11). This state will dominate the sums in (9.11) and (9.12); indeed to first approximation there would be only one term in the series (assuming only one such important collective state). If the collective state has a definite value of T, the isospin, its excitation may affect the value of α_n and α_p differently. This effect is carried by the matrix element $\langle \phi_{f,i}\psi_{c'} | Q_{jm} | \psi_c \phi_{f,i} \rangle$. Recall that Q_{jm} is composed of an isoscalar and the "3" component of an isovector. It follows that the effective charge is also composed of an isoscalar plus the "3" component of an isovector. In a pure isoscalar $c \rightarrow c'$ transition $\alpha_n = \alpha_p$ while for an isovector $c \rightarrow c'$ transition $\alpha_n = -\alpha_p$. We refer the reader back to the discussion of empirical situation.

10. INTERNAL CONVERSION AND PAIR FORMATION

An excited nucleus can also decay by emitting a virtual photon that can then be either absorbed by an atomic electron or if there is sufficient energy ($E > 2mc^2$, m = electron mass) be converted into an electron-position pair. The first is called internal conversion; the other internal pair formation (Fig. 10.1). These processes must of course be included in the calculation of the total transition probability. The internal conversion coefficient, α, is defined by

$$w_{\text{obs}} = (1 + \alpha)\, w_\gamma \tag{10.1}$$

where w_{obs} is the observed transition probability and w_γ is the transition probability if only gamma-ray decay occurs. As we shall see, α varies with multipole order j and is therefore sometimes more precisely written as α_j. The electric multipole case is denoted by α, the magnetic by β. If an atomic electron in a K orbit is ejected, the conversion coefficient is written $\alpha(K)$.

These processes are important particularly because the electrons are readily observed. In addition, because the longitudinal electromagnetic field can also

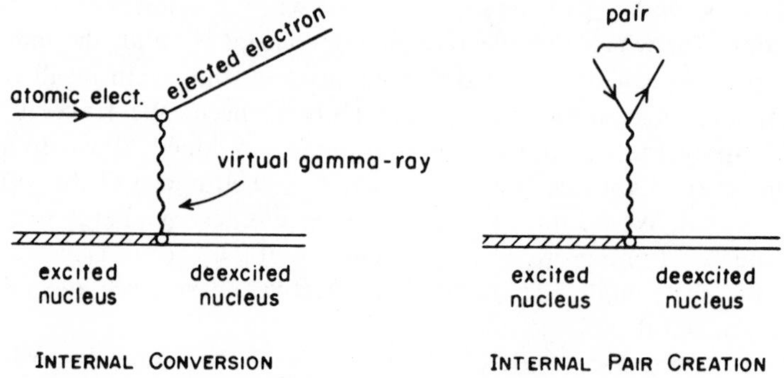

FIG. 10.1. Internal conversion and pair formation.

contribute, certain of the transitions that cannot proceed via the transverse electromagnetic field such as the $0^+ \to 0^+$, the so called $E0$ or electric monopole, can occur. They are similarly important for the decay of long-lived states.

We shall not discuss how the calculations are done as the details are not especially germane to our subject. We refer the reader to the original memoirs (Dancoff and Morrison (39), Rose (58), Sliv and Bond (56, 58), Church and Weneser (56, 60)].

What is of interest is the sensitivity of the internal-conversion coefficients to the energy, to the multipole order and parity change involved in the transition, and of course to the radius and atomic number of the nucleus involved. Most of this dependence can be qualitatively understood as the consequence of the overlap of the electronic wavefunctions and the radiation field that for a point source* has the radial dependence $h_j(kr)$ for multipole order j, where h_j is the spherical Hankel function and $k = (E_i - E_f)/\hbar c$. This function becomes more singular as either kr goes to 0 or as $j \to \infty$ being O $[(1/kr)^{j+1}]$. Hence we can expect the internal-conversion coefficients to be larger for the lower energy transitions. Since the radius of the K and L orbits decreases with increasing Z, α_j should be larger for larger Z. Since h_j is more singular for larger j, α_j should be greater for the larger multipole order. The dependence on parity, that is, the difference between α_j and β_j is not as marked but nevertheless real. These qualitative results are borne out by the following tables (Table 10.1) that have been extracted from Rose (60).

For a point nucleus, the nuclear-matrix elements cancel out in the ratio that defines α or β. This result is not appreciably changed for most transitions for finite nuclei if reasonable charge and current densities are used. This result is most important since it separates nuclear-structure dependence from the dependence on multipole order and parity. Hence by the study of either the magnitude of the internal conversion coefficient or by the ratio $\alpha(K)/\alpha(L)$ one can determine the polarity of the radiation and similarly the parity by comparing the α and β coefficients. Finally, in the case of a mixed transition such as the $(E2) + (M1)$, one can determine the ratio of the two nuclear electromagnetic transition probabilities, the so-called $(E2)/(M1)$ ratio if the absolute conversion coefficients are measured by both K and L conversion.

The independence of the conversion coefficients on the detailed structure of the radiating nucleus disappears when the gamma-ray transitions are inhibited. Then the small effects that originate in the overlap of the atomic electronic wave functions and the nucleus referred to as "the penetration effect" become important.

*Quantitatively it is necessary to take the finite size of the nucleus into account.

TABLE 10.1 K-Shell and L-Shell Conversion Coefficients [taken from M. E. Rose (60)]

k	α_1	α_2	α_3	α_4	α_5	β_1	β_2	β_3	β_4	β_5
					K Shell, $Z = 25$					
0.05	2.10 (0)	6.53 (1)	1.59 (3)	3.86 (4)	9.22 (5)	1.21 (0)	4.28 (1)	1.29 (3)	3.62 (4)	1.05 (6)
0.10	2.70(−1)	5.20 (0)	8.19 (1)	1.26 (3)	1.92 (4)	1.68(−1)	3.13 (0)	5.50 (1)	9.47 (2)	1.66 (4)
0.15	7.76(−2)	1.11 (0)	1.31 (1)	1.48 (2)	1.66 (3)	5.52(−2)	7.18(−1)	8.90 (0)	1.11 (2)	1.40 (3)
0.20	3.19(−2)	3.67(−1)	3.47 (0)	3.16 (1)	2.83 (2)	2.56(−2)	2.58(−1)	2.54 (0)	2.49 (1)	2.47 (2)
0.40	3.86(−3)	2.56(−2)	1.45(−1)	7.73(−1)	4.09 (0)	4.31(−3)	2.50(−2)	1.40(−1)	7.84(−1)	4.40 (0)
0.60	1.18(−3)	5.72(−3)	2.41(−2)	9.66(−2)	3.82(−1)	1.62(−3)	7.00(−3)	2.89(−2)	1.19(−1)	4.91(−1)
0.80	5.31(−4)	2.10(−3)	7.26(−3)	2.40(−2)	7.83(−2)	8.30(−4)	2.97(−3)	1.01(−2)	3.40(−2)	1.15(−1)
1.0	2.98(−4)	1.01(−3)	3.04(−3)	8.75(−3)	2.48(−2)	5.04(−4)	1.58(−3)	4.66(−3)	1.36(−2)	3.95(−2)
1.5	1.16(−4)	3.05(−4)	7.28(−4)	1.67(−3)	3.79(−3)	2.11(−4)	5.31(−4)	1.25(−3)	2.90(−3)	6.63(−3)
2.0	6.48(−5)	1.47(−4)	3.05(−4)	6.11(−4)	1.21(−3)	1.19(−4)	2.61(−4)	5.37(−4)	1.08(−3)	2.14(−3)
					K Shell, $Z = 35$					
0.05	3.78 (0)	7.40 (1)	1.04 (3)	1.88 (4)	1.85 (5)	5.10 (0)	1.75 (2)	3.52 (3)	6.18 (4)	1.08 (6)
0.10	5.39(−1)	8.25 (0)	1.00 (2)	1.17 (3)	1.30 (4)	6.72(−1)	1.23 (1)	1.81 (2)	2.59 (3)	3.72 (4)
0.15	1.66(−1)	1.97 (0)	1.92 (1)	1.79 (2)	1.66 (3)	2.13(−1)	2.75 (0)	3.02 (1)	3.34 (2)	3.67 (3)
0.20	7.03(−2)	6.97(−1)	5.65 (0)	4.44 (1)	3.39 (2)	9.61(−2)	9.53(−1)	8.68 (0)	7.77 (1)	7.13 (2)
0.40	9.08(−3)	5.48(−2)	2.80(−1)	1.36 (0)	6.58 (0)	1.52(−2)	8.82(−2)	4.70(−1)	2.40 (0)	1.35 (1)
0.60	2.89(−3)	1.31(−2)	5.13(−2)	1.92(−1)	7.12(−1)	5.53(−3)	2.41(−2)	9.64(−2)	3.73(−1)	1.55 (0)
0.80	1.34(−3)	4.98(−3)	1.64(−2)	5.14(−2)	1.59(−1)	2.77(−3)	1.00(−2)	3.34(−2)	1.11(−1)	3.60(−1)
1.0	7.56(−4)	2.46(−3)	7.12(−3)	1.97(−2)	5.35(−2)	1.65(−3)	5.23(−3)	1.53(−2)	4.39(−2)	1.24(−1)
1.5	3.03(−4)	7.86(−4)	1.83(−3)	4.13(−3)	9.13(−3)	6.57(−4)	1.73(−3)	4.07(−3)	9.26(−3)	2.10(−2)
2.0	1.68(−4)	3.87(−4)	7.99(−4)	1.59(−3)	3.08(−3)	3.56(−4)	8.28(−4)	1.71(−3)	3.43(−3)	6.74(−3)

					K Shell, $Z = 85$					
0.20	3.58(−1)	3.75(−1)	1.76(−1)	7.08(−2)	2.70(−2)	9.75 (0)	5.97 (1)	8.84 (1)	5.33 (1)	2.40 (1)
0.40	6.48(−2)	1.58(−1)	3.88(−1)	9.94(−1)	2.71 (0)	1.35 (0)	5.39 (0)	1.50 (1)	3.88 (1)	1.01 (2)
0.60	2.51(−2)	6.54(−2)	1.73(−1)	4.62(−1)	1.26 (0)	4.47(−1)	1.44 (0)	3.69 (0)	9.42 (0)	2.39 (1)
0.80	1.34(−2)	3.51(−2)	9.10(−2)	2.31(−1)	5.87(−1)	2.04(−1)	5.84(−1)	1.38 (0)	3.21 (0)	7.52 (0)
1.0	8.43(−3)	2.22(−2)	5.52(−2)	1.33(−2)	3.13(−1)	1.13(−1)	2.98(−1)	6.52(−1)	1.41 (0)	3.04 (0)
1.5	3.84(−3)	1.01(−2)	2.32(−2)	4.93(−2)	1.01(−1)	3.89(−2)	9.30(−2)	1.80(−1)	3.39(−1)	6.33(−1)
2.0	2.28(−3)	5.93(−3)	1.29(−2)	2.51(−2)	4.69(−2)	1.85(−2)	4.25(−2)	7.66(−2)	1.32(−1)	2.26(−1)
					K Shell, $Z = 95$					
0.40	7.79(−2)	1.43(−1)	2.86(−1)	5.77(−1)	1.17 (0)	3.25 (0)	1.05 (1)	2.07 (1)	3.69 (1)	6.48 (1)
0.60	3.22(−2)	7.65(−2)	1.92(−1)	4.72(−1)	1.03 (0)	1.05 (0)	2.86 (0)	6.05 (0)	1.25 (1)	2.59 (1)
0.80	1.78(−2)	4.65(−2)	1.19(−1)	2.87(−1)	6.84(−1)	4.78(−1)	1.18 (0)	2.39 (0)	4.82 (0)	9.78 (0)
1.0	1.15(−2)	3.14(−2)	7.85(−2)	1.80(−1)	3.99(−1)	2.62(−1)	6.08(−1)	1.18 (0)	2.24 (0)	4.32 (0)
1.5	5.50(−3)	1.55(−2)	3.64(−2)	7.55(−2)	1.49(−1)	8.85(−2)	1.91(−1)	3.38(−1)	5.84(−1)	1.01 (0)
2.0	3.34(−3)	9.52(−3)	2.11(−2)	4.06(−2)	7.33(−2)	4.12(−2)	8.79(−2)	1.47(−1)	2.37(−1)	3.79(−1)
					L-Shell, $Z = 25$					
0.05	1.93(−1)	1.09 (1)	8.14 (2)	5.43 (4)	3.16 (6)	1.12(−1)	5.67 (0)	3.00 (2)	1.61 (4)	8.46 (5)
0.10	2.38(−2)	5.85(−1)	1.63 (1)	4.83 (2)	1.39 (4)	1.54(−2)	3.51(−1)	8.33 (0)	2.07 (2)	5.24 (3)
0.15	6.81(−3)	1.12(−1)	1.88 (0)	3.41 (1)	6.29 (2)	4.98(−3)	7.45(−2)	1.15 (0)	1.86 (1)	3.08 (2)
0.20	2.80(−3)	3.53(−2)	4.29(−1)	5.56 (0)	7.46 (1)	2.30(−3)	2.58(−2)	2.98(−1)	3.59(0)	4.44 (1)
0.40	3.34(−4)	2.33(−3)	1.45(−2)	9.12(−2)	5.91(−1)	3.81(−4)	2.33(−3)	1.42(−2)	8.82(−2)	5.61(−1)
0.60	1.02(−4)	5.09(−4)	2.27(−3)	1.00(−2)	4.48(−2)	1.42(−4)	6.33(−4)	2.76(−3)	1.22(−2)	5.46(−2)
0.80	4.63(−5)	1.87(−4)	6.70(−4)	3.36(−3)	8.37(−3)	7.28(−5)	2.67(−4)	9.41(−4)	3.33(−3)	1.19(−2)
1.0	2.57(−5)	8.87(−5)	2.76(−4)	8.29(−4)	2.50(−3)	4.37(−5)	1.40(−4)	4.26(−4)	1.29(−3)	3.92(−3)
1.5	1.01(−5)	2.69(−5)	6.49(−5)	1.53(−4)	3.57(−4)	1.84(−5)	4.69(−5)	1.12(−4)	2.65(−4)	6.24(−4)
2.0	5.56(−6)	1.28(−5)	2.68(−5)	5.46(−5)	1.10(−4)	1.02(−5)	2.27(−5)	4.75(−5)	9.69(−5)	1.96(−4)

TABLE 10.1 K-Shell and L-Shell Conversion Coefficients (Continued)

k	α_1	α_2	α_3	α_4	α_5	β_1	β_2	β_3	β_4	β_5
					L-Shell, $Z = 35$					
0.05	4.43(−1)	3.98 (1)	4.08 (3)	2.76 (5)	1.53 (7)	5.57(−1)	3.52 (1)	2.30 (3)	1.37 (5)	7.34 (6)
0.10	5.75(−2)	1.73 (0)	6.77 (1)	2.32 (3)	6.89 (4)	7.32(−2)	1.91 (0)	5.15 (1)	1.41 (3)	3.79 (4)
0.15	1.70(−2)	3.08(−1)	6.83(0)	1.49 (2)	3.01 (3)	2.30(−2)	3.77(−1)	6.38 (0)	1.12 (2)	1.99 (3)
0.20	7.19(−3)	9.47(−2)	1.42 (0)	2.23 (1)	3.36 (2)	1.03(−2)	1.25(−1)	1.55 (0)	2.01 (1)	2.66 (2)
0.40	9.13(−4)	6.21(−3)	4.19(−2)	2.99(−1)	2.20 (0)	1.60(−3)	1.01(−2)	6.40(−2)	4.18(−1)	2.81 (0)
0.60	2.88(−4)	1.40(−3)	6.46(−3)	3.08(−2)	1.52(−1)	5.70(−4)	2.63(−3)	1.18(−2)	5.39(−2)	2.52(−1)
0.80	1.33(−4)	5.21(−4)	1.91(−3)	7.08(−3)	2.69(−2)	2.83(−4)	1.07(−3)	3.86(−3)	1.40(−2)	5.18(−2)
1.0	7.57(−5)	2.56(−4)	7.98(−4)	2.49(−3)	7.91(−3)	1.68(−4)	5.53(−4)	1.71(−3)	5.29(−3)	1.65(−2)
1.5	2.97(−5)	7.95(−5)	1.94(−4)	4.67(−4)	1.12(−3)	6.71(−5)	1.78(−4)	4.33(−4)	1.04(−3)	2.49(−3)
2.0	1.67(−5)	3.88(−5)	8.27(−5)	1.71(−4)	3.52(−4)	3.64(−5)	8.47(−5)	1.80(−4)	3.71(−4)	7.60(−4)
					L Shell $Z = 85$					
0.05	2.06 (0)	3.92 (3)	3.85 (5)	1.03 (7)	1.76 (8)	8.20 (4)	1.32 (4)	2.14 (6)	1.25 (8)	3.62 (9)
0.10	4.36(−1)	1.41 (2)	7.17 (3)	1.59 (5)	2.79 (6)	1.25 (1)	6.19 (2)	2.69 (4)	7.91 (5)	1.78 (7)
0.15	1.47(−1)	1.96 (1)	6.85 (2)	1.22 (4)	1.74 (5)	3.83 (0)	1.01 (2)	2.40 (3)	4.67 (4)	7.78 (5)
0.20	6.77(−2)	5.12 (0)	1.35 (2)	1.98 (3)	2.37 (4)	1.67 (0)	2.95 (1)	4.68 (2)	6.79 (3)	8.86 (4)
0.40	1.13(−2)	2.32(−1)	3.12 (0)	2.76 (1)	2.04 (2)	2.33(−1)	1.82 (0)	1.24 (1)	8.81 (1)	6.26 (2)
0.60	4.30(−3)	4.60(−2)	4.12(−1)	2.67 (0)	1.47 (1)	7.65(−2)	4.11(−1)	1.88 (0)	9.12 (0)	4.53 (1)
0.80	2.25(−3)	1.66(−2)	1.10(−1)	5.73(−1)	2.56 (0)	3.54(−2)	1.53(−1)	5.49(−1)	2.08 (0)	8.15 (0)
1.0	1.41(−3)	8.10(−3)	4.31(−2)	1.89(−1)	7.30(−1)	1.98(−2)	7.35(−2)	2.25(−1)	7.14(−1)	2.35 (0)
1.5	6.22(−4)	2.57(−3)	9.61(−3)	3.13(−2)	9.27(−2)	6.99(−3)	2.10(−2)	5.04(−2)	1.22(−1)	3.02(−1)
2.0	3.69(−4)	1.29(−3)	3.91(−3)	1.05(−2)	2.61(−2)	3.41(−3)	9.18(−3)	1.93(−2)	4.00(−2)	8.40(−2)

L-Shell $Z = 95$										
0.05	2.44 (0)	8.63 (3)	7.43 (5)	1.09 (7)	8.42 (7)	2.70 (2)	3.85 (4)	6.13 (6)	3.33 (8)	7.00 (9)
0.10	5.10(−1)	3.23 (2)	1.41 (4)	2.56 (5)	3.51 (6)	4.08 (1)	1.81 (3)	7.56 (4)	2.14 (6)	4.16 (7)
0.15	1.80(−1)	4.60 (1)	1.42 (3)	2.18 (4)	2.62 (5)	1.25 (1)	2.93 (2(	6.47 (3)	1.21 (5)	1.83 (6)
0.20	8.64(−2)	1.21 (1)	2.89 (2)	3.77 (3)	3.91 (4)	5.42 (0)	8.47 (1)	1.24 (3)	1.70 (4)	2.11 (5)
0.40	1.63(−2)	5.31(−1)	6.89 (0)	5.67 (1)	3.88 (2)	7.63(−1)	5.14 (0)	3.22 (1)	2.18 (2)	1.49 (3)
0.60	6.60(−3)	1.14(−1)	1.01 (0)	6.17 (0)	3.20 (1)	2.52(−1)	1.15 (0)	4.86 (0)	2.23 (1)	1.07 (2)
0.80	3.62(−3)	4.08(−2)	2.75(−1)	1.38 (0)	5.91 (0)	1.16(−1)	4.25(−1)	1.41 (0)	5.09 (0)	1.92 (1)
1.0	2.32(−3)	1.97(−2)	1.09(−1)	4.64(−1)	1.73 (0)	6.52(−2)	2.04(−1)	5.80(−1)	1.76 (0)	5.55 (0)
1.5	1.09(−3)	6.17(−3)	2.42(−2)	7.84(−2)	2.28(−1)	2.31(−2)	5.83(−2)	1.30(−1)	3.02(−1)	7.27(−1)
2.0	6.63(−4)	3.14(−3)	9.72(−3)	2.63(−2)	6.47(−2)	1.13(−2)	2.56(−2)	5.00(−2)	9.98(−2)	2.04(−1)

Since the atomic electron wave functions are well known, the penetration effect in principle can be exploited to probe the charge and current structure of the nucleus. One finds [Church and Weneser (60)] that the operators determining the internal-conversion transition probability are weighted averages of the multiple moment operators. The interaction Hamiltonian, after the integration over electronic coordinates have been performed so that only dependence on nuclear coordinates remain, has thus the form:

$$\int d\mathbf{r}\, \mathbf{j}_N \cdot \mathbf{A}_{jm}{}^{(E,M)} \left[1 + \text{power series in } (r/R)^2\right] \tag{10.2}$$

where R is the nuclear radius and $\mathbf{j}_N$ is the nuclear-current density. The first term gives the long wavelength-limit value for internal conversion. We also see why, in the approximation in which only the first term in (10.2) is taken into account, the internal-conversion coefficient is independent of nuclear structure. The remaining terms beginning with $(r/R)^2$ are the penetration-effect corrections. The origin of these terms can be readily seen from the familiar expression for the current-current interaction energy:*

$$\iint \mathbf{j}_N(\mathbf{r}) \cdot \frac{e^{ik|\mathbf{r}-\mathbf{r}'|}}{|\mathbf{r}-\mathbf{r}'|} \mathbf{j}_e(\mathbf{r}')\, d\mathbf{r}\, d\mathbf{r}' \tag{10.3}$$

where $\mathbf{j}_e$ is the electron current density. Form (10.2) is obtained by integration employing the orthogonality properties of the vectors $\mathbf{A}_{jm}^{(E,M)}$:

$$\left[1 + \text{power series in } \left(\frac{r}{R}\right)^2\right] \sim \iint \mathbf{A}_{jm}^{(E,M)*}(\mathbf{r}) \frac{e^{ik|\mathbf{r}-\mathbf{r}'|}}{|\mathbf{r}-\mathbf{r}'|} \cdot \mathbf{j}_e(\mathbf{r}')\, d\mathbf{r}'\, d\hat{\mathbf{r}} \tag{10.4}$$

The essential point that is exploited in the evaluation of (10.3) is the short range character of $\mathbf{j}_N(\mathbf{r})$ and the long wavelength character of $\mathbf{j}_e$. Because of these features an expansion of the right-hand side in a power series in r^2 will after integration (10.4) yield a rapidly convergent result for (10.2). Moreover the coefficients of each term in the series on the left-hand side of (10.4) are, according to the right hand side, independent of nuclear structure. Tables of these coefficients have been calculated by Green and Rose (58). As we shall see the $E1$ and $M1$ transitions do have large hindrance factors so that measured conversion coefficients tend to be anomalous for these transitions. In principle information regarding nuclear-structure information can be extracted from these anomalies. Examples have been given by Snyder and Frankel (57), Reiner (58), Nilsson and Rasmussen (58), and Asaro, Stephens, Hollander, and Perlman 60).

Internal conversion for $0^+ \to 0^+$ transitions must be a penetration effect,

*The full expression replaces $\mathbf{j}_N \cdot \mathbf{j}_e$ in (10.3) by $(\mathbf{j}_N \cdot \mathbf{j}_e - \rho_N \rho_e)$.

since there is no $E0$ gamma ray. In this case the nuclear matrix element that enters is

$$\int d\mathbf{r}\, \rho_N \left(\frac{r}{R}\right)^2 \tag{10.5}$$

where ρ_N is the nuclear-charge density. $E0$ transitions can also occur for $J^\pi \longrightarrow J^\pi$ transitions. An example is the $2^+ \longrightarrow 2^+$ transition in ^{196}Pt [Gerholm and Pettersson, (58)]. For other examples, see Church and Weneser (60). The second moment of the nucleon-charge density is also measured by electron scattering and can be determined from the spectra of μ-mesic atoms. We see that the monopole transition case also informs us with respect to this important quantity.

11. THE EXPERIMENTAL SITUATION

We shall limit the discussion to general remarks and for the most part to particle stable states. For detailed accounts and data the reader is referred to recent reviews [Perdrisat (66), Marelius, Sparrman, and Sundstrom (68), Goldhaber and Sunyar (65), Wilkinson (60), Talmi and Unna (60), Skorka, Hertel, Retz-Schmidt (66), and Warburton and Weneser (69)].

We begin by noting that there is every indication that the fundamental hypotheses employed in the evaluation of the probability of electromagnetic transitions are correct. The selection rules (1.6) for angular-momentum conservation are well satisfied. Although there is evidence of a failure of parity conservation, this effect, which originates in the weak interactions (see Section IX.16), is very small and can be neglected in all the circumstances discussed in the preceding sections. The assumption of locality, that is, that the nucleon-electromagnetic interaction is a sum of single-particle operators, (Eqs. 2.1 to 2.4) seems adequate. It is very difficult to separate the effects of model assumptions, that is, the effects of residual interactions from the effects of exchange charges and currents (Section 3). To most clearly see the effects of exchange currents we must turn to the two- and three-body systems. This will be discussed in Volume II, (see also Section 11d). The general conservation of isospin that implies $\Delta T = \pm 1, 0$ has not been until recently tested because of the experimental unavailability of the appropriate levels that, for light nuclei, are in the continuum. However through excitation of nuclei by protons a number of $T = 2$ levels have been observed in nuclei with $T = 0$ ground states. The isobar analog resonance phenomenon plays an important role here. If the $T = 2$ levels decay directly to the ground state, isospin-invariance violation is involved. As of 1968 [Snover, Riess, Hanna (68)] such transitions have not been observed. See however recent review by Hanna (73).

Generally the transitions of greatest interest are those that are either greatly enhanced or greatly retarded with respect to, say, single-particle transition rates. A large enhancement may point to importance of configuration mixing that in its extreme form leads to the vibrational and rotational collective states. The earlier discussion of effective charge in Sect. 9 gives an example. On the other hand, a large retardation may express an approximate selection rule that might be the consequence of an approximate symmetry principle or of the dynamics of the system. Whether retarded or enhanced, insight into various nuclear models that predict specific model-dependent selection rules is gained. In particular modifications in the model are suggested.

We turn now to specific transition types. We shall consider only $E0$, $E1$, and $M1$ transitions as these are sufficient to illustrate most of the points of interest. $E2$ and $M1$ transitions in the deformed region are discussed in Chapter VI.

A. Electric Monopole ($E0$) Transitions

A number of transitions of this type are listed in Table 11.1. According to (10.3) the average value of $(r/R)^2$ can be deduced from a transition. These are listed in the second column of Table 11.1. The selection rule for $E0$ transitions is $\Delta J = 0$ and no change in parity. $E0$ transitions are most easily observed in a $0^+ \to 0^+$ transition because in that case the decays can proceed only by either internal conversion or by internal pair formation. But in principle $E0$ "radiation" is possible for $J^\pi \to J^\pi$ transitions but are much more difficult to observe because of the competing $M1$ and $E2$, etc., radiation. Hence except for unusual cases such as the $2^+ \to 2^+$ transition in ^{196}Pt, $E0$ "radiation" has been observed only in $0^+ \to 0^+$ transitions.

TABLE 11.1 E0 Transitions ($0^+ \to 0^+$). ρ Equals the Average Value of $(r/R)^2$ [taken from M. Goldhaber and A. Sunyar (65)]

Nucleus	Energy (MeV)	ρ
^{12}C	7.68	≈ 0.5
^{16}O	6.04	≈ 0.5
^{70}Ge	1.21	≈ 0.09
^{72}Ge	0.69	≈ 0.11
^{214}Po	1.414	≈ 0.05
^{152}Sm	0.685	(lifetime unknown)
^{152}Gd	0.615	(lifetime unknown)
^{90}Zr	1.75	≈ 0.06
^{40}Ca	3.35	≈ 0.1
^{42}Ca	1.84	≈ 0.5

Each of the nuclear models make specific predictions about the $E0$ transition and therefore can be tested by comparison with experiment. We briefly summarize below the nature of these predictions.

a. Shell Model. Two single-particle levels with the same value of the angular momentum and parity are separated by three shells and have a very large energy difference $\sim 120/A^{1/3}$ MeV so that one of the levels is not particle stable. Hence $E0$ transitions will generally not be observed between single-particle levels. Multiparticle states can decay, but the intervention of the residual interaction is essential.

b. Vibrational model (Section VI.13). We consider the case of 2^+ phonons, which will suffice to illustrate the principle involved. Since the $E0$ radiation carries off zero angular momentum, there is no change in the angular momentum of the nuclear states. Hence the associated change in the number of phonons must even. This corresponds to the composition of the 0^+ states, the first excited state, from two 2^+ phonons, the ground state being 0^+. We recall that in the vibration model the radius R becomes a dynamical variable. See Chapter VI.13. If we keep only the quadrupole vibrations the variable is written as

$$R = R_0 \left[1 + C\Sigma \left(a_\mu Y_{2\mu}(\theta, \phi) + a_\mu^\dagger Y_{2\mu}(\theta, \phi)\right)\right] \tag{11.1}$$

where (R,θ,ϕ) is a point on the spherical surface and C is a constant. The quantities a_μ are boson-phonon operators with a_μ a destruction operator and $a_\mu^\dagger$ a creation operator. Since the $E0$ transition amplitude depends non-linearly on R (see Eq. 10.3), terms like R^2 will occur. Hence from (11.1) terms that are bilinear in a_μ will appear. The term $a_\mu\, a_{-\mu}$ will permit the destruction of the two phonons carrying a total angular momentum of zero. The transition amplitude is thus proportional to C^2 of (11.1).

c. Rotational nuclei. Since a given value of J occurs only once in a rotational band, a $J^+ \rightarrow J^+$ transition will only occur between bands. Moreover because of the K-selection rule

$$|K_f - K_i| \leqslant j$$

where j is the multipole order, *the bands must have the same value of* K for an $E0$ transition. Within the framework of the model the excited K-band is based on a vibrational state where the vibrations now are about the deformed shape of the nucleus. The selection rules with respect to the phonon number are then identical with that of the vibrational model. Calculations of the monopole nuclear-matrix element have been made for several values of the initial angular momentum. Comparison with experiment for even-even nuclei has been made with some difficulties appearing in the rare earth region. For further discussion see Davidson (68).

For odd-even nuclei, with a single-valence nucleon plus a core that may be deformed, at least one of the states of the same J^π will generally involve core-excitation for low-lying levels. Construction of such states by excitation of the valence nucleon requires a jump of two shells. The experimental evidence is rather sparse and we refer the reader to Davidson's monograph.

As a final general remark, note that there is some evidence for the existence of a giant monopole state. This would mean that most of the monopole strength is concentrated in a small energy region probably of the order of several MeV in width. The existence of such a giant state would of course seriously affect the calculation of $E0$ transition probabilities. (See the next section for a discussion of the giant electric dipole state.)

B. $E1$ Transitions

The giant dipole resonance in which photons are either absorbed or emitted with enhanced probability over an energy region generally stretching for several MeV is of great importance for the understanding of $E1$ transitions in nuclei. An example of a number of such resonances is shown in Fig. 11.1. For gamma energies ranging from roughly 15 MeV to 30 MeV and somewhat beyond, the probability of gamma-ray absorption is greatly enhanced. This phenomenon is widespread as is indicated by Fig. 11.2, which shows the mean

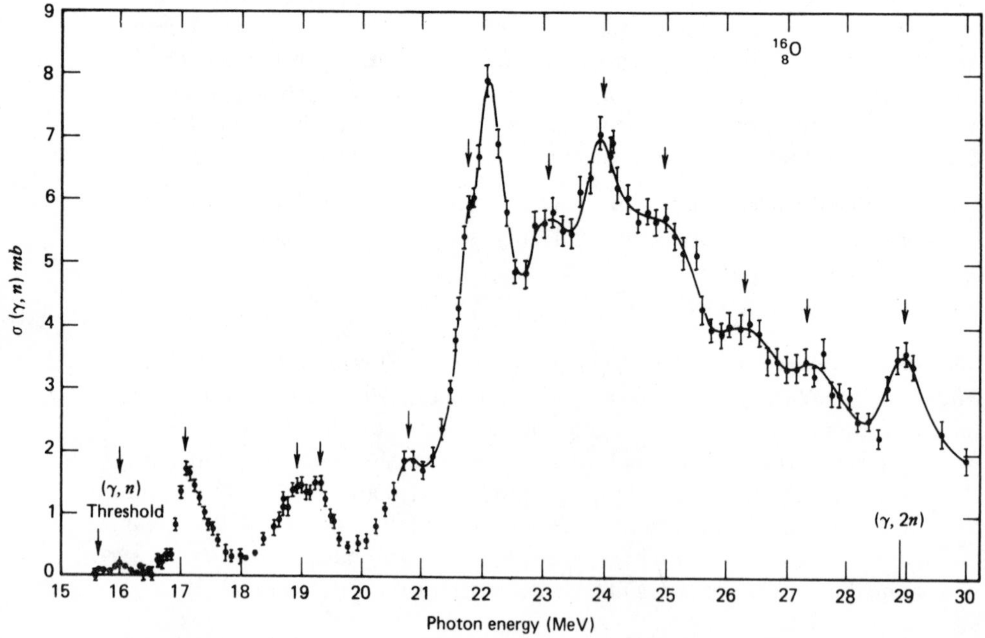

FIG. 11.1. The photoneutron across section of ^{16}O up to 30 MeV. [taken from Bramblett, Caldwell, Harvey, Fultz (64)].

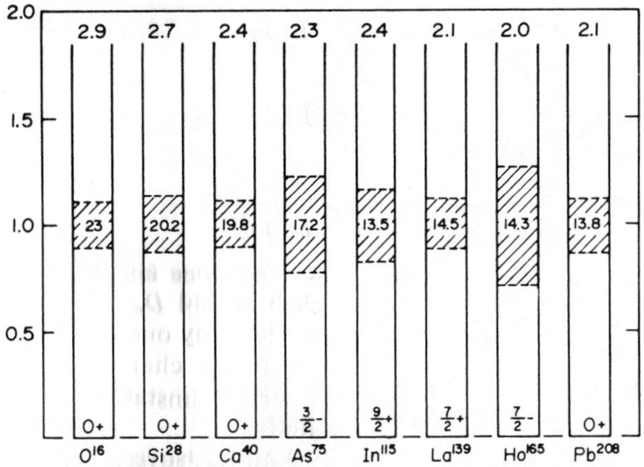

FIG. 11.2. Giant resonance summary. The resonance energy has been put equal to unity. The shaded region indicates the width of the resonance, the number inserted gives the mean energy of the resonance. The ground-state spin and parity of the target nucleus are given. The numbers at the top give the energies in units of the resonance energies at which $\lambda\!\!\!^-$ is equal to the nuclear radius [taken from E. G. Fuller (66)].

energy and width for the dipole resonance for a number of nuclei extending over the periodic table. The resonance energies are shown in Fig. 11.3. For light even-even nuclei the resonating states are found to have the quantum numbers $J^\pi = 1^-$ and $T = 1$.

We shall not review the detailed theory of these resonances, leaving that task for Chapter VI.14 and Volume II. For our present purposes a qualitative description will suffice. The essential physical point, which can be deduced from experiment, is that the state generated, when the isovector dipole moment (the isoscalar dipole moment vanishes, Eq. 7.7) is applied to the ground state of the target nucleus (assuming an incident photon beam), is an approximate

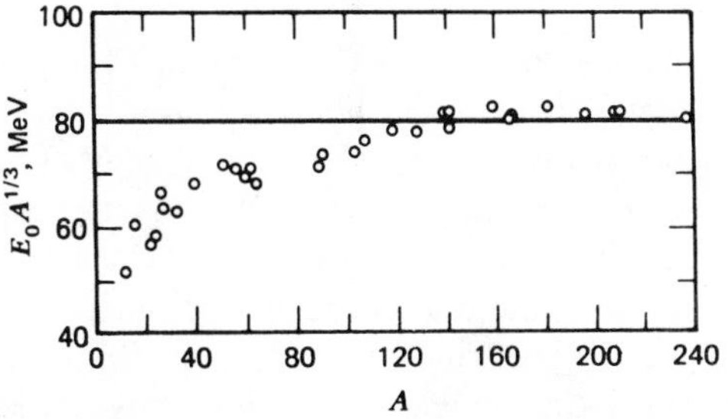

FIG. 11.3. The giant resonance energy times $A^{1/3}$ as a function of A [Taken from Hayward (65)].

eigenstate of the exact Hamiltonian of the system. Symbolically the generated wave function, ψ_{1m} is given by

$$\psi_{1m} = D_m \psi_0 \tag{11.2}$$

where

$$D_m = \sum_i \tau_3(i) r_i Y_{1m}(\hat{\mathbf{r}}_i)$$

and

$$H \psi_{1m} \approx E_1 \psi_{1m} \tag{11.3}$$

where E_1 is approximately equal to the resonance energy. Of course (11.3) cannot be exactly satisfied since the nuclear H and D_m do not commute. For one thing ψ_1 must decrease exponentially when any one of its spatial variables goes to infinity in accordance with bound-state character of ψ_0. But as is clear from experiment the exact state is particle unstable (Fig. 11.1) and the exact wave function must be finite at infinity.

For another thing, if ψ_1 were exact, the entire isovector dipole strength of the ground state would be concentrated in the transition to this one level. Only one $E1$ transition to or from the ground state would be possible and it would exhaust the dipole sum rule (8.9). However, since ψ_1 is approximately an eigenstate, transitions to the dipole resonant state may very well exhaust much of the dipole sum rule with the consequence that other $E1$ transitions to the ground state would be inhibited. In a microscopic theory of the dipole resonance, it is found that such is in fact true, as is indicated by the calculations of Elliott and Flowers (57). Here we shall be content with the empirical evidence. We compare the sum-rule value (8.9) (just the kinetic-energy contribution) with the integrated photoabsorption cross section shown in Fig. 11.4; the integration extends from the threshold for particle emission to

Problem. Suppose the ground state of ^{4}He is the solution of an harmonic-oscillator Hamiltonian. Show that ψ_1 is an exact solution of the same Hamiltonian.

about 30 MeV gamma-ray energy and thus includes the giant resonance but not the transitions not involving particle emission and not of course the particle-emission cross sections that occur above 30 MeV. All these omitted cross sections are included in the sum rule. We observe that for the heavier nuclei ($A > 50$), the kinetic energy contribution to the sum rule is completely exhausted leaving only the potential-energy contribution. The latter has been estimated to about 1.2 $\times$ the kinetic-energy contribution. But part of this strength will be present in the unmeasured region above 30 MeV. On the other hand for the light nuclei, the giant dipole empirically includes a much smaller part of the sum rule. We conclude that on the average electric dipole matrix

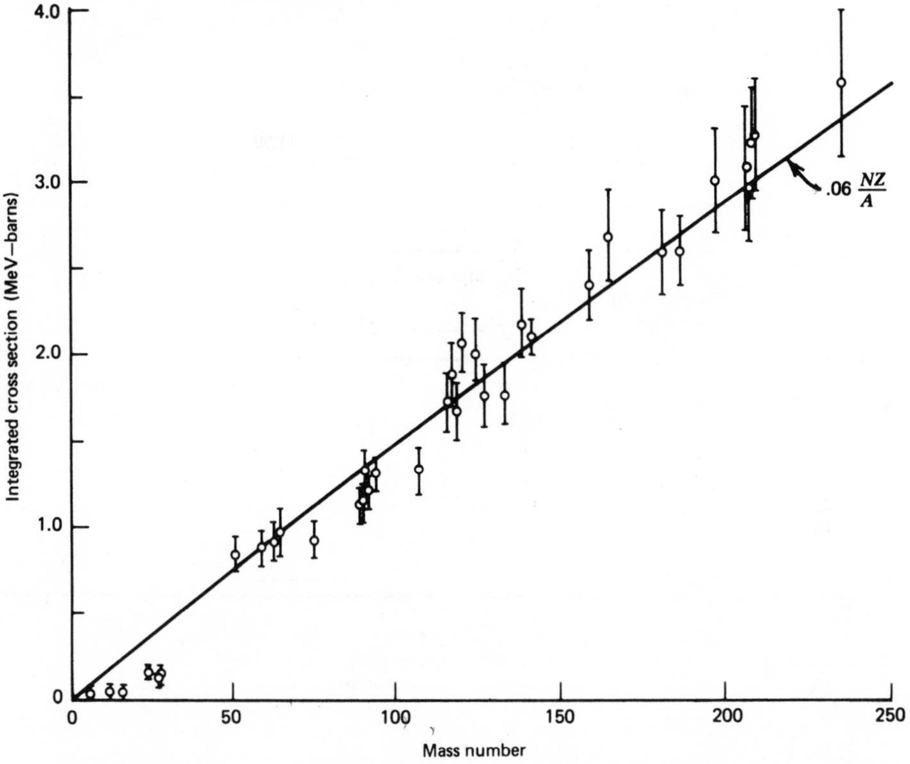

FIG. 11.4. Integrated cross sections of $\sigma(\gamma,n) + \sigma(\gamma, 2n) + \sigma(\gamma,np)$ vs. mass. [Private Communication from B. L. Berman (71).]

elements for transitions to the ground state for levels lying below the particle-emission threshold will be more hindered in the heavier elements. Since giant dipole resonances proceeding to and from the excited states also exist (see Fig. 11.5), it seems plausible that this greater inhibition should hold on the average for all $E1$ transitions in the heavier elements. This qualitative conclusion is in excellent agreement with experiment, as we shall now see.

A histogram of the transition probabilities for isospin-allowed transitions (≡ "normal") for nuclei with $A \leq 40$ is shown in Fig. 11.6 taken from Skorka et al. (66) while those for medium and heavy nuclei taken from Perdrisat (66) are shown in Fig. 11.7 and Fig. 11.8. Note that the "hindrance" factor is the factor with which the Weisskopf single-particle estimate should be multiplied to obtain the experimental value. We see that the single-particle estimate is a vast overestimate and that generally the heavier elements have larger hindrance factors; a mean for transition in light nuclei is about 2.6×10^{-3}, in the heavy $5 \times 10^{-7} - 6 \times 10^{-5}$.

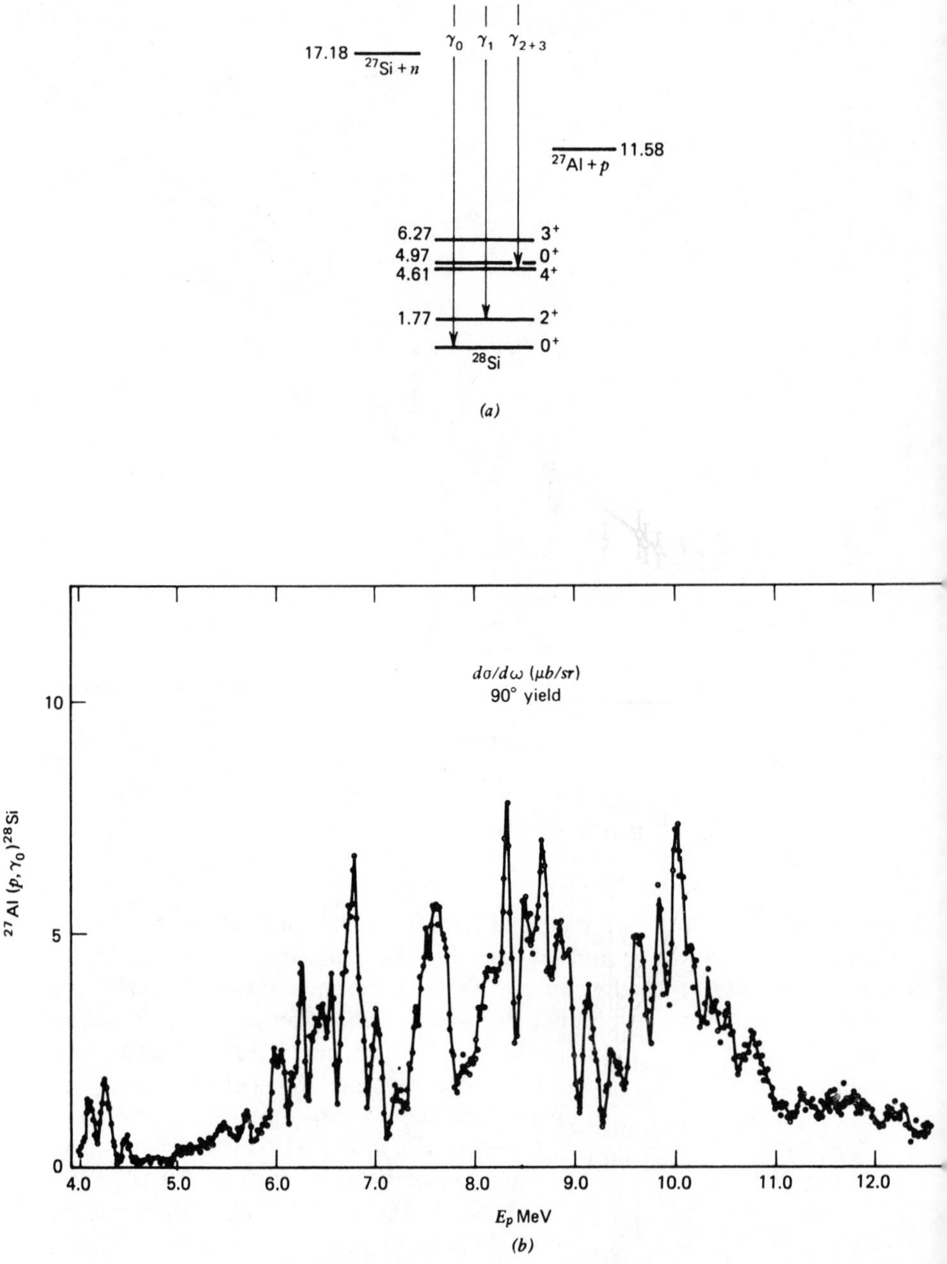

FIG. 11.5. Giant resonances and fine structure in ^{28}Si from the $^{27}Al(p,\gamma)$ ^{28}Si reaction, (*a*) Energy-level diagram of ^{28}Si, showing the transitions observed in the reaction ^{27}Al (p,γ) ^{28}Si in the giant-resonance region. (*b*) Differential yield curve for ^{27}Al (p,γ_0) ^{28}Si.

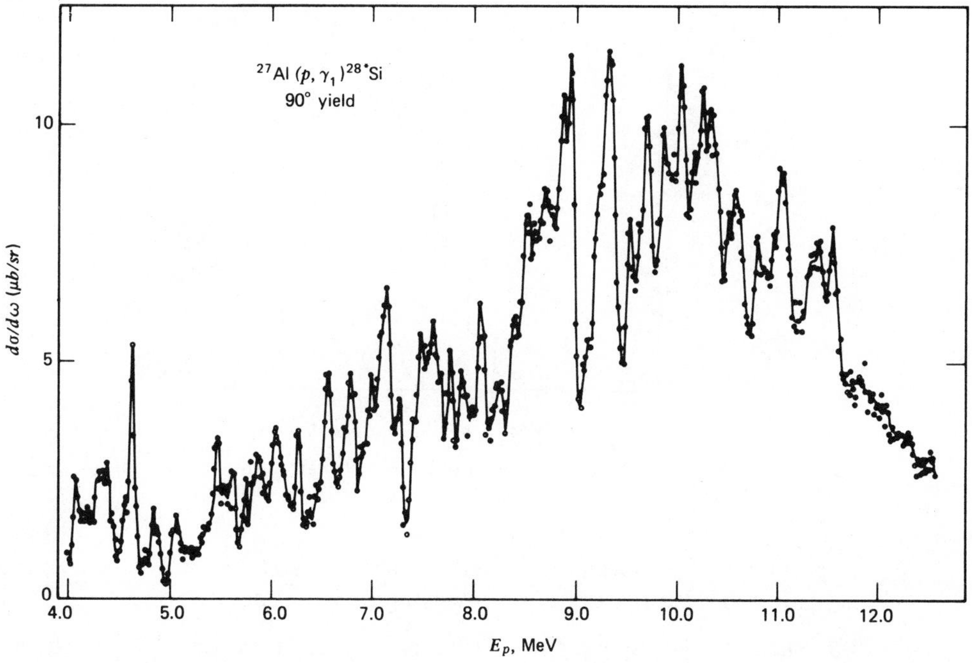

(c)

FIG. 11.5 Cont. (c) Differential yield curve for $^{27}Al\ (p,\gamma_1)\ ^{28}Si$[from Singh et. al. (65)].

Another general property of $E1$ transitions that can be readily seen in the data is the substantial agreement with the isospin-selection rules discussed earlier. These are:

(a) Corresponding $E1$ transitions in conjugate nuclei have equal strength.

(b) $T = 0$, $E1$ transitions in self-conjugate nuclei ($A = 2Z$) are forbidden. Selection rule (b) can be checked for light nuclei. A histogram of the transition probabilities for isospin forbidden $E1$ transitions is shown in Fig. 11.9. Their existence demonstrates the approximate nature of isospin invariance. Fig. 11.9 should be compared with the allowed transition shown in Fig. 11.6. For the latter the mean-transition strength is about 2.6×10^{-3} while for the forbidden group the mean-transition strength is one-seventh of this value. However there is considerable overlap.

With respect to selection rule (a), only a few cases are available among the light nuclei. We refer the reader to the article by Warburton and Weneser (69) for details. Fair agreement with rule (a) is obtained.

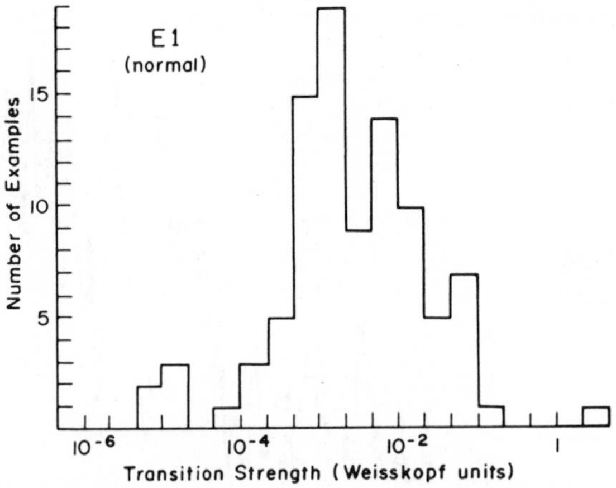

FIG. 11.6. Histogram of *E*1 transition probabilities. Taken from Skorka et al. (66).

Special selection rules prevail for the various models. Again we restrict the discussion to transitions between low-lying excited states.

a. Shell Model. Since *E*1 transitions require a parity change, single-particle transitions can only occur for light nuclei if the particle changes shell. For heavy nuclei, spin-orbit coupling does bring opposite parity orbitals down in energy. Generally spin-orbit coupling is most effective for states of large spin with the consequence that it is the high spin states that come down in energy, approaching in energy the low spin states in the neighboring shells that are not much affected by the spin-orbit potential. As a consequence a comparatively low energy gamma-ray transition between a state of very large spin and one of small spin will occur. The multipole order will be comparatively large and the lifetime of the state long. These are the isomers discussed in Chapter I. Thus the spin-orbit mechanism is not effective for single-particle E1 transitions. These transitions will in the shell model generally involve the particle-changing shells. The gamma-ray energy will be comparatively high, being about $40/A^{1/3}$ MeV. *E*1 transitions between low-lying states cannot be single-particle transitions. If an *E*1 transition between such states occurs, the initial or final state must contain many-particle configurations that are generated from the shell-model states by the residual interactions. For a single-valence nucleon nucleus this must involve core excitations. *E*1 transitions should thus be sensitive to the magnitude and nature of configuration mixing and to the residual interactions. Large hindrance factors will occur if the amplitude of these multiparticle configurations is small.

The transition probability will be reduced because of the existence of the giant electric dipole resonance. The giant dipole state, as is clear from (11.2), is composed of configurations whose transition amplitudes to the final state

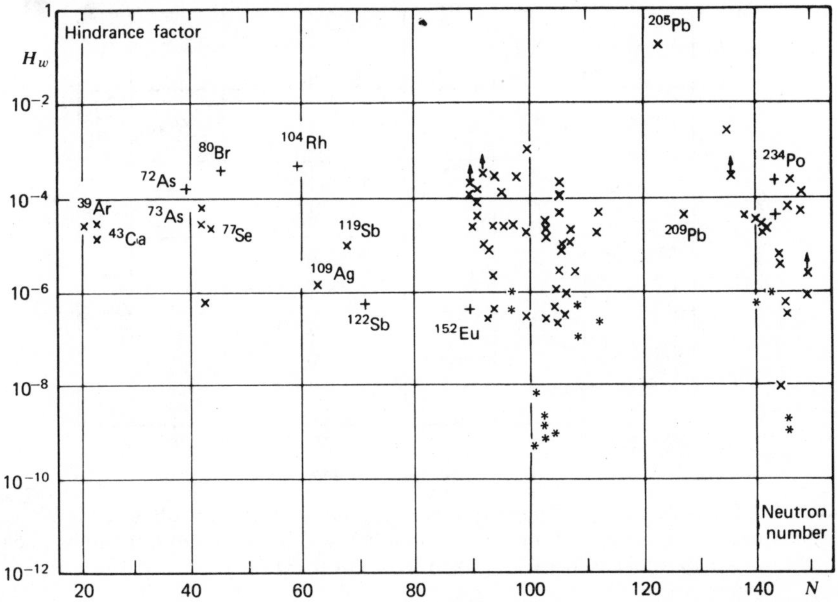

FIG. 11.7. Medium and heavy odd-A and odd-odd nuclei: Experimental hindrance factors $H_W(E1)$ relative to the Weisskopf estimate, as a function of the neutron number N. X odd-A nuclei; +, odd-odd nuclei; *K-forbidden transitions in deformed nuclei [taken from Perdrisat (66)].

interfere constructively. Other states with the same quantum numbers as the giant-dipole state must be orthogonal to it. As a consequence, the transition amplitudes from the various configurations will interfere destructively leading to large hindrance factors. This of course is a manifestation of the near exhaustion of the dipole sum rule by the giant dipole resonance. We see how the necessary reduction in the probability of other dipole transitions is accomplished. It is also clear from this discussion that states which can be mixed with the giant-dipole state by the residual interactions will have an enhanced transition probability. In fact the mixing in of the giant-dipole state through the action of the isospin symmetry breaking Coulomb potential is responsible for many of the isospin forbidden transitions noted in Fig. 11.9.

b. Vibrational Nuclei (*see Section VI.13*). If the only vibrational quantum is a 2^+ phonon no $E1$ transitions to the ground state will occur, since it is not then possible to build an excited state whose parity is opposite to that of the ground state. However, in the presence of a 3^- phonon, opposite parity excited states can obviously exist. Excited states with the same parity as the ground state may exist because of the presence of a 2^+ phonon state or two 3^- phonons

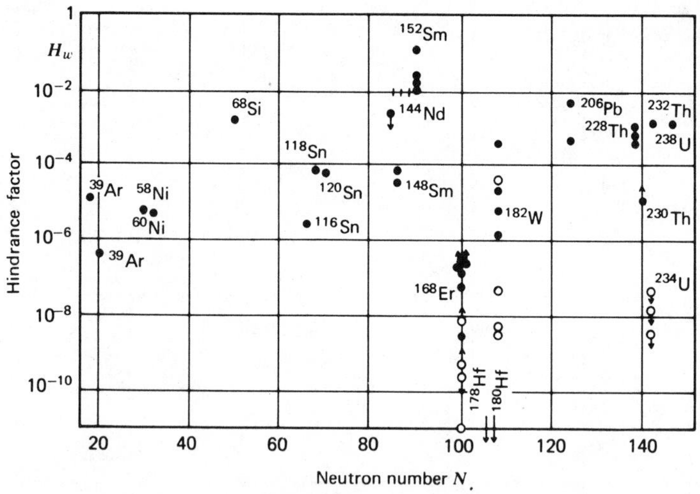

FIG. 11.8. Even-*A* nuclei; Experimental hindrance factors $H_W(E1)$ relative to the Weisskopf estimate, as a function of the neutron number *N*. O, *K*-forbidden transitions in deformed nuclei; ●, other transitions [taken from Perdrisat (66)].

can combine. Transitions such as the $3^- \rightarrow 2^+$ are then observed. For a detailed discussion see Perdrisat (66) who discusses these transitions in terms of equivalent quasi-particle language.

c. Deformed Nuclei. For deformed nuclei the following selection rule for dipole transitions

$$|\Delta K| \leq 1 \tag{11.4}$$

applies in addition to the general requirement for a parity change and $\Delta J \leq 1$. We have already discussed an important example of forbiddenness because of this rule in the $J = 8^-$ to $J = 8^+$ decay of ^{180}Hf. In this case the hindrance factor is 10^{-15}! Selection rule (11.4) is not exact because of interband mixing. Empirically one finds that even for values of ΔK that are allowed, the $\Delta K = \pm 1$ are less probable than the $\Delta K = 0$ transition. This is illustrated in Table 11.2 and Fig. 11.10 for odd nuclei. But note that in the figure the hindrance

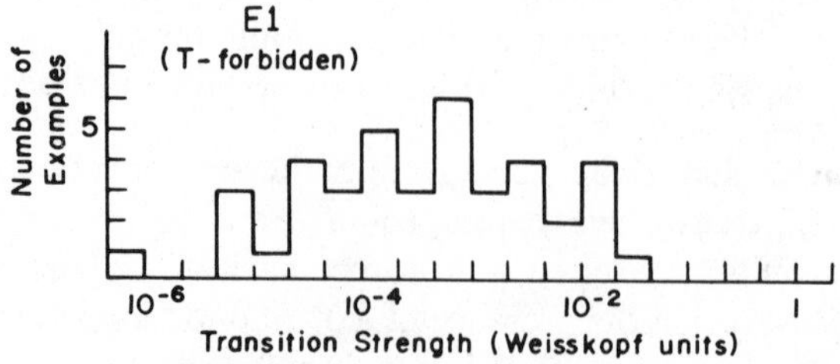

FIG. 11.9. Histograms of $E1$ transition probabilities [taken from Skorka et al. (66)].

TABLE 11.2 Reduced Transition Probabilities B(E1) for Transitions between Intrinsic States in Rare-Earth Nuclei [taken from Perdrisat (66)]

ΔK	Range of $B(E1)$	Number of Transitions Observed
0	2×10^9–1.6×10^{11}	7
+1	1×10^9–2.6×10^9	5
−1	2×10^7–4×10^9	10
≥2	6×10^4–5×10^7	11[a]

[a] None of the measured K-forbidden transitions occurs between intrinsic levels.

factors are taken with respect to the Nilsson model rather than the shell-model estimates as is appropriate for deformed nuclei. Indeed using the Weisskopf estimate does not lead to as useful classification of the observed matrix element.

$E1$ transitions in odd-A deformed nuclei are for the most part single-particle transitions. The large energy spacing between neighboring shells discussed above is no longer valid for strongly deformed nuclei. As one can see from Fig. VI. 10.1 giving the single-particle levels in a deformed potential, the spacing between neighboring shells is reduced as the deformation increases, with the consequence that single-particle $E1$ transitions of low energy occur. For transitions with $\Delta K = 0$ the Nilsson model gives the correct order of

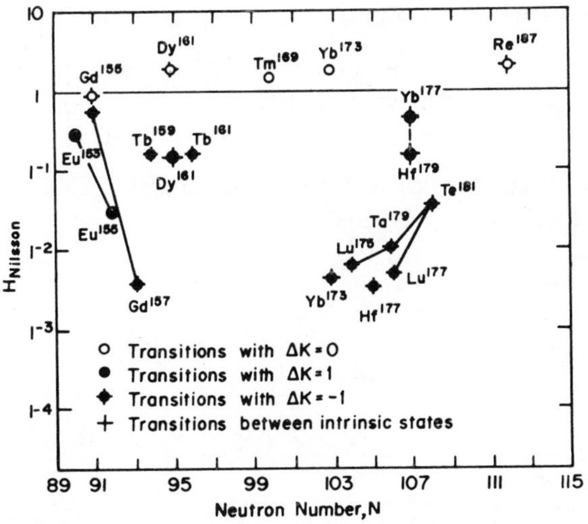

FIG. 11.10. Odd-A deformed nuclei: region of the rare earths. Experimental hindrance factor $H_N(E1)$ relative to the Nilsson estimate, as a function of the neutron number N. Transitions involving similar Nilsson states are connected by a solid line [taken from Perdrisat (66)].

magnitude as is clear from Fig. 11.10. Complete agreement for the $\Delta K = \pm 1$ transitions can be obtained if the ratio of the deformations of the initial and final states is treated as a parameter. However another effect must be taken into account that arises because the spacing of the single-particle level can very well be of the same order as that of the rotational levels. Under these circumstances small perturbations are very effective, and K will no longer be an exact quantum number, and there will be interband mixing. One perturbation that has been considered by Kerman (56) is the rotation-particle interaction (see Chapter VI) that mixes states having identical parities and total angular momentum but with values of K that differ by ± 1. Considerable improvement in predicted transition probabilities is reported [Vergnes and Rasmussen (65)] when interband mixing effects are incorporated in the analysis.

The discussion of $E1$ transitions for even-even nuclei involves bands based on octupole vibrations and is similar to that given earlier in Section 11b. For details see Perdrisat (66).

C. *M*1 Transitions

In the long wavelength limit the relevant reduced matrix element (see Eq. 5.36) is $(J_f||\boldsymbol{\mu}||J_i)$ where $\boldsymbol{\mu}$ is the magnetic moment operator. According to (5.39)*

$$\boldsymbol{\mu} = \mu_0 \sum_i \left\{ \frac{1 + \tau_3(i)}{2} \mathbf{l}_i + \frac{1}{2} [(g_p + g_n) + \tau_3(i)(g_p - g)] \mathbf{s}_i \right\} \tag{11.5}$$

where μ_0 is the nuclear magneton. The selection rules for $M1$ transitions are thus

$$\left.\begin{array}{ll} \Delta J = \pm 1, 0 & 0 \to 0 \text{ forbidden} \\ \text{no change in parity} & \\ \Delta T = \pm 1, 0 & 0 \to 0 \text{ forbidden} \end{array}\right\} \tag{11.6}$$

Approximate selection rules discovered by Morpurgo (58, 59) depend for their validity on the relatively small magnitude of

$$\frac{1}{2}(g_p + g_n) = 0.88$$

compared to

$$\frac{1}{2}(g_p - g_n) = 4.71$$

---

*The value of $B(M1)$ in terms of $\boldsymbol{\mu}$ is:

$$B(M1) = \frac{3}{4\pi} \mu_0^2 \frac{|(J_f||\boldsymbol{\mu}||J_i)|^2}{2J_i + 1}$$

For this reason the isoscalar component of $\boldsymbol{\mu}$ is relatively small. More explicitly let

$$\boldsymbol{\mu}_S = \frac{1}{2}\mu_0 \sum_i [\mathbf{l}_i + (g_p + g_n)\,\mathbf{s}_i] \tag{11.7}$$

Replace $\mathbf{l}_i$ by $\mathbf{j}_i - \mathbf{s}_i$. Then

$$\boldsymbol{\mu}_S = \frac{1}{2}\mu_0 \sum [\,\mathbf{j}_i + (g_p + g_n - 1)\,\mathbf{s}_i]$$

or

$$\boldsymbol{\mu}_S = \frac{1}{2}\mu_0\,[\mathbf{J} + (g_p + g_n - 1)\,\mathbf{S}] \tag{11.8}$$

For a transition, the matrix elements of $\mathbf{J}$ will be zero so that

$$\boldsymbol{\mu}_S \rightarrow \frac{1}{2}\mu_0\,(g_p + g_n - 1)\,\mathbf{S} \tag{11.9}$$

where $(1/2)(g_p + g_n - 1) = 0.38$. Note the contribution of the orbital angular momentum reduces the magnitude of the isoscalar magnetic moment transition operator.

The isovector contribution is

$$\boldsymbol{\mu}_V \equiv \frac{1}{2}\mu_0 \sum_i \tau_3(i)\,[\mathbf{l}_i + (g_p - g_n)\,\mathbf{s}_i] \tag{11.10}$$

or

$$\boldsymbol{\mu}_V \equiv \frac{1}{2}\mu_0 \sum_i \tau_3(i)\,[\mathbf{j}_i + (g_p - g_n - 1)\,\mathbf{s}_i] \tag{11.11}$$

It is not possible to rigorously eliminate the dependence on $\mathbf{j}_i$ as in the case of $\boldsymbol{\mu}_S$. However there are special cases in which the matrix element of this term is zero. In one example when the particles of one type form a closed shell, the transition occurs by changing the state of the other type particles. In that case

$$\sum \tau_3(i)\,\mathbf{j}_i \rightarrow \pm \sum \mathbf{j}_i = \pm\,\mathbf{J}$$

The transition matrix element of $\mathbf{J}$ is zero as in the scalar case. In another case the transition involves changing the j-value of a particle, the particle going to another orbit with the same value of l. In the limit in which the spin-orbit potential is neglected, the radial matrix element for the transition will vanish.

In both of these cases the ratio of the isoscalar transition probability to the isovector probability is

$$[(g_p + g_n - 1)/(g_p - g_n - 1)]^2 = 0.0082$$

showing the relatively low probability of the isoscalar. [For a more detailed discussion see Talmi and Unna (60).]

Recall from Section 6 page 704 and (6.7) and (6.8) the result that applies when the isoscalar matrix element is small, compared to the isovector matrix element: *corresponding* $\Delta T = 0\,(M1)$ *transition in conjugate nuclei are approximately of equal strength.* Exceptions to this rule can occur if for some reason the widths for these transitions are much below the average, for then the isovector component is reduced and the inequality may no longer hold. A second approximate selection rule states that $\Delta T = 0\,(M1)$ *transitions that proceed only in virtue of the isoscalar part of the interactions* (this will be the case of self-conjugate $A = 2Z$ nuclei) *will be weaker by a factor of 100 than the average M1 transition.* The validity of these results can be seen from Figs. 11.11, 11.12, and 11.13. In Fig. 11.11 we have the $A = 2Z$ nuclei, $|\Delta T| = 1$ transition. Figure 11.12 contains the $A = 2Z$ nuclei with $|\Delta T| = 0$. The large difference in transition strength is immediately obvious, the ratio of the average values being 0.008, very close to the rough value obtained above. In Fig. 11.13, we give the $A \neq 2Z$ transition, not attempting to distinguish between the $\Delta T = 0$ and $\Delta T = 1$ transitions. The average transition strength for these transitions is somewhat lower than the T-favored.

The average transition strength for these transitions in $A \leqslant 40$ nuclei is less than the single-particle value, averaging according to Skorka, Hertel, and Retz-Schmidt (66) 0.39 times the single-particle value for the T-favored transitions, 0.10 for the "normal" transitions, and .0048 for the T-forbidden cases. The hindrance factors are somewhat greater for heavier nuclei (Fig. 11.14). As was the case for $E1$ transitions the deviation from the single

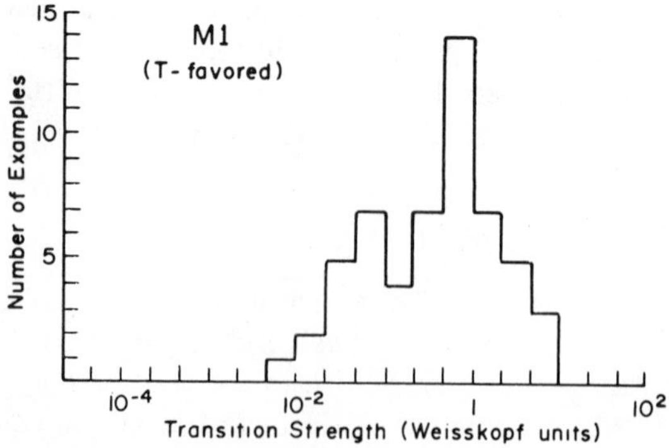

FIG. 11.11. Histogram of $M1$ transition probabilities [From Skorka et. al. (66)].

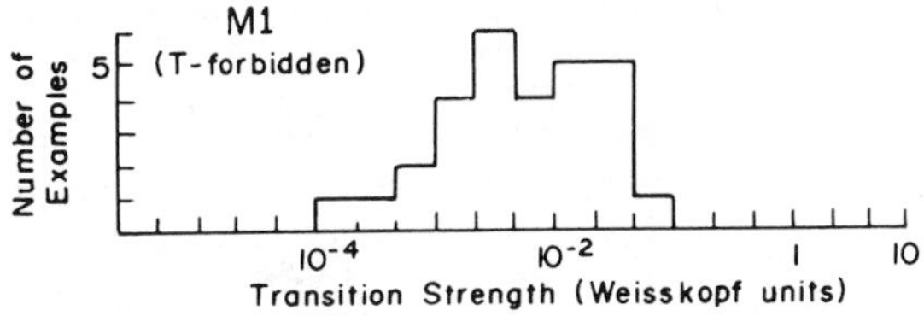

FIG. 11.12. Histogram of $M1$ transition probabilities [From Skorka et. al. (66)].

particle values can be largely explained by the influence of the residual interaction that leads to configuration mixing.

Special selection rules prevail for the various models. Those for deformed nuclei have been discussed in Chapter VI p. 429 in terms of the rotational model. In the case of the shell model we note two selection rules of interest. In the first case consider a transition between two different states formed from a j^n configuration of identical particles. In that event the matrix element of (11.5) is proportional to **J** (see Section IV.17) and the transition matrix element vanishes. As discussed in Section V.11 the nucleus ^{51}V may be a good example of an $(f_{7/2})^3$ nucleus, the neutrons forming a closed shell of 28; three of the protons going into the $f_{7/2}$ orbit, the remaining 20 also forming a closed shell. The $M1$ radiative transitions are shown in Fig. 11.15 [Bhattacherjee (71) and Horoshko, Cline, Lesser (70)]. Since the single-particle $B\ (M1)$ is about two, hindrance factors of the order of 400 and more are indicated, which is to be compared with the factor of 10 for "normal transitions." The selection rule for j^n configuration is approximately satisfied. A similar conclusion can be drawn from the relative value of the magnetic moment of the ground and first excited state. For a $(j)^n$ configuration the g-factors are identical for all

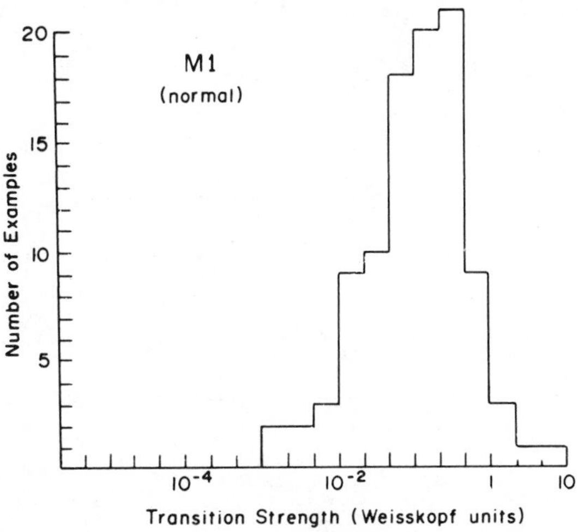

FIG. 11.13. Histogram of $M1$ transition probabilities [From Skorka et. al. (66)].

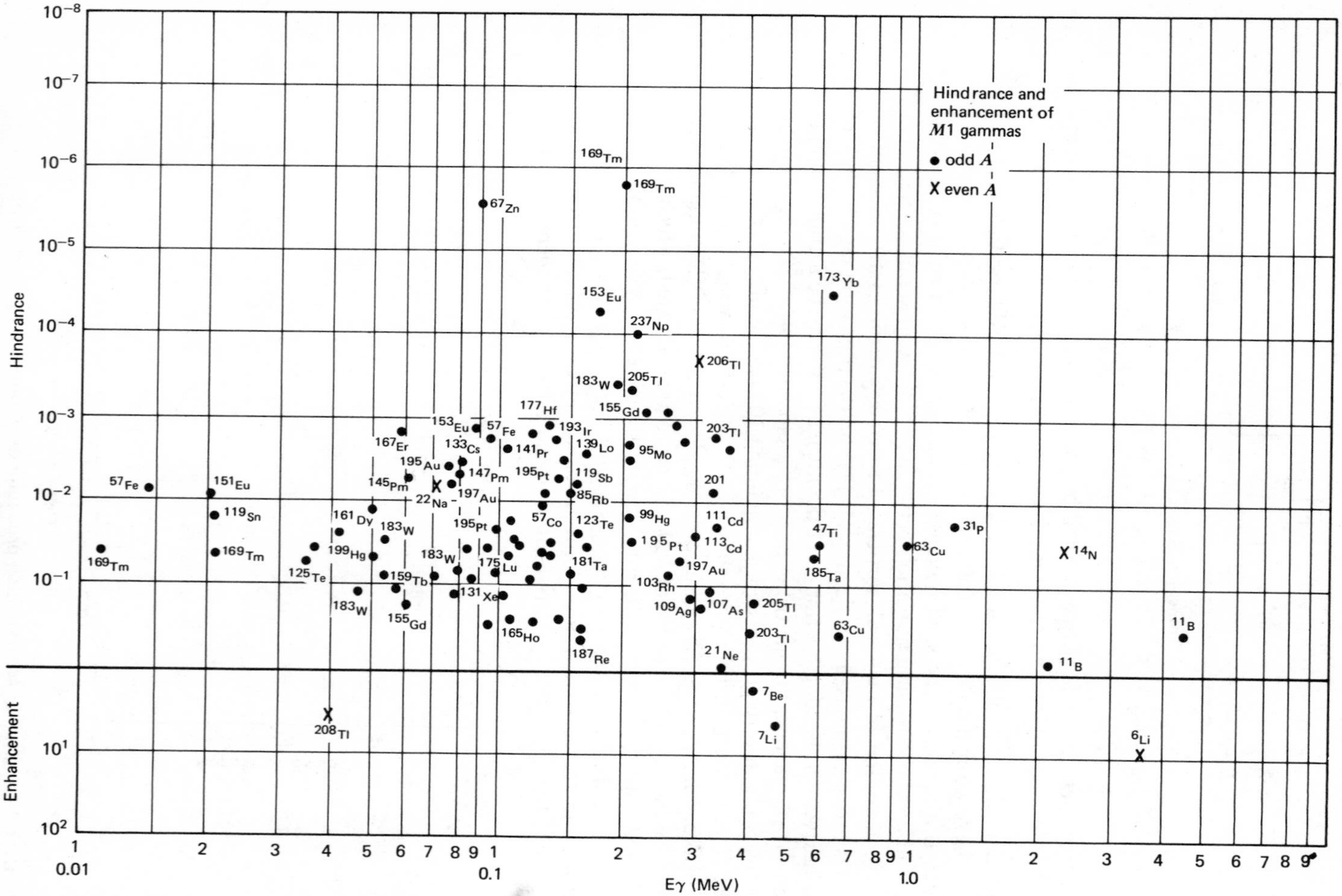

FIG. 11.14. Hindrance and enhancement factors of $M1$ gammas [taken from Gove (66)].

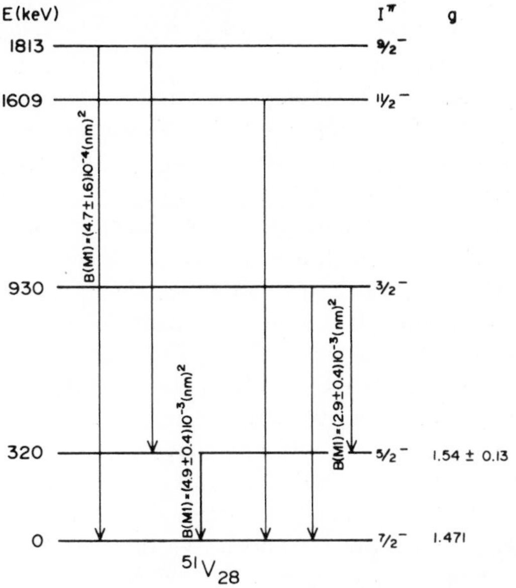

FIG. 11.15. *M*1 transitions in ^{51}V [taken from Bhattacharjee (71) and Horoshko et al (70)].

J-values (see Chapter IV p. 245). The experimental value is 1.471 for the $(7/2)^-$ ground state, which is to be compared with the experimental value of 1.54 ± 0.13 for the $(5/2)^-$ state [Varga et al. (69)]. Obviously a better experiment is needed. The existence of decays (and any possible difference between $g(7/2)$ and $g(5/2)$) demonstrates that these nuclear wave functions are not pure $(7/2)^3$. There are admixtures.

A second selection rule is concerned with the so-called "*l*-forbidden" transitions. The matrix elements of the operator $\boldsymbol{\mu}$ vanish unless the change in the angular momentum of each of the radiating particles is zero ($\Delta l = 0$). For example [Talmi and Unna (60)] in N^{14} there are no isovector *M*1 transitions from the $p_{1/2}\ s_{1/2}$ ($J = 1$) to the $p_{1/2}\ d_{5/2}$ ($J = 2$) state. More commonly hindrance factors of the order of 100 are reported [(Arima, Horie, Sano (57)].

The small violation of these shell-model selection rules indicate the presence of small admixtures of configurations other than that of the simplest shell model. These small components can be understood in terms of the effect of residual interaction. The discussion is similar to the discussion in Section 9 on effective charge. In ^{131}Xe (Arima, Horie, and Sano (57)] the radiating neutron makes an *l*-forbidden transitions from $3s_{1/2}$ orbit to a $2d_{3/2}$ orbit. Note that the unfilled shells contain $(1g_{7/2})^4$ protons, $(1h_{11/2})^{10}\ (2d_{3/2})^2\ (3s_{1/2})$ neutrons. In the presence of a tensor interaction between the $s_{1/2}$ and $h_{11/2}$ neutrons the excited state contains a term in which the $s_{1/2}$ particle moves to the $d_{3/2}$

orbit while the $h_{11/2}$ neutron goes to the $h_{9/2}$. The $h_{9/2}$ particle radiates thus returning to the $h_{11/2}$ orbit. Note that since the operator that admixes the $h_{9/2}$ state is not a magnetic moment type operator (it is of the form $[\boldsymbol{\sigma} Y_2]_\mu{}^1$, that is, the tensor combination of the spin and Y_2 that transforms under rotations like a tensor of rank 1) this effect cannot be described by simply changing the size of the neutron magnetic moment. The effective magnetic moment concept needs to be generalized to be applicable to l-forbidden transitions (See Section 11.d). From the work of Arima et al. (57) it would appear that the lifetime of many of the l-forbidden transitions can be explained by the effect of the residual interactions employing first-order perturbation theory since the admixtures are small.

Appropriate diagonal matrix elements of the magnetic moment operator (11.5) yield the magnetic moment of the ground state and excited states. The configuration mixing discussed above also plays a role in these calculations. It is thus necessary to understand not only the M1-transition probability but also the magnetic moments of the states involved in the transition as exemplified by the discussion of ^{51}V above. We turn now to a discussion of the magnetic moments.

One final word: we have not discussed the effect of giant $M1$ resonances. These have been seen with the aid of inelastic electron scattering in ^{12}C and ^{16}O. To what extent these resonances exhaust the sum rule for $M1$ is not yet clear.

d. Magnetic Moments. The magnetic moments of nuclei have been discussed to some extent in Chapter 1 where the empirical results are outlined, in Chapters IV and V where the shell-model theory is applied, and in Chapter VI where the results obtained for the rotational model for deformed nuclei are discussed. Briefly, qualitative agreement with experimental values was obtained indicating that the fundamental pictures underlying each of these models are valid. There are however considerable discrepancies as shown by the deviation of the nuclear-magnetic moments from the Schmidt lines (Fig. I.8.2). These deviations indicate the need for correcting the simple shell-model theory by configuration mixing including core excitations. For deformed nuclei the rotational model provides a simple way of doing this but again there are deviations (Figs. VI.8.1 to VI.8.4). Again configuration mixing modifying the simple deformed nucleus shell model is suggested. Clearly information on the nuclear wave function can be extracted from these results. We shall describe some of that program for the spherical shell-model case for the "near spherical" nuclei [Bhattacherjee (71), Talmi (70)]. The analogous procedures for other models can be found in Kisslinger and Sorensen (63), Bodenstedt and Rogers (64), O. Prior, F. Boehm, and S. G. Nilsson (68).

Although our discussion will be centered about the effects of the residual interaction the possible contribution of exchange currents (Section 3) [A. deShalit (51), F. Bloch (51) H. Miyazawa (51)] should not be forgotten.

Unfortunately it is not possible at the present level of understanding, to empirically distinguish between the effects of exchange currents and configuration mixing. Since exchange currents involve at least two particles, the corresponding magnetic moment operator (see Section 3) will be a two- or more body operator. But much the same can be said for the effect of configuration interaction that results from the action of the residual interaction changing the orbit of two or more particles and thereby affecting the magnetic moments. On second thought the difficulty of distinguishing between these two effects should occasion no surprise since both the exchange currents and residual interactions are carried by mesons.

There is one exception to this ambiguity. As we shall indicate below the effect of the residual interaction usually is to move the magnetic moment value into the interior lying between the two bounding Schmidt lines. Hence experimental magnetic moments lying outside the region bounded by the Schmidt line will be explained with difficulty by configuration mixing, and will require an exchange current contribution. As can be seen from Fig. I.8.2, the nuclei ^3H and ^3He are examples of this phenomenon. The magnetic moment $\mu(^3H)$ is $-2.1276\ \mu_0$ exceeding the neutron magnetic moment in magnitude by nearly 10% while $\mu\ (^3He)$ is 2.9789 μ_0 to be compared with $\mu(^1H)$ of 2.7927 μ_0. Detailed calculations demonstrate that configuration mixing will not explain these discrepancies from the Schmidt value [Delves and Phillips (69)]. As we shall see $[\mu(^3HE) + \mu\ (^3H)]$ is independent of the exchange currents, the experimental value of the sum being given by these calculations [Sachs and Schwinger (46), Sachs (47)]. On the other hand, the difference is sensitive to the exchange current; the theories of the three-body system failing to yield the experimental value [Delves and Phillips (69)]. From this example and a theoretical estimate [Drell and Walecka (60)] we can expect exchange effects of the order of 0.1 to 0.2 μ_0.

As in the case of transitions, relations exist between the magnetic moments of isospin multiplets and mirror nuclei. The operator $\boldsymbol{\mu}$ has an isoscalar and an isovector vector component. Therefore one can expect for an isospin multiplet that

$$\mu = a(T) + b(T)\ M_T \tag{11.12}$$

The values of two magnetic moments for two members of the multiplet are enough to determine the constants $a(T)$ and $b(T)$ and therefore predict μ for the other members. Unfortunately there is no case for which this rule can be tested. In one case [Talmi (70)] $A = 12, T = 1$, and $J = 1$ magnetic moments are known, $\mu\ (^{12}B) = (1.003 \pm .001)\ \mu_0$,; $\mu(^{12}N) = (\pm\ 0.457 \pm .001)\ \mu_0$ yielding the values $a = 0.730\ \mu_0$, and $b = -\ 0.273\ \mu_0$. We have taken the positive value for ^{12}N. Then relation (11.12) predicts $\mu\ (^{12}C)_{T=1,\ J=1}$ to be 0.730 μ_0.

Turning to the mirror nuclei, we can see from (11.12) that the sum $\mu\ (M_T)$

$+ \mu(-M_T)$ depends only upon the isoscalar component. Hence from (11.8) one obtains

$$\mu(M_T) + \mu(-M_T) = \mu_0 [J + (g_n + g_p - 1) \langle S_z \rangle] \tag{11.13}$$

Note that the exchange current contributions also drop out in this sum since that current is an isovector (Eq. 3.7). Only the isoscalar component contributes for $T = 0$ nuclei, the isovector component vanishing:

$$\mu(T = 0) = \frac{\mu_0}{2} [J + (g_n + g_p - 1) \langle S_z \rangle] \tag{11.14}$$

In the case of a j^n configuration [Mayer and Jensen (55)] the expectation value of S_z can be evaluated. Within this Hilbert space the expectation value of $\mathbf{s}$ (the spin operator for the ith particle) is proportional to $\mathbf{j}$ (the total spin for the ith particle). The proof of (4.10′) can be adapted to prove this statement. Then

$$\langle \mathbf{s} \rangle \to \frac{\langle \mathbf{s} \cdot \mathbf{j} \rangle \mathbf{j}}{j(j+1)} = \frac{j(j+1) + 3/4 - l(l+1)}{2j(j+1)} \mathbf{j}$$

$$= \begin{cases} \dfrac{1}{2j} \mathbf{j} & j = l + \dfrac{1}{2} \\[2ex] -\dfrac{1}{2(j+1)} \mathbf{j} & j = l - \dfrac{1}{2} \end{cases}$$

Hence

$$\mu(M_T) + \mu(-M_T) = \mu_0 J \left[1 + \begin{cases} \dfrac{g_n + g_p - 1}{2j} & j = l + \dfrac{1}{2} \\[2ex] -\dfrac{g_n + g_p - 1}{2(j+1)} & j = l - \dfrac{1}{2} \end{cases} \right] \tag{11.15}$$

while the expression for $\mu(M_T = 0)$ is just 1/2 of (11.15).

By using empirical values of μ it becomes possible to determine $\langle S_z \rangle$ (see Eq. 11.13) from experiment and by comparison with (11.15) obtain some notion as to the validity of $j - j$ coupling. The values of $\langle S_z \rangle$ obtained in this fashion are shown in Fig. 11.16 [Sugimoto (69)]. Talmi (70) has also compared these results with the jj-coupling predictions in his Table II for $M_T = 1/2$ as well as $T = 0$. As an example, experimentally the sum $\mu(^{19}F) + \mu(^{19}Ne)$ is 0.742 μ_0 while (11.15) yields 0.576 μ_0 assuming $l_j = d_{5/2}$. On the other hand, $\mu(^{21}Na) + \mu(^{21}Ne) = 1.724$ μ_0 experimentally, the assumption of $l_j = d_{5/2}$ giving 1.728 μ_0. One systematic result is obtained: except for the mirror nuclei with $A = 37$ all deviations of $\langle S_z \rangle$ from the jj model are in the same direction: the absolute value of the experimental $\langle S_z \rangle$ is smaller than the jj-model prediction. These results only check the validity of the jj-model if isospin is conserved.

The difference of the magnetic moments of mirror nuclei is proportional to the matrix elements of the isovector magnetic moment operator (11.11).

$$\mu(M_T) - \mu(-M_T) = \mu_0 [\langle \sum \tau_3(i)\, j_z(i) \rangle + (g_p - g_n - 1) \langle \sum \tau_3(i)\, s_z(i) \rangle]_{M_T} \tag{11.16}$$

Because of the large value of $g_p - g_n - 1$, (11.16) is sensitive to the value of $\langle \sum_i \tau_3(i)\, \sigma_z(i) \rangle$. It turns out that the magnitude of this matrix can be obtained from the Gamow-Teller matrix element for the β-decay between a nucleus and its mirror. We shall discuss this more fully in the next chapter, but for the present it will suffice if we note that the β-decay transition probability for the transition $M_T = 1/2 \rightarrow M_T = -(1/2)$ is proportional to

$$\frac{1}{2J+1} \sum_{MM'q} |\langle J, M, T = \tfrac{1}{2}, M_T = -\tfrac{1}{2} | \sum_i \tau_1(i)\, \sigma_q(i)\, | J, M', T = \tfrac{1}{2}, M_T = \tfrac{1}{2} \rangle|^2$$

Using the Wigner-Eckart theorem this becomes

$$\frac{1}{2}\, \frac{1}{2J+1} \begin{pmatrix} \frac{1}{2} & 1 & \frac{1}{2} \\ \frac{1}{2} & -1 & \frac{1}{2} \end{pmatrix}^2 (J, T = \tfrac{1}{2} \| \sum_i \tau(i)\, \sigma(i) \| J, T = \tfrac{1}{2})^2$$

$$= \frac{1}{6(2J+1)} (J, T = \tfrac{1}{2} \| \sum_i \tau(i)\, \sigma(i) \| J, T = \tfrac{1}{2})^2 \tag{11.17}$$

Similarly the magnetic moment matrix element

$$\langle J, M_J = J, T = \frac{1}{2}, M_T = \frac{1}{2} | \sum_i \tau_3(i)\, \sigma_z(i) | J, M_J = J, T = \frac{1}{2}, M_T = \frac{1}{2} \rangle$$

$$= \begin{pmatrix} J & 1 & J \\ -J & 0 & J \end{pmatrix} \begin{pmatrix} \frac{1}{2} & 1 & \frac{1}{2} \\ -\frac{1}{2} & 0 & \frac{1}{2} \end{pmatrix} (J, T = \tfrac{1}{2} \| \sum_i \tau(i)\, \sigma(i) \| J, T = \tfrac{1}{2})$$

$$= \sqrt{\frac{J}{6(J+1)(2J+1)}}\, (J, T = \tfrac{1}{2} \| \sum_i \tau(i)\, \sigma(i) \| J, T = \tfrac{1}{2}) \tag{11.18}$$

We thus see that the magnitude of the matrix element in the second and most important term on the right-hand side of (11.16) is given apart from known J-dependent factors by the square root of the β-decay probability. Of course this relation does not provide the sign of the matrix element.

Turning once again to *jj*-coupling and employing the relation above (11.15) to replace **j** by **s**, one obtains

$$\mu(M_T = \tfrac{1}{2}) - \mu(M_T = -\tfrac{1}{2})$$

$$= \mu_0 (g_p - g_n + 2l) \langle \sum \tau_3(i)\, s_z(i) \rangle_{M_T = 1/2} \qquad j = l + \tfrac{1}{2}$$

$$= \mu_0 (g_p - g_n - 2l - 2) \langle \sum \tau_3(i)\, s_z(i) \rangle_{M_T = 1/2} \qquad j = l - \tfrac{1}{2} \tag{11.19}$$

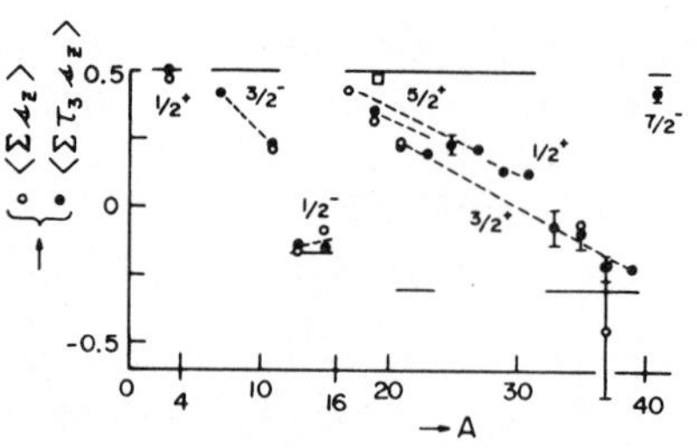

FIG. 11.16. Spin parts of the angular momentum distribution of mirror nuclei. Expectation value for the state with $T_3 = +(1/2)$ is used. The solid lines represent the single-particle values, while the broken lines connect the points of the same J^π [taken from Sugimoto (69)].

Thus by determining this difference one can obtain the expectation value $\langle \sum \tau_3 \sigma_z \rangle$ (see Fig. 11.16). It can also be obtained from β-decay using the relation indicated by (11.17) and (11.18). Two examples will indicate the level of agreement. For the (^{19}F, ^{19}Ne) pair the magnetic-moment difference yields -0.67 for the matrix element while β-decay gives excellent agreement, a *magnitude* of 0.72 ± 0.03. For the (^{21}Na, ^{21}Ne) pair the magnetic moments yield 0.45 for the matrix element that agrees fairly well with the β-decay result of 0.56 ± 0.05. For further examples see Fig. 11.C.5 and Table I of Talmi's paper (70).

We turn now to a discussion of the deviation of the magnetic moments of odd nuclei from their single-particle Schmidt values in which we shall describe the effects of the residual interaction in the spherical shell-model limit. (See earlier comment on page 751). For reference the Schmidt value of the magnetic moments are given in Table 11.3. As a first orientation, we shall estimate

TABLE 11.3 The Schmidt Values for Magnetic Moments in Units of Nuclear Magneton μ_0[a]

Odd-Proten Nuclei				Odd-Neutron Nuclei			
$j = l + 1/2$ States		$j = l - 1/2$ States		$j = l + 1/2$ States		$j - l - 1/2$ States	
$s_{1/2}$	2.793			$s_{1/2}$	−1.913		
$p_{3/2}$	3.793	$p_{1/2}$	−0.264	$p_{3/2}$	−1.913	$p_{1/2}$	0.638
$d_{5/2}$	4.793	$d_{3/2}$	0.124	$d_{5/2}$	−1.913	$d_{3/2}$	1.148
$f_{7/2}$	5.793	$f_{5/2}$	0.862	$f_{7/2}$	−1.913	$f_{5/2}$	1.366
$g_{9/2}$	6.793	$g_{7/2}$	1.717	$g_{9/2}$	−1.913	$g_{7/2}$	1.488
$h_{11/2}$	7.793	$h_{9/2}$	2.624	$h_{11/2}$	−1.913	$h_{9/2}$	1.565
$i_{13/2}$	8.793	$i_{11/2}$	3.560	$i_{13/2}$	−1.913	$i_{11/2}$	1.619

[a] Taken from Noya, Arima, Horie (58).

the relative size of the deviation for odd-N and odd-Z nuclei in the "odd-group" model [deShalit (53)]. In this model it is assumed that the contribution of the even group of nucleons to the magnetic moment as well as to the spin is zero. Second, it is assumed that the expectation value of the spin operator S_z for the odd-nucleon groups is the same as long as their number is identical. Then (11.5) becomes in nuclear magnetons

$$\mu = g_l J + (g_s - g_l)\langle S_z\rangle$$

where

$$g_l = \begin{cases} 1 & \text{protons} \\ 0 & \text{neutrons} \end{cases} \qquad g_s = \begin{cases} 4.5856 \\ -3.8262 \end{cases}$$

The Schmidt value for a ($J = j$)-nucleon is

$$\begin{aligned} \mu_{\text{Sch}} &= g_l J + (g_s - g_l)\langle s_z\rangle \\ &= j\left(g_l \pm \frac{g_s - g_l}{2l+1}\right) \qquad \text{for } j = l \pm \frac{1}{2} \end{aligned}$$

It follows that

$$\frac{(\mu - \mu_{\text{Sch}})_{Z\ \text{odd}}}{(\mu - \mu_{\text{Sch}})_{N\ \text{odd}}} = \frac{(g_s - g_l)_{\text{prot}}}{(g_s - g_l)_{\text{neut}}} = \frac{g_p - 1}{g_n} = -1.2$$

One must be circumspect in using this result, because of the approximation that the even group does not contribute directly to the magnetic moment. In fact the effects of the residual interaction will, among other things, depend upon the state of the even group, which depends importantly on the number of nucleons in the even group. Except for the light nuclei, the N for an odd-Z nucleus is much larger than the Z for an odd-N nucleus where odd N = odd Z. Under these circumstances, that is, for the heavier nuclei, the assumption that S_z for odd N equals S_z for an odd-Z nucleus (odd N = odd Z) is suspect. The above equation furnishes a rough estimate that is generally in agreement with experiment for the light mirror nuclei. The important qualitative result which follows is that the deviation of the odd-proton nuclei from the Schmidt value is slightly larger than the corresponding odd-neutron nucleus.

Turning now to more quantitative considerations we notice that even within the shell model the magnetic moment of an odd nucleus whose spin J equals j, the value of the odd nucleon is not necessarily given by the Schmidt value if the valence shell contain both neutrons and protons and if the ground state has good isospin. Of course the Schmidt value is the shell-model value if the shell contains only one particle. In particular, suppose the shell contains n-nucleons of which N_n are neutrons and N_p are protons. Each particle has a spin j, that is, jj coupling will be assumed. There are many ways to couple the spins and isospins to achieve $J = j$ and $T = \frac{1}{2}|N_n - N_p|$ for the ground state. In the "pairing" approximation (see Chapter VII), [see Section 34 and 35 in de-Shalit and Talmi (63)], the lowest state is obtained by coupling pairs of particles into units whose spin is zero and whose isospin is one. The final odd particle has spin j and isospin 1/2. For a given final T of the whole system, the wave function is unique (deShalit and Talmi, p. 444) and the magnetic moment can be calculated. [See deShalit and Talmi p. 449, Grayson Jr. and Yost (56) for the full story.] The results are:

$$N_p < N_n \qquad g - g_p(lj) = [g_n(lj) - g_p(lj)]\frac{N_n}{4(j+1)(T+1)}$$

odd-Z nuclei:

$$N_p > N_n \qquad = [g_n(lj) - g_p(lj)]\frac{2j+1-N_n}{4(j+1)(T+1)}$$

For odd-N nuclei replace $g_p \leftrightarrow g_n$ and $N_n \leftrightarrow N_p$. For most cases of interest, the first of these formulas is useful for odd-Z nuclei, the second for odd-N nuclei. Note that the Schmidt moment for odd-Z nuclei is obtained when N_n is zero or $(2j + 1)$, that is, when the neutrons form a closed shell. For the light nuclei the corrections given by the right-hand side of the above equations brings the shell-model value into closer agreement with experiment. A number of examples are given by Noya, Arima, and Horie (58) (their Table 8). There are substantial improvements in some cases but substantial differences from experiment still remain.

To approach experiment more closely the residual interactions must be taken into account. These of course modify the nuclear wave functions and thus give rise to a change in the magnetic moment. The study of the magnetic moments permits us therefore to study some aspects (which we shall see below) of the residual interaction and of its effect on the wave function (Section 9).

Suppose then that

$$\Psi = \Psi^{(0)} + \Psi^{(1)}$$

where $\Psi^{(1)}$ is the first-order correction to $\Psi^{(0)}$. From $\mu = \langle\mu_z\rangle$ we obtain

$$\mu = \mu^{(0)} + \delta\mu$$

where $\mu^{(0)}$ is the unperturbed value of the magnetic moment and $\delta\mu$ is

$$\delta\mu = 2 \operatorname{Re} \langle\Psi^{(1)}|\sum \mu_z(i)|\Psi^{(0)}\rangle_{M=J} \tag{11.20}$$

Since $\Psi^{(1)}$ is generated from $\Psi^{(0)}$ by a scalar potential, the total spin J of $\Psi^{(1)}$ will be identical to that of $\Psi^{(0)}$. The operator $\sum_i \mu_z(i)$ will generally change one of the j's of the unperturbed state $\Psi^{(0)}$ from $l \mp \frac{1}{2}$ to $l \pm 1/2$. Note that the l's are the same because $\mu_z(i)$ can be written as a sum of two terms, one proportional to $j_z(i)$ and the other proportional to $s_z(i)$, which is independent of space coordinates. In jj-coupling the possible components of $\Psi^{(1)}$ are related to $\Psi^{(0)}$ as follows [Noya, Arima, Horie (58)]. Suppose $\Psi^{(0)} = |j_1{}^{n_1}(0)\, j_2{}^{n_2}(0)\, j^n(j)\, J = j\rangle$ where the notation means n_1 particles in the j_1 orbit coupled to zero angular momentum, n_2 particles in the j_2 orbit coupled to zero angular momentum, and n particles in the j orbit coupled to j so that the total angular momentum J is j. The numbers n_1 and n_2 are even, n is odd. Then $\Psi^{(1)}$ will contain components generated as follows: (1) the orbit of a particle originally in j_1 is shifted to j_2 yielding $\Psi_1^{(1)} = |j_1^{n_1-1}(j_1)j_1^{n_2+1}(j_2)(J_1 = 1)j^n(j)J = j\rangle$. The angular momentum J_1 is obtained by combining the angular momentum of the n_1 group j_1 with that of the n_2 group, j_2. Since according to our earlier discussion this is possible only if $\mathbf{j}_1 - \mathbf{j}_2 = \mathbf{1}$, J_1 must equal one. (2) the orbit of a particle in j_1 can be moved into orbit j yielding $\Psi_2^{(1)} \equiv |j_1^{n_1-1}(j_1)j_2^{n_2}(0)$ $j^{n+1}(J_1)J = j\rangle$ where $\mathbf{J} = \mathbf{j}_1 + \mathbf{J}_1$; again $\mathbf{j}_1 + \mathbf{j} = \mathbf{1}$. (3) The orbit of a particle in j can be moved to the j_2 orbit yielding

$$\Psi_3^{(1)} \equiv |j_1^n(0)\,j_2^{n_2+1}(j_2)\,j^{n-1}(J_2)\,J = j\rangle$$

where the $(n - 1)$ particles in the orbit j couple to J_2 and J_2 couples with j_2 to yield j_1 and $\mathbf{j}_2 + \mathbf{j} = \mathbf{1}$. (4) The particles in j_1 (or j_2) couple to unity which couples with j to yield j: $\Psi_4^{(1)} = |j_1^n(1)\,j_1^{n_2}(0)\,j_n(j)\,J = j\rangle$. The amplitude of these four types of admixture is according to perturbation theory given by

$$\frac{1}{\Delta E_n}\langle\Psi_n^{(1)}|V_R|\Psi^{(0)}\rangle$$

where V_R is the residual interaction and ΔE_n is the appropriate energy difference, so that the net contribution to $\delta\mu$ is:

$$\delta\mu = 2\mathrm{Re}\sum_n \frac{\langle\Psi^{(0)}|V_R|\psi_n^{(1)}\rangle\,\langle\Psi_n^{(1)}|\sum\mu_z(i)|\Psi^{(0)}\rangle}{(\Delta E)_n} \tag{11.21}$$

This is in essence the formula used for example by Arima and his collaborators [Noya, Arima, Horie, (58), Arima and Horie (54)] for the heavier nuclei, and by Kurath for the lighter nuclei in their shell-model calculations of the magnetic moments of odd nuclei.

The overall effect of V_R can be seen qualitatively as follows. First, note that V_R must have spin dependence in order for it to be able to excite $\Psi_n^{(1)}$ from $\Psi^{(0)}$, $\Psi_n^{(1)}$ being formed by a spin flip, (the l's remain unchanged). Thus the spin-spin term $\boldsymbol{\sigma}_i\cdot\boldsymbol{\sigma}_j\,V_\sigma(r_{ij})$ as well as the tensor interaction

$$[3(\boldsymbol{\sigma}_i\cdot\mathbf{r}_{ij})(\boldsymbol{\sigma}_j\cdot\mathbf{r}_{ij}) - \boldsymbol{\sigma}_i\cdot\boldsymbol{\sigma}_j\,r_{ij}^2]\,V_T(r_{ij})$$

are relevant.* These are the commonly chosen forms for the residual interaction, chosen to conform to the form of the interaction when the particles are otherwise free.

Suppose that the odd particle in this odd nucleus is a proton. The residual interaction will then induce configurations to form $\Psi_n^{(1)}$. When interacting

---

*Note that the spin-orbit shell-model potential may, depending on its origin, give rise to an additional interaction with a static field H of the following form $[(\boldsymbol{\sigma}\cdot\mathbf{r})\mathbf{r} - (1/3)\boldsymbol{\sigma}r^2]\cdot\mathbf{H}$. The operator involved is of the form $[Y_2(\hat{\mathbf{r}})\boldsymbol{\sigma}]^{(1)}$.

with a proton the spin-spin force will dominate. Notice the potential will be strongest when the spins of the two protons are antiparallel. Hence the induced magnetic moment from the proton-proton interaction will pull the magnetic moment value away from the Schmidt line into the region between the lines. A similar result follows for the interaction of the proton with the neutrons. If the residual interaction is similar to the free-space nucleon-nucleon interaction, it follows that the $(n\text{-}p)$ potential is stronger in the triplet rather than the singlet state; in other words the neutron and proton tend to have parallel spin. But the g-factor for the neutron is negative so that the action of this term adds to the effect of the residual proton-proton interaction. To sum up: the polarization induced by the proton in the remainder of the nucleus because of the residual interaction quenches the contribution of the spin to the magnetic moment. This change is in the direction needed in order to obtain agreement with the experimental facts.

This discussion can be taken somewhat further employing the concept of the effective magnetic moment operator (deShalit 60, 63) that is suggested by the analogous effective charge or more precisely effective multipole moment of Section 9. Let us suppose that the residual potential V_R is a two-body potential. Then

$$V_R = \sum_{i \neq j} V_R\,(ij) \tag{11.22}$$

$V_R\,(ij)$ can be expanded in terms of the irreducible spherical tensors $T_p{}^{(sk)r}$ in the combined spin and coordinate space:

$$V_R\,(ij) = \sum_{ss';\,kk',r} v_{ss',kk',r}\,(r_i,\,r_j)\;T^{(sk)r}(i)\cdot T^{(s'k')r}(j) \tag{11.23}$$

The value of s gives the tensor character in spin space. Since the residual potential that acts must be spin dependent in order to affect the magnetic moment, V_R must depend on $\boldsymbol{\sigma}(i)$ and $\boldsymbol{\sigma}(j)$. Hence the $s = 1$ components are the ones of interest. The value of k gives the tensor character in coordinate space, $k = 0$ corresponding to scalar, etc. Finally r is constructed by taking the "vector sum" of s and k. From the description of the mixing configuration $\Psi_n{}^{(1)}$ above, it follows that only $r = 1$ components of $V_R\,(ij)$ will contribute to $\delta\mu$. It follows then that only the combinations of (s,k) $s = 1$, $k = 0$ and $s = 1$, $k = 2$ are significant.

Inserting (11.23) into (11.21) yields

$$\delta\mu_z = 2\,\mathrm{Re} \sum_n \sum_{ss',kk'} \sum_{i \neq j} \frac{1}{\Delta E_n} \langle \Psi_0 | T^{(sk)1}(i)\; T^{(s'k')1}(j)\; v_{zz',\,kk',\,r}\,(r_i,r_j) | \Psi_n{}^{(1)} \rangle$$

$$\cdot\, \langle \Psi_n{}^{(1)} | \sum \mu_z(i) | \Psi_0 \rangle$$

Let us separate off the dependence on the wave function ψ for the particles in the j, m_j orbital as follows:

$$\Psi_0 = \frac{1}{\sqrt{A+1}} \mathcal{A} [\psi(j, m = j) \Phi_0]$$

$$\Psi_n^{(1)} = \frac{1}{\sqrt{A+1}} \mathcal{A} [\psi (j,m = j) \Phi_n]$$

where Φ_0 and Φ_n are the wave functions for the rest of system. Substituting in the above expression for $\delta\mu$ and *neglecting exchange* terms yields

$$\delta\mu_z = \langle \Psi_0 | \sum_i \delta\mu_z(i) | \Psi_0 \rangle \tag{11.24}$$

where

$$\delta\mu_z(i) = T^{(10)1}(i)$$

$$\sum_{ns'k'}\sum_{j\neq i} \frac{1}{\Delta E_n} \langle \Phi_0 | T^{(s'k')1}(j) v_{10,s'k'1}(r_i, r_j) | \Phi_n \rangle \langle \Psi_n^{(1)} | \sum \mu_z(i) | \Psi_0 \rangle$$

$$+ T^{(12)1}(i) \cdot \sum_{ns'k'}\sum_{j\neq i} \frac{1}{\Delta E_n} \langle \Phi_0 | T^{(s'k')1}(j) v_{12,s'k'1}(r_i,r_j) | \Phi_n \rangle \langle \Psi_n^{(1)} | \sum \mu_z(i) | \Psi_0 \rangle$$

Noting that $T^{(10)1} \sim \mathbf{s}$ and $T^{(12)1} \sim [i^2 Y_2(i)\mathbf{s}]^{(1)}$ this expression is more compactly written*:

$$\delta\boldsymbol{\mu}(i) = \delta\hat{g}_s(i)\, \mathbf{s} + \delta\hat{g}_p(i)\, [i^2 Y_2(i)\, \mathbf{s}(i)]^1 \tag{11.25}$$

where $\delta\hat{g}_s(i)$ and $\delta\hat{g}_p(i)$ can be obtained from the sums multiplying $T^{(10)1}(i)$ and $T^{(12)1}(i)$, respectively. Note that both of these are functions of r_i. This result gives the induced magnetic moment that arises from the interaction of the nucleons in the valence orbitals with the rest of the nucleus. It is easy to verify that both the spin-spin and tensor potentials in the residual interaction will contribute. Presuming the accuracy of the shell-model description of the nucleus near closed shells, some aspects of the nature of these forces can be inferred from the nuclear magnetic moments.

For recent results and a critical analysis of the concept of the effective magnetic moment operator, the reader should turn to Talmi (70) and Bhattacherjee (71). A very rough qualitative comparison with experiment is made by Bodenstedt and Rogers (64). Letting δg_s and δg_p be independent of the radius coordinate and giving δg_s a magnitude $= -(1/2)\, g_s$ and adjust δg_p so that $\delta\mu$ is zero for $p_{1/2}$ nuclei, values of $(g_s)_{eff} \equiv g_s + 2\delta\mu$ were calculated. Their comparison with experiment is shown in Fig. 11.17. Although the

---

*Note: $[Y_2(\hat{r})\mathbf{s}]^{(1)} \sim 3\,(\mathbf{s}\cdot\hat{r})\hat{r} - \mathbf{s}$.

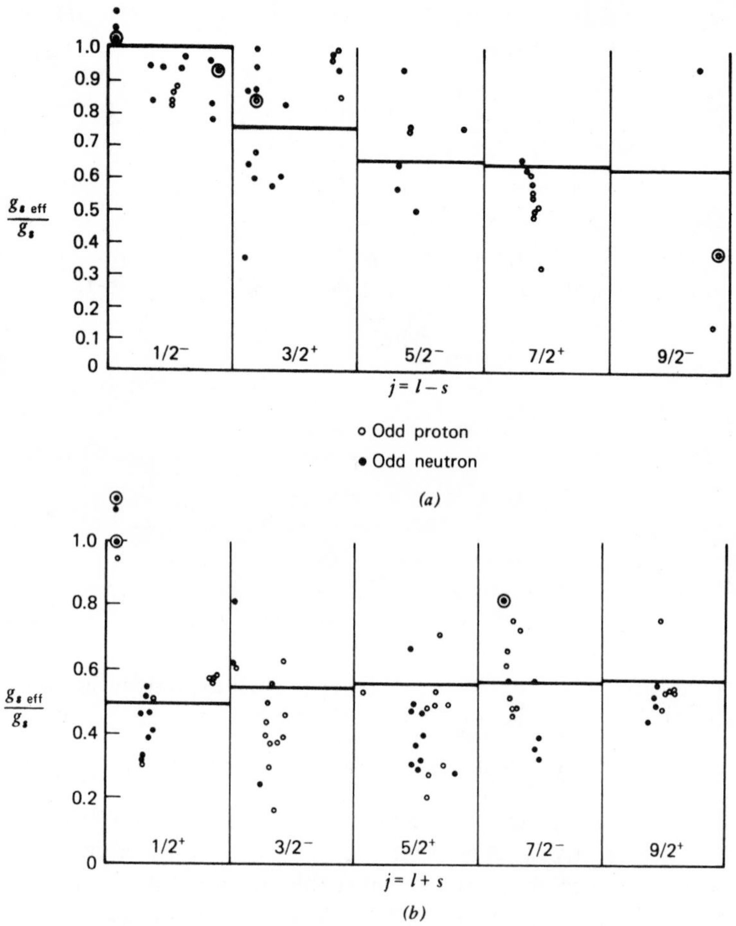

FIG. 11.17. Effective g_s values as function of j for $j = l + (1/2)$ [From Bodenstedt et. al. (64)].

individual values of the magnetic moments are of course not obtained with this crude approximation, one sees an improvement over the simple theory.

Neither the effective moment operator representation above or the more refined perturbation analysis can be expected to give accurate results until more is learned about the giant $M1$ states, or less picturesquely about the distribution of the $M1$ strength associated with a transition to (or from) a particular level. Suppose in analogy to the $E1$ giant dipole, a giant $M1$ state is built on a first approximation by operating with the magnetic dipole operator $\boldsymbol{\mu}$ on Ψ_0. Then as follows from (11.24) the principle contribution to $\delta\hat{\mu}_z$ will come from this $M1$ state. Clearly the presence or absence of such a concentration of a magnetic dipole strength and the size of the energy gap involved would make a large difference.

By the same token it is also clear that any core state Φ_n that couples preferentially to Φ_0 will be of particular importance. In the case that there is one such state (or perhaps a few) it becomes possible to construct a special model, the *core-excitation* model [deShalit (61)]. If this core state is collective its effect can be expected to be especially significant.

12. THE CORE EXCITATION MODEL

The model will be described for an odd nucleus. The odd nucleon is pictured as coupled to a core whose properties are presumed to be very much similar if not identical to the neighboring even-even nucleus. In a familiar example [Braunstein (Gal) and deShalit (62), deShalit (65), McKinley and Rinard (66)] ^{197}Au can be thought of as $d_{3/2}$ proton coupled to a core whose properties are close to those of the nucleus ^{196}Pt. The ground state of the nucleus consists then of the core in the ground state coupled to the valence $d_{3/2}$ proton. Excited states of the nucleus can be made up either of the core in the ground state and the valence nucleon excited to another orbit or with the valence nucleon unchanged but with the core excited. The second alternative is most likely when the core-excitation energy is considerably less than the energy change involved in the single-particle excitation and if the excited core state has special properties, such as collectivity, which lead to a larger matrix element for core excitation than for single-particle excitation. These matrix elements must not be too large, for then the neglect of other possible configurations will not be possible. And in addition it would be very difficult to identify those states that consist of the excited core plus valence nucleon. If the spin of the excited core state is $J_c{}^*$ and that of the valence nucleon is j, then states with spins J that lie between $|j - J_c{}^*|$ and $j + J_c{}^*$ will occur. In the absence of an interaction between the core and valence particle, these states will be degenerate. The presence of this interaction will lift the degeneracy, giving rise to $2J_c{}^* + 1$ levels if $j > J_c{}^*$ or $2j + 1$ if $j < J_c{}^*$. As an example, consider ^{197}Au once more. In this case $J_c{}^* = 2^+$ and $j = 3/2^+$, giving rise to levels with spins 7/2, 5/2, 3/2, and 1/2. These are shown in Fig. 12.1 as presented by McGowan, Milner, Robinson and Stelson (71). Including the electromagnetic properties of the ground state, μ and Q and the Q of the core nucleus ^{196}Pt, there are sixteen measured quantities listed in Table 12.1. As we shall see, the core-excitation model expresses these data in terms of five parameters. It is thus possible to test the model.

It will be useful to use the ^{197}Au case as a guide. Here we note that both the ground state and one of the excited states have a spin of 3/2. These states will generally consist of a mixture of the states formed by coupling the valence nucleon with the core in the ground state (spin $= J_c$) and with the core in an excited state (spin $= J_c{}^*$). Thus for $J_c = 0$, $J_c{}^* = 2$, $j = 3/2$, we obtain

$$\text{Ground state} = |3/2\rangle = A|0\;3/2\;3/2\rangle + \sqrt{1 - A^2}|2\;3/2\;3/2\rangle \tag{12.1}$$

$$\text{Excited state} = |(3/2)_2\rangle = -\sqrt{1 - A^2}|0\;3/2\;3/2\rangle + A|2\;3/2\;3/2\rangle \tag{12.2}$$

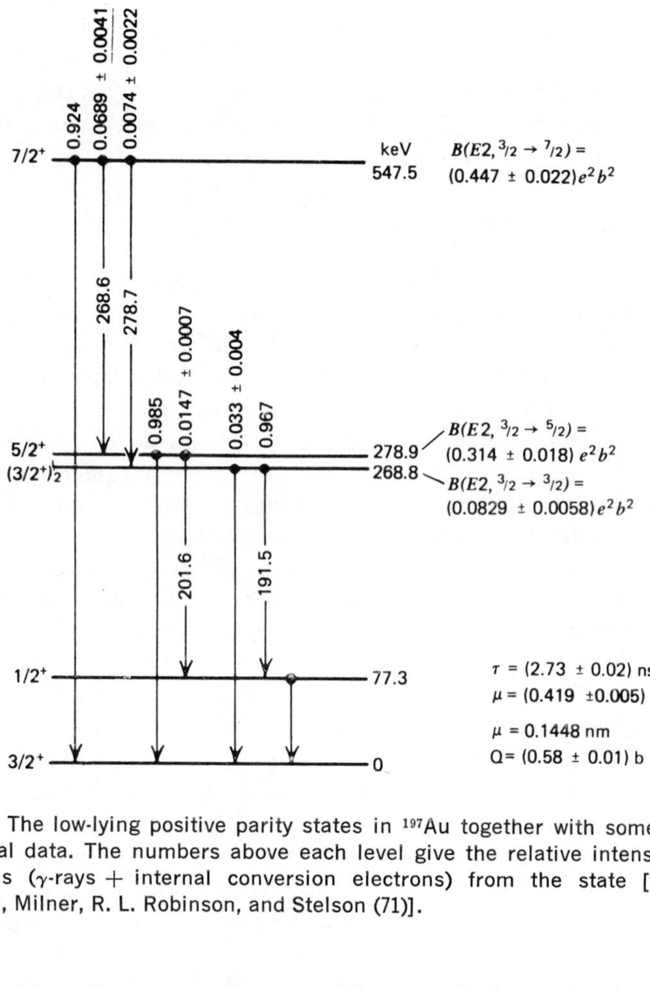

FIG. 12.1. The low-lying positive parity states in ^{197}Au together with some of the experimental data. The numbers above each level give the relative intensities of the transitions (γ-rays + internal conversion electrons) from the state [taken from McGowan, Milner, R. L. Robinson, and Stelson (71)].

The notation of the kets is $|J_c\, j\, J\rangle$, that is, the spin of the core state, of the valence particle, of the total spin in that order. The factor A is a constant that is to be determined empirically or in principle from the particle-core coupling interaction. The other excited states have the simple form

$$|\frac{1}{2}\rangle = |2\ 3/2\ 1/2\rangle \tag{12.3}$$

$$|\frac{5}{2}\rangle = |2\ 3/2\ 5/2\rangle \tag{12.4}$$

$$|\frac{7}{2}\rangle = |2\ 3/2\ 7/2\rangle \tag{12.5}$$

TABLE 12.1 Comparison between the Experimental Results and the Core-Excitation Model Predictions Relevant to Five States in ^{197}Au[a]

Quantity	Experimental Value	Core-Excitation Model Predictions $A^2 = 1$	$A^2 = .969$
$B(E2, 7/2 \rightarrow 3/2)$	32.8 ± 1.6	$[32.8 \pm 1.6]$	$[32.8 \pm 1.6]$
$B(E2, 7/2 \rightarrow (3/2)_2)$	6.9 ± 2.1	9.6 ± 1.7	4.5 ± 1.2
$B(E2, 7/2 \rightarrow 5/2)$	$.18 \pm .07$	$.26 \pm .67$	0.7 ± 1.1
$B(M1, 7/2 \rightarrow 5/2)$	$(2.1 \pm .2) \times 10^{-2}$	$(2.9 \pm .1) \times 10^{-2}$	$(2.9 \pm .1) \times 10^{-2}$
$B(E2, 5/2 \rightarrow)3/2)$	30.7 ± 1.7	32.8 ± 1.6	$[30.7 \pm 1.7]$
$B(M1, 5/2 \rightarrow 3/2)$	$(7.1 \pm 1.5) \times 10^{-2}$	0	$(1.5 \pm .2) \times 10^{-3}$
$B(E2, 5/2 \rightarrow 1/2)$	16.8 ± 3.2	$[16.8 \pm 3.2]$	$[16.8 \pm 3.2]$
$B(E2, (3/2)_2 \rightarrow 3/2)$	$12.2 \pm .8$	32.8 ± 1.6	$[12.2 \pm .8]$
$B(M1, (3/2)_2 \rightarrow 3/2)$	$< 3.7 \times 10^{-4}$	0	$(2.5 \pm .3) \times 10^{-3}$
$B(E2, (3/2)_2 \rightarrow 1/2)$	20 ± 4	17.2 ± 5.7	10.8 ± 4.3
	$9.6^{+7.2}_{-5.5}$		
$B(M1, (3/2)_2 \rightarrow 1/2)$	$(1.8 \pm .3) \times 10^{-1}$	$(5.1 \pm .2) \times 10^{-2}$	$(4.9 \pm .3) \times 10^{-2}$
	$(1.9 \pm .3) \times 10^{-1}$		
$B(E2, 1/2 \rightarrow 3/2)$	31.4 ± 3.5	32.8 ± 1.6	38.1 ± 3.1
$B(M1, 1/2 \rightarrow 3/2)$	$(7.8 \pm .2) \times 10^{-3}$	0	$(3.2 \pm .4) \times 10^{-3}$
$\mu(1/2)$	$.419 \pm .005$ nm	$[.419 \pm .005$ nm$]$	$[.419 \pm .005$ nm$]$
$\mu\ 3/2$	.1448 nm	.125 nm	.139 nm
$Q(3/2)$	$.58 \pm .01$b	$[.58 \pm .01$b$]$	$.79 \pm .05$b
$Q(2+)$ ^{196}Pt	$.58 \pm .18$b	$.65 \pm .23$b	$.60 \pm .20$b

[a] The $B(E2)$ and $B(M1)$ values are given in units of single-particle $B(E2)$ and (nuclear magnetons)^2, respectively. The values enclosed by brackets were used to adjust the parameters of the model. In some cases values obtained by different experimental groups differ. Both values are quoted. Taken from McGowan et al. (71).

From this example it is clear that we achieve sufficient generality by considering two states of the nucleus that may radiate to one another as having the forms

$$|J\rangle = A|J_c\, j\, J\rangle + \sqrt{1 - A^2}|J_c^*\, j\, J\rangle$$

and (12.6)

$$|J'\rangle = -\sqrt{1 - A^2}|J_c\, j\, J'\rangle + A|J_c^*\, j\, J'\rangle$$

We may now compute the various electromagnetic properties of the levels with the assumption that the reduced matrix element of the multipole Q_L can be split into two components, one involving the core $Q_L^{(c)}$, the other the particle $Q_L^{(p)}$:

$$(J'\|Q_L\|J) = (J'\|Q_L^{(c)}\|J) + (J'\|Q_L^{(p)}\|J) \tag{12.7}$$

Inserting (12.6) yields

$$(J'\|Q_L\|J) = A^2\,(J_c^*\, j\, J'\|Q_L\|J_c\, j\, J) + A\sqrt{1 - A^2}\,[(J_c^*\, j\, J'\|Q_L\|J_c^*\, j\, j) - (J_c\, j\, J'\|Q_L\|J_c\, j\, J)] - (1 - A^2)\,(J_c\, j\, J'\|Q_L\|J_c^*\, j\, J) \tag{12.8}$$

Each of the reduced matrix elements can now be reduced to matrix elements involving the core or particle separately by making use of (12.7) and the relations in the Appendix (Eq. A 2.55). We obtain

$$(J_c^*\, j\, J'\|Q_L\|J_c\, j\, J) = (-)^{J_c^*+j+J+L}\sqrt{(2J + 1)(2J' + 1)}\begin{Bmatrix} J_c^* & J' & j \\ J & J_c & L \end{Bmatrix}(J_c^*\|Q_L^{(c)}\|J_c) + (-)^{J_c^*+j+J'+L}\sqrt{(2j + 1)(2J' + 1)}\begin{Bmatrix} j & J' & J_c^* \\ J & j & L \end{Bmatrix}(j\|Q_L^{(p)}\|j)\,\delta(J_c, J_c^*) \tag{12.9}$$

Using these results it becomes possible to calculate the radiative transition probabilities in terms of the parameters $(J_c^*\|Q_L^{(c)}\|J_c)$ and $(j\|Q_L^{(p)}\|j)$. The The magnetic moment of the state with angular momentum J is

$$\mu\,(J) = A^2\,\langle J_c\, j\, J|\,Q_{1z}|J_c\, j\, J\rangle + (1 - A^2)\,\langle J_c^*\, j\, J|\,Q_{1z}|J_c^*\, j\, J\rangle \tag{12.10}$$

where Q_1 is the magnetic moment multipole. The matrix elements are given by

$$\langle J_c\, j\, J|\,Q_{1z}|J_c\, j\, J\rangle = \frac{J(J + 1) + J_c\,(J_c + 1) - j(j + 1)}{2(J + 1)}\,g_c(J_c) + \frac{J(J + 1) + j(j + 1) - J_c\,(J_c + 1)}{2(J + 1)}\,g_p(j) \tag{12.11}$$

This result may be related to the reduced matrix element of Q_1 by the following equation giving the g factor for the core or particle state with angular momentum $\mathcal{J}$:

$$\mu(\mathcal{J}) = g\mathcal{J} = \left(\frac{\mathcal{J}}{(\mathcal{J}+1)(2\mathcal{J}+1)}\right)^{1/2} (\mathcal{J}\| Q_1 \|\mathcal{J}) \qquad (12.12)$$

Finally we note that the quadrupole moment Q is related to the reduced matrix elements of Q_2 by the following equation

$$Q = \left(J\, M_J = J \middle| \sqrt{\frac{16\pi}{5}}\, Q_{20} \middle| J\, M_J = J\right) = \sqrt{\frac{16\pi}{5}} \begin{pmatrix} J & 2 & J \\ -J & 0 & J \end{pmatrix} (J\| Q_2 \|J)$$

or

$$Q = \sqrt{\frac{64\pi}{5}} \frac{J(2J-1)}{\sqrt{(2J-1)\, 2J\, (2J+1)(2J+2)(2J+3)}} (J\| Q_2 \|J) \qquad (12.13)$$

The reduced matrix element can then be expressed in terms of $(J_c^*\| Q_2^{(c)} \|J_c)$ etc. employing (12.8) and (12.9).

Returning to the ^{197}Au case, it is seen that the transition probabilities as well as the static electromagnetic properties of the ground and excited states can be expressed in terms of the following reduced matrix elements:

$$(j\| Q_1^{(p)} \| j),\ (j\| Q_2^{(p)} \|j),\ (J_c^*\| Q_1^{(c)} \|J_c^*),\ (J_c^*\| Q_2^{(c)} \|J_c^*)$$
$$\text{and} \qquad (J_c^*\| Q_2^{(c)} \|J_c) \qquad (12.14)$$

Note that $J_c = 0$ for this case and for that reason the other possible reduced matrix elements are zero. The first and third of (12.14) may according to (12.11) be replaced by g_p and g_c, the g factors for the valence nucleon and the excited core state, respectively. These then are the five phenomenological parameters of the model and are adjusted to fit the experimental results listed in Table 12.1 of McGowan, Milner, Robinson, and Stelson (71). A sixth parameter is the value of A, which should not differ too greatly from unity. In Table 12.1 the values in the square brackets were fitted by choosing the parameters appropriately, g_p was taken to have the Schmidt value of 0.083 μ_0 and the remaining values computed. In the $A^2 = 1$ limit, $B(M1)$ transitions from any of the excited states to the ground state are forbidden and the $B(E2)$ transitions to the ground state will have equal values. The most glaring discrepancy in that limit is the $B(E2)$ for the $(3/2)_2 \rightarrow 3/2$ transition for which the prediction is much too large. This can however be fitted by making a small adjustment of the value of A as is done in the last column of the table. Note that then A is close to unity, indicating the sensitivity of some of the transition probabilities to the details of the wave function. Generally good agreement is obtained for this very simple model. The only large discrepancy is the $B(M1)$ for the $(5/2) \to 3/2$ transition where the predicted value is 1/50 of the experimental value. The reader is referred to Braunstein and deShalit (62) for that small generalization (essentially bringing in the single-particle excitation to the $d_{5/2}$ level) needed for a cure. As a final test the parameters involving the core should be compared with their values as obtained for the core nucleus ^{196}Pt. This comparison is made in the last line of Table 12.1 for the quadrupole moment of the 2^+ excited state. The agreement is striking. The $B(E2)$ for the $2^+ \to 0^+$ transition in ^{196}Pt is related to $(J_c^* \| Q_2^{(c)} \| J_c)$. The value obtained by fitting is 32.8 ± 1.6 single-particle units while the experimental value is 44.1 ± 1.5 single-particle units.

Discussion of a number of examples in which the core excitation model has been applied with particular reference to the predicted magnetic moments is given by Bhattarcharjee (71) in his Tables III and IV. Substantial agreement is obtained in many but by no means all cases. A deeper understanding of when the model is valid has yet to be obtained.

APPENDIX

The apparatus developed in Sections 4 and 5 is needed if exact expressions valid to all orders in (kR) and for arbitrary multipole orders are to be obtained. However, if we are interested in only the first few multipoles and then only to lowest order in (kR) and accept some minor handwaving, it is possible to obtain correct results in a simple fashion. Of course one can always go back to the main text for justification of any step. The transition probability per unit time for radiative decay as given by (1.1) and (5.1) is

$$dw_{fi} = \left(\frac{2\pi}{\hbar}\right)\frac{k^2}{\hbar c}\,|H_{fi}|^2\,d\Omega \tag{A.1}$$

where

$$H_{fi} = \langle f|\ -\frac{1}{c}\int \mathbf{j}\cdot\mathbf{A}\,d\mathbf{r}\,|i\rangle \tag{A.2}$$

The vector potential $\mathbf{A}$ for emission is proportional to $(2\pi)^{-3/2}\,\hat{\boldsymbol{\varepsilon}}\,e^{-i\mathbf{k}\cdot\mathbf{r}}$ where $\hat{\boldsymbol{\varepsilon}}$ is the polarization and $\mathbf{k}$ is the direction of propagation of the emitted photon. The factor $(2\pi)^{-3/2}$ is the plane-wave normalization corresponding to the density of states $(k^2/\hbar c)$ in (A.1). For absorption the complex conjugate is used. The constant of proportionality is determined by the requirement that

$$(2\pi)^3\left|\frac{\mathbf{E}^2}{4\pi}\right| = \frac{(2\pi)^3\omega^2|\mathbf{A}|^2}{4\pi\,c^2} = \frac{\hbar\omega}{2}$$

or

$$|A|^2 = \frac{\hbar c^2}{(2\pi)^2\omega}$$

so that

$$\mathbf{A} = \sqrt{\frac{\hbar c^2}{(2\pi)^2\omega}}\,\hat{\boldsymbol{\varepsilon}}\,e^{-i\mathbf{k}\cdot\mathbf{r}}\ \text{(emission)} \tag{A.3}$$

Hence

$$H_{fi} = -\sqrt{\frac{\hbar}{(2\pi)^2\omega}}\,\langle f|\int \hat{\boldsymbol{\varepsilon}}\cdot\mathbf{j}\,e^{-i\mathbf{k}\cdot\mathbf{r}}\,d\mathbf{r}\ |i\rangle \tag{A.4}$$

The various multipoles (up to lowest order in kR) can be obtained by expanding the exponential yielding

$$\langle f|\int \hat{\boldsymbol{\varepsilon}}\cdot\mathbf{j}\,e^{-i\mathbf{k}\cdot\mathbf{r}}\,d\mathbf{r}\ |i\rangle = \langle f|\int \hat{\boldsymbol{\varepsilon}}\cdot\mathbf{j}\,d\mathbf{r}\ |f\rangle - i\,\langle f|\int(\hat{\boldsymbol{\varepsilon}}\cdot\mathbf{j})\,(\mathbf{k}\cdot\mathbf{r})\,d\mathbf{r}\ |i\rangle + \ldots \tag{A.5}$$

Replacing $\mathbf{j}$ by $\rho \mathbf{v}_i$ (see Eq. 2.3, point nucleons are assumed), the subscript numbering the particles, and $\mathbf{v}_i$ by

$$\mathbf{v}_i = \frac{i}{\hbar}[H, \mathbf{r}_i]$$

where H is the nuclear Hamiltonian, the first term becomes

$$\langle f| \int \hat{\boldsymbol{\varepsilon}} \cdot \mathbf{j}\, d\mathbf{r}\, |i\rangle = \hat{\boldsymbol{\varepsilon}} \cdot \langle f| \int \mathbf{v}\, \rho d\mathbf{r}\, |i\rangle = -i\omega\, \hat{\boldsymbol{\varepsilon}} \cdot \langle f | \int \mathbf{r}\, \rho\, d\mathbf{r}\, |i\rangle$$

Introducing the electric dipole moment by

$$\mathbf{D} = \int \rho\, \mathbf{r}\, d\mathbf{r} \tag{A.6}$$

we finally obtain

$$\langle f| \int \hat{\boldsymbol{\varepsilon}} \cdot \mathbf{j}\, d\mathbf{r}\, |i\rangle = -i\, \omega \hat{\boldsymbol{\varepsilon}} \cdot \mathbf{D}_{fi} \tag{A.7}$$

The second term of (A.5) can be broken up into a symmetric and antisymmetric part. The latter involves

$$\frac{1}{2} \hat{\boldsymbol{\varepsilon}} \cdot (\mathbf{v}\, \mathbf{r} - \mathbf{r}\, \mathbf{v}) \cdot \mathbf{k} = \frac{1}{2m} \hat{\boldsymbol{\varepsilon}} \cdot (\mathbf{p}\, \mathbf{r} - \mathbf{r}\, \mathbf{p}) \cdot \mathbf{k}$$

Introducing the vector cross product this term becomes

$$-\frac{\hbar}{2m} (\hat{\boldsymbol{\varepsilon}} \times \mathbf{k}) \cdot \mathbf{L} = \frac{\hbar\omega}{2mc} (\hat{\mathbf{k}} \times \hat{\boldsymbol{\varepsilon}}) \cdot \mathbf{L} \tag{A.8}$$

where $\hbar\mathbf{L}$ is just $(\mathbf{r} \times \mathbf{p})$.

The contribution to (A.5) is thus

$$-\frac{i\hbar\omega}{2mc} (\mathbf{k} \times \hat{\boldsymbol{\varepsilon}}) \cdot \langle f| \int d\mathbf{r}\, \mathbf{L}\, \rho |i\rangle \tag{A.9}$$

This term can also be written:

$$-i\, \omega\, (\hat{\mathbf{k}} \times \hat{\boldsymbol{\varepsilon}}) \cdot \boldsymbol{\mu}_{fi} \tag{A.10}$$

where $\boldsymbol{\mu}$ the magnetic moment operator. Equation A.9 contains only the contribution of the orbital motion of the particles. To obtain the complete expression we must add the effect of the intrinsic magnetization (see Eq. 2.4). This amounts to adding $g\mathbf{s}$ to $\mathbf{L}$. The final expression is given by (11.5) for point nucleons. In comparing (A.10) and (A.7), note that $\hat{\boldsymbol{\varepsilon}}$ is in the direction of the electric field, and $\hat{\mathbf{k}} \times \hat{\boldsymbol{\varepsilon}}$ is in the direction of the magnetic field. Hence one can obtain (A.10) from (A.7) by replacing the electric field and the electric dipole moment by the magnetic field and the magnetic dipole moment.

The remaining symmetric term is

$$\frac{1}{2}(\mathbf{r}_i\mathbf{v}_i + \mathbf{v}_i\mathbf{r}_i) = \frac{i}{2\hbar}\,\mathbf{r}_i\,[H, \mathbf{r}_i] + [H, \mathbf{r}_i]\,\mathbf{r}_i = \frac{i}{2\hbar}\,[H, \mathbf{r}_i\mathbf{r}_i]$$

Hence the matrix element yields

$$-\frac{i\omega}{2}\,\hat{\boldsymbol{\varepsilon}}\cdot\,\langle f|\int d\mathbf{r}\,\rho\,\mathbf{r}\mathbf{r}\,|i\rangle\cdot\hat{\mathbf{k}}$$

Since $\boldsymbol{\varepsilon}\cdot\mathbf{k} = 0$, this expression can be rewritten

$$-\,i\frac{\omega}{6}\,\hat{\boldsymbol{\varepsilon}}\cdot\,\langle f|\,\overleftrightarrow{Q}\,|i\rangle\,\cdot\hat{\mathbf{k}} \tag{A.11}$$

where $\overleftrightarrow{Q}$, the quadrupole moment tensor, has the form

$$Q_{ij} = 3\int d\mathbf{r}\,\rho\left(r_i r_j - \frac{1}{3}\,\delta_{ij}\,r^2\right) \tag{A.12}$$

the subscript giving the components of $\mathbf{r} = (r_1, r_2, r_3) = (x, y, z)$. Collecting all the terms we obtain

$$H_{fi} = i\sqrt{\frac{\hbar\omega}{(2\pi)^2}}\left(\hat{\boldsymbol{\varepsilon}}\cdot\mathbf{D}_{fi} + (\hat{\mathbf{k}}\times\hat{\boldsymbol{\varepsilon}})\cdot\boldsymbol{\mu}_{fi} - \frac{i}{6}\,\hat{\boldsymbol{\varepsilon}}\cdot(\overleftrightarrow{Q}_{fi})\cdot\hat{\mathbf{k}} + \ldots\right) \tag{A.13}$$

Inserting (A.13) into (A.1) yields the expressions obtained in the main body of the text.

CHAPTER IX
THE WEAK INTERACTION

1. INTRODUCTION

In even a casual examination of the history of modern physics, one is struck by the key role played by beta decay of nuclei. The discovery of natural radioactivity and the associated phenomena of β-decay as well as α- and γ-decay signaled the start of nuclear physics. The study of β-decay eventually led to the prediction of the existence of a massless, neutral, weakly interacting particle (the neutrino) [Pauli (33), Wu (60)]. The evidence for the neutrino relied on the argument that with such a particle it was possible to conserve energy and linear and angular momentum in the β-decay of nuclei [Enge (66)]. But as important as these traditional arguments was the realization that the energy distribution of the emitted electrons was that predicted for a three-particle final state (modified by the Coulomb interaction of the electron with the residual nucleus), in which only one of the particles is observed, thus providing direct evidence for a third particle in addition to the electron and residual nucleus. In order to incorporate these features Fermi formulated a field theory for β-decay [Fermi (33,34)]. Up to that time there had been only one relativistic quantum field theory, quantum electrodynamics, which was very solidly based on a deep understanding of the classical limit. The Fermi theory had no such basis. It was constructed as it were out of whole cloth with the single purpose of explaining the observed phenomena. It was the first such field theory and served as a model for other field theories developed later to describe interactions among the elementary particles.

The Fermi theory assumes that the β-decay of a nucleus proceeds via the β-decay of one of the nucleons making up the nucleus. It thus predicted the decay of the neutron. This prediction was later verified. The measured mean life of the neutron is $(0.935 \pm 0.014) \times 10^2$ sec.

The neutron was the first of the "elementary particles" that was discovered to decay with a relatively long lifetime, indicating the approximately stable nature of these particles or equivalently the weakness of the interaction producing the decay. The reader is referred to Table I.10.1, which lists some of these decays and to the much more complete table given by the Particle Data

Group (72) in which the particles of interest are listed under "Stable Particle Table." In this chapter we shall be mostly concerned with those particles whose decay involves the emission of at least one neutrino. Examples include the decay of the muon into an electron and two neutrinos, the decay of the pion into a muon and a neutrino, as well as its decay into an electron plus a neutrino. These decays are said to involve leptons—a term used to designate the muon, the electron, and the neutrino. Many of the nearly stable particles can decay by emitting pions. For example, the lamda (Λ) can decay into a negative pion and a proton, or a neutral pion and a neutron. However it can also β-decay like the neutron, by emitting an electron and neutrino or by emitting a muon and a neutrino. The neutral kaon can decay into two or three pions according to whether it is the short-lived kaon $K^{\circ}{}_{S}$ or the long-lived variety $K^{\circ}{}_{L}$. The latter can also decay by emitting a pion as well as an electron (or muon) and a neutrino. The charged kaons decay like the pions into a muon (or electron) and a neutrino.

The theory formulated by Fermi for nuclear β-decay (some simple generalizations are sometimes needed) has in very large measure been found adequate for the description of these decays. Among the first to be carefully examined was the decay of the muon. Not only was the Fermi theory found to be directly applicable but most significantly the strength of this interaction was very close to that required for the decay of the neutron, the very considerable difference in lifetimes (the muon half-life is $(2.1994 \pm 0.0006) \times 10^{-6}$ sec.) being almost completely accounted for by the difference in the energy released in the decay which is 0.78 MeV for the neutron decay and about 100 MeV for the muon decay. This near identity that was repeated when other weak decays were investigated led to the suggestion that there is a universal weak interaction that explained all the relatively long-lived decays of elementary particles, differences in the actual rates being a consequence of simple kinematical effects. Thus when the two- and three-pion decay mode of the kaon seemed to require the nonconservation of parity in the weak interactions, it was natural to turn to a nuclear β-decay for verification [Lee and Yang (56)]. In fact, parity nonconservation was observed first in the classic ^{60}Co experiment of Wu, Ambler, Hayward, Hopper, and Hudson (57). In this experiment the ^{60}Co ($J = 5$) nucleus was polarized. It was observed that the electrons from β-decay are preferentially emitted in a direction opposite to the direction of the nuclear spin. In other words the expectation value of $\mathbf{J} \cdot \mathbf{p}$ where $\mathbf{J}$ is the nuclear spin and $\mathbf{p}$ the electron momentum differs from zero and in this case is negative. Since $\mathbf{J}$ is even under reflection and $\mathbf{p}$ is odd, this expectation value can differ from zero only if parity is not conserved. A similar test in the $\pi \rightarrow \mu + e$ decay and the subsequent μ decay demonstrated the nonconservation of parity that would be expected from the universality of the weak interaction [Garwin, Lederman, and Weinrich (57); Friedman and Telegdi (57)].

During the next few years many researches directed toward the dilineation of the weak interaction were performed and will be reported later in this chapter. But equally as important was the immediate challenge to other conservation principles. It was found that charge conjugation invariance* was also violated in the weak interactions. On the other hand the only violation of time reversal invariance** and of invariance under the combined operation of charge conjugation and parity inversion found to date has been in the conversion of the short-lived neutral kaon (which decays into two pions) into the long-lived variety (which decays into three). No evidence for this phenomenon has been found in nuclear decays or for that matter in any other decay.

One result of importance was the discovery of two kinds of neutrinos. The neutrino emitted in nuclear β-decay is designated by ν_e while the neutrino emitted in pion decay, the μ-neutrino

$$\pi^- \to \mu^- + \bar{\nu}_\mu$$

is indicated by the subscript μ. This additional neutrino is presumably a not-as-yet understood clue regarding the differing natures of the electron and muon. At first glance, the muon appears to be just a more massive electron. Its interaction with the electromagnetic field differs in no essential way from that of the electron. In β-decay when it is energetically possible both electron and muon β-decay will occur. Nevertheless in spite of its greater mass, and its otherwise apparent identity with the electron, it does not decay into the latter by gamma-ray emission. This fact together with the fact that the neutrino accompanying decay by μ emission differs from the neutrino accompanying electron β-decay has suggested that there is some qualitative property of the muon that is not possessed by the electron and vice versa. This concept has been formalized by the introduction of the lepton quantum number, the electron and muon, and their associated neutrinos having different lepton numbers. This idea will be discussed in Section 12. For the present, it will suffice to say that these lepton numbers are chosen so that conservation of lepton numbers in a decay leads to muon emission being accompanied by ν_μ, electron emission by ν_e. This is of course not a "fundamental" explanation but instead a way of summarizing one interpretation of the experimental results.

As it evolved from investigations into the weak interactions the natural variables entering into the description of the weak interaction are the currents of the various interacting fields. In the case of the electromagnetic interaction

---

*For a discussion of invariance principles see Wick (58).

**See Appendix A (Sec. 3) for a discussion of this invariance.

this is a familiar description, the interaction of two charged fields being given by

$$\sum_{\kappa\lambda} \iint j_\kappa^{(1)} (\mathbf{r})\, G_{\kappa\lambda} (\mathbf{r}, \mathbf{r}')\, j_\lambda^{(2)} (\mathbf{r}')\, d\mathbf{r}\, d\mathbf{r}'$$

where $j_\mu^{(1)}$ is the current density of the first field. $G_{\kappa\lambda}$ is the tensor Green's function (Morse and Feshbach (53), Chapter XIII). It is the propagator for the photons that couple the two charged fields. A similar description is attempted for the weakly interacting fields, with however the important difference that two sorts of currents are required. The electromagnetic current is a polar four vector. But for weak interactions an axial vector designated by $J_\kappa^{(A)}$ (the polar vector by $J_\kappa^{(V)}$) is required in addition in order to account for the two types of nuclear transitions, the Fermi and the Gamow-Teller, and to permit the description of parity nonconservation. A far-reaching and profound analogy was found to exist between the polar vector current of the weak interaction and the isospin vector component of the nucleon electromagnetic current. This analogy suggests among other things that the weak four-vector current $J_\kappa^{(V)}$ is conserved:

$$\sum_\kappa \frac{\partial J_\kappa^{(V)}}{\partial x_\kappa} = 0 \tag{1.1}$$

This hypothesis known as the conserved vector current (CVC) theory has thus far been borne out by experiment; a particularly important test occurring in the β-decays of ^{12}B and ^{12}N into ^{12}C will be discussed later. An extension of the hypothesis (1.1) to the axial current is known as the partially conserved axial vector current (PCAC) theory. The PCAC hypothesis furnishes a bridge between the strong and weak interactions as exemplified by the Goldberger-Treiman (58) relation connecting the amplitude for the decay of the pion ($\pi^+ \rightarrow \mu^+ + \nu_\mu$) that proceeds via the axial vector current, the Gamow-Teller coupling constant that originates in the axial vector contribution to the neutron decay, and the coupling constant describing the strong interaction between nucleons and pions. These relations offer significant implications for nuclear physics, which we shall discuss later on in this chapter.

The question naturally arises as to whether there are particles analogous to the photons that "carry" the weak interaction. In the absence of such particles (or at least the corresponding propagator analogous to the photon $G_{\kappa\lambda}$), the weak interactions become point interactions, that is, interactions of zero range with a consequent set of divergences and failures of the theory at high energies. The experimental search for these particles has been like the quest for the Holy Grail. At this writing these particles have not yet been observed.*

We have emphasized thus far the development of our understanding of weak interactions and the role played by nuclear studies. However, one should not lose sight of the importance of the β-decay interaction for studies of nuclear structure. We have alluded to this in the preceding chapter. The very weakness of the interaction makes possible the probing of the nucleus with a minimal disturbance of the system. On the other hand that same weakness makes the cross sections very small, and we are thus some distance away from the use of neutrino beams for the study of nuclear structure. Most nuclear information is at present obtained from the β-decay of radioactive nuclei as well as from the related absorption of muons by nuclei.

2. SIMPLE THEORY OF β-DECAY†

In this section, the simpler facts and the description of β-decay as they are currently understood will be briefly summarized. For more detail see a more elementary text like Enge (66): No attempt will be made to describe here the experiments from which that theory was induced.

Nuclear β-decay involves the transition:

$$\text{Electron emission: } (Z, N) \rightarrow (Z + 1, N - 1) + e^- + \bar{\nu}_e \tag{2.1a}$$

$$\text{Positron emission: } (Z, N) \rightarrow (Z - 1, N + 1) + e^+ + \nu_e \tag{2.1b}$$

$$\text{Atomic electron capture } (Z, N) + e^- \rightarrow (Z - 1, N + 1) + \nu_e \tag{2.1c}$$

The bar on ν_e in (2.1a) signifies that an antineutrino is emitted in this process while in (2.1b) a neutrino is emitted. Capture describes a process in which an atomic electron is captured by the atomic nucleus. If the electron is captured

---

*A recent theory of these particles that puts them and the photon on comparable footing, thus unifying the electromagnetic and weak interactions, has been proposed by Weinberg (67, 72). As expected the particles that transmit the weak interactions are vector (spin 1) particles, they are massive, and in Weinberg's formulation there are neutral as well as charged particles. One of the accomplishments of the theory is that it is renormalizable in spite of the presence of charged spin 1 fields.

†Because the energies of the leptons in β-decay are large compared with their rest mass, it is necessary to employ relativistic wave functions for them. These are described in the Appendix to this chapter.

from the K-shell it is referred to as K-capture. The elementary processes involved in (2.1a) and (2.1b) are thought to be

$$\begin{aligned} n &\rightarrow p + e^- + \bar{\nu}_e \\ p &\rightarrow n + e^+ + \nu_e \end{aligned} \tag{2.2}$$

Since the neutron is more massive than the proton only the first of these reactions occur in free space with observed mean life, of $(0.935 \pm 0.014) \times 10^3$ sec. [Particle Data Group (72)]. Positron decay can occur in heavier nuclei because of the coulomb energy gained in the replacement of a proton by a neutron. The energy release E_0 in electron decay is

$$E_0 = [M(Z, N) - M(Z + 1, N - 1)]\, c^2 \text{ (electron emission)} \tag{2.3}$$

where M is the *atomic* mass. Bear in mind that the recoiling residual nucleus will have a small but finite kinetic energy. In the case of positron emission

$$E_0 = [M(Z, N) - M(Z - 1, N + 1) - 2m]\, c^2 \tag{2.4}$$

where m is the electron mass. The energy balance in atomic electron capture gives the energy release

$$E_0 \cong [M(Z, N) - M(Z - 1, N + 1)]\, c^2 \tag{2.5}$$

Comparing (2.4) and (2.5) we see that there is a range of atomic mass differences for which atomic electron capture is possible but positron emission is not. K capture is always possible when positron decay occurs.

The energy released in β-decay may be as low as 20 KeV as in the decay of ^3H or as much as 20 MeV in the decay of ^8B. The important point for us here is the associated electron and neutrino wavelength. For the electron

$$\lambda\!\!\!^{-} = \frac{386}{\sqrt{\epsilon^2 - 1}} \text{ fm} \tag{2.6}$$

where ϵ is the *total* energy of the electron including the electron rest mass in units of the electron rest mass (0.511 MeV). An electron with kinetic energy of $2\,mc^2$ (1.022 MeV) has a wavelength/2π of 137 fm; $\lambda\!\!\!^{-}$ for a $40\,mc^2$ electron is 9.5 fm. Similar numbers are obtained for the neutrino wavelength/2π. We may thus conclude that the wavelength/2π of both the electron and neutrino are much larger than the radii of the residual nuclei so that to a good approximation the corresponding wave functions are constant over the nuclear volume. Finally we comment on the recoil energy of the residual nucleus. This is given by

$$E_R = \frac{1}{2MA}(\mathbf{p}_e + \mathbf{p}_\nu)^2$$

whose maximum in the limit of large electron energy is given by

$$(E_R)_{\max} = \frac{E_0^2}{2MAc^2} = \frac{\epsilon^2}{3672A} mc^2 \tag{2.7}$$

Thus for a $2mc^2$ beta ray and a medium weight nucleus ($A \sim 100$) E_R is about 13 eV. For the most part, this recoil energy will be neglected in what follows. However it is observable and, in fact, its observation is essential for the determination of the electron-neutrino angular correlation in β-decay (see Section 8).

Experimentally the electron neutrino mass has been found to be less than 60 eV [Particle Data Group (72). Thus the zero mass for the neutrino that is used throughout this chapter is consistent with experiment.

From Volume II, Chapter 12, the probability per unit time for emitting an electron with momentum between $\hbar\,\mathbf{k}_e$ and $\hbar(\mathbf{k}_e + d\mathbf{k}_e)$ and an antineutrino with a momentum between $\hbar\mathbf{k}_\nu$ and $\hbar(\mathbf{k}_\nu + d\mathbf{k}_\nu)$ is

$$dw_{fi} = \frac{2\pi}{\hbar} |H_{fi}|^2 \frac{d\,\mathbf{k}_e}{(2\pi)^3} \frac{d\,\mathbf{k}_\nu}{(2\pi)^3} \delta(E_0 - E_e - E_\nu) \tag{2.8}$$

Sums over the spins of the final states and averages over the initial spins must be performed. Since the interaction is weak, perturbation theory may be used. The operator H in (2.8) is then just the interaction Hamiltonian density. The electron and antineutrino wave functions are normalized to unit plane waves at infinity. Because of their wavelength these wave functions will be assumed to be constant over the nuclear volume. Since they appear only in the final state, $|H_{fi}|^2$ is proportional to the electron and antineutrino density, and these can be factored out of the matrix element H_{fi}. Summing over the electron and antineutrino spin, the antineutrino density relative to the unit density at infinite is also unity. The electron density relative to unit density at infinity will depend on the nuclear charge, the radius, and the electron energy. It will be written $\rho\,(Z, R, \epsilon)$. Once these density dependences are removed from $|H_{fi}|^2$ the remainder should depend principally upon the nuclear states involved in the transition. The delta function in (2.8) insures the conservation of energy. The recoil energy has been neglected.

To obtain the energy distribution of the electrons we integrate dw_{fi} over the antineutrino momentum and over the direction of motion of the electron. Spin and isospin sums over final nucleon states and averages over the initial nuclear states are also performed. One obtains

$$\frac{dw_{fi}}{d\epsilon} = \frac{mc^2}{\hbar} \frac{\Gamma^2}{2\pi^3} \rho\,(Z, R, \epsilon)\, |M_{fi}|^2 (\epsilon_0 - \epsilon)^2 \epsilon(\epsilon^2 - 1)^{1/2} \tag{2.9}$$

where $\epsilon_0 = (E_0/mc^2)$

Here we have placed

$$\overline{|H_{fi}|^2} \equiv g^2 \rho(Z, R, \epsilon) |M_{fi}|^2 \tag{2.10}$$

where the bar denotes the spin and isospin average and sum referred to above. We have assumed that $|M_{fi}|^2$ is dimensionless, independent of $\mathbf{k}_e$ and $\mathbf{k}_\nu$ and is defined (see 2.22) so that it has unit order of magnitude. The factor g in (2.10) then indicates, explicitly the size of the interaction. The constant Γ is

$$\Gamma \equiv \frac{g}{mc^2}\left(\frac{mc}{\hbar}\right)^2 \tag{2.11}$$

As can be seen from (2.9) it is Γ that is nondimensional and therefore more properly measures the size of the interaction. Empirically the time

$$\tau_0 \equiv \frac{\hbar}{mc^2}\frac{2\pi^3}{\Gamma^2} \tag{2.12}$$

is found to be about 9000 sec by fitting (2.9) to the decay of ^{14}O and ^{14}N. The corresponding value of dimensionless Γ is 3×10^{-12}. These numbers are only meant to set the scale. We shall give more accurate values later. The use of constant electron and neutrino densities is correct only for the so-called *allowed transition* that in the present case can be defined as those for which $|M_{fi}|^2$ differs from zero.

The electron-energy distribution obtained from (2.9) is illustrated by Fig. 2.1 and 2.2 taken from Wu and Moskowski (66), showing the effects of the Coulomb field and the dependence upon the total energy ϵ_0, which is identical with the end point of the spectrum. As the energy increases ($\epsilon_0 \gg 1$) the $Z = 0$ spectrum becomes more symmetrical, the symmetry about $\epsilon_0/2$ becoming exact as $\epsilon_0 \to \infty$. The Coulomb field increases the electron density at the nucleus and decreases the positron density since the electron is attracted while the positron is repelled by the nucleus. Since the attraction is more effective and therefore the density increase larger for the slower charged particles and less effective for the faster particle, the energy distribution will be skewed toward the lower energy for electrons and toward higher energy for positron spectra. This can be seen directly if we examine the explicit form for ρ. In the nonrelativistic case it is sufficient to take the value of the density for a point nucleus:

$$\rho(Z, R = 0, \epsilon) = \left|\frac{2\pi\eta}{1 - e^{-2\pi\eta}}\right| \quad \text{nonrelativistic} \tag{2.13}$$

where

$$\eta = \pm\frac{ZE_e}{137\, cp_e} \quad \text{for} \quad e^{(\mp)} \tag{2.14}$$

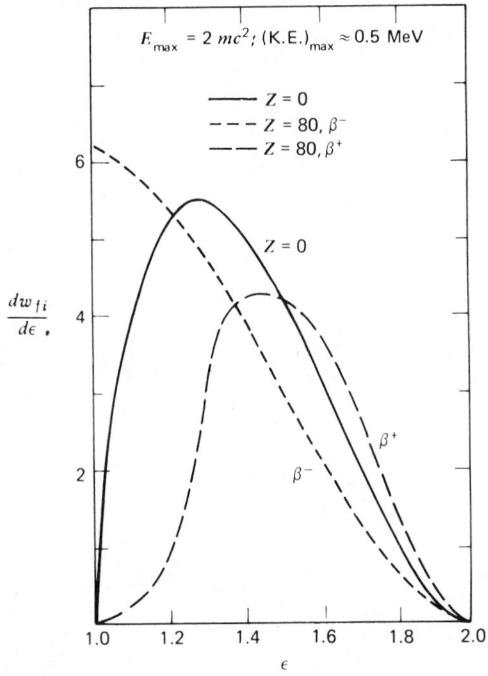

FIG. 2.1. The energy distribution $dw_{fi}/d\epsilon$ for residual nuclei $Z = 0$ and $Z = 80$. The maximum energy is $2\,mc^2$. Coulomb corrections are included [taken from Wu and Moskowski (66)].

and we have replaced $e^2/\hbar c$ by $1/137$. The relativistic density for a point nucleus is infinite at the nucleus. As a consequence Fermi (34) took the value of the density at the nuclear surface to be the ρ of (2.10). The value of ρ is then

$$\rho = 2\,(2kR)^{2(s-1)}\,\frac{(1+s)}{s^2+\eta^2}\left|\frac{e^{\pi\eta/2}\,\Gamma(s+1+i\eta)}{\Gamma(2s+1)}\right|^2 \quad \text{relativistic} \qquad (2.15)$$

$$s^2 = 1 - \left(\frac{Z}{137}\right)^2$$

Tables from which ρ can be calculated are given in a National Bureau of Standards Table (52). [The factor $|e^{\pi\eta/2}\,\Gamma(s+1+i\eta)|$ is given in *Handbook of Mathematical Functions* (64). See also Gove and Martin (71).)]

The Fermi prescription for calculating ρ has been improved by a number of authors. An example is furnished by the results of Behrens and Bühring (58) who evaluate ρ at $r = 0$ of the *finite* nucleus. The corrections to the matrix element that arise because of the variation of the wave functions over the

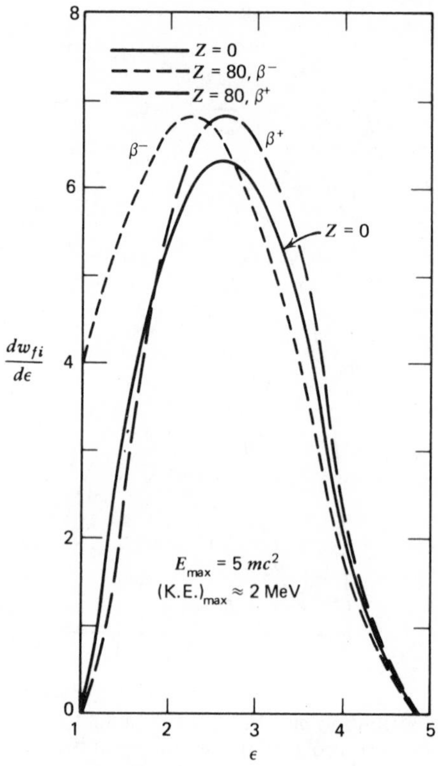

FIG. 2.2. The energy distribution $dw_{fi}/d\epsilon$ for residual nuclei $Z = 0$ and $Z = 80$. The maximum energy is $5\,mc^2$. Coulomb corrections are included [taken from Wu and Moskowski (66)].

nucleus (= error because of long wavelength approximation) are small [Blin-Stoyle (69)].

The standard method for examination of the β-ray spectrum is the so-called Kurie plot proposed by Kurie, Richardson, and Paxton (36). In this method one plots

$$K(E_e) = \left(\frac{N(E_e)}{p\epsilon\sqrt{\epsilon^2 - 1}}\right)^{1/2} \tag{2.16}$$

where $N\,dE_e$ is the number of β-particles with energy between E_e and $E_e + dE_e$. According to (2.9), assuming that $|M_{fi}|^2$ is energy independent, this quantity is proportional to $(E_0 - E_e)$. Hence the experimental $K(E_e)$ should be a straight line with slope equal to -1. In the early pre-World War II days there was considerable debate whether $K(E_e)$ was in fact a straight line. Deviations from a straight line would possibly be caused by a finite neutrino mass. The effect due to such a mass would occur, however, only near the end point.

With an upper bound of 60 eV for the neutrino mass a deviation from a straight-line dependence over a considerable portion of the spectrum would have to be ascribed to an energy dependence of $|M_{fi}|^2$. As it turned out for allowed transitions, the straight-line dependence is experimentally verified so that $|M_{fi}|^2$ is in fact energy independent for these transitions.

An example of a Kurie plot is given in Fig. 2.3. The deviation from the straight line at low energies can be ascribed to deviations from the Fermi form (2.15) used [Wu and Moszkowski (66)]. The reader should recall that (2.16) is appropriate for allowed transitions.

The total probability for decay is obtained by integrating the right-hand side of (2.9) over the electron energy to obtain

$$w_{fi} = \frac{mc^2}{\hbar}\frac{\Gamma^2}{2\pi^3}\,|M_{fi}|^2 f(Z, R, \epsilon_0) \tag{2.17}$$

where

$$f(Z, R, \epsilon_0) = \int_1^{\epsilon_0} d\epsilon\, \rho(Z, R, \epsilon)\,(\epsilon_0 - \epsilon)^2\,\epsilon(\epsilon^2 - 1)^{1/2} \tag{2.18}$$

For tables of f see Gove and Martin (71). In the limit of no Coulomb field

$$f(0, R, \epsilon_0) = (\epsilon_0^2 - 1)^{1/2}\left(\frac{\epsilon_0^4}{30} - \frac{3\epsilon_0^2}{20} - \frac{2}{15}\right)$$

$$+ \frac{\epsilon_0}{4}\log\left(\epsilon_0 + \sqrt{\epsilon_0^2 - 1}\right)\Bigg) \underset{\epsilon_0 \gg 1}{\longrightarrow} \frac{\epsilon_0^5}{30} \tag{2.19}$$

We see that the transition probability goes up very rapidly with ϵ_0. Thus the short lifetime of the μ-meson (2.2×10^{-6} sec mean life) compared to that of the neutron (.93 10^3 sec mean life) is primarily caused by the much larger energy release ($\sim$105.7 MeV) in μ-decay.*

These kinematic factors can be removed by considering the ratio f/w_{fi}. The usual procedure uses the "ft" value where $t_{1/2}$ is the half-life related to the mean life w_{fi} as follows

$$t_{1/2} = \frac{\ln 2}{w_{fi}} \tag{2.20}$$

Hence the "ft" value is

$$ft_{1/2} = \frac{\hbar}{mc^2}\frac{2\pi^3}{\Gamma^2}\frac{\ln 2}{|M_{fi}|^2} \tag{2.21}$$

---

*Of course (2.19) does not apply directly to μ-decay. However the various differences do not affect this qualitative conclusion. For the correct expression for the μ-decay probability see (12.5).

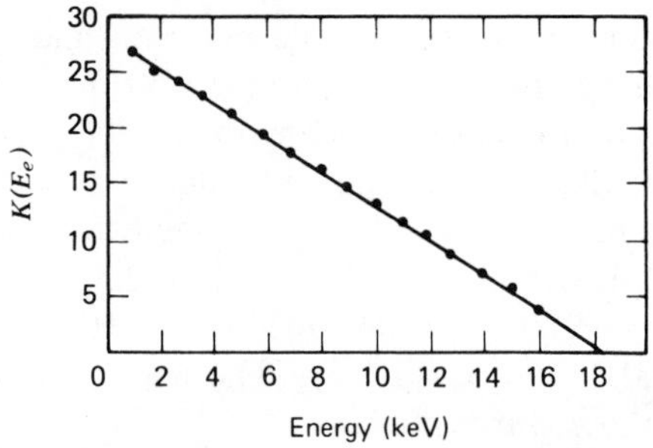

FIG. 2.3. The Kurie plot of the H^3 β^- spectrum [taken from Curran (52)].

Since f is a known function and $t_{1/2}$ an experimental number it becomes possible to determine $|M_{fi}|^2$ for each nuclear transition. The value of Γ^2 is obtained by examining transitions in which $|M_{fi}|^2$ is thought to be known. This matrix element carries the nuclear information which β-decay lifetime studies provide.

Those nuclei for which the "ft" values are smallest are the allowed transitions. Even these break up into two groups (Fig. 2.4). Those for which $\log f_0 t_{1/2}$ (f_0 is given in (2.19)) is of the order of 3.5 and those for which $\log f_0 t_{1/2}$ has the range 5.7 ± 1.1. The first group is called *superallowed*. Decays with larger

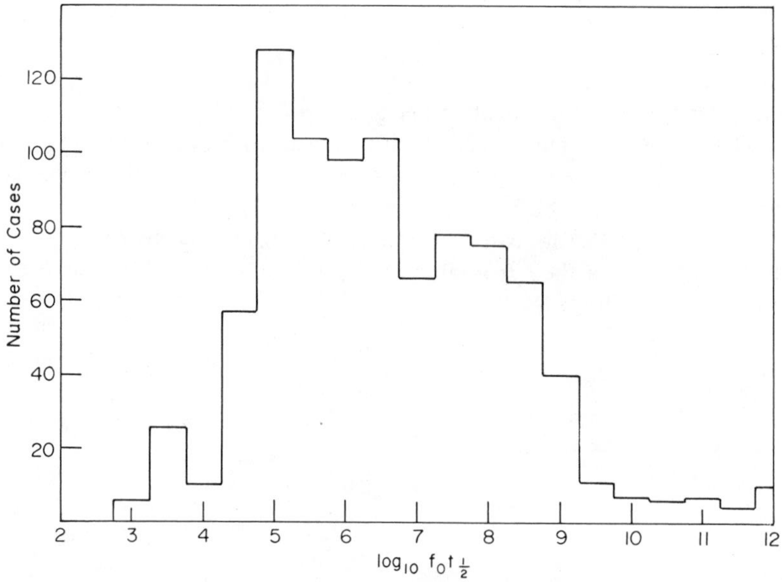

FIG. 2.4. Histogram showing the number of radioactive nuclei versus their $ft_{1/2}$ value. [taken from Gleet, Tang, and Coryell (63)].

values of log f_0t (larger than 5.7 $\pm$ 1.1) will belong to forbidden transitions of higher order. For these $|M_{fi}|^2$ vanishes. We must then return to $|H_{fi}|^2$ and calculate it more carefully. For example, one can no longer assume the electron and neutrino wave functions are constant within the nucleus. But there are other terms as well. It is necessary to have a more complete description of H_{fi} before proceeding further.

It is possible to determine the character of $|M_{fi}|^2$ for allowed transitions from general principles. We assume that the operators involved in H_{fi} are in the first approximation single-body operators. This is reasonable since, first, the free-neutron decays and, second, its lifetime is consistent with the lifetime of decaying nuclei, such as ^{3}H. Moreover, since the vectors $\mathbf{k}_e$ and $\mathbf{k}_\nu$ as well as the spin orientation of the electron and neutrino have been averaged over, M_{fi} can only involve nuclear coordinates. Third, since a neutron at rest decays, the operators in M_{fi} must be nonzero even if the decaying nucleon is at rest. Fourth, again because of neutron decay and translational invariance, the transition operator cannot depend upon spatial coordinates. There are only three such operators: the unit operator, the spin $\boldsymbol{\sigma}(i)$ and the isospin $\boldsymbol{\tau}(i)$. Finally the operators must transform a neutron into a proton for β^- decay and vice versa for β^+ decay. This is accomplished by the isospin operators

$$\tau^{(-)}\,|p\rangle = \tfrac{1}{2}(\tau_1 - i\tau_2)|p\rangle = |n\rangle$$

$$\tau^{(-)}\,|n\rangle = 0$$

$$\tau^{(+)}\,|p\rangle = \tfrac{1}{2}(\tau_1 + i\tau_2)\,|p\rangle = 0$$

$$\tau^{(+)}\,|n\rangle = |p\rangle$$

The matrix element M_{fi} must involve the isospin operator only through $\tau^{(-)}$ for positron emission or $\tau^{(+)}$ for electron emission. The dependence on $\boldsymbol{\sigma}$ is fixed by the requirement that $|M_{fi}|^2$ is rotationally invariant. These conditions determine the form $|M_{fi}|^2$ can take to be

$$|M_{fi}|^2 = \frac{|C_F|^2}{2J_i + 1} \sum_{f,i} |\langle f| \sum_k \tau^{(-)}(k)|i\rangle^2$$

$$+ \frac{|C_{GT}|^2}{2J_i + 1} \sum_{f,i} |\langle f| \sum_k \tau^{(-)}(k)\,\boldsymbol{\sigma}(k)|i\rangle|^2 \qquad (2.22)$$

where J_i is the spin of the initial state, C_F and C_{GT} are constants and

$$|\langle f| \sum_k \tau^{(-)}\boldsymbol{\sigma}|i\rangle|^2 \equiv \sum_j |\langle f| \sum_k \tau^{(-)}(k)\,\sigma_j(k)\,|i\rangle|^2$$

Here j refers to the different components of the vector $\boldsymbol{\sigma}(k)$. The sum over k is a sum over different nucleons of the decaying nucleus. A simplified notation that is often used replaces (2.22) as follows:

$$|M_{fi}|^2 = |C_F|^2\,|M_F|^2 + |C_{GT}|^2\,|M_{GT}|^2 \qquad (2.23)$$

where

$$|M_F|^2 = \frac{1}{2J_i + 1} \sum_{f,i} |\langle f| \sum_k \tau^{(\pm)}(k) |i\rangle|^2 \tag{2.24}$$

and

$$|M_{GT}|^2 \equiv \frac{1}{2J_i + 1} \sum_{f,i} |\langle f| \sum_k \tau^{(\pm)}(k)\, \boldsymbol{\sigma}(k) |i\rangle|^2 \tag{2.25}$$

For electron emission use $\tau^{(+)}$; for positron emission $\tau^{(-)}$. The first of these (2.24) is referred to as the Fermi matrix element since the interaction he employed [Fermi (34)] led directly to (2.24). Hence the subscript in C_F. The second (2.25) is the Gamow-Teller matrix element [Gamow and Teller (36)]. Each of these have characteristic selection rules:

Fermi: $\Delta J = 0$ no change in parity

$$\Delta T = 0 \qquad \Delta T_3 = 1 \quad \text{electron emission} \tag{2.26}$$

$$= -1 \quad \text{positron emission}$$

Gamow-Teller: $\Delta J = \pm 1, 0$ $(0 \to 0$ forbidden) no change in parity (2.27)

$$\Delta T = \pm 1, 0 \ (0 \to 0 \text{ forbidden})$$

$$\Delta T_3 = \pm 1$$

The Fermi selection rules follow because the Fermi matrix is a spatial scalar and because

$$\sum_k \tau^{(\pm)}(k) = T^{(\pm)} \tag{2.28}$$

commutes with T^2. The Gamow-Teller rules follow since

$$\sum \tau^{(\pm)}(k)\, \boldsymbol{\sigma}(k)$$

transforms as a vector in coordinate space and as a vector in isospin space.

The existence of these two types of matrix elements and selection rules is well substantiated by experiment. An example of Fermi type transition is the decay

$${}^{34}_{17}\text{Cl} \to {}^{24}_{16}\text{S} + e^+ + \nu \tag{2.29}$$

in which the transition is from a $J^\Pi = 0^+$ state in ^{34}Cl to a $J^\Pi = 0^+$ in ^{34}S. On the other hand any $\Delta J = 1$ transition requires the Gamow-Teller matrix element. Such is the case for the transition

$${}^{12}_{5}\text{B} \to {}^{12}_{6}\text{C} + e^- + \bar{\nu} \tag{2.30}$$

The spin and parity of ^{12}B are 1^+ while that of ^{12}C is 0^+. On the other hand in the decay of the neutron

$$_0^1n \rightarrow _1^1p + e^- + \bar{\nu} \tag{2.31}$$

contributions come from both Fermi and Gamow-Teller matrix elements.

The selection rules (2.26) and (2.27) indicate that in a Fermi transition the electron and antineutrino carry off a net zero angular momentum while in the Gamow-Teller case they carry off a unit angular momentum.

When *superallowed transitions* occur, the matrix elements M_F and M_{GT} have their maximum value. For M_F this can occur for the light elements and when the initial and final state are members of an isospin multiplet. The reason is that under these circumstances there is a maximum overlap between the initial and final nuclear wave functions, the beta transition replacing a neutron in an orbital by a proton in the identical orbital. In the transition between mirror nuclei, with $T = 1/2$ from $T_3 = -1/2$ to $T_3 = 1/2$, the wave function generated by the transition is shown in Fig. 2.5. It is identical with the $T_3 = 1/2$ state of the final nucleus.

When the initial and final state are members of an isospin multiplet, the Fermi matrix element can be immediately obtained:

$$|M_F|^2 = |\langle f|T^{(+)}|i\rangle|^2 = [T(T+1) - T_3(T_3+1)] \qquad e^- \text{ decay}$$

$$= |\langle f|T^{-}|i\rangle|^2 = [T(T+1) - T_3(T_3-1)] \qquad e^+ \text{ decay} \tag{2.32}$$

where T_3 has the value of $(Z - N)/2$ for the initial nucleus. Thus for the transition given in (2.29), $T = 1, T_3 = 0$, and $|M_F|^2 = 2$. On the other hand for the decay of a neutron into a proton (2.31), $T = 1/2, T_3 = -1/2$ so that $|M_F|^2 = 1$.

The Gamow-Teller matrix elements are not so readily obtained except for neutron decay. In that case

$$|M_F|^2 = \frac{3}{2} \sum_{m_i, m_f} |\langle m_f|\sigma_k|m_i\rangle|^2 = \frac{3}{2} \sum_{m_f} \langle m_f|m_f\rangle = 3$$

Hence for the neutron decay

$$|M_{fi}|^2 = |C_F|^2 + 3\,|C_{GT}|^2 \qquad \text{neutron decay} \tag{2.33}$$

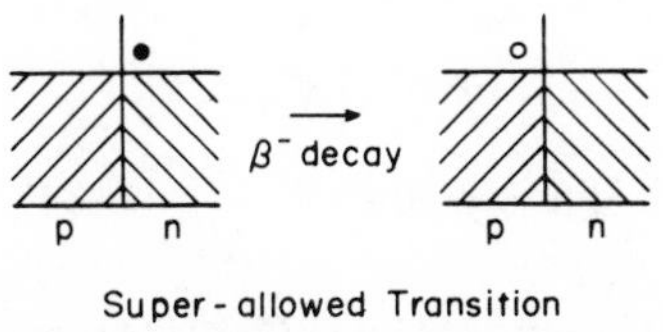

FIG. 2.5. Wave function overlap in superallowed transitions.

As in the case of the electromagnetic transitions it is possible to obtain a "single-particle" value for M_{fi} that provides an order of magnitude estimate. The jj shell model is assumed. It is also assumed that a single nucleon changes its isospin when β-decay occurs. The β decay transition operators do not depend upon space coordinates. Hence the orbital angular momentum of transforming particle will remain unchanged. Thus in the single-particle model

$$\Delta l = 0 \tag{2.34}$$

The single-particle value will assume that it is possible to satisfy this requirement. The estimate obtained in this way can be quite wrong. It may be the case that the appropriate orbital for the product particle will be occupied so that (2.34) cannot be satisfied, or the jj model may not provide a good description of the nuclear states. For example, core-excited many-particle excitations may make significant contributions to the wave function. Then the matrix element is reduced considerably from the single-particle estimate we shall now calculate.

In the single-particle limit described above the isospin operators in (2.25) can be replaced by unity. In terms of spherical tensors for $\boldsymbol{\sigma}$, $T_\kappa^{(\sigma)}$,

$$|M_{GT}|^2 = \frac{1}{2J+1} \sum_{\kappa, m_i, m_f} (-)^\kappa \left(\int T_\kappa^{(\sigma)} \right) \left(\int T_{-\kappa}^{(\sigma)} \right)^* \tag{2.35}$$

where

$$\int T_\kappa^{(\sigma)} \equiv \langle \tfrac{1}{2} l J_f m_f | T_\kappa^{(\sigma)} | \tfrac{1}{2} l J_i m_i \rangle \tag{2.36}$$

Here l gives the orbital angular momentum of the decaying nucleon. Since $\boldsymbol{\sigma}$ does not involve any spatial dependence, l is not changed in the course of the transition. From the Wigner–Eckart theorem we have

$$\int T_\kappa^{(\sigma)} = (-)^{J_f - m_f} \begin{pmatrix} J_f & 1 & J_i \\ -m_f & \kappa & m_i \end{pmatrix} (\tfrac{1}{2} l J_f || T^{(\sigma)} || \tfrac{1}{2} l J_i)$$

The reduced matrix can be further reduced according to (A.2.55):

$$(\tfrac{1}{2} l J_f || T^{(\sigma)} || \tfrac{1}{2} l J_i)$$
$$= (-)^{1/2 + l + J_i + 1} \sqrt{(2J_f + 1)(2J_i + 1)} \begin{Bmatrix} \tfrac{1}{2} & \tfrac{1}{2} & 1 \\ J_i & J_f & l \end{Bmatrix} (\tfrac{1}{2} || T^{(\sigma)} || \tfrac{1}{2})$$

Finally from (A.2.47)

$$(\tfrac{1}{2} || T^{(\sigma)} || \tfrac{1}{2}) = \sqrt{6}$$

The second factor in (2.35) is given by:

$$\left(\int T_{-\kappa}^{(\sigma)} \right)^* \equiv \langle \tfrac{1}{2} l J_i m_i | T_{-\kappa}^{(\sigma)} | \tfrac{1}{2} l J_f m_f \rangle \tag{2.37}$$

Obviously the analysis used to evaluate (2.36) is immediately applicable to (2.37).

Combining these results yields

$$|M_{GT}|^2 = 6(2J_f+1)\begin{Bmatrix} \frac{1}{2} & \frac{1}{2} & 1 \\ J_i & J_f & l \end{Bmatrix}^2 \sum \begin{pmatrix} J_f & 1 & J_i \\ -m_f & \kappa & m_i \end{pmatrix}\begin{pmatrix} J_i & 1 & J_f \\ -m_i & -\kappa & m_f \end{pmatrix}$$

or

$$|M_{GT}|^2 = 6(2J_f+1)\begin{Bmatrix} \frac{1}{2} & \frac{1}{2} & 1 \\ J_i & J_f & l \end{Bmatrix}^2 \tag{2.38}$$

Inserting explicit values for the $6j$ symbol yields the following table

TABLE 2.1 Single Particle Gamow-Teller Matrix Elements $|M_{GT}|^2$

J_f \ J_i	$l+1/2$	$l-1/2$
$l+1/2$	$\frac{J_f+1}{J_f}$	$\frac{2J_f+1}{J_f}$
$l-1/2$	$\frac{2J_f+1}{J_f+1}$	$\frac{J_f}{J_f+1}$

These results are to be substituted in (2.23) to obtain the single particle $|M_{fi}|^2$. It is to be expected that the value of $|M_{fi}|^2$ obtained in this way will give reasonable values for the light nuclei but will fail for the heavier nuclei since for these nuclei it will be more difficult to satisfy the single-particle selection rule (2.34). Some typical single-particle values obtained from Table 2.1 for transitions in the light nuclei are given in Table 2.2.

TABLE 2.2 Values of Single-Particle Gamow-Teller Matrix Elements

Decay	$J_i\ (=J_f)$	l_j	$\|M_{GT}\|^2$
$^{11}C \rightarrow {}^{11}B$	3/2	$p_{3/2}$	5/3
$^{15}O \rightarrow {}^{15}N$	1/2	$p_{1/2}$	1/3
$^{17}F \rightarrow {}^{17}O$	5/2	$d_{5/2}$	7/5
$^{31}S \rightarrow {}^{31}P$	1/2	$s_{1/2}$	3
$^{33}Cl \rightarrow {}^{33}S$	3/2	$d_{3/2}$	3/5
$^{41}SC \rightarrow {}^{41}Ca$	7/2	$f_{7/2}$	9/7

3. DETERMINATION OF THE WEAK COUPLING CONSTANT AND THE RATIO C_F/C_{GT}

The most accurate determination of the weak interaction coupling constant involves the measurement of the lifetime of $0^+ \rightarrow 0^+$, $T = 1$ superallowed β^+ decays in light nuclei, together with a careful measurement of the energy released. For β^+ decaying $T = 1$ nuclei, $|M_F|^2 = 2$ according to (2.32) so that the $ft_{1/2}$ value for such decays should be equal. The $ft_{1/2}$ values as obtained by Freeman et al. (66a, 66b, 68, 69a, 69b) and tabulated by Blin–Stoyle (69) are shown in Table 3.1.

TABLE 3.1 $ft_{1/2}$ Values and Resultant $C_F g$ for Superallowed $0^+ \rightarrow 0^+$, $T = 1$ Positron Decays

Decay	$ft_{1/2}$ (raw)	$ft_{1/2}$ (with radiative corrections)	$C_F g$(erg cm^3)10^{49}
$^{10}C \rightarrow {}^{10}B$	2973^{+49}_{-44}	3020^{+49}_{-44}	$1.439^{+0.0107}_{-0.0115}$
$^{14}O \rightarrow {}^{14}N$*	3050 ± 11	3098 ± 11	1.4089 ± 0.0022
$^{26m}Al \rightarrow {}^{26}Mg$*	3040 ± 4	3090 ± 4	1.4103 ± 0.0010
$^{34}Cl \rightarrow {}^{34}S$*	3045 ± 7	3097 ± 7	1.4092 ± 0.0015
$^{42}Sc \rightarrow {}^{42}Ca$	3064 ± 9	3111 ± 9	1.4060 ± 0020
$^{46}V \rightarrow {}^{46}Ti$	3072 ± 8	3115 ± 8	1.4051 ± 0.0018
$^{50}Mn \rightarrow {}^{50}Cr$	3059 ± 9	3102 ± 9	1.4080 ± 0.0020
$^{54}Co \rightarrow {}^{54}Fe$	3063 ± 17	3103 ± 17	1.4078 ± 0.0039

* The starred cases were recently presented by Freeman et. al. (73).

The corrections and uncertainties in the $ft_{1/2}$ value have been discussed by Blin–Stoyle (69). Of these we have included the radiative correction that is of the order of $(e^2/\hbar c) \sim (1/137)$ in the third column of the table. This correction arises because of the possible emission and absorption of virtual photons by the charged particles involved in the decay as illustrated in Fig. 3.1. This is a logarithmically divergent process requiring a cutoff and because of this requirement its evaluation remains not completely satisfactory. The corrections given in Table 3.1 are based on the calculations of Kinoshita and

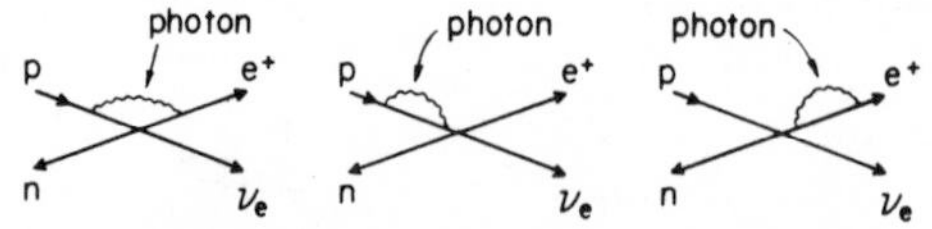

FIG. 3.1. Radiative corrections in β^+ decay.

Sirlin (59) who take the cutoff at the nucleon mass. A more recent discussion is given by Brene et al. (68). New developments, the Weinberg theory of weak interactions, may remove the need for a cutoff. An uncertainty of $\frac{1}{2}\%$, theoretical in origin, which is model dependent remains. This together with other nonstructure dependent uncertainties discussed by Blin–Stoyle (69) give rise to a total uncertainty of 1% in the absolute magnitude of the $ft_{1/2}$ values although he emphasizes the relative error is considerably smaller.

The radiative correction changes the $ft_{1/2}$ magnitudes but not their relative values especially. The agreement shown and illustrated in Fig. 3.2 demonstrates the validity of the isospin conservation and the isospin purity of the nuclear wave functions. The value of $C_F g$ can be calculated from the $ft_{1/2}$ values obtained for ^{14}O, ^{26m}Al and ^{34}Cl decays. We shall adopt the value:

$$C_F g = 1.4100\ 10^{-49}\ \text{erg cm}^3 \tag{3.1}$$

$$\simeq 10^{-5}\left(\frac{\hbar}{Mc}\right)^3 Mc^2$$

where M is the nucleon mass.

Let us turn next to the determination of the ratio $|C_{GT}/C_F|$. For this purpose note from (2.21) that

$$ft_{1/2}[|C_F|^2|M_F|^2 + |C_{GT}|^2|M_{GT}|^2]$$

equals a universal constant and must therefore be the same for all nuclei. Then by comparing two nuclei with known matrix elements it becomes possible to determine $|C_{GT}/C_F|^2$. For example, comparing ^{14}O and neutron decay yields

$$(ft_{1/2})_{^{14}O}\, 2|C_F|^2 = (ft_{1/2})_n[|C_F|^2 + 3|C_{GT}|^2]$$

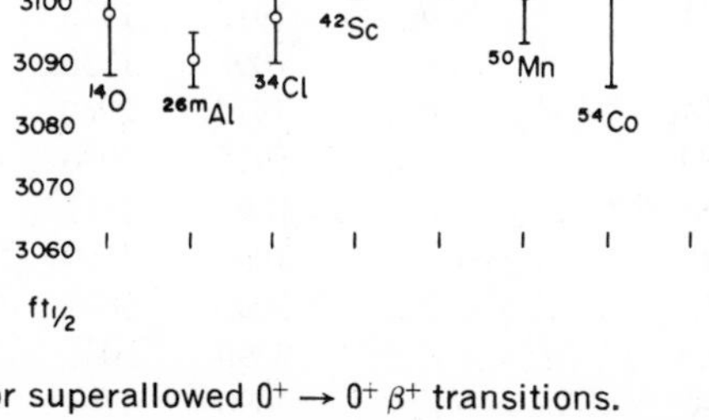

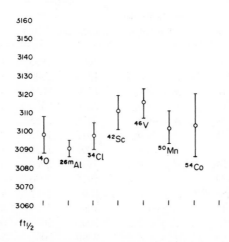

FIG. 3.2. $ft_{1/2}$ values for superallowed $0^+ \rightarrow 0^+$ β^+ transitions.

Inserting the $ft_{1/2}$ value for ^{14}O decay of 3098 from Table 3.1 and the value of $ft_{1/2}$ for neutron decay, 1092, from which $|C_{GT}/C_F| = 1.248 \pm 0.010$ [Particle Data Group (73)] indicating the close relative strength of the Fermi and Gamow–Teller interactions. This calculation assumes that C_{GT} and C_F do not depend upon the nuclear species decaying.

Once the value of $C_F g$ and the ratio $|C_{GT}/C_F|$ are determined, we can determine the combination $|M_F|^2 + |C_{GT}/C_F|^2|M_{GT}|^2$ from experiment and compare the result with model predictions such as the single-particle results in section 2.

4. ALLOWED TRANSITIONS IN LIGHT NUCLEI; *p*-SHELL NUCLEI

As a first step, let us consider β-decay of a nucleus into its mirror, an allowed transition. Some of these together with the associated single-particle transitions taken from Konopinski (66) are listed in Table 4.1. The agreement between the single-particle value of ft (the value of M_F is taken to be 1; M_{GT} is given by Table 2.1) listed in the column $(ft)_{\text{theo}}$ and $(ft)_{\text{exp}}$ is not particularly impressive. A considerable improvement is obtained if the single-particle evaluation of M_{GT} is replaced by the semiempirical value given by (Eq. VIII.11.19) obtained from the measured difference between the magnetic moments of mirror nuclei. To obtain this expression, jj coupling is assumed. It is also assumed that the isospin is a good quantum number. The resulting values are labeled $(ft)_\mu$ in Table 4.1. It can be seen that the agreement with

TABLE 4.1 Decay in Mirror Nuclei [from Konopinski (66)]

Parent	Daughter	Single-Particle Transition	$(ft)_{\text{theo}}$ (sec)	(ft_μ)	$(ft)_{\text{exp}}$
^3_1H	^3_2He	$s_{1/2} \to s_{1/2}$	1092	953	1137 ± 20
^7_4Be	^7_3Li	$p_{3/2} \to p_{3/2}$	1730	2300	2300 ± 78
$^{11}_6C$	$^{11}_5B$	$p_{3/2} \to p_{3/2}$	1730	3025	3840 ± 70
$^{13}_7N$	$^{13}_6C$	$p_{1/2} \to p_{1/2}$	3876	3515	4700 ± 80
$^{15}_8O$	$^{15}_7N$	$p_{1/2} \to p_{1/2}$	3876	3765	4475 ± 30
$^{17}_9F$	$^{17}_8O$	$d_{5/2} \to d_{5/2}$	1915	1925	2330 ± 80
$^{19}_{10}Ne$	$^{19}_9F$	$s_{1/2} \to s_{1/2}$	1092	1230	1900 ± 100
$^{25}_{13}Al$	$^{25}_{12}Mg$	$d_{5/2} \to d_{5/2}$	1915	2960	4280 ± 350
$^{27}_{14}Si$	$^{27}_{13}Al$	$d_{5/2} \to d_{5/2}$	1915	3080	4500 ± 100
$^{29}_{15}P$	$^{29}_{14}Si$	$s_{1/2} \to s_{1/2}$	1092	3247	4750 ± 200
$^{31}_{16}S$	$^{31}_{15}P$	$s_{1/2} \to s_{1/2}$	1092	4160	4820 ± 250
$^{33}_{17}Cl$	$^{33}_{16}S$	$d_{3/2} \to d_{3/2}$	3090	5642	6000 ± 500
$^{35}_{18}A$	$^{35}_{17}Cl$	$d_{3/2} \to d_{3/2}$	3090	5088	5680 ± 400
$^{39}_{20}Ca$	$^{39}_{19}K$	$d_{3/2} \to d_{3/2}$	3090	4764	4150 ± 300

experiment is greatly improved. This improvement verifies the inadequacy of the single-particle description. However, large differences still remain that may be due in part to the assumption that $M_F = 1$. In general M_F will be less than one, which would have the effect of increasing $(ft)_\mu$, which is generally the direction to improve comparison with experiment. Beyond this one must turn to deviations from isospin symmetry induced by Coulomb forces and to the effects of configuration mixing.

We turn next to β-decays of other light nuclei. A recurrent theme in this and the preceding chapter has been the value of the electromagnetic and weak interactions as probes of nuclear structure. This is demonstrated in the present context by the sensitivity of the static moments and the transition amplitudes for radiative and β-decay to the wave functions of the states involved. The excitation energies of low-lying energy levels and their quantum numbers, spin, and isospin are not in themselves sufficient. As we know from the variational principle the error in the energy is second order in the wave-function error whereas the error in moments or transition amplitudes is first order. But this is not all. Because the usual calculations have many phenomenological parameters that are to be fixed by comparison with experiment it becomes possible by suitable choices of the parameters to obtain equally good fits to the observed energy levels. A further selection among these various models may sometimes be obtained by comparison of predicted and observed transition amplitudes as well as with the static electromagnetic moments.

To illustrate let us consider the case of "p-shell" nuclei, that is, nuclei with $5 \leq A \leq 16$. In the simple independent-particle model there are four nucleons, two neutrons, and two protons in $1s_{1/2}$ orbits, forming a core. The remaining particles are in $1p_{1/2}$ or $1p_{3/2}$ orbits, the p-shell being completely filled for ^{16}O. The simplest shell model proceeds by assuming a central shell model potential including a spin-orbit force that essentially provides the splitting between the $1p_{3/2}$ and $1p_{1/2}$ single-particle energies as well as a central residual interaction. This Hamiltonian is diagonalized using states formed from single-particle $1p_{1/2}$ and $1p_{3/2}$ states. Energies are computed relative to the ^4He core, which is assumed to be quiescent. This simple model has been developed by Inglis (53, 54), Lane (53, 55), Lane and Radicati (54), and Kurath (56, 57, 60). Briefly, reasonably good fits are obtained if the strength of the central spin-orbit potential is assumed to be weak at the beginning of the shell so that LS-coupling is valid there but increases as the more nucleons are added. This model, referred to as the *intermediate-coupling* model, will fit a large majority of the low-lying excitation spectra of the p-shell nuclei [Kurath (56, 57)].

We shall discuss more fully a less ambitious approach as formulated by Cohen and Kurath (65) and applied by them as well as Poletti, Warburton, and Kurath (67) to the calculation of magnetic dipole moments, $M1$ transitions, the ratio of $E2$ amplitudes to the $M1$, and Gamow–Teller beta decays of p-shell nuclei. The Talmi method will be used. It will be recalled (Page 338) if the residual interaction V is a two-body interaction, the matrix elements of V

for a multiparticle system can be expressed in terms of the two-body matrix elements. In the Talmi method these matrix elements are treated as empirical parameters. In jj coupling there are 15 such matrix elements, $\langle j_1 j_2 JT | V | j_3 j_4 JT \rangle$ of which 10 are diagonal. In the first scheme labeled by Cohen and Kurath

Problem. Prove this statement. Remember that for two identical particles $J + T$ must be odd.

"$2BME$", these 15 matrix elements plus the single particle $p_{1/2}$ and $p_{3/2}$ energies are treated as parameters, making 17 in all, which are adjusted to give as good a match as possible to the binding energy and energies of a selected number of excited states. Nuclei with $8 \leq A \leq 16$ were used. In the second method labeled, "POT", the two-body matrix elements were calculated in the LS-scheme $\langle LSJT | V | \bar{L}\bar{S}JT \rangle$. There are 11 matrix elements, 10 diagonal. The radial wave functions for the p-states are assumed to be identical and the nucleon–nucleon interaction symmetric. Adding in the single particle energies give 13 parameters to be adjusted to give the best fits. Note that since we are dealing with p-particles only, the party of the states being considered is $(-)^A$.

A few of the nuclear spectra fitted in this way are shown in Fig. 4.1. Both models, the POT and $2BME$, provide an equally good fit for these and for the other nuclei in the range $8 \leq A \leq 16$. However the fit is by no means perfect and in one case, ^9Be, a low-lying level unobserved thus far is predicted. The magnitude of the matrix elements involved is given in Table 4.2.

We turn next to the magnetic moments (Table 4.3) and the ft values for Gamow–Teller transitions (Table 4.4) taken from Cohen and Kurath (65) and a private communication from Kurath (71). In the latter table the values predicted assuming pure jj coupling (Table 2.1) and pure LS-coupling are also given. Also given are the results obtained using the intermediate coupling model of Inglis; the parameter (a/k) measures the ratio of the strength of the spin-orbit potential to the central residual potential. In Table 4.5 $M1$ transition between the low-lying odd parity levels of ^{11}B as well as the $(11) \rightarrow (00)$ transition in ^{12}C are compared with the values predicted by the POT model as well as by the simple jj- and LS-models. Finally in Table 4.6 the ratio of the $E(2)$ to the $M1$ amplitude is given. In the column labeled $x(E2/M1)^b$ the $E2$ amplitude has been enhanced by assuming that both proton and neutron have an additional charge $\alpha_p = \alpha_n = (1/2)e$ (see Section VIII.9 and Table VIII.9.1).

From Table 4.4 listing the ft-values note that the simple jj- and LS-models differ substantially in their predictions from each other and from the more sophisticated Inglis, POT and $2MBE$ models. In Table 4.5 a similar comparison made for $\mu(^{11}B)$ and the $(11) \rightarrow (00)$ transition in ^{12}C again demonstrates the inadequacy of the simple models. Once more we see that at the very least the wave functions for the states must be linear combinations of the various states with a given J and T, which can be constructed from the two-single particle $p_{1/2}$, $p_{3/2}$ states. Or in the language of LS-coupling, various states

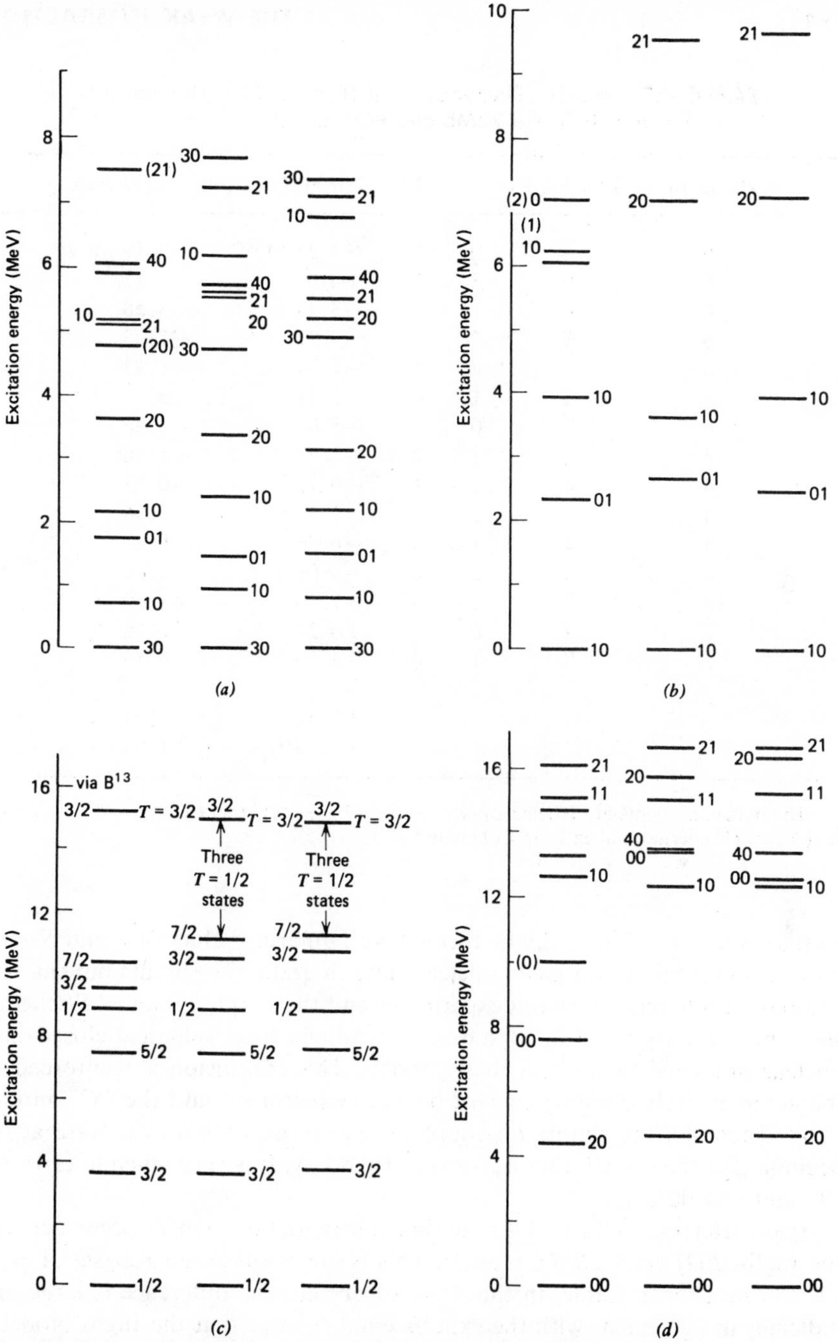

FIG. 4.1. Some representative level schemes [(*a*) of ^{10}B, (*b*) of ^{14}N, (*c*) of ^{13}N, and (*d*) of ^{12}C] obtained using the *POT* and *2BME* models. Column 1 is experimental, Column 2 results from the *POT* scheme, Column 3 from the *2BME* scheme. Each level is labeled by *J* and *T*. Levels with identification on the right were included in the energy fits while those with identification on the left were subsequently predicted from these fits [taken from Cohen and Kurath (65)].

TABLE 4.2 Matrix Elements $\langle j_1 j_2 JT | V | j_3 j_4 JT \rangle$ Determined Empirically in the 2BME and POT model[a]

$2(j_1 + j_2)$	$2(j_3 + j_4)$	J	T	POT	$2BME$
6	6	0	1	−3.33 MeV	−3.19 MeV
6	6	2	1	+0.09	−0.17
6	6	1	0	−3.44	−3.58
6	6	3	0	−7.27	−7.23
6	4	2	1	−1.74	−1.92
6	4	1	0	+3.21	+3.55
6	2	0	1	−5.05	−4.86
6	2	1	0	+1.77	+1.56
4	4	1	1	+0.73	+0.92
4	4	2	1	−1.14	−0.96
4	4	1	0	−6.56	−6.22
4	4	2	0	−4.06	−4.00
4	2	1	0	+1.20	+1.69
2	2	0	1	+0.24	−0.26
2	2	1	0	−4.29	−4.15
$\epsilon_{1/2}$				+2.42	+1.57
$\epsilon_{3/2}$				+1.13	+1.43

[a] The matrix elements are labeled by $(2j_1 + 2j_2)$, $(2j_3 + 2j_4)$, J and T. $\epsilon_{1/2}$ and $\epsilon_{3/2}$ give the single-particle energies [taken from Cohen and Kurath (65)]

with a given J and T that can be formed with differing values of L and S enter into the description of a given nuclear state. Kurath (71) points out that the substantial difference between experiment and the simple jj- and LS-schemes show how greatly the ground state of ^{12}C differs from spherical closed-shell nucleus predicted by both of these models. This conclusion is reenforced by the comparatively close agreement between experiment and the POT model. From Table 4.4 the simple LS-coupling appears to be most accurate at the beginning of the p-shell, that is, for the ^6He decay. It is very much in error for ^{14}C and ^{14}O decays.

Again referring to Table 4.4 note that substantial differences occur between the Inglis, POT, and $2BME$ models. This is not always the case (see $A = 6$, 12, 13) but, for example, in the $A = 10$ nucleus the difference is large and indicates in agreement with the experimental ft-value that the Inglis model is not valid for that nucleus. Comparing the POT and $2BME$ models is more difficult especially since the $2BME$ model contains more empirically adjusted parameters. Each model yields quite good agreement with the experimentally

TABLE 4.3 Magnetic Moments in the p-shell in nuclear magnetons

	J	EXP	Inglis (a/k)	POT	$2BME$	$(2BME)'$
^{8}Li	2	1.653	1.326(1.5)	1.367	1.377	1.556
^{9}Be	3/2	−1.177	− .975(3.0)	−1.270	−1.324	
^{10}B	3	1.801	1.768(4.5)	1.812	1.812	
^{11}B	3/2	2.688	1.910(4.5)	2.631	2.488	
^{13}C	1/2	.702	0.670(6.0)	.757	.700	
^{13}B	3/2	3.177			<u>3.150</u>	
^{14}N	1	0.404	0.347(6.0)	0.331	0.326	

[a] $(2BME)'$ refers to a fit using jj-type matrix elements including data from $6 \leq A \leq 16$ rather than only $8 \leq A \leq 16$ as in $2BME$. Taken from Cohen and Kurath (65). The underlined value is from Kurath (71).

observed magnetic moments, Table 4.3, with substantial improvements over the Inglis model. Agreement with the experimental transition strengths, Tables 4.3 and 4.4, is generally better for the strong decays as compared to the weak decays. This is not surprising. The method should work best when a description in terms of $p_{1/2}$ and $p_{3/2}$ orbitals is accurate. When these states give large transition probabilities, the contributions of the omitted small components involving for example excitations to the s and d shells can be omitted. But of course this is no longer correct when the transition probabilities are small.

A major point of interest is the decay of ^{14}C and ^{14}O both of which are remarkably long-lived. The model results obtained are still far from the experimentally observed (ft) values of 2×10^9 and 2×10^7. However, note that substantial improvement over the Inglis model is obtained in the $2BME$ model that gives 2.6×10^5 for both decays. Small changes in the wave functions and or small admixtures of s and d orbitals would provide the change needed for the almost complete cancellation to occur. This is the tenor of other attempts to understand this phenomena [Visscher and Ferrell (57), Sherr et al. (55)]. One is always suspicious of "accidental" cancellations of this kind. Is there a symmetry principle operating here?

Finally, the $E2/M1$ mixing ratios of Table 4.6 emphasizes the point made earlier (section VIII.9) that $E2$ transition probabilities agree more closely with experiment if the effective charge is increased, $\alpha_p = \alpha_n = 1/2$. This means that significant contributions to the wave function from outside the p-shell have been neglected; contributions that cannot be simulated by $p_{1/2}$ and $p_{3/2}$ wave functions and a two-body interaction.

One method for including configuration mixing effects is the use of a deformed potential, that is, the Nilsson model. Filling up the single-particle

TABLE 4.4 ft Values in Units of 10^3 sec for Allowed Gamow-Teller Transitions

Parent: Mass Number	Parent: Spin J	Parent: Isospin T	Daughter Spin	Exp	*LS*	jj	Inglis (*a*/*k*)	*POT*	*2MBE*
6	0	1	1	.81 ± .03	.73	1.32	.74(1.5)		0.79
8	2	1	2	500			467 (1.5)	58.9	336
9	3/2	3/2	3/2	316			115 (3.0)	47.9	56.3
			5/2	50.1			112 (3.0)	97.7	89.1
			1/2				138 (3.0)	46.8	83.2
			3/2*				77.9 (3.0)	53.7	214
			5/2*				725 (3.0)	115	51.3
10	0	1	1	1.0	.73	1.32	6.17(4.5)	.89	.87
			1*	63^{+26}_{-19}			1.0 (4.5)	7.4	5.25
12	1	1	0	13.1 ± 0.1		3.96	13.5 (4.5)	14.2	12
			2	115 ± 10			63 (4.5)	69	95.5
			1	3.5 ± 1.3			4.1 (4.5)	4.5	4.3
13	3/2	3/2	1/2	10.5 ± 0.5		3.3	6.2 (6.0)	9.2	7.8
			3/2	29.5 ± 3.4		2.6	24 (6.0)	52.5	38.9
			5/2	235 ± 5.0			40 (6.0)	170	955
			1/2*	42.6 ± 8.8				81	
14	0	1	1	2×10^6	.73	6.58	10 (6.0)	28.2	263
	0	1	1	$(2.1 \pm 0.3)10^4$	.73	6.58	10 (6.0)	31	263
			1*	1.20 ± 0.15				0.91	

The final isospin equals $T - 1$. Taken from Cohen and Kurath (65), the underlined values from Kurath (71).

levels yields an intrinsic wave function (see Section VI.II) that when combined with the proper rotational wave functions yield the wave functions for the states. Such a model has been developed by Kurath (59, 65), Kurath and Picman (59), and Clegg (61, 62). In this model the ground state of ^{12}C is oblately deformed, the $p_{3/2}$ levels splitting with increasing deformation, both levels decreasing in energy from their zero deformation value. By comparing this fit of the *p*-shell nuclei with that obtained from say the *POT* fit discussed above, a measure of the expected effective charge can be obtained. The details of this comparison are discussed by Poletti et al. (67) with the result that $\alpha_p = \alpha_n = 1/2$ seem quite reasonable for the transitions in Table 4.6 considering the limitations of the models.

TABLE 4.5 M1 Transitions in ^{11}B and ^{12}C Compared with Experiment (from Cohen and Kurath (65) and Kurath (71))

Spectrum	E
7/2	6.74
3/2*	5.02
5/2	4.44
1/2	2.12
3/2	0

Ground state magnetic moment, μ

Exp	*POT*	*2BME*	*jj*	*LS*
2.69 nm	2.63	2.49	3.79	3.43

Transition ($J_i \to J_f$)	B_{M1} (Exp) nm^2	B_{M1} (*POT*) nm^2
$1/2 \to 3/2$	$1.10 \pm .05$	2.00
$5/2 \to 3/2$	$0.53 \pm .05$	0.53
$3/2^* \to 3/2$	$0.99 \pm .07$	1.31
$\to 1/2$	$0.96 \pm .07$	0.85
$\to 5/2$	$<2.4 \pm .2$	2.84
$7/2 \to 5/2$	>0.005	0.01

J	E
1^+	12.73 MeV
0^+	0

*M*1 Transition in ^{12}C Compared with Experiment

Transition	B_{M1} (Exp)	*POT*	*jj*	*LS*
$1 \to 0$	0.95	0.77	3.75	0

TABLE 4.6 Experimental Values of the Mixing Ratio and Values Calculated from the Cohen and Kurath (65) Wave Function[a]

Nucleus	Transition	$x(E2/M1)^a$	$x(E2/M1)^b$	x_{exp}
^{13}C	$3.68 \to 0$	−0.054	−0.100	$-(0.096_{-0.021}^{+0.030})$
^{13}N	$3.51 \to 0$	+0.030	+0.068	$+(0.092 \pm 0.02)$
^{15}O	$6.18 \to 0$	0	−0.048	$-(0.16 \pm 0.016)$
^{15}N	$6.32 \to 0$	+0.081	+0.122	$+(0.12 \pm 0.015)$
^{11}B	$4.46 \to 0$	−0.123	−0.224	$-(0.20 \pm 0.02)$
^{11}C	$4.32 \to 0$	+0.090	+0.204	$+(0.17 \pm 0.02)$
^{14}N	$7.03 \to 0$	+0.361	+0.722	$+(0.60 \pm 0.1)$

[a] The values in the column labeled x^a were obtained without enhancement of the *E*2 rate while the values in x^b have been increased using an additional effective charge $\alpha_n = \alpha_p$ of 1/2.

5. ALLOWED UNFAVORED TRANSITIONS

When the Fermi or the Gamow–Teller matrix elements differ from zero the transition is said to be allowed. The selection rules that must be satisfied are given in (2.26) and (2.27). The parity and spin selection rules are absolute but the isospin-selection rule is not because of the presence in nuclei of isospin-breaking interactions, such as the Coulomb interaction. Empirically the allowed transitions divide roughly into two groups—those which are referred to as superallowed discussed in Sections 2, 3, and 4 and those which will be discussed in this section. The superallowed transitions occur only in light nuclei, for only in that mass region is it possible to obtain good overlap between the incident and final nuclei. This is illustrated in Fig. 2.5. However, in the β-decay of medium and heavy nuclei the overlap cannot be as complete. In these nuclei the number of neutrons is greater than the number of protons and in the shell model different particle levels are occupied. In positron decay, in which the proton converts into a neutron, the final states that give maximum overlap will generally be occupied. On the other hand in β^- decay the final proton level will not be occupied but now there will be a mismatch in energy because of the Coulomb interaction that shifts this particular proton level in the heavier nuclei toward positive energies.

The empirical evidence for these remarks is presented in Tables 5.1 and 5.2. Note that the $\log ft$ values are much larger than those which prevail for the data presented in Tables 4.1, 4.2, and 3.1. For these, except for the "accidents" of ^{14}C and ^{14}O, $\log ft$ ranges from about 3 to about 4. In terms of the single-particle value, the last column in Table 5.1 shows hindrance factors that range from 0.08 to 0.004. Table 5.2 provides a list of several nuclei for which the shell model decay is forbidden since selection rule (2.34) $\Delta l = 0$ is not satisfied. These decays are called l-forbidden decays. It is to be emphasized that these decays are all allowed insofar as selection rules (2.26) and (2.27) are concerned. The explanation for the reduced values of the matrix elements can be understood in much the same way as the hindrance factors for electromagnetic decays, discussed in Chapter 8. Take for example the case of l-forbidden decay of ^{31}Si into ^{31}P. The shell-model state of ^{31}Si is $\pi(d_{5/2})^6$, $\nu(d_{5/2})^6(s_{1/2})^2 d_{3/2}$. Through the action of the residual potential a $d_{5/2}$ proton is excited into the $s_{1/2}$ level (Fig. 5.1). The $d_{3/2}$ neutron can now decay into the resultant $d_{5/2}$ proton hole. Just as in the l forbidden electromagnetic transition (Chapter VIII, p. 756) the residual interaction changes the nature of the ground state of the decaying nucleus so that the nucleus can decay. Configuration mixing in the final nucleus state may also contribute. In any event one can expect a reduction in the matrix element for the decay from its maximum value since the state illustrated in Fig. 5.1b is only a component of the total wave function required to describe the ground state of ^{31}Si. In analogy with the discussion of magnetic moments (Chapter VIII, page 759) it is possible to introduce an effective one-body operator including in this case the combination $\tau_-[\sigma Y_2]$.

TABLE 5.1 Allowed Transitions among Odd-A Nuclei [Konopinski (66)]

Parent	Daughter	$\log_{10}(ft)_{exp}$	$\vert M_{GT}\vert^2_{exp}$	Transition	$\vert M_{GT}\vert^2_{theo}$	Hindrance Factor
$^{35}_{16}S_{19}$	$_{17}Cl_{18}$	4.98	0.048	$d_{3/2} \rightarrow d_{3/2}$	0.6	0.080
$^{37}_{18}A_{19}$	$_{17}Cl_{20}$	5.04	0.040	$d_{3/2} \rightarrow d_{3/2}$	0.6	0.067
$^{45}_{20}Ca_{25}$	$_{21}Sc_{24}$	5.94	0.005	$f_{7/2} \rightarrow f_{7/2}$	1.28	0.0039
$^{45}_{22}Ti_{23}$	$_{21}Sc_{24}$	4.59	0.110	$f_{7/2} \rightarrow f_{7/2}$	1.28	0.086
$^{49}_{21}Sc_{28}$	$_{22}Ti_{27}$	5.06	0.038	$f_{7/2} \rightarrow f_{7/2}$	1.28	0.030
$^{49}_{23}V_{26}$	$_{22}Ti_{27}$	5.10	0.035	$f_{7/2} \rightarrow f_{7/2}$	1.28	0.027
$^{61}_{27}Co_{34}$	$_{28}Ni_{33}$	5.20	0.028	$f_{7/2} \rightarrow f_{5/2}$	1.71	0.016
$^{61}_{29}Cu_{32}$	$_{28}Ni_{33}$	4.94	0.050	$p_{3/2} \rightarrow p_{3/2}$	1.67	0.030
$^{63}_{30}Zn_{35}$	$_{29}Cu_{34}$	5.40	0.017	$p_{3/2} \rightarrow p_{3/2}$	1.67	0.010
$^{69}_{30}Zn_{39}$	$_{31}Ga_{38}$	4.37	0.186	$p_{1/2} \rightarrow p_{3/2}$	2.67	0.070
$^{71}_{32}Ge_{39}$	$_{31}Ga_{40}$	4.30	0.219	$p_{1/2} \rightarrow p_{3/2}$	2.67	0.082
$^{71}_{33}As_{38}$	$_{32}Ge_{39}$	5.80	0.0069	$f_{5/2} \rightarrow f_{5/2}$	0.71	0.0097
$^{77}_{33}As_{44}$	$_{34}Se_{43}$	5.76	0.0076	$p_{3/2} \rightarrow p_{1/2}$	1.33	0.0057
$^{77}_{35}Br_{42}$	$_{34}Se_{43}$	5.36	0.019	$p_{3/2} \rightarrow p_{1/2}$	1.33	0.0014
$^{81}_{34}Se_{47}$	$_{35}Br_{46}$	4.72	0.083	$p_{1/2} \rightarrow p_{3/2}$	2.67	0.031
$_{34}Se^{*}_{49}$	$^{83}_{35}Br_{48}$	5.22	0.026	$p_{1/2} \rightarrow p_{3/2}$	2.67	0.010
$^{83}_{35}Br_{48}$	$_{36}Kr_{47}$	5.13	0.032	$p_{5/2} \rightarrow p_{1/2}$	1.33	0.024
$_{42}Mo^{*}_{49}$	$^{91}_{41}Nb_{50}$	5.72	0.0083	$p_{3/2} \rightarrow p_{1/2}$	0.33	0.025
$^{95}_{41}Nb_{54}$	$_{42}Mo_{53}$	5.08	0.036	$g_{9/2} \rightarrow g_{7/2}$	1.78	0.020
$^{97}_{41}Nb_{56}$	$_{42}Mo_{55}$	5.35	0.020	$g_{9/2} \rightarrow g_{7/2}$	1.78	0.011
$^{111}_{50}Sn_{61}$	$_{49}In_{62}$	4.69	0.089	$g_{7/2} \rightarrow g_{9/2}$	2.22	0.040
$^{121}_{50}Sn_{71}$	$_{51}Sb_{70}$	5.00	0.044	$d_{3/2} \rightarrow d_{5/2}$	2.40	0.018
$^{121}_{53}I_{68}$	$_{52}Te_{69}$	5.05	0.039	$d_{5/2} \rightarrow d_{3/2}$	1.60	0.024
$^{123}_{50}Sn_{73}$	$_{51}Sb_{72}$	5.27	0.023	$d_{3/2} \rightarrow d_{5/2}$	2.40	0.009
$^{127}_{52}Te_{75}$	$_{53}I_{74}$	5.66	0.0096	$d_{3/2} \rightarrow d_{5/2}$	2.40	0.004
$^{131}_{55}Cs_{76}$	$_{54}Xe_{77}$	5.50	0.014	$d_{3/2} \rightarrow d_{3/2}$	0.60	0.023
$^{133}_{54}Xe_{79}$	$_{55}Cs_{78}$	5.58	0.012	$d_{3/2} \rightarrow d_{5/2}$	2.40	0.005

TABLE 5.2 Apparent l-Forbidden Transitions Among Odd-A Nuclei [Konopinski (66)]

Parent	Daughter	Transition	$\log_{10}(ft)_{exp}$
$^{31}_{14}Si_{17}$	$^{31}_{15}P_{16}$	$d_{3/2} \rightarrow s_{1/2}$	5.51
$^{33}_{15}P_{18}$	$^{33}_{16}S_{17}$	$s_{1/2} \rightarrow d_{3/2}$	5.05
$^{63}_{28}Ni_{35}$	$^{63}_{29}Cu_{34}$	$f_{5/2} \rightarrow p_{3/2}$	6.56
$^{65}_{28}Ni_{37}$	$^{65}_{29}Cu_{36}$	$f_{5/2} \rightarrow p_{3/2}$	6.56
$^{65}_{30}Zn_{35}$	$^{65}_{29}Cu_{36}$	$f_{5/2} \rightarrow p_{3/2}$	7.34
$^{67}_{29}Cu_{38}$	$^{67}_{30}Zn_{37}$	$p_{3/2} \rightarrow f_{5/2}$	6.30
$^{69}_{32}Ge_{37}$	$^{69}_{31}Ga_{38}$	$f_{5/2} \rightarrow p_{3/2}$	6.4

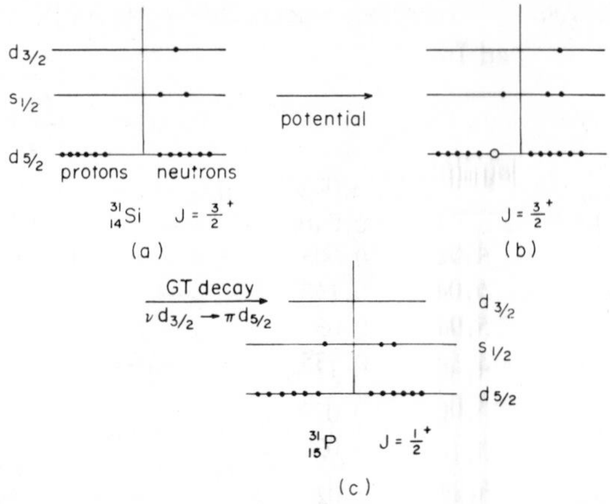

FIG. 5.1. *l*-forbidden decay of ^{31}Si into ^{31}P.

Again we emphasize the sensitivity of these matrix elements to the nature of the residual interaction.

The analogy with electromagnetic transitions can also be exploited to understand isospin forbidden β transitions. It will be recalled (Section VIII.11.B) that $E1$ transitions among low-lying nuclear states are inhibited because most of the strength of $E1$ transition is contained in the giant $E1$ resonances based on the low-lying levels. This could be deduced from the empirical fact that the sum rule for $E1$ transitions to or from the ground state was to a great extent exhausted by the giant resonance. A substantially similar situation prevails for β-decay, the role of the giant resonance being played by the isobar analog resonance; the matrix element affected is M_F [Blin–Stoyle (69)].

The empirical situation is shown in Table 5.3 taken from Bhattacherjee et al. (67) as quoted by Blin–Stoyle (69). [The last column in this table will be discussed after (5.3)]. This table is not exhaustive. Lists of values have also been given by Bloom (64, 66), Daniel and Schmitt (65), Behrens (67), and Damgärd (68). There is considerable disagreement with respect to specific values but the order of magnitudes are all similar to that given in the Table 5.3. We observe that the hindrance factor is of the order of 10^{-6}, which is smaller in magnitude than the hindrance factors for $E1$ decay (Fig. VIII.11.6). On the other hand, consider the hindrance factors for the β-decays listed in Table 5.1. These transitions can also proceed via the Gamow–Teller matrix element M_{GT} (2.25) showing that this matrix element is at least an order of magnitude larger than M_F. Daniel and Schmitt give a table of the ratio M_F/M_{GT}, which indicates that this result is fairly general.

The relation of these observations to the existence of isobar-analog states will now be discussed. We briefly review the nature of the isobar-analog reso-

TABLE 5.3 Values of M_F for Isospin-Forbidden Decays [from Bhattacherjee et al. (67)]

Parent	Daughter	T_i	T_f	$\lvert M_F \rvert \times 10^3$	$\lvert \langle \psi_A \lvert H_c \psi_f \rangle \rvert$ (keV)
$^{20}_{9}F$	$^{20}_{10}Ne$	1	0	2.0 ± 3.5	12 ± 22
$^{24}_{11}Na$	$^{24}_{12}Mg$	1	0	0.7 ± 1.0	2.6 ± 3.7
$^{24}_{13}Al$	$^{24}_{12}Mg$	1	0	0.3 ± 3.7	1.1 ± 13.5
$^{41}_{18}A$	$^{41}_{19}K$	$\frac{5}{2}$	$\frac{3}{2}$	4.0 ± 11.4	12.7 ± 3.6
$^{44}_{21}Sc$	$^{44}_{20}Ca$	1	2	8.4 ± 3.0	16.7 ± 6.0
$^{52}_{25}Mn$	$^{52}_{24}Cr$	1	2	1.0 ± 1.8	3.0 ± 5.4
$^{56}_{27}Co$	$^{56}_{26}Fe$	1	2	0.5 ± 0.1	1.4 ± 0.3
$^{67}_{28}Ni$	$^{57}_{27}Co$	$\frac{1}{2}$	$\frac{3}{2}$	14.4 ± 2.6	54 ± 10
$^{65}_{28}Ni$	$^{65}_{29}Cu$	$\frac{9}{2}$	$\frac{7}{2}$	1.3 ± 1.9	3.6 ± 5.7
$^{156}_{63}Eu$	$^{156}_{64}Gd$	15	14	1.02 ± 0.05	3.3 ± 0.2
$^{170}_{71}Lu$	$^{170}_{70}Yb$	14	15	1.03 ± 0.15	2.5 ± 0.4
$^{234}_{93}Np$	$^{234}_{92}U$	24	25	$4.2^{+1.6}_{-1.0}$	10.8 ± 3.6

nance restricting the discussion to those elements that are relevant for β-decay. A more complete discussion is given in Volume II. (See also Chapter I, page 103.) The two nuclei of interest have Z-protons, N-neutrons, and $(Z+1)$-protons, and $(N-1)$-neutrons, respectively. The first is called the *parent*. In its ground state it has an isospin $(N-Z)/2$ labeled $T_>$ and a value of $T_3 = (Z-N)/2$. In the present context we are concerned with a parent nucleus which β-decays, although this is not generally true. The ground state of the resultant nucleus will have an isospin equal to $(N-Z)/2-1$, labeled $T_<$. The energy levels of each nucleus are shown in Fig. 5.2 in the *absence of the Coulomb interaction*. Although the number of nucleons in both cases is the same and although charge independence is taken to be exact, the ground state of the smaller isospin lies lower. This is because the nuclear states with the smaller value of T will have a wave function that is more symmetric in space and spin and therefore more able to take advantage of the short ranged nuclear forces. The corresponding levels of the (Z, N)-nucleus are present in

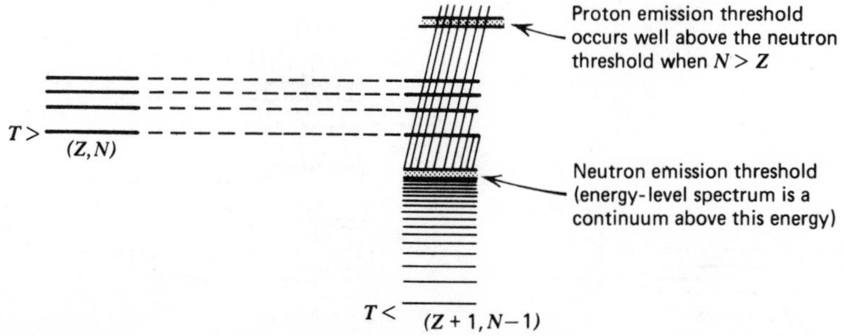

FIG. 5.2. Schematic energy-level diagram in the *absence* of the Coulomb interaction [taken from Feshbach and Kerman (67)].

the $(Z+1, N-1)$-system; for the heavier nuclei they lie in the continuum above the threshold for neutron emission. However, they remain bound states because their isospin differs from that of the isospin of the continuum levels. When isospin is conserved, there is no connection between these bound states and the neutron unstable states.

The introduction of the Coulomb force has three effects. Because of the Coulomb repulsion the entire spectrum of the $(Z+1, N-1)$-nucleus is displaced upwards (Fig. 5.3). Second, because of the repulsion, the protons are more readily emitted, and the proton threshold is lowered with respect to the neutron threshold. The $T_>$ states in the $(Z+1, N-1)$-nucleus are the *analog* of the low-lying levels in the (Z, N)-nucleus. They can be observed for example in the elastic proton scattering by the $(Z, N-1)$-nucleus where their presence results in a narrow resonance. The narrow width indicates the weak coupling between the $T_>$ and the $T_<$ states. In the light nuclei these analog levels are stable against particle emission. We restrict the discussion to the heavier nuclei where the situation illustrated in Fig. 5.3 obtains. A third effect of the Coulomb repulsion is the loss of isospin purity. The $T_>$ and $T_<$ states will no longer be exact eigenvectors of the operator T^2, that is, the Coulomb force will mix in components with other values of T so that the wave functions will have an "isospin impurity." Thus the states $T_>$ will now have $T_<$ admixtures. Since the isospin-symmetry breaking interaction is weak, a perturbative calculation of the admixture will suffice.

In the preceding discussion the situation pertaining to β^- emission has been described (Fig. 5.4*a*). A similar effect can also occur for β^+ emission from the excited state of the daughter nucleus to the parent state as illustrated in Fig. 5.4*b*. Since the Coulomb interaction is a tensor in isospin space (Chapter I, p. 33) the isospin of the decaying level can differ from the analog state by as much as two units. Of course the value of T_3 remains at $-(T-1)$. Note that the magnitude of the matrix element of the tensor component of the Coulomb interaction is smaller than that of the isovector component, which is much smaller than that of the isoscalar component.

To see the importance of this phenomenon for the Fermi matrix element,

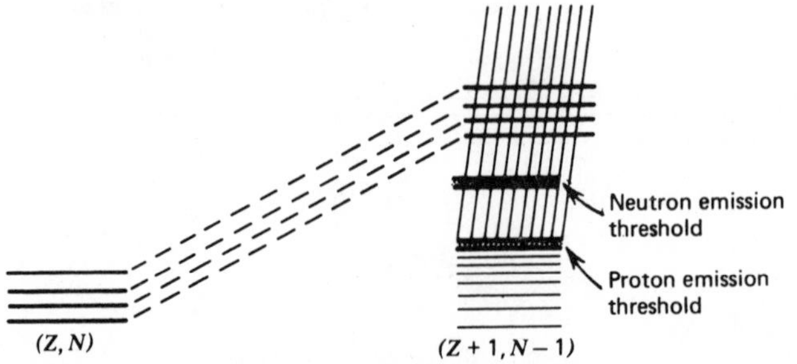

FIG. 5.3. Schematic energy-level diagram in the *presence* of Coulomb interaction [taken from Feshbach and Kerman (67)].

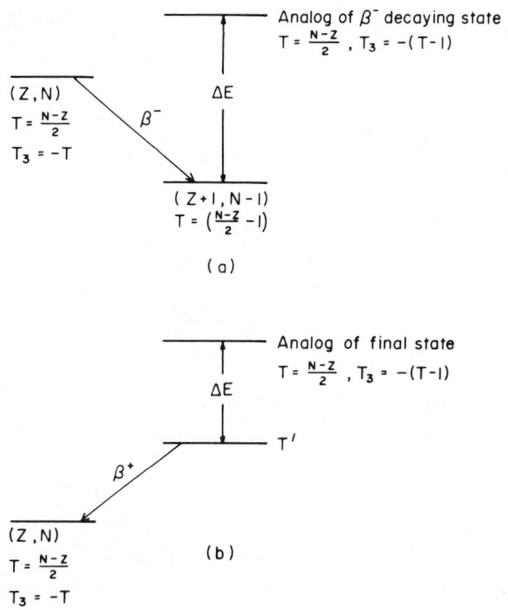

FIG. 5.4. Influence of isobar analog state on β^+ decay.

consider the β^- decay of the ground state of a (Z, N)-nucleus to the $(Z+1, N-1)$-nucleus (Fig. 5.4*a*).

$$M_F \equiv \langle \psi_f | T^{(+)} \psi_i \rangle$$

In the *absence* of the Coulomb interaction, that is, when ψ_i has a definite isospin, $T^{(+)}\psi_i$ is a state with the same isospin in the $(Z+1, N-1)$-nucleus. These are just the analog levels. These analogs because of their high energy cannot serve as final states. Energetically the transitions can only go to the $T_<$ less states for which M_F then vanishes. In another language the transition strength is, in the absence of the Coulomb interaction, entirely concentrated in the analog state $T^{(+)}\psi_i$ with the consequence that the matrix elements for transitions to other nuclear states are zero. Transitions occur only because of the isospin impurities in the nuclear states. Thus the observed decays are often referred to as "isospin-forbidden β-decays."

Let us calculate the contribution to M_F originating in the isospin impurity of the wave function given by the overlap of ψ_f with the analog state ψ_A:

$$\psi_f = \psi(T-1, T_3 = T-1) + \langle \psi_A | \psi_f \rangle \psi_A(T, T-1) \qquad (5.1)$$

The contribution of the isotensor Coulomb term is not included as it is expected to be of a much smaller magnitude. We obtain $\langle \psi_A | \psi_f \rangle$ by perturbation theory:

$$\langle \psi_A | \psi_f \rangle = -\frac{\langle \psi_A | H_c \psi_f \rangle}{\Delta E} \qquad \Delta E = E_A - E_f \qquad (5.2)$$

where H_c is the isospin symmetry breaking potential. Its principal component is the Coulomb interaction. ΔE is the energy difference between the analog state and the final state ψ_f. Then

$$|M_F|^2 = \left|\frac{\langle\psi_A|H_c\psi_f\rangle}{\Delta E}\right|^2 |\langle\psi_A|T^{(+)}\psi_i\rangle|^2$$

$$= [T(T+1) - T_3(T_3+1)]\left|\frac{\langle\psi_A|H_c\psi_f\rangle}{\Delta E}\right|^2$$

Finally,

$$|M_F|^2 = 2T\left|\frac{\langle\psi_A|H_c\psi_f\rangle}{\Delta E}\right|^2 \tag{5.3}$$

From the empirical value of $|M_F|^2$ and ΔE one can now obtain the value of the matrix element of the isospin breaking interaction H_c. These are given in Table 5.3. One word of caution. The very small values of $|M_F|^2$ may not be accurate because of the possibility that the transition will proceed via a second-forbidden transition that is present when finite wavelength of the electron and neutrino are taken into account. Since ΔE is of the order of 10 MeV it, follows that the isospin impurity as measured by $\langle\psi_A|\psi_f\rangle$ as given by (5.2) will be of the order of 10^{-3}, thus providing evidence of the isospin purity of these states. We also see that the diagonal matrix elements of H_c which give the nuclear Coulomb energy are very much larger than the nondiagonal ones. This conclusion also follows from the narrowness of the analog resonances, since the lifetime of these states is determined in part by the coupling of the analog state via H_c. This point is emphasized by Fujita and Ikeda (66a) who establish the connection of the analog width and M_F.

The theoretical calculation of the M_F from a nuclear model provides a test of these models. For a summary of the recent work in this field see Blin–Stoyle (69) as well as the papers of Fujita and Ikeda (65, 66a, 66b).

The hindrance factors for the Gamow–Teller matrix elements do not have as obvious an explanation as that for M_F. Fujita and Ikeda (66a) propose the existence of an analog resonance connected to the parent state by the application of the Gamow–Teller operator

$$\sum_i \boldsymbol{\sigma}(i)\tau^{\pm}(i)$$

to the parent state in analogy with the analog resonance state discussed above that was formed by applying $T^{(+)}$ to the parent state. In order that the effect not be as pronounced as on the Fermi case, the width of this resonance would have to be of the order of a few MeV as is clear from 5.2 and Table 5.1. In the isobar-analog case the analog involves the production of a constructively interfering proton particle–neutron hole states with $J^\pi = 0^+$ and $T = 0$. Similarly in the Gamow–Teller case the particle-hole combination would have $J^\pi = 1^+, T = 1$.

In general the hindrance factors will come from configuration mixing. The Fujita–Ikeda hypothesis is one possibility. It would be interesting to see if their analog resonance can be observed directly as in the case of the isobar-analog resonance.

6. ORBITAL ELECTRON CAPTURE

This process has been briefly mentioned at the beginning of Section 2. The elementary one-body transition involved.

$$p + e^- \to n + \nu_e \tag{6.1}$$

is closely related to positron emission

$$p \to n + e^+ + \nu_e \tag{6.2}$$

Of course the latter cannot occur in free space because the neutron's atomic mass is greater than the proton's. However both processes are possible in nuclei if E_0 the energy release as given in (2.4) and (2.5) is greater than zero:

$$E_0 = [M(Z, N) - M(Z - 1, N + 1) - 2m]c^2 \qquad \text{positron emission} \tag{6.3}$$

$$E_0 \cong [M(Z, N) - M(Z - 1, N + 1)]c^2 \qquad \text{orbital capture} \tag{6.4}$$

Since

$$E_0 \text{ (orbital capture)} = E_0 \text{ (positron emission)} + 2mc^2 - \text{electron binding energy} \tag{6.5}$$

it follows that there will be decays in which orbital capture will be energetically allowed but positron emission will be forbidden. This is the case in the decay of ^{37}A:

$$^{37}_{19}\text{A} + e^- \to {}^{37}_{18}\text{Cl} + \nu_e \tag{6.6}$$

The energy release in this process is 0.81 MeV, not sufficient for positron emission. On the other hand, when positron emission is possible, capture can always occur. The ratio of the probabilities for these two processes is, as we shall see, independent of the nuclear matrix elements M_F and M_{GT}.

A second difference occurs because reaction (6.6) involves a two-body final state whereas positron emission involves a three-body final state: the positron, the neutrino, and the daughter nucleus. The energy of the emitted positrons thus form a continuous spectrum extending up to E_0 (see Eq. 2.9). The energies of the neutrino and the recoiling nucleus in capture are unique. If the momentum of the neutrino is E_ν/c the momentum of the recoiling nucleus is also E_ν/c and the recoil energy is (neglecting the binding energy of the electron)

$$E_R = \frac{E_\nu^2}{2M_N c^2} \qquad (M_N \equiv \text{nuclear mass}) \tag{6.7}$$

while

$$E_R + E_\nu = E_0$$

Neglecting E_R in this equation so that $E_\nu \simeq E_0$ and

$$E_R \simeq \frac{E_0^2}{2M_N c^2} = 139 \frac{\epsilon_0^2}{A} \text{ eV} \tag{6.8}$$

where ϵ_0 is E_0 in units of mc^2. For the ^{37}A case of (6.6) where the energy release is 0.816 ± 0.004 MeV, E_R equals (9.67 ± 0.08) eV, which is to be compared with the experimental value of (9.63 ± 0.06) eV of Snell et al. (55a, 55b). Observing this recoil and its unique energy is a demonstration of the existence of the neutrino, assuming of course the conservation of linear momentum. For a summary of these experiments see Kofoed–Hansen (65).

The transition probability per unit time for orbital capture is given by

$$dw_c = \frac{2\pi}{\hbar} |H_{fi}|^2 \frac{d\mathbf{k}_\nu}{(2\pi)^3} \delta(E_0 - E_\nu) \tag{6.9}$$

Sums over the final spin states and averages over the initial must be performed. In comparing with positron emission, we note in that case the sum over the spin of positron was required whereas with present case an average over the spin of the capture electron is needed. However in K-capture, that is, capture of an electron in a K-orbit, note that there are two electrons in the K-orbit with opposite spin orientation. Since both of these contribute, the averages and spin sums give rise to identical factors for both cases. There are differences of course: for example, the electron density at the nucleus differs from that of the positron. Following (2.10) we write in the present case

$$|H_{fi}|^2 = g^2\rho |M_{fi}|^2$$

where ρ is the orbital electron density within the nucleus. We shall discuss it later. $|M_{fi}|^2$ is given by (2.23). Integrating (6.9) over $\mathbf{k}_\nu$ yields

$$w_K = \frac{mc^2}{\hbar} \frac{\Gamma^2}{\pi} \left(\frac{\hbar}{mc}\right)^3 \rho |M_{fi}|^2 \epsilon_\nu^2 \tag{6.10}$$

where Γ is the dimensionless coupling constants (2.11), and ϵ_ν is the neutrino energy in units of mc^2. We have used the subscript, K, to indicate (6.10) applies for capture from the K-orbit. The ratio of the K-capture probability to that for positron emission is from (2.17) given by

$$\frac{w_K}{w_{e^+}} = \frac{2\pi^2(\hbar/mc)^3\rho\epsilon_\nu^2}{f(-Z, R, \epsilon_0)} \tag{6.11}$$

This ratio is independent of the details of the weak interaction but it does depend upon properties of the nuclear systems through the density ρ. This result is reminiscent of the situation that prevails for the internal conversion coefficient of Section VIII.10. In analogy to that case, nuclear information is more easily extracted if the β decay matrix elements are unusually weak.

The density factor $(\hbar/mc)^3\rho$ determines the overall behavior of w_K with the nuclear charge. Using the nonrelativistic expression (2.13) for ρ at the nuclear center

$$\rho = \frac{Z^3}{\pi a_0^3} \tag{6.12}$$

where a_0 is the Bohr radius yields

$$\left(\frac{\hbar}{mc}\right)^3 \rho = \frac{1}{\pi}(\alpha Z)^3$$

where α is the fine structure constant. This yields the obvious result that the K-capture probability grows with increasing Z, reflecting the increasing attraction between the K-electron and the nucleus as Z increases, which leads to an increasing orbital electron density at the nucleus. The branching ratio (6.11) can also be evaluated in the limit of large ϵ_0, for then the simple limiting result (2.19) is valid for positron emission. One obtains

$$w_K/w_{e^+} \xrightarrow[\epsilon_0 \gg 1]{} 60\pi\left(\frac{\alpha Z}{\epsilon_0}\right)^3 \tag{6.13}$$

showing that for large energies positron emission becomes relatively more important. The source of this more rapid dependence of w_{e^+} on energy is in the three-body final state nature of the β^+ decay. The "phase space" for three body increases more rapidly with increasing energy than the two-body final state "phase space."

These features can be seen in the calculated branching ratios for allowed transitions shown in Figs. 6.1 and 6.2 [Feenberg and Trigg (50)]. The simple density (6.12) is replaced by its relativistic counterpart, a correction that is especially important for heavier nuclei. More precise atomic electronic wave functions that take screening and finite nuclear size into account have been in use in calculations of capture. For a critical review see the papers by Bouchez (52, 60) and Berenyi (63). The substantial agreement with theory is in-

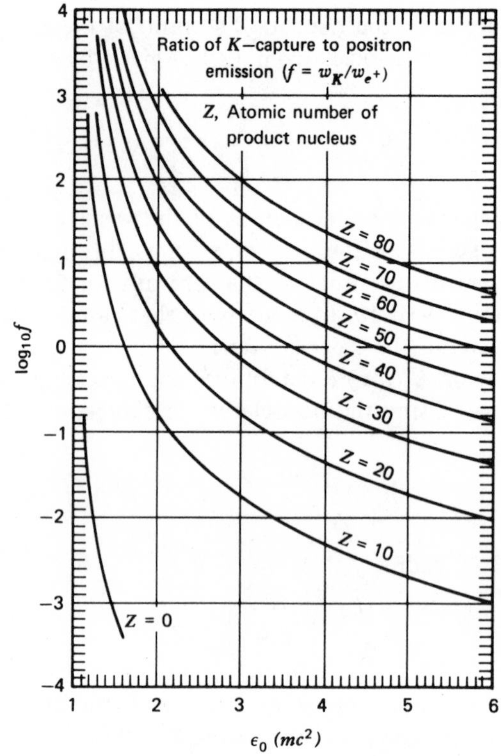

FIG. 6.1. The w_K/w_{e^+} ratio low energy. [taken from Rose (60) attributed to Feenberg and Trigg].

dicated in Table 6.1 given by Bouchez. Tables of the ratio (6.11) are given by Gove and Martin (71).

It is of course possible to capture electrons from L and higher orbits. Since electrons in these orbits are more weakly bound than the K-electrons there will be a range of energies for which K-capture will not be energetically allowed and L-capture will be. Outside of this relatively small energy range K-capture

TABLE 6.1 The Ratio of K-Capture to the Positron Emission Transition Probability [from Bouchez (60)]

Parent	(w_K/w_e $^+$) (theory)	(w_K/w^+) (experiment)
^{18}F	0.029	0.030 ± 0.002
^{48}V	1.066	0.068 ± 0.02
^{52}Mn	1.77	1.81 ± 0.07
^{107}Cd	310	320 ± 30
^{111}Sn	1.5	2.5 ± 0.25

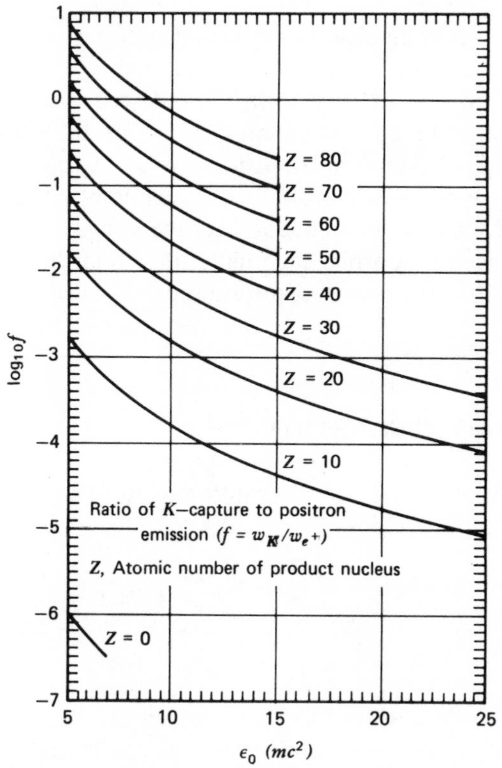

FIG. 6.2. The w_K/w_{e^+} ratio high energy. [taken from Rose (60) attributed to Feenberg and Trigg].

can be expected to be more probable than *L*-capture since the overlap of the nuclear and electronic wave functions will be more complete for the *K*-electron. For details see Brysk and Rose (58) and Gove and Martin (71).

Another phenomenon associated with orbital capture is electromagnetic in origin. There are of course x-rays that are generated when the hole left in the *K*-shell by the capture is filled by orbital electrons. But in addition radiation is generated because of the change in nuclear charge from Z to $Z + 1$. This is present in both *K*-capture and β-decay and is referred to as "inner bremsstrahlung." It is of some importance in *K*-capture because the maximum energy of the radiation equals the energy released in the decay. Measurement of this end point provides the most reliable method for determining E_0 in *K*-capture. For a review see Petterson (65).

As a final comment we note that orbital electron capture is one example of a widespread phenomenon of absorption of an orbital particle by an atomic nucleus. Other examples involve the μ^-, π^-, K^-, and the antiproton. The first of these also involves the weak interaction and will be discussed in this chapter

(Section 18). The others involve the strong interactions and will be described in Volume II. In each case the particle is captured in an outer atomic orbit and by ordinary atomic radiative transitions descends to orbits of smaller and smaller radii until because of its interaction with the nucleus is absorbed by it. In each case the absorption transition probability per unit time depends upon the overlap of the atomic orbital wave functions with the nucleus. This is true not only for the absorption but also for the x-rays emitted as the particle changes orbits, since these orbits eventually lie within the nucleus. In either event, absorption or x-ray energies, information on nuclear structure can be extracted.

7. ANTINEUTRINO ABSORPTION

An obvious variant of orbital electron capture of (6.1) is antineutrino absorption. The elementary process is

$$p + \bar{\nu}_e \rightarrow n + e^+ \tag{7.1}$$

Another example is

$$n + \nu_e \rightarrow p + e^- \tag{7.2}$$

Processes (7.1) can be observed for a hydrogen target if the energy of the antineutrino is sufficiently large. Both processes can be seen if the target is a nucleus

$$(Z, N) + \bar{\nu}_e \rightarrow (Z - 1, N + 1) + e^+ \tag{7.3}$$

while (7.2) in nuclei leads to

$$(Z, N) + \nu_e \rightarrow (Z + 1, N - 1) + e^- \tag{7.4}$$

The antineutrino needed for reaction (7.3) are produced plentifully in reactors. Here the principal β-decays are from neutron-rich elements so that the relevant decay process

$$(Z, N) \rightarrow (Z + 1, N - 1) + e^- + \bar{\nu}_e$$

Another source of electron neutrinos is the decay of μ-mesons:

$$\begin{aligned} \mu^- &\rightarrow e^- + \bar{\nu}_e + \nu_\mu \\ \mu^+ &\rightarrow e^+ + \nu_e + \bar{\nu}_\mu \end{aligned} \tag{7.5}$$

However, nuclear experiments with these neutrinos lie in the future as this is written.

Antineutrino ($\bar{\nu}_e$) absorption has been observed, the target material in one experiment being $CdCl_2$, the antineutrino source, the Savannah River high-flux reactor. The most recent experiment of this type that was initiated by Cowan, Reines et al (56) was performed in 1966 by Nezrick and Reines (66) who obtained the absorption cross section

$$\sigma_{\text{exp}} = (0.94 \pm 0.13)10^{-43} \text{ cm}^2$$

One of the problems in this measurement, and we mention it because it is common to all measurements of neutrino-induced reactions, is the determination of the incident antineutrino flux. In this case it is necessary to determine this flux as generated by the reactor in question.

It is easy to derive the cross section for antineutrino absorption. The result will suffice; the details are left as an "exercise" for the student. It is

$$\sigma = \left(\frac{\hbar}{mc}\right)^2 \frac{\Gamma^2}{\pi} |M_{fi}|^2 \epsilon\sqrt{\epsilon^2 - 1} \tag{7.6}$$

where ϵ is approximately the positron energy in units of mc^2.

In the discussion thus far it has been tacitly assumed that the antineutrino emitted in neutron decay differs from the neutrino emitted in proton decay (2.2). There is no necessity for this assumption and indeed Majorana (37) developed a theory in which the neutrino and antineutrino are identical particles. If this were the case the inverse of the reaction

$$^{37}\text{A} + e^- \rightarrow {}^{37}\text{Cl} + \nu$$

namely

$$^{37}\text{Cl} + \nu \rightarrow {}^{37}\text{A} + e^-$$

could be induced by antineutrinos as well as neutrinos. Since the reactor produces antineutrinos as discussed above, the Majorana hypothesis can be tested by seeing if ^{37}Cl would absorb reactor antineutrinos. This experiment was performed by R. Davis (55) who found a cross section of $(0.1 \pm 0.6) \times 10^{-45}$ cm^2 that is to be compared with the cross section 2×10^{-45} cm^2 calculated from (7.6). Thus, to within background errors arising from neutrinos developed from the decay of cosmic-ray muons, the neutrino and antineutrino do differ.

We shall discuss other experiments yielding the same conclusion such as double β-decay later in this chapter.

8. ELECTRON-NEUTRINO ANGULAR CORRELATION

In calculating β-decay lifetimes in the preceding sections we have summed over the spins and directions of emission of the electron and neutrino. In this section we shall investigate the dependence of the transition probability on the angle between the direction of motion of the electron and of the neutrino. We shall see that this dependence provides important information on the nature of the weak interaction. This electron–neutrino angular correlation is experimentally determined by measuring the angular correlation between the electron and the recoiling nucleus to which it is directly related because of the overall conservation of momentum and energy. For a description of these experiments and a summary of results see Wu and Moszkowski (66).

Let us begin by describing on an intuitive basis the origin of the correlation, and for simplicity let us begin with the Fermi matrix element. In that case it will be recalled (see page 785) that the electron antineutrino pair are emitted in a singlet spin state ($S = 0$), that is, their spins are antiparallel. Suppose (Fig. 8.1) that the electron is emitted in a given direction with *negative* helicity, that is, the spin of the electron is opposite to its direction of motion. If the antineutrino has *positive* helicity, its spin will cancel the spin of the electron if it moves in the same direction as the electron only as shown in Fig. 8.1*a*. Hence more generally we expect that under these circumstances that it will be more probable for the electron and neutrino to be emitted in the same direction. In terms of the canonical form for the angular correlation to be derived in Section 11

$$1 + a\mathbf{p}_e \cdot \mathbf{p}_{\bar{\nu}} \tag{8.1}$$

where $\mathbf{p}_e$ and $\mathbf{p}_{\bar{\nu}}$ are the electron and neutrino momenta, the constant a is positive. On the other hand, if the antineutrino helicity is negative the electron and neutrino go off in opposite directions as shown in Fig. 8.1*b*. The net angular correlation will thus depend upon which is more probable, the emission of a positive or negative helicity antineutrino, a matter of some importance for the fundamental β-interaction. These conclusions are not changed if the electron has the opposite, that is, positive helicity, the relative helicity being the crucial element. If it is more probable for the antineutrino to have a helicity opposite to that of the electron, the constant a is positive for the Fermi transition.

Turning next to the Gamow–Teller case in which the electron–neutrino pair are emitted in an $S = 1$ state, it is clear from Fig. 8.2 that the constant a is positive if emission of the antineutrino with the same helicity as the electron is more probable, negative if the emission with opposite helicity is more probable.

This is as far as we can go without a more detailed description of the β-interaction, which will shortly be given (Section 11). But the qualitative dis-

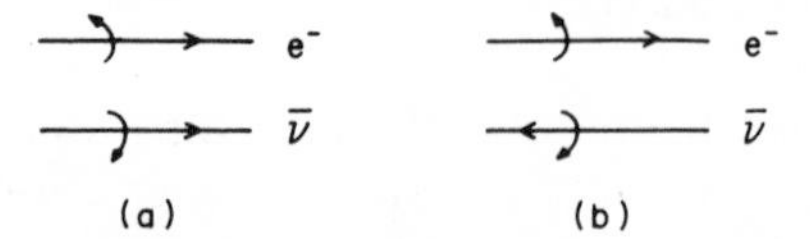

FIG. 8.1. Electron-neutrino angular correlation for a Fermi transition.

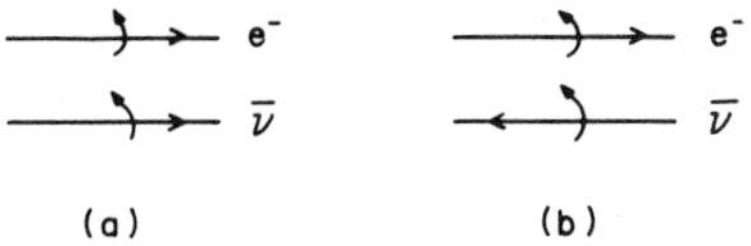

FIG. 8.2. Electron-neutrino angular correlation for a Gamow-Teller transition.

cussion given above permits us to conclude the following from the experimental results: for both Fermi and Gamow–Teller transitions it is more probable for the antineutrino and the electron to have opposite helicities; that is, the constant a in (8.1) is positive for Fermi transitions, negative for Gamow–Teller transitions. The analogous result holds for positron emission. In that case, the helicity of the positron and the neutrino are opposite.

9. THE HELICITY OF THE NEUTRINO, ELECTRON POLARIZATION

A most important property of the β-ray interaction is that the neutrino (or antineutrino) emitted in a β-decay has a definite helicity. A priori it would be expected that the antineutrino of Fig. 8.1 could have both helicities illustrated. This is not the case; the antineutrino exists in only one helicity state, that shown in Fig. 8.1*a*. It is possible to demonstrate this fact directly for the neutrino. An ingenious experiment for this purpose has been devised and performed by Goldhaber, Grodzins, and Sunyar (58). The first element is a neutrino source for which the neutrino direction of motion can be determined. This is provided by a *K*-capture neutrino, the direction of the neutrino emission being determined by observing the direction of motion of the recoiling nucleus (Section 6). A second element is the associated emission of a gamma ray, whose helicity is directly related to the helicity of the neutrino. The helicity of the gamma ray, that is, its circular polarization can be determined by its transmission through magnetized iron.

The reaction used by Goldhaber et al. is shown in Fig. 9.1. The helicity of the photon emitted in the transition $^{152}Sm^*$ to ^{152}Sm is related to that of the neutrino by the conservation of angular momentum. In this experiment the helicity of the photon whose direction of motion is opposite to that of neutrino is determined. Choosing the neutrino direction to be the "z"-direction, the z-components of the angular momentum of the *K*-electron, neutrino, and the spin of $^{152}Sm^*$ must satisfy the following equation

$$(\text{spin of } K_{\text{electron}})_z = (\text{spin of neutrino})_z + (\text{spin of } ^{152}\text{Sm}^*)_z$$

Thus if the helicity of the neutrino is -1, the component in the z direction of the spin of $^{152}Sm^*$ is either 0 or 1. The emitted gamma ray must then have a helicity of 1 in order to deexcite the $(+1)$ substate of $^{152}Sm^*$. This can be most

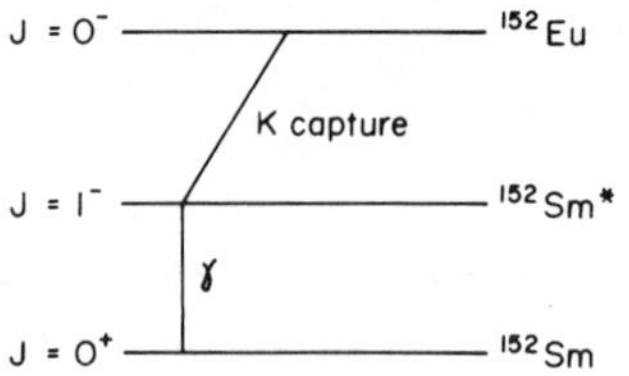

FIG. 9.1. Decay scheme for ^{152}Eu.

readily seen from the overall balance since the spins of the initial and final nuclei are zero. Hence

$$(\text{spin of } K_{el})_z = (\text{spin of neutrino})_z + (\text{spin of photon})_z \tag{9.1}$$

Thus if the neutrino has a negative helicity, the photon must have a positive helicity if (9.1) is to be satisfied (Fig. 9.2). Of course if the helicity of the neutrino is not definite, that is, if there is an equal probability for the emission of a neutrino or negative helicity, the photon will also show no definite helicity.

For a description of the experiment and the way in which the various experimental conditions were met see the original paper [Goldhaber et. al. (58)]. The results show that the neutrino has an excess of negative helicity, the amount being very close to a value that would be obtained if the neutrino helicity were exactly -1, that is, the results show that the neutrino is left circularly polarized.

An immediate consequence of the results quoted in the preceding section is that in a positron decay the positrons are polarized since in the decay the helicities of the positron and neutrino are opposite. Moreover the polarization of the positrons should be positive. Applying a similar analysis to the electrons emitted in β^- decay, the polarization of the electrons is negative as in Fig. 8.1. These electron and positron polarizations have been measured by a number of methods summarized by Wu and Moskowski (66) and these conclusions

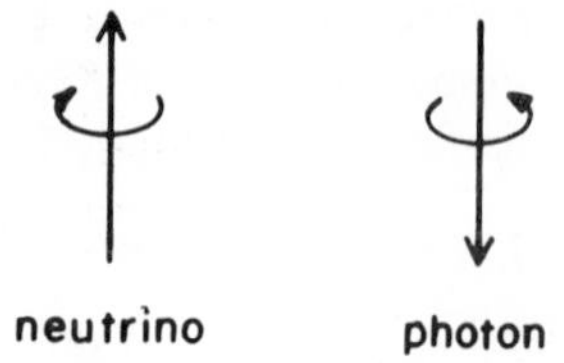

FIG. 9.2. Angular momentum conservation in the decay of ^{152}Eu.

verified. The polarizations are of the correct sign and of the order of magnitude v_e/c where v_e is the velocity of the electron or positron.*

But there is still another consequence. Our world is one in which all neutrinos are left circularly polarized. There are no right circularly polarized neutrinos. The state (e.g., right circularly polarized neutrino) obtained from one known to exist (e.g. left circularly polarized neutrino) by reflection ($\mathbf{r} \rightarrow -\mathbf{r}$) is not realized in nature. Thus states that remain unchanged under reflection cannot be prepared. In other words *parity is not conserved in the weak interactions.*

As discussed earlier, the "fall of parity" was first clearly observed in the classic experiments of Wu et al, (57). We turn now to this experiment that involved the β decay of a polarized nucleus.

10. THE "FALL OF PARITY"; β-DECAY OF A POLARIZED NUCLEUS

This writer remembers with great vividness the excitement that swept the world of physics upon the discovery that parity is not conserved in the weak interactions. The possibility had been voiced by many in connection with the τ–θ puzzle. In more modern terminology the neutral kaon was found to have two modes of decay, with two pions and three pion final states, respectively, with a short (S) and long (L) lifetime.

$$K^0 \rightarrow \begin{cases} K_S^0 & \begin{matrix} \rightarrow \pi^+ + \pi^- \\ \rightarrow \pi^0 + \pi^0 \end{matrix} \\ K_L^0 & \begin{matrix} \rightarrow \pi^+ + \pi^- + \pi^0 \\ \rightarrow \pi^0 + \pi^0 + \pi^0 \end{matrix} \end{cases}$$

But these two final states were found to be of different parity (the pions are in S-states and have intrinsic negative parities). A parity-conserving model in which there were two kaons whose masses were accidentally equal was suggested. It was the brilliant suggestion of Lee and Yang (56, 57) that by examining the weak interaction of other systems the issue might be resolved. Implicit was the concept that the weak interactions had universal properties; that is, if parity is not conserved in K-decay it is not conserved in nuclear β-decay.

Let us briefly discuss the notion of parity conservation. In an elementary form it states that, by operating on an eigenstate Ψ of a Hamiltonian H with

---

*It is interesting to note that the first evidence for the polarization of β-ray electrons is present in the early work of Cox et al. (28). However the effect was small and uncertain because of several technical problems.

the inversion operator $R(R^{-1}\mathbf{x}R = -\mathbf{x})$, a new state $R\Psi$ is formed that is also an eigenstate of H. The condition on H which must then hold is

$$[R, H] = 0 \tag{10.1}$$

From this condition it follows directly (see Volume II) that the transition operator* $\mathfrak{I}$ commutes with R and, therefore, that the transition operator is invariant under inversion if parity is conserved:

$$[R, \mathfrak{I}] = 0$$

or

$$R^{-1}\mathfrak{I}R = \mathfrak{I} \tag{10.2}$$

$\mathfrak{I}$ generally will be a function of momenta and spin operators; $\mathfrak{I}(\mathbf{p}, \boldsymbol{\sigma})$. Result (10.2) reads

$$R^{-1}\mathfrak{I}(\mathbf{p}, \boldsymbol{\sigma})R = \mathfrak{I}(-\mathbf{p}, \boldsymbol{\sigma}) = \mathfrak{I}(\mathbf{p}, \boldsymbol{\sigma}) \tag{10.3}$$

since

$$R^{-1}\mathbf{p}R = -\mathbf{p} \qquad \text{and} \qquad R^{-1}\boldsymbol{\sigma}R = \boldsymbol{\sigma} \tag{10.4}$$

Result (10.3) *holds only if parity is conserved.* We can now see formally why the experiment of Goldhaber et al. (58) demonstrates the violation of parity conservation. The nondegeneracy of the nuclear states involved is assumed. The quantity describing the spin state of the emitted neutrino is the expectation value of $(\boldsymbol{\sigma}\cdot\mathbf{p}/|p|)$ that is

$$\left\langle\frac{\boldsymbol{\sigma}\cdot\mathbf{p}}{|p|}\right\rangle = \langle\psi_f|\frac{\boldsymbol{\sigma}\cdot\mathbf{p}}{|p|}|\psi_f\rangle = \langle\mathfrak{I}\psi_i|\frac{\boldsymbol{\sigma}\cdot\mathbf{p}}{|p|}|\mathfrak{I}\psi_i\rangle = \langle\psi_i|\mathfrak{I}^\dagger\frac{\boldsymbol{\sigma}\cdot\mathbf{p}}{|p|}\mathfrak{I}|\psi_i\rangle$$

This result holds for a given initial state. Averaging over the initial state yields

$$\overline{\left\langle\frac{\boldsymbol{\sigma}\cdot\mathbf{p}}{|p|}\right\rangle} = \frac{\sum_i \langle\psi_i|\mathfrak{I}^+(\boldsymbol{\sigma}\cdot\mathbf{p}/|p|)\mathfrak{I}|\psi_i\rangle}{\sum_i \langle\psi_i|\psi_i\rangle} = \frac{\operatorname{tr}\,\mathfrak{I}^\dagger(\boldsymbol{\sigma}\cdot\mathbf{p}/|p|)}{\operatorname{tr}\,1} \tag{10.5}$$

Assuming parity conservation, that is, $\mathfrak{I}$ can be replaced by $R^{-1}\mathfrak{I}R$ according to (10.3) so the numerator of (10.5) can be written

$$\operatorname{tr}\,R^{-1}\mathfrak{I}^\dagger R\frac{\boldsymbol{\sigma}\cdot\mathbf{p}}{|p|}R^{-1}\mathfrak{I}R = \operatorname{tr}\,\mathfrak{I}^\dagger R\frac{\boldsymbol{\sigma}\cdot\mathbf{p}}{|p|}R^{-1}\mathfrak{I} = -\operatorname{tr}\,\mathfrak{I}^\dagger\frac{\boldsymbol{\sigma}\cdot\mathbf{p}}{|p|}\mathfrak{I}$$

---

*$\mathfrak{I}$ is the operator whose matrix elements between initial and final states gives the transition amplitude.

In obtaining this result the theorem tr AB = tr BA and (10.4) were used. As a result we have

$$\overline{\left\langle\frac{\boldsymbol{\sigma}\cdot\mathbf{p}}{|p|}\right\rangle} = -\overline{\left\langle\frac{\boldsymbol{\sigma}\cdot\mathbf{p}}{|p|}\right\rangle}$$

or

$$\overline{\left\langle\frac{\boldsymbol{\sigma}\cdot\mathbf{p}}{|p|}\right\rangle} = 0 \qquad \text{if parity is conserved} \tag{10.6}$$

In general for any operator $O(\mathbf{p}, \boldsymbol{\sigma})$

$$\langle\overline{O(\mathbf{p}, \boldsymbol{\sigma})}\rangle = \langle\overline{O(-\mathbf{p}, \boldsymbol{\sigma})}\rangle \qquad \text{if parity is conserved} \tag{10.7}$$

In the experiment of Goldhaber et al. $\langle\overline{(\boldsymbol{\sigma}\cdot\mathbf{p}/|p|)}\rangle$ for the emitted neutrino differed from zero, being in fact negative. Hence, parity is not conserved.

This rather heavy-handed proof furnishes the justification for saying that the *nonzero expectation of a pseudoscalar quantity demonstrates the nonconservation of parity.*

In the experiment of Wu et al (57), the pseudoscalar quantity investigated $\mathbf{p}_e\cdot\mathbf{J}$ where $\mathbf{p}_e$ is the momentum of the emitted electron and $\mathbf{J}$ is the spin of the emitting nucleus that is polarized by placing it in a suitable environment. The decay scheme of the radioactive nucleus employed ^{60}Co is shown in Fig. 10.1. The nucleus was lined up by the strong internal magnetic field of the crystal [2 Ce $(No_3)_3\cdot 3$ Mg $(NO_3)_2$; 24 H_2O]. It was found that $\langle\mathbf{p}_e\cdot\mathbf{J}\rangle$ is negative, that is, there is a preferential emission of electrons in the direction opposite to that of the nuclear spin, proving that parity is not conserved.

This result can be readily understood in terms of the angular momentum balance, the positive helicity of the antineutrino and the electron negative helicity. The ^{60}Co transition is from a $J = 5$ to a $J = 4$ state. Thus the electron antineutrino pair must have a unit angular momentum in the direction of the ^{60}Co spin. The balance is shown in Fig. 10.2. In order to obtain the total spin of unity the antineutrino must travel in a direction along the direction of the nuclear spin, the electrons in the opposite direction. Again this conclusion can be tested by measuring the electron polarization.

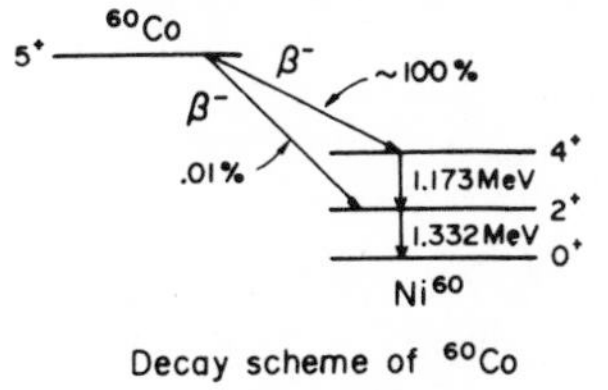

FIG. 10.1. Decay scheme for ^{60}Co.

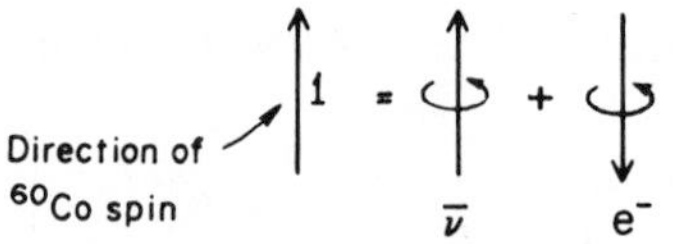

FIG. 10.2. Angular momentum balance in β^- Decay of ^{60}Co.

11. FORMAL β-DECAY THEORY

It is time to make the arguments of the preceding sections more precise and quantitative. For this purpose it is necessary to formulate a theory of β-decay. In this section the historical pattern will be followed, but in later sections it will be shown how the consequent description is part of a universal theory of weak interactions that applies not only to nuclear β-decay but also to the decay of relatively long-lived baryons (e.g., $\Lambda \rightarrow p + \pi^-$) of mesons (e.g., $\pi^+ \rightarrow \mu + \nu_\mu$) and of the muon ($\mu^- \rightarrow e^- + \nu_\mu + \bar{\nu}_e$).

The theory as first developed by Fermi (34) was patterned after the field theory of quantum electrodynamics. In analogy with that theory the starting point is a postulated interaction Hamiltonian density $\mathcal{H}$ expressed in terms of the electron–neutrino and nucleon fields. The Hamiltonian density is taken to be Lorentz invariant. (For interactions that do not involve derivative couplings, Lorentz invariance of the interaction Hamiltonian density follows directly from the Lorentz invariance of the Lagrangian density). The Lorentz invariant forms we shall employ are suggested by the electromagnetic interaction Hamiltonian density:

$$\mathcal{H} = -\frac{e}{c}\sum_\mu j_\mu A_\mu \tag{11.1}$$

where j_μ is the four-vector current and A_μ is the four-vector potential. The current itself is a bilinear form. Non-relativistically the convection current (VIII.2.3) is

$$\mathbf{j} = \frac{\hbar}{2im}(\hat{\psi}^*\nabla\hat{\psi} - \hat{\psi}\nabla\hat{\psi}^*), \qquad \rho = \hat{\psi}^*\hat{\psi}$$

Relativistically (the Dirac equation and operators are described in the appendix p. 910)

$$j_\mu = ic\hat{\bar{\psi}}\gamma_\mu\hat{\psi} \tag{11.2}$$

where

$$\hat{\bar{\psi}} = \hat{\psi}^\dagger\gamma_4$$

$$\mathbf{j} = c\hat{\psi}^\dagger\boldsymbol{\alpha}\hat{\psi} \qquad \rho = \hat{\psi}^\dagger\hat{\psi} \tag{11.3}$$

The Hamiltonian density in terms of (11.2) is $\mathcal{H}_\gamma = -ie\sum_\mu \hat{\bar{\psi}}_\mu\gamma_\mu\hat{\psi}A_\mu$. Since the operator $\hat{\psi}$ destroys (creates) an electron (positron) and $\hat{\bar{\psi}}$ does the

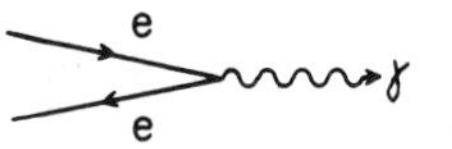

FIG. 11.1. The e-eγ electromagnetic vertex.

reverse creating (destroying) an electron (positron) the structure of $(\mathcal{H}_\gamma)$ can be represented by the diagram of this elementary vertex as illustrated in Fig. 11.1.

The corresponding diagram for β^- decay is shown in Fig. 11.2. Comparing the two figures suggests an analogy between the electron–electron combination of Fig. 11.1 and the proton–neutron lines of Fig. 11.2, with the photon line of Fig. 11.1 replaced by the electron–neutrino combination. [In principle it is possible to associate the electron and proton line and the neutrino and neutron lines, for example. However it can be shown that this produces no real change in $\mathcal{H}_\beta$ if the original form is sufficiently general. The transformation that proves the identity of the various possible pairings in Fig. 11.2 is known as the Fierz transformation (Good (55)).] This analogy suggests that forms bilinear in the nucleon fields and in the lepton fields be contracted to form the Hamiltonian density. For example one such term could be

$$\sum_\mu \hat{\bar{\psi}}_p \gamma_\mu \hat{\psi}_n \hat{\bar{\psi}}_e \gamma_\mu \hat{\psi}_\nu + \text{h.c.} \tag{11.5}$$

where h.c. ≡ hermitian adjoint. In the first term of (11.5) the fields have the following effects: $\hat{\psi}_\nu$ creates an antineutrino or destroys a neutrino, $\hat{\bar{\psi}}_e$ creates an electron or destroys a positron, $\hat{\psi}_n$ destroys a neutron or creates an antineutron, and $\hat{\bar{\psi}}_p$ creates a proton or destroys an antiproton. The processes that the first term in (11.5) can describe include

$$n \to p + \bar{\nu}_e + e^-$$

$$\nu_e + n \to p + e^-$$

$$e^+ + n \to p + \bar{\nu}_e$$

$$\bar{p} \to n + \bar{\nu}_e + e^-$$

The expansion of a spinor field in creation and destruction operators that lies behind this statement is

$$\hat{\psi} = \sum_{\mathbf{p},s} [\psi^{(+)}(\mathbf{p}, s)\hat{a}_{\mathbf{p},s} + \psi^{(-)}(\mathbf{p}, s)\hat{b}^\dagger_{-\mathbf{p},-s}] \tag{11.6}$$

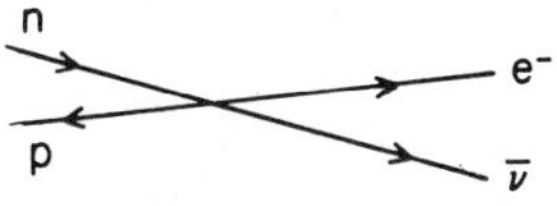

FIG. 11.2. The weak interaction vertex.

Here $\psi^{(+)}(\mathbf{p}, s)$ is the positive energy solution of the Dirac equation with momentum $\mathbf{p}$ and "z"-component of the spin, s, while $\psi^{(-)}$ is the negative energy solution. The operator $\hat{a}_{\mathbf{p},s}$ destroys an electron of momentum $\mathbf{p}$ in spin state s described by wave function $\psi^{(+)}(\mathbf{p}, s)$ while $\hat{b}^{\dagger}_{-\mathbf{p},-s}$ creates a positron of momentum $(-\mathbf{p})$ spin state $-s$, whose wave function is obtained from $\psi^{(-)}(\mathbf{p}, s)$ by charge conjugation

$$\psi_c = -\gamma_2 [\psi^{(-)}(\mathbf{p}, s)]^* \tag{11.7}$$

The abbreviation "h.c." requires the addition of the hermitian adjoint operator in order that the total expression be hermitian. In the case of (11.5)

$$\text{h.c.} = \sum_{\mu} \hat{\bar{\psi}}_n \gamma_\mu \hat{\psi}_p \hat{\bar{\psi}}_\nu \gamma_\mu \hat{\psi}_e \tag{11.8}$$

In one of the transitions such a term describes, a positron, neutrino, and neutron are created and a proton destroyed; in other words, positron β-decay.

There are many other invariants that can be formed besides (11.5), the analog of the electromagnetic case (11.4). To classify these it is useful to list the various bilinear combinations with specific behavior under a Lorentz transformation. We shall omit subscripts for the field operators since this classification will apply to both the baryonic and leptonic fields.

TABLE 11.1 Bilinear Covariants

Scalar	$\hat{\bar{\psi}}\ \hat{\psi}$
Vector	$\hat{\bar{\psi}}\ \gamma_\mu\ \hat{\psi}$
Tensor	$\hat{\bar{\psi}}\ \sigma_{\mu\nu}\ \hat{\psi}$
Axial vector	$\hat{\bar{\psi}}\ \gamma_\mu\ \gamma_5\ \hat{\psi}$
Pseudoscalar	$\hat{\bar{\psi}}\ \gamma_5\ \hat{\psi}$

The pseudoscalar behaves like a scalar under a Lorentz transformation. However under an inversion $(\mathbf{r} \to -\mathbf{r})$ it changes sign. Under an inversion

$$\hat{\psi} \to \gamma_4 \hat{\psi}$$

so that

$$\hat{\bar{\psi}}\gamma_5\hat{\psi} \to \hat{\bar{\psi}}\gamma_4\gamma_5\gamma_4\hat{\psi} = -\hat{\bar{\psi}}\gamma_5\hat{\psi}$$

Similarly the space components of the axial vector do not change while the fourth component changes sign:

$$\hat{\bar{\psi}}\gamma_k\gamma_5\hat{\psi} \to \hat{\bar{\psi}}\gamma_4\gamma_k\gamma_5\gamma_4\hat{\psi} = \hat{\bar{\psi}}\gamma_k\gamma_5\hat{\psi}$$

$$\hat{\bar{\psi}}\gamma_4\gamma_5\hat{\psi} \to \hat{\bar{\psi}}\gamma_4\gamma_4\gamma_5\gamma_4\hat{\psi} = -\hat{\bar{\psi}}\gamma_4\gamma_5\hat{\psi}$$

This behavior under inversion is opposite to that of a four vector.

A combination of terms of the form

$$\sum_{\mu} \hat{\bar{\psi}}_p \gamma_\mu \hat{\psi}_n \hat{\bar{\psi}}_e (C_V \gamma_\mu + C'_V \gamma_\mu \gamma_5) \hat{\psi}_\nu \tag{11.9}$$

where C_V and $C_{V'}$ are constants, is Lorentz invariant. It is not invariant against an inversion and thus will not conserve parity.

It will be convenient for later considerations to use nucleon wave operators instead of specifying whether one is dealing with a neutron or a proton. This is simply done by introducing the appropriate isospin operators so that in (11.5), the wave operator $\hat{\psi}$ is now being a two-element matrix in isospin space:

$$\hat{\psi} = \begin{pmatrix} \hat{\psi}_p \\ \hat{\psi}_n \end{pmatrix}$$

Then (11.9) can be written

$$\sum_{\mu} \hat{\bar{\psi}} \tau^{(+)} \gamma_\mu \hat{\psi} \hat{\bar{\psi}}_e (C_V \gamma_\mu + C'_V \gamma_\mu \gamma_5) \hat{\psi}_\nu$$

Following this as an example it is easy to see how to write down the most general Lorentz invariant interaction that does not necessarily conserve parity. And one can go systematically through the various experiments discussed earlier and finally select that interaction which yields closest agreement. This is too long a program to be carried out here. The details can be found in various monographs and review papers [Wu and Moskowski (66) and Konopinski (66)].

We restrict the general discussion in this section to allowed transitions only and even so will give only a few typical calculations rather than present an exhaustive and exhausting discussion.

For allowed transitions the full relativistic description is not needed for the nuclear matrix elements. (They do enter in forbidden transitions.) We may therefore take the nonrelativistic limits of the forms in Table 11.1 with the understanding that the $\hat{\psi}$'s refer to nucleons and not to leptons. The results are given in Table 11.2. (*Note:* $\hat{\bar{\psi}} = \hat{\psi}^\dagger \gamma_4$.)

As expected only terms that eventually lead to Fermi matrix elements, $\hat{\psi}^\dagger \hat{\psi}$, and to Gamow–Teller matrix elements, $\hat{\psi}^\dagger \boldsymbol{\sigma} \hat{\psi}$ survive. The connection with the Fermi and Gamow–Teller operators is established by the following relationships between the field operators and the isospin operators:

$$\hat{\psi}^\dagger \tau^{(-)} \hat{\psi} \rightarrow \sum_i \tau^{(-)}(i) \qquad \hat{\psi}^\dagger \tau^{(+)} \hat{\psi} \rightarrow \sum_i \tau^{(+)}(i)$$

which represent the usual relationship between field and configuration space operators. Introducing the Fermi and Gamow–Teller operators

$$\hat{M}_F{}^{(\pm)} = \sum \tau^{(\pm)}(i); \qquad \hat{\mathbf{M}}_{GT}{}^{\pm} = \sum \tau^{(\pm)}(i) \boldsymbol{\sigma}(i) \tag{11.10}$$

TABLE 11.2 Nonrelativistic Limit of Bilinear Covariants

Scalar	$\hat{\bar{\psi}}\hat{\psi} = \hat{\psi}^\dagger \gamma_4 \hat{\psi}$	$\rightarrow \hat{\psi}^\dagger \hat{\psi}$
Vector	$\hat{\bar{\psi}} \gamma_k \hat{\psi}$	$\rightarrow 0$
	$\hat{\bar{\psi}} \gamma_4 \hat{\psi}$	$\rightarrow \hat{\psi}^\dagger \hat{\psi}$
Tensor	$\hat{\bar{\psi}} \sigma_{ij} \hat{\psi}$	$\rightarrow \hat{\psi}^\dagger \sigma_k \hat{\psi}$ (i, j, k cyclical)
	$\hat{\bar{\psi}} \sigma_{4j} \hat{\psi}$	$\rightarrow 0$
Axial vector	$\hat{\bar{\psi}} \gamma_k \gamma_5 \hat{\psi}$	$\rightarrow i\hat{\psi}^\dagger \sigma_k \hat{\psi}$
	$\hat{\bar{\psi}} \gamma_4 \gamma_5 \hat{\psi}$	$\rightarrow 0$
Pseudoscalar	$\hat{\bar{\psi}} \gamma_5 \hat{\psi}$	$\rightarrow 0$

we can write the most general β-decay interaction for *allowed transitions* as

$$H_\beta = \frac{g}{\sqrt{2}} \hat{M}_F{}^{(+)} [\hat{\bar{\psi}}_e(C_S + C_S'\gamma_5 + C_V\gamma_4 + C_V'\gamma_4\gamma_5)\hat{\psi}_\nu]$$

$$+ \frac{g}{\sqrt{2}} \hat{\mathbf{M}}_{GT}{}^{(+)} \cdot [\hat{\bar{\psi}}_e\gamma_4\boldsymbol{\sigma}(C_A + C_A'\gamma_5 + C_T\gamma_4 + C_T'\gamma_4\gamma_5)\hat{\psi}_\nu]$$

$$+ \text{h.c.}$$

The values of the constants C_S, C_S', etc., are to be chosen from experiment. The final result is so remarkably simple that we restrict our discussion to it:

$$H_\beta = \frac{g}{\sqrt{2}} C_V\hat{M}_F{}^{(+)}\{\hat{\bar{\psi}}_e\gamma_4(1 + \gamma_5)\hat{\psi}_\nu\} + \frac{g}{\sqrt{2}} C_A\hat{\mathbf{M}}_{GT}{}^{(+)}$$

$$\cdot \{\hat{\bar{\psi}}_e\gamma_4\boldsymbol{\sigma}(1 + \gamma_5)\hat{\psi}_\nu\} + \text{h.c.} \qquad (11.11)$$

Comparing this with (11.10) we see first that the parity nonconserving component is maximized (i.e., $C_V = C_V'$, $C_A = C_A'$) and second that there are no tensor and scalar components (i.e., $C_S = 0$, $C_T = 0$). For this reason the interaction is often referred to the V–A interaction.

To see how these results are in accord qualitatively with the experimental results it is necessary to understand the factor $(1 + \gamma_5)\psi_\nu$. Toward this end consider the equation satisfied by ψ_ν for a massless neutrino:

$$c(\boldsymbol{\alpha}\cdot\mathbf{p})\psi_\nu = E\psi_\nu \qquad (11.12)$$

Since $|E| = cp$ for a massless particle

$$\gamma_5(\boldsymbol{\sigma}\cdot\mathbf{p})\psi_\nu = -p\psi_\nu \qquad (11.13)$$

It is easy to verify that the following four expressions are solutions with the indicated values of E. Choosing the "z" axis to be along $\mathbf{p}$ and placing

$$\psi = u \exp[(i/h)(\mathbf{p}\cdot\mathbf{r} - Et)] \qquad (11.14)$$

we obtain

$$
\begin{array}{l|l}
u_{\mathrm{I}} = \begin{pmatrix} \alpha \\ \alpha \end{pmatrix} & u_{\mathrm{III}} = \begin{pmatrix} \beta \\ \beta \end{pmatrix} \\
\sigma_z u_{\mathrm{I}} = u_{\mathrm{I}} & \sigma_z u_{\mathrm{III}} = -u_{\mathrm{III}} \\
(1+\gamma_5) u_{\mathrm{I}} = 0 & (1+\gamma_5) u_{\mathrm{III}} = 0 \\
E = |p| > 0 & E = -|p| < 0 \\
\hline
u_{\mathrm{II}} = \begin{pmatrix} \beta \\ -\beta \end{pmatrix} & u_{\mathrm{IV}} = \begin{pmatrix} \alpha \\ -\alpha \end{pmatrix} \\
\sigma_z u_{\mathrm{II}} = -u_{\mathrm{II}} & \sigma_z u_{\mathrm{IV}} = u_{\mathrm{IV}} \\
\left(\dfrac{1+\gamma_5}{2}\right) u_{\mathrm{II}} = u_{\mathrm{II}} & \left(\dfrac{1+\gamma_5}{2}\right) u_{\mathrm{IV}} = u_{\mathrm{IV}}
\end{array}
\tag{11.15}
$$

The spinors α and β are the two-element Pauli spinors corresponding to spin "up" and spin "down," respectively.* The Dirac spinors u have four elements; u_{I}, for example, is given by

$$
u_{\mathrm{I}} = \begin{pmatrix} 1 \\ 0 \\ 1 \\ 0 \end{pmatrix}
$$

Solution I describes a neutrino with positive energy whose spin is in the same direction as its direction of motion. However solution I cannot enter into the β-decay interaction (11.11) since $(1 + \gamma_5)\psi$ is zero. The positive energy solution with a finite value of $(1 + \gamma_5)\psi$ is solution II. Note that for this neutrino state the helicity is negative. In other words the direction of the spin is opposite to the direction of motion in agreement with the experimental results summarized in Section 9. The antineutrino wave function is proportional to the charge conjugate of the negative energy solution IV (see Eq. 11.7), the particle created (see Eq. 11.6) having the momentum $-\mathbf{p}$ and spin-state β as one readily determines using (11.7). For the antineutrino it follows that the states in which it is found in nature have positive helicity. Its spin and direction of motion point in the same direction.

---

*Not to be confused with the α and β Dirac matrices.

Not only do the neutrino and antineutrino have negative and positive helicity, respectively, as required by experiment but the accompanying positron and electron have the appropriate opposite helicity. This can be immediately seen from interaction (11.11). Let us take the Fermi term as an example using the term appropriate to positron emission

$$\hat{\bar{\psi}}_\nu \gamma_4 (1 + \gamma_5) \hat{\psi}_e$$

In the limit $v_e \to c$ for which case the wave functions for ψ_e approach those tabulated in (11.15), we see that again positive energy solution II and negative energy IV solution are selected by the operator $(1 + \gamma_5)$. For the limiting electron solution II, the helicity is negative; that is, the spin of the electron is opposite to that of its direction of motion. For the positron the helicity is positive, and the spin is in the same direction as the direction of motion of the positron. This perfect polarization obtains only in the limit $v_e \to c$. The actual polarization will not be unity but rather of the order of v_e/c as we shall see. The important result to be gathered from this discussion is that interaction (11.11) is in agreement qualitatively with the experimental data. The next step in the discussion will show that this agreement is also a quantitative one.

Let us consider the case of allowed β^- decay induced by the terms given explicitly in (11.11). For simplicity we shall neglect the Coulomb interaction between the electron and residual nucleus. The matrix element H_{fi} entering expression (2.8) for the transition probability is:

$$H_{fi} = \frac{g}{\sqrt{2}} C_V \langle f | \hat{M}_F{}^{(+)} | i \rangle \langle 1(p_e s_e); 1(\bar{p}_\nu, \bar{s}_\nu) | \hat{\bar{\psi}}_e \gamma_4 (1 + \gamma_5) \hat{\psi}_\nu | 0_e, 0_\nu \rangle$$

$$+ \frac{g}{\sqrt{2}} C_A \langle f | \mathbf{M}_{GT}{}^{(+)} | i \rangle \cdot \langle 1(p_e s_e); 1(\bar{p}_\nu \bar{s}_\nu) | \hat{\bar{\psi}}_e \gamma_4 \boldsymbol{\sigma} (1 + \gamma_5) \hat{\psi}_\nu | 0_e, 0_\nu \rangle$$

The states $|i\rangle$ and $|f\rangle$ on the right-hand side of this equation are the initial and final nuclear states. The leptonic matrix elements describe the creation from the vacuum state $|0_e, 0_\nu\rangle$ of a state containing an electron of four momentum p_e and spin s_e, as well as an antineutrino of momentum $\bar{p}_\nu$ and spin $\bar{s}_\nu$.

Evaluating the leptonic matrix elements by using expansion (11.6) and relation (11.15) yields:

$$H_{fi} = \frac{g}{\sqrt{2}} C_V \langle f | \hat{M}_F{}^{(+)} | i \rangle \bar{u}(p_e, s_e) \gamma_4 (1 + \gamma_5) u(-\bar{p}_\nu, -\bar{s}_\nu)$$

$$+ \frac{g}{\sqrt{2}} C_A \langle f | \mathbf{M}_{GT}{}^{(+)} | i \rangle \cdot \bar{u}(p_e, s_e) \gamma_4 \boldsymbol{\sigma} (1 + \gamma_5) u(-\bar{p}_\nu, -\bar{s}_\nu) \qquad (11.16)$$

The next step in the calculation requires the evaluation of

$$\sum_{s_e, s_\nu} |H_{fi}|^2 \tag{11.17}$$

where the sum is restricted to positive energy electrons and antineutrinos. This restriction can be lifted if a projection operator is appropriately introduced. This operator must be zero for the negative energy states and unity for the positive energy states. Let us consider a typical term in (11.17) of the form

$$S = \sum_{s_e, \bar{s}_\nu, (E_e, \bar{E}_\nu > 0)} [\bar{u}(p_e, s_e)\, A\, u(-p_\nu, -s_\nu)][\bar{u}(-\bar{p}_\nu, -\bar{s}_\nu)\gamma_4 B^\dagger \gamma_4 u(p_e, s_e)]$$

where A and B can be either $\gamma_4(1 + \gamma_5)$ or $\gamma_4\boldsymbol{\sigma}(1 + \gamma_5)$. This expression has a somewhat simpler appearance if we use $u^\dagger$ rather than $\bar{u}$:

$$S = \sum_{s_e \bar{s}_\nu, (E_e, \bar{E}_\nu > 0)} [u^\dagger(p_e, s_e)\gamma_4 A u(-\bar{p}_\nu, -\bar{s}_\nu)][u^\dagger(-\bar{p}_\nu, -\bar{s}_\nu)B^\dagger \gamma_4 u(p_e, s_e)] \tag{11.18}$$

To obtain the projection operators note that

$$[c\boldsymbol{\alpha}\cdot\mathbf{p} + \gamma_4 mc^2]u(p, s) = Eu(p, s)$$

Reversing the energy yields

$$[c\boldsymbol{\alpha}\cdot\mathbf{p} + \gamma_4 mc^2]u(p, -s) = -Eu(p, -s)$$

Hence

$$\frac{[c\boldsymbol{\alpha}\cdot\mathbf{p} + \gamma_4 mc^2 + E]}{2E} u(p, s) = u(p, s)$$

$$\frac{c\boldsymbol{\alpha}\cdot\mathbf{p} + \gamma_4 mc^2 + E}{2E} u(p, -s) = 0 \tag{11.19}$$

and

$$\frac{-c(\boldsymbol{\alpha}\cdot\mathbf{p}) - \gamma_4 mc^2 + E}{2E} u(p, s) = 0$$

$$\frac{-c(\boldsymbol{\alpha}\cdot\mathbf{p}) - \gamma_4 mc^2 + E}{2E} u(p, -s) = u(p, -s) \tag{11.20}$$

Hence S can be written

$$S = \sum_{s_e s_\nu, E_e \bar{E}_\nu} \left(u^\dagger(p_e, s_e) \frac{c(\boldsymbol{\alpha}\cdot\mathbf{p}_e) + \gamma_4 m_e c^2 + E_e}{2E_e} \gamma_4 A u(-\bar{p}_\nu, -\bar{s}_\nu) \right)$$

$$\left(u^\dagger(-\bar{p}_\nu, -\bar{s}_\nu) \frac{c(\boldsymbol{\alpha}\cdot\mathbf{p}_\nu) + \bar{E}_\nu}{2\bar{E}_\nu} B^\dagger \gamma_4 u(p_e, s_e) \right)$$

where in virtue of (11.19) and (11.20) no restrictions need be placed on the sum; both positive and negative E's are included. More compactly

$$S = \sum_{s_e s_\nu, E_e \bar{E}_\nu} \left(u^\dagger(p_e, s_e) \frac{c\gamma_\mu p_{e\mu} + im_e c^2}{2iE_e} Au(-\bar{p}_\nu, -\bar{s}_\nu) \cdot u^\dagger(-\bar{p}_\nu, -\bar{s}_\nu) \frac{c\gamma_\lambda \bar{p}_{\nu\lambda}}{2i\bar{E}_\nu} \gamma_4 B^\dagger \gamma_4 u(p_e, s_e) \right)$$

Since all four solutions of the Dirac equation are now included in the sum, S can be written as follows:

$$S = -\frac{c^2}{4E_e\bar{E}_\nu} \operatorname{tr} \{ [\gamma_\mu p_{e\mu} + im_e c] A \gamma_\lambda \bar{p}_{\nu\lambda} \gamma_4 B^\dagger \gamma_4 \} \tag{11.21}$$

where a sum over μ and λ is understood. Thus the calculation of (11.17) reduces to the calculation of traces of various assorted Dirac matrices. Note

$$\begin{aligned}
&\operatorname{tr} 1 &&= 4 \\
&\operatorname{tr} \gamma_\mu &&= 0 \\
&\operatorname{tr} \gamma_5 &&= 0 \\
&\operatorname{tr} \gamma_\mu\gamma_\nu &&= 4\delta_{\mu\nu} \\
&\operatorname{tr} \gamma_\mu\gamma_\nu\gamma_\lambda &&= 0 \\
&\operatorname{tr} \gamma_\mu\gamma_\nu\gamma_\lambda\gamma_\omega &&= 4[\delta_{\mu\nu}\delta_{\lambda\omega} - \delta_{\mu\lambda}\delta_{\nu\omega} + \delta_{\mu\omega}\delta_{\nu\lambda}] \\
&\operatorname{tr} \gamma_5 &&= 0 \\
&\operatorname{tr} \gamma_\mu\gamma_5 &&= 0 \\
&\operatorname{tr} \gamma_\mu\ \gamma_\nu\ \gamma_\lambda\ \gamma_\omega\ \gamma_5 &&= 4\epsilon_{\mu\nu\lambda\omega} = \begin{cases} 4 \text{ if } \mu\nu\lambda\omega = 1, 2, 3, 4 \text{ or an even permutation} \\ \text{thereof. } -4 \text{ if } \mu\nu\lambda\omega \text{ is an odd permutation of} \\ 1, 2, 3, 4. \end{cases}
\end{aligned} \tag{11.22}$$

To illustrate the evaluation of (11.21) let us consider the case $A = B = \gamma_4(1 + \gamma_5)$. This is the only term present in $|H_{fi}|^2$ for a pure Fermi 0—0 transition. We consider

$$S = -\frac{c^2}{4E_e\bar{E}_\nu} \operatorname{tr} [\gamma_\mu p_{e\mu} + im_e c]\gamma_4(1 + \gamma_5)\gamma_\lambda \bar{p}_{\nu\lambda}\gamma_4(1 + \gamma_5)]$$

To obtain (11.17) for this case multiply S by $|C_V|^2/2| \langle f | M_F{}^{(+)} | i \rangle |^2$. When we use the anticommutation rule $\gamma_5\gamma_\lambda = -\gamma_\lambda\gamma_5$ and $(1 + \gamma_5)^2 = 2(1 + \gamma_5)$, S becomes

$$S = -\frac{c^2}{2E_e\bar{E}_\nu} \operatorname{tr} [\gamma_\mu p_{e\mu} + im_e c]\gamma_4(1 + \gamma_5)\gamma_\lambda \bar{p}_{\nu\lambda}\gamma_4$$

Next, note that the m_e term will not contribute since its coefficient has a zero trace. Hence

$$S = -\frac{c^2}{2E_e\bar{E}_\nu}\,\mathrm{tr}\,(\gamma_\mu p_{e\mu})\gamma_4(1+\gamma_5)\gamma_\lambda\bar{p}_{\nu\lambda}\gamma_4$$

or

$$S = -\frac{c^2}{2E_e\bar{E}_\nu}\,\mathrm{tr}\,(\gamma_\mu\gamma_4\gamma_\lambda\gamma_4 p_{e\mu}\bar{p}_{\nu\lambda} + \gamma_\mu\gamma_4\gamma_\lambda\gamma_4\gamma_5 p_{e\mu}\bar{p}_{\nu\lambda})$$

We see that the term containing γ_5 does not contribute. Using (11.22) for the first term yields

$$\begin{aligned} S &= -\frac{2c^2}{E_e\bar{E}_\nu}(2\delta_{\mu4}\delta_{\lambda4} - \delta_{\mu\lambda})p_{e\mu}\bar{p}_{\nu\lambda} \\ &= -\frac{2c^2}{E_e\bar{E}_\nu}\left(-\frac{1}{c^2}E_e\bar{E}_\nu - \mathbf{p}_e\cdot\bar{\mathbf{p}}_\nu\right) \\ &= 2\left(1 + c^2\frac{\mathbf{p}_e\cdot\bar{\mathbf{p}}_\nu}{E_e\bar{E}_\nu}\right) \end{aligned} \tag{11.23}$$

Note that the nonparity conserving terms in the Hamiltonian (11.10) do not change the form of the result from that which would be obtained in its absence. The only change is the normalization factor of 2. In retrospect this is no surprise. The violation of parity would be indicated by the presence of a pseudoscalar in S. But it is not possible to form a pseudoscalar from $\mathbf{p}_e$ and $\mathbf{p}_\nu$ the only vectors available in a Fermi-type decay. Hence observing the angular correlation or transition probability in this type of decay will not shed any light on parity conservation. However the correlation between the electron and antineutrino momenta:

$$1 + c^2\frac{\mathbf{p}_e\cdot\bar{\mathbf{p}}_\nu}{E_e\bar{E}_\nu} = 1 + \frac{v_e}{c}\cos\theta \tag{11.24}$$

where θ is the angle between $\mathbf{p}_e$ and $\bar{\mathbf{p}}_\nu$, and v_e is the electron velocity is characteristic of the vector interaction, showing in this case that it is more probable for the electron and the antineutrino to go off the same direction. If the more general interaction above (11.11) is used for a $0 \rightarrow 0$ transition, (11.24) is replaced by

$$1 + \left[\frac{v_e}{c}\cos\theta \frac{|C_V|^2 + |C'_V|^2 - |C_S|^2 - |C'_S|^2}{|C_V|^2 + |C'_V|^2 + |C_S|^2 + |C_S'|^2 + (2m_e/E_e)\,\mathrm{Re}\,(C_S^*C_V + C'^*_SC'_V)}\right] \tag{11.25}$$

We note that if the interaction were scalar ($C_V = C'_V = 0$), then the probability that the electron and antineutrino would travel in opposite directions would be greater. Experiment showed the opposite and, as noted earlier, agreement (Section 8) with experiment is obtained if $C_S = C'_S = 0, C_V = C'_V$.

Let us now turn to the more general case, that is, to transitions for which both Fermi (vector) and Gamow–Teller type interactions are effective so that both terms in (11.16) contribute. One important result that can be obtained with very little effort applies to the case in which the emitting nucleus is unpolarized and the lepton polarizations are not observed. In that event the only lepton variables that can occur in (11.17) are ($\mathbf{p}_e$, E_e) and $\mathbf{p}_\nu$, $\bar{E}_\nu$) while the only nuclear terms must be $|\langle f|\hat{M}_F{}^{(+)}|i\rangle|^2$ and $\langle f|\hat{\mathbf{M}}_{GT}{}^{(+)}|i\rangle|^2$. From these variables it is not possible to form a pseudoscalar. Hence there will be no observable effect of the nonparity conserving component of (11.16) in the β-ray spectrum or in the electron-antineutrino correlation. Hence the discussion in Sections 2 to 7 of the allowed β-transitions of unpolarized nuclei remain valid in the presence of a parity nonconserving interaction. This interaction becomes observable only when the emitting nucleus is polarized or in the polarization of the emitted leptons. The pseudoscalar that can be then formed is the scalar product of the average spin direction of the emitting nucleus or emitted lepton and lepton momentum.

These results can be verified by examining the final result for 11.17:

$$\sum_{s_e\bar{s}_\nu} |H_{fi}|^2 = g^2|C_V|^2|M_F|^2\left(1 + c^2\frac{\mathbf{p}_e\cdot\bar{\mathbf{p}}_\nu}{E_e\bar{E}_\nu}\right)$$
$$+ g^2|C_A|^2\left\{\mathbf{M}_{GT}\cdot\mathbf{M}_{GT}^*\left(1 - c^2\frac{\mathbf{p}_e\cdot\bar{\mathbf{p}}_\nu}{E_e\bar{E}_\nu}\right)\right.$$
$$+ \frac{c^2}{E_e\bar{E}_\nu}\left[(\mathbf{M}_{GT}\cdot\mathbf{p}_e)(\mathbf{M}_{GT}^*\cdot\bar{\mathbf{p}}_\nu) + (\mathbf{M}_{GT}\cdot\bar{\mathbf{p}}_\nu)(\mathbf{M}_{GT}^*\cdot\mathbf{p}_e)\right] + i(\mathbf{M}_{GT}\times\mathbf{M}_{GT}^*)$$
$$\left.\left(\frac{c\bar{\mathbf{p}}_\nu}{\bar{E}_\nu} - \frac{c\mathbf{p}_e}{E_e}\right)\right\} - 4\,\mathrm{Re}\,C_AC_V^*\mathbf{M}_{GT}\cdot\left[\frac{c\bar{\mathbf{p}}_\nu}{\bar{E}_\nu} + \frac{c\mathbf{p}_e}{E_e} + \frac{ic^2(\mathbf{p}_e\times\bar{\mathbf{p}}_\nu)}{E_e\bar{E}_\nu}\right] \qquad (11.26)$$

where we have abbreviated $\langle f|\hat{M}_F{}^{(+)}|i\rangle$ by M_F and $\langle f|\hat{\mathbf{M}}_{GT}{}^{(+)}|i\rangle$ by $\mathbf{M}_{GT}$. Averaging over the spin of the target nucleus yields

$$\frac{1}{2J_i+1}\sum_{s_e s_\nu, M_i}|H_{fi}|^2 = \frac{g^2}{2J_i+1}\sum_{M_i}\left\{|C_V|^2|M_F|^2\left(1 + c^2\frac{\mathbf{p}_e\cdot\bar{\mathbf{p}}_\nu}{E_e\bar{E}_\nu}\right)\right.$$
$$\left. + |C_A|^2\left[\mathbf{M}_{GT}\cdot\mathbf{M}_{GT}^*\left(1 - \frac{1}{3}c^2\frac{\mathbf{p}_e\cdot\bar{\mathbf{p}}_\nu}{E_e\bar{E}_\nu}\right)\right]\right\} \qquad (11.27)$$

where we note that the average of $\mathbf{M}_{GT}$, and $\mathbf{M}_{GT}\times\mathbf{M}_{GT}^*$ are zero while

$$\text{Average }(\mathbf{M}_{GT}\cdot\mathbf{p}_e\mathbf{M}_{GT}^*\cdot\bar{\mathbf{p}}_\nu) = \tfrac{1}{3}\mathbf{p}_e\cdot\bar{\mathbf{p}}_\nu\text{ Average }\mathbf{M}_{GT}\cdot\mathbf{M}_{GT}^*$$

Note that the angular correlation of the electron and antineutrino in a Gamow–Teller transition is smaller and of opposite sign to that seen in a Fermi transition. This result is in agreement with the discussion of Section 8, remembering that the helicity of the electron is negative, that of the antineutrino is positive. Experimentally, this has been verified for the pure Gamow–Teller decay of $He^6(J_i^\pi = 0^+; J_f^\pi = 1^+)$ where Johnson et al. (63) obtained the value -0.334 ± 0.003 for the coefficient of the $c^2\mathbf{p}_e \cdot \bar{\mathbf{p}}_\nu / E_e\bar{E}_\nu$ on the second line of (11.27)

Averaging (11.27) over neutrino direction yields

$$g^2 \frac{1}{2J_i + 1} \sum_{M_i} (|C_V|^2 |M_F|^2 + |C_A|^2 |\mathbf{M}_{GT}|^2)$$

a result identical with (2.23) and, thereby, verifying the identity of the "C" coefficients in (11.16) with those given in (2.23) $C_V = C_F$ and $C_A = C_{GT}$.

Let us consider two of the experiments that do demonstrate the nonconservation of parity. First, consider the experiment of Wu et al. (57) that deals with the angular distribution of electrons emitted by a polarized nucleus in a Gamow–Teller transition. Since experimentally the direction of motion of the antineutrino does not concern us, we integrate (11.26) over neutrino momenta keeping only the Gamow–Teller term:

$$\frac{1}{4\pi} \int d\bar{\Omega}_\nu \sum_{s_e, s_\nu} |H_{fi}|^2 = g^2 |C_A|^2 \left(\mathbf{M}_{GT} \cdot \mathbf{M}_{GT}^* - i(\mathbf{M}_{GT} \times \mathbf{M}_{GT}^*) \cdot \frac{c\mathbf{p}_e}{E_e} \right) \tag{11.28}$$

We shall need to sum over the spin of the final nuclear states, while the properties of initial nuclear states will be inserted later. Take the first term in the brackets. Note that it will not contribute to that part of the electron angular distribution which indicates a violation of parity conservation. But it does act as a background term and it must be evaluated before the measurability of the second term in the bracket can be determined. We have

$$\sum_{M_f} \mathbf{M}_{GT} \cdot \mathbf{M}_{GT}^* = \sum \langle J_f M_f | \mathbf{M}_{GT}{}^{(+)} | J_i M_i \rangle \cdot \langle J_i M_i | \mathbf{M}_{GT}{}^{(+)} | J_f M_f \rangle$$

If we now insert the spherical tensor expressions for the components of $\mathbf{M}_{GT}$, this expression becomes

$$\sum_{M_f, \kappa} (-)^\kappa \langle J_f M_f | M_\kappa{}^{(1)} | J_i M_i \rangle \langle J_i M_i | M_{-\kappa}{}^{(1)} | J_f M_f \rangle$$

Applying the Wigner–Eckart theorem one obtains

$$\sum_{M_f} \mathbf{M}_{GT} \cdot \mathbf{M}_{GT}^* = \sum_{M_f, \kappa} (-)^\kappa (-)^{J_f - M_f - J_i + M_i} \begin{pmatrix} J_f & 1 & J_i \\ -M_f & \kappa & M_i \end{pmatrix}$$

$$\times \begin{pmatrix} J_i & 1 & J_f \\ -M_i & -\kappa & M_f \end{pmatrix} (J_i || M_{GT} || J_f)(J_f || M_{GT} || J_i)$$

Employing $M_f = \kappa + M_i$ and [see Edmonds (57), p. 78]

$$(J_i||M_{GT}^{\dagger}||J_f) = (-)^{J_i - J_f}(J_f||M_{GT}||J_i)$$

it follows that

$$\sum_{M_f} \mathbf{M}_{GT}\cdot\mathbf{M}_{GT}^* = |(J_f||M_{GT}||J_i)|^2 \sum_{M_f,\kappa} \begin{pmatrix} J_f & 1 & J_i \\ -M_f & \kappa & M_i \end{pmatrix}^2$$

or

$$\sum_{M_f} \mathbf{M}_{GT}\cdot\mathbf{M}_{GT}^* = \frac{1}{2J_i+1}|(J_f||M_{GT}||J_i)|^2 \tag{11.29}$$

Consider next the second term in (11.28). Only the component of $\mathbf{M}_{GT} \times \mathbf{M}_{GT}^*$ along $\mathbf{p}_e$ is of interest. Choosing therefore the "z" axis to be along $\mathbf{p}_e$, only the matrix elements of the z-component of $(\mathbf{M} \times \mathbf{M}^*)$ are needed. We need only evaluate

$$\sum_{M_f} [\langle J_fM_f|M_x^{(+)}|J_iM_i\rangle\langle J_iM_i|M_y^{(-)}|J_fM_f\rangle - \langle J_fM_f|M_y^{(+)}|J_iM_i\rangle$$

$$\times \langle J_iM_i|M_x^{(-)}|J_fM_f\rangle] = i\sum_{M_f} (\langle J_fM_f|M_{-1}^{(1)}|J_iM_i\rangle\langle J_iM_i|M_1^{(1)}|J_fM_f\rangle$$

$$- \langle J_fM_f|M_1^{(1)}|J_iM_i\rangle\langle J_iM_i|M_{-1}^{(1)}|J_fM_f\rangle)$$

or

$$= \sum_{M_f} i(-)^{J_f - J_i - M_f + M_i}(J_f||M_{GT}||J_i)(J_i||M_{GT}^{\dagger}||J_f)$$

$$\times\left[\begin{pmatrix} J_f & 1 & J_i \\ -M_f & -1 & M_i \end{pmatrix}\begin{pmatrix} J_i & 1 & J_f \\ -M_i & 1 & M_f \end{pmatrix} - \begin{pmatrix} J_f & 1 & J_i \\ -M_f & 1 & M_i \end{pmatrix}\begin{pmatrix} J_i & 1 & J_f \\ -M_i & -1 & M_f \end{pmatrix}\right]$$

The first pair of three j symbols require $M_f = M_i - 1$, the second pair $M_f = M_i + 1$. Both terms enter in the final expression because of the sum over M_f. After some elementary manipulations the sum becomes

$$\sum_{M_f} \langle \mathbf{M}_{GT} \times \mathbf{M}_{GT}^* \rangle_z = i|(J_f||M_{GT}||J_i)|^2$$

$$\times\left[\begin{pmatrix} J_i & J_f & 1 \\ M_i & -(M_i+1) & 1 \end{pmatrix}^2 - \begin{pmatrix} J_i & J_f & 1 \\ -M_i & M_i - 1 & 1 \end{pmatrix}^2\right] \tag{11.30}$$

Since this expression is odd under the transformation $M_i \to -M_i$, it follows that is an odd function of M_i; as we shall see, it is linear in M_i, vanishing as a consequence if the initial nucleus is unpolarized. The values of $3j$ symbols can be introduced in (11.30) to obtain

$$\frac{1}{4\pi}\int d\bar{\Omega}_\nu \sum_{s_e,\bar{s}_\nu} |H_{fi}|^2 = |C_A|^2 \frac{1}{2J_i+1} |(J_f||\mathbf{M}_{GT}||J_i)|^2$$

$$\cdot \left[1 + \frac{cp_{ez}}{E_e} M_i \begin{cases} \dfrac{1}{J_i+1} & J_f = J_i + 1 \\[2ex] -\dfrac{1}{J_i(J_i+1)} \quad \text{if} & J_f = J_i \\[2ex] -\dfrac{1}{J_i} & J_f = J_i - 1 \end{cases} \right] \tag{11.31}$$

If P_i is the probability that the initial nucleus is in the substate M_i, then the polarization of the emitting nucleus $\langle \mathbf{J} \rangle$ is given by

$$\langle \mathbf{J} \rangle_z = \sum P_i M_i$$

Hence the angular distribution of the electrons with respect to the direction of polarization of the emitting nucleus is

$$1 + \alpha(J_i)\frac{\mathbf{v}_e}{c} \cdot \langle \mathbf{J} \rangle \tag{11.32}$$

where according to (11.31)

$$\alpha(J_i) = \begin{cases} \dfrac{1}{J_i+1} & J_f = J_i + 1 \\[2ex] -\dfrac{1}{J_i(J_i+1)} & J_f = J_i \\[2ex] -\dfrac{1}{J_i} & J_f = J_i - 1 \end{cases} \tag{11.33}$$

The third possibility corresponds to the conditions of the Wu experiment since $J_f = 4$, $J_i = 5$. The electrons in that case should then be preferentially emitted in the direction opposite to the polarization of $^{60}Co^*$. This is precisely what is observed, as discussed in Section 10, on an intuitive basis.

Problem: Derive the angular distribution of the electrons emitted from a polarized nucleus when both Gamow–Teller and Fermi terms contribute

($\Delta J = 0$). Show that the interference between the two terms averaged over the antineutrino momentum and summed over final states is given by

$$-4\,\frac{\mathbf{v}_e\cdot\langle\mathbf{J}\rangle}{c}\,\mathrm{Re}\left[C_A C_V^*\,\frac{(J||\widehat{\mathscr{M}}_F||J)(J||\widehat{\mathbf{M}}_{GT}||J)}{\sqrt{J(J+1)(2J+1)}}\right] \tag{11.34}$$

The second experiment to be discussed is the polarization of the electrons emitted from *unpolarized* nuclei averaging over the antineutrino direction of emission. The polarization is specified for a given state of the electron by the expectation value of the helicity $\boldsymbol{\sigma}\cdot\hat{\mathbf{p}}_e$ for that state. If P_+ is the probability that the helicity is positive and P_- that it is negative, the expectation value of $\boldsymbol{\sigma}\cdot\hat{\mathbf{p}}_e$ is:

$$\mathrm{Pol} \equiv \langle\boldsymbol{\sigma}\cdot\hat{\mathbf{p}}_e\rangle = \frac{P_+ - P_-}{P_+ + P_-} \tag{11.35}$$

The probability P_+ is proportional to transition probability for the formation of a state with the helicity $+1$. This can be extracted from the probability for the transition without regard to helicity by inserting a projection operator that is one for positive helicity and zero for negative helicity:

$$\hat{P}_+ = \tfrac{1}{2}(1 + \boldsymbol{\sigma}\cdot\hat{\mathbf{p}}_e) \tag{11.36}$$

The corresponding operator for negative helicity is

$$\hat{P}_- = \tfrac{1}{2}(1 - \boldsymbol{\sigma}\cdot\hat{\mathbf{p}}_e) \tag{11.37}$$

Note

$$(\hat{P}_\pm)^2 = \hat{P}_\pm, \quad \hat{P}_+\hat{P}_- = 0 \tag{11.38}$$

The total transition probability can be obtained from (11.16). Since the nucleus is unpolarized, the interference between Gamow–Teller and Fermi matrix elements will not survive. Therefore the general term will involve the form:

$$|\bar{u}(p_e, s_e)\mathrm{A}u(-\bar{p}_\nu, -\bar{s}_\nu)|^2|M_A|^2$$

where A can be either $\gamma_4(1+\gamma_5)$ or $\gamma_4\boldsymbol{\sigma}(1+\gamma_5)$ and M_A, correspondingly, is the nuclear matrix element of $\widehat{\mathscr{M}}_F{}^{(+)}$ or $\widehat{\mathbf{M}}_{GT}{}^{(+)}$. Summing over antineutrino spin and momentum yields

$$\int d\bar{\Omega}_\nu \sum_{\bar{s}_\nu,\bar{E}_\nu>0} \langle\bar{u}(p_e,s_e)M_A\mathrm{A}u(-\bar{p}_\nu,-\bar{s}_\nu)\rangle\langle\bar{u}(-\bar{p}_\nu,-\bar{s}_\nu)M_A^*\gamma_4\mathrm{A}^\dagger\gamma_4u(p_e,s_e)\rangle$$

P_+ and P_- are proportional to

$$P_\pm \sim \int d\bar{\Omega}_\nu \sum_{\bar{s}_\nu\bar{E}_\nu>0} \langle\bar{u}(p_e,s_e)M_A\mathrm{A}u(-\bar{p}_\nu,-\bar{s}_\nu)\rangle\langle\bar{u}(-\bar{p}_\nu,-\bar{s}_\nu) \cdot M_A^*\gamma_4\mathrm{A}^\dagger\gamma_4\hat{P}_\pm u(p_e,s_e)\rangle$$

Taking for example P_+, $\hat{P}_+ u(p_e, s_e)$ will vanish for that component of $u(p_e, s_e)$ which has negative helicity and will give unity for the positive helicity component. We will be able to perform the sum over $\bar{s}_\nu$ by following the technique described below (11.17). The result is (see Eq. 11.21)

$$P_\pm \sim \int d\bar{\Omega}_\nu \, \langle \bar{u}(p_e, s_e) M_A \mathrm{A} \frac{\gamma_\lambda \bar{p}_{\nu\lambda}}{2i\bar{E}_\nu} \gamma_4 \mathrm{A}^\dagger M_A^* \gamma_4 \hat{P}_\pm u(p_e, s_e) \rangle$$

The antineutrino momentum occurs only in the factor $p_{\nu\lambda}$. Only the $\lambda = 4$ component will survive so that (dropping a 4π factor)

$$P_\pm \sim \tfrac{1}{2} \langle \bar{u}(p_e, s_e) M_A \mathrm{A} \mathrm{A}^\dagger M_A^* \gamma_4 \hat{P}_\pm u(p_e, s_e) \rangle \tag{11.39}$$

This gives the result for a particular electron spin state $u(p_e, s_e)$. To obtain the total P_+ it is now necessary to sum (11.39) over the spin states for positive energy. The final result (see Eq. 11.21) is (dropping numerical factors)

$$P_\pm \sim \mathrm{tr}\ [(\gamma_\mu p_{e\mu} + i m_e c) M_A \mathrm{A} \mathrm{A}^\dagger M_A^* \gamma_4 \hat{P}_\pm]$$

Substituting this result in (11.35) one obtains

$$\mathrm{Pol} = \frac{\mathrm{tr}\ [(\gamma_\mu p_{e\mu} + i m_e c) M_A \mathrm{A} \mathrm{A}^\dagger M_A^* \gamma_4 \boldsymbol{\sigma} \cdot \hat{\mathbf{p}}_e]}{\mathrm{tr}\ [\gamma_\mu p_{e\mu} + i m_e c) M_A \mathrm{A} \mathrm{A}^\dagger M_A^* \gamma_4]} \tag{11.40}$$

or

$$\mathrm{Pol} = \frac{\mathrm{tr}\ [\gamma_\mu p_{e\mu} M_A \mathrm{A} \mathrm{A}^\dagger M_A^* \gamma_4 \boldsymbol{\sigma} \cdot \hat{\mathbf{p}}_e]}{\mathrm{tr}\ [\gamma_\mu p_{e\mu} M_A \mathrm{A} \mathrm{A}^\dagger M_A^* \gamma_4]} \tag{11.41}$$

The value of this expression can be readily evaluated. If the initial nucleus is unpolarized, (11.41) yields the identical result for both Fermi and Gamow–Teller interactions:

$$\mathrm{Pol} = -\frac{v_e}{c} \qquad \text{unpolarized nucleus; } \beta^- \text{ emission} \tag{11.42}$$

The spin of the electron thus points in a direction opposite to the direction of motion. The result for position emission has the magnitude of v_e/c but the helicity is positive

$$\mathrm{Pol} = v_e/c \qquad \text{unpolarized nucleus, } \beta^+ \text{ emission} \tag{11.43}$$

This essentially maximum polarization is a consequence of the equal magnitude of the parity and nonparity-conserving components in the interaction as symbolized by the $(1 + \gamma_5)$ factor.

One corrollary shall be needed later. The polarization (11.42) is identical to that which would be obtained if the electron wave function were of the form

$$u_- + \sqrt{\frac{1 - v_e/c}{1 + v_e/c}}\, u_+$$

where u_- is a state of negative helicity, u_+ one of positive helicity. This is one member of the complete set for positive energies; the other is

$$u_+ - \sqrt{\frac{1 - v_e/c}{1 + v_e/c}}\, u_-$$

One may regard the first of these as the electron state developed by the β-ray interaction choosing, however, an appropriate $u(p_e, s_e)$ upon which to operate. We note that the probability of the electron having a positive helicity is

$$\tfrac{1}{2}(1 - v_e/c) \tag{11.44}$$

To conclude this section we write down the β-decay Hamiltonian that is consistent with all these experiments including of course the β-neutrino correlation, the angular distribution of electrons emitted from polarized nuclei, and the polarization of electrons emitted from unpolarized nuclei. In (11.10) we take $C_S = C_S' = 0$, $C_T = C_T' = 0$, $C_V = C_V'$, $C_A = C_A'$. Then returning to the relativistic form (see Eq. 11.9) we have

$$H_\beta = \sum_\mu \frac{g}{\sqrt{2}} [C_V [\hat{\bar{\psi}}_e \gamma_\mu (1 + \gamma_5)\hat{\psi}_\nu] \hat{\bar{\psi}} \gamma_\mu \tau^{(+)} \hat{\psi} \\ - C_A [\hat{\bar{\psi}}_e \gamma_\mu (1 + \gamma_5)\hat{\psi}_\nu] \hat{\bar{\psi}} \gamma_\mu \gamma_5 \tau^{(+)} \psi] + \text{h.c.} \tag{11.45}$$

Note $\gamma_\mu \gamma_5 (1 + \gamma_5) = \gamma_\mu (1 + \gamma_5)$ and

$$gC_V = 1.4100 \times 10^{-49} \text{ erg cm}^3$$

and

$$C_A/C_V = -1.248 \pm 0.010 \tag{11.46}$$

These last equations summarize the present understanding of nuclear β-decay. Using perturbation theory the various phenomena listed above are correctly described. To this must be added the result that this agreement holds not only for allowed transitions (the only transitions discussed thus far) but also for forbidden transitions to be discussed in Section 17. However it should not be forgotten that there is some experimental uncertainty remaining. The elegant form (11.45) can be modified by the addition of small terms without contradicting experiment.

12. MU-MESON DECAY; THE UNIVERSAL WEAK INTERACTION

It is a remarkable fact that weak interactions, of which β-decay is the prototype, act between all elementary particles. It is even more remarkable that the weak interaction has identical properties for each set of interacting particles. The word "identical" will be defined more precisely below. For the present it will suffice to say that the strength of the interaction as well as the effective Hamiltonian corresponding to (11.16), including its consequences such as nonconservation of parity and the maximal polarization of (11.42), are the same to first approximation for all. It is possible to construct a formalism that makes this universality of the weak interaction explicit. It is our intent to describe this theory in this section.

We shall begin by discussing briefly muon decay and muon capture:

$$\mu^- \to e^- + \nu_\mu + \bar{\nu}_e \qquad \mu\text{-decay} \tag{12.1a}$$

$$\mu^- + p \to n + \nu_\mu \qquad \mu\text{-capture} \tag{12.1b}$$

Some features of these reactions should be noted. The μ-decay involves only leptons. The strongest interaction that can exist among these particles is electromagnetic. The μ-capture involves the strongly interacting neutron and proton as does ordinary β-decay or orbital capture:

$$n \to p + e^- + \bar{\nu}_e \tag{12.2a}$$

$$e^- + p \to n + \nu_e \tag{12.2b}$$

Note that there are two kinds of neutrinos distinguished by the subscripts μ and e. The μ-neutrino is also produced in pion decay

$$\pi^- \to \mu^- + \bar{\nu}_\mu \tag{12.3}$$

The helicity of the μ-neutrino is observed to be the same as that of the electron neutrino ν_e, that is, negative, assuming the V–A interaction. This differentiation of various kinds of neutrinos is based on experimental evidence [Danby et al. (62), Bienlein et al. (64)] that show that ν_μ neutrinos cannot replace the electron neutrino ν_e in the capture reaction

$$n + \nu_e \to p + e^-$$

In fact the absorption of ν_μ results in the emission of muons:

$$n + \nu_\mu \to p + \mu^-$$

There are thus two kinds of leptons that can be involved in the weak interactions. The electron mode (ν_e, e^-) of (12.2) and the muon mode of (12.1b). The description of electron capture is identical to muon capture except for the important difference between the electron and muon mass. The connection between the two lepton sets provided by μ-decay (12.1a), is again weak.

The most quantitative justification for the universality of the weak interaction comes from comparing the strength of interaction for nuclear β-decay and μ^--decay. The former was discussed in Section 3 with the favored value for $C_V g$ being

$$C_V g = 1.4100 \times 10^{-49} \text{ erg cm}^3 \tag{12.4}$$

The calculation of the (transition probability/time) for μ-decay is a straightforward generalization of the procedures of the preceding section. The V–A interaction form is used placing $C_V = C_A$ (a point to be discussed later in Section 14). The analog of the neutron is the μ^-, of the proton, the e^-, etc. [compare Eqs. 12.1a and (12.2a)]. The calculation is however lengthier because of the high energy of all the emerging particles in μ-decay, while in (12.2a) one can in first approximation neglect the energy of. the recoiling proton. We shall therefore quote only the result [see Wu and Moskowski (66)]. The transition probability/time for μ-decay is

$$w = \frac{G_\mu^2}{192\pi^3\hbar^7} m_\mu^5 c^4 \tag{12.5}$$

where G_μ is analogous to $C_V g$ for β-decay. The experimental lifetime is 2.198×10^{-6} sec yielding [Lee and Wu (65)]:

$$G_\mu = (1.4350 \pm 0.0011) \times 10^{-49} \text{ erg cm}^3 \tag{12.6}$$

some 2% larger than $C_V g$.

Other features of the V–A interaction are demonstrated by the observed helicities of the ν_μ. We shall cite three examples. In the first, pionic decay (12.3), since the spin of the pion is zero, the spin of the μ^- and the spin of the $\bar{\nu}_\mu$ must be opposite and because of momentum conservation, they move in opposite directions. If the helicity of the $\bar{\nu}_\mu$ is positive (opposite to the helicity of the ν_μ), the reaction can proceed only because the μ^- has some probability of having positive helicity, the probability being given by $(1/2)\cdot(1 - v_\mu/c)$ according to (11.44). In other words the observed μ^- have positive helicity. In the second, suppose we discuss μ^- decay assuming that the μ^- are produced in the π–μ decay of (12.3) so that they have positive helicity. Then as we shall now show the angular distribution and polarization of the e^- in the μ-decay will characteristically depend upon its energy. When the electron is of high energy, both neutrinos must go off in a direction opposite to that of the electron in order to conserve momentum. Since they have opposite helicity, the spin of the electron must be the same as the helicity of the μ^-, that is, positive (Fig. 12.1). Hence it must move in a direction opposite to that of μ^- spin direction. On the other hand for soft electrons, the neutrinos will go off antiparallel. Since the helicity of ν_μ is negative and $\bar{\nu}_e$ positive, the the $\bar{\nu}_e$ must move in a direction along the direction of the spin of the μ^-. The electron moves in the same direction as the $\bar{\nu}_e$, its negative helicity serves to

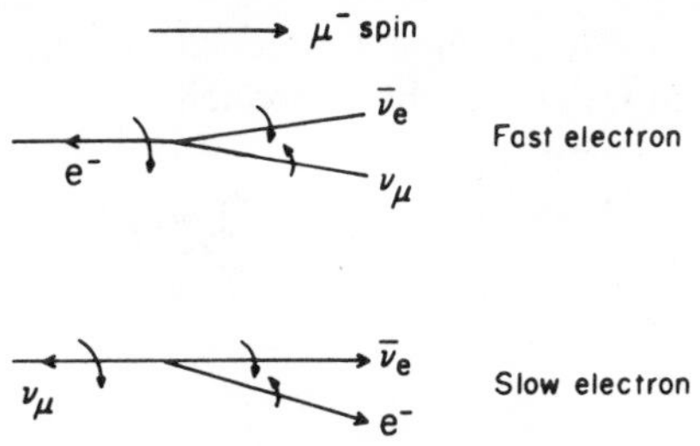

FIG. 12.1. Helicities and direction of motion in the decay of $\mu^- \rightarrow e^- + \bar{\nu}_e + \nu_\mu$ for fast and slow electrons.

cancel out the helicity of the one of the neutrinos. Hence it moves in the forward direction, that is, in the direction parallel to that of the μ^- spin (Fig. 12.1).

This description is in agreement with the experimental evidence. The polarization of the muons has been measured by Alikhanov et al. (60), Backenstoss et al. (61), and Plano (60). The angular distribution of the electrons in μ-decay has been observed by Garwin, Lederman and Weinrich (57) and Friedman and Telegdi (57).

As a third and final indication of the universality of the V–A interaction, that is, that the neutrino has a well-defined helicity and that e^- is paired with $\bar{\nu}_e$, consider π–e decay:

$$\pi^- \rightarrow e^- + \bar{\nu}_e \tag{12.7}$$

As in the case of π–μ decay this reaction can proceed only because the e^- has a small probability

$$\tfrac{1}{2}(1 - v_e/c)$$

of having a positive helicity. This is much smaller than the corresponding probability for the μ^- to be right-handed since

$$\frac{1 - v_e/c}{1 - v_\mu/c} = \frac{2m_e^2}{m_\pi^2 + m_e^2} \cdot \frac{m_\pi^2 + m_\mu^2}{2m_\mu^2} = \frac{m_e^2(m_\pi^2 + m_\mu^2)}{m_\mu^2(m_\pi^2 + m_e^2)}$$

This has the consequence that π–e decay is considerably retarded compared to π–μ decay. A detailed calculation gives the ratio of the transition probabilities to be 1.23×10^{-4}, which is to be compared with the experimental value of $(1.24 \pm 0.03)\ 10^{-4}$. The close correspondence furnishes another experimental proof of the universality of the weak interaction.

There are many more examples of this universality; but the above should be sufficiently convincing. We now proceed to formalize the concept of a universal weak interaction by developing a Hamiltonian that incorporates this feature. We recall the discussion following (11.4) where the analogy between the β-decay and electromagnetic interaction was first mentioned. Let us make the analogy more explicit by introducing the currents

$$\hat{J}_\lambda^{(N)\star} = i\hat{\bar{\psi}}_p\gamma_\lambda\left(1 + \frac{C_A}{C_V}\gamma_5\right)\hat{\psi}_n \tag{12.8}$$

$$\hat{j}_\lambda^{(e)} = -i\hat{\bar{\psi}}_e\gamma_\lambda(1 + \gamma_5)\hat{\psi}_{\nu_e} \tag{12.9}$$

The β-ray interaction can then be written

$$H_\beta = \frac{C_V g}{\sqrt{2}}\sum_\lambda \hat{J}_\lambda^{(N)\star}\hat{j}_\lambda^{(e)} + \text{h.c.} \tag{12.10}$$

We now ask how this interaction can be generalized to include the μ-decay and μ-capture (12.1) as well as nuclear β-decay. Their relationship is summarized in Fig. 12.2, the so-called Puppi triangle. In analogy with (12.8) we now define a muon-neutrino "current":

$$\hat{j}_\lambda^{(\mu)} = -i\hat{\bar{\psi}}_\mu\gamma_\lambda(1 + \gamma_5)\hat{\psi}_{\nu_\mu} \tag{12.11}$$

Then all the features of the three interactions of the Puppi triangle are obtained with the Hamiltonian

$$H = \frac{C_V g}{\sqrt{2}}\sum_\lambda \{\hat{J}_\lambda^{(N)\star}[\hat{j}_\lambda^{(\mu)} + \hat{j}_\lambda^{(e)}] + \hat{j}_\lambda^{(\mu)\star}\hat{j}_\lambda^{(e)}\} + \text{h.c.} \tag{12.12}$$

where

$$\hat{j}_\lambda^\star = \begin{cases} \hat{j}_\lambda^\dagger & \text{for} \quad \lambda = 1, 2, 3 \\ -\hat{j}_\lambda^\dagger & \text{for} \quad \lambda = 4 \end{cases}$$

A further generalization that is referred to as the current-current hypothesis employs the total current

$$\hat{J}_\lambda = \hat{J}_\lambda^{(N)} + \hat{j}_\lambda^{(\mu)} + \hat{j}_\lambda^{(e)} + \ldots \tag{12.13}$$

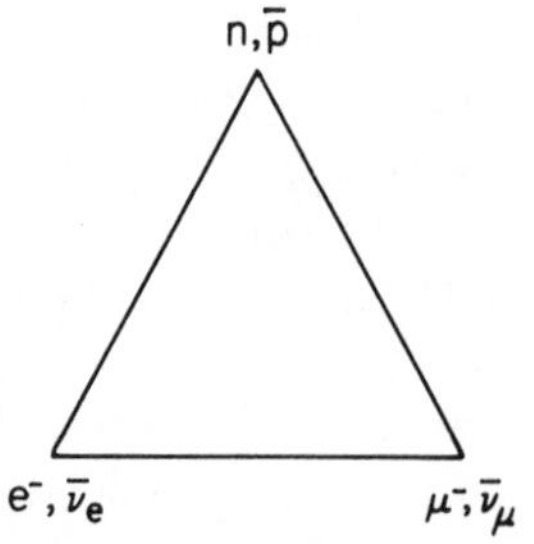

FIG. 12.2. The Puppi triangle.

where the dots refer to currents for other particles, such as other baryons and mesons. We shall discuss their form later. In any event, using (12.13) the current–current weak interaction has the form

$$H_{\text{weak}} = \frac{C_V g}{\sqrt{2}} \sum_{\lambda} (\hat{J}_\lambda{}^{\star}\hat{J}_\lambda + \text{h.c.}) \tag{12.14}$$

This Hamiltonian certainly contains all the terms in (12.12). But it contains a good deal more. The new terms that are added are called the diagonal terms (Gell–Mann et al. 69). In the present case they include

$$\sum J_\lambda{}^{(N)\star}J_\lambda{}^{(N)}, \qquad \sum j_\lambda{}^{(e)\star}j_\lambda{}^{(e)}, \qquad \sum j_\lambda{}^{(\mu)\star}j_\lambda{}^{(\mu)} \tag{12.15}$$

The last two lead to such processes as the scattering of electrons and muons by the appropriate neutrinos:

$$\begin{aligned} e^- + \nu_e &\rightarrow e^- + \nu_e \\ \mu^- + \nu_\mu &\rightarrow \mu^- + \nu_\mu \end{aligned} \tag{12.16}$$

processes that have yet to be observed. The first term in (12.15) leads to additional nucleon–nucleon scattering that will not conserve parity. That is, nuclear forces will have a very small component that violates parity conservation. This has been recently seen and will be discussed in more detail later in this chapter (Section 16).

Another result that is implicit in the currents (12.8), (12.9), (12.11) is *lepton conservation.* As one can see from these formal expressions or from the Puppi triangle, the process

$$\mu^- \rightarrow e^- + \bar{\nu}_\mu + \nu_e \tag{12.17}$$

is not allowed. If e^- is emitted it can, according to (12.9), be accompanied only by $\bar{\nu}_e$; similarly μ^- and $\bar{\nu}_\mu$ are coupled. We have remarked earlier that at least in nuclear β-decay, this correlation has been demonstrated experimentally by R. Davis (55) (Section 7). The existence or nonexistence of reaction (12.17) still remains to be investigated.

This pairing (e^-, $\bar{\nu}_e$), (μ^-, $\bar{\nu}_\mu$), etc. can be formalized by introducing a μ- and e-lepton number l_μ and l_e and by requiring conservation of lepton number of each type, the total lepton number being obtained by *addition* of the individual lepton numbers. The lepton numbers assigned each particle is listed in Table 12.1.

TABLE 12.1 Lepton Quantum Numbers

Particle	l_e	Particle	l_μ
e^-	1	μ^-	1
ν_e	1	ν_μ	1
e^+	-1	μ^+	-1
$\bar{\nu}_e$	-1	$\bar{\nu}_\mu$	-1

The nucleons and other hadrons are presumed to have a zero lepton number. The particles at each vertex of the Puppi triangle have a total lepton number, obtained by addition of their individual lepton numbers, of zero. Hence, reactions in which the particles at one vertex are transformed into the pair at any other vertex are allowed. On the other hand reaction (12.17) is not allowed since the muon lepton number of the right side of the equation is (-1), on the left-hand side it is $(+1)$. Similarly the reaction

$$p + \nu_e \rightarrow n + e^+$$

is not allowed. A reaction in which an electron or e-neutrino converts into a muon is not allowed. Thus

$$p + \bar{\nu}_e \rightarrow n + \mu^+$$

is not allowed, as described earlier in this section.

The conservation of lepton number is not at this writing a settled question. However in nuclear β-decay, because of the Davis experiment (see p. 811) and the double β-decay results to be discussed it seems to hold to within present experimental accuracy that would permit deviations of the order of several percent.

It is convenient at this point to mention a further modification of (12.13) that enters because of the existence of strange particles. This requires a change in the hadron current. There is only one component of that current in (12.13): $\hat{\mathcal{J}}_\lambda^{(N)}$. When strange particles such as Λ, Σ, Ξ particles are included a strange-particle current must be defined, $\hat{\mathcal{J}}_\lambda^{(S)}$. Then $\hat{\mathcal{J}}_\lambda^{(N)}$ in (12.13) is replaced by

$$\hat{\mathcal{J}}_\lambda^{(N)} \rightarrow \cos\theta_c \hat{\mathcal{J}}_\lambda^{(N)} + \sin\theta_c \hat{\mathcal{J}}_\lambda^{(S)} \tag{12.18}$$

where θ_c is the Cabibbo angle. We note that the nuclear decay transition amplitude will be proportional to $\cos\theta_c$ (strange particle β-decay to $\sin\theta_c$) while μ-decay involving only lepton currents will be unaffected. Thus the difference noted earlier between $C_V g$ and G_μ, (12.4) and (12.6), can be used to give a measurement of $\cos\theta_c$. The resulting value of the Cabibbo angle is then about 0.20, which is close to the value 0.21 obtained in the comparison between K and μ-decay. However, as can be seen from Table 3.1, the value of θ_c obtained from nuclear β-decay is uncertain in the second significant figure.

13. CONSERVED VECTOR CURRENT (CVC)

The preceding discussion has been based on first-order perturbation theory with H_{weak} as the perturbation. However the nucleons are strongly interacting particles, and the question naturally arises as to the modifications that can be introduced because of their coupling to, say, the pion field (others might be the ρ, ω, etc. fields). The diagrams of first order in the meson-nucleon coupling are shown in Fig. 13.1. The μ-decay, however, does not involve particles that couple strongly with other particles. Nevertheless the coupling constants G_μ and gC_V, except for the small Cabibbo effect, are equal.

A solution to this problem is suggested by the solution to a similar problem for the electric charge. Although a proton and an electron have very different couplings to, for example pions, their couplings to the electromagnetic field, as described by their electric charge, are identical. The analogy to the weak interaction is really quite complete. This is indicated in Fig. 13.2 where we see a one-to-one relationship with the diagrams of Fig. 13.1, the single-photon line taking the place of the $(e^-, \bar{\nu}_e)$ combination, the incident and final particles are now protons and both are charged. One can of course build further structures on each of the pion lines, but it is clear that identical features can be built into the corresponding diagrams of Fig. 13.1. If we recall the original considerations described in Section 11 that led to the formal β-decay theory this one-by-one correspondence should occasion no surprise. But it has important consequences: that if we organize the weak interaction theory so that it has the same formal elements as electrodynamics, then the equality of the muon and nucleon weak-coupling constants would immediately follow.

In electrodynamics the essential element is charge conservation. The introduction of virtual particles such as pions will not change the net charge of the system since charge conservation occurs at each vertex, that is, at each interaction. The effect then of these couplings is to surround the bare particle

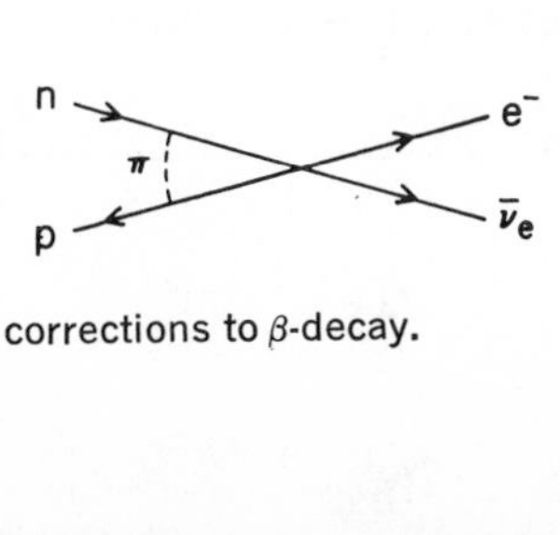

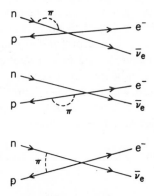

FIG. 13.1. First-order pionic corrections to β-decay.

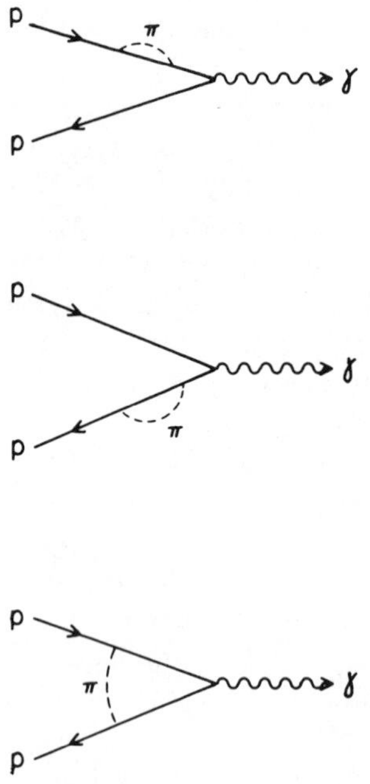

FIG. 13.2. First order pionic corrections to the electromagnetic p-p-γ vertex.

with a cloud of charged and neutral particles whose net charge is zero. This cloud may very well have a size and a structure (e.g., the electron–proton scattering probes the proton structure). But at infinite wavelength only the total charge counts and, thus, the effect of the cloud of virtual particles vanishes. The electric charge is defined in this infinite wavelength limit.

This argument although substantially correct is not rigorous because of the divergences of the theory. For example it is known that there is a renormalization of the charge in quantum electrodynamics. The rigorous argument proceeds as follows. The conservation of charge has the Ward identity as one of its consequences. This identity [Bjorken and Drell (65)] relates the electromagnetic interaction to the propagator for the charged particle so that we need only discuss the propagator. The first correction to the propagator is a photon-electron bubble as indicated by Fig. 13.3. The photon line has of course corrections but these will be independent of the particle which generated the virtual photon in the first instance. Hence, if the bare charge of each of the charged particles were the same, the interaction effects will preserve their ratio.

These results follow from conservation of charge. If $\hat{\rho}$ is the charge density operator then

$$\hat{Q} = \int \hat{\rho}\, d\mathbf{r} \tag{13.1}$$

FIG. 13.3. Corrections to the electron propagator.

is a constant of the motion. In differential form the conservation of charge becomes

$$\sum_\lambda \frac{\partial}{\partial x_\lambda} \hat{j}_\lambda{}^{(el)} = 0 \tag{13.2}$$

To carry over this result—the independence of the ratio of charges to interaction effects—to the weak interactions we shall postulate that the four-vector component of the currents $\hat{J}_\lambda{}^{(N)}$, $\hat{V}_\lambda{}^{(N)}$, satisfy the continuity equation satisfied by the four-vector $\hat{j}_\lambda{}^{(el)}$. This four-vector component is

$$\hat{V}_\lambda{}^{(N)} = i\hat{\bar{\psi}}\gamma_\lambda\tau^{(+)}\hat{\psi} \tag{13.3}$$

The same also holds for the hermitian conjugate vectors. For completeness, the remainder of $\hat{J}_\lambda$ is referred to as the axial vector component, the symbol $\hat{A}_\lambda$ now being employed. Thus

$$\hat{J}_\lambda{}^{(N)\bigstar} \equiv \hat{V}_\lambda{}^{(N)} + \hat{A}_\lambda{}^{(N)} \tag{13.4}$$

where

$$\hat{A}_\lambda{}^{(N)} = i\hat{\bar{\psi}}\gamma_\lambda\gamma_5\tau^{(+)}\hat{\psi} \tag{13.5}$$

Note that conservation of the axial current is *not* postulated, the electrodynamic model being slavishly followed. Actually, as we shall see, exact conservation for the axial current would result in the absence of the decay $\pi \to \mu + \nu$, and thus would contradict experiment (see p. 854).

Returning to the vector component of the current, we now postulate current conservation [R. P. Feynman and M. Gell-Mann (58), Gershtein and Zel'dovich (55)]

$$\sum_\lambda \frac{\partial}{\partial x_\lambda} \hat{V}_\lambda = 0 \tag{13.6}$$

and*

$$\int \hat{V}_0 \, d\mathbf{r} \qquad \text{is a constant of motion} \tag{13.7}$$

where

$$\hat{V}_4 \equiv i\hat{V}_0 \tag{13.8}$$

---

*In the presence of an electromagnetic field $\mathcal{A}_\lambda$, (13.6) becomes

$$\sum_\lambda \frac{\partial \hat{V}_\lambda}{\partial x_\lambda} = \frac{ie}{\hbar c} \sum_\lambda \hat{\mathcal{A}}_\lambda \hat{V}_\lambda$$

With this postulate the ratio of the observed coupling constants gC_V and G_μ equals its "bare" value. And these are now assumed to be related by

$$gC_V = G_\mu \cos \theta_c \tag{13.9}$$

Postulate (13.6) is referred to as the weak conserved vector-current hypothesis.

The strong CVC is obtained by comparing the nucleon-vector current (13.3) with the nucleon electromagnetic current. The latter can be written as linear combination of an isoscalar and an isovector (see Eq. VIII.2.4). In field theoretic notation

$$\hat{j}_\lambda{}^{(el)} = \frac{i}{2}\hat{\bar{\psi}}\gamma_\lambda(1 + \tau_3)\hat{\psi} \tag{13.10}$$

The charge density is

$$\hat{\rho} = \tfrac{1}{2}\hat{\psi}^\dagger(1 + \tau_3)\hat{\psi} \tag{13.11}$$

Note that

$$\int \hat{\psi}^\dagger\hat{\psi}\, d\mathbf{r}$$

is proportional to the total number of nucleons and, according to the principle of the conservation of baryon number, is conserved. Hence both the isoscalar and isovector components of (13.10) separately satisfy a continuity equation:

$$\sum_\lambda \frac{\partial}{\partial x_\lambda}\hat{\bar{\psi}}\gamma_\lambda\hat{\psi} = 0 \tag{13.12}$$

$$\sum_\lambda \frac{\partial}{\partial x_\lambda}\hat{\bar{\psi}}\gamma_\lambda\tau_3\hat{\psi} = 0 \tag{13.13}$$

Compare this last equation with the conservation conditions satisfied by the nucleon current (13.6):

$$\sum_\lambda \frac{\partial}{\partial x_\lambda}\hat{\bar{\psi}}\gamma_\lambda\tau^{(\pm)}\hat{\psi} = 0$$

Similarly

$$\frac{1}{2}\int \hat{\psi}^\dagger\tau_3\hat{\psi}\, d\mathbf{r} \tag{13.14}$$

as well as

$$\int \hat{\psi}^\dagger\tau^{(\pm)}\hat{\psi}\, d\mathbf{r} \tag{13.15}$$

are conserved. The strong form of the conserved vector current hypothesis suggests itself: *The hadron vector current of the weak interaction is related to the isovector component of the electromagnetic current by a rotation in isospin space.* This is true for the nucleon current considered here

$$i\bar{\psi}\gamma_\lambda\boldsymbol{\tau}\psi$$

The important point is that this relation between the weak current and the isovector component of the electromagnetic current holds for *all* hadron electromagnetic currents. For example, the isovector component of the current for the nucleon plus pion system is

$$i\{\tfrac{1}{2}\hat{\bar{\psi}}\tau_3\gamma_\lambda\hat{\psi} - [\hat{\pi}^*T_3\partial_\lambda\hat{\pi} - (\partial_\lambda\hat{\pi})^*T_3\hat{\pi}]\} \tag{13.16}$$

where $\hat{\pi}$ is the pion field and T_i is the isospin operator for unit isospin. According to the CVC hypothesis the weak current $\hat{\mathcal{J}}_\lambda$ will contain the term

$$\hat{\mathcal{J}}_\lambda^\star = i[\tfrac{1}{2}\hat{\bar{\psi}}\tau^{(+)}\gamma_\lambda\hat{\psi} - (\hat{\pi}^*T^{(+)}\partial_\lambda\hat{\pi} - (\partial_\lambda\hat{\pi})^*T^{(+)}\hat{\pi}] + \ldots \tag{13.17}$$

Combining this with the lepton current we see that the CVC hypothesis predicts π-decay

$$\pi^+ \longrightarrow \pi^0 + e^+ + \nu \tag{13.18}$$

One of the triumphs of the CVC theory is the prediction of the lifetime of this decay.

Let us now consider the application of CVC to nuclear β-decay. Because of the relation between (13.16) and (13.17) it becomes possible to relate electromagnetic transition matrix elements and β-decay matrix elements for members of an isospin multiplet. There are two ways to proceed. One is to evaluate the matrix element of the nucleon current between neutron and proton and then to simply generalize it to finite nuclei that is, to a many-body system. Another procedure that has been pioneered by Kim and Primakoff (65, 66) treats the transitions between the nuclear states directly. Let us discuss the first of these now.

The matrix element of interest is

$$M_\beta \equiv \sum_\lambda \langle e\bar{\nu}p| \int d\mathbf{r}\, \hat{V}_\lambda{}^{(N)}(\mathbf{r})\hat{j}_\lambda{}^{(e)}(\mathbf{r})|n\rangle \tag{13.19}$$

Because of translational invariance (there is no unique origin)

$$\exp(-i\hat{\mathbf{P}}\cdot\mathbf{r})\hat{V}_\lambda{}^{(N)}(\mathrm{o})\exp(i\hat{\mathbf{P}}\cdot\mathbf{r}) = \hat{V}_\lambda{}^{(N)}(\mathbf{r})$$

and similarly for $\hat{j}_\lambda{}^{(e)}$. We obtain

$$\hat{V}_\lambda{}^{(N)}(\mathbf{r})\hat{j}_\lambda{}^{(e)}(\mathbf{r}) = \exp(-i\hat{\mathbf{P}}\cdot\mathbf{r})\hat{V}_\lambda{}^{(N)}(\mathrm{o})\hat{j}_\lambda{}^{(e)}(\mathrm{o})\exp(i\hat{\mathbf{P}}\cdot\mathbf{r})$$

Here $\hat{\mathbf{P}}$ is the total momentum operator. Then

$$M_\beta \equiv \sum_\lambda \int d\mathbf{r}\,\langle e\bar{\nu}p|\,\exp\,(-i\hat{\mathbf{P}}\cdot\mathbf{r})\hat{V}_\lambda{}^{(N)}(\mathrm{o})\hat{\jmath}_\lambda{}^{(e)}(\mathrm{o})\exp\,(i\hat{\mathbf{P}}\cdot\mathbf{r})|n\rangle$$

Evaluating yields

$$\begin{aligned} M_\beta &= \sum_\lambda \int d\mathbf{r}\,\exp\,[i(\mathbf{p}_n - \mathbf{p}_e - \mathbf{p}_\nu - \mathbf{p}_p)\cdot\mathbf{r}]\langle e\bar{\nu}p|\hat{V}_\lambda{}^{(N)}(\mathrm{o})\hat{\jmath}_\lambda{}^{(e)}(\mathrm{o})|n\rangle \\ &= (2\pi)^3\delta(\mathbf{p}_n - \mathbf{p}_e - \mathbf{p}_\nu - \mathbf{p}_p)\sum_\lambda\ \langle e\bar{\nu}|\hat{\jmath}_\lambda{}^{(e)}(\mathrm{o})|0\rangle\langle p|\hat{V}_\lambda{}^{(N)}(\mathrm{o})|n\rangle \end{aligned}$$

The delta function provides for conservation of momentum, which is thus a consequence of the homogeneity of space. The next two factors give rise, in the nonrelativistic limit for nucleon motion, to the first term of (11.16), the lepton current matrix element yielding the bilinear term in u while the nucleon current matrix element becomes the Fermi matrix element. Let us concentrate on this term.

From general covariance it follows that

$$\langle p|\hat{V}_\lambda{}^{(n)}(\mathrm{o})|n\rangle = \bar{u}_p\mathrm{O}_\lambda u_n \tag{13.20}$$

where u are the Dirac spinors discussed earlier. O_λ is an operator in spinor space (there can be no dependence on $\mathbf{r}$) that transforms as a four vector.* It must involve the Dirac operators γ_λ, $\sigma_{\lambda\nu}$, etc. and the momenta of the neutron and proton. Since $p_p{}^2 = \mathbf{p}_p{}^2 - E^2/c^2 = -\ M_p{}^2c^2$ with a similar relation for the neutron momentum, there is only one four-dimensional scalar, which we choose as

$$k^2 \equiv (p_n - p_p)^2 \tag{13.21}$$

For momentum variables we choose

$$\begin{aligned} k_\lambda &\equiv (p_n)_\lambda - (p_p)_\lambda \\ K_\lambda &\equiv (p_n)_\lambda + (p_p)_\lambda \end{aligned} \tag{13.22}$$

The most general four vector involving k_λ, K_λ, and the Dirac matrices is then

$$\mathrm{O}_\lambda = a\gamma_\lambda + b\sum_\nu \sigma_{\lambda\nu}k_\nu + ck_\lambda + d\sum_\nu \sigma_{\lambda\nu}K_\nu + eK_\lambda \tag{13.23}$$

where the coefficients a, b, etc. are functions only of k^2. The matrix elements

*To be completely independent of representation, it is more accurate to say that $\bar{u}_p\mathrm{O}_\lambda u_n$ transforms as a four vector. However it is convenient to use the simplified statement.

of these operators are not independent because u_p and u_n satisfy the Dirac equation**

$$
\begin{aligned}
[(\gamma p_n) - iM_n c]u_n &= 0 \\
\bar{u}_p[(\gamma p_p) - iM_p c] &= 0
\end{aligned}
\qquad (13.24)
$$

where $(\gamma p) \equiv \sum_\lambda \gamma_\lambda p_\lambda$

The consequence is that all dependence on K_λ can be eliminated as we shall now show. Multiply the first equation in (13.24) by $\bar{u}_p\gamma_\nu$ from the left and the second equation from the right by $\gamma_\nu u_n$. Subtract the two equations and express p_n and p_p in terms of k and K to obtain

$$\bar{u}_p[\tfrac{1}{2}\gamma_\nu(\gamma K) - \tfrac{1}{2}(\gamma K)\gamma_\nu + \tfrac{1}{2}\gamma_\nu(\gamma k) + \tfrac{1}{2}(\gamma k)\gamma_\nu - i(M_n - M_p)c\gamma_\nu]u_n = 0$$

or

$$\sum_\lambda \bar{u}_p\sigma_{\nu\lambda}K_\lambda u_n = i\bar{u}_p k_\lambda u_n + (M_n - M_p)c\bar{u}_p\gamma_\lambda u_n$$

By adding the two equations (13.24) one can similarly show that

$$\bar{u}_p K_\nu u_n = i\bar{u}_p(M_n + M_p)c\gamma_\nu u_n - i\bar{u}_p\sigma_{\nu\lambda}k_\lambda u_n$$

Hence the K_λ terms in (13.23) are not independent of the other terms. Returning to (13.20) the most general form of the matrix element of $\hat{V}_\lambda^{(N)}(\mathrm{o})$ is thus

$$\langle p|\hat{V}_\lambda^{(N)}(\mathrm{o})|n\rangle = i\bar{u}_p[g_V\gamma_\lambda + g_W\frac{1}{2Mc}\sum_\nu \sigma_{\lambda\nu}k_\nu + ig_S\frac{1}{2Mc}k_\lambda]u_n \qquad (13.25)$$

where the factors Mc have been introduced so that the factors g_W and g_S are dimensionless. The coefficients g are functions of k^2 and are real if time-reversal invariance holds. Note that this form is not identical with the form that would have been obtained if zero-order perturbation theory had been applied to $\hat{V}_\lambda^{(N)}$. The differences should then be ascribed to interaction effects.

To obtain a measure of these, the CVC hypothesis will be applied. First suppose the coupling constant $C_V g$ is given its empirical value

$$g_V(\mathrm{o}) = 1 \qquad (13.26)$$

the argument preceding (13.9) applying to infinitely long wavelength for the leptons. To proceed further the strong form of CVC is needed. We must compare (13.25) with the electomagnetic form factor (see Eq. VIII. 3.8)

$$\langle p_1|V_\lambda^{(el)}(\mathrm{o})|p_2\rangle = i\langle\bar{u}(p_1)|F_1^{(V)}(k^2)\gamma_\lambda + F_2^{(V)}(k^2)\sum \sigma_{\lambda\nu}k_\nu|u(p_2)\rangle \quad (13.27)$$

**This is true for free protons and neutrons. For protons and neutrons inside nuclei (13.24) must be modified, leading to two-body, three-body, etc. corrections to the dominant one-body term.

where $F_1^{(V)}$ and $F_2^{(V)}$ are the isovector form factors (the coefficient of $\tau_3/2$). In the limit of zero momentum transfer

$$F_1^{(V)}(0) = 1 \qquad F_2^{(V)}(0) = \frac{\kappa_p - \kappa_n}{2Mc} \tag{13.28}$$

where κ_p and κ_n are the proton and neutron anomalous magnetic moments in units of the nuclear magneton μ_0. Comparing (13.25) and (13.27) yields

$$g_V(k^2) = F_1^{(V)}(k^2) \qquad g_V(0) = 1 \tag{13.29}$$

$$\frac{1}{2Mc} g_W(k^2) = F_2^{(V)}(k^2) \qquad g_W(0) = \kappa_p - \kappa_n = 3.706 \tag{13.30}$$

We note that no term of the scalar type, g_S, appears. This is a consequence in the electromagnetic case of the application of the electromagnetic current conservation. The same proof does not obtain for form (13.25) because the mass of the proton does not equal that of the neutron, as can be seen from the following argument.

The matrix element of $\Sigma \partial \hat{V}_\lambda^{(N)}/\partial x_\lambda$ according to the conservation law must zero while according to (13.25)

$$\sum_\lambda \langle p| \frac{\partial \hat{V}_\lambda^{(N)}}{\partial x_\lambda} |n\rangle = 0 = -\frac{1}{\hbar} \bar{u}_p \left[g_V(k\gamma) + \frac{ig_S}{2Mc} k^2 \right] u_n \tag{13.30'}$$

Employing the Dirac equation (13.24), (13.30′) becomes

$$0 = g_V(M_n - M_p)c + \frac{g_S}{2Mc} k^2$$

Recalling that g_V and g_S are functions of k^2 it follows that g_S is zero to within the isospin symmetry breaking interaction responsible for the mass difference between the neutron and proton. Recall that the conservation of the vector current is valid only in the limit in which this interaction vanishes.

The induced scalar can also be eliminated if one invokes *G*-parity conservation. The strong interactions that do give rise to the induced terms such as g_S and g_W conserve *G*-parity to within the electromagnetic isospin breaking interaction and, therefore, to this approximation one might ask that the form (13.25) have a unique *G*-parity. Let us examine this argument now reviewing first the definition of *G*-parity.

We first recall the behavior of the electromagnetic current under charge conjugation. Let C be the charge conjugation operator with the property that the charge conjugate spinor corresponding to a spinor u is given by Cu^* (Appendix A.C.6). The current for this charge conjugate spinor is

$$\overline{Cu^*}\gamma_\lambda Cu^* = -\bar{u}\gamma_\lambda u \tag{13.31}$$

so as expected the current changes sign. The same result does not hold for the weak hadron current, $\bar{u}_p\gamma_\lambda u_n$, since the two spinors do not refer to the same particle. This difficulty can however be remedied by including a rotation about the "2" axis in isospin space of 180° so that G is given by

$$G \equiv e^{i\pi T_2}C \tag{13.32}$$

which for isospinors is

$$G \equiv i\tau_2 C$$

Thus the effect of τ_2 and C is to change not only the particle into an antiparticle but in addition to change it from a proton to a neutron or vice versa. Then it can be shown that

$$\overline{i\tau_2 C u_p{}^*}\gamma_\lambda i\tau_2 C u_n = -\bar{u}_p\gamma_\lambda u_n$$

On the other hand the scalar term that modifies g_S, $\bar{u}_p\, u_n$ will not change sign under this transformation and thus has a differing G-parity from the leading term. It is thus plausible that this term is not present in (13.25).*

Problem: Using $C = -\gamma_2$ (11.7) show that

$$\overline{i\tau_2 C u_p{}^*} A i\tau_2 C u_n = -\bar{u}_p\gamma_4\gamma_2 A^T\gamma_2\gamma_4 u_n$$
$$= (\bar{u}_p A u_n)\cdot\begin{cases} -1 & \text{if} \quad A = 1,\ \gamma_5,\ \gamma_\lambda\gamma_5 \\ 1 & \text{if} \quad A = \gamma_\lambda,\ \sigma_{\nu\lambda} \end{cases}$$

Note that $\gamma_2\gamma_k{}^T\gamma_2 = \gamma_k$ and $\gamma_2\gamma_4\gamma_2 = -\gamma_4$

Finally, with $g_S = 0$

$$\langle p|\hat{V}_\lambda{}^{(N)}(\text{o})|n\rangle = i\bar{u}_p\left[F_1{}^{(V)}(k^2)\gamma_\lambda + F_2{}^{(V)}(k^2)\frac{1}{2Mc}\sum \sigma_{\lambda\nu}k_\nu\right]u_n \tag{13.33}$$

This is a most remarkable formula, the nucleon structure effects in β-decay being determined from proton and neutron electron scattering. Of course the usual momentum changes that occur in β-decay are small so that the only values of k^2 close to zero are involved. (Somewhat greater momenta changes occur in μ-capture.) In that limit

$$\langle p|\hat{V}_\lambda{}^{(N)}(\text{o})|n\rangle = i\bar{u}_p\left(\gamma_\lambda + \frac{\kappa_p - \kappa_n}{2Mc}\sum \sigma_{\lambda\nu}k_\nu\right)u_n \tag{13.34}$$

The second term was referred to as "weak magnetism" by Gell–Mann because of its relation to the anomalous magnetic moments of the nucleons. In both

---

*Some objection may be made to this argument since it involves a comparison between particle and antiparticle currents, the latter being hardly accessible. However the same argument is valid when C is replaced by the product of time reversal and parity inversion.

cases, the anomaly as well as the weak magnetism are the consequences of the strong interactions the nucleons have with other particles. Weak magnetism is a specific prediction of CVC with no undetermined constants. The extraordinary thing is that it agrees with experiment verifying the strong CVC hypothesis.

Before describing the experiment it is useful to obtain the nonrelativistic reduction of (13.34). For the $\lambda = 4$ component the result up to $O(1/M^2)$ remains

$$\bar{u}_p \gamma_\lambda u_n \to u_p^\dagger u_n$$

The spatial components can also be readily obtained by noting that the operator in (13.34) is the electromagnetic current operator. Therefore the reduction will yield the same result as in that case for which γ_k contributes the "normal" Dirac value of one magneton which when added on to the anomalous term gives the total magnetic moment.

$$\langle p\bar{\nu}_e e | H^V | n \rangle \to \frac{g C_V}{\sqrt{2}} \left\{ u_p^\dagger u_n \bar{u}_e \gamma_4 (1 + \gamma_5) u_{\bar{\nu}} - \frac{\mu_p - \mu_n}{2Mc} u_p^\dagger \boldsymbol{\sigma} u_n \cdot \bar{u}_e (\mathbf{k} \times \boldsymbol{\gamma}) \times (1 + \gamma_5) u_{\bar{\nu}} \right\} + \text{h.c.} \qquad (13.35)$$

where $\boldsymbol{\gamma} = (\gamma_1, \gamma_2, \gamma_3)$. Note that $\mathbf{k}$ when expressed in lepton variables is given as a consequence of the conservation of momentum by

$$\mathbf{k} = \mathbf{p}_e + \bar{\mathbf{p}}_\nu \qquad (13.36)$$

The induced term is of the Gamow–Teller type and represents a small correction to the principal term of that type that comes from the axial vector component of $J_\mu^{(N)}$. However it is momentum dependent, vanishing as expected as the momentum transfer $\mathbf{k}$ goes to zero. It is this momentum dependence that permits this effect to be seen in the presence of the dominant momentum independent axial interaction. The electron energy distribution will be modified, and the Kurie plot of Fig. 2.3 will no longer be a straight line.

The effect can be amplified by comparing positron and electron emitters since, as we shall now show, the deviation from the Kurie plot for β^+ emitters will be exactly opposite to that for β^- emitters. The weak magnetism term must be added to the axial vector term. The essential point is that the relative sign between the matrix elements for these two contributions will be different for the two kinds of decay.

This can be seen from the following argument. When the electron-antineutrino operator

$$\hat{\psi}_e^\dagger A \hat{\psi}_\nu$$

is rewritten in terms of the charge conjugate field operators, that is, $\hat{\psi}_{e^-}$ is replaced according to the following

$$\hat{\psi}_{e^-} = -\gamma_2 \hat{\psi}_{e^+}^*$$

it becomes

$$(\hat{\psi}^{\dagger}{}_{e^+})^* \gamma_2 A \gamma_2 \hat{\psi}_{\bar{\nu}}{}^* = -\hat{\psi}_{\bar{\nu}}{}^{\dagger} \gamma_2 A^T \gamma_2 \hat{\psi}_{e^+}$$

where we have included the change of sign that comes from the anticommutation of spin 1/2 wave operators. This is not yet the term we want. It still gives the β^- emission. To obtain the appropriate term for β^+ emission we need to take its hermitan conjugate which yields finally

$$-\hat{\psi}_{e^+}{}^{\dagger} \gamma_2 A \gamma_2 \hat{\psi}_{\bar{\nu}e}$$

giving the operator for positron decay. We now note that A is of the form $(A' + A'\gamma_5)$, the second term being the axial part and

$$\gamma_2(A' + A'\gamma_5)\gamma_2 = \gamma_2 A' \gamma_2 (1 - \gamma_5)$$

proving that the relative phase between the vector and axial vector for β^- decay is opposite to that for β^+ decay.

Gell–Mann (58) suggested an experiment exploiting these features. The nuclear levels are the members of a $T = 1$ triad all decaying to a $T = 0$ level (Fig. 13.4). The spin of the $T = 1$ levels are $J^\pi = 1^+$ decaying by a pure Gamow–Teller transition β^- and β^+, and electromagnetically (for the transition involving the corresponding nuclear states) by a magnetic dipole transition. Thus we can compare the two β-decays as suggested in the above paragraph. By using the isovector part of the transition magnetic dipole moment that dominates the electromagnetic transition, it becomes possible to calculate

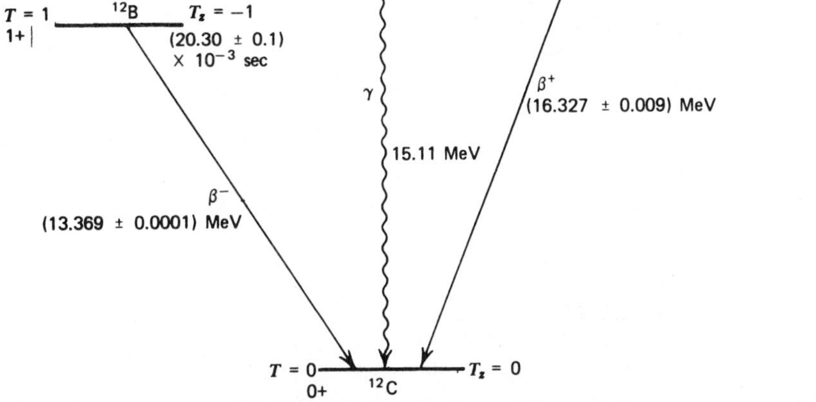

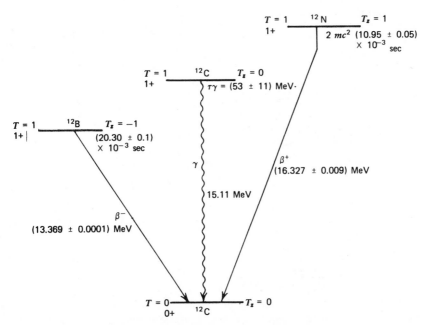

FIG. 13.4. The $A = 12$ triad [taken from Wu and Moskowski (66)].

the magnitude of the matrix element of the weak magnetism term for β-decay and, therefore, the magnitude of the deviation from the Kurie plot. The nuclei in question are the $A = 12$ triad.

We shall not go through the details of calculating the new energy distribution as they are straightforward. The comparison between CVC theory and the experimental electron and positron spectra is shown in Fig. 13.5. The agreement is excellent. The theoretical predictions to be compared with the figures in parenthesis are $(0.55 \pm 0.12)\%$ and $(-0.55 \pm 0.12)\%$.

We have not discussed several possible corrections. These are contributions coming from the finite wavelength of the electron and neutrino wave functions, electromagnetic effects that the CVC hypothesis neglects, and second-forbidden matrix elements (See Wu (64)). A recent experiment (Garvey and Tribble (73)) avoids these difficulties.

In the preceding discussion the CVC theory has been applied to the elementary transition of the nucleon. Kim and Primakoff in a series of papers [Kim and Primakoff (65a), (65b), (66), Primakoff (69)] have applied CVC directly to the nuclei. The electromagnetic form factors of (13.33) for the nucleon are replaced by the corresponding form factors for the nucleus. This has the advantage that these form factors can often be directly determined from electron nuclear scattering. The more traditional method, from this point of view, is essentially a calculation of these form factors from the form factors of the elementary nucleons.

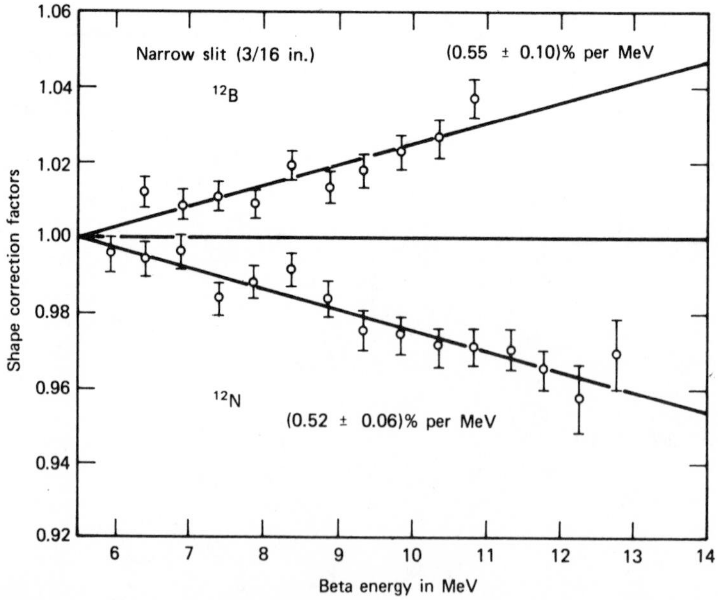

FIG. 13.5. Comparison between experimental electron and positron spectra and predictions of CVC [Lee, Mo, and Wu (63), Wu (64)].

14. PARTIALLY CONSERVED AXIAL VECTOR CURRENT (PCAC)

Let us now apply some of the reasoning of the preceding section to the axial currents. The analogs of (13.20) and (13.23) are

$$\langle p|\hat{A}_\lambda^{(N)}(o)|n\rangle = \bar{u}_p P_\lambda u_n \tag{14.1}$$

where P_λ is a pseudovector operator.* After eliminating K_λ, P_λ has the form

$$P_\lambda = i\left[g_A\gamma_\lambda\gamma_5 + \frac{i}{2Mc}g_P\gamma_5 k_\lambda + \frac{1}{2Mc}g_T\sum_\nu \sigma_{\lambda\nu}k_\nu\gamma_5\right] \tag{14.2}$$

In addition to the original pseudovector term, there are induced pseudoscalar and tensor terms. The G parity associated with the tensor term is opposite to that of the other two that are even, the tensor term being odd. We shall discuss the experimental evidence that bears on the existence of this term later. We have already alluded twice (pages 753 and 790) to the relation between the first term in (14.2) and the spin component of the magnetic moment operator

$$\boldsymbol{\mu}_s = \left(\frac{\mu_p - \mu_n}{2}\right)\tau_3\boldsymbol{\sigma}$$

The nonrelativistic approximation for the first term in (14.2) (see Table 11.2) involves the operator for β^- decay $\tau^{(+)}\boldsymbol{\sigma}$ so that in this limit the principal axial vector term and $\boldsymbol{\mu}_s$ are members of a vector in isospin space. This relationship has been exploited in the discussion following (Eq. VIII.11.19) and on p. 790.

The value of g_A is 1.25 according to (11.46). The universal weak interaction (12.12) postulates this factor to be unity in the original unrenormalized interaction. The renormalized value of (1.25) (which does not occur in the lepton terms) must then be a consequence of the strong interactions. One of the triumphs of weak interaction theory is the prediction of this number.

We begin by discussing the possibility that the axial current also satisfies a conservation condition, always in the context of universality. Conservation of the axial current is not possible because it would result in a zero probability for the decay of the pion into a muon and neutrino [Taylor (58)]. Since the pion is a pseudoscalar particle the weak Hamiltonian for this decay must be of the form

$$\sum_\lambda \hat{j}_\lambda^{(\mu)*}\hat{A}_\lambda \tag{14.3}$$

where $\hat{j}_\lambda^{(\mu)}$ involves the muon and ν_μ fields. The matrix element for the decay will involve the matrix element

$$\langle o|\hat{A}_\lambda|\pi\rangle$$

Since the pion is a pseudoscalar it follows that

$$\langle o|\hat{A}_\lambda|\pi\rangle \simeq ik_\lambda F(k^2)$$

*See footnote p. 846.

where k_λ is the pion momentum and F is a scalar function of k^2. It is convenient to describe pion decay in a system in which the pion is at rest. Then $k^2 = -m_\pi^2$. Suppose now that

$$\sum \frac{\partial \hat{A}_\lambda}{\partial x_\lambda} = 0$$

Taking the matrix element of this equation we have

$$\langle \text{o}| \sum \frac{\partial \hat{A}_\lambda}{\partial x_\lambda} |\pi\rangle \sim ik_\lambda \langle \text{o}|\hat{A}_\lambda|\pi\rangle \sim -m_\pi^2 F(-m_\pi^2) = 0$$

proving the point that conservation of the axial current leads to zero probability for π-decay.

It was suggested by several authors [Gell–Mann, Levy (60), Bernstein, Fubini, Gell–Mann, and Thirring (60a), Bernstein, Gell–Mann, and Michel (60b), Nambu (60) and Chou (61)] that $\hat{A}_\lambda$ is partially conserved, that is, it is conserved in some limit. Since $\hat{A}_\lambda$ is an axial vector its four divergence must be proportional to a pseudoscalar field. The one of lowest mass is the pion field and was thought to be the most important possibility. Let us then initially assume that

$$\sum \frac{\partial}{\partial x_\lambda} \hat{A}_\lambda = a\phi_\pi \tag{14.4}$$

where a is a constant. To determine a we follow the procedure of Adler (65).

We shall need the matrix element (14.1)

$$\langle p|\hat{A}_\lambda^{(N)}|n\rangle = i\bar{u}_p \left(g_A\gamma_\lambda\gamma_5 + \frac{i}{2Mc} g_P\gamma_5 k_\lambda + \frac{g_T}{2Mc} \sum_\lambda \sigma_{\lambda\nu}k_\nu\gamma_5 \right) u_n$$

If we take the matrix element of the left hand side of (14.4) evaluating it at $k^2 = 0$ we obtain

$$\langle p| \sum_\lambda \frac{\partial \hat{A}_\lambda^{(N)}}{\partial x_\lambda} |n\rangle_{k^2=0} = -(1/\hbar)\bar{u}_p \sum_\lambda (g_A k_\lambda\gamma_\lambda\gamma_5)u_n|_{k^2=0}$$

$$= 2i\frac{Mc}{\hbar} g_A(\text{o})\bar{u}_p\gamma_5 u_n \tag{14.5}$$

On the other hand the matrix element of the right-hand side of (14.4) $\langle p|\hat{\phi}_\pi|n\rangle$ may be evaluated as follows. The pseudoscalar pion field is generated in the presence of nucleons according to the inhomogeneous Klein–Gordon equation (Volume II)

$$\left[\square^2 - \left(\frac{m_\pi c}{\hbar}\right)^2\right] \langle p|\hat{\phi}_\pi|n\rangle = i\sqrt{2}g_{pp\pi}(k^2)\bar{u}_p\gamma_5 u_n$$

where $g_{pp\pi}$ is the coupling constant between the nucleon and pion fields. Empirically from studies of the one pion exchange part of nuclear forces, and from pion–nucleon scattering [Lomon and Feshbach (68)]

$$\frac{g^2{}_{pp\pi}(-m_\pi{}^2)}{4\pi\hbar c} = 14.9 \pm 0.3$$

Then

$$\langle p|\hat{\phi}_\pi|n\rangle = -\frac{i\sqrt{2}\hbar^2}{k^2 + m_\pi{}^2c^2} g_{pp\pi}\bar{u}_p\gamma_5 u_n \tag{14.6}$$

Comparing (14.6) and (14.5) at $k^2 = 0$, according to (14.4)

$$a = -\frac{\sqrt{2}(Mc\, m_\pi{}^2c^2)}{\hbar^3}\frac{g_A(\text{o})}{g_{pp\pi}(\text{o})} \tag{14.7}$$

Relation (14.7) is known as the Goldberger–Trieman relation [Goldberger and Treiman (58)].

Assuming the universality of relation (14.4), we can use it to compute the lifetime of the pion against μ-decay. One may then evaluate a from the measured lifetime and compare it with the value calculated from (14.7). Or one can eliminate a and determine $g_A(\text{o})$. This is the procedure we shall follow.

The π-decay transition probability w is given by

$$dw = \frac{2\pi}{\hbar}\frac{G_\mu{}^2}{2}\sum_{s_\mu s_{\bar{\nu}}} \Big|\sum_\lambda \langle \text{o}|\hat{A}_\lambda|\pi\rangle \bar{u}_\mu\gamma_\lambda(1+\gamma_5)u_{\bar{\nu}}\Big|^2\rho_f \tag{14.8}$$

where ρ_f, the density of final states, is given by

$$\rho_f = \frac{1}{(2\pi\hbar c)^3}\frac{E_\nu{}^2E_\mu}{E_\pi}\, d\bar{\Omega}_\nu$$

In the reference frame in which the pion is at rest, $E_\pi = m_\pi c^2$. To get the total probability of decay it is necessary to integrate over $\bar{\Omega}_\nu$.

The matrix element $\langle \text{o}|\hat{A}_\lambda|\pi\rangle$ is related to constant a. To exhibit this relation take the matrix element of both sides of (14.4)

$$\langle \text{o}|\sum_\lambda \frac{\partial}{\partial x_\lambda}\hat{A}_\lambda|\pi\rangle = a\langle \text{o}|\hat{\phi}_\pi|\pi\rangle \tag{14.9}$$

The right-hand side can be directly evaluated

$$a\hbar c/\sqrt{2E_\pi}$$

The factor multiplying a is the normalization typical of a boson field. We have encountered it in our discussion of the electromagnetic field in the Appendix to Chapter VIII. The left-hand side of (14.9) can be evaluated as follows. We first observe

$$\langle o|\sum_\lambda \frac{\partial}{\partial x_\lambda}\hat{A}_\lambda|\pi\rangle = i\sum k_\lambda\langle o|\hat{A}_\lambda|\pi\rangle$$

where $\hbar k_\lambda$ is the four momentum of the pion. Second the matrix element $\langle o|\hat{A}_\lambda|\pi\rangle$ must be proportional to k_λ, the only vector in the field. We therefore write

$$\langle o|\hat{A}_\lambda|\pi\rangle = ik_\lambda\frac{\hbar c}{\sqrt{2E_\pi}}F(-m_\pi{}^2) \tag{14.10}$$

where F, the axial pion form factor, is a function of k^2. Employing (14.10)

$$\langle o|\sum \frac{\partial}{\partial x_\lambda}\hat{A}_\lambda|\pi\rangle = -\sum_\lambda k_\lambda{}^2\frac{\hbar c}{\sqrt{2E_\pi}}F(-m_\pi{}^2) = a\frac{\hbar c}{\sqrt{2E_\pi}}$$

It follows that

$$a = -\sum k_\lambda{}^2F(-m_\pi{}^2) = \left(\frac{m_\pi c}{\hbar}\right)^2F(-m_\pi{}^2)$$

Returning to (14.10) one obtains

$$\langle o|\hat{A}_\lambda|\pi\rangle = ik_\lambda a\left(\frac{\hbar}{m_\pi c}\right)^2\frac{\hbar c}{\sqrt{2E_\pi}} \tag{14.11}$$

It is now a straightforward matter to perform the spin sums in (14.8) and integrate over $d\Omega_\nu$. The final answer is

$$w(\pi\to\mu\nu_\mu) = \frac{G_\mu{}^2\,m_\mu{}^2}{8\pi c\,m_\pi{}^3}\left(1-\frac{m_\mu{}^2}{m_\pi{}^2}\right)^2a^2 \tag{14.12}$$

Finally we can introduce the Goldberger–Treiman relation, with the result

$$w(\pi\to\mu\nu_\mu) = g_A{}^2(o)\frac{G_\mu{}^2}{4\pi\hbar^2c}\left(\frac{m_\mu c}{\hbar}\right)^2\left(\frac{Mc}{\hbar}\right)^2\left(\frac{m_\pi c}{\hbar}\right)\frac{\hbar c}{g_{pp\pi}^2}\left(1-\frac{m_\mu{}^2}{m_\pi{}^2}\right)^2 \tag{14.13}$$

Introducing the masses, the pion-nucleon coupling constant and the empirical value for w [the mean life is $(2.6024 \pm .0024) \times 10^{-8}$ sec] yields a value of $g_A(o)G_\mu$, which is 3% greater than the experimental value. This is well within the experimental error both of $g_A(o)$ and of the pion-nucleon coupling constant to which must be added the error associated with (14.13). This is a striking confirmation of the PCAC hypothesis.

In the evaluation of (14.13), the value of $g_{pp\pi}$ at $k^2 = -m_\pi{}^2$ is employed rather than its value at zero k^2. It is thus assumed that k^2 dependence of $g^2{}_{pp\pi}$ is weak in the range $-m_\pi{}^2 \leq k^2 \leq 0$. Perturbative investigation of the source of such an energy dependence indicates that it might arise from the corrections to the unrenormalized nucleon-nucleon-pion vertex. These corrections involve intermediate states generated by the virtual emission and absorption of the heavier bosons such as the ρ. This suggests that the "small parameter" measuring the energy dependence of $g^2{}_{pp\pi}$ is $(m_\pi/m_\rho)^2 \sim 0.02$.

As a second consequence of PCAC, it is possible to obtain an estimate of the size of the coefficient g_P in (14.2). The physical process in which g_P, the induced pseudoscalar term, originates is illustrated in Fig. 14.1. In a perturbation theory, this diagram would give a contribution proportional to

$$(g_{pp\pi}\bar{u}_p\gamma_5 u_n)\left(\frac{1}{k^2 + m_\pi{}^2c^2}\right)\langle\pi|\hat{A}_\lambda|\text{o}\rangle\bar{u}_e\gamma_\lambda(1 + \gamma_5)u_{\bar{\nu}} \tag{14.14}$$

the first term in parentheses coming from vertex labeled 1 in the figure, the second the pion propagator going from vertex 1 to vertex 2 where the pion decays according to the process $\pi^- \rightarrow e^- + \bar{\nu}_e$. Since $\langle\pi|\hat{A}_\lambda|\text{o}\rangle$ is proportional to k_λ we see that the form of the resulting term is identical with that of the pseudoscalar term in (14.2), $\bar{u}_p\gamma_5 k_\lambda u_n$. This is a perturbation calculation. With the PCAC hypothesis it is possible to make a nonperturbative calculation that is outlined as follows.

For this purpose we evaluate $\langle p|\sum_\lambda \partial A_\lambda{}^{(N)}/\partial x_\lambda|n\rangle$ but this time we shall evaluate it at a finite value of k^2. In that event (14.5) is replaced by

$$\langle p|\sum_\lambda \frac{\partial\hat{A}_\lambda{}^{(N)}}{\partial x_\lambda}|n\rangle = \frac{i}{\hbar}\sum_\lambda k_\lambda\langle p|\hat{A}_\lambda{}^{(N)}(\text{o})|n\rangle$$

or using (14.2)

$$\langle p|\sum_\lambda \frac{\partial\hat{A}_\lambda{}^{(N)}}{\partial x_\lambda}|n\rangle = -\frac{1}{\hbar}\bar{u}_p\left[g_A(k\gamma)\gamma_5 + \frac{ik^2}{2Mc}g_P\gamma_5\right]u_n$$

$$= -\frac{1}{\hbar}\bar{u}_p\left(-2iMcg_A + \frac{ik^2}{2Mc}g_P\right)\gamma_5 u_n$$

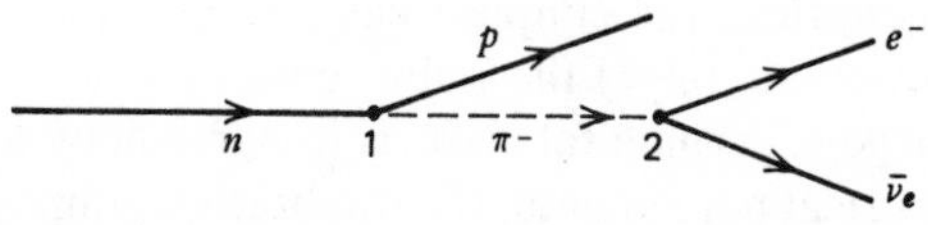

FIG. 14.1. Pionic correction to β-decay.

Employing (14.4) and (14.6)

$$\langle p|\sum_\lambda \frac{\partial \hat{A}_\lambda}{\partial x_\lambda}|n\rangle = \frac{-ia\sqrt{2}\hbar^2}{k^2 + m_\pi{}^2c^2} g_{pp\pi}\bar{u}_p\gamma_5 u_n$$

so that

$$-\frac{\sqrt{2}a\hbar^3}{k^2 + m_\pi{}^2c^2} g_{pp\pi} = 2Mcg_A - \frac{k^2}{2Mc} g_P \tag{14.15}$$

Since g_A and $g_{pp\pi}$ are assumed to be slowly varying, it immediately follows that near $k^2 = -m_\pi{}^2$

$$g_P \cong \frac{4M^2c^2}{k^2 + m_\pi{}^2c^2} g_A(0) \tag{14.16}$$

where a has been replaced according to (14.7). The next approximation to this equation for g_P is a constant term proportional to the first derivative of $g_{pp\pi}(k^2)$ evaluated at $k^2 = -m_\pi{}^2c^2$. Employing (14.16) is thus consistent with the assumption made above that $g_{pp\pi}$ and g_A are constant in the range of k^2 of interest, $-m_\pi{}^2c^2 < k^2 < 0$. Of course (14.16) becomes more accurate the closer k^2 is to $-m_\pi{}^2c^2$. Returning to (14.2) we see that the magnitude of the term modifying γ_5 is

$$\frac{2Mc}{k^2 + m_\pi{}^2c^2} k_\lambda g_A$$

For nucleon β-decay $k_\lambda \sim m_e c$, while for μ capture $k_\lambda \sim m_\mu c$. Hence the magnitude of the induced pseudoscalar terms for nucleon β-decay is

$$\frac{2(Mc)(m_e c)}{(m_e c)^2 + m_\pi{}^2c^2} g_A = \frac{2Mm_e}{m_e{}^2 + m_\pi{}^2} g_A \qquad \text{(nucleon } \beta\text{-decay)}$$

$$\sim 0.05 g_A \tag{14.17}$$

For μ-capture the magnitude of the pseudoscalar term is

$$\frac{2Mm_\mu}{m_\pi{}^2 + m_\mu{}^2} g_A \sim 6.4 g_A \tag{14.18}$$

Thus at least for the allowed nuclear β-decay the effect of the induced pseudoscalar term is not of much importance. On the other hand it is expected to be of importance in μ-capture, the simplest case being capture by a proton. In that case [Quaranta et. al. (69)] the value obtained is $(5 \pm 2.5)g_A$, which within the rather large experimental error is in agreement with (14.18).

Beside these numerical agreements, the qualitative point should be stressed that the PCAC condition (14.4) provides a connection between strong interactions and the weak interaction currents. As a consequence, properties of the strong interactions can be investigated by examining the connected weak interaction whose form can be more readily deduced from weak decays, at least for the range in momentum transfer involved. Such applications have proved of importance in recent developments in modern physics. It is not appropriate to discuss them here. However there is one connection between strong and weak interactions, the Adler–Weisberger relation, [Adler (65), Weisberger (65)] which is of importance for our subject. It will not be possible to give all details of the calculation. But we shall carry it far enough so that the elements that enter into the final result are visible. Essentially Adler's method is used.

For this purpose the analog of the charge operator for the vector current is defined for the axial vectors. Let

$$\hat{Q}_3^{(V)}(t) = \frac{1}{2}\int \Psi^\dagger \tau_3 \hat{\psi}\, d\mathbf{r} = \int \hat{V}_0\, d\mathbf{r} \tag{14.19}$$

and

$$\hat{Q}_\pm^{(A)}(t) = \int \hat{\psi}^\dagger \gamma_5 \tau^{(\pm)} \hat{\psi}\, d\mathbf{r} = \int \hat{A}_0^{(\pm)}\, d\mathbf{r} \tag{14.20}$$

where $\hat{Q}$ and $\hat{\psi}$ are Heisenberg operators. If $\hat{\psi}^\dagger$ and $\hat{\psi}$ satisfy the free field anticommutation rules these charge operators satisfy the commutation rule

$$[\hat{Q}_+^{(A)}(t),\ \hat{Q}_-^{(A)}(t)] = 2\hat{Q}_3^{(V)} \tag{14.21}$$

We shall assume this result to hold generally, one of the hypotheses of the current algebra approach. $\hat{Q}_3^{(V)}$ is familiar since it is the isovector component of the electric charge operator.

To exploit (14.21) we shall first use PCAC to connect the charges $Q_\pm^{(A)}$ which are intimately related via the weak processes to the strong processes. This is essential since we shall use only the equality between the first and third quantities of (14.20) to define $Q_\pm^{(A)}$. Starting from this definition we have

$$\frac{d}{dt}\hat{Q}^{(A)} = \int \frac{\partial \hat{A}_0}{\partial t}\, d\mathbf{r}$$

Upon assuming that $\mathbf{A}$ vanishes at large distances we can add in div $\mathbf{A}$ to $\partial \hat{A}_0/\partial t$ to obtain

$$\frac{d\hat{Q}^{(A)}}{dt} = \int \sum_\lambda \frac{\partial \hat{A}}{\partial x_\lambda}\, d\mathbf{r}$$

Hence from (14.4)

$$\frac{d\hat{Q}^{(A)}}{dt} = a\int \hat{\phi}_\pi\, d\mathbf{r} \tag{14.22}$$

Since $\hat{\phi}_\pi$ is a pion field operator, it is clear that, as a consequence of PCAC, $\hat{Q}^{(A)}$ connect states that differ in pion occupation number by unity. For example, if $|p\rangle$ is a state containing one nucleon and zero pions, and $|k\rangle$ is a one nucleon plus one pion state then

$$\langle p|\,\hat{Q}^{(A)}\,|k\rangle = \frac{\hbar a}{i\,(E_p - E_k)}\,\langle p|\int \hat{\phi}_\pi\, d\mathbf{r}\,|k\rangle \tag{14.23}$$

If we are considering the matrix element of $\hat{Q}_+^{(A)}$, then $|k\rangle$ is a state with a neutron and a positive pion. On the other hand $\hat{Q}_-^{(A)}$ connects a state $|k\rangle$ involving a proton and a negative pion with a neutron.

Relation (14.23) holds in all momentum frames. Fubini and Furlan (65) exploited this fact by considering (14.23) in the frame in which the momentum $\mathbf{p}$ is infinite. In this limit the four-momentum transfer

$$(p - k)^2 = 2\left(\frac{E_pE_k}{c^2} - \mathbf{p}\cdot\mathbf{k}\right) - M_p{}^2c^2 - M_k{}^2c^2$$

tends to zero. M_k is the invariant mass of the intermediate system in state $|k\rangle$ while $\mathbf{k}$ is its three momentum. To prove this statement we observe that $\mathbf{p} = \mathbf{k}$. [Prove it by employing translational invariance to show that the matrix element on the right-hand side of (14.23) is proportional to $\delta(\mathbf{p}\text{–}\mathbf{k})$.] Then

$$\frac{E_pE_k}{c^2} = [M_p{}^2c^2 + p^2]^{1/2}[M_k{}^2c^2 + p^2)^{1/2}$$

$$\xrightarrow[p\to\infty]{} p^2\left[1 + \frac{1}{2}\frac{M_p{}^2c^2}{p^2} + \frac{1}{2}\frac{M_k{}^2c^2}{p^2} + 0\left(\frac{1}{p^4}\right)\right]$$

Hence

$$(p - k)^2 = 0\left(\frac{1}{p^2}\right) \to 0 \qquad \text{as} \qquad p \to \infty$$

Thus, if the mass of the pion described by $\hat{\phi}_\pi$ were zero, the matrix element in (14.23) would conserve not only momentum but also energy. It can then be shown that it is proportional to the pion–nucleon forward scattering amplitude for zero mass pions. We refer the reader to Adler (65) and to Chapter IV in Adler and Dashen (68) for the details.

Returning to (14.21), take the matrix element with respect to a proton

$$\langle p|\,\hat{Q}_+^{(A)}\hat{Q}_-^{(A)} - \hat{Q}_-^{(A)}\hat{Q}_+^{(A)}\,|p'\rangle = \langle p|\,2\hat{Q}_3^{(V)}\,|p'\rangle \tag{14.24}$$

Expand the left-hand side in a complete set of intermediate states. These will include the one neutron state $|n\rangle$ as well as the nucleon + one-pion state $|k\rangle$. Then the left-hand side of the above equation becomes

$$\sum_n \langle p|\hat{Q}_+^{(A)}|n\rangle\langle n|\hat{Q}_-^{(A)}|p'\rangle - \langle p|\hat{Q}_-^{(A)}|n\rangle\langle n|\hat{Q}_+^{(A)}|p'\rangle$$
$$+\sum_k (\langle p|\hat{Q}_+^{(A)}|k\rangle\langle k|\hat{Q}_-^{(A)}|p'\rangle - \langle p|\hat{Q}_-^{(A)}|k\rangle\langle k|\hat{Q}_+^{(A)}|p'\rangle)$$

where the sum over n goes over the spin and momentum of the neutron. The one-nucleon intermediate state terms can be evaluated in terms of the form factors (14.2), (see Eq. 14.20) and will therefore involve g_A^2. The matrix elements in the remaining terms, as we have indicated above, can be related to the pion–nucleon scattering amplitude for zero mass pions. It is thus not surprising that it can be shown [Adler (65)] that the sum over k can be related to the zero mass pion–nucleon scattering cross section. Thus the left-hand side of (14.24) will involve a term proportional to g_A^2 and another term involving an integral over E_K (see Eq. 14.23), or equivalently over the intermediate state pion energy, of the zero mass pion–nucleon scattering cross sections. The right-hand side is easily evaluated from (14.19)

$$\langle p|2\hat{Q}_3^{(V)}|p'\rangle = 2\langle p|T_3|p'\rangle = \delta(\mathbf{p} - \mathbf{p}')$$

These remarks outline the procedures that are used to obtain the Adler–Weisberger relation. [For further details see the original papers by Weisberger (65), Adler (65), Fubini et al. (65), the account given by Adler and Dashen (68), and by Gasiorowicz (67).] Considering (14.21) in the infinite momentum frame and placing the pion mass equal to zero, one obtains

$$1 = g_A^2 + \frac{1}{\pi} a^2 \int_{m_\pi}^{\infty} \frac{dE_\pi}{E_\pi} [\sigma^{(0)}(\pi^-, p; E_\pi) - \sigma^{(0)}(\pi^+, p; E_\pi)] \tag{14.25}$$

It is of course no surprise that the renormalization of g_A for nucleon β-decay is related to the nucleon–pion interaction but it is a surprise that the relation should involve the *zero pion mass* total cross section $\sigma^{(0)}$, which can be estimated by extrapolation from the cross sections for finite mass pions. There are uncertainties because of this extrapolation and this as well as other uncertainties result in an uncertain value of g_A. However the results obtained from (14.25) are within a *few percent* of the empirical value.

With the astonishing agreements, (14.14), (14.18), and (14.25) one is left with the mysteries, the justification of the PCAC and charge commutator postulates (14.4) and (14.21). But perhaps "justification" is the wrong word to use here. Perhaps these currents, vector and axial, are the appropriate dynamical variables for these systems, while PCAC and (14.4) must then be regarded as fundamental postulates in analogy to the dynamical variables **r** and **p** and the commutation relation between these variables of ordinary quantum mechanics.

The Adler–Weisberger rotations have been generalized to finite nuclei by Kim and Primakoff [Kim and Primakoff (66)] where now the pion–nucleon cross sections are replaced by the pion–nucleus cross sections and the nucleon axial form factor g_A is replaced by the nuclear axial form factor. The reader is referred to the original papers for details.

15. SECOND-CLASS CURRENTS

The discussion thus far has not touched upon the induced tensor form factor g_T of (14.2). As pointed out by Weinberg (58) the G-parity of this term differs from the other terms, such as the axial vector term in (14.2), and thus originates from currents whose G-parity differs from that of $\bar{u}_p\gamma_\lambda\gamma_5 u_n$. Weinberg refers to these currents as *second-class currents*. The existence or absence of second-class currents is important for our understanding of the nature of the weak interactions. According to Weinberg (72) their presence would pose a serious problem for the modern theory.

With respect to nuclear β-decay the argument is made that since the strong interactions satisfy G-parity invariance (as well as parity invariance) only those terms are induced that have the same G-parity as the original term. Hence the observation of a term similar to the induced tensor term would indicate the existence of G-violating terms. In interpreting the experimental values it should be realized that G-parity invariance is broken when isospin symmetry is broken by interactions such as the Coulomb interaction.

The induced tensor term can be distinguished from the other terms by comparing the ft-values for the decay of mirror nuclei [Huffaker and Greuling (63)]. We have already noted that the spectrum of a Gamow–Teller decay will differ for mirror nuclei because of the weak magnetism term (page 850). However it can be shown that the weak magnetism term in a first approximation when Coulomb effects are neglected [Wu and Moskowski (66)] does not affect the ratio of the ft values for mirror nuclei. The β-spectra are changed but the electrons that are removed from the distribution for $E_e > E_0/2$ are compensated by the electrons that are added for $E_e < E_0/2$. This compensation does not occur for the induced tensor term. Therefore, by comparing the ft-values for corresponding decays of mirror nuclei the presence of this term can be detected. The dependence of the ft-ratio on the energy release E_0^+ and E_0^- in $\beta^\pm$ decay is of the form

$$\delta = \frac{(ft)_{\beta^+}}{(ft)_{\beta^-}} = 1 + b(E_0^+ + E_0^-) \tag{15.1}$$

where b is proportional to $g_A g_T$. Hence the ft ratio will be a linear function of the end-point energy. Recent experiments [Wilkinson (70)] yield an upper bound to g_T

$$|g_T| \leq 2.5 \tag{15.2}$$

There is then still some distance to go before we arrive at the limit set by isospin-breaking interactions.

It has been recently pointed out by Delorme and Rho (71) that additional second-class currents can be present when the interaction between the decaying nucleon and the other nuclear nucleons are taken into account. *G*-parity can be carried off by the rest of the nucleus so that it need not be conserved in the elementary decay. According to these authors off-the-mass shell contributions might be as important as the induced tensor terms. Further experiments are required. According to Greuling and Kim (64) the electron–neutrino and electron–nuclear spin correlation are also sensitive to the presence of second-class currents. For further discussion see Delorme and Rho.

16. OTHER TESTS OF THE THEORY OF WEAK INTERACTIONS

A. Double β-Decay

In the process of double β-decay a nucleus (Z, N) decays spontaneously to the nucleus $(Z \pm 2, N \mp 2)$ by the emission of two electrons (or positrons) accompanied by two antineutrinos (or neutrinos) according to the weak interaction theory developed above. Double β-decay is described by second-order perturbation theory. The first step is a transition from the original (Z, N) nucleus to a $(Z + 1, N - 1)$ nucleus plus an electron and antineutrino. This intermediate state then decays into the final $(Z + 2, N - 2)$ nucleus by the emission of another electron–antineutrino pair. This decay is known as the *two-neutrino decay*. In the theory described above it is not possible to make the transition to the final nucleus by absorbing the antineutrino emitted in the first leg, since a neutron can transform into a proton only by neutrino absorption. Hence the *no-neutrino decay* is not allowed. Clearly, observing double β-decay and particularly the extent to which no neutrino decay occurs will provide a test of the theory.

Only a few nuclei are stable except for double β-decay. In particular, single β-decay to the intermediate nucleus must be forbidden energetically or, if allowed energetically, strongly inhibited because of the large angular momentum change involved. Nuclei that satisfy these conditions are shown in Table 16.1. The column labeled "two neutrino" gives the value of the lifetime as obtained with the standard theory while the last column gives the lifetime if no-neutrino decay is allowed. We observe that the no-neutrino theory lifetimes are a factor of 10^6 shorter. The reason for this large ratio lies in the virtual character of the neutrinos emitted and absorbed in the no-neutrino process, the available phase space being thus larger than in the two-neutrino case. Another and related difference between the two possibilities lies in their electron spectrum. In the no-neutrino case the sum of the energies of the two electrons is a constant (neglecting as usual the recoil energy of the decaying nu-

TABLE 16.1 Double β-decay [from Rosen and Primakoff (65)]

Decay Process	Energy Release (MeV)	Half-life (years): Experimental	Half-life (years): Theoretical, Two-Neutrino	Half-life (years): Theoretical, No-Neutrino
$^{130}Te_{52} \rightarrow {}^{130}Xe_{54}$	3.0	$>10 \pm 0.12^{21.34}$	$8 \times 10^{21\pm2}$	$8 \times 10^{15\pm2}$
$^{238}U_{92} \rightarrow {}^{238}Pu_{94}$	1.0	$> 6 \times 10^{18}$	$1 \times 10^{26\pm2}$	$2 \times 10^{18\pm2}$
$^{124}Sn_{50} \rightarrow {}^{124}Te_{52}$	2.2	$>1 \times 10^{16}$	$2 \times 10^{23\pm2}$	$5 \times 10^{16\pm2}$
		$>1 \times 10^{17}$		
		$>2 \times 10^{17}$		
		$>5 \times 10^{16}$		
		$>1.5 \times 10^{17}$		
$^{116}Cd_{48} \rightarrow {}^{116}Sn_{50}$	2.7	$>1 \times 10^{17}$	$1 \times 10^{22\pm2}$	$2 \times 10^{16\pm2}$
		$>6 \times 10^{16}$		
$^{100}Mo_{42} \rightarrow {}^{100}Ru_{44}$	2.3	$=2 \times 10^{16}$	$2 \times 10^{23\pm2}$	$5 \times 10^{16\pm2}$
		$>3 \times 10^{17}$		
$^{106}Cd_{48} \rightarrow {}^{106}Pd_{46}$	0.9	$>6 \times 10^{16}$	$8 \times 10^{28\pm2}$	$6 \times 10^{20\pm2}$
$^{96}Zr_{40} \rightarrow {}^{96}Mo_{42}$	3.3	$>2 \times 10^{16}$	$4 \times 10^{21\pm2}$	$6 \times 10^{15\pm2}$
		$>5 \times 10^{17}$		
$^{48}Ca_{20} \rightarrow {}^{48}Ti_{22}$	4.3	$=2 \times 10^{17}$	$1 \times 10^{21\pm2a}$	$3 \times 10^{15\pm2}$
		$>2 \times 10^{18}$		
$^{150}Nd_{60} \rightarrow {}^{150}Sm_{62}$	3.7	$>4 \times 10^{18}$	$8 \times 10^{20\pm2}$	$2 \times 10^{15\pm2}$
$^{b}\,^{64}Zn_{30} \rightarrow {}^{64}Ni_{28}$	1.1	$>1 \times 10^{16}$	$1 \times 10^{30\pm2}$	$3 \times 10^{26\pm2}$

[a] On the basis of shell-model type estimates of the relevant nuclear matrix elements, Belyaev and Zaharyev and also Meichsner obtain a half-life ≈ 10^{19} years for the two-neutrino $\beta^-\beta^-$-decay of $^{48}Ca_{20} \rightarrow {}^{48}Ti_{22}$.

[b] KK-capture.

cleus) while in the two-neutrino case only the maximum of the energy sum equals the energy released, since the two neutrinos can carry off energy.

These two differences, lifetime and electron spectrum have formed the main bases of the various experimental attempts that have been made to observe the double β-decay process. It was recently seen by Kirsten et al. (68) by examining the tellurium ores of a known age. The ore was found to contain ^{130}Xe in an abundance far greater than that contained in atmospheric Xenon. These authors show that this excess could only have resulted from the double β-decay of ^{130}Te. The half-life was found to be $10^{21.34\pm0.12}$. Comparing with Table 16.1, we see that the two-neutrino theory fits experiment and that the pure no-neutrino theory is incorrect. A small component of the no-neutrino process cannot be excluded because of experimental error. This possibility has been investigated by Primakoff and Rosen (69) who were able to set a rather small upper limit on this possibility.

B. Parity Nonconserving Nuclear Forces

Thus far in our discussion only particular terms of the weak interaction Hamiltonian (12.14) have been considered: those involving cross terms between different kinds of currents. Nuclear β-decay involves the nucleon and electron-neutrino currents; μ- decay this last current and the muon and μ^- neutrino current. There are, however, "diagonal terms" (see Eq. 12.15). There is, for example, the product

$$\frac{C_V g}{\sqrt{2}} \sum \hat{\jmath}_\lambda{}^{(e)\star} \hat{\jmath}_\lambda{}^{(e)} + \text{h.c.}$$

which leads to the scattering of electrons by neutrinos and antineutrinos:

$$\bar{\nu}_e + e^- \rightarrow \bar{\nu}_e + e^- \tag{16.1a}$$

$$\nu_e + e^- \rightarrow \nu_e + e^- \tag{16.1b}$$

The predicted cross sections for these processes are very small. For the first process it is of the order of

$$(0.56\ 10^{-44}\ \text{cm}^2) \times E_\nu$$

where E_ν is neutrino energy in MeV. The second process (16.1b) cross section is three times this number. To this date only an upper limit for (16.1a), which is 2.5 times the above estimate, has been established [Reines and Gurr (70)] by scattering the anti-neutrinos available at reactors. Electron neutrinos are generated in $\mu^{(+)}$ decay and thus experiment (16.1a) may be eventually performed at accelerators that produce muons with sufficient intensity. The importance of these processes for the understanding of weak interactions has been recently emphasized by Gell–Mann et al. (69).

There are of course other "diagonal terms." One of interest for us is

$$\frac{C_V g}{\sqrt{2}} \sum_\lambda \hat{J}_\lambda{}^{\star(N)} \hat{J}_\lambda{}^{(N)} + \text{h.c.} \tag{16.2}$$

The diagram for this term is shown in Fig. 16.1. It clearly leads to a nucleon–nucleon interaction. The important point is that since (16.2) does not conserve parity there is a component of the nuclear force that does not conserve parity. One immediate effect is that nuclear states have mixed parity. The wave function may be written

$$\psi = \psi_{\text{norm}} + F\chi \tag{16.3}$$

where ψ_{normal} is the wave function in the absence of the parity nonconserving potential. The amplitude $F\chi$ is the addition generated by that potential. The parity of χ is opposite to that of ψ_{norm}. If χ is normalized $|F|$ measures the size of the effect.

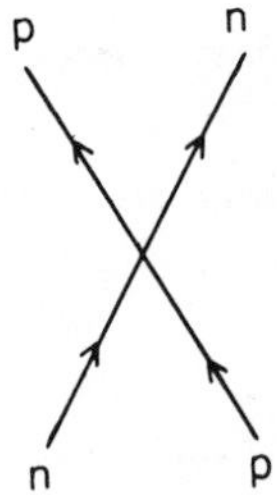

FIG. 16.1. Weak proton-neutron interaction according to Eq (16.2).

Experimental approaches directed toward the determination of F have been of two types. In one, a transition that would be forbidden if parity is conserved is examined. A case of this kind that has received considerable attention is [Segal et. al. (61)] the α-decay of 8.88 MeV 2^- state in ^{16}O to the 0^+ ground state of ^{12}C (Fig. 16.2). If parity is conserved the width for this decay is zero. By observing the number of alphas of the appropriate energy and comparing to the known decays, an estimate of the width of this state can be obtained. Wäffler (72) obtained a parity-forbidden alpha width of

$$\Gamma_\alpha = (1.0 \pm 0.2) \times 10^{-10} \text{ eV}$$

from which the estimate

$$|F|^2 < 3 \times 10^{-14} \tag{16.4}$$

was obtained. Theoretical estimates have been made by a number of authors. These calculations all predict $\Gamma_\alpha \sim 10^{-10}$ in agreement with experiment [Gari (70)]. This experiment is of some importance since it will tell us if the nonparity conserving potential has an isoscalar component, the isospin of the 2^- level, and the ground state of ^{12}C being identical.

Successful observation of nonparity conserving transitions have been made by examining the character of electromagnetic transitions. The advantages of this method were first described by Wilkinson (58). For example, if the principal mode of decay of a state is by an electric dipole transition, the normal amplitude, the presence of the small $F\chi$ component would add a small magnetic dipole component, the abnormal amplitude. Since this magnetic dipole radiation amplitude is opposite in parity to that of the electric dipole, the interference between the two amplitudes will produce a component of the intensity that is odd under space inversion, indicating the presence of a non-parity-conserving interaction. The strength of this odd term is proportional to F not to $|F|^2$ as in the experiment just discussed and should therefore be much more easily observed.

Several methods of observing this effect have been employed. If the gamma rays emitted by an unoriented nucleus are partially circularly polarized,

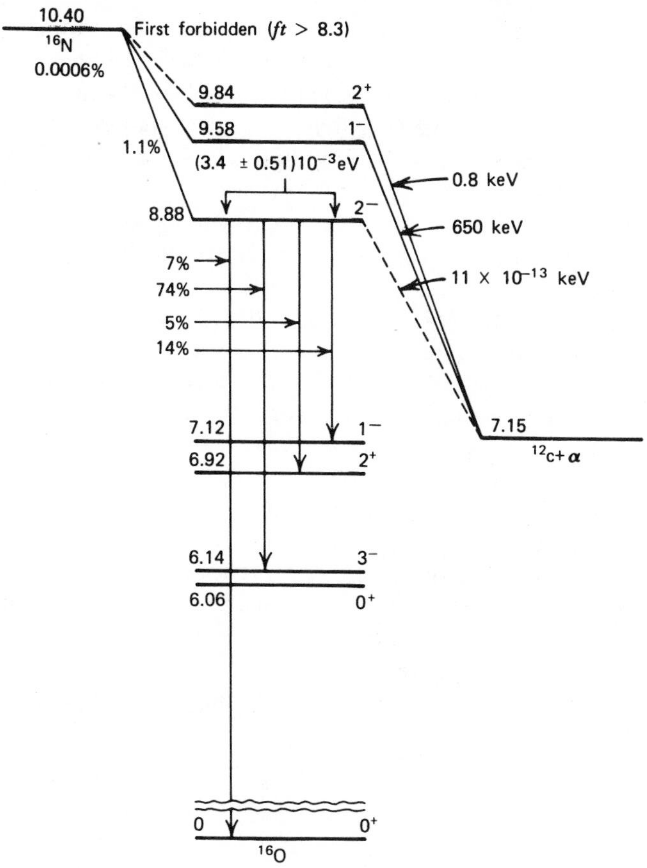

FIG. 16.2. Energy level diagram, branching ratios and approximate absolute widths of relevant $T = 0$ states in ^{16}O produced by the β-decay ^{16}N [taken from Henley (69)].

parity is not conserved in the transition since, as noted in Chapter VIII, the state of a photon with a given helicity is of mixed parity, the vector potential having the form $\mathbf{e} \pm i(\mathbf{e} \times \mathbf{k})$, the sum of a polar and axial vector. The vector $\mathbf{k}$ is in the direction of propagation of the photon. In another experiment, the asymmetry of the gamma rays emitted in the capture of polarized thermal neutrons with respect to the direction of polarization is evidence of parity nonconservation. If $\mathbf{P}$ is the polarization direction, then in this experiment the expectation value of $\mathbf{P}\cdot\mathbf{k}$ where $\mathbf{k}$ is the direction of motion of the photon is determined. This expectation value is zero if parity is conserved since $\mathbf{P}\cdot\mathbf{k}$ is a pseudoscalar, polarization being a pseudovector. (For the spin 1/2 case $\mathbf{P} = \langle\boldsymbol{\sigma}\rangle$.) The asymmetry is a consequence of the interference between the normal and abnormal amplitudes. If only a single multipole is involved, the asymmetry is absent (see Eq. VIII.5.13).

Because of the small value of F, these experiments, even though the effect depends linearly on F, are still very difficult. The "background" out of which one wishes to extract the effect involves at the very least the normal transitions so that one way to improve the visibility of the effect is to search for it in cases where the normal transition probability happens to be reduced. As we have noted in Chapter VIII that is the case for electric dipole transition and magnetic dipole transitions because of the action of isospin-selection rules (Chapter VIII p. 734). In rotational nuclei, highly K forbidden transitions present another possibility.

Several such measurements have been made. These are listed in Table 16.2 taken mostly from Gari's review article (72). In several of these experiments, the decays in ^{175}Lu, ^{181}Ta, ^{114}Cd, and ^{180}Hf the uncertainty is sufficiently small and there is agreement among several experiments so that there is no question that the presence of nonparity-conserving nuclear forces has been observed. However, substantial quantitative differences do still remain and more experimental work is essential. The large value obtained for ^2H is unexpected theoretically. [Hadjimichael and Fischbach (71), F. Partovi (64)].

Let us now return to the theory of these processes. We shall not attempt to derive the nonparity-conserving forces but just to point to significant features. Particular attention will be paid to their isospin character and range. Let us first consider the nonstrangeness changing part of the hadron current–current interaction arising from the non-strangeness changing currents. This will have the form (16.2)

$$H_{W1}{}^{(N)} = \frac{G_\mu}{\sqrt{2}} \cos^2 \theta_c \sum_\lambda \left[\hat{J}_\lambda{}^{(N)(+)} \hat{J}_\lambda{}^{(N)(-)} + \hat{J}_\lambda{}^{(N)(-)} \hat{J}_\lambda{}^{(N)(+)} \right] \tag{16.5}$$

where $\hat{J}_\lambda{}^{(\pm)}$ has the same isospin-transformation properties as the isospin operators, $T^{(\pm)}$. Thus the transformation properties of $H_{W1}{}^{(N)}$ are those of $T^{(+)}T^{(-)} + T^{(-)}T^{(+)}$ where T_i transforms like a vector in isospin space. $H_{W1}{}^{(N)}$, since it is bilinear, might behave like a linear combination of an isoscalar, isovector, and an isotensor. However, since $H_{W1}{}^{(N)}$ is symmetric, there can be no isovector component so that $H_{W1}{}^{(N)}$ can induce $T = 0$ and $T = 2$ transitions, both of which can occur in the nucleon–nucleon systems. In terms of the nucleon–nucleon isospin variables the nucleon–nucleon interaction will involve terms like $\tau_1{}^{(+)}\tau_2{}^{(-)} + \tau_1{}^{(-)}\tau_2{}^{(+)}$ where the subscripts number the nucleons.

Turning now to the range of this interaction we observe since the various currents entering into (16.5) are evaluated at one point, that the range of the force is zero. This is however not correct since the effect of the strong interactions are important as we know from the renormalization and form-factor effects that are present in β-decay (Fig. 13.1 and p. 841). In the present process the bosons can be exchanged between all four particles (the incident nucleons and the final nucleons). An estimate of the maximal range can be obtained by asking for the minimum mass of bosons with isospin $T = 0$ and $T = 2$

that can be exchanged between the nucleons. The minimum mass is provided by two pions corresponding to a range of $(\hbar/2m_\pi c)$ or about 0.7 fm. *Single pion exchange is not allowed* in this theory.

Single pion exchange is permitted in the contribution to the potential of the strangeness-changing part of the hadron currents. The Hamiltonian involved is (12.18)

$$H_{W2}{}^{(N)} = \frac{G\mu}{\sqrt{2}} \sin^2 \theta_c \sum_\lambda \hat{J}_\lambda{}^{(S)*} \hat{J}_\lambda{}^{(S)} \qquad (16.6)$$

Recall that $\hat{J}_\lambda{}^{(S)}$ is responsible for the decay of the Λ into a proton plus a negative pion. Since Λ has zero isospin the current might involve a $T = 1/2$ component and possibly a 3/2 one. However, experimental evidence very strongly suggests that only the $T = 1/2$ component is present in $\hat{J}_\lambda{}^{(S)}$. From this it follows that a component of $H_{W2}{}^{(N)}$ given by (16.6) will transform like a vector in isospin space. In terms of the nucleon–nucleon isospin variable, $\boldsymbol{\tau}_1$ and $\boldsymbol{\tau}_2$, the only vector is $\boldsymbol{\tau}_1 \times \boldsymbol{\tau}_2$. In this case, since the exchanged boson can have $T = 1$, a pion is the candidate with a minimum mass with the consequent range of $(\hbar/m_\pi c)$ or 1.4 fm. We leave it as an exercise for the reader to devise a process in which according to (16.6) a nucleon can weakly decay into a nucleon plus a pion. $H_{W2}{}^{(N)}$ appears to be considerably smaller than $H_{W1}{}^{(N)}$ by the factor $\tan^2 \theta_c$. However the much larger range of $H_{W2}{}^{(N)}$ plus the small probability of finding two nucleons close together because of the hard core in the parity-conserving part of the nucleon–nucleon interaction makes the potentials obtained from $H_{W1}{}^{(N)}$ and $H_{W2}{}^{(N)}$ have about the same importance for the determination of F.

Some remarks on the procedure involved in obtaining finally the nonparity-conserving potentials follow. [For details see, for example, Blin–Stoyle and Herczeg (68).] It is necessary to evaluate the matrix element

$$\langle N'_1 N'_2 | H_{W1}{}^{(N)} | N_1 N_2 \rangle$$

where $N_{1,2}$ and $N'_{1,2}$ refer to the initial and final states of the two-nucleon system.

The assumption is made that the only intermediate states of consequence are single-nucleon states. The consequences as shown for the first term in (16.5) is

$$\langle N'_1 N'_2 | \hat{J}_\lambda{}^{(N)(+)} \hat{J}_\lambda{}^{(N)(-)} + \hat{J}_\lambda{}^{(N)(-)} \hat{J}_\lambda{}^{(N)(+)} | N_1 N_2 \rangle \simeq \langle N'_1 | \hat{J}_\lambda{}^{(N)(+)} | N_1 \rangle$$
$$\times \langle N'_2 | \hat{J}_\lambda{}^{(N)(-)} | N_2 \rangle + \langle N'_1 | \hat{J}_\lambda{}^{(N)(-)} | N_1 \rangle \langle N'_2 | \hat{J}_\lambda{}^{(N)(+)} | N_2 \rangle \qquad (16.7)$$

To obtain the parity-nonconserving component (PNC) we write $\hat{J}_\lambda = \hat{V}_\lambda + \hat{A}_\lambda$ and keep only the cross terms between $\hat{V}_\lambda$ and $\hat{A}_\lambda$. The diagonal terms $\hat{V}_\lambda \hat{V}_\lambda$ and $\hat{A}_\lambda \hat{A}_\lambda$ conserve parity and are completely negligible compared to the terms generated by the strong interactions. One may then employ (13.25) for

TABLE 16.2 Experimental Measurements of the Circular Polarization of Photons, P_γ, Emitted in Nuclear Transitions and the Asymetry A_γ for Photons Emitted After Capture of Polarized Neutrons [from Gari (72) mostly].

Nucleus	Transition Energy	Normal Transitions	Abnormal Transitions	Experimental Results	References
^{2}H	2.2 keV	$^1S_0 \xrightarrow{M1} {}^3S_1 + {}^3D_1$	$\tilde{E}1$	$P_\gamma = \underline{-(1.7 \pm 0.7)10^{-6}}$	V. M. Lobashov et al. (71)
		$(n + p \xrightarrow{\text{thermal}} d + \gamma)$		$= -(1.5 \pm 1.5)10^{-6}$	V. M. Lobashov et al. (70)
^{19}F	110 keV	$1/2^- \xrightarrow{E1} 1/2^+$	$\tilde{M}1$	$A_\gamma = (4.3 \pm 5.2)10^{-4}$	Moline, Morris, Dyer, and Barnes (70)
^{41}K	1291 keV	$7/2^- \xrightarrow{M2} 3/2^+$	$\tilde{E}2$	$P_\gamma = \underline{(1.9 \pm 0.3)10^{-5}}$	V. M. Lobashov et al. (69)
				$A_\gamma = (2.4 \pm 3.6)10^{-3}$	Boehm and Hauser (60)
^{57}Fe	14.4 keV	$3/2^- \xrightarrow{M1} 1/2^-$	$\tilde{E}1$	$P_\gamma = (2.0 \pm 6.0)10^{-5}$	E. Kankeleit (64)
^{75}As	401 keV	$5/2^+ \xrightarrow{E1} 3/2^-$	$\tilde{M}1$	$P_\gamma = \underline{-(6.0 \pm 2.0)10^{-5}}$	Vanderleeden, Boehm, and Lipson (72)
^{114}Cd	~9.05 MeV	$1^+ \xrightarrow{M1} 0^+$	$\tilde{E}1$	$A_\gamma = \underline{-(3.5 \pm 1.2)10^{-4}}$	Yu. G. Abov et al. (68)
				$= \underline{-(3.7 \pm 0.9)10^{-4}}$	
					Yu. G. Abov et al. (65)
				$= -(2.5 \pm 2.2)10^{-4}$	
					E. Warming et al. (67)
				$= -(6.0 \pm 1.5)10^{-4}$	J. L. Alberi et al. (72)
				$= -(0.6 \pm 1.7)10^{-4}$	E. Warming (69)

^{133}Ca	81 keV	$5/2^+ \xrightarrow{M1} 7/2^+$	$\tilde{E}1$	$A_\gamma = (1.5 \pm 1.5)10^{-3}$	Boehm and Hauser (60)
^{159}Tb	363 keV	$5/2^- \xrightarrow{E1} 3/2^+$	$\tilde{M}1$	$P_\gamma = -(1.1 \pm 2.8)10^{-4}$	K. S. Krane et al. (71)
				$P_\gamma = -(1.0 \pm 5.0)10^{-4}$	Lipson, Boehm, and Vanderleeden (71)
	348 keV	$5/2^+ \xrightarrow{M1} 3/2^+$	$\tilde{E}1$	$A_\gamma = (1.7 \pm 5.0)10^{-3}$	K. S. Krane et al. (71)
	226 keV	$5/2^+ \xrightarrow{M1} 3/2^+$	$\tilde{E}1$	$A_\gamma = -(1.2 \pm 4.0)10^{-3}$	
^{161}Dy	49 keV	$3/2^- \xrightarrow{M1} 5/2^-$	$\tilde{E}1$	$A_\gamma = -(3.0 \pm 3.2)10^{-3}$	Krane, Olsen, Sites, and Steyert (71)
	75 keV	$3/2^- \xrightarrow{E1} 5/2^+$	$\tilde{M}1$	$A_\gamma = -(2.0 \pm 2.0)10^{-4}$	
^{175}Lu	396 keV	$9/2^- \xrightarrow{E1+M2} 7/2^+$	$\tilde{M}1$	$P_\gamma = (4.0 \pm 1.0)10^{-5}$	V. M. Lobashov et al. (67)
				$P_\gamma = (6.2 \pm 0.8)10^{-5}$	Vanderleeden and Boehm (69)
				$A_\gamma = (4.5 \pm 3.0)10^{-3}$	W. Pratt et al. (70)
	343 keV	$5/2^+ \xrightarrow{M1+E2} 7/2^+$	$\tilde{E}1$	$P_\gamma = (2.0 \pm 3.0)10^{-5}$	Boehm and Kankeleit (68)

TABLE 16.2 (Continued)

Nucleus	Transition Energy	Normal Transitions	Abnormal Transitions	Experimental Results	References
^{177}Hf	208 keV	$9/2^+ \xrightarrow{E1} 9/2^-$	$\tilde{M}1$	$P_\gamma = (3.0 \pm 13.0)10^{-4}$	Boehm and Hauser (60)
^{180}Hf	501 keV	$8^- \xrightarrow{(M2+E3)} 6^+$	$\tilde{E}2$	$P_\gamma = -(2.8 \pm 0.45)10^{-3}$	Jenschke and Bock (70)
				$= -(2.3 \pm 0.6)10^{-3}$	Lipson, Boehm, and Vanderleeden (71)
				$= -(1.8 \pm 0.6)10^{-3}$	Kuphal, Daum, and Kankeleit (71)
				$A_\gamma = -(1.6 \pm 0.18)10^{-2}$	Krane, Olsen, Sites, and Steyert (71)
				$P_\gamma \leq 2\%$	H. Blumberg et al. (66)
				$P_\gamma = -(1.4 \pm 0.7)10^{-3}$	Povel and Bock (68)
	57.5 keV	$8^-8 \xrightarrow{E1} 8^+0$	$\tilde{M}1$	$P_\gamma = (5.9 \pm 6.1)\ \%$	Paul, McKeown, and Scharff-Goldhaber (67)
				$= -(2.3 \pm 3)\ \%$	Bock, Jenschke, and Schopper (66)
^{181}Ta	482 keV	$5/2^+ \xrightarrow{M1+E2} 7/2^+$	$\tilde{E}1$	$P_\gamma = -(3.1 \pm 2.5)10^{-6}$	Lipson, Boehm, and Vanderleeden (71)
				$= -(3.9 \pm 1.2)10^{-6}$	Vanderleeden and Boehm (70)
				$= -(6.0 \pm 1.0)10^{-6}$	V. M. Lobashov et al. (67)
				$= -(4.1 \pm 1.3)10^{-6}$	Bock and Jenschke (71)
				$P_\gamma = -(9.0 \pm 6.0)10^{-5}$	Cruse and Hamilton (69)
				$= -(2.1 \pm 1.1)10^{-5}$	P. DeSaintignon et al. (71)
				$= -(2.8 \pm 0.6)10^{-5}$	Bodenstedt, Ley, Schlenz, and Wehman (69)

^{181}Ta (cont'd)				$P_\gamma = (3.0 \pm 21.0)10^{-5}$	Bock and Schopper (65)
				$= (2.0 \pm 4.0)10^{-6}$	Kuphal, Daum, and Kankeleit (71)
				$= -(1.0 \pm 4.0)10^{-5}$	Boehm and Kankeleit (68)
				$= (0.7 \pm 7.0)10^{-5}$	J. van Roojen et al. (67)
^{182}W	1189 keV	$22^- \xrightarrow{E1+M2} 20^+$	$\tilde{M}1 + \tilde{E}2$	$P_\gamma = -(2.5 \pm 4.0)10^{-5}$	Lipson, Boehm, and Vanderleeden (71)
				$A_\gamma = -(2.8 \pm 1.7)10^{-4}$	K. S. Krane et al. (71)
^{203}Tl	279 keV	$3/2^+ \xrightarrow{M1+E2} 1/2^+$	$\tilde{E}1$	$P_\gamma = -(3.0 \pm 1.8)10^{-5}$	P. De Saintignon and Chabre (70)
				$= (3.0 \pm 8.0)10^{-6}$	Kuphal, Daum, and Kankeleit (71)
				$= -(2.0 \pm 3.0)10^{-6}$	Boehm and Kankeleit (68)
				$= (2.0 \pm 5.0)10^{-6}$	Vanderleeden, Boehm, and Lipson (72)
				$= -(4.0 \pm 10.0)10^{-6}$	Lipson, Boehm, and Vanderleeden (71)
				$A_\gamma = (6.0 \pm 8.0)10^{-5}$	Baker and Hamilton (70)

the matrix element of the vector current and (14.2) for the axial current. Finally to obtain the potential in coordinate space it is necessary to take the Fourier transform from momentum space to coordinate space.

The result as given by Blin–Stoyle and Herczeg (68) is instructive*

$$V_{NPC}^{(1)} = \left[(\boldsymbol{\sigma}_1 \times \boldsymbol{\sigma}_2)\cdot \mathbf{r} f_A(r) + (\boldsymbol{\sigma}_1 - \boldsymbol{\sigma}_2)\cdot \frac{\mathbf{p}_1 - \mathbf{p}_2}{Mc} h_A(r) \right] \times [\boldsymbol{\tau}_1 \cdot \boldsymbol{\tau}_2 - (\tau_1)_3(\tau_2)_3] \tag{16.8}$$

where $\mathbf{r} = \mathbf{r}_1 - \mathbf{r}_2$. Since $\boldsymbol{\sigma}_i$ does not change sign on parity inversion and $\mathbf{r}$ and $\mathbf{p}_i$ do, it is clear that $V_{\text{NPC}}^{(1)}$ is odd under this transformation. Note that $V_{\text{NPC}}^{(1)}$ is symmetric under the exchange $1 \leftrightarrow 2$ and is time reversal invariant. The factors f_A and h_A are directly related to the form factors of (13.25) and (14.2)

$$f_A = -\frac{G_\mu}{\sqrt{2}} \cos^2\theta_c \frac{\hbar}{2Mc} \operatorname{Re}\left\{ \left(\frac{1}{2\pi}\right)^3 \frac{d}{dr} \int [g_V(k^2) + g_W(k^2)] g_A(k^2) \times \exp[i(\mathbf{k}\cdot\mathbf{r})]\, d\mathbf{k} \right\} \tag{16.9}$$

$$h_A = -\frac{G_\mu}{\sqrt{2}} \cos^2\theta_c \frac{1}{2} \operatorname{Re}\left[\left(\frac{1}{2\pi}\right)^3 \int g_V(k^2) g_A(k^2) \exp(i\mathbf{k}\cdot\mathbf{r})\, d\mathbf{k} \right] \tag{16.10}$$

We have dropped the g_S and g_T terms. The vector form factors g_V and g_W can be obtained by CVC from electron–nucleon scattering. The momentum dependence of g_A is not so nearly well determined and only reasonable guesses have been made so far. The range of f_A and h_A are given by $(1/k)$ where k is roughly the value of k beyond which the form factor products in (16.10) are not appreciable. Figure 16.3 shows h_A and f_A as obtained by Blin–Stoyle and Herczeg.

The evaluation of the matrix element $\langle N'_1N'_2|H_{W2}^{(N)}|N_1N_2\rangle$ and the determination of the consequent nuclear potential [McKellar (67), Fischbach and Trabert (68)] cannot be readily understood without substantial discussion of $SU(3)$ classification of particles. This will take us too far afield. But the form of the answer can be made reasonable in terms of Fig. 16.4 which corresponds perturbatively to the actual evaluation which relies on PCAC as well as $SU(3)$ theories. In this diagram weak interaction theory enters at the weak interaction vertex where the *weak* decay $N_1 \to N'_1 + \pi$ occurs. The emitted pion is then absorbed by N_2 via the strong interaction involving then the strong

*Note that for two-particle systems $\sigma_1 \times \sigma_2 \to -i(\sigma_1 - \sigma_2)$.

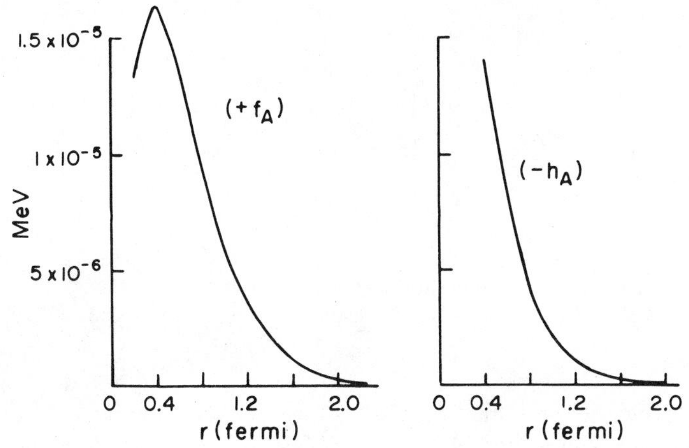

FIG. 16.3. Form factors h_A and f_A obtained by Blin-Stoyle and Herczeg (68).

coupling constant $g_{pp\pi}$. The weak decay probability is determined from the pionic decays of the Λ, Σ, and Ξ hyperons making use of properties of the $SU(3)$ group that relates these decays to the weak nucleon decay. It is found that the matrix element γ for the process $n \rightarrow p + \pi^-$ is $\pm(4.2 \pm 0.8) \times 10^{-8}$, the sign not being determined by this method. The corresponding nucleon–nucleon potential

$$V_{\text{NPC}}^{(2)} = \pm \frac{\gamma g_{pp\pi}}{8\pi\sqrt{2}} \left(\frac{m_\pi}{M}\right) m_\pi c^2 [(\boldsymbol{\sigma}_1 + \boldsymbol{\sigma}_2)\cdot\hat{\mathbf{x}}] e^{-x} \left(\frac{1}{x^2} + \frac{1}{x^3}\right) (\boldsymbol{\tau}_1 \times \boldsymbol{\tau}_2)_3 \tag{16.11}$$

where

$$\mathbf{x} = \frac{m_\pi c}{\hbar} (\mathbf{r}_1 - \mathbf{r}_2)$$

Note that $V_{\text{NPC}}^{(2)}$ conserves the spin of the two-body system while $V_{\text{NPC}}^{(1)}$ does not. At the pion compton wavelength both $V_{\text{NPC}}^{(1)}$ and $V_{\text{NPC}}^{(2)}$ are of the order 10^{-7} MeV while the parity-conserving nucleon–nucleon potential is of the order of several MeV, at the same internuclear distance. As x gets smaller $V_{\text{NPC}}^{(1)}$ will eventually dominate; at $x = 0.4$, $V_{\text{NPC}}^{(1)} \sim 10\ V_{\text{NPC}}^{(2)}$.

Observation of such a small interaction, one of the order of electron volts, in the presence of the much stronger parity-conserving nucleon–nucleon potential is indeed remarkable. However, in comparing theory with experiment one must consider that there are several approximations involved in deriving this force. In the derivation we have pointed to the approximations involved in obtaining (16.7) and to the unknown momentum dependence of g_A. Other uncertainties are present in $V_{\text{NPC}}^{(2)}$. But in addition one must now use these

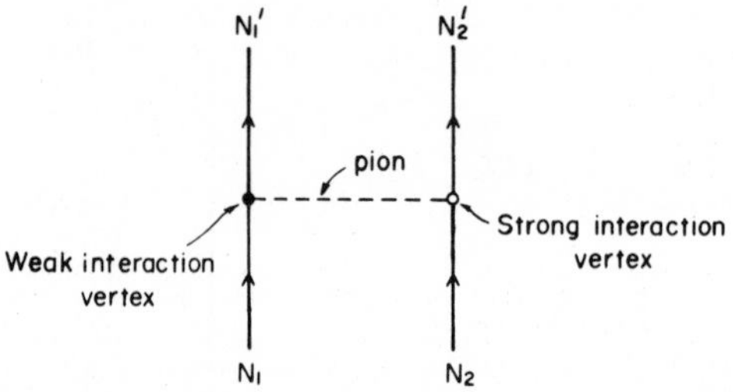

FIG. 16.4. One-pion exchange weak interaction originating in the strange current—strange-current interaction.

potentials together with nuclear models for the nuclei involved. At the present moment there is considerable difference in the results of these calculations. In the case of ^{181}Ta [the experimental value for circular polarization is $-(0.6 \pm 0.1) \times 10^{-5}$] McKellar (68) obtains $(-2 \pm 1) \times 10^{-5}$ if positive γ is used, (-3 ± 1) for negative γ. Fischbach and Trabert (68) obtain similar values but Desplanques and Vinh Mau (71) obtain much smaller values. Such a large difference is partially caused by differing treatments of the hard core of the parity-conserving potential.

Further study of nuclear structure effects are obviously in order. But as was suggested some time ago, Blin–Stoyle and Feshbach (61), many of these difficulties might be avoided if the simpler nuclei such as ^{2}H, ^{3}H, or ^{3}He were involved. These authors and F. Partovi (64) have considered the effects of V_{NPC} on the photo disintegration of ^{2}H as well as neutron capture by protons and ^{2}H. In experiments involving the deuteron asymmetry and polarization effects of the order of $10^{-6} - 10^{-7}$ were predicted but in (n, d) capture larger effects are predicted. Danilov (65) and Tadic (68) have shown that the polarization in (n, p) capture depends only upon the $T = 0, 2$ components of V_{NPC}, that is, on $V_{\text{NPC}}^{(1)}$, while the asymmetry in the capture of polarized neutrons by protons depends only on $V_{\text{NPC}}^{(2)}$. The deuteron experiments have been performed recently by Loboshov (70, 71). For a theoretical discussion see the papers by Partovi (64) and Hadjimichael and Fischbach (71)]. The (n, d) capture process seems more promising but of course the interpretation will be more difficult [Moskalev (68), Hadjimichael, Harms, Newton (71)].

Light nuclei also present possibilities because the wave functions are better known. In addition one can choose levels for which $E1$ is retarded. Recall (see page 739) the approximate selection rule that states that $E1$ transitions in self-conjugate nuclei are forbidden (Fig. VIII.11.9). Similarly isoscalar $M1$ transitions (Chapter VIII, page 746 and Fig. VIII.11.12) are approximately forbidden. This suggestion that was made by Dashen et al. (64) has been recently investigated by Henley (68) who finds that the transition

$2^-(5.105\text{ MeV}) \xrightarrow{E1} 1^+ (0.717\text{ MeV})$ in ^{10}B and $0^-(1.08\text{ MeV}) \xrightarrow{E1} 1^+$ (g.s.) in ^{18}F are promising.*

TABLE 16.3 Isospin Analysis of Experiments in Parity Violation[a]

Experiment	$\lvert\Delta T\rvert = 0$	$\lvert\Delta T\rvert = 1$	$\lvert\Delta T\rvert = 2$	Performed
α-decay $^{16}O(8.88\text{ MeV}) \to {}^{12}C + \alpha$	X			Yes
$n + p \rightleftarrows d + \gamma$ Circular polarization	X			Yes
$n + p \rightleftarrows d + \gamma$ Angular asymmetry		X		No
$n + d \rightleftarrows t + \gamma$ Circular polarization	X	X		In progress
Complex nuclei (^{181}Ta, etc.)	X	X	X	Yes
$\alpha + d \to {}^{6}Li + \gamma$		X		Yes
Self-conjugate nuclei (^{18}F, ^{10}B, etc.)		X		No
$e + d \to e + d$	X			No

[a] The cross (X) indicate the isospin terms in V_{NPC} to which the experiment is sensitive [from Fischbach and Tadic (72)].

C. Neutral Currents; Time-Reversal Invariance; Nonparity-Conserving Potential

The discovery that kaon decay is not time-reversal or CP invariant (C = charge conjugation, P = space inversion) [Christenson, Cronin, Fitch, Turlay (64)] has stimulated much theoretical and experimental activity. On the experimental side besides attempting to elucidate the kaon decay, a search has been made for other examples of time-reversal invariance violating transi-

*In a recent excellent paper, Fischbach and Tadic (71) review the current (1972) status of the origin of nonparity-conserving potentials and their effect on nuclear transitions. Unfortunately it arrived well after the material in this section had been written so that it was not possible to include relevant aspects of their discussion. We shall however include one of their tables that gives the sensitivity of the various experiments to the isospin components of the parity-nonconserving potential (Table 16.3). Another even more recent review has been made by Gari (72).

tions without success. Because of the association of the one observed case with weak decay, the nuclear β-decay has been subjected to considerable scrutiny. Possible experiments of this type were first discussed by Jackson, Treiman, and Wyld (57). The principal involved is analogous to that used in the investigation of nonparity-conserving β-decay—to observe the expectation value in the final state of a quantity that is not time-reversal invariant. One example involves the measurement of the antineutrino momentum (obtained from the recoiling nucleus), and the electron momentum as emitted from a polarized nucleus. One then asks for the expectation value of $\mathbf{P} \cdot (\mathbf{p}_e \times \mathbf{p}_{\bar{\nu}})$ where $\mathbf{P}$ is the polarization vector for the emitting nucleus. Since $\mathbf{P}$ transforms like a spin, it is odd under the time-reversal transition as are $\mathbf{p}_e$ and $\mathbf{p}_\nu$. The reader can easily form other examples of such noninvariant quantities. For a recent discussion of these experiments as well as those involving electromagnetic decay of nuclear systems and nuclear reactions, see Henley's article (69). An example of the latter, nuclear reactions, is shown in Fig. 16.5 where the detailed balance between the reactions $^{24}\text{Mg}(\alpha, p)\ ^{27}\text{Al}$ and $^{27}\text{Al}(p, \alpha)\ ^{24}\text{Mg}$ is examined. The authors claim an upper bound of 3×10^{-3} for the ratio of reaction amplitude odd under time-reversal compared to the amplitude even under time reversal. The discussion of electromagnetic transitions takes advantage of the fact that if time-reversal invariance holds it is possible to choose the phases of the nuclear wave functions so that the matrix elements for the transition are real. Hence a violation of time-reversal invariance is implied if the relative phase of say an $E2$ matrix element and an $M1$ matrix element is observed to be other than 0 or π. For a discussion of these experiments see Henley (69).

Problem. Discuss other β-decay measurements that would test time-reversal invariants. An example is $\langle \boldsymbol{\sigma}_e \cdot (\mathbf{p}_e \times \mathbf{p}_{\bar{\nu}}) \rangle$ where $\boldsymbol{\sigma}_e$ refers to the electron spin.

From a theoretical point of view one can insert a lack of time-reversal invariance phenomenologically by making the form factors in the effective nuclear β-decay Hamiltonian g_V, g_M, g_A, etc. complex. From a more fundamental point of view, one procedure involves keeping the current–current interaction form but adding currents that do not preserve CP- or T-invariance. One such theory [Oakes (68)] adds neutral currents, suggested by the $SU(3)$ classification, to the isospin current used in the theory, as discussed up to this point. The latter currents are charged, that is, one of the particles involved is charged, the other neutral (see Eqs. 12.8, 12.9, and 12.11). Neutral currents were introduced much earlier by d'Espagnat (63) in order to guarantee the $\Delta T = 1/2$ rule in nonleptonic weak decays [Gasiorowicz (66)]. For the discussion here it is important to note that these currents have a large effect on the nonparity-conserving potential discussed in Section 16c. The new feature is that a component of the neutral currents as introduced by Oakes transforms

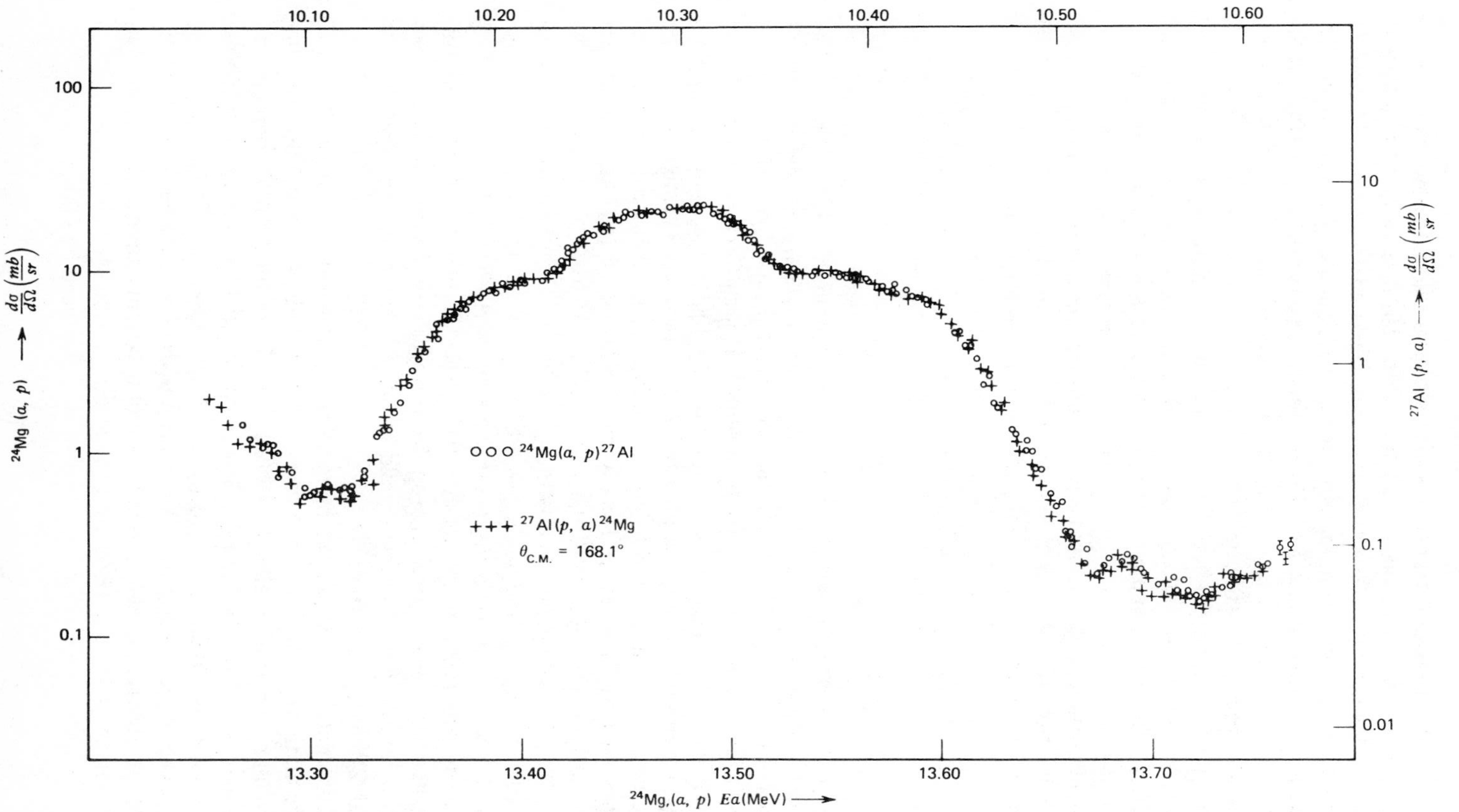

FIG. 16.5. Detailed balance between $\alpha + {}^{24}Mg \rightarrow p + {}^{27}Al$ and $p + {}^{27}Al \rightarrow \alpha + {}^{24}Mg$ Excitation functions at $\theta_{cm} = 168.1°$. The errors at the high energy end are statistical [taken from Von Witsch, Richter, and von Brentano (66,67,68) and Henley (69)].

like T_3 (and also is nonstrangeness changing) breaking the symmetry of the $T^{(+)}T^{(-)} + T^{(-)}T^{(+)}$ (see discussion below Eq. 16.5). The T_3 combines with $T^{(+)}$ and $T^{(-)}$ to obtain, in addition to the isoscalar and isotensor of the charged current–current interaction, an isovector as well. Because the current–current interaction will now have a $T = 1$ component, it follows that the range of the resultant nucleon–nucleon force can be the order of *one* pion Compton wavelength and, therefore, will contribute to $V_{\text{NPC}}^{(2)}$; moreover the factor of $\sin^2 \theta_c$ of that term will be replaced by $\cos^2 \theta_c$. The consequent enhancement of $V_{\text{NPC}}^{(2)}$ according to McKellar (68) is a factor of 30, making $V_{\text{NPC}}^{(2)}$ dominant and increasing correspondingly the effects in nuclear transition of nonparity-conserving potentials.

It is important to realize that such a pion term can only occur if the effective interaction H_{eff} does *not* conserve CP. To prove this consider the most general effective local $\Delta S = 0$ interaction that does *not* conserve parity

$$H_{\text{eff}} \sim \int d\mathbf{r} \left\{ c_1 \hat{\bar{\psi}}_N \hat{\psi}_N \hat{\phi}_\pi + c_2 \sum_\lambda \hat{\bar{\psi}}_N \gamma_\lambda \hat{\psi}_N \frac{\partial \hat{\phi}_\pi}{\partial x_\lambda} + c_3 \sum_{\lambda\nu} \hat{\bar{\psi}}_N \sigma_{\lambda\nu} \hat{\psi}_N \frac{\partial^2 \hat{\phi}_\pi}{\partial x_\lambda \partial x_\nu} \right\} \tag{16.12}$$

where c_i are constants, $\hat{\psi}_N$ is the nucleon field, and $\hat{\phi}_\pi$ is a pseudoscalar neutral field like that of the π° (but it could also be the η°). Recall that under charge conjugation $\hat{\phi}_\pi$ is even and that $\hat{\bar{\psi}}\hat{\psi}$ is even under P and C. If H_{eff} is to conserve CP, the first term must be dropped. It is odd under P since $\hat{\phi}_\pi$ is a pseudoscalar but even under C. If we integrate the second term by parts we see that it vanishes when we employ (13.25) for the matrix element of $\bar{\psi}_N \gamma_\mu \psi_N$ (a vector current) and realize that the initial and final nucleon are identical and thus have the same mass. Finally the c_3 term in (16.11) vanishes because of the antisymmetry of $\sigma_{\lambda\nu}$ and the symmetry of $\partial^2/\partial x_\lambda \partial x_\nu$ with respect to the indices λ and ν. The conclusion is that the effective Hamiltonian cannot conserve CP, strangeness and at the same allow for single neutral pion exchange. The addition of a CP violating neutral current is required.

Thus by observing nonparity-conserving transitions we not only confirm the existence of the diagonal baryon–baryon weak interaction, but also from quantitative interpretation one should be able to determine the existence of a neutral CP violating current.

Unfortunately at the present moment there seems to be a disagreement in the calculations: McKellar (68) finds, when including neutral currents, that the circular polarization effect for ^{181}Ta is $(-10 \pm 5) \times 10^{-5}$ compared to his no-neutral current value of -3×10^{-5}. Vinh Mau and Bruneau's (69) value rises from -0.08×10^{-5} to -1.55×10^{-5}. Both calculations reflect the strong effect of the neutral current. Since the experimental value is $-(0.6 \pm 0.1) \times 10^{-5}$ McKellar claims the absence of a neutral current while Vinh Mau and Bruneau predict its existence. This disagreement, which is partially a consequence of different treatments of the hard core of the parity-conserving nucleon–nucelon interaction, is undoubtedly transitory. As soon as it is resolved (or further experiments on better known nuclei are performed) the existence of neutral currents will be established. The reverse would also be true; if these experiments ever were to become more routine (and V_{NPC} better known), they would permit further studies of aspects of nuclear wave functions. For a recent review of the status of these calculations [see Gari (72)].

17. FORBIDDEN β-DECAY

In discussing nuclear β-decay we have made a number of approximations, which limit the validity of the results obtained, to the allowed transitions of Section 2. These approximations need to be improved upon when (1) the Fermi and/or Gamow–Teller selection rules of (2.26) and (2.27) are not satisfied, and/or (2) when the electron or neutrino wavelength is so small that the approximation that these lepton wave functions are constant over the nuclear volume is invalid. Under these conditions it is no longer correct to take the nonrelativistic limit (Table 11.2) of the nucleon operators. In addition one must also include the effects of the strong interactions, as given by the momentum dependence of the form factors g_V, g_W, g_A, and g_P. We shall place the second-class current form factors g_S, g_T equal to zero. We shall now improve upon the earlier simpler theory in order to take these effects into account.

When these effects are taken into account, finite values are obtained for the nuclear matrix elements for transitions that are not allowed according to the selection rules (2.26) and (2.27). These transitions are referred to as *forbidden transitions*. As will be discussed the spatial variation of the leptonic wave functions is describable in terms of a multipole expansion, so that the forbidden transitions can generally be classified according to the angular momentum and parity carried off by the leptonic fields. This is completely analogous to the various electromagnetic multipole transitions that arise because the photon field similarly contains components of various angular momentum and parity.

The selection rules and degree of forbiddeness are also affected by the relativistic corrections to the nonrelativistic nucleon operators that are carried out below to first order in v_N/C where v_N is the nucleon velocity. As we shall see the consequent *form* for the effective β-decay Hamiltonian is not sensitive to the errors in this first-order approximation. The quantitative conclusions are correct to the extent that the nuclear wave functions used in the evaluation of the matrix elements of this Hamiltonian already implicitly contain higher-order relativistic effects.

Let us begin by first developing this effective Hamiltonian.

We shall first generalize the one-body nucleon amplitude (see Eqs. 13.25 and 14.2)

$$\langle p|\hat{V}_4^{(N)}(0) + \hat{A}_4^{(N)}(0)|n\rangle L_4(0) + \sum_a \langle p|\hat{V}_a^{(N)}(0) + \hat{A}_a^{(N)}(0)|n\rangle L_a(0)$$

$$= L_4(0)\bar{u}_p\left\{\gamma_4[g_V(k^2) + \gamma_5 g_A(k^2)] + \frac{i}{2Mc}g_P\gamma_5 k_4 + \frac{1}{2Mc}g_W\sum_\nu \sigma_{4\nu}k_\nu\right\}u_n$$

$$+ \sum_a L_a(0)\bar{u}_p\left\{\gamma_a[g_V(k^2) + \gamma_5 g_A(k^2)]\right.$$

$$\left. + \frac{i}{2Mc}g_P\gamma_5 k_a + \frac{1}{2Mc}g_W\sum_\nu \sigma_{a\nu}k_\nu\right\}u_n \qquad (17.1)$$

The factor $L_\lambda(0)$ involves the lepton wave functions:

$$L_\lambda(0) = \bar{\psi}_e\gamma_\lambda(1 + \gamma_5)\psi_{\bar{\nu}}$$

while k_λ is the momentum transfer

$$k_\lambda \equiv (p_n)_\lambda - (p_p)_\lambda \qquad (13.22)$$

Equation 17.1 applies to β^- decay. For positron decay, the form of (17.1) (assuming time-reversal invariance) is unchanged. Of course p is exchanged for n in the initial and vice versa in the final states while L_λ is replaced by

$$\tilde{L}_\lambda = \bar{\psi}_\nu\gamma_\lambda(1 + \gamma_5)\psi_e$$

In obtaining the matrix element for the nuclear transition several approximations will be made. First, we assume that (17.1), which is valid for the free nucleon, is not modified inside the nucleus. This assumption can in part be described as neglecting the influence of the strong interactions with the other nucleons in the nucleus upon the β-decay interaction. Since these effects will depend not only upon the coordinates of the decaying nucleon but also upon those of the other interacting nucleons, the resulting corrections will add in many-body terms, involving the coordinates of two or more particles. The mesonic exchange considered by Bell and Blin–Stoyle (57) provide an example of this phenomena. It is closely related to the mesonic exchange currents that were discussed in Section VIII.3 for electromagnetic transitions. The results to be obtained below will thus contain only the one-body part of the nuclear β-decay interaction. (See however p. 894).

Effective Nuclear β-Decay Hamiltonian

We are now left with the problem of putting the nucleon matrix element into a form suitable for nuclear transitions. Coordinate space representation seems most convenient. To determine its form we recall that H is given by

$$H = \frac{C_V g}{\sqrt{2}} \sum_{\lambda} \int d\mathbf{r} \langle f | \hat{\mathcal{J}}_{\lambda}^{(N)\star}(\mathbf{r}) \hat{\jmath}_{\lambda}^{(l)}(\mathbf{r}) | i \rangle \tag{17.2}$$

Factoring out the leptonic contribution yields

$$H = \frac{C_V g}{\sqrt{2}} \sum_{\lambda} \int d\mathbf{r} \langle f_N | \hat{\mathcal{J}}_{\lambda}^{(N)\star}(\mathbf{r}) | i_N \rangle \langle e\bar{\nu} | \hat{\jmath}_{\lambda}^{(l)}(\mathbf{r}) | o \rangle$$

The second bracket is proportional (there is a factor of $-i$) to $L_{\lambda}(\mathbf{r})$. The first factor is rewritten as follows:

$$\langle f_N | \hat{\mathcal{J}}_{\lambda}^{(N)\star}(\mathbf{r}) | i_N \rangle = \int d\mathbf{r}' \int d\mathbf{r}'' \int \frac{d\mathbf{p}'}{(2\pi\hbar)^3} \int \frac{d\mathbf{p}''}{(2\pi\hbar)^3}$$
$$\cdot \langle f_N | \mathbf{r}' \rangle \langle \mathbf{r}' | \mathbf{p}' \rangle \langle \mathbf{p} | \hat{\mathcal{J}}_{\lambda}^{(N)\star}(\mathbf{r}) | \mathbf{p}'' \rangle \langle \mathbf{p}'' | \mathbf{r}'' \rangle \langle \mathbf{r}'' | i_N \rangle$$

But

$$\langle \mathbf{p}' | J_{\lambda}^{(N)\star}(\mathbf{r}) | \mathbf{p}'' \rangle = \exp\left[- i/\hbar (\mathbf{p}' - \mathbf{p}'') \cdot \mathbf{r} \right] \langle \mathbf{p}' | J_{\lambda}^{(N)\star}(o) | \mathbf{p}'' \rangle$$

and

$$\langle \mathbf{p}'' | \mathbf{r}'' \rangle = \exp\left(i\mathbf{p}'' \cdot \mathbf{r}'' / \hbar \right)$$

so that

$$\langle f_N | \hat{\mathcal{J}}_{\lambda}^{(N)\star}(\mathbf{r}) | i_N \rangle = \int d\mathbf{r}' \int d\mathbf{r}'' \int \frac{d\mathbf{p}'}{(2\pi\hbar)^3} \int \frac{d\mathbf{p}''}{(2\pi\hbar)^3} < f_N | \mathbf{r}' \rangle \langle \mathbf{r}'' | i_N \rangle$$
$$\times F_{\lambda}\left(\frac{\mathbf{p}''}{\hbar} - \frac{\mathbf{p}'}{\hbar} \right) \cdot \exp\left[-i/\hbar \left[(\mathbf{p}' - \mathbf{p}'') \cdot \mathbf{r} + \mathbf{p}'' \cdot \mathbf{r}'' - \mathbf{p}' \cdot \mathbf{r}' \right] \right]$$

where

$$F_{\lambda}\left(\frac{1}{\hbar} (\mathbf{p}'' - \mathbf{p}') \right) \equiv \langle \mathbf{p}' | \hat{\mathcal{J}}_{\lambda}^{(N)\star}(o) | \mathbf{p}'' \rangle \tag{17.3}$$

Integrating over $(\mathbf{p}' + \mathbf{p}'')/2$ yields as a factor $\delta(\mathbf{r}' - \mathbf{r}'')$. Integrating over $\mathbf{r}''$ one finally obtains

$$\langle f_N | \hat{\mathcal{J}}_{\lambda}^{(N)\star}(\mathbf{r}) | i_N \rangle = \int \langle f_N | \mathbf{r}' \rangle \, d\mathbf{r}' \int \frac{d\mathbf{k}}{(2\pi)^3} F_{\lambda}(\mathbf{k}) \exp\left[-i\mathbf{k} \cdot (\mathbf{r} - \mathbf{r}') \right] \langle \mathbf{r}' | i_N \rangle$$
$$= \int d\mathbf{r}' \langle f_N | \mathbf{r}' \rangle I_{\lambda}(\mathbf{r} - \mathbf{r}') \langle \mathbf{r}' | i_N \rangle \tag{17.4}$$

where I_{λ} is the Fourier transform of F_{λ}.

In this equation $\mathbf{r}'$ is the nucleon coordinate. In going to many nucleons we must sum over the various nucleons in the decaying nucleus. Hence the effective Hamiltonian is

$$H = \frac{C_V g}{\sqrt{2}} \int d\mathbf{r} \left[\sum_i I_4(\mathbf{r} - \mathbf{r}_i) L_4(\mathbf{r}) + \sum_{i,k} I_k(\mathbf{r} - \mathbf{r}_i) L_k(\mathbf{r}) \right] \tag{17.5}$$

where

$$I_\lambda(\mathbf{r}) = \frac{1}{(2\pi)^3} \int d\mathbf{k} \exp - (i\mathbf{k}\cdot\mathbf{r}) \Big\{ \gamma_\lambda [g_V(k^2) + \gamma_5 g_A(k^2)] + \frac{i}{2Mc} g_P \gamma_5 k_\lambda + \frac{1}{2Mc} g_W \sum_\nu \sigma_{\lambda\nu} k_\nu \Big\} \tag{17.6}$$

Or for the spatial components* of I_λ:

$$\mathbf{I}(\mathbf{r}) = \boldsymbol{\gamma}[G_V(\mathbf{r}) + \gamma_5 G_A(\mathbf{r})] + \gamma_5 \frac{i\hbar}{2Mc} \nabla G_P(\mathbf{r}) - \frac{i\hbar}{2Mc} (\boldsymbol{\sigma} \times \nabla) G_W(\mathbf{r}) + \frac{iE_0}{2Mc^2} \boldsymbol{\alpha} G_W \tag{17.7}$$

where for example

$$G_V(\mathbf{r}) = \frac{1}{(2\pi)^3} \int \exp(-i\mathbf{k}\cdot\mathbf{r}) g_V(k^2)\, d\mathbf{k} \tag{17.8}$$

For the fourth component

$$I_4(\mathbf{r}) = \gamma_4 [G_V(\mathbf{r}) + \gamma_5 G_A(\mathbf{r})] - \frac{E_0}{2Mc^2} \gamma_5 G_P(\mathbf{r}) - \frac{i\hbar \boldsymbol{\alpha}\cdot\nabla}{2Mc} G_W(\mathbf{r}) \tag{17.9}$$

In the limit of zero momentum for the leptonic field g_V, g_A constant, g_P, g_A, zero,

$$G_V(\mathbf{r}) = g_V \delta(\mathbf{r}) \qquad G_A(\mathbf{r}) = g_A \delta(\mathbf{r}) \tag{17.10}$$

Nonrelativistic Reduction

To convert these expressions to forms convenient for nuclear transitions it is necessary, in virtue of the nonrelativistic description of the nuclear many-body problem commonly used, to approximate the nuclear Dirac matrices by nonrelativistic forms.

---

*The fourth component of k, k_4, equals $i\,(E_e + \bar{E})\,\hbar c$ which by energy conservation equals $(iE_0/\hbar c)$. E_0 is the energy release in the decay.

This is simple because of the low momentum transfer to the nucleus. The general problem is the evaluation of $\bar{u}_p\, \mathrm{O}\, u_n$ where O is a Dirac matrix. However u_n and u_p may be written to first order in the recoil momenta as follows:*

$$u \simeq \begin{pmatrix} u_L \\ \dfrac{\boldsymbol{\sigma}\cdot\mathbf{p}}{2Mc}\, u_L \end{pmatrix} = \begin{pmatrix} 1 & 0 \\ \dfrac{\boldsymbol{\sigma}\cdot\mathbf{p}}{2Mc} & 0 \end{pmatrix} \begin{pmatrix} u_L \\ 0 \end{pmatrix}$$

where u_L is a two-component wave function, the subscript L denoting the "large" component. The small component is given to the order of v_N/c where v_N is the nucleon velocity. The normalization condition involves a second-order correction. To first order

$$\bar{u}_p O u_n = u_p^{\dagger}\gamma_4 O u_n \rightarrow (u^*_{pL} 0) \begin{pmatrix} 1 & (\boldsymbol{\sigma}\cdot\mathbf{p})/2Mc \\ 0 & 0 \end{pmatrix} \begin{pmatrix} 1 & 0 \\ 0 & -1 \end{pmatrix} O \begin{pmatrix} 1 & 0 \\ \dfrac{\boldsymbol{\sigma}\cdot\mathbf{p}}{2Mc} & 0 \end{pmatrix} \begin{pmatrix} u_{nL} \\ 0 \end{pmatrix}$$

$$= u^*_{pL} \left\{ O_{LL} - \frac{\boldsymbol{\sigma}\cdot\overleftarrow{\mathbf{p}}}{2Mc} O_{SL} + O_{LS} \frac{\boldsymbol{\sigma}\cdot\mathbf{p}}{2Mc} \right\} u_{nL} \tag{17.11}$$

where

$$O \equiv \begin{pmatrix} O_{LL} & O_{LS} \\ O_{SL} & O_{SS} \end{pmatrix} \tag{17.12}$$

and the left-facing arrow on $\overleftarrow{\mathbf{p}}$ implies that this operator acts on u^*_{pL}. Thus the relativistic operator O may be replaced by

$$O' \rightarrow \tau^{(+)} \left(O_{LL} - \frac{\boldsymbol{\sigma}\cdot\overleftarrow{\mathbf{p}}}{2Mc} O_{SL} + O_{LS} \frac{\boldsymbol{\sigma}\cdot\mathbf{p}}{2Mc} \right) \tag{17.13}$$

where O' is to be taken between the nonrelativistic Pauli spinor wave functions employed in nuclear theory. Various O and their corresponding O' are listed in Table 17.1.

We now apply these results to the β-ray Hamiltonian below

$$H_{\beta-} = \frac{C_V g}{\sqrt{2}} \left[L_4\gamma_4(G_V + \gamma_5 G_A) + \boldsymbol{\gamma}\cdot\mathbf{L}(G_V + \gamma_5 G_A) - \frac{i\hbar}{2Mc} (\boldsymbol{\sigma} \times \boldsymbol{\nabla})\cdot\mathbf{L} G_W \right] \tag{17.14}$$

*More precisely, if the nucleon Dirac equation describes a particle moving in a four-vector potential then $(\boldsymbol{\sigma}\cdot\mathbf{p}/2Mc)$ in the above equation is replaced by $c/(E - V + Mc^2)\, \boldsymbol{\sigma}\cdot[\mathbf{p} - (1/c)\, \mathbf{A}]$.

TABLE 17.1 Nonrelativistic Limits of Dirac Operators

O	O'
$1 = \begin{pmatrix} 1 & 0 \\ 0 & 1 \end{pmatrix}$	1
$\gamma_4 = \begin{pmatrix} 1 & 0 \\ 0 & -1 \end{pmatrix}$	1
$\gamma_5 = -\begin{pmatrix} 0 & 1 \\ 1 & 0 \end{pmatrix}$	$\frac{\boldsymbol{\sigma}\cdot\overleftarrow{\mathbf{p}}}{2Mc} - \frac{\boldsymbol{\sigma}\cdot\mathbf{p}}{2Mc}$
$\gamma_4\gamma_5 = \begin{pmatrix} 0 & -1 \\ 1 & 0 \end{pmatrix}$	$-\frac{\boldsymbol{\sigma}\cdot\overleftarrow{\mathbf{p}}}{2Mc} - \frac{\boldsymbol{\sigma}\cdot\mathbf{p}}{2Mc}$
$\boldsymbol{\gamma} = \begin{pmatrix} 0 & -i\boldsymbol{\sigma} \\ i\boldsymbol{\sigma} & 0 \end{pmatrix}$	$-\frac{i}{2Mc}[(\boldsymbol{\sigma}\cdot\overleftarrow{\mathbf{p}})\boldsymbol{\sigma} + \boldsymbol{\sigma}(\boldsymbol{\sigma}\cdot\mathbf{p})]$
	$= -\frac{i}{2Mc}\{\overleftarrow{\mathbf{p}} + \overrightarrow{\mathbf{p}} + i[\boldsymbol{\sigma}\times(\overleftarrow{\mathbf{p}} - \overrightarrow{\mathbf{p}})]\}$
$\boldsymbol{\gamma}\gamma_5 = \begin{pmatrix} i\boldsymbol{\sigma} & 0 \\ 0 & -i\boldsymbol{\sigma} \end{pmatrix}$	$i\boldsymbol{\sigma}$
$\sigma_{4k} \rightarrow -\boldsymbol{\alpha} = -\begin{pmatrix} 0 & \boldsymbol{\sigma} \\ \boldsymbol{\sigma} & 0 \end{pmatrix}$	$\frac{\boldsymbol{\sigma}\cdot\overleftarrow{\mathbf{p}}}{2Mc}\boldsymbol{\sigma} - \boldsymbol{\sigma}\frac{\boldsymbol{\sigma}\cdot\mathbf{p}}{2Mc}$
	$= \frac{1}{2Mc}[\overleftarrow{\mathbf{p}} - \overrightarrow{\mathbf{p}} + i\boldsymbol{\sigma}\times(\overleftarrow{\mathbf{p}} + \overrightarrow{\mathbf{p}})]$

The term proportional to G_P has been dropped because of its small effect for β-decay (see Eq. 14.17). Separating out the first order (in nuclear v_n/c) terms we have:

$$H_{\beta^-}(\mathbf{r} - \mathbf{r}_i) = \frac{C_V g}{\sqrt{2}}\,[h_{\beta^-} - \Delta(\mathbf{r} - \mathbf{r}_i)\tau^+] \tag{17.15}$$

where

$$h_{\beta^-} \equiv (L_4 G_V + i\boldsymbol{\sigma}_i\cdot\mathbf{L}G_A)\tau^{(+)} \tag{17.16}$$

and

$$\Delta = \frac{L_4}{2Mc}[\boldsymbol{\sigma}_i \cdot \overleftarrow{\mathbf{p}}_i G_A + G_A \boldsymbol{\sigma}_i \cdot \mathbf{p}_i] + \frac{i\hbar}{2Mc}(\boldsymbol{\sigma}_i \times \nabla_i G_W) \cdot \mathbf{L}$$

$$+ \frac{i}{2Mc}\{(\overleftarrow{\mathbf{p}}_i \cdot \mathbf{L})G_V + G_V(\mathbf{L} \cdot \mathbf{p}_i) + [\boldsymbol{\sigma}_i \times (\overleftarrow{\mathbf{p}}_i G_V - G_V \mathbf{p}_i)] \cdot \mathbf{L}\} \qquad (17.17)$$

Note that the operator $\mathbf{p}_i$ can operate on G_A, G_V, and G_W since they are functions of the nuclear coordinates $\mathbf{r}_i$ but not on L_λ since it is a function of $\mathbf{r}$. H_{β^-} reduces in the limit of constant g_V and g_A (see 17.10) to the combination of Fermi and Gamow–Teller terms employed in discussing the allowed transitions with the difference that L_4 and $\mathbf{L}$ in (17.16) are not necessarily constants. Note that the weak magnetism term involving G_W is new in the sense that the pre-CVC theories did not include this term. It combines with a similar term involving G_V. We rewrite Δ as follows:

$$\Delta = \frac{i\hbar}{2Mc}(\boldsymbol{\sigma}_i \times \nabla_i) \cdot \mathbf{L}(G_W + G_V) + \frac{i}{2Mc}[\mathbf{p}_i \cdot (\mathbf{L}G_V - i\boldsymbol{\sigma}_i G_A L_4)$$

$$+ (G_V \mathbf{L} - iG_A L_4 \boldsymbol{\sigma}_i) \cdot \mathbf{p}_i] \qquad (17.18)$$

where the direction of the arrow on $\mathbf{p}_i$ has been reversed by assuming that the wave functions it operates on describe bound states and assuming that eventually an integration over $\mathbf{r}_i$ will be performed. Introducing the explicit forms for L_4 and $\mathbf{L}$ yields

$$h_{\beta^-} = [G_V \bar{\psi}_e \gamma_4 (1 + \gamma_5) \psi_\nu - G_A \boldsymbol{\sigma}_i \cdot \bar{\psi}_e \gamma_4 (1 + \gamma_5) \boldsymbol{\sigma} \psi_\nu] \tau^{(+)} \qquad (17.19)$$

and

$$\Delta = -\frac{\hbar}{2Mc}[\boldsymbol{\sigma}_i \times \nabla_i (G_W + G_V)] \cdot \bar{\psi}_e \gamma_4 (1 + \gamma_5) \boldsymbol{\sigma} \psi_\nu$$

$$- \frac{1}{2Mc}[\bar{\psi}_e \gamma_4 (1 + \gamma_5) \boldsymbol{\sigma} \psi_\nu \cdot (\mathbf{p}_i G_V + G_V \mathbf{p}_i) \qquad (17.20)$$

$$- \bar{\psi}_e \gamma_4 (1 + \gamma_5) \psi_\nu (\boldsymbol{\sigma}_i \cdot \mathbf{p}_i G_A + G_A \boldsymbol{\sigma}_i \cdot \mathbf{p}_i)]$$

This is as far as one can go formally in reducing the nuclear decay amplitude to order $(1/M)$. The functions G_V and G_W are "known" via electron scattering and CVC. Very much less is known about G_A, although some information in principle should become available from an analysis of nuclear β-decays. In using CVC for G_V and G_W one should not forget that the simple CVC relationships hold only when the isospin symmetry breaking parts of the hadron interactions are neglected.

Selection Rules—Effects of the Spatial Dependence of Lepton Fields

Thus far, we have made explicit those contributions that are generated by the form factors and by the relativistic components (the "small" wave functions) of the nucleon Dirac wave functions. We turn now to the spatial dependence of the lepton fields that is neglected in the calculation of allowed transitions. If we use plane-wave descriptions of the electron and antineutrino, the lepton functions will be proportional to

$$\exp\,[-(i/\hbar)(\mathbf{p}_e + \bar{\mathbf{p}}_\nu)\cdot\mathbf{r}] = \exp\,[-(i/\hbar)(\mathbf{p}_e + \bar{\mathbf{p}}_\nu)\cdot\mathbf{r}_i] \times \exp\,[-(i/\hbar)(\mathbf{p}_e + \bar{\mathbf{p}}_\nu)\cdot(\mathbf{r} - \mathbf{r}_i)]$$

The second factor will be integrated over $\mathbf{r}$. The first factor can be expanded in a series involving $Y_{lm}(\hat{\mathbf{r}}_i)$, the first term $(l = 0)$ giving the allowed transition. The order of magnitude of each term is

$$\left(\left|\frac{\mathbf{p}_e + \bar{\mathbf{p}}_\nu}{\hbar}\right| R\right)^l$$

where R is the nuclear radius. The maximum value of this factor is approximately

$$\left(\frac{E_0 R}{\hbar c}\right)^l$$

where E_0 is the energy release in the reaction. This factor is very small compared to one for nuclear β-decay. (See Eq. (2.6).

The contributions to a given order of forbiddenness are then a combination of Y_{lm} and $Y_{lm}\boldsymbol{\sigma}$ from (17.19); $Y_{lm}\boldsymbol{\sigma}$, $Y_{lm}\mathbf{p}$, $Y_{lm}\mathbf{p}\cdot\boldsymbol{\sigma}$ from (17.20). The corresponding selection rules are given in Table 17.2.

TABLE 17.2 Selection Rules

Transition Type		Nuclear Angular Momentum Change ΔJ	Nuclear Parity Change
Allowed	$l = 0$	$\Delta J = 0$	No
	$l = 0$	$\Delta J = \pm 1, 0$	No
First forbidden	$(l = 1)$	$\Delta J = \pm 1, 0$	Yes
	$(l = 1)$	$\Delta J = \pm 2, 1, 0$	Yes
	$(l = 0)$	$\Delta J = \pm 1, 0$	Yes
Second forbidden	$(l = 2)$	$\Delta J = \pm 2, 1, 0$	No
	$(l = 2)$	$\Delta J = \pm 3, 2, 1$	No
	$(l = 1)$	$\Delta J = \pm 2, 1, 0$	No
	$(l = 1)$	$\Delta J = \pm 1, 0$	No

The $l = 0$ terms in the first forbidden and the $l = 1$ in the second forbidden class come from the Δ term (17.18). The magnitude of the *amplitude* for the nth order of forbiddenness in the plane-wave approximation is $(E_0R/\hbar c)^n$ if the term is generated from (17.19). If its source is Δ, the magnitude is $(E_0R/\hbar c)^{n-1}(3\hbar/McR)$ or $(E_0R/\hbar c)^{n-1}(v_N/c)$ depending upon whether the source is in the first term in Δ or the other two. The factor of three modifying $(\hbar/McR)$ takes the anomalous magnetic moment contribution coming from g_W into account. (See Eq. VIII.5.46). Since $(E_0R/\hbar c) \sim R/\lambdabar$ is a small number the nth order amplitude is significant only if the $(n - 1)$th amplitude is zero or anomolously small.

The factor $(E_0R/\hbar c)^l$ is a measure of the effect of the angular momentum barrier for the electrons and neutrino. However, because of the Coulomb field between the electron and the nucleus, this factor may be modified, the effect being of importance for low l for which the "centripetal" potential and the Coulomb potential may be comparable. The Coulomb field introduces a "momentum" (Ze^2/Rc) where R is the nuclear radius. The Coulomb parameter that can replace $E_0R/\hbar c$ is obtained by replacing the lepton momentum E_0/c by this momentum. The result is $(Ze^2/\hbar c)$ and can be used in place of $(E_0R/\hbar c)$ in the estimates given above when the Coulomb potential dominates (p. 892). Of course for high l the angular momentum barrier will dominate.

Electron Polarization in Forbidden Decays

It will be observed that the leptonic factors in (17.19) and (17.20) are identical with those which occur in the allowed transition matrix element. In the case where the spatial dependence of both the electron and neutrino amplitudes are given by plane waves, the spinor lepton factors are $\bar{u}_e\gamma_4\cdot(1 + \gamma_5)u_{\bar{\nu}}$ and $\bar{u}_e\gamma_4(1 + \gamma_5)\boldsymbol{\sigma}u_{\bar{\nu}}$ identical with the results obtained in the allowed case. The nuclear matrix elements are no longer the simple M_F and M_{GT} but are modified. We define an $\mathfrak{M}_F$ and $\boldsymbol{\mathfrak{M}}_{GT}$ where

$$M_{fi} = \langle f| \sum_i \int d\mathbf{r} H_{\beta^-}(\mathbf{r} - \mathbf{r}_i)|i\rangle$$

$$\equiv \frac{C_V g}{\sqrt{2}} [\mathfrak{M}_F \bar{u}_e\gamma_4(1 + \gamma_5)u_{\bar{\nu}} + \boldsymbol{\mathfrak{M}}_{GT}\cdot\bar{u}_e\gamma_4(1 + \gamma_5)\boldsymbol{\sigma}u_{\bar{\nu}}] \tag{17.21}$$

where the nuclear matrix elements $\boldsymbol{\mathfrak{M}}_F$ and $\boldsymbol{\mathfrak{M}}_{GT}$ are defined by

$$\mathfrak{M}_F \equiv \sum_i \langle f_N|\tau^{(+)}(i) \int d\mathbf{r} \exp - (i\mathbf{k}\cdot\mathbf{r}) \times \left(G_V - \frac{1}{2Mc}(\boldsymbol{\sigma}_i\cdot\mathbf{p}_i G_A + G_A\boldsymbol{\sigma}_i\cdot\mathbf{p}_i)\right)|i_N\rangle \tag{17.22}$$

and

$$\mathfrak{M}_{GT} \equiv \sum_i \langle f_N | \tau^{(+)}(i) \int d\mathbf{r} \exp(-i\mathbf{k}\cdot\mathbf{r})$$

$$\times \left\{ -G_A \boldsymbol{\sigma}_i + \frac{\hbar}{2Mc} [\boldsymbol{\sigma}_i \times \nabla_i (G_W + G_V)] + \frac{1}{2Mc} (\overleftarrow{\mathbf{p}}_i G_V + G_V \mathbf{p}_i) \right\} | i_N \rangle \tag{17.23}$$

where $\mathbf{k}$ is the momentum carried off by the neutrino-electron pair

$$\hbar\mathbf{k} = \mathbf{p}_e + \bar{\mathbf{p}}_\nu \tag{17.24}$$

Equation 17.21 provides a completely general form for M_{fi} in the plane-wave approximation for the leptons while (17.22) and (17.23) provide approximate expressions for M_F and $\mathfrak{M}_{GT}$ to first order in (v_N/c). *We see that any result obtained in Section 11 for allowed transitions will hold generally if it is independent of M_F and* $\mathbf{M}_{GT}$. The polarization of electrons emitted from an unpolarized nucleus is such an example, the result given by (11.42) being valid to all orders in the plane-wave approximation.

Unique Forbidden Spectra

The electron energy spectrum for a given order of forbiddenness and change of angular momentum (ΔJ) and parity is generally dependent upon nuclear matrix elements simply because several terms in (17.22) and (17.23) can contribute. There is one exception to this rule, when ΔJ takes on its maximum value of $(n + 1)$ for the nth-forbidden transition. There is only one term that contributes: $G_A \boldsymbol{\sigma}_i (\mathbf{k}\cdot\mathbf{r})^n$ with the consequence that there is a *unique forbidden spectrum* in these cases. As an example, consider the first forbidden transition with $\Delta J = \pm 2$ and with a change in the parity of the nuclear state. The nuclear matrix element including the dependence upon $\mathbf{k}$ is

$$\sum_a k_a \langle f_N | r_a \boldsymbol{\sigma} | i_N \rangle$$

This vector should then be inserted in place of $(\mathbf{M}_{GT})$ in (11.26) after which the appropriate averages over spins and lepton momenta can be performed in order to obtain the electron-energy spectrum. The term that contributes to the $\Delta J = 2$ transition *only* is the traceless tensor part of $r_a \sigma_b$. We therefore define*

$$\mathfrak{B}_{ab} \equiv \langle f_n | \tfrac{1}{2}(r_a \sigma_b + \sigma_b r_a) - \tfrac{1}{3}\delta_{ab} \mathbf{r}\cdot\boldsymbol{\sigma} | i_N \rangle \tag{17.25}$$

---

*$\langle f_N | r_a \sigma_b | i_N \rangle = \frac{1}{3} \langle f_N | \mathbf{r}\cdot\boldsymbol{\sigma}\, \delta_{ab} | i_N \rangle + \frac{1}{2} \langle f_N | r_a \sigma_b - r_b \sigma_a | i_N \rangle + \mathfrak{B}_{ab}$

The first term δ_{ab} transforms as a scalar, the second term as an axial vector $(r \times \boldsymbol{\sigma})_c$ where c is cyclically related to (a, b).

Thus for this transition with $\Delta J = 2$

$$\mathfrak{M}_{GT} = \mathbf{k}\cdot\overleftrightarrow{\mathfrak{B}} \tag{17.26}$$

Inserting into (11.26) and averaging over the direction of the neutrino emission it can be verified that the average of the cross term and that the average of the combination

$$\mathfrak{M}_{GT}\cdot\mathbf{p}_e\overset{*}{\mathfrak{M}}_{GT}\cdot\mathbf{p}_\nu + \mathfrak{M}_{GT}\cdot\mathbf{p}_\nu\overset{*}{\mathfrak{M}}_{GT}\cdot\mathbf{p}_e - \mathfrak{M}_{GT}\cdot\overset{*}{\mathfrak{M}}_{GT}\mathbf{p}_e\cdot\mathbf{p}_\nu$$

vanish leaving only the average of $\mathfrak{M}_{GT}\cdot\overset{*}{\mathfrak{M}}_{GT}$. If we now also average over the direction of the electron motion, we finally obtain for the average

$$\hbar^2\langle\mathfrak{M}_{GT}\cdot\overset{*}{\mathfrak{M}}_{GT}\rangle = \tfrac{1}{3}(p_e{}^2 + \bar{p}_\nu{}^2)\sum_{ab}\overset{*}{\mathfrak{B}}_{ab}\mathfrak{B}_{ab} \tag{17.27}$$

The important point here is the additional energy dependence contained in the factor

$$a_1 \equiv p_e{}^2 + \bar{p}_\nu{}^2 = [(\epsilon^2 - 1) + (\epsilon_0 - \epsilon)^2]m^2c^2 \tag{17.28}$$

which modifies the phase-space factor $(\epsilon_0 - \epsilon)^2(\epsilon)(\epsilon^2 - 1)^{1/2}$ of (2.9). Bear in mind that we are dealing with the $Z = 0$ limit. The energy dependence of (17.28) weights the high energy end of the electron spectrum and therefore is readily observable. The Kurie plot (see Eq. 2.16) will now be a straight line only if the Kurie function for allowed transitions (2.16) is replaced by

$$K_{\text{allowed}}/\sqrt{a_1}$$

The effect on the electron spectrum has been observed in a number of nuclei. The first clearly observed example, that of Langer and Price (49) is shown in Fig. 17.1.

Unique spectra with n greater than one have also been observed. The factor $k^2 = (p_e{}^2 + \bar{p}_\nu{}^2)$ obtained for $n = 1$ is just $(\mathbf{p}_e + \bar{\mathbf{p}}_\nu)^2$ averaged over the neutrino emission angle. For larger n it turns out [Konopinski (66)] that the factor modifying the allowed electron energy distribution is simply the similarly averaged value of

$$(\hbar k)^{2n} = (\mathbf{p}_e + \bar{\mathbf{p}}_\nu)^{2n}$$

Labeling the resulting averages by a_n one obtains

$$\begin{aligned} a_1 &= p_e{}^2 + \bar{p}_\nu{}^2 & \Delta J &= 2 \quad \text{(yes)} \\ a_2 &= p_e{}^4 + \tfrac{10}{3}p_e{}^2\bar{p}_\nu{}^2 + \bar{p}_\nu{}^4 & \Delta J &= 3 \quad \text{(no)} \\ a_3 &= p_e{}^6 + 7p_e{}^4\bar{p}_\nu{}^2 + 7p_e{}^2\bar{p}_\nu{}^4 + \bar{p}_\nu{}^6 & \Delta J &= 4 \quad \text{(yes)} \end{aligned} \tag{17.29}$$

For a further discussion of the empirical evidence for unique forbidden spectra, see Wu and Moskowski (66), Konopinski (66), and Weidenmuller (61). These authors also include the effect of the Coulomb electron–nuclear interaction on these spectra.

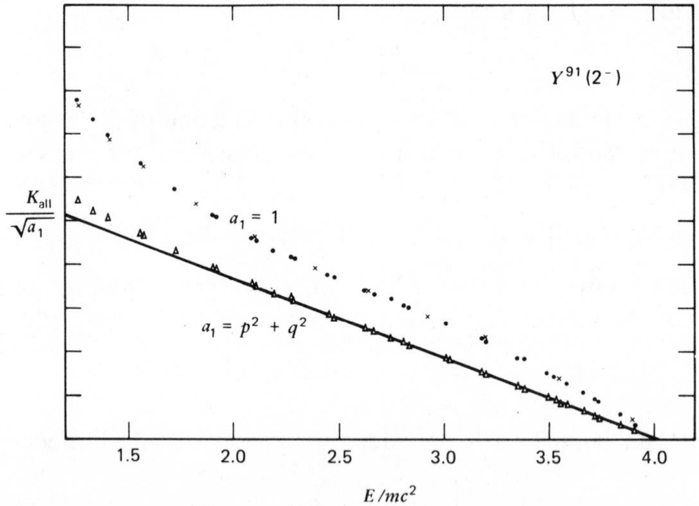

FIG. 17.1. Comparison between the predicted spectrum for a unique $\Delta J = 2$ (yes) transition in ^{91}Y and experiment [Langer and Price (49)]. The ordinate is K allowed/ $\sqrt{a_1}$ where a_1 is given by (17.29).

Coulomb Effect for First Forbidden Transitions

It is convenient at this point to mention another Coulomb effect that is important for the first forbidden $l = 1 \Delta J < 2$ decays. As was discussed earlier (p. 889) the estimate for $l = 1$ transition probability, $(E_0R/\hbar c)^2$ times the allowed transition probability, is correct only if the height of the angular momentum barrier is large compared to the Coulomb energy at approximately the nuclear radius. If the opposite condition holds: (this is the relativistic version for $l = 1$)

$$\frac{Ze^2}{R} \gg E_0$$

or

$$\frac{Ze^2}{\hbar c} \gg \frac{E_0R}{\hbar c} \tag{17.30}$$

the rough order of magnitude estimate for the transition probability mentioned above is no longer valid, the factor $(E_0R/\hbar c)$ being replaced by $Ze^2/\hbar c$ as discussed above (p. 889). In that event it can be shown that the electron spectrum for the first forbidden transition under discussion will have an allowed shape. Condition (17.30) is more likely to hold for large Z nuclei where for example the energy spectrum of electrons emitted in the decay of ^{198}Au is close to that of an allowed transition, the Kurie plot being nearly a straight line. For a large order of forbiddenness this effect is not important since the angular momentum barrier becomes eventually dominant. The Coulomb effect is important for $s_{1/2}$ and $p_{1/2}$ electrons but is already much reduced in importance for $p_{3/2}$ electrons, the principal type involved in the $n = 1$ unique first forbidden transition.

Forbidden β-Transitions and Electromagnetic Multipole Moments

A second point of interest calls attention to the contributions of the nuclear current $\hat{V}_\lambda$: the terms in (17.22) and (17.23) involving G_V and G_W. These terms are, except for the replacement of τ_3 by $\tau^{(+)}$, identical with the isovector electromagnetic current of Chapter VIII. This can be readily seen if one considers the case that g_V and g_W are constants equal to unity and $\frac{1}{2}(g_p - g_n)$, respectively. Then

$$G_V(\mathbf{r} - \mathbf{r}_i) = \delta(\mathbf{r} - \mathbf{r}_i)$$

$$G_W(\mathbf{r} - \mathbf{r}_i) = \tfrac{1}{2}(g_p - g_n)\,\delta(\mathbf{r} - \mathbf{r}_i)$$

Comparison can now be made with (VIII.2.3) and (VIII.2.4). First this weak interaction nucleon current satisfies the continuity equation. Second its reduced nuclear matrix elements are simply related to those of the electromagnetic current by a rotation in isospin space that rotates τ_3 into $\tau^{(+)}$. However the multiplying vector

$$\mathbf{L} = -\bar{u}_e\gamma_4(1 + \gamma_5)\boldsymbol{\sigma}u_{\bar{\nu}} \exp(-i\mathbf{k}\cdot\mathbf{r})$$

the analog of $\mathbf{A}$, the vector potential does not satisfy a condition analogous to the radiation gauge condition satisfied by $\mathbf{A}$, div $\mathbf{A} = 0$ (Eq. VIII.4.1). There are some longitudinal contributions. Hence it is not possible to take over entirely the results of Chapter VIII. However the $(\boldsymbol{\sigma}_i \times \boldsymbol{\nabla}_i)(G_W + G_V)$ term is divergenceless. For this term, only the divergenceless part of $\mathbf{L}$ can contribute to the transition amplitude. In this case it is thus possible to obtain the contributing components of the nuclear matrix element from the corresponding electromagnetic matrix elements after replacing τ_3 by $\tau^{(+)}$. Finally the transition probability involving these terms, summed over the electron and neutrino spins is similarly related to the square of the reduced matrix element of the isovector component of the magnetic multipole operator (Eq. VIII.5.38).

The remaining terms in (17.22) and (17.23) combined with divergenceless part of $\mathbf{L}$

$$-\bar{u}_e\gamma_4(1 + \gamma_5)[\boldsymbol{\sigma} - \hat{\mathbf{k}}(\boldsymbol{\sigma}\cdot\hat{\mathbf{k}})]u_{\bar{\nu}} \exp(-i\mathbf{k}\cdot\mathbf{r})$$

will yield results analogous (i.e., $\tau_3 \rightarrow \tau^{(+)}$) to those obtained in the electromagnetic case as given by the isovector term of the electric multipole operator (8.5.37). However the longitudinal component

$$-\bar{u}_e\gamma_4(1 + \gamma_5)\hat{\mathbf{k}}(\sigma\cdot\hat{\mathbf{k}})u_{\bar{\nu}} \exp(-i\mathbf{k}\cdot\mathbf{r})$$

will remain. This term will then need to be expanded in the longitudinal modes

$$\mathbf{A}_{lm}{}^{(L)} \equiv \frac{1}{k}\,\mathrm{grad}\, j_l(kr)Y_{lm}(\hat{\mathbf{k}}) \tag{17.31}$$

which together with $\mathbf{A}_{lm}{}^{(M)}$ and $\mathbf{A}_{lm}{}^{(E)}$ of (VIII.4.4a) and (VIII.4.4b) form a complete set of vectors. We shall not carry out this analysis here. Detailed results for forbidden transitions are given by Konopinski (66).

Siegert's Theorem for the Weak Interactions

An analog of Siegert's theorem (see discussion following Eq. VIII.3.1) can be formulated, with an important and interesting difference from the electromagnetic case because of the differing isospin dependence. We recall the argument of Chapter VIII. In that discussion, the approximation

$$\boldsymbol{\varepsilon} \exp(i\mathbf{k}\cdot\mathbf{r}) \simeq \mathrm{grad}\,[(\boldsymbol{\varepsilon}\cdot\mathbf{r})\exp(i\mathbf{k}\cdot\mathbf{r})] + O(kR) \tag{17.32}$$

was made. In the present circumstance $\boldsymbol{\varepsilon}$ is replaced by

$$\mathbf{l} \equiv \bar{u}_n\gamma_4(1 + \gamma_5)\boldsymbol{\sigma}u_p \tag{17.33}$$

so that

$$\mathbf{L} = \mathbf{l}\exp(-i\mathbf{k}\cdot\mathbf{r})$$

Replacing the $\tau^{+}(i)$ dependence by using the commutation relation

$$\tau^{(+)}(i) = -\tfrac{1}{2}[T^{(+)}, \tau_3(i)] \tag{17.34}$$

permits the direct introduction of the isovector part of the *electromagnetic* nucleon current $\mathbf{j}_{el}{}^{(N)}$ including the τ_3 dependence. Hence the $\mathfrak{M}_{GT}$ term in (17.21) becomes

$$-\tfrac{1}{2}\sum_i \langle f_N| \left[T^{(+)}, \int d\mathbf{r}[\mathbf{l}\cdot\mathbf{j}_{el}{}^{(N)} - \sum_i G_A\mathbf{l}\cdot\boldsymbol{\sigma}_i\tau_3(i)]\exp(-i\mathbf{k}\cdot\mathbf{r})\right]|i_N\rangle \tag{17.35}$$

Consider only the term depending on $\mathbf{j}_{el}{}^{(N)}$:

$$M_{fi} \equiv -\tfrac{1}{2}\langle f_N| \left[T^{(+)}, \int d\mathbf{r}\,\mathbf{l}\cdot\mathbf{j}_{el}{}^{(N)}\exp(-i\mathbf{k}\cdot\mathbf{r})\right]|i_N\rangle \tag{17.36}$$

Employing approximation (17.32) in the form

$$\mathbf{l} \exp(-i\mathbf{k}\cdot\mathbf{r}) \simeq \text{grad } [\mathbf{l}\cdot\mathbf{r} \exp(-i\mathbf{k}\cdot\mathbf{r})]$$

integrating by parts and using the continuity equation yields

$$M_{fi} = -\tfrac{1}{2}\langle f_N | \left[T^{(+)}, \int d\mathbf{r} \left(\frac{\partial}{\partial t}\rho_{el}{}^{(N)}\right) \mathbf{l}\cdot\mathbf{r} \exp(-i\mathbf{k}\cdot\mathbf{r}) \right] | i_N \rangle \tag{17.37}$$

Inserting

$$\frac{\partial \rho}{\partial t} = \frac{i}{\hbar}[H_N, \rho]$$

where H_N is the nuclear Hamiltonian yields

$$M_{fi} = -\frac{i}{2\hbar}\langle f_N | \int d\mathbf{r}\, [T^{(+)}, [H_N, \rho_{el}{}^{(N)}]]\mathbf{l}\cdot\mathbf{r} \exp(-i\mathbf{k}\cdot\mathbf{r}) | i_N \rangle$$

To evaluate the double commutator we first employ the Jacobi identity:

$$[T^+, [H_N, \rho_{el}{}^{(N)}]] = -[H_N, [\rho_{el}{}^{(N)}, T^+]] - [\rho_{el}{}^{(N)}, [T^+, H_N]] \tag{17.38}$$

The matrix element of the first term on the right-hand side of (17.38) can be reduced to the matrix element of the commutator $[\rho_{el}{}^{(N)}, T^{(+)}]$:

$$\langle f_N | -[H_N, [\rho_{el}{}^{(N)}, T^+]] | i_N \rangle = (E_i - E_f)\langle f_N | [\rho_{el}{}^{(N)}, T^+] | i_N \rangle \tag{17.39}$$

This term is in the form expected from the electromagnetic Siegert's theorem. The second term on the right-hand side of (17.38) is new and is present only in virtue of the isospin symmetry breaking terms in the nuclear Hamiltonian such as the Coulomb interaction and the neutron–proton mass difference. Denoting this interaction by H_c we have

$$[H_N, T^{(+)}] = [H_c, T^{(+)}] \tag{17.40}$$

a commutator that is familiar from the theory of isobar analog resonances. In order to write the second term as a correction to the first it has been customary to make the following approximation. Using identity (17.38) once again we have

$$[\rho_{el}{}^{(N)}, [T^+, H_c]] = -[T^+, [H_c, \rho_{el}{}^{(N)}]] - [H_c, [\rho_{el}{}^{(N)}, T^+]]$$

The commutator for $[\rho, H_c]$ is usually taken to vanish since both H_c and ρ are assumed to depend only upon $\tau_3(i)$. The matrix element of the first term is then customarily [Ahrens and Feenberg (52)] approximated by assuming that H_c is diagonal. Hence

$$\begin{aligned}\langle f_N | [\rho_{el}{}^{(N)}, [T^{(+)}, H_c]] | i_N \rangle &\simeq -\langle f_N | [H_c, [\rho_{el}{}^{(N)}, T^{(+)}]] | i_N \rangle \\ &\simeq [\langle i_N | H_c | i_N \rangle - \langle f_N | H_c | f_N \rangle] \\ &\qquad \times \langle f_N | [\rho_{el}{}^{(N)}, T^{(+)}] | i_N \rangle\end{aligned}$$

Finally

$$M_{fi} = -\frac{i}{2\hbar}\int d\mathbf{r}\,\langle f_N|\,[\rho_{el}, T^{(+)}]\,|i_N\rangle\{(E_i - E_f) - \langle i_N|H_c|i_N\rangle + \langle f_N|H_c|f_N\rangle\} \tag{17.41}$$

The curly bracket can be easily evaluated in terms of the neutron–proton mass difference and the Coulomb energy to yield

$$-\frac{i}{2\hbar}\{E_0 - [M_N - M_P] + \Delta E_{\text{coul}}\} \equiv -\frac{i\delta}{2\hbar} \qquad \beta^- \text{ decay} \tag{17.42}$$

defining δ and where ΔE_{coul} is the increase in Coulomb energy that occurs because of the replacement of neutron by a proton. The main difference term and this Coulomb term will have an opposite sign for positron emission.

With this approximation the β-decay analog of Siegert's theorem becomes: the matrix element M_{fi} defined by (17.36) can be calculated to $O(kR)$ by making the following replacement

$$-[T^{(+)}, \mathbf{l}\cdot\mathbf{j}_{el}^{(N)}] \to -\frac{i\delta}{\hbar}[T^{(+)}, \rho_{el}]\mathbf{l}\cdot\mathbf{r}$$

Inserting the current that contributes to $\partial\rho/\partial t$ via the continuity equation we obtain

$$\mathbf{l}\cdot\sum_i \langle f_N|\tau^{(+)}(i)\int d\mathbf{r}\,\frac{1}{2Mc}(\mathbf{p}_i G_V + G_V\mathbf{p}_i)\exp(-i\mathbf{k}\cdot\mathbf{r})|i_N\rangle$$

$$\to -\frac{\delta}{\hbar}\mathbf{l}\cdot\sum_i \langle f_N|\tau^{(+)}(i)\int d\mathbf{r}G_V\mathbf{r}\exp(-i\mathbf{k}\cdot\mathbf{r})|i_N\rangle \qquad \beta^- \text{ decay} \tag{17.43}$$

The deviation from the CVC result ($\delta = E_0$) is a consequence of the breaking of isospin symmetry. In the case of β-decay it is one of these corrections, that resulting from the nuclear coulomb interaction, which dominates δ. This point and others are discussed by Wu and Moskowski (66) and Konopinski (66).

We shall stop our discussion of forbidden nuclear β-decay at this point. We have chosen to discuss the specially important unique forbidden transitions and those terms that are significantly related to electromagnetic transitions. The latter transitions do not have unique nuclear matrix elements. Contributions originate not only in the vector and weak magnetism terms but also in the axial vector interactions indicated by the form factor G_A. To obtain a complete expansion for $\mathfrak{M}_{GT}$ in multipole moments is not difficult. The results for the "magnetization" term ($\sim\boldsymbol{\sigma}_i \times \boldsymbol{\nabla}_i$) have been given in terms of the isovector component of the magnetic multipole operator. The remaining dependence on G_V has been transformed via Siegert's theorem (17.43) into a form that can be readily expanded in terms of spherical harmonics $Y^*_{lm}(\hat{\mathbf{r}})$. The axial component of $\mathfrak{M}_{GT}$ is also simply treated requiring only the decomposition of $\boldsymbol{\sigma} Y^*_{lm}(\hat{\mathbf{r}})$ into spherical tensors. Turning now to $\mathfrak{M}_F$ the term in G_V will upon expansion of the exponential involve simply the spherical harmonics $Y^*_{lm}(\hat{\mathbf{r}})$. It is most convenient to use the same expansion for the remaining terms in $\mathfrak{M}_F$ since $\boldsymbol{\sigma}_i \cdot \mathbf{p}_i$ is a scalar. We shall not give the rather complicated results here. Detailed expressions are given by Konopinski (66) who includes the effects of the electron–nucleus Coulomb interaction that have been omitted in this section's discussion.

The reader is referred to Konopinski (66, Chapter 8), Wu and Moskowski (66, Chapter 3) and Weidenmuller (61) for a discussion of the experimental material and the insight into the validity of nuclear models it provides.

18. MU ABSORPTION BY NUCLEI

When a μ^- is stopped in matter it is possible for the muon to be captured into an atomic orbit forming a μ-mesic atom. Then it undergoes a series of transitions to more deeply bound orbits accompanied either by x-ray emission or ejection of atomic electrons, this process stopping when the muon is in the K-orbit.* During this process it becomes increasingly possible for the muon to be captured by the nucleus via the elementary process

$$\mu^- + p \to n + \nu_\mu \tag{18.1}$$

which for nuclear capture becomes

$$\mu^- + A(Z, N) \to A(Z - 1, N + 1) + \nu_\mu \tag{18.2}$$

The probability for this process reaches its maximum when the muon is in the K-orbit. Actually this orbit is within the interior of a medium weight and heavier nuclei.† It is the probability for absorption from the K-orbit that we

*This process will be described in more detail in Volume II where the relation between the observed energies of the μ-mesic atoms and the charge distribution of the nucleus will be described and comparison made with results obtained from the elastic scattering of electrons by nuclei.

†The Bohr radius for a muon moving in the Coulomb field of a nucleus of atomic nucleus Z is

$$\frac{\hbar^2}{Zm_\mu e^2} = \frac{m_e}{Zm_\mu} \cdot (0.53 \times 10^{-8}\ \text{cm}) \sim \frac{2.6}{Z} \times 10^{-11}\ \text{cm}$$

Taking the nuclear radius to be $\sim(1.2\ 10^{-13}\ \text{cm})\ A^{1/3}$ the ratio of the muon Bohr radius to the nuclear radius is $\sim 200/ZA^{1/3}$ which equals one for medium weight nuclei becoming less than unity for heavier nuclei.

will consider here, although capture from larger orbits when compared to K-orbit capture can be related to the relative matter distribution at the outer edges of the nucleus.

The elementary decay process

$$\mu^- \rightarrow e^- + \bar{\nu}_e + \nu_\mu \tag{18.3}$$

competes with nuclear capture. The issue as to which occurs first depends on the ratio of the lifetime of the free muon (2.2×10^{-6} sec) to the sum of the times for the atomic cascade process and the lifetime against capture by the nucleus. This sum decreases rapidly with Z so that nuclear capture dominates for $Z \gtrsim 11$ while decay (18.3) is the more important process for $Z < 11$. However even capture in hydrogen is present and has been observed, the capture rate being 1/1000 of the spontaneous decay rate.

Nuclear μ^- capture is analogous to K capture of atomic electrons (Section 6). There are however important energetic differences because of the greater mass of muon (106.6599 ± 0.0014) MeV. As a consequence there is a large energy release so that the residual nucleus can be left in a variety of excited states. In the elementary process (18.1) we find from momentum and energy balance that the neutrino has an energy of about 100 MeV while the neutron carries off a kinetic energy of about 5 MeV. In a nucleus, the average excitation energy can be as much as $\sim$15 to 20 MeV, an estimate obtained using the Fermi-gas model of the nucleus. Not only are more nuclear levels excited than in the case of electron orbital capture but also the excitation energy range includes the giant dipole state (see Fig. VIII.11.2), a matter of considerable importance as we shall see.

Qualitative Considerations for Heavy Nuclei

The capture rate particularly by heavy nuclei has been studied extensively by Primakoff (59) and collaborators [e.g., Fujii (59)], by Tolhoek [e.g., Luyten et. al. (63)] and Telegdi (62). For a recent review see Walecka (73). The Primakoff formula as far as its dependence on atomic and mass number can be made plausible as follows. The capture rate (see Eq. 6.10, for the capture rate for orbital electrons) will be proportional to the muon density within the nucleus times the square of the matrix element describing the nuclear transition. The first of these factors is for a hydrogenic wave function proportional to Z^3. We now estimate the Z- and A-dependence of the nuclear factor. Note that the momentum transfer to the nucleus is approximately equal to the neutrino momentum 100 MeV/c since the momentum of the captured muon is relatively small. The corresponding wavelength/2π, $\lambda\!\!\!^{-}$, is of the order of 2 fm. Because this distance is comparable to the distance between the nucleons in the nucleus, it is unlikely that any appreciable part of the transition probability involves a major fraction of the nucleus acting coherently. Most of the transition probability originates in the interaction of the muon with a single proton. Hence the transition probability will simply be proportional to, Z, the number of protons.

This result must however be modified by the effects of the Pauli principle. We have discussed this point earlier in Section 5 in connection with nuclear β-decay. In the present context some of the possible states for the neutron produced by muon absorption are not allowed because they are occupied. Roughly speaking we would expect that this Pauli effect would be proportional to the fraction, number of neutrons in the nucleus, $(A - Z)/A$ so that the nuclear factor in the capture rate instead of being proportional to Z as far as its dependence on A and Z is concerned is proportional to

$$Z\left(1 - \delta\frac{A - Z}{2A}\right)$$

Here δ is a constant of proportionality for the Pauli effect. Roughly speaking, it measures the degree of overlap of the produced neutron wave function and the original nucleon neutrons. It is assumed to be relatively insensitive to Z and A for the heavy nuclei. Combining leads to the following for the transition probability per unit time:

$$w \sim Z^4\left(1 - \frac{A - Z}{2A}\,\delta\right)$$

Finally we recall that for heavy nuclei, the muon K-orbit is inside the nucleus so that the Coulomb potential it sees is generated by a charge Z_{eff}, which is less than Z. The final formula is then

$$w \sim Z_{\text{eff}}^3 Z\left(1 - \frac{A - Z}{2A}\,\delta\right) \tag{18.4}$$

Plausible values for Z_{eff}^3 have been calculated by Sens (58). This result can be tested by plotting the experimental capture rate/$Z_{\text{eff}}^3\,Z$ against $(A - Z)/2A$. The result should be a straight line. This plot is shown in Fig. 18.1. We see that the major effects in the dependence of w on Z and A are well represented by Primakoff's formula (18.4). Primakoff also gives the constant of proportionality in (18.4) and calculates δ, values that compare favorably with the values obtained from Fig. 18.1. From the figure $\delta \sim 3$ and the constant of proportionality is $\sim 10^{-28}$ cm^3/sec.

Quantitative Theory of Muon Capture

We turn now to more formal considerations with the view in mind of constructing a theory for muon capture by atomic nuclei. It will differ from the theory of nuclear β-decay described in Section 17 in the nature of the approxi-

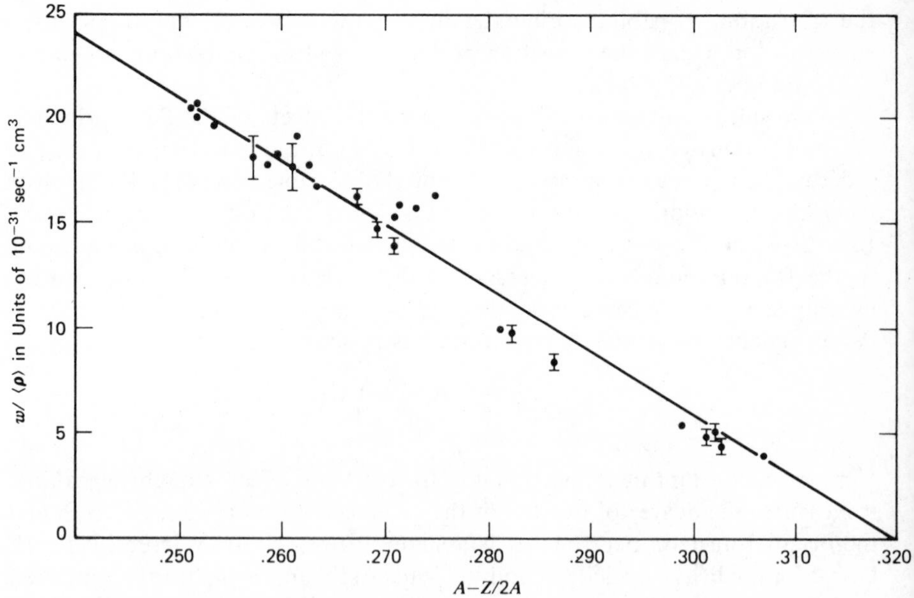

FIG. 18.1. Comparison between experimental muon (capture rate)/$Z^3_{\text{eff}}Z$ and Primakoff theory [Taken from Telegdi (63) Telegdi's $\langle\rho\rangle = Z|\phi|^2{}_{AV}$ where ϕ is the atomic muon wave function; $\langle\rho\rangle \sim Z^3_{\text{eff}}Z$.]

mations employed. In both cases the motion of the nucleons is nonrelativistic so that the reduction carried out in Section 17 to nonrelativistic variables can be used here. There is one exception: the induced pseudoscalar term is to be kept for the present problem. The reader will recall from Section 14, (see discussion following Eq. 14.16) that the induced pseudoscalar coupling constant g_P is more than six times the axial coupling constant g_A. This combined with the relatively large energy release makes this term appreciable. When we turn to the lepton amplitudes, it is of course necessary to use the appropriate atomic wave function for the muon rather than the plane wave employed in the expressions for $\mathfrak{M}_F$ and $\mathfrak{M}_{GT}$ of (17.22) and (17.23). It is however possible for most nuclei to use a nonrelativistic muon wavefunction, a procedure we shall sometimes adopt.

There is one other approximation which is traditionally employed in this field. Instead of carrying out the transformations outlined on page 883, effective coupling constants are used by evaluating g_V, g_A, etc., at a particular value of the nucleon momentum transfer, k^2. In μ-capture, if one neglects the initial muon momentum, k^2 is given by

$$k^2 = 2m_\mu c p_\nu - (m_\mu c)^2$$

Using an average neutrino momentum estimated to be about $0.8m_\mu c$ by Primakoff (59) and Luyten et al. (63)

$$k^2 \approx 0.6(m_\mu c)^2 \tag{18.5}$$

The various constants in (17.1) are then evaluated at this value of k^2

$$g'_V = 1 - \tfrac{1}{6}k^2\langle r_p^{\,2}\rangle = 0.98$$

$$g'_A \simeq g_A(0) = 1.2$$

$$\frac{1}{2Mc}g'_P = \frac{2Mc}{k^2 + m_\pi^{\,2}c^2}g_A = \frac{7.5}{m_\mu c}g_A(0) \tag{18.6}$$

$$\frac{g'_W}{2Mc} = \frac{\kappa_P - \kappa_n}{2Mc}g'_V = \frac{3.7}{2Mc}$$

Because of this approximation the forms (17.7) and (17.9) are no longer appropriate. Returning to (17.11) and employing Table 17.1 one obtains

$$I_4(\mathbf{r}) = \delta(\mathbf{r})\left\{g'_V - \frac{1}{2Mc}g'_A\boldsymbol{\sigma}\cdot(\overleftarrow{\mathbf{p}} + \overrightarrow{\mathbf{p}}) + \frac{i}{(2Mc)^2}g_Pk_4\boldsymbol{\sigma}\cdot(\overleftarrow{\mathbf{p}} - \overrightarrow{\mathbf{p}})\right.$$

$$\left. + \frac{1}{(2Mc)^2}g'_W\mathbf{k}\cdot[\overleftarrow{\mathbf{p}} - \overrightarrow{\mathbf{p}} + i\boldsymbol{\sigma}\times(\overleftarrow{\mathbf{p}} + \overrightarrow{\mathbf{p}})]\right\} \tag{18.7}$$

and

$$\mathbf{I}(\mathbf{r}) = \delta(\mathbf{r})\left\{\frac{-ig'_V}{2Mc}[\overleftarrow{\mathbf{p}} + \overrightarrow{\mathbf{p}} + i\boldsymbol{\sigma}\times(\overleftarrow{\mathbf{p}} - \overrightarrow{\mathbf{p}})] + ig'_A\boldsymbol{\sigma}\right.$$

$$+ \frac{i}{(2Mc)^2}g'_P\mathbf{k}[\boldsymbol{\sigma}\cdot(\overleftarrow{\mathbf{p}} - \overrightarrow{\mathbf{p}})] - \frac{g'_W}{2Mc}(\boldsymbol{\sigma}\times\mathbf{k})$$

$$\left. - \frac{g'_Wk_4}{(2Mc)^2}[\overleftarrow{\mathbf{p}} - \overrightarrow{\mathbf{p}} + i\boldsymbol{\sigma}\times(\overleftarrow{\mathbf{p}} + \overrightarrow{\mathbf{p}})]\right\} \tag{18.8}$$

The operator $\overleftarrow{\mathbf{p}}$ is replaced by $\mathbf{p}_n$ and $\overrightarrow{\mathbf{p}}$ by $\mathbf{p}_p$. Remember we are considering μ^- capture so that the final nucleon is a neutron, the original a proton. Neglecting the momentum of the muon

$$\overleftarrow{\mathbf{p}} - \overrightarrow{\mathbf{p}} = \mathbf{p}_n - \mathbf{p}_p = -\mathbf{p}_\nu \tag{18.9}$$

while

$$\overleftarrow{\mathbf{p}} + \overrightarrow{\mathbf{p}} = 2\mathbf{p}_p - \mathbf{p}_\nu \tag{18.10}$$

Note that $k_\lambda = (p_\nu)_\lambda$. The nonrelativistic approximation may also be used for the muon wave function. If we write

$$\psi_\mu = \begin{pmatrix} u_\mu \\ 0 \end{pmatrix} \tag{18.11}$$

and

$$\psi_\nu = \frac{1}{\sqrt{2}}\begin{pmatrix} u_\nu \\ \boldsymbol{\sigma}\cdot\mathbf{p}_\nu u_\nu \end{pmatrix} \tag{18.12}$$

it follows that

$$L_4 \equiv \bar{\psi}_\nu\gamma_4(1+\gamma_5)\psi_\mu \rightarrow \frac{1}{\sqrt{2}}u_\nu^*(1-\boldsymbol{\sigma}\cdot\hat{\mathbf{p}}_\nu)u_\mu \tag{18.13}$$

and

$$\mathbf{L} \equiv \bar{\psi}_\nu\boldsymbol{\gamma}(1+\gamma_5)\psi_\mu \rightarrow \frac{i}{\sqrt{2}}u_\nu^*(1-\boldsymbol{\sigma}\cdot\hat{\mathbf{p}}_\nu)\vec{\boldsymbol{\sigma}}u_\mu \tag{18.14}$$

These then are to be combined with (18.7) and (18.8) according the formula for H given by (17.5). The result to be given below is obtained after a number of simple manipulations. One example is furnished by combining the two terms

$$g'_V L_4 + i\frac{g'_V}{2Mc}\mathbf{p}_\nu\cdot\mathbf{L}$$

The $\mathbf{p}_\nu\cdot\mathbf{L}$ contribution comes from the $\overleftarrow{\mathbf{p}} + \overrightarrow{\mathbf{p}}$ terms in (18.8) and (18.10). Inserting (18.14) this term becomes

$$-\frac{g'_V}{2Mc}u_\nu^*\frac{(1-\boldsymbol{\sigma}\cdot\hat{\mathbf{p}}_\nu)}{\sqrt{2}}\boldsymbol{\sigma}\cdot\mathbf{p}_\nu u_\mu = \frac{g'_V}{2Mc}p_\nu u_\nu^*\frac{(1-\boldsymbol{\sigma}\cdot\hat{\mathbf{p}}_\nu)}{\sqrt{2}}u_\mu$$

so that

$$g'_V L_4 + i\frac{g'_V}{2Mc}\mathbf{p}_\nu\cdot\mathbf{L} \rightarrow g'_V\left(1+\frac{p_\nu}{2Mc}\right)L_4$$

The total result obtained by Foldy and Walecka (64) is:

$$H_{\text{eff}} = \frac{G_\mu}{\sqrt{2}}\sum_j \tau^{(-)}(j)\left\{L_4(r_j)\Gamma_V + i\mathbf{L}(\mathbf{r}_j)\cdot\left[\Gamma_A\boldsymbol{\sigma}(j) - \Gamma_P\hat{\mathbf{p}}_\nu(\boldsymbol{\sigma}(j)\cdot\hat{\mathbf{p}}_\nu)\right.\right.$$
$$\left.\left. + g'_V\mathbf{p}_j/Mc - g'_A\hat{\mathbf{p}}_\nu\frac{\boldsymbol{\sigma}(j)\cdot\mathbf{p}_j}{Mc}\right]\right\} \tag{18.15}$$

The matrix elements of this Hamiltonian taken between the initial and final states gives the matrix element in a non-relativistic approximation defined above for the absorption of a muon by a nucleus with the emission of a μ-neutrino. The spatial dependence of $L_\lambda(\mathbf{r}_j)$ is given by

$$\exp(-i\mathbf{p}_\nu\cdot\mathbf{r}_j)\phi_\mu(\mathbf{r}_j)$$

when ϕ_μ is the muon atomic wave function evaluated as the position of the absorbing nucleon, $\mathbf{r}_j$. The constants Γ_V etc. are

$$\Gamma_V = g'_V\left(1 + \frac{p_\nu}{2Mc}\right)$$

$$\Gamma_A = \left\{g'_A + \frac{p_\nu}{2Mc}\,[g'_V + g'_W]\right\}$$

$$\Gamma_P \simeq \frac{p_\nu}{2Mc}\,\{7.5g'_A - g'_A + g'_V + g'_W\} \tag{18.16}$$

It is now relatively simple to calculate the transition rate in which the capturing nucleus goes from state $|i_N\rangle$ to state $|f_N\rangle$. Two simplifications are often made (i) the average value of the atomic muonic density over the volume of the nucleus is used to take this quantity out of the matrix element and (ii) nucleon recoil terms of the order of p/Mc are neglected. The total transition rate to all possible final states is then

$$w \simeq \frac{|\phi_\mu|^2{}_{AV}G_\mu{}^2}{2\pi\hbar^2 c\lambda\!\!\!^{-}_\mu{}^2}\,[\Gamma_V{}^2M_V{}^2 + 3\Gamma_A{}^2M_A{}^2 + (\Gamma_P{}^2 - 2\Gamma_P\Gamma_A)M_P{}^2] \tag{18.17}$$

where

$$\lambda\!\!\!^{-}_\mu = \frac{\hbar}{m_\mu c}$$

and from (3.1)

$$\frac{G_\mu{}^2}{2\pi\hbar^2 c\lambda\!\!\!^{-}_\mu{}^2} = \frac{10^{-10}}{2\pi}\left(\frac{m_\mu}{M}\right)^2\left(\frac{\hbar}{Mc}\right)^2 c$$

The squared matrix elements $M_V{}^2$, $M_A{}^2$, and $M_P{}^2$ are

$$M_V{}^2 = \frac{1}{2J_i + 1}\sum_{M_i}\sum_f (k_{if}\lambda\!\!\!^{-}_\mu)^2\int\frac{d\hat{\mathbf{k}}_{if}}{4\pi}\,|\langle f_N|\sum_j \tau^{(\mp)}_j \exp(-i\mathbf{k}_{if}\cdot\mathbf{r}_j)|i_N\rangle|^2 \tag{18.18}$$

$$M_A{}^2 = \frac{1}{3(2J_i + 1)}\sum_{M_i}\sum_f (k_{if}\lambda\!\!\!^{-}_\mu)^2$$

$$\times\int\frac{d\hat{\mathbf{k}}_{if}}{4\pi}\,|\langle f_N|\sum_j \tau^{(\mp)}_j\boldsymbol{\sigma}_j \exp(-i\mathbf{k}_{if}\cdot\mathbf{r}_j)|i_N\rangle|^2 \tag{18.19}$$

$$M_P{}^2 = \frac{1}{2J_i + 1}\sum_{M_i}\sum_f (k_{if}\lambda\!\!\!^{-}_\mu)^2\int\frac{d\hat{\mathbf{k}}_{if}}{4\pi}\,|\langle f_N|\sum_j \tau^{(\mp)}_j\hat{\mathbf{p}}_\nu\cdot\boldsymbol{\sigma}_j \exp(-i\mathbf{k}_{if}\cdot\mathbf{r}_j)|i_N|\rangle^2 \tag{18.20}$$

Here

$$\mathbf{k}_{if} = \mathbf{p}_\nu/\hbar; \qquad k_{if} = \frac{1}{\hbar c}\,[m_\mu c^2 - (E_f - E_i)] \tag{18.21}$$

where the binding energy of atomic muon has been neglected.

Because of the large number of final states that are excited by muon capture, one can in a very approximate way consider $\boldsymbol{\sigma}_j$ in these formulas to be classical and that all directions of these vectors are equally probable. Under these circumstances

$$M_V{}^2 \simeq M_A{}^2 \simeq M_P{}^2 \tag{18.22}$$

The proof of these results for doubly closed-shell nuclei in the shell model or in the closure approximation in which the set $|f_N\rangle$ is assumed to be complete has been given by Luyten et al. (63) [for additional references see Walecka's review (73)]. Inserting this rough relation into (18.17) yields

$$w \simeq \frac{|\phi_\mu|^2{}_{AV} G_\mu{}^2}{2\pi\hbar^2 c \lambda\!\!\!^-{}_\mu{}^2}\,[\Gamma_V{}^2 + 2\Gamma_A{}^2 + (\Gamma_P - \Gamma_A)^2]M_V{}^2 \tag{18.23}$$

From this point on calculations, for example those of Tolhoek et al. (64), have for the most part proceeded through the use of models. By comparing (18.23) with experiment the induced pseudoscalar coupling constant g_P can be obtained and then compared with the predictions of PCAC. This comparison [Quaranta et. al. (69)] was referred to in Section 14.

The Matrix Element $\mathbf{M}_V$ and the Giant Dipole Resonance

The preceding discussion is appropriate for the heavier nuclei. At the average excitation energy the density of levels for these nuclei is sufficiently great that it becomes possible to use the statistical arguments leading to (18.22). It is thus not expected for the detailed properties of the levels to play an important or particularly visible role. This is not quite so true for the light nuclei for which the levels are considerably sparser. It may then become possible to observe transitions to particular final states and thus obtain the averaged momentum transfer ($\mathbf{k}_{if}$) dependence of the vector, axial and pseudoscalar square matrix elements. Allowed β-decay gives these values for nearly zero momentum transfer. In addition the excitation of levels with "unnatural" parity, 0^-, 1^+, 2^-, etc., becomes possible via the axial current, as is seen from expression (18.19) for $|M_A|^2$. [For recent reviews see Uberall (66) and Walecka (73).]

Observation of the transition to a given level is however difficult because generally these levels are not only unstable against gamma-ray emission but also against the emission of particles such as neutrons. The levels are thus quite broad and overlap so that it becomes a nontrivial problem to disentangle the observed particle or gamma-ray spectrum. However as T. Ericson pointed out [the first published works are those of Foldy and Walecka (64) and of Balashov, Beliaev, Eramjian and Kabachnik (64); for later references see Uberall (66) and Walecka (73)] transitions to giant resonance should dominate and may be more visible. This suggestion was stimulated by calculations of Luyten, Rood and Tolhoek (63) for the muon capture rate in ^{16}O and ^{40}Ca. These authors used a shell-model description of these nuclei adjusting the single-particle potential so as to obtain the charge radius as given by electron scattering and the average separation energy of the last nucleon. Their results gave a muon capture rate by ^{16}O which is approximately twice the experimental value. In addition, and this is most important, they found that the capture process in ^{16}O was dominated by the transition to the 1^- states in ^{16}N. This is expected since these states occur in the excitation region of roughly 20 MeV, the average excitation energy following μ^- capture. But these are just the giant dipole states (the isospin partner of the well known state in ^{16}O). These are collective states whose properties cannot be predicted by the simple shell model. It is necessary to take the residual nucleon–nucleon interaction into account (Section VI.13). Indeed, Barlow et. al. (64) showed by shifting the energy of the dipole states calculated by Luytens et al. (63) to their greater experimental value that the discrepancy between the calculated and observed capture rates could be removed.

As Foldy and Walecka (64) point out these results strongly suggest that it would be useful to relate the muon capture matrix elements from the ground state of ^{16}O to the giant dipole states in ^{16}N to the dipole transition from the ground state in ^{16}O to the corresponding giant dipole state in ^{16}O. In the following we shall sketch how these authors obtained that relation and the consequent comparison with experiment.

Their starting point is (18.23). They consider nuclei with $T = 0$, $T_3 = 0$ and assume isospin conservation. The relation to an electro-magnetic process involves the same device (17.34) as that used in the derivation of Siegert's theorem for β-decay. In this case we use

$$\tau_j^{(-)} = \tfrac{1}{2}[T^{(-)}, \tau_{3j}] \tag{18.24}$$

Inserting this result into (18.18), one obtains

$$M_V^2 = \frac{1}{2J_i + 1} \sum_{M_i} \frac{1}{4} \sum_f (k_{if}\lambda\!\!\!{}^-_\mu)^2 \int \frac{d\hat{\mathbf{k}}_{if}}{4\pi} \left| \langle f_N | \sum_j [T^{(-)}, \tau_{3j}] \right.$$
$$\left. \times \exp(-i\mathbf{k}_{if}\cdot\mathbf{r}_j) | i_N \rangle \right|^2$$

Writing the commutator out, inserting a complete set of states, and noting that ($T = 0$ for $|i_N\rangle$)

$$T^{(\pm)} | i_N \rangle = 0$$

we obtain

$$M_V^2 = \frac{1}{2J_i + 1} \sum_{M_i} \sum_{ff'f''} \frac{1}{4} (k_{if}\lambda_\mu)^2 \int \frac{d\hat{\mathbf{k}}_{if}}{4\pi} \langle i_N | \sum_j \tau_{3j} \exp (i\mathbf{k}_{if}\cdot\mathbf{r}_j) | f' \rangle$$

$$\cdot \langle f' | T^{(+)} | f_N \rangle \langle f_N | T^{(-)} | f'' \rangle \langle f'' | \sum_j \tau_{3j} \exp (-i\mathbf{k}_{if}\cdot\mathbf{r}_j) | i_N \rangle$$

From the properties of $T^{(\pm)}$, $|f'\rangle$, $|f_N\rangle$, $|f''\rangle$ must belong to the same T-state. Since the sums in the first and last factor transform like a vector under rotations in isospin space and the isospin state of i_N is $T = 0$, $T_3 = 0$, it follows that the isospins of $|f'\rangle$ and $|f''\rangle$ are unity. Moreover the isospin sum over the product

$$\langle f' | T^{(+)} | f_N \rangle \langle f_N | T^{(-)} | f'' \rangle$$

yields $2\delta_{f'f''}$. Hence

$$M_V^2 = \frac{1}{2J_i + 1} \sum_{M_i} \sum_{f'} \tfrac{1}{2} (k_{if}\lambda_\mu)^2 \int \frac{d\hat{\mathbf{k}}_{if}}{4\pi} |\langle f' | \sum_j \tau_{3j} \exp (-i\mathbf{k}_{if}\cdot\mathbf{r}_j) | i_N \rangle|^2 \tag{18.25}$$

The quantity k_{if} is still given by (18.21) but it is now more convenient to refer it to the state $|f'\rangle$ as follows

$$k_{if} = \frac{1}{\hbar c} [E_m - (E_{f'} - E_i)] \tag{18.26}$$

where E_m is defined as the neutrino energy when $E_{f'} = E_i$. The states $|f'\rangle$ are excited states of the initial nucleus $|i_N\rangle$ with $T = 1$. They are the isobar analogs of the states excited in the final nucleus. Inserting now the assumption that only the dipole states are important we find that the matrix element in (18.25) becomes

$$|M_D|^2 = |\langle f' | \sum_j \tau_{3j}\mathbf{k}_{if}\cdot\mathbf{r}_j \frac{3j_1(k_{if}r_j)}{k_{if}r_j} | i \rangle|^2$$

where D refers to dipole.

Since the operator involved in M_D is a single-particle operator and since $|f'\rangle$ and $|i\rangle$ are antisymmetrized

$$|M_D|^2 = A^2 | \langle f' | \tau_{3j}\mathbf{k}_{if}\cdot\mathbf{r}_j \frac{3j_1(k_{if}r_j)}{k_{if}r_j} | i \rangle|^2$$

where the subscript j refers to any one nucleon coordinate. If now the assumption that $|f'\rangle$ is a giant dipole state is inserted by assuming that the only nonzero matrix element of $\tau_{3j}\mathbf{k}_{if}\cdot\mathbf{r}_j$ is $\langle f' | \tau_{3j}\mathbf{k}_{if}\cdot\mathbf{r}_j | i \rangle$, $|M_D|^2$ becomes

$$|M_D|^2 = |\langle f' | 3 \sum_j \tau_{3j}\mathbf{k}_{if}\cdot\mathbf{r}_j | i \rangle|^2 [G(k_{if})]^2 \tag{18.27}$$

where

$$G^2(k_{if}) = \frac{1}{A^2} |\langle f' | \sum_j \frac{j_1(k_{if}r_j)}{k_{if}r_j} |f'\rangle|^2$$

Relating $|f'\rangle$ and $|i\rangle$ according to the assumption leading to (18.27) the expression for G can be replaced by a diagonal matrix element with respect to $|i\rangle$. Assuming harmonic-oscillator functions for $|i\rangle$ it follows that G^2 is proportional to F^2 where

$$F \equiv \langle i| \frac{1}{A} \sum_j \exp(i\mathbf{k}_{if}\cdot\mathbf{r}_j)|i\rangle$$

$$= \int \rho(\mathbf{r}) \exp(i\mathbf{k}_{if}\cdot\mathbf{r})\, d\mathbf{r}$$

where ρ is the ground state density. F is the elastic scattering form factor. [For other derivations of this result see Fallieros, Ferrell, and Pal (60) and Goldemberg et al. (63)]. F is evaluated at the neutrino momentum ν_R required for excitation of the giant dipole. The first factor in (18.27) is just the required squared electromagnetic dipole matrix element and can therefore be related to the photoabsorption cross section. The final result is

$$(M_V{}^2) = \frac{F^2(\nu_R)}{2\pi^2(e^2/\hbar c)\lambda\!\!\!^{-}_\mu{}^2} \left(\frac{E_m}{m_\mu c^2}\right)^4 \int_0^{E_m} \frac{(E_m - E)^4}{E_m{}^4} \frac{\sigma_\gamma(D)(E)}{E}\, dE \tag{18.31}$$

With this result one finally relates mu meson capture with the dipole photoabsorption by nuclei. It is a result that permits a quick albeit somewhat rough evaluation of $|M_V|^2$ and thereby of the muon capture rate.

The results obtained by Foldy and Walecka corrected by contributions from other multipoles as well as from nucleon recoil terms neglected in (18.17) are shown in Table 18.1. The substantial agreement with experiment verifies that muon capture is dominated by transitions to dipole giant resonance. This positive result encourages the search for other types of resonances by examination of the γ and neutron decay spectra. We shall make no attempt to describe the results so far obtained. Much more data on giant resonance states obtained in this fashion can be expected once the more intense muon sources to be provided by accelerators now being built are operational. For the present we refer the reader to the review paper of Walecka (73) for references. A most recent treatment for closed-shell nuclei is given by Donnelly and Walker (70).

Finally we mention the relation of these studies to the β-decay of very highly excited nuclei. These nuclei that have an excitation energy of several MeV can β-decay to many levels of the daughter nucleus. The branching ratios to these levels have an interesting structure that seems to be connected with

TABLE 18.1 Total Muon Capture Rates [from Walecka (73)]

Element	ω_{tot}^{theory} in sec^{-1}	ω_{tot}^{exp} in sec^{-1}	References[c]
^{40}Ca	31.8×10^5	$25.5 \pm .5 \times 10^5$	Sens (1958)
^{16}O	1.07×10^5	$.98 \pm 0.05 \times 10^5$	Barlow et al. (1964)
		$.97 \pm 0.03 \times 10^5$	Eckhouse (1962)
		$.39 \pm 0.01 \times 10^5$	Eckhouse (1962)
^{12}C	$.36 \times 10^5$[b]	$.37 \pm 0.01 \times 10^5$	Reiter et al. (1960)
		$.36 \pm 0.01 \times 10^5$	Lanthrop et al. (1961)
^{4}He	278	364 ± 46	Block et al. (1968)
		336 ± 75	Bizzarri et al. (1964)

[b] Includes the allowed contribution to the ground state of B^{12}.
[c] For complete references cited in this table, see Walecka (73).

groups of levels for which the transitions are enhanced or inhibited. These phenomena have become accessible only recently. See for example Duke et al. (70).

19. CONCLUDING REMARKS

In this chapter we have seen how the weak interaction can be used to probe the structure of the nucleus and how the nucleus can be used to uncover properties of the weak interactions. This is a continuing process. One of the more fundamental questions of weak interaction physics relates to postulated neutral currents. Experiments that measure the strength of the parity-nonconserving nuclear forces will help to reveal the presence or absence of these currents. But as we have emphasized elucidation of this process will not only require a superb experimental effort but also very solid theoretical interpretation.

This complementarity between probing and investigation of nuclear structure and the study of the probing interaction is characteristic of all of nuclear physics. And of course it applies most directly to the nuclear force itself. In studying nuclear structure we are at the same time attempting to discover the characteristic modes of nuclear excitation, and to uncover the essential elements in the nuclear forces that are their origin.

In Volume II we shall ask what is the nature of nuclear dynamics as revealed by nuclear reactions. Again the first question will be: what are the outstanding features? Again it will be necessary to study systematically a number of cases before the salient aspects can be discovered. And each time the range of experimental variables or accuracy is extended new significant features will be revealed. Again the question will be: how are these related to nuclear forces? And finally what do we learn about nuclear forces from these experiments? As the energy of the probes increase other forces will become of interest—the pion–nucleon, the hyperon–nucleon, the kaon–nucleon, and the various unstable bosons and the nucleon. Indeed for many cases it will be only inside the nucleus that these forces will become accessible to investigation. The understanding of nuclear structure and nuclear dynamics will be essential for these tasks; yet at the same time their study may reveal new facets of nuclear properties.

APPENDIX

DIRAC EQUATION

1. NOTATION

Energy-Momentum Four Vector. p_μ ($\mu = 1, 2, 3, 4$) consists of a spatial three vector $p_k (k = 1, 2, 3)$ and a fourth component $p_4 = iE/c$.

Coordinate Four Vector. x_μ ($\mu = 1, 2, 3, 4$) consists of a spatial three vector $x_k (k = 1, 2, 3) \equiv \mathbf{r}$ and a fourth component $x_4 = ict$.

Gradient four Vector. $\partial/\partial x_\mu$ ($\mu = 1, 2, 3, 4$) consists of a spatial three vector $\partial/\partial x_k (k = 1, 2, 3) \equiv \nabla$ and a fourth component $\partial/\partial x_4 = \partial/\partial(ict)$.

Four-vector potential. A_μ ($\mu = 1, 2, 3, 4$) consists of a spatial three vector $A_k (k = 1, 2, 3) \equiv \mathbf{A}$, the vector potential and $A_4 = i\phi$ where ϕ is the scalar potential.
The invariant sum $\Sigma A_\mu B_\mu$ where A_μ and B_μ are four vectors is sometimes written as (AB).

2. DIRAC MATRICES

These are 4 x 4 matrices that we shall write as two by two matrices each element of which is a two by two matrix. For example

$$\rho_1 = \begin{pmatrix} 0 & 1 \\ 1 & 0 \end{pmatrix} = \begin{pmatrix} 0 & 0 & 1 & 0 \\ 0 & 0 & 0 & 1 \\ 1 & 0 & 0 & 0 \\ 0 & 1 & 0 & 0 \end{pmatrix} \tag{2.1}$$

Define also

$$\rho_2 = \begin{pmatrix} 0 & -i \\ i & 0 \end{pmatrix}, \qquad \beta \equiv \rho_3 = \begin{pmatrix} 1 & 0 \\ 0 & -1 \end{pmatrix} \quad 1 = \begin{pmatrix} 1 & 0 \\ 0 & 1 \end{pmatrix} \tag{2.2}$$

where

$$\rho_1\rho_2 = i\rho_3 \qquad \rho_1\rho_2 + \rho_2\rho_1 = 0 \qquad \rho_i^2 = 1 \tag{2.3}$$

Define

$$\boldsymbol{\sigma} = \begin{pmatrix} \boldsymbol{\sigma} & 0 \\ 0 & \boldsymbol{\sigma} \end{pmatrix} \tag{2.4}$$

In terms of ρ_i and $\boldsymbol{\sigma}$:

$$\boldsymbol{\alpha} = \rho_1 \boldsymbol{\sigma} = \begin{pmatrix} 0 & \boldsymbol{\sigma} \\ \boldsymbol{\sigma} & 0 \end{pmatrix}, \qquad \alpha_4 = \beta \tag{2.5}$$

$$\alpha_\mu \alpha_\nu + \alpha_\nu \alpha_\mu = 2\delta_{\mu\nu} \tag{2.6}$$

$$\gamma_k = \rho_2 \boldsymbol{\sigma} = \begin{pmatrix} 0 & -i\boldsymbol{\sigma} \\ i\boldsymbol{\sigma} & 0 \end{pmatrix}, \qquad \gamma_4 = \beta \tag{2.7}$$

$$\gamma_\mu \gamma_\nu + \gamma_\nu \gamma_\mu = 2\delta_{\mu\nu} \tag{2.8}$$

$$\gamma_5 = \gamma_1 \gamma_2 \gamma_3 \gamma_4 = -\rho_1 \tag{2.9}$$

$$\gamma_5 \gamma_\mu + \gamma_\mu \gamma_5 = 0 \tag{2.10}$$

$$\sigma_{\mu\nu} = \frac{1}{2i} (\gamma_\mu \gamma_\nu - \gamma_\nu \gamma_\mu) \tag{2.11}$$

$$= \frac{1}{i} \gamma_\mu \gamma_\nu \qquad \mu \neq \nu$$

$$\sigma_{ij} = \frac{1}{i} \sigma_i \sigma_j = \sigma_k \qquad i, j, k, \text{ cyclical} \tag{2.12}$$

$$\sigma_{k4} = -\sigma_{4k} = \alpha_k$$

$$\alpha_k = i\gamma_4 \gamma_k \qquad \text{(2.13}$$

3. DIRAC EQUATION: HAMILTONIAN FORM

$$H\psi = i\hbar \partial\psi/\partial t \tag{3.1}$$

$$H = c\boldsymbol{\alpha} \cdot \left(\mathbf{p} - \frac{e}{c}\mathbf{A}\right) + mc^2\beta + e\phi \tag{3.2}$$

where $\mathbf{A}$ and ϕ are the electromagnetic vector and scalar potentials respectively. If

$$\psi = \begin{pmatrix} \psi_1 \\ \psi_2 \\ \psi_3 \\ \psi_4 \end{pmatrix} \tag{3.3}$$

then

$$\psi^\dagger \equiv (\psi_1^*, \psi_2^*, \psi_3^*, \psi_4^*)$$

satisfies

$$\psi^\dagger\left[c\boldsymbol{\alpha}\cdot\left(\overleftarrow{p} + \frac{e}{c}\mathbf{A}\right) - mc^2\beta - e\phi\right] = i\hbar\partial\psi^\dagger/\partial t \tag{3.4}$$

$$\text{current density } \mathbf{J} = ec\psi^\dagger\boldsymbol{\alpha}\psi \tag{3.5}$$

$$\rho = e\psi^\dagger\psi \tag{3.6}$$

4. DIRAC EQUATION: COVARIANT FORM

$$\left[\sum_\mu \gamma_\mu\left(p_\mu - \frac{e}{c}A_\mu\right) - imc\right]\psi = 0 \tag{4.1}$$

$$\text{adjoint of } \psi \equiv \overline{\psi} = \psi^\dagger\gamma_4 \tag{4.2}$$

$$\overline{\psi}\left[\sum_\mu \gamma_\mu\left(p_\mu + \frac{e}{c}A_\mu\right) + imc\right] = 0 \tag{4.3}$$

four vector current density: $j_\mu = iec\overline{\psi}\gamma_\mu\psi$, $j_4 = ic\rho$

$$\sum_\mu \frac{\partial j_\mu}{\partial x_\mu} = 0 \tag{4.4}$$

5. FREE PARTICLE SOLUTIONS OF THE DIRAC EQUATION ($A_\mu = 0$)

(a) Zero momentum. The four solutions of the Dirac equation are:

$$u^{(1)} = \begin{pmatrix}1\\0\\0\\0\end{pmatrix} \quad u^{(2)} = \begin{pmatrix}0\\1\\0\\0\end{pmatrix} \quad u^{(3)} = \begin{pmatrix}0\\0\\1\\0\end{pmatrix} \quad u^{(4)} = \begin{pmatrix}0\\0\\0\\1\end{pmatrix} \tag{5.1}$$

or

$$u^{(1)} = \begin{pmatrix}\alpha\\0\end{pmatrix} \quad u^{(2)} = \begin{pmatrix}\beta\\0\end{pmatrix} \quad u^{(3)} = \begin{pmatrix}0\\\alpha\end{pmatrix} \quad u^{(4)} = \begin{pmatrix}0\\\beta\end{pmatrix} \tag{5.2}$$

where

$$\alpha = \begin{pmatrix} 1 \\ 0 \end{pmatrix}, \qquad \beta = \begin{pmatrix} 0 \\ 1 \end{pmatrix} \tag{5.3}$$

The corresponding energies and spin directions are

	E	m_s
$u^{(1)}$	mc^2	$+$
$u^{(2)}$	mc^2	$-$
$u^{(3)}$	$-mc^2$	$+$
$u^{(4)}$	$-mc^2$	$-$

(5.4)

(b) Finite momentum **p**. Let

$$E(p) \equiv \sqrt{m^2c^4 + c^2p^2} \tag{5.5}$$

and $H(p)$

$$H(p) \equiv c(\boldsymbol{\alpha}\cdot\mathbf{p}) + \beta mc^2 \tag{5.6}$$

then the four normalized solutions of the Dirac equation are

$$\psi^{(i)} = \frac{1}{\sqrt{2E(p)[E(p) + mc^2]}}\,[H(p)\beta + E(p)]u^{(i)} \exp\,[(i/\hbar)\mathbf{p}\cdot\mathbf{r}] \tag{5.7}$$

Explicitly,

$$E = E(p){:}\quad \sqrt{\frac{2E(p)}{E(p) + mc^2}}\,\psi^{(1)} = \begin{pmatrix} 1 \\ 0 \\ cp_z/[E(p) + mc^2] \\ c(p_x + ip_y)/[E(p) + mc^2] \end{pmatrix} \times \exp\,(i/\hbar\mathbf{p}\cdot\mathbf{r}) \tag{5.8}$$

$$E = E(p){:}\quad \sqrt{\frac{2E(p)}{E(p) + mc^2}}\,\psi^{(2)} = \begin{pmatrix} 0 \\ 1 \\ c(p_x - ip_y)/[E(p) + mc^2] \\ -cp_z/[E(p) + mc^2] \end{pmatrix} \times \exp\,(i/\hbar\mathbf{p}\cdot\mathbf{r}) \tag{5.9}$$

$$E = -E(p): \sqrt{\frac{2E(p)}{E(p)+mc^2}}\,\psi^{(3)} = \begin{pmatrix} -cp_z/[E(p)+mc^2] \\ -c(p_x+ip_y)/[E(p)+mc^2] \\ 1 \\ 0 \end{pmatrix} \times \exp(i/\hbar\mathbf{p}\cdot\mathbf{r}) \tag{5.10}$$

$$E = -E(p): \sqrt{\frac{2E(p)}{E(p)+mc^2}}\,\psi^{(4)} = \begin{pmatrix} -c(p_x-ip_y)/[E(p)+mc^2] \\ cp_z/[E(p)+mc^2] \\ 0 \\ 1 \end{pmatrix} \times \exp(i/\hbar\mathbf{p}\cdot\mathbf{r}) \tag{5.11}$$

6. CHARGE CONJUGATE SOLUTION ψ_c

If ψ_c satisfies

$$\left[c\alpha\cdot\left(\mathbf{p}+\frac{e}{c}\mathbf{A}\right)+mc^2\beta-e\phi\right]\psi_c = \hbar i\frac{\partial\psi_c}{\partial t} \tag{6.1}$$

then

$$\psi_c = -\gamma_2\psi^* \tag{6.2}$$

where ψ satisfies (3.1).

Example. If $\psi = \psi^{(4)}$ of (5.11) then

$$\psi_c^{(1)} = -\gamma_2\psi^{(4)*} = \sqrt{\frac{E(p)+mc^2}{2E(p)}}\begin{pmatrix} 1 \\ 0 \\ -cp_z/[E(p)+mc^2] \\ c(p_x+ip_y)/[E(p)+mc^2] \end{pmatrix} \times \exp(-i/\hbar\mathbf{p}\cdot\mathbf{r}) \tag{6.3}$$

$\psi_c^{(1)}$ describes a particle not only of opposite charge to that of $\psi^{(4)}$ but also of opposite spin, momentum, and energy.

7. SOLUTION OF DIRAC EQUATION IN COULOMB FIELD OF NUCLEUS

$$[c\boldsymbol{\alpha}\cdot\mathbf{p} + \beta mc^2 + V(r)]\psi = E\psi \tag{7.1}$$

$$V(r) = -Ze^2/r \qquad \text{electron}$$

$$= Ze^2/r \qquad \text{positron}$$

Let

$$\psi = \begin{pmatrix} u \\ v \end{pmatrix} \tag{7.2}$$

Then

$$c(\boldsymbol{\sigma}\cdot\mathbf{p})v = (E - mc^2 - V)u \tag{7.3}$$

$$c(\boldsymbol{\sigma}\cdot\mathbf{p})u = (E + mc^2 - V)v \tag{7.4}$$

Let

$$\mathcal{Y}_m(j, l) = \sum (\tfrac{1}{2}m_s l m_l | jm)\chi_{1/2}(m_s)\mathcal{Y}_{lm_l} \tag{7.5}$$

where $\chi_{1/2}(1/2) = \alpha$, $\chi_{1/2}(-1/2) = \beta$, and $\mathcal{Y}_{lm}$ is defined by appendix A, (A.2.37).

Let

$$u = \frac{G(r)}{r}\mathcal{Y}_m(j, l)$$

$$v = \frac{f(r)}{r}(-i\boldsymbol{\sigma}\cdot\hat{\mathbf{r}})\mathcal{Y}_m(j, l) \tag{7.7}$$

Then

$$\frac{1}{c}(E - mc^2 - V)G + \frac{\partial f}{\partial r} + \kappa\frac{f}{r} = 0 \tag{7.8}$$

$$-\frac{1}{c}(E + mc^2 - V)f + \frac{\partial G}{\partial r} - \kappa\frac{G}{r} = 0 \tag{7.9}$$

where

$$\kappa = 1 + \boldsymbol{\sigma}\cdot\mathbf{L} = \begin{cases} -(j + 1/2) & \text{for} \quad l = j + 1/2 \\ j + 1/2 & \text{for} \quad l = j - 1/2 \end{cases}$$

Solutions.

$$G = (\epsilon + 1)^{1/2} N_s \operatorname{Re} [\exp (i\phi_\kappa) \exp (-ikr)(2kr)^s F(s + i\eta; 2s + 1; 2ikr)] \tag{7.10}$$

$$f = (\epsilon - 1)^{1/2} N_s \operatorname{Im} [\exp (i\phi_\kappa) \exp (-ikr)(2kr)^s F(s + i\eta; 2s + 1; 2ikr)] \tag{7.11}$$

where

$$s = \sqrt{\kappa^2 - \alpha^2} = \sqrt{(j + 1/2)^2 - \alpha^2}$$

$$\alpha = \frac{Z}{137}, \qquad \eta = \frac{\alpha \epsilon mc}{p}, \qquad \epsilon = \frac{E}{mc^2}$$

$$N_s = \frac{|\Gamma(s + 1 + i\eta)|}{\Gamma(2s + 1) \exp (-\pi\eta/2)} \qquad k = p/\hbar$$

$$\exp (2i\phi_\kappa) = \frac{\kappa - i\alpha mc/p}{s + i\eta} \qquad F(a; c; x) = 1 + \frac{a}{c} x + \frac{a(a + 1)}{c(c + 1)} \frac{x^2}{2!} + \ldots$$

$$G(r \to 0) \to (\epsilon + 1)^{1/2} N_s \cos \phi_\kappa (2kr)^s \tag{7.12}$$

$$f(r \to 0) \to (\epsilon - 1)^{1/2} N_s \sin \phi_\kappa (2kr)^s \tag{7.13}$$

$$G(r \to \infty) \to (\epsilon + 1)^{1/2} \cos \left[kr + \eta \log 2kr - \frac{\pi s}{2} + \delta \right] \tag{7.14}$$

$$f(r \to \infty) \to -(\epsilon - 1)^{1)2} \sin \left[kr + \eta \log 2kr - \frac{\pi s}{2} + \delta \right] \tag{7.15}$$

$$\delta = \frac{k + i\alpha mc/p}{s - i\eta} \frac{\Gamma(s + 1 - i\eta)}{\Gamma(s + 1 + i\eta)} \tag{7.16}$$

APPENDIX A

It is assumed that the reader is familiar with the principles of quantum mechanics as expounded in Gottfried's (66) and Messiah's (63) texts. This appendix will be used to fix notation (Section 1) and to collect some of the results involving rotations, angular momenta and coupling coefficients (Section 2) and time reversal (Section 3) that are frequently encountered in these volumes.

1. NOTATION

(a) Dirac notation: this is primarily used when a state vector is described in terms of its quantum numbers α.

State vector: "ket" $\equiv |\alpha\rangle$

"bra" $\equiv \langle\alpha|$

Scalar product: $\langle\alpha'|\alpha\rangle$

Matrix element of an operator O: $\langle\alpha'|O|\alpha\rangle$

(b) In another notation capital Greek letters will be used to denote a state vector. The scalar product is then

Scalar product: $\langle\Phi|\Psi\rangle$

Matrix element of an operator O: $\langle\Phi|O\Psi\rangle$

(c) The *hermitian adjoint* of an operator O is: $O^\dagger$. It is defined by

$$\langle\Phi|O\Psi\rangle \equiv \langle O^\dagger\Phi|\Psi\rangle$$

or

$$\langle\alpha'|O^\dagger|\alpha\rangle \equiv \langle\alpha|O|\alpha'\rangle^*$$

where the * denotes complex conjugation.
The *transpose* of O is: O^T

Definition: $\langle\alpha'|O^T|\alpha\rangle = \langle\alpha|O|\alpha'\rangle$

The *complex conjugate* of O is O^*

Definition: $\langle\alpha'|O^*|\alpha\rangle = \langle\alpha'|O|\alpha\rangle^*$

Hence $O^\dagger = (O^T)^* = (O^*)^T$

It will be *occasionally* necessary in order to avoid confusion to make explicit the operator nature of a quantity. In that event a "hat" will be used so that $\hat{p}$ is the momentum operator, $\hat{\psi}$ is the field operator in field theory and so on.

(d) Lower case Greek letters ψ, ϕ, χ will generally be used to designate *wave functions*. A mixed notation is used to relate wave functions and state vectors:

$$\psi(\alpha) = \langle\alpha|\Psi\rangle$$

The scalar product of two wave functions is sometimes written

$$\int \psi^*(\alpha)\phi(\alpha)$$

2. ROTATIONS, ANGULAR MOMENTA, AND COUPLING COEFFICIENTS

For a discussion of the origin and significance of the following results the reader is referred to Gottfried (66, p. 264f) and deShalit and Talmi (63, p. 63f). Other important references are Rose (57), Rotenberg et al. (59) Edmonds (60), Brink and Satchler (62).

Three-dimensional rotations R are specified by the Euler angles α, β, γ shown in Fig. A.1. The corresponding unitary operator* D(R) given by

$$D(R) \equiv D(\alpha, \beta, \gamma) = \exp(-i\alpha J_z) \exp(-i\beta J_y) \exp(-i\gamma J_z) \tag{2.1}$$

relates $\psi(\mathbf{r})$ and $\psi(R\mathbf{r})$:

$$\psi(R\mathbf{r}) = D\psi(\mathbf{r}) \tag{2.2}$$

with D in the coordinate representation
More explicitly, for a rotation of the coordinate system about the z-axis through an angle α, $\mathbf{r}' = R\mathbf{r}$ is

$$x' = x\cos\alpha + y\sin\alpha$$

$$y' = -x\sin\alpha + y\cos\alpha$$

$$z' = z$$

*Unfortunately there are different conventions for the D operator. In the older literature one may even find D's that refer to the rotation of left-handed rather than right-handed reference frames. Our convention agrees with that of Rose (57), Jacob and Wick (59), deShalit and Talmi (63), and Gottfried (66). It is the convention that is used in particle physics. Much of the literature on collective motion in nuclei uses a D that is the complex conjugate of the one used here [see, for example, Bohr and Mottelson (69)].

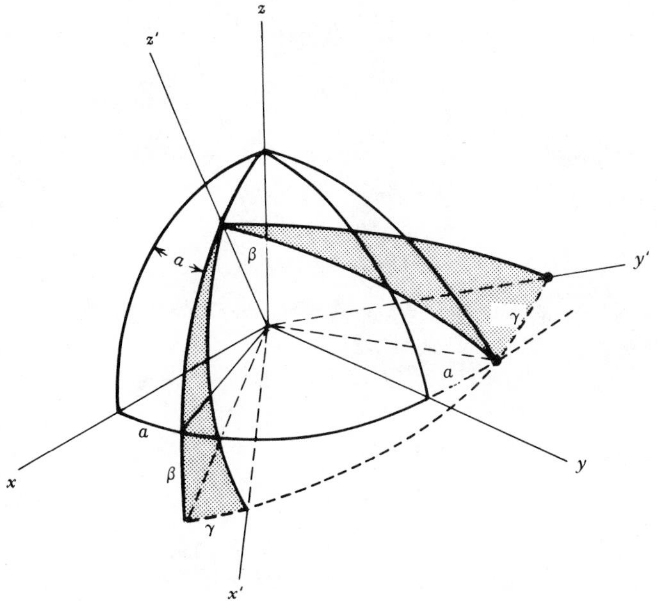

FIG. A.1.

The D-matrix is the matrix formed by taking matrix elements of the D-operator with respect to an irreducible representation of the rotation group of dimension $2j + 1$ corresponding to the angular momentum j. Each element of the representation is identified by two eigenvalues j and m. Thus the D-matrix, $D_{mm'}^{(j)}(\alpha, \beta, \gamma)$ is defined by

$$D_{mm'}^{(j)}(\alpha, \beta, \gamma) = \langle jm | D(\alpha, \beta, \gamma) | jm' \rangle$$
$$= \langle jm | \exp(-i\alpha J_z) \exp(-i\beta J_y) \exp(-i\gamma J_z) | jm' \rangle \quad (2.3)$$

where $\mathbf{J}$, the generator of an infinitesimal rotation, equals, except for a factor of $\hbar$, the angular momentum operator. The components of $\mathbf{J}$ satisfy the commutation rule $[J_x, J_y] = iJ_z$. Below are some properties of D that we have found useful.

Unitarity

$$D^\dagger(\alpha, \beta, \gamma) = D^{-1}(\alpha, \beta, \gamma) = D(-\gamma, -\beta, -\alpha) \quad (2.4)$$

$$[D_{mm'}^{(j)}(\alpha, \beta, \gamma)]^* = D_{m'm}^{(j)}(-\gamma, -\beta, -\alpha) \quad (2.5)$$

$$\sum_{\mu} D_{m\mu}^{(j)}(R) D_{m'\mu}^{(j)}(R) = \delta_{mm'} \tag{2.6}$$

$$\sum_{\mu} D_{\mu m}^{(j)}(R) D_{\mu m'}^{(j)}(R) = \delta_{mm'} \tag{2.7}$$

Group Property

$$D(R_2)D(R_1) = D(R_2R_1) \tag{2.8}$$

$$\sum_{m'} D_{mm'}^{(j)}(R_2) D_{m'm''}^{(j)}(R_1) = D_{mm''}^{(j)}(R_2R_1) \tag{2.9}$$

$$D(R)D(R^{-1}) = D(R)D^{-1}(R) = 1 \tag{2.10}$$

Orthogonality

$$\int_0^{2\pi} d\alpha \int_0^{2\pi} d\gamma \int_0^{\pi} \sin\beta \, d\beta \, [D_{mm'}^{(j)}(\alpha, \beta, \gamma)]^* D_{\mu\mu'}^{(j')}(\alpha, \beta, \gamma)$$

$$= \frac{8\pi^2}{2j+1} \delta_{m\mu} \, \delta_{m'\mu'} \, \delta_{jj'} \tag{2.11}$$

Explicit Evaluation of $D_{mm'}^{(j)}(\alpha, \beta, \gamma)$

$$D_{mm'}^{(j)}(\alpha\beta\gamma) = \exp(-im\alpha) d_{mm'}^{(j)}(\beta) \exp(-im\gamma) \tag{2.12}$$

$$d_{mm'}^{(j)}(\beta) = (-)^{m-m'} \left[\frac{(j+m)!(j-m)!}{(j+m')!(j-m')!} \right]^{1/2} (\cos\tfrac{1}{2}\beta)^{m+m'} (\sin\tfrac{1}{2}\beta)^{m-m'}$$

$$\cdot \, P_{j-m}^{(m-m', m+m')}(\cos\beta) \tag{2.13}$$

$$P_n^{(\mu,\nu)}(\cos\beta) = \sum_{\alpha} \binom{\mu+n}{\alpha} \binom{\nu+n}{n-\alpha} (-)^{n-\alpha} (\sin^2\tfrac{1}{2}\beta)^{n-\alpha} (\cos^2\tfrac{1}{2}\beta)^{\alpha} \tag{2.14}$$

$$= \binom{n+\mu}{n} F(-n, n+\mu+\nu; \mu+1; (\sin\tfrac{1}{2}\beta)^2) \tag{2.15}$$

$$= \binom{n+\mu}{n} (\cos\tfrac{1}{2}\beta)^{2n} F(-n, -n-\nu; \mu+1; -\tan^2\tfrac{1}{2}\beta) \tag{2.16}$$

F is the hypergeometric function

$$F(a, b; c; x) = 1 + \frac{ab}{c} x + \frac{a(a+1)b(b+1)}{c(c+1)} \frac{x^2}{2!} + \ldots$$

Symmetry Properties of $d_{mm'}{}^{(j)}$

$$d_{mm'}{}^{(j)}(\beta) = d_{-m',-m}{}^{(j)}(\beta) \tag{2.17}$$

$$d_{mm'}{}^{(j)}(\beta) = d_{m'm}{}^{(j)}(-\beta)$$

$$= (-)^{m-m'} d_{m'm}{}^{(j)}(\beta) \tag{2.18}$$

$$d_{m'm}{}^{(j)}(\pi - \beta) = (-)^{j-m'} d_{m,-m'}{}^{(j)}(\beta)$$

Special Values and Symmetries of $D_{mm'}{}^{(j)}$

$$D_{mm'}{}^{(j)*}(R) = (-)^{m-m'} D_{-m,-m'}{}^{(j)}(R) = D_{m',m}{}^{(j)}(R^{-1}) \tag{2.19}$$

When $R = R_z$, a rotation about the z-axis through the angle α

$$D_{mm'}{}^{(j)}(R_z) = \exp(-im\alpha)\delta_{mm'} \tag{2.20}$$

When R is a rotation about the y-axis through the angle π:

$$D_{mm'}{}^{(j)} = (-)^{j-m}\delta_{m,-m'} \tag{2.21}$$

When R is a rotation about the x-axis through an angle π:

$$D_{mm'}{}^{(j)} = (-)^{j}\delta_{m,-m'} \tag{2.22}$$

Relation to Spherical Harmonics

$$D_{m0}{}^{(l)}(\alpha\beta\gamma) = \sqrt{\frac{4\pi}{2l+1}}\, Y_{lm}^{*}(\beta, \alpha) \tag{2.23}$$

$$D_{0m}{}^{(l)}(\alpha\beta\gamma) = (-)^{m}\sqrt{\frac{4\pi}{2l+1}}\, Y_{lm}^{*}(\beta\gamma) \tag{2.24}$$

$$D_{00}{}^{(l)}(\alpha\beta\gamma) = P_l(\cos\beta) \tag{2.25}$$

$$Y_{lm'}(\hat{\mathbf{r}}') = \sum_m Y_{lm}(\hat{\mathbf{r}})\, D_{mm'}{}^{(l)}(R) \tag{2.26}$$

where $R\mathbf{r} = \mathbf{r}'$.

Properties of Spherical Harmonics

$$Y_{lm}^{*}(\hat{\mathbf{r}}) \equiv Y_{lm}(\theta, \phi) = (-)^{m}\left[\frac{(2l+1)(l-|m|)!}{4\pi(l+|m|)!}\right]^{1/2} P_l{}^{|m|}(\cos\theta) \times \exp(im\phi) \tag{2.27}$$

$$P_l{}^{m}(x) = (1-x^2)^{m/2}\frac{d^m P_l(x)}{dx} \tag{2.28}$$

$$\int Y^*_{l'm'}(\hat{\mathbf{r}})Y_{lm}(\hat{\mathbf{r}})\,d\hat{\mathbf{r}} = \delta_{ll'}\,\delta_{mm'} \tag{2.29}$$

$$P_l(x) = \frac{1}{2^l l!}\frac{d}{dx^l}(x^2-1)^l \tag{2.30}$$

$$P_l{}^m(-x) = (-)^{l+m}P_l{}^m(x);\; P_l{}^{-m}(x) = (-)^m\frac{(l-m)!}{(l+m)!}P_l{}^m(x) \tag{2.31}$$

$$\begin{aligned} Y^*_{lm}(\theta,\phi) &= (-)^m Y_{l,-m}(\theta,\phi);\; Y_{lm}(-\hat{\mathbf{r}}) = Y_{lm}(\pi-\theta,\pi+\phi) \\ &= (-)^l Y_{lm}(\hat{\mathbf{r}}) \end{aligned} \tag{2.32}$$

$$Y_{l0}(\theta,\phi) = \sqrt{\frac{2l+1}{4\pi}}\,P_l(\cos\theta) \tag{2.33}$$

$$P_l(\cos\theta) = \frac{4\pi}{2l+1}\sum_m Y^*_{lm}(\hat{\mathbf{r}}')Y_{lm}(\hat{\mathbf{r}}) \qquad \hat{\mathbf{r}}'\cdot\hat{\mathbf{r}} = \cos\theta \tag{2.34}$$

$$Y_{l_1m_1}(\hat{\mathbf{r}})Y_{l_2m_2}(\hat{\mathbf{r}}) = \sum_{lm}\left[\frac{(2l_1+1)(2l_2+1)(2l+1)}{4\pi}\right]^{1/2} \times \begin{pmatrix} l_1 & l_2 & l \\ m_1 & m_2 & m\end{pmatrix}\begin{pmatrix} l_1 & l_2 & l \\ 0 & 0 & 0\end{pmatrix} Y_{lm}(\hat{\mathbf{r}}) \tag{2.35}$$

$$\sqrt{\frac{(2l_1+1)(2l_2+1)(2l+1)}{4\pi}}\begin{pmatrix} l_1 & l_2 & l \\ 0 & 0 & 0\end{pmatrix} Y_{lm}(\hat{\mathbf{r}}) = \sum_{m_1m_2}\begin{pmatrix} l_1 & l_2 & l \\ m_1 & m_2 & m\end{pmatrix} \times Y_{l_1m_1}(\hat{\mathbf{r}})Y_{l_2m_2}(\hat{\mathbf{r}}) \tag{2.36}$$

When discussing time reversal invariance it will be useful to introduce

$$\mathcal{Y}_{lm}(\mathbf{r}) = i^l Y_{lm}(\hat{\mathbf{r}}) \tag{2.37}$$

with the property

$$\mathcal{Y}^*_{lm}(\hat{\mathbf{r}}) = (-)^{l+m}\mathcal{Y}_{l,-m}(\hat{\mathbf{r}}) \tag{2.38}$$

Spherical Tensors

An irreducible spherical tensor field of degree k, $T_\kappa{}^{(k)}$ $(-k < \kappa < k)$, has $2k+1$ components that transform under rotations according to (2.26)

$$T_\kappa{}^{(k)\prime}(\mathbf{r}') = \sum_{\kappa'} T_{\kappa'}{}^{(k)}(\mathbf{r})D_{\kappa'\kappa}{}^{(k)}(R); \qquad \mathbf{r}' = R\mathbf{r} \tag{2.39}$$

where $T_\kappa{}^{(k)\prime}$ are the components of $T_\kappa{}^{(k)}$ referred to the rotated coordinate system.

If $\mathbf{J}$ is the infinitesimal rotation operator then

$$\delta T_\kappa{}^{(k)} = -i\delta\theta\hat{\mathbf{n}}\cdot\mathbf{J}T_\kappa{}^{(k)}$$

where δT is the change in T because of a rotation of the coordinate system by an angle $\delta\theta$ about a direction $\hat{\mathbf{n}}$ and

$$J_z T_\kappa{}^{(k)} = \kappa T_\kappa{}^{(k)} \tag{2.40}$$

$$(J_x \pm iJ_y)T_\kappa{}^{(k)} = \sqrt{k(k+1) - \kappa(\kappa \pm 1)}\,T_{\kappa\pm 1}{}^{(k)} \tag{2.41}$$

Spherical Tensor Operators: Commutation Rules An irreducible tensor *operator* $T_\kappa{}^{(k)}$ satisfies

$$[J_z, T_\kappa{}^{(k)}] = \kappa T_\kappa{}^{(k)} \tag{2.42}$$

$$[J_x \pm iJ_y, T_\kappa{}^{(k)}] = \sqrt{k(k+1) - \kappa(\kappa \pm 1)}\,T_{\kappa\pm 1}{}^{(k)} \tag{2.43}$$

Spherical Tensors: Wigner-Eckart Theorem

$$\langle jm|T_\kappa{}^{(k)}|j'm'\rangle = (-)^{j-m}\begin{pmatrix} j & k & j' \\ -m & \kappa & m' \end{pmatrix}(j||T^{(k)}||j') \tag{2.44}$$

where the double-barred quantity is called the *reduced matrix element.* It is independent of the magnetic quantum numbers m, κ, and m'. The symbol

$$\begin{pmatrix} j & k & j' \\ -m & \kappa & m' \end{pmatrix}$$

is the Wigner $3j$ symbol defined in (2.66). The Wigner–Eckart theorem implies the *selection rules*

$$m = \kappa + m' \qquad \text{and} \qquad j + k \geq j' \geq |j - k| \tag{2.45}$$

For scalar tensors ($k = 0$). Some special reduced matrix elements:

$$(j||1||j') = \sqrt{2j+1}\,\delta_{jj'} \tag{2.46}$$

$$(j||\mathbf{J}||j) = \sqrt{j(j+1)(2j+1)} \tag{2.47}$$

$$(l||Y_k||l') = (-)^l\begin{pmatrix} l & k & l' \\ 0 & 0 & 0 \end{pmatrix}\sqrt{\frac{(2l+1)(2k+1)(2l'+1)}{4\pi}} \tag{2.48}$$

$$(\tfrac{1}{2}lj||Y_k||\tfrac{1}{2}l'j') = \tfrac{1}{2}(-)^{j-1/2}[1 + (-)^{l+k+l'}]\begin{pmatrix} j & k & j' \\ 1/2 & 0 & 1/2 \end{pmatrix} \cdot\sqrt{\frac{(2j+1)(2k+1)(2j'+1)}{4\pi}} \tag{2.49}$$

Tensor Operators: Matrix Elements of Tensor Products

The tensor product $T^{(k)}$ of $T^{(k_1)}$ and $T^{(k_2)}$ is defined by

$$T_\kappa^{(k)}(k_1k_2) = \sum_{\kappa_1\kappa_2} (k_1\kappa_1, k_2\kappa_2|k\kappa)T_{\kappa_1}^{(k_1)}T_{\kappa_2}^{(k_2)} \tag{2.50}$$

where the coefficient $(k_1\,\kappa_1, k_2\,\kappa_2|k\kappa)$ is defined by (2.57). $T_\kappa^{(k)}$ transforms under rotations like an irreducible spherical tensor or order k. The scalar product of two tensors $T^{(k)}$ and $U^{(k)}$ is defined as

$$\mathbf{T}^{(k)}\cdot\mathbf{U}^{(k)} = (-)^k\sqrt{2k+1}\,T_0^{(0)}(k,k) \tag{2.51}$$

$$= \Sigma\,(-)^\kappa\,T_{-\kappa}^{(k)}\,T_\kappa^{(k)} \tag{2.51a}$$

Reduced Matrix Elements of Tensor Products

If $T^{(k_1)}$ and $T^{(k_2)}$ operate on the same set of coordinates:

$$(\alpha j||T^{(k)}||\alpha' j') = (-)^{j+k+j'}\sqrt{2k+1}\sum_{\alpha''j''}\begin{Bmatrix} k_1 & k_2 & k \\ j' & j & j'' \end{Bmatrix} \cdot (\alpha j||T^{(k_1)}||\alpha'' j'')(\alpha'' j''||T^{(k_2)}||\alpha' j') \tag{2.52}$$

where { } is the Wigner 6j coefficient defined by (2.84)

If $T_1^{(k_1)}$ and $T_2^{(k_2)}$ operate on two differing systems denoted by subscripts 1 and 2:

$$(\alpha_1 j_1\alpha_2 j_2 J||T^{(k)}||\alpha_1' j_1'\alpha'_2 j'_2 J') = \sqrt{(2J+1)(2k+1)(2J'+1)} \cdot \begin{Bmatrix} j_1 & j_2 & J \\ j_1' & j_2' & J' \\ k_1 & k_2 & k \end{Bmatrix} (\alpha_1 j_1||T_1^{(k_1)}||\alpha'_1 j'_1)(\alpha_2 j_2||T_2^{(k_2)}||\alpha'_2 j'_2) \tag{2.53}$$

where { } is the Wigner 9j coefficient defined by (2.100)

$$(\alpha_1 j_1\alpha_2 j_2 J||\mathbf{T}_1^{(k)}\cdot\mathbf{T}_2^{(k)}||\alpha'_1 j'_1\alpha'_2 j_2' J') = (-)^{j_2+J+j_1'}\sqrt{2J+1}\begin{pmatrix} j_1 & j_2 & J \\ j_2' & j_1' & k \end{pmatrix} \cdot \delta_{JJ'}(\alpha_1 j_1||T_1^{(k)}||\alpha_1' j_1')(\alpha_2 j_2||T_2^{(k)}||\alpha_2' j_2') \tag{2.54}$$

$$(\alpha_1 j_1\alpha_2 j_2 J||T_1^{(k)}||\alpha_1' j_1'\alpha_2' j_2' J') = (-)^{j_1+j_2+J'+k}\sqrt{(2J+1)(2J'+1)} \cdot \delta_{\alpha_2\alpha_2'}\,\delta_{j_2' j_2'}\begin{Bmatrix} j_1 & J & j_2 \\ J' & j' & k \end{Bmatrix}(\alpha_1 j_1||T_1^{(k)}||\alpha'_1 j'_1) \tag{2.55}$$

$$(\alpha_1 j_1\alpha_2 j_2 J||T_2^{(k)}||\alpha_1' j_1'\alpha_2' j_2' J') = (-)^{j_1'+j_2'+J+k}\sqrt{(2J+1)(2J'+1)} \cdot \delta_{\alpha_1\alpha'_1}\,\delta_{j_1,j_1'}\begin{Bmatrix} j_2 & J & j_1 \\ J' & j'_2 & k \end{Bmatrix}(\alpha_2 j_2||T_2^{(k)}||\alpha_2' j_2') \tag{2.56}$$

Clebsch-Gordan Coefficients

The state of a system with a given angular momentum J, "z-component" M constructed from two systems each in a state with angular momentum, j_i, m_i $(i = 1, 2)$ is:

$$\psi(j_1 j_2 JM) = \sum_{m_1 m_2} (j_1 m_1 j_2 m_2 | JM) \psi(j_1 m_1) \psi(j_2 m_2) \tag{2.57}$$

The Clebsch–Gordan coefficient $(j_1 m_1 j_2 m_2 | JM)$ is real.

$$(j_1 m_1 j_2 m_2 | JM) = \langle \psi(j_1 j_2 JM) | \psi(j_1 m_1) \psi(j_2 m_2) \rangle \tag{2.58}$$

is a matrix of a unitary transformation:

$$\psi(j_1 m_1) \psi(j_2 m_2) = \sum_{JM} (j_1 m_1 j_2 m_2 | JM) \psi(j_1 j_2 JM) \tag{2.59}$$

Unitarity:

$$\sum_{m_1 m_2} (j_1 m_1 j_2 m_2 | JM)(j_1 m_1 j_2 m_2 | J'M') = \delta_{JJ'} \, \delta_{MM'} \tag{2.60}$$

$$\sum_{JM} (j_1 m_1 j_2 m_2 | JM)(j_1 m'_1 j_2 m'_2 | JM) = \delta_{m_1 m'_1} \, \delta_{m_2 m'_2} \tag{2.61}$$

Symmetries:

$$(j_1 m_1 j_2 m_2 | j_3 m_3) = (-)^{j_2 + m_2} \left[\frac{2j_3 + 1}{2j_1 + 1} \right]^{1/2} (j_2 - m_2 j_3 m_3 | j_1 m_1) \tag{2.62}$$

$$= (-)^{j_1 - m_1} \left[\frac{2j_3 + 1}{2j_1 + 1} \right]^{1/2} (j_3 m_3 j_1 - m_1 | j_2 m_2) \tag{2.63}$$

$$= (-)^{j_1 + j_2 - j_3} (j_1 - m_1 j_2 - m_2 | j_3 - m_3) \tag{2.64}$$

$$= (-)^{j_1 + j_2 - j_3} (j_2 m_2 j_1 m_1 | j_3 m_3) \tag{2.65}$$

Wigner 3-j Symbol

$$\begin{pmatrix} j_1 & j_2 & j_3 \\ m_1 & m_2 & m_3 \end{pmatrix} \equiv \frac{(-)^{j_1 - j_2 - m_3}}{\sqrt{2j_3 + 1}} (j_1 m_1 j_2 m_2 | j_3 - m_3) \tag{2.66}$$

$$= 0 \quad \text{if} \quad m_1 + m_2 + m_3 \neq 0 \quad \text{and} \quad \mathbf{j}_1 + \mathbf{j}_2 \neq \mathbf{j}_3 \tag{2.67}$$

(See Eq. 2.87 for definition of $\mathbf{j}_1 + \mathbf{j}_2 = \mathbf{j}_3$)

$$(j_1 m_1 j_2 m_2 | j_3 m_3) = (-)^{j_1 - j_2 - m_3} \sqrt{2j_3 + 1} \begin{pmatrix} j_1 & j_2 & j_3 \\ m_1 & m_2 & -m_3 \end{pmatrix} \tag{2.68}$$

Symmetries:

$$\begin{pmatrix} j_1 & j_2 & j_3 \\ m_1 & m_2 & m_3 \end{pmatrix} = \begin{pmatrix} j_2 & j_3 & j_1 \\ m_2 & m_3 & m_1 \end{pmatrix}$$

$$= \begin{pmatrix} j_3 & j_1 & j_2 \\ m_3 & m_1 & m_2 \end{pmatrix}$$

$$= (-)^{j_1+j_2+j_3} \begin{pmatrix} j_1 & j_3 & j_2 \\ m_1 & m_3 & m_2 \end{pmatrix}$$

$$= (-)^{j_1+j_2+j_3} \begin{pmatrix} j_1 & j_2 & j_3 \\ -m_1 & -m_2 & -m_3 \end{pmatrix} \tag{2.69}$$

Unitarity:

$$\sum_{m_1 m_2} \begin{pmatrix} j_1 & j_2 & j_3 \\ m_1 & m_2 & m_3 \end{pmatrix} \begin{pmatrix} j_1 & j_2 & j'_3 \\ m_1 & m_2 & m'_3 \end{pmatrix} = \frac{1}{2j_3 + 1} \delta_{j_3 j_3'} \delta_{m_3 m'_3} \tag{2.70}$$

$$\sum_{j_3 m_3} (2j_3 + 1) \begin{pmatrix} j_1 & j_2 & j_3 \\ m_1 & m_2 & m_3 \end{pmatrix} \begin{pmatrix} j_1 & j_2 & j_3 \\ m'_1 & m'_2 & m'_3 \end{pmatrix} = \delta_{m_1 m'_1} \delta_{m_2 m'_2} \tag{2.71}$$

3-*j* Coefficients and the D Matrices

$$\begin{pmatrix} j_1 & j_2 & j_3 \\ m_1 & m_2 & m_3 \end{pmatrix} = \sum_{m'_1 m'_2 m'_3} \begin{pmatrix} j_1 & j_2 & j_3 \\ m'_1 & m'_2 & m'_3 \end{pmatrix} D^{(j_1)}_{m'_1 m_1}(R) D^{(j_2)}_{m'_2 m_2}(R) D^{(j_3)}_{m'_3 m_3}(R) \tag{2.72}$$

$$\int D^{(j_3)}_{m'_3 m_3} D^{(j_1)}_{m_1 m'_1} D^{(j_2)}_{m_2 m'_2} \sin\beta\, d\beta\, d\alpha\, d\gamma = 8\pi^2 \begin{pmatrix} j_1 & j_2 & j_3 \\ m_1 & m_2 & m_3 \end{pmatrix} \begin{pmatrix} j_1 & j_2 & j_3 \\ m_1' & m_2' & m_3' \end{pmatrix} \tag{2.73}$$

$$D^{(j_3)*}_{m_3 m'_3} = (2j + 1) \sum_{m_1 m'_1 m_2 m'_2} \begin{pmatrix} j_1 & j_2 & j_3 \\ m_1 & m_2 & m_3 \end{pmatrix} \begin{pmatrix} j_1 & j_2 & j_3 \\ m'_1 & m'_2 & m'_3 \end{pmatrix} D^{(j_1)}_{m_1 m'_1} D^{(j_2)}_{m_2 m'_2} \tag{2.74}$$

$$D^{(j_1)}_{m_1 m'_1} D^{(j_2)}_{m_2 m'_2} = \sum_{j_3} (2j_3 + 1) \begin{pmatrix} j_1 & j_2 & j_3 \\ m_1 & m_2 & m_3 \end{pmatrix} \begin{pmatrix} j_1 & j_2 & j_3 \\ m'_1 & m'_2 & m'_3 \end{pmatrix} D^{(j_3)}_{m_3 m'_3} \tag{2.75}$$

Special Values of the Clebsch-Gordan and 3-j Coefficients

$$(jm\,j' - m'|00) = \frac{(-)^{j-m}}{\sqrt{2j+1}}\,\delta_{jj'}\,\delta_{mm'} \tag{2.76}$$

$$(jm\,00|j'm') = \delta_{jj'}\,\delta_{mm'} \tag{2.77}$$

$$\begin{pmatrix} j_1 & j_2 & j_3 \\ 0 & 0 & 0 \end{pmatrix} = 0 \quad \text{if} \quad j_1 + j_2 + j_3 \quad \text{is odd} \tag{2.78}$$

$$\begin{pmatrix} j & 1 & j \\ -m & 0 & m \end{pmatrix} = (-)^{j-m}\frac{m}{\sqrt{j(2j+1)(j+1)}} \tag{2.79}$$

$$\begin{pmatrix} j & 2 & j \\ -m & 0 & m \end{pmatrix} = (-)^{j-m}\frac{3m^2 - j(j+1)}{\sqrt{(2j-1)(2j+1)(j+1)(2j+3)}} \tag{2.80}$$

$$\begin{pmatrix} j_1 & j_2 & J \\ \frac{1}{2} & -\frac{1}{2} & 0 \end{pmatrix} = -\sqrt{(2l_1+1)(2l_2+1)}\begin{pmatrix} l_1 & l_2 & J \\ 0 & 0 & 0 \end{pmatrix}\begin{Bmatrix} j_1 & j_2 & J \\ l_2 & l_1 & \frac{1}{2} \end{Bmatrix} \tag{2.81}$$

where $l_i = j_i \pm \frac{1}{2}$, $l_1 + l_2 + J$ even

$(\frac{1}{2}m_s l m_l|jm)$

j \ m_s	$1/2$	$-1/2$
$l + \frac{1}{2}$	$\sqrt{\frac{l + (1/2) + m}{2l+1}}$	$\sqrt{\frac{l + (1/2) - m}{2l+1}}$
$l - \frac{1}{2}$	$\sqrt{\frac{l + (1/2) - m}{2l+1}}$	$-\sqrt{\frac{l + (1/2) + m}{2l+1}}$

(2.82)

$(1m_s l m_l|jm)$

j \ m_s	1	0	1
$l + 1$	$\sqrt{\frac{(l+m)(l+m+1)}{(2l+1)(2l+2)}}$	$\sqrt{\frac{(l-m+1)(l+m+1)}{(2l+1)(l+1)}}$	$\sqrt{\frac{(l-m)(l-m+1)}{(2l+1)(2l+2)}}$
l	$\sqrt{\frac{(l+m)(l-m+1)}{2l(l+1)}}$	$-\frac{m}{\sqrt{l(l+1)}}$	$-\sqrt{\frac{(l-m)(l+m+1)}{2l(l+1)}}$
$l - 1$	$\sqrt{\frac{(l-m)(l-m+1)}{2l(2l+1)}}$	$-\sqrt{\frac{(l-m)(l+m)}{l(2l+1)}}$	$\sqrt{\frac{(l+m+1)(l+m)}{2l(2l+1)}}$

(2.83)

Recoupling Coefficients: 6*j* Symbols

Coupling of three angular moments $\mathbf{j}_1$, $\mathbf{j}_2$, $\mathbf{j}_3$. There are two orders of coupling $\mathbf{j}_1 + \mathbf{j}_2$ to get $\mathbf{j}_{12}$ that is then coupled to $\mathbf{j}_3$ to give $\mathbf{J}$. Or $\mathbf{j}_2$ and $\mathbf{j}_3$ can be coupled to $\mathbf{j}_{23}$ which when coupled to $\mathbf{j}_1$ gives $\mathbf{J}$. The unitary transformation that connects these two schemes is:

$$\langle j_1 j_2 (j_{12}) j_3;\, J | j_1, j_2 j_3 (j_{23});\, J\rangle \equiv (-)^{j_1+j_2+j_3+J} \sqrt{(2j_{12}+1)(2j_{23}+1)} \cdot \begin{Bmatrix} j_1 & j_2 & j_{12} \\ j_3 & J & j_{23} \end{Bmatrix} \qquad (2.84)$$

$$\langle j_1 j_3 (j_{13}) j_2;\, J | j_1 j_2 (j_{12}) j_3;\, J\rangle = (-)^{j_2+j_3+j_{13}+j_{12}} \sqrt{(2j_{12}+1)(2j_{13}+1)} \cdot \begin{Bmatrix} j_1 & j_2 & j_{12} \\ J & j_3 & j_{13} \end{Bmatrix} \qquad (2.85)$$

where the curly bracketed factors are *Wigner 6j symbols.*

$$\begin{Bmatrix} j_1 & j_2 & j_{12} \\ j_3 & J & j_{23} \end{Bmatrix} \text{ vanishes unless } \begin{matrix} \mathbf{j}_1 + \mathbf{j}_2 = \mathbf{j}_{12} \\ \mathbf{j}_3 + \mathbf{j}_2 = \mathbf{j}_{23} \\ \mathbf{j}_1 + \mathbf{J} = \mathbf{j}_{23} \\ \mathbf{j}_3 + \mathbf{J} = \mathbf{j}_{12} \end{matrix} \qquad (2.86)$$

The notation $\mathbf{a} + \mathbf{b} = \mathbf{c}$ is shorthand for $a + b \geqslant c \geqslant |a - b|$ (2.87)

Symmetry:

$$\begin{Bmatrix} j_1 & j_2 & j_3 \\ l_1 & l_2 & l_3 \end{Bmatrix} = \begin{Bmatrix} j_2 & j_3 & j_1 \\ l_2 & l_3 & l_1 \end{Bmatrix} = \begin{Bmatrix} j_3 & j_1 & j_2 \\ l_3 & l_1 & l_2 \end{Bmatrix} = \begin{Bmatrix} j_2 & j_1 & j_3 \\ l_2 & l_1 & l_3 \end{Bmatrix} = \begin{Bmatrix} l_1 & l_2 & j_3 \\ j_1 & j_2 & l_3 \end{Bmatrix} \qquad (2.88)$$

Unitarity:

$$\sum_{j_3} (2j_3+1)(2l_3+1) \begin{Bmatrix} j_1 & j_2 & j_3 \\ l_1 & l_2 & l_3 \end{Bmatrix} \begin{Bmatrix} j_1 & j_2 & j_3 \\ l_1 & l_2 & l'_3 \end{Bmatrix} = \delta_{l_3 l'_3} \qquad (2.89)$$

Composition of recoupling transformations:

$$\sum_{j} (-)^{j+j'+j''} (2j+1) \begin{Bmatrix} j_1 & j_2 & j' \\ j_3 & j_4 & j \end{Bmatrix} \begin{Bmatrix} j_1 & j_3 & j'' \\ j_2 & j_4 & j \end{Bmatrix} = \begin{Bmatrix} j_1 & j_2 & j' \\ j_4 & j_3 & j'' \end{Bmatrix} \qquad (2.90)$$

$$\sum_k (-)^{k_1+k_2+k}(2k+1)\begin{Bmatrix} j_1 & j'_1 & k \\ j'_2 & j_2 & j \end{Bmatrix}\begin{Bmatrix} k_1 & k_2 & k \\ j_1' & j_1 & j_2'' \end{Bmatrix}\begin{Bmatrix} k_1 & k_2 & k \\ j_2' & j_2 & j_1'' \end{Bmatrix}$$

$$= (-)^{j_1+j_2+j_1'+j_2'+j''_2+j''_1+j}\begin{Bmatrix} j_1 & j_2 & j \\ j_1'' & j_2'' & k_1 \end{Bmatrix}\begin{Bmatrix} j_1' & j_2' & j \\ j_1'' & j_2'' & k_2 \end{Bmatrix} \tag{2.91}$$

Regge relation:

$$\begin{Bmatrix} j_1 & j_2 & j_3 \\ l_1 & l_2 & l_3 \end{Bmatrix} = \begin{Bmatrix} \frac{1}{2}(j_1+j_2+l_1-l_2) & \frac{1}{2}(j_1+j_2+l_2-l_1) & j_3 \\ \frac{1}{2}(l_1+l_2+j_1-j_2) & \frac{1}{2}(l_1+l_2+j_2-j_1) & l_3 \end{Bmatrix} \tag{2.92}$$

Relation to $3j$ symbols:

$$\begin{pmatrix} j_1 & j_2 & j_3 \\ m_1 & m_2 & m_3 \end{pmatrix}\begin{Bmatrix} j_1 & j_2 & j_3 \\ l_1 & l_2 & l_3 \end{Bmatrix} = \sum_{m_1' m_2' m_3'} (-)^{l_1+l_2+l_3+m'_1+m'_2+m'_3}$$

$$\cdot \begin{pmatrix} j_1 & l_2 & l_3 \\ m_1 & m'_2 & -m'_3 \end{pmatrix}\begin{pmatrix} l_1 & j_2 & l_3 \\ -m'_1 & m_2 & m'_3 \end{pmatrix}\begin{pmatrix} l_1 & l_2 & j_3 \\ m'_1 & -m'_2 & m_3 \end{pmatrix} \tag{2.93}$$

Further relations can be obtained from unitarity properties of the $3j$ and $6j$ symbols.

$$\begin{Bmatrix} j_1 & j_2 & j_3 \\ l_1 & l_2 & l_3 \end{Bmatrix} = \sum (-)^{l_3-j_3-j_1-j_2-l_1-l_2-m_1-m'_1}\begin{pmatrix} j_1 & j_2 & j_3 \\ m_1 & m_2 & -m_3 \end{pmatrix}$$

$$\cdot \begin{pmatrix} l_1 & l_2 & j_3 \\ m'_1 & m'_2 & m_3 \end{pmatrix}\begin{pmatrix} j_2 & l_1 & l_3 \\ m_2 & m'_1 & -m'_3 \end{pmatrix}\begin{pmatrix} l_2 \cdot & j_1 & l_3 \\ m'_2 & m_1 & m'_3 \end{pmatrix} \tag{2.94}$$

$$\begin{pmatrix} j_1 & j_2 & j_3 \\ m_1 & m_2 & -m_3 \end{pmatrix}\begin{pmatrix} l_1 & l_2 & j_3 \\ m'_1 & m'_2 & m_3 \end{pmatrix} = \sum (2l_3+1)(-)^{j_1+j_2-j_3+l_1+l_2+l_3-m_1-m_1'}$$

$$\cdot \begin{Bmatrix} j_1 & j_2 & j_3 \\ l_1 & l_2 & l_3 \end{Bmatrix}\begin{pmatrix} l_2 & j_1 & l_3 \\ m_2' & m_1 & m_3' \end{pmatrix}\begin{pmatrix} j_2 & l_1 & l_3 \\ m_2 & m_1' & -m_3' \end{pmatrix} \tag{2.95}$$

Special value:

$$\begin{Bmatrix} j_1 & j_1' & 0 \\ j_2 & j_2' & j_3 \end{Bmatrix} = \frac{(-)^{j_1+j_2+j_3}}{\sqrt{(2j_1+1)(2j_2+1)}}\,\delta_{j_1j_1'}\,\delta_{j_2j_2'} \tag{2.96}$$

Other notation:

$$\begin{Bmatrix} j_1 & j_2 & j_3 \\ l_1 & l_2 & l_3 \end{Bmatrix} = (-)^{j_1+j_2+l_1+l_2} W(j_1 j_2 l_2 l_1; j_3 l_3) \tag{2.97}$$

Coupling of Four Angular Momenta: Wigner 9j Symbols

Four angular momenta can be coupled as follows to achieve a final value of J;

$$\mathbf{j}_1 + \mathbf{j}_3 = \mathbf{j}_{13},\ \mathbf{j}_2 + \mathbf{j}_4 = \mathbf{j}_{24},\ \mathbf{j}_{13} + \mathbf{j}_{24} = \mathbf{J} \tag{2.98}$$

or

$$\mathbf{j}_1 + \mathbf{j}_2 = \mathbf{j}_{12},\ \mathbf{j}_3 + \mathbf{j}_4 = \mathbf{j}_{34},\ \mathbf{j}_{12} + \mathbf{j}_{34} = \mathbf{J} \tag{2.99}$$

The unitarity transformation connecting these two schemes is

$$\langle j_1 j_3 (j_{13}) j_2 j_4 (j_{24}); J | j_1 j_2 (j_{12}) j_3 j_4 (j_{34}); J \rangle$$

$$= \sqrt{(2j_{13}+1)(2j_{24}+1)(2j_{12}+1)(2j_{34}+1)} \begin{Bmatrix} j_1 & j_2 & j_{12} \\ j_3 & j_4 & j_{34} \\ j_{13} & j_{24} & J \end{Bmatrix} \tag{2.100}$$

where the curly bracketed factor is the Wigner $9j$ symbol.

Unitarity:

$$\sum (2j_{13}+1)(2j_{24}+1) \begin{Bmatrix} j_1 & j_2 & j_{12} \\ j_3 & j_4 & j_{34} \\ j_{13} & j_{24} & J \end{Bmatrix} \begin{Bmatrix} j_1 & j_2 & j_{12}' \\ j_3 & j_4 & j_{34}' \\ j_{13} & j_{24} & J \end{Bmatrix}$$

$$= \frac{1}{(2j_{12}+1)(2j_{34}+1)} \delta_{j_{12}' j_{12}} \delta_{j_{34}' j_{34}} \tag{2.101}$$

Composition:

$$\sum_{j_{13}\, j_{24}} (-)^{(j_2+j_4+j_{24})+(j_2+j_3+j_{23})} (2j_{13}+1)(2j_{24}+1)$$

$$\cdot \begin{Bmatrix} j_1 & j_3 & j_{13} \\ j_2 & j_4 & j_{24} \\ j_{12} & j_{34} & J \end{Bmatrix} \begin{Bmatrix} j_1 & j_4 & j_{14} \\ j_3 & j_2 & j_{23} \\ j_{13} & j_{24} & J \end{Bmatrix} = (-)^{j_3+j_4+j_{34}} \begin{Bmatrix} j_1 & j_4 & j_{14} \\ j_2 & j_3 & j_{23} \\ j_{12} & j_{34} & J \end{Bmatrix} \tag{2.102}$$

$$\sum_{jj'j''} (2j+1)(2j'+1)(2j''+1) \begin{Bmatrix} j_1 & j_2 & j_{12} \\ j_3 & j_4 & j \\ j_{13} & j' & j'' \end{Bmatrix} \begin{Bmatrix} j & j'' & j_{12} \\ j_3 & j_5 & j_{35} \\ j_4 & j_6 & j_{46} \end{Bmatrix} \begin{Bmatrix} j' & j_2 & j_4 \\ j'' & j_5 & j_6 \\ j_{13} & j_{25} & j_{46}' \end{Bmatrix}$$

$$= \frac{\delta_{j_{46},j_{46}'}}{2j_{46}+1} \begin{Bmatrix} j_{46} & j_{13} & j_{25} \\ j_{12} & j_1 & j_2 \\ j_{35} & j_3 & j_5 \end{Bmatrix} \qquad (2.103)$$

Symmetries: an odd permutation of the rows or columns changes the numerical values by the factor $(-)^{j_1+j_2+j_3+j_4+j_{12}+j_{34}+j_{13}+j_{24}+J}$.

An even permutation or a transposition leaves the value unchanged.

$$\begin{Bmatrix} j_1 & j_2 & j_{12} \\ j_1 & j_2 & j_{12} \\ l_1 & l_2 & l_3 \end{Bmatrix} = 0 \quad \text{if} \quad 2(j_1+j_2+j_{12}) + (l_1+l_2+l_3) \quad \text{is odd.}$$

(2.104)

Relation to $3j$ coefficients:

$$\begin{Bmatrix} j_1 & j_2 & j_{12} \\ j_3 & j_4 & j_{34} \\ j_{13} & j_{24} & J \end{Bmatrix} = \sum_{m's} \begin{pmatrix} j_1 & j_2 & j_{12} \\ m_1 & m_2 & m_{12} \end{pmatrix} \begin{pmatrix} j_3 & j_4 & j_{34} \\ m_3 & m_4 & m_{34} \end{pmatrix} \begin{pmatrix} j_{13} & j_{24} & J \\ m_{13} & m_{24} & M \end{pmatrix}$$

$$\cdot \begin{pmatrix} j_1 & j_3 & j_{13} \\ m_1 & m_3 & m_{13} \end{pmatrix} \begin{pmatrix} j_2 & j_4 & j_{24} \\ m_2 & m_4 & m_{24} \end{pmatrix} \begin{pmatrix} j_{12} & j_{34} & J \\ m_{12} & m_{34} & M \end{pmatrix} \qquad (2.105)$$

Relation to $6j$ coefficients:

$$\begin{Bmatrix} j_1 & j_2 & J \\ j_1' & j_2' & J \\ k & k & 0 \end{Bmatrix} = (-)^{j_2+j'_1+k+J} \frac{\begin{Bmatrix} j_1 & j_2 & J \\ j_2' & j_1' & k \end{Bmatrix}}{\sqrt{(2J+1)(2k+1)}} \qquad (2.106)$$

$$\begin{Bmatrix} j_1 & j_2 & j_{12} \\ j_3 & j_4 & j_{34} \\ j_{13} & j_{24} & J \end{Bmatrix} = \sum (-)^{2j}(2j+1) \begin{Bmatrix} j_1 & j_3 & j_{13} \\ j_{24} & J & j \end{Bmatrix} \begin{Bmatrix} j_2 & j_4 & j_{24} \\ j_3 & j & j_{34} \end{Bmatrix} \begin{Bmatrix} j_{12} & j_{34} & J \\ j & j_1 & j_2 \end{Bmatrix}$$

(2.107)

$$(-)^{2j}\begin{Bmatrix} j_3 & j_4 & j_{34} \\ j_2 & j & j_{24} \end{Bmatrix}\begin{Bmatrix} j_{13} & j_{24} & J \\ j & j_1 & j_3 \end{Bmatrix} = \sum_{j_{12}} (2j_{12}+1)\begin{Bmatrix} j_1 & j_2 & j_{12} \\ j_3 & j_4 & j_{34} \\ j_{13} & j_{24} & J \end{Bmatrix} \cdot \begin{Bmatrix} j_1 & j_2 & j_{12} \\ j_{34} & J & j \end{Bmatrix} \qquad (2.108)$$

Special value:

$$\begin{Bmatrix} j_1 & j_2 & J \\ j_3 & j_4 & J \\ L & L & 1 \end{Bmatrix} = \frac{j_1(j_1+1) - j_3(j_3+1) - j_2(j_2+1) + j_4(j_4+1)}{\sqrt{4J(J+1)(2J+1)^2L(L+1)(2L+1)^2}} \cdot (-)^{L+J+j_1+j_4}\begin{Bmatrix} j_1 & j_2 & J \\ j_4 & j_3 & L \end{Bmatrix} \qquad (2.109)$$

3. TIME REVERSAL

The formulation of time reversal that will be presented here will generalize and modify the usual textbook discussion in several respects. (For a summary of the resultant changes see p. 937) First the concept of time reversal is extended so that it can be applied to systems with nonhermitian Hamiltonians.* The complex Hamiltonian of the optical model is an example of where such an extension is needed in nuclear physics. In the standard formulation, this Hamiltonian is said to be not time-reversal invariant; but one hastens to add that the symmetry of the scattering amplitude is still preserved and detailed balance is maintained. The customary definition thus sets up a distinction of no observable significance. In the formulation of this appendix the choice of definition of time reversal is such that whenever the S-matrix is symmetric, the corresponding Hamiltonian is time-reversal invariant and vice versa. For hermitian Hamiltonians the two formulations, the traditional one and the one proposed here, are identical. As a by-product it becomes necessary to discuss the behavior of the time reverse of systems whose Hamiltonian is not time-reversal invariant. Such systems do exist as has been shown by experiments on the decay of neutral K-mesons [Christiansen, Cronin, Fitch, and Turlay (64)]. In the main text we shall be mostly concerned with systems that are time-reversal invariant. For a description of the various possible tests of time-reversal invariance of nuclear systems that have been considered see

*These for physical reasons are absorptive.

Henley and Jacobson (59) and a recent review article by Henley (69). As of this writing, no conclusive evidence for a violation of time-reversal invariance, other than in the neutral kaon decay, has been found.

Following Messiah (63) let us begin with a discussion of time reversal in classical mechanics. Consider the simple classical particle trajectory shown in Fig. A.2 in which the particle position from $t = 0$ to $t = T$ is recorded. The classical time-reversed path is defined to be the path obtained by reversing the velocity, without changing its magnitude, at every point in the trajectory. The new trajectory obtained in this way coincides spatially with the original one but is described in the opposite direction by the particle. It is important to realize that such a reversed path can always be achieved dynamically. If the forces are conservative, then we need only require that at $t = T$, the particle's velocity be reversed, and the same forces be allowed to act. If the forces are not conservative, for example, if the particle moves in a viscous medium, the time-reversed path will be described by the particle only if the forces are changed appropriately. We need to change the Hamiltonian or, as we shall say, the time-reversed system is not the same as the original one. In the case when viscous frictional forces act on the original system, $0 < t < T$, the new forces, operating while the time-reversed trajectory is described, must be accelerating rather than decelerating, being equal at every point in magnitude to the forces that are frictional but opposite in sign. If the frictional force is $\mathbf{F}(\mathbf{r}, \mathbf{v})$ then the force acting in the time-reversed system is $F(\mathbf{r}, -\mathbf{v})$. If these forces, $\mathbf{F}(\mathbf{r}, -\mathbf{v})$, act once the particle's velocity is reversed at $t = T$, the time-reversed trajectory will be described by the particle.

The trajectory in the interval $0 < t < T$ will be referred to as $\mathbf{r}(t)$. The time t in this expression is that given by a "laboratory clock." The position of the particle on the time-reversed trajectory can also be written as a function of t. However another possibility is to use the first trajectory to define a clock and refer the reversed motion to it. Turning to Fig. A.2 we can see how this can be done. For each point on the original trajectory $(0 < t < T)$, there is a

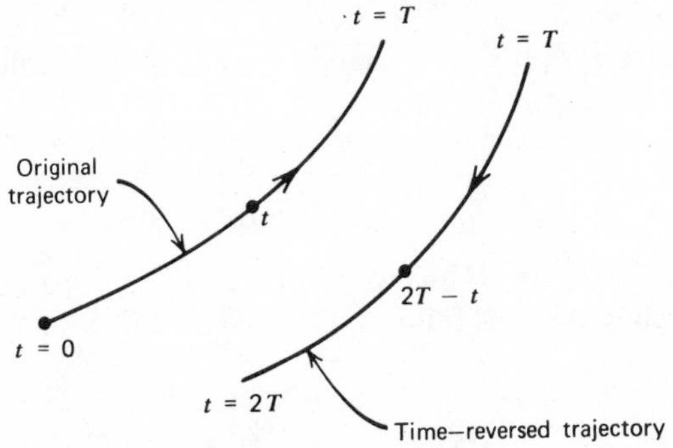

FIG. A.2.

corresponding point on the reversed trajectory ($T < t < 2T$). The position at t is identical with the position on the reversed trajectory at a time equal to $2T - t$. We may write the position of the particle on the reversed trajectory as a function of the laboratory clock time t', or *equivalently* as a function of t where t is defined by

$$t' \equiv 2T - t \qquad t < T \tag{3.1}$$

referring the time t' along the reversed orbit to the time t at which the particle arrives at the corresponding point on the original trajectory. Note that t' increases from T to $2T$ as t goes from T to 0. The t-clock thus effectively runs counterclockwise, and it is essentially for this reason that the phrase "time reversed" is used. Of course time is not reversed; rather the motion of the hands of the clock is reversed.

We turn now to the development of these ideas in quantum mechanics. Let $\mathcal{S}$ be a system whose time evolution is governed by the Hamiltonian H so that if the state of $\mathcal{S}$ at $t = 0$ is described by $\Psi(0)$, at a time t later:

$$\Psi(t) = \exp(-i/\hbar Ht)\Psi(0) \tag{3.2}$$

Time reversal is connected with the use of t defined by (3.1) as the independent variable; that is, through the introduction of a clock that runs counterclockwise. Let us first investigate the effects of using such a description of time and later return to the problem of defining the time-reversed system and state. Suppose then that $\Phi(t')$ is the state of a system $\tilde{\mathcal{S}}$ (which is not necessarily the same as $\mathcal{S}$) at a time t'. In terms of t

$$\Phi(t') = \Phi(2T - t) \equiv \tilde{\Phi}(t) \tag{3.3}$$

defining $\tilde{\Phi}(t)$. The time development of Φ for $t < T$, that is, for $T < t' < 2T$ is governed by a Hamiltonian operator $\tilde{H}^\dagger$ whose nature depends upon $\tilde{\mathcal{S}}$:

$$\Phi(t') = \exp[-i/\hbar\tilde{H}^\dagger(t' - T)]\Phi(T) \tag{3.4}$$

or

$$\tilde{\Phi}(t) = \exp[-i/\hbar\tilde{H}^\dagger(T - t)]\Phi(T) \tag{3.5}$$

From this equation it is possible to obtain the equation satisfied by $\tilde{\Phi}$ by differentiating with respect to t. We find

$$\tilde{H}^\dagger\tilde{\Phi}(t) = -i\hbar\frac{\partial\tilde{\Phi}}{\partial t} \tag{3.6}$$

Assuming that H, or better H^T as we shall see, and $\tilde{H}^\dagger$ span the same Hilbert space, one can relate them as follows. Compare (3.6) with the equation satisfied by Ψ^*:

$$H^*\Psi^* = -i\hbar\frac{\partial\Psi^*}{\partial t} \tag{3.7}$$

Thus if

$$\tilde{\Phi} = K\Psi^*(t) \equiv \Theta\Psi(t) \tag{3.8}$$

where K is a unitary operator then

$$K^\dagger \tilde{H}^\dagger K = H^*$$

or

$$\tilde{H} = KH^T K^\dagger \tag{3.9}$$

The operator Θ consists in taking the complex conjugate and then transforming with K. Θ is referred to as *anitunitary* operator. In contrast with a unitary operator it does not conserve the inner product in Hilbert space but only its magnitude. To show this consider $\langle\Psi|\Psi'\rangle$. Note

$$\langle\Psi|\Psi'\rangle = \langle\Psi^*|\Psi'^*\rangle^* = \langle\Psi'^*|\Psi^*\rangle = \langle K^\dagger\tilde{\Phi}'|K^\dagger\tilde{\Phi}\rangle$$

or

$$\langle\Psi|\Psi'\rangle = \langle\tilde{\Phi}'|\tilde{\Phi}\rangle \tag{3.10}$$

proving the assertion.

To determine how an operator transforms consider $\langle\Psi|\mathrm{O}\Psi'\rangle$. Following the steps that led to (3.10) yields

$$\langle\Psi|\mathrm{O}\Psi'\rangle = \langle\mathrm{O}^*\Psi'^*|\Psi^*\rangle = \langle\mathrm{O}^*K^\dagger\tilde{\Phi}'|K^\dagger\tilde{\Phi}\rangle$$

or

$$\langle\Psi|\mathrm{O}\Psi'\rangle = \langle\tilde{\Phi}'|K\mathrm{O}^T K^\dagger\tilde{\Phi}\rangle$$

Comparing with (3.9) suggests the definition

$$\tilde{\mathrm{O}} \equiv K\mathrm{O}^T K^\dagger \tag{3.11}$$

Under the transformation Θ, $\mathrm{O} \to \tilde{\mathrm{O}}$, and

$$\langle\Psi|\mathrm{O}\Psi'\rangle = \langle\tilde{\Phi}'|\tilde{\mathrm{O}}\tilde{\Phi}\rangle \tag{3.12}$$

The definition of the time-reversed state $\tilde{\Psi}(t)$ corresponding to $\Psi(t)$ of (3.2) requires (i) fixing a relation between the two states at a given time, in this case T, and (ii) determining K from which $\tilde{H}$ can be obtained from H (3.9) and the time development of $\tilde{\Psi}$ can be calculated according to (3.5). Recalling the classical discussion we require that at $t = T$ and for a spinless particle

$$|\langle\mathbf{p}|\Psi(T)\rangle| = |\langle-\mathbf{p}|\tilde{\Psi}(T)| \qquad \text{that is} \qquad |\tilde{\mathbf{p}}\rangle = |-\mathbf{p}\rangle \tag{3.13}$$

so that the time-reversed state has the same probability density in momentum space at $-\mathbf{p}$ as the original state at $\mathbf{p}$ at a time T. This is the quantum analog to the classical condition that at a time T the velocity of the particle is reversed.

Only the magnitudes of $\langle \mathbf{p}|\Psi\rangle$ and $\langle -\mathbf{p}|\tilde{\Psi}\rangle$ can be compared because the antiunitary operator Θ relates $\tilde{\Psi}$ with Ψ^* not with Ψ. In coordinate space the relation between Ψ and $\tilde{\Psi}$ is

$$|\langle \mathbf{r}|\Psi(T)\rangle| = |\langle \mathbf{r}|\tilde{\Psi}(T)\rangle| \qquad \text{that is} \qquad |\tilde{\mathbf{r}}\rangle = |\mathbf{r}\rangle \tag{3.14}$$

To determine $\tilde{\mathbf{p}}$ from (3.13) consider the relation

$$\langle \mathbf{p}_0|\mathbf{p}|\mathbf{p}_1\rangle = \mathbf{p}_1\,\delta(\mathbf{p}_0 - \mathbf{p}_1)$$

But according to (3.12) and (3.13) the left-hand side can be rewritten to yield

$$\langle -\mathbf{p}_1|\tilde{\mathbf{p}}| - \mathbf{p}_0\rangle = \mathbf{p}_1\,\delta(\mathbf{p}_0 - \mathbf{p}_1)$$

This is possible if $\tilde{\mathbf{p}} = -\mathbf{p}$. The relation of $\tilde{\mathbf{r}}$ and $\mathbf{r}$ can be obtained in the same way. Hence

$$\tilde{\mathbf{r}} = K\mathbf{r}^T K^\dagger = \mathbf{r}$$

$$\tilde{\mathbf{p}} = K\mathbf{p}^T K^\dagger = -\mathbf{p} \tag{3.15}$$

When K satisfies these relations for a spinless particle (the additional requirements for particles with spin will be discussed later) then

$$\tilde{\Psi} = K\Psi^* \tag{3.16}$$

is the "time-reversed" state and

$$\tilde{O} = KO^T K^\dagger \tag{3.17}$$

is the "time reverse" of operator O.

From (3.9) we see that with this definition the time reverse of the Hamiltonian operator H is $\tilde{H}$ in agreement with (3.17). Moreover if H depends upon the momentum operator, so that $H = H(\mathbf{r}, \mathbf{p})$

$$\tilde{H} = H^T(\tilde{\mathbf{r}}, \tilde{\mathbf{p}}) = H^T(\mathbf{r}, -\mathbf{p}) \tag{3.18}$$

Note that the time dependence of $\tilde{\Psi}(t)$ is, according to (3.4), determined by $\tilde{H}^\dagger$ rather than $\tilde{H}$. This is important only when H is nonhermitian. If H is absorptive, as one would expect from the classical analog discussed earlier, the time-reversed system should involve a regenerative potential that corresponds in the classical case to changing a frictional force into an accelerating one. Hence the Hamiltonian $\tilde{H}^\dagger$ describing the time-reversed system $\tilde{\mathbb{S}}$ is obtained by time-reversing H according to (3.9) and then taking the hermitian adjoint to convert from an absorptive to a regenerative potential.

We now turn to the question of time-reversal invariance. We say that an operator O is time-reversal invariant if

$$O = \tilde{O}$$

for then the matrix element of O between the states Ψ and Φ and between the corresponding time-reversed states $\tilde{\Phi}$ and $\tilde{\Psi}$ are equal; that is, from (3.12)

$$\langle \Psi|O\Phi\rangle = \langle \tilde{\Phi}|O\tilde{\Psi}\rangle \qquad \text{time-reversal invariance} \tag{3.19}$$

When H is a function of $\mathbf{r}$ only, H according to (3.18) is time-reversal invariant. Thus if the complex potential of the optical model is local, the corresponding H is time-reversal invariant. More generally for a spin independent H we shall show that H need only be symmetric, that is, $H = H^T$ for time-reversal invariance to hold.

To prove this last result and to discuss more general situations it is necessary to specify the properties of K, a task to which we now turn.

Both the traditional discussion of time reversal and the discussion in this appendix employ the identical relation between Ψ and $\tilde{\Psi}$; that is the transformation Θ between the two is given by (3.16) in both treatments. However the traditional definition of time-reversal invariance requires O^* to equal $\tilde{O}$ rather than (3.19). For hermitian operators the two definitions are identical. For nonhermitian operators, they differ. But as we can see from the discussion just above (3.19) and comparing with the classical case, it makes more sense physically to use $\tilde{H}^\dagger$ as the effective Hamiltonian to describe the time dependence of $\tilde{\Psi}$. And, although we shall not prove it here, such use also makes it possible to relate the symmetry of the scattering amplitude with the time-reversal invariance of H without requiring the introduction of additional concepts.

Properties of K

A condition is put on K by the requirement that application of two time reversals on an operator must yield the original operator. Hence

$$(\tilde{\tilde{O}}) = O \qquad \text{or} \qquad K(KO^TK^\dagger)^TK^{-1} = 0$$

This is identical with

$$[KK^*, \mathrm{O}] = 0 \tag{3.20}$$

This equation states that KK^* commutes with any operator. It must therefore be proportional to the unit operator. Since KK^* is unitary it follows that

$$KK^* = \exp(i\eta)$$

or

$$K = \exp(i\eta)K^T$$

Taking the transpose of both sides we immediately obtain

$$\exp(2i\eta) = 1$$

so that $\eta = \mathrm{o}$ or π. If $\eta = \mathrm{o}$, K is symmetric; if $\eta = \pi$, K is antisymmetric.

Examples

(a) We begin with position operator $\mathbf{r}$ and its conjugate momentum $\mathbf{p}$. According to (3.15) K must satisfy the two equations

$$\mathbf{r} = K\mathbf{r}^TK^\dagger \qquad \mathbf{p} = -K\mathbf{p}^TK^\dagger$$

In the representation in which $\mathbf{r}$ is diagonal,

$$\mathbf{r}^T = \mathbf{r}$$

The operator $\mathbf{p}$ is given by $(\hbar/i)\nabla$ so that

$$\mathbf{p}^T = -\mathbf{p}$$

Comparing with (3.15) we see that K can be taken to be a constant say ϵ, where $|\epsilon| = 1$ as far as its dependence on $\mathbf{r}$ and $\mathbf{p}$ is concerned.

With the time reversal of the two fundamental vectors $\mathbf{r}$ and $\mathbf{p}$ determined, the time-reversal properties of operators such as $\mathbf{L}$, the angular momentum, constructed from $\mathbf{r}$ and $\mathbf{p}$ can be readily obtained. From

$$\mathbf{L} = \mathbf{r} \times \mathbf{p}$$

we find

$$\tilde{\mathbf{L}} = (\mathbf{r} \times \mathbf{p})^T = (\mathbf{r}^T \times \mathbf{p}^T) = -(\mathbf{r} \times \mathbf{p})$$

or

$$\tilde{\mathbf{L}} = -\mathbf{L} \tag{3.21}$$

Returning to the Hamiltonian, we see that when H is a function of $\mathbf{r}$ and $\mathbf{p}$ only,

$$\tilde{H} = H^T$$

Then the time-reversal invariance becomes the condition $H^T = H$, or H must be symmetric. In a one-channel theory, for example, any local or symmetric potential is time reversal invariant.

(b) When a particle has internal degrees of freedom, the corresponding operators are not functions of $\mathbf{r}$ and $\mathbf{p}$ and their behavior under the operator of time reversal must be determined by other considerations. In these volumes we will be concerned with three such internal degrees of freedom, spin, charge, and isospin. As a first example, consider spin. The spin of a particle is an angular momentum and will combine additively with orbital angular momenta to give the total angular momentum J. In order that the properties of the sum J be preserved under time reversal, $\boldsymbol{\sigma}$ like $\mathbf{L}$ must be odd under time reversal:

$$\tilde{\boldsymbol{\sigma}} = -\boldsymbol{\sigma} \tag{3.22}$$

Somewhat more physically we may envisage a more detailed description of the spin degree of freedom. Such a model will always involve a rotation of some kind and therefore will give rise to time-reversal properties identical with those of the orbital angular momentum.

In the usual representation in which σ_x and σ_y are symmetric and σ_y is antisymmetric, the unitary operator K must satisfy

$$K\sigma_x K^\dagger = -\sigma_x$$

$$K\sigma_y K^\dagger = \sigma_y$$

$$K\sigma_z K^\dagger = -\sigma_z$$

These conditions are met by

$$K = -i\sigma_y = \begin{pmatrix} 0 & -1 \\ 1 & 0 \end{pmatrix} \tag{3.23}$$

where again it is possible to multiply this K by a complex number of unit magnitude. The time-reversed wave function for a positive spin state

$$\psi = \begin{pmatrix} 1 \\ 0 \end{pmatrix}$$

is

$$\tilde{\psi} = -i\sigma_y \begin{pmatrix} 1 \\ 0 \end{pmatrix} = \begin{pmatrix} 0 \\ 1 \end{pmatrix}$$

In other words the time-reversed state has the opposite spin orientation.

It is also possible to obtain the transformation operator K for spin 1/2 particles by exploiting the properties of the Dirac equation for such particles. The equation implicitly treats the spin and orbital angular momentum on the same footing so that there is no surprise that the conclusion remains unchanged in the relativistic situation. The Dirac Hamiltonian for a free particle of mass m is

$$H = c(\boldsymbol{\alpha}\cdot\mathbf{p}) + \beta mc^2 \tag{3.24}$$

Assuming H to be time-reversal invariant we write

$$\tilde{H} = KH^TK^\dagger = H \qquad \text{for time-reversal invariance}$$

Inserting (3.24) on both sides one obtains

$$K[-c(\boldsymbol{\alpha}^*\cdot\mathbf{p}) + \beta mc^2]K^\dagger = c(\boldsymbol{\alpha}\cdot\mathbf{p}) + \beta mc^2$$

The consequent conditions on $\boldsymbol{\alpha}$ and β are

$$\tilde{\boldsymbol{\alpha}} = K\boldsymbol{\alpha}^*K^\dagger = -\boldsymbol{\alpha}$$

$$\tilde{\beta} = K\beta K^\dagger = \beta$$

It can be readily verified that these equations are satisfied with the K given by (3.23). To illustrate consider the effect of time reversal on a plane-wave solution of the Dirac equation for a particle of mass m energy E and momentum $\mathbf{p}$ with positive spin orientation when viewed in the rest frame. Then

$$\psi = \begin{pmatrix} 1 \\ 0 \\ cp_z/(mc^2 + E) \\ c(p_x + ip_y)/(mc^2 + E) \end{pmatrix} \exp\,(i\mathbf{p}/\hbar\cdot\mathbf{r})$$

The time-reversed solution ψ is

$$\tilde{\psi} = K\psi^* = -i\sigma_y\psi^*$$

or

$$\tilde{\psi} = \begin{pmatrix} 0 \\ 1 \\ -c(p_x - ip_y)/(mc^2 + E) \\ cp_z/(mc^2 + E) \end{pmatrix} \exp(-i\mathbf{p}/\hbar\cdot\mathbf{r})$$

Thus $\tilde{\psi}$ corresponds to a particle with reversed spin and momentum.

Problems. 1. Prove that ρ and $\mathbf{j}$ defined by

$$\rho = \psi^*\psi \qquad \mathbf{j} = \psi^*\boldsymbol{\alpha}\psi$$

transform as follows

$$\tilde{\rho} = \rho \qquad \tilde{\mathbf{j}} = -\mathbf{j} \tag{3.25}$$

Note that $\psi^*\psi$ is a shorthand for $\Sigma\psi_\alpha^*\psi_\alpha$.

2. Show that under time reversal, the Dirac γ_μ transform as follows:

$$\tilde{\gamma}_k = -\gamma_k$$
$$\tilde{\gamma}_4 = \gamma_4 \tag{3.26}$$

(c) We turn next to the electromagnetic field. *Presuming time-reversal invariance* of the interaction term $(-e/c\mathbf{j}\cdot\mathbf{A})$, $e\mathbf{A}$ must be odd under time reversal since $\mathbf{j}$ is odd. It will be convenient to choose e to be invariant so that $\mathbf{A}$ itself is odd:

$$\tilde{\mathbf{A}} = -\mathbf{A} \tag{3.27}$$

It follows from the relations between $\mathbf{A}$ and the electromagnetic fields that

$$\tilde{\mathbf{H}} = -\mathbf{H} \tag{3.28}$$

and

$$\tilde{\mathbf{E}} = \mathbf{E} \tag{3.29}$$

These results may also be verified directly from the Maxwell equations.

(d) The time reverse of the isospin operators is given by (3.17). Thus

$$\tilde{\tau}_x = K\tau_x{}^T K^\dagger = K\tau_x K^\dagger$$
$$\tilde{\tau}_y = -K\tau_y K^\dagger$$
$$\tilde{\tau}_z = K\tau_z K^\dagger$$

The choice of K is dictated by the requirement that the charge remain invariant so that τ_z should not be changed by time reversal. Physically this means that time reversal cannot change a proton into a neutron. K must then be of the form $a + b\tau_z$. It causes no loss of generality to take K to be a constant so that

$$\begin{aligned} \tilde{\tau}_x &= \tau_x \\ \tilde{\tau}_y &= -\tau_y \\ \tilde{\tau}_z &= \tau_z \end{aligned} \tag{3.30}$$

One can verify that with this choice the β-ray Hamiltonian is time-reversal invariant. But for such verification, we must discuss time reversal for quantum field theories.

Time Reversal in Quantum Field Theory

The results (3.16) and (3.17) obtained earlier in this appendix apply as well to the operators and states of quantum field theory. In this domain, the procedure we have adopted corresponds to Gasiorowicz's (67) "second definition of the antiunitary operation," and is the one he also employs. This differs from the more traditional treatments as given for example by Schweber (61); and again the differences are of no practical consequence as long as matrix elements of hermitian operators are involved.

Let $\hat{\psi}(\mathbf{r})$ be the nonrelativistic operator (see Eq. VII.1.69) that destroys a particle at $\mathbf{r}$. Then according to (3.17) the time-reversed operator $\tilde{\psi}(\mathbf{r})$ is

$$\hat{\tilde{\psi}}(\mathbf{r}) = \hat{K}\hat{\psi}^T\hat{K}^\dagger \tag{3.31}$$

where the hat on $\hat{K}$ indicates that this quantity will generally operate on a many-body state vector.* To obtain $\hat{\psi}^T$ we shall use a representation

$$\tilde{\psi} = K\psi^* \tag{3.32}$$

in which $\mathbf{r}_i$ is diagonal. The possible state vectors in addition to the vacuum state $|0\rangle$ include $|\mathbf{r}_1\rangle$, $|\mathbf{r}_1, \mathbf{r}_2\rangle$, . . . , $|\mathbf{r}_1, \mathbf{r}_2, \ldots, \mathbf{r}_N\rangle$ . . . eigenvectors of the operators $\hat{\mathbf{r}}_1$, $\hat{\mathbf{r}}_1$ and $\hat{\mathbf{r}}_2$, $\hat{\mathbf{r}}_1$, $\hat{\mathbf{r}}_2$, . . . , $\hat{\mathbf{r}}_N$ respectively. Following the techniques given in Section VII.1, or using (VIII.1.69) directly, one finds that

$$\hat{\psi}(\mathbf{r}) = |0\rangle\langle\mathbf{r}| + \int d\mathbf{r}_1 |0, \mathbf{r}_1\rangle\langle\mathbf{r},\mathbf{r}_1| + \int d\mathbf{r}_1 \int d\mathbf{r}_2 |0, \mathbf{r}_1, \mathbf{r}_2\rangle\langle\mathbf{r}, \mathbf{r}_1, \mathbf{r}_2| + \ldots \tag{3.33}$$

The operator $\hat{\psi}(\mathbf{r})$ thus destroys a particle in the volume element $d\mathbf{r}$ located at $\mathbf{r}$.

---

*In this section, K without a "hat" will operate on a single-particle wave function so that according to (3.16)

$$\tilde{\psi} = K\psi^*$$

The single-particle wave function $\psi(\mathbf{r})$ is given by

$$\psi(\mathbf{r}) = \langle 0|\hat{\psi}(\mathbf{r})|\mathbf{r}\rangle \tag{3.34}$$

If its time reverse is

$$\tilde{\psi}(\mathbf{r}) = \langle \mathbf{r}|\hat{\tilde{\psi}}|0\rangle$$

then

$$\hat{\tilde{\psi}} = \hat{\psi}^{\dagger}K^{\dagger} = \hat{K}\hat{\psi}^{T}\hat{K}^{\dagger} \tag{3.35}$$

and

$$K\hat{\psi} = \hat{K}\hat{\psi}^{*}\hat{K}^{\dagger}$$

where K operates on $\hat{\psi}$. Time reversal thus transforms a destruction operator $\hat{\psi}$ into a creation operator.

To obtain the behavior under time reversal in momentum space we employ the expansion (VII.1.69)

$$\hat{\psi} = \sum a_k\phi_k(\mathbf{r}) \tag{3.36}$$

According to (3.35)

$$\hat{\tilde{\psi}} = \sum a_k^{\dagger}\phi_k^{*}K^{\dagger} = \sum a_k^{\dagger}\tilde{\phi}_k$$

For spinless free particles

$$\phi_k = \frac{1}{\sqrt{\Omega}}\exp(i\mathbf{k}\cdot\mathbf{r}) \qquad \text{and} \qquad K = 1$$

where Ω is a normalization volume. Expanding $\hat{\tilde{\psi}}$ as follows

$$\hat{\tilde{\psi}} = \frac{1}{\sqrt{\Omega}}\sum_k \tilde{a}_k \exp(i\mathbf{k}\cdot\mathbf{r})$$

where $\tilde{a}_k$ is defined to be time reverse of a_k we find upon comparing with (3.36) that

$$\tilde{a}_k = a_{-k}^{\dagger} \tag{3.37}$$

Thus the time-reversal transformation converts a destruction operator for a particle of momentum $\hbar\mathbf{k}$ into a creation operator for a particle of momentum $(-\hbar\mathbf{k})$. If the particle has spin, the spin orientation would have been reversed upon time reversal.

Let us examine the behavior of some typical field theory operators. The charge density operator

$$\hat{\rho} = \hat{\psi}^{\dagger}\hat{\psi}$$

becomes upon time reversal (see Eq. 3.35)

$$\begin{aligned}\hat{\tilde{\rho}} &= \hat{K}(\hat{\psi}^{\dagger}\hat{\psi})^{T}\hat{K}^{\dagger}\\ &= \hat{K}\hat{\psi}^{T}\hat{\psi}^{*}\hat{K}^{\dagger} = (\hat{K}\hat{\psi}^{T}\hat{K}^{\dagger})(\hat{K}\hat{\psi}^{*}\hat{K}^{\dagger})\\ &= \hat{\psi}^{\dagger}K^{\dagger}K\hat{\psi} = \hat{\psi}^{\dagger}\hat{\psi}\end{aligned}$$

or

$$\hat{\hat{\rho}} = \hat{\rho} \tag{3.38}$$

Hence $\hat{\rho}$ is invariant under time reversal.

The momentum operator $\hat{\mathbf{p}}$ for a field is defined by

$$\hat{\mathbf{p}} = \frac{\hbar}{i} \int \hat{\psi}^\dagger \nabla \hat{\psi} \, d\mathbf{r} \tag{3.39}$$

Then following the steps that lead to (3.38) one finds that

$$\hat{\hat{\mathbf{p}}} = \frac{\hbar}{i} \int (\nabla \hat{\psi}^\dagger) \hat{\psi} d\mathbf{r}$$

Integrating by parts, dropping contributions from "infinity" one obtains the expected result

$$\hat{\hat{\mathbf{p}}} = -\hat{\mathbf{p}} \tag{3.40}$$

Under time reversal the field momentum changes sign.

As a final example consider the β-decay Hamiltonian density in the non-relativistic limit for the nucleons:

$$\hat{\mathscr{H}} = C_F g [\tau_+ \hat{\psi}_e^\dagger \hat{\psi}_\nu + \tau_- \hat{\psi}_\nu^\dagger \hat{\psi}_e] \tag{3.41}$$

Only the Fermi term (Chapter IX) is given. Then using (3.30)

$$\tau_+ \hat{\psi}_e^\dagger \hat{\psi}_\nu \leftrightarrow \tau_- \hat{\psi}_\nu^\dagger \hat{\psi}_e \qquad \text{under time reversal}$$

Hence

$$\hat{\hat{\mathscr{H}}} = \hat{\mathscr{H}}$$

and the Hamiltonian density (3.41) is time-reversal invariant.

4. Shell-Model States

(a) Single nucleon states in the shell model

Shell	State
1	$1s_{1/2}$
2	$1p_{3/2}, 1p_{1/2}$
3	$1d_{5/2}, 2s_{1/2}, 1d_{3/2}$
4	$1f_{7/2}$
5	$2p_{1/2}, 1f_{5/2}, 2p_{1/2}, 1g_{9/2}$
6	$2d_{3/2}, 1g_{7/2}, 3s_{1/2}, 2d_{3/2}, 1h_{11/2}$
7	$1h_{9/2}, 2f_{7/2}, 3p_{3/2}, 2f_{5/2}, 3s_{1/2}, 1i_{13/2}$

(b) Possible total spin J for various configurations $(j)^k$ [taken from Mayer and Jensen (55).]

$j = \frac{3}{2}$
$k = 1$ $\frac{3}{2}$.
$= 2$ 0, 2.

$j = \frac{5}{2}$
$k = 1$ $\frac{5}{2}$.
$= 2$ 0, 2, 4.
$= 3$ $\frac{3}{2}, \frac{5}{2}, \frac{9}{2}$.

$j = \frac{7}{2}$
$k = 1$ $\frac{7}{2}$.
$= 2$ 0, 2, 4, 6.
$= 3$ $\frac{3}{2}, \frac{5}{2}, \frac{7}{2}, \frac{9}{2}, \frac{11}{2}, \frac{15}{2}$.
$= 4$ 0, 2 (twice), 4 (twice), 5, 6, 8.

$j = \frac{9}{2}$
$k = 1$ $\frac{9}{2}$.
$= 2$ 0, 2, 4, 6, 8.
$= 3$ $\frac{3}{2}, \frac{5}{2}, \frac{7}{2}, \frac{9}{2}$ (twice), $\frac{11}{2}, \frac{15}{2}, \frac{17}{2}, \frac{21}{2}$.
$= 4$ 0 (twice), 2 (twice), 3, 4, (3 times), 5, 6 (3 times), 7, 8, 9, 10, 12.
$= 5$ $\frac{1}{2}, \frac{3}{2}, \frac{5}{2}$ (twice), $\frac{7}{2}$ (twice), $\frac{9}{2}$ (3 times), $\frac{11}{2}$ (twice), $\frac{13}{2}$ (twice), $\frac{15}{2}$ (twice), $\frac{17}{2}$ (twice), $\frac{19}{2}, \frac{21}{2}, \frac{25}{2}$.

$j = \frac{11}{2}$
$k = 1$ $\frac{11}{2}$.
$= 2$ 0, 2, 4, 6, 8, 10.
$= 3$ $\frac{3}{2}, \frac{5}{2}, \frac{7}{2}, \frac{9}{2}$ (twice), $\frac{11}{2}$ (twice), $\frac{13}{2}, \frac{15}{2}$ (twice), $\frac{17}{2}, \frac{19}{2}, \frac{21}{2}, \frac{23}{2}, \frac{27}{2}$.
$= 4$ 0 (twice), 2 (3 times), 3, 4 (4 times), 5 (twice), 6 (4 times), 8 (4 times), 9 (twice), 10 (3 times), 11, 12 (twice), 13, 14, 16.
$= 5$ $\frac{1}{2}, \frac{3}{2}$ (twice), $\frac{5}{2}$ (3 times), $\frac{7}{2}$ (4 times), $\frac{9}{2}$ (4 times), $\frac{11}{2}$ (5 times), $\frac{13}{2}$ (4 times), $\frac{15}{2}$ (5 times), $\frac{17}{2}$ (4 times), $\frac{19}{2}$ (4 times), $\frac{21}{2}$ (3 times), $\frac{23}{2}$ (3 times), $\frac{25}{2}$ (twice), $\frac{27}{2}$ (twice), $\frac{29}{2}, \frac{31}{2}, \frac{35}{2}$.
$= 6$ 0 (3 times), 2 (4 times), 3 (3 times), 4 (6 times), 5 (3 times), 6 (7 times), 7 (4 times), 8 (6 times), 9 (4 times), 10 (5 times), 11 (twice) 12 (4 times), 13 (twice), 14 (twice), 15, 16, 18.

BIBLIOGRAPHY

GENERAL REFERENCES

Below are listed several books, review articles and, occasionally, original papers that we have found useful or that supplement our discussion in an important way. Good reviews that present the most recent description of various fields are found in review journals such as *Advances in Physics*, *Advances in Nuclear Physics*, *Annual Review of Nuclear Science*, *Comments on Nuclear and Particle Physics*, *Physics Reports*, *Progress in Nuclear Physics*, *Reports on Progress in Physics*, *Revista del Nuovo Cimento*, and *Reviews of Modern Physics*. In addition, there are reports from schools, such as the *Proceedings of the International School of Physics*, "*Enrico Fermi*," and the *Cargese Lectures in Physics*. The reader will also find it useful to consult conference reports. There are several of these reports each year on special subfields. The "International Conferences on Nuclear Physics" in which the entire field is reviewed have been held in 1960 (Kingston), 1964 (Paris), 1966 (Gatlinburg), 1968 (Dubna), 1970 (Tokyo), and 1973 (Munich).

Gottfried, K., *Quantum Mechanics*, Vol. I, W. A. Benjamin, New York (1966).

Messiah, A., *Quantum Mechanics*, North Holland, Amsterdam (1961, 1963).

Blatt, J. M., and **V. F. Weisskopf,** *Theoretical Nuclear Physics*, John Wiley and Sons, New York, (1962).

Bohr, Aage, and **Ben R. Mottelson,** *Single Particle Motion*, Vol. I, W. A. Benjamin, New York (1969).

Brown, G. E., *Unified Theory of Nuclear Models and Forces*, North Holland, Amsterdam, (1967).

Eisenberg, J. M., and **W. Greiner,**
Vol. 1 *Nuclear Models*
Vol. 2 *Excitation Mechanisms of the Nucleus*
Vol. 3 *Microscopic Theory of the Nucleus*, North Holland, Amsterdam (1970).

Enge, H., *Introduction to Nuclear Physics*, Addison-Wesley, Reading, Mass. (1966).

Flugge, S., editor, *Handbach der Physik*, Springer-Verlag, Berlin, the volumes in this series were published in different years.

Irvine, J. M., *Nuclear Structure Theory*, Pergamon Press, Oxford, (1972).

Marmier, P., and **E. Sheldon,** *Physics of Nuclei and Particles*, Academic Press, New York (1969).

Meyerhof, Walter E., *Elements of Nuclear Physics*, McGraw-Hill, New York, (1967).

Preston, M. A., *Physics of the Nucleus*, Addison-Wesley, Reading, Mass., (1962).

Proceedings International School of Physics, "Enrico Fermi" Course 15, (G. Racah, editor). Academic Press, New York (1962).

Proceedings International School of Physics, "Enrico Fermi" Course 23, (V. F. Weisskopf, editor). Academic Press, New York, (1963).

Proceedings of the International School of Physics "Enrico Fermi," Course 36, (C. Bloch, editor). Academic Press, New York, (1966).

Proceedings International School of Physics "Enrico Fermi," Course 40, (M. Jean, editor). Academic Press, New York (1969).

Rowe, D. J., L. E. H. Trainer, S. S. M. Wong, and **T. W. Donnelly,** *Dynamic Structure of Nuclear States*, University of Totonto Press, (1972).

Roy, R. R., and **B. P. Nigam,** *Nuclear Physics*, John Wiley and Sons, New York, (1967).

Siegbahn, K., editor, *Alpha, Beta and Gamma Ray Spectroscopy*, North Holland, Amsterdam, (1965).

Chapters 3 and 7

Baranger, M., "Theory of Finite Nuclei: in *Cargese Lectures in Theoretical Physics*, (M. Levy, editor). W. A. Benjamin, New York, (1963).

Baranger, M., in Proceedings International School of Physics "Enrico Fermi," Course 40, (M. Jean, editor). Academic Press, New York, (1969).

Barrett, B. R., and **M. W. Kirson,** *Advances in Nuclear Physics*, *6* to be published (1973).

Bes, D. R., and **R. A. Sorensen,** *Advances in Nuclear Physics*, 2 (1969) 129.

Bethe, H. A., *Ann. Rev. Nucl. Sci.21* (1971) 93.

Brandow, B. H., *Rev. Mod. Phys. 39* (1967) 771.

Brown, G. E., *Many-Body Problems*, North Holland, Amsterdam, (1973).

Cargese Lectures in Physics, Vol. 3, (M. Jean, editor). Gordon and Breach, New York, (1969).

Day, R. B., *Rev. Mod. Phys. 39* (1967) 719.

Fetter, A. L. and **J. D. Walecka,** *Quantum Theory of Many Particle Systems*, McGraw-Hill Company, New York, (1971).

Ichimura, M., *Progress in Nucl. Phys. 10*, 309 (1969).

Macfarlane, M. H., "Reaction Matrix in Nuclear Shell Theory" in *Proceedings International School of Physics*, Course 40 (M. Jean, editor). Academic Press, New York, (1969).

Marshalek, E., and **J. Weneser,** *Ann. Phys.* (New York) *53* (1969) 569.

Migdal, A. B., *Nuclear Theory—The Quasi-Particle Method* W. A. Benjamin, New York, (1968).

Proceedings of the Gull Lake Symposium on Two Body Forces in Nuclei (1971).

The Many Body Problem, (C. de Witt and Phillipe Nozieres, editors) Les Houches Summer School, John Wiley and Sons, New York (1959).

Thouless, D. J., *The Quantum Mechanics of Many Body Systems*, Academic Press, New York (1961).

Chapters 4 and 5

Arima, A., and **I. Hamamoto,** *Ann. Rev. Nucl. Sci. 21* (1971) 55.

deShalit, A., and **I. Talmi**, *Nuclear Shell Theory*, Academic Press, New York (1963).

Elliott, J. P., and **A. M. Lane,** *The Nuclear Shell Model*, *Handbuch der Physik*, *39*, 241 (S. Flugge, editor) Springer-Verlag, Berlin (1957).

Grodzins, L. *Ann. Rev. Nucl. Sci. 18* (1968) 291.

Mayer, M. G., and **J. H. D. Jensen,** *Elementary Theory of Nuclear Shell Structure*, John Wiley and Sons, New York, (1955).

Mayer, M. G., J. H. D. Jensen, and **D. Kurath,** in *Alpha, Beta, and Gamma Ray Spectroscopy*, (K. Siegbahn, editor) North Holland, Amsterdam (1965).

Talmi, I., and **I. Unna,** *Ann. Rev. Nucl. Sci. 10* (1960) 353.

Chapter 6

Alder, K., A. Bohr, T. Huus, B. Mottelson, and **A. Winther,** *Rev. Mod. Phys. 28* (1956) 452.

Brink, D. M., in *Progress in Nuclear Physics* (O. R. Frisch, editor), Pergamon Press, Oxford (1960).

Danos, M., and **E. G. Fuller,** *Ann. Rev. Nucl. Sci. 15* (1965) 29.

Davidson, J. P., *Collective Models of the Nucleus*, Academic Press, New York (1968).

Green, A. M., *Rep. Prog. Phys. 28* (1965) 113.

Johnson, A., and **Z. Szymanski,** *Phys. Rep. 7C* (1973) 183.

Kerman, A. K., in *Nuclear Reactions*, (P. M. Endt and M. Demeur editors). North Holland, Amsterdam (1959).

Pauli, H. C., *Phys. Rep. 7C* (1973) 37.

Rogers, J. D., *Ann. Rev. Nucl. Sci. 15* (1965) 241.

Rowe, D. J., *Nuclear Collective Motion*, Methuen, London (1970).

Spicer, B. M., *Adv. Nucl. Phys. 2* (1969) 1.

Chapter 8

Arima, A., and **I. Hamamoto,** *Ann. Rev. Nucl. Sci. 21* (1971) 55.

Alder, R., and **R. M. Steffen,** *Ann. Rev. Nucl. Sci. 14* (1964) 403.

Bodenstedt, E., and **J. D. Rogers,** "The Magnetic Moments of Excited Nuclear States' in *Perturbed Angular Correlations*, (E. Karlsson, E. Matthias, and K. Siegbahn, editors) North Holland, Amsterdam (1964).

Church, E. L., and **J. Weneser,** *Ann. Rev. Nucl. Sci. 10* (1960) 193.

Grodzins, L., *Ann. Rev. Nucl. Sci. 18* (1968) 291.

Hyperfine Structure and Nuclear Radiations (E. Matthias and D. A. Shirley, editors) North Holland, Amsterdam (1968).

Perdrisat, C. F., *Rev. Mod. Phys. 38* (1966) 41.

Warburton, E. K., and **J. Weneser,** in *Isopin in Nuclear Physics* (D. H. Wilkinson, editor) North Holland, Amsterdam (1969).

Yoshida, S., and **L. Zamick,** *Ann. Rev. of Nucl. Sci. 22* (1972) 121.

Chapter 9

Adler, S. L., and **R. F. Dashen,** *Current Algebras and Applications to Particle Physics* W. A. Benjamin, New York (1968).

Blin-Stoyle, R. J., in *Isopsin in Nuclear Physics.* (D. H. Wilkinson, editor) North Holland Amsterdam (1969).

Blin-Stoyle, R. J., and **S. C. K. Nair,** *Adv. Physics 15* (1966) 493.

Blin-Stoyle, R. J., *Fundamental Interactions and the Nucleus*, North Holland, Amsterdam (1973).

Bressani, T., *Revista del Nuovo Cimento 1* (1971) 268.

Fiorini, E., *Revista del Nuovo Cimento 2* (1972) 1.

Fischbach, E., and **D. Tadic,** *Physics Reports 6C* (1973) 123.

Gari, M., "Parity Non-Conservation in Nuclei," *Phys. Reports 6C* (1973) 317.

Gasiorowicz, S., *Elementary Particle Physics*, John Wiley and Sons, New York, 1966.

Hamilton, W. D., *Progress Nucl. Phys. 10* (1969) 1.

Henley, E. M., *Ann. Rev. Nucl. Sci. 19* (1969) 367.

Kofoed-Hansen, O. M., and **C. J. Christensen,** "Experiments on Beta Decay, Vol. 41/2, (1962) 1. *Handbuch der Physik* (S. Flugge, editor) Springer-Verlag, Berlin (1962).

Konopinski, E. J., *The Theory of Beta Radioactivity*, Oxford University Press, London (1966).

Lee, T. D., and **C. S. Wu,** "Weak Interactions," *Ann. Rev. Nucl. Sci. 15* (1965) 381, *16* (1966) 471; *17* (1967) 513.

Lipkin, H. J., *Beta Decay for Pedestrians*, North Holland, Amsterdam (1962).

Schopper, H. F., *Weak Interactions and Nuclear Beta Decay*, North Holland, Amsterdam (1966).

Wu, C. S., and **S. A. Moszkowski,** *Beta Decay*, John Wiley and Sons, New York (1966).

REFERENCES

A

Abov, Yu. O., P. A. Krupchitsky, M. I. Bulgakov, O. N. Yermakov, and **I. L. Karpiklin,** *Phys. Letters 27B* (1968) 16.

Abov, Yu. O., P. A. Krupchitsky, and **Y. A. Oratovsky,** *Yadern Fiz. 1* (1965) 479 [translation: *Soviet J. Nucl. Phys. 1* (1965) 341].

Adelberger, E. G., C. L. Cocke, C. H. Davids, and **A. B. McDonald,** *Phys. Rev. Letters 22* (1969) 352.

Adler, S. L., *Phys. Rev. 139B*, 1638 (1965).

Adler, S. L., and **R. F. Dashen** *Current Algebras and Applications to Particle Physics*, W. A. Benjamin, New York (1968) Chapter IV.

Ahrens, T. and **E. Feenberg,** *Phys. Rev. 86* (1952), 64.

Ajzenberg-Selove, F., *Nucl. Phys A190* (1972), 1.

Ajzenberg-Selove, F., and **T. Lauritsen**, *Ann. Rev. Nucl. Sci. 10*, (1960), 409.

Alburger, D. E., and **D. H. Wilkinson,** *Phys. Letters 32B* (1970) 190.

Alder, K., A. Bohr, T. Huus, B. Mottelson, and **A. Winther,** *Rev. Mod. Phys. 28*, (1956) 432.

Alikhanov, Galaktionov, Gorodkov, Eliseyev and **Lyubunov,** *Proceedings International Conference on High Energy Physics*, Rochester 1960.

Anderson, J. D., C. Wong, and **J. W. McClure,** *Phys. Rev. 126* (1962) 2170.

Araujo, J. M., *Nuclear Reactions*, Vol. II, (P. M. Endt and P. B. Smith, editors) North Holland, Amsterdam, 1962, p. 195.

Arima, A., and **M. Horie,** *Prog. Theor. Phys.* (Kyoto), *12* (1954) 623.

Arima, A., H. Horie, and **H. Sano.** *Prog. Theor. Phys.* (Kyoto) *17* (1957) 367.

Arima, A., S. Cohen, R. D. Lawson, and **M. H. McFarlane,** *Nucl. Phys. A108* (1968) 94.

Asaro, F., F. S. Stephens, J. M. Hollander, and **I. Perlman,** *Phys. Rev. 117* (1960) 492.

Auerbach, N., J. Hüfner, A. K. Kerman, and C. M. Shakin, *Rev. Mod. Phys.44* (1972) 48.

Auerbach, N., *Nucl. Phys. 76*(1966) 321.

Auerbach, N., and **I. Talmi,** *Nucl. Phys. 64* (1965) 458.

Austern, N., Argonne Heavy Ion Conference (1972).

Austern, N., *Direct Nuclear Reaction Theories*, John Wiley and Sons, New York (1970).

Austern, N., and **R. G. Sachs,** *Phys. Rev. 81* (1951) 710.

Avida, R., J. Burde, and **A. Molchadzki,** *Nucl. Phys. A115* (1968) 405.

Avida, R., M. B. Goldberg, G. Goldring, and **A. Sporinzak,** *Nucl. Phys. A135* (1969) 678.

B

Backenstoss, G., B. D. Hyams, G. Knop, P. C. Martin, and **V. Stierlin,** *Phys. Rev. Letters 6*, (1961) 415.

Baker, G. A., Jr., *Phys Rev. 128*, 1485 (1962).

Baker, G. A., Jr., *Phys. Rev. B140* (1965) 9.

Baker, G. A., Jr., and **J. Kahane,** *J. Math. Phys. 10* (1969) 1967.

Baker, G. A., Jr., M. G. Hind, and **J. Kahane,** *Phys. Rev. C2* (1970) 841.

Baker, K. O., and **W. D. Hamilton,** *Phys. Letters 31B* (1970) 557.

Balashov, V. V., V. B. Beliaev, R. N. Eramjian, and **N. M. Kabachnick,** *Phys. Letters 9* (1964) 168.

Balian, R., and **C. Bloch,** *Ann. Phys. 60* (1970) 401.

Balian, R., and **C. Bloch,** *Ann. Phys. 63* (1971) 592.

Baranger, M., in *1962 Cargese Lectures in Theoretical Physics*, M. Levy, (editor), W. A. Benjamin, New York, (1963).

Baranger, M., "Recent Progress in the Understanding of Finite Nuclei from the Two Nucleon Interaction" in *Proceedings of the International School of Physics: Nuclear Structure and Nuclear Reactions* (M. Jean, editor) Academic Press, New York (1969).

Baranger, M., and **K. Kumar,** *Nucl. Phys. A122* (1968) 241, *A110* (1968) 490.

Bardeen, J., L. N. Cooper, and **J. R. Schrieffer,** *Phys. Rev. 108* (1957) 1175.

Bardeen, J., *1962 Cargese Lectures in Theoretical Physics* (M. Levy, editor), W. A. Benjamin Press, (1963).

Barlow, J., J. C. Sens, P. J. Duke, M. A. R. Kemp, *Phys. Letters 9* (1964) 84.

Barnett, A. R., in *Isobaric Spin in Nuclear Physics* (J. D. Fox and D. Robson, editors) Academic Press, New York (1966).

Barrett, B. R., *Phys. Rev. 154* (1967) 955.

Barrett, Bruce R., and **Michael W. Kirson,** "The Microscopic Theory of Nuclear Effective Interactions and Operators" to appear in *Advances in Nuclear Physics* Vol. 6.

Barrett, Bruce R., *Bull. Am. Phys. Soc. 16* (1971) 623.

Barschall, H. H., *Phys. Rev. 86* (1952) 431.

Bassel, R. H., and **C. Wilkin,** *Phys. Rev. 174* (1968) 1179.

Bassichis, W. H., B. Giraud, and **G. Ripka,** *Phys. Rev. Letters 15* (1965) 980.

Bassichis, W. H., A. K. Kerman, and **J. P. Svenne,** *Phys. Rev. 160* (1967) 746.

Bassichis, W. H., B. A. Pohl, and **A. K. Kerman,** *Nucl. Phys. A112* (1968) 360.

Bassichis, W. H., and **L. Wilets,** *Phys. Rev. Letters 22* (1969) 799.

Bayman, B. F., and **L. Silverberg,** *Nucl. Phys. 16* (1960) 625.

Becker, J. A., J. W. Olness, and **D. H. Wilkinson,** *Phys. Rev. 155* (1967) 1089.

Becker, J. A., and **D. H. Wilkinson,** *Phys. Rev. 134B* (1964) 1200.

Becker, R. L., A. D. MacKellar, and **A. D. Morris,** *Phys. Rev. 174* (1968) 1264.

Behrens, H., *Zeit. f. Phys. 201* (1967) 153.

Behrens, H., and **W. Buhring,** *Nucl. Phys. A106* (1958) 433.

Bell, J. A., and **R. J. Blin-Stoyle,** *Nucl. Phys. 6* (1957) 87.

Bellicard, J. P., P. Bounin, R. F. Frosch, R. Hofstadter, J. S. McCarthy, F. J. Urhane, M. R. Yearian, B. C. Clark, R. Herman, and **D. G. Ravenhall,** *Phys. Rev. Letters 19* (1967) 527.

Bellicard, J. B., and **K. J. van Oostrum,** *Phys. Rev. Letters 19* (1967) 242.

Belyaev, S. T., *Mat. Fys. Medd. Dan. Vid. Selsk 31* #11.

Berenyi, D., *Nucl. Phys. 48* (1963) 121.

Berman, B. L., private communication.

Bernstein, A., *Adv. Nucl. Phys. 3* (1970) 325.

Bernstein J., M. Gell-Mann, and **L. Michel,** *Nuovo Cimento 16* (1960) 560.

Bernstein, J., S. Fubini, M. Gell-Mann, and **W. Thirring,** *Nuovo Cimento 17* (1960) 757.

Bertozzi, W., and **S. Kowalski,** editors, *Medium Energy Nuclear Physics with Electron Linear Accelerators*, M.I.T. 1967 Summer Study. USAEC Div. Tech. Info. Tid 24667.

Bertozzi, W., J. Friar, J. Heisenberg, J. W. Negele, *Phys. Letters 41B* (1972) 408.

Bertsch, G. F., *Nucl. Phys. 74* (1965) 234.

Bethe, H. A., "Theory of Nuclear Matter" in *Ann. Rev. Nucl. Sci. 21* (1971) 93.

Bethe, H. A., *Phys. Rev. 167* (1968) 879.

Bethe, H. A., B. H. Brandow, and **A. G. Petschek,** *Phys. Rev. 129* (1963) 225.

Bethe, H. A., and **J. Goldstone,** *Proc. Roy. Soc. A238* (1957) 551.

Bhaduri, R. K., and **P. Van Leuven,** *Phys. Letters 20* (1966) 182.

Bhattacherjee, S. K., *Ann. Phys.* (New York) *63* (1971) 613.

Bhattacharjee, S. K., S. K. Mitra, and **H. C. Padhi,** *Nucl. Phys. A96* (1967) 81.

Bienlein H. et al. *Phys. Letters 13* (1964) 80.

Bilpuch, E. G., in *Isobaric Spin in Nuclear Physics*, (J. D. Fox and D. Robson editors) Academic Press, New York (1966) p. 235.

Bizzari, R., E. D. Capua, U. Dore, G. Gialanella, P. Guidoni, and **I. Laaka,** *Nuovo Cimento 33* (1964) 1497.

Bjorken, J. D., and **S. D. Drell,** *Relativistic Quantum Fields*, McGraw Hill, New York (1965).

Bjørnholm, *Nuclear Excitations in Even Isotopes of the Heaviest Elements*, Munksgaard, Copenhagen (1965).

Blair, J. S., in *Nuclear Spin-Parity Assignments* (N. B. Gove editor), Academic Press, New York (1962) p. 327.

Blair, J. S., "The Scattering of Strongly Absorbed Particles" in *Lectures in Theoretical Physics*, Vol. VIII, (C. Kunz, D. Lind, W. Britten, editors) University of Colorado Press, Boulder (1966).

Blatt, J. M., and **V. F. Weisskopf,** *Theoretical Nuclear Physics*, John Wiley and Sons, New York (1952).

Bleck, J., D.W. Haag, W. Leitz, R. Michaelsen, W. Ribbe, and **F. Sechelschmidt,** *Nucl. Phys. A123* (1969) 65.

Blin-Stoyle, R. J., *Proc. Phys. Soc.* (London) *66* (1953) 1158.

Blin-Stoyle, R. J., "Isospin in Nuclear Beta Decay," in *Isopin in Nuclear Physics* (D. H. Wilkinson, editor), North Holland, Amsterdam (1969) p. 115-172.

Blin-Stoyle, R. J., *Rev. Mod. Phys. 28* (1956) 75.

Blin-Stoyle, R. J., and **H. Feshbach,** *Nucl. Phys. 27* (1961) 395.

Blin-Stoyle, R. J., and **P. Herczeg,** *Nucl. Phys. B5* (1968) 291.

Bloch, F., *Phys. Rev. 83* (1951) 839.

Bloch, C., *Nucl. Phys. 7* (1958) 451.

Bloch, C. and **J. Horowitz,** *Nucl. Phys. 8* (1958) 91

Block, B. and **H. Feshbach,** *Ann. Phys.* (New York) *23* (1963) 47.

Block, M. M., T. Kikuchi, S. Koetke, C. R. Sun, R. Walker, G. Culligan, V. L. Telegdi, and **R. Winston,** *Nuovo Cimento 55A* (1968) 501.

Bloom, S. D., *Nuovo Cimento 32* (1964), 1023.

Bloom, S. D., in *Isobaric Spin in Nuclear Physics*, (J. D. Fox and D. Robson editors). Academic Press, New York, (1966).

Bloom, S. D., L. G. Mann, R. Polizhar, J. R. Richardson, and **A. Scott,** *Phys. Rev. 134B* (1964) 481.

BNL-325. *Neutron Cross-Sections* (J. Stehn, G. Goldberg, R. W. Chasman, S. Mughabghab, B. Magurno and V. Max, editors).

Bock, P., B. Jenschke, and **H. Schopper,** *Phys. Letters 22* (1966) 316.

Bock, P., and **B. Jenschke,** *Nucl. Phys. A160* (1971) 550.
Bock, P., and **H. Schopper,** *Phys. Letters 16* (1965) 284.
Bodenstedt, E., L. Ley, H. O. Schlenz, and **U. Wehmann,** *Nucl. Phys. A137* (1969) 33, *Phys. Letters 29B* (1969) 165.
Bodenstedt, E., and **J. D. Rogers,** in *Perturbed Angular Correlations* (E. Karlsson, F. Matthias, and K. Siegbahn, editors) North Holland Amsterdam, (1964) p. 91.
Boehm, F., and U. Hauser, *Nucl. Phys. 14* (1960) 615.
Boehm, F., and **E. Kankeleit,** *Nucl. Phys. A109* (1968) 457.
Boer, J. de, and **J. Eichler,** in *Advances in Nuclear Physics, 1* (1968).
Bogolyubov, N. N., *Nuovo Cimento 7* (ser. 10) (1958) 794.
Bohr, A., *Kgl. Dan. Mat- Fys. Medd. 26*, no. 14 (1952).
Bohr, A., and **B. R. Mottelson,** *Dan. Mat-Fys. Medd. 27*, no. 16 (1953).
Bohr, A., and **B. R. Mottelson,** *Dan. Mat-Fys. Medd, 30* (1955) no. 1.
Bohr, A., and **B. R. Mottelson,** in *Beta and Gamma Ray Spectroscopy*, (K. Siegbahn, editor), North Holland, Amsterdam, (1955), p. 468.
Bohr, A., and **B. R. Mottelson,** *Nuclear Structure*, Vol. I. W. A. Benjamin, New York, (1969).
Bouchez, R., *Physica 18* (1952) 1171.
Bouchez, R., and **P. Depommier,** *Rep. Prog. Phys. 23* (1960) 395.
Bramblett, R. J., J. T. Caldwell, R. R. Harvey, and **S. C. Fultz,** *Phys. Rev. 133B* (1964) 869.
Brandow, B. H., *Rev. Mod. Phys. 39* (1967) 771.
Braunstein (Gal) A. and **A. deShalit,** *Phys. Letters 1* (1962) 264.
Brene, N., M. Roos, and **A. Sirlin,** *Nucl. Phys 136* (1968) 255.
Brenig, W., *Nucl. Phys. 4* (1957) 363.
Bressel, C. N., A. K. Kerman, and **B. Rouben,** *Nucl. Phys. A124* (1969) 624.
Brink, D. M., and **A. K. Kerman,** *Nucol. Phys.* 12 (1959) 314.
Brink, D. M., and **G. R. Satchler,** *Angular Momentum*, Clarendon Press, Oxford (1962).
Brody, T. A., and **M. Moshinsky,** *Table of Transformation Brackets*, Monografias Del Instituto De Fisica, Mexico (1960).
Broglia, R., (to be published in *Annals of Physics*, 1973).
Bromley, D. A., and **J. Weneser,** *Comments on Nuclear and Particle Physics, 2* (1968) 151.
Bromley, D. A., and **J. Weneser,** *Comments on Nuclear and Particle Physics, 1* (1967) 80.
Bromley, D. A., H. E. Gove, E. B. Paul, A. E. Litherland, and **E. Almquist,** *Can. J. Phys. 35* (1957) 1042.
Bromley, D. A., H. E. Gove, and **A. E. Litherland,** *Can. J. Phys 35* (1957) 1057.
Bromley, D. A., H. E. Gove, E. B. Paul, A. E. Litherland, and **E. Almquist,** *Can. J. Phys. 35* (1957) 1042.
Brown, G. E., *Unified Theory of Nuclear Models*, North Holland, 1st ed. (1964), 2nd ed. (1967).
Brown, G. E., and **M. Bolsterli,** *Phys. Rev. Letters 3* (1959) 472.
Brown, G. E., J. D. Jackson, and **T. T. S. Kuo,** *Nucl. Phys. A133* (1969) 481.
Brown, G. E., *Proceedings Int. Conf. on Photonuclear Reactions* (B. L. Berman, editor), Lawrence Livermore Lab. (1973).
Brueckner, K. A., *Phys. Rev. 96* (1954) 508.
Brueckner, K. A., and **C. A. Levinson,** *Phys. Rev. 97* (1955) 1344.
Brueckner, K. A., A. Lockett, and **A. M. Rotenberg,** *Phys. Rev. 121* (1961) 255.

Brueckner, K. A., J. Gammel, and J. L. Weitzner, *Phys. Rev. 110* (1958), 431.
Brysk, H., and M. E. Rose, *Rev. Mod. Phys. 30* (1958) 1169.
Buck, B., and F. G. Perey, *Phys. Rev. Letters 8* (1962) 444.
Bugg, D. V., D. C. Salter, G. H. Stafford, R. F. George, K. F. Riley, and M. R. Yearian, *Phys. Rev. 164* (1966) 980.
Burde, J., L. P., G. Engler, A. Ginsburg, A. A. Jaffe, and A. Marinov, in *Isobaric Spin in Nuclear Physics*, (J. D. Fox and D. Robson, editors), Academic Press, New York, (1966) p. 63.
Butler, G., J. Cerny, S. Cosper, and R. McGrath, *Phys. Rev. 166* (1968) 1096.

C

Calogero, F., *Problem Symposium in Nuclear Physics*, 2nd. ed., Novosibirsk (1970).
Campbell, E. J., H. Feshbach, C. E. Porter, and V. F. Weisskopf, *Laboratory of Nuclear Science Technical Report No. 75*, M.I.T., Cambridge (1960) p. 132.
Carlson, B. T., and I. Talmi, *Phys. Rev. 96* (1954) 436.
Castel, B., *Can. J. of Phys. 46* (1968) 2571.
Chase, D. M., L. Wilets, and A. R. Edmonds, *Phys. Rev. 110* (1958) 1080.
Chou Kuang-Chao, *Sov. Phys. JETP 12* (1961) 492.
Christensen, J. H., J. W. Cronin, V. L. Fitch, and R. Turlay, *Phys. Rev Letters 13* (1964) 138.
Church, E., and J. Weneser, *Phys. Rev. 104* (1956) 1382.
Church, E. L., and J. Weneser, *Ann. Rev. Nucl. Sci, 10* (1960) 193.
Clark, J. M., and J. P. Elliott, *Phys. Letters 19* (1965) 294.
Clegg, A. B., *Nucl. Phys. 38* (1962) 353.
Clegg, A. B., *Phil. Mag. 6* (1961) 1207.
Clement, D. M., and E. U. Baranger, *Nucl. Phys. A108* (1968) 27.
Coester, F., *Nucl. Phys. 7* (1958) 421.
Coester, F., S. Cohen, B. Day, and C. M. Vincent, *Phys. Rev. C1* (1970) 769.
Cohen, B. L., and R. E. Price, *Phys. Rev. 121* (1961) 1441.
Cohen, E. R., and J. W. M. DuMond, *Rev. Mod. Phys. 37* (1965) 537.
Cohen, S., and D. Kurath, *Nucl. Phys. 73* (1965) 1.
Condon, E. U., and G. H. Shortley, *The Theory of Atomic Spectra* Cambridge University Press (1945).
Conlon, T. W., *Nucl. Phys. A100* (1967) 545.
Cook, C. F., and T. W. Bonner, *Phys. Rev. 94* (1954) 651.
Cooper, L. N., and E. M. Henley, *Phys. Rev. 92* (1953) 801.
Cosman, E. R., D. N. Schramm, H. A. Enge, A. Sperduto, and C. H. Paris, *Phys. Rev. 163* (1967) 1134
Cowan, C. L., Jr., F. Reines, F. B. Harrison, H. W. Kruse, and A. D. McGuire, *Science 154* (1956) 103.
Cox, R. T., C. G. McIlwraith, and B. Kurrelmeyer, *Proc. Nat. Acad. Sci., U.S. 14* (1928) 544.
Cruse, D. W., and W. D. Hamilton, *Nucl. Phys. 125* (1969) 241.
Curran, S. C., *Physica 18* (1952) 1161.

D

Danby, G., J. M. Gaillard, K. Goulianos, L. M. Lederman, N. Mistry, M. Schwartz, and **J. Steinberger,** *Phys. Rev. Letters 9* (1962) 36.
Dancoff, S. M., and **P. Morrison,** *Phys. Rev. 55* (1939) 122.
Daniell, H., and **H. Schmitt,** *Nucl. Phys. 65* (1965) 481.
Danilov, G. S., *Phys. Letters 18* (1965) 40.
Danos, M. *Nucl. Phys. 5* (1957) 23.
Danos, M., and **E. G. Fuller,** *Ann. Rev. Nucl. Sci. 15* (1965) 29.
Dar, A., *Phys. Rev. 139B* (1965) 1193.
Dashen, R. F., S. C. Frautschi, M. Gell-Mann, and **Y. Hara** *The Eight-Fold Way*, W. A. Benjamin, New York (1964). p. 253.
Davidson, J. P., *Rev. Mod. Phys. 37* (1965) 105.
Davidson, J. P., in *Nuclear Spin-Parity Assignments* (N. B. Gove, editor), Academic Press, New York (1966), p. 446.
Davidson, J. P., *Collective Models of the Nucleus*, Academic Press, New York (1968).
Davies, K. T. R., S. J. Krieger, and **M. Baranger,** *Nucl. Phys. 84* (1966) 545.
Davies, K. T. R., and **R. J. McCarthy,** *Phys. Rev. C4* (1971) 81.
Davis, R., Proceedings *International Conference Radioisotopes*, Paris, 1957, Pergamon Press, London (1958).
Davydov, A. S., and **G. F. Filippov,** *Nucl. Phys. 8* (1958) 237.
Dawson, J. F., I. Talmi, J. D. Walecka, *Ann. Phys.* (New York) *18* (1962) 339.
Day, B. D., *Rev. Mod. Phys. 39* (1967) 719.
Deforest, T., and **J. D. Walecka,** *Adv. Phys. 15* (1966) 1.
Delorme, J., and **M. Rho,** *Mirror Asymmetry, Meson Exchanges and Second Class Currents*, Preprint, (May 1971).
Delves, L. M., and **A. C. Phillips,** *Rev. Mod. Phys. 41* (1969) 497.
deShalit, A., *Helv. Phys. Acta 24* (1951) 296.
deShalit, A., *Phys. Rev. 91* (1953) 1479.
deShalit, A., *Phys. Rev. 90* (1953), 83.
deShalit, A., *Selected Topics in Nuclear Theory*, I.A.E.A., (F. Janovich, editor). Vienna (1963).
deShalit, A. *Rendiconti della Scuoloa Int. di Physica*, Varenna 1960, Zanchelli, Bologna (1962)
deShalit, A., *Phys. Letters 15* (1965) 170.
deShalit, A., *Phys. Rev. 122* (1969) 1530.
deShalit, A., and **I. Talmi,** *Nuclear Shell Theory*, Academic Press, New York, 1963.
deShalit, A., and **J. D. Walecka,** *Phys. Rev. 147* (1966) 763.
deShalit, A., and **V. F. Weisskopf,** *Ann. Phys.*, (New York) *5* (1958) 282.
Desplanques, B., and **N. Vinh Mau,** *Phys. Letters 35B* (1971) 28.
Devons, S., and **I. Duerdoth,** "Mounic Atoms" in *Adv. Nucl. Phys. 2* (1969) 295.
Diamond, R. H., and **F. S. Stephens,** *Nucl. Phys. 45* (1963) 632.
Dickens, J. K., E. Eichler, and **G. R. Satchler,** *Phys. Rev. 168* (1967) 1355.
Donnelly, T. W., and **G. E. Walker,** *Ann. Phys.* (New York) *60* (1970) 209.
Drell, S. D., and **J D. Walecka,** *Phys. Rev. 120* (1960) 1069.
Drozdov, S. I., *JETP 38* (1960) 499 [translation *Soviet Phys. JETP 11* (1960) 362].

Duke, C. L., P. G. Hansen, O. B. Nielsen, G. Rudstam and **Isolde Collaboration,** C.E.R.N. *Nucl. Phys. A151* 609 (1970).
Dutton, L. M. C., *Phys. Letters B25* (1967) 245.

E

Eckhouse, M., Thesis, Carnegie Institute of Technology, Pittsburgh, (1962).
Eden, R. J., and **V. J. Emery,** *Proc. Roy. Soc. A248* (1958) 266.
Eden, R. J., V. J. Emery, and **S. Sampanthar,** *Proc. Roy. Soc, A253* (1959), 177, 186.
Eden, R. J., and N. C. Francis, *Phys. Rev. 97* (1955) 1366.
Edmonds, A. R., *Angular Momentum in Quantum Mechanics*, Princeton University Press, Princeton, N.J., (1957).
Ehrenberg, H. F., R. Hofstadter, U. Meyer-Berkhout, D. G. Ravenhall, and **S. E. Sobottka,** *Phys. Rev.113* (1959) 666.
Eisenberg, J. M., and **W. Greiner,** *Nuclear Theory*, North Holland, Amsterdam (1970).
Ekstein, H., *Phys. Rev. 117* (1960) 1590.
Elliott, J P., *Proc. Roy. Soc. 245* (1958) 128.
Elliott, J. P., *Proc. Roy. Soc. 245* (1958) 562.
Elliott, J. P., and **A. M. Lane,** *Encyclopedia of Physics*, XXXIX, Springer-Verlag, Berlin (1957) p. 337.
Elliott, J. P., and **B. H. Flowers,** *Proc. Roy. Soc.* (London) *A242* (1957) 57.
Elliott, J. P., and **C. E. Wilsdon,** quoted by Wilkinson (1967).
Elliott, J. P., and **T. H. R. Skyrme,** *Proc. Roy. Soc. A232* (1955) 561.
Enge, Harald, *Introduction to Nuclear Physics*, Addison-Wesley, Reading, Mass. (1966).
Ericson, T. E. O., *Phys. Rev. Letters 5* (1960) 430.
Ericson, T. E. O., *Ann. Phys.* (New York) *32*(1963) 390.
d'Espagnat, B., *Phys. Letters 7* (1963) 209.
Estrada, L., and **H. Feshbach,** *Ann. Phys.* (New York) *23* (1963) 123.

F

Fallieros, S., R. A. Ferrell, and **M. K. Pal.,** *Nucl. Phys. 15* (1960) 363.
Fallieros, S., B. Goulard, and **R. H. Venter,** *Phys. Letters 19* (1965) 398.
Feenberg, E., and **G. Trigg,** *Rev. Mod. Phys. 22* (1950) 399.
B. T. Feld, *Models of Elementary Particles*, Blaisdell, Waltham (1969).
Fermi, E. *Ric. Sci 2* (1933) part 12.
Fermi, E. *Zf. Physik 88*, (1934) 161.
Fernbach, S., R. Serber, and **T. B. Taylor,** *Phys. Rev. 75* (1949) 1352.
Feshbach, H., *Ann. Phys.* (New York) *5* (1958) 357.
Feshbach, H., *Ann. Phys.* (New York) *19* (1962) 287.
Feshbach, H. "Reaction Mechanisms—Retarded Interactions and Compound Nucleus, Statistical Theories," Plenary Session #4b, Congres International de Physique Nucleaire Vol. I (1964) Centre National de la Recherche Scientifique.
Feshbach, H., in *Nuclear Structure Study with Neutrons* (M. Neve de Mevergnies, P. VanAssche and J. Vervier editors,) North Holland, Amsterdam (1966), p. 257.
Feshbach, H., and **A. Kerman,** Comments on Nuclear and Particle Physics *1* (1967) 69.
Feshbach, H., A. K. Kerman, and **R. H. Lemmer,** *Ann. Phys.* (New York) *41* (1967) 230.
Feshbach, H., and **E. L. Lomon,** *Ann. Phys.* (New York) *29* (1964) 19.

Feshbach, H., C. E. Porter, and **V. F. Weisskopf.** *Phys. Rev. 96* (1954) 448.
Feynman, R. P., and **M. Gell-Mann,** *Phys. Rev. 109* (1958) 193
Firk, F. W. K , "Low Energy Photonuclear Reactions", *Ann. Rev. Nucl. Sci., 39* (1970)
Fischbach, E., and **Trabert K.,** *Phys. Rev. 174* (1968) 1843.
Fischbach, Ephraim and **Dubravko Tadic,** "Parity Violating Nuclear Interactions and Models of the Weak Hamiltonian," *Physics Reports 6C* (1973) 125.
Fischer, T. R., S. S. Hanna, D. D. Healey, and **P. Paul,** *Phys. Rev. 176* (1968) 1130.
Foldy, L. L., *Phys. Rev. 83* (1951) 397.
Foldy, L. L., and **J. D. Walecka,** *Nuovo Cimento, Suppl. 4* (1966) 781.
Ford, K. W., and C. Levinson, *Phys. Rev. 100* (1955) 1.
Fox, J. D., and **D. Robson,** *Isobaric Spin in Nuclear Physics,* Academic Press, New York (1966).
Frauenfelder and R. M. Steffen, "Angular Correlation," in *Alpha, Beta and Gamma Ray Spectroscopy,* Vol. *2* (K. Siegbahn, editor) North Holland, Amsterdam (1965), p. 997.
Freeman, J. M., J. G. Jenkin, and **G. Murray,** *Phys. Letters 22* (1966) 177.
Freeman, J. M., J. G. Jenkin, G. Murray, and **W. E. Burcham** *Phys. Letters 16* (1966) 959.
Freeman J. M., J. G. Jenkin, D. C. Robinson, G. Murray, and **W. E. Burcham,** *Phys. Letters B27* (1968) 156.
Freeman, J. M., J. G. Jenkin, and **G. Murray,** *Nucl. Phys. A124* (1969) 393.
Freeman, J. M., G. J. Clark, J. E. Draper, D. C. Robinson, J. S. Rider, W. E. Burcham, and **G. T. A. Squier,** *Proceedings International Conf. on Nuclear Physics,* Munich (1973) (J. de Boer and H. J. Mang, editors).
Friedman, F. L., and **V. F. Weisskopf,** in *Niels Bohr and the Development of Physics,* (W. Pauli, L. Rosenfeld, V. F. Weisskopf editors), Pergamon Press, London (1955).
Friedman, J. I., and **V. L. Telegdi,** *Phys. Rev. 105* (1957) 1681.
Frosch, R. F., R. Hofstadter, J. S. McCarthy, G. K. Nöldeke, K. J. van Oostrum, and **M R. Yearian,** *Phys. Rev. 174* (1968) 1380.
Fubini, S., and **G. Furlan,** *Physics 1* (1965) 229.
Fubini, S., G. Furlan, and **C. Rossetti,** *Nuovo Cimento 40* (1965) 1171.
Fujii, A. L., and **H. Primakoff,** *Nuovo Cimento 12* (1959) 327.
Fujita, J. I., and **K. Ikeda,** *Nucl. Phys. 67* (1965) 145.
Fujita, J. I., and **Ikeda, K.,** *Prog. Theor. Phys. 36* (1966) 288.
Fujita, J. I., and **K. Ikeda,** *Prog. Theor. Phys. 36* (1966) 530.
Fuller, E. G., "Photoneutron Reactions" *Nuclear Structure Study with Neutrons* M. N. de Mevergnies, P. Van Assche, and J. Vervier editors) North Holland, Amsterdam (1966) p. 359.
Fuller, E. G., and **E. Hayward,** *Nuclear Reactions,* Vol. II, p. 113 (P. M. Endt and P. B. Smith, editors) North Holland, Amsterdam (1962).
Fuller, E. G., and **M. S. Weiss,** *Phys. Rev. 112* (1958) 560
Fuller, G. H., and **V. W. Cohen** "Nuclear Moments," Appendix 1 to *Nuclear Data Sheets,* National Academy of Sciences—National Research Council, Oak Ridge National Laboratory (May 1965).
Fultz, S. C., R. L. Bramblett, J. T. Caldwell, and **N. A. Kerr,** *Phys. Rev. 137* (1962) 1273.

G

Gal, A., *Phys. Letters 20* (1966) 414.
Gallagher, C. J., and **S. A. Mozkowski,** *Phys. Rev. 111* (1958) 1282.

Gamow, G. and **E. Teller,** *Phys. Rev. 49* (1936) 895.
Gari, Manfred, "Parity Non-Conservation in Nuclei," *Phys. Rep. 6C* (1973) 317.
Gari, M., *Phys. Letters 31B* (1970) 627.
Gartenhaus, S., and **C. Schwartz,** *Phys. Rev. 108* (1957) 482.
Garvey, G. T., W. J. Gerace, R. L. Jaffe, I. Talmi, and **I. Kelson,** *Rev. Mod. Phys. 41* (1969) S1.
Garvey, G. T., and **I. Kelson,** *Phys. Rev. Letters 16* (1966) 1967.
Garvey, G. T., W. T. Gerace, R. L. Jaffe, I. Talmi, and **I. Kelson,** *PUC-937-331* (1968).
Garvey, G. T., and **R. Tribble,** Princeton thesis (1973).
Garwin, R. L., L. M. Lederman, and **M. Weinrich,** *Phys. Rev., 105* (1957) 1415.
Gasioriowicz, S., *Elementary Particle Physics,* J. Wiley and Sons, Inc., New York (1967).
Gell-Mann, M., *Phys. Rev. 111* (1958) 362.
Gell-Mann, M., M. Goldberger, N. Kroll, and **F. Low,** *Phys. Rev. 179* (1969) 1518.
Gell-Mann, M., M. L. Goldberger, W. E. Thirring, *Phys. Rev. 95* (1954) 1612.
Gell-Mann, M., and **M. Levy,** *Nuovo Cimento 16* (1960) 705.
Gell-Mann, M., and **V. L. Telegdi,** *Phys. Rev. 91* (1953) 169.
Gerholm, T. R., and **B. G. Petterson,** *Phys. Rev. 110* (1958) 1119.
Gershtein, S. S., and **Ia. B. Zel'dovich,** *JETP* (USSR) *29* (1955) 698.
Gillet, V., A. M. Green, and **E. A. Sanderson,** *Nucl. Phys. 88* (1966) 321.
Glassgold, A., W. Heckrotte, and **K. M. Watson,** *Ann. Phys.* (New York) *6* (1959) 1.
Glauber, R. J., "High Energy Collision Theory," in *Lectures in Theoretical Physics,* Vol. I (W. E. Britten and L. G. Dunham, editors). Interscience Publishers, New York (1959).
Glauber, R. J. "Theory of High Energy Hardron-Nucleus Collisions" in *High Energy Physics and Nuclear Structure,* (S. Devons, editor). Plenum Press New York, (1970).
Glauber, R. J., and **P. C. Martin,** *Phys. Rev. 95* (1954) 272.
Glauber, R. J., and **P. C. Martin,** *Phys. Rev. 104* (1956) 158.
Gleit, C. E., C. W. Tang, and **C. D. Coryell,** Beta Decay Transition Probabilities, *Nuclear Data Sheets,* Appendix 5 (1963).
Goldberger, M., and **S. Treiman,** *Phys. Rev. 111* (1958) 354.
Goldemberg, J., Y. Torizuka, W. C. Barber, and **J. D. Walecka,** *Nucl. Phys. 43* (1963) 242.
Goldhaber, M., L. Grodzins, and **A. M. Sunyar,** *Phys. Rev. 109* (1958) 1015.
Goldhaber, M., and **A. Sunyar,** Chapter XVIII **in** *Alpha, Beta and Gamma Ray Spectroscopy,* (K. Siegbahn, editor), North Holland, Amsterdam (1965).
Goldhaber, M., and **E. Teller,** *Phys. Rev. 74* (1948) 1046.
Goldstone, J., *Proc. Roy. Soc.* (London) *A293* (1957) 265.
Gomes, L. C., J. D. Walecka, and **V. F. Weisskopf,** *Ann. Phys.* (New York) *3* (1958) 251.
Good, R. H., *Rev. Mod. Phys. 27* (1955) 187.
Gottfried, K., *Quantum Mechanics,* Vols. I., Benjamin, New York (1966).
Gottfried, K., and **D. Yennie,** *Phys. Rev. 182* (1969) 1595.
Goulard, B., C. Goulard, and **H. Primakoff,** *Phys. Rev. 133B* (1964) 186.
Gove, H. E., in *Proceedings of the International Conference on Nuclear Structure* at Kingston, Canada. (D. A. Bromley and E. W. Vogt, editors) Univ. of Toronto Press (1960).
Gove, N. B., and **M. J. Martin,** *Nuclear Data Tables 10* (1971) 206.
Gove, N. B., *Nuclear Spin Parity Assignments,* (N. B. Gove editor), Academic Press, New York (1966).

Gove, N. B., and **M. Yamada,** *Nuc. Data A4* (1968) 237.
Grayson, W. C., Jr., and **L. W. Nordheim,** *Phys. Rev. 102* (1956) 1084.
Green, T. A., and **M. E. Rose,** *Phys. Rev. 110* (1958) 105.
Green, I. M., and **S. A. Mozkowski,** *Phys. Rev. 139B* (1965) 790.
Griffin, J. J., *Phys. Rev. Lett. 17* (1966) 478.
Greuling E., and **J. K. Kim.** (unpublished) (1964).
Gunye, M. R., *Phys Letters 27B* (1968) 136.
Gunye, M. R., and **C. S. Warke,** *Phys. Rev. 156* (1967) 1087
Gustafson, T., *Dan. Mat-Fys Medd. 30* #5 (1955).

H

Hadjimichael, E., and **E. Fischbach,** *Phys. Rev. D3* (1971) 755.
Hadjimichael, E., E. Harms, and **V. Newton,** *Phys. Rev. Letters 27* (1971) 1322.
Hafele, J. C., and R. Woods, *Phys. Letters 23* (1966) 579.
Haftel, M., and **F. Tabakin,** *Nucl. Phys. A158* (1970) 1.
Haftel, M., and **F. Tabakin,** *Phys. Rev. C3* (1971) 921.
Halbert, E. C., J. B. McGrory, B. H. Wildenthal and **S. P. Pandya,** *Adv. Nucl. Phys. 4* (1971) 316.
Hamada, T., and **I. D. Johnston,** *Nucl. Phys. 34* (1962) 382.
Handbook of Mathematical Functions (M. Abramowitz and I. A. Stegun, editors) National Bureau of Standards, Applied Mathematics Series *55* (1964).
Hanna, S. S., "Analog Resonances Formed in Isospin Forbidden Reactions, in *Nuclear Isospin* (Anderson, Bloom, Cerny and True, editors) Academic Press New York and London (1969).
Hanna, S. S., *Proceedings Int. Conf. Photonuclear Reactions* (B. L. Berman, editor) Lawrence Livermore Lab. (1973).
Harmatz, B., T. Handley, and **J. W. Mihelich,** *Phys. Rev. 128* (1962) 1186.
Harvey, M. *Adv. Nucl. Phys., 1* (1968) 67
Hättig, H., K. Hunchen and, **H. Waffler,** *Phys. Rev. Letters 25* (1970) 941.
Haxel, O., T. H. D. Jensen and **H. E. Suess,** *Phys. Rev. 75* (1949) 1766.
Hayward, Evans, ' Photonuclear Reactions," *Nuclear Structure and Electromagnetic Transitions* (N. MacDonald, editor) Plenum Press, New York (1965).
Heller, Leon, "Some Basic Questions in Nucleon-Nucleon Bremsstrahlung" in *Proceedings of the Gull Lake Symposium on the Two-Body Forces in Nuclei* (1971).
Henley, Ernest M. *Ann. Rev. Nucl. Sci. 19* (1969) 367.
Henley, E. M., and **B. A. Jacobsohn,** *Phys. Rev. 113* (1959) 225.
Herzberg, G., *Atomic Spectra and Atomic Structure*, Dover Publications, New York (1949).
Hill, D. L., and **J. A. Wheeler,** *Phys. Rev. 89* (1953) 1102.
Hill, D. L., in *Handbuch der Physik* (S. Flugge, editor) Springer-Verlag, Berlin (1957) p. 178.
Hitlen, D., S, Bernan, S. Devons, I. Duerdoth, J. W. Kast, E. R. Macagno, J. Rainwater, and **C. S. Wu,** *Phys. Rev. C1* (1970) 1184.
Hodgson, P. E. *The Optical Model of Elastic Scattering*, Oxford University Press (1963).
Hofstadter, R., *Electron Scattering and Nuclear and Nucleon Structure*, (R. Hofstadter, editor) W. A. Benjamin, New York (1963).

Horoshko, R. N., D. Cline, and **P. M. S. Lesser,** "Measurement and Interpretation of Electromagnetic Transitions in ^{51}V" *Rep: UR-NSRL-32* (February 1970).
Huang, K., and **C. N. Yang,** *Phys. Rev. 105* (1957) 767.
Hubbard, J. *Proc. Roy. Soc. A240* (1957) 539.
Huber, M. G., M. Danos, H. J. Weber, and **W. Greiner,** *Phys. Rev. 155* (1967) 1073.
Huffaker, J. N., and **E. Greuling,** *Phys. Rev. 132* (1963) 738.
Hugenholtz, N. M., *Physica 23* (1957) 481.
Hulthen, L., and **M. Sagawara,** *Handbuch der Physik*, Vol. XXXIX (1957)
Hyde, E. K., I. Perlman, and **G. T. Seaborg,** *The Nuclear Properties of Heavy Elements,* Prentice Hall, Englewood Cliffs N.J. (1964).

I

Inglis, D. R., *Phys. Rev. 55* (1939) 988.
Inglis, D. R., *Rev. Mod. Phys. 25* (1953) 390.
Inglis, D. R., *Phys. Rev. 96* (1954) 1059
Inglis, D. R.. *Phys. Rev. 97* (1955) 701.

J

Jackson, D., S. Treiman, and **H. Wyld,** *Nucl. Phys.4* (1957) 206.
Jackson, D., S. Treiman, and **H. Wyld,** *Phys. Rev. 106* (1957) 517.
Jacob, M., and **G. C. Wick,** *Ann. Phys.* (New York) *7* (1959) 404.
James, A. N. in *High Energy Physics and Nuclear Structure,* (S. Devons, editor) Plenum Press, New York (1970).
James, A. N., P. T. Andrews, P. Kirkby, and **B. G. Lowe,** *Nucl. Phys. A138* (1969) 145.
Janecke, J., in *Isopsin in Nuclear Physics,* (D. H. Wilkinson, editor) North Holland, Amsterdam (1969), Ch. 8.
Jastrow, R., *Phys. Rev. 98* (1955) 1479.
Jean, M., and **L. Wilets,** *Phys. Rev. 102* (1956) 788.
Jenschke, P., and P. Bock, *Phys. Letters 31B* (1970) 65.
Jensen, J. H. D., and **M. G. Mayer,** *Phys. Rev. 85* (1952) 1040 (L).
Johnson, M. B., and **M. Baranger,** *Ann. Phys.* (New York) *62* (1971) 172.

K

Kallio, A., and K. Kollveit, *Nucl. Phys. 53* (1964) 87
Kallio, A., and **B. D. Day,** *Phys. Letters 25B* (1967) 72.
Kallio, A., and **B. D. Day,** *Nucl. Phys. A124* (1969) 177.
Kankeleit, E., Congres International de Physique Nucleaire, Paris, Vol. II (1964) p. 1206.
Kao, E. I., and **S. Fallieros,** *Phys. Rev. Letters 25* (1970) 827.
Kelson, I., and **C. A. Levinson,** *Phys. Rev. 134B* (1964) 269.
Kerman, A. K., *Mat. Fys. Medd. Dan. Vid. Selsk 30* (1956) no. 15.
Kerman, A. K., in *Nuclear Reactions* (P. M. Endt and M. Demeur, editors, North Holland, Amsterdam, (1959), p. 427.
Kerman, A. K., and **M. K. Pal,** *Phys. Rev. 162* (1967) 970.

Kerman, A. K., "Nuclear Forces and Hartree-Fock Calculations," *Cargese Lectures* (M. Jean, editor), Gordon and Breach (1969).

Kerman, A. K. J. P. Svenne, and **F. M. H. Villars,** *Phys. Rev. 147* (1966) 710.

Kim, C. K., *Nucl. Phys. 49* (1963) 383.

Kim, C. W., and H. Primakoff, *Phys. Rev. 139B* (1965) 1447.

Kim, C. W., and **H. Primakoff,** *Phys. Rev. 140B* (1965) 566.

Kim, C. W., and **H. Primakoff,** *Phys. Rev. 147* (1966) 1034.

Kinoshita, T., and **A. Sirlin,** *Phys. Rev. 113* (1959) 1652.

Kirson, M. W., *Ann Phys.* (New York) *66* (1971) 624.

Kirson, M. W., and **L. Zamick,** *Ann. Phys.* New York *60* (1970) 188.

Kirsten, T., O. A. Schaeffer, E. Norton, and **R. W. Stoenner,** *Phys. Rev. Letters 20* (1968) 1300.

Kisslinger, L. S., and **R. A. Sorensen,** *Rev. Mod. Phys. 35* (1963) 853.

Kisslinger, L., and **R. A. Sorensen,** *Mat. Fys. Medd. Dan. Vid. Selsk. 32*, no. 9 (1960).

Klein, A., *Theories of Collective Motion in Dynamic Structure of Nuclear States* (Rowe, Trainor, Wong and Donnelly, editors) University of Toronto Press (Toronto) 1972.

Klein, A., and **A. Kerman** *Phys. Rev. 138B* (1965) 1323.

Kofoed-Hansen, O., "Neutrino Recoil Experiments" in *Alpha, Beta, and Gamma Ray Spectroscopy*, (K. Siegbahn, editor) North Holland, Amsterdam (1965).

Kofoed-Hansen, O., and **C. J. Christensen,** "Experiments on Beta Decay," *Handbuch der Physik* (S. Flügge, editor) Vol. 41, part 2. Springer-Verlag, Berlin (1962).

Köhler, H. S., and **R. J. McCarthy,** *Nucl. Phys. 106* (1967) 313.

Konopinski, E. J., *The Theory of Beta Radioactivity*, Oxford University Press, (1966).

Krane, K. S., C. E. Olsen, J. R. Sites, and **W. A. Steyert,** *Phys. Rev. C4* (1971) 1942.

Krane, K. S., C. E. Olsen, J. R. Sites, and **W. A. Steyert,** *Phys. Rev. Letters 26* (1971) 1579.

Krane, K. S., C. E. Olsen, J. R. Sites, and W. A. Steyert, Phys. Rev. *C4* (1972) 1907.

Krause, I. Y., *Phys. Rev. 129* (1963) 1330.

Kravtsov, V. A., and **N. N. Skachkov,** *Nucl. Data A1* (1966) 491.

Kumar, K., and **M. Baranger,** *Nucl. Phys. A92* (1967) 608.

Kumar, K., and **M. Baranger,** *Nucl. Phys. A122* (1968) 273.

Kuo, T. T. S., *Nucl. Phys. A103* (1967) 71.

Kuo, T. T. S., *Nucl. Phys. A122* (1968) 325.

Kuo, T. T. S., and **G. E. Brown,** *Nucl. Phys. 85* (1966) 40.

Kuphal, E., M. Daum, and **E. Kankeleit,** preprint (1971).

Kuphal, E., M. Daum, and **E. Kankeleit,** *High Energy Physics and Nuclear Structure* (S. Devons, editor) Plenum Press, New York (1970), p. 770.

Kurath, D., *Phys. Rev. 91* (1953) 1430.

Kurath, D., *Phys. Rev. 101* (1956) 216.

Kurath, D., *Phys. Rev. 106* (1957) 975.

Kurath, D., and **L. Picman,** *Nucl. Phys. 10* (1959) 313.

Kurath, D., *Nucl. Phys. 14* (1959) 398.

Kurath, D., "Nuclear Coupiing Schemes" in *Nuclear Spectroscopy* (Fay-Ajzenberg-Selove, editor) Academic Press, New York (1960).

Kurath, D., *Phys. Rev. 140B* (1965) 1190.

Kurath, D., private communication (1971).

Kurie, F. N. D., J. R. Richardson, and **H. C. Paxton,** *Phys. Rev. 49* (1936) 368.

L

Lambert, E., and **H. Feshbach,** *Phys. Letters 38B* (1972) 487, *Ann. of Phys.* (New York) *76* (1973) 80.

Lane, A. M., *Nuclear Theory*, W. A. Benjamin, New York (1964).

Lane, A. M., *Proc. of Phys. Soc.* (London) *A66* (1953) 977.

Lane, A. M., *Proc. of Phys. Soc.* (London) *A68* (1955) 189, 197

Lane, A. M., R. G. Thomas, and **E. P. Wigner**, *Phys. Rev. 98* (1955) 693

Lane, A. M., and **L. A. Radicati,** *Proc. Phys. Soc.* (London) *A67* (1954) 167.

Langer, L. M., and **H. C. Price, Jr.,** *Phys. Rev. 76* (1949) 641.

Lanthrop, J. L., R. A. Lundy, V. L. Telegdi, R. Winston, and **D. D. Yavanovitch,** *Phys. Rev. Letters 7* (1961) 107.

Laukien, G., Kernmagnetische Hockfrequenz-Spectroscopie. "External Properties of Atomic Nuclei," Vol. 38/1 *Handbuch der Physik* (S. Flügge, editor) Springer-Verlag, Berlin (1958).

Lawson, R. D., and **J. L. Uretsky,** *Phys. Rev. 108* (1957) 1300.

Lederer, C. M., J. M. Hollander, and **I. Perlman.** *Table of Isotopes*, 6th edit., John Wiley and Sons, New York (1967).

Lee, L. L., J. P. Schiffer, B. Zeidman, G. R. Satchler, R. M. Drisko, and **R. H. Bassel,** *Phys. Rev. 136B* (1964) 971.

Lee, Y. K., L. W. Mo, and **C. S. Wu,** *Phys. Rev. Letters 10* (1963) 253.

Lee, T. D., and **C. S. Wu.** *Ann. Rev. Nucl. Sci. 15* (1965) 381.

Lee, T. D., and **C. N. Yang,** *Phys. Rev. 104* (1956) 254.

Lee, T. D., and **C. N. Yang** *Phys. Rev. 105* (1957) 1671

Levinger, J. S., *Phys. Rev. 91* (1955) 122

Levinger, J. S., *Phys. Rev. 101* (1956) 733.

Levinger, J. S., and **D. C Kent,** *Phys. Rev. 95* (1950) 418.

Levinger, J. S., and **H. A. Bethe,** *Phys. Rev. 78* (1950) 115.

Levinger, J. S. *Nuclear Photodisintegration*, Oxford University Press, 1960.

Lindgren, Richard A., "A Study of Light Rigid Rotor Nuclei," Yale Ph.D. Thesis. Research Advisor: D. A. Bromley (1969).

Lipkin, H. J., *Nucl. Phys. 26* (1961) 147.

Lipkin, H. J., *Ann. Phys.* (New York) *9* (1960) 272.

Lipkin, H. J. *Beta Decay for Pedestrians* North Holland, Amsterdam (1962).

Lipson, E. D., F. Boehm, and **J. C. Vanderleeden,** *Phys. Letters 35B* (1971) 307.

Litherland, A. E., E. B. Paul, G. A. Bartholomew, and **H. E. Gove,** *Phys. Rev. 102* (1956) 208.

Litherland, A. E., H. McManus, E. B. Paul, D. A. Bromley, and **H. E. Gove,** *Can. J. Phys. 36* (1958) 378.

Lobashov, V. M., V. A. Nazarenko, L. F. Saenko, L. M. Smotritskii, and **G. I. Kharkovitch.** *Zh Eksp. Teor. Fiz. Pis'ma 3* (1966) 268 [translation: *Soviet Phys. JETP Letters 3* (1966) 173].

Lobashov, V. M., V. A. Nazarenko, L. F. Saenko, and **L. M. Smotritskii,** *Zh. Eksp. Teor. Fiz. Pis'ma 5* (1967) 73 [translation *5*, (1967) 59]. *Phys. Letters 25B* (1967) 104.

Lobashov, V. M., N. A. Lozovoy, V. A. Nazarenko, L. M. Smotritskii, and **O. I. Kharkevitch,** *Phys. Letters 30B* (1969) 39.

Lobashov, V. M. et al., quoted by O. S. Danilov, *Phys. Letters 35B* (1971) 579.

Lobashov, V. M., A. E. Egrov, D. M. Kaminker, V. A. Nazarenko, L F. Saenko, L. M. Smotritskii, O. I. Kharkevich, and **V. A. Knyaz'kov,** *JETP Letters 11* (1970) 76.

Lobashov, V. M., *High Energy Physics and Nuclear Structure* (S. Devons, editor) Plenum Press, New York (1970), p. 713; *Soviet J. Nucl. Phys.* 2 (1965) 683.

Lobashov, V. M., V. A. Nazarenko, L. F. Saenko, L. M. Smotritskii, and **O. I. Kharkevitch,** *Proceedings International Conference on Nuclear Structure, Phys. Soc. of Japan* Tokyo (1968).

Lomon, E. L., submitted for publication, (1972).

Lomon, E. L., and **H. Feshbach,** *Ann. Phys.* (New York) *48* (1968) 94.

Luyten, J. R., H. P. C. Rood, and **H. A. Tolhoek,** *Nucl. Phys. 41* (1963) 236.

M

McCarthy, R. J., and **H. S. Köhler,** *Nucl. Phys. 99* (1967) 65.

McCarthy, R. J., *Nucl. Phys. A130* (1969) 305.

McFadden, L. and **G. R. Satchler,** *Nucl. Phys. 84* (1966) 177.

McGowan, F. K., and **P. H. Stelson,** *Phys. Rev. 122* (1961) 1274.

McGowan, F. K., W. T. Milner, R. L. Robinson, and **P. H. Stelson,** *Ann. Phys.* (New York) *63* (1971) 549.

McGrath, R. L., J. Cerny, J. C. Hardy, G. Goth, and **A. Arima,** *Phys. Rev. C1* (1970) 184.

McIntyre, J. A., T. L. Watts, and **F. C. Jobes,** *Phys. Rev. 119* (1960) 1331.

McKellar, B. H. J., *Phys. Rev. Letters 21* (1968) 1822.

McKinley, J. M., and P. M. Rinard, *Nucl. Phys. 79* (1966) 159.

MacFarlane, M. H., "The Reaction Matrix in Nuclear Shell Theory: Int'l School of Physics," *Nuclear Structure and Nuclear Reactions*, (M. Jean, editor) Academic Press (New York) 1969.

Majorana, E., *Nuovo Cimento 14* (1937) 171.

Malik, F. B., and **W. Scholz,** *Phys. Rev. 150* (1966) 919.

Malik, F. B., and **W. Scholz,** in *Nuclear Structure* (A. Hossain, Editor) North Holland Amsterdam (1967).

Marelius, A., P. Sparrman, and **T. Sundstrom,** in *Hyperfine Structure and Nuclear Radiations* (E. Matthias and D. A. Shirley, editors) North Holland, Amsterdam (1968).

Marshalek, E. R., and **J. O. Rasmussen,** *Nucl. Phys. 43* (1963) 438.

Marshalek, E. R., and **J. Weneser** *Ann. Phys.* (New York) *53* (1969) 569.

Matthias, E., and **D. A. Shirley** (editors) *Hyperfine Structure and Nuclear Radiations*, North Holland, Amsterdam (1968).

Mattauch, J. H. E., W. Thiele, and **A. H. Wapstra,** *Nucl. Phys. 67* (1965) 1, 32.

Mayer, M. G., *Phys. Rev. 75* (1949) 1969.

Mayer, M. G., and **J. H. D. Jensen,** *Elementary Theory of Nuclear Shell Structure.* John Wiley and Sons, New York (1955).

Mayer, M. G., and **J. H. D. Jensen,** in *Alpha, Beta, Gamma Ray Spectroscopy* (K. Siegbahn, editor) North Holland, Amsterdam (1965).

Messiah, A., *Quantum Mechanics*, North Holland, Amsterdam (1961).

Meyerhof, W. E., and T. A. Tombrello, *Nucl. Phys. A109* (1968) 1.

Migneco, E., and **J. P. Theobold,** *Nucl. Phys. A112* (1968) 603.

Miller, D. W., R. K. Adair, C K. Bockelman, and **S. E. Darden,** *Phys. Rev. 88* (1952) 83.

Mittelstaedt, P., *Acta. Phys. Hung. 19* (1965) 303.

Miyazawa, H., *Prog. Theor. Phys. 5* (1951) 801.

Moinester, M., J. P. Schiffer, and **W. P. Alford,** *Phys. Rev. 179* (1969) 984.

Moline, A., J. Morris, P. Dyer, and **C. A. Barnes,** unpublished report (1970).

Moniz, E., *Phys. Rev. 184* (1969) 1154.

Moniz, E. J., I. Sick, R. R. Whitney, J. R. Ficenec, R. D. Kephart, and **W. P. Trower,** *Phys. Rev. Letters 26* (1971) 444.

Morinaga, H., and **N. L. Lark,** *Nucl. Phys. 67* (1965) 315.

Morita, T., *Prog. Theor. Phys. 29* (1963) 351.

Morpurgo, G., *Phys. Rev. 110* (1958) 721.

Morpurgo, G., *Phys. Rev. 114* (1959) 1075.

Morse, P. M., and **H. Feshbach,** *Methods of Theoretical Physics*, McGraw-Hill, New York (1953).

Moshinsky, M., Group Theory and the Many Body Problem, Gordon and Breach, New York (1968).

Moskalev, A. N., *JETP Letter 8* (1968) 27.

Moszkowski, S. A., *Handbuch der Physik*, XXXIX (S. Flügge, editor) Springer-Verlag, Berlin (1957) p. 411.

Moszkowski, S. A., and **B. L. Scott,** *Ann. Phys.* (New York) *11* (1960) 65.

Moszkowski, S. A., and **B. L. Scott,** *Ann. Phys.* (New York) *14* (1961) 107.

Moszowski, S. A., *Phys. Rev. C2* (1970) 402.

Mottelson, B. R., and **S. G. Nilsson,** *Phys. Rev. 99* (1955) 1615.

Mottelson, B. R., and **S. G. Nilsson,** *Mat. Fys. Skr. Dan. Vid. Selsk. 1* (1959) No. 8.

Mottelson, B., *Proceedings of the Enrico Fermi School of Physics*, Academic Press, New York (1960)

Myers, W. D., and **W. J. Swiatecki,** *Nucl. Phys. 81* (1966) 1.

N

Nambu, Y., *Phys. Rev. Letters 4* (1960) 380.

Nathan, O., *Studies of Nucleon Quadrupole and Octupole Vibrations*, Munksgard, Copenhagen (1964).

Nathan, O., and **S. G. Nilsson,** Chap. X in *Alpha, Beta, and Gamma Ray Spectroscopy* (K. Siegbahn, editor). North Holland, Amsterdam (1965).

Negele, J. W., *Phys. Rev. C1* (1970) 1260.

Nemeth, J., and **D. Vautherin,** *Phys. Letters B32* (1970) 561.

Nerenson, N., and **S. Darden,** *Phys. Rev. 89* (1953) 775.

Nerenson N., and **S. Darden,** *Phys. Rev. 94* (1954) 1678.

Nestor, C. W., K. T. R. Davies, S. J. Krieger, and **M. Baranger** *Nucl. Phys. A113* (1968) 14.

Nezrick, F. A., and **F. Reines,** *Phys. Rev. 142* (1966) 852.

Nilsson, S. G., *Dan. Mat. Fys. Medd. 29* (1955) No. 16.

Nilsson, S. G., in *Perturbed Angular Correlations*, (E. Karlsson, E. Matthias, and K. Siegbahn, editors). North Holland, Amsterdam (1964), p. 163.

Nilsson, S. G., and **J. O. Rasmussen,** *Nucl. Phys. 5* (1958) 617.

Noble, J. V., *Ann. Phys.* (New York) *67* (1971) 98.

Noya, H., A. Arima, and **H. Horie,** *Prog. Theo. Phys.*, Suppl. *8* (1958) 33.

O

Oakes, R. J., *Phys. Rev. Letters 20* (1968) 1359.
Oertzen, W., von, H. H. Gutbrod, M. Muller, O. Voos, and **R. Bock,** *Phys. Letters 26B* (1968) 291.
Okamoto, K., *Phys. Rev. 110* (1958) 143.
Okazaki, A., S. E. Darden, and **R. B. Walton,** *Phys. Rev. 93* (1954) 461.
Oquidam, B., and **B. Jancovici,** *Nuovo Cimento 11* (1959) 578.
Obsorn, R. K., and **L. L. Foldy,** *Phys. Rev. 79* (1950) 795.

P

Palevsky, H., J. L. Friedes, R. J. Sutter, G. W. Bennet, G. J. Igo, W. D. Simpson, G. C. Phillips, D. M. Corley, N. S. Wall, R. L. Stearns, and **B. Gottschalk,** *Phys. Rev. Lett. 18* (1967) 1200.
Pandya, S. P., *Phys. Rev. 103* (1956) 956.
Particle Data Group N. Barash-Schmidt, A. Barbaro-Gaiteri, C. Briemen, T. Lasinski, A. Rittenberg, M. Roos, A. H. Rosenfeld, P. Söding, T. G. Trippe, and **C. G. Wohl,** *Rev. Mod. Phys. 43*, (April 1971).
Particle Data Group, *Physics Letter* (April 1972). Chrman,. A. H. Rosenfeld.
Partovi, F., *Ann. Phys.* (New York) *27* (1964) 114.
Partovi, F., and **E. L. Lomon,** *Phys. Rev. D5* (1972) 1192.
Partovi, M. H., and **E. L. Lomon,** *Phys. Rev. D2* (1970) 1999.
Paul, E. B., *Phil. Mag. 2* (1957) 311.
Paul, H., O. McKeown, and **G. Scharff-Goldhaber,** *Phys. Rev. 158* (1967) 1112.
Pauli, W., *Handbruch der Physik*, Vol. *24/1.* Springer-Verlag, Berlin (1933).
Pauli, W., *Proc. Solvay Congress*, Brussels (1933).
Pease R. L., and **H. Feshbach,** *Phys. Rev. 88* (1952) 945.
Peierls, R. E., and **D. J. Thouless,** *Nucl. Phys. 38* (1962) 154.
Peierls, R. E., and **J. Yoccoz,** *Proc. Roy. Soc.* (London) *A70* (1957) 381.
Perdrisat, Charles F., *Rev. Mod. Phys. 38* (1966) 41.
Perey, F. G., in *Nuclear Spin and Parity Assignments*, (N. B. Gove, editor) Academic Press, New York (1966).
Peterson, G. A., *Phys. Letters 25B* (1968) 549.
Petterson, B. G., "Internal Bremsstrahlung" in *Alpha, Beta and Gamma-Ray Spectroscopy* (K. Siegbahn, editor). North Holland, Amsterdam (1965).
Plano, R. J., *Phys. Rev. 119* (1960) 1400.
Poletti, A. R., E. K. Warburton, and **D. Kurath,** *Phys. Rev. 155* (1967) 1096.
Povel, H. P. Diplomarbeit, Universität Karlsruhe (1968).
Pratt, W., R. I Schermer, J. R. Sites, and **W. A. Steyert,** *Phys. Rev. C2* (1970) 1499.
Primakoff, H. *Rev. Mod. Phys. 31* (1959) 819.
Primakoff, H., *Proceedings of International School of Physics "Enrico Fermi,"* Course 32 (T. D. Lee, editor) Academic Press, New York (1966).
Primakoff, H., *Weak Interactions in Nuclear Physics*, Karlsruhe Summer Institute Lecture (1969).
Primakoff, H., and **S. P. Rosen,** *Phys. Rev. 184* (1969) 1925.
Prior, O., F. Boehm, and **S. G. Nilsson, Nucl. Phys.** A*110* (1968) 257.

Q-R

Quaranta, A., et al., *Phys. Rev. 127* (1969) 2118.

Radicati, L. A., *Phys. Rev. 87* (1952) 521.

Rakavy, G., *Nucl. Phys. 4* (1957) 375.

Rajaraman, R., and **H. A. Bethe**, *Rev. Mod. Phys. 39* (1967) 745.

Rasmussen, J. O., *Nucl. Phys. 19* (1960) 85.

Ravenhall, D. G., in *Medium Energy Nuclear Physics* (W. Bertozzi and S. Kowalski, editors) M.I.T. Summer Study (1967) USAEC Div. Tech. Info. TID 24667.

Redlich, M., *Phys. Rev. 110* (1958) 468.

Reiner, A. S., *Nucl. Phys. 27* (1961) 115.

Reiner, A. S., (Rinat) *Nucl. Phys. 5* (1958) 544.

Reines, F., and **C. L. Cowan, Jr.**, *Phys. Rev. 113* (1959) 273.

Reines, F., and **H. Gurr**, *Phys. Rev. Letters 24* (1970) 1448.

Ricci, R. A., *Proceedings of the International School of Physics "Enrico Fermi,"* Course XXXVI (C. Bloch, editor) Academic Press, New York (1966) p. 566.

Richard, P., **C. F. Moore**, **D. Robson**, and **J. D. Fox**, *Phys. Rev. Letters 13* (1964) 343.

Riihimaki, E., M.I.T. Thesis (1970).

Ripka, G., *Advances in Nuclear Physics, 1* (1968) 183

Robson, D., *Ann. Rev. Nucl. Sci., 16* (1966) 119.

Roijen, J. J. van., **P. Pronk**, **S. U. Ottevangers**, and **J. Block**, *Physica 37* (1967) 32.

Rose, M. E., **Multipole Fields**, John Wiley and Sons, New York (1955).

Rose, M. E., *Elementary Theory of Angular Momentum*, John Wiley and Sons, New York (1957).

Rose, M. E., "Analysis of Internal Conversion Data" in *Nuclear Spectroscopy* (Fay Ajzenberg-Selove, editor) Academic Press, New York (1960) part B.

Rose, M. E., "Analysis of Beta Decay Data" in *Nuclear Spectroscopy* (Fay Ajzenberg-Selove, editor), Academic Press, New York (1960) part B.

Rose, M. E., *Internal Conversion Coefficients*, North Holland, Amsterdam (1958).

Rosen, L., **J. G. Beery**, **A. S. Goldhaber**, and **E. H. Auerbach**, *Ann. Phys. 34* (1965) 96.

Rosen, L., in *Nuclear Structure Study with Neutrons*, (M. Neve de Mevergnies, P. Van Assche, and J. Vervier, editors), North Holland Amsterdam, (1966), p. 379.

Rosen, S. P., and **H. Primakoff**, in *Alpha, Beta and Gamma Ray Spectroscopy*, (K. Siegbahn, editor) North Holland, Amsterdam (1965).

Rotenberg, M., **R. Bivins**, **N. Metropolis**, and **J. K. Wooten, Jr.**, '*The 3-j and 6-j Symbols*," The Technology Press, M.I.T., Cambridge, Mass. (1959).

Rowe, D. J., *Nucl. Phys. 80* (1966) 209.

Roy, R. R. and **B. P. Nigam**, *Nuclear Physics, Theory and Experiment*. John Wiley and Sons, New York (1967).

S

Sachs, R. G., *Phys. Rev. 126* (1962) 2256.

Sachs, R. G., *Phys. Rev. 72* (1947) 312.

Sachs, R. G., and **N. Austern**, *Phys. Rev. 81* (1951) 705.

Sachs, R. G., and **J. Schwinger**, *Phys. Rev. 70* (1946) 41.

Saintignon, P. De and **M. Chabre**, *Phys. Letters 33B* (1970) 463.

Saintignon, P. De, J. J. Lucas, J. B. Viano, M. Chabre, and **D. Depommier,** *Nucl. Phys. A160* (1971) 53.

Satchler, G. R., "Proton Inelastic Scattering" in *Polarization Phenomena* in *Nuclear Reactions*, (H. H. Barschall and W. Haeberli, editors) University of Wisconsin Press, Madison, Wis., (1971).

Satchler, R. G., in *Lectures in Theoretical Physics*, Volume VIII C. University of Colorado Press, Boulder (1966) p. 73-176.

Sawicki, J., "Microscopic Theories of Vibrational States of Even-Even Nuclei, including 4 Quasi-Particle Excitations." *International School of Physics, Nuclear Structure and Nuclear Reactions* (M. Jean, editor) Academic Press, New York, (1969).

Scholz, W., and **F. B. Malik,** *Phys. Rev. 147* (1966) 836.

Scholz, W. and **F. B. Malik,** *Phys. Rev. 153* (1967) 1071.

Schwager, J. E., *Phys. Rev. 121* (1961) 569.

Schwartz, C., and **A. deShalit,** *Phys. Rev. 94* (1954) 1257.

Seeger, P. A., in *Proceedings Third Intl. Conference on Atomic Masses* (R. C. Barber, editor), University of Manitoba Press, Winnepeg, Manitoba (1967).

Segel, R. E., J. W. Olness, and **E. L. Sprenkel,** *Phys. Rev. 123* (1961) 1382.

Segre, E., *Nuclei and Particles*, W. A. Benjamin, New York (1964).

Sens, J. C., *Phys. Rev. 113* (1958) 679.

Sens, J. C., in *Second International Conference on High Energy and Nuclear Structure, Rehovot 1967.* North Holland, Amsterdam (1967).

Seth, K. K., *Nucl. Data A2* (1966) 299.

Shao, J., W. H Bassichis, and **E. L. Lomon,** *Phys. Rev. 6C* (1972) 758.

Shao, J., and **E. L. Lomon,** *Phys. Rev.* C (to be published).

Sherr, R., J. Gerhart, H. Horie, and **W. Hornyak,** *Phys. Rev. 100* (1955) 945.

Shirley, V. S., "Table of Nuclear Magnetic Moments," *Proceedings of International Conference of Hyperfine Structure and Nuclear Radiation*, Asilomar 1967. North-Holland, Amsterdam (1968).

Siegbahn, K., Editor, *Alpha, Beta, and Gamma Ray Spectroscopy*, North Holland, Amsterdam (1965).

Siegert, A. J. F. *Phys. Rev. 52* (1937) 787.

Singh, P. P., R. E. Segel, L. Meyer Schützmeister, S. S. Hanna and R. G. Allen, *Nucl. Phys 65* (1965) 577.

Skorka, S. J., J. Hertel, and **T. W. Ritz-Schmidt,** *Nuclear Data*, Section A, *2* (1966) 347.

Skyrme, T. H. R., *Phil. Mag. 1* (1956) 1043.

Skyrme, T. H. R., *Nucl. Phys. 9* (1959) 615.

Sliv, L. A., and **I. M. Band.** *Coefficients of Internal Conversion of Gamma Radiation.* Part I, "K Shell" (1956), Part II, "L Shell" Physico-Technical Institute, Academy of Sciences, Leningrad, U.S.S.R., (1958).

Sliv, L. A., and **I. M. Band,** "Tables of Internal Conversion Coefficients, Appendix J" in *Alpha, Beta, and Gamma Ray Spectroscopy* (K. Siegbahn, editor) North Holland, Amsterdam (1965).

Snell, A. H., and **F. Pleasanton,** *Phys. Rev. 97* (1955) 246.

Snell, A. H., and **F. Pleasanton,** *Phys. Rev. 100* (1955) 1396

Snyder, E. S., and **S. Frankel,** *Phys. Rev. 106* (1957) 755.

Soper, J. M., "Isospin Purity of Low-Lying States" in *Isospin in Nuclear Physics.* (D. H. Wilkinson, editor) North Holland, Amsterdam (1969).

Spicer, B. M., H. H. Theis, J. E. E. Baglin, and **F. R. Allum,** *Australian J. Phys. 11* (1958) 298.

Spicer, B. M., in *Adv. Nucl. Phys. 2* (1969) 1.

Sprung, D. W. L., and **P. K. Banerjee,** *Nucl. Phys. A168* (1971) 273.

Sprung, D. W. L., *Adv. Nucl. Phys. 5* (1972) 225.

Steinwedel, H., and **J. H. D. Jensen,** *Zeit. fur Naturforsch 5* (1950) 413.

Stelson, P. H., and **L. Grodzins,** *Nucl. Data 1* (1965) 21.

Stelson, P. H., and **F. K. McGowan,** *Phys. Rev. 121* (1961) 209.

Stephens, F. S., N. L. Lark, and **R. M. Diamond** *Nucl. Phys. 63* (1965) 82.

Stratton, Julius Adams, *Electromagnetic Theory*, McGraw-Hill, New York (1941).

Strutinsky, V. M., *Nucl. Phys. A95* (1967) 420.

Sugimoto, K., *Phys. Rev. 182* (1969) 1051.

T

Tabakin, F., *Ann. Phys.* (New York) *30* (1964) 51.

Tadić, D. *Phys. Rev. 174* (1968) 1694.

Talmi, I., *Helv. Phys. Acta 25* (1952) 185.

Talmi, I., and **I. Unna.,** *Ann. Rev. Nucl. Sci. 10* (1960) 353.

Talmi, I., *Lecture in Theoretical Physics*, Vol. VIII, University of Colorado Press, Boulder (1965) p. 39.

Talmi, I., "Nuclear Magnetic Moments—Shell Model Interpretation," *Lecture at International Conference on Hyperfine Interactions Detected by Nuclear Radations*, Rehovot and Jerusalem (Sept. 1970).

Tamura, T., and **L. G. Komai,** *Phys. Rev. Letters 3* (1959) 344.

Taylor, J. C., *Phys. Rev. 110* (1958) 1216.

Telegdi, V. L., *Weak Interactions and Topics in Dispersion Physics*, W. A. Benjamin New York (1962).

Telegdi, V. L., *Phys. Rev. Letters 8* (1963) 327.

Temmer, G. M., and **N. P. Heydenburg,** *Phys. Rev. 104* (1956) 967.

Theiberger, P., A. W. Sunyar, P. C. Rogers, N. Lark, O. C. Kistner, E. der Mateosian, S. Cochavi, and E. H. Auerbach, "Symposium on Heavy Ion Reactions and Many Particle Excitations." Sept. 1971. *Jour de Physique*, Colloque C-6 (1971).

Thouless, D. J., *The Quantum Mechanics of Many Body Systems*, Academic Press, New York (1961).

Tolhoek, H. A., "The Induced Pseudoscalar Interaction in Muon Capture and Nuclear Structure" in *Selected Topics in Nuclear Spectroscopy*. (B. J. Verhaar, editor) North Holland, Amsterdam (1964).

U-V

Uberall, H., *Nuovo Cimento Suppl 4* (1966) 781.

Valatin, J. G., *Nuovo Cimento 7* (1958) 843.

Vanderleeden, J. C., and **F. Boehm,** *Phys. Letters 30B* (1969) 467.

Vanderleeden, J. C., and **F. Boehm,** *Phys. Rev. C2* (1970) 748.

Vanderleeden, J. C., F. Boehm, snd **E. Lipson,** *Phys. Rev. C4* (1972) 2218.

Van Patter, D. M., *Nucl. Phys. 14* (1959) 42.

Varga, L., I. Demeter, L. Keszthelyi, Z. Szokefalvi-Nagy, and **Z. Zamori,** *Phys. Rev. 177* (1969) 1783.

Vautherin, D., and **D. M Brink,** *Phys. Rev. C5* (1972) 626.
Vautherin, D., M. Vénéroni and **D. M. Brink,** *Phys. Letters 33B* (1970) 381.
Vergnes, M. N., and **J. O. Rasmussen,** *Nucl. Phys. 62* (1965) 233.
Verhaar, B. J., editor, *Selected Topics in Nuclear Spectroscopy* North Holland, Amsterdam, (1964).
Villars, F., *Proceedings of the International School of Physics "Enrico Fermi,"* Course XXXVI (C. Bloch, editor), Academic Press, New York (1966).
Villars, F., and **G. Cooper,** *Ann. Phys.* (New York) *56* (1970) 224.
Villars, F., and **N. S. Rogerson,** *Ann. Phys.* (New York) *63* (1971) 443.
Vinh Mau, N., and **A. M. Bruneau,** *Phys. Letters, 29B* (1969) 408.
Visscher, V. M., and **R. A. Ferrell,** *Phys. Rev. 107* (1957) 781.

W

Wäffler, H. (1972), quoted by Gari (73).
Walecka, J. D., "Semi-Leptonic Weak Interactions in Nuclei" in *Muon Physics.* (V. W. Hughes and C. S. Wu, editors) (1973).
Walt, M., R. L. Becker, A. Okazaki, and **R. E. Fields,** *Phys. Rev. 89* (1953) 1271.
Wang, W. L. and **C. M. Shakin,** *Phy. Rev. C5* (1972) 1898.
Wapstra, A. H., and **N. B. Gove** *Nuclear Data Tables 9* (1971).
Warburton, E., and **J. Weneser** in *Isospin in Nuclear Physics,* (D. H. Wilkinson, editor). North Holland Amsterdam (1969).
Warburton, E. K., D. E. Alburger, D. H. Wilkinson, and **J. M. Soper,** *Phys. Rev. 129* (1963) 2191.
Warburton, E. K., in *Isobaric Spin in Nuclear Physics* (J. D. Fox and D. Robson, editors) Academic Press, New York (1966) p. 90-112.
Ward, D., R. M. Diamond, and **F. S. Stephens,** *Nucl. Phys. A117* (1968) 309.
Warming, E., *Phys. Letters 29B* (1969) 564.
Warming, E., F. Stecher-Rasmussen, W. Ratymski, and **J. Kopecky,** *Phys. Letters 25B* (1967) 200.
Way, K., and F. W. Hurley, *Nuclear Data A1* (1966) 473.
Weidenmuller, H., *Rev. Mod. Phys. 33* (1961) 574.
Weisberger, W. L. *Phys. Rev. Letters 14* (1964) 1047.
Weise, W., *Proceedings Int. Conf. on Photonuclear Reactions* (B. L. Berman, editor), Lawrence Livermore Lab. (1973).
Weisskopf V. F., *Atti del Convengo Mendeleeviano,* Torino-Roma, 1969 [(M. Verde, editor) Accademia delle Scienze de Torino and Accademi Nazionale dei Lincie, Torino (1971)].
Werntz, C., and **W. E. Meyerhof,** *Nucl. Phys A121* (1968) 38.
Wick, Gian Carlo, *Ann. Rev. Nucl. Sci. 8* (1958) 1.
Wildenthal, B., J. B. McGrory, E. C. Halbert, and **P. W. M. Glaudemans,** *Phys. Letters 26B* (1968) 692.
Wilkinson, D. H., in *Nuclear Spectroscopy Part B.* (F. Ajzenberg-Selove, editor) Academic Press, New York, London (1960)
Wilkinson, D. H., "Comments on Nuclear and Particle Physics," *1*, (1967) 139.
Wilkinson, D. H., *Phys. Rev. 109* (1958), 1603, 1611.
Wilkinson, D. H., and **D. E. Alburger,** *Phys. Rev. Letters 24* (1970) 1134.
Wilkinson, D. H., *Phys. Letters 31B* (1970) 447.

Wilson, R., *The Nucleon Nucleon Interaction*, Interscience Publishers, New York (1963).

Winter, C. von, *Physica 20* (1954) 274.

Witsch, W. von, A. Richter, and **P. von Brentano,** *Phys. Letters 22* (1966) 631.

Witsch, W. von A. Richter and **P. von Brentano,** *Phys. Rev. Letters, 19* (1967) 524.

Witsch, W. von, A. Richter, and **P. von Brentano,** *Phys. Rev. 169* (1968) 923.

Wong, C. W., *Nucl. Phys. A91* (1967) 399.

Wong, C. W , *Nucl. Phys. A104* (1967) 417.

Wu, C. S., E. Ambler, R. W. Hayward, D. D. Hoppes, and **R. F. Hudson,** *Phys. Rev. 105* (1957) 1413.

Wu, C. S., "The Neutrino," in *Theoretical Physics in the Twentieth Century* (M. Fierz and V. F. Weisskopf, editors) Interscience Publishers, London (1960).

Wu, C. S., and **S. A. Moszkowski,** *Beta Decay*, Interscience Publishers, New York (1966).

Wu, C. S., *Rev. Mod. Phys.36* (1964) 618.

Wu, C. S., and **L. Wi!ets** *Ann. Rev. Nucl. Sci. 19* (1969) 527.

X-Y-Z

Yamazaki, T and **G. T. Ewan** *Phys. Letters 24B* (1967) 278.

d'Yasevich, A., and **V. G. Antonenko** *Sov. J. Nucl. Phys. 10* (1970) 278.

Zaidi, S. A. A., J. L. Parish, J. G. Kulleck, C. F. Moore, and **P. von Brentano,** *Phys. Rev. 165* (1968) 1312.

Zamick L., and **J. McCullen,** *Bull. Am. Phys. Soc. 10* (1965) 485.

INDEX

**Italicized numbers refer to more detailed discussion of the subject.*